SMITH AND WILLIAMS'

INTRODUCTION TO THE PRINCIPLES OF DRUG DESIGN AND ACTION

FOURTH EDITION

SMITH AND WILLIAMS'

INTRODUCTION TO THE PRINCIPLES OF DRUG DESIGN AND ACTION

FOURTH EDITION

EDITED BY

H. John Smith
University of Cardiff
Cardiff, Wales

A CRC title, part of the Taylor & Francis imprint, a member of the Taylor & Francis Group, the academic division of T&F Informa plc.

Published in 2006 by
CRC Press
Taylor & Francis Group
6000 Broken Sound Parkway NW, Suite 300
Boca Raton, FL 33487-2742

Printed in the United States of America on acid-free paper
10 9 8 7 6 5 4 3 2 1

International Standard Book Number-10: 0-415-28877-0 (Hardcover)
International Standard Book Number-13: 978-0-415-28877-4 (Hardcover)
Library of Congress Card Number 2005041802

Library of Congress Cataloging-in-Publication Data

Smith and Williams' introduction to the principles of drug design and action / edited by H. John Smith.--4th ed.
p. ; cm.
Includes bibliographical references and index.
ISBN 0-415-28877-0
1. Drugs--Design. 2. Drugs--Structure-activity relationships. I. Title: Introduction to the principles of drug design and action. II. Smith, H. J., 1930- III. Williams, Hywel.
[DNLM: 1. Drug Design. 2. Pharmacologic Actions. QV 744 S642 2005]

RS420.S64 2005
615'.7--dc22 2005041802

Taylor & Francis Group
is the Academic Division of T&F Informa plc.

Visit the Taylor & Francis Web site at
http://www.taylorandfrancis.com

and the CRC Press Web site at
http://www.crcpress.com

Preface

This book, *Smith and Williams' Introduction to the Principles of Drug Design and Action, Fourth Edition*, is intended for use in undergraduate pharmacy courses in medicinal chemistry and as an aid in similar courses in pharmacology and biochemistry where there is a need to appreciate the rationales behind the design of drugs. It provides a suitable background for graduates in chemistry who are just entering the pharmaceutical industry.

The emphasis in this book is on principles, which are appropriately illustrated by groups of drugs in current (or even future) use. It is not the intention of this book to deal comprehensively with all conceivable groups of drugs, or to consider drugs grouped on the basis of particular pharmacological actions. This would require repeated descriptions of a range of design aspects relevant to each group so that design considerations would become subservient to the biologically observable actions. Instead the aim is to provide a framework of basic drug design and principles into which current drugs, and more importantly future drugs based on new developments, may be fitted. This approach should provide the newly qualified graduates with an understanding of new developments as they take place in future years.

The time is now right for the third edition to be revised and updated in view of the availability of many new drugs and the entrenchment of design techniques, which were in their infancy at the time of preparation of that edition.

Many new sections to existing chapters as well as several new chapters have been included: Pharmacokinetics, as an expansion of drug handling by the body; Peptide Drug design, in view of increasing importance of this category of drugs; the Human Genome and Its Impact on Drug Discovery and Therapy following insight into relationships between diseases and genetic makeup, and the possibility of finding novel targets for drug action based on new sites of action; and Combinatorial Chemistry, as a means of increasing the rate of discovery of new drugs with novel structures.

H.J.S.
Cardiff

About the Editor

H. John Smith received his B. Pharm and Ph.D. (Medicinal Chemistry) degrees at the School of Pharmacy, University of London where he was also an assistant lecturer in Pharmaceutical Chemistry. After short periods in the Chemical Defence Experimental Establishment at Porton Down working on an antidote to organophosphorous poisoning, and with the Pyrethrum Board of Kenya on insecticide research, he joined the Welsh School of Pharmacy, Cardiff University, eventually becoming Reader in Medicinal Chemistry. He is Editor-in-Chief of the *Journal of Enzyme Inhibition and Medicinal Chemistry* and co-editor of four authoritative texts on the design of enzyme inhibitors including, recently, *Proteinase and Peptidase Inhibition: Recent Potential Targets for Drug Development and Enzymes and Their Inhibition: Drug Development.* He is a Fellow of the Royal Society of Chemistry and Fellow of the Royal Pharmaceutical Society of Great Britain. In 1995 he received the D.S.C. (London).

Contributors

Dr. David J. Barlow
Department of Pharmacy
King's College London
London, United Kingdom

Dr. Mark T.D. Cronin
School of Pharmacy and Chemistry
John Moores University
Liverpool, United Kingdom

Dr. Robin H. Davies
Welsh School of Pharmacy
Cardiff University
Cardiff, United Kingdom

Professor John C. Dearden
School of Pharmacy and Chemistry
John Moores University
Liverpool, United Kingdom

Dr. Philip N. Edwards
School of Pharmacy & Pharmaceutical Sciences
University of Manchester
Manchester, United Kingdom

Dr. Mark Gumbleton
Welsh School of Pharmacy
Cardiff University
Cardiff, United Kingdom

Dr. Andrew J. Hutt
Department of Pharmacy
King's College London
London, United Kingdom

Dr. Barrie Kellam
School of Pharmacy
University of Nottingham
Nottingham, United Kingdom

Professor Ian W. Kellaway
School of Pharmacy
University of London
London, United Kingdom

Professor Andrew W. Lloyd
Faculty of Science and Engineering
University of Brighton
Brighton, United Kingdom

Dr. Anjana Patel
Independent Pharmaceutical Consultant
Harrow, Middlesex
United Kingdom

Professor Frederick J. Rowell
School of Health, Natural & Social Sciences
University of Sunderland
Sunderland, United Kingdom

Professor A Denver Russell*
Welsh School of Pharmacy
Cardiff University
Cardiff, United Kingdom

Professor Walter Schunack
Freie Universität Berlin
Institut für Pharmazie
Berlin, Germany

Dr. Robert D.E. Sewell
Welsh School of Pharmacy
Cardiff University
Cardiff, United Kingdom

Dr. Claire Simons
Welsh School of Pharmacy
Cardiff University
Cardiff, United Kingdom

Dr. H. John Smith
Welsh School of Pharmacy
Cardiff University
Cardiff, United Kingdom

Professor Holger Stark
Johann Wolfgang Goethe-universitaet
Institut für Pharmazeutische Chemie
Frankfurt am Main, Germany

Dr. Torsten Steinmetzer
Curacyte Chemistry Gmbh
Jena, Germany

Professor Philip G. Strange
School of Animal and
Microbial Sciences
University of Reading
Reading, United Kingdom

Professor David M. Taylor
School of Chemistry
Cardiff University
Cardiff, United Kingdom

Dr. Glyn Taylor
Welsh School of Pharmacy
Cardiff University
Cardiff, United Kingdom

Professor David E. Thurston
School of Pharmacy
University of London
London, United Kingdom

Professor David R. Williams
Chemistry Department
Cardiff University
Cardiff, United Kingdom

* Deceased

Abbreviations

τ	dosing interval
AADC	aromatic amino acid decarboxylase
ACE	angiotensin 1-converting enzyme
AD	Alzheimer's disease
ADME	Absorption, distribution, metabolism, and excretion
AG	aminoglutethimide
AGP	$\alpha_1{}^N$-acid glycoprotein
AIDS	Acquired immune deficiency syndrome
c-AMP	adenine 3′,5′-cyclic phosphate
AMP	adenosine monophosphate
ANP	atrial natriuretic peptide
2-APAs	2-arylpropionic acids
ATP	adenosine triphosphate
AUC	area under the plasma concentration curve
BA	bioavailability
BCRP	breast cancer resistance protein
Boc	*t*-butyloxycarbonyl
BP	British Pharmacopoeia
CL	clearance
CLogP	calculated log *P*
CME	1-cyano-1-methyl-ethyl group
CNS	central nervous system
CoA	coenzyme A
COMT	catechol-*O*-methyltransferase
COX	cycloxygenase
CSCC	cholesterol side chain cleavage enzyme
CSF	cerebrospinal fluid
CYP	cytochrome P450 enzyme
CYP 19	aromatase
DAG	diacylglycerol
DHEA	dehydroepiandrosterone
DHEAS	dehydroepiandrosterone sulphate
DHFR	dihydrofolate reductase
DICR	dosage interval concentration
D-Lac	D-alanine-D-lactate
DPI	dry powder inhaler

DTPA	diethylenetriaminepentaacetate
E	extraction ratio
EMs	extensive metabolisers
EMATE	oestrone-3-sulfamate
ER	oestrogen receptor
EU	European Union
FDA	Food and Drug Administration
F_{oral}	oral bioavailability
FRE	fibrinogen recognition site
FTIR	Fourier Transform Infra Red
F_u	unbound fraction in plasma
F_{ut}	unbound fraction in tissues
GABA	γ-aminobutyric acid
GABA-T	γ-aminobutyric acid transaminase
GC	gas chromatography
GDP	guanosine diphosphate
GI	gastrointestinal
GlcNAc	*N*-acetylglucosamine
Glu	L-glutamate
GPCRs	G-protein coupled receptors
GSH	glutathione
GTP	guanosine triphosphate
HA	H-bond acceptor
HCS	high content screening
HD	H-bond donor
HDL	high density lipoprotein
HF	hydrofluoric acid
HIV	human immunodeficiency virus
^{1}H NMR	proton nuclear magnetic resonance
HPLC	high performance liquid chromatography
HPMNL	human polymorphonuclear leukocytes
HSA	human serum albumin
17β-HSD	17β-hydroxysteroid dehydrogenase
5-HT	5-hydroxytryptamine
HTS	High Throughput Screening
IC_{50}	concentration required for 50% inhibition
IgG	immunoglobulin
IP3	1,4,5-inositol triphosphate
IPA	inhibitors of platelet aggregation
IPGS	inhibitor of prostaglandin synthesis
IV	intravenous
JAM	junction associated membrane protein
K_i	inhibition constant
LDL	low density lipoprotein
LFERs	linear free energy relationships
LH	luteinizing hormone
LH-RH	luteinizing hormone releasing hormone
LPs	lipopolysaccharide stimulation

MaLDI-TOF-MS	matrix associated laser desorption ionisation-time of flight-mass spectrometry
MDI	metered dose inhaler
MDR	multidrug resistance
MDT	mean disposition time
MEC	minimum effective plasma concentration
MEP	membrane metalloendopeptidase
MIC	minimum inhibitory concentration
MIT	mean input time
MLR	multiple linear regression
MMC	migrating myoelectric complex
MMP	matrix metalloproteinase
MR	molar refractivity
MRPs	multidrug resistant proteins
MRSA	methicillin-resistant *Staphylococcus aureus*
MRT	mean residence time
MS	mass spectrometry
MSC	maximum safe concentration
α-MSH	α-melanocyte stimulating hormone
MurNAc	*N*-acetylmuramic acid
NMDA	*N*-methyl D-aspartate
NMR	nuclear magnetic resonance
NO	nitric oxide
NSAIDs	non-steroidal anti-inflammatory drugs
N_{voc}	nitroveratryloxycarbonyl
ODC	ornithine decarboxylase
$P450_{AROM}$	aromatase
$P450_{scc}$	cholesterol side chain cleavage enzyme
PAGE	polyacrylamide gel
PC	principal components
PD	pharmacodynamics
PGE_2	prostaglandin E_2
P-gp	P-glycoprotein
PIP2	phosphatidylinositol biphosphate
PK	pharmacokinetics
PKC	protein kinase C
PLS	partial least squares
PMs	poor metabolisers
PMF	protonmotive force
PPIs	proton pump inhibitors
PVDF	polyvinylidene difluoride
Q	blood flow
QSAR	quantitative structure-activity relationships
QSPKR	quantitative structure-pharmacokinetic relationships
Ro	rate of drug input
R & D	research and development
m-RNA	messenger RNA
RT	reverse transcriptase
SAR	structure–activity relationships
S.C.	subcutaneous injection
SERMS	selective estrogen receptor modulators

SNPs	single nucleotide polymorphisms
SPS	solid phase synthesis
SR	sustained release
SSRIs	selective serotonin re-uptake inhibitors
$t_{1/2}$	half life
TCR	therapeutic concentration range
TFA	trifluoracetic acid
TJ	tight junction
TNF-α	tumour necrosis factor-alpha
TRH	thyrotropin releasing hormone
TXB_2	thromboxane B_2
UV	ultraviolet
UVE	ultraviolet-induced erythema
V_p	plasma volume
V_{TW}	aqueous volume outside plasma
ZO	zonula occluden

Contents

*Deceased.

1

Processes of Drug Handling by the Body

Mark Gumbleton

CONTENTS

1.1 INTRODUCTION

To be useful as a medicine, a drug must be capable of being delivered to its site of action achieving concentrations sufficient to initiate and maintain the appropriate pharmacological response. The concentration of drug at the site of action will depend upon the amount of drug administered, the rate and extent of its absorption and distribution in the body, and simply the rate at which the drug is eliminated from the body. The processes by which the body handles drugs and those that determine the temporal profile of drug concentrations in the body are categorized into: (i) *absorption*, the process of drug transport from the site of release of drug from the delivery system (e.g., tablet, ointment or depot injection) into the systemic blood circulation; (ii) *distribution*, the process of reversible transport of drug from the site of absorption to those tissues of the body into which the drug is able to distribute. Whether a drug will be able to distribute into a particular tissue is dependent upon the drug's physicochemical properties and the nature of the tissue barrier itself; (iii) *elimination*, the process of irreversible removal of drug from the body. Elimination of drug from the body will occur by metabolism, which is the chemical modification of drug, or by excretion, which is the physical removal of drug from the body, e.g., renal or biliary excretion. From the above is derived the acronym ADME, which stands for absorption, distribution, metabolism, and excretion, describing the qualitative processes by which the body handles drugs. The term disposition is used to describe collectively the processes of drug distribution and elimination, i.e., the processes exclusive of drug input, for example, an intravenous bolus dose of drug, but not an intravenous infusion where an input process exists. The processes of ADME will occur concurrently in that as soon as drug molecules begin to be absorbed and enter the blood circulation, they will also be subjected to the processes of distribution and elimination.

Qualitative considerations of drug handling by the body are most often considered synonymous with pharmacokinetics (PK), which quantitatively defines the temporal relationship between administered drug dose and drug concentration in the body (Figure 1.1) through pharmacokinetic parameters such as clearance, volume of distribution, etc.; the handling of drug by the body means that the relationship between drug dose and concentration varies with time (Figure 1.2). The

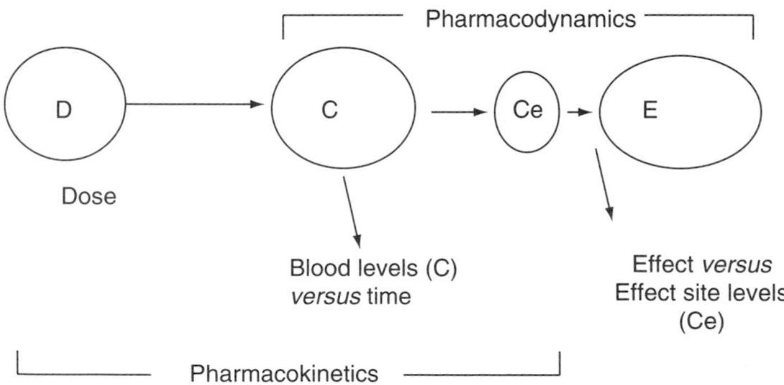

Figure 1.1 The inter-relationship between pharmacokinetics, a relationship between dose and concentration that varies with time, and pharmacodynamics, a relationship between concentration and effect.

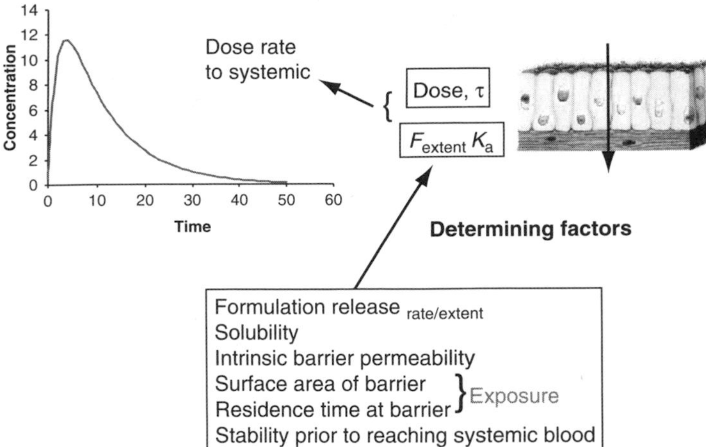

Figure 1.2 A range of factors influences the rate (K_a) and extent (F) of drug bioavailability following absorption across an epithelial barrier. Together with dose and the frequency of dosing (τ), the pharmacokinetic parameters K_a and F determine the dose rate of drug to the systemic blood.

relationship between drug concentration at the pharmacological receptor and pharmacological response is termed pharmacodynamics (Figure 1.1). However, measured drug concentrations are generally determined from peripheral venous plasma sampled from the saphenous vein in the arm. The definition of pharmacodynamics is therefore extended to include the relationship between plasma drug concentration and effect, with the assumption that once a drug has reached equilibrium within the body, the drug venous plasma concentration–time profile will parallel the tissue drug concentration–time profile. This does not mean that the absolute magnitude of drug concentrations in plasma and tissue will be the same, but merely that at equilibrium the relative kinetic profiles with respect to time are indistinguishable. In establishing a pharmacokinetic–pharmacodynamic relationship, it is important that the issue of the time required to achieve equilibrium is taken into account, as a pharmacodynamic relationship is independent of time. This simply means that a given concentration should give the same level of pharmacological response irrespective of whether the drug concentrations in the body are rising during the absorption phase or they are falling as absorption is nearing completion. The usefulness of pharmacokinetics is that for a given drug and patient profile, and with pharmacokinetic and pharmacodynamic parameters remaining constant independent of time and drug dose, such parameters will provide a quantitative framework for the prospective optimization of therapeutic dosage regimens. Knowledge of ADME provides a mechanistic platform to understand pharmacokinetic and pharmacodyamic relationships.

1.2 ABSORPTION

1.2.1 Mammalian Cell Membranes

A plasma membrane encloses every cell of the body, defining the cell's extent and maintaining the essential differences between the cell's interior and its environment. Consideration of the composition and structure of biological membranes is fundamental to understanding the relationship between a drug's physicochemical properties and its membrane transport, and as a corollary, the understanding of the processes of drug absorption, distribution, and elimination.

Eukaryote cell membranes have a common general structure as assemblies of lipid and protein molecules held together mainly by noncovalent forces. Lipid molecules are arranged as bilayers (~5 nm thick) with protein molecules that are surface bound or are traversing the lipid bilayer mediating a significant element of the membrane's functioning. The three major types of lipids in cell membranes are phospholipids, cholesterol, and glycolipids, with the ''fluidity'' (viscosity) of the lipid bilayer depending upon the nature and composition of its lipid components. A phospholipid molecule, e.g., phosphatidylcholine, possesses a polar head group, e.g., choline–phosphate–glycerol, and two hydrophobic fatty acid tails or chains. The shorter the hydrophobic tails and the higher the degree of unsaturated *cis*-double bonds in the fatty acid chain, the less is the tendency for hydrophobic chains of adjacent phospholipids to interact and pack together, and the more ''fluid'' is the membrane. The plasma membrane of most eukaryotic cells contains a variety of phospholipids, for example, those based on glycerol include phosphatidylcholine, phosphatidylserine (possesses a net negative charge at physiological pH), and phosphatidylethanolamine. Sphingomyelin is a membrane phospholipid based upon ceramide. Cholesterol is a major component of plasma membranes (15 to 20% of total lipid by weight) and a key determinant of membrane ''fluidity.'' Glycolipids are oligosaccharide-containing lipid molecules which are located exclusively in the outer leaflet of the membrane bilayer, i.e., the leaflet exposed to interstitial fluid that bathes all cells, with the polar sugar groups exposed at the surface. In animal cells, almost all glycolipids are based on ceramide (cf. sphingomyelin). These glycosphingolipids have a general structure comprising a polar head group and two hydrophobic fatty acid chains. Glycolipids are distinguished from one another by their polar head group, which consists of one or more sugar residues. The most widely distributed glycolipids in the plasma membranes of eukaryotes are the neutral glycolipids whose polar head groups consist of 1 to 15 or more uncharged sugar residues.

Most of the specific functions of biological membranes are carried out by proteins. The amounts and types of protein in a membrane are highly variable. Membrane proteins associate with a lipid bilayer in many different ways. Transmembrane proteins extend across the bilayer as a single α-helix or multiple α-helices. For example, the xenobiotic efflux transporter, P-glycoprotein (P-gp) consists of 12 transmembrane spanning domains and two adenosine 5′-triphosphate (ATP) binding domains. The homology of the latter leading to the classification of P-gp as member of the ATP-binding cassette (ABC) superfamily of transporter proteins, whose other members also include the family of efflux transporters constituted by the multidrug resistance proteins (MRPs) and the efflux transporter, breast cancer resistance protein (BCRP). The great majority of transmembrane proteins are glycosylated with the oligosaccharide chains exposed to the extracellular environment; P-gp is N-glycosylated, which probably has important implications for membrane targeting, insertion, and stability of this protein. Some integral membrane proteins are attached to the bilayer only by means of a fatty acid chain, while others are attached covalently via a specific oligosaccharide. Integral membrane proteins can be released only by disrupting the bilayer with detergents or organic solvents.

All eukaryotic cells have carbohydrate on their surface membranes both as oligosaccharide chains covalently bound to proteins (glycoproteins) and those that are covalently bound to lipids (glycolipids). The term glycocalyx is used to describe the carbohydrate-rich peripheral zone on the externally orientated surface of cell membranes. The glycocalyx is characterized by a net negative charge, which results from the presence of sialic acid or sulfate groups at the nonreducing termini of the glycosylated molecules. This ''blanket'' of negative charge serves to protect the underlying membranes and has clear implications with respect to xenobiotic–membrane interactions as well as cell–cell communication.

1.2.2 Overview of Epithelial Barriers and Tight Junctions

Epithelial barriers can be subclassified morphologically as *squamous* epithelium, e.g., alveolar epithelium; *simple columnar* epithelium, e.g., those lining the conducting airways of the lung or gastrointestinal tract; *stratified* epithelium, e.g., non-keratinized epithelium lining the buccal cavity (Figure 1.3).

Multiple type of epithelia

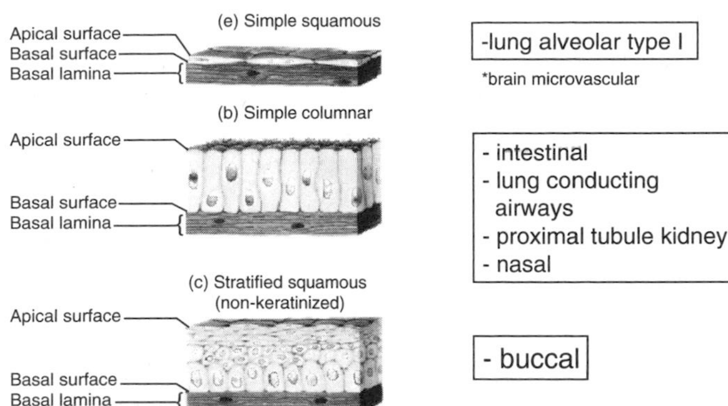

Figure 1.3 Different types of epithelia exist at the protective barriers of the body. The blood–brain barrier is constituted anatomically by the brain microvascular endothelium, which nevertheless is a highly restrictive barrier possessing a squamous morphology.

Epithelial and endothelial cells form cellular barriers separating compartments of different composition. In forming such barriers the cells need to form intercellular junctions. Tight junctions (TJ) (or zonula occludens) are the most apical of the intercellular junctions selectively restricting the intercellular diffusion of solutes on the basis of solute molecular size, shape, and charge, and in doing so minimizing the transfer of potentially harmful solutes while maximizing the functional significance of the cell's plasma membrane active transport systems. The TJ complexes also fulfil a role as a membrane ''fence'' restricting the intermixing of apical and basolateral membrane lipids and proteins. This confinement of specific proteins and lipids to specific membranes leads to the polarization of the cell, i.e., the cell possesses two distinctively different membrane surfaces and, by inference, different capabilities for interacting with drug molecules. For example P-gp is localized to the apical (i.e., luminal) membranes in kidney and small intestinal epithelial cells.

By freeze-fracture electron microscopy, the TJs appear as a set of continuous, anastomozing intramembraneous strands which contact similar strands on the adjacent cells and thus seal the intercellular space (Figure 1.4). The transmembrane protein occludin is one of the major constituents of the TJ strands or fibrils, and was the first candidate protein considered to fulfil the functional restrictive properties of the TJ fibril network. More recently other proteins have been identified, including the claudin protein family, the members of which are integral membrane proteins localizing to TJ strands and which bind homotypically to the TJ strands of adjacent cells. In addition, a junction-associated membrane protein (JAM), a member of the immunoglobulin superfamily, has been localized to TJ complexes although less is known about the functional role of this protein.

In addition, there are several accessory proteins localized to the cytoplasmic surface of TJs. The zonula occluden (ZO) proteins, ZO-1, ZO-2, and ZO-3 form heterodimers potentially serving as the major molecular scaffold for the TJ network. These ZO proteins form crosslinks between the TJ strands and actin filaments. Cingulin is a double-stranded myosin-like protein that associates with the cytoplasmic face of the TJ complex and apparently directly with ZO proteins. Cingulin may function in linking the TJ strands with the actomyosin cytoskeleton.

The Ca^{2+} ion has a key role to play in the maintenance of TJ paracellular restrictiveness. Ca^{2+} ions act primarily on the extracellular side of the cell interacting with the extracellular part of E-cadherin, a critical cell–cell adhesion molecule in the zonula adherens junction that lies underneath the zonula occludens. Extracellular Ca^{2+} activates E-cadherin, which is then able to aggregate with other E-cadherin molecules on the same cell, an arrangement that favors binding

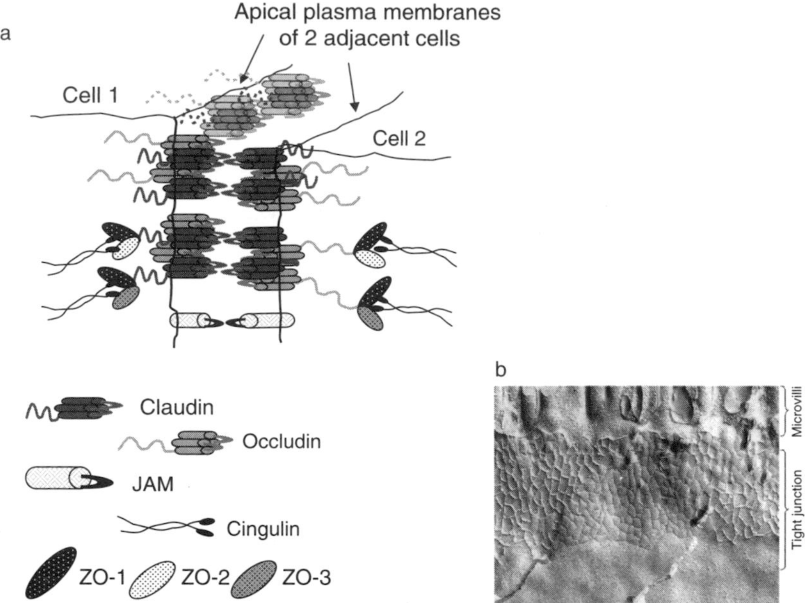

Figure 1.4 (a) Schematic of the protein–protein interactions involved in tight junctional formation between two adjacent cells. The tight junctional complex is termed the zonula occludens and involves intercellular homotypic interactions between the proteins occludins, claudins, and JAM. Cytoplasmically located proteins ZO-1, ZO-2, and ZO-3 that lie in close proximity to the ZO complex link the extracellular orientated proteins to the cell's cytoskeleton. (b) Freeze-fracture image to show the tortuous fibril nature of the tight junctional strands formed between two adjacent cells.

to E-cadherin of an adjacent cell. In the absence of Ca^{2+}, E-cadherins are inactive and cell–cell adhesion is lost leading to functional impairment of the TJ complexes; the divalent ion chelator, EDTA, is often used in laboratory studies of drug transport to chelate free Ca^{2+} and disrupt the restrictive properties of the TJs. Ca^{2+} ions also promote binding of E-cadherin with intracellular-located catenins, which in turn binds to vinculin, and actinin and indirectly to the cytoskeleton of actin. The cytoskeleton appears to fulfil a key role in delivering signals from the adherens junctions to the TJ; inhibitors of microfilaments and microtubules disrupt TJ junction formation.

1.2.3 Membrane Permeability and Drug Physicochemical Properties

During the process of drug absorption into the systemic circulation, the drug molecules traverse an epithelial barrier, e.g., the single layer of columnar epithelium lining the small intestine. If we focus our discussion on oral absorption then the rate at which drug molecules reach the systemic circulation will (assuming rate of drug release from the formulation is not rate-limiting) depend upon: (i) the rate of drug dosing which takes into account the administered dose (mass) and the dosing interval (t; time); (ii) the extent of oral bioavailability (F_{oral}; %), and (iii) the apparent oral absorption rate constant for the drug (K_a; time^{-1}) (Figure 1.2). The parameters F_{oral} and K_a will themselves depend upon the rate and extent of release of drug from the formulation, the solubility of drug in the intestinal lining fluid, the intestinal surface area available for absorption, and the residence time of the drug in solution at the absorption site. Also, the stability of drug during the absorption process and the intrinsic permeability of the intestinal barrier to the drug are critical factors in determining F_{oral} and K_a.

It is on the permeability of the absorption barrier that our discussion now focuses.

Firstly, if we consider only the transport of drug molecules across an epithelial barrier via passive diffusional processes then the overall flux (J) of a drug in one dimension, i.e., the net mass of drug that diffuses through a unit area per unit time, can be described by the following equation:

$$J = -\frac{DK_{\mathrm{p}}A}{x}\left(\frac{\mathrm{d}C}{\mathrm{d}x}\right)_t \tag{1.1}$$

where J is the flux of the drug, D is the diffusion coefficient of the drug across the cellular barrier, K_{p} is a global partition coefficient (cell membrane/aqueous fluid), A is the surface area of the barrier available for absorption, x is the thickness of the absorption barrier, and $(\mathrm{d}C/\mathrm{d}x)_t$ is the concentration gradient of drug across the absorption barrier. The negative sign in Equation (1.1) indicates that diffusion proceeds from high to low concentration and hence the flux is a positive quantity. The greater the concentration gradient, the greater the rate of diffusion of a drug across the cell membrane.

The apparent permeability coefficient (ρ) of an epithelial barrier to a given drug will approximate $D \cdot K_{\mathrm{p}}/x$. The processes of drug partitioning within the cell (including partitioning between extracellular fluid and plasma membrane, partitioning between plasma membrane and cell cytosol, and other organelle interactions, etc.) and of drug diffusion across the cell (including a range of organelle and macromolecule interactions that will influence the diffusion process) will themselves depend upon the molecular properties of the drug. These can be categorized as *steric* properties (i.e., molecular size, shape, volume), *ionic* properties (i.e., hydrogen bonding potential, pK_{a}), and *hydrophobic* properties. These molecular properties will determine if passive diffusional transport across an epithelial barrier will involve either a predominantly paracellular (between cells) pathway negotiating a tortuous intercellular route via the aqueous channels formed by the anastomozing TJ fibrils between adjacent cells, or predominantly (but not exclusively) transcellular (across the cell) pathway requiring partitioning of drug into the plasma membrane bilayer.

The drug's partitioning between the lipid cell membrane and the aqueous extracellular fluid is a major factor. Most drugs are weak acids or weak bases, existing in aqueous solution as an equilibrium mixture of nonionized and ionized species. The nonionized species if sufficiently hydrophobic in nature will readily partition into cell membranes. In contrast, ionized compounds partition poorly and as a result will only be slowly transported across biological membranes. The ratio of nonionized to ionized drug when in aqueous solution depends upon the pK_{a} of the drug and the pH of the environment, and can be calculated from the Henderson–Hasselbach equation:

$$\mathrm{pH} = \mathrm{p}K_{\mathrm{a}} + \log\frac{\text{conjugate base}}{\text{conjugate acid}} \tag{1.2}$$

where pK_{a} is the dissociation constant and a conjugate acid refers to the H^+ ion donor, and a conjugate base refers to the H^+ ion acceptor. A simple *aide memoire* for determining the extent of ionization of a weak acid is that when the pH equals the pK_{a} then the molecule will be 50% ionized. At 1 pH unit above the pK_{a} (i.e., more alkaline conditions), a weak acid is 90% ionized and at 2 pH units above the pK_{a} it is 99% ionized. Similarly, at 1 pH unit below the pK_{a} (i.e., more acid conditions), a weak acid is 90% nonionized and at 2 pH units below the pK_{a} it is 99% nonionized. The same principle would apply for weak bases except that the nonionized form predominates as the pH is increased above the pK_{a} of the weak base (i.e., in more alkaline conditions), and the ionized form predominates as the pH is reduced below the pK_{a} (i.e., in more acidic conditions).

Plainly put, if a drug's molecular properties afford partitioning into cellular membranes (i.e., nonionized form of the drug predominates and is of a sufficient hydrophobic nature), then the membrane surface area available for transcellular diffusion will be considerably greater by many orders of magnitude than the surface area available for diffusion via the paracellular route. As a corollary, transcellular diffusion will potentially lead to a higher epithelial permeability and a higher rate and extent of absorption. However, even for drugs displaying unfavorable membrane partitioning properties their steric properties may allow for a significant extent of absorption via the paracellular route, e.g., atenolol with a log D of -1.9 displays an extent of oral bioavailability of 50% in humans.

The linkage between molecular properties and membrane permeability underlies the significant effort now devoted in drug discovery programs to the physicochemical and computational assessment of the molecular properties of drug candidates with the aim of predicting which ones are likely to display poor *in vivo* absorption characteristics. These physicochemical and computational assessments are aimed at addressing, among others, solubility, permeability, pK_a, and hydrophobicity issues through a range of profiling strategies. In 1995 Christopher Lipinkski from Pfizer Inc. presented a molecular property based postsynthetic alert derived from the analysis of 2245 USAN (United States Adopted Name) or INN (International Nonproprietary Name) named compounds that had undergone some form of clinical exposure. This alert was seen as a guide for medicinal chemists in aiding the prediction of which drug candidates may display oral absorption problems. The recommendations of Lipinski and colleagues were subsequently published (Lipinski et al. 1997) as a set of simple rules to be used as an indicator that poor oral absorption is more likely when a compound possesses:

- More than five H-bond donors
- More than ten H-bond acceptors — the presence of hydrogen-bond acceptors and donors increases the desolvation energy necessary to partition from an aqueous environment into the hydrophobic environment of the inner cell membrane
- $C \log P$ (calculated $\log P$) greater than five (or $M \log P$ greater than 4.15) — high $C \log P$ is generally associated with poor aqueous solubility
- Molecular mass over 500 Da — increasing molecular mass will hinder not only the passage through the paracellular pathway of hydrophilic molecules but also the rate of transcellular diffusion.

These rules should not be considered as absolute predictors for poor absorption. It is merely that when a compound displays the above mentioned properties that it is more likely to display absorption problems. They serve as a useful guide to be used in conjunction with *in vitro* physicochemical and computational assessments of a compound's molecular properties, and the correlation of these properties to *a priori* determined absorption parameters.

1.2.4 Gastrointestinal Absorption

The most common route of drug administration is oral. It provides a convenient, relatively safe, and economical method of dosing that in general meets the needs of both patient and the pharmaceutical industry. The epithelium of the stomach is principally concerned with secretion, although water, ethanol, and other nonionized solutes of low molecular weight may show appreciable absorption from this site. The pH of gastric juice is normally 1 to 2 on a fasting stomach, increasing up to 4 following the ingestion of food. Clearly the pH of the stomach will impact upon drug stability; for example, benzylpenicillin is degraded by acid pH and is required to be administered by injection, whereas the synthetic analogue ampicillin is acid stable. The use of enteric-coated capsules or tablets triggered in a pH-sensitive manner to disintegrate and dissolve in the more alkaline environment of the small intestine is a way in which acid labile drugs may be protected from degradation in the stomach. Nevertheless, whether the drug is formulated as a tablet, capsule, or liquid preparation, the most important site for drug absorption is the small intestine because it offers a far greater epithelial surface area for drug absorption than other parts of the gastrointestinal tract.

The small intestine has a luminal diameter of 2.5 cm and a length of about 700 cm, and comprises the distinct zones of duodenum, jejunum, and ileum. The luminal foldings of the small intestinal mucosa and submucosa, the villi bearing enterocytes at their apical surface, which are the intestinal epithelial absorptive cells, and the microvilli present on the surface of each enterocyte all increase the potentially available drug absorption surface area across the small intestine to approximately 200 m^2. The first part of the small intestine, the duodenum, functions to neutralize the gastric acid and initiate further digestive processes. Pancreatic secretions and bile duct secretions enter at the level of the duodenum. Beyond the duodenum is the jejunum and ileum, where the majority of food absorption occurs. The pH of the small intestine increases from about 5.5 in the duodenum to about 6 to 7 in the jejunum and ileum.

The rate and extent of drug absorption from the small intestine are related to the release of the active ingredient from a dosage form, its solubility in the liquid phase of gastrointestinal contents, and the transport of the dissolved compound or the intact dosage form from the stomach into the duodenum. Further, in dynamic systems such as dosing to the gastrointestinal tract, the rate and extent of absorption cannot be dissociated from the residence time of the drug in solution at an absorption surface. As such, gastric emptying and intestinal motility can be critical determinants for the absorption of drugs.

The stomach has its own motility pattern dependent on the presence of foodstuffs. In the fasting state, the motility pattern of the stomach has four different phases defined by the interdigestive myoelectric cycle, or migrating myoelectric complex (MMC), which are bursts of smooth muscle contraction that move from the stomach toward the ileocecal valve at regular frequency during the interdigestive period. Each phase lasts for a different period of time and possesses different contraction strength. Phase I lasts for 40 to 60 min with rare contractions. Phase II lasts for a similar period of time with increasing contraction strength and frequency. Phase III lasts for 4 to 6 min with the highest contraction strength, which is necessary to empty large indigestible particles (e.g., enteric-coated tablet) from the stomach. Phase IV is a transition period between phase III and phase I, and lasts typically for 15 to 30 min. The whole cycle is repeated approximately every 2 h until a meal is ingested and the stomach contractures change to those of the ''fed state.''

Gastric emptying of liquids in the fasting state appears to be a function of volume, with administration of small volumes (approximately <150 mL) not changing the gastric emptying pattern. Most of the liquid in the stomach is emptied as a bolus upon arrival of phase II activity and completely by the start of phase III. This may take 40 to 50 min from ingestion if administration occurs at the onset of phase I. Administration of large fluid volumes converts the stomach to a ''fed state'' bringing about continuous and immediate emptying which follows first-order kinetics. The time required for discharge of 50% of a large volume of liquid from the stomach is about 12 to 15 min.

The motility pattern of the stomach during the ''fed state'' is completely different from that in the fasted state. There are no distinctive phases and the contraction is only moderate in strength. There are two types of contraction: peristaltic and systolic. These two contraction patterns are coordinated to achieve the diminution of the size of the stomach by mixing and grinding. The systolic contraction is mainly the result of circular muscle contraction, whereas the peristaltic contraction is mainly the result of longitudinal contraction. Greater volume of food, its solid nature, high fat, carbohydrate, and protein content decrease gastric emptying.

Intestinal motility is a continuation of the gastric motility, and displays motility patterns similar to that of the stomach. The main difference is that while the whole stomach is always in the same phase, in the intestine different segments are likely to be in different phases at any given time. Gastric contraction in the fasted state also tends to be stronger than that of small intestinal contraction in the fasted state where each motility cycle takes about 140 to 150 min. In the ''fed state'' the contractions serve to mix food and promote contact between food ingredients and the absorption surface. An increase in the solid food content of a meal tends to increase the number of contractions. Generally, most of the digestive processes and absorption of beneficial nutrients occur within the first 12 h of consumption of food. Total intestinal transit times are highly variable but can be as long as 48 h.

The effect of food upon the absorption of oral medications is an issue that needs comment. The most obvious is that delayed gastric emptying following ingestion of a meal will also delay the intestinal delivery of any co-administered tablet or capsule. Although when the pharmaceutical preparation releases the active compound within the stomach, or for example, an enteric formulation comprising micropellets, the gastric emptying of the drug may not be significantly delayed. In the extreme, a delay in gastric emptying will directly impact the onset of drug absorption from solid dose enteric-coated tablet or capsule formulations designed to disintegrate only in the more alkaline conditions of the intestine. If a solid enteric-coated formulation is taken as part of a multiple dose regimen then it is certainly possible that the first dose would still exist in the stomach as an intact tablet or capsule when a second dose is taken. Certain drugs, e.g., nonsteroidal antiinflammatories, irritate the gastric epithelium and are, therefore, advised to be given with food. Griseofulvin is widely recognized to have dissolution problems and its extent of absorption is increased when given with a meal, particularly one of a high fat content. The meal not only delays gastric emptying enabling a longer time for griseofulvin dissolution, but also the high fat increases bile acid production which promotes the dissolution of poorly water-soluble drugs. In contrast, certain drugs may interact with food, e.g., beta-lactam antibiotics, resulting in decreased extents of absorption.

1.2.5 Absorption from the Oral Cavity

Absorption from the oral cavity includes: (i) sublingual drug delivery across the epithelial lining of the ventral surface of the tongue and of a region of the mouth itself underlying the tongue and (ii) buccal drug delivery across the epithelial lining of the cheeks. Both sublingual and buccal regions of the oral cavity pertinent to drug delivery comprise stratified squamous non-keratinized epithelium, with a thickness of the sublingual mucosal barrier ~200 μm and a thickness of the buccal mucosal barrier 500 to 800 μm. Both the regions receive a relatively good blood supply, with molecules absorbed from these sites not subjected to first-pass metabolism of the gut mucosa and liver. The buccal route of delivery affords the opportunity for application of adhesive drug delivery systems, able to release drug over a prolonged period of time; this is not applicable to the sublingual route where system retention and flow of saliva make such systems less amenable. In contrast, the sublingual route can provide for extremely rapid absorption, for example the sublingual absorption of glyceryl trinitrate for immediate relief from anginal pain.

1.2.6 Pulmonary Absorption

While oral dosing via the gastrointestinal tract remains the optimal route for patient compliance and for a majority of drug developers, inhalational drug delivery possesses distinct advantages in certain circumstances. Firstly, for the treatment of local lung disease such as asthma, the local respiratory aerosol administration of bronchodilators or of corticosteroids can be undertaken at much lower doses than those used for the equivalent oral dosing of these agents. For example, the oral tablet dose for the β_2-adrenoreceptor agonist salbutamol is 2 to 4 mg three to four times daily while the inhalational dose is 100 to 200 μg three to four times daily. In this way systemic side effects are negated, or at least significantly minimized, and the drug is more efficiently delivered to its target tissue. A second reason for the increasing interest in pulmonary drug delivery is that proteins and peptides appear to show good extents and rates of bioavailability across the epithelium of the deep lung compared to other barriers such as in the gastrointestinal tract or across the skin. For example, the absolute extent of absorption for insulin via inhalation drug delivery approximates 10% or the relative bioavailability compared to s.c. injection, which approximates 20 to 25%. The predominant mechanism of absorption for proteins from lung airspace to blood is still unclear, although a range of processes from passive paracellular diffusion to active vesicular trafficking have been considered.

The respiratory system itself comprises two distinct regions, the upper and lower respiratory tracts. The upper respiratory tract consists of the nasal and paranasal passages, and the pharynx (collectively termed the nasopharyngeal region). The nasopharyngeal region serves to warm and humidify inhaled air and prevents the passage of particulates and microorganisms to the membranes of the lower respiratory tract to a large extent.

The trachea is the first conducting airway and connects the nasopharyngeal region to the lower respiratory tract lung tissue. The main function of the conducting airways is to carry inspired air through to the gaseous exchange region of the alveoli. The conducting airways are a series of bifurcations, the trachea bifurcates into the bronchi which subsequently bifurcate into bronchioles, etc. After about 12 generations of airway bifurcation, cartilage disappears from the airway walls and the airways (now termed bronchioles) rely on lung volume to maintain airway calibre. Bronchodilators by acting upon the submucosal smooth muscle are able to influence such airway diameters. The epithelium of the proximal bronchi is lined with pseudostratified ciliated columnar epithelium, with the epithelial cells (cell depth ~50 μm) interspersed with secretory and basal cells. With increasing distance inside the conducting airways, the epithelium becomes progressively cuboidal (cell depth ~10 μm), and the ciliated cells become more sparse in number. Approximately 40% of the epithelial cells lining the conducting airways of mammals are ciliated. These cilia beat in one direction only, propelling the airway surface liquid upward toward the pharyngeal region. The airway surface liquid comprises an aqueous sol phase, which lies proximal to the surface epithelium surrounding the cilia, and a higher viscosity gel phase, which sits on top of the cilia. The principal macromolecular structural component of the gel phase is mucin glycoproteins, secreted from goblet cells of the surface epithelium and serous cells of the submucosal tissue. It is the cilia-driven upward movement of this gel toward the pharyngeal region which constitutes the mucociliary clearance mechanism of the lung, the function of which is to clear deposited particulate matter trapped within the mucus barrier.

After about 16 generations of airway bifurcation the conducting airway to respiratory transitional zone begins, which eventually gives rise to the alveolar sacs. Each alveolus (diameter 250 μm) is lined by alveolar epithelium and supported by a thin basement membrane which interfaces with numerous blood capillaries in such an arrangement that presents an extensive air to blood network with separation by only a minimal tissue barrier and therefore allowing for optimal diffusion of gases. The alveolar epithelium is comprised of up to 93% of its surface area by the squamous alveolar epithelial type I cells. These cells are extremely thin having an average cell thickness of 0.35 μm ranging from 2 to 3 μm in the perinuclear region of the cell to approximately 0.2 μm in the peripheral attenuated regions of the cell. Using electron microscopy techniques, the total alveolar epithelial surface area within an average adult human lung has been estimated to be as large as 140 m^2; this contrasts with an approximate 2 to 4 m^2 surface area for the tracheobronchiole conducting airways. The alveolar epithelial surface is covered with a surface film of surfactant that lowers the surface tension in lungs and is essential if the alveolar sacs are to expand during inspiration. The volume of this alveolar film has been calculated to be 7 to 20 mL per 100 m^2 alveolar surface area.

Macrophages are also present in the alveolus and are able to migrate across the alveolar epithelium; they are found at all the lung epithelial surfaces. Since there are no cilia in the alveoli, the macrophages are the first line of defence and engulf exogenous and endogenous particles that have escaped from the mucus trap and the ciliary escalator of the upper respiratory tract.

Some of the above morphometric features of the alveolar epithelial–pulmonary capillary barrier (e.g., large surface area, thin cellular barrier, absorptive surface beyond the mucociliary escalator, high tissue blood perfusion) exemplify the ''*favorable*'' anatomical determinants to be considered in delivering proteins and peptides to the systemic circulation and have driven the challenge to improve the delivery of therapeutic aerosols to the lung periphery, and in particular the alveolar epithelium.

For inhalational drug delivery a therapeutic aerosol must be generated. There are three principal categories of aerosol generator: (i) pressurized metered-dose inhalers (pMDIs) which convert a drug solution or suspension in volatile propellant into an aerosol; (ii) dry powder inhalers (DPI) which permit the drug to be delivered to the airways as a dry powder aerosol; (iii) nebulizers which are devices that convert aqueous drug solutions or micronized suspensions into aerosols by either high velocity airstream dispersion (air-jet nebulizers) or by ultrasonic energy dispersion (ultrasonic nebulizers).

Aerosol particles may vary in size, shape, and density, which impact upon the aerodynamic behavior of the aerosol particle within the respiratory airstream. The preferred method of describing an aerosol particle is through the term "aerodynamic equivalent diameter" (d_{ae}), which describes the diameter of a perfect sphere of unit density (1 g/cm^3) that has the same settling velocity in still air as the particle that is being described. Mass median aerodynamic diameter (MMAD) is used to describe the characteristics of an aerosol population and is the hypothetical aerodynamic diameter below which 50% of the aerosol mass is contained.

The therapeutic benefit from a medical aerosol-containing drug depends upon the aerosol particles depositing on the lung epithelial surface at the appropriate part of the respiratory tract. The aerodynamic behavior of aerosol particles within the respiratory tract is described by the distinct processes of inertial impaction, sedimentation, and diffusion. Significant device and aerosol technology has been committed to obtaining optimal deposition patterns for therapeutic aerosols.

1.2.7 Nasal Absorption

Drugs may be delivered to the nasal cavity for local action, such as nasal decongestants, or for the purposes of achieving systemic drug absorption. Examples of prescription available drugs delivered nasally for the latter objective include not only low molecular weight organic molecules but also peptides such as buserelin, desmopressin, oxytocin, and calcitonin. The physicochemical properties of a molecule which affect its absorption across the nasal epithelium are broadly the same as those affecting transepithelial absorption at any site, although the rate of degradative drug metabolism within the nasal cavity and mucosa is certainly less than that seen in the gastrointestinal tract; absorption via the nose avoids hepatic first-pass metabolism.

The nasal cavity itself extends from the nostrils to the nasopharynx and is divided laterally by the nasal septum. The total surface area of the nasal cavity is about 150 cm^2. The nostrils lead into the nasal vestibule, the most proximal zone of the nasal cavity, which is lined by squamous epithelium. From the vesitibule, the cavity extends toward the turbinates (inferior, middle, superior), which constitute the main part of the nasal passage. The turbinates are delicate spiral bones found in the nasal passages covered with highly vascularized pseudostratified ciliated columnar secretory epithelium. Microvilli are found on the columnar cells, which increase the surface area available for absorption of drugs. The main function of the turbinates is in humidification and dehumidification of inspired or expired air. The olfactory region of the nose is located at the roof of the nasal cavity close to the superior turbinate and is lined with nonciliated neuro-epithelium.

The ciliated epithelium overlying the turbinates takes part in the mucocillary clearance of particulate matter, such as dust and microorganisms, that deposit upon the mucus blanket lining this region of the nasal cavity. The mucus with entrapped particulates is propelled by the cilia towards the nasopharynx to be either swallowed or expectorated. The clearance of the bulk of the mucus from the nose to the nasopharynx occurs over 10 to 20 min.

Drugs are administered to the nasal cavity as aqueous drops or aerosol sprays. Nasal drops disperse a drug solution throughout the length of the nasal cavity from the vestibule to the nasopharynx, offering a relatively large area for immediate absorption. Nasal sprays produce an aerosol of drug containing droplets analogous to the aerosols produced in pulmonary drug delivery, although the device technology is of a much simpler design. Aerosol droplets produced by nasal sprays tend to deposit drug at the front of the nasal cavity with little quantity reaching the turbinates. Upon absorption, the significance of the above deposition patterns is that nasal drops while instantly spreading throughout the full nasal cavity will also be subjected to the immediate effects of mucocilliary clearance, thus limiting the residence time the drug remains in contact with the absorption surface. In contrast, aerosols, while depositing initially in the nonciliated regions of the anterior nasal passage, will have a much longer residence time at the absorption surface, although with time an increasing proportion of the drug aerosol will be subject to clearance by mucociliary transport.

A range of strategies have been explored to increase the absorption of drugs from the nasal cavity including the use of penetration enhancers to alter nasal epithelium permeability, and the use of bioadhesives to improve contact, and prolong residence time, with the nasal mucosa. There is increasing evidence that drugs administered to the roof of the nasal cavity proximal to the olfactory region may gain direct access to the central nervous system, without having to cross the blood–brain barrier.

1.2.8 Transdermal Absorption

The skin is an extremely efficient barrier that minimizes water loss from the body. As such drug absorption across the skin is potentially more limiting and the relationship between a drug's physicochemical properties and epidermal permeability are less ambiguous than with mucosal barriers. The body's skin surface area is approximately 1.5 m^2 and represents a readily accessible surface for application of drug delivery systems. The outermost region of the skin is the epidermis (100 to 250 μm thickness) which is a stratified, squamous, keratinizing epithelium with keratinocytes constituting the major cellular component (>90% of the cells). Underlying the epidermis is the dermis, which comprises primarily connective tissue and provides support to the epidermis. It contains blood and lymphatic vessels, and nerve endings. The dermis also bears the skin's appendageal structures, specifically the hair follicles and sweat glands. The epidermis is avascular and for drugs to gain access to the capillary network, they must traverse the full thickness of the epidermis to reach the underlying vascularized dermis. Drugs absorbed across the skin avoid hepatic first-pass metabolism.

The epidermis is divided histologically into five distinct layers corresponding to the sequential nature of keratinocyte cell differentiation from the basal layer, *stratum basale*, which bears keratinocyte stem cells and is the site for proliferation of new keratinocytes, to the outermost layer, *stratum corneum*, bearing terminally differentiated keratinocytes; keratinocyte differentiation and migration from *stratum basale* to *stratum corneum* is a continuous process taking 20 to 30 days in duration. However, it is the *stratum corneum* comprising approximately 20 cell layers in depth that provides the principal barrier to skin. An often used analogy for the *stratum corneum* is that of a ''brick wall'' with the fully differentiated *stratum corneum* keratinocytes, or corneocytes as they are alternatively known, comprising the ''bricks,'' embedded in a ''mortar'' constituted by intercellular lipids which include ceramides, cholesterol, and free fatty acids. An obvious distinction should now be apparent in that the intercellular (paracellular) pathway in mucosal barriers is aqueous in nature, while the intercellular pathway in the *stratum corneum* barrier is lipid in nature. It is the convoluted lipid intercellular pathway that is considered the primary route for drug permeability across the skin barrier.

From the discussion above, it should be clear that lipophilicity is a key physicochemical drug property for *stratum corneum* permeability with optimum log[octanol – water] partition coefficients for transport in the range of 1 to 3. For very lipophilic compounds, e.g., log[octanol – water] partition coefficients >4, however, then the rate-limiting step in absorption may indeed be the partitioning of the drug from the *stratum corneum* into the underlying more aqueous viable epidermis. The potential with this kind of molecule is for significant lag times in absorption and for the *stratum corneum* serving as a reservoir for drug even after removal of the delivery system. Partitioning of charged or very polar molecules into the *stratum corneum* is essentially nonexistent without the use of some form of penetration enhancement.

Chemical penetration enhancers will work by either: (i) facilitating the partitioning of drug from the vehicle into the epidermis; (ii) reducing the diffusional barrier of the *stratum corneum*

an apparent and not an absolute volume is that the parameter also incorporates the balance in a drug's binding characteristics between the plasma and tissue compartments. For example, some drugs have a volume of distribution that is in excess of total body water (42 L); for example nortriptyline has an apparent volume of distribution of 22 to 27 L/kg. Other drugs display a volume of distribution that is surprisingly smaller than that predicted on the basis of their membrane partitioning potential. This is best seen from the following equation:

$$V = V_{\mathrm{p}} + V_{\mathrm{TW}} * F_{\mathrm{U}}/F_{\mathrm{UT}} \tag{1.3}$$

where V represents the apparent volume of distribution, V_{p} is the plasma volume (~3 L), V_{TW} is the aqueous volume outside plasma (~29 L), F_{U} is the fraction unbound in plasma, and F_{UT} is the fraction unbound in tissues. When a higher proportion of a given drug in the tissues is in the bound state (i.e., lower F_{UT}), for example binding to membrane or soluble proteins, or DNA, or even sequestration within coalesced globules of triglyceride, the higher is its volume of distribution. Conversely, increased binding in the plasma (i.e., lower F_{U}) will lead to a decrease in its apparent volume of distribution. The clinical impact of drug displacement from plasma protein binding depends firstly upon the extent of the fraction of drug in plasma that is bound (greater than 90%) and secondly, that it is not widely distributed throughout the body, e.g. warfarin. It is important to view the degree of binding to plasma proteins in relation to a drug's apparent volume of distribution. Although a drug may display high plasma protein binding, its volume of distribution may be large so that displacement by another drug will be clinically insignificant.

In addition to the extent of a drug's distribution in the body, consideration should also be given to the rate at which a drug achieves equilibrium between the plasma and various tissues of the body. The entry rate of a drug into a tissue depends on the rate of blood flow to the tissue, on tissue mass, and on partition characteristics of the drug between the blood and the tissue. For a given drug, distribution equilibrium (when entry and exit rates are the same) between blood and tissue is reached more rapidly in richly perfused tissues (e.g., lungs, kidney, thyroid, and adrenal glands, and liver) than in poorly perfused areas (e.g., bone, cool skin, inactive muscle, and fat) unless diffusion across membrane barriers is the rate-limiting step. For any given tissue, a higher tissue to plasma drug partitioning ratio at equilibrium implies that for a given blood perfusion rate the equilibrium will take longer to achieve. After equilibrium is attained the temporal profiles (not absolute concentrations) of drug behavior in tissues and in extracellular fluids are reflected by the plasma concentration.

1.4 ELIMINATION

Drugs are eliminated from the body either by chemical modification (i.e., metabolism) to form metabolites, or by the process of excretion from the body, i.e., drug is physically removed from the body without undergoing chemical modification. Through metabolism the physicochemical properties of a drug are modified to better enable excretion of the resulting metabolites.

1.4.1 Phase I Metabolism

Drug metabolism involves a wide range of chemical reactions, including oxidation, reduction, hydrolysis, hydration, conjugation, condensation, and isomerization. The enzymes involved are present in many tissues but generally are more concentrated in the liver. Drug metabolism is generally defined as occurring in two apparent phases. Phase I reactions involve the formation of a new or modified functional group or a cleavage (oxidation, reduction, hydrolysis). Phase II reactions involve conjugation with an endogenous compound (e.g., glucuronic acid, sulfate, glutathione). Metabolites formed in the phase II reactions tend to be more polar and more readily excreted by the kidneys (in urine) and the liver (in bile) than those formed in the phase I reactions. Drugs undergo either phase I or phase II reactions or indeed be subject to both of these phases; therefore the phase numbers reflect functional rather than sequential classification.

The most important enzyme system of phase I metabolism is cytochrome P-450, an integral membrane protein primarily located within the cell's endoplasmic reticulum (although cytochrome P405[rg5] can also be found within the mitochondrial membrane), and comprising a superfamily of isoenzymes that transfer electrons to catalyze the oxidation of many drugs. The cytochrome P-450 system catalyzes a variety of reactions that may be grouped into the categories of carbon hydroxylation, heteroatom oxygenation, heteroatom release, dehydrogenation, epoxidation, and oxidative group transfer. Table 1.1 shows the nature and range of such changes. Cytochrome P-450s are also known to catalyze some reductive reactions dependent upon the tissue oxygen tension. A typical cytochrome P-450 catalyzed reaction is

$$\mathrm{NADPH} + \mathrm{H}^{+} + \mathrm{O}_2 + \mathrm{RH} \Rightarrow \mathrm{NADP}^{+} + \mathrm{H}_2\mathrm{O} + \mathrm{R{-}OH}$$

The electrons are supplied by NADPH-cytochrome P-450 reductase, a flavoprotein that transfers electrons from NADPH (the reduced form of nicotinamide-adenine dinucleotide phosphate) to cytochrome P-450. The name cytochrome P-450 derives from the identification of this enzyme within a microsomal suspension through difference spectroscopy at 450 nm following first its reduction and then its exposure to carbon monoxide; the P stands for ''pigment.''

Phase I oxidations can also be undertaken by the flavin monoxygenase system (Cashman 2003) present also within the membranes of the endoplasmic reticulum. The human flavin-containing monooxygenases (FMO) catalyze the oxygenation of nucleophilic heteroatom-containing drugs to more polar materials that are more efficiently excreted in the urine. Evidence for six forms of the FMO gene exist although FMO form 3 (FMO3) is the prominent form in adult human liver and is likely associated with the bulk of FMO-mediated metabolism. Human FMO3 N-oxygenates primary, secondary, and tertiary amines. Human FMO1 is only highly efficient at N-oxygenating tertiary amines. Both human FMO1 and FMO3 S-oxygenate a number of nucleophilic sulfur-containing substrates, such as cimetidine. FMO2 is the major FMO in the lung. Other oxidations by, for example cytosolic alcohol and aldehyde dehyrogenases, amine oxidases, xanthine oxidases, or reductions, for example azo cleavage, or hydrolysis of ester and amide linkages also contribute to the phase I drug metabolism reactions.

Cytochrome P-450 gene family

Cytochrome P-450 enzymes are grouped into a large number of gene families, subfamilies, and isoforms. Cytochrome P-450 enzymes are designated by a symbol CYP, followed by: (i) a number for the particular P450 family to which the enzyme belongs; (ii) a letter for the subfamily; and (iii) another number for the specific gene or isoform itself, e.g., CYP3A4 relates to gene or isoform 4 belonging to family 3, subfamily A. In humans there are 17 cytochrome P-450 families providing up to 50 different cytochrome P-450 genes or isoforms. A CYP *family* contains genes that have at least a 40% sequence homology with each other; members of a subfamily must display at least 55% identity. Enzymes within the 1A, 2B, 2C, 2D, and 3A subfamilies appear the most important in mammalian metabolism with CYP1A2, CYP2C9, CYP2C19, CYP2D6, CYP2E1, and CYP3A4 quantitatively (in terms of drug substrates) the most important in human drug metabolism. Many drug interactions are a result of inhibition or induction of CYP450 enzymes. Enzyme inhibition usually involves competition with another drug for the enzyme-binding site. An interaction will begin within the first dosing phase of the inhibitor with the onset and reversal of inhibition correlating with the half-lives of the drugs involved. Enzyme induction occurs when a drug stimulates the synthesis of more CYP450 protein, enhancing the enzyme's metabolizing capacity.

Table 1.1 Some microsomal oxidations

Aromatic oxidation

$$\text{C}_6\text{H}_5\text{-R} \longrightarrow \text{R-C}_6\text{H}_4\text{-OH}$$

(phenobarbitone → 5-ethyl-5(4-hydroxyphenyl)barbituric acid)

Aliphatic (side-chain) oxidation

$$\text{R-CH}_3 \rightarrow \text{RCH}_2\text{OH}$$

(pentobarbitone → 5-ethyl-5(3-hydroxy-1-methylbutyl)barbituric acid)

Epoxidtion

$$\text{R–CH=CH}_2 \longrightarrow \text{R–CH–CH}_2 \text{ (epoxide, bridged by O)}$$

(carbamazepine → carbamazepine-10,11-epoxide)

Dealkylation

$$\text{R-X-CH}_3 \rightarrow \text{R-X-CH}_2\text{OH} \rightarrow \text{R-XH} + \text{HCHO}$$

(X = NH or NCH_3, imipramine → desmethylimipramine)
(X = O, phenacetin → paracetamol)
(X = S, 6-methylthiopurine → 6-mercaptopurine)

Oxidative deamination

$$\text{R–CH(NH}_2\text{)–CH}_3 \longrightarrow \left[\text{R–C(OH)(NH}_2\text{)-CH}_3\right] \longrightarrow \text{R–COCH}_3 + \text{NH}_3$$

(amphetamine → phenylacetone)

N-hydroxylation

$$\text{C}_6\text{H}_5\text{-NH}_2 \longrightarrow \text{C}_6\text{H}_5\text{-NHOH}$$

(aminoglutethimide → *N*-hydroxyaminoglutethimide)

N-oxidation

$$\text{R}_3\text{N} \rightarrow \text{R}_3\text{N} \rightarrow \text{O}$$

(trimethylamine trimethylamine *N*-oxide)

Sulfoxidation

$$\text{R}_2\text{S} \rightarrow \text{R}_2\text{S} \rightarrow \text{O}$$

(chlorpromazine → chlorpromazine sulfoxide)

Desulfurization

$$\text{R}_2\text{C}{=}\text{S} \rightarrow \text{R}_2\text{C}{=}\text{O}$$

(thiopentone → pentobarbitone)

The clinical impact of induction is likely to be more prolonged as it requires increased synthesis of a CYP450 enzyme generally (but not always) following multiple administrations of the inducer. Table 1.2 provides a list of the substrates, inducers, and inhibitors for the above mentioned CYP450

Table 1.2 Table of substrates, inducers and inhibitors for the most pharmaceutically significant of cytochrome P-450 enyzmes

CYP450 ISOFORM	Substrate	Inducer	Inhibitor
CYP1A2	Amitriptyline	Omeprazole	Fluvoxamine
	Clomipramine	Lansoprazole	Cimetidine
	Clozapine	Phenobarbital	Grapefruit juice
	Imipramine	Phenytoin	Ticlopidine
	Propranolol	Rifampacin	
	R-warfarin	Polycyclic	Quinolones
	Theophylline	Hydrocarbons	Ciprofloxacin
	Tacrine	(smoking, charcoal-	Enoxacin
	Cyclobenzaprine	Broiled meat)	Norfloxacin
	Mexillitene		Ofloxacin
	Naproxen		Lomefloxacin
	Riluzole		
	Caffeine		
	Erythromycin		
	Haloperidol		
	Ropivacaine		
CYP2C9	Celecoxib	Rifampacin	Ritonavir
	Fluvastatin	Secobarbital	Amiodarone
	Phenytoin	Carbamazepine	Isoniazid
	Sulfamethoxazole		Ticlopidine
	Fluoxetine		Sulfaphenazole
	Dextromethorphan		Sulphinpyrazone
	Tamoxifen		Amiodarone
	Tolbutamide		Cimetidine
	Torsemide		
	S-warfarin		Antifungals
			Fluconazole
	NSAIDs		Ketoconazole
	Ibuprofen		Metronidazole
	Piroxicam		Itraconazole
	Naproxen		
	Oral Hypoglycemic Agents:		
	Tolbutamide		
	Glipizide		
	Angiotensin II Blockers:		
	Irbesartan		
	Losartan		
CYP2C19	Clomipramine	Phenobarbitone	Fluoxetine
	Imipramine		Fluvoxamine
	Amitriptyline		Ketoconazole
	Diazepam		Lansoprazole
	Meprazole		Omeprazole
	Lansoprazole		Ticlopidine
	Pantoprazole		Sertraline
	Propranolol		Ritonavir
	(S) mephenytoin		Sulfaphenazole
	Phenytoin		
	Phenobarbitone		Oral contraceptives
	Cyclophosphamide		
	Progesterone,		
	Dextromethorphan, sertraline		
	Aminopyrine		
CYP2D6	Debrisoquine		Paroxetine
	Ondansetron		Fluoxetine
	Tamoxifen		Sertraline
	Dexfenfluramine		Fluvoxamine
	Tolterodine		Nefazodone
			Venlafaxine
	Antidepressants		Clomipramine
	Amitriptyline		Amitriptyline
	Clomipramine		Cimetidine

(*continued*)

Table 1.2 Table of substrates, inducers and inhibitors for the most pharmaceutically significant of cytochrome P-450 enyzmes — *continued*

CYP450 ISOFORM	Substrate	Inducer	Inhibitor
	Desipramine		Fluphenazine
	Doxepin		Haloperidol
	Fluoxetine		Perphenazine
	Imipramine		Thioridazine
	Nortriptyline		Methadone
	Paroxetine		Quinidine
	Venlafaxine		Ritonavir
	S-mianserin		Amiodarone
	Trazadone		Chlorpheniramine
	Antipsychotics		
	Haloperidol		
	Chlorpromazine		
	Perphenazine		
	Risperidone		
	Thioridazine		
	zuclopenthixol		
	Beta blockers		
	S-Metoprolol		
	Penbutolol		
	Propranolol		
	Timolol		
	Antiarrhythmics		
	Propafenone		
	Flecainide		
	Mexiletine		
	Flecainide,		
	Procainamide		
	Narcotics		
	Codeine, tramadol		
	Dextromethorphan		
	Fentanyl, pethidine		
CYP2E1	Paracetamol	Ethanol	Disulfiram
	Ethanol	Isoniazid	
	Pentobarbitone,		
	Tolbutamide,		
	Propranolol,		
	Rifampicin		
	Chlorzoxazone		
	Volatile anaesthetics: isoflurane		
	Sevoflurane		
	Enflurane		
CYP3A4	Amitriptyline	Carbamazepine	Amiodarone
	Carbamazepine	Phenobarbital	Cimetidine
	Dexamethasone	Phenytoin	Clarithromycin
	Erythromycin	Rifampacin	Diltiazem
	Ethinyl estradiol	Troglitazone	Erythromycin
	Glyburide	Ramactane	Fluoxetine
	Imipramine	Rifabutin	Fluvoxamine
	Ketoconazole	Artemisinin	Grapefruit juice
	Nefazodone		Mibefradil
	Terfenadine		Troleandomycin
	Astemizole		Verapamil
	Sertraline		Cyclosporine
	Testosterone		Propofol
	Theophylline		Vinblastine vincristine
	Venlafaxine		Ergotamine
	Azelastine		Progesterone
	Tirilazad		Dexamethasone
	Pimozide		Quinidine

(*continued*)

Table 1.2 Table of substrates, inducers and inhibitors for the most pharmaceutically significant of cytochrome P-450 enyzmes — *continued*

CYP450 ISOFORM	Substrate	Inducer	Inhibitor
	Alfentanil		
	Sufentanil		
	Fentanyl		HIV protease inhibitors
	Quinidine		Indinavir
	Dextromethorphan		Nelfinavir
	Cisapride		Ritonavir
	Codeine		Saquinavir
	Granisetron		Azole antifungals
	Lignocaine		Keoconazole
	Ropivacaine		Itraconazole
	Hydrocortisone		Fluconazole
			Antidepressants
	Macrolide antibiotics		Nefazodone
	Clarithromycin		Fluvoxamine
	Erythromycin		Fluoxetine
	Benzodiazepines		Sertraline
	Alprazolam		Paroxetine
	Diazepam		Venlafaxine
	Midazolam		
	Triazolam		
	Immune modulators		
	Cyclosporine		
	Tacrolimus		
	HIV protease inhibitors		
	Indinavir		
	Ritonavir		
	Saquinavir		
	Nelfinavir		
	Antihistamines		
	Astemizole		
	Chlorpheniramine		
	Calcium channel blockers		
	Amlodipine		
	Diltiazem		
	Felodipine		
	Nifedipine		
	Nisoldipine		
	Nitrendipine		
	Verapamil		
	HMG CoA reductase INH		
	Atorvastatin		
	Cerivastatin		
	Lovastatin		
	Pravastatin		
	Simvastatin		

Adapted from reference texts and the following CYP450 webpages:
http://medicine.iupui.edu/flockhart/table.htm,
http://www.anaesthetist.com/physiol/basics/metabol/cyp/cyp.htm,
http://www.aafp.org/afp/980101ap/cupp.html
http://www.hospitalist.net/highligh.htm

enzymes. It should be noted that not all the drugs listed in the table as inhibitors or inducers are recognized to result in clinically significant interactions with the respective substrates.

Cytochrome P-450 enzymes are found throughout the body, with some isoforms distributed widely while others are limited to a particular tissue, e.g., CYP11B2 in the glomerulosa zone of the adrenal gland. The liver as the main organ involved in xenobiotic elimination possesses quantitatively the widest range of P450 isoforms at relatively high expression levels, although significant amounts, CYP3A4 in particular, are also found in the intestinal mucosa. Most of the cytochrome

P-450 isoforms (a notable exception being CYP2D6) can be induced with the appropriate agent to higher expression levels, which will hence alter susbtrate turnover. Variable expression of cytochrome P-450 enzymes may lead to substantial clinical consequences, not only between different individuals and different race groups, but also in individuals as they progress from infancy to old age. For example, CYP1A2 is not expressed in neonates, making them particularly susceptible to toxicity from caffeine. Differences between individuals and between different race groups may arise due to genetic polymorphism at a particular cytochrome P-450 gene loci. Polymorphism is the genetic variation in a population with both gene variants existing at a frequency of at least 1%. Of the important cytochrome P-450s in humans, CYP2C9, CYP2C19, and CYP2D6 exist in polymorphic forms. Where the principal pathway in the metabolism of a new drug candidate involves a polymorphic CYP450, there is a strong likelihood that the further development of that molecule is aborted.

CYP1A2

CYP1A2 is the only isoform known to be affected by tobacco smoking. For example, cigarette smoking can result in an increase of as much as threefold in CYP1A2 activity with smokers requiring higher doses of theophylline than nonsmokers. It is the polycyclic aromatic hydrocarbons within tobacco that serve as inducers, and the levels of these agents are also increased in charbroiled foods. The interaction between theophylline and quinolone antibiotics, particularly ciprofloxacin and enoxacin, is of documented clinical significance.

CYP2C9

CYP2C9 gene is found on chromosome 10 and is subject to polymorphism with two polymorphic phenotypes, poor and extensive metabolizers; individuals with normal CYP2C9 activity are termed extensive metabolizers. Approximately 1 to 3% of the Caucasian population are poor metabolizers. One of the most recognized substrates subject to drug–drug interactions through CYP2C9 is warfarin. Although CYP2C9 is not the only CYP450 family member to be involved in warfarin metabolism, CYP2C9 inhibitors such as fluconazole, ketoconazole, metronodazole, and itraconazole can significantly reduce warfarin metabolism causing marked elevations of prothrombin time and the potential for serious bleeding.

Warfarin is produced as a racemic mixture of *R*-warfarin and *S*-warfarin, but the main pharmacologic activity resides in the *S*-enantiomer. Most of the metabolism of *S*-warfarin is through CYP2C9. The patients expressing the ''poor metabolizer'' phenotype will be less efficient in eliminating *S*-warfarin and will be fully anticoagulated at much lower doses than the majority of the population. Poor metabolizers will also be relatively poor at drug activation through CYP2C9, for example the prodrug losartan will be poorly activated. The metabolism of phenytoin is primarily through CYP2C9, and is another good example of a potentially clinically significant drug interaction arising through CYP2C9.

CYP2C19

CYP2C19 is found in the duodenum with few other extrahepatic sites of expression. Its expression within the liver is lower than that of CYP2C9.

CYP2C19 gene is found on chromosome 10 with expression of the phenotype for poor metabolizers found in 3 to 5% of the Caucasian population, 8% Africans, 19% African-American, 15 to 20% of the Asian population, and 71% of Pacific islanders. Clinical examples of excessive or adverse drug effects in people who are CYP2C19-deficient are lacking. However, cure rates for peptic ulcer treated with omeprazole are substantially greater in individuals with defective CYP2C19.

CYP2D6

CYP2D6 gene is found on chromosome 22 with the gene product also known as debrisoquine hydroxylase after the substrate used in the initial identification of CYP2D6 polymorphism. CYP2D6 represents about 2% of all liver CYP450s with little expression of this isoform in intestinal tissue. CYP2D6 is expressed in two polymorphic phenotypes, poor and extensive metabolizers, with approximately 5 to 10% of Caucasians poor metabolizers. CYP2D6 appears not to be inducible as are many of the other CYP450 family members.

Numerous psychoactive medications are metabolized via CYP2D6, and in poor metabolizers, subjects will be predisposed to drug toxicities caused by antidepressants or neuroleptics. Other drugs that have caused problems in those lacking 2D6 include dexfenfluramine, propafenone, and mexiletine. Conversely, when formation of an active metabolite is essential for drug action, poor metabolizers of CYP2D6 can exhibit less response to drug therapy compared with extensive metabolizers. Codeine is *O*-demethylated by CYP2D6 to the active morphine species, which accounts at least partially for its analgesic effect.

With respect to drugs inhibiting CYP2D6, cimetidine, some antidepressants (tricyclic and selective serotonin reuptake inhibitors) appear clinically important. Of the antidepressants, paroxetine appears to have the greatest ability to inhibit the metabolism of CYP2D6 substrates.

CYP2E1

CYP2E1 represents about 7% of liver CYP450 expression. This isoform is inducible by ethanol and isoniazid and is responsible in part for the metabolism of paracetamol. The product of paracetomol's cytochrome P-450 metabolism is a highly reactive intermediate that must be detoxified by conjugation with glutathione. Patients with alcohol dependence may be at increased risk from paracetamol hepatotoxicity because ethanol induction of CYP2E1 increases formation of this reactive intermediate, and glutathione concentrations are decreased in these patients.

CYP3A4

The CPY3A enzyme subfamily is the most abundant of the human cytochrome enzymes. This subfamily metabolizes a large number of drug substrates, and is involved in many clinically significant drug interactions. The CYP3A4 enzyme is the most important member of the subfamily and represents about 30% of hepatic CYP450 expression and 70% of small intestinal CYP450 expression.

The most important inducers of CYP3A4 are antimicrobials such as rifampicin, and anticonvulsants like carbamazepine and phenytoin, but potent steroids such as dexamethasone may also serve as inducing agents. The long catalogue of agents metabolized by CYP3A4 can be seen from Table 1.2 and includes, among others, opioids, benzodiazepines, and local anaesthetics, as well as erythromycin, cyclosporine, haloperidol, calcium channel blockers, cisapride, and pimozide. Oral contraceptives are also metabolized and their efficacy may be impaired when an inducer such as rifampicin is taken. Perhaps of even more clinical significance are the inhibitors of CYPA4 which include, among others, the azole antifungals — in particular ketoconazole and itraconazole, HIV protease inhibitors, calcium channel blockers, some macrolides like troleandromycin and erythromycin, and selective serotonin reuptake inhibitor antidepressants — in particular nefazodone, fluvoxamine, norfluoxetine, and fluoxetine. Inhibition of the metabolism of cisapride through CYP3A4 by azole antifungals, resulting in the development of ventricular arrhythmias is a particularly well recognized clinically important drug–drug interaction. Azole antifungals by inhibiting CYP3A4 also limit the activation of the nonsedating antihistamine product, terfenadine (a prodrug that undergoes complete first-pass metabolism) to its active carboxymetabolite. Of more consequence, however, is that it is the parent terfenadine that is associated with cardiotoxicity. The active metabolite of terfenadine is now marketed, fexofenadine, as a noncardiotoxic alternative to terfenadine.

Exemplified best by CYP3A4 is the association between CYP450 and P-glycoprotein activities. CYP3A4 and P-glycoprotein certainly share a number of common substrates, and it is perceived that CYP3A4 functions synergistically with P-glycoprotein to maximize the removal of xenobiotics, i.e., two barriers acting in series. For example, in the intestinal enterocyte P-glycoprotein expressed in the apical membrane effluxes drug molecules from the apical membrane back into the intestinal lumen. However, it is inevitable that some of the drug molecules escape efflux and upon further transport into the cell interior they then face the second sequential barrier represented by CYP3A4. The fact that P-glycoprotein continually effluxes drug back into the intestinal lumen also limits the mass of the drug that at any single time is presented to the CYP3A4, and as such minimizes the possibility of negotiating this barrier through enzyme saturation.

1.4.2 Phase II Metabolism

Major phase II reactions include (i) glucuronide conjugation catalyzed by UDP-glucuronyl transferase isoenzymes and (ii) glutathioneconjugation catalyzed by glutathione-*S*-transferase isoenzymes and sulfation, methylation, and acetylation reactions catalyzed by the respective transferase enzymes. Glucuronidation is the most common phase II reaction and is the only one that occurs in the liver microsomal enzyme system. Glucuronides are secreted in bile and eliminated in urine. Amino acid conjugation with glutamine or glycine produces conjugates that are readily excreted in urine but are not extensively secreted in bile. Acetylation is the primary metabolic pathway occurring for sulfonamides for example. Sulfate conjugation is the reaction between phenolic or alcoholic groups and inorganic sulfate, partially derived from sulfur-containing amino acids, e.g., cysteine. The sulfate esters formed are polar and readily excreted in urine. Thyroxine is an example of a molecule that gives rise to sulfate conjugates. Methylation is a major metabolic pathway for inactivation of some catecholamines.

1.4.3 Excretion

The main route for excretion is via the kidneys in the urine. Some drug molecules may be excreted in the bile, with subsequent elimination with the faeces or alternatively the molecules that are excreted into the intestine with bile may subsequently undergo enterohepatic recycling, i.e., reabsorption across the intestinal epithelium back into blood. Volatile substances may be excreted via the lungs with expired air.

The functional unit of the kidney is the nephron, with each kidney containing approximately 10^6 nephrons. In the nephron blood, is ultrafiltrated through the glomerular capillaries into the renal tubules, i.e., glomerular filtration. As the glomerular filtrate passes down the tubules its volume is reduced and its composition altered by the processes of tubular reabsorption (transport of some water and certain solutes across the tubule epithelium back to the blood) and tubular secretion (secretion of solutes across the tubular epithelium from blood to tubule lumen to be excreted in the urine).

The kidneys receive about 22% of cardiac output (~1.1 L/min blood flow) and in an individual with healthy kidneys approximately 10% of this is filtered at the glomerulus. Specifically, approximately 125 mL of plasma water passes into the glomerular tubule each minute, i.e., glomerular filtration rate (GFR) is about 125 mL/min. The glomerular capillaries are freely permeable to water, electrolytes, and most constituents of plasma, including drug molecules. However, under normal conditions in the healthy kidney there is a molecular size restriction on filtration at the glomerulus represented by the lack of filtration of the plasma protein albumin (67 kDa). Since the glomeruli effectively restrict the passage of drug bound to the plasma proteins it is only the free or unbound fraction (F_U) of any drug molecules in plasma that are available for filtration. Under equilibrium conditions the elimination of any unbound drug will also lead to a net decrease in the mass of drug that is plasma protein bound, and hence ultimately glomerular filtration alone would eventually lead to the complete removal of drug from the body but at a relatively slow rate dependent solely upon GFR and F_U.

The principles of transmembrane passage governing passive renal tubular reabsorption of drugs are the same as those for drug passage across any biological barrier. Polar compounds and ions cannot diffuse back into the circulation and are excreted unless a specific transport mechanism for their reabsorption exists (e.g., as for glucose, ascorbic acid, and B vitamins). The glomerular filtrate that enters the proximal tubule has the same pH as plasma, but the pH of voided urine varies from 4.5 to 8.0. This variation in pH may markedly affect the rate of drug reabsorption and hence drug excretion in the urine. Unionized forms of nonpolar weak acids and weak bases tend to be reabsorbed readily from tubular fluids, acidification of urine increases reabsorption (i.e., decreases excretion) of weak acids and decreases reabsorption (i.e., increases excretion) of weak bases. The opposite occurs after alkalinization of urine. In some cases of overdose, these principles may be applied to enhance the excretion of weak acids or bases. For example, alkalinization of urine increases the excretion of the weak acids phenobarbital and aspirin, and acidification may accelerate the excretion of bases, such as methamphetamine. The extent to which changes in urinary pH alter the rate of drug elimination depends on the contribution of the renal route to total elimination as well as on the polarity of the unionized form and the degree of ionization of the molecule.

Mechanisms for active tubular secretion in the proximal tubule are important in the elimination of many drugs (e.g., penicillin, mecamylamine, salicylic acid). This energy-dependent process may be blocked by metabolic inhibitors. When drug concentration is high, an upper limit for secretory transport can be reached; each substance has a characteristic maximum secretion rate (transport maximum). Anions and cations are handled by separate transport mechanisms. Normally, the anion secretory system eliminates metabolites conjugated with glycine, sulfate, or glucuronic acid. Anionic compounds compete with one another for secretion, as do cations.

Arising within the interstitium of the liver parenchyma are the biliary canaliculi, which drain via the bile ductules into the gall bladder. From the gall bladder, biliary fluid can be secreted via the common bile duct into the lumen of the small intestine. Drugs and their metabolites that are extensively excreted in bile are transported across the biliary epithelium against a concentration gradient, requiring active secretory transport. Secretory transport may approach an upper limit at high plasma concentrations of a drug (transport maximum), and substances with similar physicochemical properties may compete for excretion via the same mechanism. As a general rule increasing the molecular weight of a drug (>300 Da), for example by phase II conjugation particularly glucuronidation, increases its potential for biliary excretion; the glucuronide moiety has a molecular weight of 180 Da. In the enterohepatic cycle, a drug secreted in bile is reabsorbed from the intestine. Drug conjugates secreted into the intestine also undergo enterohepatic cycling when they are hydrolyzed and the drug is reabsorbed. Biliary excretion eliminates substances from the body only to the extent that enterohepatic cycling is incomplete, i.e., when some of the secreted drug is not reabsorbed from the intestine.

1.5 SPECIALIZED TOPICS

1.5.1 Efflux Transporters

In the sections above we have discussed the processes of drug transport across biological membranes involving passive diffusional mechanisms. The permeability of an epithelial barrier to a drug may, however, also be subject to membrane transporters which recognize drugs as substrates and serve to either restrict or promote drug transport across the barrier.

In the last decade considerable attention has been directed to membrane transporters that restrict drug transport. These transporters have been termed 'efflux' transporters and the most documented ones belong to the ABC superfamily of transporters, including P-gp, MRPs, and BCRP (see Section 2.1).

P-glycoprotein

P-glycoprotein (P-gp) is probably one of the most studied ABC transporters. It consists of two very similar halves, each containing six transmembrane domains and an intracellular ATP binding site. In humans there are two gene products MDR-1 and MDR-3. P-gp MDR-3 has been shown to transport membrane components such as phosphatidylcholine, while the P-gp MDR-1 is associated with the transport of a wide variety of xenobiotics including, among others, HIV protease inhibitors, corticosteroids, antibiotics, and a range of cytotoxic agents and as such contributes to the multidrug resistance (MDR) phenotype in cancer. P-gp MDR-1 is constitutively expressed at high levels in the bile cannicular membrane of hepatocytes and the villus tips of the enterocytes of the gastrointestinal tract. In this way xenobiotics may be extruded from blood into the bile for excretion into the gastrointestinal tract, and respectively, prevented from crossing the gastrointestinal epithelium to be absorbed into the mesenteric blood supply. P-gp MDR-1 is also highly expressed within the luminal capillary membranes of the brain microvascular cells that constitute the blood–brain barrier. Additionally P-gp is expressed at the blood–testis barrier, within the apical (luminal) membranes of renal proximal tubule epithelial cells, and within apical (luminal) membranes of lung epithelial cells (both bronchiolar and alveolar). As such the constitutive expression of P-gp represents a significant barrier in epithelial and endothelial drug transport.

The issue of P-gp serving as a barrier to the absorption and tissue distribution of drugs is further compounded for the pharmaceutical industry by the fact that P-gp displays an enormous diversity in the structure of the substrates that it transports. Substrates for P-gp vary in size from 150 Da to approximately 2000 Da, many contain aromatic groups although nonaromatic linear or circular molecules are also transported. Many of the substrates that are most efficiently transported are uncharged or weakly basic in nature, but acidic compounds can also be transported, although generally at a lower rate. It is this wide diversity in substrate structure that has made it difficult to generate structure—activity relationships (SAR) for P-gp. A common physicochemical property of P-gp substrates is however their tendency to be hydrophobic in nature, consistent with the fact that the P-gp substrate binding sites are buried within the lipid bilayer. In principle therefore P-gp substrates would, in the absence of P-gp, generally display relatively rapid transmembrane passive transport. When such a substrate is faced with P-gp its net transport reflects a balance between the driving force for drug efflux, i.e., expression level of P-gp within the membrane, the concentration of drug substrate relative to the binding affinity of P-gp versus the driving force for inward passive diffusional transport down the drug's electrochemical gradient.

There is considerable overlap between the substrate selectivity and tissue localization of P-gp-MDR1 and CYP3A4 which has led to the view (Benet and Cummins 2001) that this enzyme and transporter pair function in a coordinated manner, synergistically maximizing the removal of xenobiotics, i.e., two barriers acting in series. For example, in the intestinal enterocyte P-gp expressed in the apical membrane will efflux drug molecules from the apical membrane back into the intestinal lumen. However, it is inevitable that some drug molecules escape efflux and upon further transport into the cell interior they will then face the second sequential barrier represented by CYP3A4. That P-glycoprotein continually effluxes drug back into the intestinal lumen also limits the mass of the drug that at any single time is presented to the CYP3A4, and as such minimizes the possibility of overcoming this metabolic barrier through enzyme saturation.

Multidrug resistance proteins

The family of multidrug resistance proteins includes to date MRP-1 through MRP-9, with MRP-1 and MRP-2 being the most investigated with respect to pharmaceutical barriers (Schinkel and Jonker 2003). MRP-4 and MRP-5 proteins possess a similar structure to that of P-gp, whereas MRP-1, MRP-2, and MRP-3 possess an additional N-linked segment comprising a five-transmembrane domain.

MRP-1 displays quite a striking overlap with P-gp in respect to the transport of cytotoxic agents conferring resistance to antracyclines, vinca alkaloids, epipodophyllotoxins, camptothecins, and methotrexate, but not to taxanes that are an important component of the P-gp anticancer substrate profile. However, the general substrate selectivities of P-gp and MRP-1 do nevertheless show marked differences in that while P-gp substrates are neutral or weakly positively charged lipophilic compounds, MRP-1 is able to transport lipophilic anions including a structurally diverse group of molecules conjugated with glutathione, glucuronide, or sulfate. Further, it appears that transport of neutral compounds such as some of the anticancer agents by MRP-1 are also highly dependent upon cellular glutathione, with drug export involving a combination of cotransport with glutathione. As such MRP-1-conferred resistance will be subject to inhibition with agents that block glutathione synthesis.

In epithelial cells MRP-1 is localized on the basolateral membrane, and is expressed in the choroid epithelium effluxing substrate from cerebral spinal fluid (CSF) to blood. It is also expressed at high levels within intestinal mucosa, lung mucosa, and within the kidney. Low expression is seen in the liver.

MRP-2 (also termed cMOAT) like MRP-1 is essentially an organic anion transporter displaying similar substrate selectivity with respect to glutathione and glucuronide conjugates. The pattern of expression compared to MRP-1 is, however, different with MRP-2 expressed in the apical membranes of polarized cells, with particularly high expression in the liver canaliculi, and with lower levels in the renal proximal tubules and intestinal enterocytes.

The substrate selectivity of MRP-3 appears to overlap with that of MRP-1 and MRP-2 with respect to the transport of glutathione and glucuronide conjugates, although the affinity for conjugates is reported to be less than that of MRP-1 and MRP-2. Its spectrum of anticancer agents transported is more limited and may not require glutathione for cotransport. MRP-3 is expressed in the liver, intestine, adrenal gland, and to a lower extent in the pancreas and kidney. Like MRP-1, MRP-3 is expressed on the basolateral membranes of epithelia.

MRP-4 and MRP-5 are organic anion transporters with the capacity to transport substrates such as oestradiol-17-β-glucuronide, methotrexate, and reduced folates. In addition MRP-4 and MRP-5 are able to mediate the transport of cAMP and cGMP, with the ability to confer resistance to certain nucleotide analogues, for example 6-mercaptopurine, 6-thioguanine, and azidothymidine. MRP-4 has been reported to be expressed in a number of tissues including lung, kidney, and small intestine among others. It can be found in either apical or basolateral membranes depending upon the tissue in question. MRP-5 shows highest expression in the brain and skeletal muscle and is expressed in the basolateral membranes.

Breast cancer resistant protein

BCRP was first cloned from the doxorubicin-resistant breast cancer cell line, MCF7. However, its expression is not specific for breast cancer cells and indeed this transporter may not necessarily fulfil a critical role in chemotherapy resistance in breast cancer. BCRP can be considered structurally as a half-transporter comprising a single six-transmembrane domain segment with a single N-terminal ATP binding site. In polarized epithelial cells BCRP localizes to the apical membrane. It is expressed in kidney, in the bile canalicular membrane of liver hepatocytes and the luminal

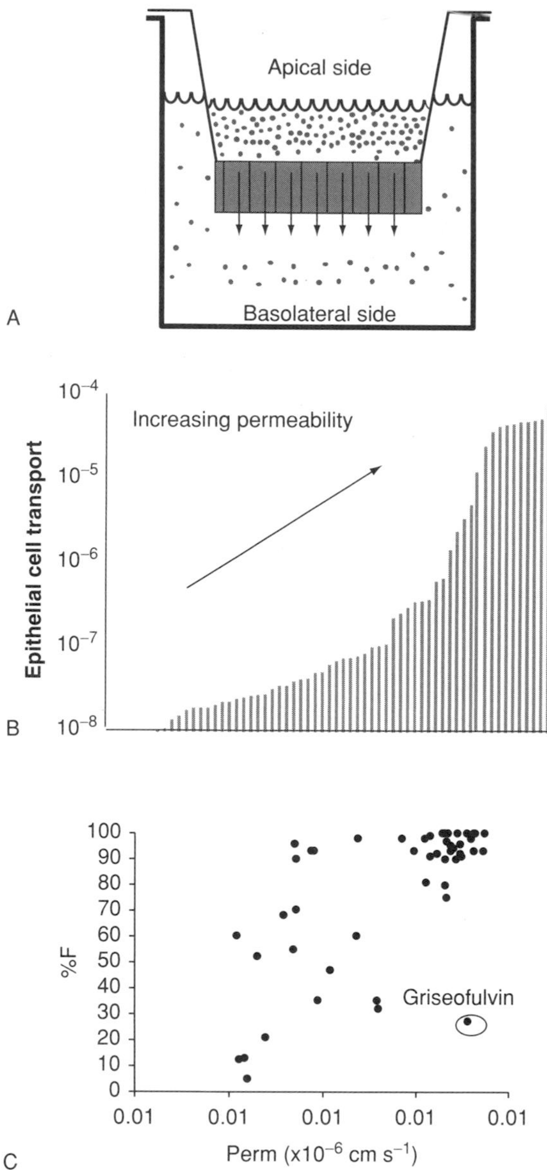

Figure 1.6 The permeability to drug of a cellular barrier representing the rate-limiting barrier *in vivo*, for example intestinal epithelium, may be assessed by culturing a monolayer model of enterocytes upon a semipermeable membrane with both apical and basolateral membrane surfaces exposed to culture medium. Drug in solution is placed in the apical (donor) chamber and will transport down its electrochemical gradient to the basolateral (receiver) chamber (A). Permeability of the biological model will also reflect any active transport mechanisms and metabolism pathways that may perceive the drug as substrate. Based upon *in vitro* permeability determinations a ranking of a large number of candidate drug molecules will highlight those that are likely to show poor absorption (B). With a significantly large *in vitro* data set correlations between *in vitro* permeability (using data from a single laboratory) may be established to literature determinations of the drugs' *in vivo* extent of bioavailability. Some anomalies will inevitably appear, for example griseofulvin whose *in vivo* absorption is so dependent upon dissolution and gastric emptying rate (C).

membrane of the epithelial cells of the small and large intestine. BCRP is also found at the apical surface of the ducts and lobules in the breast, and in many microvascular beds.

There is considerable but varying overlap in the anticancer drug substrate specificity between BCRP, P-gp, MRP-1, and MRP-2, with likely substrates including mitoxantrone, topptecan, and doxorubicin. Little or no resistance is seen for vincristine, paclitaxel, or cisplatin.

ABC drug efflux transporter inhibitors

An area of research that has attracted much attention within the pharmaceutical industry is the identification of P-gp inhibitors that could be used in the management of cytotoxic-resistant tumors expressing P-gp MDR-1. The second and third generation inhibitors have been specifically designed to inhibit P-gp with associated low toxicities and high selectivities. PSC833 (valspodar) is a structural analogue of cyclosporine A (a clinically effective immunosuppressant and first generation P-gp inhibitor) that also inhibits MRP-2. Another third generation inhibitor, GF120918 (elacridar) inhibits both P-gp and BCRP, while LY335979 (zosuquidar) is able to inhibit P-gp without apparent effects upon MRP-1, MRP-2, MRP-3, or BCRP. Both valspodar and zosuquidar produce effective P-gp inhibition without effect upon CYP3A4.

1.5.2 Caco-2 Cell Monolayers for Oral Absorption Prediction

Section 2.3 commented upon the use of *in vitro* physicochemical and computational assessments of a drug's molecular properties and the application of this information in the prediction of potential drug absorption problems. Clearly the drug properties examined by these assessments will mainly relate to their potential for passive diffusional transport. These assessments would not account for the role of cellular metabolism or active transport in the overall permeability of a drug. In particular, the role of active drug transporters are recognized, including not only carrier-mediated pathways facilitating drug passage across a barrier, but more importantly (since a little problematic from a predictive point of view) the major drug efflux mechanisms such as the P-gp transporter, whose substrate specificity is clearly very broad and not well defined. Therefore, in addition to transcellular and paracellular diffusional pathways, a cell-based model system, such as Caco-2, offers the potential to account for metabolism and active transport processes as well as nondefined interactions between a drug and cellular material (e.g., drug–protein interactions) that may impact upon the overall epithelial permeability to a drug. Caco-2 cells are the most well characterized cellular system used in academic and commercial settings for the experimental study of drug transport mechanisms and for the purposes of characterizing the potential epithelial, enterocyte, permeability of drugs.

Caco-2 cells were first isolated and established as a growing culture in 1974 from a colon carcinoma at the Sloan Kettering Memorial Cancer Centre, New York (Fogh and Trempe 1975). In culture, 100% of the cells undergo terminal differentiation to a columnar absorptive cell type, which although showing characteristics of colonic fetal type cells, also share a number of differentiation features to that of small intestinal enterocytes. The differentiated cell monolayer culture exhibits a typical intestinal cell apical brush border and displays intercellular tight junctional complexes at the apical domains. Furthermore, the cells possess a number of enzymes and transporters that are present in the corresponding *in vivo* cell type.

The growth of Caco-2 cell monolayers upon semipermeable membrane supports, such as the Transwell system, has afforded their wide use in transepithelial transport studies where drug in solution bathing the apical membrane surface of the cell monolayer may cross the epithelial barrier to be sampled from fluid bathing the basolateral membrane surface of the cells (Figure 1.6A). From such transport experiments a quantitative measure of the permeability (permeability coefficient; see Equation (1.1)) of the barrier to drug molecules can be determined, a measure which when used in a ranking comparative analyses (Figure 1.6B) may indicate if *in vivo* oral absorption problems are likely. Figure 1.6C shows a typical plot of Caco-2 permeability ($\times 10^{-6}$ cm s^{-1}) versus extent of oral bioavailability in humans (% *F*) using a sample of the data of pharmaceutical compounds where % *F* is documented. The typical relationship seen for a large group of compounds of varying physicochemical properties is that of a sigmoidal relationship between Caco-2 permeability and % *F in vivo*, where the rising phase of the relationship can be steep, i.e., little discrimination in Caco-2 permeability for quite marked changes in % *F*. This occurs because the paracellular pathway in the Caco-2 monolayer is much more restrictive than *in vivo* intestinal epithelium. Nevertheless a ranking of compound permeability is still possible, a ranking which may reveal potential drug absorption problems of individual molecules. In Figure 1.6C, it is noteworthy to highlight that doxorubicin Caco-2 permeability is low and the *in vivo* % *F* is indeed low, both despite physicochemical properties for doxorubicin that would predict much better membrane partitioning. This reflects the presence of P-glycoprotein expression in the Caco-2 model as well as the *in vivo* enterocyte. The Caco-2 model would appear to overestimate the *in vivo* % *F*, reflecting the reason for low *in vivo* bioavailability being one of formulation dissolution rather than intrinsic membrane partitioning; in the Caco-2 system drugs are invariably added in solution form.

Clearly for a robust prospective prediction of barrier permeability, a number of methodologies should be exploited and used appropriately such as the *in vitro* physicochemical or computational assessment for high throughput screening and the *in vitro* cell-based models together with the *in vivo* studies for medium to low throughput candidate selection.

REFERENCES

Cited in Text

Benet, L.Z. and Cummins, C.L. (2001) The drug efflux-metabolism alliance: biochemical aspects. *Adv. Drug Deliv. Rev.* **50**, S3–S11.

Fogh, J. and Trempe, G. (1975) New Human Tumour Cell Lines. In *Human Tumour Cells In-Vitro*, edited by Fogh, J., pp. 115–141. New York: Plenum Press.

Lipinski, C.A., Lombardo, F., Dominy, B.W. and Feeney, P.J. (1997) Experimental and computational approaches to estimate solubility and permeability in drug discovery and development settings. *Adv. Drug Deliv. Rev.* **23**, 3–25.

ABC Efflux Transporters

Schinkel, A.H. and Jonker J.W. (2003) Mammalian drug efflux transporters of the ATP cassette (ABC) family: an overview. *Adv. Drug Deliv. Rev.* **55**, 3–29.

Tight Junctions

Balda, M.S. and Matter, K. (2000) Transmembrane proteins of tight junctions. *Sem. Cell Dev. Biol.* **11**, 281–289.

Cereijido, M., Shoshani, L. and Contreras, R.G. (2000) Molecular physiology and pathophysiology of tight junctions I: biogenesis of tight junctions and epithelial polarity. *Am. J. Physiol.* **279**, G477–G482.

Lapierre, L.A. (2000) The molecular structure of the tight junction. *Adv. Drug Deliv. Rev.* **41**, 255–264.

Drug Metabolism

Cashman, J.R. (2003) The role of flavin-containing monooxygenases in drug metabolism and development. *Curr. Opin. Drug Discov. Devel.* **6**, 486–493.

Ingelman-Sundberg, M. (2002) Polymorphism of cytochrome P450 and xenobiotic toxicity. *Toxicology* **182**, 447–452.

Chang, G.W. and Kam, P.C.A. (1999) The physiological and pharmacological roles of cytochrome P450 isoenzymes. *Anaesthesia* **54**, 42–50.

Thummel, K.E. and Wilkinson, G.R. (1998) *In vitro* and *in vivo* drug interactions involving human CYP3A. *Ann. Rev. Pharmacol. Toxicol.* **38**, 389–430.

Predicting Drug Absorption Using *in vitro* Approaches

Artursson, P., Palm, K. and Luthman, K. (2001) Caco-2 monolayers in experimental and theoretical predictions of drug transport. *Adv. Drug Deliv. Rev.* **46**, 27–43.

Butina, D., Segall, M.D. and Frankcombe, K. (2002) Predicting ADME properties in-silico: methods and models. *Drug Discov. Today* **7**, S83–S88.

Kerns, E.H. (2001) High throughput physico-chemical profiling for drug discovery. *J. Pharm. Sci.* **90**, 1838–1858.

Kramer, S.D. (1999) Absorption prediction from physico-chemical parameters. *Pharm. Sci. Tech. Today* **2**, 373–380.

van de Waterbeemd, H., Smith, D.A., Beumont, K. and Walker, D.K. (2001) Property-based design: Optimisation of drug absorption and pharmacokinetics. *J. Med. Chem.* **44**, 1314–1330.

2

The Design of Drug Delivery Systems

Ian W. Kellaway

CONTENTS

2.1 INTRODUCTION

Drugs are rarely, if ever, administered to patients in an unformulated state. The vast majority of the available medicinal compounds, which are potent at the milligram or microgram levels, could not be presented in a form providing an accurate and reproducible dosage unless mixed with a variety of excipients and converted by controlled technological processes into medicines. Indeed, the primary skills of the pharmacist lie in the design, production, and evaluation of a wide range of dosage forms, each providing an optimized delivery of drug by the selected route of administration. The aims of this chapter, therefore, are to outline mechanisms by which the onset, duration, and magnitude of the therapeutic responses can be controlled by the designer of the drug delivery system.

It has been appreciated for a considerable time that dosage forms possessing the same amount of an active compound (chemically equivalent) do not necessarily elicit the same therapeutic response. The rate at which the drug is released from the dosage form and the subsequent absorption, distribution, metabolism, and excretion kinetics will determine the availability of the active species at the receptor site.

The majority of systemically acting drugs are administered by the oral route and therefore must traverse certain physiological barriers including one or more cell membranes. Prodrugs may alter this part of the overall rate process (see Chapter 7) although generally, control of plasma levels is achieved by modulation of the drug release process from the dosage form. The critical drug activity at the receptor site is usually related to blood and other distribution fluid levels, as well as elimination rates. Other factors affecting activity include deposition sites, biotransformation processes, protein binding, and the rate of appearance in the blood. Hence in order to obtain the desired response, the drug must be absorbed both in sufficient quantity and at a sufficient rate.

The term bioavailability is used to express the rate and extent of absorption from a drug delivery system into the systemic circulation. The crucial influence of rate as well as extent of absorption in considerations of bioavailability can be seen in Figure 2.1.

The plasma levels are illustrated following a single oral administration of three chemically equivalent delivery systems (A, B, and C) but with different drug release rates (A $>$ B $>$ C). Formulation A has a shorter duration of activity but results in a more rapid onset of activity compared with formulation B. The magnitude of the therapeutic response is also greater for A than B. Formulation C is therapeutically inactive, as the minimum effective plasma concentration

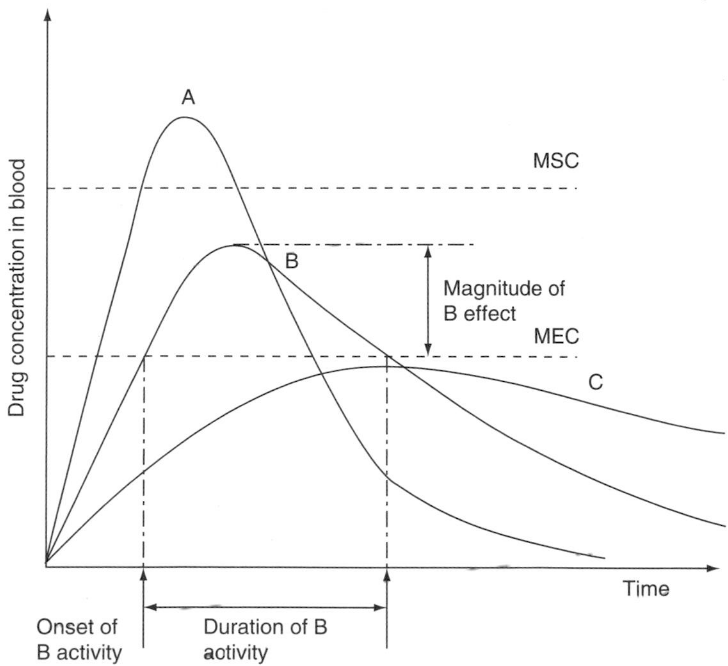

Figure 2.1 The influence of drug release rate on the blood level–time profile following the oral administration of three chemically equivalent formulations. MSC, maximum safe concentration; MEC, minimum effective concentration.

(MEC) is not achieved. Therefore, unless a multiple dosing regimen is to be considered, C has no clinical value.

It should also be noted that the plasma concentrations from A exceed the maximum safe concentration (MSC) and some toxic side effects will be observed. Unless rapidity of action is of paramount importance and the toxic effects can be tolerated, B therefore becomes the formulation of choice. Generally, however, a rapid and complete absorption profile is required to eliminate variation in response due to physiological variables, which include gastric emptying rate and gut motility. Bioavailability can also therefore be influenced by physiological and pathological factors, although in this chapter only the pharmaceutical or formulation aspects will be considered.

Bioavailability may be assessed by the determination of the induced clinical response, which makes quantitation difficult because it often involves an element of subjective assessment. Measurement of drug concentrations at the receptor site is not feasible; therefore, the usual approach is the determination of plasma or blood levels as a function of time, making the implicit assumption that these concentrations correlate directly with the clinical response. Areas under the concentration–time profiles give the amount of drug absorbed and hence (if related to those of an intravenous solution of the same drug) permit an absolute bioavailability to be determined, while if related to a ''standard'' formulation (often the original or formula of the patent holder) then the term ''relative bioavailability'' is employed.

The constraints of space dictate the limitation of both discussion and examples to the oral route, which is the most widely used route for systemically active agents, and the pulmonary route for which there is an interdependence between the device and the formulation in order to optimize drug efficiency. However, there are alternative nonparenteral routes to be considered including nasal, ocular, transdermal, buccal, vaginal, and rectal; details are available from specialist textbooks (see Further Reading).

2.2 FORMULATION AIMS

Formulation aims, in the light of bioavailability considerations, are to produce a drug delivery system such that:

a. A unit dose contains the intended quantity of drug. This is achieved by homogeneity during the manufacturing process and a suitable choice of excipients, stabilizers, and manufacturing conditions to ensure both drug and product stability over the expected shelf-life.
b. The drug is usually totally released but always in a *controlled* manner, in order to achieve the required onset, intensity, and duration of clinical response as previously outlined. Most dosage forms can be designed to give a rapid response; if, however, a long duration of response is required then it is easier to achieve sustained release using solid rather than liquid formulations.

2.3 ORAL DRUG DELIVERY

This is the most convenient route of drug administration, is patient acceptable, and generally affords good compliance. However, it is not without limitations, for example drugs that are poorly absorbed and/or degraded. Some limitations may be overcome by advances in drug delivery system design.

2.3.1 The Gastrointestinal Tract

The digestive system serves to process ingested food into simple molecules, which are absorbed into the blood or the lymph. This process occurs as transit takes place from the oral cavity to the stomach, and into the small intestine, and finally the colon. Drugs may be absorbed from any of these four regions, although for most drugs the principal absorption site is the small intestine comprising the duodenum, jejunum, and ileum. The absorption area of the small intestine is approximately 200 m^2 in humans, which is achieved by the plica circulares (circularly arranged folds of the mucosa and submucosa), the villi (finger-like projections of the mucosa), microvilli (present on the luminal surface of each epithelial cell), and crypts of Lieberkuhn (mucosal invaginations at the base of the villi).

The epithelia of the gastrointestinal (GI) tract show considerable variation in different regions of the tract according to functional requirements. Mucosae can be classified as protective, secretory, absorptive, and absorptive/protective. In addition, there is gut-associated lymphoidal tissue (GALT) which when present as discrete, nonencapsulated aggregates of lymphoid follicles is referred to as Peyer's patches. Occurring largely in the distal ileum, they participate in antigen sampling. M-cells cover the surface of the patches and are capable of extensive uptake of macromolecules and microparticles. Vaccination by the oral route is therefore possible by employing microparticle formulations.

GI transit of dosage forms

The motility of the GI tract is influenced by many factors, food intake being the most important. The transit of dosage forms is greatly influenced by whether the fed or fasted state exists. When food enters the stomach, contractions in the antrum serve to mix and grind the contents. Solid material is periodically moved to the distal antrum. When the pylorus contracts, liquids and small (<5 mm) suspended solids pass into the duodenum. Larger material is returned to be subjected to further grinding and mixing; a process which continues until the stomach is empty of food. Various feedback mechanisms operate to control the emptying process with receptors sensitive to acid, fat, osmotic pressure, and amino acids. Dosage forms, which are nondisintegrating or disintegrated fragments >5 mm, normally remain within the stomach while it is in the fed state.

In contrast, in the fasted state (or postdigestive phase), the stomach empties undigested material by the migrating myoelectric complex mechanism. This comprises four phases, the third of which (the "housekeeper wave") consists of a series of rapid contractions occurring approximately every 2 h, causing the undigested material to be swept through the open pylorus into the duodenum. A disintegrating or pelletized dosage form will be emptied from the stomach, while a nondisintegrating single unit administered in the fed state or any dosage form administered in the fasted state will be cleared by the myoelectric motor complex.

When fasted, both single units and pellets are emptied from the stomach quite rapidly. From a lightly fed stomach, the emptying of single units will be delayed, while for pellets their rate of emptying and their spreading in the small intestine will depend on the quantity of food ingested. Following a heavy meal, single units will be retained in the stomach as long as it remains in the fed state and for pellets a slow, steady emptying will occur with an appreciable degree of spreading in the small intestine. Hence the dependence of emptying on the nature of the dosage form and on food intake will have an appreciable bearing on the design of controlled release dosage forms. For example, if a drug is absorbed only from the small intestine or has a window of absorption in the duodenum, then gastric retention of the dosage form would ensure that drug can access the absorption site over a prolonged period of time. Less intersubject variability in plasma concentrations would therefore be expected from pelleted formulations compared with single unit dosage forms, especially if there is no attempt to regulate diet and where the dosage forms show pH-dependent release profiles.

Whereas gastric emptying is a highly variable process, the transit time in the small intestine is relatively constant taking 3–4 h. It is independent of formulation, fasted state, age, pathological condition, and exercise. Propulsion of the dosage form along the small intestine occurs by peristalsis, which is a sequential annular contraction of the gut.

pH in the GI tract

The pH in the fasted stomach is between 0.8 and 2.0, which transiently rises to 4–5 on ingestion of food, only to fall again as acid is secreted. The duodenal contents are normally in the pH range 5–7, the jejunum 6–7, and the ileum 6–7.5. In the colon, the pH range is 5.5–7.0. Dissolution, solubilization, and absorption processes of ionizable drugs are generally influenced by the pH of the surrounding media.

Influence of food on drug absorption

Food generally reduces the rate and/or extent of drug absorption by (a) slowing gastric emptying, (b) increasing the viscosity of luminal contents (hence decreasing dissolution rate and drug diffusion to the gut wall), (c) drug complexation with food components, and (d) stimulation of GI fluid secretion which may degrade the drug.

Influence of mucus on drug absorption

Mucus may form an additional barrier to drug absorption. This viscoelastic gel, although containing approximately 95% water, prevents the diffusion of large molecules (>1 kDa) to the epithelial surface. Small drug molecules are generally able to diffuse through the gel interstices and it is only for those drugs that bind specifically with the glycoprotein network that reduced bioavailability may be expected.

Drug metabolism and active secretion in the GI tract

Cytochrome P450 3A4 is the major phase I drug metabolizing enzyme in humans, which together with the multidrug efflux pump, P-glycoprotein, is present in the enterocytes of the villi in the small intestine. P-glycoprotein is a major route for the elimination of anticancer drugs, e.g., vinca alkaloids, taxol, and etoposide. Oral bioavailability can be enhanced by the inhibition of cytochrome P450 3A4.

Drugs absorbed from the GI tract are transported to the liver where they may be metabolized and hence lost from the circulation, i.e., the first-pass effect. Only for prodrugs requiring metabolism for activation is such an effect desirable.

2.3.2 Physicochemical Factors Influencing Drug Bioavailability

Drug concentrations in the blood are controlled either by the rate of drug release from the dosage form or by the rate of absorption. In many cases it is the drug dissolution rate that is the rate-determining step in the process. Dissolution is encountered in all solid dosage forms, i.e., tablets and hard gelatin capsules as well as suspensions, whether intended for oral use or administration via the intramuscular or subcutaneous routes. If absorption is rapid, then it is almost inevitable that drug dissolution will be the rate-determining step in the overall process and hence any factor which affects the solution process will result in changes in the plasma–time profile. Hence the formulator has the opportunity of controlling the onset, duration, and intensity of the clinical response by controlling the dissolution process.

Rate of solution

Dissolution of a drug from a primary particle in a nonreacting solvent can be described by the Noyes–Whitney equation

$$\mathrm{d}w/\mathrm{d}t = k(c_\mathrm{s} - c) = DA/h(c_\mathrm{s} - c), \tag{2.1}$$

where $\mathrm{d}w/\mathrm{d}t$ is the rate of increase of the amount of drug dissolved, k is the rate constant of dissolution, c_s the saturation solubility of the drug in the dissolution media, c the concentration of drug at time t, A is the surface area of drug undergoing dissolution, D the diffusion coefficient of the dissolved drug molecules, and h the thickness of the diffusion layer. Hence it can be readily appreciated that the dissolution rate is dependent on the diffusion of molecules through the diffusion layer of thickness h. Closer examination of this equation will demonstrate some of the mechanisms for controlling solution rate.

(1) $\mathrm{d}w/\mathrm{d}t \propto A$. Reduction in the particle size of the primary particle will result in an increase in surface area and hence more rapid dissolution will be achieved. A change in the shape of the plasma–time profile will result and it is possible also to increase the area under this curve, which of course means an increase in bioavailability. It is therefore possible to achieve a reduction in the time necessary for the attainment of maximum plasma levels, an increase in the intensity of the response, and an increase in the percentage of the dose absorbed. Griseofulvin is one of the most widely studied drugs in relation to bioavailability, as this poorly water-soluble, antifungal drug exhibits a striking example of dissolution rate-limited absorption. Plasma levels have been shown to increase linearly with an increase in specific surface area and thus, despite the cost of micronization, griseofulvin is marketed as a preparation in this form because identical blood levels can be achieved by using half the amount of drug present in the unmicronized formulation. Micronization, however, is not the only solution to the griseofulvin bioavailability problem. For example, microcrystalline dispersions have been formed in a water-soluble solid matrix in which the dispersion state is

determined by the preparative procedures, some of which result in true solid solutions. The two most widely accepted approaches are (a) crystallization of a melt, resulting from fusing of drug and carrier and (b) co-precipitation of drug and carrier from a common organic solvent. In the latter case a griseofulvin–polyvinylpyrrolidone dispersion resulted in a tenfold increase in solution rate, compared with a micronized preparation. It should be emphasized, however, that griseofulvin is at the extreme end of the bioavailability spectrum. For drugs exhibiting good aqueous solubility, little is to be gained by reducing the particle size of the drug, as plasma levels are unlikely to be dissolution rate-limited. Indeed, if enzymatic or acid degradation of the drug occurs in the stomach, then increasing dissolution rates by reducing particle size can result in reduced bioavailability.

(2) $dw/dt \propto c_s$. Many drugs are weak acids or bases and hence exhibit pH-dependent solubility. It is therefore possible to increase c_s in the diffusion layer by adjustment of pH in either (a) the whole dissolution medium or (b) the microenvironment of the dissolving particle. The pH of the whole medium can be changed by the co-administration of an antacid. This raises the pH of the gastric juices and hence enhances the dissolution rate of a weak acid. However, this is rarely a practical proposition and therefore most pH adjustments are made within the very localized environment of the dissolving drug particles. Solid basic substances may be added to a weakly acidic drug, which raises the pH of the microenvironment. Probably the best known example is that of buffered aspirin products which use the basic substances such as sodium bicarbonate, sodium citrate, or magnesium carbonate. Rather than employ another agent to alter the pH, a highly water-soluble salt of the drug can be equally, if not more, effective. The dissolving salt raises the pH of the gastric fluids immediately surrounding the dissolving particle. On mixing with the bulk of the gastric fluids the free acid form of the drug will be precipitated, but in a microdispersed state with a large surface area to volume ratio, it will rapidly redissolve. The process is represented diagrammatically in Figure 2.2.

Many examples exist to illustrate the importance of salt formation on bioavailability. One such example is provided by the antibiotic novobiocin, where the bioavailability was found to decrease in the order, sodium salt > calcium salt > free acid. The dissolution rate of weak bases can be similarly changed by salt formation; however, dissolution rate-limited absorption is less important for bases than acids. This is because little absorption occurs in the stomach where the bases are ionized, most of the drug being absorbed by postgastric emptying and this delay compensates any benefits accruing from more rapid solution rates. However, basic drugs are often administered as

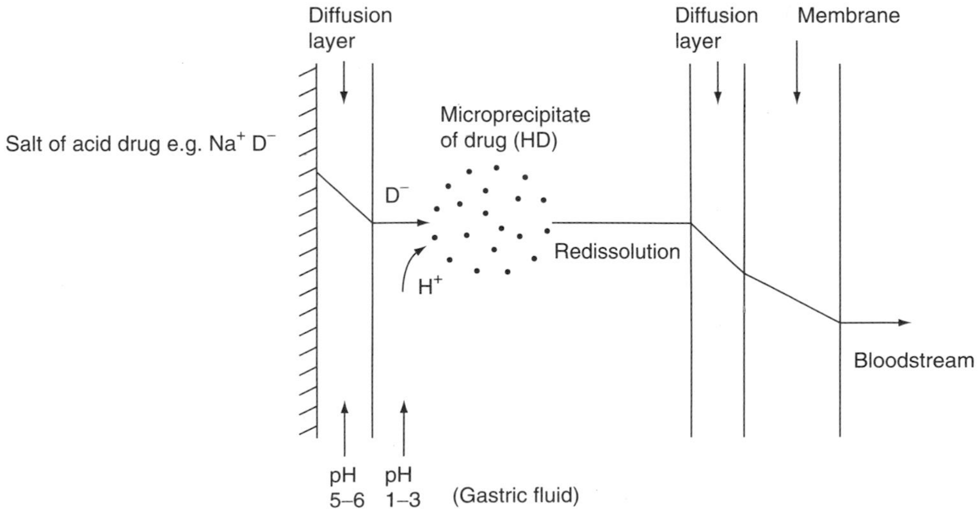

Figure 2.2 The dissolution of a highly water-soluble salt of a weak acid in the stomach.

salts, e.g., phenothiazines and tetracyclines, to ensure that gastric emptying, and not dissolution, will be the rate-limiting factor in the absorption process.

A large number of drugs exhibit polymorphism, that is, they exist in more than one crystalline form. Polymorphs exhibit different physical properties including solubility, although only one polymorph will be stable at any given temperature and pressure. Others may exist in a metastable condition, reverting to the stable form at rates that may permit their use in drug delivery systems. The most desirable property of the metastable forms is their inherently higher solubility rates, which arise from the lower crystal lattice energies.

Amorphous or noncrystalline drugs are always more soluble than the corresponding crystalline form because of the lower energy requirements in transference of a molecule from the solid to the solution phase. Crystalline novobiocin dissolves slowly *in vitro* compared with the amorphous form, the kinetics of which correlate well with bioavailability data. Amorphous chloramphenicol stearate is hydrolyzed in the GI tract to yield the absorbable acid, while the crystalline form is of such low solubility that an insufficient quantity is hydrolyzed to give effective plasma levels.

Solvates are formed by some drugs: when the solvent is water, the hydrates dissolve more slowly in aqueous solutions than the anhydrous forms, e.g., caffeine and glutethimide. For ampicillin, greater bioavailability has been shown for the higher energy form anhydrate than the trihydrate, which illustrates the dependence of solubility and dissolution rates on the free energy of the molecules within the crystal lattice. Conversely, organic solvates such as alkanoates dissolve more rapidly in aqueous solvents than the desolvated forms.

Complexation

Increased solubility or protection against degradation may be achieved by complex formation between the drug and a suitable agent. Complexes may also arise unintentionally as a result of drug interaction with an excipient or with substances occurring in the body. Complex formation is a reversible process and the effect on bioavailability is often dependent on the magnitude of the association constant. As most complexes are nonabsorbable, dissociation must therefore precede absorption.

The formation of lipid-soluble ion-pairs between a drug ion and an organic ion of opposite charge should result in greater drug bioavailability. Rarely have such results been achieved, presumably due to the dissociating influence of the mucosa and the poor membrane partitioning of the bulky ion-pair.

Surfactants are used in a wide range of dosage forms often to increase particle wetting, control the stability of dispersed particles, and to increase both solution rates and the equilibrium solubility by the process of solubilization. Bioavailability may, however, be enhanced or retarded and often exhibits surfactant concentration-dependent effects. Below the critical micelle concentration (CMC), enhanced absorption may be encountered due to partition of the surfactant into the membrane, which results in increased membrane permeability. At post-CMC levels, the dominant effect is the ''partitioning'' of the drug into the micelle, a lower drug thermodynamic activity results and absorption is reduced. Micellar solubilization of membrane components with a loss of membrane integrity can also occur. Thus it is not easy to predict the effect of surfactants on bioavailability for, although dissolution rates will be increased by high concentrations of surfactant, the effect on the absorption phase may be complex.

Drug stability

Drug stability, in addition to being of paramount importance to product shelf-life, can also affect bioavailability. Some therapeutic substances are degraded by the acid conditions of the stomach or by enzymes encountered in the GI tract. Reduced or zero therapeutic effectiveness will result. Penicillin G is an example of a drug rapidly degraded in the stomach and for which enteric coating is not a solution to the problem, as the drug is poorly absorbed from the small intestine. The semisynthetic penicillins such as ampicillin and amoxacillin show much greater acid stability. Improved bioavailability of acid-labile drugs can sometimes be achieved by reducing the rate of drug release from the dosage form.

2.3.3 Influence of Type of Dosage Form

Bioavailability, in addition to being dependent on the route of administration, will also be influenced by the dosage form selected. Although it is not possible to generalize completely regarding the relative drug release rates and hence bioavailabilities from different dosage forms, Table 2.1 attempts to provide guidelines. It is however possible, for example, to produce a tablet with bioavailability equivalent to an aqueous solution.

Aqueous solutions are rarely used due to solubility, stability, taste, and nonunit dosing problems. The use of oils as drug carriers either as an emulsion, in which homogeneity and flavor masking are important, or in a soft gelatin capsule, provides efficient oral dosage forms. The release of the oil from the soft gelatin capsule shell is rapid but the surface area of the oil–water interface is lower than in an emulsion and hence partitioning of the drug is slower. Suspensions are suited to drugs of low solubility and high stability. Although a large surface area is provided, a dissolution stage nevertheless exists. On proceeding along the sequence from powders to hard gelatin capsules to tablets (see Table 2.1), the particles become more compacted and hence the deaggregation/dissolution phase becomes longer (see Figure 2.3).

2.3.4 Formulation Factors

It should by now be appreciated that, by design, it is possible to formulate a potent, well-absorbed drug in such a manner that it is essentially nonabsorbable. Hence the formulating scientist can significantly influence the therapeutic efficacy of a drug. In most cases, the formulator can only influence bioavailability if the drug release phase is the rate-controlling step in the overall process.

Solutions

As the drug is in a form readily available for absorption, few problems should exist. However, if the drug is a weak acid or a cosolvent is employed, then precipitation of the drug in the stomach may take place. Rapid redissolution of these ''microprecipitates'' normally occurs. Aqueous solutions will require the addition of a suitable selection of colors and flavors to minimize patient noncompliance, and preservatives and perhaps buffers to optimize stability. Such factors would be elucidated in preformulation studies.

Table 2.1 The ranking of dosage forms for oral administration with respect to the rate of drug release

		Aqueous solutions
		Emulsions
Increasing release rates and bioavailability	↑	Soft gelatin capsules
		Suspensions
		Powders
		Granules
		Hard gelatin capsules
		Tablets
		Coated tablets

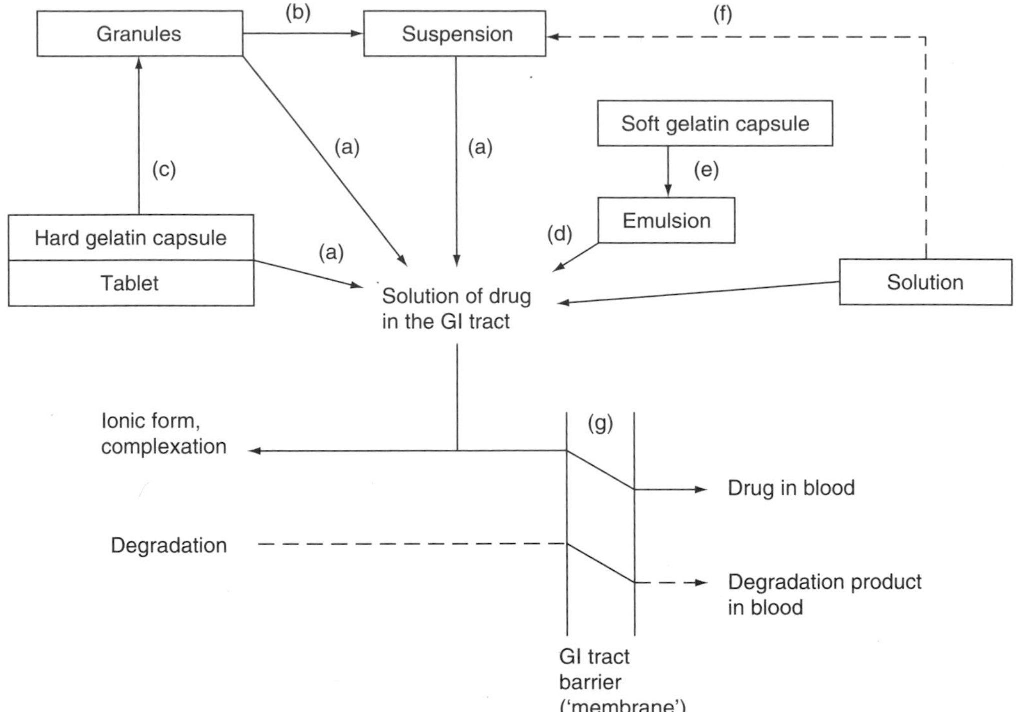

Figure 2.3 Summary of the processes following oral administration of dosage forms. Processes (a) dissolution; (b) deaggregation; (c) disintegration; (d) partitioning; (e) dispersion; (f) precipitation; (g) absorption.

Emulsions

The use of oral emulsions is on the decline. Most oils are unpalatable and an emulsion is an inherently unstable system. The choice of carrier oil dictates the extent and rate of drug partitioning between the oil and water. Emulsifying agents are either a mixture of surfactants or a polymer. Polymers may also be present to control the rheological properties of the emulsion and achieve an acceptable rate of creaming. The effect of surfactants on bioavailability has been previously discussed. Polymers can form nonabsorbable complexes with drugs and an increase in viscosity brought about by ''thickening agents'' can delay gastric emptying, which in turn may affect absorption. Viscosity effects, however, are not likely to be encountered with small dose volumes (5–10 mL).

Soft gelatin capsules

After rupture of the glycero-gelatin shell, a crude emulsion is formed when the oil containing the drug is dispersed in the aqueous contents of the GI tract. Oils are not always used to fill soft gelatin capsules; indeed occasionally water-miscible compounds such as polyethylene glycol 400 are used as vehicles. Soft gelatin capsules are a convenient unit dosage form generally exhibiting good bio-availability.

Suspensions

A high surface area of the dispersed particles ensures that the dissolution process begins immediately after the administered dose is diluted with the fluids of the GI tract. Most pharmaceutical suspensions may be described as coarse, that is they have particles in the size range 1–50 μm. Colloidal dispersions are expensive to produce and the theoretically faster solution rates arising from increased surface area are often offset by the spontaneous aggregation of the particles due to the possession of high surface energy. Particles >50 μm result in poor suspensions with rapid sedimentation, slower solution rates, and poor reproducibility of the unit dose. In order to achieve desirable settling rates and ease of redispersion of the resulting sediments, controlled flocculation of the suspension is necessary. This is normally achieved by the use of surfactant or polymers, both of which may significantly influence drug bioavailability for reasons previously discussed. Polymers are also used, as with emulsions, as thickening agents to achieve the desired bulk rheological properties. On storage, the particle size distribution of suspensions may change with the growth of large particles at the expense of small particles. Hence solution properties and bioavailability may well be altered on storage.

Hard gelatin capsules

It might be assumed that powders distributed into loosely packed beds within a rapidly dissolving hard gelatin capsule would not provide bioavailability problems. However, in practice, this is not true. One of the classic bioavailability cases in the pharmaceutical literature arose when the primary excipient in phenytoin capsules, calcium sulphate dihydrate, was substituted by lactose by the manufacturing company in Australia. Minor adjustments were also made to the magnesium silicate and magnesium stearate levels. The overall effect was that previously stabilized epileptic patients suddenly developed the symptoms associated with phenytoin overdose. It is now generally accepted that the calcium ions form a poorly absorbable complex with phenytoin.

Another study demonstrated the reduced bioavailability of tetracycline from capsules in which calcium sulphate and dicalcium phosphate were used as fillers. The calcium–tetracycline complex formed in such formulations is poorly absorbed from the GI tract.

The choice and quantity of lubricant employed can greatly influence bioavailability. Even with a water-soluble drug it is possible to vary the drug release patterns from rapid and complete to slow and incomplete. With hydrophobic drugs, the problems can be even more acute. Hence, hydrophilic diluents should be employed to aid the permeation of aqueous fluids throughout the powder mass, reduce particle clumping, and hence increase solution rates.

Tablets

For economic reasons as well as for the convenience of the patient, the compressed tablet is the most widely used dosage form. However, by virtue of the relatively high compression forces used in tablet manufacture, together with the inevitable need of a range of excipients (including fillers, disintegrants, lubricants, glidants, and binders), tabletting of drugs can give rise to serious bioavailability problems. As was seen in Figure 2.3, the active ingredient is released from the tablet by the processes of disintegration, deaggregation, and dissolution, the latter occurring, however, at all stages in the overall release process. The rate-limiting step is normally dissolution, although, by the use of insufficient or an inappropriate type of disintegrant, disintegration may become the all-important rate-limiting step. Division of the disintegrant between the granule interior and the intragranular void spaces can accelerate the disintegration process. Several interdependent factors determine disintegration rates, including concentration and type of drug, the nature of diluent, binder, and disintegrant as well as the compaction force. High compression forces will often result in the retardation of disintegration due to reduced fluid penetration and extensive interparticulate bonding. Soluble drugs and excipients may lead to a decrease in disintegration due to the local formation of viscous solutions.

The effect of hydrophobic lubricants is similar to that observed for capsules. The method by which the lubricant is incorporated, as well as the efficiency of mixing, has also been shown to influence drug dissolution rate from tablets. When the excipient–drug ratio is increased, thus increasing tablet size, solution rates of poorly water-soluble drugs also increase.

2.3.5 Biopharmaceutical Drug Classification

A classification scheme was proposed in 1995 for correlating *in vitro* dosage form dissolution and *in vivo* bioavailability following oral dosing. This scheme recognized that dissolution and GI permeability are the principal factors controlling the rate and extent of drug absorption. The drug classes identified were:

Case 1, High solubility–high permeability drugs: *In vitro–in vivo* correlation is to be expected only if the dissolution rate is slower than gastric emptying. The drug is well absorbed and dissolution or gastric emptying (if dissolution is rapid) become the rate-limiting step. For immediate release dosage forms, bioequivalence is probable if 85% of the dose dissolves in <15 min.

Case 2, Low solubility–high permeability drugs: *In vitro–in vivo* correlation is expected if the *in vitro* dissolution rate is similar to *in vivo* dissolution rate. The dissolution profile must be well defined (a number of time points and a minimum of 85% dissolution at several physiological pH values) and reproducible.

Case 3, High solubility–low permeability drugs: Limited or no *in vitro–in vivo* correlation is expected as permeability is the rate-limiting step. As for Class 1 drugs, the dissolution profile should be well defined but with a simplified dissolution specification for immediate release dosage forms. Both rate and extent of absorption may be variable but with fast dissolution any variation will be due to GI transit, luminal contents, and epithelial permeability.

Case 4, Low solubility–low permeability drugs: Limited or no *in vitro–in vivo* correlation is expected. Drugs of this class are expected to show significant problems for effective oral delivery.

> The alveolar region, together with reduced extracellular enzyme levels compared with the GI tract, ensures that pulmonary administration is a potentially attractive route for the delivery of systemically active agents including the new generation of biotechnology molecules.

2.4 DRUG DELIVERY INTO THE LUNG

2.4.1 Therapeutic Aerosol Generation and Particle Fate

There are three principal types of aerosol generators currently used in inhalation therapy, viz. the pressurized pack (metered dose) inhaler (MDI), the nebulizer for continuous administration and the unit-dose dry powder inhaler (DPI). The pharmaceutical formulator is not only concerned with the drug formulation but also the selection of the appropriate device as it is the intimate relationship between device and formulation that leads to optimal drug deposition within the lower respiratory tract. The latter consists of the bronchial and pulmonary regions and in order to deliver drug to these regions, the polydispersed therapeutic aerosol containing particles/droplets of the drug should ideally be in the size range of 2–5 μm in diameter.

The influence of particle diameter in determining deposition site is illustrated in Figure 2.4, where the fraction deposited in the alveolar and tracheobronchial regions of the lung is shown as a function of aerodynamic particle diameter. Tracheobronchial deposition may occur by various mechanisms but inertial impaction, sedimentation, and Brownian diffusion predominate. Mouth breathing — the normal route of pulmonary delivery of medicinal agents — bypasses the nasal removal of large particles, which are therefore deposited in the throat and part of the tracheobronchial region. In the bronchioles, ciliated cells are dominant, and in conjunction with mucus secreted

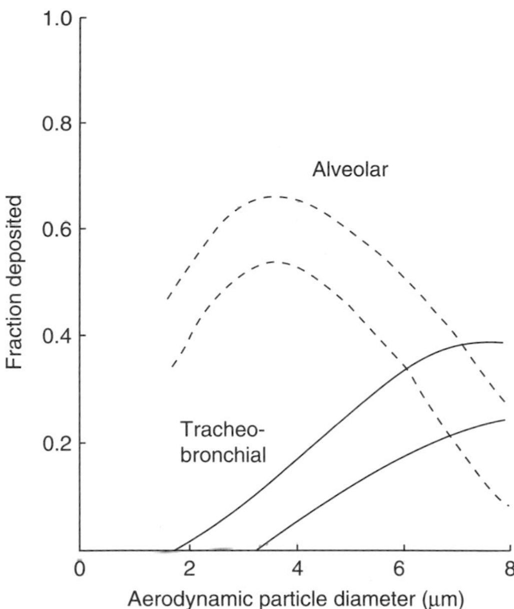

Figure 2.4 Particle diameter dependence of alveolar and tracheobronchial deposition for mouth breathing. Tidal volume 1 L, breathing frequency 7.5/min, mean flow rate 250 cm^3/s, inspiration/expiration times 4 s each. (Reproduced from A.T. Florence and E.G. Salole (eds.), *Routes of Drug Administration*, 1990, p. 53. London: Wright.)

by goblet cells and submucosal glands, constitutes the "mucociliary escalator," which ensures rapid (within hours) removal of insoluble or slowly soluble deposited particles by transport to the mouth for subsequent swallowing. Soluble particles, in contrast, dissolve and may enter the bloodstream. Particles penetrating to the pulmonary compartment may be retained on the pulmonary surfaces as a result of settling, diffusion, and interception processes. Several mechanisms ensure clearance, including dissolution with absorption, phagocytosis of particles by macrophages with translocation to the ciliated airways, and lymphatic uptake. Aerosol characteristics will therefore determine the depth of penetration within the airways and hence particle fate.

2.4.2 Metered Dose Inhalers

This is a sprayable product in which the propellant force is a liquified or compressed gas (Figure 2.5). They are currently the major device used by domiciliary patients and consist of a container hermetically sealed by a metering valve and composed of aluminum or glass protected with a plastic outer casing. As most drugs are of low propellant solubility, they are frequently formulated as micronized suspensions. Stability is achieved by the addition of surfactants, which also serve as a lubricant of the metering valve assembly. Solution formulations may be achieved by the addition of a cosolvent such as ethanol or by solubilization in the added surfactant. Hydrofluorocarbons HFA–134a and HFA–227 are currently the propellants of choice, replacing the previously employed chlorofluorocarbons in response to the requirements of the Montreal Protocol on Substances that Deplete the Ozone Layer.

2.4.3 Nebulizers

Nebulizers are devices for converting aqueous solutions or micronized suspensions of drug into an aerosol. This is effected by two principal mechanisms, either high-velocity airstream dispersion (the air-jet nebulizers) or by ultrasonic energy dispersion (the ultrasonic nebulizers). The former

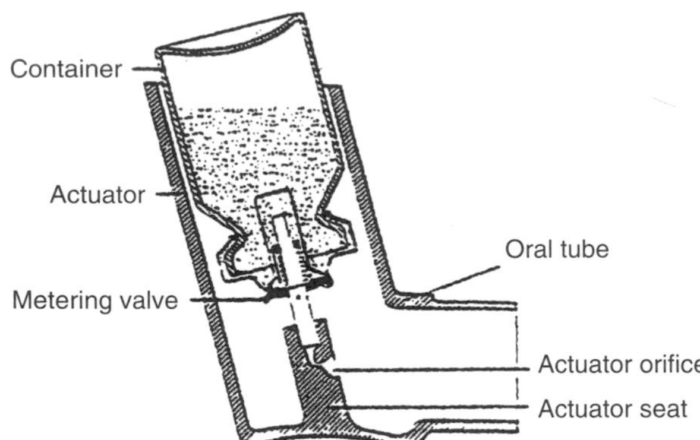

Figure 2.5 Diagram of a metered dose inhaler. Within the container is the drug formulation which typically comprises micronized drug suspended in the propellant and stabilized by a surfactant. (Reproduced from Morén, F. (1981) *Int. J. Pharm.* **8**, 1–10. With permission.)

requires a source of compressed gas (cylinder or air compressors) and hence tend to be more frequently encountered in hospitals than the domiciliary environment. Ultrasonic nebulizers are, in contrast, easily portable. Although these nebulizers produce a dense aerosol plume, often the population of droplets have a higher mass median aerodynamic diameter compared with those generated by the air-jet nebulizers.

Drug formulations for use in nebulizers are, wherever possible, aqueous solutions. Selection of appropriate salts and pH adjustment will usually permit the desired concentration to be achieved. If this is not feasible, then the use of cosolvents such as ethanol and/or propylene glycol can be considered. However, such solvents change both the surface tension and viscosity of the solvent system, which, in turn, influences aerosol output and droplet size. Water-insoluble drugs can be formulated by either micellar solubilization or by forming a micronized suspension.

Nebulizer solutions are often presented as concentrated solutions from which aliquots are withdrawn for dilution before administration. Such solutions require the addition of preservatives, e.g., benzalkonium chloride and antioxidants (e.g., sulphites). Both excipient types have been implicated with paradoxical bronchospasm and hence the current tendency to use small unit-dose solutions that are isotonic and free from preservatives and antioxidants.

Nebulizers of different design produce aerosols of different output and particle size of droplets. For maximum efficacy, the drug-loaded droplets need to be less than 5 μm. In the treatment of prophylaxis of *Pneumocystis carinii* pneumonia with nebulized pentamidine and where the target is the alveolar space, it is desirable to use nebulizers capable of generating droplets of less than 2 μm. During the nebulization from air-jet nebulizers, cooling of the reservoir solution occurs which, together with vapour loss, results in concentration of the drug solution. This can lead to drug recrystallization with subsequent blockage within the device or variation in aerosol droplet size. In contrast, ultrasonic nebulization results in a rise in solution temperature and a decrease in aerosol size.

Although aerosol size distributions are a critical determinant of effective pulmonary drug delivery, it is also desirable to consider output in selection of a nebulizer. For most applications drug administration should occur over a maximum of 10–15 min to optimize patient compliance.

2.4.4 Dry Powder Inhalers

These breath-activated devices aerosolize a set dose of micronized drug on an airstream. The earliest devices consisted of the micronized drug contained within a single-dose capsule, which often contained lactose as an inert drug carrier and diluent. On rapid inhalation, mechanical deaggregation of the powder occurs but the high inertia ensures a significant deposition of the powder on the back of the throat. DPIs tend to be even less efficient than MDIs but because of the higher doses employed, an equivalent therapeutic effect can be achieved. Multidose systems are now available, e.g., Diskhaler® and Turbuhaler®, the latter functioning at low inspiratory flow rates with the capability of delivering, for example, 200 × 1 mg doses of terbutaline sulphate.

2.4.5 Pulmonary Drug Selectivity and Prolongation of Therapeutic Effects

Prodrugs

In addition to improved selectivity of action in the lung relative to other organs, it is possible to obtain prolongation of therapeutic effects and enhancement of pulmonary activity by the design of appropriate prodrugs. Lung accumulation from the blood pool is achieved by many drugs, which are both highly lipophilic and strongly basic amines. Such drugs exhibit very slowly effluxable lung pools.

Lung tissue exhibits high nonspecific esterase activity, which is species dependent and capable of cleaving carboxylate or carbonate ester linkages. *In vivo* prodrug conversion to active drug moiety can be controlled by use of different aliphatic or aromatic coupling agents, together with stereochemical modifications.

Terbutaline (**2.1**) is an example of a bronchodilator drug for which a number of prodrugs exist. Terbutaline exhibits little affinity for lung tissue being rapidly absorbed following inhalation with peak plasma concentrations occurring within 0.5 h. The diisobutyryl ester (ibuterol) (**2.2**) results in an increased bioavailability of 1.6-fold over terbutaline following oral administration. However, it is three times as effective as terbutaline postinhalation in inhibiting bronchospasm. Enhanced effects are attributable to more rapid absorption and better tissue penetration. Bambuterol (**2.3**) is the bis-*N, N*-dimethylcarbonate of terbutaline and as such is well absorbed from the GI tract and is relatively resistant to hydrolysis, leading to a sustained release oral product. However, it is not readily metabolized in the lung, which precludes its administration by the pulmonary route.

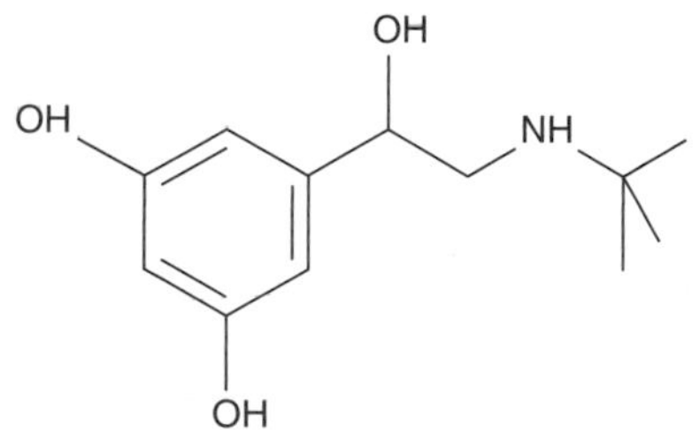

OH
OH
NH
OH

(2.1); terbutaline

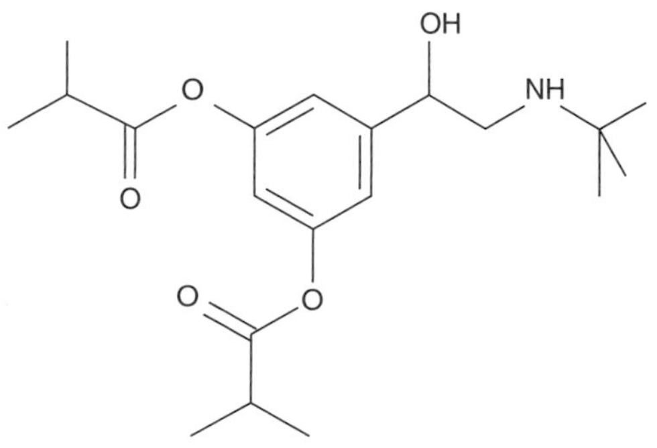

OH
O
NH
O
O
O

(2.2); ibuterol

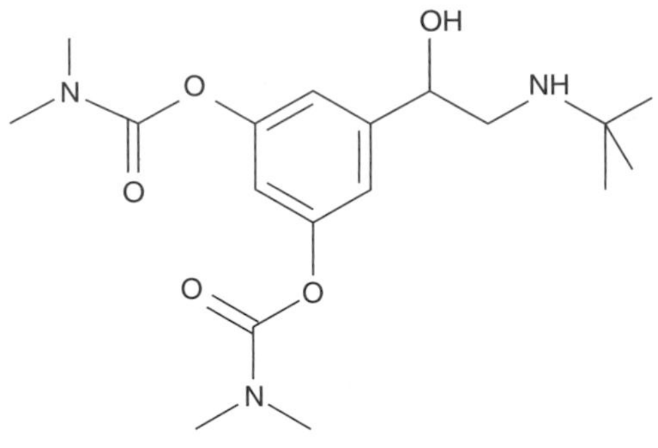

(2.3); bambuterol

Polyamine active transport system

The cell types, which accumulate polyamines such as endogenous putrescine, spermidine, and spermine, together with compounds such as paraquat, are the Clara cells and the alveolar Type I and Type II cells.

Rate control achievable by employing colloidal drug carriers

Control of the duration of local drug activity and of the plasma levels of systemically active agents may be achievable by employing a colloidal carrier possessing appropriate drug release characteristics. Tracheobronchial deposition of such carriers may not be desirable as their clearance will occur in a relatively short time period on the mucociliary escalator. Pulmonary deposition will, in contrast, result in extended clearance times, which may be dependent upon the composition of the colloid. The mechanism by which clearance is effected will also vary, but will involve alveolar macrophage uptake, with subsequent metabolism or deposition on to the mucus blanket in the ciliated regions or lymphatic uptake. Colloidal carriers, of which liposomes are an example, can therefore control both drug delivery rates and availability. Technological problems, however, exist such as the design of delivery devices to ensure deposition in the appropriate regions of the lung without degradation or loss of entrapped drug. Toxicological considerations, foremost amongst which is the processing of the colloid, also require to be addressed.

2.4.6 Delivery of Drugs to the Systemic Circulation by the Pulmonary Route

The large surface area, thin epithelial membrane provided by Type I cells, and a rich blood supply ensure that many compounds are readily transported from the airways into the systemic circulation. Gaseous anaesthesia and oxygen therapy are examples of efficient clinical utilization of the pulmonary absorption process. Compounds are absorbed by different processes including active transport and passive diffusion through both aqueous pores and lipophilic regions of the epithelial membranes. Absorption can be both rapid and efficient; for example, sodium cromoglycate is well absorbed from the lung whereas less than 5% is absorbed from the GI tract.

Small lipophilic molecules, such as the gaseous anaesthetics, are absorbed by a nonsaturable passive diffusion process. Hydrophilic compounds are absorbed more slowly and generally by a paracellular route. Aqueous pores are, by virtue of their size, capable of controlling the rate and extent of hydrophilic compound absorption. Sodium cromoglycate is absorbed by both active and passive (paracellular) mechanisms. The rates of absorption by the paracellular route decrease as the molecular weight of the compound increases.

The efficiency of absorption from the lung is species dependent. For example, insulin is absorbed from the human lung but less efficiently than in the rat or rabbit. Human growth hormone (molecular weight 22 kDa) is absorbed from the lungs of hypophysectomized rats with an estimated bioequivalence of 40% relative to the subcutaneous route and an absolute bioavailability of 10%, sufficient to induce growth. A nonapeptide (leuoprolide acetate) has been shown to have an absolute bioavailability following aerosolization to healthy male volunteers of between 4 and 18% which, when corrected for respirable fraction, corresponds to 35–55%.

Protein absorption, however, is postulated to occur through the extremely thin Type I cells by the vesicular process of transcytosis. The passage from lung to blood of proteins in the rat has recently been shown to increase during inflammatory conditions with the observed transport correlating to the severity of the lung injury. The pulmonary route therefore warrants further investigation for the systemic delivery of peptides and proteins.

2.5 SUSTAINED AND CONTROLLED RELEASE DOSAGE FORMS

Figure 2.6 illustrates the differences between three distinct drug release profiles achieved by the use of (A) the usual single-dose preparation, (B) a sustained release preparation, and (C) a prolonged release preparation. Sustained release products are rarely achieved in practice, although in many respects they represent an ideal delivery system. Initially a loading dose is rapidly released from the sustained action delivery system to provide the necessary blood levels to elicit the desired pharmacological response. The remaining fraction of the dose (maintenance dose) is then released from the preparation at rates that ensure the maintenance of a constant blood level. Prolonged action delivery systems merely extend the duration of the pharmacological response compared with the usual single-dose preparation. Not all drugs are suitable candidates for prolonged action medication as (a) the drug must be absorbed efficiently over a substantial portion of the GI tract, (b) the drug must possess a reasonably short biological half-life (<12 h) (c) the size of the prolonged dosage form must not be too large for ease of swallowing, i.e., the drug must be effective at a "reasonable" dose level, and (d) the pharmacological activity of the drug should be clinically desirable. In some instances, the latter has been questioned if tolerance to the drug may result.

It should be noted that various terms have been employed to describe oral dosage forms, which provide long-term therapeutic action. These include "sustained," "prolonged," "slow,"

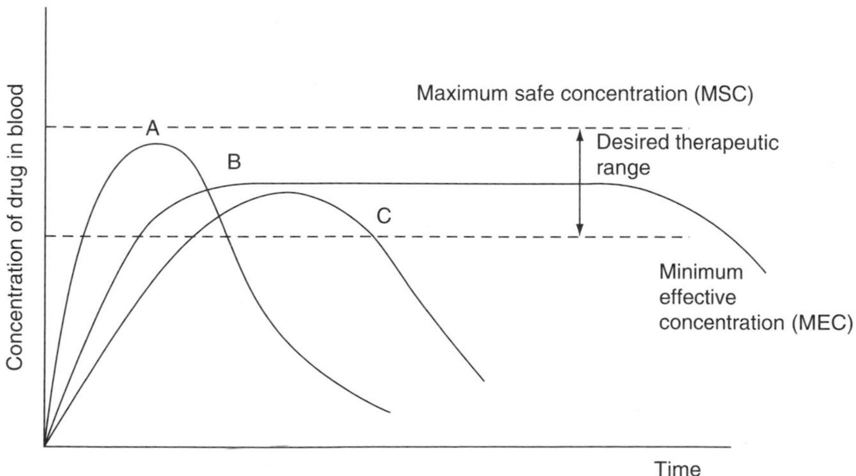

Figure 2.6 The difference between sustained and prolonged release dosage forms as illustrated by the blood concentration–time profiles.

"gradual," "timed," "extended," and "controlled." Often such terms are used interchangeably, although "controlled" should be reserved for drug delivery systems where the rate of drug release is determined solely by the device and is therefore independent of any anatomical or physiological constraints.

2.5.1 Potential Advantages of Sustained Controlled Release Products

Prolonged drug absorption and reduced peak blood concentrations are two obvious advantages of effectively designed sustained release products. As a consequence of the prolonged absorption phase, therapeutic effects should also be extended and a more regular and even pattern established. It has been claimed that a reduction in dosing frequency to once daily will lead to improved compliance and a concomitant reduction in unwanted side effects from high peak blood levels. Irritant drugs such as nonsteroidal antiinflammatory agents, which are slowly released within the gut, should result in reduced inflammatory responses in the gastric mucosa.

2.5.2 Therapeutic Concentration Ranges and Ratios

Rapid diffusion of the drug across capillary walls will result in equilibrium drug concentrations at the target site equivalent to the free serum concentration. Under these conditions, concentration–effect relationships can be established. Often it is difficult to define both MSCs and MECs due to variability in reported values (Figure 2.6). Therapeutic ranges are often related to patient age, disease state, and concomitant therapy. The presence of active metabolites and intersubject variation in plasma–protein binding often complicates both the highest tolerable and the minimum therapeutic concentrations; the ratio of which is the therapeutic concentration ratio (TCR). For drugs exhibiting a low TCR, it is critical to minimize variations in peak and trough plasma concentrations. Hence, a controlled release dosage form becomes highly desirable because, in addition, minimization of variations in serum concentration between doses will achieve both increased therapeutic effectiveness and safety.

2.5.3 Dosage Interval Concentration Ratio and Rate of Elimination

The dosage interval concentration (DICR) is the ratio of the peak to the minimum plasma concentration achieved during a single dosing interval. It is dependent on and will increase with dosing frequency and absorption and elimination rates (see Chapter 1). Elimination rate is an intrinsic property of the drug molecule and therefore, unlike absorption rate, cannot be controlled by formulation factors. Minimization of the DICR for rapidly cleared drugs can be achieved by frequent administration, resulting in patient inconvenience and hence poor compliance, or by the more pragmatic approach, the design of sustained release formulations in order to prolong the absorption phase.

2.5.4 Mechanisms of Achieving Sustained Release by the Oral Route

Oral sustained release products have been produced employing various drug release mechanisms. Unfortunately, no single approach is universally acceptable and selection is inevitably related to drug properties. For *bona fide* controlled release, *in vitro* release profiles are superimposable on those achieved *in vivo*. If drug release is pH-dependent, then greater variability in *in vivo* performance is to be expected. Excluding molecular modification in order to change drug solubility (prodrug, salt, or complex formation) and the use of the limited applicability of ion exchange resins, the design of sustained release products is often accompanied by employing one or more of the following approaches.

Hydrophilic gel tablets or capsules

Hydrophilic gums are used to form a gel layer surrounding the tablet upon introduction to an aqueous medium. Diffusion across this layer constitutes the rate-determining release step. Nitroglycerin (glyceryl trinitrate) dispersed in hydroxypropylmethylcellulose for buccal administration permits sustained release for over 4 h. Capsules, the contents of which swell within the stomach to produce a plug which is buoyant on the gastric contents, provide a further example of this type of technology.

Matrix tablets

The eroding variety of matrix tablets are generally slowly disintegrating tablets, although similar release systems can be achieved by using semisolid lipophilic materials in hard gelatin capsules. This approach generally leads to poorer control of *in vivo* performance. Drugs dispersed in inert matrices can also be employed for sustained release tablets, and better *in vivo* reproducibility generally results, as drug release rates are not dependent on enzyme levels or gastric intestinal transit rates. Zero-order release kinetics are not obtained from these tablets; cumulative drug release is often proportional to the square root of time.

Capsules containing pellets with disintegrating coatings

By employing differentially coated pellets, i.e., pellets with varying thickness of a slowly dissolving or eroding polymer, it is possible to provide an extended dissolution profile of a drug. Such a coating may be pH sensitive or insensitive, the former to provide positional release (e.g. classic enteric coatings), the latter to achieve drug release rates independent of transit profiles within the GI tract.

Pellets or tablets coated with diffusion-controlling membranes

Pellets or a compressed tablet may be coated with a rate-controlling membrane (nondisintegrating) across which the drug may diffuse. Pellets are normally presented in hard gelatin capsules. By encapsulating drugs in excess of their solubility, a constant concentration gradient will be maintained as long as the saturation state exists and zero-order kinetics will prevail.

The OROS® elementary osmotic pump comprises a central core of a salt to provide an osmotic gradient together with drug particles. Surrounding this core is a semipermeable membrane. Fluid is drawn into the core at a rate controlled by both the membrane characteristics and the osmotic gradient. Saturated drug solution is then forced from the device through a small orifice into the surrounding membrane at a constant rate.

2.5.5 Positional Controlled Release

Considerable benefits may ensue from drug delivery to a specific region of the GI tract. An example of buccal absorption to eliminate first-pass metabolism has previously been described (Section 2.5.4). Gastric retention of the drug delivery system may be required to achieve localized drug concentrations or to delay passage of the dosage form past the absorbing membranes of the small intestine, which may lead to improved bioavailability. Prolonged gastric retention may be achieved by the use of gel rafts, which ''float'' on the gastric contents (as with the hydrodynamically balanced capsule).

It is often necessary to prevent drug release in the stomach to avoid gastric degradation, reduce gastric side effects such as inflammatory responses, or provide localized drug concentrations in the small intestine or colon. Enteric coating of both tablets and capsules provides the most widely used approach to avoid drug release in the stomach and is achieved by the use of film-forming ionizing polymers of suitable pK_a and degree of substitution.

Alternative, more sophisticated technologies exist to achieve colon-specific drug delivery. For example, the Pulsincap® utilizes a novel hydrogel plug fitted into the neck of a water-insoluble capsule. A water-soluble cap fits over the capsule body and dissolves within the gastric juice to expose the underlying plug. The hydrogel swells at a controlled rate and ejects when it can no longer be contained, thus giving rise to pulsed delivery of the capsule contents within the colon approximately 5 h after administration. Applications include treatment of local disorders, e.g., irritable bowel disease, and peptide delivery for absorption at what is the preferred GI site.

2.6 SITE-SPECIFIC DRUG DELIVERY

Having briefly reviewed the principal approaches to the temporal control of drug from the dosage form, it is pertinent now to examine the subject of site-specific delivery or drug targeting. Originally driven out of a desire for therapeutical optimal delivery of cytotoxic agents, the interest in site-specific delivery has expanded. Molecular biology advances have led to a better understanding of a number of states and the identification of potential target sites (receptors) for drugs. Currently witnessing a revolution in availability of biotechnological drug products (e.g., peptidergic mediators, antisense oligonucleotides), their successful clinical utility requires significant application of drug delivery science. In essence, site-specific drug delivery attempts to optimize drug activity by insuring exclusive availability specific receptors, i.e., differential accessibility and in so doing provide protection drug and body.

2.6.1 Carrier Systems

Target selectivity by differential sensitivity (drug distributed throughout the body acting exclusively on target) is virtually impossible to achieve. Therefore, in selectively delivering drugs, it is necessary to utilize a drug carrier system (Table 2.2) A site-specific carrier is required, in addition, to being a guiding device, to protect drug excretion or inactivation, prevent drugs from eliciting adverse immune reactions, provide for site recognition and being retained at that site to achieve drug release at appropriate timescale, and finally to be degraded/excreted.

Parenteral administration is the simplest and most efficient approach for any of the carrier systems, although considerable efforts have been expended to develop systems that will be capable of reaching the blood pool via mucosal epithelia (nasal, pulmonary, buccal, intestinal, vaginal, rectal). Transport mechanisms include fluid-phase pinocytosis, paracellular and transcellular

Table 2.2 Classification of drug carriers

Macromolecular carriers	Microparticulate carriers
Proteinaceous carriers (e.g., antibodies, albumin, glycoproteins, lipoproteins, gelatin, polypeptides)	Cellular carriers (e.g., erythrocytes, leucocytes, lymphoid cells, fibroblasts)
Lectins Hormones	Vesicular carriers (e.g., liposomes, niosomes)
Polysaccharides (e.g., dextran)	Lipid carriers (e.g., emulsions, waxes, lipoproteins, chylomicrons)
Deoxyribonucleic acid	Microspheres/nanoparticles (e.g., albumin, starch, dextran, polyalkylcyano-acryl polyamide, polyanhydrides, poly(lactic glycolic) acid
	Viruses and viral envelope products

diffusion, receptor-mediated trans- and endo-cytosis, nutrient carrier processes, and lymphatic translocation.

2.6.2 Fate of Site-specific Delivery Systems

The target site may be reached by either active or passive events. Microparticulate carriers (Table 2.2) injected into the circulation are usually recognized as foreign and undergo opsonization. This process consists of adsorption of serum components or opsonins (mainly proteins), which trigger the microparticle uptake by the cells of the mononuclear phagocytic system (MPS). Depending on surface properties of the carrier particles, some 50–90% of the injected dose will distribute to the liver where the main target cells are the Kupffer cells with a smaller portion taken up by hepatocytes. The spleen, lungs, and bone marrow will also accumulate the opsonized particles but to a lesser degree. This passive targeting process can be controlled by selecting particles with different surface properties leading to a different spectrum and mass of adsorbed opsonins. Attempts to produce microparticles with long circulation times have focused on preventing and reducing opsonization by preadsorption or surface grafting of surfactants or polymers on the microparticle surface. By utilizing surfactants or polymers which contain hydrophilic moieties such as polyoxyethylene chains, a steric barrier is created preventing serum protein adsorption, which is predominantly a hydrophobic interaction. ''Stealth'' liposomes are examples of sterically stabilized microparticulate drug carriers.

Active targeting, in contrast, involves some cell-specific recognition event achieved by incorporating a target recognition moiety at the surface of the carrier. Immunoliposomes, for example, may be prepared by covalent coupling of Fab$'$ fragments to the preformed liposome surface. The Fab$'$ is prepared from IgG raised against a tumor-specific cell surface antigen.

Although microparticulate carriers have the advantage of providing a high, well-protected drug payload, there are two principal disadvantages to their usage. The first already discussed is their potential capture by cells of the MPS before reaching their targets. The second is the barrier afforded by the endothelia to extravasation. There are three main types of endothelia:

1. HContinuous — where the cells are close together and no gaps exist. The cells overlie a continuous basement membrane. This is the most frequently occurring form and its morphology prevents colloid extravasation. Molecules of 5–10 kDa can pass into the tissue space and subsequently into the lymphatic system.
2. Fenestrated endothelia occur in the exocrine glands, possess a continuous basement membrane, but differ from the continuous type in that the endothelial surface contains fenestrae (50–60 nm holes), which are covered by a diaphragm. Although colloids might penetrate the fenestrae, transport would not occur across the basement membrane.
3. Sinusoidal endothelia are a discontinuous type that is found in the liver, spleen, and bone marrow. Gaps are present in the wall and the basement membrane is absent. In the liver, sieve plates exist through which colloids of less than 100 nm can pass into the space of Disse and hence into the parenchymal cells. Sinusoidal endothelia therefore provide the only opportunity for the extravasation of microparticulate drug carriers. The opportunity therefore exists for treating abnormal cells, such as cancerous or virally infected through macrophage activation, by incorporating an immunomodulatory drug such as muramyl dipeptide in a colloidal carrier. In rheumatoid arthritis, damaged endothelia may provide the opportunity for colloidal carrier accumulation.

Intravascular targets, for example, diseased macrophages (fungal, parasitic and viral, autoimmune diseases and enzyme storage diseases) and other bood cells (cancer, gene, and antiviral therapy) are readily accessible to sub-1.0 μm colloids. Also, the opportunity exists of providing long-circulating microparticulates for the slow release of a range of pharmacological agents.

2.7 BIOEQUIVALENCE

There are many instances reported in the pharmaceutical literature of chemically equivalent products, which have been shown to be bioinequivalent. Any change in a formulation can potentially change the bioavailability. Even with identical formulae, changes in the many process variables can occur in the production of a generic product undertaken by different manufacturers, giving rise to bioinequivalent products. In a granulation process, for example, the nature, quantity, and method of addition of the granulating fluid, the drying process, the ageing and storage conditions of the granules prior to tabletting can all influence the drug release characteristics of the final product. The choice of granulating agent can lead to differences in tablet hardness upon storage, which of course in turn will lead to prolonged disintegration and dissolution times. However, some drugs are known to provide greater bioinequivalence problems than others. The Food and Drug Administration , in response to this situation, published a list of 115 drug substances for which *in vitro* or *in vivo* bioequivalence data are required.

FURTHER READING

Chien, Y.W. (1992) *Novel Drug Delivery Systems*, 2nd edn. New York: Marcel Dekker.

Florence, A.T. and Salole, E.G. (eds.) (1990) *Routes of Drug Administration*. London: Wright.

Junginger, H.E. (ed.) (1992) *Drug Targeting and Delivery: Concepts in Dosage Form Design*. Chichester, England: Ellis Horwood.

Kreuter, J. (ed.) (1994) *Colloidal Drug Delivery Systems*. New York: Marcel Dekker.

Lee, V.H.L. (ed.) (1991) *Peptide and Protein Drug Delivery*. New York: Marcel Dekker.

Robinson, J.R. and Lee, V.H.L. (eds.) (1987) *Controlled Drug Delivery: Fundamentals and Applications*, 2nd edn. New York: Marcel Dekker.

3

Fundamental Pharmacokinetics

Glyn Taylor

CONTENTS

3.1 INTRODUCTION

Pharmacokinetics (PK) is the scientific discipline concerned with the time course of drug absorption, distribution, metabolism, and excretion. Defining the time course involves measuring drug concentrations in plasma, blood, saliva, urine, or a combination of these. Pharmacokinetic analysis then allows us to predict drug concentrations in fluids and tissues, which would result from administering different doses, multiple doses, doses given by different routes, and doses given to different groups of patients. Moreover a study of pharmacokinetics helps us better understand how drugs are subject to the processes of absorption, distribution, and elimination by the body and use this knowledge to design new drugs with better pharmacokinetic profiles.

Pharmacodynamics (PD) is the scientific discipline concerned with the relationships between drug concentrations at the site of action and drug effects.

There are a number of common features between PK and PD but this chapter will focus on describing the fundamentals of PK analysis and highlight some of the areas of particular relevance to drug design.

3.2 BASIC KINETICS ASSOCIATED WITH DRUG HANDLING BY THE BODY

The processes determining the fate of drug molecules in the body can be categorized in a number of ways. Figure 3.1 is a schematic illustrating drug liberation from the dosage form, absorption into the bloodstream, reversible distribution into tissues, elimination by metabolism or excretion, and delivery of drug to the effect sites where the response is elicited, either directly or indirectly. The response may be the requisite pharmacological effect or a toxic side effect. The schematic is a simplistic overview of the elements involved but does serve to illustrate that the response elicited by a particular drug is determined by a number of biophysical processes and gaining an understanding of the kinetics of these processes is crucial to optimizing drug usage.

3.2.1 Liberation

The influence of formulation on drug release is fully discussed in Chapter 2. From the PK perspective it is important to appreciate that for the overwhelming majority of drugs, they will need to be in solution before transport across the absorption barrier can occur.

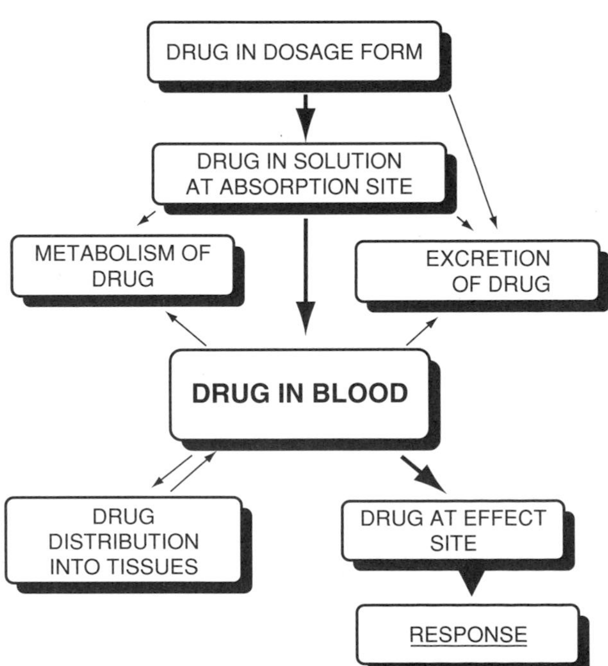

Figure 3.1 Schematic of liberation, absorption, distribution, metabolism, excretion, and response (LADMER).

3.2.2 Absorption

As discussed in Chapter 1 the absorption of drugs after oral dosing is determined by a number of factors and predominantly occurs in the small intestine rather than the stomach. The small intestine has a much larger surface area and has specialized transport processes and thus is the main site of absorption for orally administered drugs. Drug absorption kinetics after oral dosing are multifaceted and depend upon various physicochemical and molecular properties of the drug. A number of processes including dissolution, gastric emptying, diffusion, partitioning, degradation, intestinal metabolism, intestinal motility, and binding to luminal contents may all be contributory factors for different drugs. The observed kinetics resulting from the combination of these processes are determined by the rate-limiting process. Many of the factors, including dissolution, gastric emptying, degradation, and diffusion, under certain conditions, can be manifest as first-order processes and in most cases the kinetics of oral absorption are approximated as a first-order process and an absorption half-life ($t_{1/2,\mathrm{a}}$) can be determined as described later (Section 3.3.3). It should however be recognized that the multiple processes involved in oral absorption kinetics may not always be accurately described by a first-order process, another approach is to use a ''model independent'' approach and determine, for example mean input time (MIT) (Section 3.5).

Absorption kinetics from routes of administration other than oral, such as rectal, pulmonary, and nasal, are also most often ascribed to first-order processes, whereas subcutaneous and intramuscular injections can result in either first- or zero-order kinetics, depending upon the drug's physicochemical properties.

Absorption from these alternative routes is often manifest as absorption rate limited pharmacokinetics, due to the drugs being absorbed slowly or because the drugs (in particular proteins) have very short elimination half-lives in plasma. Thus, administration by these routes offers a means of attaining sustained blood concentrations and effect for a number of drugs.

3.2.3 Distribution

The extent to which drugs are distributed throughout the body varies widely from those few drugs which almost solely reside within blood (e.g., warfarin) to drugs such as chloroquine for which less than 0.1% of the total amount in the body will be circulating in blood at any given time. The extent of distribution is represented by a ''volume of distribution,'' which is the theoretical volume of blood required to account for the total amount of drug in the body. Thus the volume of distribution of warfarin is only around 5 L but is approximately 20,000 L for chloroquine. The volume of distribution of a drug does not usually relate to any physiological space but is a very useful parameter which we use to relate blood concentrations and amounts of drug in the body, using the following equation:

$$V = \frac{\text{Amount in body}}{\text{Concentration in blood}} \tag{3.1}$$

Distribution of drugs within the blood pool occurs very rapidly after intravenous (IV) dosing and for almost all practical purposes can be regarded as instantaneous. A number of factors including drug lipophilicity, $\mathrm{p}K_\mathrm{a}$ and molecular size may each influence affinities for plasma proteins and tissue constituents. Consequently these factors determine the partitioning and transport kinetics of a drug from the blood into a particular tissue. Distribution from the blood to other tissues is generally very rapid for highly perfused tissues but can take up to several hours for poorly perfused tissues. Blood perfusion rates for different tissues varies greatly, for example lung tissue is perfused at approximately 300 times the rate for peripheral fat. The net result is that both the extent and time course of distribution show wide variations for different drugs. The kinetics of drug distribution are usually ascribed to first-order, since most of the processes involve diffusion or partitioning. Kinetically distribution is usually described by a first-order distribution rate constant (α).

3.2.4 Elimination

The processes of both elimination and distribution result in drug removal from the blood, whereas distribution is a reversible process, elimination is non-reversible. Drug elimination may involve two mechanisms: (i) metabolism, which occurs primarily in the liver, lung, and kidneys and (ii) excretion of unchanged drug into urine, bile (and breath for volatile drugs) from kidney and liver (and lung). Thus liver, kidney, and lung are the primary eliminating organs for drugs.

The most important PK term used to quantify drug elimination is clearance (CL). As blood flows through the eliminating organs in the body, a specific proportion of the drug present in the blood will be eliminated. Clearance is determined by blood flow (Q) to the eliminating organ and the extraction ratio (E) by the following equation:

$$\mathrm{CL} = QE \tag{3.2}$$

From Equation (3.2) we can see that blood flow to the organ is important in determining CL and thus the major organs for drug elimination have high blood flows. The extraction ratio (E) is the fraction of drug molecules eliminated (by metabolism or excretion) during a single passage through that organ. Alternatively, E can be considered as the fraction by which the arterial drug concentration perfusing the organ is reduced (i.e., 1 – output/input concentration). The extraction ratio can vary between 0 and 1 for different drugs and different organs. For example, promethazine has an E value of approximately 0.75 in the liver but less than 0.05 in kidney, whereas *para*-aminohippurate has an E value of 0 in liver but virtually 1 in kidney. By virtue of the latter, kidney blood flow can be estimated from *para*-aminohippurate PK.

Clearance can be defined as the volume of blood which is totally cleared of drug in unit time and can vary from 0 to the blood flow for the organ. For example, promethazine has a hepatic clearance of approximately 68 L h^{-1} (since Q for liver blood flow is about 90 L h^{-1} and E is 0.75). In reality clearance does not result in drug being totally removed from a specific volume of blood but it does cause a consistent reduction in the drug concentration in blood across a specific organ. Clearance can also be considered at the whole body level, where the whole body clearance will reflect the sum of clearances for each of the eliminating organs. Since the rate of elimination of most drugs follows first-order kinetics that rate is determined by the amount of drug presented to the eliminating organs. The amount presented to the eliminating organs is blood flow multiplied by concentration and it follows that rate of elimination is proportional to blood concentration. Clearance is the parameter which relates rate to concentration

$$\mathrm{CL} = \frac{\text{Rate of elimination}}{\text{Concentration in blood}} \tag{3.3}$$

Describing elimination by a first-order process may at first seem surprising considering that drug metabolism and other elimination processes often involve Michaelis–Menten-type kinetics. If, however, we consider drug elimination by metabolism in the liver, during the drug's passage through that organ a number of processes including diffusion, partitioning, and drug–enzyme interaction will occur. The former two are driven by first-order processes and although Michaelis–Menten kinetics may dictate the processes of metabolite formation from the drug–enzyme complex; for almost all drugs their therapeutic blood concentrations are much lower than their half-saturation concentration, and thus apparent (or pseudo-) first-order kinetics are seen.

Similarly in considering drug elimination by excretion, the main processes involved are filtration, active transport, and re-absorption (by partitioning). The filtration and partitioning processes are first-order and the transport processes are usually far from being saturated and are manifest as pseudo-first-order kinetics.

In a few cases, the elimination processes above may start to become saturated and there will be deviation from first-order kinetics towards zero-order kinetics and is one category of non-linear PK. This situation occurs for example with the metabolism of phenytoin and alcohol by the liver. However, for the overwhelming majority of drugs, the elimination kinetics are first-order. Where absorption, distribution, and elimination are all first-order processes then the drug is deemed to show ''linear PK,'' since doubling the dose administered will double the resulting concentrations.

While the most important PK parameter used to quantify elimination is clearance, another often cited parameter is the elimination half-life ($t_{1/2}$), which is the time taken for half of the drug to be eliminated from the body. Drug half-lives vary from a few minutes for a drug such as glyceryl trinitrate to several days for a drug such as phenobarbital. For drugs which show first-order kinetics, their half-life will be constant but for those showing non-linear elimination kinetics, the time taken to eliminate half of the drug will increase with higher doses thus doubling the dose will produce more than a two-fold increase in the resulting blood concentrations.

3.2.5 Pre-systemic Elimination

After oral administration, there are a number of barriers which a drug must transverse before it reaches the systemic circulation, as illustrated in Figure 3.2. Thus some of the drug molecules may be eliminated from the body before they have the opportunity to elicit an effect. Hence factors involved in pre-systemic elimination, which is one element of drug bioavailability (BA), need to be considered. The Food and Drug Administration (FDA) definition of BA is ''the rate and extent to which the active ingredient or active moiety is absorbed from a drug product and becomes available

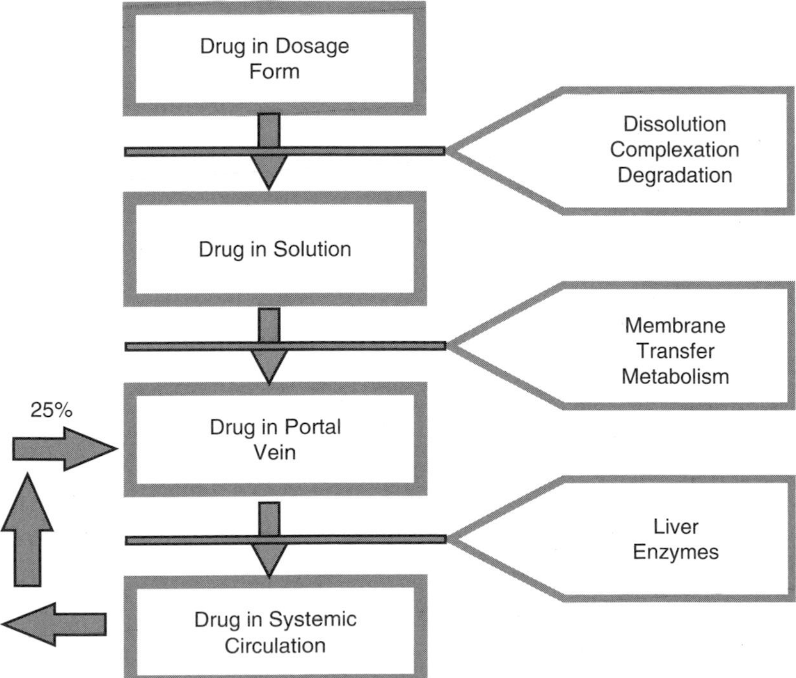

Figure 3.2 Factors determining bioavailable fraction.

at the site of action.'' For drugs which act systemically, this concept of BA translates to the rate and relative amount of drug reaching the systemic blood circulation. The rate is usually determined from the peak blood concentration (C_{max}) and the time to reach the peak blood concentration (t_{max}). This is discussed in Section 3.3.3. The relative amount is determined by the bioavailable fraction (F), i.e., the fraction of the drug dose which reaches the systemic circulation as intact drug (i.e., non-metabolized).

Since clearance relates rate of elimination to blood concentration, then from Equation (3.3) we can express the rate of elimination mathematically as a decrease in the amount of drug in the body with time ($-\mathrm{d}A/\mathrm{d}t$)

$$-\frac{\mathrm{d}A}{\mathrm{d}t} = \mathrm{CL}\ C \tag{3.4}$$

Resolving Equation (3.4) results in ($-\mathrm{d}A/\mathrm{d}t$) integrated to the total amount of drug reaching the blood at some time after drug administration (i.e., F dose); CL is a constant for drugs showing linear PK; and concentration integrates to $C\mathrm{d}t$, which is the area under the concentration vs. time profile (AUC) from time zero to infinity

$$\mathrm{CL} = \frac{F\ \mathrm{Dose}}{\mathrm{AUC}} \tag{3.5}$$

F is usually regarded as 1 for IV administered drugs and thus CL is calculated after IV dosing. Equation (3.5) and its variants are used in many different aspects of PK and are regarded as a model independent calculation of CL, since few assumptions are made about the drug's PK.

The value of F for a drug given by different routes of administration can thus be calculated from Equation (3.6) by comparing the dose normalized AUC after (e.g., oral) dosing with that seen after IV dosing

$$F = \frac{\mathrm{AUC}_{\mathrm{PO}}\ \mathrm{Dose}_{\mathrm{IV}}}{\mathrm{AUC}_{\mathrm{IV}}\ \mathrm{Dose}_{\mathrm{PO}}} \tag{3.6}$$

One important pharmacokinetic contributor to F for orally administered drugs is that of pre-systemic elimination, which may occur by metabolism in the intestinal lumen for drugs such as insulin, in the intestinal wall for example with midazolam, cyclosporine, and morphine, and finally during first passage through the liver. Most drugs are subject to some degree of first-pass metabolism in the liver after oral administration and for some drugs, such as alprenolol, terfenadine, and tricyclic antidepressants, they are extensively metabolized on first-pass, with as little as 5–10% of the dose reaching the systemic circulation as intact drug.

First-pass metabolism in the liver is determined by the hepatic clearance and extraction ratio of the drug across the liver. First-pass metabolism of drugs given orally occurs because after the drug is absorbed from the intestine, it is transported to the liver where the whole of the absorbed dose is vulnerable to metabolism by the liver enzymes. Once the drug has traversed the liver and entered the systemic circulation, it will be re-circulated through the liver but since only about 25% of the cardiac output re-circulates through the liver then each molecule will be statistically less vulnerable to metabolism (see Figure 3.2). Thus for a drug whose extraction ratio across the liver is 0.9, only 10% of the absorbed dose will appear in the blood (i.e., F will be 0.1 if all of the dose is absorbed intact across the intestinal membrane). Thus when comparing the same dose of drug given orally and IV, lower concentrations will be observed after oral dosing due to pre-systemic elimination.

Drugs given by the rectal route are also subjected to liver first-pass metabolism, depending upon the site of absorption within the rectum (de Boer and Breimer, 1997). Some inhaled drugs are also subject to first-pass metabolism by the lung (Dickinson and Taylor, 1996).

Formulation factors can also influence a drug's BA and it is often important to consider bioequivalence between different formulations of the same drug. This subject is discussed in Chapter 2.

3.3 SINGLE-DOSE PHARMACOKINETIC MODELS

One of the purposes of using PK models is to aid our understanding of the kinetics associated with how the body handles drugs. Models, or at least empirical equations, are necessary for predicting and summarizing PK data. Modeling techniques range from the use of very simplistic models to exquisite physiologically based models such as illustrated in Chapter 1 (Figure 1.5).

The most commonly used modeling technique in PK is to represent the body as containing one, two, or three compartments. These compartments bear little relation to any physiological spaces but can be used to give useful predictions. The physiologically based PK models are built using elements which represent specific tissues and organs in the body and are particularly useful where *ex vivo* studies in isolated tissues can be used to predict how drugs might be handled by the body. A detailed description of this type of modeling is beyond the scope of this chapter but is reviewed elsewhere (Oliver et al., 2001).

In the following sections we will look at one- and two-compartment models.

3.3.1 Elimination Model (IV Bolus 1-Compartment Open Model)

This is the simplest model which is in common use in PK and is illustrated in Figure 3.3. It is sometimes referred to as the one-compartment (open) model and may be used to predict PK for some drugs which are administered as an IV bolus, i.e., the drug dose is injected intravenously within a short period of time (less than a minute). The body is viewed as a single homogeneous compartment into which the drug is instantaneously administered. After injection, elimination is the only kinetic process affecting the drug and this is a first-order process defined by the first-order elimination rate constant (k). It is an ''open model'' since the eliminated drug is not recorded. We could add another compartment to the end of the arrow to represent the amount of drug eliminated. This would then be a two-compartment ''closed model'' and this approach might be useful if we were particularly interested in predicting the production of metabolite or the amount of drug excreted in urine.

In this model the drug is assumed to instantaneously distribute from the site of injection (venous blood) to all other tissues in the body for which the drug has affinity. As discussed in Section 3.2.3 there are a number of factors which determine the distribution kinetics. The model has limited applications but for some drugs such as warfarin and gentamicin, which have restricted distribution, the model is very useful

$$C = C_{(0)}e^{-kt} \tag{3.7}$$

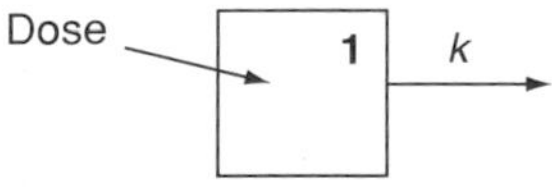

Figure 3.3 A one-compartment open disposition model.

Typical blood concentrations associated with this model are illustrated in Figure 3.4. The concentrations decline exponentially (in accord with Equation (3.7)) and thus a straight line is given when concentrations are plotted on a logarithmic scale against time. In plotting this type of blood concentration–time profile, if the base of natural logarithms, e, is chosen for the log scale and $\log_e$ concentrations are plotted against time, then the slope of line is given by $-k$ and an intercept of $\log_e C_{(0)}$, since taking logs of Equation (3.7) reveals:

$$\log_e C = \log_e C_{(0)} - kt \tag{3.8}$$

In Figure 3.4 a log scale has been used, however it is not the $\log_e$ scale but one from which the blood concentrations can be directly read. Using this approach we can fit Equation (3.7) to experimental blood concentration data and $C_{(0)}$ (the theoretical concentration at time zero) can be read directly from the Y-axis (as shown in Figure 3.4). We know that immediately after injection (at $t = 0$) the amount of drug in the body is the injected dose and thus we can use Equation (3.1) to determine the drug's volume of distribution

$$V = \frac{\text{Dose}}{C_{(0)}} \tag{3.9}$$

In the one-compartment model the drug has only one volume of distribution, i.e., the V is constant over time (cf. Section 3.3.2).

We can also use Equation (3.7) to predict the drug concentrations (C) at any time after injection. For example we can calculate the time taken for the concentration to decrease by a half and this period of time is known as the elimination half-life ($t_{1/2}$). The time taken for any two-fold decrease in concentration (from C to $1/2C_{(0)}$; or $1/2C_{(0)}$ to $1/4C_{(0)}$; or $2/3C_{(0)}$ to $1/3C_{(0)}$; etc.) will be the same, that is one half-life. Substituting $t = t_{1/2}$ and $C = 1/2C_{(0)}$ in Equation (3.7) and solving for $t_{1/2}$ (by taking natural logs, $\log_e$), we find the relationship between the elimination rate constant (k) and half-life and thus we can calculate k from the half-life:

$$k = \frac{\log_e 2}{t_{1/2}} = \frac{0.693}{t_{1/2}} \tag{3.10}$$

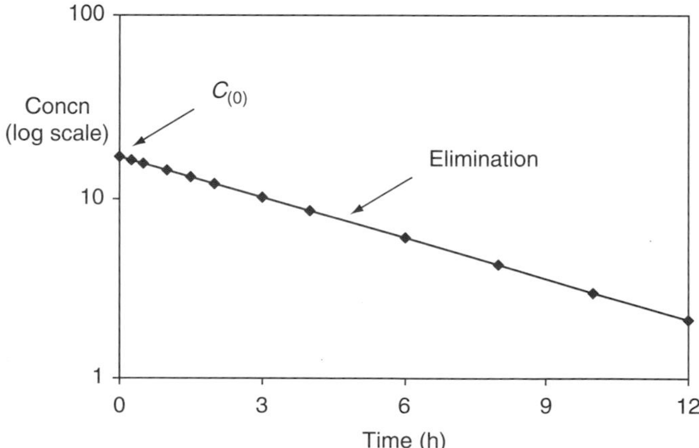

Figure 3.4 A typical concentration–time profile of a drug showing one-compartment PK after IV dosing.

The elimination rate constant k relates rate of elimination to amounts of drug in the body:

$$k = \frac{\text{Rate of elimination}}{\text{Amount in body}} \tag{3.11}$$

Equation (3.11) has limited practical applications since amounts of drug in the body at various times after administration are not readily determined, hence we use clearance, introduced in Section 3.2.4, which relates rate of elimination to blood concentrations (Equation (3.3)). Since blood concentrations are related to amounts in the body by volume of distribution (Equation (3.1)) then we can derive the relationship between CL and k:

$$k = \frac{\text{CL}}{V} \tag{3.12}$$

Thus using Equation (3.12) to substitute for k in Equation (3.10) and solving for $t_{1/2}$ gives:

$$t_{1/2} = \frac{0.693V}{\text{CL}} \tag{3.13}$$

Equation (3.13) is important in concept since the drug half-life is dependent upon both CL and V, i.e., both elimination and distribution will influence the observed elimination half-life. Thus a drug may have a long half-life because the eliminating organs are not efficient at removing the drug (i.e., the drug has a low CL) or because the drug has a large V and thus only a very small fraction of the amount of drug in the body is passing through the eliminating organs at any one time. For example the drug chlorpromazine has a fairly long half-life of around 12–30 h yet it is very efficiently eliminated by the liver with the chlorpromazine molecules having about a 80% probability of being metabolized each time they pass through the liver (i.e., hepatic E is approximately 0.8).

The clearance of the drug can be calculated experimentally using Equation (3.13) from a prior calculation of V and $t_{1/2}$, however a more useful approach, which has general applicability to many different compartment models is to derive clearance from the area under the blood concentration–time curve (AUC) using Equation (3.5) where F is 1 for IV-administered drugs.

3.3.2 Distribution Plus Elimination Model (IV Bolus 2-Compartment Open Model)

For most drugs, their distribution throughout tissues in the body is not instantaneous and thus the one-compartment model has restricted uses. The model shown in Figure 3.5, the two-compartment open model, incorporates the phenomenon of drug distributing into tissues over a finite amount of time. In this model, Compartment 1 comprises the blood and ''highly perfused'' tissues. Thus the IV dose is injected and is instantaneously distributed throughout blood and all other tissues in this compartment. Compartment 2 contains the ''poorly perfused'' tissues, into which drug slowly distributes. Compartment 1 is often described as the ''central compartment'' and Compartment 2 as the ''peripheral compartment.'' As discussed previously, compartment modeling is primarily

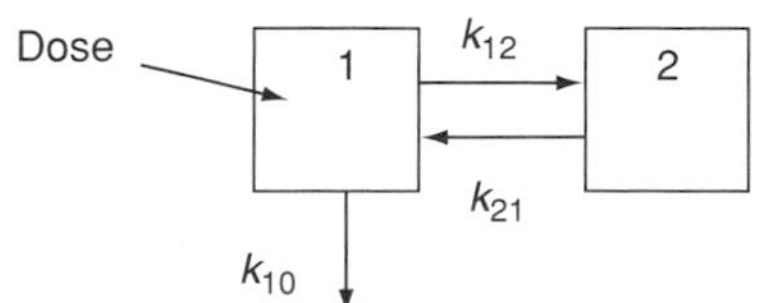

Figure 3.5 A two-compartment open disposition model.

used to make predictions and aid in our understanding of PK. The terms ''poorly'' and ''highly'' perfused and ''central'' and ''peripheral'' should not be interpreted too literally. In the two-compartment model the distribution kinetics are determined by the first-order rate constants k_{12} and k_{21}, which define drug transfer from Compartment 1 to Compartment 2 and vice versa. As discussed previously, distribution kinetics are likely to be manifest as first-order processes. Drug elimination only occurs from Compartment 1, since (as described in Section 3.2.4) the major eliminating organs are usually those with high blood flows.

In a concentration–time profile resulting from this model shown in Figure 3.6, a number of differences with Figure 3.4 can be observed. The blood concentrations immediately after IV injection of the drug decline rapidly, which is due to a combination of distribution (from Compartment 1 to Compartment 2) and elimination (from Compartment 1). During this initial phase, concentrations in Compartment 1 decline rapidly while those in Compartment 2 will increase. After a certain period of time, approximately 3 h in the example shown in Figure 3.6, a ''distribution equilibrium'' is reached and concentrations in Compartment 1 and Compartment 2 will decline in parallel during this terminal phase where the decline is determined by the elimination half-life $t_{1/2,\ \beta}$.

Equation (3.14) is used to determine blood concentrations in this model

$$C = A\mathrm{e}^{-\alpha t} + B\mathrm{e}^{-\beta t} \tag{3.14}$$

There are similarities between this equation and Equation (3.7). Equation (3.14) has two exponential elements, the $B\mathrm{e}^{-\beta t}$ component is similar to $C_{(0)}\mathrm{e}^{-kt}$ in Equation (3.7) insofar as these both reflect elimination of the drug. Thus β is the elimination rate constant in the two-compartment model. The $A\mathrm{e}^{-\alpha t}$ component reflects distribution and α is the distribution rate constant. The distribution and elimination rate constants (α and β) are related to the respective distribution and elimination half-lives ($t_{1/2,\alpha}$ and $t_{1/2,\beta}$) as shown in Equations (3.15) and (3.16), which are analogous to Equation (3.10)

$$\alpha = \frac{0.693}{t_{1/2,\alpha}} \tag{3.15}$$

$$\beta = \frac{0.693}{t_{1/2,\beta}} \tag{3.16}$$

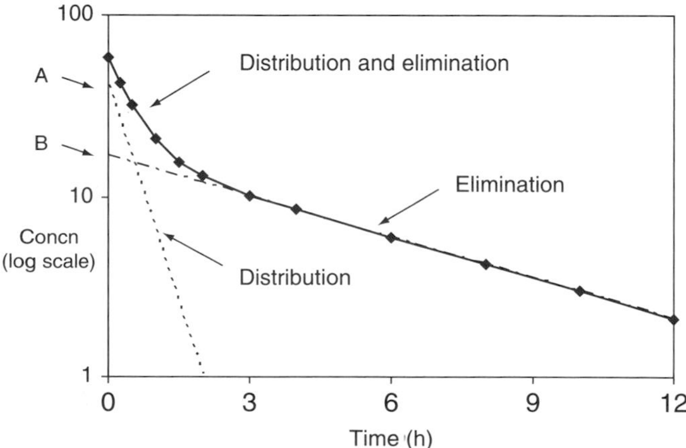

Figure 3.6 A typical concentration–time profile for a drug showing two-compartment PK after IV dosing.

The elimination half-life $t_{1/2,\beta}$ can be readily determined from the time taken for concentrations to decline by one half during the elimination phase, which is the period after distribution has finished (i.e., after approximately 3 h in the example shown in Figure 3.6). Determination of the distribution half-life $t_{1/2,\alpha}$ is more complicated, and it is not simply the time taken for initial blood concentrations to decline by one half, since both distribution and elimination influence this initial decline. Techniques such as curve stripping or non-linear regression are required for calculating the distribution half-life. The curve stripping technique involves subtracting concentrations which lie on the elimination line (shown dashed in Figure 3.6) from the observed blood concentrations during the initial phase. This results in a distribution line (shown dotted in Figure 3.6) from which the distribution half-life is calculated as the time taken for the stripped (or residual) concentrations on the distribution line to decrease by one half. An alternative technique for calculating rate constants and half-lives is to use non-linear regression analysis on computer using programs such as Winlonlin, Modfit, and Minim.

In Figure 3.6, A and B are the intercepts on the Y-axis of the distribution and elimination lines — as with Figure 3.4 note that a logarithmic scale has been used where the blood concentrations, and thus A and B can be read directly from the scale. If a $\log_e$ scale was used then the intercepts would be $\log_e(A)$ and $\log_e(B)$.

For similar reasons to those discussed for the one-compartment model (Section 3.3.1) a volume of distribution can be calculated using the following equation:

$$V_1 = \frac{\text{Dose}}{C_{(0)}} = \frac{\text{Dose}}{A + B} \tag{3.17}$$

Note that the volume calculated here is V_1, the ''initial volume of distribution'' or the ''volume of Compartment 1.'' As with other volume terms it relates concentration to amount in the body, but it is only valid immediately after injection. In Figure 3.6 the steep decline in blood concentrations reflects drug distribution (i.e., from Compartment 1 to Compartment 2 in Figure 3.5) plus drug elimination. Thus the blood concentration decreases much faster than the decrease in amount in the body, as a consequence the volume of distribution, which relates concentration to amount, increases. Hence the volume will increase with time from an initial value of V_1 until the distribution equilibrium is reached (at around 3 h for the example in Figure 3.6). After the distribution equilibrium, the volume is constant and can be calculated from the drug clearance (CL) and elimination rate constant (β) using Equation (3.18). Note the similarity between Equation (3.18) and Equation (3.12)

$$\beta = \frac{\text{CL}}{V} \tag{3.18}$$

The clearance of drugs showing two-compartment PK after intravenous dosing is calculated in exactly the same manner as for drugs showing one-compartment PK, i.e., using Equation (3.5).

The steepness of decline between the initial and terminal phases, and the time to reach the distribution equilibrium, will vary considerably between different drugs and may also vary for the same drug administered to different subjects. In Figure 3.7, three concentration profiles are shown, all of which are derived from the two-compartment model. Drug X shows a rapid distribution phase and short time to distribution, while Drug Y illustrates a slower but less pronounced distribution and Drug Z highlights a slow distribution which is difficult to distinguish from elimination. All drugs have the same elimination half-life and hence their terminal phases decline in parallel. Drug X has large A/B and α/β ratios, Drug Y has a small A/B ratio and Drug Z has a small α/β ratio. In practice there is substantial biological variability in PK and it is not unusual to find a range of profiles similar to those shown in Figure 3.7 for the same drug administered to different subjects.

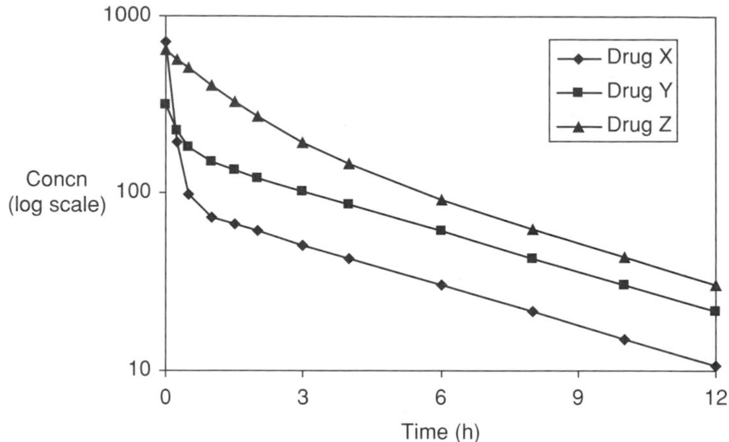

Figure 3.7 Profiles for drugs showing different distribution kinetics.

3.3.3 Absorption Plus Elimination Model (Oral 1-Compartment Open Model)

As discussed in Section 3.2.2, the absorption kinetics of many drugs can be approximated as a first-order process and the model shown in Figure 3.8 is commonly used for orally administered drugs.

In this model the drug dose is initially present in the absorption compartment (A) and the drug's transit into the blood (Compartment 1) is determined by the first-order absorption rate constant (k_a). Not all of the administered drug dose will appear in the blood, since as discussed in Section 3.2.5, the bioavailable fraction of most orally administered drugs is less than 1. Consequently there is a non-specific route for drug elimination from Compartment A, labeled ''other.'' The kinetic processes associated with this ''other'' route may influence the calculated value of k_a and it should be borne in mind that k_a will not always accurately, or solely, reflect the transport kinetics of drug from the absorption site into blood.

This model has general applicability to orally administered drugs and also other routes of administration where absorption may be approximated by a first-order process, such as rectal, vaginal, pulmonary, nasal, and buccal drug administration.

A typical blood-concentration–time profile after oral dosing is shown in Figure 3.9. Immediately after drug ingestion the blood concentration is zero since a finite amount of time is required for the drug to dissolve from the dosage form, empty from the stomach, and be absorbed from the intestine. During the pre-peak phase after drug administration, blood concentrations increase since the rate of drug absorption into the blood (which is the product of k_a and the amount in Compartment A) is greater than the rate of drug elimination (i.e., the product of k and amount in Compartment 1). As time progresses drug is absorbed into the blood and thus the amount in Compartment A decreases whilst the amount in Compartment 1 increases. Consequently the rate of absorption decreases while the rate of elimination increases. When the two rates are equal, the blood concentration is at a peak (C_{max}) and this occurs at the peak time (t_{max}) which is around 2 h in the example shown in Figure 3.9. After the peak, the amount at the absorption site is small, the rate of absorption becomes negligible,

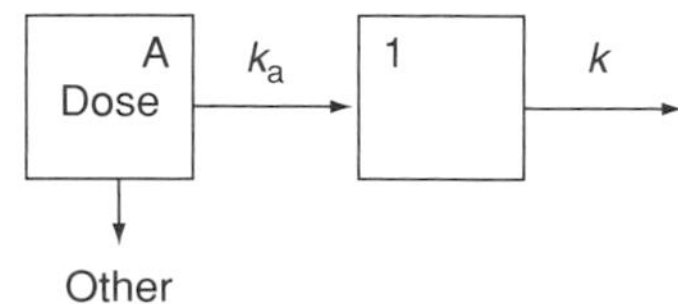

Figure 3.8 A one-compartment open disposition model with absorption.

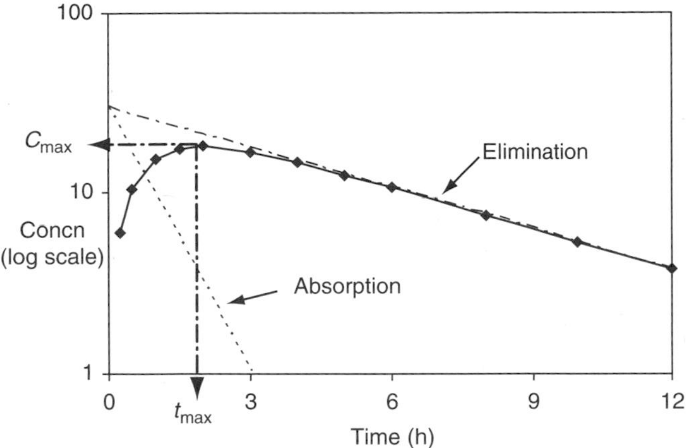

Figure 3.9 A typical concentration–time profile of a drug showing one-compartment disposition PK after oral dosing.

and consequently the post-peak concentrations decline due to elimination of drug from the blood. Eventually a linear decline is seen in the log concentration–time plot (from 4 h onwards in Figure 3.9) and the elimination half-life ($t_{1/2}$) of the drug is given by the time taken for blood concentrations to decrease by a half in this phase.

Determination of the absorption half-life ($t_{1/2,a}$) is more complex and requires the application of curve stripping or non-linear regression. The curve stripping technique involves subtracting the pre-peak concentrations from the backward extrapolated elimination line (shown as a dashed line on Figure 3.9). This results in an absorption line (shown dotted in Figure 3.9) from which the absorption half-life is calculated as the time taken for the stripped (or residual concentrations) on this absorption line to decrease by one half.

The equation describing blood concentrations after oral dosing is given below and has the format of one exponential function (associated with elimination) minus another exponential function (associated with absorption)

$$C = \frac{FD}{V}\frac{k_a}{(k_a - k)}\left(e^{-kt} - e^{-k_a t}\right) \tag{3.19}$$

In Equation (3.19) the bioavailable fraction F multiplied by the dose (D) gives the amount of drug delivered to the blood. The peak concentration is influenced by all of the parameters in the equation, thus increasing F, D, or k_a will all increase C_{max}. Conversely t_{max} is only influenced by k_a and k.

The effect of slowing absorption, resulting in decreased k_a values, on blood-concentration profiles is illustrated in Figure 3.10. Case A represents a rapid absorption example, k_a/k is 6 and the resulting C_{max} is around 70% of the $C_{(0)}$ of an equivalent IV dose and occurs at a t_{max} of 1 h. Decreasing the k_a/k ratio to 3, 1.1, and 0.5 (Cases B, C, and D) results in decreases in C_{max} to 58%, 38%, and 25% of those from an IV dose, together with increases in t_{max} to 1.5, 3, and 4 h (for B, C, and D, respectively). The rate of drug absorption into the bloodstream may be slow due to a number of factors including poor physicochemical properties, causing the drug to dissolve slowly, slow release from the dosage form, or slow transit across the absorbing membrane.

A further PK phenomenon is observed when the absorption rate constant is less than the elimination rate constant as illustrated in Case D in Figure 3.10, where in addition to the lower C_{max} and longer t_{max}, the decline in blood concentrations after the peak does not parallel that after IV dosing (as seen with Cases A–C) but shows a much slower decline. This is a situation known as absorption rate-limited kinetics where the concentrations after the peak are determined by

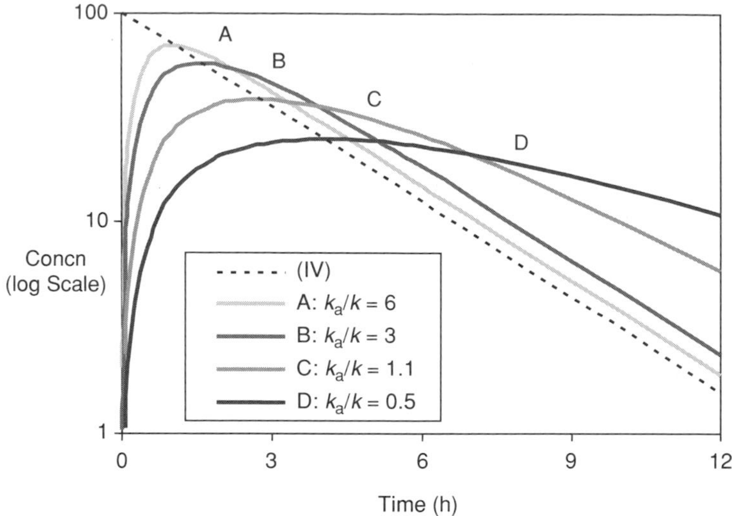

Figure 3.10 The effect of changing absorption rate constant on blood-concentration profiles.

absorption, not elimination, and the half-life measured after the peak is the absorption half-life. This type of kinetics is sometimes referred to as ''flip–flop'' kinetics since the elimination half-life is determined from the pre-peak concentrations and absorption half-life is calculated from the post-peak concentrations (i.e., the opposite of the normal case). This phenomenon is commonly seen when drugs are given as first-order sustained release dosage forms, with the aim of reducing the decline in post-peak blood concentrations, providing a longer duration of drug action, and producing smaller fluctuations in peak and trough concentrations after multiple dosing (see Section 3.4). Note that controlled or sustained release dosage forms do not alter elimination of the drug from the blood, they merely slow drug input into the blood such that absorption is the rate-limiting step and absorption rate-limited kinetics are seen.

3.3.4 Absorption Plus Distribution Plus Elimination Model (Oral 2-Compartment Open Model)

The PK model shown in Figure 3.11 shows an absorption element coupled to the two-compartment disposition model introduced in Section 3.3.2. As discussed previously, the distribution of most drugs does not occur instantaneously and the two-compartment model is needed for IV-administered drugs. The situation is however somewhat more complicated for oral administration of the same drugs. It is fairly evident from Figure 3.11 that the model is more complex and this is reflected in Equation (3.20), which is used to describe concentrations after oral dosing in the two-compartment model

$$C = \frac{FD}{V_1} k_a \left(\frac{(k_{21} - \alpha)}{(k_a - \alpha)(\beta - \alpha)} e^{-\alpha t} + \frac{(k_{21} - \beta)}{(k_a - \beta)(\alpha - \beta)} e^{-\beta t} + \frac{(k_{21} - k_a)}{(\beta - k_a)(\alpha - k_a)} e^{-k_a t} \right) \tag{3.20}$$

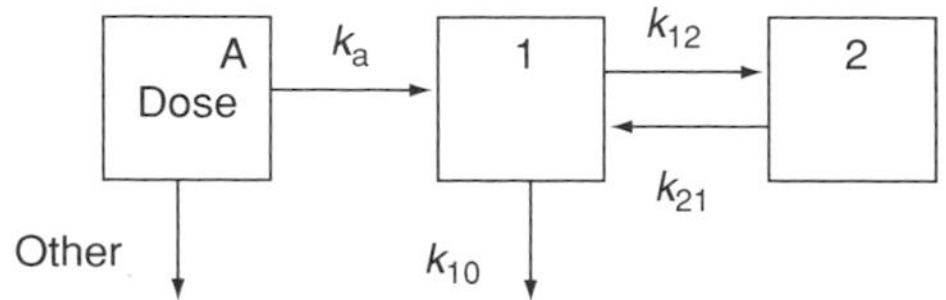

Figure 3.11 A two-compartment open disposition model with absorption.

Note that there are six unknowns in Equation 3.20 (or strictly speaking five, since the ratio F/V_1 cannot be resolved from oral data alone). This number of unknown variables creates some practical problems since a typical PK study involves collecting only 10–15 samples and thus the individual parameters cannot be determined with a high degree of precision. Increasing the number of samples could improve the confidence in the calculated PK values; however, the kinetics of most drugs is such that their absorption and distribution occur at similar rates and thus the resultant PK profile is often practically indistinguishable from that shown in Figure 3.9, where there is no distribution element. Therefore for most oral absorption situations, the one-compartment model described in Section 3.3.3 is used, even though we know that the drug displays two-compartment PK after IV dosing. In using this approach we should recognize that the calculated k_a is not a pure measure of absorption, but encompasses both absorption and distribution. It is possible to analyze oral PK data by non-linear regression using Equation (3.20) and calculate separate values for absorption and distribution, however we usually do not have a high degree of confidence in the estimates unless absorption of the drug occurs at a much faster rate than distribution.

3.4 MULTIPLE DOSE PHARMACOKINETIC PROFILES

3.4.1 Plasma Concentrations after Intravenous Infusion Dosing

When a drug is administered at a constant (zero-order) infusion rate (R_0), plasma concentrations of the drug will increase. The increase in concentration will result in an increase in rate of elimination (from Equation (3.3)) since CL is constant for most drugs. Eventually an equilibrium, known as steady-state, will be reached when the rate of elimination equals the rate of drug input (rate out = rate in) and concentrations reach a plateau, as illustrated in Figure 3.12.

The concentration seen at steady-state (C_{ss}) is solely determined by the drug's clearance (CL) and the rate of drug input (R_0):

$$C_{ss} = \frac{R_0}{\mathrm{CL}} \tag{3.21}$$

Therefore, if CL for the drug is known, the infusion rate (R_0) required to produce a target plasma concentration can be readily determined. Thus in the example shown in Figure 3.12 the drug has a clearance of 6.4 L/h and a 0.8 mg/h intravenous infusion results in a steady-state concentration of

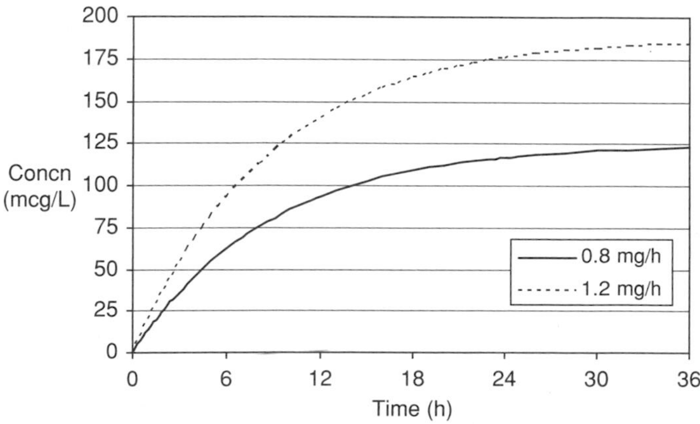

Figure 3.12 A typical multiple dosing blood concentration–time profiles after intravenous infusion dosing.

125 mcg/L. Equation (3.21) is also useful for other zero-order drug delivery situations, including certain subcutaneous implants and depot intramuscular injections. In these cases, R_0 represents the appearance rate of drug in the blood, since not all of the drug released from these types of formulation may appear in the blood. In all cases, increasing the rate of drug input produces a proportionately higher steady-state concentration as illustrated in Figure 3.12, where increasing R_0 to 1.2 mg/h results in a steady-state concentration of 188 mcg/L.

The time taken for plasma concentrations to reach steady-state is solely dependent upon the half-life of the drug. The example drug in Figure 3.12 has a half-life of 6 h and the concentrations seen after one half-life (at 6 h after starting the infusion) are 50% of the respective steady-state values for both the 0.8 and 1.2 mg/h infusion rates. After two half-lives, 75% of the steady-state concentrations are achieved then 90%, 95%, and 99% values are reached after 3.3, 4.3, and 6.6 half-lives, respectively.

3.4.2 Steady-State Plasma Concentrations after Oral Dosing

Multiple dosing of drugs most often occurs orally, with the drug given at regular intervals. The principles described in Section 3.4.1 will apply to regimens of regular oral dosing and a ''steady-state situation'' will be achieved where the average rate of drug administration will be balanced by an ''average'' rate of drug elimination. Clearly, however the blood concentrations will not plateau in the same manner as for infusion dosing, but will rise and fall in a pulsatile manner. In oral dosing regimens and other regimens where discrete doses are regularly administered (e.g., multiple IV bolus dosing) the blood concentrations will not be constant but rise from a trough (C_{min}) immediately prior to dosing, to a maximum (C_{max}) and then decline again to a trough concentration. Steady-state is achieved when there is no difference in C_{max} between subsequent doses and this is the maximum steady-state concentration ($C_{max,ss}$). At steady-state, there is similarly no difference in the trough ($C_{min,ss}$) between subsequent doses or in the ''average'' steady-state concentration ($C_{av,ss}$). The $C_{av,ss}$ is not the arithmetic mean of $C_{max,ss}$ and $C_{min,ss}$ but is an ''area average'' determined from AUC. This type of steady-state situation is shown in Figure 3.13, where for a drug with a half-life of 6 h, there is practically no change in the maximum blood concentrations for any of the doses given more than 24 h after starting the dosing regimen.

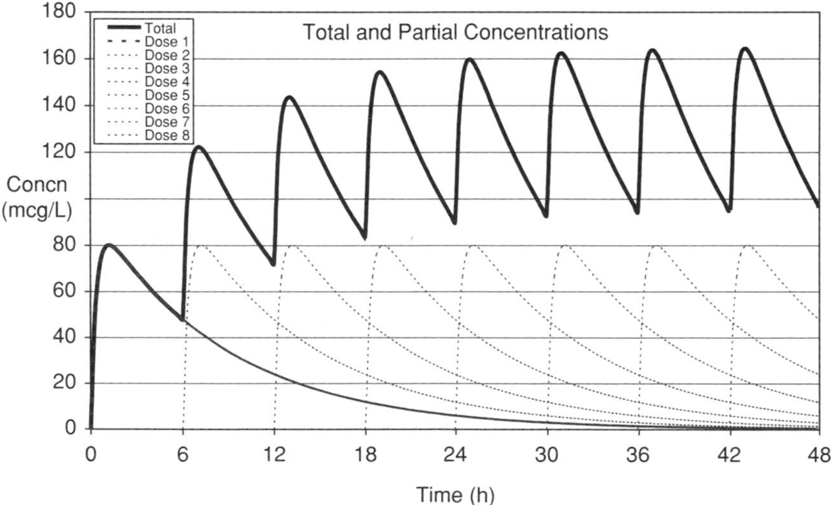

Figure 3.13 A typical multiple dosing blood concentration–time profile after oral dosing.

Calculation of $C_{av,ss}$ can be determined using Equation (3.22), which arises from the same ''rate out = rate in'' principles used in Equation (3.21):

$$C_{av,ss} = \frac{F\,Dose}{\tau\,CL} \quad (3.22)$$

The ''average'' rate of drug elimination is the product of CL and $C_{av,ss}$ whilst the average rate of drug input is the available dose (F Dose) divided by dosing interval (τ). Equation (3.22) can also be simplified by substituting CL from Equation (3.5) to highlight the ''area average'' basis of $C_{av,ss}$. In Equation (3.23), AUC is the area under the curve, between time zero and infinity, seen after a single dose of the drug, and the same AUC is also given by the area under the curve during one dosing interval at steady-state:

$$C_{av,ss} = \frac{AUC}{\tau} \quad (3.23)$$

The $C_{max,ss}$ and $C_{min,ss}$ can be determined from the concentrations seen after a single dose using a method of ''superposition.'' This is illustrated in Figure 3.13, where for example, the maximum concentration seen at around 32 h after starting the dosing regimen is derived from the concentration resulting from the sixth dose (80 mcg/L) plus the sum of those concentrations remaining at 32 h from the fifth dose (42 mcg/L), fourth dose (21 mcg/L), third dose (10.5 mcg/L), etc. This method can be used for other situations of regular discrete dosing, such as multiple IV bolus dosing.

The $C_{max,ss}$ and $C_{min,ss}$ values are dependent upon CL, the available dose (F Dose), and the ratio of dosing interval (τ) to post-peak half-life. A longer dosing interval will result in a greater difference between $C_{max,ss}$ and $C_{min,ss}$. If the drug is fairly rapidly absorbed then the $C_{max,ss}/C_{min,ss}$ ratios will be approximately 2, 4, and 8 if dosing intervals equal to 1, 2, and 3 half-lives are used. Additionally, if a rapidly absorbed drug is regularly dosed at intervals of one drug half-life, then $C_{max,ss}$ will be approximately twice the C_{max} seen after a single dose. This is the case illustrated in Figure 3.13, where the drug is dosed every 6 h and has a half-life of 6 h. If the post-peak half-life is increased, for example by using a sustained release dosage form to produce absorption rate limited kinetics, then this will result in smaller differences between $C_{max,ss}$ and $C_{min,ss}$.

In common with intravenous infusion dosing, the time taken to reach steady-state is solely dependent upon the drug's half-life and 90% of the steady-state concentrations will be reached after about 3.3 half-lives.

3.5 NON-COMPARTMENTAL METHODS

As indicated in Section 3.3, there is often complexity in assigning models to particular PK data sets and a number of different models may mathematically give equally valid descriptions of the experimental data. The same drug given intravenously to a group of individuals may show one-compartment PK in some subjects, yet two-compartment PK in others. Similarly in an oral dosing study, some data sets may be best described by first-order absorption while others indicate zero-order. Even with the sophisticated non-linear regression computer programs currently available it can be difficult to decide which equation gives the best fit to a particular set of data. Hence a number of different non-compartmental methods, which make fewer assumptions about the kinetic processes have been developed. Equation (3.5) has been discussed previously and is one example of a non-compartmental approach. Another commonly used (and related method) is based upon residence times. Mean residence time (MRT) is the average lifetime of a drug molecule in the body or in a particular physiological space. MRT can be calculated for a drug administered after intravenous, oral, or other routes of administration using the following equation:

$$\mathrm{MRT} = \frac{\mathrm{AUMC}}{\mathrm{AUC}} \tag{3.24}$$

In Equation (3.24) the term AUMC is the area under the first moment curve which is calculated from the area under the concentration×time vs. time plot from time zero to infinity. If the MRT is calculated after IV bolus administration, then it represents the average time for a drug molecule to be disposed (by metabolism or elimination) from the body, i.e., it is the mean disposition time (MDT). MDT is numerically equivalent to $1/k$ calculated from the one-compartment model in Section 3.3.1 and a function of both the distribution and elimination rate constants in the two-compartment model of Section 3.3.2.

If MRT is calculated from oral PK data and MDT for the drug is known from IV bolus studies, then the mean input time (MIT) of the bioavailable drug can be determined using the following equation:

$$\mathrm{MIT} = \mathrm{MRT} - \mathrm{MDT} \tag{3.25}$$

MIT is then numerically equivalent to $1/k_a$ calculated from the one-compartment model of Section 3.3.3. If the input is characterized by zero-order kinetics, for example by IV infusion or controlled release absorption, then MIT indicates the time when half of the bioavailable drug has been delivered to the bloodstream. This approach offers a simple method to resolve the problem discussed in Section 3.3.4 where the classical compartmental method results in a complexity which is not readily applicable to most practical situations. MRT can also be compared for drugs administered orally as a solution and a solid dosage form in order to determine an *in vivo* mean dissolution time (MST) of the bioavailable drug using the following equation:

$$\mathrm{MST} = \mathrm{MRT}_{\mathrm{Tablet}} - \mathrm{MRT}_{\mathrm{Solution}} \tag{3.26}$$

As with any PK analytical method, there are limitations to this approach. Determination of both AUC and AUMC requires estimation of the extrapolated areas beyond the last sampling point to infinity. Hence blood samples may need to be taken for prolonged periods in order to minimize the influence of errors in these extrapolated areas on the overall AUC and AUMC values.

3.6 PREDICTIVE PHARMACOKINETICS

There is clearly a need during drug development to make some predictions concerning the pharmacokinetics of a new drug in humans prior to designing and undertaking clinical trials. A number of techniques are used which may provide data useful in this scaling up from *in vitro* or laboratory animal PK. These techniques include using immortalized cell lines such as Caco–2 cells to predict absorption as described in Chapter 1. Microsomes, isolated from hepatocytes and cultured human hepatocytes, have been used to predict hepatic clearance. The role of transporters may however be underestimated in such systems. Isolated perfused tissues, such as gut, liver, lung, and kidney, have also provided much useful information on absorption, distribution, and elimination of drugs via these organs.

3.6.1 Allometric Scaling

Allometric scaling entails predicting the interspecies differences in pharmacokinetics based upon body size (allometry). This is an important element of drug design since it may allow prediction from animal studies conducted early in the drug development program to the likely clinical dosage regimens and an assessment of their viability. The relationships between pharmacokinetic parameters such as V and CL and body size, determined using either body weight or surface area, have been studied for many different classes of drugs. These parameters are often more closely related by body surface area but pragmatically body weight is most often measured and relationships of the general form shown below have been derived:

$$\text{PK parameter} = A(\text{Weight})^{B} \tag{3.27}$$

Thus a log–log plot of a pharmacokinetic parameter such V or CL vs. body weight may reveal a linear relationship. The values of A and B in Equation (3.27) can be calculated from the intercept and slope of the log–log plots. The parameter A varies greatly between drugs and drug classes, however the power function (B) is more constant for a particular PK parameter. Thus B for a number of drugs has typically reported values of 0.9–1 for V and 0.6–0.8 for CL. Values of $B < 1$ indicate that on body weight basis, the parameter is higher in smaller animals, thus for example CL, expressed in terms of l/h/kg will be higher in smaller animals. This occurs because the eliminating organs of smaller animals occupy a greater proportion of body mass and have proportionately higher blood flows than in humans.

3.6.2 Quantitative Structure–Pharmacokinetic Relationships

In addition to trying to predict human pharmacokinetics from *in vitro* or laboratory animal models, some research efforts have been made into using molecular and physicochemical parameters of drug molecules for such predictions. These studies share some commonalities with quantitative structure–activity relationships (QSAR), which are described in Chapter 6. Clearly predicting the whole time course of events which determine a drug's fate is complex and quantitative structure–pharmacokinetic relationship (QSPKR) studies have tended to focus on specific elements such as drug absorption, hepatic clearance, renal clearance, protein binding, and volume of distribution. Moreover, where relationships have been identified, they tend to be stronger for drugs within congeneric series. The physicochemical parameters, which have been commonly used are lipophilicity, assessed using $\log P$ or chromatographic techniques, adsorption, and aqueous diffusivity. The molecular parameters studied include van der Waal's dimensions and connectivity. More recently a few studies with artificial neural network techniques have been applied to elucidate patterns of PK behavior.

3.7 SUMMARY

The PK properties of a drug will determine its concentration and duration of residence at sites of action and toxicity. Understanding the principles of PK allows us to recognize the factors which provide optimal drug concentrations and indicate where the selection of drug formulation, route of administration, and dosage regimen may all be optimized to ensure that the selected drug is used to its maximum clinical potential. Additionally, predictive PK techniques aid in drug design to ensure that lead compounds will have PK profiles that are consistent with practicable dosing regimens.

FURTHER READING

Boxenbaum, H. (1982). Interspecies scaling, allometry, physiological time, and the ground plan of pharmacokinetics. *Journal of Pharmacokinetics and Biopharmaceutics* **10**, 201–227.

de Boer, A.G. and Breimer, D.D. (1997) Hepatic first-pass effect and controlled drug delivery following rectal administration. *Advanced Drug Delivery Reviews* **28**, 229–237.

Dickinson, P.A. and Taylor, G. (1996) Pulmonary first-pass and steady-state metabolism of phenols. *Pharmaceutical Research* **13**, 744–748.

Houston, J.B. and Galetin A. (2003) Progress towards prediction of human pharmacokinetic parameters from *in vitro* technologies. *Drug Metabolism Reviews* **35**, 393–415.

Oliver, R.E., Jones, A.F. and Rowland M. (2001) A whole-body physiologically based pharmacokinetic model incorporating dispersion concepts: short and long time characteristics. *Journal of Pharmacokinetics and Biopharmaceutics* **28**, 27–55.

Rowland, M. and Tozer, T. (1995) *Clinical Pharmacokinetics: Concepts and Applications*, 3rd ed. Williams and Wilkins, Media, USA.

Schoenwald, R.D. (Ed.) (2002) *Pharmacokinetics in Drug Discovery and Development*. CRC Press, Boca Raton, USA.

Testa, B., Crivori, P., Resit, M. and Carrupt, P.A. (2000) The influence of lipophilicity on the pharmacokinetic behaviour of drugs: Concepts and examples. *Perspectives in Drug Discovery and Design* **19**, 179–211.

Thummel, K.E., Kunze, K.L. and Shen D.D. (1997) Enzyme-catalyzed processes of first-pass hepatic and intestinal drug extraction. *Advanced Drug Delivery Reviews* **27**, 99–127.

Turner, J.V., Maddalena, D.J. and Cutler D.J. (2004) Pharmacokinetic parameter prediction from drug structure using artificial neural networks. *International Journal of Pharmaceutics* **270**, 209–219.

GLOSSARY OF PHARMACOKINETIC TERMS

Commonly used Terms and Symbols in PK are Given Below:

***Absorption Half-Life* ($t_{1/2,a}$, min):** The half-life associated with absorption. This half-life is associated with the pre-peak plasma concentrations, except in conditions which result in absorption rate-limited pharmacokinetics (q.v.).

***Absorption Rate-Limited Pharmacokinetics*:** This occurs when the rate constant for drug input is much slower than the rate constant for drug elimination. This results in the terminal half-life reflecting absorption kinetics (rather than elimination kinetics).

***Bioavailability*:** The rate and relative amount of drug reaching the systemic circulation.

***Bioequivalence*:** Comparable bioavailability.

***Clearance* (CL, L/h):** A measure of the capacity of an organ or the whole body to eliminate drug.

***Compartment*:** The composite of tissues and fluids within the body whose rates of uptake and elimination of a specific drug appear to be the same. Compartments do not usually correspond to anatomical entities.

***Disposition*:** The processes which determine the fate of absorbed drug, i.e., those of distribution and elimination.

***Distribution*:** Transfer of drug from one compartment to another in the body. Distribution processes are reversible (cf. elimination).

***Distribution Equilibrium*:** This is achieved when drug concentrations in all tissues and fluids in the body decline in parallel.

***Distribution Half-Life* ($t_{1/2},\alpha$ h):** The half-life of drug distribution. Usually the half-life associated with the first exponential phase seen after intravenous bolus dosing.

***Elimination*:** Removal of drug from the body. Elimination processes are irreversible (cf. distribution). Elimination is the sum of metabolic and excretory processes.

***Elimination Half-Life* ($t_{1/2}$, h):** The half-life associated with the elimination phase.

Extraction Ratio **(E, –):** The fraction of drug molecules elminated during a single passage through an organ. Extraction ratio is also the ratio of; blood clearance by a specific organ/blood flow through that organ.

Excretion**:** Elimination of drug without altering its chemical form (cf. ''metabolism'').

First-Pass Metabolism **(first-pass effect, pre-systemic elimination):** Chemical conversion of a drug occurring after drug administration but prior to entering the systemic (general) circulation. This commonly occurs after oral administration of drugs during their first passage through the liver (and the GI mucosa).

Flip–Flop Pharmacokinetics**:** This occurs when the rate constant for drug input is much slower than the rate constant for drug elimination. This results in the terminal half-life reflecting absorption kinetics (rather than elimination kinetics).

Half-Life **(half-time, h):** The time taken for plasma (or blood) concentrations to decline by a half. (See also distribution half-life, elimination half-life.)

Hepatic Clearance **(CL_H, L/h):** The clearance associated with drug elimination by the liver.

Initial Volume of Distribution **(V_l, l):** The (extrapolated) volume of distribution measured at time zero after an intravenous bolus dose. This is equivalent to the volume of the central compartment (Compartment 1).

Linear Pharmacokinetics**:** This occurs when the processes of drug disposition are controlled by first-order kinetics. Under these conditions plasma concentrations are linearly related to dose (cf. non-linear pharmacokinetics).

Mean Absorption Time **(MAT, h):** The average time a drug molecule is resident in the body, or part of the body.

Mean Residence Time **(MRT, h):** The average time taken for a drug molecule to be absorbed into the bloodstream.

Metabolism**:** Elimination of drug by conversion to active or inactive metabolites (cf. ''excretion'').

Non-linear Pharmacokinetics **(Saturation Pharmacokinetics):** This occurs when disposition processes are not controlled by first-order kinetics, but by zero-order or Michaelis–Menten kinetics. This results in a non-linear relationship between dose and plasma concentrations: larger doses resulting in higher than proportionate plasma concentrations.

Renal Clearance **(CL_R, L/h):** The clearance associated with drug elimination by the kidney.

Terminal Half-Life **($t_{1/2,z}$, h):** The half-life associated with the terminal plasma concentrations. The terminal half-life is usually the elimination half-life.

Total Clearance **(CL, L/h):** The overall clearance of a drug by the body. This is equal to the sum of individual clearances by eliminating organs e.g., liver, kidney.

Volume of Distribution **(V, V_1, l):** A hypothetical volume relating concentration of drug in plasma (or blood) to the total amount of drug in the body. Most drugs have more than one volume of distribution since the blood:tissue ratio changes during the period before the distribution equilibrium (q.v.) is reached.

4

Intermolecular Forces and Molecular Modeling

Robin H. Davies

CONTENTS

4.1 INTRODUCTION

Structural specificity in biological interactions involving macromolecules is generally dominated by noncovalent intermolecular forces. While localized biological interactions may be dependent on proton or electron transfer between groups or, in the case of an enzyme action, on the transfer of a functional group between molecules, the orientation and complementarity of the interacting groups for fast, efficient action usually dictate a much larger structure. The resulting interacting surface is dominated by the noncovalent intermolecular binding forces of the two molecules. Often a biological reaction may require a signal to be transmitted over a distance to a site of action, messages, for example, from periplasmic to cytoplasmic molecules having to traverse the cell membrane through receptor proteins.

Apart from the minimal structures essential to allow a given mechanism to occur, the evolutionary selectivity of biological interactions has produced additional structure to ensure relatively unique complementarity between the noncovalently interacting molecules. However, even the localized interaction may be dominated by noncovalent forces. Attractive noncovalent hydrogen bond forces may be able to dominate the activation of proton transfer from a tyrosine residue by excitation through specific hydrogen bond proton donor interactions on the residue's phenolic oxygen atom. If energy is not to be wasted in such excitation, accurate directionality in hydrogen bonding is required. In the case of an enzyme interaction, noncovalent forces can dominate the formation of the Michaelis complex prior to activation of the reaction. Again for biological efficiency, the favorable binding must be such as to eliminate further energetic input, the binding being such as to prepare for the reaction with the substrate close to the transition state.

Further structure may have evolved on interacting macromolecules as the efficient deployment of excess energy involved in a specific mechanism may be utilized in an associated reaction to absorb the surplus energy. Signals may have evolved between the two mechanisms for efficient deployment of the energy resource. The wider aspect of signal control and stability may again have utilized further additional structure to allow interactions for feedback mechanisms to switch off the action or reaction concerned. In modeling such reactions, it is useful to bear in mind such problems and, in view of their complexity, to remember that we are usually experimentally led.

We therefore use crystallographic and other experimental evidence whenever available to assist prediction. Over 6000 biological structures are now available and readily accessible. The theoretical determination of protein structure from first principles based on the intramolecular interactions of the individual amino acids would have high significance in the design of inhibitory or stimulatory ligands in many areas of drug therapy, but the search for a full three-dimensional structure by seeking a global energy minimum in any sizeable protein is of very high dimensionality. This so-called multiple minima problem has meant, in practice, a recourse to a variety of empiric assumptions to reduce the scale of the problem although theoretical effort in this field is intense. As the size of the macromolecule increases and attempts to compute free energies of small molecules interacting with macromolecules gain ground, it becomes essential to base the intermolecular forces on fast methods using interatomic potentials. These may be conveniently developed from more rigorous theoretical methods and we review later the current scale of attack on these problems.

Focus will be given in this chapter to the underlying factors contributing to the free energies of binding of small ligands to large biomolecules. We define first the fundamental interactions in molecular recognition.

4.2 MOLECULAR INTERACTIONS

4.2.1 Electrostatic Interactions

The electrostatic potential energy between two isolated charge distributions A and R is given by

$$E = \sum_i \sum_j \frac{q_{A_i} q_{R_j}}{r_{ij}} \tag{4.1}$$

where q_i, q_j are the charge on atoms i of A and j of R respectively and r is the interatomic distance.

Scaling electronic units of charge and distance to chemical energies in kcal/mol, the interaction energy associated with two units of charge sited 1 Å apart is 332.0. Thus, for an ion pair interacting over 3.0 Å in the gas phase, the interaction energy is over 100 kcal/mol and electrostatic interactions are large. On the other hand, in aqueous solution, the energies of ion hydration must be overcome for the ions to interact. The addition of a proton to an ammonia molecule in the gas phase is calculated to be over 200 kcal/mol while in aqueous solution the net interaction is 4.0 kcal/mol. Thus the binding of a strong polar group involves a relatively small competitive difference between hydration and the specific interaction, making the likelihood of entropic effects relevant to the overall equilibrium.

An equivalent description of the potential energy of two charge distributions as given in Equation (4.1) can be made by expanding the $1/r_{ij}$ term to give an expression of the energy in terms of all the electric multipole moments. For polar molecules with no net charges, the first term is a dipole–dipole interaction where the individual polar charges (the interactions of which, as stated, can be summed separately) are conveniently replaced by the dipoles. Nonvanishing higher terms, such as the dipole–quadrupole and quadrupole–quadrupole, etc., may be necessary for the equivalent accurate description. To exemplify the size of dipole–dipole interactions in simple systems, for charges of 0.15 electrostatic units taken over a length of 2.0 Å, the effective dipole moment μ is 1.4 Debyes. For two such dipoles aligned 4.0 Å apart, the energy of interaction is ~1 kcal/mol, the energy contribution being proportional to μ/r^3 where r is the distance between the dipole centers. An isolated charge interacting with this dipole, on the other hand, would produce an equivalent interaction of over 6 kcal/mol. Here, the interaction is proportional to μ/r^2.

The electrostatic interactions are, for modeling purposes, scaled by a dielectric factor. The dielectric constant is 80 in an aqueous medium, but this factor reduces to 3.0 for interactions in a hydrocarbon liquid. An enzyme reaction may have a heterogeneous environment but for membrane and many other proteins, there is evidence that the environment is close to a hydrocarbon-like medium. A favored empiric factor used in modeling interactions within proteins is a distance-dependent $1/4r$. For interactions in close proximity, the relevance of a general dielectric effect may be questioned.

The presence of strong charge creates two further effects. Strong charge may induce a further moment in the second molecule, producing an induced moment, a polarizing effect while an actual transfer of charge within the interacting molecules may take place. Polarization may be regarded as the influence of one set of charges upon the second molecule. Charge transfer is a quantum mechanical effect and as it now emerges over the last decade is a dominant feature of the hydrogen bond. When occupied orbitals overlap, there is repulsion between the orbitals. If, on the other hand, an occupied orbital overlaps with an unoccupied antibonding orbital, there is some charge transfer to the unoccupied orbital and increased stabilization. The closer the two orbitals are together in energy, the greater the stabilization. The largest effect is found from the interaction of the highest energy occupied molecular orbital (HOMO), usually a lone electron pair (n) with that of the lowest energy unoccupied orbital (LUMO), often a σ^* orbital, spectroscopically observed, for example, as a n–σ^* transition. The contribution of these terms to the overall energy may be determined theoretically.

4.2.2 The Hydrogen Bond

It was first noted by Coulson that although hydrogen bonding may appear to be dominantly electrostatic, the distance of approach of the proton donor and acceptor produces an almost equal repulsive effect. Decomposition of the terms contributing to the hydrogen bond, at first suggested that the main component contribution to the hydrogen bond was electrostatic, contrasting with early spectroscopically based theories of charge exchange. It has been subsequently shown that the dominant feature of the hydrogen bond does indeed involve charge exchange and that the form of the earlier decomposition underestimated this charge transfer effect. The size of these charge transfer terms can be used to distinguish hydrogen bond and nonhydrogen bond interactions and are exemplified in Table 4.1. Even where the net change in charge transfer is less than 0.01e, the net attractive energy is still around 6 kcal/mol. The more favorable the charge transfer interactions, the more repulsion can be overcome and the greater penetration of the van der Waals radius of the atom.

The degree of charge transfer, in general, follows simple electronegativity rules. Thus, for example, for complexes involving NH_3 as the electron donor (Lewis base), one can rank Lewis acid strength:

$$\mathrm{HF} > \mathrm{HOH} > \mathrm{HNH_2} > \mathrm{CO_2}(\pi^*) > \mathrm{CO}(\pi^*) \sim \mathrm{F_2}(\sigma)$$

Electron donor (Lewis base) strength decreases in the order:

$$\mathrm{H_3N} > \mathrm{H_2O} > \mathrm{HF} > \mathrm{OC} > \mathrm{OCO} \sim \mathrm{N_2} \sim \mathrm{F_2} > \mathrm{O_2}$$

There is an exception to the ranking of F_2 and O_2, which can be understood in terms of the hybrid p character of the donor lone pairs where the mixing of π-type lone pairs with the σ-type lone pair increases the charge transfer. On the other hand in the nonhydrogen bond complexes, the electrostatic interaction is more dominant, the overall energy being greater than the charge transfer. For the T-shaped complexes, the dominant charge transfer interaction is of the n–π^* type.

4.2.3 van Der Waals Interactions

Electrons at any moment in time are in arbitrary positions around an atom and may give rise to instantaneous local induced moments, although the net average will be zero. The main effect will be an induced dipole which, in turn, will polarize a second atom or molecule to induce a further moment. The dominant induced dipole-induced dipole interaction will produce a net attraction even though the atoms interacting have no net charge. Such interactions are termed nonpolar or van der Waals interactions and their energies vary inversely as the sixth power of the interatomic distance. The induced interaction is proportional to the polarizabilities of the atoms and hence the van der Waals interaction increases with the extent of the outer electron shell of the atom. Although individual van der Waals interactions are quite small, since they are common to all atoms, the net interaction effect may become dominant in the overall drug–receptor interaction.

Table 4.1 Natural bond orbital charge transfer analysis of some H-bonded and non-H-bonded complexes (calculations at HF/6–31G* level) involving HF, H_2O, and NH_3 (energies in kcal mol^{-1})

Complex (A–B)	ΔE	ΔE_{CT}	$\Delta E_{A\to B}$	$\Delta E_{B\to A}$	q_A[a]	d[b]
$H_3N\cdots HF$	−12.19	−21.96	−21.20	−0.82	+0.0339 (+0.0340)	+0.85
$H_3N\cdots HOH$	−6.47	−11.42	−11.00	−0.44	+0.0176 (+0.0217)	+0.61
$H_3N\cdots HNH_2$	−2.94	−4.96	−4.75	−0.21	+0.0078 (+0.0152)	+0.27
$H_2O\cdots HF$	−9.21	−15.17	−14.58	−0.66	+0.0206 (+0.0305)	+0.78
$H_2O\cdots HOH$	−5.64	−9.17	−8.88	−0.31	+0.0130 (+0.0253)	+0.57
$H_2O\cdots HNH_2$	−2.86	−4.10	−3.99	−0.11	+0.0065 (+0.0184)	+0.21
$HF\cdots HF$	−5.85	−10.47	−10.18	−0.34	+0.0146 (+0.0318)	+0.69
$HF\cdots HOH$	−3.99	−5.15	−5.03	−0.13	+0.0078 (+0.0238)	+0.40
$HF\cdots HNH_2$	−2.65	−3.45	−3.40	−0.06	+0.0056 (+0.0184)	+0.18
$CO_2\cdots FH$ (T)[c]	−1.58	−1.42	−0.13	−1.30	−0.0019 (−0.0069)	+0.17
$CO_2\cdots OH_2$ (T)	−3.14	−2.67	−0.33	−2.34	−0.0037 (−0.0053)	+0.31
$N_2\cdots FH$ (T)	−0.35	−0.24	−0.07	−0.17	−0.0001 (−0.0007)	−0.41
$N_2\cdots OH_2$ (T)	−0.51	−0.21	−0.09	−0.12	+0.0000 (−0.0007)	−0.54
$N_2\cdots NH_3$ (T)	−0.52	−0.23	−0.08	−0.16	−0.0001 (−0.0009)	−0.68
$O_2\cdots FH$ (T)	−0.29	−0.30	−0.16	−0.14	+0.0000 (−0.0003)	−0.37
$O_2\cdots OH_2$ (T)	−0.40	−0.26	−0.15	−0.11	+0.0001 (−0.0002)	−0.51
$O_2\cdots NH_3$ (T)	−0.36	−0.25	−0.11	−0.14	−0.0000 (−0.0005)	−0.67
$F_2\cdots FH$ (T)	−0.12	−0.34	−0.21	−0.13	+0.0001 (+0.0001)	−0.28
$F_2\cdots NH_3$ (T)	−0.02	−0.17	−0.06	−0.11	−0.0000 (−0.0002)	−0.72

[a]Charge on monomer A by natural bond orbital analysis. Values in parentheses are the corresponding Mulliken charges, shown for comparison purposes only.
[b]Distance of penetration of van der Waals radius in Å.
[c]T-shaped complex.
Source: From Reed et al. (1986). With permission.

4.2.4 Exchange Repulsion

The overlap between closed electron shells on too close an approach between atoms is strongly repulsive. This ''exchange repulsion'' is exponential in character, i.e., of the form Ae^{-br} where A and b are constants. For speed of computation of summed energies over all interacting atoms, these terms are usually replaced by an inverse 12th power of the interatomic distance or by an effective cutoff at the van der Waals contact distance.

4.3 FREE ENERGIES OF INTERACTION — GAS PHASE AND SOLUTION

4.3.1 Entropy and Free Energy Contributions in the Gas Phase

Typical entropy and free energy contributions from translations, rotations, and vibrations for simple molecules in the gas phase at 298 K are given in Table 4.2(a).

Vibrational entropies due to the ''hard'' modes of vibration (from bond stretching and large bond angle opening) are relatively unimportant for normal small rigid molecules lacking low-frequency vibrational modes but on binding and coupling there is spreading of the perturbed frequency vibrational modes to both higher and lower frequency modes. The latter will dominate the entropy contributions and may become of significant importance. For the isolated molecule, however, the internal rotations of the molecule and the ''soft'' modes of vibration due to dihedral angle variation are generally regarded as more entropy-rich than all but the lowest frequency ''hard'' vibrations, and for this reason, the latter contributions are usually neglected. These entropic and enthalpic vibrational contributions may be readily calculated using a parabolic approximation for a vibration where, at the lowest frequencies most of the contributions to the thermodynamic functions occur. The fact remains that most of the significant effects in vibration may arise from the

Table 4.2(a) Typical entropy and free energy contributions from translations, rotations, and vibrations at 298 K

Motion		S^0 (cal deg^{-1} mol^{-1})	H^0–H_0^0 (kcal mol^{-1})	G^0–H_0^0 (kcal mol^{-1})
Three degrees of translational freedom for molecular weights 20–200, standard state 1 *M*[a]		29–36	1.48	−7.2 to −9.1
Three degrees of rotational freedom[a]				
	Moments of inertia[b]			
Water	5.8×10^{-120}	10.5	0.89	−2.24
n-Propane	5.0×10^{-116}	21.5	0.89	−5.53
endo-Dicyclopentadiene	3.8×10^{-113}	27.2	0.89	−7.21
Internal rotation		3–5	0.3[c]	−0.6 to −1.2
Vibration				
	ν (cm^{-1})			
	1000	0.1	0.03	0.0
	800	0.2	0.05	−0.01
	400	1.0	0.20	−0.10
	200	2.2	0.35	−0.31
	100	3.4	0.46	−0.56

[a]Calculated.
[b]Product of three principal moments of inertia g^3 cm^6.
[c]Typical value; this quantity is a function of the barrier to rotation and the partition function.
Source: From Page and Jencks (1971). With permission.

coupling contributions between the molecules whether in the gas phase or in solution. Thus, for example in a distorted ligand–receptor three-point interaction where most of the energetic contributions arise, on average, from predominantly two out of the three points of primary interaction and some 12 degrees of freedom may contribute to balancing the unfavorable distortion against slightly more favorable attraction, a dominant entropic $T\Delta S$ contribution of over 5 kcal mol^{-1} can contribute to the binding. Such contributions only arise from the direct interaction of the two molecules and should be treated explicitly. Experimental data on receptor ligand binding to a guanine nucleotide-coupled receptor are given in Section 4.3.5.

4.3.2 Entropy and Free Energy Contributions in Solution

In the case of ligand–receptor binding the net effect is that the free energy of the binding is greater than that of the sum of solvated ligand and solvated receptor free energies. Using a simple liquid–gas thermodynamic cycle to relate gas and liquid phase interactions,

$$\begin{array}{ccccccccc} & A(g) & + & & R(g) & = & AR(g) & & \\ \Delta G_A & \downarrow\uparrow & \Delta G_R & & \downarrow\uparrow & & \downarrow\uparrow & \Delta G_{AR} & \\ & A(l) & + & & R(l) & = & AR(l) & & \end{array} \quad (4.2)$$

it is seen that liquid–gas partitioning of the appropriate species of drug (A), receptor (R), and complex (AR) may be incorporated to understand the predicted solvational behavior. The effects of solvation on the electrostatic and van der Waals interactions, as discussed earlier, give rise to competitive effects of hydration both on electrostatic and on hydrophobic interactions. We consider first the effect of solvation on translational and rotational entropies.

There will be some loss in translational and rotational entropies on solution. For a nonpolar molecule, the magnitude of this loss is ~10 e.u. or a $T\Delta S$ contribution of ~3 kcal/mol at blood temperature when the same reference concentration is taken in the gas and liquid phases. The translational and rotational entropy loss on binding to a receptor site is thus expected to be not very different between binding in solution and in the gas phase.

4.3.3 Electrostatic Interactions in Solution

The dramatic effect of hydration on electrostatic interactions has been mentioned in Section 4.2.1. Thus net ion–ion interactions are greatly reduced on hydration or may become unfavorable and only the consequent liberation of solvent molecules may create a significant entropy effect and favorable interaction. Except in close comparison of related molecules, gas-phase comparisons are of limited utility. Weaker interactions such as ion–dipole interactions may similarly be largely suppressed by competitive solvent dipoles and contribute little to the overall free energy of interaction in a polar phase. While electrostatic effects will be largely suppressed by hydration, the final binding site of the ligand molecule will usually be within a macromolecule surrounded by mobile regions of polar and nonpolar phases. Ligand concentration is usually referenced to aqueous solution. To estimate the energetics of hydration that must be lost for ligands to interact with a membrane or related protein, it can, therefore, be informative to reference the concentration to a non-aqueous hydrocarbon environment, when van der Waals interactions are automatically taken into account. The best reference model is obviously some hydrocarbon phase, but the partitioning model which has received the most attention has been the solvent octanol with its attendant problem of interaction with strong hydrogen bond acceptor solutes. This latter model is discussed in Chapter 6. An example of the insight gained on applying partitioning data to the thermodynamics of ligand–receptor protein binding is given in the next section.

4.3.4 van Der Waals Interactions in Solution and the Hydrophobic Effect

There is a driving force for nonpolar molecules to interact in aqueous solution which is termed the hydrophobic effect. This force at the macroscopic level causes aggregation of lipids in solution and the folding of proteins in self assembly. The source of this important effect has been disputed. Earlier theories had interpreted this effect as being entropically driven due to the ordering of water molecules around nonpolar solutes with their resultant liberation on nonpolar interatomic contact. More recent evidence has shown that this effect is predominantly or equally enthalpic in character.

Table 4.2(b) shows the incremental thermodynamics of partitioning of a methylene group in a homologous series between an aqueous and a hydrocarbon phase. There are relatively weak favorable enthalpic and large unfavorable entropic components in aqueous solution while in the hydrocarbon environment there is marked favorable enthalpy and weaker unfavorable incremental entropy. The thermodynamic contributions for transfer from the aqueous to the hydrocarbon phase then show the entropic and enthalpic components to produce similarly different contributions to the free energy transfer.

Further insight into the cause of the hydrophobic effect comes from cavity models of solution. The unfavorable entropic effect on aqueous solvation appears to arise from the high number of

Table 4.2(b) Incremental thermodynamics of partitioning of the $-CH_2$ group

Partitioning phases	ΔG^0	310 K ΔH^0	kcal mol^{-1} $-T\Delta S^0$
1. Cyclohexane/gas	−0.76	−1.12	+0.36
2. H_2O/gas	+0.18	−0.67	+0.85
3. Cyclohexane/H_2O	−0.94	−0.45	−0.49

Source: From Abraham (1982). With permission.

states available to water and the resultant loss in entropy on forming a cavity to adapt the solvent. The free energy of solvation may be written as a sum of the free energies of formation of the cavity formation and of the solute–solvent interaction. Although the free energies of formation of the cavity are relatively similar in the aqueous and non-aqueous phases, the enthalpic and entropic components are quite different. In the non-aqueous phase, most of the work of cavity formation goes to the enthalpic maintenance of the excluded volume and only a small contribution to the entropy or configurational exclusion of volume. For water, the reverse is the case.

As molecules become progressively larger and structures more ordered, it is not possible to be categoric in the relation of forces to their resultant effects in solution, particularly when structural reorganization becomes critical. Thus as more states become available, there is usually a weakening of enthalpy changes but a compensation in entropy effects. These compensatory effects can be shown to be large. The concept of molar concentration applied to thermodynamic changes in ordered structures such as liposomes is also less certain.

4.3.5 Some Experimental Observations — Thermodynamics of Ligand Binding to Receptor Proteins

The thermodynamics of binding of small ligand molecules within known protein sites should be computable to a good degree of accuracy. The difficulty lies not with the flexibility of the small ligand but with the uncertainty in accommodating the potential flexibility of the macromolecular structure when based on a crystal structure. The position may be exemplified by data on ligand binding to guanine nucleotide-coupled receptor proteins (GCPRs, refer to Color Figure 4.12), where an x-ray structure of the common class A defined by rhodopsin is now available (Palczewski et al., 2000). The GCPRs are a dominant class of hepta-helical membrane-spanning proteins linking cytoplasmic events through a heterotrimeric G$\alpha\beta\gamma$-protein on the cytoplasmic side of the cell to a signaling hormone binding to the receptor. Typical data for binding of related phenoxypropanolamine ligands to a turkey erythrocyte β-adrenergic receptor are given in Table 4.3, a receptor closely related to the mammalian β_1-adrenergic receptor. The prediction for the binding of the antagonist, propranolol from the weak partial agonist, practolol may be made with good accuracy. Practolol possesses a *p*-$NHCOCH_3$ group, and data at the free energy level on the mammalian receptor are concordant with an –NH hydrogen bond proton donor interaction with the receptor, not of particular strength and which can be modeled empirically using data from a long-chain ester solvent. The main enthalpic difference between the binding of the two compounds is due to loss of hydration on the amidic C=O moiety of practolol. The structure–activity relationships have indicated high flexibility in the hydrophobic residues surrounding given regions of the ligand and those residues surrounding the bound 2-substituents in phenoxy ring compounds show a receptor environment akin to a hydrophobic liquid accommodating even large substituents. Agonists on the other hand such as isoprenaline show a strong enthalpic binding with marked negative entropy. Despite ~12 kcal/mol enthalpic difference in the binding, the free energies of binding of the agonist and antagonists are quite similar. Without some indication of the regions of flexibility of the protein residues and an understanding of the energetics controlling activation of the signal, small empiric perturbations about the structure of the known ligand might still offer the best way of achieving partial agonism. Given the bovine rhodopsin crystal structure now available, on the other hand, an adapted aligned β_1-adrenergic receptor structure allows a free energy exploration of the ligand binding at some level of approximation using molecular dynamics methods or a Monte Carlo simulation and an assessment of the thermodynamic prediction.

Table 4.3 Flexibility of the protein and ligand hydration effects in the thermodynamics of binding of phenoxypropanolamine (a) and phenethanolamine (b) ligands to the turkey erythrocyte β-adrenoceptor. (a) Prediction of the binding of propranolol from practolol[a] and (b) adrenaline from isoprenaline[b]

	310°K $\Delta G°$	Kcal mol^{-1} $\Delta H°$	$-T\Delta S°$
a) Practolol	−7.46	+3.92	−11.40
−*p*-NHCOCH$_3$	−2.72	−6.6[a]	+3.9[c]
$OCH_2CHOHCH_2NHR_1$ (phenyl) Prediction (1)	−10.2	−2.7	−7.5
Entropy correction of specific-NHCO conformation	0.42		0.42
$OCH_2CHOHCH_2NHR_1$ (phenyl) Prediction (2)	−10.6	−2.7	−7.9
(benzo ring)	−2.06	−1.03	−1.03
$OCH_2CHOHCH_2NHR_1$ (naphthyl) Prediction	−12.7	−3.7	−8.9
Experiment	−12.5	−3.85	−8.65
b) $CHOHCH_2NHR_1$ (3,4-dihydroxyphenyl; HO, OH) Isoprenaline	−9.39	−13.39	−4.00
$\delta(CH(CH_3)_2—CH_3)^2$	−1.70	−0.85	−0.85
Adrenaline prediction	−7.69	−12.54	+4.85
Experiment	−7.50	−12.75	+5.24

$R_1 = CH(CH_3)_2$.

[a]Practolol, 3-(4-methylacylamino-phenoxy)–1-isopropylamino-propan–2-ol; Propranolol, 3-(α-naphthoxy)–1-isopropylamino-propan–2-ol.

[b]The ethanolamine derivatives are 2-(3,4-dihydroxy phenyl)–1-isopropylaminoethan–2-ol (isoprenaline) — with the 1-methylamino analogue being the natural hormone adrenaline. The noradrenaline lacking the methyl group on the amino moiety is the natural hormone of the β_1-adrenoceptor.

Contributions to binding based on simple phase change to the ligand:

(1) Cyclohexane/H_2O partitioning.

(2) Long-chain ester/H_2O partitioning.

[c]The error in the enthalpic estimate for the propylene glycol di-pelargonate solvent is considered to be less than ± 0.5 kcal mol^{-1}.

Source: From Davies (1987). With permission.

4.4 INTRAMOLECULAR FORCES AND CONFORMATION

4.4.1 Conformation in the Gas Phase: Intrinsic Conformation

Intrinsic conformational preference in small molecules is a guide to interpretation in larger systems. Conformational preference in the gas phase is very largely dictated by the net outcome of electrostatic and bond orbital interactions. We have already commented on the repulsive or destabilizing (''four electron'') interaction, the ''exchange repulsion,'' of occupied bond orbitals in the overlap of closed shells of electrons and the stabilizing (''two electron'') interaction of an occupied bond orbital with an unoccupied antibonding orbital, giving rise to some charge transfer. All chemists are familiar with the concept of delocalized π molecular orbitals arising from overlap of the atomic π orbitals, allowing reactivity at a site remote from the site of substitution. A set of molecular orbitals can be given an equivalent representation in terms of local bond or group orbitals of the molecule. In the case of π orbitals, the resultant interaction may extend over several atomic centers. For singly bonded flexible systems, there are more localized bond orbital interactions from vicinal orbitals about the bond which can dictate or contribute to structural preference. A knowledge of bond or group orbital interactions can thus give insight into the resultant preferred conformation of flexible systems. The term hyperconjugation has been used to define the favorable interaction of these orbitals and in view of their importance in conformational studies, a wider simple introduction to bond orbitals and their interaction is given, following Jorgensen and Salem (1973).

4.4.2 General Rules for the Interaction Between Orbitals of Different Energy

1. When two orbitals interact, they yield a lower energy bonding combination and a higher energy antibonding combination.

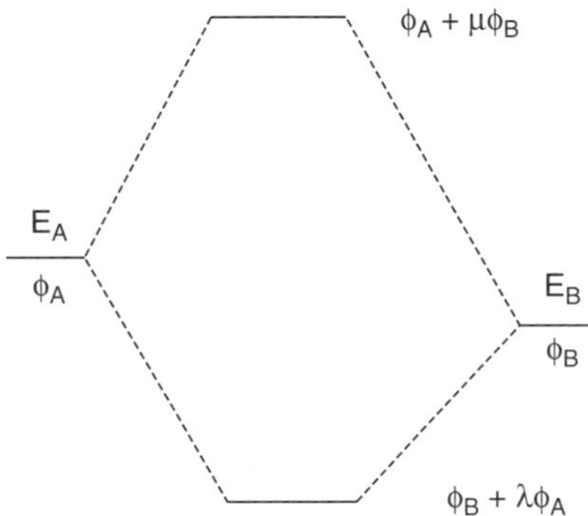

2. The destabilization of orbital φ_A (energy E_A) is always slightly larger than the stabilization of orbital φ_B (energy E_B and $E_B < E_A$).
3. Only energy levels that are close together interact strongly, the closer the better.
4. Only orbitals which overlap significantly interact.
5. If a given energy level interacts with several others of significantly different energy, the interactions are pairwise additive.

4.4.3 Examples of Orbital Interaction, e.g., C–C σ Bonds, C–C π Bonds

Two carbon p atomic orbitals interacting ''end on'' (in this interaction there is zero angular momentum about the bond which is defined as a σ interaction) are shown diagrammatically (Figure 4.1) to produce a σ C–C orbital of lower energy and a σ^* C–C antibonding orbital. The atomic contributions are out-of-phase in the antibonding orbital and characterized by a node (where there

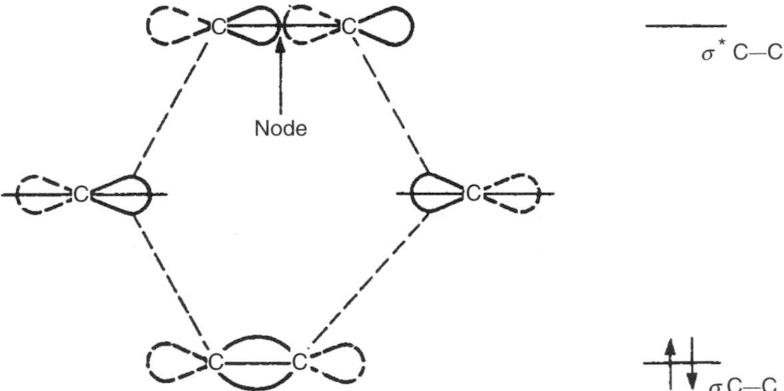

Figure 4.1 Two carbon p orbitals interacting to produce a bonding σ C–C orbital of lower energy and a higher σ* antibonding orbital. (By permission from Jorgensen, W.L. and Salem, L. (1973) *The Organic Chemist's Book of Orbitals.* New York and London: Academic Press.)

is zero charge density). The two electrons occupy the lower bonding orbital and there is a net energy stabilization on interaction.

Figure 4.2 shows the interaction of 2p orbitals to produce a π orbital overlap, in the case of forming the second bond in a double bond. Three carbon orbitals lie in the plane at right angles to the paper, and the 2p carbon orbitals are perpendicular to them. On interaction, the higher energy π* antibonding orbital has the atomic contributions out-of-phase and there is now a nodal plane perpendicular to the C–C bond. The two electrons occupy the lower energy π bonding orbital and there is net energy stabilization. The carbon double bond is thus seen to consist of a σ C–C bond and a π C–C bond.

4.4.4 Electron Donor–Acceptor Interaction

In the case of azoborane the relevant atomic orbitals are the 2p nitrogen orbital containing the electron lone pair, and the vacant 2p boron orbital. On interaction there is net stabilization in energy, the two electrons of the nitrogen atom occupying the N–B bonding orbital and a planar structure is formed (Figure 4.3).

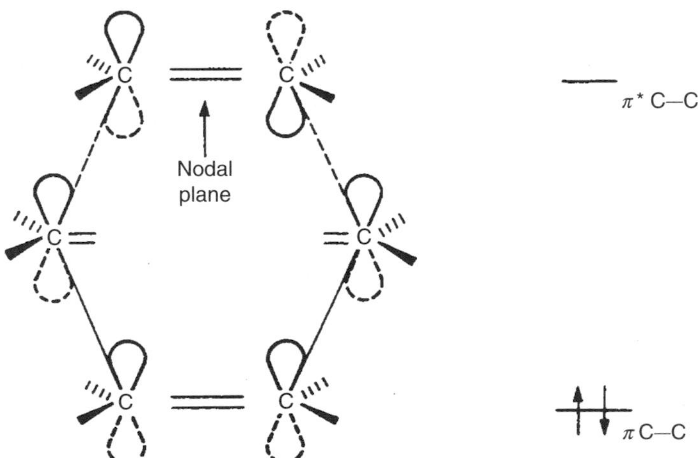

Figure 4.2 Interaction of 2p carbon atomic orbitals to produce π-orbital overlap for a double bond. (By permission from Jorgensen, W.L. and Salem, L. (1973) *The Organic Chemist's Book of Orbitals.* New York and London: Academic Press.)

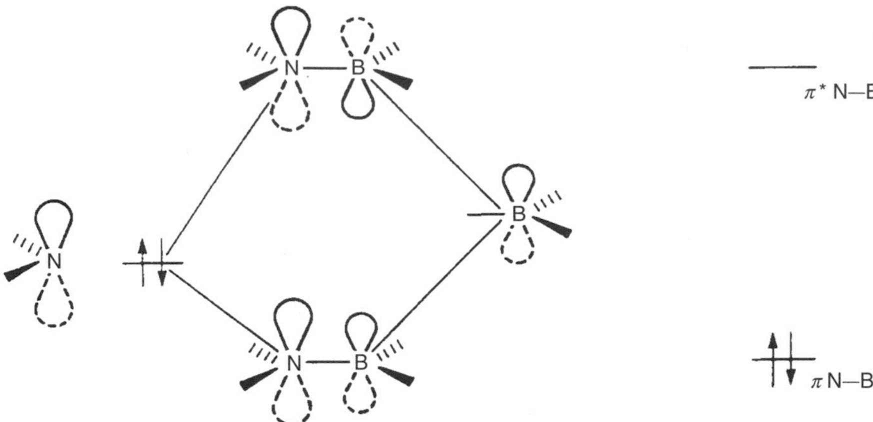

Figure 4.3 Overlap of 2p nitrogen orbital containing the electron lone pair and the vacant 2p boron orbital in azaborane. (By permission from Radom, L. (1982) In Csizmadia, I.G. (ed.), *Molecular Structure and Conformation: Recent Advances.* Amsterdam, Oxford, New York: Elsevier.)

When electron lone pairs are present in both orbitals as in hydrazine (**4.1**), the additional electrons would have to enter the π* antibonding orbital and from rule 2 (Section 4.4.2) the net energy would be destabilizing. The hydrazine structure is thus staggered, with the electron pairs lying in a *gauche* position.

For the same reason the azadipeptide (**4.2**) would be expected to show no tendency to delocalize across the N−N bond and calculation shows the amidic groups to lie preferentially at 90° to one another. In the *N*, *N*′-dialkyl hydrazino group in the ring system (**4.3**) the x-ray structure shows the amidic groups to lie in a similar orientation (Figure 4.4) although other forces in this cyclic system may be acting.

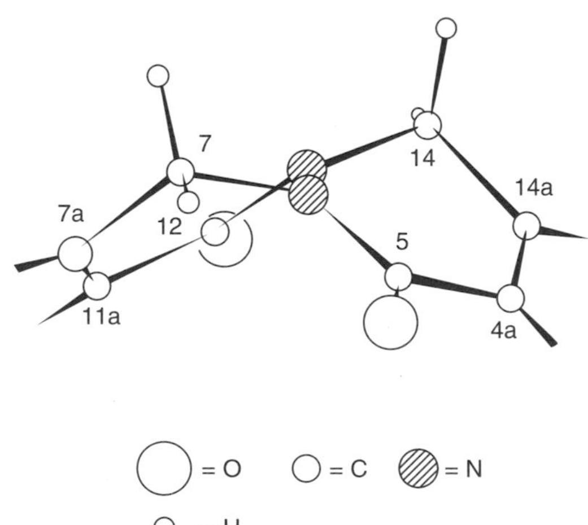

Figure 4.4 Structure of a fraction of the phthalazino (2,3-*b*)phthalazine-5,12-dione molecule. (By permission from Cariati, F., Cauletti, C., Ganadu, M.L., Piancastelli, M.N., and Sgamellotti, A. (1980) *Spectrochimica Acta* **36A**, 1037–1043.)

(4.1)

(4.2)

(4.3)

4.4.5 Hyperconjugation

For single bonds involving a heteroatom, the possibility exists that a vicinal atom may have a vacant antibonding orbital to produce two-electron stabilization. The strength of this interaction will be dependent on the energy difference between these orbitals and the extent of their overlap. In principle, any bonding–antibonding interaction will produce some effect but the highest occupied molecular orbital is usually that occupied by a heteroatom lone pair and the high energy of the localized orbital will have a dominant effect on structural preference. In the case of the vacant antibonding orbital, the more electronegative the neighboring atom or its substituent the lower will be its energy. The typical shapes of bonding and antibonding hybrid C−X orbitals are shown in Figure 4.5.

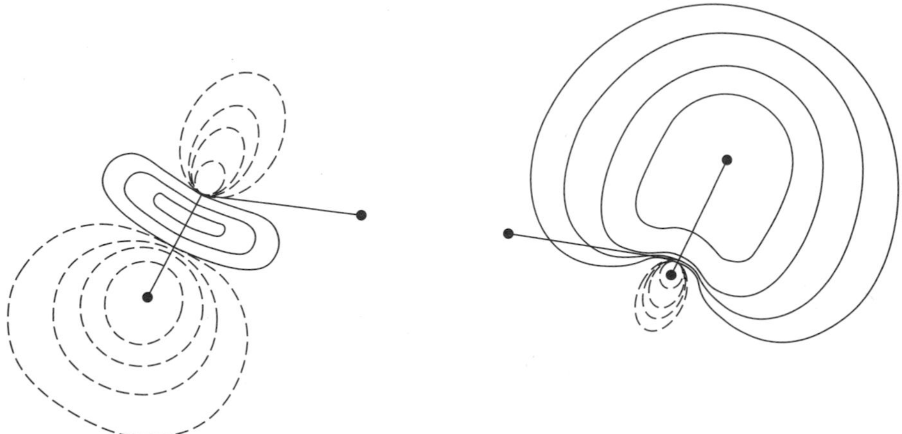

Figure 4.5 Bonding and antibonding C–H hybrid sp^3 orbitals. The solid (dashed) lines represent orbital amplitude contours of positive (negative) phase. The position of the C–C bond in the fragments is indicated. The corresponding overlap of the orbitals (with the appropriate phase) may be judged by superposition of the two C–C bonds. Each contour corresponds to half the amplitude of the preceding one. (By permission from Brunck, T.K. and Weinhold, F. (1979) *Journal of the American Chemical Society* **101**, 1700–1709.)

Maximum overlap of the bonding and antibonding orbitals tends to occur when the bonds of the neighboring groups are antiperiplanar (*trans*), or in the case of the lone pair when it is similarly antiperiplanar to the C−X bond. The strongest hyperconjugative interaction will thus tend to occur when a heteroatom lone pair is antiperiplanar to an electronegative substituent. The anomeric effect in sugars or in substituted pyranose or dioxan rings where an electronegative substituent lies preferentially axial is an example of an effect where bond orbital interaction dominates the conformer preference. The effect even so is not large being of the order 1−1.5 kcal mol^{-1} at blood temperature, giving an axial to equatorial preference of 5–15:1.

Possible acetal conformations (**4.4 a–f**) are shown where R,R′ are alkyl substituents (Deslongchamps, 1983). The antibonding orbitals of interest will lie on the bond with the electronegative oxygen heteroatom and be preferentially antiperiplanar to an oxygen atom electron lone pair. Conformers **d, e,** and **f** have two anomeric effects, **a** and **b** have only one, while conformer **c** has no suitable bond orbital interaction. However, conformers **e** and **f** have steric repulsion from alkyl R′ substituents and the order of stability is found to be **d, a, b, c** with estimated energies of 0, +1.0, +1.9 and +2.9 kcal mol^{-1}, respectively.

(**4.4**)

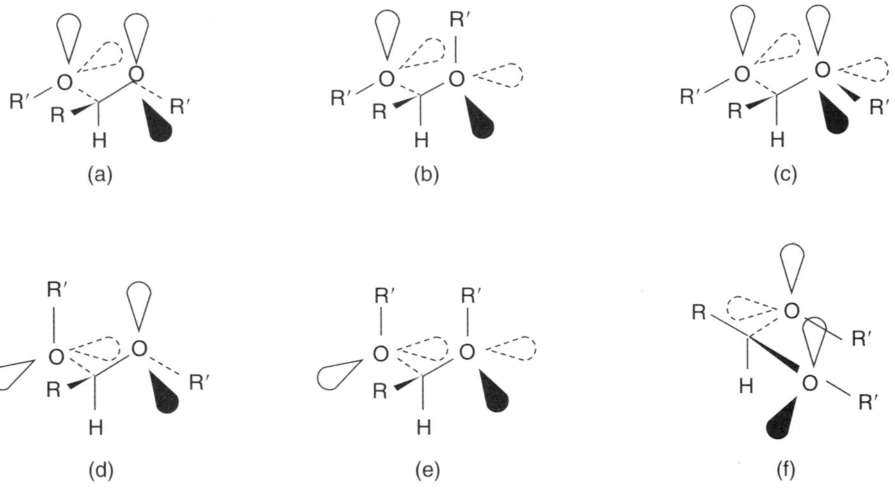

(a) (b) (c)

(d) (e) (f)

An example involving the nitrogen lone pair is shown in the conformer preference of (**4.5a**) and (**4.5b**) where conformer (**4.5b**) has a 500-fold population preference.

(**4.5a**) (**4.5b**)

4.4.6 General Remarks

Conformer preference in flexible σ-bonded systems is more usually a balance between electrostatic, exchange repulsion, and bond orbital effects. The favorable ''two-electron'' interaction has been emphasized here to give some insight into the structure of conformer preference. More detailed reading may be cited (Csizmadia, 1982; Deslongchamps, 1983).

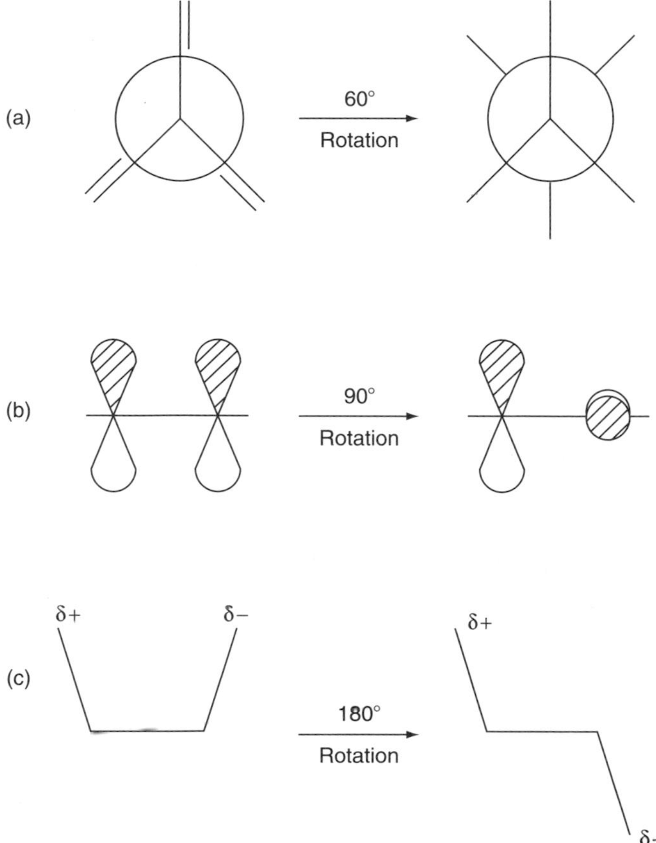

Figure 4.6 Relative components contributing to the conformer preference in a saturated aliphatic system: (a) steric (b) bond orbital, and (c) electrostatic.

It is possible to estimate the relative components contributing to the conformer preference in saturated systems by the following considerations. Figure 4.6a shows eclipsed and staggered forms of an aliphatic system using Newman projections. On rotating about the bond there is an energy well or barrier every 60° due to the exchange repulsion and the rotation is threefold symmetric. In the case of hyperconjugation, the rotation of the antibonding orbital through 90° minimizes the interaction and there is a twofold interaction on rotation about the bond through 360° (Figure 4.6b). For an electrostatic interaction, on the other hand, there is a 1-fold interaction on bond rotation though 360° (Figure 4.6c). The components and their resultant interaction may thus be separated and are shown experimentally in Figure 4.7.

4.5 MOLECULAR MODELING

4.5.1 Introduction

Over the last 10 years, the scale of machine development has continued to double each year and not too expensive machines within the budgets of most small departments can now have five twin processors in parallel, which can compute at up to 50 gigaflops/sec while spare disk capacity has become very inexpensive with some 500 gigabytes of disk space being unexceptional. Grid computing may have 100 times more computing power. The scale of problems ''*in silico*'' whether at the electronic or atomic level of description has become less formidable and yet, as with most

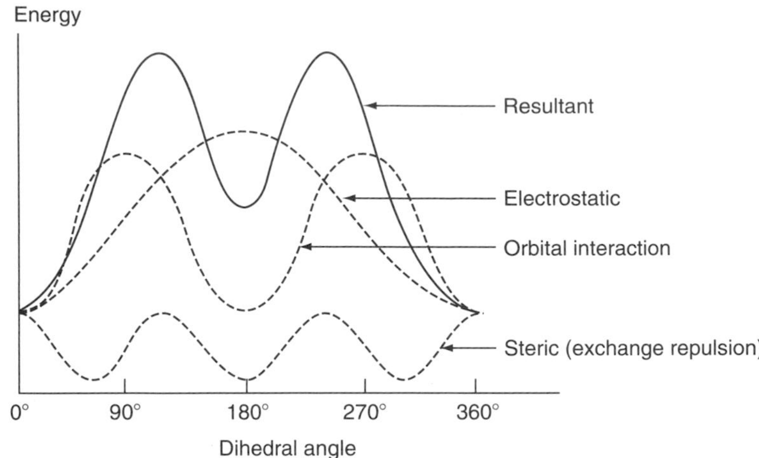

Figure 4.7 Resultant interaction of components contributing to conformer preference illustrated in Figure 4.6. (By permission from Radom, L. (1982) In Csizmadia, I.G. (ed.), *Molecular Structure and Conformation: Recent Advances.* Amsterdam, Oxford, New York: Elsevier.)

problems in science, it is best to define the limits of the accuracy and the assumptions present in attempting to gain insight into a given problem. Even where the target structure is totally unknown it should be possible to determine structural information from the target site ligand either by selective synthetic ligand constraints or by analysis of the available pharmacological data. Early development in the pharmaceutical industry relied on such methods utilizing small perturbations about the structure of a target hormone and such methods continue to have strong utility. While the structural information obtained from such approaches is not independent of the mode of binding of the particular set of ligands, even here, there are indications of efficiency from the overall gross ligand potency. Synthetically based identification of the bioactive conformers using constrained molecules aided by temperature studies on receptors using isolated membranes or intact cells *in vitro* yield thermodynamic conformer binding data and quantitative information on the mode of binding which should allow determination of the geometry around localized bonds of the ligand in many instances and, importantly, allow for some decomposition of the energetics of the binding in closely related ligands. While considerable localized information on the target site can be obtained from such methods, selective binding to sites remote from the biological action remain elusive without wide-scale random screening.

There is an increasing database on target macromolecules. The three-dimensional structures of over 6000 macromolecules now exist in the Brookhaven Protein Data bank and amino acid sequences of a further 150,000 are available. The determination of an x-ray structure for a mammalian G protein-coupled receptor bovine rhodopsin (Palczewski et al., 2000) has transformed the facility for designing useful ligand molecules within this class of receptors where economic development of a molecule with high selectivity of action against its receptor subtypes becomes feasible. The selectivity relies on exploiting the residue differences in receptor subtypes. As almost half the current therapeutic drugs utilize regulation of such receptors — almost one in a hundred pieces of mammalian DNA consist of this class of receptor — progress in the design of useful pharmacological molecules should greatly accelerate over the next few years. Structure-based ligand design is, therefore, a reality in a number of therapeutic areas. Figure 4.12 shows an adapted β_1-adrenoceptor segment of the crystallographic structure of the seven *trans*-membrane α-helices of bovine rhodopsin. Serine and aspartate proteases are another protein area where much is known. Figure 4.8 shows the localized structure of a typical serine protease containing the characteristic Asp^{102}-His^{57}-Ser^{195} catalytic triad involved in the peptide bond rupture. The catalytic site of trypsin in the presence of the bovine pancreatic trypsin inhibitor shows the presence of an

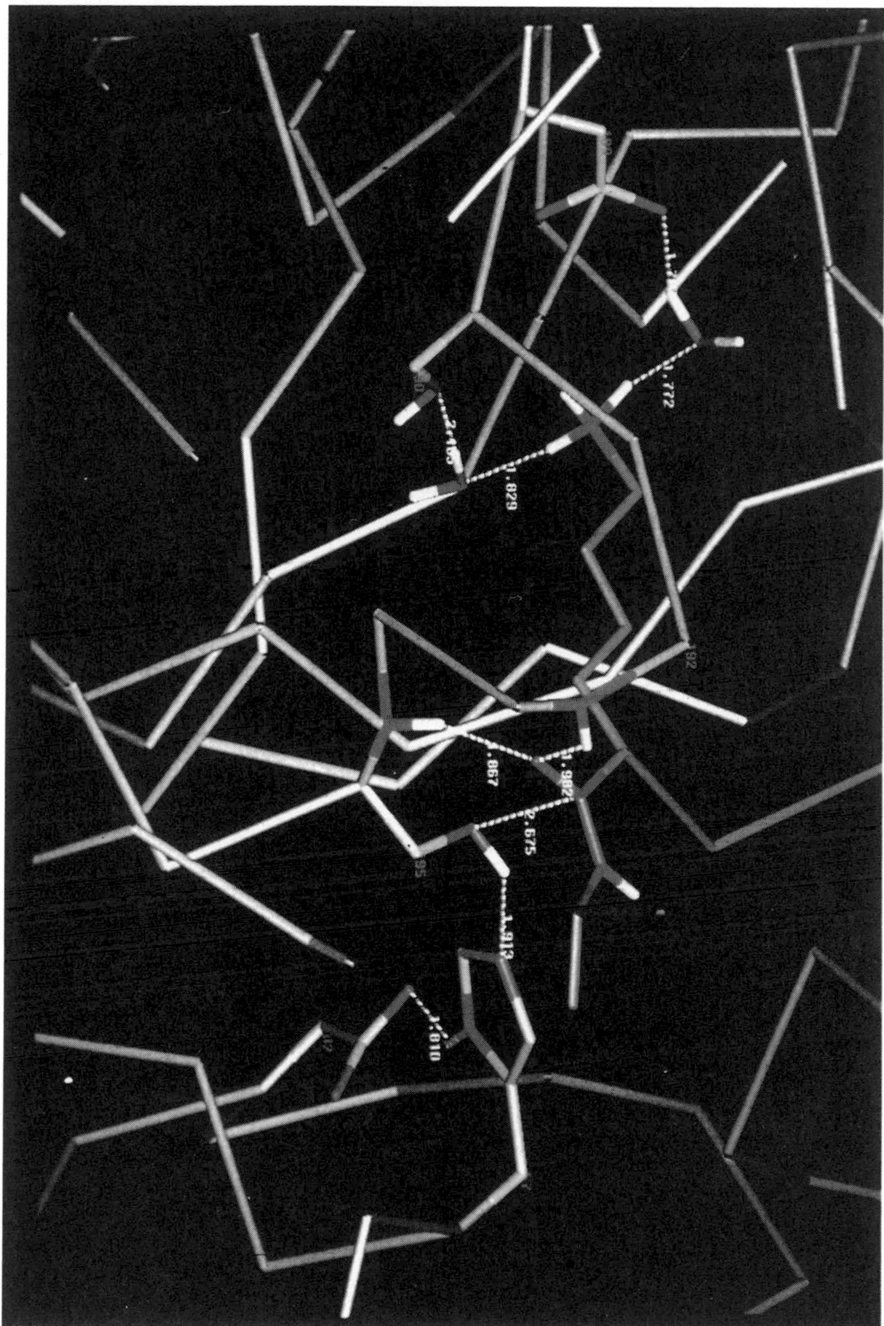

Figure 4.8 (See color insert after p. 368) Serine proteases. Proton movement and enzymatic cleavage of the peptide bond. The serine proteases are characterized by an Asp^{102}–His^{57}–Ser^{195} catalytic triad. Experimental (NMR) and theoretical results have indicated that the histidine residue remains neutral throughout the course of the reaction. The initiating attack of Ser^{195} on the peptide carbonyl carbon atom is facilitated by the abstraction of the hydroxyl proton by His^{57}. The proton originally residing on His^{57} is transferred to Asp^{102} and the incipient negative charge developing on the peptide carbonyl oxygen is stabilized by hydrogen bonding from the main chain –NH groups of residues 193 and 195. The tetrahedral intermediate collapses to an acylated enzyme with the delivery of a proton to the leaving amino group. This proton originates from His^{57} but delivery may be mediated by a water molecule. Concomitantly, the histidine accepts the proton from Asp^{102} to regenerate the initial protonation state. Deacylation follows an analogous cycle of proton transfers with a water molecule replacing Ser^{195} as the nucleophile and with the serine becoming the leaving group.

anionic site for preferential binding to basic residues involved in the peptide bond rupture. Figure 4.9 shows the inhibition of this site in the enzyme α-thrombin by the natural ligand inhibitor Hirudin where selective binding to remote ''exo-sites'' is exemplified. A variety of simple logical approaches can be deployed to examine the possibilities of occupying the binding site efficiently. Their disadvantage in accurate prediction, as mentioned earlier, is the inability of the crystal structure to convey the varying degrees of flexibility within the site. The selective binding to remote sites, however, should be efficient and of great therapeutic advantage. Figure 4.10 shows the x-ray structure of pepsin. A wider classification of proteinase and peptide classes of enzymes may be cited (Gerhartz et al., 2002). Inhibition and stimulation of signaling mechanisms involving tyrosine kinases and phosphatases are other examples of structure-based drug design targets.

As we are concerned with ligand design here, the main emphasis in this chapter will be to concentrate on the strategies available to rational ligand design both when the macromolecular structure is known and unknown. Examples of the scale of some interactions involving protein–protein, protein–single-stranded DNA, and protein–double-stranded DNA are given in Figures 4.11, 4.15, and 4.16, respectively. Figure 4.15 shows the binding of a zinc finger domain to a single strand of DNA while protein occupancy of the major and minor grooves of a piece of double-stranded DNA is shown in Figure 4.16.

4.5.2 Thermodynamics of Ligand Binding and Conformer Identification

When a ligand binds to a receptor or its target enzyme, often the energy of hydration is lost from most regions of the molecule and the interaction becomes essentially non-aqueous in character. It can therefore be useful to change the reference phase for binding to that of a model hydrocarbon liquid when simple correlations of potency and change in reference phase indicate the inherent flexibility of the target macromolecule in given regions of the molecule. Some consequences are exemplified in Section 4.3.5. Such correlations at the free energy level automatically introduce a good approximation to the van der Waals forces operating in the binding in closely related molecules. Often a 2–3 kcal/mol variation in observed binding is reduced to little more than 0.15 kcal/mol when introducing this reference change, providing a useful base line for exploring other effects within a given mode of binding. As a major target is to maximize efficient binding it is important to identify whether the binding conformation is dominant or whether only a small fraction of the ligand productively binds to the macromolecule. It is useful, therefore, to represent

Figure 4.8 Continued

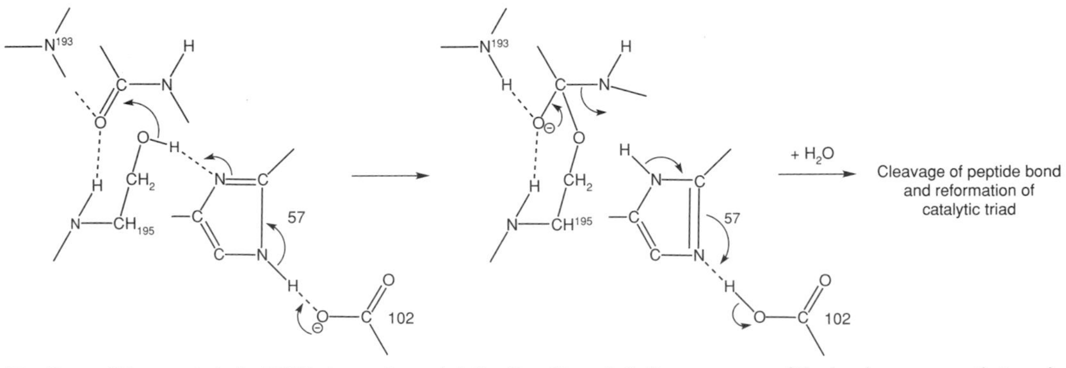

+ H_2O

Cleavage of peptide bond and reformation of catalytic triad

The figure (Marquart et al., 1983) shows the catalytic site of trypsin in the presence of the bovine pancreatic trypsin inhibitor (BPTI). The C_α trace of the enzyme (pink) shows the catalytic triad to the right. The scissile carbonyl carbon atom is shown in green. The primary recognition of the peptide bond to be cleaved results from a binding pocket for the substrate side chain in the vicinity of residue 189. The nature of the residues in this pocket predicates the particular specificity of the protease. In the case of trypsin, this residue is an aspartate and specificity is for basic side chains. In the left of the figure, a lysine side chain is shown interacting with Asp^{189} and Thr^{190} via two water molecules.

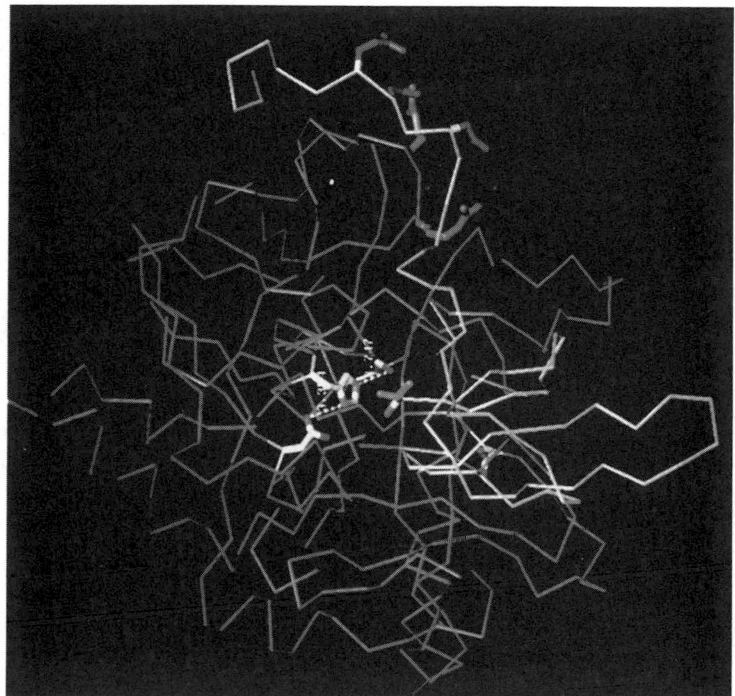

Figure 4.9 (See color insert after p. 368) Inhibition of a serine protease and protein–protein recognition. The natural ligand inhibitor, Hirudin binding to the catalytic Asp-His-Ser triad within the serine protease α-Thrombin (Vitali et al., 1992). α-Thrombin has a high specificity for peptide bonds associated with arginine residues and plays a central role in thrombosis and haemostasis. It is the product of prothrombin cleavage by factor Xα in the final step of the blood clotting cascade, and consists of two polypeptide chains, A and B, connected through a single disulfide bond. During clotting, α-thrombin converts fibrinogen into fibrin by removing fibrinopeptide A from the Aα-chain and fibrinopeptide B from the β-chains of fibrinogen. Hirudin is a small protein of 65 residues and three disulfide bonds that is isolated from the glandular secretions of the leech *Hirudo medicinalis* and is a potent natural inhibitor of thrombin.

The figure shows the large surface area of contact of the Hirudin inhibitor (blue) with the serine-protease, bovine α-thrombin (brown). The Asp^{102}-His^{57}-Ser^{195} catalytic triad of the enzyme (elemental coloring) is blocked by the first three residues of the N-terminal chain of Hirudin (refer to the Hirudin N-terminus in green). In human thrombin, two hydrogen bonds from the amino terminal group exist. In the crystal developed at pH 4.7, one is to the carbonyl group of Ser^{214} and the second is to the catalytic serine residue 195. For the crystal developed at pH 7, this second bond is to His^{57}. Neither bond is formed in this bovine complex at pH 4.7, indicating that a second bond may not be essential for Hirudin binding. Specific binding to the associated binding site for arginine residues does not occur (compare the bovine pancreatic trypsin inhibitor in Plate 3.1) but a number of exo-sites on the surface of the thrombin can interact with the inhibitor. The last 16 residues of hirudin are in an open conformation and bind between the two loops of the enzyme surface formed by Phe^{34} to Leu^{41} and by Lys^{70} to Glu^{80}. This region of the enzyme is marked by positively charged side chains, and interaction with Hirudin's anionic residues Asp^{53}, Asp^{55}, Glu^{56}, Glu^{57}. The latter three residues are shown in red. Salt bridges are formed by Asp^{55} and Glu^{57} interacting with the enzyme residues Arg^{73} and Arg^{75}, respectively.

the gross ligand binding constant in a conformer representation and to examine possible relations between the phase environment and the conformer representation (Davies, 1987).

In terms of standard partial free energies, μ° the gross binding constant may be written as

$$\mu^\circ_{AR} - \mu^\circ_{A} - \mu^\circ_{R} = kT \log K \tag{4.3}$$

where the subscripts AR, A, and R refer to the complex, drug, and receptor, respectively, k is the Boltzmann constant, and T the absolute temperature.

Using second indices to identify the conformer i of the drug A engaged in binding, with $j*$ its receptor counterpart. For the $ij*$ conformer interaction, Equation (4.3) may be written

$$\mu^\circ_{A_iR_{j*}} + (\mu^\circ_{AR} - \mu^\circ_{A_iR_{j*}}) - \mu^\circ_{A_i} - (\mu^\circ_{A} - \mu^\circ_{A_i}) - \mu_{R_{j*}} - (\mu^\circ_{R} - \mu^\circ_{R_{j*}}) = kT \log K \tag{4.4}$$

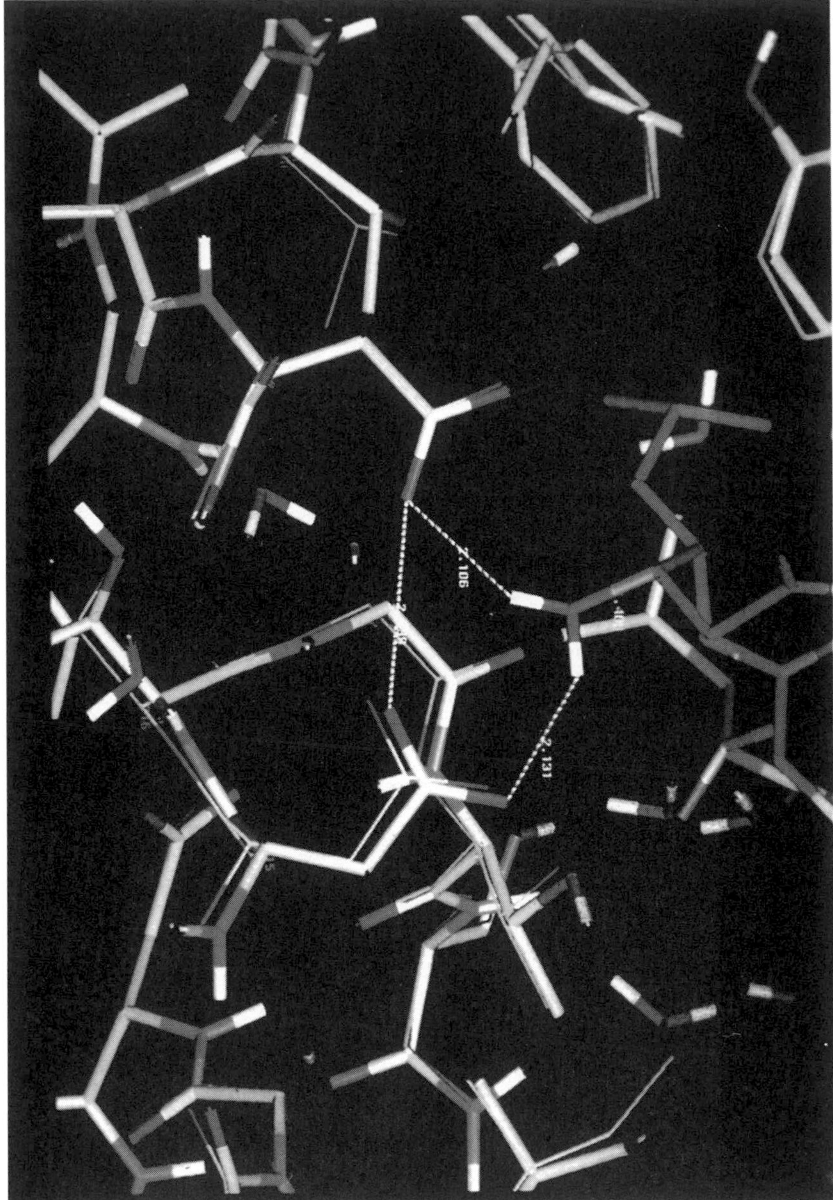

Figure 4.10 (See color insert after p. 368) Aspartate proteases. As for the serine proteases, electron reorganization coupled to proton movement is critical to the cleavage of the peptide bond. In this case, the catalytic site consists of two adjacent aspartate residues, Asp^{32} and Asp^{215} (pepsin numbering) which localize a solvent water molecule between their carboxyl groups. This highly polarized water molecule is the initiator of the peptide bond hydrolysis. Studies of the pH dependence of catalysis by porcine pepsin leads to estimates of two pK_a values of 1.2 and 4.7 and, hence, one of the apartate residues is thought to be protonated in the resting state. High refinement (1.8 Å, Sielecki et al., 1990) of the pepsin structure shows that the oxygen atom of the catalytic water molecule lies in the plane of Asp^{215} whereas the carboxylate group of Asp^{32} is twisted by some 22° with repect to this common plane. In the resting state, the carboxylate oxygen atoms are arranged such that one from each residue is within hydrogen bonding distance of the water molecule and the two adjacent "inner" carboxylate oxygen atoms from each residue are also hydrogen bonded together. From the interatomic distances, the "inner" oxygen atom of Asp^{32} and the "outer" oxygen of Asp^{215} are hydrogen bonded to the water. Hence the probable location of the proton that forms the hydrogen bond between the carboxylate groups is on the "inner" oxygen atom of Asp^{215}. The shorter contact distance from the water molecule is also to Asp^{32} (2.6 Å) rather than to Asp^{215} (2.9 Å) suggesting that in the resting state, it is Asp^{215} that is protonated. The proposed hydrogen bond length between Asp^{32} and Asp^{215} is 2.8 Å. The precise mechanism of catalysis is not known and the following description represents one possible hypothesis. Nucleophilic attack on the peptide carbonyl carbon atom is probably facilitated by movement of a proton from the water molecule to Asp^{215}

Using conformer populations f^{i} of A, and f^{j*} of R, and the relations

$$\mu^{\circ}_{\mathrm{A}} - \mu^{\circ}_{\mathrm{A}_i} = kT \log f^{i} \tag{4.5}$$

$$\mu^{\circ}_{\mathrm{R}} - \mu^{\circ}_{\mathrm{R}_{j*}} = kT \log f^{j*}$$

and summing over the bound states

$$\sum_{i}\sum_{j*} K^{ij*} f^{i} f^{j*} = K^{ij} \tag{4.6}$$

where K^{ij*}, the conformer binding constant is given by

$$\mu^{\circ}_{\mathrm{A}_i\mathrm{R}_{j*}} - \mu^{\circ}_{\mathrm{A}_i} - \mu^{\circ}_{\mathrm{R}_{j*}} = -kT \log K^{ij*} \tag{4.7}$$

The binding constant is a sum of the conformer binding constants weighted by their appropriate conformer fractions.

It is more convenient to define the conformer binding constant $K^{ij*}_{\square}$ referenced to the average states of A and R.

$$\mu^{\circ}_{\mathrm{A}_i\mathrm{R}_{j*}} - \mu^{\circ}_{\mathrm{A}} - \mu^{\circ}_{\mathrm{R}} = -kT \log K^{ij*} f^{i} f^{j*} = -kT \log K^{ij*} \tag{4.8}$$

It is often helpful to consider comparative drug binding with a change of reference to a hydrocarbon lipid phase L. The standard free energy change of A can then be written

$$\mu^{\circ}_{\mathrm{AiL}} - (\mu^{\circ}_{\mathrm{A}} - \mu^{\circ}_{\mathrm{AL}}) - (\mu^{\circ}_{\mathrm{AL}} - \mu^{\circ}_{\mathrm{AiL}}) \tag{4.9}$$

$$\mu^{\circ}_{\mathrm{AiL}} - (\mu^{\circ}_{\mathrm{A}} - \mu^{\circ}_{\mathrm{Ai}}) - (\mu^{\circ}_{\mathrm{Ai}} - \mu^{\circ}_{\mathrm{AiL}})$$

Figure 4.10 Continued

along with proton transfer from Asp^{215} to Asp^{32}. The "inner" oxygen atoms are located via hydrogen bonds from the main chain –NH groups of Gly^{34} and Gly^{217} and two residues, Ser^{35} and Thr^{218} may hydrogen bond to the "outer" carboxylate oxygen atoms. Ser^{35} may help to stabilize the incipient oxyanion of the tetrahedral intermediate and Thr^{218} may position a second water molecule in order to mediate the transfer of the proton from Asp^{215} to the leaving amino group on breakdown of the tetrahedral intermediate. As the proton is transferred from the "outer" oxygen atom of Asp^{215}, the proton on Asp^{32} is transferred to the "inner" carboxylate of Asp^{215} so restoring the initial state.

The hypothetical proton and electron reorganizations are shown in the scheme below:

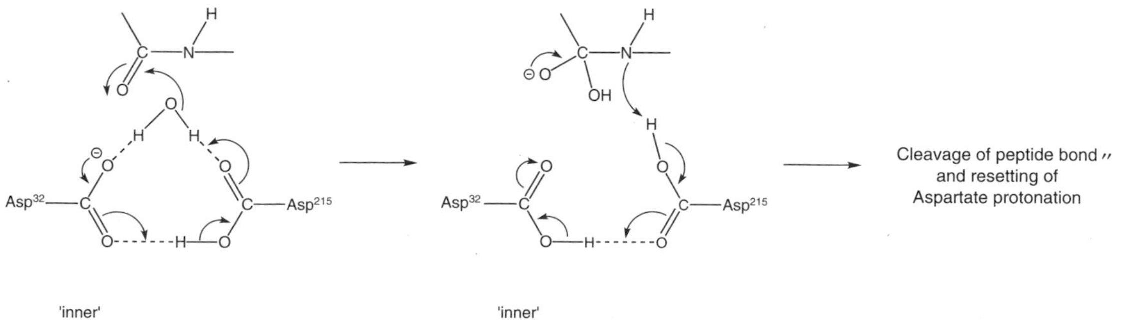

Cleavage of peptide bond and resetting of Aspartate protonation

'inner' 'inner'

Inhibitors of aspartate proteases, such as pepstatin, displace the catalytic water molecule by an appropriately orientated hydroxyl group. The figure shows the pepsin catalytic site with the resident water molecule superimposed on a second structure determined in the presence of pepstatin (green). In the center of the figure are the hydroxyl group of pepstatin and the catalytic water molecule with Asp^{32} located to the left.

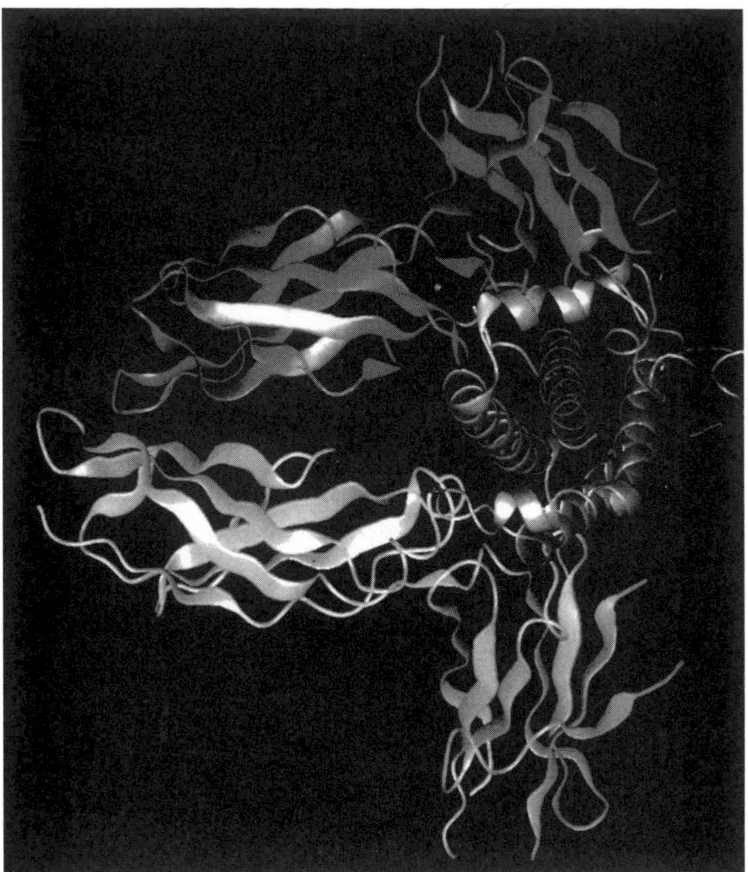

Figure 4.11 (See color insert after p. 368) Protein–protein recognition. The influence of a hormone on protein dimerization. Human growth hormone (hGH) binding to the extracellular domain of its receptor (de Vos et al., 1992). The binding of hGH to its receptor is required for regulation of normal human growth and development. The extracellular domain of the receptor (hGHbp) complex, here shown as a ribbon structure, consists of one molecule of growth hormone per two molecules of receptor (orange and blue, respectively). The hormone (lilac) is a four helix bundle. The binding protein consists of two distinct domains which have some similarity to immunoglobulin domains. In the complex, both receptors donate essentially the same residues to interact with the hormone even though the two binding sites on hGH have no structural similarity. In addition to the hormone–receptor interfaces, there is also substantial contact between the carboxyl-terminal domains of the receptors.

The core of the helix bundle is made up of primarily hydrophobic residues. The extracellular part of the receptor consists of the two domains linked by a four residue segment of polypeptide chain. Each domain contains seven β-strands that together form a sandwich of two antiparallel β-sheets, one with four strands and one with three with the same topology in each domain. The 30 residues of the receptor's amino terminal domain show conformational flexibility and are not given in the crystal structure. The carboxy-terminal domains are closely parallel, the termini pointing away from the hormone in the expected direction of the membrane. Intact receptors would have an additional eight residues at the end of the seventh strand (bottom right), which form the putative membrane-spanning helix.

and since

$$\mu^{\circ}_{AL} - \mu^{\circ}_{A_iL} = -kT \log f^i_{L} \tag{4.10a}$$

$$\mu^{\circ}_{A_i} - \mu^{\circ}_{A_iL} = -kT \log P^i \tag{4.10b}$$

$$\mu^{\circ}_{A} - \mu^{\circ}_{L} = -kT \log P \tag{4.10c}$$

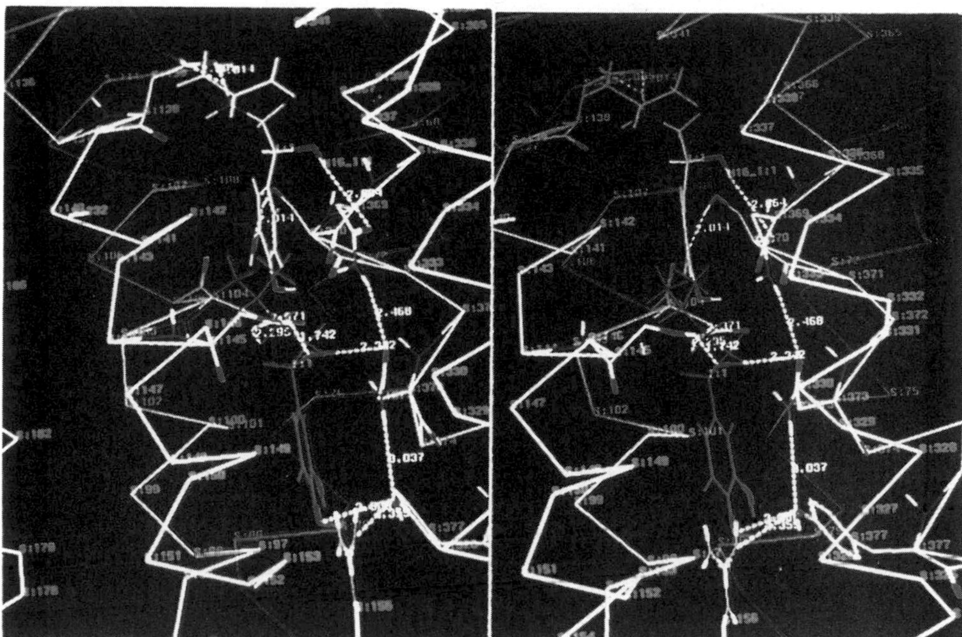

Figure 4.12 (See color insert after p. 368) A segment of the β_1-adrenergic receptor adapted from the crystallographic structure of the mammalian G protein-coupled receptor, bovine rhodopsin (Palczewski et al., 2000), receptors characteristically represented by seven *trans*-membrane α-helices and activated by small ligand hormones. The known agonist conformer of the activating hormone, noradrenaline, here represented by the isopropylamino analogue, isoprenaline, is shown attached to two aspartate residues, 138 (helix III) and 104 (helix II) in upper (pink) and lower (mauve) positions relative to the membrane periplasmic interface. The agonist conformation is conserved in both potential positions of the ligand. A model for the activating mechanism of a ligand–receptor–G protein ternary complex acting as a monocation driven proton pump developed using a bacteriorhodopsin-adapted structure for the β_1-adrenoceptor (Nederkoorn et al., 1998) can be developed further using such a representation. The proton may be used either to assist exchange of guanosine triphosphate (GTP) for the diphosphate (GDP) or to activate synthesis of GTP from GDP resident within the G protein. It is not possible to separate the two mechanisms by any steady-state representation of such ligand receptor interaction (Broadley et al., 2000).

where f_{L}^{i} is the conformer fraction of i in a non-aqueous medium and P^{i} is the conformer- or micropartition coefficient of the species i which, often, is easily estimated (Davies et al., 1981). It follows that

$$
\begin{aligned}
&\sum_{i}\sum_{j^*} K_{\mathrm{L}}^{ij^*} f_{\mathrm{L}}^{i} f_{\mathrm{L}}^{j^*} P = \bar{K} \quad \text{(a)} \\
&\sum_{i}\sum_{j^*} K_{\mathrm{L}}^{ij^*} f^{i} f_{\mathrm{L}}^{j^*} P^{i} = \bar{K} \quad \text{(b)}
\end{aligned}
\qquad (4.11)
$$

These relations may be observed from the free energy diagram in Figure 4.17. The appropriate thermodynamic relations may be similarly expressed. The two equations show the relationship between conformer populations in aqueous and non-aqueous phases.

For a set of close analogues which bind to the receptor in the same way, $K^{ij^*}f^{j^*}$ is often invariant and the binding constant will vary directly with the relevant conformer fraction of the ligand. For a rotation about a single bond, the conformer population can be readily calculated from the rotamer energetics by use of the Boltzmann distribution. The reason that classical statistics can be applied to rotamer energetics is that, unless the barrier to rotation is very high, conformer interchange is very fast. The number of molecules with energy E_i is given by

$$n_i = \frac{e^{-E_i/kT}}{\sum_i e^{-E_i/kT}} \tag{4.12}$$

The relative population between two states 1 and 2 is given by

$$n_2/n_1 = e^{-(E_2-E_1)}/kT \tag{4.13}$$

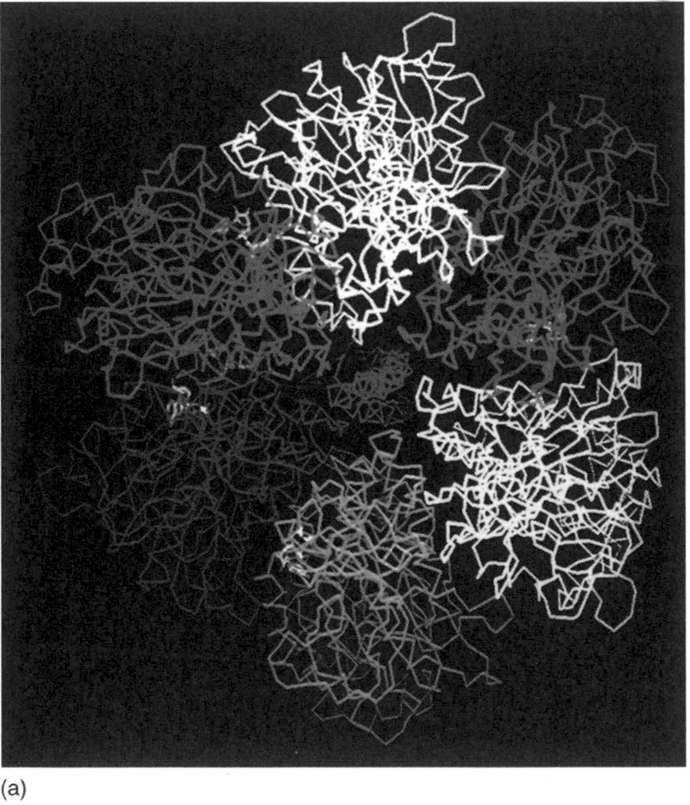

(a)

Figure 4.13 (See color insert after p. 368) (a) Adenosine triphosphate synthase (ATP synthase, F_1 F_0 synthase) is the central enzyme in energy conversion in mitochondria, chloroplasts, and bacteria. and uses a proton gradient across the membrane to synthesis ATP from the diphosphate, ADP, and inorganic phosphate. The multisubunit assembly consists of a globular domain, F_1, and an intrinsic membrane domain, F_0, linked by a slender stalk about 45 Å long. The F_1 domain is an approximate sphere 90–100 Å in diameter and contains the catalytic binding sites for the substrates ADP and inorganic phosphate. About three protons flow through the membrane per ATP synthesized but the mechanism of synthesis is not known. The F_1 structure is a complex of five different proteins with the stoichiometry 3α:3β:1γ:1δ:1ε. The sequences of the α- and β-subunits are homologous (~20% identical), including the P-loop nucleotide-binding motif. The catalytic sites are in the β-subunits while the function of the α-subunits is obscure. It has been suggested that the structures of the three catalytic sites are always different, but each passes through a cycle of "open," "loose," and "tight" states. In this respect crystals developed with AMP-PNP (where the nitrogen atom defines the analogue of ATP) show occupancy of the nucleotide sites in different states of phosphorylation. The α- and β-subunits are arranged alternatively like the segments of an orange around a central α-helical domain containing both the N- and C-terminals of the γ-subunit. As the three β-subunits vary in nucleotide occupancy (ADP, AMP-PNP, and empty) and have different conformations, the structure as found in the crystal (2.8 Å resolution) is compatible with one of the states to be expected in the cyclical binding change mechanism (Abrahams et al., 1996). The figure shows the arrangement of the three α- (A, pink; B, blue; C, green) and β- (D, purple; E, yellow; F, white), around the central F_0 stalk (orange). The positions of nucleotides are given in elemental coloring.

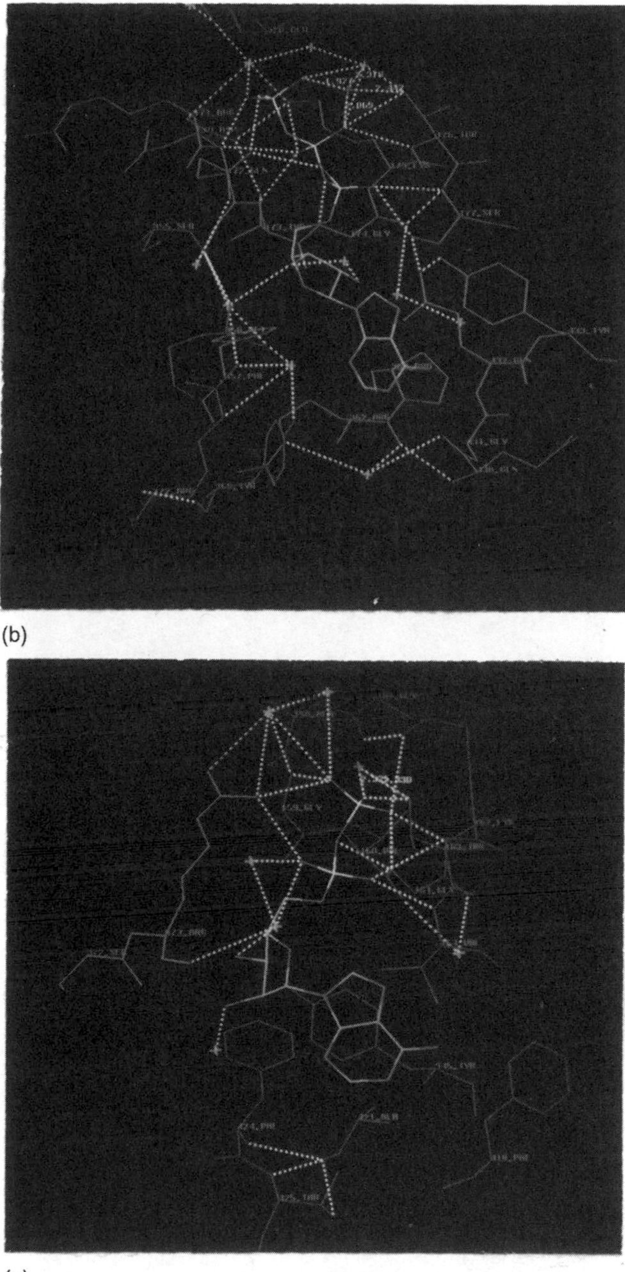

(b)

(c)

Figure 4.13 Continued
Plates (b) and (c) show the similarity of the binding sites of the nucleotides in the α- and β-subunits. (b) The nucleotide AMP-PNP is between the A,α- and D,β-subunits. All the nucleotide-binding sites are in the α- (A, pink) except for those indicated (β-, D (purple)). The magnesium ion assisting the phosphorylation is shown in red between the two terminal phosphate groups. (c) The ADP is bound very predominantly to the β- (D, purple) subunit. The relations to the α- (C, green) subunit are indicated.

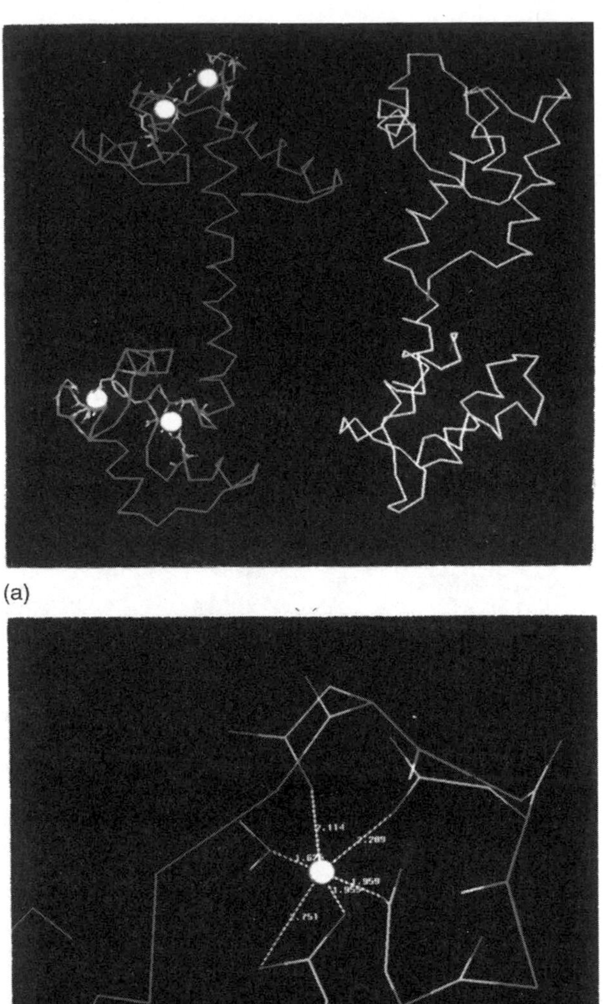

(a)

(b)

Figure 4.14 (See color insert after p. 368) The influence of strong charge on conformation. The structure of calmodulin with and without the interaction of four calcium ions. Calmodulin is the principal calcium-dependent regulator of a variety of intracellular processes. The 148-residue protein has four Ca^{2+} sites and a number of acidic residues. It is a ubiquitous protein in eukaryotes and plays a critical role in coupling transient Ca^{2+} influx, caused by a stimulation at the cell surface, to events in the cytosol. The Ca^{2+}-binding sites have the "EF hand" configuration also identified in other Ca^{2+} binding proteins such as intestinal calcium binding protein and troponin C. The "EF hand" comprises a helix–loop–helix structure which can be identified from the sequence homology alone. The basic structural unit of the globular domain consists of a pair of EF-hands rather than a single binding site.
Figure (a) *Top*: Calcium-bound calmodulin from *Drosophila melanogaster* (2.2 Å resolution; Taylor et al., 1991) has a seven-turn α-helix connecting the two calcium-binding domains. The dumb-bell shaped molecule contains seven α-helices and four "EF" calcium-binding sites and closely resembles the mammalian structure.

More strictly, writing ΔG_{2-1} for the free energy difference between the two rotamers and taking logarithms

$$\Delta G_{2-1} = -NkT\log n_2/n_1 \tag{4.14}$$

where N is Avogadro's number. Using the $\log_{10}$ scale, $NkT = 1.418$ kcal mol^{-1} at blood temperature (37°C). Thus for two conformers differing in energy by 1 kcal mol^{-1}, $\log_{10} n_2/n_1 \sim 0.7$ (where conformer 2 is the more favorable) giving a conformer population ratio n_2/n_1 of 5:1 at this temperature.

The utility of Equation (4.10) is shown by a simple early example in Figure 4.18 where the bioactive conformer with the basic side chain perpendicular to the aromatic ring (calculated on the intrinsic conformer preference in Table 4.4) of the CNS agent viloxazine (**4.6**, R = 2-OCH_3), an inhibitor of biogenic amine release, is plotted as a function of the potency component which has been referenced to a non-polar phase environment. The relation is of unit slope. (In this early example, octanol has been used as the reference non-polar solvent. As the relatively weak hydrogen bond proton acceptor properties of the solute phenoxy oxygen atom are weaker than those of the polar reference solvent octanol, little error is introduced in this set of data by the use of this solvent compared with that of a hydrocarbon.)

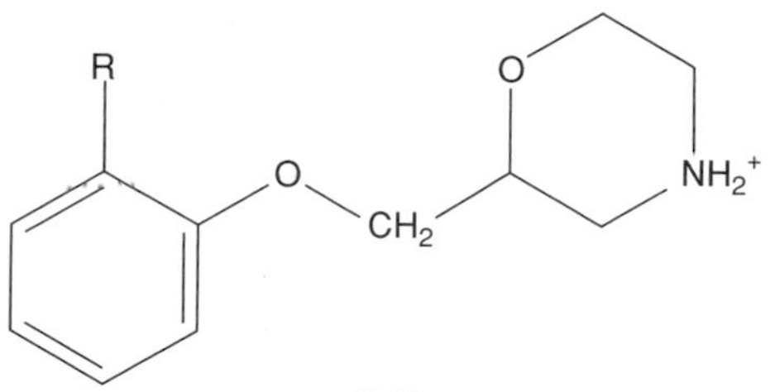

R O O CH_2 NH_2^+

(**4.6**)

While most data of this type are very much related to details of ligand conformation and of localized energetics in the target site, an advantage of such information can be an understanding of the detailed thermodynamics of binding of closely related molecules as exemplified in Section 4.3.5. A model alignment of residues of the β_1-adrenergic receptor on the bacteriorhodopsin model (Nederkoorn et al., 1998) based on the known bound conformation of the natural agonist noradrenaline showed that the α-carbon interatomic distance between residues defining the alignments of helices III and VII on which a suggested proton pumping mechanism was based were within 1 Å of the bovine rhodopsin crystal distance. Other helices were much less well defined. Refinement of the proton pumping mechanism may now be possible and open to more detailed interpretation.

Figure 4.14 Continued
The six-coordination octahedral form of a binding site is shown in plate (b) where the Ca^{2+} ion is held by four acidic residues. In each site, the coordination (one shared) comes from five side-chain oxygen atoms, a carboxyl oxygen (not shown) and one water molecule. *Bottom*: The NMR-determined calcium-free structure of calmodulin (Kuboniwa et al., 1995). Each calmodulin domain consists of a strongly twisted but tightly packed bundle of four helices. Upon binding of Ca^{2+} most of the change occurs within each of the "EF hands" with interhelix angle changes. The structural rearrangement on binding Ca^{2+} ion results in a pronounced hydrophobic pocket on the surface of each domain. These pockets appear to be of importance from structure studies on Ca^{2+}-bound complexes with different synthetic target peptides. The accuracy of NMR-determined structures is highest at the center of the protein and decreases as one moves towards the surface. The accuracy in the determination of the Ca^{2+} binding loops requires, in principle, further refinement. The conformation of the long central helix in the crystal structure was not previuosly consistent with extensive biochemical data on these proteins. The Ca^{2+}-free structure shows increased flexibility and this "connecting spacer" can be viewed as a flexible tether between the two domains. This is confirmed by x-ray structures on calmodulin complexed with peptide fragments of its intracellular receptors, e.g., myosin light-chain kinase where the two domains of cadmodulin swing round and envelope the target peptide.

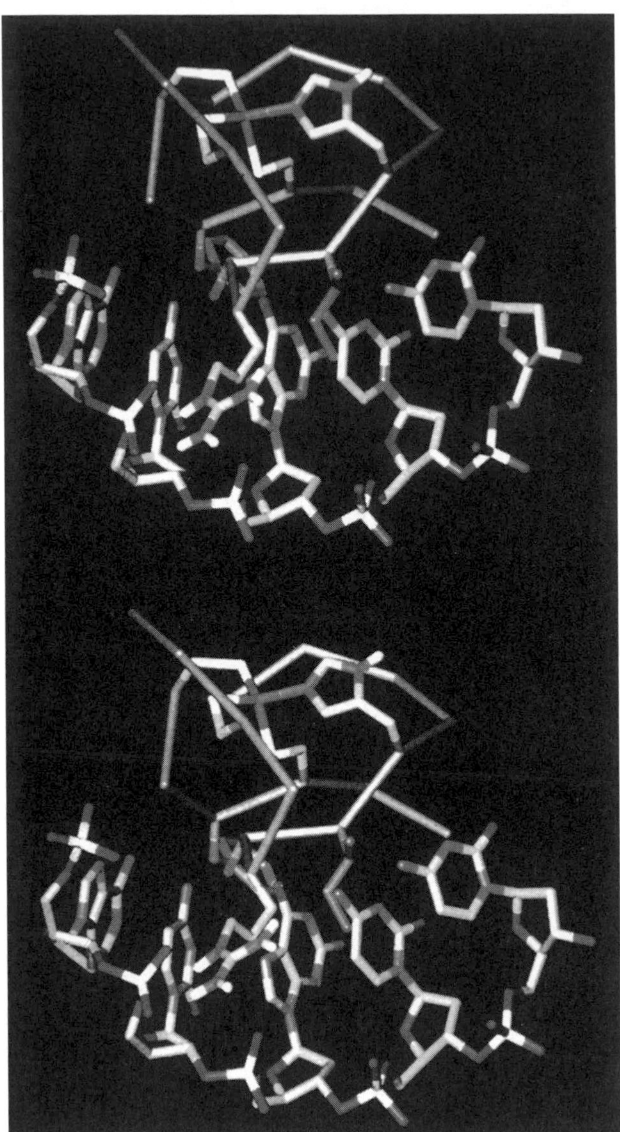

Figure 4.15 (See color insert after p. 368) Protein–single strand DNA recognition. A zinc finger domain binding to a single-stranded DNA sequence. Interaction of an NMR-determined zinc finger domain in the HIV–1 nucleocapsid protein (South et al., 1991). A common feature of proteins containing the "retroviral-type" (r.t.) zinc finger domain (Cys-X_2-Cys-X_4-His-X_4-Cys) is that they appear to be involved at some stage in sequence-specific single-stranded nucleic acid binding analogous to the zinc finger motif found widely in duplex–DNA binding proteins. Zinc finger r.t. domains are found both in the N-terminal and C-terminal chains of the intact HIV–1 nucleocapsid protein isolated from virus particles. The sequences have been shown to bind zinc stoichiometrically and with high affinity. The figure shows an 18 amino acid HIV1-F1 peptide C_α sequence (Val-Lys-Cys-Phe-Asn-Cys-Gly-Lys-Glu-Gly-His-Ile-Ala-Arg-Asn-Cys-Arg-Ala in pink) bound to a single-strand DNA sequence A-C-G-C-C). The tetrahedral coordination of the Zn ion with the three cysteine residues and His_{11} is shown bonded schematically on the right of the figure. The hydrophobic interactions of the peptide residues (Phe_4, Ile_{12}, Ala_{13}) are shown in green while the strong polar interaction of Arg_{14} with DNA backbone phosphate groups is seen at the end of the finger.

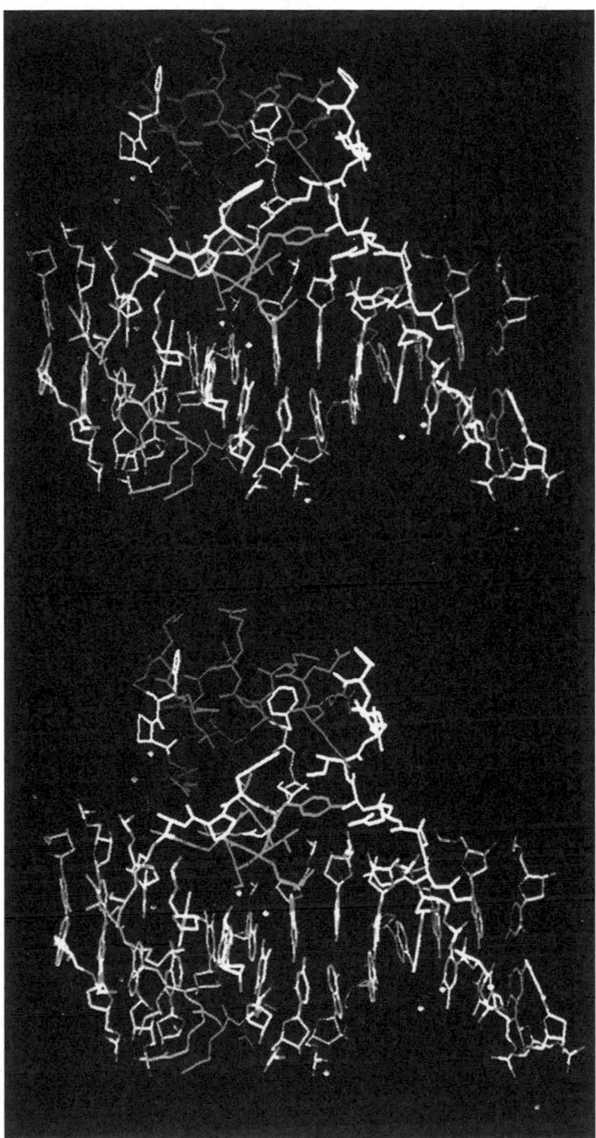

Figure 4.16 (See color insert after p. 368) Protein–double strand DNA recognition. The selectivity of protein binding in the major and minor grooves of the DNA. The binding of the prokaryotic enzyme Hin recombinase to DNA in the *Salmonella* chromosome (Feng et al., 1994). This site-specific recombination reaction controls the alternate expression of two flagellin genes by reversibly switching the action of a promoter. During the process of inverting the extended segment of DNA, two Hin proteins in the form of a dimer bind to the the left and right recombination sites located at the boundaries of the invertible DNA segment. Through interaction with a third interacting site (held by an additional protein) the overall complex aligns the two recombination sites correctly and the Hin protein is activated to initiate the exchange of DNA strands leading to inversion of the intervening DNA.

The recombination half-site of the double helical sugar–phosphate backbone of the DNA (elemental coloring) linked by the heterocyclic base pairs (blue) is shown occupied by the helix–loop–helix–loop–helix of the Hin protein.The third Hin helix (green) sits in the major groove of the DNA where the residues Arg 178, Thr 175, and Tyr 179 are shown on the lower side of this helix. Helices 1 and 2 (purple) are approximately orthogonal to helix 3. The amino terminal loop (white) at the bottom right of the picture attached to Helix 1 lies in the minor groove with two arginine residues (140 and 142) interacting with the helical backbone of the DNA. The carboxyl terminal chain extending from helix 3 (white) leads again into the minor groove at the upper left of the figure where the portion of the chain interacting with the DNA is shown in pink. The short loops joining helices 1 and 2 (top right) and helices 2 and 3 (middle right) are also indicated in white. Water molecules within the x-ray crystal structure (determination at 2.3 Å resolution) are shown with a white cross.

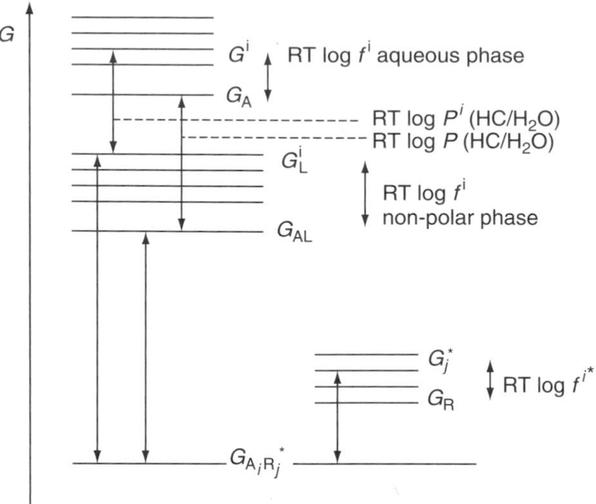

Figure 4.17 Schematic representation of the free energy relations for the conformer I of the drug (A) interacting with the relevant receptor conformer j^x of the receptor protein complex and possible pathways for relating the bound conformer free energy $G_{A_iR_j}{}^x$ to the reference energy G_A. The standard free energy of the conformer *i* of the drug is related to the average free energy G_A by the conformer fraction or population f^i. A change of reference phase from aqueous to hydrocarbon is shown by the subscript L. The partition coefficient *P* defines the average free energy difference of A between the two phases and individual conformers in the different phase environments may be related similarly by conformer partition coefficients P^i. (By permission from Davies R.H. (1987) *International Journal of Quantum Chemistry, Quantum Biology Symposium* **14**, 221–243.)

4.5.3 Ligand Design — Macromolecular Structure Known

There are a number of simple ligand modeling strategies that have evolved to take advantage of the structural information on target proteins derived from x-ray crystallography or NMR spectroscopy. Given that the structure of the site is known, the strategies resolve to devising efficient schemes for the logical exploration of the space of the target site and the housing of the ligand's appropriate interacting groups. Whether to build upon interacting groups to probe obvious target sites and link these probes back to some representative molecule or whether to fill the volume of the site with nominal atoms and then to choose viable subsets for efficient interaction, the choice is perhaps dependent on the degree of understanding of the mechanism involved. It should be remembered that binding is a free energy process and those methods which incorporate the statistics of the binding both in macromolecule and in ligand should prove the most powerful. The limiting problem is likely to be computational effort but there is no substitute for knowledge of molecular structure.

Multiple fragment probes: locate and link methods

In the probe approaches, the so-called locate and link methods, a site-specific small probe can be geometrically constructed or better its interaction calculated and the orientation for the best localized orientation of the small probe molecule determined. Variants on optimizing the location of the probe can be generalized. One may place a set of small groups randomly on a coarse grid (0.5 Å) and optimize the translational and orientational variables using search methods based on the rate of change in energy as a function of the variables or by stochastic methods such as Monte Carlo. Similar approaches using the protein–fragment interaction forces and employing molecular dynamics for locating probes on a large number of polar fragments (e.g., 1000) randomly distributed within the binding site are used to calculate, via Newton's laws, the independent motion of each

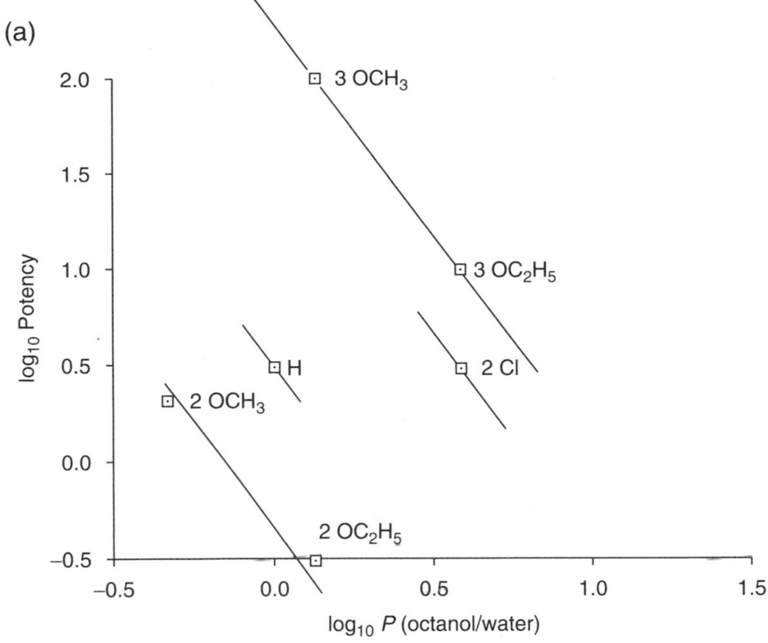

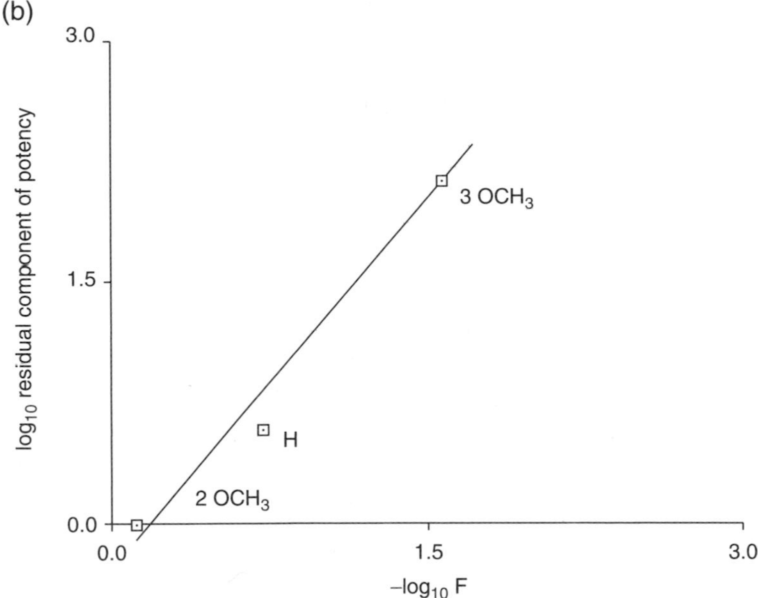

Figure 4.18 (a) Potency *in vivo* of viloxazine analogues plotted against a partitioning effect using the octanol–water model on the $\log_{10}$ scale. (b) Residual variation in potency of viloxazine analogues after allowance for a partitioning effect plotted on the $\log_{10}$ scale against the fraction of the conformers having the side chain perpendicular to the aromatic ring. (By permission from Davies R.H. (1987) *International Journal of Quantum Chemistry, Quantum Biology Symposium* **14**, 221–243.)

Table 4.4 Intrinsic conformer preference of substituted anisoles and related molecules at 37°C. *Ab initio* estimates and NMR data

	Conformational energy preference planar/perpendicular	Concentration difference 37°C
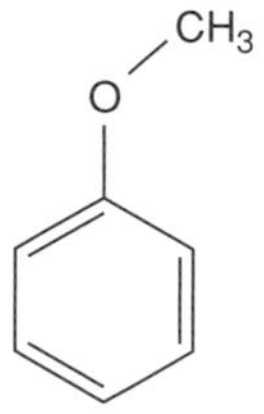	1.2 kcal[a] (STO-3*G*) 0.7 kcal (4-31*G*)[b]	5:1
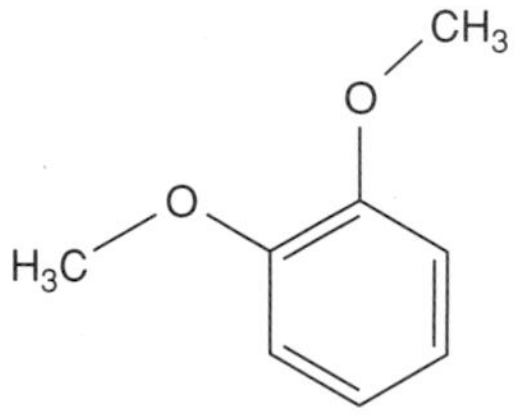	0.0 kcal (STO-3*G*)	1:1
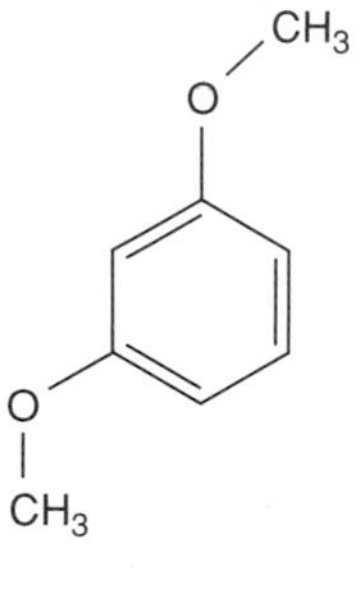	2.0 kcal (STO-3*G*)	25:1
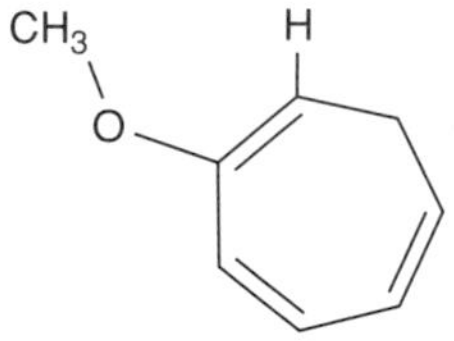	1.3 kcal[c] Synperiplanar to H Relative to antiperiplanar (in $CDCl_3$, NMR)	8:1
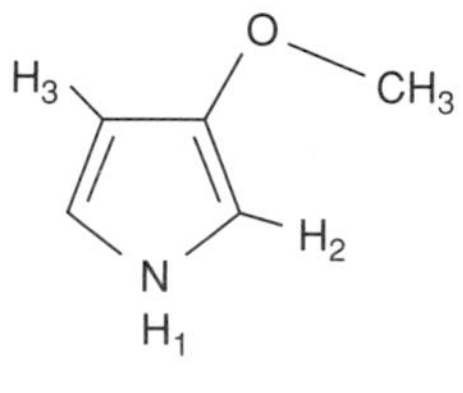	1.2 kcal [b](4-31*G*) synperiplanar to H_2 or ⊥[r] antiperiplanar less favored	7:7:1

Davies (1987) and references therein.

fragment. By slowly cooling the system to absolute zero, optimal binding positions for the probe groups can be determined.

Steric features of a site can be exploited by constructing spheres in contact with the protein surface such that the centroids represent positions for locating interacting atoms. In place of calculation, optimal positions for groups to partner hydrogen bonding moieties in the protein can be derived from data surveys of small-molecule x-ray structures and via microwave spectroscopy and quantum mechanical calculations. The resulting positions for donor hydrogen or acceptor atoms and their connected atoms form a set of vectors on which candidate probe hydrogen bond groups can be overlayed.

A special case in refined x-ray structures is given by bound solvent water molecules, which represent experimentally located probe fragments. Such water molecules can indicate opportunities for the location of hydrogen bond groups, although depending on their degree of interaction, not all can be replaced in an energetically favorable manner. Water is potentially tetracoordinate in hydrogen bonding though its two hydrogen bond proton donors and its two-electron lone pairs as proton acceptors. Whether to treat a located water molecule as a candidate for replacement or as strongly held by the protein depends on the number of potential interactions made with the macromolecule. The following simple table on the categories of bound water and their implications for substitution may be constructed:

H_1	LP_1	H_2	LP_2	Category of water	Implication
Protein	Protein	Protein	Protein	Sequestered	Not available
Protein	Protein	protein	available	Structural	Locate a donor
Available	Protein	Protein	Protein	Structural	Locate an acceptor
Available	Protein	Available	Protein	Ligand-like	Replace by e.g., –CO
Protein	Available	Protein	Available	Ligand-like	Replace by e.g., $-NH_2$
Protein	Protein	Available	Available	Ligand-like	Replace by e.g., –OH

Where only one interaction with the protein occurs, then replacement should be possible unless the water molecule under consideration forms a link in a chain of interacting hydrogen bonded groups from the protein or other water molecules. In this case preservation of the chain may be an important consideration. For two or less interactions with the protein or surrounding system, the bound water may be described as ligand-like and it should be possible to displace it with a favorable energetic outcome provided that there is no degradation in the quality of the replacement interactions. This approach is rather simplistic and takes no allowance for more subtle competitive displacement of the water molecule and the resultant energetics.

Linking the probes

The construction of potential intramolecular links between two probe groups is a straightforward if tedious problem of determining the possible spans by constructing a series of bonds with standard lengths, angles, and torsions and elucidating those links which do not clash with the protein. A number of methods have been developed to address this problem. For up to six bonds, given the bond lengths and angles, the required torsion angles can be solved analytically. Beyond this, connection can be achieved by constrained optimization of the torsion angles introducing constraints using the method of Lagrange multipliers. If torsion angles are sampled at appropriate minima, combinations of bond geometries (tetrahedral and trigonal) can be assembled into a growing network which terminate when a connection between fragments is made. Generally, several linker chains of varying length and composition will connect the probes and often these can be combined to give cyclic structures, eliminating unwanted conformational freedom and associated entropic effects. Similarly when the linkers show certain patterns of torsion angles, for example, a series of planar torsions, they may be reinforced by constructing rings incorporating those torsions.

Alternatively, the spatial arrangement of functional groups within the binding site allows this geometric structure to interrogate a database of small-molecule three-dimensional structures such as the Cambridge structure database. In pharmaceutical companies, such databases contain up to 1 million molecules. Molecules matching the required criteria can be tested for their ability to bind to the protein. The fastest searches assume a single conformation for each small molecule, but multiple conformations can be sampled if pre-stored in the database. A method requiring less data storage at the expense of description of strain within the molecule and of computer time attempts to fit the spatial constraints of the search query using distance geometry methods (Blumenthal, 1970).

Further database methods utilize the vector nature of the probe to its potential link with the putatative ligand. The geometric relationship between the vectors can be defined in terms of distances, angles, and torsions. A searchable vector database can be generated from any set of molecules by identifying templates with a number of connector bond vectors and then tabulating the geometric relationships between them. These templates are often rigid ring systems and the connectors, C—H bonds. Starting from commercially available compounds, a database of more than 30,000 templates can be derived. Searching a vector database for templates capable of connecting the localized functional groups simply correlates matching the appropriate distances, angles, and torsions within given tolerances. An ability to synthesize the appropriate template is usually an overriding choice amongst the matching templates.

Finally, it may be noted that the whole process of matching a ligand to its site can be machine based without recourse to an experimental database. If a decision is taken on the basis of the synthetic chemistry to be exploited, for example that of substituted benzdiazepines, then the most promising substitution patterns can be identified. Since the chemical reactions are specified, the reagents that are available commercially can be used as input to the computation and the output can be exploited using robotic methods in multiple parallel syntheses to generate libraries of candidate compounds (see Chapter 11).

Single fragment probes and ligand evolution

As the name implies, an initial target binding site is selected and an initial fragment probe developed from which the ligand is allowed to grow. This growth can be done by successive addition of atoms using a correlated acceptance or rejection procedure on each addition, the choice being dependent on their fitness to the protein environment. At a geometric level, the quality of the ground rules and the range of atom types considered are critical to the validity of the method. A variant of the method allows atoms to be "mutated" and segments of one molecule to be exchanged for a second. These evolutionary steps of addition, mutation, and crossover form the basis of a "genetic" algorithm. The number of structures evolved is controlled by assessing the fitness of the protein environment. Further criteria are required for realistic segments to be be identified and to ensure that the consequences of mutating atoms on surrounding atoms are transmitted into the next generation.

Again a fragment database with connector bonds can be attached to candidate "hooks" within the seed. Selection procedures based on some "protein binding" score catering for the interaction and the degree of distortion involved with the new link are used in such procedures. The quality of the criteria and the potentially regressive effect of the enlargement on previous substitutions, where a net favorable gross interaction may occur, highlight difficulties with these automated procedures. As further steps become involved, the enlargement may lead to combinatorial explosion in the number of candidates.

Filling the target site

The final approach is to fill the target site with nominal atoms and then choose viable subsets and determine the chemical nature of the atoms constituting the candidate ligand. Again much of these automated procedures are based on simple logical procedures. A regular lattice such as the diamond, tetrahedral, or planar hexagonal is positioned in the binding site using interactive graphics or by calculating minimal steric clashes with the protein surface. Complementary ligand–receptor interatomic interactions may be assigned and viable subsets of atoms selected. It is, however, difficult to mix different regular lattices to form realistic molecules.

Alternatively one may place a set of small acyclic and cyclic fragments to fill the site with all possible combination frameworks. It is then necessary to select candidate subgraphs and assign atom types via the protein environment. This method allows different geometries to be used together and overcomes the combinatorial explosion by using a small set of fragments involving only carbon atoms. Here the main difficulties lie in the selection of viable subgraphs and the assignment of atom types. A much simpler approach is to characterize the shape of the protein binding site as a defined ellipsoid and search a database of small molecules for identifying suitable ligands which fill the site approximately. The structure may then be substituted to adapt and complement the target site.

The last approach in this category is to fill the site with atoms whose individual nature is randomly assigned. The system is equilibrated using molecular dynamics with a force field that allows for ''soft'' repulsion between the atoms. A ''mother'' atom is randomly selected and attempts are made to form bonds with neighboring atoms using probabilistic rules. If accepted, the system is then relaxed using molecular dynamics and a new ''mother'' atom selected. The process is repeated for a specified number of selections, resulting in the emergence of a candidate ligand from the initial aggregate of atoms. The process is thus stochastic and may take many repeats to arrive at a synthetically useful ligand. The rules for bond formation and the associated acceptance criteria are crucial to this approach.

4.5.4 Accommodation of The Protein to Ligand Binding: Estimating Interaction Free Energies

In the previous section, the structure of the protein was taken to be fixed at the average determined by x-ray crystallography or NMR spectroscopy. By comparing native protein structure with those of complexes, it is apparent that some degree of accommodation to the ligand always occurs on binding. Indeed, in many cases, significant conformational changes accompany the ligand binding. Given the choice of a native protein structure or the structure of the protein partner from a ligand complex, experience has indicated that the latter is the better starting point for ligand design. The problem here, as with basing new design on an active ligand conformer when the structure of the protein binding site is unknown, is the inherent bias of the bound ligand conformation. Clearly one should design much better, if the accommodation of the protein to novel structure were taken into account.

The local fluctuations in protein and ligand structures can be introduced in a given mode of binding to yield a free energy. Using Monte Carlo or molecular dynamics methods (see for example, Beveridge and DiCapua, 1989; Allen and Tildesley, 1987; Valleau and Whittington, 1977), an ensemble of local fluctuations within the ligand and the protein are calculated to yield the thermodynamic functions of binding. In the former method, the sample space is efficiently explored using an algorithm based on Boltzmann weighting while in the latter the dynamics of the interactions are explored over a period of nanoseconds. Both methods thus allow for an ensemble of protein structures to be explored and replace the single rigid structure used hitherto. As indicated earlier in the context of building fragments, one problem is the expansion in the number of protein ''structures,'' which are associated with individually designed ligands and the limits on computing time. Again restriction on the variables undergoing change in the fluctuations to torsional angle subsets may alleviate the problems to some extent. If large conformational changes occur on binding, then the changes are difficult to simulate in any predictive way.

Some experimental information on restricting the scale of the structure to be relaxed can be given by the x-ray or NMR structure. NMR-determined structures are defined by an ensemble of structures that meet the NMR structural criteria. This ensemble can be used instead of a single structure. The mobility of atoms in structure determined by x-ray crystallography is often represented by an associated temperature factor, and these data could be incorporated into the design process. There is a structural hierarchy in relation to the protein's accommodation to the ligand, from side-chain reorientation, then local main-chain adjustments and finally large hinge-bending movements of whole regions of the protein structure.

Although many of these decision making processes may be introduced into automated regimes, the introduction of specific constraints removes some of the objective character of the procedures involved, and all methods are limited by the adequacy of the physical descriptions of the interactions defined in Sections 4.2 to 4.4. Specific polarizing effects of strong charge interactions inducing changes in the charge distribution both in ligand and in protein are not introduced into standard fast potential routines unless potentials are specifically developed over the sets of ligand and protein atoms for the particular interaction concerned using more fundamental quantum mechanical calculations. There is a case for doing this in any area of detailed study. The difficulties of determining accurate free energies of binding should not be underestimated. It would, of course, be desirable to calculate all interactions by fundamental quantum mechanical methods but the physical constraint on machine time becomes quickly rate limiting. The scale of the problem with current machine capabilities is summarized in Section 4.7. The philosophy with the current state of computing power should be (once one is approaching likely candidate structures of interest)

1. to determine calibrated intra- and interatomic potentials from rigorous quantum mechanical methods on local residue interactions
2. to check the influence of longer range interactions possibly by introducing specific local charges into the potential, and
3. to determine the minimum energies of the ligand–protein interaction followed by calculating the statistics of the ligand–protein interaction allowing local variations in the protein environment.

4.6 PROTEINS

The theoretical determination of protein structure from first principles based on the intramolecular interactions of the individual amino acids, as we remarked earlier, would have high significance in the design of inhibitory or stimulatory ligands in many areas of drug therapy. This is a large subject and we refer to more specialized treatments. The possible number of sequences in an average sized protein of some 400 amino acids is 20^{400} based on the 20 amino acids and the question as to why only a very small fraction occurs in nature may resolve to structures that have unique and stable native states. A paper (Li et al., 1996) which avoids most details of the chemistry of the amino acid interactions examines a polymer of 27 amino acids occupying all sites of a $3\times3\times3$ cube employing simple interactions on a lattice (hydrogen bonding or otherwise). The great majority of sequences have multiple ground states and hence may fold into different structures assuming no inherent large kinetic barrier. Thus ''foldability'' focuses on the sequence selecting potentially functional ones while ''designability'' is based on the structure of the resulting protein, which is quantified by measuring the number of sequences that uniquely fold into a particular structure. In evaluating the 2^{27} structure in the simple amino acid scheme, the distribution gives a number of patterns. At the tail of the distribution, there are structures that are highly desirable and are also more stable. The number of sequences (N_S) associated with a given structure (S) differs from structure to structure but preferred structures emerge with N_S values much larger than the average. Analysis of the mutation patterns of the homologous sequences for highly designable structures revealed phenomena similar to those observed in real proteins, some sites being highly mutable while others are highly conserved. Although the initial categorization is elementary, such an approach may offer a pathway to introducing constraints on the multiple minima problem in addition to the already established methods. One view is that an ensemble of states exist as the components of the protein assembly interact to find their minimum, in which probably the van der Waals forces play the important role (Dill and Chan, 1997). Thus the contacts may be viewed as a form of ''hydrophobic zipping'' where similar energies of assemblage allow further localized specific interactions to occur, the association of these specific proximal interactions being facilitated by multiconformational starting structures.

4.7 ACCURATE CALCULATION OF INTERMOLECULAR INTERACTIONS

Again, we refer to more specialized treatments for quantum mechanics methods and limit ourselves to a brief perspective of the scale of the machine problem for calculating noncovalent molecular interactions. In fundamental or *ab initio* quantum mechanical calculations, each electron's interaction with the nuclei is not strictly independent of the position of other electrons in the system. To simplify the problem, the initial approximation is made that each electron interacts with the average field of the other electrons, i.e., the motions of the electrons are uncorrelated (Hartree–Fock approximation). An electron will thus have kinetic energy while its potential energy consists of its interaction with other nuclei and with the average field of the other electrons, so that the problem is reduced to a set of one-electron equations. Molecular geometries, dipole moments, and electrostatic effects may be calculated to good accuracy with this approximation. The neglect of electron correlation, however, means that dispersive or van der Waals interactions are not present, while in situations where electron correlation is important, for example in transition states with molecules near dissociation limits, the approximation is completely invalid.

The electrons on each atom are characterized by molecular orbitals and a molecular orbital is constructed from a linear combination of the atomic orbitals. A variational procedure to minimize the energy is employed to determine the relevant contributions of each orbital to the molecular wave functions. The set of vector functions defining the atomic orbitals is known as a basis set. If one function is used to characterize the atomic orbitals, the set is known as a minimal basis, and broadly viewed, a minimal basis has insufficient flexibility to enable the valence electrons to spread themselves out satisfactorily and such conditions can have different consequences depending on the occupied and unoccupied orbitals. Providing two functions to characterize each orbital (doubling the basis set) allows much more flexibility in the wave functions but can produce exaggerated properties. As interactions become stronger, the introduction of d orbitals in atoms in the first row of the periodic table becomes significant contributing to polarization effects but leading to some 15 basis functions to describe a first row atom and the size of the basis set rapidly expands even with relatively small molecules. In Hartree–Fock theory, the number of two-electron integrals rises as n^4 where n is the number of basis functions. Geometry optimization requires an n^5 calculation. An alternative approach which has gained ground in recent years is based on the Kohn–Sham theory that an exact solution to the Schrodinger equation exists which leads to self-consistent equations as in Hartree–Fock theory. The many-electron problem can be replaced by an exactly equivalent set of one-electron equations with an effective one-particle potential. This effective potential will reproduce the exact density and the exact total energy if the definition of this potential can be defined. The problem is in its definition. The advantage of this density functional approach, which has required considerable development, is that the scale of this approach rises as n^2. Thus for larger problems of chemical interest, the potential becomes high. But how good is the initial description?

A decade ago, the cutting edge of computation was a machine with a speed of some 100 mflops/sec but as stated earlier, this is now available to most quite small operators. Utilizing the benefits of parallelization of machines applies only to certain calculations where the problem can be dismembered satisfactorily to run time-limiting sections in parallel as with the calculation of two-electron integrals and the basic time-dependent problem in *ab intio* calculations. Using 64 node machines, the practical limit on basis functions using *ab initio* methods is approximately 4000. Grid computing in 2004 could, in principle, take this set to 20,000. This allows, depending on accuracy, an interaction of some 1250 to 5000 atoms. The basis set limit using density functional theory is perhaps 20,000. Semiempirical quantum mechanical methods cannot utilize the benefit of parallelization beyond about eight nodes and the practical limit of scale is again of this order. For free energy calculations using empiric potentials, molecular dynamics methods can handle up to 2,500,000 atoms for 1 nanosec timescale. Vibrations and rotations with timescales of 10^{-15} and 10^{-12} sec, respectively, can be handled by such calculations. However docking in molecular recognition (of the order of 10^{-6}) and translational motion in liquids are on longer timescales.

REFERENCES

Allen, M.P. and Tildesley, D.J. (1987) *Computer Simulations of Liquids*. Oxford: Clarendon.

Abraham, M.H. (1982) Free energies, enthalpies and entropies of solution of gaseous nonpolar nonelectrolytes in water and nonaqueous solvents. The hydrophobic effect. *Journal of the American Chemical Society* **104**, 2085–2094.

Abrahams, J.P., Buchanan, S.K., van Raaij, M.J., Fearnley, I.M., Leslie, A.G.W. and Walker J.E. (1996) The structure of bovine F_1-ATPase complexed with the peptide antibiotic efrapeptin. *Proceedings of the National Academy of Science* **93**, 9420–9424.

Beveridge, D.L. and DiCapua, F.M. (1989) Free energy via molecular simulation: a primer. In van Gunsteren, W.F. and Weiner, P.K. (eds.), *Computer Simulations of Biomolecular Systems*. Leiden: ESCOM, pp. 1–26.

Blumenthal, L.M. (1970) *Theory and Applications of Distance Geometry*, 2nd edn. Bronx, New York: Chelsea.

Broadley, K.J., Nederkoorn, P.H.J., Timmerman, H., Timms, D. and Davies, R.H. (2000) The ligand–receptor–G protein ternary complex as a GTP-synthase. Steady-state proton pumping and dose–response relationships for β-adrenoceptors. *Journal of Theoretical Biology* **205**, 297–320.

Brunck, T.K. and Weinhold, F. (1979) Quantum mechanical studies on the origin of barriers to internal rotation about single bonds. *Journal of the American Chemical Society* **101**, 1700–1709.

Cariati, F., Cauletti, C., Ganadu, M.L., Piancastelli, M.N. and Sgamellotti, A. (1980) Spectroscopic investigations on phthalazino(2,3-b)phthalazine–5,12-dione and some of its mono and di-substituted derivatives. *Spectrochimica Acta* **36A**, 1037–1043.

Csizmadia, I.G. (ed.) (1982) *Molecular Structure and Conformation*. Amsterdam: Elsevier.

Davies, R.H. (1987) Drug and receptors in molecular biology. *International Journal of Quantum Chemistry, Quantum Biology Symposium* **14**, 221–243.

Davies, R.H., Sheard, B. and Taylor, P.J. (1981) Conformation, partition and drug design. *Journal of Pharmaceutical Sciences* **68**, 396–397.

Deslongchamps, P. (1983) *Stereoelectronic Effects in Organic Chemistry*. Oxford: Pergamon.

de Vos, A.M., Ultsch, M. and Kossiakoff, A.A. (1992) Human growth hormone and extracellular domain of its receptor: crystal structure of the complex. *Science* **255**, 306–312.

Dill, K.A. and Chan, H.S. (1997) From levinthal to pathways to funnels. *Nature Structural Biology* **4**, 10–19.

Feng, J.-A., Johnson, R.C. and Dickerson, R.E. (1994) Hin recombinase bound to DNA: The origin of specificity in major and minor groove interactions. *Science* **263**, 348–355.

Gerhartz, B., Niestroj, A.J. and Demuth, H.-U. (2002) Enzyme classes and mechanisms. In Smith, H. J. and Simons, C. (eds.) *Proteinase and Peptidase Inhibition*. London: Taylor and Francis, pp. 1–34.

Jorgensen, W.L. and Salem, L. (1973) *The Organic Chemist's Book of Orbitals*. New York and London: Academic Press.

Kuboniwa, H., Tjandra, N., Grzesiek, S., Ren, H., Klee, C.B. and Bax, A. (1995) Solution structure of calcium-free calmodulin. *Nature Structural Biology* **2**, 768–776.

Li, H., Helling, R., Tang, C. and Wingreen, N. (1996) Emergence of preferred structures in a simple model of protein folding. *Science* **273**, 666–669.

Marquart, M., Walter, J., Deisenhofer, J., Bode, W. and Huber, R. (1983) The geometry of the active site and of the peptide groups in trypsin, trypsinogen and its complexes with inhibitors. *Acta Crystallographica B* **39**, 480–490.

Nederkoorn, P.H.J., Timmerman, H., Timms, D., Wilkinson, A.J., Kelly, D.R., Broadley, K.J. and Davies, R.H. (1998) Stepwise phosphorylation mechanisms and signal transmission within a ligand–receptor–Gαβγ–protein complex. *Journal of Molecular Structure (Theochem)* **452**, 25–47.

Page, M.I. and Jencks, W.P. (1971) Entropic contributions to rate accelerations in enzymic and intramolecular reactions and the chelate effect. *Proceedings of the National Academy of Science* **68**, 1678–1683.

Palczewski, K., Kumaska, T., Hori, T., Behnke, C.A., Motoshima, H., Fax, B.A., Le Trong, I., Teller, D.C., Okada, T., Stenkamp, R.E., Yamamoto, M. and Miyano, M. (2000) Crustal structure of rhodopsin. A G protein-coupled receptor. *Science* **289**, 739–745.

Radom, L. (1982) Structural consequences of hyperconjugation. In I.G. Csizmadia (ed.), *Molecular Structure and Conformation: Recent Advances.* Amsterdam, Oxford, New York: Elsevier, pp. 1–64.

Reed, A.E., Weinhold, F., Curtiss, L.A. and Potachko, D.J. (1986) Natural bond orbital analysis of molecular interactions: the theoretical studies of binary complexes of HF, H_2O, NH_3, N_2, O_2, F_2, CO and CO_2 with HF, H_2O and NH_3. *Journal of Chemical Physics* **84,** 5687–5705.

Sielecki, A.R., Fedorov, A.A., Boodhoo, A., Andreeva, N.S. and James, M.N.G. (1990) Molecular and crystal structures of monoclinic porcine pepsin refined at 1.8 Å resolution. *Journal of Molecular Biology* **214**, 143–170.

South, T.L., Blake, P.R., Hare, D.R. and Summers, M.F. (1991) C-Terminal retroviral-type zinc finger domain from the HIV–1 nucleocapsid protein is structurally similar to the N-terminal zinc finger domain. *Biochemistry* **30**, 6342–6349.

Taylor, D.A., Sack, J.S., Maune, J.F., Beckingham, K. and Quiocho, F.A. (1991) Structure of a recombinant calmodulin from *Drosophila melanogaster* refined at 2.2 Å resolution. *Journal of Biological Chemistry* **266**, 21375–21380.

Valleau, J.P. and Whittington, S.G. (1977) A guide to Monte Carlo for statistical mechanics; 1. HighWays. In Berne, B.J. (ed.), *Statistical Mechanics Part A: Equilibrium Techniques.* New York and London: Plenum, pp. 137–168.

Vitali, J., Martin, P.D., Malkowski, M.G., Robertson, W.D., Lazar, J.B., Winant, R.C., Johnson, P.H. and Edwards, B.F.P. (1992) The structure of a complex of bovine α-thrombin and recombinant hirudin at 2.8 Å resolution. *Journal of Biological Chemistry* **267**, 17670–17678.

5

Drug Chirality and its Pharmacological Consequences

Andrew J. Hutt

CONTENTS

5.1 INTRODUCTION

One in four therapeutic agents are marketed and administered to humans as mixtures. These mixtures are not drug combinations in the accepted meaning of the term, i.e., two or more coformulated therapeutic agents, but combinations of isomeric substances, the biological activity of which may vary markedly. The majority of these mixed formulations arise as a result of the use of racemates, an equal parts mixture of enantiomers, of synthetic chiral drugs and, less frequently, mixtures of diastereoisomers. A survey of 1675 drug structures carried out in the 1980s revealed the extent of the problem. Of the 1200 (72%) agents classified as synthetic, 422 (25%), and 58 (3.5%) were marketed as racemates and single enantiomers, respectively.

That the individual enantiomers present in a racemate may exhibit differential biological properties has been known for over a century. However, only relatively recently with advances in the chemical technologies associated with the synthesis, analysis, and preparative scale separation of chiral molecules has the potential significance of stereochemical considerations in pharmacology and therapeutics been appreciated and, in some instances exploited, to a great extent. These new technologies have facilitated both the pharmacological evaluation of single stereoisomers and their production on a commercial scale. Such biological evaluation has resulted in an increased awareness of the potential significance of the differential pharmacodynamic and pharmacokinetic properties of the enantiomers present in a racemate, particularly with respect to safety issues, and the use of such mixtures has become a cause of concern.

The interaction of a drug with its target site involves interactions between functionalities on the drug molecule and complementary sites, or groups, on the target. Such interactions may have considerable steric constraints in terms of interatomic distance and bulk and, in the case of stereoisomers, the three-dimensional spatial arrangement of such functionalities is of considerable significance. At the molecular level biological environments are highly chiral being composed of chiral biopolymers, e.g., proteins, glycolipids, and polynucleotides, from the chiral building blocks of L-amino acids and D-carbohydrates. Additionally, many of the natural ligands at drug target sites, e.g., neurotransmitters, autocoids, hormones, endogenous opioids, etc. are chiral single enantiomer molecules. As nature has expressed a preference in terms of its stereochemistry it should not be surprising that receptors, enzyme active sites, ion channels, etc. frequently exhibit a preference for one of a pair of enantiomers. Indeed, Lehmann has stated that ''the stereoselectivity displayed by pharmacological systems constitutes the best evidence that receptors exist and that they incorporate concrete molecular entities as integral components of their active sites.''

5.2 DEFINITIONS AND NOMENCLATURE

Stereochemistry is concerned with the three-dimensional spatial arrangement of the atoms within a molecule; the prefix stereo originating from the Greek *stereos* meaning solid or volume. Stereoisomers are compounds which differ in the three-dimensional spatial arrangement of their constituent atoms and may be divided into two groups namely enantiomers and diastereoisomers. Enantiomers are stereoisomers which are nonsuperimposable mirror images of one another and are therefore pairs of compounds related as an object to its mirror image in the same way that an individual's left and right hands (or feet, or ears) are related. Such molecules are said to be chiral, from the Greek *chiros* meaning handed. Stereoisomers of this type are also referred to as optical isomers, due to their ability to rotate the plane of plane polarized light, which for a pair of enantiomers is equal in magnitude but opposite in direction, and also as enantiomorphs, from the Greek *enantios* opposite, *morph* form.

Stereoisomers that are not enantiomeric, i.e., are not related as nonsuperimposable mirror images, are said to be diastereomeric. The term diastereoisomer therefore refers to all other stereoisomeric compounds regardless of their ability to rotate the plane of plane polarized light, and the definition also includes geometrical, i.e., *cis*/*trans*, isomers.

In a pair of enantiomers the relative positions and interactions between the individual atoms are identical, as are their energy contents and, other than the direction of rotation of plane polarized light, their physicochemical properties are also identical. In contrast, the relative positions of the individual atoms, their interactions, and hence the energy content of a pair of diastereoisomers differ, as do their physicochemical properties. This fundamental difference in the properties of the two types of stereoisomer has considerable significance. As a result of their identical properties the separation, or resolution, of a pair of enantiomers cannot be readily achieved by standard chemical techniques, whereas a pair of diastereoisomers may, in principle at least, be separated by distillation, crystallization, and chromatography.

In terms of the compounds of interest in medicinal chemistry and pharmacology the most frequent cause of chirality arises from the presence of an sp^3 hybridized tetrahedral carbon atom in a molecule to which four different atoms or groups are attached (**5.1**). Such atoms are known as stereogenic centers, centers of chirality or, in older texts, asymmetric centers. The presence of one such center in a molecule gives rise to a pair of enantiomers (e.g., ibuprofen **5.2**), the presence of n such different centers yields 2^n stereoisomers and half that number of pairs of enantiomers. Those stereoisomers which are not enantiomeric are diastereoisomeric. For example the antibiotic chloramphenicol contains two stereogenic carbon atoms (numbered 1 and 2 in structure **5.3**) and therefore four stereoisomers are possible, two pairs of enantiomers (compounds labeled (1*R*,2*R*)- and (1*S*,2*S*)-**5.3**; (1*R*,2*S*)- and (1*S*,2*R*)-**5.4**), those stereoisomers which are not enantiomeric being diastereomeric. Thus, in the diagram presented those compounds related horizontally, i.e., the upper and lower pairs of structures are enantiomeric, whereas those related vertically, e.g., (1*R*,2*R*)-**5.3** with either of the structures **5.4** are diastereomeric. Diastereoisomers, which differ in configuration about one stereogenic center only, are termed epimers. Thus, (1*R*,2*R*)-chloramphenicol (**5.3**) is epimeric with structure (1*S*,2*R*)-**5.4** at carbon atom 1 and with (1*R*,2*S*)-**5.4** at carbon atom 2.

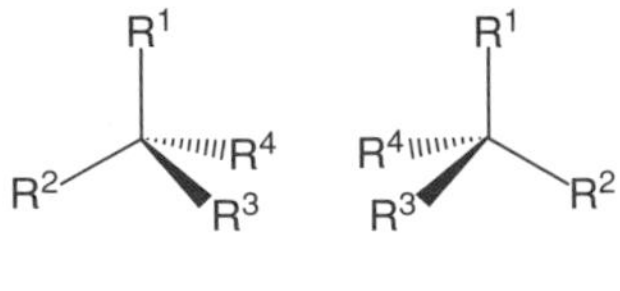

(**5.1**)

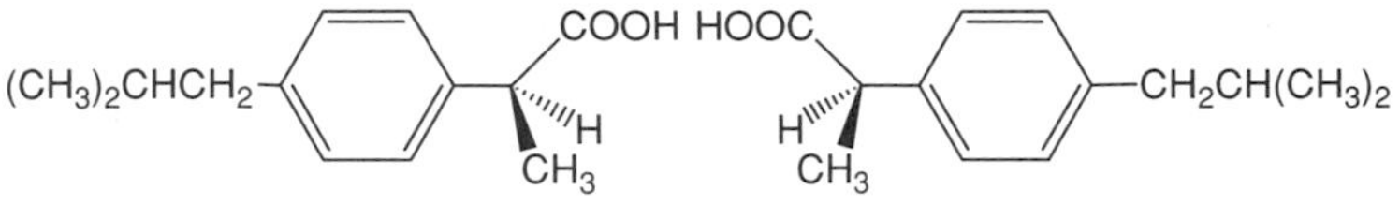

(+)-(*S*)-ibuprofen (**5.2**) (–)-(*R*)-ibuprofen (**5.2**)

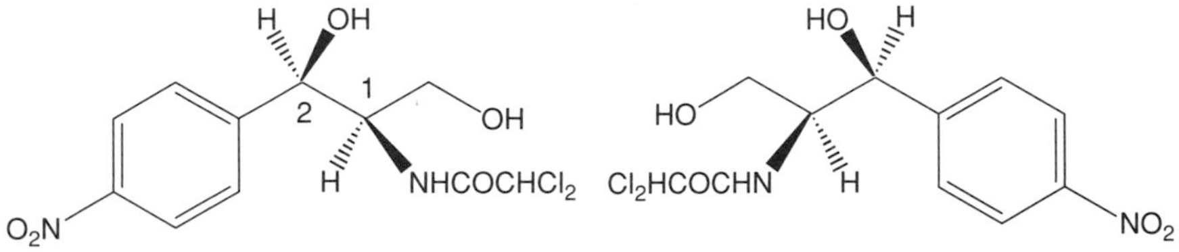

1*R*,2*R*-(**5.3**) 1*S*,2*S*-(**5.3**)

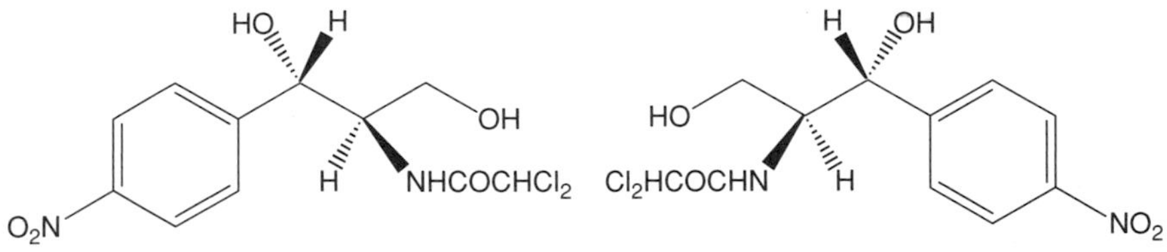

1*R*,2*S*-(**5.4**) 1*S*,2*R*-(**5.4**)

In addition to carbon, other atoms frequently found in organic molecules have a tetrahedral arrangement of the attached ligands, e.g., nitrogen, phosphorus, and sulfur, and chiral molecules with these elements as the stereogenic center are also known. In the case of trivalent derivatives of nitrogen the lone pair of electrons may be considered to be the fourth ligand. However, rapid inversion of the pyramidal forms occurs, through a planar transition state (**5.5**→ **5.6**→ **5.7**); the energy barrier for inversion is very low, so that separation of enantiomers is not possible. However, chiral compounds in which the nitrogen atom is part of a rigid cage structure, preventing inversion, are also known, e.g., the alkaloid quinine (**5.8**).

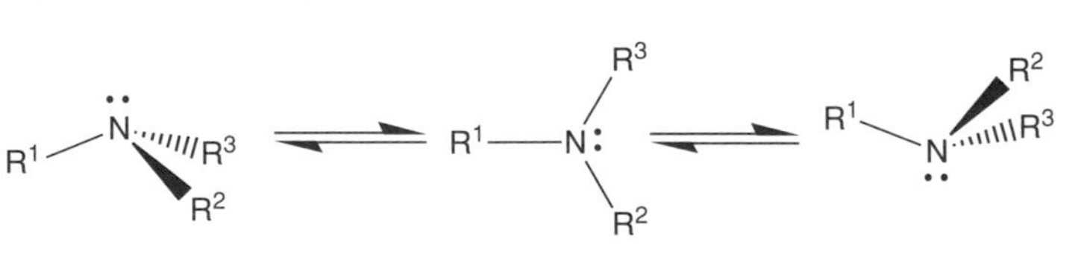

(**5.5**) (**5.6**) (**5.7**)

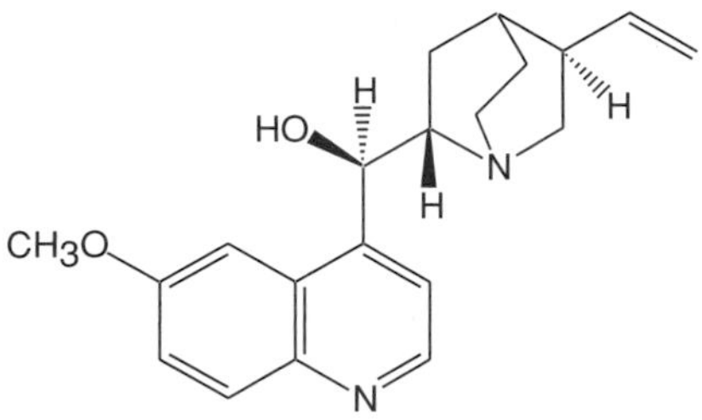

Quinine (**5.8**)

The formation of quaternary ammonium compounds, e.g., the neuromuscular blocking agent atracurium besylate (**5.9**), or the formation of an amine oxide, e.g., pargyline *N*-oxide (**5.10**), a metabolite of pargyline, results in the formation of a stereogenic nitrogen center and such compounds may be resolved into their enantiomers.

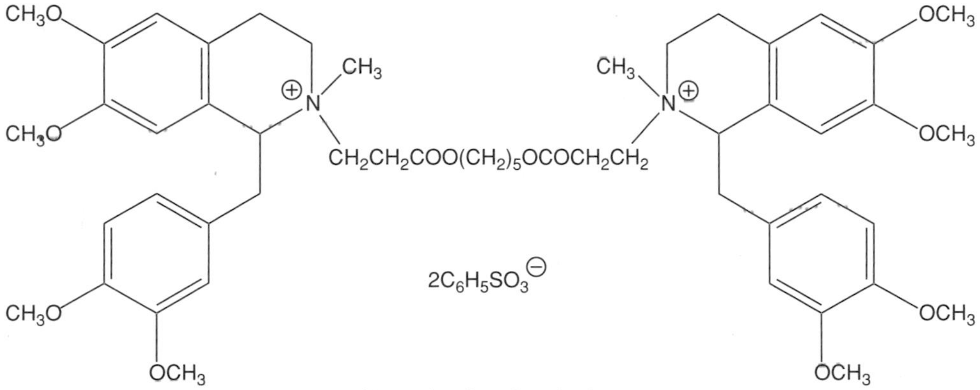

Atracurium besylate (**5.9**)

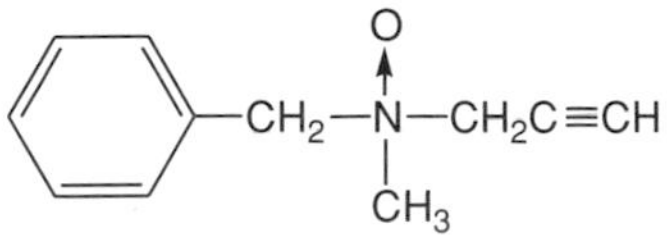

Pargyline *N*-oxide (**5.10**)

In contrast to trivalent derivatives of nitrogen, trivalent pyramidal sulfur derivatives have a higher energy of activation for inversion and the rate is slow enough that the individual enantiomers are relatively stable. Examples of drug molecules containing stereogenic sulfur and phosphorus centers include the proton pump inhibitors, e.g., omeprazole (**5.11**) and related benzimidazole derivatives, and the phosphamide mustard prodrugs, cyclophosphamide (**5.12**) and ifosfamide (**5.13**).

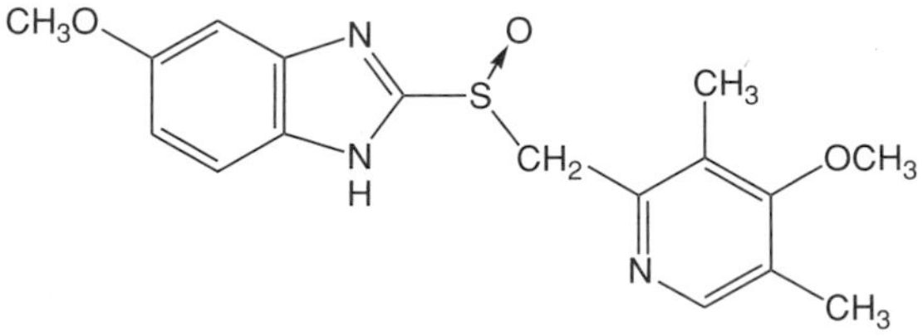

Omeprazole **(5.11)**

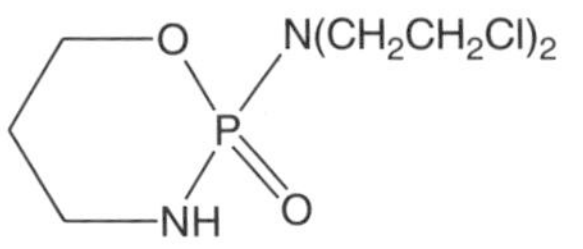

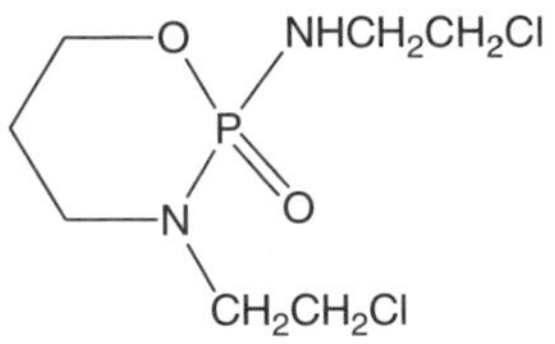

Cyclophosphamide (**5.12**) Ifosfamide (**5.13**)

Molecules that do not possess a stereogenic center as part of their structure may also exist in enantiomeric forms as a result of an axis or plane of chirality. Such structures occur less frequently in compounds of pharmaceutical interest. Atropoisomerism (Greek, *atropos* inflexible) is a term used to characterize stereoisomers which are chiral due to hindered rotation about a single bond, e.g., *ortho*-substituted biphenyl derivatives with two different substituents on each ring (**5.14**). In this case rotation about the carbon–carbon bond linking the two phenyl rings is restricted by the steric bulk of the *ortho*-substituents resulting in configurational stability. Examples of interest include the hypnotic methaqualone (**5.15**) and the male antifertility agent gossypol (**5.16**).

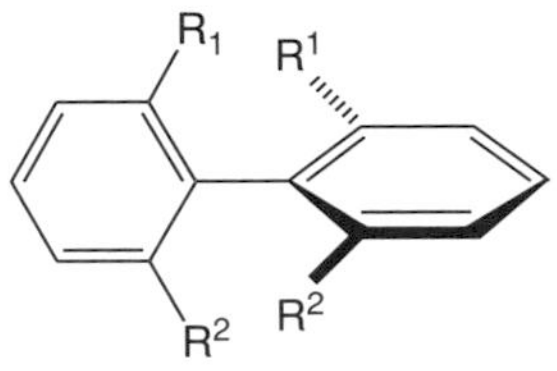

ortho-Substituted biphenyl (**5.14**)

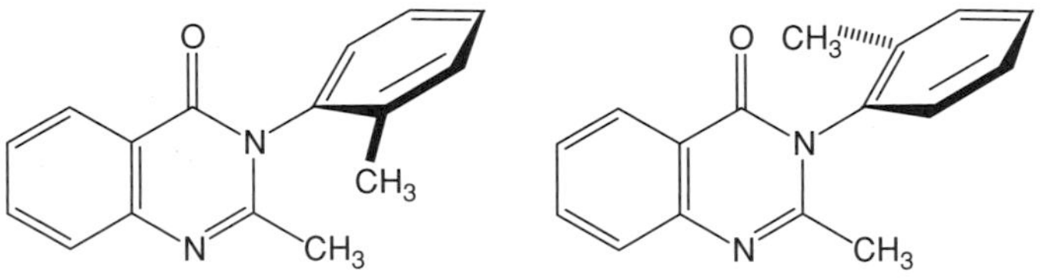

Methaqualone enantiomers (**5.15**)

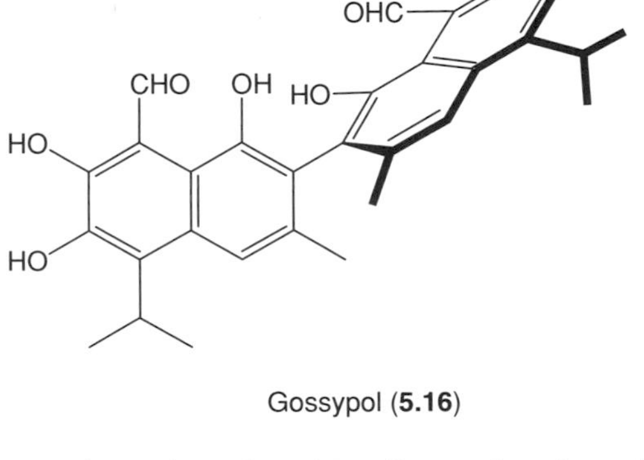

Gossypol (**5.16**)

The presence of adjacent double bonds as found in allenes also gives rise to enantiomerism, e.g., structures (**5.17**) and (**5.18**). In the case of these compounds the substituents R^1 and R^2 lie in intersecting planes and the two structures are nonsuperimposable. This type of isomerism is found in the naturally occurring antibiotics mycomycin (**5.19**) and nemotinic acid (**5.20**).

$R^1R^2C{=}C{=}CR^1R^2$ (**5.17**) $R^1R^2C{=}C{=}CR^1R^2$ (**5.18**)

$$HC{\equiv}C{-}C{\equiv}C{-}CH{=}C{=}CH{-}CH{=}CH{-}CH{=}CHCH_2COOH$$

Mycomycin (**5.19**)

$$HC{\equiv}C{-}C{\equiv}C{-}CH{=}C{=}CH{-}CH(OH){-}CH_2CH_2COOH$$

Nemotinic acid (**5.20**)

The macromolecular structures of biopolymers also give rise to chirality as a result of helicity. Helical structures may have either a left- or right-handed turn in the same way that a corkscrew, or spiral staircase, may be either left (**5.21**) or right (**5.22**) handed. For example the α-helix of proteins, composed of L-α-amino acids, is right-handed and the two polynucleotide strands of the DNA double helix wind around a common axis with a right-handed twist. In the case of these biopolymers not only are the individual building blocks, i.e., the amino acids and nucleotides chiral, but also the macromolecular structures of these biopolymers themselves exhibit chirality as a result of helicity.

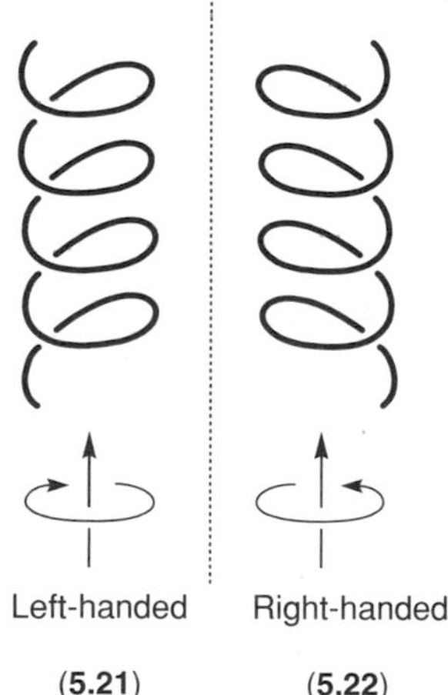

(**5.21**) (**5.22**)

5.2.1 Nomenclature and Designation of Stereoisomers

The classical method of distinguishing between a pair of optical isomers makes use of their unique property of rotation of the plane of plane polarized light. Those isomers which rotate the plane to the right are termed dextrorotatory, indicated by a (+)-sign before the name of the compound while those which rotate the plane to the left are termed levorotatory indicated by a (−)-sign. In the older literature the letters *d*- and *l*-are also used to indicate (+)- and (−)-enantiomers, respectively. The use of these lower case letters gives rise to confusion as the upper case D and L are used for the designation of configuration, and their use to indicate direction of rotation should be avoided. A racemic mixture, a 1:1 mixture of enantiomers, is indicated by a (±)-sign before the name of the compound. It is important to appreciate that this form of designation yields information concerning a physical property of the material and does indicate that a single enantiomer or racemate is present. It does not provide information concerning the three-dimensional spatial arrangement, or absolute configuration, of the molecule, the significant feature with respect to the biological activity of the drug. Also considerable care is required when using the direction of rotation as a stereochemical descriptor as both the magnitude and direction of rotation may vary with the conditions used to make the determination, e.g., temperature, solvent, analyte concentration, salt form, etc. For example the *S*- and *R*-enantiomers of the selective serotonin reuptake inhibitor fluoxetine, as the hydrochloride salt, are dextrorotatory and levorotatory, respectively, when the determination is carried out in methanol but reversed using water as the solvent. The active isomer of chloramphenicol (**5.3**) has the 1*R*,2*R*-absolute configuration, but this stereoisomer is dextrorotatory when the measurement is carried out in ethanol and levorotatory in ethyl acetate. Similarly, the *R*- and *S*-enantiomers of the antiarrhythmic agent propafenone are levorotatory and dextrorotatory as the free bases, but dextrorotatory and levorotatory as their hydrochloride salts; the active *S*-enantiomer of the 2-arylpropionic acid nonsteroidal anti-inflammatory drug fenoprofen, is dextrorotatory as the free acid and levorotatory as the anion.

Additional complications arise if the drug material is a mixture of two diastereoisomers, e.g., the β-lactam antimicrobial agent moxalactam (**5.23**) is a mixture of two epimers both of which are levorotatory. Their designations, based on the configuration of the side chain stereogenic center and optical rotation, are (−)-(*R*)- and (−)-(*S*)-moxalactam. In this case the designation of the material by optical rotation alone is meaningless and provides no information concerning the stereochemical composition of the material, i.e., single isomer or mixture.

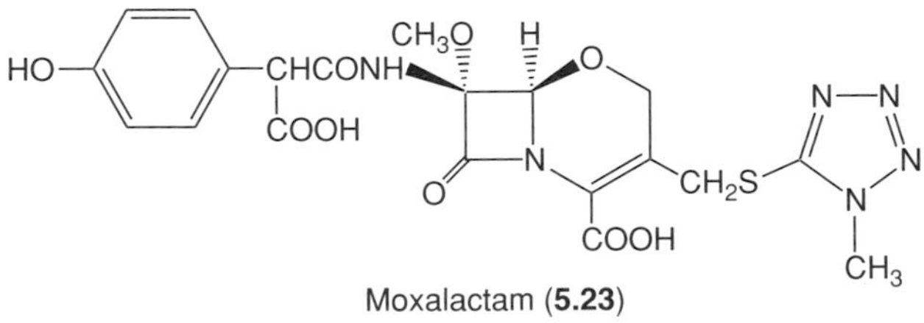

Moxalactam (**5.23**)

Once the three-dimensional structure of a stereoisomer has been determined, by for example x-ray crystallography, then the absolute configuration of a molecule may be indicated by the use of a prefix letter to the name of the compound. Two systems are currently used, the *R*/*S* or sequence rule nomenclature of Cahn, Ingold, and Prelog and the older D/L system of Fischer and Rosanoff.

One of the major problems in organic chemistry is the representation of three-dimensional structures on two-dimensional sheets of paper; the relationships between stereoisomers can best be seen and understood by the use of molecular models. The Fischer projection, devised by the carbohydrate chemist Emil Fischer, is a common method for two-dimensional representations of three-dimensional structures. In Fischer projections the structure is drawn in a vertical rather than a horizontal form with the lowest numbered carbon atom, in standard nomenclature terms, or the most highly oxidized end of the chain, drawn at the top. At each stereogenic center along the main axis of the molecule the vertical bonds project back away from the reader (into or behind the plane of the paper) while the horizontal bonds project up towards the reader (out of or above the plane of the paper). In the case of glyceraldehyde (2,3-dihydroxypropanal) the simplest carbohydrate, containing one stereogenic carbon atom, the individual enantiomers are drawn as represented by structures (**5.24**).

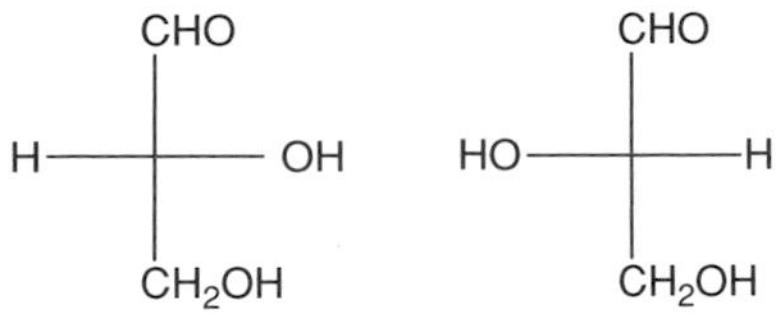

CHO / H — — OH / CH_2OH CHO / HO — — H / CH_2OH

D-(**5.24**)   L-(**5.24**)

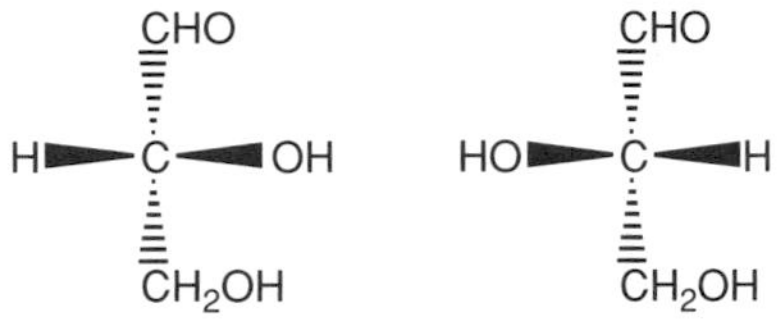

CHO / H — C — OH / CH_2OH CHO / HO — C — H / CH_2OH

D-(**5.24**)   L-(**5.24**)

The stereogenic carbon atom is regarded as being in the plane of the paper and those groups which are bonded horizontally, i.e., the H and OH project up towards the reader, and those bonded vertically, i.e., the CHO and CH_2OH project back away from the reader. In the diagram drawn the upper pair of structures are Fischer projections of the enantiomers of glyceraldehyde and the lower pair indicates what these projections represent in terms of spatial arrangement.

The structure of glyceraldehyde with the secondary hydroxyl group drawn on the right in the Fischer projection was designated as having the D-configuration D-(**5.24**) and that with the secondary hydroxyl on the left the L-configuration L-(**5.24**). At the time this representation of structure was developed it was not possible to determine the three-dimensional structures of molecules and the observed optical rotations of the two enantiomers were arbitrarily assigned as D-(+) and L-(−). At this time the letters *d*- and *l*- were used to indicate the direction of rotation rather than (+) and (−), and the combination of both upper and lower case letters to define both the shape of the molecule and the physical property continues to add to the confusion associated with the study of stereochemistry. It was not until the 1950s that it was possible to show that the optical rotation assignment in fact corresponded to the structures drawn, which was highly fortuitous. Stereoisomers of compounds, which can be related to D-glyceraldehyde by synthesis, are given the D-configuration, irrespective of the observed direction of rotation of polarized light and compounds related to L-glyceraldehyde are given the L-configuration. For example (+)-glucose (**5.25**), (−)-2-deoxyribose (**5.26**), and (−)-fructose (**5.27**) having the same configuration as D-(+)-glyceraldehyde, at the highest numbered stereogenic center (i.e., at the penultimate carbon atom) are assigned to the D-series. In the case of the amino acids the reference compounds used are D-(+)- and L-(−)-serine (**5.28**).

The use of this system presents a number of problems particularly if there is more than one stereogenic center in the molecule. Thus the amino acid L-threonine (**5.29**) may be related to L-serine at carbon-2 and D-glyceraldehyde at carbon-3. In the case of the α-amino acids the α-carbon atom is used to define the stereochemistry and the majority of naturally occurring amino acids have the L-configuration at this center. D-Amino acids are however found in a number of peptide antibiotics, e.g., bacitracin, the penicillins, etc.

In an attempt to overcome the difficulties associated with the D/L designation Cahn, Ingold, and Prelog devised their sequence rule system. Using this method the substituent atoms bonded to a stereogenic center are ranked in an order of priority based on their atomic number. The higher the atomic number the greater the priority. If a decision on priority cannot be made on the basis of the atoms directly bonded to the stereogenic center, then the atoms two bonds away are considered.

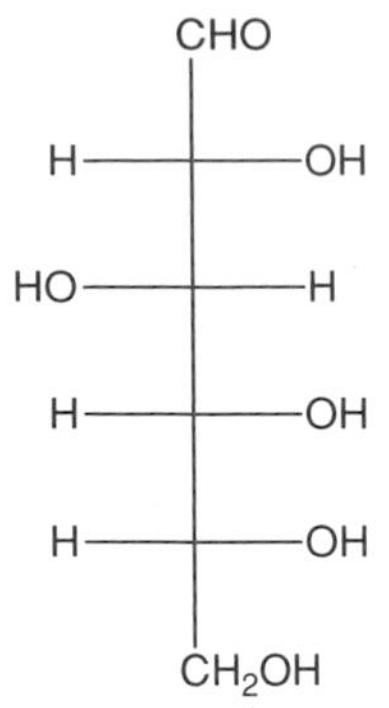

D-(**5.25**) 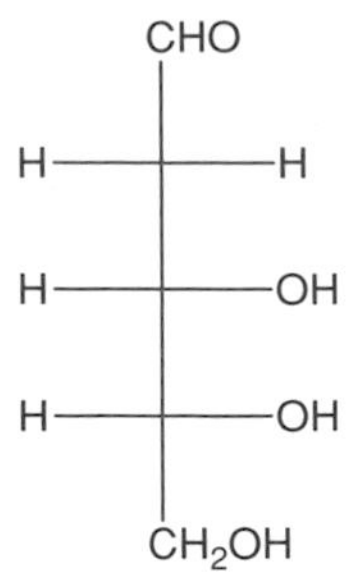 D-(**5.26**) 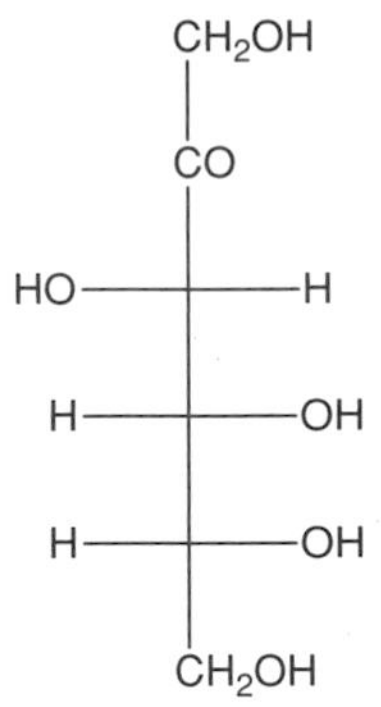 D-(**5.27**)

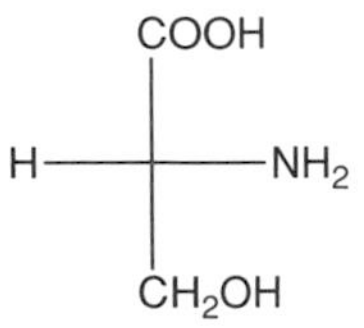

D-(**5.28**) 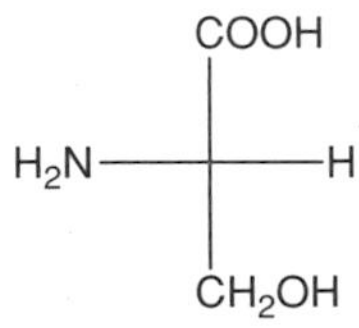 L-(**5.28**) 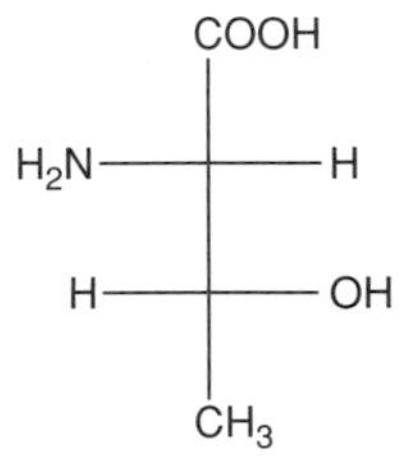 L-(**5.29**)

This process is continued along a substituent until all the priorities have been assigned. The molecule under examination is then viewed from the side opposite to the group of lowest priority. If the priority sequence, highest to lowest, is to the right (i.e., clockwise) then the center is of the *rectus* or *R*-absolute configuration (Latin *rectus*, right) and if to the left (i.e., anticlockwise) the *sinister* or *S*-absolute configuration (Latin *sinister*, left).

In the case of glyceraldehyde (**5.24**) the priority order of the substituents is: HO−(highest), −CHO, $-CH_2OH$, −H (lowest). The carbonyl group has a higher priority than the primary alcohol as the carbonyl carbon atom is considered to be bonded to two oxygen atoms, one ''real'' and one ''ghost'' or ''phantom'' oxygen so that the carbon–oxygen double bond is taken into account. The application of these rules to the enantiomers of glyceraldehyde is illustrated below (**5.24**).

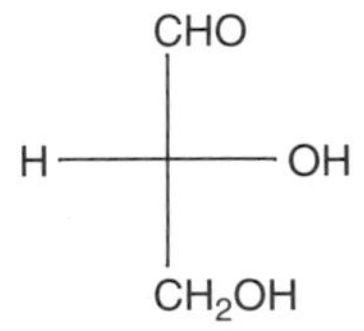

D-(**5.24**) 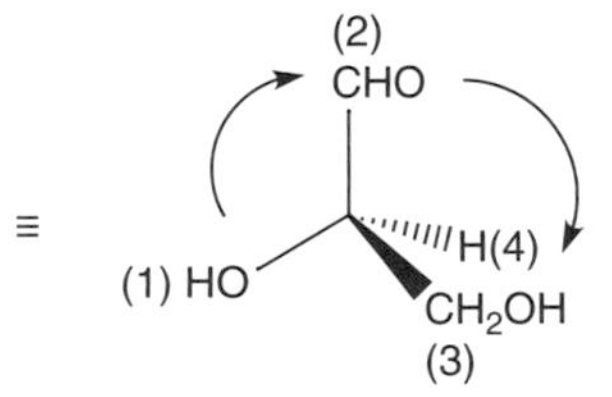 ≡ *R*-(**5.24**)

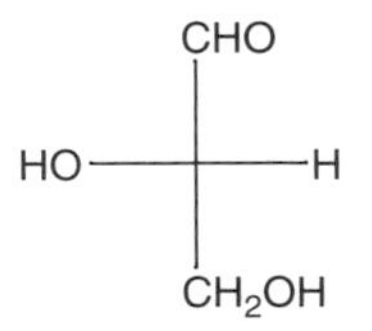

L-(**5.24**) 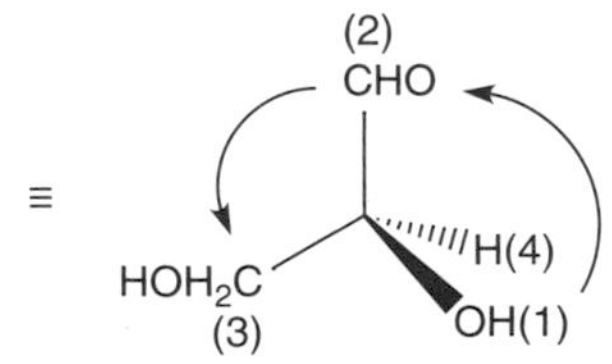 ≡ *S*-(**5.24**)

Thus, D-(+)-glyceraldehyde has the *R*-absolute configuration using the Cahn, Ingold, and Prelog sequence rules and L-(−)-glyceraldehyde has the *S*-absolute configuration.

The naturally occurring catecholamines, (−)-noradrenaline (**5.30**), and (−)-adrenaline (**5.31**) have been stereochemically related, by chemical degradation studies, to D-(−)-mandelic acid (**5.32**) and therefore these two compounds are assigned the D-configuration. In the case of noradrenaline (**5.30**) and adrenaline (**5.31**), and related chiral derivatives of phenylethylamine, the convention regarding the presentation of Fischer projections with the lowest numbered carbon atom at the top is not strictly applied. These agents are conventionally drawn ''upside down'' as Fischer projections as shown in the structures (**5.30**) and (**5.31**).

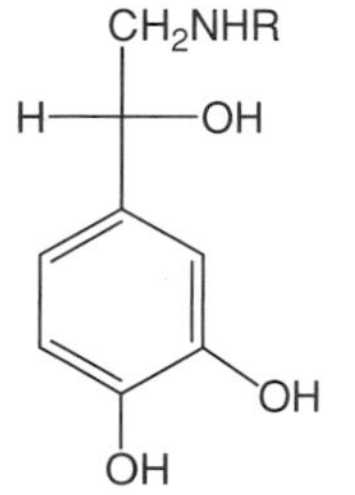

CH_2NHR H— —OH OH OH

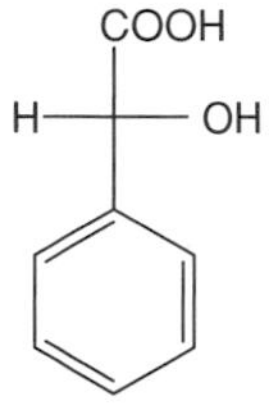

COOH H— —OH

D-(−)-Noradrenaline R=H (**5.30**) D-(**5.32**)

D-(−)-Adrenaline R=CH_3 (**5.31**)

Redrawing the Fischer projections of (**5.30**) and (**5.31**) to a form suitable for assigning the priority sequence yields structure (**5.33**) and examination of the sequence indicates that the D-enantiomers of both catecholamines correspond to the *R*-absolute configuration.

(2) CH_2NHR (1) HO H(4) (3) OH OH

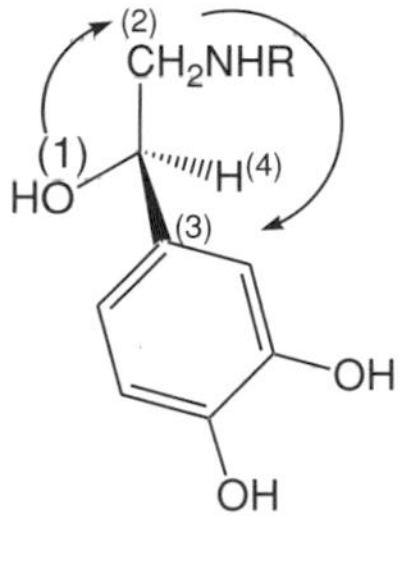

(**5.33**)

One of the major problems with stereochemical nomenclature is the continued use of both the above systems for designation of configuration and also the use of the physical descriptors (+) and (−). The potential problems associated with the use of the physical descriptors have been presented above. The reason the D/L system continues to be used is essentially biochemical. For example D-(+)-glucose (**5.25**) could be known as (2*R*, 3*S*, 4*R*, 5*R*)-2,3,4,5,6-pentahydroxyhexanal or (2*R*, 3*S*, 4*R*, 5*R*)-aldohexose, which does not take into account the cyclic structure of the molecule and the two possible anomeric forms. It is obviously simpler to refer to D-(+)-glucose. Also the naturally occurring chiral amino acids are of the L-configuration and the application of the *R*/*S* system results in a lack of consistency within the series. For example L-serine (**5.28**) has the *S*-absolute configuration while L-cysteine (**5.34**) has the *R*-configuration as a result of the presence of the sulfur atom.

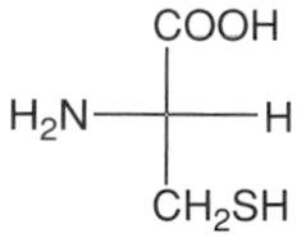

COOH H_2N— —H CH_2SH

L-(**5.34**)

Additional complexities may also arise in the nomenclature of semisynthetic products, as in some cases both systems are used to designate the stereochemistry of the molecule. In the case of the β-lactam antibiotics the absolute stereochemistry of the 6-aminopenicillanic acid and 7-aminocephalosporanic acid nucleii have been determined and defined in terms of the *R*/*S* system. The addition of a side chain, e.g., ampicillin (**5.35**) and cefalexin (**5.36**), may result in the introduction of an additional stereogenic center and within the older literature the two possible epimeric diastereoisomers of such compounds are frequently defined in terms of D/L.

It is important to appreciate that the stereochemical designations, *R* and *S*, are defined by a set of arbitrary rules and that with respect to biological activity the relevant feature is the three-dimensional spatial arrangement of the functionalities within the molecule. A change in one functional group may result in an alteration of the configurational designation but have no influence

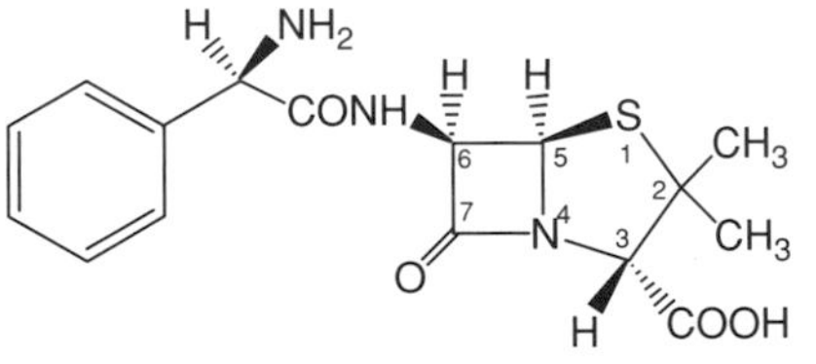

Ampicillin (**5.35**)

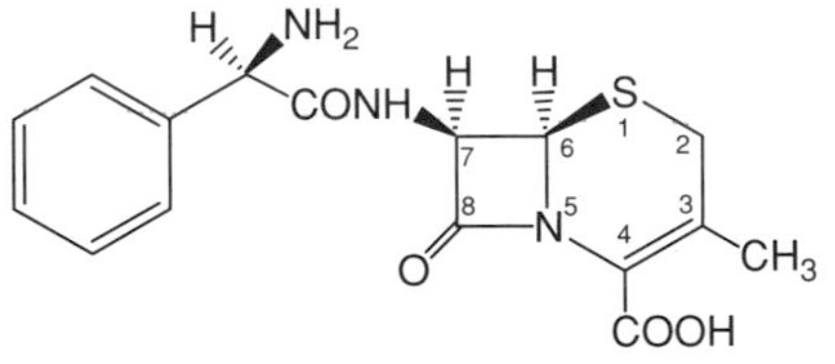

Cefalexin (**5.36**)

on the relative orientation of the functionalities required for biological activity with respect to one another. For example the active enantiomers of the 2-arylpropionic acid nonsteroidal anti-inflammatory drugs (NSAIDs) have the *S*-configuration (**5.37**) which corresponds to the *R*-configuration of the 2-aryloxypropionic acid herbicides (**5.38**). An appreciation of this reversal in configurational designation is of significance for an understanding of the stereoselectivity of metabolism within the two series of compounds (see Section 5.4.3). Similarly in the case of the β-adrenoceptor antagonists the active agents of the arylethanolamine series have the *R*-configuration (**5.39**) whereas those of aryloxypropanolamine series have the *S*-configurational (**5.40**) designation. Without an appreciation of the sequence rules and their application it could be assumed that the stereochemical requirements for activity within the two series of compounds were for some reason reversed.

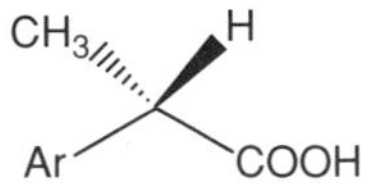

S-(**5.37**) 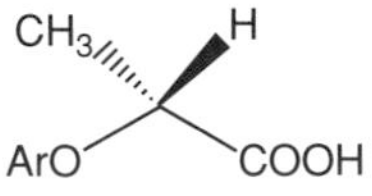 ***R***-(**5.38**)

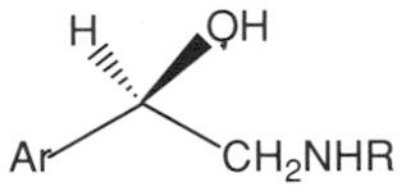

R-(**5.39**) 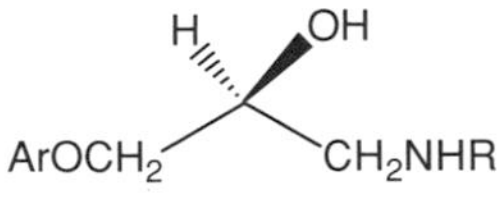 *S*-(**5.40**)

The metabolism of a drug may also result in an alteration of configurational designation with no change in the spatial arrangement of the functionalities. For example, fonofos (**5.41***), a cholinesterase inhibitor, undergoes oxidation to yield fonofos-oxon (**5.42**), which is also active. As a result of the sequence rule designations the *R*-enantiomer of fonofos yields the *S*-enantiomer of fonofos-oxon and vice versa. In the case of fonofos this change in designation is important as the activity and toxicity of the *R*-enantiomer is greater than that of the *S*-isomer, whereas the situation is reversed for fonofos-oxon, i.e., $S > R$. Without an appreciation of the structures of the individual enantiomers it would appear that the activity of the oxygen derivatives showed the reverse stereoselectivity to the sulfur series which is obviously not the case.

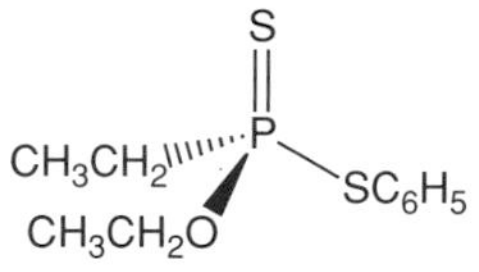

S-(**5.41**) 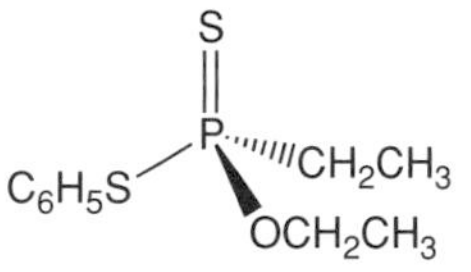 *R*-(**5.41**)

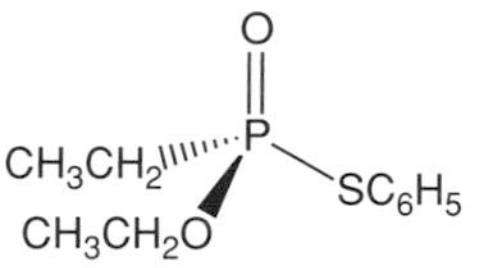

R-(**5.42**) 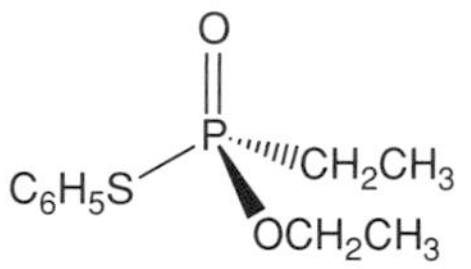 *S*-(**5.42**)

5.2.2 The Nomenclature Problem in Generic Names

A major problem in therapeutics is the lack of readily available information on the stereochemical identity or composition of a chiral drug in the majority of standard reference works. In a number of cases, it is impossible to determine if the material used is a single enantiomer, a racemic mixture, a mixture of diastereoisomers, or some other possibility. It is frequently the case that the (±)-prefix is used to indicate that the material is a racemic mixture, but if the compound in question contains two stereogenic centers in its structure then four stereoisomeric forms are possible, i.e., two pairs of enantiomers and hence two racemic mixtures. Which of the two possible racemates is the drug or is it a mixture of all four stereoisomers? The use of the (±)-prefix in this case does not specify the composition of the material. There is therefore a need within drug nomenclature to provide a system of generic names which will indicate if a compound may exist in more than one stereoisomeric form, and also the nature of the material used, i.e., single isomer or mixture. This situation has attained increased significance in recent years with the advent of the chiral switch (see Section 5.8.1) and the possibility that both single enantiomer and racemic mixture products of some drugs either are, or will be, available at the same time.

As pointed out above the direction of rotation of the plane of plane polarized light is frequently used in the designation of a pair of enantiomers and the abbreviations dex or dextro, and lev or levo, have been adopted as a prefix to the approved names of a number of single enantiomer drugs. This approach to nomenclature does provide information with respect to a physical property and the nature of the material, i.e., single enantiomer rather than racemic mixture. However, as pointed out above the direction of rotation may vary with experimental conditions, which require specification. The current editions of the British National Formulary (BNF No. 49, March 2005) and British

*The designation applied to structure (**5.41**) may appear to be incorrect, but in the sequence rules the participation of *d*-orbitals in bonding is neglected for assignment of designation, thus the phosphorus–sulfur and phosphorus–oxygen double bonds in these structures are regarded as single for the assignment of configuration.

Table 5.1 Single stereoisomer compounds indexed using the dex/lev prefix in the British National Formulary and British pharmacopoeia

Levamisole (hydrochloride)[a]	Dexamethasone (plus esters)
Levetiracetam[b]	Dexamfetamine (dexamphetamine)
Levobunolol (hydrochloride)[a]	Dexchlorpheniramine maleate[c]
Levobupivacaine[b]	Dexfenfluramine[b]
Levocabastine[b]	Dexketoprofen[b]
Levocarnitine[c]	Dexpanthenol[c]
Levocetirizine[b]	Dextromethorphan (hydrobromide)[a]
Levodopa	Dextromoramide (tartrate)[a]
Levodropropizine[c]	Dextropropoxyphene (hydrochloride, napsilate)[a]
Levofloxacin[b]	
Levofolinic acid[b] (calcium levofolinate)	
Levomenthol[c]	
Levomepromazine	
Levomethadone hydrochloride[c]	
Levonorgestrel	
Levothyroxine (sodium)[a]	

[a]BP, salt form indexed.
[b]BNF (No. 49, March 2005) only.
[c]BP (2004) only.

Pharmacopoeia (BP 2004) list a number of agents using this approach (Table 5.1). The current edition of the BP also uses the prefix *Rac* or *Race* to indicate a racemic mixture for a limited number of compounds. Thus monographs for racementhol (decongestant), racemic camphor (counter-irritant), and racephedrine hydrochloride (β-adrenoceptor agonist) are included in the current BP. In the case of these compounds monographs are also included for the single enantiomers, levomenthol, natural camphor (D-camphor), and ephedrine hydrochloride, the 1*R*,2*S*-stereoisomer.

An alternative approach recently introduced incorporates the configurational designation into the names of some agents previously available as racemates, the prefixes "es" and "ar" being used to designate the single *S*- and *R*-enantiomers, respectively. Thus, the single enantiomer forms of the proton pump inhibitor (*S*)-omeprazole and the selective serotonin reuptake inhibitor (*S*)-citalopram have been named esomeprazole (**5.43**) and escitalopram (**5.44**), respectively. However, this approach to nomenclature is not without problems. The above agents have only one stereogenic center in their structures and the application of this system to agents with more than one center may prove problematic. There are also a number of agents whose names begin with "es" or "ar" which have no association with their stereochemical designation, e.g., esmolol a racemic short acting β-adrenoceptor antagonist; articaine a racemic local anesthetic agent; the amino acid arginine used as the single *S*-enantiomer, in addition to a number of nonchiral agents. In some instances a completely different name has been used for a single isomer product, e.g., dilevalol for the β-adrenoceptor antagonist stereoisomer of the combined α- and β-antagonist labetalol (see Section 5.6.1).

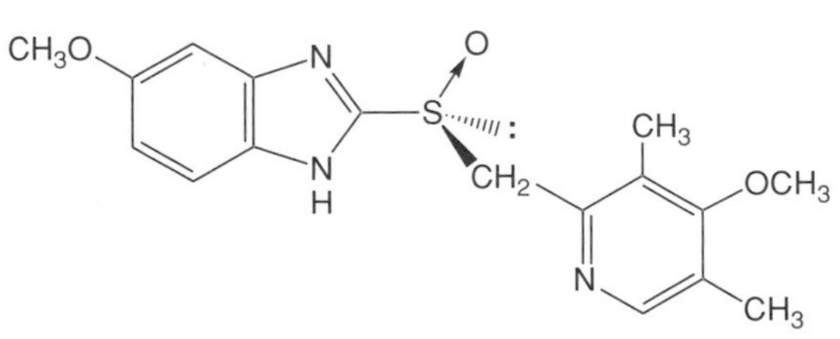

Esomeprazole (**5.43**)

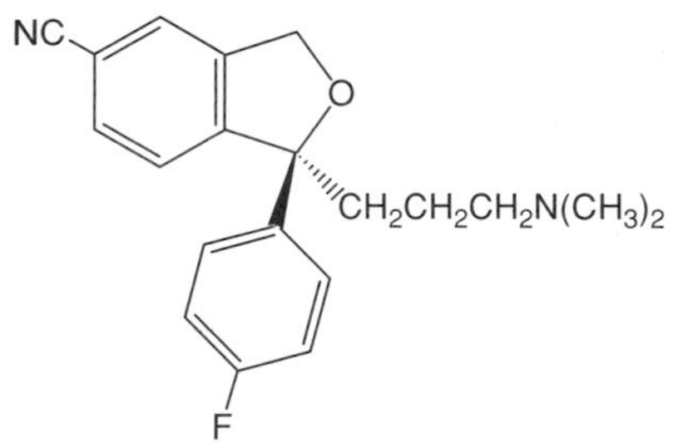

Escitalopram (**5.44**)

5.2.3 *Meso* Compounds

In molecules with two or more stereogenic centers in which the stereogenic atoms are bonded to identical substituents the number of possible stereoisomers is less than that obtained by application of the $\mathbf{2}^n$ rule. In the simplest case of a molecule with two stereogenic centers, e.g., tartaric acid, structures (**5.45**) and (**5.46**) are nonsuperimposable and are therefore a pair of enantiomers with the 2*S*,3*S*- and 2*R*,3*R*-configurations, respectively. In contrast structures (**5.47**) and (**5.48**) are identical, are superimposable on one another, and represent the same structure as rotation through 180° in the plane of the paper and superimposition will show. These structures possess a plane of symmetry between the two stereogenic carbon atoms 2 and 3, the "top" half of the molecule is a reflection of the "lower" half. Similarly, the configurational designation of the "top" stereogenic carbon, number 2 in structure (**5.47**), is the opposite of that of the "lower" carbon-3, i.e., is 2*R*,3*S*-stereoisomer, whereas that in structure (**5.48**) as drawn, is "top," "lower" 2*S*,3*R*-. Such molecules are not optically active as the effects of the two opposite stereogenic centers are "self-canceling." In the older literature such molecules are described as "internally compensated." Thus in the case of tartaric acid three stereoisomeric forms are possible, a pair of enantiomers (**5.45** and **5.46**) and an optically inactive form which is diastereoisomeric with the enantiomeric pair. Such optically inactive forms are known as *meso* compounds and are achiral, even though they contain stereogenic centers as part of their structure. Examples of compounds used in therapeutics that may exist as *meso* forms include the antitubercular agent ethambutol (see Section 5.5.6), the β-adrenoceptor antagonist nebivolol (see Section 5.6.1), and the neuromuscular blocking drug atracurium (**5.9**).

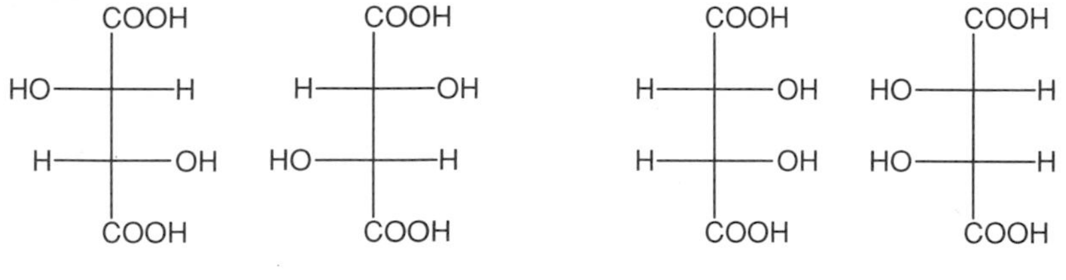

(–)-2*S*,3*S*-(**5.45**) (+)-2*R*,3*R*-(**5.46**) 2*R*,3*S*-(**5.47**) 2*S*,3*R*-(**5.48**)

5.2.4 Prochirality

Atoms that are bonded to two identical groups and to two other different groups are said to be prochiral. For example if either of the two methylene group hydrogen atoms in ethanol (**5.49**) were replaced by another group, e.g., deuterium, then the carbon atom (C_1) becomes chiral and two enantiomeric forms are possible (**5.50**). If ethanol (**5.49**) is viewed from the side opposite to the hydrogen atom indicated** then the sequence of groups about C_1, i.e., HO, CH_3, H, is anticlockwise. If the molecule is viewed from the side opposite to the hydrogen indicated* then the sequence of groups is reversed, i.e., clockwise. In terms of their molecular environments these two hydrogen atoms are not equivalent, the carbon atom C_1 is prochiral and the two hydrogen atoms are said to be enantiotopic. If H** is arbitrarily preferred over H* then an *R*-designation is obtained and H** is designated pro-*R* and H* as pro-*S* (**5.51**). Differentiation of enantiotopic groups may be of considerable significance in biochemistry and metabolism (see Section 5.3).

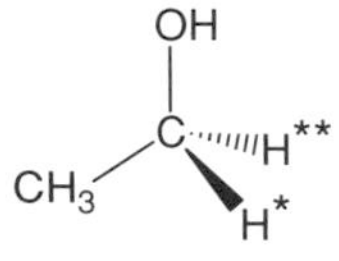

Ethanol (**5.49**)

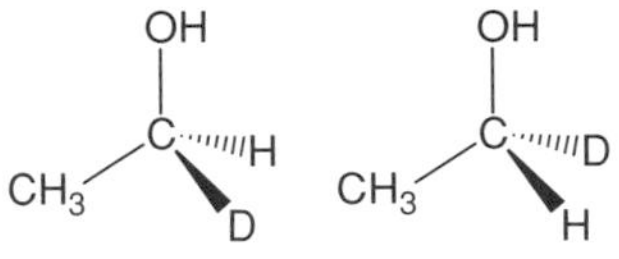

(**5.50**)

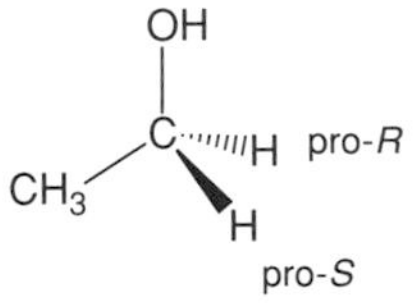

(**5.51**)

5.3 BIOLOGICAL ACTIVITY

That enantiomers should be regarded as different compounds, rather than different forms of the same compound and that in some instances, a racemate may be regarded as a "third compound," is particularly emphasized on examination of their biological properties. As pointed out in Section 5.1 the fact that enantiomers may exhibit different biological activities has been appreciated for over a century. One of the first reported observations of the differential physiological actions of stereoisomers was that of Piutti, who in 1886 isolated the enantiomers of the amino acid asparagine (**5.52**) and reported that the (+)-enantiomer tasted sweet whereas the (−)-enantiomer was bland. Similar observations have been reported for other amino acids; those of the D-series taste sweet, whereas the L-series are either tasteless or bitter. Enantiomers may also exhibit different odors the (−)-enantiomer of carvone (**5.53**) smells of spearmint whereas (+)-carvone has an odor of caraway; the (+)-enantiomer of the related terpene limonene (**5.54**) smells of orange and the (−)-enantiomer of lemon.

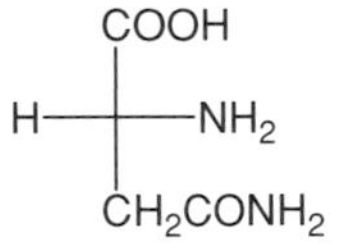

D-(+)-(**5.52**)

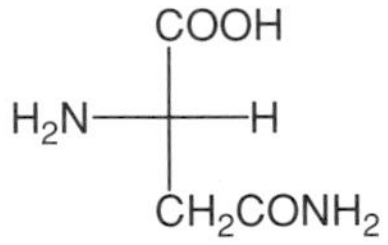

L-(−)-(**5.52**)

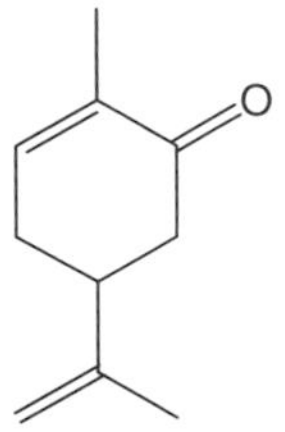

Carvone (**5.53**)

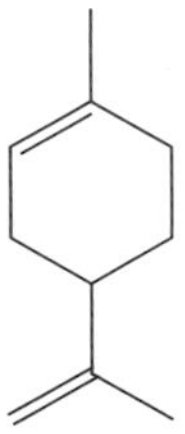

Limonene (**5.54**)

The differential pharmacological activity of drug enantiomers was shown in the early years of the last century when the British pharmacologist Cushny demonstrated differences in the activity of (−)-hyoscyamine and atropine (racemic hyoscyamine) and (+)- and (−)-adrenaline. In order to rationalize the observed differences in pharmacological activity between enantiomers Easson and Stedman, in 1933, suggested a ''three point fit'' model between the more active enantiomer and its receptor (Figure 5.1). In Figure 5.1 the enantiomer on the left is involved with three simultaneous bonding interactions with complementary sites on the receptor surface, whereas that on the right may take part in two such interactions. Alternative orientations of the enantiomer on the right to the receptor surface are possible but only two interactions may take place at any one time. According to the Easson–Stedman model the more potent enantiomer is involved with a minimum of three intermolecular interactions with the receptor surface whereas the less potent isomer may interact at two sites only. Thus the ''fit'' of the enantiomers to the receptor are different, as are their binding affinities. Similarly, an achiral analog of the drug should also interact at two sites with an affinity and/or activity similar to that of the less potent enantiomer.

The Easson–Stedman model was supported by an examination of the activity of the enantiomers of adrenaline and the achiral desoxy analog *N*-methyldopamine. The three functionalities involved in the drug receptor interaction are postulated to be the methylamino group, the catechol ring system, and the secondary alcohol. Only in (−)-(*R*)-adrenaline (**5.31**) are these functionalities appropriately configured to take part in three simultaneous interactions with the receptor. In the case of (*S*)-adrenaline the hydroxyl group is orientated in an unfavorable position to interact with the receptor and only a two-point interaction is possible. Similarly, *N*-methyldopamine may interact at two sites, with the result that the activity is similar to that of the *S*-enantiomer and much less than that of (*R*)-adrenaline. Similar data have been obtained for the corresponding enantiomers and achiral derivatives of (*R*)-noradrenaline (**5.30**) and (*R*)-isoprenaline for both α- and β-adrenoceptor activity.

On examination of related chiral and desoxy achiral adrenergic agents the Easson–Stedman model was found not to hold always. In some instances the achiral analogs were found to be more active than the ''less active'' enantiomers. These anomalies were subsequently found to be associated with variable direct and indirect actions of the compounds. The ''active'' (−)-isomers were found to be more potent than their (+)-enantiomers and achiral analogs in both normal and catecholamine depleted, reserpine-pretreated tissues, whereas the (+)-enantiomers and achiral analogs were equipotent in catecholamine-depleted tissues and of variable potency in normal tissue. These observations resulted in the conclusion that the Easson–Stedman model only applies at sites of direct drug action. Thus, an examination of the stereoselectivity of drug action also provided additional insight into the mechanism of action.

Additional support for the Easson–Stedman three-point interaction model for the catecholamines has recently become available. The elucidation of the amino acid sequences of many

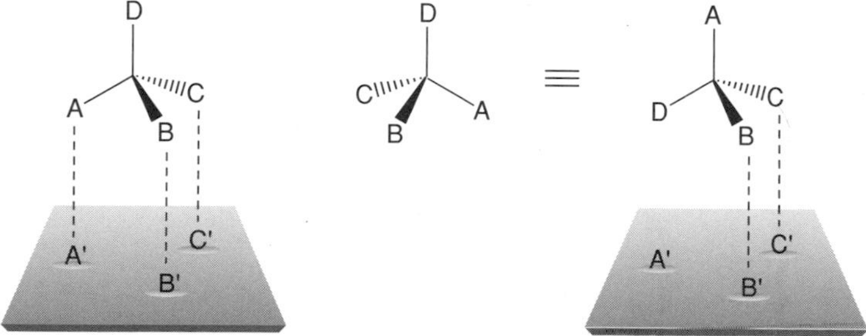

Figure 5.1 Stereochemical discrimination on interaction of drug enantiomers with a chiral biological macromolecule. The enantiomer on the left is involved in three simultaneous bonding interactions with complementary functionalities on the "receptor" surface, whereas that on the right can interact at two sites only. Alternative orientations of the enantiomer on the right to the receptor surface are possible but only two interactions may take place at anytime.

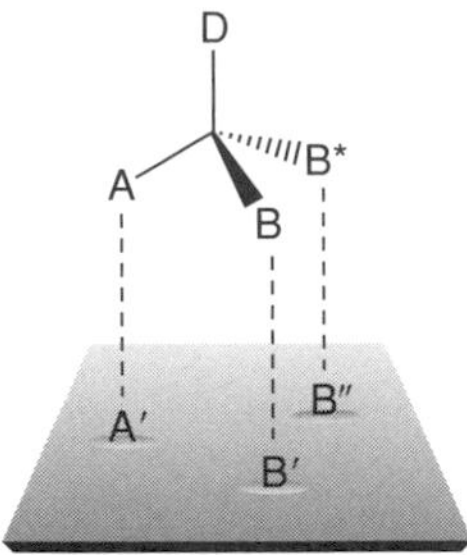

Figure 5.2 Three-point interaction model of a prochiral substrate at an enzyme active site. The groups B and B* are identical and enantiotopic. If B″ is the enzyme catalytic site then group B* in the substrate, and not B, will undergo biotransformation to yield a chiral product.

G-protein-coupled receptors, including adrenoceptors has resulted in the development of models of their organization and ligand interactions. The β_2-adrenoceptor is thought to contain seven transmembrane α-helices and site-directed mutagensis has resulted in the identification of an aspartate residue (Asp 113 in domain III) and two serine residues (Ser 204 and 207 in domain V) as the binding sites for the protonated methylamino and catechol functionalities of (*R*)-adrenaline, respectively. Identification of the third binding site, associated with the secondary alcohol, proved more problematic. A recent report, involving site-directed mutagenesis of the human β_2-adrenergic receptor, has indicated the potential significance of an asparagine residue (Asn 293) in transmembrane domain VI. Replacement of Asn 293 by leucine (Leu) in the mutant receptors resulted in a reduction in the stereoselectivity of isoprenaline, adrenaline, and noradrenaline binding and receptor activation, which was predominantly associated with a reduction in affinity of the more active *R*-enantiomers. In contrast the affinity of achiral dopamine and racemic dobutamine, which lack the secondary hydroxyl groups were not affected. As replacement of Asn 293 with Leu did not completely abolish the stereoselectivity of drug action the data were interpreted as indicating that additional interaction sites may also contribute to the chiral recognition process. Interestingly, examination of the activity of three chiral antagonists (propranolol, alprenolol, and metoprolol) indicated a modest reduction in affinity for the mutated receptor but no alteration in their stereoselectivity of action.

In 1948 Ogston, unaware of the Easson–Stedman hypothesis, proposed a similar three-point attachment model in order to rationalize the results from enzymatic studies using prochiral substrates. In the case of the prochiral compound CABB*D, the identical enantiotopic groups B/B* may be differentiated on interaction with an enzyme active site such that only one of the groups undergoes transformation. Ogston proposed that the substrate interacts with three sites on the enzyme but that only one of the complimentary sites to the enantiotopic groups B/B* is involved with the biochemical transformation (Figure 5.2). If reaction can only occur at site B″ then group B* in the substrate, but not group B, is converted in the product, i.e., the groups B and B* are not sterically equivalent.

Transformations of this type are relatively common in biochemistry and drug metabolism. For example, the synthesis of (−)-(*R*)-noradrenaline (**5.30**) from dopamine (**5.55**), mediated by dopamine-β-hydroxylase, proceeds with total stereoselectivity, i.e., is stereospecific.

HO, HO, H, H, CH_2NH_2 → HO, HO, H, OH, CH_2NH_2

(**5.55**) *R*-(**5.30**)

Similar specificity is shown by this enzyme in the metabolism of other substrates, e.g., (+)-α-methyldopamine to (−)-(1*R*,2*S*)-α-methylnoradrenaline. The antihypertensive agent α-methyldopa (**5.56**) is marketed as the single L- enantiomer corresponding to the *S*-configuration using the sequence rule designation. This agent undergoes decarboxylation, mediated by dopa decarboxylase, to yield (+)-(*S*)-α-methyldopamine (**5.57**), which then undergoes dopamine β-hydroxylase-mediated oxidation to (1*R*,2*S*)-α-methylnoradrenaline (**5.58**), the active agent. As (+)-(*S*)-α-methyldopamine is chiral the two hydrogen atoms on the β-carbon atom are said to be diastereotopic rather than enantiotopic.

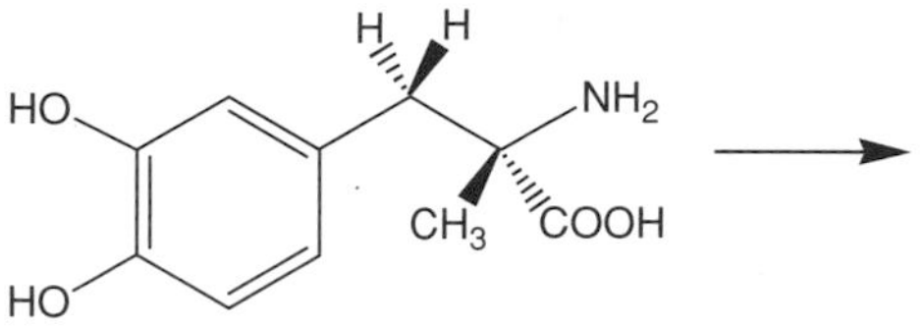

α-Methyldopa (**5.56**)

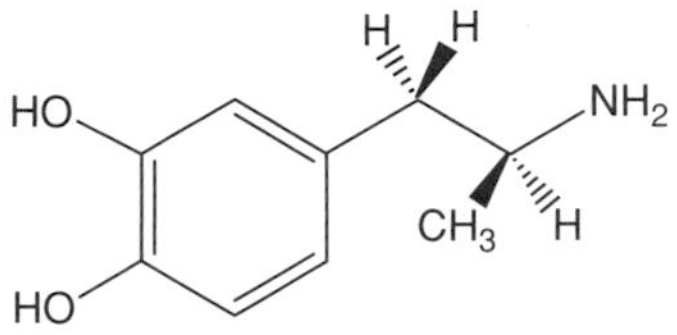

(*S*)-α-Methyldopamine (**5.57**)

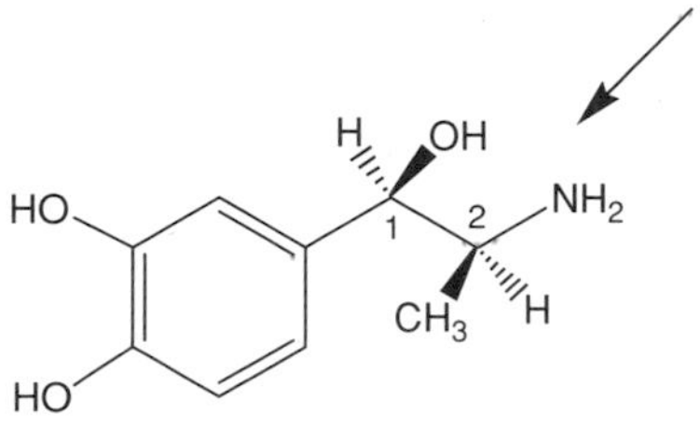

(1*R*,2*S*)-α-Methylnoradrenaline (**5.58**)

The above models are very useful but relatively simplistic representations of what may in fact occur during the drug, or substrate, interaction with a receptor, or enzyme, and assume that the ligand has to adopt a particular orientation in relation to the target site. It is also feasible to propose that the interactions do not necessarily need to be attractive and that both attractive and repulsive, e.g., steric and/or electrostatic, interactions may also be involved. In such instances the less active, or less potent, enantiomer may be involved in three intermolecular interactions, which do not enhance ligand binding. The less active enantiomer could also interact with the target at three sites resulting from additional interactions with the biomolecule, which do not occur with the more active enantiomer. In addition, the interaction between the drug and the receptor/enzyme target may result in conformational changes in both the target macromolecule and the ligand. Thus, the final interaction model may be fairly complex and both the stereochemistry and conformational flexibility of the ligand need to be taken into account.

The chiral recognition process continues to be a topic of considerable interest and alternative models, and refinements to existing models have been proposed. A recent investigation, concerned with the interaction of the enantiomers of isocitrate with the enzyme isocitrate dehydrogenase, has indicated that a three-point attachment model is not sufficient to explain the observed enantioselectivity. Crystallographic analysis of enzyme–substrate complexes indicated that both enantiomers of the substrate interact with three common binding sites, located in a cavity within the enzyme structure, and enantioselectivity was determined by an additional fourth site. As a result of these observations a so-called four-location model has been proposed in which either four interaction sites, or alternatively three interaction sites together with a specific direction/orientation are required for enantioselectivity. The latter being essentially the Easson–Stedman model.

Similarly, a conformationally flexible model has been proposed to explain enzyme enantioselectivity with respect to prochiral substrates. In this model the substrate is envisaged to bind to the enzyme via two interaction sites and once bound has conformational flexibility. The enantioselectivity of the process/transformation is determined by the orientation and flexibility of the substrate target groups in relation to the catalytic site of the enzyme. Whereas in the Ogston model the substrate is required to be specifically orientated before binding to the enzyme, in this more dynamic model binding takes place followed by conformational alterations in the enzyme–substrate complex.

5.3.1 Terminology Used in the Pharmacological Evaluation of Stereoisomers

The differential biological activity of a pair of stereoisomers has given rise to additional terminology. Thus, the stereoisomer with the greater receptor affinity, or activity, is termed the *eutomer* and that with the lower affinity, or activity, the *distomer*. The ratio of affinities, or activities, eutomer to distomer is known as the *eudismic ratio* and its logarithm as the *eudismic index*.

It is important to appreciate that such terminology applies to a particular activity of a drug. For example in the case of a dual action drug the eutomer for one activity may be the distomer for another, or the enantiomers may be equal in activity. In the case of the β-adrenoceptor antagonist propranolol the eutomer for β-blocking activity is the enantiomer of the *S*-absolute configuration, which is between 40- and 100-fold more potent than (*R*)-propranolol, depending on the test system used. In contrast both enantiomers of propranolol have similar activities with respect to their membrane-stabilizing properties.

The significance of stereochemistry with respect to drug action is dependent on the position of the stereogenic center within the molecule. Is the stereogenic center located in a position that will influence the interaction of the drug with the target receptor? A number of situations are possible:

a. The stereogenic center is located at a critical position within the molecule such that the enantiomer, or an achiral analog exhibits a marked reduction in activity, e.g., the situation with (*R*)-adrenaline (**5.31**) referred to previously.

b. The stereogenic center is located in a critical position within the molecule but the eutomer has enhanced, or the same activity, as an achiral analog, the distomer reduced in activity compared to the achiral compound. For example, examination of the activity of the acetylcholine analog (*S*)-β-methacholine (**5.59**) on isolated rat intestine yields a pD_2 value of 6.8, compared to the value of 7.0 obtained with acetylcholine, whereas, the *R*-enantiomer, the distomer, yields a value of 4.1. In this case, it appears that a two-point interaction is required for activity but that the orientation of the methyl group at the stereogenic center is critical for activity. In the *S*-enantiomer, the eutomer, the methyl group is presumably orientated in a noncritical-binding region of the receptor, whereas in the *R*-enantiomer the orientation results in steric repulsion.

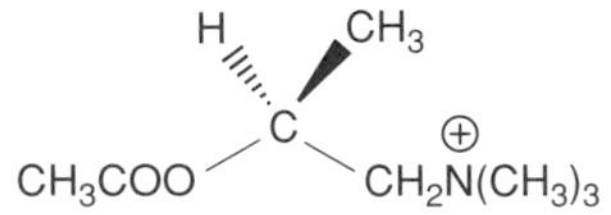

H, CH_3, C, CH_3COO, $CH_2\overset{\oplus}{N}(CH_3)_3$

(*S*)-β-Methacholine (**5.59**)

c. The chiral center is in a noncritical position in the molecule such that both enantiomers and the achiral analog have the same, or similar, activities. Examination of the properties of the H_1-antihistamine terfenadine (**5.60**), in either pharmacological or biochemical assay systems, indicates no difference in activity between the enantiomers. Replacement of the hydroxyl group at the stereogenic carbon atom by hydrogen yields an achiral derivative that has activity similar to that of the enantiomers of terfenadine. Thus, the hydroxyl group is located in a noncritical position for receptor binding.

If the stereogenic center is located in a critical region of the molecule then differences in activity between isomers are expected and such differences would be greater for stereoisomers than for

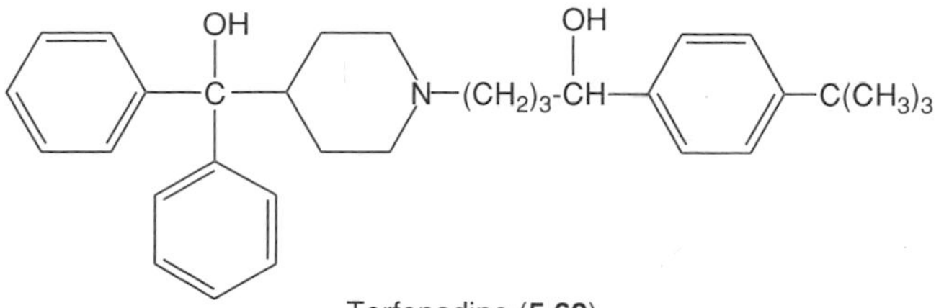

Terfenadine (**5.60**)

homologs, or analogs resulting from relatively simple isosteric replacements. In order to derive useful data from quantitative structure–activity relationships (QSAR) of chiral compounds each series of stereoisomers should be examined independently.

As a general rule, the eudismic index is a function of the affinity of the eutomer, the higher the affinity of the drug the greater the degree of complementarity between the drug and its receptor site. Whereas for low-affinity compounds the complementarity between the drug and receptor site will be lower and the extent of chiral discrimination, i.e., the eudismic index, will be reduced. For drugs such as terfenadine, i.e., those in which the chirality is not critical for activity, a similarly low ratio would be expected.

A plot of eudismic index versus logarithm of the affinity of the eutomer in a homologous, or congeneric, series of compounds frequently yields a straight line, the slope of which is positive and is known as the eudismic affinity quotient (EAQ). EAQ is therefore a measure of chiral discrimination with increase in affinity for a particular biological effect. This relationship, the greater the affinity of the eutomer the greater the eudismic ratio, appears to be common for many series of drugs and is known as Pfeiffer's rule. Examples of compounds, which do not follow this generalization, are known. In these cases, the stereogenic center may be in a noncritical site in the molecule; two of the four groups attached to the stereogenic center are bioisosteric and therefore, in biological terms at least, are not distinguished; the increased affinity of the distomers is due to additional interactions with the biomolecule which do not occur with the eutomer.

5.3.2 Receptor Selectivity

As pointed out above eudismic ratios are only meaningful for a particular biological activity of a drug. For drugs which act at two or more sites differences in the eudismic ratio between the sites provides useful information in terms of the stereochemical demands and geometry of each site, a means of comparison between receptors in different tissues and may be used as a method of distinguishing receptor subtypes. Obviously such comparisons must be made with caution to ensure that potentially misleading factors, e.g., diffusion barriers, tissue uptake, and metabolism, are taken into account or controlled as such factors may vary markedly between tissues.

The activities of the enantiomers of the neuroleptic agent butaclamol have been investigated with tissue preparations containing D_2-dopaminergic, α-adrenergic, $5HT_2$ and $5HT_1$ serotoninergic, and opioid receptors. The eudismic ratio, (+)/(−), varied markedly with receptor system, (+)-(3*S*,4a*S*,13b*S*)-butaclamol (**5.61**) being 1250 times more active than the (−)-enantiomer in displacing haloperidol at D_2-receptors, 143 times more active at displacing LSD at 5-HT_2-receptors and equally active at displacing nalorphine at opioid receptors. The greater eudismic ratio was observed for actions in which the compound showed the greatest potency.

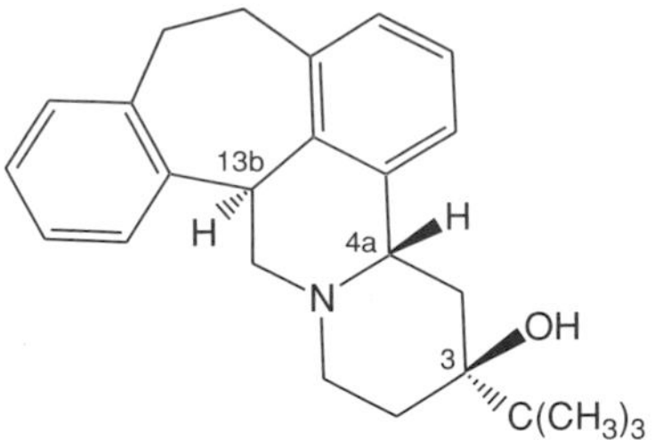

(+)-(3*S*,4a*S*,13b*S*)-Butaclamol (**5.61**)

Comparison of the stereochemical discrimination of the enantiomers of noradrenaline by α_1- and α_2-adrenoceptors indicates basic differences between the two receptor subtypes. The eudismic ratios (*R*/*S*) obtained are 107- and 480-fold for α_1- and α_2-receptors, respectively. Similar differences are also observed for α-methylnoradrenaline, the eudismic ratios for the 1*R*,2*S*/1*S*,2*R* enantiomeric pair being α_1, 60 and α_2, 550. Thus, for phenylethylamine derivatives the steric demands of the α_2-receptor subtype are more stringent than those of the α_1-receptor.

Differential stereoselectivity has also been observed with agonists at the histamine receptor subtypes. The introduction of α-methyl group in histamine results in the chiral molecule α-methylhistamine (**5.62**), examination of the activity of the enantiomers of this compound at the three histamine receptor subtypes yields eudismic ratios (*R*/*S*) of 1, 0.6, and approximately 100 at H_1, H_2, and H_3 receptors, respectively. The H_1-receptor showing no stereoselectivity, the H_2-receptor limited selectivity for the *S*-enantiomer, and the H_3-receptor showing marked stereoselectivity. Examination of pD_2 values for the *R*-enantiomer at the three-receptor subtypes yields values of 4.54 for H_1, 3.96 for H_2, and 8.40 for H_3. Thus, (*R*)-α-methylhistamine is a highly selective H_3-agonist and stimulation at H_3-receptors would be expected to occur at concentrations 10^4 times lower than those required for H_1- or H_2-receptor stimulation. Similar stereoselectivity for the H_3-receptor is also observed with α,β-dimethylhistamine. In this case the α*R*,β*S*-stereoisomer (**5.63**) is 100-fold more active than its α*S*,β*R*-enantiomer and shows 130,000-fold greater selectivity for the H_3-receptor than the other two subtypes. These two examples illustrate that the introduction of chirality into a critical site in a molecule may result in significant receptor subtype selectivity.

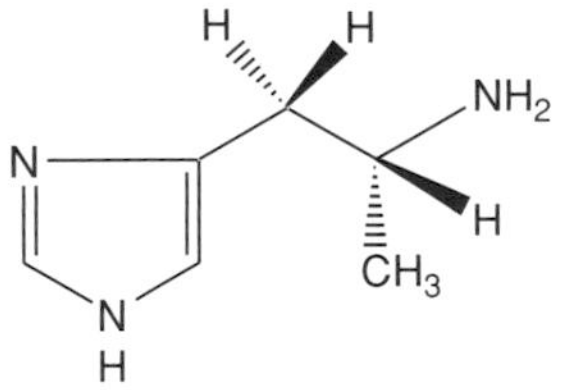

(*R*)-α-Methylhistamine (**5.62**)

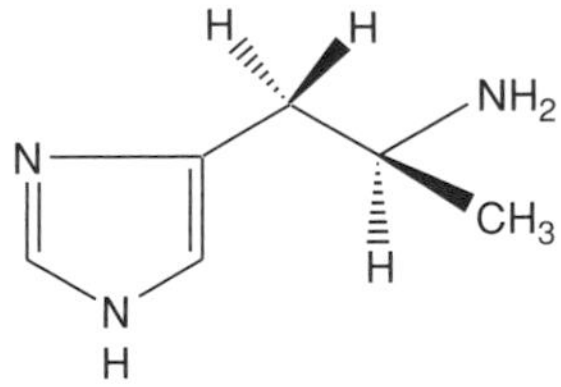

(*S*)-α-Methylhistamine (**5.62**)

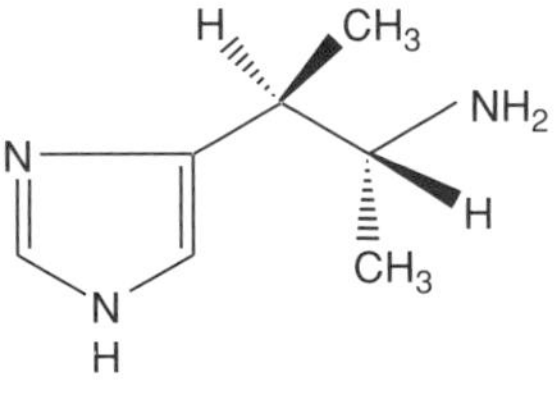

(α*R*, β*S*)-Dimethylhistamine (**5.63**)

5.3.3 "Purity" of Enantiomerically Pure Drugs

Determination of the eudismic ratio obviously depends on the availability of enantiomerically pure compounds and reported eudismic ratios for a particular compound may vary widely within the literature. While data of this type would be expected to vary from one laboratory to another, an important contributory factor is associated with the enantiomeric purity of the materials examined. As the eudismic ratio increases then the significance of a small quantity of the eutomer as an impurity of the "inactive" distomer also increases. For example initial evaluation of the enantiomers of isoprenaline on cat blood pressure yielded a ratio (*R*/*S*) of approximately 12; further experimentation and improved resolution, in this case repeated fractional crystallization of the diastereoisomeric (+)-bitartrate salts, resulted in a 1000-fold difference in activity.

The influence of relatively small quantities of stereoisomeric impurities on eudismic ratio may be illustrated by a report of the activity of the stereoisomers of formoterol a β_2-selective adrenoceptor agonist. Formoterol (**5.64**) has two stereogenic centers and therefore exists in four stereoisomeric forms, two pairs of enantiomers, the two stereogenic centers being positioned α and β to the aliphatic nitrogen atom. An examination of the activity of the four stereoisomers on the relaxation of airway smooth muscle indicated a relative order of potency of $\alpha R,\beta R > \alpha S,\beta R \approx \alpha S,\beta S > \alpha R,\beta S$. In an initial report the eudismic ratio for the enantiomeric pair $\alpha R,\beta R/\alpha S,\beta S$ was determined to be 14. A more recent investigation reported the same relative order of isomeric potency but a eudismic ratio $\alpha R,\beta R/\alpha S,\beta S$ of 50. In the latter study the distomer, the $\alpha S,\beta S$-enantiomer, was contaminated with 1.5% of the active $\alpha R,\beta R$-stereoisomer. Reduction in the "active impurity," to less than 0.1%, resulted in an increase in eudismic ratio $\alpha R,\beta R/\alpha S,\beta S$ to 850, and similar reductions of the "impurity" in the $\alpha S,\beta R$- and $\alpha R,\beta S$-stereoisomers resulted in an alteration in the order of potency to $\alpha R,\beta R > \alpha S,\beta R \approx \alpha R,\beta S > \alpha S,\beta S$.

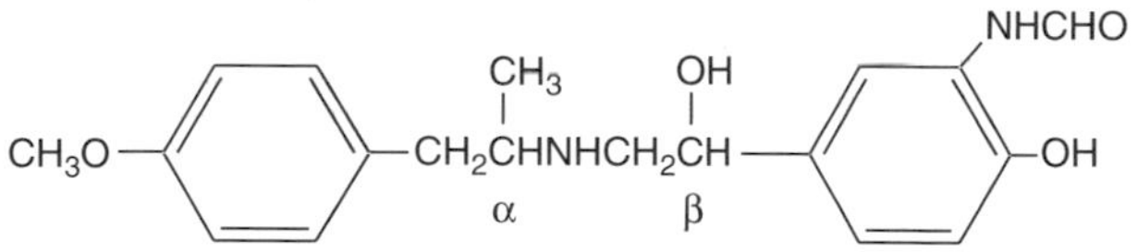

CH_3O–C_6H_4–$CH_2CH(CH_3)NHCH_2CH(OH)$–C_6H_3(NHCHO)(OH)
(α at CH(CH3), β at CH(OH))

Formoterol (**5.64**)

The degree of enantiomeric purity is frequently not specified in the pharmacological literature, or alternatively, is presented in terms of optical rotation, which is not a particularly sensitive technique at levels of contamination of a few percent. Indeed at one time it was proposed that purification to constant biological activity was a better criterion of enantiomeric purity than constant specific rotation. However, as a result of developments in stereospecific analytical methodology, particularly the chromatographic techniques using chiral stationary phases, it should now be considered unacceptable to present pharmacological data on individual enantiomers without quoting their enantiomeric purity.

5.4 PHARMACOKINETIC CONSIDERATIONS

As many of the processes of drug absorption and disposition involve an interaction between the enantiomers of a drug and a chiral biological macromolecule, it is hardly surprising that stereoselectivity is observed during these processes.

5.4.1 Absorption

The most important mechanism of drug absorption is passive diffusion through biological membranes, a process that is dependent upon the physicochemical properties of the molecule, e.g., lipid solubility, pK_a, molecular size, etc. If a chiral drug is absorbed by a passive process then differences between enantiomers would not be expected. In contrast, diastereoisomers may show differences in absorption as a result of the differences in their solubility. For example, the aqueous solubility of ampicillin (with 2*R*-stereochemistry in the acylated side chain, corresponding to the D-configuration, **5.35**) is greater than that of the 2*S*-epimer (L-configuration in the side chain). Differences between enantiomers may occur if the drug is a substrate for an active transport or carrier-mediated transport system. Such processes require the reversible combination of a substrate with a biological macromolecule and involve movement against a concentration gradient, expenditure of metabolic energy, and may be saturated. Such systems exhibit substrate specificity and therefore would be expected to exhibit stereoselectivity. Stereospecific transport systems are known to exist in the gastrointestinal tract for L-amino acids, dipeptides, and D-carbohydrates, etc. and drugs which are similar in structure to such naturally occurring substrates may be expected to be actively transported. Thus, L-dopa (**5.65**), L-penicillamine (**5.66**), and L-methotrexate (**5.67**) have been shown to be more rapidly absorbed from the gastrointestinal tract than their D-enantiomers, which are not substrates and are absorbed by passive diffusion. Such active processes may be expected, in theory at least, to increase the rate rather than the extent of absorption. In fact the bioavailability of D-methotrexate is only 2.5% that of the L-isomer.

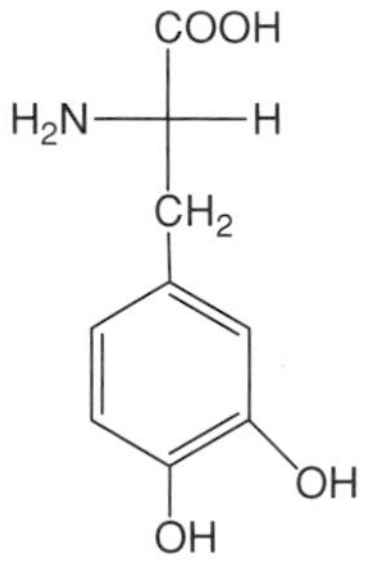

L-Dopa (**5.65**) 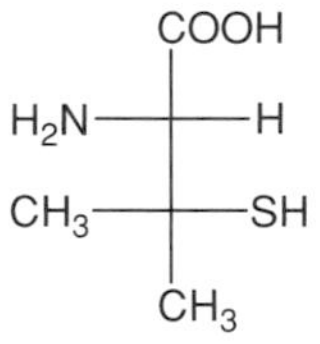 L-Penicillamine (**5.66**)

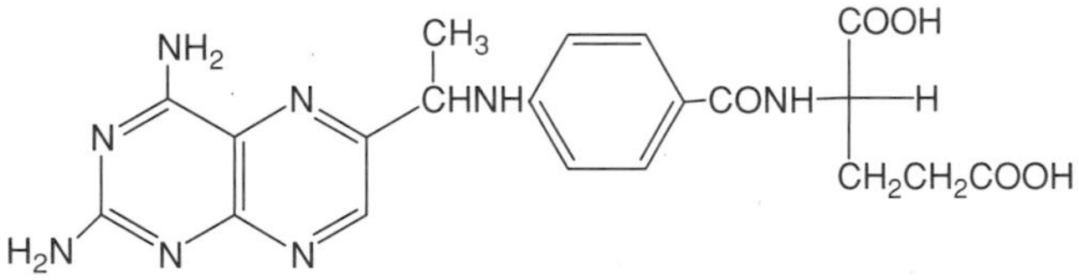

L-Methotrexate (**5.67**)

Many of the β-lactam antibiotics are substrates for the gut dipeptide transport system and as such their absorption would be expected to be stereoselective. The influence of the stereochemistry of the 7-acyl side chain on the absorption of the diastereoisomers of cefalexin has been investigated in the rat. Both diastereoisomers are substrates for the carrier-mediated transport system with the L-epimer showing a higher affinity than, and acting as a competitive inhibitor for, D-cefalexin (**5.36**) transport. However, the L-epimer is also more susceptible to the intestinal wall peptidases and cannot be detected in serum, whereas the D-isomer is well absorbed. The drug is marketed as the single D-epimer.

Additional biochemical or pharmacological factors may also influence the stereoselectivity of drug absorption. For example, the greater oral bioavailability of (−)-(*R*)-terbutaline compared to the less active (+)-*S*-enantiomer arises as a result of stereoselectivity in first pass metabolism and possibly due to the (−)-enantiomer increasing membrane permeability.

Differences in absorption may also be observed if the individual enantiomers differ in their effects on local blood flow. For example, (−)-bupivacaine has a longer duration of action than (+)-bupivacaine following intradermal injection. This difference in activity is due to the vasoconstrictor effects of the (−)-enantiomer reducing blood flow and hence systemic absorption.

5.4.2 Distribution

Protein binding

The majority of drugs undergo reversible binding to plasma proteins. In the case of chiral drugs the drug enantiomer–protein complexes are diastereoisomeric and individual enantiomers would be expected to exhibit differences in binding affinity to the circulating proteins. Such differences in binding affinity result in differences between enantiomers in the free, or unbound, fraction that is able to distribute into tissue (Table 5.2). The two most important plasma proteins with respect to drug binding are human serum albumin (HSA) and α_1-acid glycoprotein (AGP). In general acidic drugs bind predominantly to HSA, whereas basic drugs bind predominantly to AGP.

Differences between enantiomers in plasma protein binding may be relatively small (Table 5.2) and in some cases less than 1%. However, such low stereoselectivity in binding may result in much larger differences in the enantiomeric composition of the free, or unbound, fraction particularly for highly protein-bound drugs. For example, the free fractions of (−)-(*R*)- and (+)-(*S*)-indacrinone are 0.9% and 0.3%, respectively, i.e., a threefold difference. An extreme example of stereoselectivity in binding is the amino acid tryptophan, the L-enantiomer binding to HSA with an affinity approximately 100 times greater than that of the D-isomer. In terms of drugs, (*S*)-oxazepam hemisuccinate (**5.68**) binds to HSA with an affinity 40 times than that of the *R*-enantiomer. However, using bovine

Table 5.2 Stereoselectivity in plasma protein binding

	Unbound (%)		
	S-enantiomer	*R*-enantiomer	Ratio (*S*/*R*)
Acidic drugs			
Acenocoumarol	2.0	1.8	1.1
Etodolac	0.85	0.47	1.8
Flurbiprofen	0.048	0.082	0.59
Ibuprofen	0.64	0.42	1.5
Indacrinone	0.3	0.9	0.33
Mephobarbitone	53	66	0.80
Moxalactam	32	47	0.68
Pentobarbitone	26.5	36.6	0.72
Phenprocoumon	0.72	1.07	0.67
Warfarin	0.9	1.2	0.75
Basic drugs			
Bupivacaine	4.5	6.6	0.68
Carvedilol	0.63	0.45	1.4
Chloroquine	33.4	51.5	0.64
Disopyramide	22.2	34	0.64
Fenfluramine	2.8	2.9	0.96
Gallopamil	5.7	4.0	1.4
Methadone	9.2	12.4	0.74
Mexiletine	28.3	19.8	1.4
Propafenone	2.5	3.9	0.64
Sotalol	62	65	0.95
Tocainide	83–89	86–91	~1.0
Verapamil	11	6.4	1.7

serum albumin as a protein source, the difference in affinity is only threefold. Such species variation in the enantioselectivity of plasma protein binding has also been reported for phenprocoumon and disopyramide.

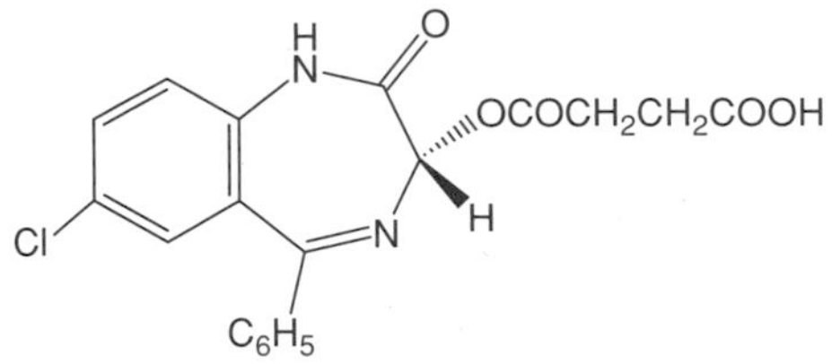

Oxazepam hemisuccinate (**5.68**)

Enantioselectivity in binding may also vary between HSA and AGP. For example, the binding of propranolol to AGP is stereoselective for the *S*-enantiomer, whereas binding to HSA is selective for the *R*-enantiomer. In whole plasma the binding to AGP predominates and the fraction unbound of the *R*-enantiomer exceeds that of (*S*)-propranolol (Table 5.3).

Stereoselectivity in plasma protein binding also influences clearance for drugs with a low hepatic extraction ratio, total clearance being proportional to fraction unbound. In addition stereoselective displacement of drug enantiomers from plasma protein-binding sites may give rise to complexities in drug interactions. Interactions between enantiomers for plasma protein-binding sites may also result in pharmacokinetic complications. For example, the protein binding of disopyramide is stereoselective, concentration-dependent, and competitive and as a result the pharmacokinetic parameters of the individual enantiomers differ depending if the drug is administered as the racemate or single isomer.

Tissue distribution

The extent of tissue distribution of a drug depends on both its lipid solubility and relative plasma to tissue protein binding. In a number of instances differences in calculated volumes of distribution between enantiomers are lost when plasma protein binding is taken into account and unbound volumes of distribution are compared. Similarly, apparent stereoselective distribution of some drugs into various tissues and fluids may be rationalized by differences between enantiomers in protein binding, e.g., the stereoselective distribution of (*S*)-ibuprofen into synovial fluid may be explained by differences in protein binding.

Lipid solubility is obviously an important factor for drug transfer across biological membranes and it would appear that lipophilicity is of greater significance than chiral drug–lipid interactions. However, recent evidence has indicated that some basic drugs preferentially accumulate in tissues containing acidic phospholipids, e.g., phosphatidylserine. Stereoselective interactions, assumed to be electrostatic, between phosphatidylserine and morphine have been reported and there is evidence that other basic chiral drugs, e.g., disopyramide and verapamil, undergo preferential and stereoselective distribution in tissues containing a high content of phosphatidylserine. Thus,

Table 5.3 Stereoselectivity of the plasma protein binding of propranolol enantiomers

	Enantiomer free fraction		
Protein source	***R***	***S***	**Ratio (*R*/*S*)**
Whole plasma	0.203	0.176	1.15
HSA	0.607	0.647	0.94
AGP	0.162	0.127	1.28

both stereoselective protein binding and phosphatidylserine content of tissue may influence the distribution of basic drugs.

Stereoselective distribution may also occur as a result of interactions with tissue uptake transporter and efflux processes, and storage mechanisms. The skeletal muscle relaxant baclofen (4-amino-3-(4-chlorophenyl)butyric acid), the activity of which resides in the *R*-enantiomer, the eudismic ratio (*R*/*S*) for the mimetic effect on the $GABA_B$ receptor being approximately 100, undergoes stereoselective transport across the blood–brain barrier (BBB). The BBB clearance of the *R*-enantiomer is fourfold greater than that of either (*S*)- or racemic baclofen. Such data are indicative of an enantiomeric interaction in carrier-based transport, which is thought to be the large neutral amino acid carrier system. The stereoselective efflux of the enantiomers of (*E*)-10-hydroxynortriptyline, a metabolite of both amitriptyline and nortriptyline, has been reported from the cerebrospinal fluid in depressed patients; the secretion of the (−)-enantiomer being greater than that of the (+)-metabolite. The *S*-enantiomers of the β-adrenoceptor antagonists propranolol and atenolol undergo selective storage and secretion by adrenergic nerve terminals in cardiac and other tissue. In the case of (−)-atenolol, uptake into storage granules has been reported to be fivefold that of the (+)-enantiomer, a value in good agreement with the fourfold (*R*/*S*) selectivity reported for the enantiomers of noradrenaline.

Stereoselective distribution may also be associated with metabolism. The *R*-enantiomers of some of the 2-arylpropionic acid NSAIDs, e.g., ibuprofen and fenoprofen, undergo selective incorporation into lipid resulting in the formation of ''hybrid'' triglycerides. The mechanism of this will be discussed below (see Section 5.6.2), but the selective deposition results in the accumulation of these agents into adipose tissue the toxicological significance of which is unknown.

5.4.3 Metabolism

In contrast to other processes involved in drug absorption and disposition, drug metabolism frequently exhibits marked stereoselectivity. Stereoselectivity in metabolism may be associated with the binding of the substrate to the enzyme, and therefore associated with the chirality of the enzyme-binding site. Alternatively, selectivity may be associated with catalysis due to differential reactivity and/or orientation of potential target groups with respect to the enzyme catalytic site.

An examination of the stereochemistry of drug metabolism is of importance as the individual enantiomers of a racemic drug may be metabolized by different routes to yield different products and they are frequently metabolized at different rates. In addition, species differences may occur in the metabolism of individual enantiomers and as data derived from animal studies are used to assess potential toxic hazard to man the information may have little relevance.

The stereoselectivity of the reactions of drug metabolism may be associated with:

1. substrate stereoselectivity, i.e., the selective metabolism of one enantiomer compared to the other in either rate and/or route of metabolism;
2. product stereoselectivity, i.e., the preferential formation of a particular stereoisomer rather than other possible stereoisomers;
3. substrate–product stereoselectivity, i.e., the selective metabolism of one enantiomer resulting in the preferential formation of one of a number of possible diastereoisomeric products.

In terms of the stereochemical outcome metabolic transformations may be divided into five groups as indicated below.

Prochiral to chiral transformations

In reactions of this type the molecule acquires chirality by metabolism, which may take place at either a prochiral center or on an enantiotopic group bonded to it. The antiepileptic drug phenytoin (**5.69**) has a prochiral center at carbon-5 of the hydantoin ring system and the two phenyl rings are enantiotopic as indicated by pro-*S* and pro-*R* (**5.69**). The major route of metabolism of phenytoin in both animals and humans involves aromatic oxidation, which in humans shows product stereoselectivity for formation of (*S*)-4′-hydroxyphenytoin (**5.70**). In contrast, in the dog oxidation takes place in the pro-*R* ring to yield (*R*)-3′-hydroxyphenytoin the reaction showing species selectivity in both stereochemistry and regiochemistry (position).

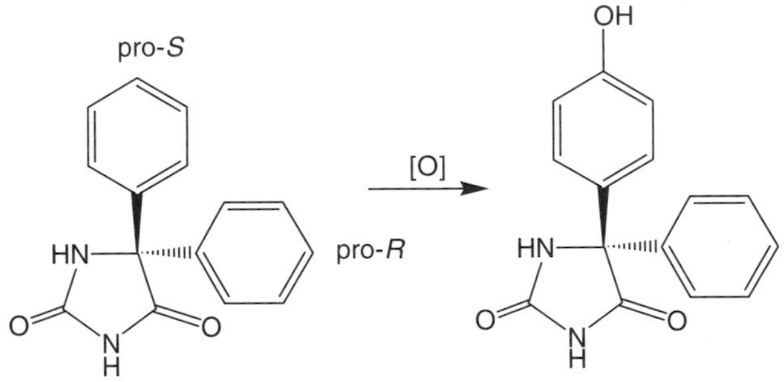

Phenytoin (**5.69**) (*S*)-4-Hydroxyphenytoin (**5.70**)

It has been pointed out above that sulfoxides may be chiral and therefore the metabolic oxidation of sulfides to sulfoxides will produce chiral metabolites. Cimetidine (**5.71**) undergoes oxidation at sulfur to yield a chiral sulfoxide (**5.72**) as a urinary metabolite. The reaction is product stereoselective for the formation of the (+)-enantiomer, the enantiomeric composition of the material in urine is approximately (+/−) 3:1.

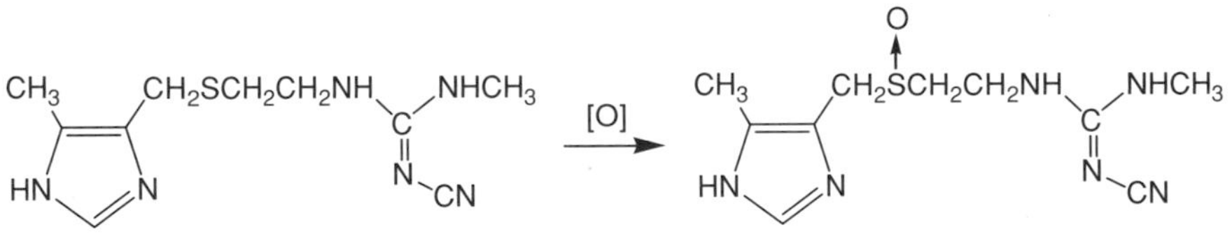

Cimetidine (**5.71**) Cimetidine *S*-oxide (**5.72**)

Chiral to chiral transformations

In transformations of this type metabolism takes place at a site in the molecule that does not alter the chirality of the metabolite relative to that of the drug.

Esmolol (**5.73**) is an ultrashort acting, relatively cardioselective β-adrenoceptor antagonist, which is administered intravenously for the short-term treatment of supraventricular arrhythmias and sinus tachycardia. The drug is used as a racemate but the pharmacological activity resides in the enantiomer of the *S*-configuration; the *R*-enantiomer is pharmacologically inactive. The basis of the short duration of action, 10–15 min, is the rapid hydrolysis (**5.73**→**5.74**) of the ester functionality by blood esterases. The stereoselective hydrolysis of this agent shows considerable species variability, e.g., the hydrolysis of (*S*)-esmolol is faster than that of the *R*-enantiomer in the rhesus monkey, rabbit, and guinea pig and shows reversed stereoselectivity in rat and dog. In man the hydrolysis of both enantiomers occurs at similar rates.

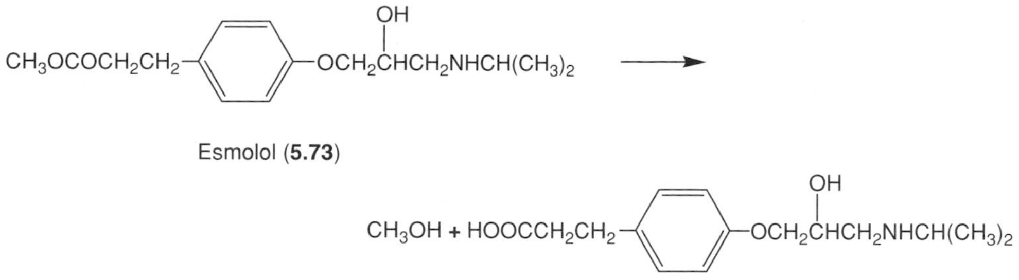

$CH_3OCOCH_2CH_2$–C_6H_4–$OCH_2CH(OH)CH_2NHCH(CH_3)_2$ ⟶

Esmolol (**5.73**)

CH_3OH + $HOOCCH_2CH_2$–C_6H_4–$OCH_2CH(OH)CH_2NHCH(CH_3)_2$

(**5.74**)

Aromatic oxidation of warfarin (**5.75**) yields the 7-hydroxy metabolite a reaction which is stereoselective for the more active *S*-enantiomer (urinary recovery ratio *S*/*R* ~14) of the drug in man. In contrast, oxidation at the 6-position of the coumarin ring system shows no stereoselectivity in humans. In the rat 7-hydroxywarfarin is a major metabolite but for the *R*-enantiomer, i.e., the oxidation shows the reverse stereoselectivity compared to humans. Studies using human expressed cytochrome P450 (CYP) isoenzymes have indicated that the isoform CYP2C9 is primarily responsible for the oxidation of (*S*)-warfarin to the 6- and 7-hydroxy compounds whereas isoform CYP1A2 is involved in the formation of (*R*)-6-hydroxywarfarin.

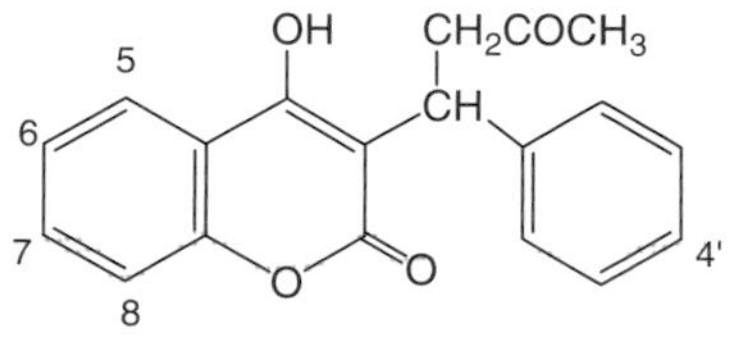

Warfarin (**5.75**)

Chiral to diastereoisomer transformations

Transformations of this type involve the introduction of an additional stereogenic center into a chiral molecule. Such centers may arise by a Phase I, or functionalization, metabolic reaction at a prochiral center or by a Phase II, or conjugation, process by reaction with a chiral-conjugating agent.

Reactions of the first type include reduction of the prochiral ketone group in warfarin (**5.75**) to yield a pair of diastereoisomeric warfarin alcohols. In both rat and human, the reduction is substrate selective for (*R*)-warfarin (**5.75**) and the predominantly formed isomer of the alcohol (**5.76**) has the

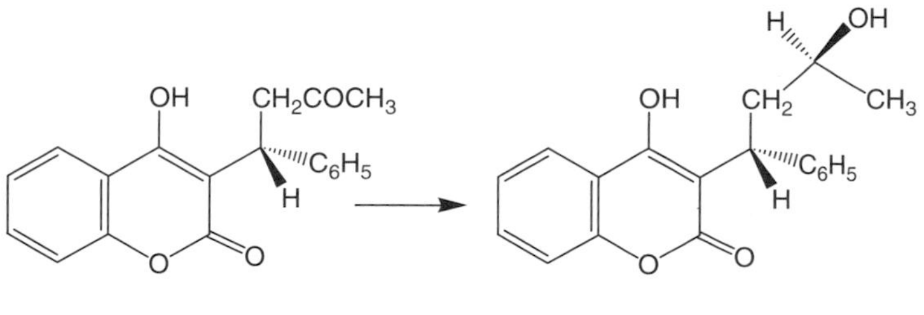

(*R*)-Warfarin (**5.75**) ⟶ (*S,R*)-Warfarin alcohol (**5.76**)

S-configuration at the new center. The Phase II or conjugation reactions of drug metabolism are biosynthetic and involve the combination of the drug, or a Phase I metabolite of the drug, with an endogenous molecule. Many of the endogenous molecules involved in the conjugation reactions are chiral, e.g., D-glucuronic acid, the amino acid L-glutamine, and the tripeptide glutathione, and hence chiral drugs which undergo conjugation with these agents will produce diastereoisomeric products.

Oxazepam (**5.77**) is a chiral benzodiazepine which is used as a racemic mixture. The individual enantiomers of oxazepam are stereochemically unstable and readily undergo racemization in aqueous media. Both enantiomers of oxazepam undergo conjugation with D-glucuronic acid to yield a pair of stereochemically stable diastereoisomeric conjugates the proportions of which may vary between species. In humans, dog, and rabbit, the diastereoisomer produced from (*S*)-oxazepam (**5.78**) predominates, *S*/*R* ratios varying between 2 and 3.4, whereas in the rhesus monkey (*R*)-oxazepam glucuronide (**5.78**) is preferentially formed (ratio $S/R = 0.5$). The formation of the stereochemically stable glucuronides has facilitated the examination of the stereoselective aspects of oxazepam disposition. It is of interest to note that hydrolysis of either conjugate diastereoisomer results in the formation of the racemic drug.

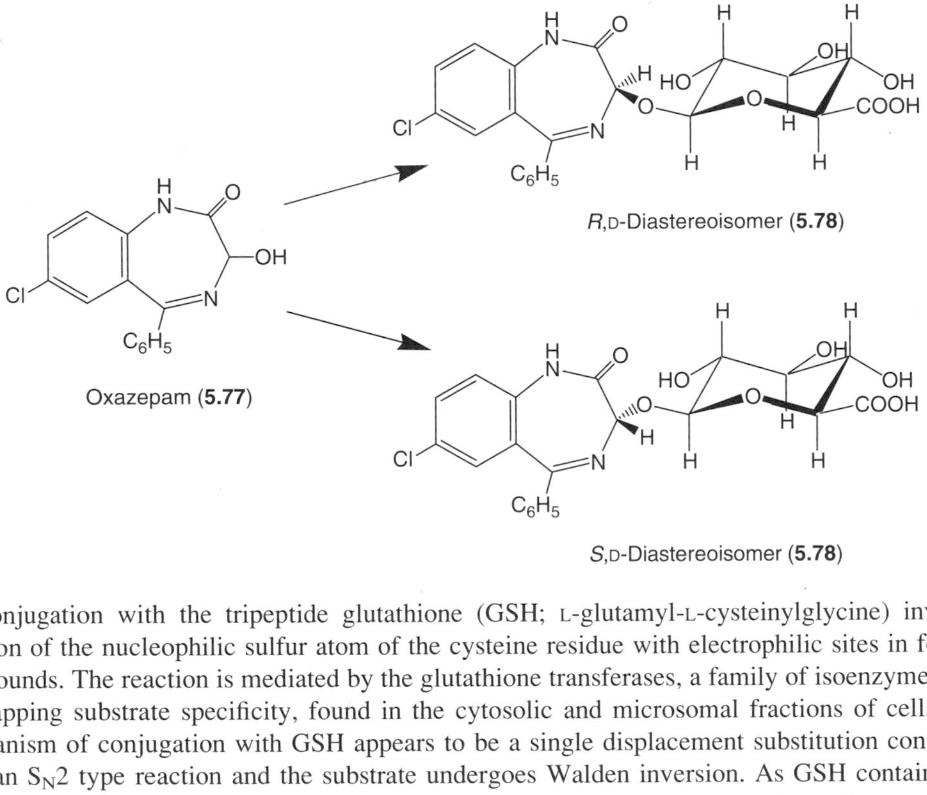

Oxazepam (**5.77**)

R,D-Diastereoisomer (**5.78**)

S,D-Diastereoisomer (**5.78**)

Conjugation with the tripeptide glutathione (GSH; L-glutamyl-L-cysteinylglycine) involves reaction of the nucleophilic sulfur atom of the cysteine residue with electrophilic sites in foreign compounds. The reaction is mediated by the glutathione transferases, a family of isoenzymes with overlapping substrate specificity, found in the cytosolic and microsomal fractions of cells. The mechanism of conjugation with GSH appears to be a single displacement substitution consistent with an S_N2 type reaction and the substrate undergoes Walden inversion. As GSH contains two L-amino acid residues in its structure, reaction with a racemic substrate results in diastereomeric glutathione conjugates. The conjugation of the obsolete chiral hypnotic agent α-bromoisovalerylurea (**5.79**) with GSH involves nucleophilic displacement of the bromine atom at the stereogenic center and the glutathione conjugates (**5.80**) have the reverse configurational designation to those in the drug. In the case of α-bromoisovalerylurea the reaction is stereoselective for the *R*-enantiomer of the drug, the cytosolic enzymes showing a threefold greater activity for the *R*-enantiomer compared to the S-enantiomer. The stereoselectivity of the reaction varies with isoenzyme such that examination of purified enzyme systems indicates that the isoenzymes of the mu-family show a stereopreference for conjugation of (R)-α-bromoisovalerylurea, whereas with those of the α-family show a preference for the S-enantiomer.

$H_2NCONHCO$ / $(CH_3)_2CH$ / H / Br —GSH→ GS / $CONHCONH_2$ / $CH(CH_3)_2$ / H

S-(**5.79**) *R*-(**5.80**)

$H_2NCONHCO$ / H / $(CH_3)_2CH$ / Br —GSH→ GS / $CONHCONH_2$ / H / $CH(CH_3)_2$

R-(**5.79**) *S*-(**5.80**)

Chiral to achiral transformations

In reactions of this type the biotransformation results in a loss of chirality, the reaction taking place at the stereogenic center. Examples of interest are provided by the 1,4-dihydropyridine calcium-channel blocking agents, e.g., nitrendipine and nilvadipine (**5.81**). These agents undergo cytochrome P450-mediated oxidation to yield the corresponding achiral pyridine analogs (**5.82**). In the case of nilvadipine this reaction is stereoselective for the (+)-enantiomer in the rat, but for the (−)-enantiomer in dog and humans. Reduction of the chiral sulfoxide moiety in the NSAID prodrug sulindac (**5.83**) results in the formation of the active cyclooxygenase inhibiting achiral sulfide (**5.84**) metabolite. Similarly, sulfoxidation of sulindac results in the loss of chirality yielding the sulfone metabolite (**5.85**), which has recently been shown to have antiproliferative activity and may prove to be useful in the treatment of some forms of colon cancer. Thus, both achiral metabolites of sulindac appear to possess useful biological activity.

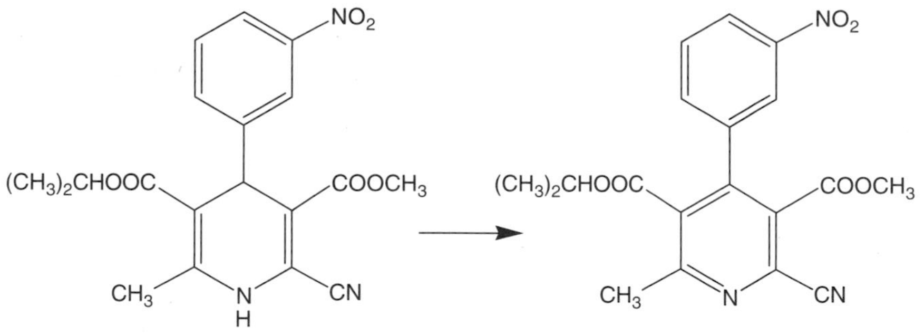

NO_2 / $(CH_3)_2CHOOC$ / $COOCH_3$ / CH_3 / N / H / CN → NO_2 / $(CH_3)_2CHOOC$ / $COOCH_3$ / CH_3 / N / CN

Nilvadipine (**5.81**) (**5.82**)

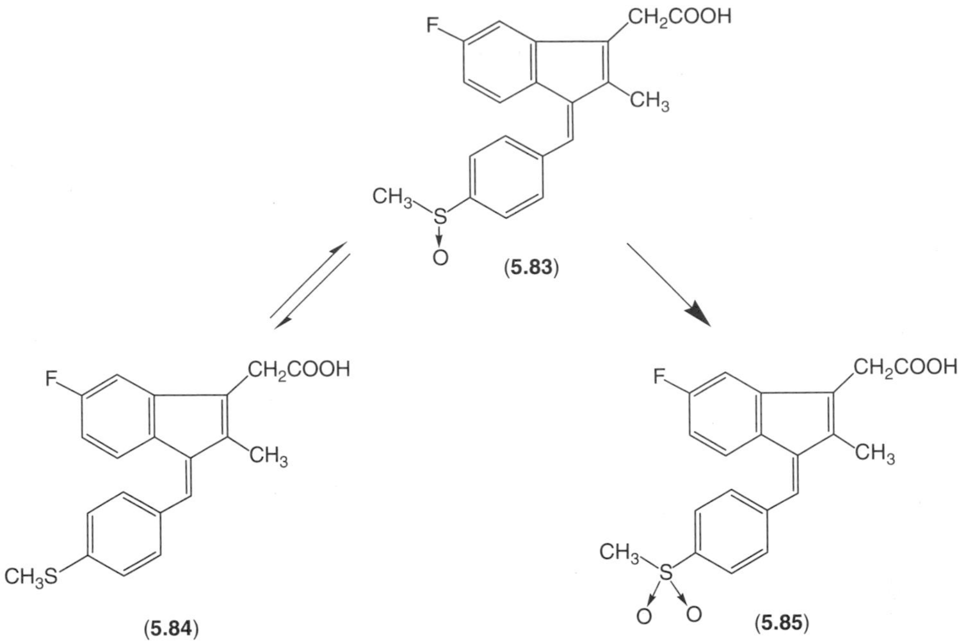

(5.83)

(5.84)

(5.85)

Chiral inversion

Chiral inversion is a relatively rare metabolic transformation and involves the conversion of a stereoisomer to its enantiomer with no other chemical change to the molecule. The reaction was initially observed with the 2-arylpropionic acid NSAIDs, e.g., ibuprofen (**5.2**), and has since been found to occur with the chemically related 2-aryloxypropionates, which are used as herbicides, e.g., haloxyfop (**5.86**) and more recently with compound (**5.87**) a selective topoisomerase IIβ inhibitor with antitumor activity. In the case of the 2-arylpropionic acids the reaction involves inversion of the relatively inactive, with respect to inhibition of cyclooxygenase, *R*-enantiomers to their *S*-eutomers. Whereas with the 2-aryloxypropionates the reaction appears to be reversed, i.e., the transformation is from the *S*- to the *R*-enantiomers. This difference in the stereochemistry of the inversion reaction is apparent and arises as a result of the sequence rule designation, the three-dimensional spatial arrangement of the *R*-2-arylpropionic acids corresponding to that of an *S*-2-aryloxypropionate. The mechanism of this reaction will be examined in Section 5.6.2.

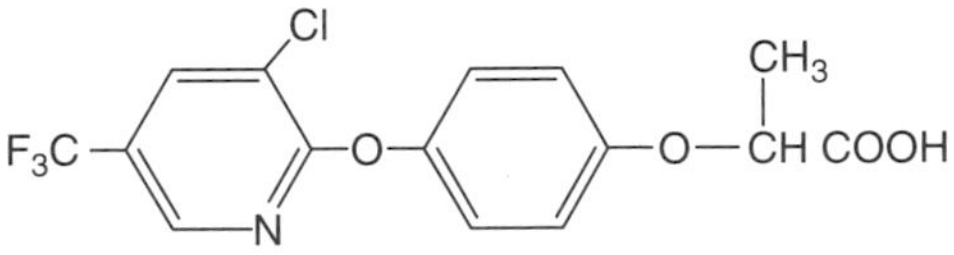

Haloxyfop (**5.86**)

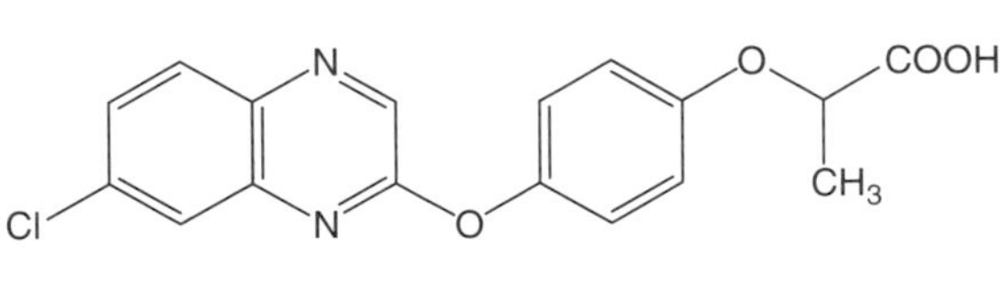

(**5.87**)

5.4.4 Excretion

Renal excretion is the net result of glomerular filtration, active secretion, and passive and active reabsorption. Since glomerular filtration is a passive process differences between enantiomers would not be expected. However, apparent stereoselectivity in renal clearance may arise as a consequence of stereoselectivity in protein binding. Stereoselectivity in renal clearance may be observed as a result of active secretion; however, active reabsorption and renal metabolism may also be significant. Active renal tubular secretion is thought to be responsible for the differential clearance of the enantiomers of a number of basic drugs with stereoselectivities in the range of 1.1 to 3.0 (Table 5.4). The renal clearance of quinidine has been reported to be four times greater than that of its diastereoisomer quinine.

The renal clearance of the diastereoisomeric glucuronide conjugates of both ketoprofen and propranolol has also been reported to show stereoselectivity. In both cases renal clearance is selective for the *S*-enantiomer conjugate of the drug with selectivities of 3.2- and 1.3-fold for propranolol and ketoprofen, respectively.

Relatively little is known regarding the stereoselectivity of the active processes involved in the biliary secretion of drugs. Differences in the biliary recovery of enantiomers have been reported, e.g., acenocoumarol in the rat, however it is not clear if this is due to stereoselectivity in biliary clearance or as a result of other stereoselective processes.

5.4.5 Pharmacokinetic Parameters

As a result of the above processes compounds administered as racemates rarely exist as 1:1 mixtures of enantiomers in biofluids and tissues, and do not reach their sites of action in equal concentration. The pharmacokinetic profiles of the enantiomers of a racemic drug may differ markedly and hence an estimation of pharmacokinetic parameters, or an examination of drug concentration–effect relationships based on ''total,'' i.e., the sum of the two enantiomer, concentrations present in biological samples may at best yield data of limited value and are potentially highly misleading.

In comparison to the differences observed between enantiomers in terms of their receptor-mediated pharmacodynamic activities, the magnitude of the differences in their pharmacokinetic parameters tend to be relatively modest, frequently one- to threefold (Table 5.5). However, the degree of stereoselectivity observed for a particular pharmacokinetic parameter is also influenced by the organizational level that the parameter represents. Pharmacokinetic parameters may be divided into three levels of organization, i.e., the whole body (systemic clearance, volume of distribution, elimination half-life); organ (hepatic and renal clearance), and macromolecular (intrinsic metabolite formation clearance, fraction unbound). Parameters representing the whole body level of organization are determined by multiple organ parameters, which in turn are a reflection of

Table 5.4 Stereoselectivity in the renal clearance of basic drugs in man

Drug	Stereochemistry	Ratio
Chloroquine	(+) > (−)	1.6
Disopyramide	(+) > (−)	1.3
Flecainide	*S* > *R*	1.1
Metoprolol	*S* > *R*	1.1
Mexiletine	*S* > *R*	1.2
Pindolol	*R* > *S*	1.2
Prenylamine	*S* > *R*	3.0
Salbutamol	*R* > *S*	1.6
Sotalol	*R* > *S*	1.1
Terbutaline	*S* > *R*	1.8

Table 5.5 Stereoselectivity in pharmacokinetic parameters following administration of racemic drugs to man

Drug	Route of administration	Enantiomer	Clearance (units)	Volume of distribution (units)	Half-life (h)
Bupivacaine	IV	*R*	0.40 L min^{-1}	84 L	3.5
		S	0.32 L min^{-1}	54 L	2.6
		R[a]	7.3 L min^{-1}	1576 L	—
		S[a]	8.7 L min^{-1}	1498 L	—
Carvedilol	Oral	*R*	0.87 L min^{-1}	302 L	5.3
		S	1.26 L min^{-1}	487 L	5.1
Etodolac	Oral	*R*	22 mL h^{-1} kg^{-1}	0.21 L kg^{-1}	6.6
		S	288 mL h^{-1} kg^{-1}	1.6 L kg^{-1}	4.3
Ifosfamide	IVI	*R*	0.060 L h^{-1} kg^{-1}	0.61 L kg^{-1}	7.1
		S	0.072 L h^{-1} kg^{-1}	0.63 L kg^{-1}	6.0
Hexobarbitone	Oral	*R*	136 L h^{-1}	—	6.7
		S	21 L h^{-1}	—	2.8
Ketorolac	IM	*R*	19.0 mL h^{-1} kg^{-1}	0.075 L kg^{-1}	3.6
		S	45.9 mL h^{-1} kg^{-1}	0.135 L kg^{-1}	2.4
Mephobarbitone	Oral	*R*	170 L h^{-1}	716 L	3.1
		S	1.5 L h^{-1}	105 L	50.5
Metoprolol	Oral	*R*	1.7 L h^{-1} kg^{-1}	7.6 L kg^{-1}	2.7
		S	1.2 L h^{-1} kg^{-1}	5.5 L kg^{-1}	3.0
Mexiletine	Oral	*R*	8.6 mL min^{-1} kg^{-1}	6.6 L kg^{-1}	9.1
		S	8.1 mL min^{-1} kg^{-1}	7.3 L kg^{-1}	11.0
	Oral	*R*	7.9 mL min^{-1} kg^{-1}	5.3 L kg^{-1}	8.1
		S	8.8 mL min^{-1} kg^{-1}	6.0 L kg^{-1}	8.4
Nitrendipine	IV	*R*	1.6 L min^{-1}	3.7 L kg^{-1}	4.0
		S	1.5 L min^{-1}	3.9 L kg^{-1}	4.3
	Oral	*R*	6.6 L min^{-1}	—	7.5
		S	3.1 L min^{-1}	—	7.7
Nivaldipine	Oral	*R*	110 mL min^{-1} kg^{-1}	—	2.1
		S	39.5 mL min^{-1} kg^{-1}	—	1.5
Prenylamine	Oral	*R*	4.0 L min^{-1}	—	8.2
		S	20.5 L min^{-1}	—	24
Propranolol	IV	*R*	1.21 L min^{-1}	4.82 L kg^{-1}	3.5
		S	1.03 L min^{-1}	4.08 L kg^{-1}	3.6
	Oral	*R*	6.9 L min^{-1}	—	4.3
		S	4.6 L min^{-1}	—	4.8
Reboxetine	IV	*R,R*	0.027 L h^{-1} kg^{-1}	0.39 L kg^{-1}	10.6
		S,S	0.071 L h^{-1} kg^{-1}	0.92 L kg^{-1}	9.42
	Oral	*R,R*	41.6 mL min^{-1}	50.9 L	14.8
		S,S	99.3 mL min^{-1}	114 L	14.4
Salbutamol	IV	*R*	0.62 L h^{-1} kg^{-1}	2.0 L kg^{-1}	2.0
		S	0.39 L h^{-1} kg^{-1}	1.8 L kg^{-1}	2.9
	Oral	*R*	0.17 L h^{-1} kg^{-1}	—	—
		S	0.19 L h^{-1} kg^{-1}	—	—
	IV	*R*	46.8 L h^{-1}	—	2.5
		S	14.7 L h^{-1}		4.7
	Oral	*R*	—	—	2.9
		S			6.0
	Inhalation	*R*	—	—	2.0
		S			4.5
Sotalol	Oral	*R*	12.4 L h^{-1}	2.0 L kg^{-1}	7.9
		S	11.7 L h^{-1}	2.0 L kg^{-1}	8.2
Terodiline	Oral	*R*	—	391 L	98
		S	59 mL h^{-1} kg^{-1}	443 L	86
Thiopental[b]	IV	*R*	0.30 L min^{-1}	139 L	9.6
		S	0.23 L min^{-1}	114 L	9.0
	IVI	*R*	0.10 L min^{-1}	313 L	14.6
		S	0.08 L min^{-1}	273 L	14.7

(*continued*)

Table 5.5 Stereoselectivity in pharmacokinetic parameters following administration of racemic drugs to man — continued

Drug	Route of administration	Enantiomer	Clearance (units)	Volume of distribution (units)	Half-life (h)
Tocainide	IVI	*R*	11.1 L h^{-1}	136 L	9.3
		S	6.3 L h^{-1}	134 L	17.1
Verapamil	IV	*R*	0.80 L min^{-1}	2.74 L kg^{-1}	4.1
		S	1.40 L min^{-1}	6.42 L kg^{-1}	4.8
	Oral	*R*	1.72 L min^{-1}	—	—
		S	7.46 L min^{-1}	—	—
Warfarin	Oral	*R*	1.9 mL h^{-1} kg^{-1}	129 mL kg^{-1}	47.1
		S	2.0 mL h^{-1} kg^{-1}	70.5 mL kg^{-1}	24.4

Average values of the reported parameters presented. IV, intravenous injection; IVI, intravenous infusion; IM, intramuscular injection.
[a]Values of unbound enantiomer clearance and volume of distribution.
[b]Thiopental doses: IV, 0.25–0.5 g; IVI, between 12.5 and 86.9 g, duration of infusion between 31 and 285 h.

multiple macromolecular interactions. Differences between enantiomers are potentially greatest in these latter parameters which are associated with a direct interaction with a chiral biological macromolecule. Thus, differences in pharmacokinetic parameters between a pair of enantiomers may be amplified or attenuated with each level of organization, and it is therefore possible that comparison of parameters that reflect the whole body level of organization may mask stereoselectivity at the level of the organ or macromolecule.

In the case of the antiarrhythmic agent verapamil, the ratio (*S*/*R*) of the half-lives of the individual enantiomers is relatively modest at approximately 1.2, reflecting the whole body level of organization and dependence on volume of distribution (*S*/*R* 2.34) and clearance (*S*/*R* 1.77). However, examination of metabolite formation clearance for demethylation, a macromolecular parameter, yields a ratio (*S*/*R*) of 33. Thus, the modest ratio in enantiomer half-life of verapamil masks the significant enantioselectivity of the demethylation metabolic pathway.

For drugs which are subject to extensive stereoselective first pass, or presystemic metabolism, the differential bioavailability of the individual enantiomers may give rise to apparent anomalies in drug–concentration effect relationships with route of administration if the enantiomeric composition of material in plasma is not taken into account. Thus, based on measurements of ''total'' plasma concentrations verapamil appears to be more effective when given intravenously than orally, whereas propranolol shows the opposite effect. In both cases the explanation for the observed effect is stereoselective presystemic metabolism, which in the case of verapamil is selective for the more active *S*-enantiomer and for propranolol the less active *R*-enantiomer (Table 5.6).

Care should be taken in therapeutic drug monitoring of chiral drugs administered as racemic mixtures. The determination of the plasma concentrations of the individual enantiomers of chiral drugs would be advantageous to define the ''real'' therapeutic range of such compounds. The therapeutic concentrations of the racemic antiarrhythmic agent tocainide, based on ''total'' drug covers a threefold range. However, the plasma half-life of the more active *R*-enantiomer, at approximately 10 h, is shorter than that of (*S*)-tocainide ($t_{1/2}$, 15–17 h) with the result that following intravenous infusion the enantiomeric ratio (*S*/*R*) in plasma concentration increases from ca. 1 at 2 min to ca. 1.7 after 48 h. Hence ''total'' drug plasma concentrations will increase progressively during the infusion but with relatively small changes in pharmacological effect. There is also considerable interpatient variability in the enantiomeric composition of the drug in plasma, the *S*/*R* ratio varying between 1.3 and 3.8, which is probably associated with variability in metabolism.

Stereochemical considerations may also be of significance for understanding drug interactions between both the enantiomers of chiral drugs and a second agent and also to rationalize differences in the disposition of chiral drugs when given as racemic mixtures or single isomers (Table 5.6).

Table 5.6 Factors influencing the stereoselectivity of drug action and disposition in man

Factor	Drug	Comment
Route of administration		
	Nitrendipine	Stereoselective availability following oral administration, fractions absorbed 8% and 13.5% for the *R*- and *S*-enantiomers, respectively; results in a twofold greater AUC of the more active *S*-enantiomer following oral administration compared to a ~1.2-fold ratio following intravenous administration.
	Verapamil	Clearance of the more active *S*-enantiomer greater than that of (*R*)-verapamil resulting in a twofold ratio (*R*/*S*) in plasma concentration following intravenous administration; stereoselective first pass metabolism, oral bioavailability ~50% and ~20% for the *R*- and *S*-enantiomers, respectively, results in a fivefold ratio (*R*/*S*) in plasma concentrations. Examination of concentration–effect relationships based on "total" drug plasma concentrations indicates an enhanced effect following intravenous compared to oral administration.
	Propranolol	Appears to be more potent following oral compared to intravenous administration when concentration effect relationships are based on "total" drug; due to stereoselective first pass metabolism of the less active *R*-enantiomer.
Formulation		
	Verapamil	Enantiomeric ratio (*R*/*S*) of the maximum plasma concentrations (C_{max}) and area under the plasma concentration versus time curves (AUC) significantly lower following immediate release (IR) compared to sustained release (SR) formulations (C_{max}, IR, 4·52; SR, 5·83; AUC, IR, 5·04; SR, 7·75); variation associated with concentration and/or input rate related saturable first pass metabolism of (*S*)–verapamil.
Drug interactions		
	Warfarin	Most extensively investigated drug with respect to stereoselectivity in drug interactions. Some agents (e.g., metronidazole, cotrimoxazole, ticrynafen) are selective for the more active *S*-enantiomer, a reduction in clearance resulting in an enhanced effect; others are selective for the *R*-enantiomer (e.g., cimetidine, enoxacin, omeprazole) decrease clearance with no effect on activity; others either decrease the clearance of both enantiomers (e.g., amiodarone, miconazole) resulting in an enhanced effect, or increase the clearance of both (e.g., rifampicin, secobarbital) resulting in a reduced effect. Some agents (e.g., phenylbutazone, sulfinpyrazone) increase the clearance of the *R*-enantiomer, as a result of selective displacement from plasma protein-binding sites, but reduce clearance of (*S*)-warfarin by inhibition of metabolism resulting in an increased effect.
	Verapamil	Stereoselective reduction in clearance of the *S*-enantiomer following administration with cimetidine, resulting in an increased negative dromotropic effect on atrioventricular conduction.
	Hexobarbitone	Stereoselective increase in clearance following administration with rifampicin; *S*-enantiomer, sixfold increase in both young and elderly subjects; *R*-enantiomer, 89-fold increase in young subjects but only 19-fold increase in the elderly.
Enantiomeric interactions		
	Propafenone	Oral clearance of the *S*-enantiomer reduced following administration of the racemate compared to administration as a single enantiomer, due to inhibition of CYP2D6 mediated metabolism by (*R*)-propafenone. Plasma concentrations of the *S*-enantiomer following administration of the racemate are similar to those obtained following a double dose of the single enantiomer, whereas those of the *R*-enantiomer are unaffected. The *S*- but not the *R*-enantiomer causes β-blockade which is observed following administration of the racemate.
	Disopyramide	Following administration of the individual enantiomers there are no significant differences in pharmacokinetic parameters; on administration of the racemate the *S*-enantiomer has a lower total clearance, renal clearance, volume of distribution, and shorter half-life compared to (*R*)-disopyramide. The differences arise as a result of enantiomeric interactions in plasma protein binding which is also concentration dependent.

Table 5.6 Factors influencing the stereoselectivity of drug action and disposition in man — Continued

Factor	Drug	Comment
Aging		
	Warfarin	Significant positive correlation between unbound clearance of both enantiomers with age between 1 and 18 years; increased clearance of *S*-enantiomer greater than that of (*R*)-warfarin with age.
	Verapamil	Differential reduction in oral clearance in elderly (*R*-enantiomer, 0.53 L h^{-1} kg^{-1}; *S*-enantiomer, 4.8 L h^{-1} kg^{-1}) compared to young subjects (*R*-enantiomer, 1.3 L h^{-1} kg^{-1}; *S*-enantiomer, 22.5 L h^{-1} kg^{-1}); negative chronotropic and dromotropic effects observed in the elderly.
	Hexobarbitone	Stereoselective decrease in clearance with age; *S*-enantiomer twofold greater oral clearance in young (16.9 mL min^{-1} kg^{-1}) compared to elderly (8.2 mL min^{-1} kg^{-1}) volunteers; *R*-enantiomer, no age-related effect (oral clearance, young 1.9 mL min^{-1} kg^{-1}; elderly, 1.7 mL min^{-1} kg^{-1}).
Disease		
	Nimodipine	Bioavailability in patients with liver cirrhosis increased 3 to 4- and 17-fold for the *R*- and *S*-enantiomers, respectively compared to healthy volunteers due to stereoselective reduction in first pass metabolism.
	Ibuprofen	Plasma concentrations of (*S*)-ibuprofen lower than those of the *R*-enantiomer in patients with liver cirrhosis; ratio of area under the plasma concentration time curve (*S*/*R*) 0·94 in patients compared to 1.3 in healthy volunteers.
Gender		
	Metoprolol	Oral clearance and volume of distribution of both enantiomers lower in women than men (enantiomeric ratio *R*/*S*: clearance: male 1.5; female, 1.2; volume of distribution: male 1.5; female 1.2) resulting in significantly greater AUCs in women; concentration–effect relationships the same in both sexes, differences in observed effects due to gender specific pharmacokinetics.
	Mephobarbital	Oral clearance of *R*-enantiomer significantly lower in young women (45 L h^{-1}) compared to young men (170 L h^{-1}); *S*-enantiomer no significant difference (women, 1.14 L h^{-1}; men, 1.46 L h^{-1}).
Pharmacogenetics		
	Fluoxetine	Metabolism mediated by CYP2D6; oral clearance of both enantiomers similar in extensive metabolizers (EMs) at 40 and 36 L h^{-1} for the *R*- and *S*-enantiomers respectively; reduced to 17 and 3 L h^{-1} in poor metabolizers (PMs). Plasma concentrations of (*R*)- and (*S*)-fluoxetine increased ~2.5- and ~11-fold, respectively, in PMs compared to EMs.
	Hexobarbitone	Metabolism mediated by CYP2C19; enantiomeric ratio (*R*/*S*) in oral clearance 6 and 0.5 in EMs and PMs, respectively.
	Metoprolol	Metabolism mediated by CYP2D6; enantiomeric ratio (*S*/*R*) of the AUC decreases from 1·37 in EMs to 0·90 in PMs; the "total" plasma concentration–effect relationship shifts to the right in PMs compared to EMs.
	Mephenytoin	Metabolism mediated by CYP2C19; oral clearance of (*S*)-mephenytoin reduced from 4.7 L min^{-1} in EMs to 0.029 L min^{-1} in PMs; *R*-enantiomer clearance 0.03 and 0.02 L min^{-1} in EMs and PMs, respectively.

5.5 PHARMACODYNAMIC CONSIDERATIONS

As pointed out previously the greatest differences between enantiomers occur at the level of receptor interactions, and eudismic ratios of the order of 100- to 1000-fold, or even larger are not uncommon. However, both enantiomers may contribute to the observed activity of a racemate and a number of possible scenarios may arise on comparison of their pharmacodynamic properties as indicated below.

5.5.1 Pharmacodynamic Activity Resides in One Enantiomer the Other Being Biologically Inert

There are relatively few examples of drugs that possess one or two stereogenic centers as part of their structure in which the pharmacological activity is restricted to a single enantiomer the other being totally devoid of activity. In the case of α-methyldopa the antihypertensive activity resides exclusively in the *S*-enantiomer and this agent is marketed as a single isomer. For compounds with more than two stereogenic centers it is frequently found that the configurations of all such centers are fixed requirements or activity/specificity in action is either lost or considerably reduced. For example, the angiotensin converting enzyme inhibitor imidapril, containing three stereogenic centers, all of the *S*-configuration, is greater than a million fold more potent than its enantiomer and 10,000- fold more potent than its *SRS*- and *RSS*-diastereoisomers.

5.5.2 Both Enantiomers Have Similar Activities

Both enantiomers of the antihistamine promethazine (**5.88**) have similar pharmacological and toxicological properties, and the introduction of the chiral center in the dimethylaminoethyl side chain results in a 100% increase in antihistaminic potency compared to the nonchiral analog. Similarly, the enantiomers of flecainide (**5.89**) are equipotent with respect to their antiarrhythmic activity, effect on cardiac sodium channels and show no significant differences with respect to their pharmacokinetic properties. In the case of flecainide little information is available with respect to the toxicity of the individual isomers but the use of a single enantiomer would appear not to offer a therapeutic advantage.

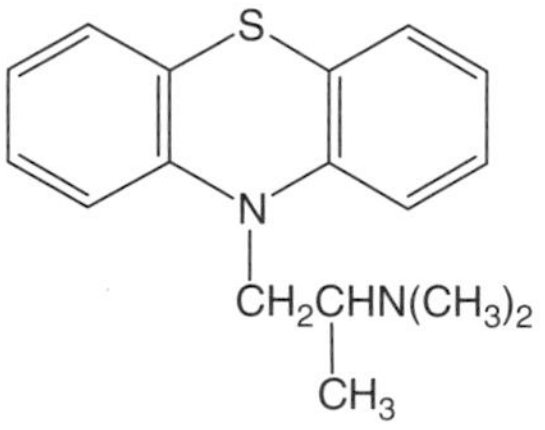

S
N
$CH_2CHN(CH_3)_2$
CH_3

Promethazine (**5.88**)

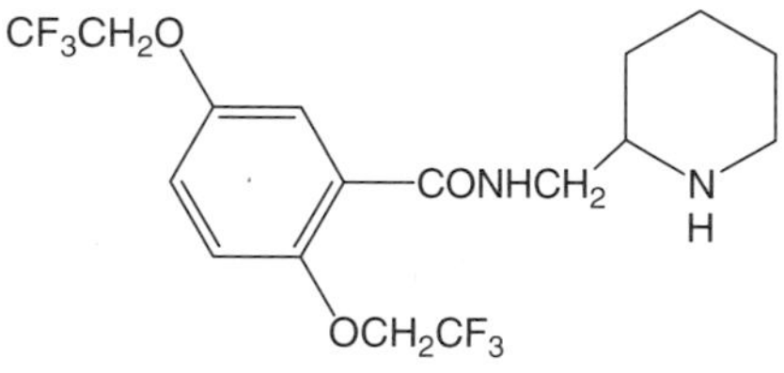

CF_3CH_2O
—CONHCH$_2$
N
H
OCH_2CF_3

Flecainide (**5.89**)

5.5.3 Both Enantiomers are marketed with Different Indications

In some instances the biological activities of a pair of enantiomers are so different that both are marketed with different therapeutic indications. Both enantiomers of propoxyphene (1-benzyl-3-dimethylamino-2-methyl-1-phenylpropyl propionate) are available, the dextrorotatory 1*S*,2*R*-enantiomer as the analgesic dextropropoxyphene (**5.90**) and levopropoxyphene (**5.91**), with the 1*R*,2*S*-configuration, as an antitussive. In the case of this example not only are the molecules mirror image related, but also are their trade names Darvon (dextropropoxyphene) and Novrad (levopropoxyphene). Similar differences in activity are found with related opiate derivatives, e.g., dextromethorphan, (+)-3-methoxy-*N*-methylmorphinan, is an antitussive agent, virtually free from analgesic, sedative, or other morphine-like effects, whereas the enantiomer, levomethorphan, is a potent opioid with antitussive activity and is addictive.

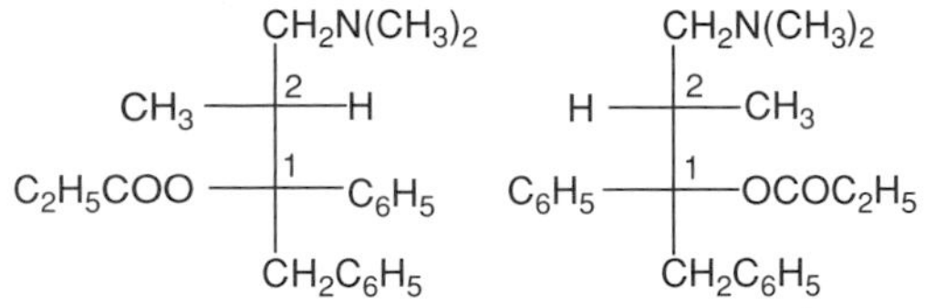

(+)-1*S*,2*R*-(**5.90**) (−)-1*R*,2*S*-(**5.91**)

5.5.4 Enantiomers Have Opposite Effects

Picenadol (**5.92**) is a phenylpiperidine analgesic, the racemate of which exhibits the properties of a partial agonist at the μ-opioid receptor. Examination of the properties of the individual enantiomers indicates that the analgesic activity resides in the (+)-3*S*,4*R*-stereoisomer, whereas the (−)-3*R*,4*S*-enantiomer is an antagonist, the partial agonist activity of the racemate arising due to the greater agonist potency of (+)-picenadol. A similar situation occurs with racemic sopromidine (**5.93**), the *R*-enantiomer being an agonist at H_2-receptors whereas (*S*)-sopromidine is an antagonist; the racemate exhibits the properties of a partial agonist on guinea pig atrium preparations.

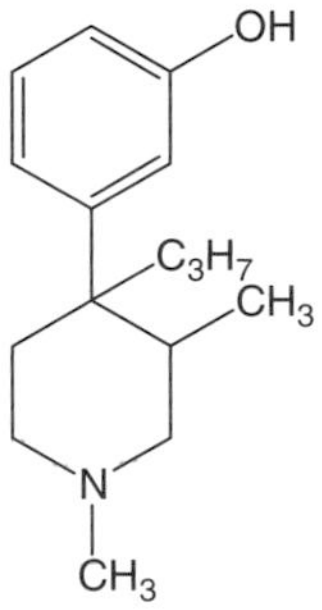

Picenadol (**5.92**)

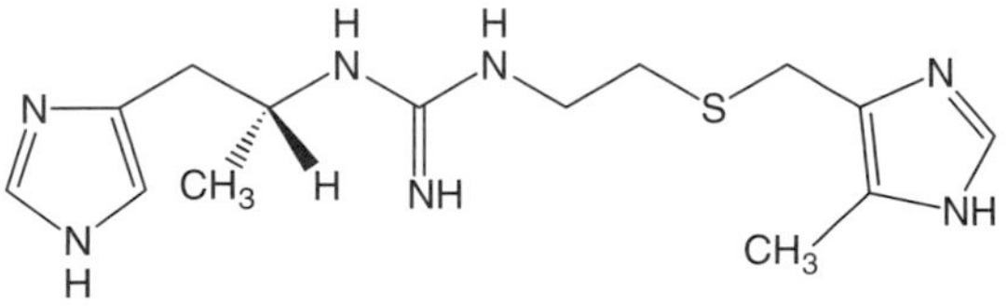

(*R*)-Sopromidine (**5.93**)

The pharmacological activities of the enantiomers of several derivatives of aporphine have been examined and in each case the *S*-enantiomers appear to be antagonists of their *R*-enantiomers. (*R*)-11-Hydroxy-10-methylaporphine (**5.94**) is a highly selective 5-HT_{1A} agonist, whereas its *S*-enantiomer is an antagonist at the same receptor. Similarly, (*R*)-11-hydroxyaporphine activates dopamine receptors and the *S*-enantiomer is an antagonist.

A more complex situation arises with 3-(3-hydroxyphenyl)-*N*-propylpiperidine (3-PPP; **5.95**). The initial pharmacological evaluation of this compound was carried out using the racemate and 3-PPP was described as a highly selective presynaptic dopaminergic agonist. Resolution and pharmacological evaluation of the individual enantiomers indicated that the situation was more complex. (*R*)-3-PPP acts as an agonist at both pre- and postsynaptic dopamine receptors, whereas (*S*)-3-PPP stimulates presynaptic and blocks postsynaptic receptors. The pharmacological profile observed with the racemate arises from the sum of the activities of the individual enantiomers.

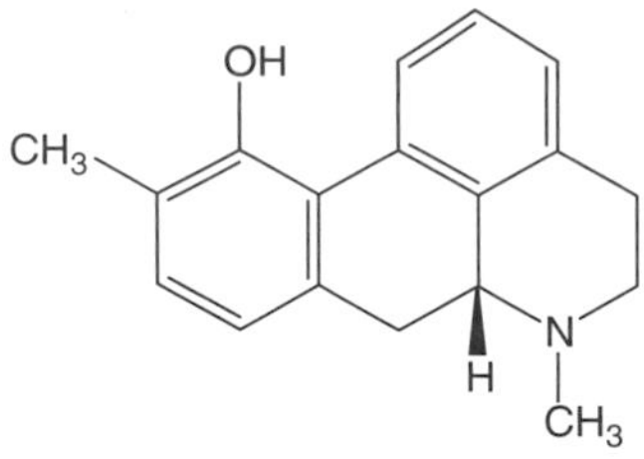

R-(**5.94**) 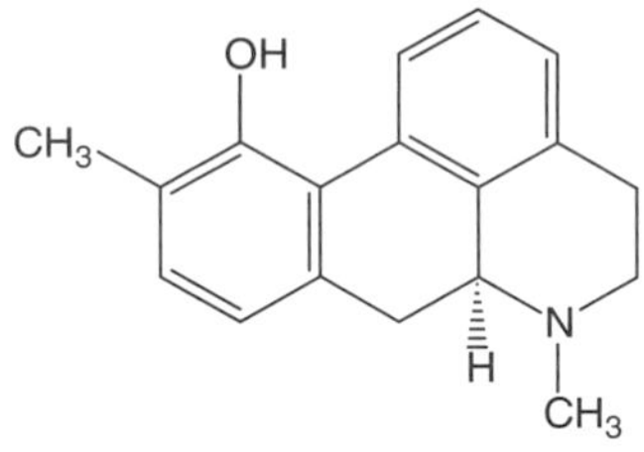 S-(**5.94**)

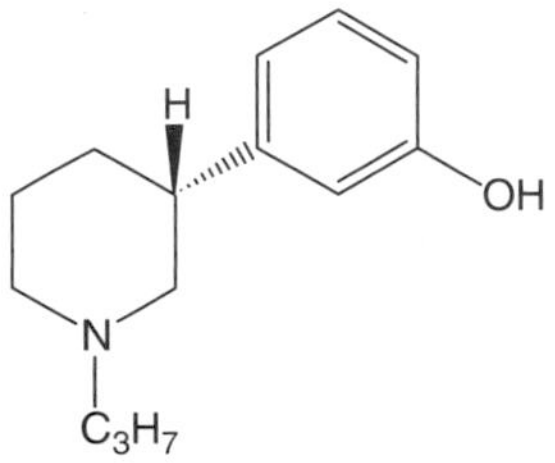

(R)-PPP (**5.95**) 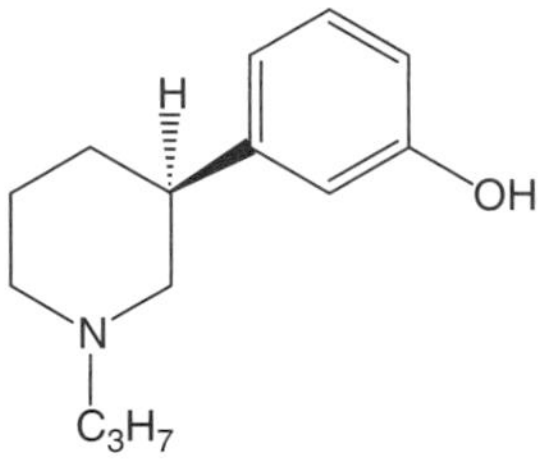 (S)-PPP (**5.95**)

The 1,4-dihydropyridines are calcium-channel blockers used for the treatment of angina and hypertension. A number of these agents possess a stereogenic center at the 4-position of the dihydropyridine ring system and examples are known in which the enantiomers have opposing actions on channel function, e.g., compounds (**5.96**), (**5.97**), and (**5.98**). The *S*-enantiomers act as potent activators, whereas the *R*-enantiomers are antagonists at L-type voltage-dependent calcium channels. It was thought that the observed effects of the enantiomers of these agents were due to interactions at different binding sites. However, it appears that the enantiomers interact with different channel *states*, open and inactivated, the drug-binding sites of which have opposite steric requirements. The situation is further complicated as the *S*-enantiomers of (**5.96**) and (**5.98**) are activators at polarized membrane potentials but become antagonists under depolarizing conditions. Indeed one author has described these agents as "molecular chameleons."

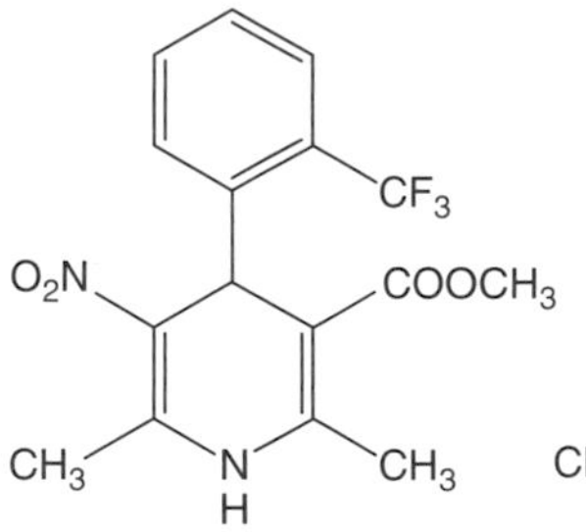

(**5.96**) 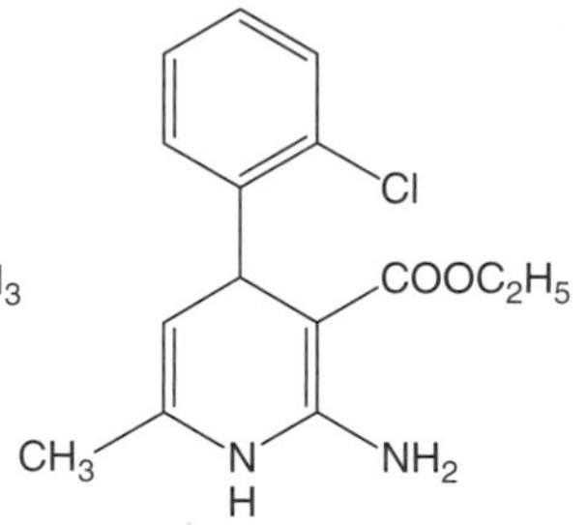 (**5.97**) 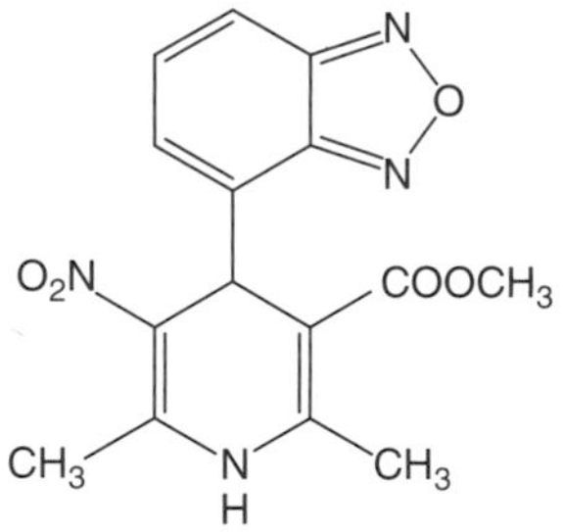 (**5.98**)

5.5.5 One Enantiomer May Antagonize the Side Effects of the Other

Indacrinone (**5.99**), a *m*-indanyloxyacetic acid derivative, is a loop diuretic with uricosuric activity, which has been evaluated for the treatment of hypertension and congestive heart failure. However, following administration of the racemate to humans serum urate levels increase. Resolution and pharmacological evaluation of the individual enantiomers indicates that the diuretic and natriuretic activity resides in the (−)-*R*-enantiomer and the uricosuric effects reside in (+)-(*S*)-indacrinone.

Following administration of the racemate to humans the plasma half-life of the *S*-enantiomer ($t_{1/2}$, 2–5 h) compared to that of (*R*)-indacrinone ($t_{1/2}$, 10–12 h) and uricosuric activity was found to be too short to prevent the increase in serum uric acid. Alteration of the enantiomeric composition of the drug from the 1:1 ratio of the racemate by increasing the proportion of the (+)-*S*-enantiomer resulted in a mixture (*S*:*R*: 4:1), which was isouricemic and further increases (*S*:*R*:8:1) resulted in a mixture which caused hypouricemia. Thus, evaluation of the pharmacodynamic and pharmacokinetic properties of the individual enantiomers of the drug and subsequent manipulation of the enantiomeric composition resulted in an agent with potential for an improved therapeutic profile. The development of indacrinone was stopped in the mid-1980s, but the concept of variation in enantiomeric composition for an improved therapeutic profile was established.

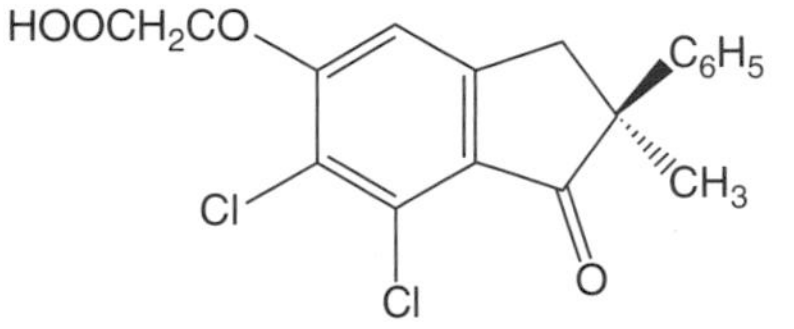

HOOCH$_2$CO, Cl, Cl, O, C_6H_5, CH_3

(*S*)-Indacrinone (**5.99**)

HOOCH$_2$CO, Cl, Cl, O, CH_3, C_6H_5

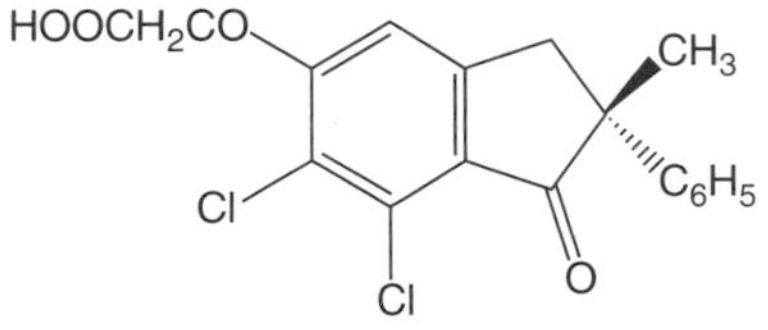

(*R*)-Indacrinone (**5.99**)

5.5.6 The Required Activity Resides in One or Both Enantiomers but the Adverse Effects are Predominantly Associated with One Enantiomer

Ketamine (**5.100**) is a general anesthetic agent with analgesic properties. The use of the drug is complicated by postanesthesia "emergence reactions," including hallucinations, vivid dreams, and agitation; the drug is also subject to abuse. Stereoselectivity in the pharmacological activity of ketamine has been known since the 1970s when the greater analgesic and hypnotic activity, and reduced locomotor activity, of the *S*-enantiomer was observed following administration to animals.

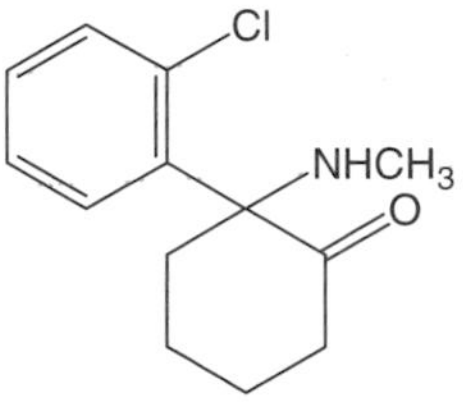

Cl, NHCH$_3$, O

Ketamine (**5.100**)

Ketamine interacts with multiple-binding sites including *N*-methyl D-aspartate (NMDA) receptors and non-NMDA glutamate receptors, nicotinic and muscarinic cholinergic, monoaminergic, and opioid receptors. The NMDA receptor, considered to be the main site of action, shows stereoselectivity with (*S*)-ketamine having a threefold greater affinity for the phencyclidine-binding site compared to the *R*-enantiomer. Similarly, enantiomeric binding to μ- and κ-opioid receptors shows a two- to fourfold selectivity for the *S*-enantiomer, but with a 10- to 20-fold reduction in affinity compared to the NMDA receptor.

Studies in surgical patients, following administration of equianesthetic doses of the enantiomers of ketamine, indicate a reduced dose requirement of (*S*)-compared to (*R*)-ketamine, associated with a potency ratio (*S*/*R*) of 3.4. The *S*-enantiomer was reported to produce more effective anesthesia, less emergence reactions, and agitated behavior than either (*R*)-ketamine or the racemate. The drug has recently undergone the chiral switch process (see Section 5.8.1), with the single *S*-enantiomer marketed in Germany. The potential advantages of the single enantiomer are a reduction in dose, more rapid recovery, and fewer psychotomimetic emergence reactions compared to the racemate.

The antitubercular drug ethambutol (**5.101**) contains two identically substituted stereogenic centers in its structure and exists in three stereoisomeric forms, the enantiomeric pair (+)-(*S,S*)- and (−)-(*R,R*)-ethambutol, together with the optically inactive diastereomeric *meso* form. The activity of the drug resides in the (+)-enantiomer which is 500- and 12-fold more potent than (−)-ethambutol and the *meso* form, respectively. The drug was initially introduced for clinical use as the racemate but was rapidly changed to the (+)-enantiomer, as a result of ocular neuropathy. The toxicity is related to both dose and treatment duration, and in the majority of cases is reversible on termination of therapy. All three stereoisomeric forms appear to be equipotent with respect to the adverse effect; thus the use of the single enantiomer provided a much improved risk–benefit ratio.

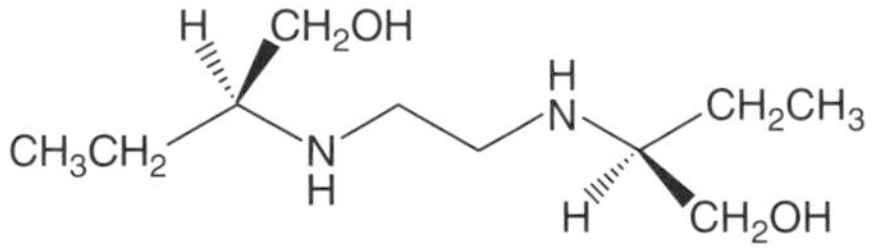

(*S,S*)-Ethambutol (**5.101**)

The β_2-selective adrenoceptor agonist salbutamol (**5.102**), also known as albuterol in the USA, is the most widely used bronchodilator for the treatment of asthma. Regular use of the drug is associated with some loss of bronchodilator potency, decreased protection against bronchoprovacation, and increased sensitivity to allergen challenge and also to some bronchoconstrictor stimuli. Studies in animals have indicated that the drug also produces bronchial hyperresponsiveness, which may also be induced by (*S*)-salbutamol.

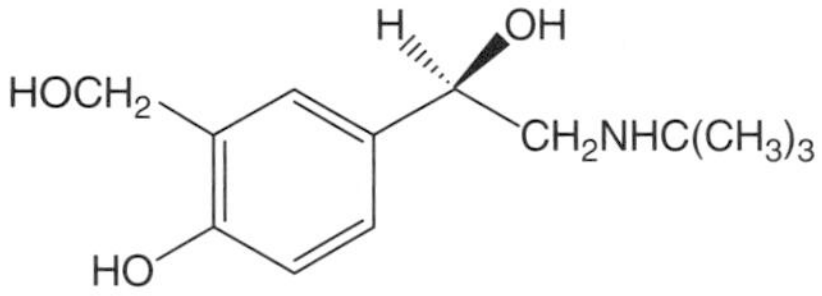

(*R*)-Salbutamol (**5.102**)

The drug is used as the racemate and initial studies indicated that the *R*-enantiomer was 68-fold more active than (*S*)-salbutamol as a β_2-agonist. More recent investigations, using transfected cells expressing human adrenoceptors, have shown the 90- to 100-fold greater binding affinity of the *R*-enantiomer for either β_1- or β_2-receptors, in comparison to (*S*)-salbutamol. In addition, (*R*)-salbutamol causes an increase in intracellular cAMP and intrinsic activity identical to that produced from twice the concentration of the racemate, and inhibits activation of mast cells and eosinophils. In contrast, the *S*-enantiomer intensifies bronchoconstrictor responses of sensitized animals, induces hypersensitivity of asthmatic airways, and promotes the activation of human eosinophils *in vitro*. Studies in healthy volunteers have indicated that single doses of (*R*)-salbutamol result in prolonged protection in the methacholine-induced bronchoconstrictor challenge test compared to the racemate, whereas the *S*-enantiomer significantly increased sensitivity to methacholine.

The drug also exhibits stereoselectivity in disposition (Table 5.5), the oral bioavailability of the ''inactive'' *S*-enantiomer ranging between 2.4- and sevenfold greater than that of (*R*)-salbutamol as a result of presystemic metabolism. In contrast, following inhalation both enantiomers are absorbed to a similar extent (~20%). The systemic clearance of the *R*-enantiomer is also greater (1.6- to 3-fold) than that of (*S*)-salbutamol resulting in a significantly longer plasma half-life and therefore greater exposure to the less active enantiomer.

As a result of the problems associated with the use of salbutamol, the single *R*-enantiomer has undergone clinical evaluation and studies in both children and adults have demonstrated that inhalation of the single enantiomer produced significantly greater bronchodilation than the racemate. The single enantiomer, levalbuterol ((*R*)-salbutamol) has been marketed in the USA, with potential for an improvement in maintenance therapy. Similarly, other β_2-adrenoceptor agonists, e.g., (*R*,*R*)-formoterol, are presently undergoing evaluation as potential single enantiomer products.

5.5.7 A Racemic Mixture Provides a Superior Therapeutic Response than Either Individual Enantiomer

Dobutamine (**5.103**) is a racemic inotropic sympathomimetic agent reported to increase the force of myocardial contraction without increasing either heart rate or blood pressure. Examination of the pharmacodynamic properties of the individual enantiomers indicates that they are both active but at different receptors. (+)-Dobutamine acts as a relatively potent agonist at both β_1- and β_2-adrenoceptors and has weak α-antagonist properties, in contrast the (−)-enantiomer is a potent α_1-adrenoceptor agonist. Thus both enantiomers contribute to the positive inotropic effects of the drug and the peripheral vasoconstrictor and vasodilator effects of the (−)- and (+)-enantiomers, respectively, essentially cancel each other out. In the case of dobutamine it would appear that the use of the racemate has advantages over either individual enantiomer.

HO HO H N CH$_3$ OH

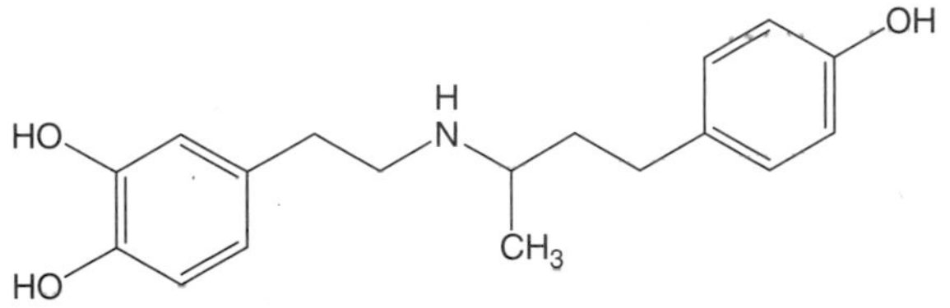

Dobutamine (**5.103**)

5.6 SELECTED THERAPEUTIC GROUPS

As pointed out in Section 5.1, approximately 25% of drugs are marketed as racemates and therefore the issues associated with drug chirality are not restricted to particular groups of compounds but extend across all therapeutic areas. Stereoselectivity of drug action *in vivo* may arise as the net result of both pharmacodynamic and pharmacokinetic processes, the relative significance of which may be difficult to discern. Some of the factors influencing disposition and action in relation to drug stereochemistry are summarized in Table 5.6. In this section the stereochemistry of some selected therapeutic agents will be examined in an attempt to illustrate some of the complexities which may arise. This is not intended to be an exhaustive compilation but merely to serve as an indication of the potential advantages of stereochemical considerations in pharmacology. Some alternative therapeutic groups, namely antiarrhythmic agents, anticoagulants, H_1-antihistamines, and antimicrobial agents, have been examined in the previous edition of this book.

5.6.1 β-Adrenoceptor Antagonists

The β-adrenoceptor antagonists may be divided into two chemical groups, the arylethanolamine and aryloxypropanolamine derivatives. These agents show a high degree of stereoselectivity with respect to their action at β-receptors with the pharmacodynamic activity residing in the enantiomers of the *R*-configuration of the arylethanolamine series and the *S*-enantiomers of the aryloxypropanolamine group. Examination of the general structures of the active enantiomers of the two series, (**5.39**) and (**5.40**), indicates that the three-dimensional spatial arrangements of the functionalities of the active enantiomers are identical in spite of their opposite configurational designations.

The stereoselectivity exhibited by these agents may vary markedly, the eudismic ratio for the binding affinity of atenolol enantiomers to the β-receptor is as low as 10 whereas that of pindolol is 1000. Differences in eudismic ratio between β-receptor subtypes have also been observed which indicate that β_1-receptors are more sterically demanding than β_2-receptors, i.e., higher endismic ratios are observed at β_1-receptors than at the β_2-subtype. This should not be surprising as there are known to be structural differences between the receptor subtypes. Examination of QSARs has indicated that the differences in eudismic index between the two receptor subtypes are associated with the higher affinity of the distomers for the β_2-receptor compared with the β_1-subtype. An additional physicochemical property of significance in determining the binding affinity of these agents is their lipophilicity. The addition of the lipophilicity parameter ($\log P$) to the QSAR correlation equations indicated that hydrophobic parameters are of greater significance for drug binding to the β_2-compared to the β_1-receptor, particularly for the distomers for which the binding ''fit'' would obviously not be expected to be as good as for the eutomers. Thus the stereochemical differences observed between the receptor subtypes may arise as a result of additional hydrophobic interactions of the distomers to the β_2-receptor, which the eutomers are unable to participate in.

For those β-antagonists which exhibit additional pharmacodynamic properties, e.g., the membrane-stabilizing effects of propranolol, the class III antiarrhythmic properties of sotalol, and the vasodilator effects of carvedilol, the individual enantiomers are frequently, but not always equipotent.

Of the β-antagonists currently available three timolol (**5.104**), penbutolol (**5.105**), and levobunolol (**5.106**), are marketed as single isomers and belonging to the aryloxypropanolamine series, are available as the *S*-enantiomers. The remainder are marketed as racemates and in the case of one compound, labetalol, as a mixture of four stereoisomeric forms. In terms of their use in the treatment of hypertension and angina there appears to be relatively little advantage in using single enantiomers particularly as the majority of the adverse effects are related to their pharmacological action and therefore a significant reduction in side effects is unlikely. There are, however, a number of reasons why the stereochemistry of the β-antagonists should not be neglected as indicated in the examples cited below.

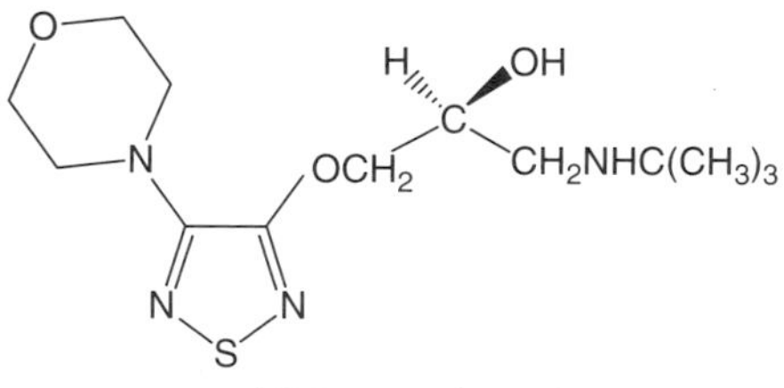

(*S*)-Timolol (**5.104**)

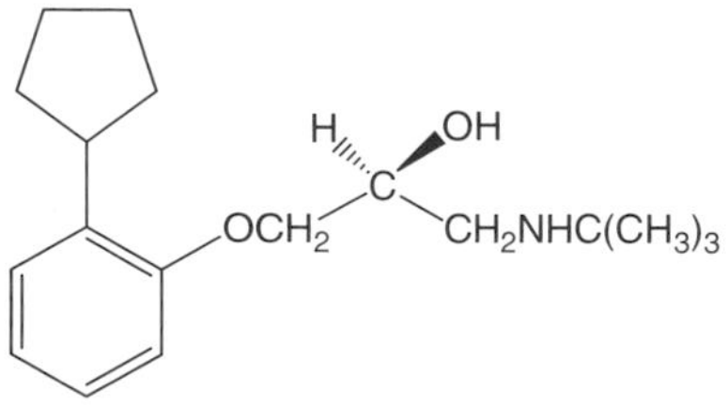

(*S*)-Penbutolol (**5.105**)

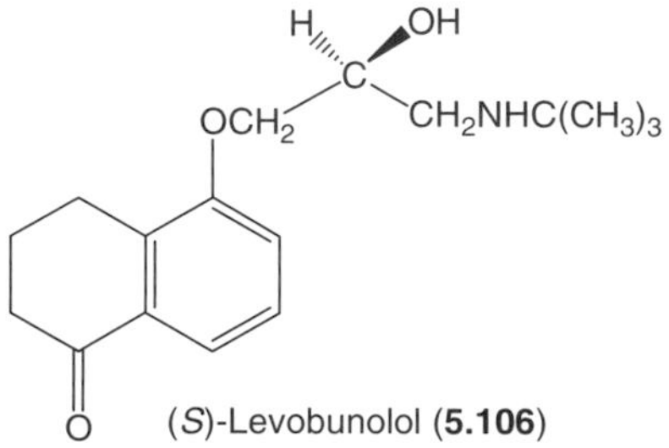

(*S*)-Levobunolol (**5.106**)

Propranolol (**5.107**) is a lipophilic nonselective β-antagonist used as the racemate, the *S*-enantiomer being between 40- and 100-fold more potent than the *R*-enantiomer. The enantiomers show no differences in activity with respect to their membrane-stabilizing properties. (*R*)-Propranolol also inhibits the conversion of thyroxine to triiodothyronine, the inhibition is highly stereoselective, if not stereospecific. Following administration of the racemic drug to man the mean plasma concentrations of thyrotropin and total thyroxine were found to significantly increase, whereas the ratio of triiodothyronine/total thyroxine decreased. Administration of half the dose of the single *S*-enantiomer resulted in no alteration in any of the above parameters. As a result of these observations, it has been suggested that (*R*)-propranolol could be used in the treatment of hyperthyroidism; the perceived benefits are a reduction in dose without β-blockade, particularly for patients with impaired cardiac function.

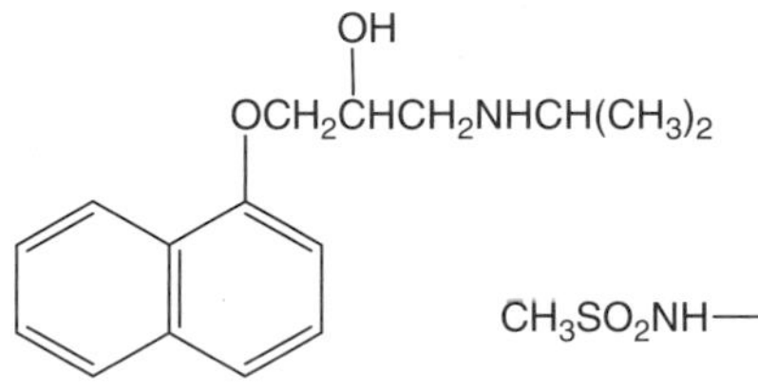

OH
$OCH_2CHCH_2NHCH(CH_3)_2$

OH
CH_3SO_2NH—$CHCH_2NHCH(CH_3)_2$

Propranolol (**5.107**) 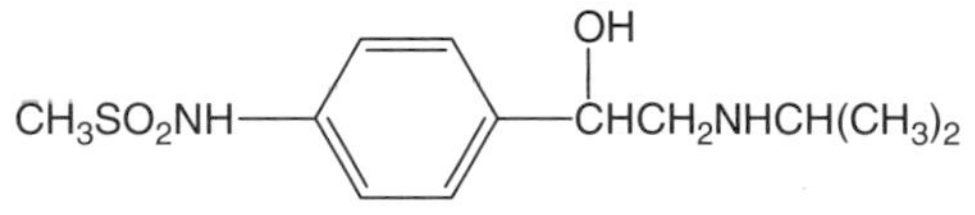 Sotalol (**5.108**)

As pointed out above timolol (**5.104**) is available as the single *S*-enantiomer. In addition to its use in the treatment of hypertension and angina timolol is also used for the treatment of wide-angle glaucoma. Following administration to the eye significant amounts of the drug are systemically absorbed and cardiovascular and pulmonary side effects have been reported. This systemic absorption is of particular significance for the use of the drug in patients for whom β-antagonists are contraindicated, e.g., those with respiratory disease states, and a number of deaths have been reported following the use of timolol eye drops in asthmatic patients.

Using pharmacological test systems for the evaluation of β-blockade (*S*)-timolol shows marked stereoselectivity in action with eudismic ratios (*S*/*R*) of between 50 and 90, depending on the test system used. These large differences in activity reduce to ca. threefold when the ocular properties of the drug are examined, e.g., reduction in aqueous humor recovery rate and inhibition of dihydroalprenolol binding in the iris-ciliary body. (*R*)-Timolol has also been shown to reduce intraocular pressure in patients with glaucoma with fewer systemic effects than (*S*)-timolol. In addition, the *R*-enantiomer has been reported to increase retinal/choroidal blood flow, whereas the (*S*)-timolol decreases it, an unrequired effect. As a result of these differences in ocular activity it has been suggested that timolol possibly represents a drug where both enantiomers could be marketed for specific therapeutic indications. However, examination of the systemic effects of both enantiomers, following ocular administration to mild asthmatics, resulted in similar reductions in respiratory function with an enantiomeric ratio similar to that reported for reduction in intraocular pressure. It is therefore unlikely that (*R*)-timolol would offer a significant improvement in risk-benefit ratio in the ''at risk'' population group.

Sotalol (**5.108**), an arylethanolamine derivative, is a nonselective β-antagonist with class III antiarrhythmic activity. The drug is used as the racemate, the β-antagonist activity residing in the (−)-enantiomer, which is 14- to 50-fold more potent than (+)-sotalol, whereas the enantiomers are equipotent with respect to their antiarrhythmic properties. (+)-Sotalol therefore provides an antiarrhythmic agent without β-blockade, and dexsotalol was evaluated in patients with reduced left ventricular function following myocardial infarction in the Survival With Oral d-Sotalol (SWORD) trial. The investigation was terminated prematurely, following recruitment of approximately half the intended number of patients, as a result of increased mortality in the treatment compared to the control group. It has been suggested that the combination of both β-antagonism and class III activity present in the racemate provides a more effective combination than antiarrhythmic activity alone, and dexsotalol represents an example where the supposed advantages of a single enantiomer drug have not been realized.

Carvedilol (**5.109**) is a racemic nonselective β-antagonist with vasodilator properties mediated via α_1-adrenoceptor antagonism. The β-antagonist activity resides in the (−)-*S*-enantiomer, whereas both enantiomers are essentially equipotent with respect to α_1-antagonism. Thus in the case of carvedilol both enantiomers contribute to the antihypertensive activity of the drug.

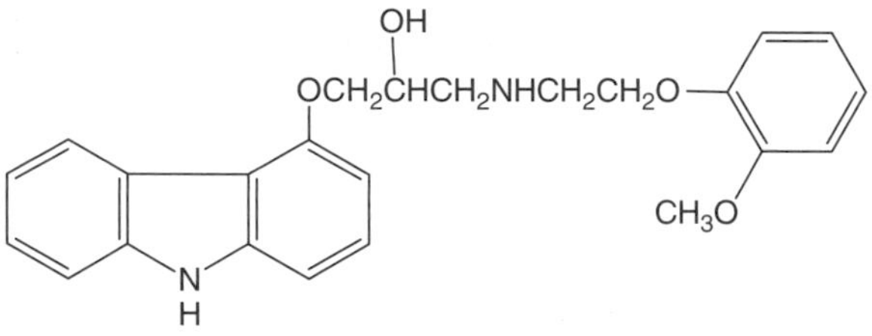

Carvedilol (**5.109**)

Nebivolol (**5.110**) has four stereogenic centers in its structure and, as a result of its symmetrical nature, ten possible stereoisomeric forms are possible, four enantiomeric pairs, and two *meso* forms. The drug is marketed as the (+)-*S*,*R*,*R*,*R*- and (−)-*R*,*S*,*S*,*S*- racemate for the treatment of hypertension. The β_1-antagonist activity resides in (+)-nebivolol, the (−)-enantiomer is essentially devoid of activity; the drug also has vasodilator properties with the (−)-enantiomer being more potent than (+)-nebivolol. This latter activity is thought to be mediated via the endothelial L-arginine/NO mechanism. (+)-Nebivolol has a depressant effect on left ventricular function, which is in part counterbalanced by the action of the (−)-enantiomer, resulting in a racemate with a beneficial cardiovascular activity profile.

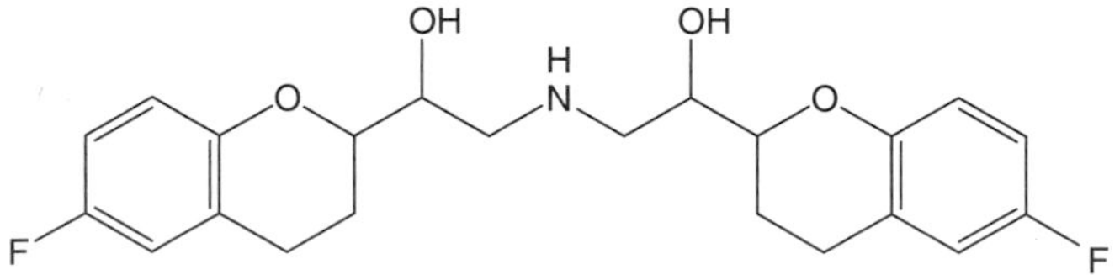

Nebivolol (**5.110**)

Labetalol (**5.111**, Table 5.7), an arylethanolamine derivative, is a dual action drug with combined α- and β-antagonist activity. Labetalol contains two stereogenic centers in its structure and is marketed as an equal parts mixture of four stereoisomers. Examination of the pharmacodynamic activity (pA_2 values) of the four stereoisomers (Table 5.7) indicates that the α_1- and β-antagonist activity reside in the *S*,*R*- and *R*,*R*-stereoisomers, respectively, with the remaining pair being essentially inactive. Labetalol is certainly not ''one drug'' with two actions. The *R*,*R*-stereoisomer, named dilevalol, was evaluated as a potential single isomer β-antagonist. However, development of the drug was terminated due to adverse effects associated with hepatotoxicity in a small number of patients. This toxicity appears to be of minor significance with respect to labetalol and the reason why the single isomer should show increased toxicity is unknown. Labetalol/dilevalol represents an example where drug development was stopped as a result of an unexpected adverse reaction, indicating that removal of isomeric ''impurities'' may not be a trivial matter.

Table 5.7 Pharmacological activity of the stereoisomers of labetalol (5.111)

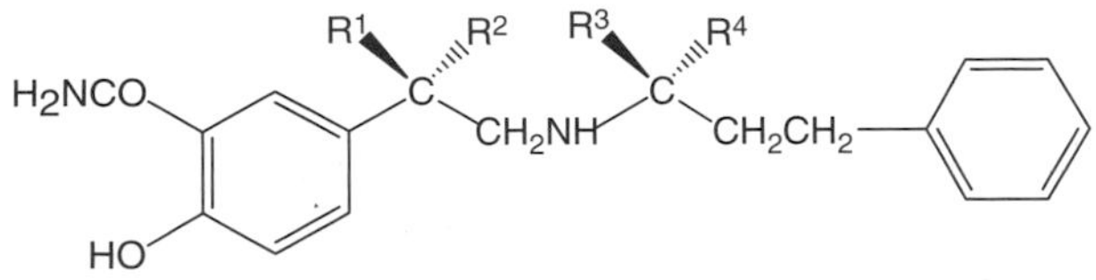

	R^1	R^2	R^3	R^4	Activity (pA2 values) α_1	β_1	β_2
R,R-	HO	H	H	CH_3	5.87	8.26	8.52
S,S -	H	OH	CH_3	H	5.98	6.43	<6.0
R,S -	HO	H	CH_3	H	5.5	6.97	6.33
S,R -	HO	OH	H	CH_3	7.2	6.37	<6.0

5.6.2 Nonsteroidal Anti-inflammatory Drugs

2-Arylpropionic acids

The 2-arylpropionic acids (2-APAs) are an important group of NSAIDs. The main pharmacodynamic action of these agents, inhibition of cyclooxygenase (COX), resides in the enantiomers of the *S*-configuration (**5.37**), the *R*-enantiomers are either inactive, or only weakly active, in *in vitro* test systems (Tables 5.8 and 5.9). The majority of these drugs, with the exception of naproxen (**5.112**) and flunoxaprofen (**5.113**), were introduced as racemic mixtures. However, the single *S*-enantiomers dexketoprofen (**5.114**) and dexibuprofen (**5.2**) have been marketed; the racemates having undergone the chiral switch process (see Section 5.8.1).

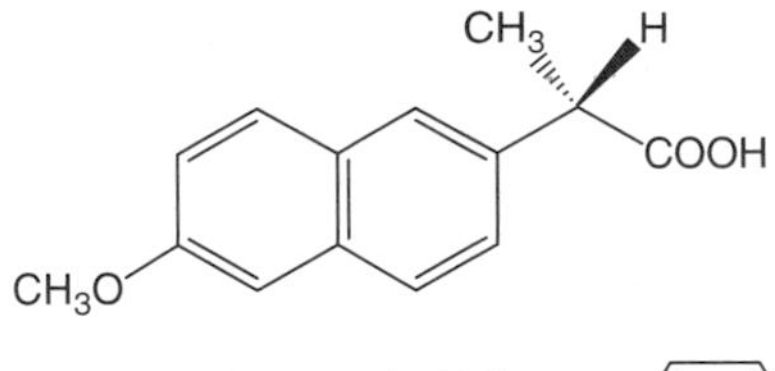

Naproxen (**5.112**)

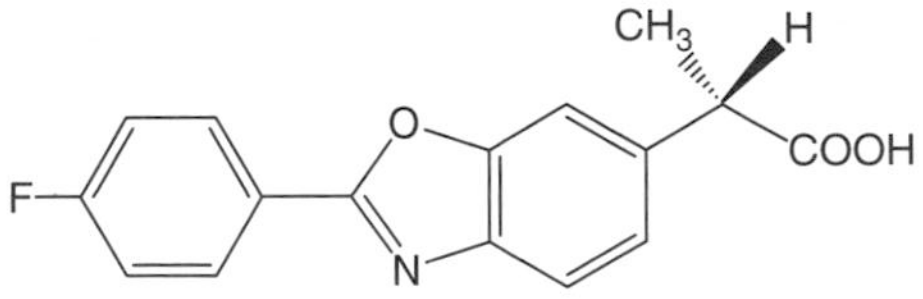

Flunoxaprofen (**5.113**)

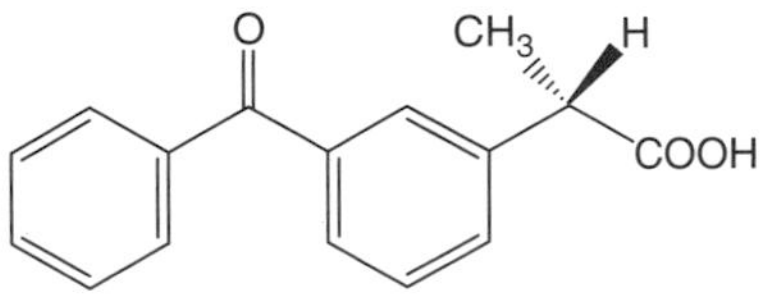

(+)-(*S*)-Ketoprofen (**5.114**)

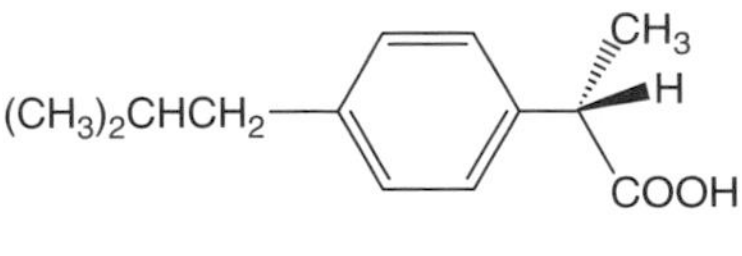

(+)-(*S*)-Ibuprofen (**5.2**)

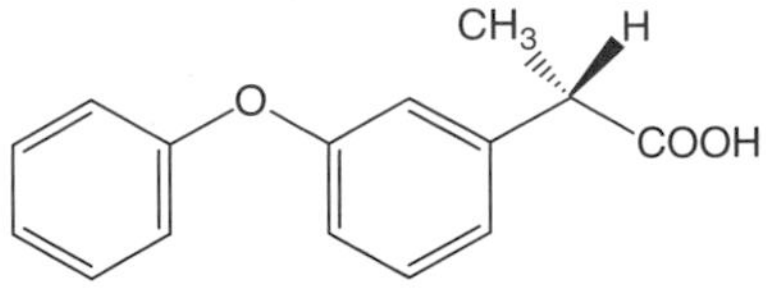

(*S*)-Fenoprofen (**5.115**)

The large differences observed in the *in vitro* activity of the enantiomers of these agents decrease markedly in *in vivo* test systems and in some cases, e.g., fenoprofen (**5.115**) and ibuprofen, the enantiomers appear to be essentially equipotent (Table 5.8). This difference in enantiomeric activity *in vivo* and *in vitro* is in part due to the metabolic chiral inversion of the inactive *R*- to the active *S*-enantiomers *in vivo*. The observed activity of these agents, when administered as racemates, are therefore closely linked with their stereoselective metabolism.

The mechanism of the chiral inversion involves the stereoselective formation of an acyl-coenzyme A (CoA) thioester of the *R*-2-APAs. In the case of ibuprofen, the most extensively investigated compound of the group, the enzyme mediating this reaction is thought to be long-chain acyl-CoA synthetase involved in the metabolism of endogenous fatty acids. Once formed the (*R*)-2-arylpropionyl-CoA thioester (**5.116**) undergoes epimerization of the 2-arylpropionyl moiety to yield a mixture of both possible epimeric 2-arylpropionyl-CoA derivatives, which may undergo hydrolysis to liberate both enantiomers of the 2-arylpropionate. The enzyme mediating the epimerization step is thought to be α-methylacyl-CoA racemase, which is involved in the metabolism of a number of α-methyl-substituted fatty acids and bile acid intermediates. The enzymes involved in the hydrolytic step have not been identified yet.

H CH$_3$ Ar COOH *R*-(**5.37**) ⇌ (AcylCoA synthetase / Thioesterase) H CH$_3$ Ar COSCoA *R*-(**5.116**)

⇅ "Epimerase"

CH$_3$ H Ar COOH *S*-(**5.37**) ← (Thioesterase) CH$_3$ H Ar COSCoA *S*-(**5.116**)

Alternative pathways to hydrolysis of the 2-arylpropionyl-CoA thioesters are acyl transfer of the 2-arylpropionyl moiety to an amino acid, resulting in the formation of an amino acid conjugate in some species, or to glycerol resulting in the formation of a ''hybrid'' triglyceride (**5.117**; R = long-chain fatty acid alkyl groups). The formation of the ''hybrid'' triglyceride results in the accumulation of drug residues in adipose tissue. The stereochemistry of the 2-arylpropionic acid moiety found in adipose tissue has been investigated in the rat following administration of both the individual enantiomers and racemic ibuprofen. Administration of both racemic and (*R*)-ibuprofen resulted in the incorporation of the drug into triglycerides, the enantiomeric composition of the material in both cases is $R > S$. Total drug lipid levels, following administration of equal doses, were approximately twice as high following the administration of (*R*)-ibuprofen compared to the racemate, whereas following the administration of (*S*)-ibuprofen only trace quantities of the drug could be detected. *In vitro* studies using (*R*)- and (*S*)-fenoprofen (**5.115**) as substrates and rat hepatocyte and adipocyte preparations, indicated stereoselective incorporation of (*R*)-fenoprofen into triglycerides and also that *R*- but not (*S*)-fenoprofen inhibited endogenous triglyceride synthesis *in vitro*.

CH_2O-COR
|
CHO-COR
|
CH_2O-COCH-Ar
|
CH_3

(**5.117**)

Table 5.8 Relative activity of the enantiomers of 2-arylpropionic acid NSAIDs in *in vitro* and *in vivo* test systems

	In vitro		*In vivo*	
Compound	**Ratio *S*/*R***	**Test**	**Ratio *S*/*R***	**Test**
Carprofen	>16	IPGS	14	Acute adjuvant induced arthritis
	>24	IPA		
Fenoprofen	35	IPA	1	Carrageenin paw edema; UVE
Flurbiprofen	200	IPA	2–16	Guinea pig anaphylaxis
	880	Antagonism of SRS-A		
Ibuprofen	160	IPGS	1.4	Toxin-induced writhing; pain threshold
			1.1	UVE
Indoprofen	100	IPGS	20	Carrageenin paw edema
			31	Granuloma pouch
			25	Toxin-induced writhing
Naproxen	130	IPGS	28	Carrageenin paw edema
	70	IPGS	15	Antipyretic activity
Pirprofen	6.4	IPGS		

IPGS, inhibition of prostaglandin synthesis; IPA, inhibition of platelet aggregation; UVE, ultraviolet induced erythema.

The above investigations indicate that both the (*R*)- and (*S*)-2-arylpropionyl moiety may be transferred from the acyl-CoA thioesters to glycerol but that the incorporation of drug depends upon the presence of the *R*-enantiomer, as would be expected from the stereoselectivity of acyl-CoA thioester formation. The toxicological significance of drug incorporation into lipid is not known, but the formation of hybrid triglycerides may result in the accumulation of these agents and possible toxicity due to their effects on normal lipid metabolism and membrane function.

Chiral inversion has been reported for a number of the 2-arylpropionic acids in both animals and humans, the rate and extent of the reaction appears to be both substrate and species dependent, for example (*R*)-flurbiprofen undergoes inversion in the dog, guinea pig, and mouse but not in rat or humans. Human pharmacokinetic studies have shown that the *R*-enantiomers of ibuprofen, fenoprofen, and benoxaprofen undergo significant inversion, whereas the reaction either does not occur, or is of minor significance, for indoprofen, flurbiprofen, ketoprofen, and carprofen. In addition to chiral inversion a number of these agents show stereoselectivity in plasma protein binding (e.g., the fraction unbound of the enantiomers of ibuprofen is $S > R$) and in other routes of metabolism, e.g., glucuronidation and oxidation. In the majority of cases following administration of the racemic drug to humans the plasma concentrations and areas under the plasma concentration versus time curves (AUC), of the active *S*-2-APAs exceed those of the *R*-enantiomers (e.g., benoxaprofen, carprofen, fenoprofen, flurbiprofen, ibuprofen, and indoprofen), but others, e.g., ketoprofen and tiaprofenic acid, show similar plasma concentrations and AUCs. The dispositional properties of these agents are also complicated by enantiomer–enantiomer interactions. For example, both enantiomers of ibuprofen show concentration-dependent plasma protein binding and compete for binding sites, which may explain the differences observed in their pharmacokinetic parameters when administered as single enantiomers or as the racemic mixture.

For those 2-arylpropionates marketed as racemic mixtures and for which chiral inversion is a significant route of metabolism, the effective dose of the active agent is unknown. In the case of these agents the *R*-enantiomers act essentially as prodrugs for the active *S*-isomers. The extent of the inversion reaction would also be expected to vary within the population, possibly with disease state and thus any attempt to relate plasma concentrations to clinical effect must take the stereochemistry of the circulating drug into account. The use of the single *S*-enantiomers of these agents offers a number of advantages: accurate dosing, simplification of pharmacokinetics and concentration–effect relationships, together with the avoidance of potential problems due to hybrid triglyceride formation and inhibition of fatty acid metabolism.

Evaluation of the pharmacological properties of the enantiomers of the 2-APAs, particularly for those agents which do not undergo chiral inversion, has indicated additional actions which are of potential clinical relevance. The antinociceptive activity of flurbiprofen is associated with both enantiomers, the activity of (*R*)-flurbiprofen being within the central nervous system (CNS). (*R*)-Flurbiprofen is also reported to be slightly more potent than the *S*-enantiomer in the inhibition of nuclear factor κB. Flurbiprofen, in addition to other NSAIDs, has also been shown to possess antiproliferative activity. Following investigations in animal models of prostate and colon cancer, in which the drug was shown to reduce both tumor formation and progression, (*R*)-flurbiprofen has undergone clinical evaluation for the treatment of late-stage prostate cancer. In addition, recent data have indicated the potential of NSAIDs for the prevention of Alzheimer's disease, which is thought to be by suppression of β-amyloid (Aβ) peptide accumulation. *In vitro* studies have indicated that the *R*-enantiomers of both ibuprofen and flurbiprofen inhibit the production of Aβ42, with (*R*)-flurbiprofen showing the greater potency. As (*R*)-flurbiprofen is essentially inactive with respect to inhibition of COX, it has been suggested that this enantiomer may be a useful candidate for clinical development for the treatment and prevention of Alzheimer's disease, without gastro-intestinal side effects, and the drug is reported to be undergoing clinical evaluation.

Thus, in the case of flurbiprofen, examination of the biological activity of the individual enantiomers has the potential to provide ''new'' indications for an ''old'' drug.

As pointed out above two of these agents have undergone the chiral switch process and are available as their single *S*-enantiomers, dexketoprofen and dexibuprofen. The latter agent is available in Austria and Switzerland. Dexketoprofen has been formulated as the trometamol salt, resulting in more rapid absorption and reduced potential for gastric ulceration in comparison to the racemic free acid, together with a reduction in dose requirement. Similarly, the use of dexibuprofen has resulted in dose reduction, a number of clinical studies indicating 1200 mg of the single enantiomer to be equivalent to double the dose of the racemate per day. However, other studies in patients with rheumatoid arthritis have indicated a dose reduction of about one third in comparison to the racemate, which would as a result of chiral inversion and metabolic activation seem more realistic.

Other chiral NSAIDs

Chiral NSAIDs are also found in other chemical groups and the situation with the chiral sulfoxide containing prodrug sulindac (**5.83**) has been outlined in Section 5.4.3. Two additional agents worthy of note are etodolac (**5.118**) and ketorolac (**5.119**); inhibition of prostaglandin synthesis is shown to reside in the *S*-enantiomers of both compounds. In the case of etodolac, a selective COX 2 inhibitor, *in vivo* studies in rats with adjuvant-induced polyarthritis indicated that the (+)-*S*-enantiomer was 2.6-fold more potent than the racemate. The (−)-*R*-enantiomer was found to be inactive at the dose level used, a 100-fold increase in dose is required to produce the same effect as (*S*)-etodolac. Similarly, the concentration of (*S*)-etodolac for inhibition of prostaglandin synthesis by 50% (IC_{50}) was half that of the racemate, the *R*-enantiomer having no inhibitory effect at a threefold greater concentration. Ketorolac, used in the UK for postoperative analgesia, also exhibits stereoselectivity for the inhibition of both COX 1 and COX 2 (Table 5.9).

Table 5.9 Eudismic ratios (*S/R*) for the inhibition of cyclooxygenase 1 and 2 by NSAIDs

Test system	Enzyme	Eudismic ratio (*S*/*R*)		
		Ketoprofen	Flurbiprofen	Ketorolac
Guinea pig whole blood				
Inhibition of TXB_2 synthesis	COX 1	87	313	7.5
LPS; inhibition of PGE_2 synthesis	COX 2	158	123	25
Intact human cells				
Inhibition of TXB_2 synthesis; HPMNL	COX 1	285	148	93
Inhibition of PGE_2 synthesis; LPS monocytes	COX 2	107	560	120
Isolated enzymes from				
Ram seminal vesicles	COX 1	46	11,000	9.5
Sheep placenta	COX 2	>19	>200	92

LPS, lipopolysaccharide stimulation; HPMNL, human polymorphonuclear leukocytes; TXB_2, thromboxane B_2; PGE_2, prostaglandin E_2.

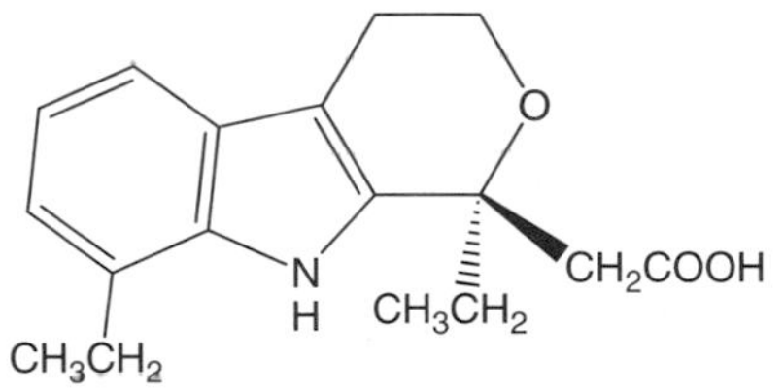

(*S*)-Etodolac (**5.118**)

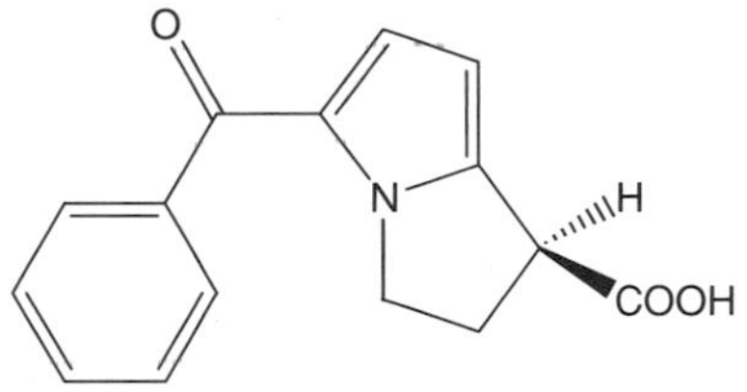

(*S*)-Ketorolac (**5.119**)

5.6.3 Proton Pump Inhibitors

The currently available proton pump inhibitors (PPIs) are benzimidiazole derivatives containing a chiral sulfoxide moiety as part of their structure (**5.11**, **5.120**, and **5.121**). These agents are marketed as racemates, with the exception of omeprazole (**5.11**) that is available as the *S*-enantiomer, esomeprazole (**5.43**), and the racemate. The PPIs inhibit gastric acid secretion by blocking the hydrogen–potassium adenosine triphosphatase enzyme (H^+/K^+-ATPase) system (the ''proton pump'') of gastric parietal cells and are effective in the treatment of gastric and duodenal ulcers and gastroesophageal reflux disease.

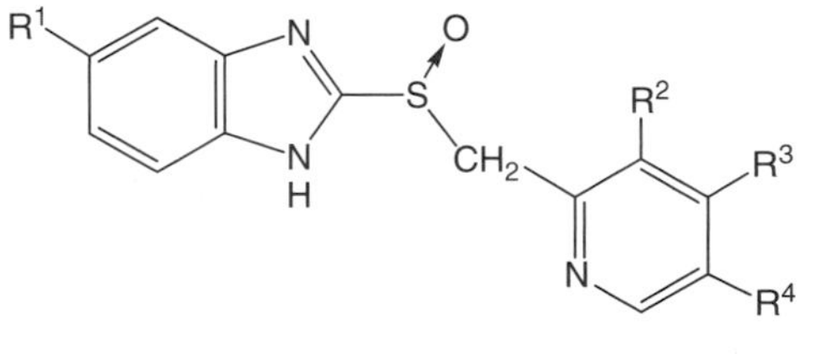

	R^1	R^2	R^3	R^4
Omeprazole **(5.11)**	CH_3O	CH_3	CH_3O	CH_3
Lansoprazole **(5.120)**	H	CH_3	CF_3CH_2O	H
Pantoprazole **(5.121)**	CHF_2O	CH_3O	CH_3O	H

The PPIs are prodrugs and, following absorption, as a result of their weakly basic properties concentrate in the secretory cannaliculi of the hydrochloric acid secreting parietal cells. In acidic environments these compounds undergo transformation to yield an achiral sulfenamide derivative which reacts with thiol groups of H^+/K^+-ATPase to yield a mixed disulfide, resulting in inactivation of the enzyme and a reduction in acid secretion (Figure 5.3). Studies *in vitro*, using gastric microsomal preparations and isolated parietal cells, together with *in vivo* investigations, have indicated no differences between the enantiomers in the inhibition of H^+/K^+-ATPase or in acid secretion.

The metabolism of these drugs involves sulfoxidation, mediated by cytochrome P450 (CYP) 3A4, to yield the achiral sulfone derivatives, together with aliphatic oxidation and *O*-demethylation mediated by CYP2C19. In the case of omeprazole these transformations exhibit stereoselectivity with enantiomeric ratios (*S*/*R*) in intrinsic clearance of: aliphatic oxidation, 0.1; 5-*O*-demethylation, 3.7 and sulfoxidation, 4.6 (Figure 5.4). The overall oxidative metabolism of the drug shows a threefold preference for the *R*-enantiomer. CYP2C19 is polymorphically expressed with some 3% of the Caucasian and 18% to 22% of the Asian population unable to express a functional form of the enzyme. These individuals are known as poor metabolizers (PMs), the remainder of the population are known as extensive metabolizers (EMs). As a result of this genetic polymorphism the pharmacokinetic properties of omeprazole, and related PPIs also metabolized by CYP2C19, show variation between the two phenotypes, which also exhibit differences in stereoselectivity.

Area under the plasma concentration versus time curve (AUC) has been reported to be the pharmacokinetic parameter best correlated to the suppression of acid secretion. Following oral administration of racemic omeprazole exposure to the (+)-*R*- and (−)-*S*-enantiomers, as measured by AUC, are approximately 7.5 and threefold greater, respectively, in PMs compared to EMs.

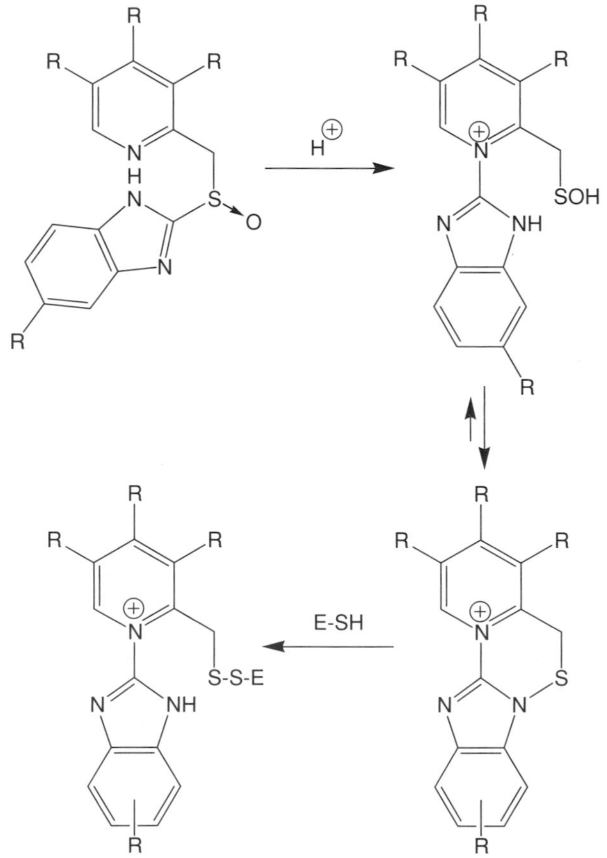

Figure 5.3 Mechanism of inactivation of H^+/K^+-ATPase by the proton pump inhibitors. The chiral drug undergoes acid catalyzed transformation to yield an achiral sulfenamide, which reacts with thiol groups on H^+/K^+-ATPase (E-SH) to yield a mixed disulfide.

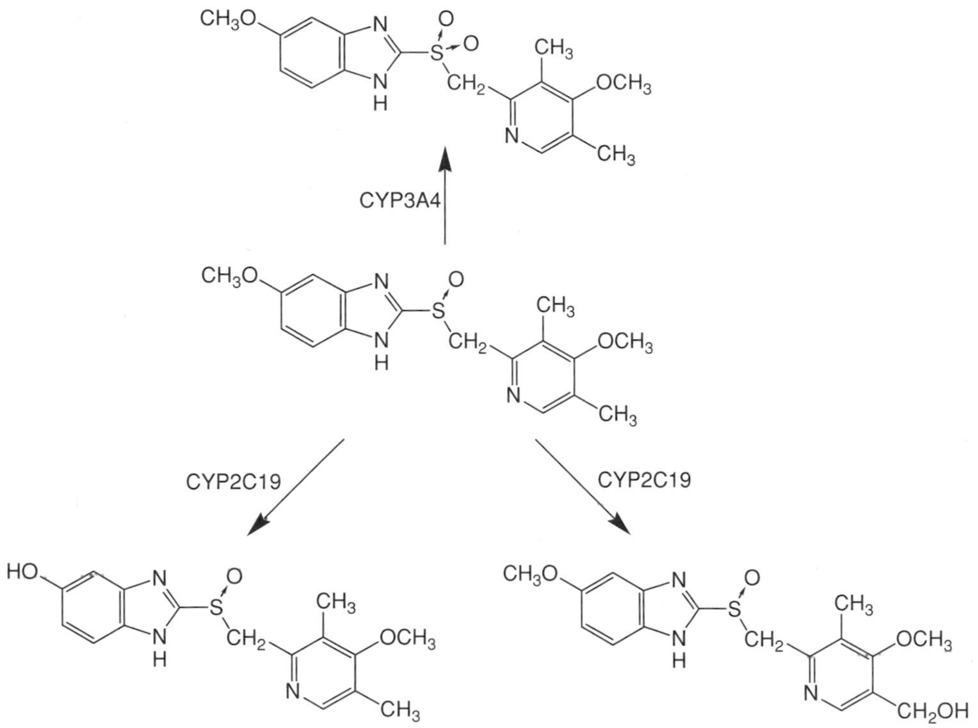

Figure 5.4 Oxidative metabolism of omeprazole by cytochrome P450 isoforms CYP3A4, to yield the sulfone, and CYP2C19, to yield the demethylated and alcohol metabolites.

Similarly, the AUC of (+)-5-hydroxyomeprazole is 3.8-fold less in PMs compared to EMs and that of the (−)-enantiomer is only 1.2-fold greater in EMs compared to PMs. In extensive metabolizers, metabolism is stereoselective for (*R*)-omeprazole to yield the (+)-5-hydroxy metabolite (enantiomeric ratios *R*/*S* of the AUCs of the drug and metabolite are 0.62 and 5.3, respectively). Whereas in PMs, metabolism is stereoselective for the (−)-*S*-enantiomer (enantiomeric ratios *R*/*S* of the AUCs of the drug and metabolite are 1.5 and 1.6, respectively); thus, in PMs not only is metabolism reduced, but also the stereoselectivity is reversed.

As pointed out above, with the exception of omeprazole, which has undergone the chiral switch (see Section 5.8.1), these agents are marketed as racemates. (*S*)-Lansoprazole (**5.120**) and (−)-pantoprazole (**5.121**) are also reported to be undergoing evaluation as potential single enantiomer products. However, as the individual enantiomers of the PPIs are equipotent in a number of pharmacodynamic test systems and require acid-mediated transformation to an achiral intermediate, there is some controversy concerning the relative merits of single enantiomers of these agents in comparison to the racemates. In the case of omeprazole, esomeprazole has been shown to have a reduced clearance and increased systemic availability compared to the *R*-enantiomer or an equal dose of the racemate. Clinical investigations indicate that the single enantiomer maintains intragastric pH above 4 in patients with gastroesophageal reflux disease longer, with a 24-h median pH greater than an equal dose of the racemate. In addition, interpatient variability in intragastric pH and AUC are reported to be less following administration of esomeprazole in comparison to the racemate, resulting in more effective acid control. Thus, the advantageous pharmacokinetic profile of esomeprazole in comparison to omeprazole has the potential to result in superior therapeutic efficacy (in theory the *S*-enantiomer of omeprazole should be less vulnerable to genetic variation in CYP 2C19).

5.6.4 Antidepressants

A number of compounds used in the treatment of depression, including members of several different pharmacological classes, possess one or more stereogenic centers in their structures. Some of these drugs were originally introduced into therapeutics as single enantiomers whereas others are used as racemates, two of which have undergone reevaluation as single enantiomer products with very different outcomes.

Monoamine oxidase inhibitors

Tranylcypromine is marketed as a racemic mixture of (+)-(1*S*,2*R*)- and (−)-(1*R*,2*S*)-*trans*-2-phenylcyclopropylamine. The (+)-enantiomer is reported to be a ca. 100- and 10-fold more potent inhibitor of MAO (mainly MAO-B) in *in vitro* and *in vivo* test systems, respectively, compared to (−)-tranylcypromine. In contrast, the (−)-enantiomer is a two- to threefold more potent inhibitor of presynaptic catecholamine uptake. It is therefore possible that both enantiomers contribute to the antidepressant effects of the drug. However, pharmacokinetic studies in man, following oral administration of both the racemate and the individual enantiomers, indicate marked differences, the area under the plasma concentration versus time curves being ca. tenfold greater for the (−)-1*R*,2*S*-enantiomer following administration of the racemate. Similarly, the oral clearance of the (+)-enantiomer is fivefold greater than that of (−)-tranylcypromine, whereas the difference in plasma half-life is 1.5-fold longer. It has been proposed that such differences may be associated with stereoselective first pass metabolism.

Selegiline (**5.122**), also known as deprenyl, is the *N*-propargyl derivative of (−)-(*R*)-methamphetamine and is a selective inhibitor of MAO-B. The (+)-*S*-enantiomer is only a weak inhibitor of MAO-B but undergoes metabolism to (*S*)-amphetamine derivatives resulting in stimulant, undesired side effects, whereas (*R*)-selegiline yields the corresponding nonstimulant (*R*)-amphetamine derivatives.

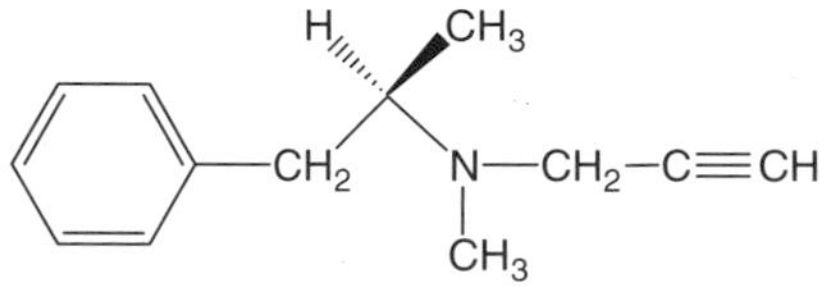

(*R*)-Selegiline (**5.122**)

Tetracyclic compounds

Mianserin (**5.123**) is a tetracyclic antidepressant, used as the racemate. (+)-(*S*)-Mianserin has been shown to possess greater activity than the (−)-*R*-enantiomer in a number of antidepressant screening tests. The (+)-*S*-enantiomer has also been shown to be between 200- to 300-fold more active than (−)-(*R*)-mianserin in the inhibition of noradrenaline uptake and a more potent inhibitor of presynaptic α_2-adrenoceptor blockade. In contrast, (*R*)-mianserin has a greater affinity for 5-HT_3 receptor subtypes, whereas the *S*-enantiomer has a greater affinity, and selectivity, for the 5-HT_2 subtype. Both enantiomers are of similar potency in terms of their sedative properties, which are thought to be associated with their antihistaminic activity.

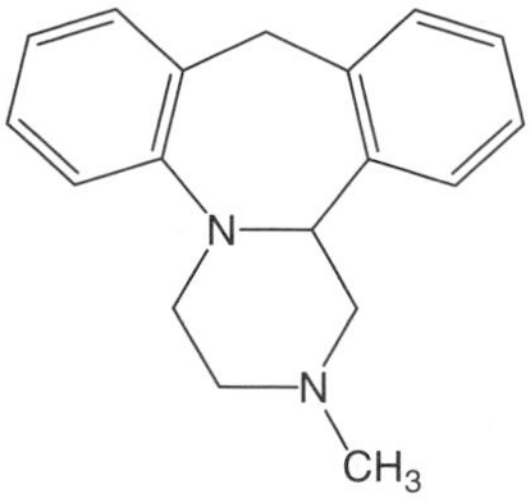

Mianserine (**5.123**)

The metabolism of mianserin involves aromatic oxidation, *N*-oxidation, and *N*-demethylation, together with the formation of cytotoxic metabolites. Studies *in vitro*, using human liver microsomal preparations, have indicated that the extent of metabolism of the individual enantiomers is similar but that the pathways differ with respect to their enantioselectivity. Thus, formation of the aromatic oxidation product, 8-hydroxymianserin, and the *N*-oxide is stereoselective for the *S*-enantiomer, whereas *N*-demethylation is selective for (*R*)-mianserin. More importantly the formation of cytotoxic metabolites was fourfold greater following incubation of (*R*)-compared to (*S*)-mianserin, whereas the formation of products undergoing irreversible protein binding was moderately selective for the *S*-enantiomer. Thus, the antidepressant activity of the drug resides in (*S*)-mianserin, while other pharmacological and toxicological properties either show no stereoselectivity or are predominantly associated with the *R*-enantiomer.

Mirtazapine (**5.124**) is chemically related to mianserin but differs markedly in pharmacological properties having a negligible effect on noradrenaline uptake and a lower affinity for α_1-adrenoceptors. The action of the drug is due to preferential blockade of α_2-adrenoceptors, resulting in an increase in noradrenaline and 5-HT neurotransmission, together with blockade of postsynaptic 5-HT_2 and 5-HT_3 receptors, such that mirtazapine only enhances 5-HT_1-mediated transmission. As a result of its mode of action mirtazapine has been classified as a noradrenergic and specific serotonergic antidepressant. The individual enantiomers exhibit considerable differences in their pharmacological profile, the (+)-enantiomer showing 10- and 37-fold greater antagonist activity than (−)-mirtazapine at α_2-auto- and heteroreceptors, respectively. Whereas, the (−)-enantiomer is approximately a 140-fold more potent antagonist at 5-HT_3 receptors. It would appear therefore that both enantiomers contribute to the observed antidepressant activity of mirtazapine.

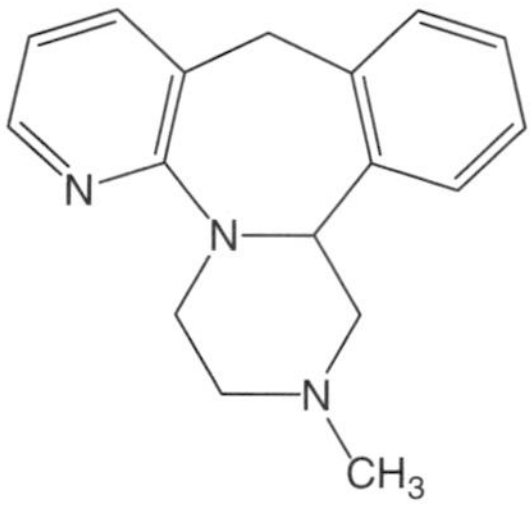

Mirtazapine (**5.124**)

Noradrenaline reuptake inhibitor

Reboxetine (**5.125**), a specific noradrenaline reuptake inhibitor, contains two stereogenic centers in its structure and is used as a racemic mixture of the (−)-*R*,*R*- and (+)-*S*,*S*-enantiomers. Studies both *in vivo* and *in vitro* have indicated the greater potency of the (+)-*S*,*S*-enantiomer in the inhibition of noradrenaline reuptake. The drug undergoes stereoselective disposition, the plasma concentrations of the more active (+)-enantiomer being ca. twofold lower than those of

(−)-(*R*,*R*)-reboxetine following either intravenous or oral administration. This difference in plasma concentration is not due to stereoselective first pass metabolism as the bioavailability of the individual enantiomers is 0.92 and 1.02 for (*R*,*R*)- and (*S*,*S*)-reboxetine, respectively. The enantiomers exhibit stereoselective pharmacokinetics; the values of clearance and volume of distribution of the (+)-*S*,*S*-enantiomer are 2.5- and 2.3-fold greater than that of (−)-(*R*,*R*)-reboxetine, respectively. The similarity of these enantiomeric differences in the two parameters results in the plasma elimination half-lives being essentially equal, indicative of the hybrid nature of this pharmacokinetic parameter.

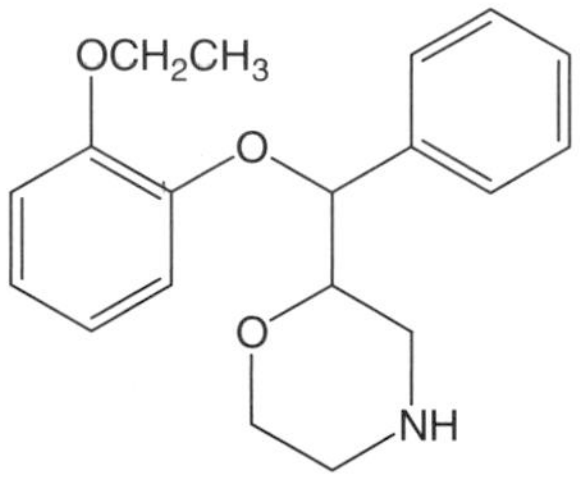

OCH_2CH_3 O O NH

Reboxetine (**5.125**)

Coadministration of reboxetine with ketoconazole, an antifungal agent and a specific inhibitor of cytochrome P450 3A4, results in a reduction in clearance of both enantiomers, with no change in volume of distribution. As a result of this metabolic interaction the half-life of both enantiomers increases, as does the area under the plasma concentration versus time curve with no alteration in the maximum observed plasma concentration. The interaction shows modest stereoselectivity, clearance of the (−)-*R*,*R*-enantiomer decreasing by 34% (from 41.6 to 27.5 mL min^{-1}) whereas that of (+)-(*S*,*S*)-reboxetine decreases by 24% (from 99.3 to 75.6 mL min^{-1}).

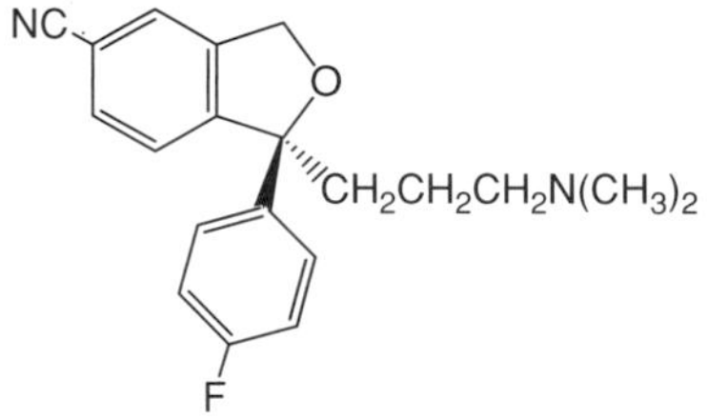

NC O $CH_2CH_2CH_2N(CH_3)_2$ F

Escitalopram (**5.44**)

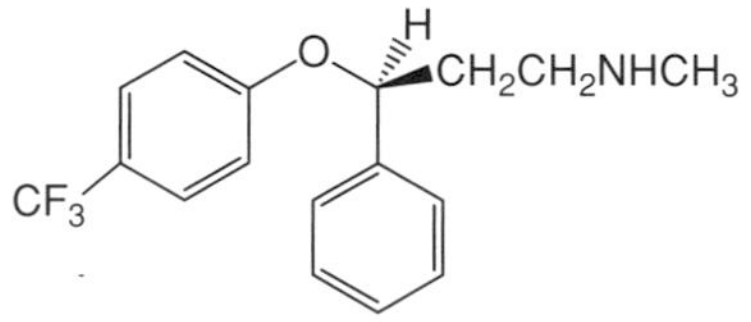

O H $CH_2CH_2NHCH_3$ CF_3

(*S*)-Fluoxetine (**5.126**)

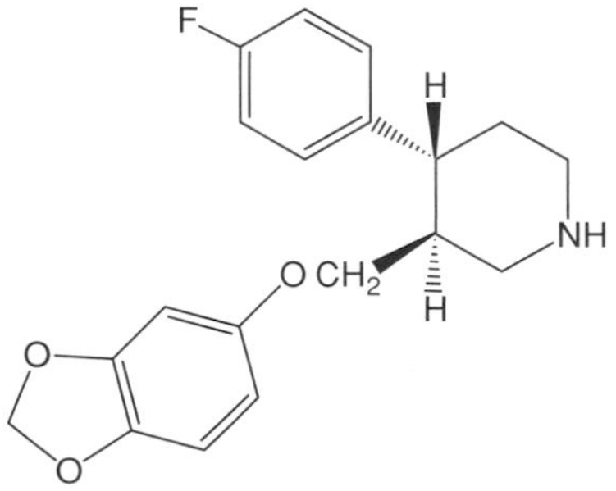

F H NH O CH_2 H O O

(3*S*,4*R*)-Paroxetine (**5.127**)

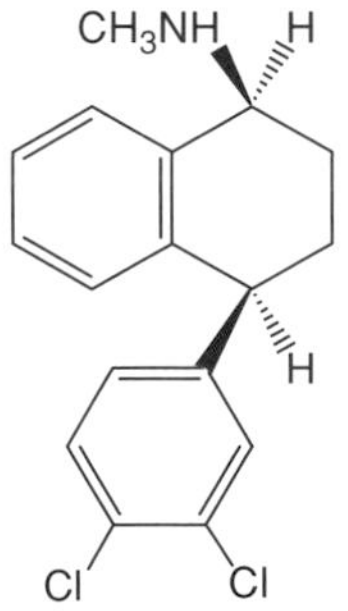

CH_3NH H H Cl Cl

(1*S*,4*S*)-Sertraline (**5.128**)

Selective serotonin reuptake inhibitors

The selective serotonin reuptake inhibitors (SSRIs) are an important group of antidepressant agents and include compounds with either one, fluoxetine and citalopram, or two, sertraline and paroxetine, stereogenic centers in their structures. In the case of the former pair both agents were originally introduced as racemates and have subsequently undergone reevaluation as single enantiomer products with very different results, whereas the latter pair were both introduced as single enantiomer products.

The enantiomers of fluoxetine have similar potencies in *in vitro* test systems for inhibition of 5-HT uptake, the eudismic ratios (*S*/*R*) varying between 1.1 and 1.9 depending on the test system used. The demethylated metabolite, norfluoxetine is also pharmacologically active with (*S*)-norfluoxetine 15- to 20-fold more potent than the *R*-enantiomer, and 1.5-fold more potent than (*S*)-fluoxetine (**5.126**), and has been considered for development as a drug in its own right. Following administration of the individual enantiomers to animals the duration of action also differs, (*S*)- and (*R*)-fluoxetine acting for 24 and 8 h, respectively, which is thought to be associated with the formation of (*S*)-norfluoxetine.

Fluoxetine also exhibits stereoselectivity in the inhibition of cytochrome P450 2D6, the ratio of the inhibition constants (*S*/*R*) being 6.3 and 4.8 for the drug and metabolite enantiomers, respectively. Fluoxetine also undergoes metabolism mediated by CYP2D6 and the pharmacokinetic parameters of the individual enantiomers vary between extensive and poor metabolizers, i.e., increased half-life and area under the plasma concentration versus time curve of both enantiomers in poor compared to extensive metabolizers.

As a result of the shorter washout period, the reduced activity of the metabolite, reduced accumulation and inhibition of CYP2D6, and increased flexibility for the treatment of depression (*R*)-fluoxetine has been evaluated as a potential single enantiomer product. However, development of the drug was stopped due to a small but significant increase in QT_c prolongation at the highest dose examined. (*S*)-Fluoxetine (**5.126**) has also been evaluated for the prophylaxis of migraine in a placebo-controlled clinical trial, the results of which indicated a reduction in attack frequency earlier and greater in the treatment group. Thus (*S*)-fluoxetine may have potential as a single enantiomer product with a new therapeutic indication.

Citalopram is one of the most selective SSRIs available and a number of biochemical studies have indicated the greater potency of the *S*-enantiomer in comparison to (*R*)-citalopram in the inhibition of 5-HT uptake, eudismic ratios (*S*/*R*) varying between 130 and 160 depending on the test system used. Examination of both the individual enantiomers and the racemate, in animal models of depression, have also indicated the twofold potency of (*S*)-citalopram in comparison to the racemate and the lack of activity of the *R*-enantiomer. Citalopram undergoes *N*-demethylation to yield an active metabolite desmethylcitalopram, the *S*-enantiomer of which is 6.7-fold less active than (*S*)-citalopram but 6.5-fold more active than (*R*)-desmethylcitalopram.

Citalopram has recently undergone the chiral switch process (see Section 5.8.1) and the single *S*-enantiomer, escitalopram (**5.44**), is commercially available in addition to the racemate. When single enantiomers are developed from previously marketed racemates the regulatory bodies permit bridging studies between the original and the new submission. As part of these investigations a comparison of the pharmacokinetic profile of the single enantiomer following administration as such and as a component of the racemate is required to ensure that interactions between enantiomers do not occur. In the case of escitalopram both the drug and the active demethylated metabolite have been shown to be bioequivalent following oral administration of the racemate and an equivalent dose of the single enantiomer. Clinical studies with escitalopram indicate a faster onset of action, a reduction in side effects, and an improvement in tolerability profile in comparison to the racemate. A recent study, in an animal model for anxiolytic activity, has indicated the greater potency of escitalopram in comparison to the racemate and also that the *R*-enantiomer attenuates the action of escitalopram. However, the mechanism of this attenuation is unknown.

Table 5.10 Stereoselectivity of sertraline (5.128) isomers for the inhibition of amine uptake

	Inhibitory concentration (IC_{50} μM)		
Stereoisomer	**Serotonin**	**Noradrenaline**	**Dopamine**
Trans-(+)-1*R*,4*S*	0.033	0.011	0.033
Trans-(−)-1*S*,4*R*	0.45	0.05	0.23
Cis-(+)-1*S*,4*S*[a]	0.06	1.2	1.1
Cis-(−)-1*R*,4*R*	0.46	0.38	0.29

[a]Sertraline.

The related agents paroxetine (**5.127**) and sertraline (**5.128**) contain two stereogenic centers in their structures and both compounds were introduced originally as single stereoisomer products. Pharmacological evaluation of the four stereoisomers of sertraline indicated the significance of stereochemical considerations in the selection of a particular isomer in terms of selectivity of action (Table 5.10). The *trans*-(+)-1*R*,4*S*-stereoisomer was found to be the most potent with respect to inhibition of uptake of the three biogenic amines investigated, serotonin, dopamine, and noradrenaline. The *cis*-(+)-1*S*,4*S*-stereoisomer was found to be the least potent with respect to inhibition of dopamine and noradrenaline uptake, but only twofold less potent than the *trans*-(+)-diastereoisomer in terms of inhibition of 5-HT uptake. Thus, the stereoisomer selected for subsequent development, as sertraline (Table 5.10) was that with the greatest selectivity of action.

5.6.5 Local Anesthetic Agents

A number of local anesthetic agents, including prilocaine (**5.129**), mepivacaine (**5.130**), ropivacaine, and bupivacaine, are chiral and while the majority are marketed as racemates, single enantiomers, e.g., (*S*)-ropivacaine (**5.131**), and in the case of bupivacaine both the racemate and single *S*-enantiomer, levobupivacaine (**5.132**), are commercially available. It has been known for a number of years that the enantiomers of these agents may differ in their duration of action, disposition, and acute toxicity following administration to animals.

These drugs act by inhibition of nerve impulse in the peripheral nervous system by blockade of sodium and potassium ion channels. Voltage-gated sodium channels exist in three conformational states resting (closed), open (activated), and inactivated. The affinity of the open and inactivated channel states for the local anesthetics are greater than that of the resting state and compounds that bind with a higher affinity, or dissociate more slowly, exhibit a greater potency of blockade in comparison to those which dissociate faster. Potassium channels, in contrast to sodium channels, are a diverse family of membrane proteins with a number of subtypes. Inhibition of potassium channels may potentiate, or antagonize, the impulse blockade primarily caused by inhibition of sodium channels.

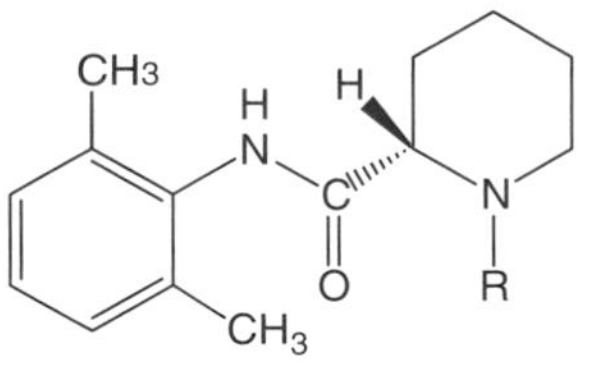

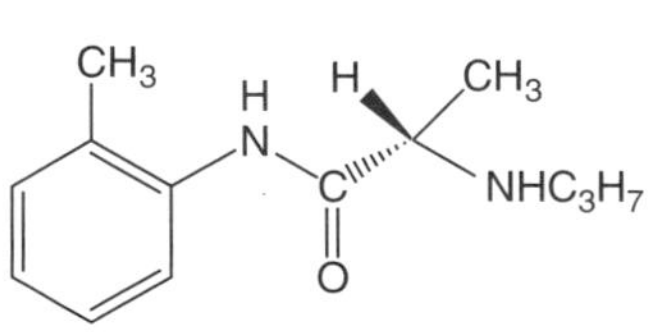

	R
(*S*)-Mepivacaine (**5.130**)	CH_3
(*S*)-Ropivacaine (**5.131**)	C_3H_7
(*S*)-Bupivacaine (**5.132**)	C_4H_9

(*S*)-Prilocaine (**5.129**)

The enantiomers of bupivacaine have been shown to exhibit relatively modest stereoselectivity of action using a variety of *in vitro* test systems including human cardiac sodium channels, (*R*)-bupivacaine exhibiting selectivity of action with eudismic ratios (*R*/*S*) varying between 1.1- and 3.0-fold depending on the methodology employed. Interestingly, the resting sodium channel conformation appears to show a slightly greater affinity for (*S*)-bupivacaine (ratio *S*/*R* 1.2). In contrast, the inactivated channel binds both enantiomers with greater affinity (greater than tenfold) and reversed stereoselectivity (ratio *R*/*S* 1.2), confirming the greater potency of the *R*-enantiomer for inhibition of both the neuronal action potential and sodium currents. The rates of dissociation of the enantiomers from inactivated sodium channels also differ and that of the *R*-enantiomer is slower than (*S*)-bupivacaine.

Inhibition of human ventricle delayed rectifier potassium channels (hKv1.5) also indicates the greater potency of the *R*-enantiomers, the enantiomers of bupivacaine and ropivacaine yielding eudismic ratios (*R*/*S*) of 7 and 2.5, respectively. In addition bupivacaine exhibits marked stereoselectivity on the flicker potassium ion channel with a eudismic ratio (*R*/*S*) of 73, which appears to be associated with the kinetics of dissociation from the binding site, the enantiomeric ratio of the dissociation rate constants (*S*/*R*) is 64, i.e., the dissociation of the *R*-enantiomer is much slower than that of (*S*)-bupivacaine.

The toxicity and adverse effects of these agents, including CNS and cardiovascular reactions, and death following accidental intravenous injection are obviously a cause of concern. The cardiovascular effects are associated with blockade of cardiac sodium channels, a property utilized with some of these agents as antiarrhythmics. The local anesthetics show a slightly greater potency for the blockade of cardiac sodium channels in comparison to those in neuronal tissue. This differential activity may be associated with either greater affinity for cardiac sodium channels or different inactivation gating properties. In addition, drugs that bind with greater affinity or dissociate slower, resulting in a longer recovery time, e.g., bupivacaine, exhibit a greater potency of blockade in comparison to agents that dissociate more rapidly, e.g., lignocaine. Such differences in dissociation are thought to account for the cardiotoxicity of bupivacaine, as a result of accumulation of blockade during diastole and slowed conduction resulting in arrhythmias, and the use of lignocaine as an antiarrhythmic agent.

The cardiotoxicity of bupivacaine is predominantly associated with the *R*-enantiomer, studies in animals indicating a greater change in conduction following (*R*)- or racemic bupivacaine compared to the *S*-enantiomer. This may be explained by the slightly greater affinity of the *R*-enantiomer for cardiac sodium channels together with its longer dissociation times.

As a result of the greater risk of cardiotoxicity of (*R*)-bupivacaine, the drug was reevaluated as the single *S*-enantiomer, levobupivacaine (**5.132**), and was the subject of the chiral switch (see Section 5.8.1). Similar considerations, with respect to the stereoselectivity of the adverse effect, resulted in the development of the *N*-propyl analog, ropivacaine (**5.131**), as the *S*-enantiomer. Comparisons of levobupivacaine with the racemate in patients have indicated that sensory block and the clinical profile resulting from the single enantiomer and an equal dose of the racemate are essentially the same. The cardiovascular effects of levobupivacaine, following intravenous administration to healthy volunteers, have also been investigated. The negative inotropic effects of levobupivacaine were found to be significantly less, approximately half, in comparison to the racemate.

Bupivacaine also exhibits stereoselectivity in its pharmacokinetic properties. The individual enantiomers of the local anesthetics differ in their effects on local blood flow and differences in their duration of action may be accounted for by differences in their systemic absorption. For example the *S*-enantiomers of mepivacaine, ropivacaine, and bupivacaine cause vasoconstriction, which may result in a longer duration of action in comparison to the *R*-enantiomers, as observed for bupivacaine following intradermal injection. Thus, stereoselectivity in pharmacodynamic activity indirectly influences the stereoselectivity of absorption. Pharmacokinetic parameters of the enantiomers of bupivacaine have been reported following intravenous administration of the racemate to man. The systemic clearance, volume of distribution at steady state, and half-life of (*S*)-bupivacaine are significantly less than those of the *R*-enantiomer (Table 5.5). The enantiomers also show differences in fraction unbound, that of the *R*-enantiomer being greater than that of (*S*)-bupivacaine (Table 5.2) with the result that there are no differences in the unbound volume of distribution at steady state, but the unbound clearance of the *S*-enantiomer is greater than that of (*R*)-bupivacaine.

Thus, from the available data, it would appear that the single *S*-enantiomer of bupivacaine results in a product with a similar clinical profile to the racemate, with a reduction in the cardiovascular adverse effects.

5.7 TOXICOLOGY

The process of drug safety evaluation is complex, expensive, and time-consuming involving acute and chronic toxicity testing, mutagenicity and genetic toxicology, reproductive toxicology, carcinogenicity, and clinical safety evaluation both pre- and postmarketing. There is also a need to carry out mechanistic and toxicokinetic studies in order to determine animal exposure to both the drug and metabolites, and to aid in the extrapolation of animal data to man. There are relatively little published data on the comparative toxicity of single enantiomers versus racemic drugs and even less information arising from clinical studies. However, examples may be cited which are illustrative of aspects of stereochemical considerations in drug safety evaluation.

Fenvalerate (**5.133**) is a synthetic pyrethroid insecticide that contains two stereogenic centers, and thus four stereoisomers are possible. Administration of the compound in the diet to a range of animal species resulted in granulomatous changes in the liver, lymph nodes, and spleen. Separation and toxicological evaluation of the individual stereoisomers indicated that the toxicity was associated with only one of the four isomers. Subsequent metabolic studies indicated that the toxicity was associated with the formation and disposition of a cholesteryl ester, (*R*)-2-(4-chlorophenyl) isovaleric acid cholesterylester (**5.134**), formed by transesterification of the single toxic stereoisomer of fenvalerate. Fortunately the active isomer of fenvalerate may be synthesized stereospecifically. While not a drug, this example does indicate that stereochemical considerations may prevent a compound being discarded following an adverse toxicological evaluation.

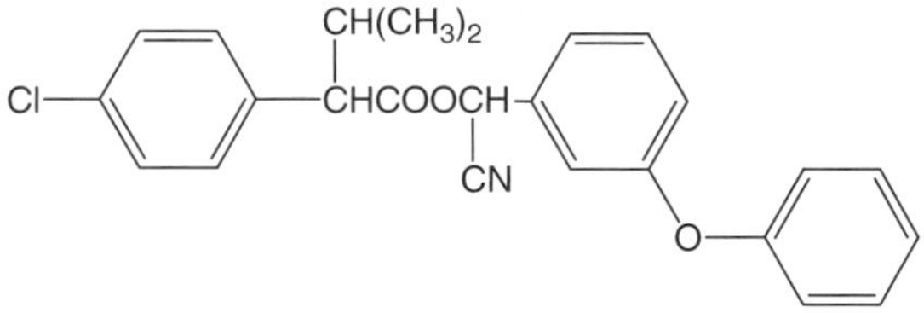

Cl — C6H4 — CH(CH(CH3)2)COOCH(CN) — C6H4 — O — C6H5

Fenvalerate (**5.133**)

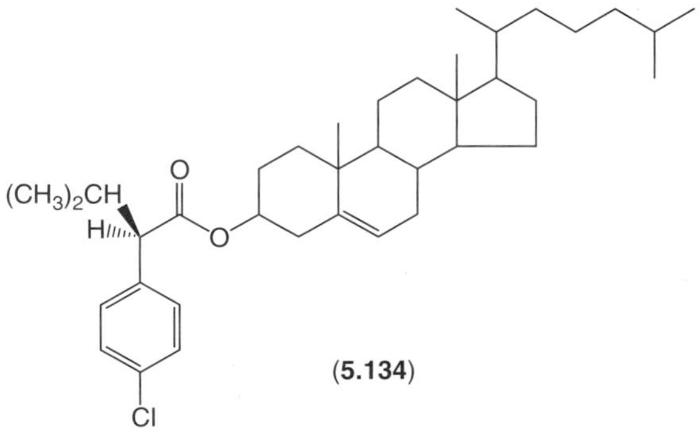

(CH3)2CH, H, O, O, Cl

(**5.134**)

Thalidomide (**5.135**) is a compound frequently cited, particularly in the popular press, to support arguments for the development of single isomer drugs. Thalidomide was introduced, as the racemate, into therapeutics in the early 1960s as a sedative-hypnotic agent and was used in pregnant women for the relief of morning sickness. However, the drug was withdrawn when it became apparent that its use in pregnancy was associated with malformations, particularly phocomelia (shortening of the limbs), in the offspring. Investigations in the late 1970s using SWS mice indicated that both isomeric forms of the drug are hypnotic agents but that the teratogenic properties of the drug reside in the *S*-enantiomer. Thus, the argument goes: if the drug had been used as the single *R*-enantiomer then the tragedy of the early 1960s could have been avoided.

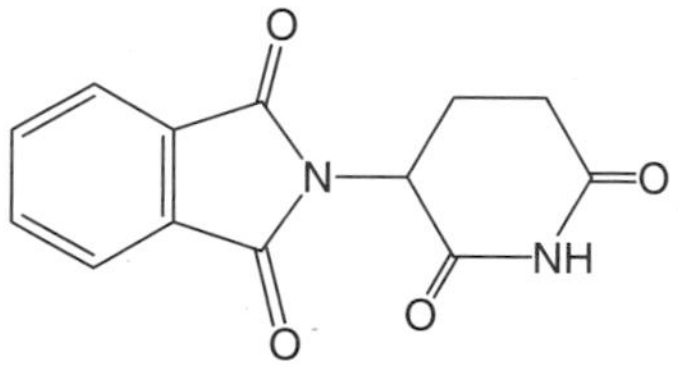

Thalidomide (**5.135**)

However, the situation with thalidomide is much more complex. Rodents are resistant to the teratogenic toxicity of the drug and the mouse is a poor model for teratogenicity testing. Data obtained in a more sensitive test species, New Zealand white rabbits, indicate that both the enantiomers of thalidomide are teratogenic. An additional problem with the drug is its stereochemical stability since the single isomers undergo rapid racemization in biological media. Thus, even if a single isomer was administered to an experimental animal the other would be formed relatively rapidly. The acute toxicity of thalidomide, as determined by the LD_{50} (lethal dose for 50%) test, also presents a complex problem. The individual enantiomers have similar reported LD_{50} values of approximately 1.0–1.2 g kg^{-1} in mice, but the value for the racemate is greater than 5 g kg^{-1}, i.e., the racemate is nontoxic. In this case it would appear that the administration of the racemic mixture is exerting a protective effect, and the mechanism is unknown.

Taken together the above information indicates that the situation with thalidomide is by no means as clear as sometimes implied and the drug is certainly not a good example to cite in support of arguments for single isomer drugs.

A similar situation arises with the thalidomide analog EM12 (**5.136**) the enantiomers of which also undergo racemization, both *in vitro* and *in vivo* following administration to animals. Additionally, administration of the individual enantiomers to pregnant marmosets indicated modest stereoselectivity in the teratogenic potency of the *S*-enantiomer. However, the facile racemization makes interpretation of the significance of such data difficult. Stereoselectivity with respect to teratogenicity has been shown with the enantiomers of 2-ethylhexanoic acid, a metabolite of the plasticizer di-(2-ethylhexyl)phthalate, the *R*-enantiomer being teratogenic and embryotoxic, following administration to mice, whereas (*S*)-2-ethylhexanoic acid was nontoxic.

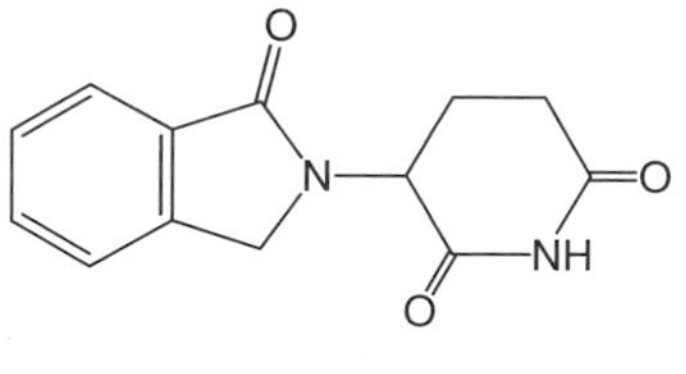

EM12 (**5.136**)

A number of chiral drugs administered as racemates have been withdrawn over the years, including the β-adrenoceptor antagonist practolol, the NSAID benoxaprofen, the H_1-antihistamine terfenadine, the antianginal agent prenylamine, and the anticholinergic calcium antagonist terodiline, as a result of adverse reactions. In some instances a possible stereochemical association with the adverse event has been examined. For example racemic terfenadine was withdrawn as a result of cardiotoxicity, the drug induced ventricular arrhythmias. The enantiomers were found to be essentially equipotent with respect to their antihistamine activity and also in their blockade of human cardiac potassium channels (hKv1.5). Thus, neither action of the drug exhibits stereoselectivity. In contrast, the anticholinergic calcium antagonist terodiline, used in the treatment of urinary incontinence was similarly withdrawn due to cardiac arrhythmias associated with electrocardiogram QT interval prolongation. The pharmacodynamic effects of the drug are enantioselective, (*S*)-terodiline being approximately tenfold more potent as a calcium antagonist, and ca. tenfold less potent as an anticholinergic than the *R*-enantiomer. Thus both enantiomers were thought to contribute to the effects on the urinary bladder, the major effect on the detrusor muscle is associated with (*R*)-terodiline. Examination of the cardiac effects of the two enantiomers in healthy volunteers indicated that the racemate and *R*-enantiomer caused significant prolongation of the QT interval with the conclusion that (*R*)-terodiline is responsible for the ventricular arrhythmias caused by the drug. In this instance the stereoselectivity of action, the drug effects on the detrusor muscle, and toxicity are selective for the same enantiomer. The antianginal drug prenylamine, also withdrawn as a result of ventricular arrhythmias, is thought to exhibit stereoselective toxicity associated with the *S*-enantiomer.

In the majority of cases, unlike some of the examples cited above, the significance of stereochemistry to the observed toxicity is difficult to assess from the available data. However, the use of single enantiomer would have halved the required dose and the adverse effects may have been reduced as a consequence.

5.8 RACEMATES VERSUS ENANTIOMERS AND REGULATION OF CHIRAL DRUGS

The realization of the potential significance of the pharmacological differences between enantiomers, in the late 1980s and early 1990s, resulted in drug stereochemistry and racemates versus enantiomers becoming a topic of debate. Advocates of the use of single enantiomers regarded racemates as compounds containing 50% impurity, while others stated that their use is essentially polypharmacy with the proportions in the mixture dictated by chemical rather than therapeutic or pharmacological criteria.

There are a number of potential advantages associated with the use of single enantiomer drugs including:

- Less complex more selective pharmacodynamic profile
- Removal of a potentially interacting impurity
- Potential for an improved therapeutic index
- Less complex pharmacokinetic profile
- Reduced potential for complex drug interactions
- Less complex relationship between plasma concentration and effect

The major regulatory authorities have examined the issues associated with drug chirality and have published policy statements or issued guidelines. It is obvious that all pharmaceutical, preclinical, and clinical regulatory requirements applicable to nonchiral drugs apply equally to chiral compounds, irrespective of their use as racemates or single enantiomers, and the guidelines concerning stereochemistry emphasize the additional information required for the development of chiral compounds. To date none of the regulatory agencies have an absolute requirement for the development of single stereoisomer products. Chiral drugs may be developed as single enantiomers, racemic, or nonracemic mixtures, and for compounds with more than one stereogenic center, mixtures of diastereoisomers. The composition of the material to be developed is left to the compound sponsor, however the decision as to which form is to be used, enantiomer, racemate, or some alternative possibility, requires scientific justification based on quality, safety, and efficacy criteria, together with the risk–benefit ratio. There are a number of arguments that could be used to support the submission of a racemate including:

- The individual enantiomers are stereochemically unstable and readily racemize *in vitro* and/or *in vivo*.
- The preparation of the drug as a single enantiomer on a commercial scale is not technically feasible.
- The individual enantiomers have similar pharmacological and toxicological profiles.
- One enantiomer is totally devoid of activity and does not provide an additional body burden or influence the pharmacokinetic properties of the other.
- The use of the racemate produces a superior therapeutic effect to either individual enantiomer.
- The therapeutic significance of the compound in relation to the disease state and adverse reaction profile.

In terms of regulatory requirements stereochemical considerations start with the chemical development process, proof of structure, and configuration are required. The final product must be characterized, as for any drug, with the additional requirement to establish the stereochemical purity of the material. With single stereoisomer products, the unrequired stereoisomers arising either during manufacture or storage are regarded as impurities and there is an additional requirement to show that unacceptable changes in stereochemical composition do not occur. Additionally, the use and stereochemical purity of individual batches of material must be known so that they may be related to safety and clinical investigations.

Preclinical and clinical investigations on single stereoisomer products are carried out as for any other new chemical entity, with the additional requirement to examine the stereochemical stability of the material *in vivo*, i.e., to establish if inversion of configuration, either chemically or biochemically mediated, takes place *in vivo*. In instances where this does occur the stereoisomer formed is treated as a metabolite, and an appreciation of such interconversions will obviously contribute to the decision to develop either a single enantiomer or racemate. Preclinical evaluation of a chiral drug should include pharmacodynamic, pharmacokinetic, and appropriate toxicological investigations on both enantiomers and the racemate. In some instances, clinical investigations on the three forms may also be required.

As a result of regulatory attitudes, together with the considerable advances in synthetic and separation methodologies, the number of single enantiomer new chemical entities submitted for approval over the last 10 years appears to have increased. Of 95 new agents evaluated by the UK Medicines and Healthcare products Regulatory Agency over the period 1996–2000, 76 were classified as synthetic origin of which 45 were chiral, with 30 (67%) submitted as single stereoisomers, and only 15 (33%) as racemic mixtures. While the total number of agents is considerably smaller than the initial survey, carried out in the mid-1980s, the trends for future drug development are obvious.

5.8.1 The Chiral Switch

In addition to new drug development a number of established agents originally marked as racemates have been reevaluated and reintroduced as single enantiomers. Several of these agents, together with their reported advantages, e.g., esomeprazole (Section 5.6.3), escitalopram (Section 5.6.4), levobupivacaine (Section 5.6.5), and ketamine (Section 5.5.6), have been discussed in the text, additional compounds and other drugs reported to be undergoing evaluation as single enantiomer products are listed in Table 5.11. The chiral switch has resulted in a number of drugs becoming commercially available as both single enantiomer and racemic mixture products at the same time. There is therefore an obvious requirement for both pharmacists and physicians to have

Table 5.11 Chiral switch marketed agents and compounds reported to be under evaluation

Marketed (availability)	Under evaluation
Dexketoprofen	(*S*)-Amlodipine
Dexibuprofen (Austria, Switzerland)	(*S*)-Doxazosin
Esomeprazole	Eszopiclone
Levofloxacin	(*R*,*R*)-Formoterol
Levobupivacaine	(*S*)-Fluoxetine
Escitalopram	(*R*)-Flurbiprofen
(*S*)-Ketamine (Germany)	(*S*)-Lansoprazole
Levocetirizine	(−)-Norcisapride
Cisatracurium	(*S*)-Oxybutinin
(*R*)-Salbutamol [Levalbuterol] (USA)	(−)-Pantoprazole
(*R*,*R*)-Methylphenidate (USA)	Sibutramine metabolite[a]

[a]Neither the structure of the metabolite nor its stereochemistry has been specified.

an appreciation of stereochemical issues in therapeutics so that appropriate medication can be provided to the patient.

The marketed single enantiomers frequently have the same, or very similar, therapeutic indications as the original racemate. However, this may not always be the case and novel indications for ''old'' compounds have been reported, e.g., (*S*)-fluoxetine for migraine prophylaxis, (*R*)-flurbiprofen in prostate cancer therapy.

The idea of investigating single enantiomers following observation of adverse effects with the racemate, or developments in technology for the production of single enantiomers is not new. D-Penicillamine, introduced originally for the treatment of Wilson's disease, has been used in rheumatology for a number of years. Initial clinical evaluation of the synthetic racemate in USA resulted in optic neuritis and the drug was withdrawn. In the UK, D-penicillamine was obtained as the single enantiomer, from the hydrolysis of penicillin, and the adverse effect was not observed. Similarly, the initial use of racemic dopa for the treatment of Parkinson's disease resulted in a number of adverse effects including nausea, vomiting, anorexia, involuntary movements, and granulocytopenia. The use of L-dopa resulted in halving the dose, a reduction in adverse effects, the granulocytopenia was not observed, and lead to an increased number of improved patients. The progestogen, norgestrel, the activity of which resides in the levorotatory enantiomer, used as an oral contraceptive and in hormone replacement therapy was initially marketed as a racemate. However, developments in the synthetic methodology resulted in the introduction of the single enantiomer in the late 1970s.

When single enantiomers are developed from previously approved racemates the regulatory bodies permit appropriate ''bridging'' studies such that data from the new single enantiomer submission may be linked to the original racemate submission; the advantage of the approach is a reduction in the number of investigations that need to be carried out. The extent of the ''bridging'' studies will depend on the compound under evaluation, but will obviously involve a comparison of the pharmacodynamic, pharmacokinetic, and toxicological profile between the single enantiomer and racemate. Should unexpected results be obtained then additional studies may be required. It may also be possible, depending on the preclinical studies, to extrapolate some of the clinical investigations with the racemate to the single enantiomer. This approach may present problems if the sponsor of the single enantiomer was not responsible for the development of the original racemate.

Such reevaluations of single enantiomers are not without problems and removal of the so-called ''isomeric ballast'' may not be a trivial matter, with the supposed advantages not being realized. The examples of (*R*,*R*)-labetalol (dilevalol), dexsotalol, and (*R*)-fluoxetine may be cited. An additional problem associated with these chiral switch ''failures'' is that the mixture products are obviously still available and widely used. Had the single enantiomer versions of these agents been selected for development originally and ''failed'' it is unlikely that they would have been redeveloped as racemates and therefore useful compounds would have been lost. Similarly, in the past racemates may have been thought unsuitable for development, whereas their individual enantiomers may have had potentially useful properties. One of the first compounds to undergo the chiral switch was the anoretic agent dexfenfluramine, the *S*-enantiomer of racemic fenfluramine. Both the racemate and single enantiomer were withdrawn following an association of the drug with valvular heart disease, and the rare but serious risk of pulmonary hypertension.

In addition to the proposed therapeutic advantages of single enantiomers, the chiral switch process has been argued to have commercial benefits in terms of extending patent life and providing some market share protection against generic competition. As a result with some agents the chiral switch has been a matter of controversy. However, the single enantiomer failures outlined above are not without considerable financial consequences and while the cost of developing a switch compound may be regarded as relatively cost-effective, in comparison to a new chemical entity, failure is still extremely expensive.

The examples cited above, and listed in Table 5.11, are concerned with the reevaluation and development of single stereoisomers from racemates or mixtures of diastereoisomers. However, it is possible, although in the present regulatory climate unlikely, that a previously marketed single enantiomer could be reevaluated as a racemic mixture. An isolated example of this unlikely scenario has been reported in the literature. The analgesic methadone is used as maintenance therapy in the management of opioid dependence. In the majority of countries the drug is used as the racemate even though the *R*-enantiomer is reported to be approximately 50-fold more potent as an analgesic in man compared to (*S*)-methadone. In Germany, the *R*-enantiomer has been available but as a result of higher costs the drug has been progressively replaced by a double dose of the racemate. This ''reversed'' chiral switch resulted in a decrease in the serum concentration/dose ratio of the *R*-enantiomer in patients undergoing maintenance therapy with the racemate and some patients experienced withdrawal symptoms, possibly as a result of enzyme induction. Thus, a change from the single enantiomer to the racemate resulted in therapeutic implications requiring in some instances adjustment of the maintenance dose.

5.9 CONCLUDING COMMENT

The material presented in this chapter was selected to provide a background to the biological significance of drug chirality and to highlight the advantages of stereochemical considerations in pharmacology. In the past such quotes as ''Warfarin enantiomers should be treated as two drugs'' and ''(*S*)- and (*R*)-propranolol are essentially two distinct entities pharmacologically'' have appeared in the literature. In the future, as a result of regulatory considerations and advances in chemical technology, the majority of chiral drugs will be introduced as single isomers. It is also to be expected that additional compounds currently available as racemates will be reintroduced as single isomers with, in some instances, novel indications. However, for many drugs, currently in use as racemates, relatively little is known regarding the pharmacodynamic or toxicological activities or pharmacokinetic properties of the individual enantiomers. The results of additional pharmacological and pharmacokinetic investigations on the enantiomers of marketed racemates may result in new indications for ''old'' drugs, improve the clinical use of these agents and hence result in increased safety and efficacy. Probably the best take home message for the budding medicinal chemist would be: ''if you make a chiral compound, finish the job and separate the isomers yourself don't expect the patient to do it for you.'' Future drug development is literally in your hands.

FURTHER READING

Aboul-Enein, H.Y. and Wainer, I.W. (eds) (1997) *The Impact of Stereochemistry on Drug Development and Use*. New York: John Wiley.

Agranat, I., Caner, H. and Caldwell, J. (2002) Putting chirality to work: the strategy of chiral switches. *Nature Reviews Drug Discovery* **1**, 753–768.

Ariens, E.J. (1984) Stereochemistry, a basis for sophisticated nonsense in pharmacokinetics and clinical pharmacology. *European Journal of Clinical Pharmacology* **26**, 663–668.

Ariëns, E.J., Soudijn, W. and Timmermans, P.B.M.W.M. (eds) (1983) *Stereochemistry and Biological Activity of Drugs*. Oxford: Blackwell.

Baumann, P., Zullino, D.F. and Eap, C.B. (2002) Enantiomers potential in psychopharmacology—a critical analysis with special emphasis on the antidepressant escitalopram. *European Neuropsychopharmacology* **12**, 433–444.

Branch, S. (2001) International regulation of chiral drugs. In: Subramanian, G. (ed.), *Chiral Separation Techniques: A Practical Approach*, 2nd edn. Weinheim: Wiley-VCH, pp. 319–342.

Burke, D. and Bannister, J. (1999) Left-handed local anaesthetics. *Current Anaesthesia and Critical Care* **10**, 262–269.

Cahn, R.S., Ingold, C.K. and Prelog, V. (1956) The specification of asymmetric configuration in organic chemistry. *Experimentia* **12**, 81–94.

Caldwell, J. and Leonard, B.E. (eds) (2001) The enantiomer debate: realising the potential of enantiomers in psychopharmacology. *Human Psychopharmacology. Clinical and Experimental* **16** (Suppl. 2), S65–S107.

Caldwell, J., Winter, S.M. and Hutt, A.J. (1988) The pharmacological and toxicological significance of the stereochemistry of drug disposition. *Xenobiotica* **18** (Suppl. 1), 59–70.

Easson, L.H. and Stedman, E. (1933) Studies on the relationship between chemical constitution and physiological action. V. Molecular dissymmetry and physiological activity. *Biochemical Journal* **27**, 1257–1266.

Eichelbaum, M., Testa, B. and Somogyi, A. (eds) (2003) *Stereochemical Aspects of Drug Action and Disposition*. Berlin: Springer-Verlag.

Eliel, E.L. and Wilen, S.H. (1994) *Stereochemistry of Organic Compounds*. New York: John Wiley.

Evans, A.M. (1992) Enantioselective pharmacodynamics and pharmacokinetics of chiral non-steroidal anti-inflammatory drugs. *European Journal of Clinical Pharmacology* **42**, 237–256.

Hutt, A.J. (2003) Drug chirality: stereoselectivity in the action and disposition of anaesthetic agents. In: Adams, A.P., Cashman, J.N. and Grounds, R.M. (eds) *Recent Advances in Anaesthesia and Intensive Care*, Vol. 22. London: Greenwich Medical Media, pp. 31–64.

Hutt, A.J. and Caldwell, J. (1983) The metabolic chiral inversion of 2-arylpropionic acids—a novel route with pharmacological consequences. *Journal of Pharmacy and Pharmacology* **35**, 693–704.

Hutt, A.J. and O'Grady, J. (1996) Drug chirality: a consideration of the significance of the stereochemistry of antimicrobial agents. *Journal of Antimicrobial Chemotherapy* **37**, 7–32.

Lane, R.M. and Baker, G.B. (1999) Chirality and drugs used in psychiatry: nice to know or need to know? *Cellular and Molecular Neurobiology* **19**, 355–372.

Lehmann, P.A.F. (1982) Quantifying stereoselectivity or how to choose a pair of shoes when you have two left feet. *Trends in Pharmacological Sciences* **3**, 103–106.

Lough, W.J. and Wainer, I.W. (eds) (2002) *Chirality in Natural and Applied Science*. Oxford: Blackwell.

McManus, C. (2002) *Right Hand, Left Hand: The Origins of Asymmetry in Brains, Bodies, Atoms and Cultures*. London: Weidenfeld & Nicolson (Winner of the Aventis Prize, 2003).

Mescar, A.D. and Koshland, D.E. (2000) A new model for protein stereospecificity. *Nature* **403**, 614–615.

Morris, D.G. (2001) *Stereochemistry*. Cambridge: The Royal Society of Chemistry.

Nau, C. and Strichartz, G.R. (2002) Drug chirality in anesthesia. *Anesthesiology* **97**, 497–502.

Patil, P.N., Miller, D.D. and Trendelenburg, U. (1975) Molecular geometry and adrenergic drug activity. *Pharmacology Reviews* **26**, 323–392.

Pfeiffer, C.C. (1956) Optical isomerism and pharmacological action a generalisation. *Science* **124**, 29–31.

Reddy, I.K. and Mehvar, R. (eds) (2004) Chirality in Drug Design and Development. New York: Marcel Dekker.

Ruffolo, R.R. (1991) Chirality in α- and β-adrenoceptor agonists and antagonists. *Tetrahedron* **47**, 9953–9980.

Shah, R.R., Midgley, J.M. and Branch, S.K. (1998) Stereochemical origin of some clinically significant drug safety concerns: lessons for future drug development. *Adverse Drug Reactions Toxicological Reviews* **17**, 145–190.

Smith, D.F. (ed.) (1989) *Handbook of Stereoisomers: Therapeutic Drugs*. Boca Raton: CRC Press.

Tucker, G.T. (2000) Chiral switches. *Lancet* **355**, 1085–1087.

Tucker, G.T. and Lennard, M.S. (1990) Enantiomer specific pharmacokinetics. *Pharmacology and Therapeutics* **45**, 309–329.

Wainer, I.W. (ed.) (1993) *Drug Stereochemistry. Analytical Methods and Pharmacology*, 2nd edn. New York: Marcel Dekker.

Walle, T., Webb, J.G., Bagwell, E.E., Walle, U.K., Daniell, H.B. and Gaffney, T.E. (1988) Stereoselective delivery and actions of beta receptor antagonists. *Biochemical Pharmacology* **37**, 115–124.

6

Quantitative Structure–Activity Relationships (QSAR) in Drug Design

John C. Dearden and Mark T.D. Cronin

CONTENTS

Upon this gifted age rains from the sky
A meteoric shower of facts . . .
They lie unquestioned, uncombined.
Wisdom enough to leech us of our ills
Is daily spun, but there exists
No loom to weave it into fabric.
Edna St. Vincent Millay

6.1 INTRODUCTION

Over 130 years ago it was recognized[1] that ''a relationship exists between the physiological action of a substance and its chemical composition and constitution.'' In fact, any change in molecular structure, however small, will alter the physical, chemical, and biological properties of a molecule. Since each of these properties is a direct consequence of the electron distribution within the molecule, it should be possible to describe any molecular property in terms of the behavior of its electrons. However, at the present time molecular orbital (MO) theory does not allow us to compute other than the simplest properties of the simplest molecules; if, therefore, we wish to model, say, a biological activity, we must look not at its absolute value but at how this activity is altered by changes in molecular structure within a series of compounds. These changes can be represented as changes in fundamental electronic properties such as atomic charge, or as changes in physicochemical or structural properties, which themselves are a reflection of electronic behavior. In either case, the important point is that a *quantitative* (numerical) description of the change of biological activity can be given; this can be used to predict other changes in biological activity as a consequence of other changes in molecular structure. This quantitative structure–activity relationship (QSAR) can be used to design more potent or less toxic compounds in the series, and this is generally considered to be the prime function of QSAR. In addition, the correlation can be used to give some indication of the mechanism involved in the biological activity, since those properties controlling the activity should be most likely to correlate with it. It should be remembered, however, that a correlation does not necessarily imply a causal relationship.

The development of a QSAR thus requires biological endpoint values for a set of chemicals, one or more physicochemical or structural descriptors for the chemicals, and a statistical method for correlating the biological data with the descriptors. The QSAR then needs to be validated; that is, its ability to predict endpoint values for compounds other than those used to develop the QSAR has to be established.

6.2 THE DEVELOPMENT OF QSAR

The first quantitative studies of the variation of biological activity with a molecular property were those of Overton and Meyer, who showed, just before 1900, that nonspecific and reversible narcosis of tadpoles by simple compounds like alcohols and ketones was directly related to the compounds' partition coefficients, which reflect their hydrophobicity. Hansch[2] subsequently expressed Overton's results as a regression equation, i.e., a QSAR:

$$\begin{aligned} \log 1/C &= 0.94 \log P + 0.87 \\ n = 51;\ & r^2 = 0.943;\ s = 0.280 \end{aligned} \tag{6.1}$$

where C is the concentration of the compound producing narcosis in 50% of tadpoles; P is the octanol–water partition coefficient (the octanol–water system was chosen partly because of octanol's lipid-like characteristics and is generally found to be the best model); n is the number of compounds studied, r is the correlation coefficient, and s is the standard error of the estimate; r^2 is the fraction of the variance described by the descriptors, so that in Equation (6.1) $\log P$ models 94.3% of the variation of narcosis. Comment is made later on the statistics of regression and other forms of correlation analysis.

Note that Equation (6.1) is in logarithmic form, as is common with QSAR equations. There are two reasons for this. Firstly, P is an equilibrium constant, so that (from the van't Hoff isotherm) $\log P$ is proportional to free energy; QSARs are sometimes called linear free energy relationships (LFERs). Secondly, descriptor and biological activity values often range over several orders of magnitude, and it is thus preferable to use their logarithmic forms. Note also that the biological activity is, as is usual in QSAR, expressed as log $1/C$. This is because a high value of C indicates a low activity, and *vice versa*; hence $1/C$ increases as activity increases, which makes for easier comprehension of the correlation.

Despite the significant work of Overton and Meyer, it was not until 1962 that QSAR as a scientific discipline could be said to have started in earnest. In that year Hansch and coworkers[3] showed that the growth-altering effect of phenoxyacetic acid herbicides could be described by physicochemical descriptors:

$$\log 1/C = 4.08\pi - 2.14\pi^2 + 2.78\sigma + 3.36 \tag{6.2}$$

This work was remarkable for five reasons. Firstly, it showed that variation in biological activity could be described by more than one molecular descriptor (multiple linear regression, MLR); secondly, it introduced the hydrophobic substituent constant π, defined as ($\log P_{\text{derivative}} - \log P_{\text{parent}}$); thirdly, it demonstrated that within a congeneric series of compounds, substituent constants such as π and σ (the Hammett constant, a measure of the electron-directing ability of a substituent) could be used as descriptors to model the variation of biological activity; fourthly, it introduced the use of the octanol–water partition coefficient as a descriptor of hydrophobicity; and finally, it pioneered the use of the quadratic equation (parabola) to describe the variation of biological activity with hydrophobicity. Such biphasic variation is quite common, and its prime cause can be explained as follows. Compounds of low partition coefficient do not partition well into lipid membranes, and thus reach the site of action only at a low rate; on the other hand, compounds of high partition coefficient, while partitioning well into lipid membranes, do not partition well from there to the next aqueous compartment and so again reach the site of action at a low rate. Compounds of intermediate partition coefficient, having the ablility to partition reasonably well both into and out of lipid membranes, are thus more active. It should be noted that computer modeling has shown that the above explanation is tenable only for situations in which a single dose of xenobiotic is administered to the organism. Figure 6.1 illustrates the biphasic dependence of anti-inflammatory activity of aspirin derivatives on $\log P$. The lack of dependence of activity on other properties (e.g., those controlling receptor binding) does not necessarily mean that they are not important, but rather that their effect is constant throughout the series of compounds.

For situations in which a constant supply of xenobiotic is available to the organism (e.g., a fish in polluted water) it has been found that there is generally a rectilinear correlation between biological activity and $\log P$, and that the prime cause of a decrease (if any) in bioactivity with increasing hydrophobicity is low aqueous solubility, since the latter and partition coefficient are inversely related. It is for this reason that most QSARs of toxicity to aquatic species show no biphasic dependence on $\log P$, unlike those concerned with, say, mammalian toxicity. For example, the oral toxicity of saturated alcohols to the rat shows a biphasic dependence on hydrophobicity[4]:

$$\begin{gathered}\log(1/\mathrm{LD}_{50}) = 0.663\log P - 0.800\log(0.076P + 1) + 1.132 \\ n = 57;\ r^2 = 0.956;\ s = 0.241\end{gathered} \tag{6.3}$$

This is an example of the bilinear equation, developed by Kubinyi, which often gives a better fit to biphasic data than does the parabolic equation. (Note that biphasic equations pass through a maximum, indicating that an optimal $\log P$ value exists for the biological endpoint being considered.) On the other hand, the toxicity of alcohols to barnacle larvæ is rectilinearly dependent on hydrophobicity[5]:

$$\begin{gathered}\log(1/\mathrm{C}) = 0.976\log P + 0.584 \\ n = 14;\ r^2 = 0.980;\ s = 0.149\end{gathered} \tag{6.4}$$

One of the main tenets of QSAR is that all the compounds used in a study should exert their biological effect by the same mechanism, otherwise poor correlations will be observed. Since it is

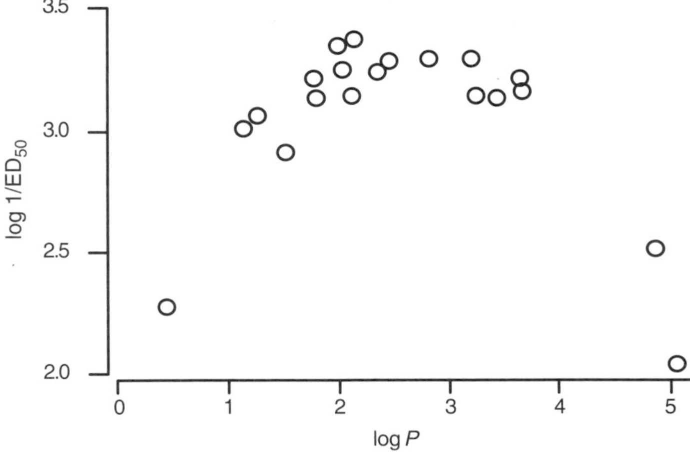

Figure 6.1 Biphasic dependence of anti-inflammatory activity of aspirin derivatives on hydrophobicity (Dearden, J.C. and George, E.; unpublished).

usually extremely difficult, if not impossible, to determine precise mechanisms of action, the assumption is usually made that members of a congeneric series act by the same mechanism, and hence QSAR studies are usually confined to congeneric series. Even so, outliers may occur, and for that reason a visual inspection of the results of a correlation analysis is often more enlightening than simply relying on the statistics of an equation.

The emphasis on congeneric series means that QSAR is essentially a lead-optimizing rather than a lead-generating method of drug design. It may be mentioned, however, that QSAR correlations that yield substituent-specific information (such as Equation (6.5)) allow some receptor-mapping to be performed, and thus can be regarded as contributing to lead generation.

Inhibition of human dihydrofolate reductase by pyrimidines[6]:

$$\begin{aligned} \log(1/K_i) &= 0.59\pi_{3,5} - 0.63\,\log(\beta \times 10^{\pi_{3,5}} + 1) \\ &\quad + 0.19\pi_4 + 0.19\,\mathrm{MR}_3 + 0.30\,\sigma + 4.03 \\ n = 38;\ r^2 &= 0.773;\ s = 0.266;\ \log\beta = -0.82 \end{aligned} \tag{6.5}$$

where MR is molar refractivity, which has the dimensions of molar volume.

In this example, different hydrophobic effects are identified at positions 3 and 5 compared with position 4, and a bulk size effect is evident only for position 3.

6.3 THE BIOLOGICAL DATA

Biological data are notorious for their variability, and clearly a good QSAR correlation will not be found if there is excessive error in these data. It is, therefore, essential that the biological data are as accurate as possible and of a consistent form; for example, a response to one dose or concentration cannot be compared with a response to a different dose or concentration.

Biological data usually take the form of a measured response to a fixed dose or concentration of xenobiotic, or a fixed response (e.g., ED_{50}, the dose that produces a 50% response in the biological test) to a range of doses or concentrations. The latter form is generally preferable, since it allows a greater numerical range of results. It should also be noted that doses or concentrations must always be in molar (e.g. mmol/kg) rather than weight terms, otherwise comparison is impossible.

The least variability is usually found with simple *in vitro* interactions such as enzyme binding, while *in vivo* data are inherently the most variable. It is therefore possible to place the greatest reliance on, and thus to draw the firmest conclusions from, QSARs relating to the simplest systems. For example, it is often possible to distinguish position-dependent effects of substitution in enzyme-binding studies, which can throw light on receptor requirements.

6.4 THE DESCRIPTORS

When a xenobiotic enters an organism, it must firstly be transported to the appropriate site of action; it must then interact with that site in order to trigger a biological response. *In vivo* transport usually involves partitioning through lipid membranes, and so can be modeled by the partition coefficient. In addition there may be metabolism of the compound, which is controlled largely by electronic factors governing bond order and also by steric effects such as shielding of a susceptible substituent. Further factors affecting transport are protein-binding and uptake into adipose tissue; both of these are a function largely of hydrophobicity, and thus can be modeled by the partition coefficient. Indeed, hydrophobicity is a key factor in the distribution and binding of xenobiotics within organisms, and some 70% of all published QSARs include a hydrophobicity term.

Table 6.1 Some commonly used QSAR descriptors

Hydrophobic descriptors	
log P	Common logarithm of the partition coefficient (usually octanol–water)
π	Hydrophobic substituent constant (log P (derivative) − log P (parent))
log k'	Capacity factor from HPLC
log S_{aq}	log (aqueous solubility)
Electronic descriptors	
σ	Hammett substituent constant
F, R	Swain–Lupton field and resonance substituent constants
HA, HD	Indicator variables for hydrogen bond acceptor and donor ability
δ	Dipole moment
q_n	Atomic charge on atom n
E_{HOMO}, E_{LUMO}	Energies of highest occupied and lowest unoccupied molecular orbitals
Steric descriptors	
MV	Molar volume
MR	Molar refractivity
ASA	Accessible surface area
E_s	Taft steric constant
L, B_1–B_5	Sterimol shape parameters
Topological descriptors	
$^n\chi$	nth order molecular connectivity
κ_n	nth order shape descriptor
S(A)	Electrotopological state index for atom A

Binding to a receptor site involves a range of electronic interactions, which can be modeled by a wide variety of electronic properties. Molecular size and shape can also influence the ability of a molecule to interact with the receptor. Mention must also be made of topological descriptors, which are derived from the connectivities (atom–atom connections) within a molecule. Table 6.1 lists some of the descriptors most widely used in QSAR studies. All of the descriptors listed, with the exception of log k', can be obtained by calculation.

It is recommended that calculated descriptors be used in QSAR modeling in preference to those requiring experimental measurement (such as NMR chemical shifts). This has two main advantages: it is much quicker to calculate descriptors than to measure them, and they can be calculated for compounds that have not been synthesized. There are now many software packages available for the calculation of descriptor values: Dragon (www.disat.unimb.it/chm/), CODESSA (www.semichem.com), and MDL QSAR (www.mdli.com) calculate hundreds of descriptors of all types; MOLCONN-Z (www.eslc.vabiotech.com/molconn) calculates topological and electrotopological descriptors; ClogP (www.biobyte.com), KOWWIN (www.epa.gov/oppt/exposure/docs/episuitedl.htm), and ChemSilico (www.logp.com), among others, calculate log P values. Dearden et al.[7] have compared the performance of a number of programs for the calculation of log P.

6.4.1 Hydrophobicity Descriptors

Partition coefficient

The most common hydrophobic descriptor is log P, where P is the n-octanol–water partition coefficient; log P values from other solvent pairs are also sometimes used. However, octanol–water log P values are by far the most readily available; thousands of such values have been measured, and software is available for their calculation (*vide ultra*). For investigations involving congeneric series, the hydrophobic substituent constant π can be used; π values are available for numerous substituents.[8]

Aqueous solubility

Aqueous solubility (S_{aq}) is inversely related to P and can be used in its place. Again, many thousands of solubility values have been measured, and software such as ChemSilico (www.logp.com) and WSKOWWIN (www.epa.gov/oppt/exposure/docs/episuitedl.htm) is available for their calculation. It should be mentioned, however, that aqueous solubilities cannot be measured or calculated as accurately as can partition coefficients.

Chromatographic parameters

Chromatographic parameters, such as R_m from reversed-phase thin layer chromatography and the capacity factor $\log k'$ from HPLC, can also be used as hydrophobic descriptors. An advantage of chromatographic methods is that they can be used for compounds whose partition coefficients are very difficult to measure, such as surfactants. Their disadvantages are that they operate only over a restricted range of $\log P$ (typically $\log P < 5$) and that they have to be obtained experimentally.

6.4.2 Electronic Descriptors

There is a vast range of electronic descriptors available, many of which are difficult to interpret physicochemically. It is best to use those whose chemical significance is understandable.

The Hammett constant

The Hammett substituent constant σ, derived from pK_a values, is a measure of the electron-directing effect of an aromatic substituent; σ values are available for many substituents.[8] Swain and Lupton's F and R values, which are the field and resonance components of the Hammett constant, are also available.[8]

Molecular orbital descriptors

Dipole moment is a whole-molecule descriptor, and can be calculated from MO theory. Atomic charges, also calculated from MO theory, can give an indication of which atoms are important in, for example, drug–receptor binding, as can frontier electron densities and superdelocalizabilities. The energy of the highest occupied molecular orbital (E_{HOMO}) and that of the lowest unoccupied molecular orbital (E_{LUMO}) are measures of electron-donating and electron-accepting ability, respectively, and have been found to be of great use in QSAR studies.

Hydrogen bonding

Hydrogen bonding is often of prime importance in drug–receptor interactions. The simplest hydrogen bonding descriptor is an indicator variable, taking the value of 1 when a molecule or substituent is capable of hydrogen bonding, and of 0 when it is not. Hydrogen bonding ability is often split into H-bond donor (HD) and H-bond acceptor (HA) ability. Thus $-OH$, NH_2 and $-COOH$ have HD values of 1 and HA values of 1, while OCH_3, NMe_2 and -COOMe have HD values of 0, but have HA values of 1. It is generally accepted that halogen atoms are not capable of significant hydrogen bonding. HD and HA values are listed by Hansch and Leo[8] for many substituents. Quantitative measures of hydrogen bonding ability are also available, for example from the Absolv (www.ap-algorithms.com) and HYBOT (www.ibmh.msk.su/qsar) software.

6.4.3 Steric Descriptors

Steric descriptors fall into two groups — those that model size and those that model shape. Size is much the easier to model and simple descriptors such as relative molecular mass (molecular weight) and molecular volume are often used.

Molecular volume

Molecular volume can be calculated by summing the van der Waals volumes of the constituent atoms, or by the use of software that rolls a water molecule over the molecular surface. Probably the simplest way of calculating molecular volume accurately is to use the McGowan characteristic volume method, which simply sums atomic and bond contributions as follows: C 16.35, H 8.71, O 12.43, N 14.39, F 10.48, Cl 20.95, Br 26.21, I 34.53, S 22.91, P 24.87; for each bond, irrespective of type, subtract 6.56. Thus for NH_2COCH_3 the value is $(2 \times 16.35) + 12.43 + 14.39 + (5 \times 8.71) - (8 \times 6.56) = 50.59\ cm^3\ mol^{-1}$.

Molecular surface area

Molecular surface area, especially accessible surface area (that surface accessible to solvent and receptor), is often a useful descriptor. It can be particularly effective if the contributions of hydrophobic and hydrophilic (polar) surface areas can be distinguished.

Molar refractivity

Molar refractivity (MR) has the units of molar volume, and is frequently used as a size descriptor. However, it is derived from refractive index, which means that it has a polarizability component, and it is sometimes used and interpreted as a polarizability term. Substituent MR values are listed by Hansch and Leo.[8]

Shape descriptors

There are only a few shape descriptors in use. Sterimol descriptors represent the length of a substituent and its widths in different directions (see Figure 6.2); they are listed by Hansch and Leo.[8] Kappa (κ) values, introduced by Kier,[9] are derived from the number of two-bond fragments (e.g., C–C–C) in a nonhydrogen molecular skeleton. Linear molecules tend to have higher kappa values, as is shown by the values for the isomeric hexanes:

n-Hexane	5.000
2-Methylpentane	3.200
3-Methylpentane	3.200
2,3-Dimethylbutane	2.222
2,2-Dimethylbutane	1.633

Kappa values are readily calculated, or can be obtained using the MOLCONN-Z software (www.eslc.vabiotech.com/molconn/).

6.4.4 3-D Descriptors

Most of the classical QSAR descriptors, with the exception of those that model shape, take no account of conformation or of the fact that most molecules are three-dimensional. Nevertheless, since a significant contribution to a molecule's biological activity arises from its fit and binding to a receptor, molecular three-dimensionality is clearly important. In recent years, therefore, much effort has gone into the examination and development of descriptors that reflect that three-

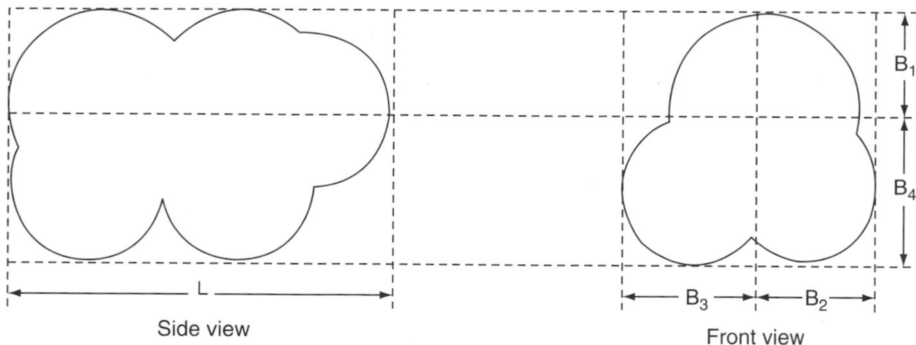

Figure 6.2 Sterimol descriptors.

dimensionality. Such descriptors can be as simple as interatomic distances or torsion angles or as complex as the distribution of electrostatic potential around a molecule. One approach that has aroused much interest is that known as comparative molecular field analysis (CoMFA) (see Section 6.7.5).

Similarity descriptors

A set of descriptors using the concept of molecular similarity has been developed by Richards, and is incorporated in the tools for structure–activity relationships (TSAR) software (www.accelrys.com). Based on equations derived by Carbo and by Hodgkin, it allows comparison of the similarity of a set of molecules to a standard (e.g., the most active in a series) on the basis of either electrostatic potential or steric parameters. Although relatively new in concept, it is finding wide application in QSAR analysis. A number of other methods of determining molecular similarity have recently been developed, and are finding use in, for example, the searching of databases for the screening of compounds for specified types of drug action.

6.4.5 Topological Descriptors

Graph theory is that branch of chemistry dealing with molecular topology, since a molecular structure is described as a graph. Graph theory is particularly concerned with the way that atoms are connected in a molecule, and many attempts have been made to relate topology to molecular properties.

Molecular connectivities

Of the topological approaches, the most successful is undoubtedly that of Kier and Hall,[10] who developed a series of topological descriptors called molecular connectivities (${}^{m}\chi$) from an original concept of Randić. The superscript m denotes the order of the descriptor. Zero-order connectivity (${}^{0}\chi$) is the simplest and is defined by Equation (6.6), where δ_i is a number assigned to each nonhydrogen atom, reflecting the number of nonhydrogen atoms bonded to it

$$ {}^{0}\chi = \sum (\delta_i)^{-(1/2)} \tag{6.6} $$

Thus for 1-butane ($C_a{-}C_b{-}C_c{-}C_d$),
$\delta_{Ca} = 1$ (because C_a is attached to C_b only),
$\delta_{Cb} = 2$ (because C_b is attached to C_a and C_c).

$$\begin{aligned}\therefore {}^0\chi &= \left(1/\sqrt{\delta_{C_a}}\right) + \left(1/\sqrt{\delta_{C_b}}\right) + \left(1/\sqrt{\delta_{C_c}}\right) + \left(1/\sqrt{\delta_{C_d}}\right)\\ &= \left(1/\sqrt{1}\right) + \left(1/\sqrt{2}\right) + \left(1/\sqrt{2}\right) + \left(1/\sqrt{1}\right)\\ &= 3.414\end{aligned}$$

The first-order connectivity (${}^1\chi$) is derived for each bond by calculating the product of the numbers associated with the two atoms of the bond. The reciprocal of the square root of this number is the bond value. Bond values are summed to give the first-order connectivity for the molecule, so that the value for 1-butane is

$${}^0\chi = \left(1/\sqrt{2}\right) + \left(1/\sqrt{4}\right) + \left(1/\sqrt{2}\right) = 1.914.$$

The ${}^1\chi$ value for 2-butane is similarly calculated to be 1.732.

Higher order connectivities are calculated by multiplying δ_i values across appropriate numbers of bonds. Heteroatoms are accounted for by use of the so-called valence molecular connectivities (${}^m\chi^v$).

The physical significance of molecular connectivities is not easy to comprehend. They contain much steric information, as the above examples of 1-butane and 2-butane show, but are perhaps best regarded as indicators of molecular complexity (e.g., branching).

A further problem with the use of molecular connectivities in drug design QSAR is that there is no easy way of translating a given ${}^m\chi$ value into a molecular feature, as one can readily do with, say, $\log P$. Recent published work is, however, addressing this problem.

Information content

Another type of topological parameter is information content. Given a molecular graph (i.e., typically the hydrogen-suppressed skeletal molecular structure), an appropriate set A of n elements is derived, based upon certain structural features. The set A is then partitioned into disjoint subsets A_i of order n_i; p_i is the probability that a selected element of A will occur in the ith subset, and is equal to n_i/n. The mean information content (IC) of an element of A is then defined by Shannon's relationship:

$$\text{IC} = \sum_{i=1}^{h} p_i \log_2 p_i \tag{6.7}$$

The total information content of the set A is then n times IC. In effect, it is a measure of molecular complexity, and although its derivation may seem somewhat abstruse, it has been found in many instances to be a useful parameter for the correlation of biological activity. Other means of calculating information content have been derived by Basak and coworkers[11] from the Shannon relationship. They have been found useful in, for example, the prediction of molecular similarity.

All of the above indices are calculated by the MOLCONN-Z software (www.eslc.vabiotech.com/molconn/).

Electrotopological state indices

The electrotopological state indices (S), developed by Kier and Hall,[12] are atom-level indices that combine the electronic character and the topological environment for each skeletal atom in a molecule. The index is formulated as an intrinsic value I_i plus a perturbation term ΔI_j arising from the electronic interaction of each atom j in the molecule. I_i is defined as $\delta(\delta^v + 1)$ for first quantum level atoms, where δ and δ^{v} are the counts of sigma and valence electrons, respectively, for a given atom, whilst ΔI_j is defined as $\sum(I_i - I_j)/r_{ij}^2$, where r_{ij} is the graph separation between atoms i and j, counted as number of atoms (including i and j). S values have been shown to correlate well with both physicochemical properties and biological activities of compounds, including those of pharmaceutical and medicinal importance.

Electrotopological state indices are calculated by the MOLCONN-Z software (www.eslc.vabiotech.com/molconn/).

6.5 STATISTICAL ANALYSIS

MLR analysis is usually used to correlate a given bioactivity with molecular descriptors. If one selects a few descriptors for correlation, it is a relatively easy matter to decide which combination of them gives the best correlation. But what is the ''best'' correlation? Standard procedure is to rely on the correlation coefficient and the standard error of the estimate, but this is not adequate, since the inclusion of virtually any additional descriptor will raise the correlation coefficient and lower the standard error. Current recommendations are to use the square of the correlation coefficient adjusted for degrees of freedom (r^2_{adj}), and to include the variance ratio, F. The significance of an individual descriptor can also be gauged from its standard error, which should be considerably smaller than its regression coefficient. The following example illustrates these points.

Suppose that 17 compounds were submitted to a pharmacological test, and that Equations (6.8–6.12) were obtained when the biological responses were correlated with various combinations of Hammett constants, Taft steric parameters (E_s), and $\log P$ values. The numbers in brackets in the equations, immediately preceding $\log P$, σ, and E_S are the standard errors of the coefficients; their significance is explained in Section 6.5.4.

Equation	n	s	r	
$\log 1/C = 5.816 + 2.342(0.105)\log P - 0.731(0.0413)(\log P)^2 + 0.0361(0.0190)\sigma + 0.195(0.1762)E_S$	17	0.126	0.957	(6.8)
$\log 1/C = 6.303 + 3.416(0.0961)\log P - 0.942(0.0114)(\log P)^2$	17	0.141	0.938	(6.9)

$F_{2,14} = 11.9;\ F_{2,14}\ \alpha,\ 0.05 = 3.74;$
$F_{2,14}\ \alpha,\ 0.001 = 11.78$

Equation	n	s	r	
$\log 1/C = 2.416 + 4.981(0.994)\log P$	17	0.314	0.895	(6.10)
$\log 1/C = 2.002 + 0.0714(0.0697)\sigma$	17	0.280	0.246	(6.11)
$\log 1/C = 5.972 + 0.0146(0.0209)E_S$	17	0.303	0.209	(6.12)

Much information can be derived from these equations, as explained below.

6.5.1 Correlation Coefficient

The correlation coefficient of Equation (6.8), because it is close to 1.00, indicates that the relationship represents the experimental results reasonably well, and explains $0.957^2 \times 100 = 91.6\%$ of the variation. However, if the steric and electronic parameters are omitted, to give Equation (6.9), the new equation still has a good correlation coefficient. It is doubtful whether the correlation coefficient of Equation (6.8) is significantly better, since it contains two extra variables. If there were 17 variables for example, r would equal 1.000, irrespective of the data. Further evidence comes from the correlation coefficients of Equation (6.11) and Equation (6.12), which are very low, indicating that neither σ nor E_S contributes to biological activity. Furthermore, the significantly lower correlation coefficient of Equation (6.10) in comparison with that of Equation (6.9) suggests that the relationship between $\log 1/C$ and $\log P$ is biphasic.

One of the problems of using the correlation coefficient r as a measure of goodness-of-fit is that inclusion of more parameters in the equation, be they relevant or not, will always increase r. However, most statistical packages now include the calculation of an r^2 value (r^2(adj)) adjusted for the degrees of freedom (see later) in the correlation. In the example above, r^2(adj) for Equation (6.8) is 0.930, whereas that for Equation (6.9) is 0.934. This clearly indicates that the two additional terms in Equation (6.8) do not improve the correlation.

6.5.2 Regression Coefficient

The coefficients of the variables give support to the evidence given by the correlation coefficients. The coefficients of σ and E_S in Equation (6.11) and (6.12) are small, and in comparison with the intercepts and coefficients of $\log P$ and $(\log P)^2$, suggest that Equation (6.11) and Equation (6.12) represent plots in which the regression lines would be almost parallel with the E_S axes. The larger coefficients in $\log P$ in Equation (6.8) and Equation (6.9) support the conclusions given by the correlation coefficients that biological activity is dependent on the hydrophobic nature of the compounds under test.

The lower coefficients of $(\log P)^2$ might give the impression that the squared term is not important; for example, the coefficient of $(\log P)^2$ in Equation (6.8) is only 0.731, in comparison with 2.342 with $\log P$. However, it must be remembered that $(\log P)^2$ is usually bigger than $\log P$. Thus if P is of the order of 1000, $(\log P)^2 = 3 \log P$, and one would anticipate a correspondingly larger coefficient for $\log P$, as is the case in Equation (6.8) and Equation (6.9). A similar pitfall occurs when parameters of considerably different magnitude are compared; for example Equation (6.13), in which V is molar volume (molecular weight/density) in $\text{cm}^3\,\text{mol}^{-1}$, suggests that molecular size is not a controlling factor. However, molar volumes are of the order of $10^2\,\text{cm}^3$, while σ has a value less than 1.00. In this light the coefficients are comparable. It is preferable that parameters should be scaled to roughly the same numerical values. This has the advantage that coefficients can be readily compared, and also improves the stability of the statistical analysis

$$\log 1/C = 1.313 + 2.456\sigma + 0.02456V \tag{6.13}$$

6.5.3 Standard Error of the Estimate

The standard error of the estimate should be as low as possible. Generally standard error decreases as the correlation coefficient increases. The values given in Equations (6.8)–(6.12) support the conclusion drawn from the correlation coefficients and regression coefficients.

Standard error of the coefficient

The figure in brackets following each regression coefficient represents the standard error of the coefficient, which means that if the experiment is repeated the coefficient should lie between these limits; for example, the coefficient for σ in Equation (6.8) should be 0.0361 $\pm$ 0.0190. Obviously the higher the standard error, the less reliable is the coefficient, and the less is the likelihood that the variable it represents is related to the biological response.

The confidence in the term can be assessed by dividing the coefficient by the standard error. Thus for the second term on the right hand side of Equation (6.8), the ratio is 2.342/0.105 = 22.3,

Table 6.2 Student *t* values

Degrees of freedom	Probability	
ϕ	0.05	0.01
12	2.179	3.055
13	2.160	3.012
14	2.145	2.977
15	2.132	2.947

which, because it is a high number, suggests that the term is important. When there is doubt whether the ratio can be considered sufficiently high, it may be compared with the limiting Student's *t* value. Most statistical text books give tables of these. Some limiting *t* values are given in Table 6.2, and it can be seen that they depend on the probability level and on the number of degrees of freedom. The probability level is generally taken as 0.05 for QSAR studies. The number of degrees of freedom (ϕ) is $(n-m-1)$, where n is the number of sets of data (17 in the example) and m is the number of variables (4 in Equation (6.8)), is therefore 12, which from Table 6.2 gives a t value of 2.179 for a probability of 0.05. Since this is less than 22.3, the term in $\log P$ is significant. The same test rejects the σ and E_s terms in Equation (6.8).

*Fisher statistic (*F*) values*

For convenience, F distribution results are given only for Equation (6.9). The two numbers in the subscript (2 and 14) following the letter F are m and $n-m-1$, as defined in the previous paragraph, and the 11.9 following is the experimental F value which fits the data. The F value indicates the probability that the equation is a true relationship between the results, or is merely coincidence. If the experimental figure exceeds the limiting value, the relationship is a true one, within the given probability level. Limiting F values can be obtained from statistical tables, from which Table 6.3 has been abstracted. The numbers running along the top represent the first number in the subscript following F, and those running down the left hand side, the second number in the subscript. Thus for Equation (6.9), $v_1 = 2$ and $v_2 = 14$ in Table 6.3, giving $F = 3.74$. Table 6.3 is based on the probability of 0.05; therefore there is less than a 1 in 20 chance that the relationship is a coincidence, and hence better than a 19 in 20 chance that the results are truly related in the manner given. $F = 11.9$ is obviously much greater than 3.74, and it would be of interest to know precisely how good it is. Consultation of a table for 0.001 points of the F distribution gives $F_{2,14}\ \alpha$, $0.001 = 11.78$, so that the probability of Equation (6.9) representing a chance relationship is less than 1 in 1000.

The mechanism of calculating the statistical parameters of regression, used above, is considered to be outside the scope of this chapter, which seeks to explain the interpretation, rather than the preparation, of QSAR data. The necessary arithmetic is usually built into the computer program.

Table 6.3 0.05 Probability points of the *F*-distribution

v_1	1	2	3	4	5
$v_2 = 1$	161.4	199.5	215.7	224.6	230.2
2	18.5	19.0	19.2	19.2	19.3
12	4.75	3.89	3.49	3.26	3.11
13	4.67	3.81	3.41	3.18	3.03
14	4.60	3.74	3.34	3.11	2.96

6.6 SOME PUBLISHED QSARs

A selection of published MLR QSAR correlations is given below, to exemplify the range of biological endpoints and descriptors used.

Toxicity of barbiturates to the mouse[13]:

$$\begin{aligned} \log(1/\mathrm{LD}_{50}) &= 1.02\ \log P - 0.27\,(\log P)^2 + 1.86 \\ n = 13;\ r^2 &= 0.852;\ s = 0.113 \end{aligned} \tag{6.14}$$

Blood levels of drugs and poisons causing human death[14]:

$$\log(1/C) = -0.896 \log S_{\mathrm{aq}} + 1.75 \tag{6.15}$$

where S_{aq} is aqueous solubility.

COX–2 inhibition by diarylspiro[2,4]heptenes[15]:

$$\begin{aligned} \log(1/C_{50}) &= -0.207\ \mathrm{MR}_{3,5} - 1.472\ \pi_4 - 1.604\ \mathrm{HA}_4 \\ &\quad - 1.360\ I_4 + 0.437 I_{\mathrm{R}} + 9.858 \\ n = 23;\ r^2 &= 0.893;\ s = 0.219 \end{aligned} \tag{6.16}$$

where I_4 and I_{R} are indicator variables for the presence of certain molecular features.

Antiemetic activity of 4-substituted 5-nitro–2-methoxy-*N*-(2-diethyl-aminoethyl)benzamides[16]:

$$\begin{aligned} \log A &= 0.914\ L - 0.140\ L^2 - 0.514 \\ n = 15;\ r^2 &= 0.846;\ s = 0.279 \end{aligned} \tag{6.17}$$

where L is Sterimol substituent length.

Olfactory threshold concentration of alkanes[17]:

$$\begin{aligned} \log(1/C) &= 2.57\ \log P - 0.24(\log P)^2 + 1.36 \\ n = 7;\ r^2 &= 0.941;\ s = 0.39 \end{aligned} \tag{6.18}$$

Odor character (benzaldehyde-likeness) of benzaldehydes and nitrobenzenes[18]:

$$\begin{aligned} \mathrm{BL} &= -8.08\,{}^3\chi^{\mathrm{v}}_{\mathrm{p}} + 2.19\,{}^4\chi^{\mathrm{v}}_{\mathrm{pc}} + 13.6 \\ n = 15;\ r^2 &= 0.926;\ s = 0.545 \end{aligned} \tag{6.19}$$

where ${}^3\chi^{\mathrm{v}}_{\mathrm{p}}$ and ${}^4\chi^{\mathrm{v}}_{\mathrm{pc}}$ are third-order valence path molecular connectivity and fourth-order valence path–cluster molecular connectivity.

Inhibition of influenza virus by benzimidazoles[19]:

$$\begin{aligned} \log(1/K_i) &= 4.27\ S(\text{N-1},3) + 0.79\ S(\text{C-2}) - 16.00 \\ n = 15;\ r^2 &= 0.91;\ s = 0.16 \end{aligned} \tag{6.20}$$

where S(N-1,3) and S(C-2) are electrotopological state indices for nitrogen in the 1- and 3-positions and carbon in the 2-position.

Physicochemical and other properties can also be correlated in this way; these are generally termed quantitative structure–property relationships (QSPRs). For example, Dearden et al.[20] were able to model Henry's law constant H (the air–water partition coefficient) using seven descriptors:

$$\begin{aligned} \log H = & -0.294\ \mathrm{HB_N} - 0.957\ \mathrm{HB_I} - 1.86\ \Delta\mathrm{MR} + 0.998\ \log P \\ & - 1.11\mathrm{MR} + 0.356\mathrm{BI_{dw}}/100 + 0.229\,{}^4\chi_\mathrm{p}^\mathrm{v} + 0.579 \\ n = 294;\ & r^2_\mathrm{adj} = 0.874;\ s = 0.769;\ F = 292.5 \end{aligned} \tag{6.21}$$

where $\mathrm{HB_N}$ is the number of H-bonds that a molecule can form, $\mathrm{HB_I}$ is a H-bond indicator variable, ΔMR is excess molar refractivity (molar refractivity of compound minus that of a straight-chain alkane of the same molecular weight) and $\mathrm{BI_{dw}}$ is the Bonchev index.

6.7 MULTIVARIATE ANALYSIS

A drawback of multiple regression analysis is that in order to minimize the risk of chance correlations, the ratio of observations to independent variables should be kept reasonably high. Topliss and Costello[21] recommended in 1972 that the ratio be kept at 5:1 or above. This ''rule'' is still too often broken (see Equation (6.18)).

A method of checking for chance correlations is to randomize the biological activity values and develop a QSAR for the randomized data. This procedure can then be repeated up to, say, 100 times. For a 1% risk that a correlation has occurred by chance, only one correlation out of the 100 should be acceptably high.

If one knows the factors influencing a given activity, one needs to utilize only the molecular descriptors that model those factors. However, there is usually no way of telling *a priori* which descriptors will best correlate with the bioactivities of a set of compounds. It is therefore common practice in QSAR analysis to generate, through the use of computational chemistry and the availability of topological indices, very large numbers of descriptors for each compound studied. One then needs to select from among these the descriptors that will best model the biological activity; this is usually done by the use of stepwise regression, best subsets regression, or preferably a genetic algorithm.

6.7.1 Data Reduction

Hyde[22] has emphasized the necessity of data reduction in so-called ''over-square'' matrices with many descriptors per compound; such reduction reduces computer time and also, through elimination of highly collinear descriptors, increases the stability of the statistical analysis and reduces the risk of chance correlations. There are several ways to carry out data reduction.

Cluster analysis

One data reduction method is cluster analysis, which groups similar objects (in this case, descriptors) together in a dendrogram, and thus indicates which descriptors can be deleted without significant loss of information. The level of similarity required can be selected from the dendrogram.

Principal components analysis

Another method is principal components (PC) analysis, in which the descriptors are combined into a smaller number of terms, called principal components, each of which is orthogonal to (i.e., uncorrelated with) the others. It is usually found, in such analysis, that a small number of principal components describes a large proportion of the variation in the biological data. For example, Cronin[23] found that a data set of 49 highly collinear variables could be reduced to five principal components describing 93.7% of the variation of the original variables. The principal components themselves have no physical significance, but can be correlated with the original variables to see which they best represent. Thus, for example, it might be found that the first PC represented largely hydrophobic terms, while the second reflected steric effects. A variation of principal components analysis is factor analysis, in which the components are rotated in space in order to aid interpretation.

6.7.2 Partial Least Squares Analysis

A refinement of principal components analysis is partial least squares (PLS) analysis. This technique carries out the formation of principal components and the multiple regression in a single step, and in addition is designed to give maximal correlation between the principal components and the dependent variable. PLS can accommodate large numbers of descriptors per compound, and is not adversely affected by high collinearities between descriptors. There is thus no need for data reduction.

6.7.3 Canonical Correlation Analysis

Canonical correlation analysis is a statistical technique that can be employed to correlate simultaneously two or more different biological effects of the same set of compounds. For example, Szydlo et al.[24] were able to distinguish the factors contributing to knockdown activity and toxicity to mustard beetles of some benzyl cyclopropane-1-carboxylate esters.

6.7.4 Artificial Neural Networks

A recently developed methodology for QSAR correlation is that of artificial neural networks (ANNs). ANNs have the ability to model nonrectilinear correlations, and thus can often model a data set better than can MLR. However, ANNs need careful training to give valid results, and they do not yield a regression equation. Furthermore, they are not transparent, in that the descriptors that have contributed to the correlation are not indicated. Maddelena[25] has reviewed the application of ANNs to QSAR.

6.7.5 Comparative Molecular Field Analysis

Most molecules are three-dimensional (3-D) and many receptor-binding sites are also 3-D. It is therefore not surprising that descriptors that model this three-dimensionality are often important. The most widely used 3-D QSAR approach is CoMFA (www.tripos.com), in which a probe atom is used to calculate the electronic and steric fields at many points in a three-dimensional lattice. This involves firstly aligning the molecules to be studied, within a three-dimensional grid or lattice. This is a simple procedure for most congeneric series, whereby the common features of the molecules can readily be superimposed. For noncongeneric series, with no obvious common features, alignment is much more difficult and more subjective. Hence most CoMFA studies to date have been concerned with congeneric series. A probe atom is then placed at each lattice point in turn, and the steric (Lennard–Jones) and electrostatic (Coulombic) fields exerted by each molecule at each lattice point are then calculated. This results in a large number of data points, and PLS statistics is used to determine the minimal set of data points necessary to model the set of compounds according to their biological activities. The PLS model then has to be cross-validated, for example by the leave-one-out (LOO) method (see Section 6.10). If necessary and appropriate, re-alignment of the most poorly modeled compounds can then be carried out, and the above steps repeated. The contoured QSAR coefficients can then be displayed to allow visualization of regions where electrostatic or steric fields have the greatest effect on activity. Martin[26] has recently reviewed the scope and applications of 3-D QSAR.

6.7.6 Molecular Similarity

It is widely accepted that similar compounds possess similar properties. Hence measures of similarity can be used to correlate biological activities. Bartlett *et al.*[27] have used shape similarity to model the incidence of cutaneous rash from oral penicillins (%ROP):

$$\sqrt{\%ROP} = 3.824\ \text{Sim.BP}\ -1.754$$
$$n = 14;\quad r^2 = 0.823;\quad s = 0.181 \qquad (6.22)$$

where Sim.BP represents shape similarity to benzylpenicillin.

Johnson and Maggiora[28] and Dean[29] have discussed similarity approaches to drug design in detail. A very recent approach is comparative molecular similarity index analysis (CoMSIA).

6.8 FREE–WILSON ANALYSIS

This is an alternative procedure to MLR analysis, in that so-called *de novo* substituent constants based on biological activities are used, rather than physical properties. As one of their examples, Free and Wilson[30] used the antimicrobial activities of some 6-deoxytetracyclines (**6.1**) against *Staphylococcus aureus*. The compounds they examined are summarized in Table 6.4, together with their antimicrobial activities.

(**6.1**)

Biological activities can be expressed in terms of the constituent groups in the molecules; for example, Equation (6.23) can be used to describe the antimicrobial activity of the first compound in Table 6.4.

$$\mu - a[\text{H}] + b[\text{NO}_2] + c[\text{NO}_2] = 60 \qquad (6.23)$$

where μ is the overall average antimicrobial activity for the whole series, and a, b, and c the contributions of the groups –R, –X, and –Y, respectively. The identity of the terms prefixed by the letters a, b, and c can best be explained if it is imagined that the contributions to the total antimicrobial activities made by the groups in position R can be determined experimentally, and are

Table 6.4 Antimicrobial activities of 6-deoxytetracyclines against *Staphylococcus aureus*

Compound	R		Supposed partial biological activity	X			Y			Experimental biological activity
	H	CH_3		Br	Cl	NO_2	NO_2	NH_2	CH_3CONH	
I	✓	—	5.9	—	—	✓	✓	—	—	60
II	✓	—	2.2	—	✓	—	✓	—	—	21
III	✓	—	1.4	✓	—	—	✓	—	—	15
IV	✓	—	50.4	—	✓	—	—	✓	—	525
V	✓	—	32.5	✓	—	—	—	✓	—	320
VI	✓	—	29.2	—	—	✓	—	✓	—	275
VII	—	✓	13.2	—	—	✓	—	✓	—	160
VIII	—	✓	5.2	—	—	✓	—	—	✓	15
IX	—	✓	14.6	✓	—	—	—	✓	—	140
X	—	✓	6.0	✓	—	—	—	—	✓	75

given in column 4 of Table 6.4. a[H] will then be defined by Free–Wilson analysis as the mean of the figures in column 4 involving R = H, minus the means of all the figures in column 4, i.e.,

$$a[\mathrm{H}] = \frac{(5.9 + 2.2 + 1.4 + 50.4 + 32.5 + 29.2)}{6} - \frac{160.6}{10} = 4.2 \qquad (6.24)$$

Similarly $a[\mathrm{CH_3}] = -6.3$.

Obviously it is not possible experimentally to determine partial biological activities of this sort, but the calculation given above serves to show that:

$$6a[\mathrm{H}] + 4a[\mathrm{CH_3}] = 6 \times 4.2 - 4 \times 6.3 = 0 \quad \text{or} \quad 4a[\mathrm{CH_3}] = 6a[\mathrm{H}] \qquad (6.25)$$

and this relationship applies irrespective of the numbers displayed in column 4 of Table 6.4. Similarly,

$$4b[\mathrm{Br}] + 2b[\mathrm{Cl}] + 4b[\mathrm{NO_2}] = 0 \qquad (6.26)$$

and

$$3c[\mathrm{NO_2}] + 5c[\mathrm{NH_2}] + 2c[\mathrm{CH_3\,CONH}] = 0 \qquad (6.27)$$

Table 6.4 yields ten equations analogous to Equation (6.23), with nine unknowns, μ, and the contributions of R = $-$H or $-CH_3$, X = $-NO_2$, or $-$Br or $-$Cl and Y = $-NO_2$ or $-NH_2$ or CH_3CONH-. μ can be equated to the mean experimental response, and three of the remainder can be eliminated through Equations (6.25–6.27), leaving five unknowns. Calculation of the best values to fit the ten equations can be carried out using a computer, and gives the substituent constants

Table 6.5 Calculated substituent constants for antimicrobial activities of 6-deoxytetracyclines against *Staphylococcus aureus*[a]

R		X		Y	
a[H]	75	b[Cl]	84	$c[NH_2]$	123
$a[CH_2]$	−112	b[Br]	−16	$c[CH_3CONH]$	18
		$b[NO_2]$	−26	$c[NO_2]$	−218

[a]Free, S.M. and Wilson, J.W. (1964) A mathematical contribution to structure–activity studies. *Journal of Medicinal Chemistry* 7, 395–399.

shown in Table 6.5, from which the antimicrobial properties of new compounds can be predicted. Thus for example, if R $= -CH_3$, X $= -Cl$ and Y $= -NH_2$, the predicted antimicrobial activity will be $1606/10 - 112 + 84 + 123 = 256$. 1606 is the total experimental biological activity.

The major weakness of this approach is that it can be used only for relationships which are rectilinear. However, the technique has been extended to parabolic relationships by introducing terms representing interactions between substituent groups, and by using equations involving both Hansch and Free–Wilson descriptors. In recent investigations, activity has been replaced by log activity, which is related to free energy, and therefore additive. Another innovation is that the activity of the unsubstituted compound (in which the substituent is hydrogen) is used as standard, thereby eliminating the need for restricting equations.

6.9 SERIES DESIGN

It is good practice to select the training set carefully (the set of compounds from which the QSAR is to be developed), so as to encompass a good range of values of hydrophobic, electronic, and steric properties. Too often this is not done; the training set is selected on other criteria such as availability or ease of synthesis. As a consequence, any QSAR that is obtained can give only limited information. To take a very commonly occurring example, if a series of compounds consists simply of homologues with different alkyl chain lengths, there will be virtually no difference in many electronic properties between the compounds and hence a QSAR will provide no information on the importance of electronic effects on bioactivity. Furthermore, since $\log P$ and molecular volume are collinear in a homologous series, correlation analysis would not be able to distinguish between hydrophobic and steric effects. Equation (6.18) falls into this category. It is therefore important not to incorporate descriptors that are highly collinear into a QSAR.

One simple approach to the design of a training set is based on cluster analysis. Hansch et al.[31] subjected a large number of substituents to this procedure, which clusters them according to similarity of properties. Selection of substituents from differing clusters thus ensures a good range of molecular properties. Other workers[32,33] have proposed various other approaches to series design. A good review of series design has been given by Pleiss and Unger.[34]

6.10 VALIDATION OF A QSAR

The prime purpose of developing a QSAR is usually to enable prediction of more active or less toxic compounds to be made. It is often assumed that provided the correlation is a ''good'' one (as indicated typically by a high correlation coefficient), then the QSAR can be used to give reliable predictions of bioactivity. However, this is by no means always true.

Firstly, a high correlation coefficient can be obtained from a poorly distributed set of training compounds, for example from two clusters of compounds (see Figure 6.3). Secondly, a QSAR cannot be expected to give reliable prediction outside the descriptor range of the training set; thus if compounds of $\log P$ range 0–6 were used to develop the QSAR, it cannot be expected to predict the bioactivity of a compound with $\log P = 10$. Thirdly, a QSAR cannot be expected reliably to predict the activity of a compound that is unlikely to act by the same mechanism; for example a QSAR developed for the toxicity of simple phenols would probably not be able to predict accurately the toxicity of a phenolic steroid.

It is thus important to validate a QSAR before using it for predictive purposes. The best way in which this is done is to leave out, typically by random selection, a number of compounds from the training set and to use those as the test set. Up to 50% of compounds can be left out, provided that the remaining compounds form a reasonable training set, in terms of both the number of compounds and the range of descriptor values spanned. This process can then be repeated a number of times. If the QSAR has good predictivity, the activities of the test set compounds should be well predicted.

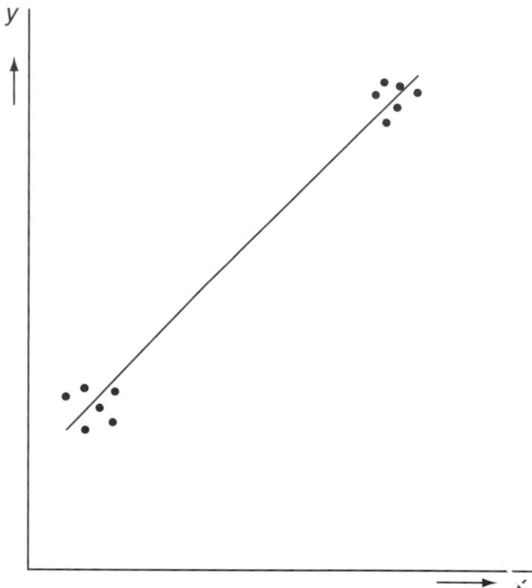

Figure 6.3 Distribution consisting of two clusters.

Another way to assess predictability, especially if a data set contains relatively few compounds, is to use a cross-validation technique (available in many statistics software packages) to derive a cross-validated r^2 value. A common way in which this is done is by the leave-one-out (LOO) method: one compound is removed from the data set and a QSAR correlation obtained for the remaining compounds is used to predict its activity. The compound is then returned to the training set, another one is removed, and a second QSAR correlation is obtained. This procedure is repeated until all the compounds in turn have been removed. The cross-validated r^2 value (r^2(CV) or Q^2) is then computed. It will be lower than the ordinary r^2 value, but should not be too much lower if the QSAR correlation is to be valid for predictive purposes. Opinions differ on how much lower is acceptable; Walker et al.[35] have suggested that for good predictability, ($r^2 - Q^2$) should not exceed 0.3. However, it should be noted that the LOO technique can overestimate the predictivity of a QSAR correlation.

6.11 TREATMENT OF NONCONTINUOUS DATA

Not infrequently, bioactivity is determined on a noncontinuous scale. For example, a compound may be reported as being carcinogenic or noncarcinogenic, or test results may be given as +++, ++, +, −. Such data cannot be handled by regression analysis, although Moriguchi and Komatsu[36] have developed an approach known as adaptive least squares analysis for noncontinuous data.

The more usual approach with noncontinuous data is to use some form of pattern recognition, whereby molecular features (which can include structural features and physicochemical properties) are used to place the compounds into the appropriate class of bioactivity. For example, Barratt[37] used principal components to distinguish between skin-corrosive and non-skin-corrosive compounds (Figure 6.4).

In practice, there is rarely complete discrimination, so the percentage of correct classifications is used as a measure of the predictive ability of the discriminant analysis. Rose and Jurs[38] used a total of 22 topological and MO-based descriptors to give 97% correct classification of the carcinogenicity of 150 nitrosamines. Lewis et al.[39] have developed a carcinogenicity discrimination model, COMPACT (based on the ability of compounds to induce cytochrome P450–1A) that uses only two molecular properties, a shape descriptor and the energy of the lowest unoccupied MO. Such procedures should, as with regression analysis, be subjected to cross-validation. Livingstone[40] has briefly reviewed pattern recognition methods in drug design.

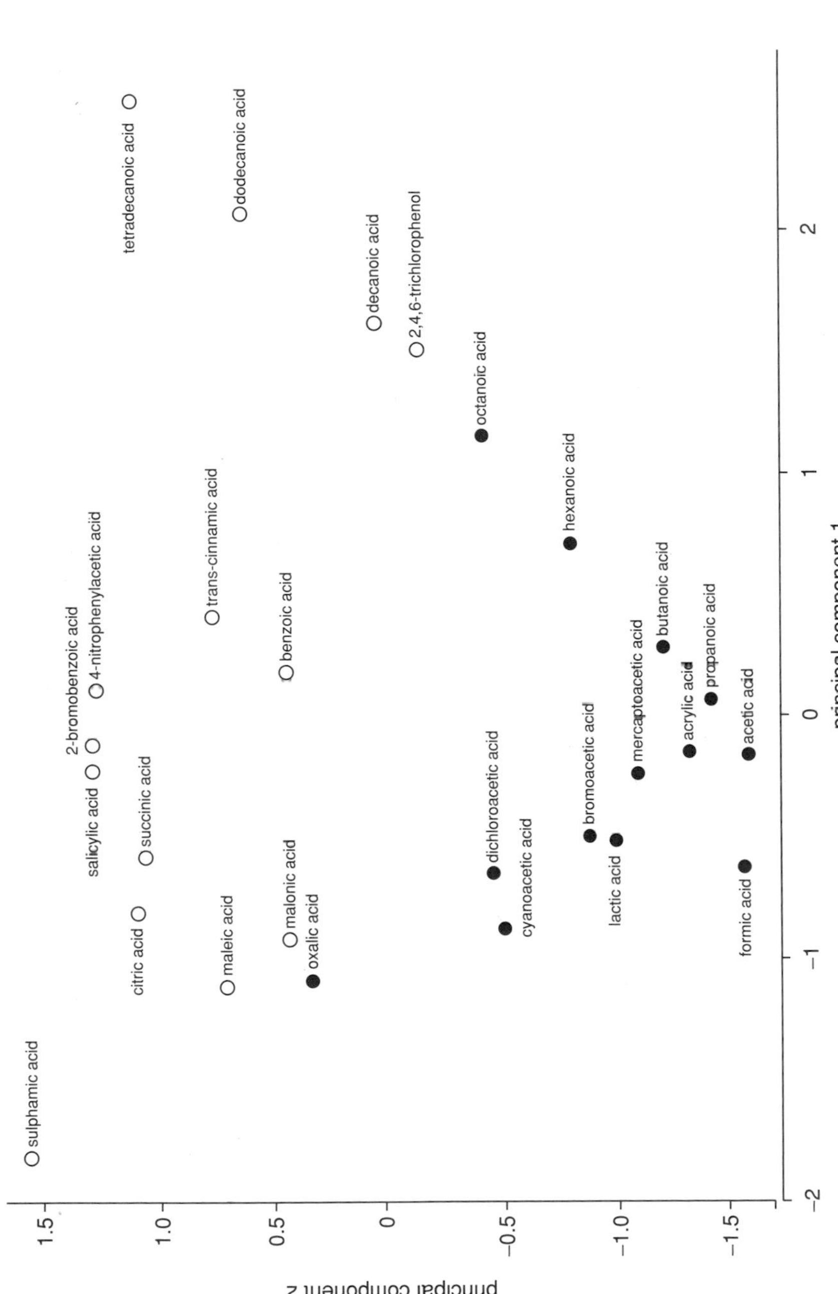

Figure 6.4 Principal components analysis of skin corrosivity of organic acids (reproduced with permission from Barratt, M.D. (1995) *Alternatives to Laboratory Animals* 23, 111–122). The principal components were derived from log *P* values, molecular volumes, melting points, and pK_a values. ● = corrosive; ○ = non-corrosive.

6.12 SOME LIMITATIONS AND PITFALLS OF QSAR

The concept of QSAR has been continually under fire since it was first introduced, but most of the shots have been directed against the ways in which the technique has been used, rather than against the overall idea. QSAR is dependent on the accuracy of the biological results which, by their nature, are susceptible to considerable experimental error. There is therefore a built-in scatter which cannot be explained mathematically. The true relationships can be hidden within this scatter (or alternatively, false correlations can evolve) and their failure to fit as closely as is desirable is blamed on biological variation. Accurate biological data are essential for this technique to work well. Concentration and dose data should always be presented in molar units (e.g., mmol kg^{-1}), otherwise they are not comparable.

The success of QSAR predictions is highly dependent on the number of results from which they are derived; the greater the number, the more reliable the correlation. Five biological results for every variable on the right hand side of the correlation equation are generally regarded as a minimum acceptable level because of the risk of chance correlations,[21] and the more this figure is exceeded the better. Correlations derived from the first results emanating from a structure–activity analysis can change considerably when more results come to hand. Coefficients can change, and physicochemical descriptors that were originally considered significant can cease to be important. Many of the earlier publications on QSAR were based on too few results.

It is also important to have a good distribution of compounds in the training set. Figure 6.3 shows an example of a training set that would yield a QSAR with a high correlation coefficient, but which gives no useful information, since only two very small regions of descriptor space are covered by the training set compounds. The line joining the two clusters of points could in fact be of any form.

A further concern is that, even though a QSAR correlation appears to be extremely good, it may not be able accurately to predict the activities of other compounds. This is because the QSAR has been developed using a certain number of compounds (the training set) having some molecular characteristics. Unless the compounds whose activities are to be predicted by the QSAR (the test set) have similar molecular characteristics, the prediction will not be accurate. This brings us to two basic tenets of QSAR development and application.

Firstly, the compounds used in the training set should span a sufficiently wide range of descriptor space; this means (i) that a reasonable range of values of any one parameter should be covered (e.g., $\log P$ values from 0 to 5) and (ii) there should be, in the training set, compounds showing variation in all the different types of descriptor — hydrophobic, electronic, and steric. An example of a poor training set in this respect would be a homologous series, with compounds differing only in the length of an alkyl chain. (Sadly, this is still sometimes seen; see Equation (6.18).) Such a training set would have two grave faults: (a) there is little or no variation in electronic descriptors as alkyl chain length increases and (b) hydrophobicity and size of alkyl groups are highly collinear, and so hydrophobic and steric effects could not be differentiated.

The second basic tenet arising from the above is that the test set compounds should cover the same range of descriptor space as do those of the training set. For example, if a training set covered a $\log P$ range of 0 to 5, the QSAR should not be used to predict the activity of a compound with, say, $\log P = 8$.

6.13 QSAR SOFTWARE

There are numerous software packages commercially available for generating descriptors or carrying out QSAR analysis; a detailed list is given by Grover et al.[41]. Some widely used packages are SYBYL (www.tripos.com), CAChe (www.cachesoftware.com), TSAR (www.accelrys.com),

CODESSA (www.semichem.com), ClogP (www.biobyte.com), KOWWIN (www.epa.gov/oppt/exposure/docs/episuitedl.htm), MOLCONN-Z (www.eslc.vabiotech.com/molconn/), MDL QSAR (www.mdli.com), MOPAC (www.schrodinger.com), HYBOT-PLUS (www.ibmh.msk.su/qsar), and Absolv (www.ap-algorithms.com).

A Russian software package, PASS, predicts the likelihood that a compound will possess one or more of about 900 pharmacological and toxicological activities. It can be used on the internet free of charge (www.ibmh.msk.su/PASS/). There are also a number of commercially available toxicity prediction packages, such as TOPKAT (www.accelrys.com), CASETOX (www.multicase.com), HAZARDEXPERT (www.compudrug.hu), and DEREK (www.lhasalimited.org), which can predict whether a compound is likely to be toxic (various toxicity endpoints being available) from a consideration of substructural features or from QSAR correlations.

6.14 CONCLUSIONS

QSARs are increasingly used to predict a wide range of activities and toxicities of drugs, pesticides, food additives, and environmental pollutants. They require, for successful application, reliable bioactivity data for a carefully selected training set of compounds from which to construct the QSAR.

QSARs need to be validated if they are to have good predictive ability. Even so, current knowledge and expertise are insufficient to guarantee fully the reliability of the prediction. They are, nevertheless, slowly being accepted for regulatory purposes and they do serve as a valuable guide for further testing and for priority setting. They find wide application in the prediction of drug potency, and are increasingly used for prediction of drug toxicity.

Used wisely, QSAR is the loom that weaves the scattered threads of our wisdom into fabric.

FURTHER READING

Cronin, M.T.D. and Livingstone, D.J. (Eds.) (2004) *Predicting Chemical Toxicity and Fate*. CRC Press, Boca Raton, FL.

Devillers, J. (Ed.) (1996) *Neural Networks in QSAR and Drug Design*. Academic Press, London.

Hansch, C. and Leo, A. (1995) *Exploring QSAR: Fundamentals and Applications in Chemistry and Biology*. American Chemical Society, Washington DC.

Karcher, W. and Devillers, J. (Eds.) (1990) *Practical Applications of Quantitative Structure–Activity Relationships* (QSAR) *in Environmental Chemistry and Toxicology*. Kluwer Academic Publishers, Dordrecht.

Kier, L.B. and Hall, L.H. (1986) *Molecular Connectivity in Structure–Activity Analysis*. John Wiley & Sons, Inc., New York.

Kier, L.B. and Hall, L.H. (1999) *Molecular Structure Description: the Electrotopological State*. Academic Press, San Diego, CA.

Kubinyi, H. (1993) *QSAR: Hansch Analysis and Related Approaches*. VCH: Weinheim.

Kubinyi, H. (Ed.) (2000) *3D QSAR in Drug Design: Theory, Method and Applications*. Kluwer Academic Publishers/ESCOM, Dordrecht.

Livingstone, D. (1995) *Data Analysis for Chemists: Applications to QSAR and Chemical Product Design*. Oxford University Press, Oxford.

Mason, J.S., (Ed.) (2005) *Comprehensive Medicinal Chemistry* II, Vol. 4: *Computer-Aided Drug Design*. Elsevier, St. Louis, MO.

van de Waterbeemd, H. (Ed.) (1996) *Structure–Property Correlations in Drug Research*. R.G. Landes Company, Georgetown, TX.

REFERENCES

1. Crum Brown, A. and Fraser, T. R. (1868–69) On the connection between chemical constitution and physiological action I. On the physiological action of the salts of the ammonium bases, derived from strychnia, brucia, thebaia, codeia, morphia, and nicotia. *Transactions of the Royal Society of Edinburgh* **25**, 151–203.
2. Hansch, C. (1971) Quantitative structure–activity relationships in drug design. In Ariëns, E. J. (Ed.), *Drug Design*; Vol. 1. Academic Press, New York, NY, pp. 271–342.
3. Hansch, C., Maloney, P.P., Fujita, T. and Muir, R.M. (1962) Correlation of biological activity of phenoxyacetic acids with Hammett substituent constants and partition coefficients. *Nature* **194**, 178–180.
4. Lipnick, R.L., Pritzker, C.S. and Bentley, D.L. (1985) A QSAR study of the rat LD_{50} for alcohols. In Seydel, J.K. (Ed.), *QSAR and Strategies in the Design of Bioactive Compounds*. VCH, Weinheim, pp. 420–423.
5. Hansch, C. and Dunn, W.J. (1972) Linear relationships between lipophilic character and biological activity of drugs. *Journal of Pharmaceutical Sciences* **61**, 1–19.
6. Li, R., Hansch, C., Matthews, D., Blaney, J.'M., Langridge, R., Delcamp, T.J., Susten, S.S. and Freisheim, J.H. (1982) A comparison by QSAR, crystallography, and computer graphics of the inhibition of various dihydrofolate reductases by 5-(X-benzyl)-2,4-diamino-pyrimidines. *Quantitative Structure–Activity Relationships* **1**, 1–7.
7. Dearden, J.C., Netzeva, T.I. and Bibby, R. (2003) A comparison of commercially available software for the prediction of partition coefficient. In Ford, M., Livingstone, D., Dearden, J. and van de Waterbeemd, H. (Eds.), *Designing Drugs and Crop Protectants: Processes, Problems and Solutions*. Blackwell Publishing, Oxford, pp. 168–169.
8. Hansch, C. and Leo, A. (1995) *Exploring QSAR: Fundamentals and Applications in Chemistry and Biology*; American Chemical Society, Washington DC.
9. Kier, L.B. (1985) A shape index from molecular graphs. *Quantitative Structure–Activity Relationships* **4**, 109–116.
10. Kier, L.B. and Hall, L.H. (1986) *Molecular Connectivity in Structure–Activity Analysis*. John Wiley & Sons Inc., New York.
11. Basak, S.C., Balaban, A.T., Grunwald, G.D. and Gute, B.D. (2000) Topological indices: their nature and mutual relatedness. *Journal of Chemical Information and Computer Sciences* **40**, 891–898.
12. Kier, L.B. and Hall, L.H. (1999) *Molecular Structure Description: the Electrotopological State*. Academic Press, San Diego, CA.
13. Hansch, C. and Clayton, J.M. (1973) Lipophilic character and activity of drugs II: the parabolic case. *Journal of Pharmaceutical Sciences* **62**, 1–23.
14. King, L. A. (1987) The use of drug solubility data in forensic toxicology. In Brandenberger, H. and Brandenberger, R.T. (Eds.), *Reports on Forensic Toxicology*. Branson Research, Männedorf, pp. 45–56.
15. Kumar, R. and Singh, P. (1997) Diarylspiro[2,4]heptenes as selective cyclooxygenase–2 inhibitors: a quantitative structure-activity relationship analysis. *Indian Journal of Chemistry B* **36**, 1164–1168.
16. Mukhomorov, V.K., Semenova, G.K. and Shagoyan, M.G. (1988) Antiemetic activity and structural features of 4-substituted 5-nitro–2-methoxy-*N*-(2-diethylaminoethyl)benzamide. *Khimiko-Farmatsevticheskii Zhurnal* **22**, 1108–1111.
17. Greenberg, M.J. (1979) Dependence of odor intensity on the hydrophobic properties of molecules. A quantitative structure-odor intensity relationship. *Journal of Agricultural and Food Chemistry* **22**, 347–352.
18. Dearden, J.C. (1994) Quantitative structure–activity relationships (QSAR) and odour. *Food Quality and Preference* **5**, 81–86.
19. Hall, L.H., Mohney, B.K. and Kier, L.B. (1991) The electrotopological state: structure information at the atomic level for molecular graphs. *Journal of Chemical Information and Computer Sciences* **31**, 76–83.
20. Dearden, J.C., Cronin, M.T.D., Sharra, J.A., Higgins, C., Boxall, A.B.A. and Watts, C.D. (1997) The prediction of Henry's law constant: a QSPR from fundamental considerations. In Chen, F. and

Schüürmann, G. (Eds.), *Quantitative Structure–Activity Relationships in Environmental Sciences — VII*. SETAC Press, Pensacola, FL, pp. 135–142.

21. Topliss, J.G. and Costello, R.J. (1972) Chance correlations in structure–activity studies using multiple regression analysis. *Journal of Medicinal Chemistry* **15**, 1066–1069.
22. Hyde, R. (1989) QSAR parameters à la carte from computer chemistry. In Fauchère, J.-L. (Ed.), *QSAR — Quantitative Structure–Activity Relationships in Drug Design*. A.R. Liss, New York, NY, pp. 91–95.
23. Cronin, M.T.D. (1990) Quantitative Structure–Activity Relationships of Comparative Toxicity to Aquatic Organisms. PhD Thesis; Liverpool Polytechnic.
24. Szydlo, R.M., Ford, M.G., Greenword, R. and Salt, D.W. (1983) The relationship between the physicochemical properties of substituted benzyl cyclopropane-1-carboxylate esters and their pharmacokinetic, pharmacodynamic and toxicological parameters. In Dearden, J.C. (Ed.), *Quantitative Approaches to Drug Design*. Elsevier, Amsterdam, pp. 203–214.
25. Maddelena, D.J. (1996) Applications of artificial neural networks to quantitative structure–activity relationships. *Expert Opinion on Therapeutic Patents* **6**, 239–251.
26. Martin, Y.C. (1998) 3D QSAR: current state, scope and limitations. In Kubinyi, H., Folkers, G. and Martin, Y.C. (Eds.), *3D QSAR in Drug Design: Recent Advances*. Kluwer Academic Publishers, Dordrecht, pp. 3–23.
27. Bartlett, A., Dearden, J.C. and Sibley, P.R. (1995) Quantitative structure–activity relationships in the prediction of penicillin immunotoxicity. *Quantitative Structure–Activity Relationships* **14**, 258–263.
28. Johnson, M.A. and Maggiora, G.M. (Eds.) (1990) *Concept and Applications of Molecular Similarity*. John Wiley & Sons Inc., New York, NY.
29. Dean, P.M. (Ed.) (1994) *Molecular Similarity in Drug Design*. Chapman & Hall, London.
30. Free, S.M. and Wilson, J.W. (1964) A mathematical contribution to structure–activity studies. *Journal of Medicinal Chemistry* **7**, 395–399.
31. Hansch, C., Unger, S.H. and Forsythe, A.B. (1973) Strategy in drug design. Cluster analysis as an aid in the selection of substituents. *Journal of Medicinal Chemistry* **16**, 1217–1222.
32. Austel, V. (1983) 2^n-Factorial schemes in drug design. Extensions increasing versatility. *Quantitative Structure–Activity Relationships* **2**, 59–65.
33. Eriksson, L. and Johansson, E. (1996) Multivariate design and modelling in QSAR. *Chemometrics and Intelligent Laboratory Systems* **34**, 1–19.
34. Pleiss, M.A. and Unger, S.H. (1990) The design of test series and the significance of QSAR relationships. In Ramsden, C.A. (Ed.), *Comprehensive Medicinal Chemistry*, Vol. 4; *Quantitative Drug Design*. Pergamon Press, Oxford, pp. 561–587.
35. Walker, J.D., Jaworska, J., Comber, M.H.I., Schultz, T.W. and Dearden, J.C. (2003) Guidelines for developing and using quantitative structure–activity relationships. *Environmental Toxicology and Chemistry* **22**, 1653–1665.
36. Moriguchi, I. and Komatsu, K. (1977) Adaptive least-squares classification method applied to structure–activity correlation of antitumor mitomycin derivatives. *Chemical and Pharmaceutical Bulletin* **25**, 2800–2802.
37. Barratt, M.D. (1995) The role of structure–activity relationships and expert systems in alternative strategies for the determination of skin sensitisation, skin corrosivity and eye irritation. *Alternatives to Laboratory Animals* **23**, 111–122.
38. Rose, S.L. and Jurs, P.C. (1981) Computer-assisted studies of structure–activity relationships of *N*-nitroso compounds using pattern recognition. *Journal of Medicinal Chemistry* **25**, 769–776.
39. Lewis, D.F.V., Ionnides, C. and Parke, D. (1989) Prediction of chemical carcinogenicity from molecular and electronic structure: a comparison of MNDO/3 and CNDO/2 molecular orbital methods. *Toxicology Letters* **49**, 1–13.
40. Livingstone, D.J. (1991) Pattern recognition methods in rational drug design. In Largone, J.J. (Ed.), *Methods in Enzymology*; Vol. 203. Academic Press, San Diego, CA, pp. 613–638.
41. Grover, M., Singh, B., Bakshi, M. and Singh, S. (2000) Quantitative structure–property relationships in pharmaceutical research — part 1. *Pharmaceutical Science and Technology Today* **3**, 28–35.

7

Prodrugs

Andrew W. Lloyd

CONTENTS

7.1 INTRODUCTION

Although pharmaceutical companies attempt to design and develop new chemical entities using rational and logical processes, very few of these compounds become clinically useful drugs because unpredictable interactions with biological systems reduce therapeutic efficacy and in many cases lead to undesirable toxicity. An alternative approach to enhance therapeutic activity relies on the chemical modification of known compounds to overcome the undesirable physical and chemical properties using prodrug design.[1–4]

A prodrug is a pharmacologically inactive compound which is metabolized to the active drug by either a chemical or enzymatic process. Some of the early pharmaceuticals were found to be prodrugs and this has lead to the subsequent introduction of the metabolite itself into therapy, particularly in cases where the active metabolite is less toxic or has fewer side effects than the parent prodrug. The administration of the active metabolite may also reduce variability in clinical response between individuals, which is attributed to differences in pharmacogenetics, particularly in disease states.

The earliest example of a prodrug is arsphenamine (**7.1**) used by Ehrlich for the treatment of syphilis. Later Voegtlin demonstrated that the activity of this compound against the syphilis organism was attributable to the metabolite oxophenarsine (**7.2**). Arsphenamine was later replaced by oxophenarsine in therapy as the metabolite was less toxic at the dose required for effective therapy.

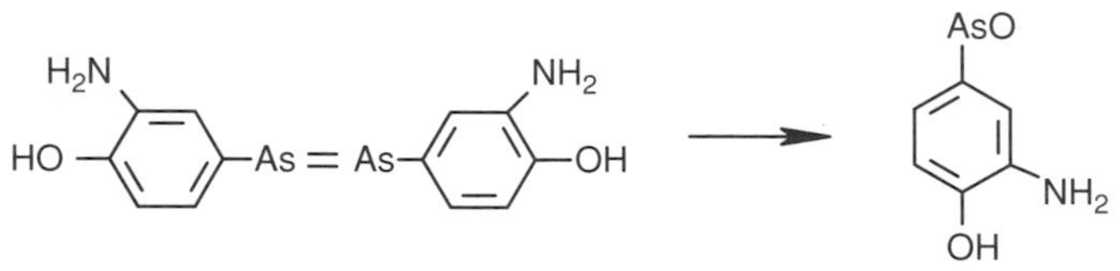

(7.1) **(7.2)**

Other such discoveries have led to the development of complete classes of drug compounds. For example the development of present day sulfonamide therapy evolved from the discovery by Domagk in 1935 that the azo dye prontosil (**7.3**) had antibacterial activity. Prontosil was subsequently shown to be a precursor which was metabolized to the active agent, *p*-aminobenzenesulfonamide (**7.4**), *in vivo*. This led to the subsequent development of a wide range of therapeutically superior sulfonamides through modification of the aminobenzenesulfonamide molecule.

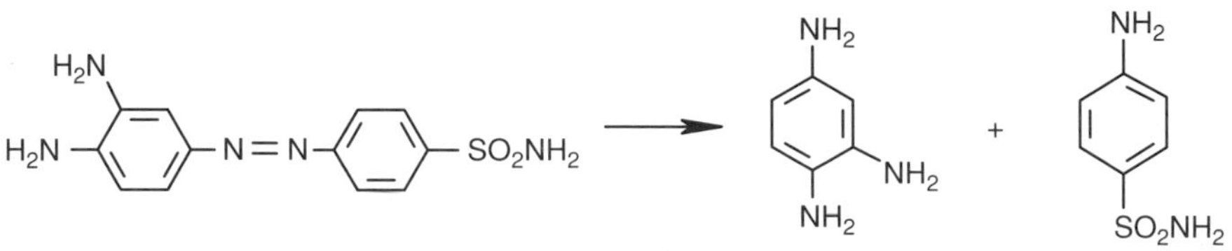

(7.3) **(7.4)**

The antimalarial drugs pamaquin (**7.5**) and paludrine (**7.7**) are also both converted to active metabolites by the body. Pamaquin is dealkylated and oxidized to the quinone (**7.6**), which is 16 times more active *in vivo* than the parent compound whereas paludrine cyclizes to give the active dihydrotriazine (**7.8**), which has structural similarities to the active antimalarial pyrimethamine (**7.9**).

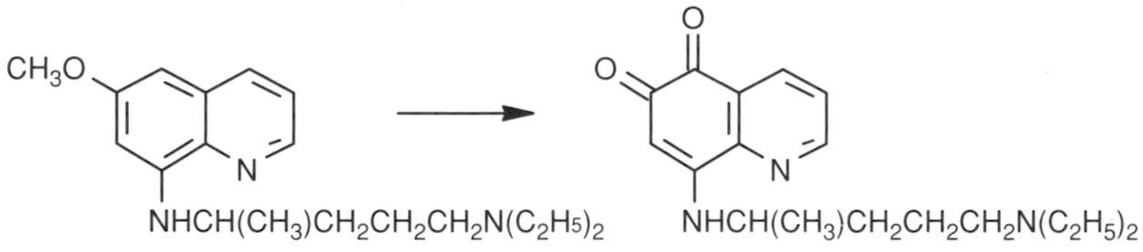

(7.5) **(7.6)**

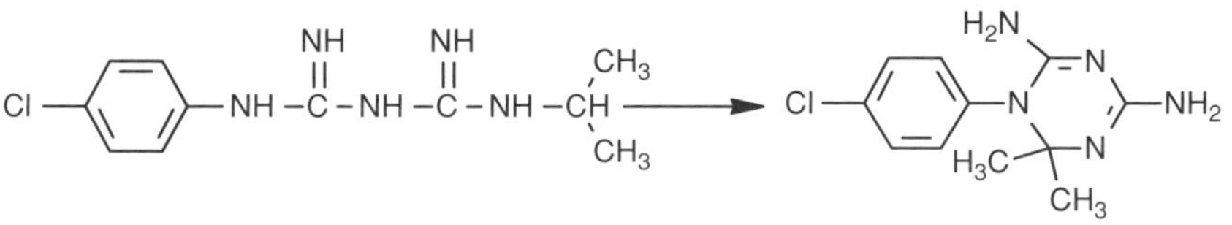

(7.7) **(7.8)**

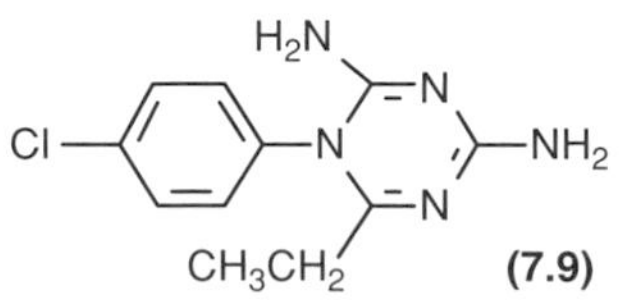

(7.9)

The dihydrotriazine metabolite, cycloguanil (**7.8**), has been administered as the insoluble pamoate salt in an oily base through a single intramuscular injection to provide malarial protection for up to several months depending on the particular particle size of the drug substance.

The development of depressants based on trichloroethanol (**7.11**) were shown to be the active metabolite of the once used hypnotic chloral hydrate (Noctec®) (**7.10**). This led to the use of trichloroethanol acid phosphate (**7.12**) for patients where choral hydrate was found to be either unpalatable or caused gastric irritation.

$Cl_3C{-}CH(OH)_2 \longrightarrow Cl_3C{-}CH_2OH$

(7.10) **(7.11)**

$Cl_3CH_2OPO(OH)ONa$

(7.12)

The antiepileptic activities of methylphenobarbitone (Prominal®) (**7.13**), primidone (Mysoline®) (**7.14**), and methsuximide (**7.15**) have also been shown to be related to the plasma levels of active metabolites. The active metabolites are obtained on demethylation of methylphenobarbitone and oxidation of primidone. Methsuximide is also demethylated and at steady state the metabolite of this compound has been shown to be present at 700-fold greater concentrations than the parent drug.

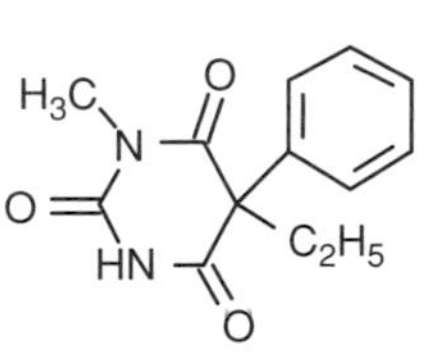

(7.13) 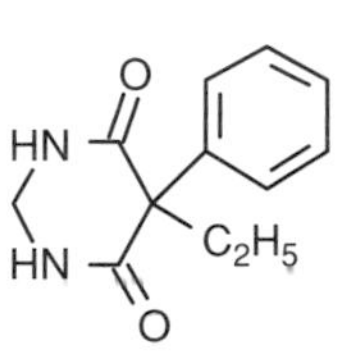 **(7.14)** 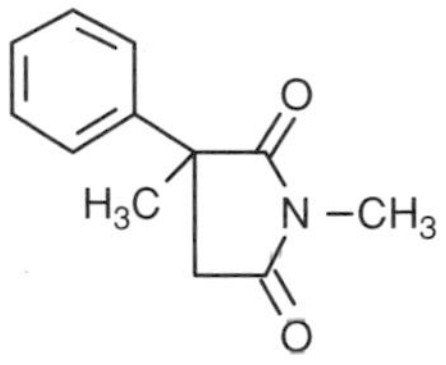 **(7.15)**

The nonsteroidal anti-inflammatory drug sulindac (Clinoril®) (**7.16**) is also a prodrug, which is reduced to the active metabolite (**7.17**), although some of the inactive sulfone (**7.18**) is formed on oxidation.

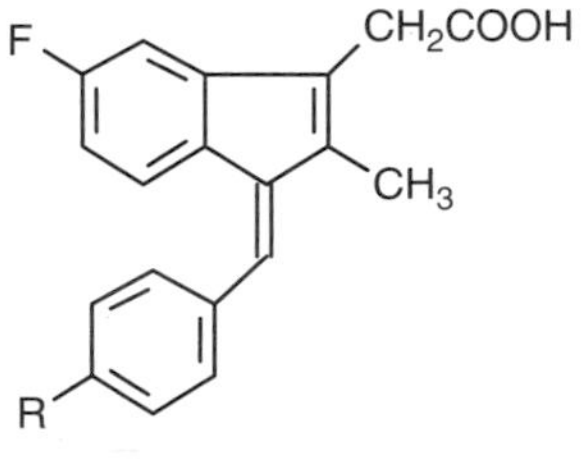

(7.16) R = —S(=O)CH$_3$

(7.17) R = —SCH$_3$

(7.18) R = —SO$_2$CH$_3$

The *in vivo* hydrolysis of aspirin (**7.19**) to salicylic acid (**7.20**) by esterases allows the administration of aspirin in preference to salicylic acid, which is more corrosive to the gastrointestinal mucosa.

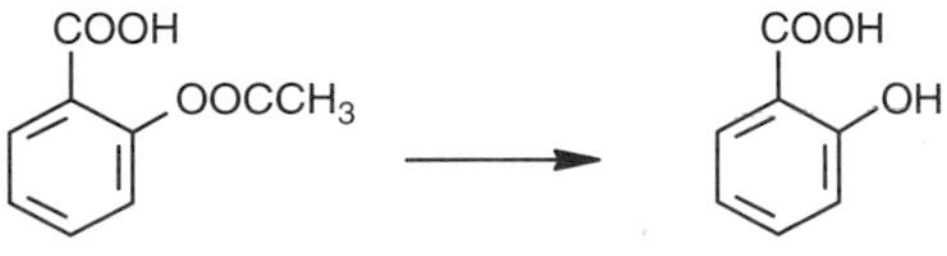

(7.19) **(7.20)**

Hexamine (Hiprex®, Mandelamine®) (**7.21**) is administered as a prodrug of formaldehyde (**7.22**) for the treatment of urinary tract infections although it was initially used to dissolve renal stones. An enteric coat is used to protect the prodrug from stomach acid; however on reaching the acidic environment of the urine the formaldehyde is released and exerts its antiseptic action.

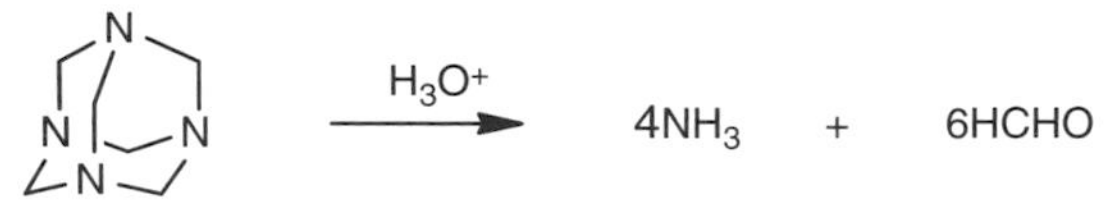

$$\xrightarrow{H_3O^+} 4NH_3 + 6HCHO$$

(7.21) **(7.22)**

Phenylbutazone (Butozolidine®) (**7.23**) is converted by the body into the two hydroxylated forms, oxyphenbutazone (**7.24**) and (**7.25**). The drug is used in therapy under hospital supervision, mainly as an anti-inflammatory agent, and this activity resides in form (**7.24**). However, another use of the drug is as a uricosuric agent, in the treatment of gout, and this action is attributable to the form (**7.25**). The observation that substitution in the side chain of phenylbutazone results in enhanced uricosuric action has led to the discovery of several other agents which have this action, in particular sulfinpyrazone (**7.26**).

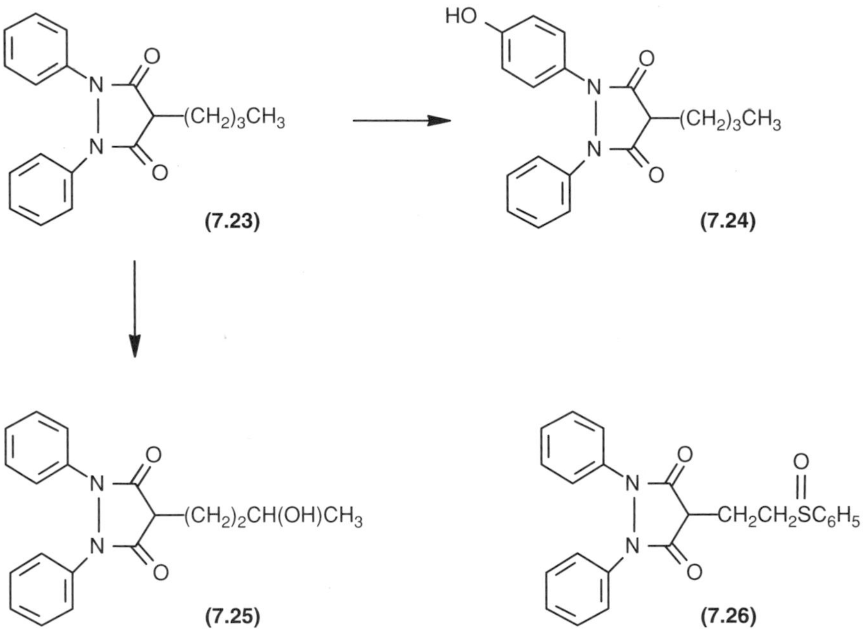

$(CH_2)_3CH_3$ **(7.23)** → HO $(CH_2)_3CH_3$ **(7.24)**

$(CH_2)_2CH(OH)CH_3$ **(7.25)**

$CH_2CH_2S(O)C_6H_5$ **(7.26)**

In addition to those drugs detailed above several drugs which were metabolized to active compounds were initially considered to be prodrugs but later shown to possess activity themselves. For example, phenacetin (**7.27**), an analgesic and antipyretic agent, is mainly metabolized in the body to an active metabolite, *N*-acetyl-*p*-aminophenol (paracetamol) (**7.28**), as well as to an inactive metabolite, the glucuronide of 2-hydroxyl phenacetin (**7.29**), in small amounts.

OCH_2CH_3 / $NHCOCH_3$ **(7.27)**

OH / $NHCOCH_3$ **(7.28)**

OCH_2CH_3 / OH / $NHCOCH_3$ **(7.29)**

Paracetamol has replaced phenacetin in therapy, because it is usually free from toxic effects associated with phenacetin, e.g., methemoglobin formation. However, extensive hepatic necrosis may occur when overdoses are ingested since the normal biotransformation pathway (conjugation with glutathione) is then saturated and a highly reactive metabolite is formed which binds irreversibly to hepatic tissue. More recent work has shown that phenacetin itself possesses antipyretic activity and that this activity is not dependent on metabolism to paracetamol.

7.2 PRODRUG DESIGN

Most chemically designed prodrugs are composed of two parts in which the active drug is linked to a pharmacologically inert molecule. The chemical bond between the two parts of the prodrug must be sufficiently stable to withstand the pharmaceutical formulation of the prodrug while permitting chemical or enzymatic cleavage at the appropriate time or site. After administration or absorption of the prodrug, the active drug is usually released either by catalyzed hydrolysis by the liver or intestinal enzymes or simply by hydrolysis although reductive processes have also been utilized. Prodrugs are most commonly used to overcome the biological and pharmaceutical barriers, which separate the site of administration of the drug from the site of action (Figure 7.1).

7.3 APPLICATION TO PHARMACEUTICAL PROBLEMS

The pharmaceutical problems that have been addressed using prodrug design include unpalatability, gastric irritation, pain on injection, insolubility, and drug instability.

7.3.1 Patient Acceptability

Unpleasant tastes and odors may often affect patient compliance. For example very young children generally require liquid medication since they are usually not amenable to swallowing capsules or coated tablets. Despite the life-threatening toxicity the antibiotic chloramphenicol (**7.30**) it is still administered orally for the treatment of typhoid fever and salmonella infections. However, the drug has an extremely bitter taste and is entirely unsuitable for administration as a suspension to such patients. To overcome this problem orally administered chloramphenicol is usually formulated as the inactive tasteless palmitate (**7.31**) or cinnamate (**7.32**) esters. The active parent drug is released from these compounds by esterases present in the small intestine.

$$O_2N{-}C_6H_4{-}CH(OH){-}CH(NHCOCHCl_2){-}CH_2OR$$

(7.30) R = H
(7.31) R = $CH_3(CH_2)_{14}CO{-}$
(7.32) R = $C_6H_5CH{=}CHCO{-}$

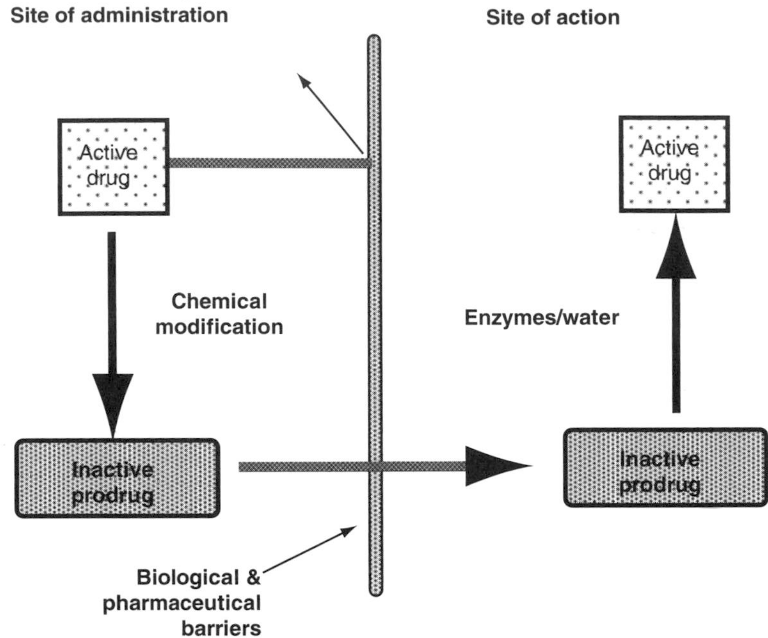

Figure 7.1 The prodrug concept. A diagrammatic representation of the prodrug concept where a pharmaceutically active drug is converted to an inactive compound to overcome pharmaceutical and biological barriers between the site of administration and the site of action.

The bitter taste of the antibiotics clindamycin and erythromycin has been similarly masked using the palmitate ester and hemisuccinate ester prodrugs, respectively. The antimicrobial metronidazole (**7.33**) is another example of a drug with an unacceptably bitter taste. To overcome this problem, it is administered as a suspension of benzoylmetronidazole (Flagel S®) (**7.34**). Likewise, ethyl dithiolisophthalate (Ditophal®) has replaced the foul-smelling liquid ethyl mercaptan for the treatment of leprosy. The odorless inactive diisophthalyl thioester is metabolized to the active parent drug by thioesterases.

O_2N … CH_3 … CH_2CH_2OR

(7.33) R = H
(7.34) R = C_6H_5CO—

7.3.2 Drug Solubility

The formulation of insoluble compounds for parenteral delivery represents a major problem as the insoluble drug will have a tendency to precipitate on injection in an organic solvent. The solubility of such compounds may be improved by the use of phosphate or hemisuccinate prodrugs. For example the insoluble glucocorticoids such as betamethasone, prednisolone, methylprednisolone, hydrocortisone, and dexamethasone are available for injection as the water-soluble prodrug in the form of the disodium phosphate ($RO.PO_3^{2-}\ 2Na^+$) or sodium hemisuccinate ($RO.CO.CH_2CH_2COO^-\ Na^+$) salts. The phosphate esters are rapidly hydrolyzed to the active steroid by phosphatases, whereas the hemisuccinate salts are less efficiently hydrolyzed by esterases, possibly due to the presence of an anionic center (COO^-) near the hydrolyzable ester bond. The poorly water-soluble anti-inflammatory steroidal alcohol dexamethasone has been shown to rapidly ($t_{1/2} = 10$ min) liberate the active steroid *in vivo* when injected as the water-soluble phosphate (**7.35**).

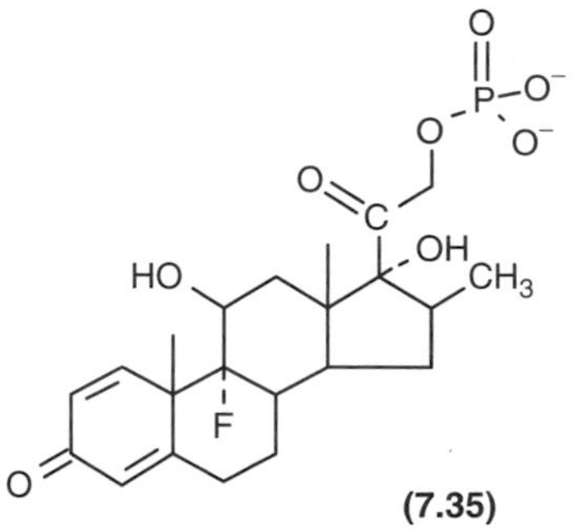

(7.35) 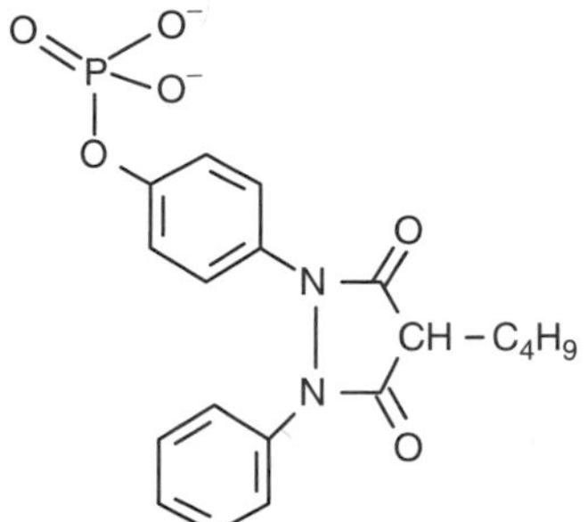 **(7.36)**

The water-soluble phosphate ester of the anti-inflammatory agent oxyphenbutazone (**7.36**) is rapidly hydrolyzed *in vivo* and gives higher blood levels of oxyphenbutazone on oral or intramuscular administration than attained on administration of the same doses of the parent drug.

Difficulties in the formulation of the anticonvulsant drug phenytoin (**7.37**) as a soluble injectable dosage form have led to the development of water-soluble prodrugs which have been shown to have a superior *in vivo* performance in rats. The prodrug is prepared by the reaction of phenytoin with an excess of formaldehyde to give the 3-hydroxymethyl intermediate (**7.38**), which is unstable in the absence of excess reagent. Conversion of the intermediate (**7.38**) to the disodium phosphate ester prodrug (**7.39**) gives a water-soluble derivative. This is metabolized *in vivo* by phosphatases to (**7.38**), which rapidly breaks down ($t_{1/2} = 2$ s) at 37°C (pH 7.4) to give the active drug, phenytoin.

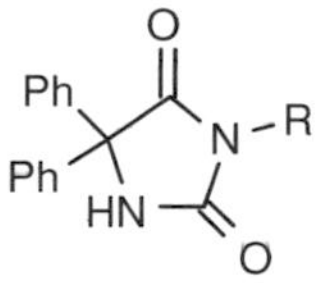

(7.37) R = H

(7.38) R = CH_2OH

(7.39) R = $CH_2OPO_3^{2-}\ Na_2^+$

7.3.3 Drug Stability

Many drugs are unstable and may either breakdown on prolonged storage or are degraded rapidly on administration. This is a particular problem on oral administration as drugs are often unstable in gastric acid. Although enteric coatings may be used, it is also possible to utilize prodrug design to overcome this problem.

For example, the antibiotic erythromycin is destroyed by gastric acid and, as an alternative to enteric-coated tablets, it is administered orally as a more stable ester. The inactive erythromycin estolate (laurylsulfate salt of the propionyl ester), when administered as a suspension, is rapidly absorbed and the propionyl ester converted by body esterases to the active erythromycin. The propionyl ester gives higher blood levels after oral administration on an equidose basis than the acetate or butyrate esters. The ethyl succinate ester has also been used.

5-Aminosalicylic acid (mesalazine) is useful in the treatment of ulcerative colitis and to a lesser degree in the management of Crohn's disease. It cannot be administered orally since firstly, it is unstable in gastric acid and secondly, it would not reach its site of action in the ileum/colon since it would be absorbed in the small intestine. Sulfasalazine (**7.40**), where mesalazine is covalently linked with sulfapyridine, is broken down in the colon by bacteria to the two components and in this way 5-aminosalicylic acid is delivered to the required site of action.

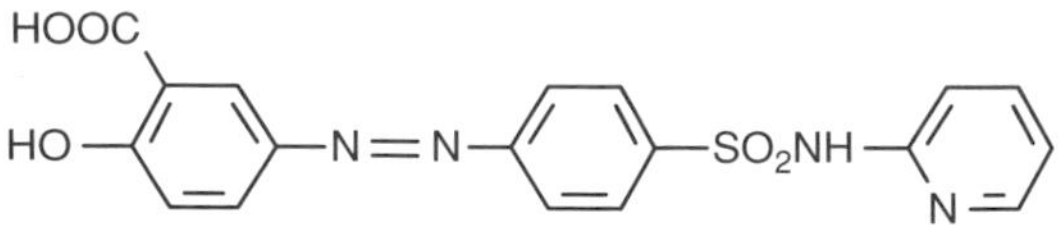

(7.40)

However, sulfapyridine is responsible for the majority of side effects attributable to this combination and is thought to have little therapeutic activity. An alternative prodrug, osalazine (**7.41**), consisting of two molecules of 5-aminosalicylic acid has been developed to overcome this problem. Reduction of the azo bond by the colonic microflora therefore liberates two molecules of 5-aminosalicylic acid. Mesalazine has also been administered orally as tablets coated with a pH-dependent acrylic-based resin, which disintegrates in the terminal ileum/colon as the environment pH rises above pH 7.

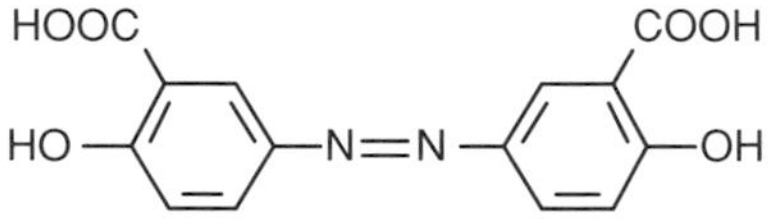

(7.41)

Microbial metabolism of prodrugs has also been utilized in the delivery of corticosteroids to the colon. Such compounds are generally readily absorbed from the upper gastrointestinal tract and therefore delivered ineffectively to the colon. Administration of corticosteroids, such as dexamethasone (**7.42**), as glycoside prodrugs overcomes these problems by reducing systemic uptake in the small intestine. Prodrugs, such as dexamethasone-β-D-glucoside (**7.43**), are hydrolyzed by the specific glycosidases produced by the colonic bacteria and the parent corticosteroid absorbed from the lumen of the large intestine, resulting in much higher concentrations in the colonic tissues.

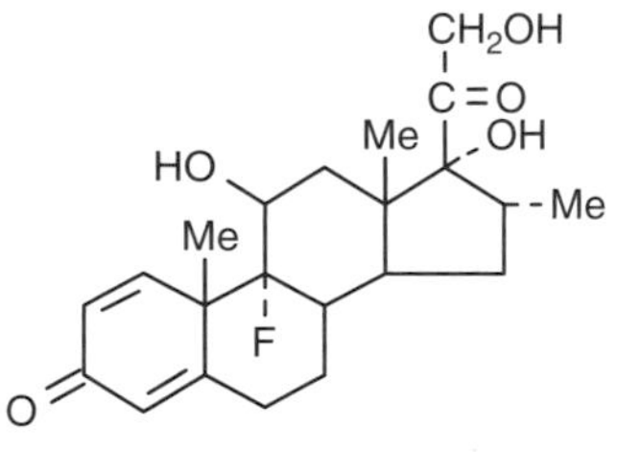

(7.42) 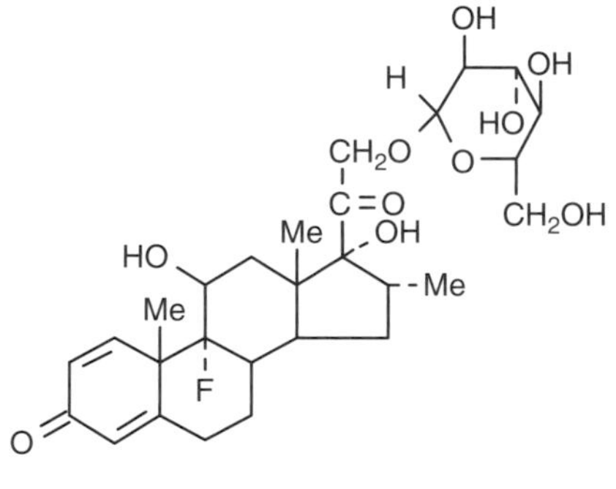 **(7.43)**

More recently macromolecular prodrugs have been investigated as means of overcoming instability and undesirable systemic uptake. For example, 5-aminosalicyclic acid has been linked to poly(sulfonamidoethylene) to give another mesalazine prodrug known as polyasa (**7.44**), which has been shown to have less side effects than sulfasalazine and is therefore better tolerated by patients found to be allergic to or intolerant to sulfasalazine (**7.40**).

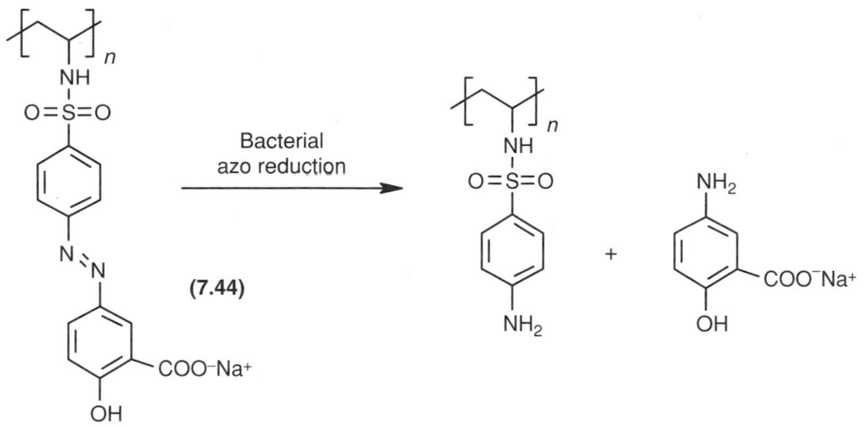

(7.44)

Macromolecular prodrugs have also been investigated as a means of reducing degradation of drugs by gastrointestinal enzymes. For example, the coupling of the B chain of insulin to water-soluble copolymers such as *N*-(2-hydroxypropyl)methacrylamide or poly(*N*-vinylpyrrolidone-co-maleic acid) appears to reduce the susceptibility of the insulin B chain to degradation by brush border peptidases *in vitro*.

7.4 PHARMACOLOGICAL PROBLEMS

There are a number of pharmacological problems which may be addressed by prodrug design. These problems may be either related to pharmacokinetic, pharmacodynamic, or toxic properties of the drug. Inappropriate pharmacokinetics may result in an undesirable rate of onset or duration of action of a drug. Poor pharmacodynamics may be a consequence of inefficient or unpredictable drug adsorption from the gastrointestinal tract, inappropriate distribution, and variable bioavailability as a consequence of presystemic metabolism or the inability to reach the site of action from the systemic circulation, e.g., penetration of the blood–brain barrier. Toxic side effects may be due to nonspecific drug delivery to the site of action.

7.4.1 Drug Absorption

Many drugs are either poorly or unpredictably adsorped from the gastrointestinal tract resulting in variation in efficacy between patients. Prodrug design has been utilized in a number of cases to optimize the adsorption of such drugs thereby improving their bioavailability.

Many penicillins are not absorbed efficiently when administered orally and their lipophilic esters have been used to improve absorption. However, simple aliphatic esters of penicillins are not active *in vivo* and therefore activated esters are necessary for release of the active penicillin from the inactive prodrug. Ampicillin (**7.45**), a wide-spectrum antibiotic, is readily absorbed orally as the inactive prodrugs, pivampicillin (**7.46**), bacampicillin (**7.47**), and talampicillin (**7.48**) which are then converted by enzymic hydrolysis to ampicillin.

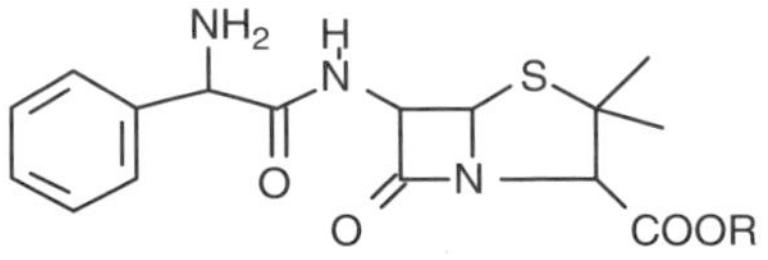

(7.45) Ampicillin R = H

(7.46) Pivampicillin R = $—CH_2OC(O)C(CH_3)_3$

(7.47) Bacampicillin R = $—CH(CH_3)OC(O)OC_2H_5$

(7.48) Talampicillin R =

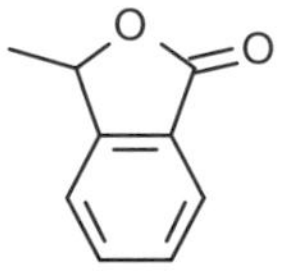

The preferred prodrug is pivampicillin since minimal hydrolysis occurs in the intestine before absorption into the systemic circulation. Pivampicillin, the pivaloyloxymethyl ester, contains an acyloxymethyl function which is rapidly hydrolyzed by enzymes to the hydroxymethyl ester. This hemi-ester of formaldehyde, spontaneously cleaves with release of ampicillin and formaldehyde. In a similar manner, bacampicillin and talampicillin are cleaved and decompose to give ampicillin together with acetaldehyde and 2-carboxybenzaldehyde, respectively.

Acyclovir (**7.49**) has been widely used for the treatment of herpes simplex and herpes-zoster infections. This prodrug is activated through phosphorylation by the viral thymidine kinase to acyclovir monophosphate, which is then converted to the triphosphate, which inhibits DNA polymerase, by host cellular enzymes. However, the use of this drug has been limited to some extent by low oral absorption; only 20% of a 200-mg dose is absorbed and little improvement is seen with doses above 800 mg. This has led to the development of a range of acyclovir prodrugs including ''6-deoxyacyclovir'' (BW A515U; (**7.50**)) which has been used for prophylaxis of herpes-virus infections in patients with hematological malignancies.[5] It is well absorbed orally and produces plasma concentrations of the drug which are much higher than those obtained by oral administration of acyclovir. The drug (**7.50**) is converted to acyclovir *in vitro* by xanthine oxidase.

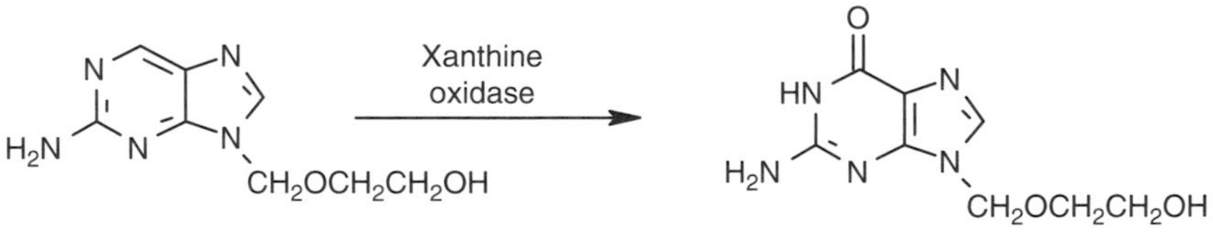

Xanthine oxidase

$CH_2OCH_2CH_2OH$

(7.50) 6-deoxyacyclovir **(7.49) Acyclovir**

An alternative orally active prodrug is valacyclovir (**7.51**), the L-valyl ester of acyclovir, which is rapidly hydrolyzed by first pass intestinal and hepatic metabolism. The mechanism of this biotransformation has yet to be fully elucidated but is thought to be enzymatic in nature.

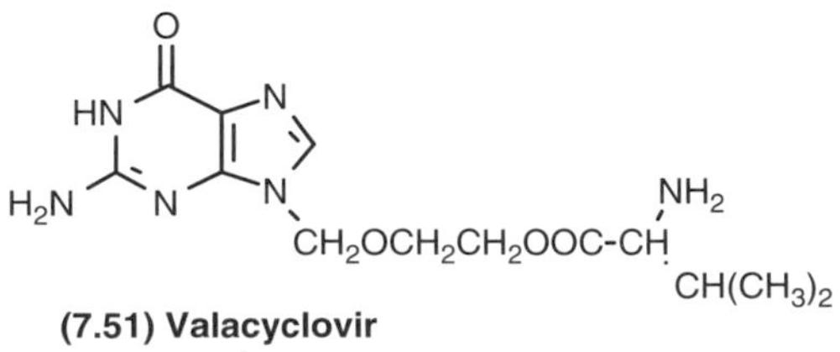

$CH_2OCH_2CH_2OOC\text{-}CH(NH_2)CH(CH_3)_2$

(7.51) Valacyclovir

More recently famciclovir (**7.52**) has been licensed in the United Kingdom for the treatment of herpes-zoster infections. Famciclovir is an orally absorbed 6-deoxy, diacetyl ester prodrug of penciclovir (**7.53**). This prodrug is rapidly deacetylated and oxidized in the intestinal wall and liver to give a systemic availability of penciclovir from famciclovir of 77% on oral administration. *In vitro* studies suggest that aldehyde oxidase, rather than xanthine oxidase, is involved in the conversion of famciclovir to penciclovir in the human liver.

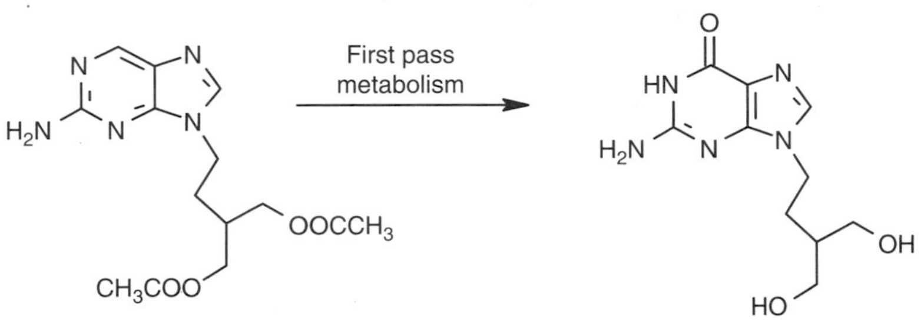

First pass metabolism

(7.52) Famciclovir **(7.53) Penciclovir**

Penciclovir is selectively phosphorylated by viral thymidine kinase in the same way as acyclovir. The penciclovir triphosphate, generated by phosphorylation of the monophosphate by cellular enzymes, is 100 times less efficient at inhibiting the DNA polymerase from herpes virus but has 10- to 20-times greater intracellular stability than acyclovir triphosphate.

Several 2′,3′-dideoxynucleoside analogs such as zidovudine (azidothymidine, AZT) (**7.54**) and 2′,3′-didehydro-3′-deoxythymidine (D4T) (**7.55**) have potent antiviral activity against human immunodeficiency virus (HIV). These compounds are phosphorylated intracellularly to the 5′-triphosphate derivatives which inhibits the viral reverse transcriptase. To achieve effective metabolic antagonism against reverse transcriptase the plasma concentration of these compounds must be maintained. However, this has proved difficult because of the rapid elimination and metabolism of these compounds. Furthermore, the undesirable side effects associated with such compounds have been attributed to elevated plasma concentrations of these drugs. In an attempt to overcome these problems, and to improve oral bioavailability, a number of workers have recently investigated the potential of ester prodrugs of these compounds. These studies have demonstrated that such prodrugs increase the circulating half-life while limiting the elevation of the plasma concentration of the parent nucleoside. Some of the ester prodrugs were also shown to have higher absolute oral bioavailabilities than the parent nucleoside drug.

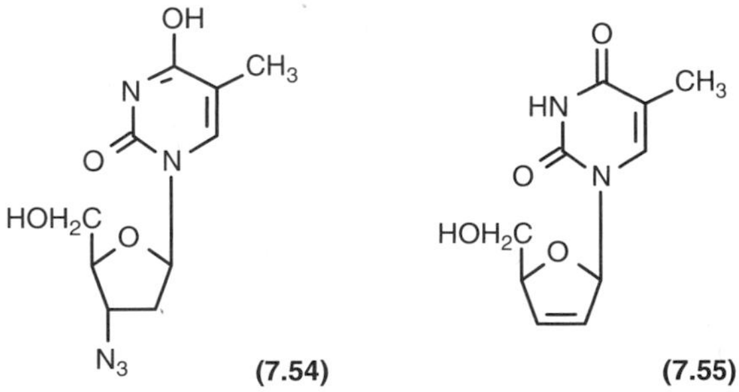

(7.54) **(7.55)**

The use of these nucleoside analogs as antiviral and antineoplastic agents is also limited by their absolute requirement for kinase-mediated intracellular phosphorylation. Nucleotide phosphates are unable to readily penetrate membranes and therefore have little therapeutic utility. This has led to the development of masked-phosphate prodrugs of anti-HIV nucleoside analogs, such as (**7.56**), which facilitate intracellular delivery of the bioactive free phosphate.[6] These compounds have been shown to be 25 times more potent and 100 times more selective than the parent nucleosides. Unlike the parent drugs they also retain good activity against kinase-deficient cells. Such strategies also have important implications for the development of much wider ranges of compounds to combat the emergence of resistance to certain nucleoside analogs.

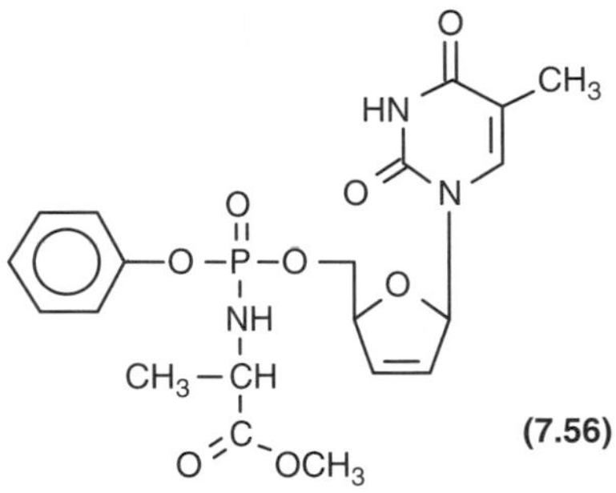

(7.56)

In another example, the antihypertensive effects after oral administration of the angiotensin-converting enzyme inhibitor enalaprilat (**7.57**) have been improved by conversion to the more efficiently absorbed ethyl ester, enalapril (**7.58**). In the active form, less than 12% is adsorbed whereas the inactive derivative has an improved adsorption of between 50% and 75%. The prodrug enalapril is converted *in vivo* to the active enalaprilat by hydrolysis in the liver following absorption from the gastrointestinal tract.

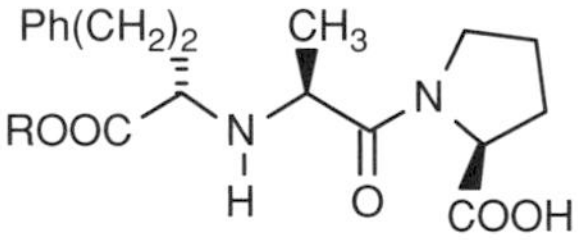

(7.57) R = H

(7.58) R = $-CH_2CH_3$

Animal studies have shown that the oral absorption of certain basic drugs may be increased by the preparation of ''soft'' quaternary salts. The ''soft'' quaternary salt is formed by reaction between α-chloromethyl ester (**7.59**) and the amino group of the drug. The quaternary salt formed is termed as ''soft'' quaternary salt since, unlike normal quaternary salts, it can release the active basic drug on hydrolysis.

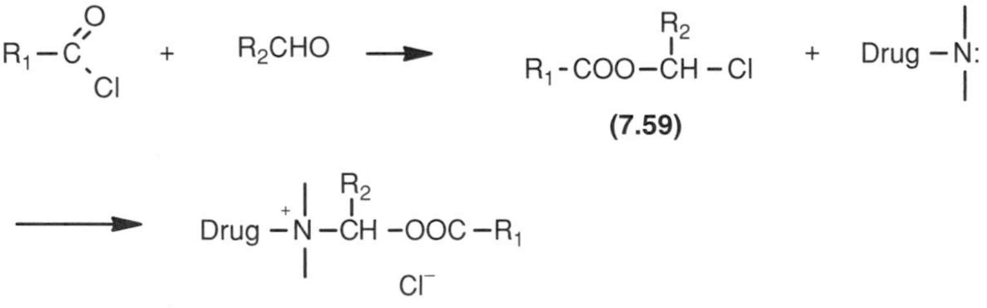

$$R_1-C(=O)Cl + R_2CHO \longrightarrow R_1\text{-COO-}CH(R_2)-Cl \;(\mathbf{7.59}) + \text{Drug}-N: \longrightarrow \text{Drug}-\overset{+}{N}-CH(R_2)-OOC-R_1 \; Cl^-$$

''Soft'' quaternary salts have useful physical properties compared with the basic drug or its salts. Water solubility may be increased compared with other salts, such as the hydrochloride, but more importantly there may be an increased absorption of the drug from the intestine. Increased absorption is probably due to the fact that the ''soft'' quaternary salts have surfactant properties and are capable of forming micelles and unionized ion pairs with bile acids, etc., which are able to penetrate the intestinal epithelium more effectively. The prodrug, after absorption, is rapidly hydrolyzed with release of the active parent drug as illustrated below.

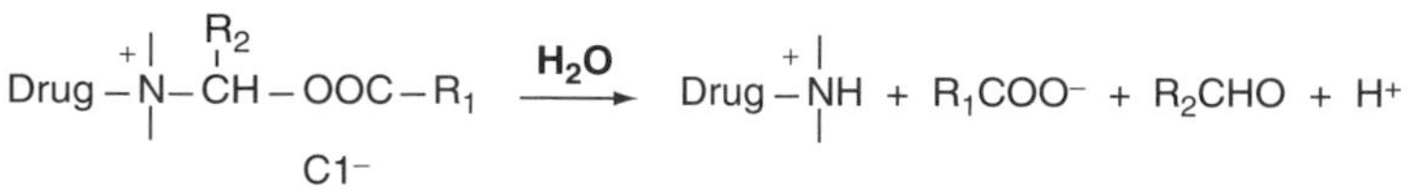

$$\text{Drug}-\overset{+}{\underset{|}{\text{N}}}-\overset{\text{R}_2}{\text{CH}}-\text{OOC}-\text{R}_1 \xrightarrow{\text{H}_2\text{O}} \text{Drug}-\overset{+}{\underset{|}{\text{NH}}} + \text{R}_1\text{COO}^- + \text{R}_2\text{CHO} + \text{H}^+$$

$$\text{C1}^-$$

Such an approach has also been utilized to achieve improved bioavailability of pilocarpine on ocular administration.[7] Pilocarpine is rapidly drained from the eye resulting in a short duration of action. The ''soft'' quaternary salt (**7.60**) has a lipophilic side-chain which has been shown to improve absorption in rabbits and gives a more prolonged effect at one tenth of the concentration of pilocarpine. The action of this compound has been shown to be due to the release of pilocarpine on hydrolytic cleavage of the ester followed by release of formaldehyde.

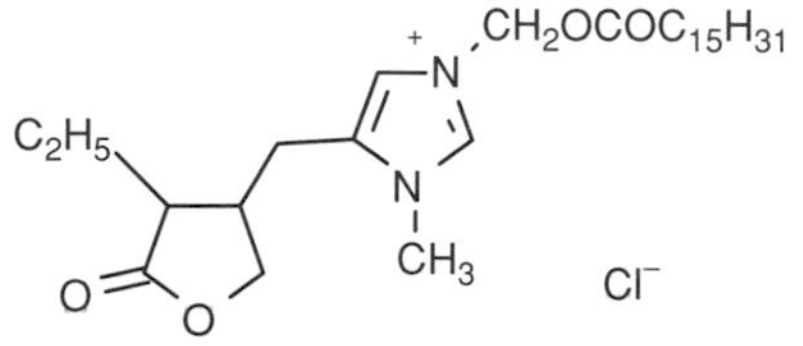

C_2H_5 … $\overset{+}{N}$–$CH_2OCOC_{15}H_{31}$ … N–CH_3 … Cl^- (**7.60**)

Topical administration is also used in the treatment of glaucoma with adrenaline (**7.61**), which lowers the intraocular pressure. Enhanced therapeutic efficacy may be achieved using a more lipophilic prodrug dipivefrin (**7.62**) which is 100 times more active than adrenaline as a consequence of more efficient corneal transport, followed by de-esterification by the corneal tissue and release of adrenaline in the aqueous humor. Consequently lower doses of dipivefrin than adrenaline can be administered to achieve the same therapeutic effect. This offers advantages in reducing the side effects associated with the use of adrenaline including cardiac effects due to systemic absorption and the accumulation of melanin deposits in the eye.

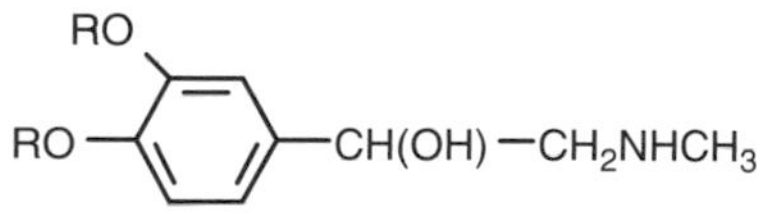

RO, RO – C_6H_3 – $CH(OH)-CH_2NHCH_3$

(**7.61**) R = H

(**7.62**) R = $COC(CH_3)_3$

7.4.2 Drug Distribution

The modification of a drug to a prodrug may lead to enhanced efficacy for the drug by differential distribution of the prodrug in body tissues before the release of the active form.

For example, more extensive distribution of ampicillin occurs in the body tissues when the methoxymethyl ester of hetacillin (a 6-side-chain derivative of ampicillin) is administered, than is obtained with ampicillin itself. Conversely, decreased tissue distribution of a drug may occur, as was observed when adriamycin as its DNA-complex was administered as a prodrug. Decreased tissue distribution restricts the action of a drug to a specific target site in the body and may therefore decrease its toxic side effects, resulting from its reaction at other sites. Anticancer drugs can suppress growth in normal as well as neoplastic tissue. Improved selective localization has been achieved using nontoxic prodrugs which release the active drug within the cancer cell as a result of either the enhanced enzyme activity in the cell or enhancement of reductase activity in the absence of molecular oxygen in hypoxic cells.

The prodrug cyclophosphamide (**7.63**) is used for the treatment of certain forms of cancer and as an immunosuppressant after organ transplant. It does not possess alkylating properties and consequently is not a tissue vesicant since the electron withdrawing properties of the adjacent phosphono-function decrease the nucleophilic properties of the β-chloroethylamino-nitrogen atom and prevent the formation of the reactive alkylating ethyleniminium ion. The prodrug requires hepatic mixed-function oxidase-mediated metabolic activation to generate 4-hydroxycyclophosphamide (**7.64**). The 4-hydroxycyclophosphamide exists in equilibrium with its open ring tautomer aldophosphamide (**7.65**), which undergoes β-elimination to produce the alkylating cytotoxic phosphoramide mustard (**7.66**) in the target cells.

Liver

(7.63)

(7.64)

Acrolein

Target cells

Active agent (7.66)

(7.65)

Aldehyde dehydrogenase

Inactive

(7.67)

Cyclophosphamide is also metabolized by aldehyde dehydrogenase to the inactive carboxyphosphoramide (**7.67**). Since this reaction provides a detoxification pathway, the effectiveness of cyclophosphamide is found to inversely correlate with the dehydrogenase activity of the target cells. The action of this alkylating species would be expected to be restricted to the target tissue but unfortunately in practice the action of the drug is more widespread and it shows toxicity to normal tissue, one of the apparent effects being alopecia.

Recently the organic thiophosphate prodrug amifostine (**7.68**) ($H_2N(CH_2)_3NH(CH_2)_2$ S-PO $(OH)_2$) has been introduced as a cytoprotective agent to reduce the toxic effects of cyclophosphamide on bone marrow. Amifostine uptake into normal cells occurs by facilitated diffusion and is therefore more rapid than the uptake into tumor cells by passive diffusion. As tumor cells are often hypoxic, poorly vascularized, and have a low pH environment they also have reduced alkaline phosphatase activity. Amifostine exploits these differences in uptake and enzyme activity to ensure that the prodrug is only dephosphorylated to the active drug in healthy tissues. The active drug ($H_2N(CH_2)_3NH(CH_2)_2SH$) therefore selectively deactivates the reactive cytotoxic species produced by cyclophosphamide in nontumor tissue without compromising the efficacy of the chemotherapy.

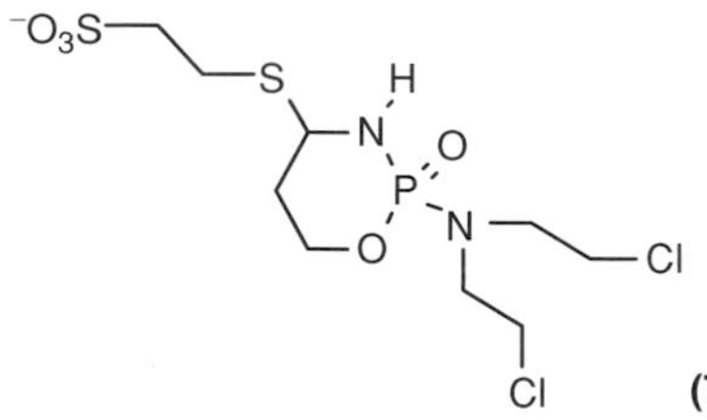

(7.68)

In addition, the acrolein produced during ring opening of (**7.64**) was initially found to cause bladder trouble. This problem has been overcome by either administration of cyclophosphamide together with an alkyl sulfide (sodium 2-mercaptoethanesulfonate, mesna, Uromitexan®) to remove acrolein as it is formed by addition to the β-carbon atom by a Michael's reaction, or use of a modified cyclophosphamide (**7.68**), which does not form acrolein after ring opening.

The anticancer effect of the prodrug procarbazine (**7.69**) has also been attributed to the formation of a cytotoxic species in the target cells. In this case, procarbazine is metabolized by the mixed function oxidase to azoprocarbazine (**7.70**) which undergoes further cytochrome P450-mediated oxidation to azoxy procarbazine isomers (**7.71**, **7.72**) which liberate the diazomethane alkylating agent in the target cells.

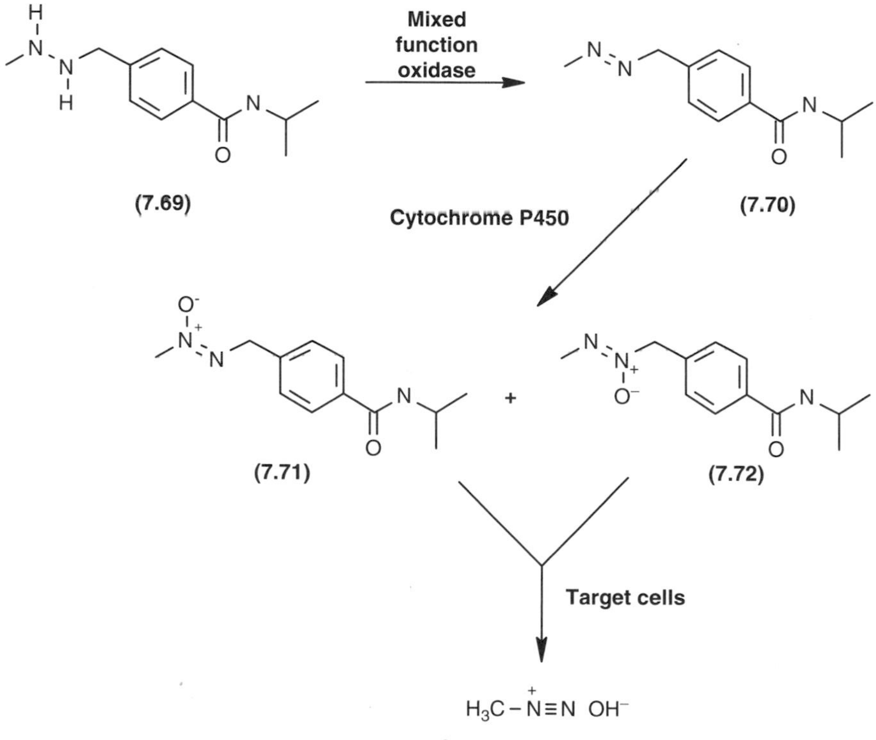

(7.69) Mixed function oxidase (7.70)

Cytochrome P450

(7.71) + (7.72)

Target cells

$H_3C-\overset{+}{N}\equiv N\ OH^-$

Cytotoxic agent

A series of other nontoxic nitrogen mustard prodrugs have also been designed to regenerate the parent alkylating agent in neoplastic tissues by taking advantage of the difference in the level of enzymatic amidase between normal and neoplastic cells. *N*, *N*-Diallyl-3-(1-aziridino)propionamide (DAAP) is active against certain forms of leukemia but does not cause leukopenia, a common toxic side effect observed with other bifunctional alkylating agents. This observation suggests that DAAP is selective in its action against dividing (neoplastic) cells where a high amidase level occurs.

7.4.3 Site-Specific Drug Delivery

Prodrugs have more recently been used to achieve site-specific drug delivery to various tissues. Such prodrugs are designed to ensure that the release of the active drug only occurs at its site of action thereby reducing toxic side effects due to high plasma concentrations of the drug or nonspecific uptake by other body tissues. This has led to the development of systems for site-specific delivery to the brain and to cancer cells.

The blood–brain barrier is impenetrable to lipid insoluble and highly polar drugs. Although lipophilic prodrugs may be used to overcome this physiological barrier, the increased lipid solubility may enhance uptake in other tissues with a resultant increase in toxicity. Furthermore therapeutic levels of such lipophilic prodrugs can only be maintained if there is a constant plasma concentration. These problems may be overcome by utilizing a dihydropyridine — pyridinium salt type redox system. This approach was first used to enhance the penetration of the nerve gas antagonist pralidoxine into the CNS using (**7.73**) as a nonpolar prodrug, which can cross the barrier and is then rapidly oxidized to the active form and trapped in the CNS.

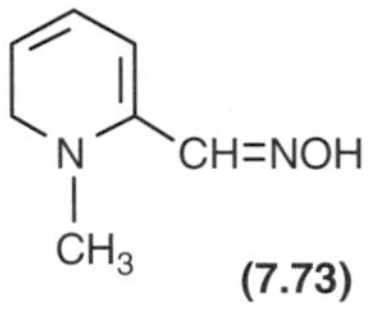

N CH=NOH
CH_3
(7.73)

More recently this approach has been developed as a general rationale for the site-specific and sustained delivery of drugs, which either do not cross the blood–brain barrier readily or are rapidly metabolized. Phenylethylamine and dopamine have been used to illustrate the principles involved and *in vivo* work has been described in animal experiments.[8]

The delivery system is prepared by condensing phenylethylamine with nicotinic acid to give (**7.74**) which is then quaternized to give (**7.75**). The quaternary ammonium salt (**7.75**) is then reduced to the 1,4-dihydro-derivative (**7.76**). The prodrug (**7.76**) is delivered directly to the brain, where it is oxidized and trapped as the prodrug (**7.75**). The quaternary ammonium salt (**7.75**) is slowly cleaved by enzymic action with sustained release of the biologically active phenylethylamine and the facile elimination of the carrier molecule. Elimination of the drug from the general circulation is by comparison accelerated, either as (**7.75**) or (**7.76**) or as cleavage products. This rationale removes excess drug and metabolic products during or after onset of the required action. This is in contrast to normal penetration of the brain by a drug from plasma, where plasma levels must be maintained to produce the required effect and can thus cause systemic side effects.

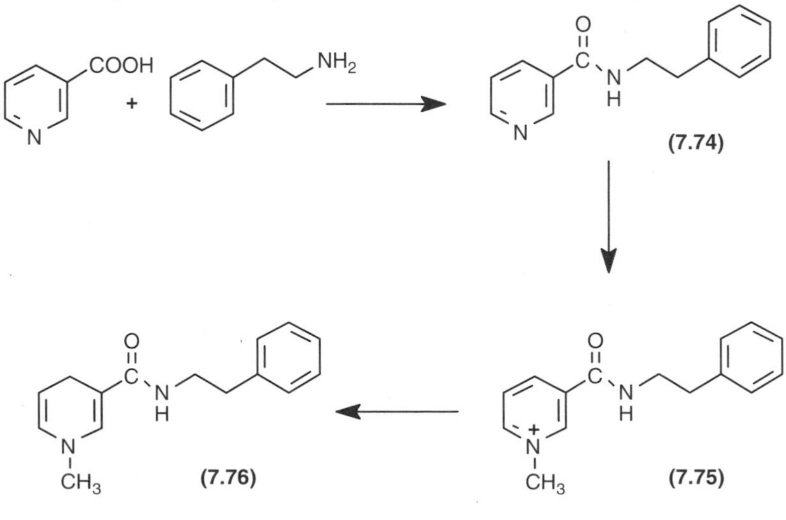

COOH + NH_2 → O C N H N **(7.74)**

O C N H N CH_3 **(7.76)** ← O C N H $\overset{+}{N}$ CH_3 **(7.75)**

In animal experiments the anti-inflammatory effect of topically applied hydrocortisone has been increased, and its systemic effects after absorption decreased, by use of the prodrug spirothiazolidine derivative (**7.77**). These beneficial effects are due to restriction of the action of hydrocortisone within the skin. After absorption, (**7.77**) is hydrolyzed in a stepwise manner with eventual release of hydrocortisone within the skin from the accumulated prodrug, resulting in a more intense anti-inflammatory effect and a decrease in its rate of leaching into the blood stream to produce systemic effects. The sustained release of hydrocortisone is due to retardation of the intermediate hydrolytic product (**7.78**) by disulfide formation (**7.79**) between its thiol group and a thiol group of the skin, followed by a slow breakdown of (**7.79**) to give hydrocortisone.

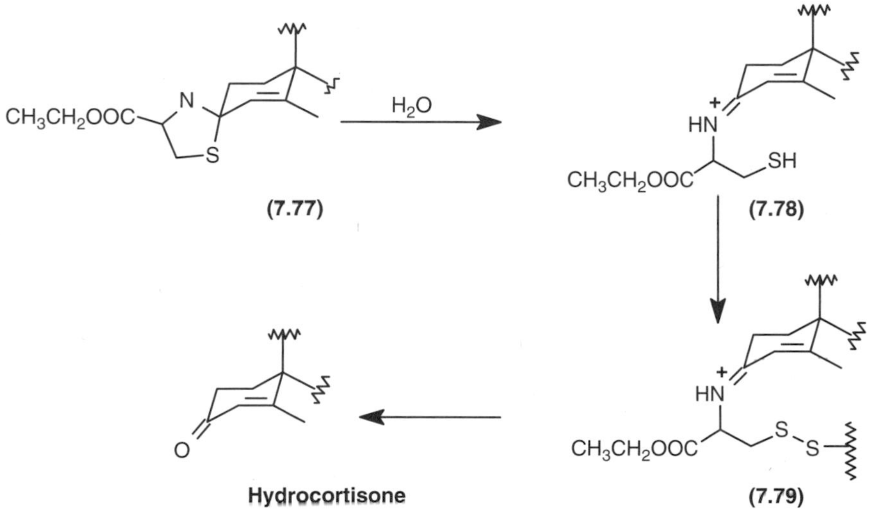

CH_3CH_2OOC N S **(7.77)** $\xrightarrow{H_2O}$ HN+ SH CH_3CH_2OOC **(7.78)** → HN+ S S CH_3CH_2OOC **(7.79)** → O **Hydrocortisone**

Success in cancer chemotherapy probably lies in utilizing differences in rates of growth between the rapidly dividing tumor cells and the slower noncycling normal tissue cells, as evidenced by responsiveness to chemotherapy of leukemia and the high-growth solid tumors. However, a different approach is needed for low-growth solid tumors.

The blood supply to large solid tumors is disorganized and internal regions may be nonvasculated and the cells, termed hypoxic, deprived of oxygen. Hypoxic tissues are known to have greater powers of reduction than oxygenated areas and the reduced species are expected to be stable in the absence of molecular oxygen, which could theoretically reverse the reduction process. This knowledge has been used in the development of a rationale for targeting drugs to the internal hypoxic regions of solid tumors, these regions are relatively inaccessible to drugs that are rapidly metabolized or strongly bound to tissue components. This approach could provide a selective chemical drug-delivery system when used in combination with treatments likely to be limited by the presence of hypoxic cells (see Chapter 14).

Certain aromatic or heterocyclic nitro-containing compounds can be reduced in a hypoxic (oxygen-deficient) environment to produce intermediates which then fragment into alkylating species. The 2-nitro-imidazole compound misonidazole (**7.80**) is selectively cytotoxic to cultured hypoxic cells. Reduction of the nitro group to the hydroxylamine (R NH_2OH) probably occurs, with further fragmentation occurring to give the DNA-alkylating species including glyoxal ($(CHO)_2$).

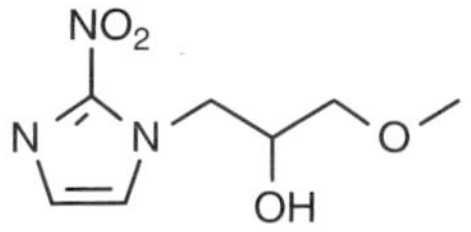

NO_2 N N O OH **(7.80)**

Nitracine (**7.81**) is another selective alkylating agent for hypoxic mammalian cells in culture after reduction, although the identity of the active species is unknown. Although nitracine is 10^5 times more potent than misonidazole in this system, it lacks activity in murine or human xenografted tumors.

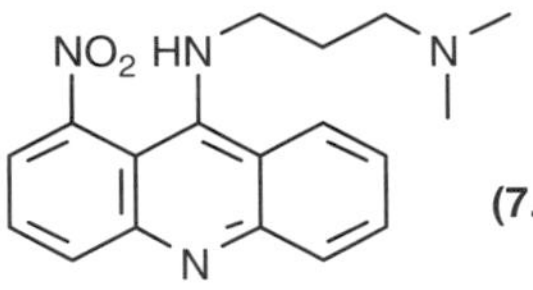

(7.81)

Research has also been directed toward the bioactivation of aromatic nitrogen mustards, where the mechanism of action is predictable and the activation step occurs by reduction of a substituent group in the aryl ring.[9] The alkylating ability of the β-chloroethylamine side-chain is dependent on the electron density on the nitrogen. The *p*-nitro substituent in (**7.82**), by exerting an electron withdrawing effect, reduces the electron density on the nitrogen thereby inhibiting the formation of the alkylating carbonium ion. Reduction of the nitro group in (**7.82**) in a hypoxic environment removes its electron withdrawing effect and restores the ability of the compound to form the alkylating species via an S_N1 reaction pathway. Whether reduction gives the hydroxylamine (**7.83**) or the amine (**7.84**) is uncertain, but both species have been calculated to have greater activity than the nitro compound.

(7.82) R = -NO$_2$
(7.83) R = -NHOH
(7.84) R = -NH$_2$

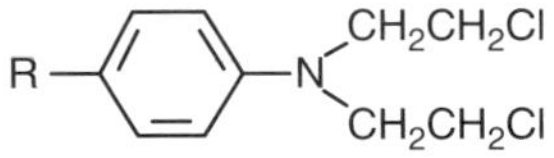

The aziridine (**7.85**) may be activated in a similar manner and has been shown to be selectively toxic to hypoxic cells. It should be noted that the presence of additional groups in the aryl ring may affect the actual electron density on the nitrogen atom, and hence the reactivity of the alkylating species generated, despite the activation process occurring on reduction.

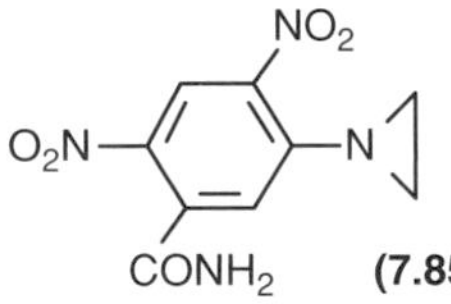

(7.85)

Soluble macromolecular prodrug delivery systems have also been developed to improve the pharmacokinetic profile of pharmaceutical agents by the controlled release of the active agent.[10] It has been suggested that such soluble polymeric carriers have the potential to improve the activity of conventional antitumor agents. Recently the potential of *N*-(2-hydroxypropyl)-methacrylamide (HPMA) copolymers as carriers for the antitumor agent doxorubicin (DOX) has been investigated.[11] Doxorubicin was linked to the polymeric carrier by peptidyl spacers designed to be cleaved by lysosomal thiol-dependent proteases, which are known to have increased activity in metastatic tumors. Such conjugates have been shown to have a broad range of antitumor activities against leukemic, solid tumor, and metastatic models. Fluorescein-labeled HPMA copolymers have been shown to accumulate in vascularized stromal regions, particularly in new growth sites in the tumor periphery. Treatment of C57 mice bearing subcutaneous B16F10 melanomas with DOX–HPMA copolymer conjugate improved the treated to control lifespan by threefold with respect to that obtained on aggressive treatment with free doxorubicin. It has been suggested that these macromolecular prodrugs reduce toxicity by controlled drug release following passive accumulation and retention within solid tumors.

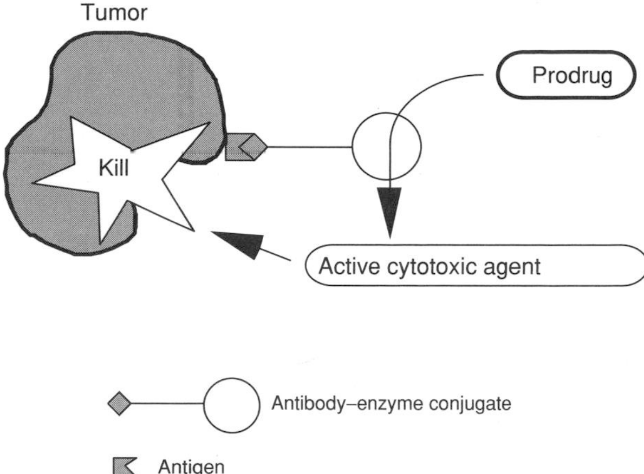

Figure 7.2 Antibody-directed enzyme prodrug therapy (ADEPT). A diagrammatic representation of the ADEPT approach to cancer chemotherapy which employs an antitumor antibody conjugated to an enzyme. The conjugate is localized at the tumor site via an antibody–antigen interaction and converts a subsequently administered prodrug into a cytotoxic agent, which attacks the tumor.

Recent research (see Chapter 14) has been directed toward alternative approaches to obtain site-specific activation of prodrugs for cancer chemotherapy using antibody-directed enzyme prodrug therapy (ADEPT) (Figure 7.2).[12,13] The ADEPT approach employs an enzyme, not normally present in the extracellular fluid or on cell membranes, conjugated to an antitumor antibody which localizes in the tumor via an antibody–antigen interaction on administration. Once any unbound antibody conjugate has been cleared from the systemic circulation, a prodrug, which is specifically activated by the enzyme conjugate, is administered. The bound enzyme–antibody conjugate ensures that the prodrug is only converted to the cytotoxic parent compound at the tumor site thereby reducing systemic toxicity. It has been shown that in systems utilizing cytosine deaminase to generate 5-fluorouracil from the 5-fluorocytosine prodrug at tumor sites, 17 times more drug can be delivered within a tumor than on administration of 5-fluorouracil alone.

The ADEPT approach has been recently investigated as a means of overcoming the side effects of taxol, which is an effective treatment for breast cancer but also attacks healthy tissues. The system utilizes a β-lactamase enzyme antitumor antibody conjugate and a prodrug (PROTAX) which consists of taxol linked via a short chain to cepham sulfoxide. Taxol is selectively released at the tumor site by the localized β-lactamase enzyme, which is not normally found in any other tissues. In studies on cultured human breast cells it has been shown that the prodrug is almost as effective as taxol on cells which have been treated with the enzyme-bound antibody, however PROTAX alone is only a tenth as toxic to cancer cells as taxol and is therefore less likely to harm healthy cells.

More recently advances in molecular biology (see Chapter 14) have led to the development of a virus-directed enzyme prodrug therapy (VDEPT) using suicide genes.[14,15] Suicide genes encode for nonmammalian enzymes which can convert a prodrug into a cytotoxic agent. Cells which are genetically transduced to express such genes essentially commit metabolic suicide in the presence of the appropriate prodrug. Typical suicide genes include herpes simplex thymidine kinase and *Escherichia coli* cytosine deaminase. Viral vectors are used to carry the gene into both tumor and normal cells. Tumor-specific transcription of the suicide gene is achieved by linking the foreign gene downstream of a tumor-specific transcription unit such as the proximal ERBB2 promoter. The ERBB2 oncogene is overexpressed in approximately a third of all breast and pancreatic tumors by transcriptional upregulation of the ERBB2 gene with or without gene amplification. In recent studies a chimeric minigene consisting of the proximal ERBB2 promoter linked to a gene coding for cytosine deaminase has been constructed and incorporated into a double-copy recombinant retrovirus. *In vitro* studies using pancreatic and breast cell lines have been used to demonstrate significant cell death on treatment of cells which expressed ERBB2 with the viral vector and 5-fluorocytosine, whereas cells which did not express ERBB2 were not affected.

7.4.4 Sustaining Drug Action

Prodrug design has also been successfully used to modify the duration of action of the parent drug by either reducing the clearance of the drug or by providing a depot of the parent drug.

The prodrug bitolterol (**7.86**), which is the di-*p*-toluate ester of *N*-*t*-butyl noradrenaline (**7.87**), has been shown in dogs to provide a longer duration of bronchodilator activity than the parent drug. Furthermore, the prodrug is preferentially distributed in lung tissues rather than plasma or heart so that the bronchodilator effect, following subsequent biotransformation of the prodrug, is not associated with undesirable cardiovascular effects and is slow and prolonged.

(7.86) R = *p*-toluoyl
(7.87) R = H

The phenothiazine group of drugs, acting as tranquillizers, have been converted to long acting prodrugs which are administered by intramuscular injection. Not only is the frequency of administration reduced, but also the problem associated with patient compliance is also eliminated. Flupenthixol (**7.88**) when administered as the decanoate ester (**7.89**) in an oily vehicle for the treatment of schizophrenia, is released intact from the depot and subsequently hydrolyzed to the parent drug, possibly after penetration of the blood–brain barrier. Maximum blood levels are observed within 11–17 days after injection and the plateau serum levels averaged 2–3 weeks in duration.

(7.88) R = -H
(7.89) R = -CO$(CH_2)_8CH_3$

Similarly, perphenazine (**7.90**) has been used as the enanthrate ester (**7.91**) and pipothiazine (**7.92**) as the undecanoate (**7.93**) and palmitate (**7.94**) esters.

(7.90) R = -H
(7.91) R = -CO$(CH_2)_5CH_3$

(7.92) R = -H
(7.93) R = $-CO(CH_2)_9CH_3$
(7.94) R = $-CO(CH_2)_{14}CH_3$

Vasopressin has been used for the treatment of bleeding varicose veins in the lower end of the esophagus (esophageal varices), a condition that affects about 1000 individuals annually. The vasoconstrictor action of the drug stops the bleeding, but the action is of short duration and cannot be prolonged by the use of higher doses due to the development of toxic side effects. Glypressin, Gly-Gly-Gly-Lys-vasopressin, is an inactive prodrug of vasopressin and after injection the glycyl residues are steadily cleaved off by enzymic action to release the active drug. A sustained low level of vasopressin is obtained in this manner, which is sufficient to produce the required vasoconstriction effect on portal blood pressure while minimizing the possibility of unwanted effects caused by high blood pressure.

The examples given in this chapter illustrate the importance of the prodrug concept as a means of overcoming pharmaceutical and pharmacological problems encountered during drug development. In addition, recent advances in biotechnology have made it possible to utilize prodrug design to develop chemical drug delivery systems which provide various means of targeting the delivery of parent drugs to specific sites within the body. Clearly, the increasing demands for more efficacious and less toxic drugs will ensure that prodrug approaches continue to be exploited in the development of future drug substances.

REFERENCES

1. Stella V.J., Charman W.N.A., and Naringrekar V.H. (1985) Prodrugs. Do they have advantages in clinical practice? *Drugs* **29**, 455–473 [see references to other reviews cited therein].
2. Riley T.N. (1988) The prodrug concept and new drug design and development. *Journal of Chemical Education* **65**, 947–953.
3. Waller D.G. and George C.F. (1989) Prodrugs. *British Journal of Clinical Pharmacology* **28**, 497–507.
4. Sinkula A.A. and Yalkowsky S.H. (1975) Rationale for design of biologically reversible drug derivatives: prodrugs. *Journal of Pharmaceutical Sciences* **64**, 181–210.
5. Easterbrook P. and Wood M.J. (1994) Successors to acyclovir. *Journal of Antimicrobial Chemotherapy* **34**, 307–311.
6. McGuigan C., Sheeka H.M., Mahmood N., and Hay A. (1993) Phosphate derivatives of d4T as inhibitors of HIV. *Bioorganic & Medicinal Chemistry Letters* **3**, 1203–1206.
7. Druzgala P., Winwood D., Drewniak-Deyrup M., Smith S., Bodor N., and Kaminski J.J. (1992) New water-soluble pilocarpine derivatives with enhanced and sustained muscarinic activity. *Pharmaceutical Research* **9**, 372–377.
8. Bodor N. and Farag H.H. (1983) Improved delivery through biological membranes. 11. A redox chemical drug-delivery system and its use in brain-specific delivery of phenylethylamine. *Journal of Medicinal Chemistry* **26**, 313–318.
9. Denny W.A. and Wilson W.R. (1986) Considerations for the design of nitrophenyl mustards as agents with selective toxicity for hypoxic tumour cells. *Journal of Medicinal Chemistry* **29**, 879–887.
10. Duncan R. (1992) Drug polymer conjugates — potential for improved chemotherapy. *Anti-Cancer Drugs* **3**, 175–210.

11. Seymour L.W., Ulbrich K., Steyger P.S., Brereton M., Subr V., Strohalm, J., and Duncan R. (1994) Tumor tropism and anticancer efficacy of polymer-based doxorubicin prodrugs in the treatment of subcutaneous murine B16F10 melanoma. *British Journal of Cancer* **70**, 636–641.
12. Bagshawe K.D., Sharma S.K., Springer C.J., and Rogers G.T. (1994) Antibody directed enzyme prodrug therapy (ADEPT). *Annals of Oncology* **5**, 879–891.
13. Huennekens F.M. (1994) Tumor targeting: activation of prodrugs by enzyme-monoclonal antibody conjugates. *Trends in Biotechnology* **12**, 234–239.
14. Harris J.D., Gutierrez A.A., Hurst H.C., Sikora K., and Lemoine N.R. (1994) Gene therapy for cancer using tumour-specific prodrug activation. *Gene Therapy* **1**, 170–175.
15. Huber B.E., Richards C.A., and Austin E.A. (1994) Virus-directed enzyme/prodrug therapy-selectively engineering drug sensitivity into tumors. *Annals of the New York Academy of Science* **716**, 104–114.

FURTHER READING

Anderson B.D. (1996) Prodrugs for improved CNS delivery. *Advanced Drug Delivery Reviews* **19**, 171–202.

Chari R.V.J. (1998) Targeted delivery of chemotherapeutics: tumor activated prodrug therapy. *Advanced Drug Delivery Reviews* **31**, 89–104.

Charman W.N. and Porter C.J.H. (1996) Lipophilic prodrugs designed for intestinal lymphatic transport. *Advanced Drug Delivery Reviews* **19**, 149–169.

Fleisher D., Bong R., and Stewart B.H. (1996) Improved oral drug delivery: solubility limitations overcome by the use of prodrugs. *Advanced Drug Delivery Reviews* **19**, 115–130.

Hoste K., De Winne K., and Schacht E. (2004) Polymeric prodrugs. *International Journal of Pharmaceutics* **277**, 119–131.

Huber B.E., Richards C.A., and Austin E.A. (1995) VDEPT: an enzyme/prodrug gene therapy approach for the treatment of metastatic colorectal cancer. *Advanced Drug Delivery Reviews* **17**, 279–292.

Jones R.J. and Bischofberger N. (1995) Minireview: nucleotide prodrugs. *Antiviral Research* **27**, 1–17.

Kearney A.S. (1996) Prodrugs and targeted drug delivery. *Advanced Drug Delivery Reviews* **19**, 225–239.

Krise J.P. and Stella V.J.(1996) Prodrugs of phosphates, phosphonates, and phosphinates. *Advanced Drug Delivery Reviews* **19**, 287–310.

Ouchi T. and Ohya Y. (1995) Macromolecular prodrugs. *Progress in Polymer Science* **20**, 211–257.

Riley T.N. (1988) The prodrug concept and new drug design and development. *Journal of Chemical Education* **65**, 947–953.

Sandborn W.J. (2002) Rational selection of oral 5-aminosalicylate formulations and prodrugs for the treatment of ulcerative colitis. *The American Journal of Gastroenterology* **97**, 2939–2941.

Senter P.D. and Springer C.J. (2001) Selective activation of anticancer prodrugs by monoclonal antibody–enzyme conjugates. *Advanced Drug Delivery Reviews* **53**, 247–264.

8

From Program Sanction to Clinical Trials: A Partial View of the Quest for Arimidex, A Potent, Selective Inhibitor of Aromatase

Philip N. Edwards

CONTENTS

8.1 INTRODUCTION

Breast cancer is the commonest cancer in women and, despite continuing advances in treatment, each year worldwide an increasing number die from the disease: in Japan the incidence has been increasing by an alarming 10% per annum. In the early stages of the disease, 30–40% of patients respond to hormonal or antihormonal therapy. One way to deprive hormone-dependent cancer of its primary mitogens, estrogens, is to prevent their synthesis — preferably by inhibition of aromatase, the ultimate and biochemically unique enzyme that converts androgens such as testosterone to mitogenic estradiol.

This account provides some of the background to the author's and ICI's involvement with hormonal modulation, but attempts mainly to cover the cytochrome P450-dependent enzyme, aromatase (estrogen synthase, P450arom/NADPH cytochrome P450 reductase), its inhibition, and the way in which the ICI Aromatase Team selected a development compound, was forced by long-term toxicity to abandon it, but was more fortunate in its second choice with the compound ICI 207658, numbered D1033 during early development and later given the name anastrozole. During the development phase of D1033, ICI Pharmaceuticals became Zeneca Pharmaceuticals and the designation ZD1033 was used for the drug which is now called Arimidex.*

The medicinal chemistry coverage, in focusing mainly on the author's team contributions to the program, is a partial account of work that involved several chemistry teams.

8.2 BACKGROUND

An undergraduate course in organic chemistry in the mid-1950s typically made good use of the inspiring work of many groups in the fields of steroid structure determination, conformational analysis, reactivity, and synthesis. The potent and multifarious biological properties of such molecules, along with the synthetic challenges presented by extremely complex structures for those times, made them synthetic targets for many eminent chemists of the day. The first formal *total* synthesis of cortisone (**8.1**), then thought to be a miracle drug, had been briefly reported by Fieser and Woodward in August 1951.

(8.1)

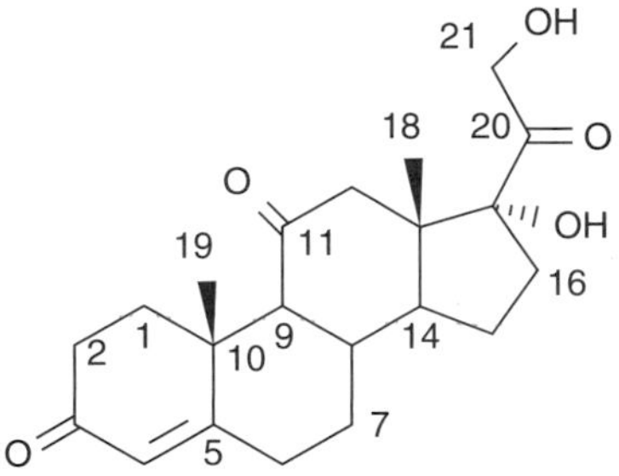

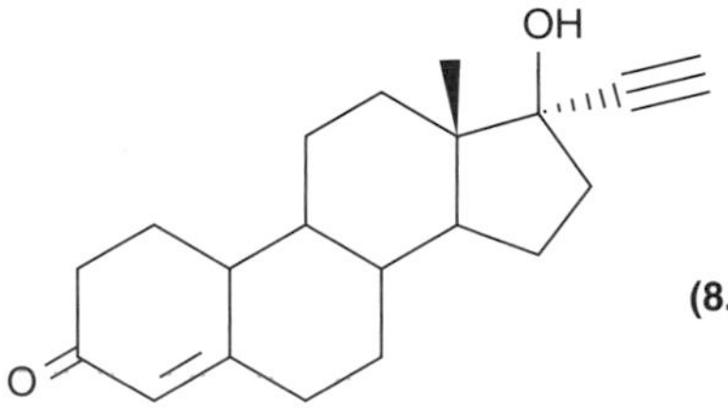

(8.2)

However, the race to achieve the first nontrivial synthesis of cortisone had unexpectedly been won by a group of young chemists in Mexico City who were employed by a small, recently formed company called Syntex, Inc. That synthesis, starting with readily available diosgenin — from Mexican yams — had commercial potential from the sale of intermediates as well as the final product and its dihydroderivative, hydrocortisone or cortisol. Cortisone was in great demand for the treatment of severe inflammatory and immunologically related conditions, as well as for treating Addison's disease, a previously life-threatening condition caused by a deficiency of cortisol synthesis in the adrenal gland of an afflicted individual.

Clearly, our target aromatase inhibitor would have to avoid *in vivo* inhibition of the cytochrome P450-dependent enzymes involved in cortisol/cortisone synthesis. Indeed it ideally had to avoid inhibiting any cytochrome P450-dependent enzyme except aromatase. Schenkman and Greim (1993) recently edited a wideranging multiauthor review of such enzymes.

Carl Djerassi was a leading member of that early Syntex group and he was soon hailed by many in academia as one of the promising synthetic chemists of the day. Such limited accolades were eventually dwarfed when he achieved broadly proclaimed ''immortality'' as the ''Inventor of The Pill.'' As so often even in those days, the title is largely the result of the mass media requirement for (over)simplification. As Djerassi (1992) emphasizes in his autobiography, his part in the invention of ''The Pill'' was that of main contributor to the search for and discovery of the orally potent progestagen, norethindrone (norethisterone) (**8.2**); this was achieved during his brief time at Syntex and overlapped the cortisone work. Others had foreseen the potential of such agents and many others were involved in the development and exploitation of this and related compounds — indeed it took more than a decade for such hormonal modulation to gain even limited acceptance as a method of contraception.

*Arimidex is a trademark, the property of AstraZeneca plc.

It is of interest in the present context that norethindrone and other terminal acetylenes such as ethynyloestradiol may owe some of their improved oral activity to irreversible, mechanism-based, covalent inhibition of drug-metabolizing cytochrome P450-dependent enzymes in the liver. This possibility and the nature of such enzymes were unknown at the time of those drugs' discovery, but with the advent of that understanding it is now possible to design inhibitors of P450s based on terminal acetylenes. Potent inhibitors of aromatase have been generated by modifying natural substrates, and close analogues, through the addition of an ethynyl group to C19 — the initial site of substrate oxidation by aromatase. Amazingly, it was only in 1982 that norethindrone was shown *in vitro* to be a rather weak (~2 μM), irreversible inhibitor of aromatase. It is unlikely, however, that this has relevance to its contraceptive use.

By the late 1950s many research groups were involved in hormonal modulation. One of those groups was in ICI Pharmaceuticals and its efforts were rewarded with the discovery and successful development of the estrogen *antagonist* tamoxifen, or Nolvadex (**8.4**) as ICI named it (Nolvadex is a trademark, the property of AstraZeneca plc.). A number of structurally related estrogen *agonists* had been discovered in the 1930s, one of which, diethylstilboestrol (**8.3**), remains, somewhat controversially, in use to this day.

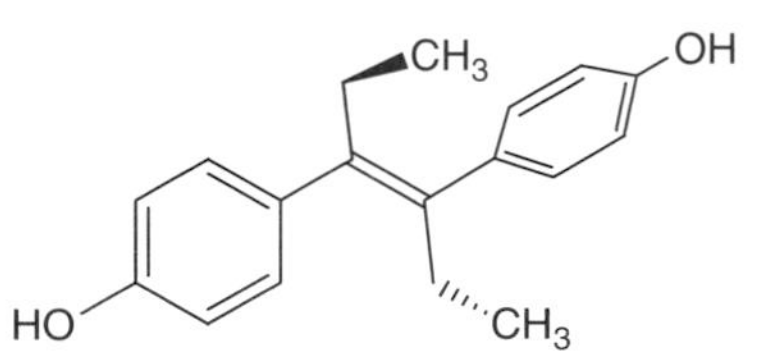

(8.3) 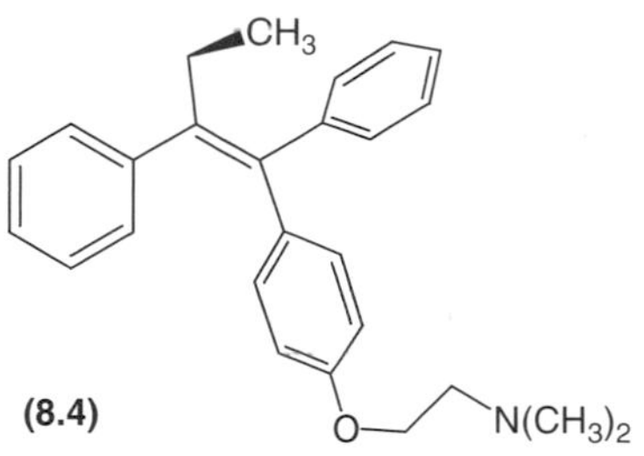 **(8.4)**

Nolvadex (**8.4**) was found to be an effective treatment for a substantial proportion of post-menopausal patients with estrogen-dependent breast cancer. It has in recent years been the largest selling chemically defined anticancer drug of all time.

The author's first year in research, 1957–1958, coincided with three events important to this discourse. First, M. Klingenberg and later D. Garfinkel independently reported the generation of a new absorption peak at 450 nm when homogenized liver supernatant was exposed to carbon monoxide: pigment 450 (P450) was born, but the function, if any, of this pigment was unknown. Second, K.J. Ryan reported that androgens incubated with human placental microsomes were converted to estrogens. This amazing process involves the removal, by then unknown chemical steps, of a very hindered, nonactivated methyl group from C10 and a nonactivated hydrogen atom from C1 of the androgenic precursors testosterone or androstenedione. What agent or agents were at work was again unknown. Third, the Swiss pharmaceutical giant, Ciba, a major force in steroid chemistry, pharmacology, and drugs, started clinical trials with aminoglutethimide (AG) (**8.5**), as a prospective anticonvulsant drug. Those trials and subsequent use under the name Elipten (later Orimeten: Ciba, and now Cytadren: Ciba–Geigy) revealed multiple adverse side effects, one of the most serious being adrenal insufficiency. A few years after launch it was withdrawn from sale, but as so often in chemotherapy, one person sees a side effect while another sees an opportunity: a medical adrenalectomy might be useful in various adrenal hormone-dependent diseases — including breast cancer where surgical adrenalectomy was an established hormonal maneuver.

(8.5) **(8.6)**

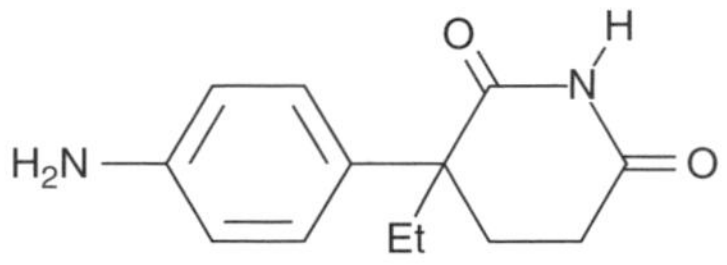

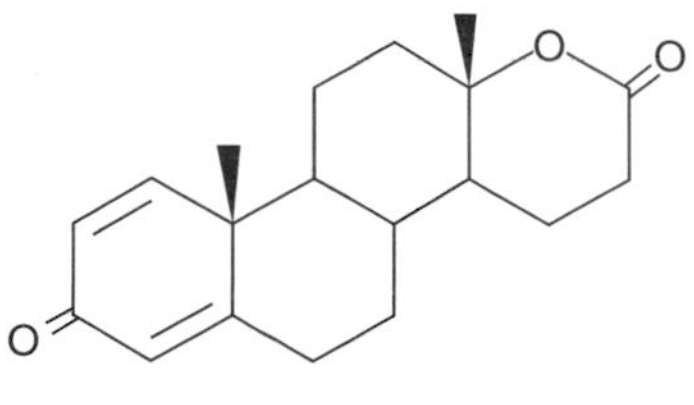

Some years later, that possibility became a reality — AG was shown to be useful in several conditions, including advanced breast carcinoma. It originally was assumed that efficacy flowed from suppression of adrenal pregnenolone synthesis. Rather low-potency (26 μ*M*) inhibition *in vitro* of an adrenal-derived enzyme, P450scc, that converts cholesterol via side chain cleavage to pregnenolone, had long since been demonstrated and adrenal hypertrophy in various species dosed with AG is ascribed to that gland's attempt to maintain steroidogenic homeostasis. Inhibition of P450scc would in turn limit synthesis of the many other steroids, including estrogens, which have pregnenolone as a precursor. Scheme 8.1 shows a selection of steroidogenic pathways — unidirectional arrows indicate that one or more of the steps in that pathway involves a cytochrome P450 enzyme. These pathways operate in differing degrees according to tissue, species, sex, age, and in pregnancy and disease and their products elicit a wide variety of responses depending on the target cell type and its environment (Castagnetta *et al.*, 1990).

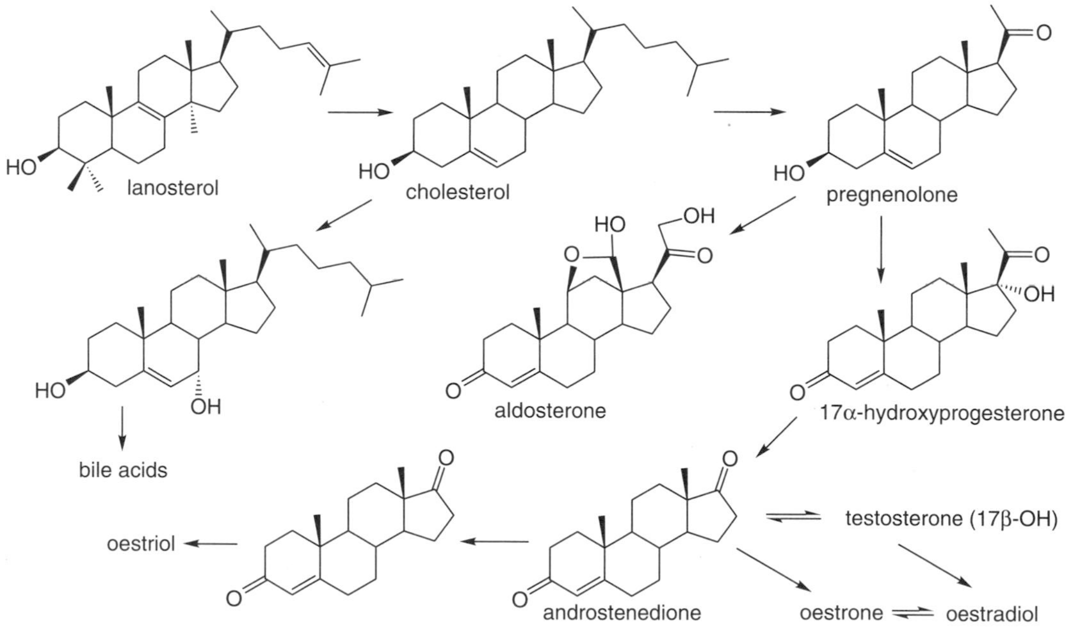

lanosterol → cholesterol → pregnenolone

bile acids

aldosterone

17α-hydroxyprogesterone

testosterone (17β-OH)

oestriol ← ← androstenedione

oestrone ⇌ oestradiol

Scheme 8.1

Replacement glucocorticoid was administered with AG during breast cancer therapy in part because of the above findings. Later quantitative studies however showed reduced estrogen levels but normal or even increased levels of androstenedione in AG-treated patients' plasma: androstenedione is the main androgenic precursor of oestrone. These new findings appeared inconsistent with substantial P450scc involvement. Subsequent *in vitro* studies, 25 years after Ryan's discovery, suggested that efficacy in postmenopausal breast cancer patients resulted mainly from inhibition of the enzyme (or enzymes) that converts androgens to estrogens. By then the placental enzyme activity was widely known, but because the enzyme is embedded in microsomal phospholipid membranes, and is functionally dependent on that association, it was not yet well characterized. Crystallization is likely to be extremely difficult or impossible.

We now know that P450arom is a single enzyme — the product of the CYP19 gene on chromosome 15 in humans. Gene expression is subject to complex and multifactorial regulation. The enzyme is widespread in humans, males and females, in the brain and the periphery, but it is much less widespread in most other species. It belongs to the cytochrome P450-dependent class of enzymes and is commonly called aromatase or P450arom, but the latter designation refers specifically to the heme-binding protein of the two-component enzyme: the second component, a flavoprotein, is reduced nicotinamide adenine dinucleotide diphosphate (NADPH)–cytochrome P450 reductase; the reductase is common to all P450-dependent enzymes. Use *in vitro* of oxidants other than air, for example Ph−I=O, or H_2O_2, allows most P450s to function in the absence of the reductase. Interestingly, and relevant to the mechanism (Scheme 8.2), iodosobenzene is ineffective in the final step of aromatization while hydrogen peroxide allows all three steps to proceed.

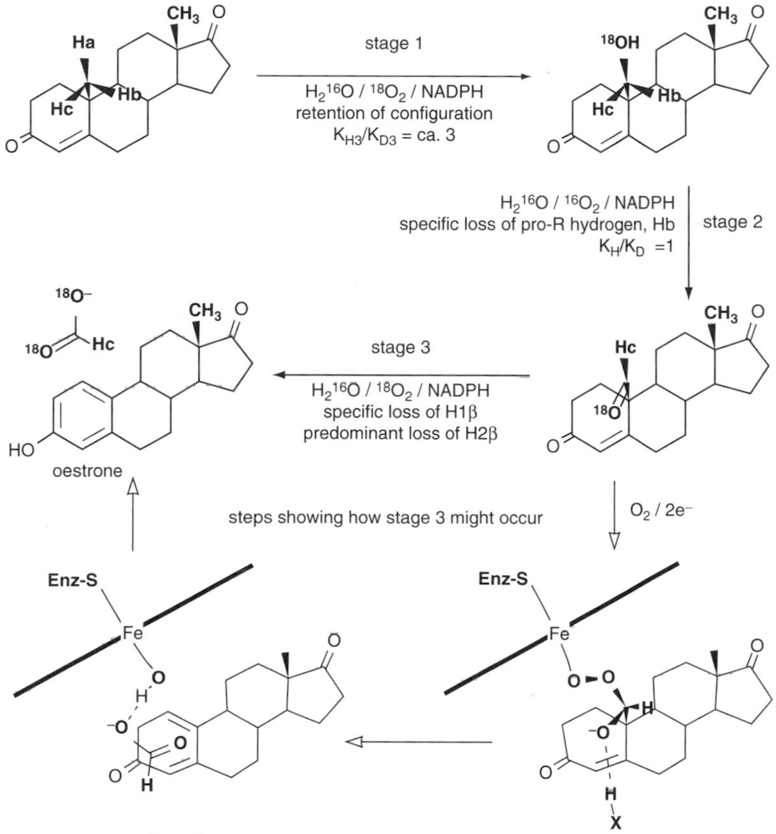

Scheme 8.2 Peroxy compound (**8.9**), as the oxy-anion, is predicted to undergo concerted exothermic fragmentation of O−O and C−C bonds to generate the much more stable formate anion and a delocalized radical. Rapid abstraction, by the jointly formed Fe^{IV}=O, of H1β from the carbon adjacent to the π-radical, generates the dieneone (**8.10**), which, being a very high-energy tautomer of a phenol, can reliably be predicted to be a much stronger acid than formic acid ($>10^3$-fold). Proton transfer of H2β to formate anion or Fe^{III}−OH is expected to be a fast, exothermic process which does not require proton transfer to the developing phenolate anion.

P450arom, or rather the estrogens it produces, has differing roles according to the sex of the animal and cell type in which it is expressed. Importantly it is expressed and is functional in producing trophic effects in many breast cancer cell lines — a blood-born estrogen supply is not always necessary for growth.

Testololactone (Teslac, Squibb) (**8.6**), used in breast cancer and originally thought to act via androgenicity, may also owe much of its efficacy to aromatase inhibition.

8.3 ICI START ON AROMATASE INHIBITION

In late 1970s the Fertility Project Team in ICI was again testing compounds for antifertility potential. Some of that effort was devoted to random screening with the end point being prevention of pregnancy in rats. One of the more potent compounds discovered, the *N*-‘benzyl’ imidazole (**8.7**), was considered to be worthy of further investigation since its structure and overall biological effects did not point to any known mode of action. There was some concern that its fragmentation to a quinone methide might be involved — if that was so, the generation of such a reactive species would make it and any analogues unattractive. The postulated quinone methide is implicated in the lung toxicity of butylated hydroxytoluene (BHT; 3,5-ditertiarybutyl-4-hydroxy-toluene), a widely used antioxidant.

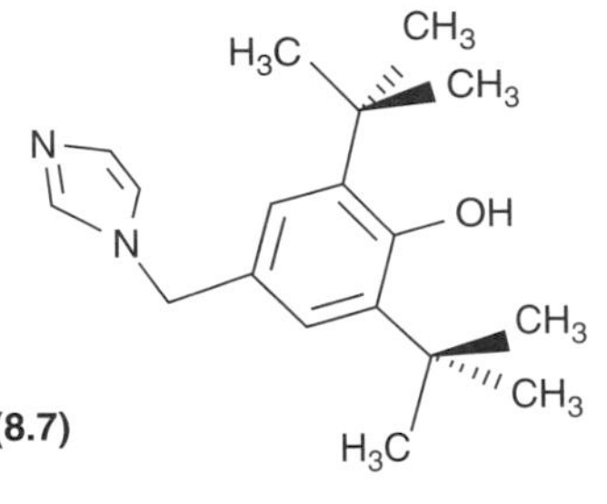

(8.7)

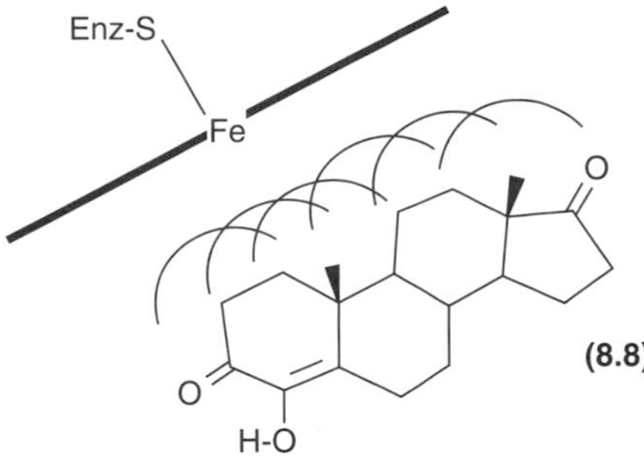

(8.8)

Also at that time, the Team's interest in aromatase had been heightened by the results of clinical and biochemical studies in patients receiving AG. Preclinical results, obtained by Angela Brodie and coworkers with 4-acetoxy-androst-4-ene-3,17-dione, were also encouraging. This steroid was active *in vivo*, especially in the estrogen-dependent dimethylbenzanthracene (DMBA) rat tumor model, but interest focused later on the 4-hydroxy compound, 4-OHA (**8.8**), (formestane), a potent, $K_i = 10$ n*M*, and time-dependent aromatase inhibitor since licensed and named Lentaron by Ciba–Geigy. Structure (**8.8**) is shown with partial van der Waals' radii for some ''atoms'' (actually CH_2 and CH_3 groups); those ''atoms'' in the enzyme-bound state are postulated to be in contact with the large, extensively planar protoporphyrin-IX prosthetic group which is depicted, edge-on, as a thick line. Partial van der Waals' radii for atoms in the porphyrin are not shown but extend to contact those shown for the steroid.

The official start of the Aromatase program, just a few weeks into the new decade, was contemporaneous with the beginning of another team's attempt to find an antiestrogen working through inhibition of translocation of the estrogen receptor from cytosol to nucleus (more details of this work are given below). Chemistry started in two main directions, steroid-based (naturally) and *azole*-based: the *N*-''benzyl'' imid*azole* (**8.7**) had by now been shown likely to be an aromatase inhibitor. Our compound collection, together with some standard antifungal agents from ICI Plant Protection, generated a structurally diverse set of leads with some remarkably simple azoles, e.g., *N*-(*m*-pentanoylbenzyl)imidazole ($IC_{50} = 2$ ng/ml), being very potent inhibitors of human placental microsomal aromatization, e.g., of testosterone to estradiol or of androstenedione to estrone.

The literature evidence at the time was consistent with all the aromatase chemistry being carried out by a single enzyme, but this became certain only much later, during the second part of the program. The multiple steps involved in this conversion were already broadly established from a vast body of work by many academic groups (Brodie et al., 1993) (see Scheme 8.2), but some of the finer mechanistic detail remains controversial (Akhtar et al., 1993). The lower part of the scheme and footnote commentary represent the author's view of how the final steps might proceed.

At this stage we knew from ICI work on antifungal agents which potently inhibit fungal lanosterol-14-methyl demethylase, a P450-mediated reaction very closely related to that performed by aromatase, as well as from literature reports of azoles inhibiting various P450 enzymes, and already rather extensive studies with the *N*-''benzyl'' imidazole (**8.7**), that superior selectivity versus AG could be the key to a successful drug. Considering how ''dirty'' AG is by modern standards, this seemed at first an easy target. We soon thought otherwise: the *in vivo* effective azoles then to hand were all clearly deficient in one or more respects. Surprisingly to us because we were not aware of any connection with P450 enzymes, all the *in vivo* more potent (but still weakly potent) azoles, including (**8.7**), caused unacceptable elevation of liver triglycerides at modest multiples of their aromatase-effective doses.

Another frequently observed effect *in vivo* — adrenal enlargement — indicated unwanted inhibition of nonestrogenic steroid production. No pattern of selectivity could be discerned. Yet another indicator of potentially inadequate selectivity was the increased sleeping times observed after co-administration of hexobarbital to mice with each of the few azoles tested: such effects are probably due to inhibition of liver P450-mediated oxidative clearance of this xenobiotic sedative. Multiple high doses of all azoles examined caused increases in liver weight to body weight ratios and elevation of some liver mixed function oxidase (P450) enzymes. These elevated P450 levels can cause increased clearance rates and modify metabolite patterns of hormones, drugs, and other natural products and xenobiotics. Both these effects and enzyme-inhibitory effects are present in AG-treated patients, but we decided that none of these effects would be acceptable at the therapeutic dose of our target molecule. Increased P450-mediated production of toxic and particularly mutagenic metabolites is one of the consequences of smoking and is implicated in the increased incidence of cancer in smokers. Smoking differs from most drug therapy in causing different P450s to become elevated, but obviously everyone would wish to minimize such risks, even in long-term drug treatment of cancer patients.

So how might one achieve selectivity? There are several possibilities: set up screens and throw everything you have at them; or, ideally with the help of precision models, Dreiding, etc., and computer modeling, try to use substrate structures and inhibitor structures as a guide to drug design; or, again using modeling, try to understand the reactions using as much detailed information as exists, make educated guesses about the three-dimensional interactions needed for recognition and mechanism, then design the drug around as many hopefully unique features as possible.

Mainly in part two of the program, we did some of each of these and developed other ideas that will be discussed later. Modeling at various levels should be, and was, an on-going process. The amino acid sequence of human aromatase was not known at the time of our work but became so soon thereafter: its sequence of 503 amino acids shows only about 30% homology with other known mammalian P450s; the latter group is highly homologous. This puts P450arom in a unique category. Despite the implications from the foregoing, several groups have published models of the enzyme-based on lipid-free, water-soluble, bacterial enzymes, e.g., P450cam, that have been crystallized and the structures determined at high resolution by x-ray diffraction methods. In the author's view none of these models is satisfactory and *any* model is at present highly speculative. Speculative hydrogen bonds indicated in Scheme 8.2 and in Figure 8.1 are those used during our work. The peroxy intermediate (**8.9**), bound to a partial enzyme model, is shown as a stereoscopic pair in Figure 8.1. This binding mode was the basis of essentially all our modeling, despite the (still) speculative nature of such a species. It is shown essentially as we used it except that the amino acid side chains on the α-helical protein fragment have been updated: we used those in P450cam. Some inhibitors throughout the chapter are drawn as they would appear in such a model, but with the partial α-helix removed and viewed edge on to the porphyrin multiring system: these changes allow easier comparisons of our suggested binding modes.

Steroidal inhibitors might seem intrinsically to hold better prospects for selectivity, but, as shown in part in Scheme 8.1, Nature's wide use of P450 enzymes in chopping, trimming, and oxidatively modifying this skeleton argues against overconfidence in this intuitive position. Furthermore, steroids typically have other problems such as rapid clearance and poor efficacy by the oral route, especially in the rat, our preferred test species. Effects through receptor interactions are

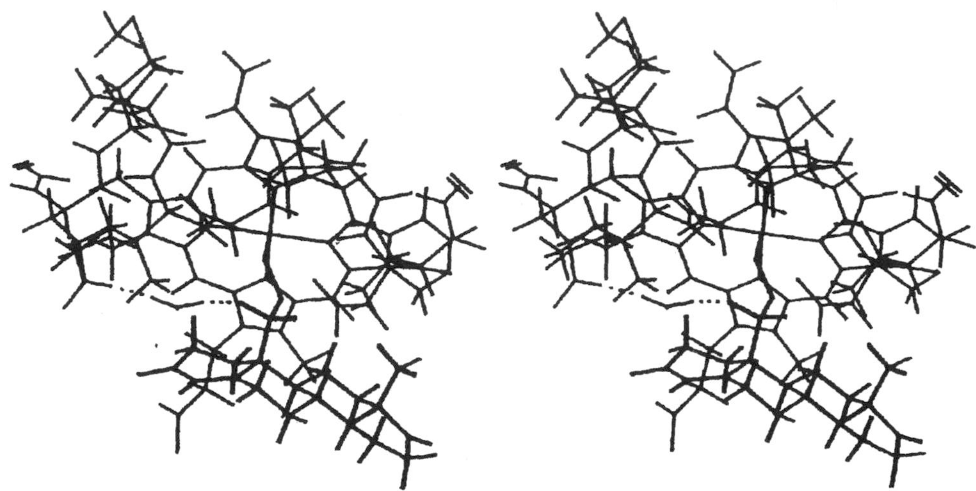

Figure 8.1 Stereo-pair of proposed (partial) active site of $P450_{arom}$.

another concern. At a practical level, synthesis of new compounds can be demanding and slow and several other groups, industrial and academic, were known to have a substantial start on us. Counterbalancing this, we had a steroid expert in the team and we thought it was worth a try.

Probably none of these considerations counted for much in the light of the excitement generated by the ''translocation'' work yielding some extremely interesting compounds. This new lead, irrespective of mechanism, was antiestrogenic in every test that was applied. It is now considered unlikely that effects on translocation, as such, are important but this idea nonetheless led to the discovery of the first ''pure'' antiestrogens (Wakeling, 1990). There is recent evidence to show that these agents are not equivalent to the total absence of estrogen, the potential outcome of aromatase inhibition. Control of gene transcription is a complex multifactorial process in which the occupied but apparently estrogenically inactive receptor still has a role. Ongoing clinical studies may start to tease out some of the therapeutic implications of this complex and still incompletely explored biochemistry.

Unsurprisingly, chemistry effort was switched from aromatase. Soon still more effort was required: it was then that the author came to work for the first time on hormonal modulation.

8.4 AROMATASE RESUMED

8.4.1 The Legacy from Antiestrogens

A ''pure'' antiestrogen development candidate, ICI 182780 (**8.11**), was chosen in 1985, but it was 2002 before it was marketed by AstraZeneca under the trade name ''Faslodex''. ''Faslodex'' is a 7α-(long side chain) substituted estradiol derivative, but many nonsteroidal frameworks were investigated during the program and most, with appropriate side chains, yielded potent, ''pure'' antiestrogens.

Generally these frameworks, linked to azole rings via short side chains, yielded moderate- to high-potency aromatase inhibitors in the ensuing second phase of that program. Because of its limited conformational freedom and very high inhibitory potency against human placental aromatase, one such azole, the (racemic) triazole derivative (**8.12**), is particularly relevant to computer modeling of the enzyme active site.

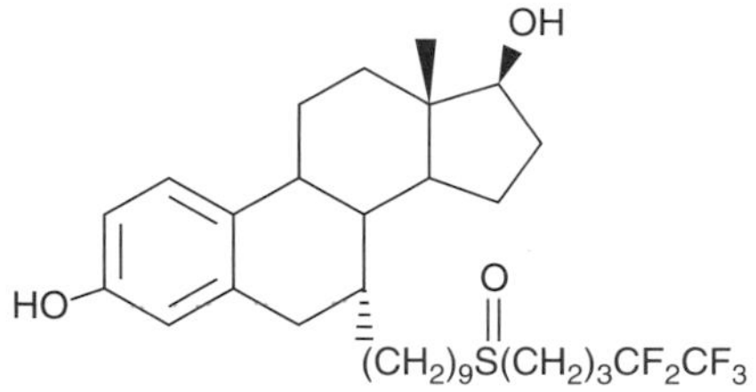

(8.11)

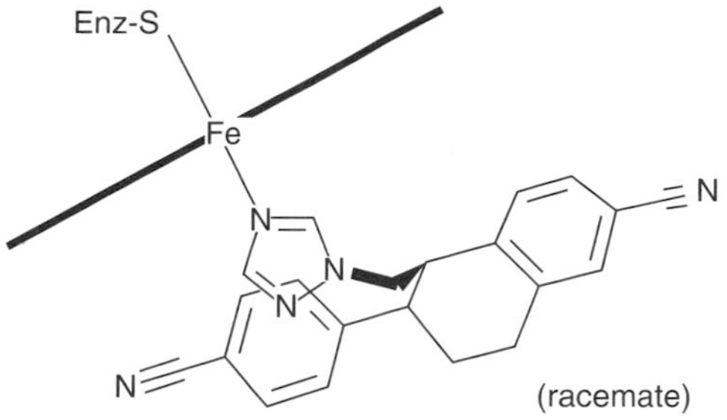

(racemate)

(8.12)

Several highly potent aromatase inhibitors arose inadvertently during the final stages of the antiestrogen work. We were attempting to find a replacement for the potency-enhancing but metabolically sensitive phenolic hydroxyl group in our pure antiestrogens: phenolic compounds bind to the receptor *in vitro* about 100-fold tighter than nonphenolic analogues. Many alternatives to the phenolic OH group had been tried, but none came even close to matching its ''magical'' effect. One possible explanation for these dramatic findings is that strong interactions occur between the receptor protein and *both* the in-plane acidic hydrogen and the in-plane oxygen lone-pair of the aromatic OH group.

With this hypothesis, no benzenoid derivative was likely to match the phenol, but that conclusion need not apply to planar heterocyclic systems: the 4-substituted pyrazole (**8.13**) was designed to interact with just such a hypothetical phenol-binding site, as shown, minus double bonds for clarity, in (**8.14**).

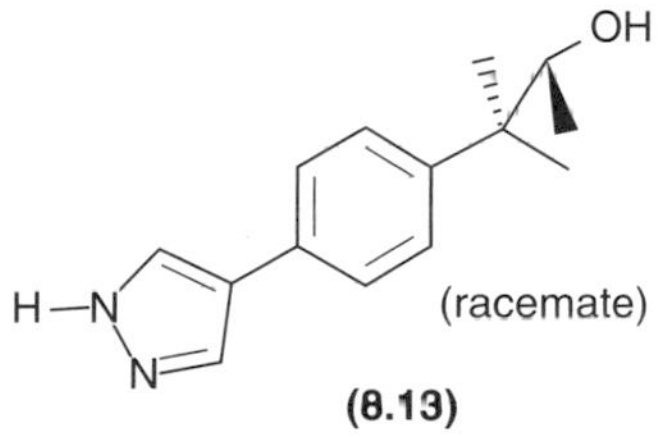

(racemate)

(8.13)

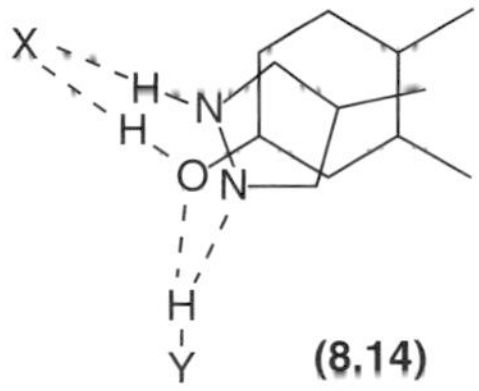

(8.14)

Because there is no positional correspondence of atoms in the two differently interacting rings, the design of the pyrazole 4-substituent could not be based on the normal steroid structure. Instead it was designed such that, overall, the molecule possesses a similar outline shape to the steroid skeleton and the hydroxyl group could be positioned roughly to correspond to that in testosterone. The design was a miserable failure: inhibition of radiolabeled estrogen binding to the receptor was undetectable; a substantial volume deficit in the steroid ring B and C regions may contribute to this result, and recent x-ray diffraction derived receptor structures show ligand phenolic OH groups involved in three potential hydrogen-bonding interactions. Inhibition of aromatase in contrast was among the best we had then seen: AR1: $IC_{50} = 2$ ng/ml (see below). Unfortunately, activity *in vivo* was not detected at the highest dose examined, 20 mg/kg, and none of several pyrazoles was better, even after intraperitoneal dosing.

These results illustrate a common problem in chemotherapy — good activity *in vitro* all too often fails to manifest itself *in vivo*. Sometimes this can be rationalized, in part, in terms of competitive phase effects: the highly lipophilic *N*-''benzyl'' imidazole (**8.7**) binds to albumin and some other macromolecules and is extracted into fat deposits, phospholipid bilayers, and other fatty body components. This drastically reduces the free aqueous concentration of the drug *in vivo* relative to that *in vitro*, and binding to the enzyme suffers in proportion. More usually poor bioavailability and rapid clearance are the most relevant parameters. The latter effect is undoubtedly relevant to potency in our OI3 test, but much more so in the OI2 test that we relied on increasingly throughout this second phase. These tests involved oral dosing of compound at 12:00 noon on day 3 (OI3) or 4:00 pm on day 2 (OI2) of the light-synchronized ovarian cycles of female rats; it had been previously determined that suppression of ovarian estrogen production from mid-afternoon until midnight of day 3 of the 4-day cycles, prevents priming of the hypothalamus for the ovulation-triggering surge of luteinizing hormone (LH) on day 4. Ovulation inhibition is the observed endpoint.

8.4.2 The Potential Importance of Uninterrupted Drug Cover

On theoretical and practical grounds (occasional noncompliance) there are good reasons for wanting a cytostatic anticancer drug to have a long half-life ($t_{1/2}$): even the transient presence each day of growth-promoting levels of estrogen may reduce response rates, quality, or duration of effect. Tumor cells can express aromatase and synthesize estrogens locally so plasma drug and estrogen levels might give an incomplete picture of intratumor drug effectiveness. On the other hand, too long a half-life may result in serious consequences in the event of a severe adverse reaction. This analysis led us, in this case, to aim for an average $t_{1/2}$ of our target drug (assuming the simplest possible kinetics) of at least 12–16 h in patients and preferably not greater than 2 days. Because of competitive pressures, this criterion was made the dominant factor at one stage of the program, despite the fact that predicting $t_{1/2}$ values in humans from data in other species was known to be little better than guesswork.

8.4.3 Increasing Concerns About Timeliness

By the autumn of 1985, as we restarted the program, the competitive situation in aromatase was intense. Many analogues of AG had been revealed and even *p*-cyclohexylaniline had been shown to possess good *in vitro* potency, equal to AG with respect to human placental aromatase but substantially less so versus rat ovarian enzyme. We had during the initial phase of the program tested a few compounds in parallel against rat and human enzymes: AG was sevenfold more potent against rat enzyme, while 1-nonylimidazole was threefold selective in the reverse sense. Almost no other comparative tests were performed as our limited resources were needed elsewhere. So our interpretation of *in vitro* (human) to *in vivo* (rat) potency ratios was always potentially flawed by species differences in enzyme binding; we had to hope, and still hope, that we were not seriously misled, but the reader needs to bear this in mind during apposite parts of structure–activity relationship (SAR) discussions.

Some of these new AG analogues showed much improved selectivity for aromatase — sometimes through improvements against the target enzyme, but often through reduced potency against other enzymes, typically P450scc. Potency in rats however, where reported, remained disappointing.

Another recently reported analogue of AG had the 4-aminophenyl group changed to 4-pyridyl and it was reported to be more selective opposite P450scc. We decided to make a sample for in-house investigation. While we were doing that, and making a number of analogues, we roughly derived the necessary parameters for imides (at that time they were not available from published lists of Allinger's MM2 force-field parameters) so that we could perform molecular mechanics calculations on such systems: we were thus able to predict that both this new analogue and AG exist very largely with the aromatic rings axially disposed. This interesting prediction led us to perform a slightly modified synthesis aimed at the analogue (**8.15**), which calculations predicted would exist overwhelmingly in the axial pyridyl conformation, and (**8.16**) which should have a moderate preference for the equatorial pyridyl conformation. The latter was only two to six times less potent in AR1 than either the parent or (**8.15**). No compound in this pyridyl series had sufficient potency *in vivo* to warrant further interest from us.

(8.15) (8.16)

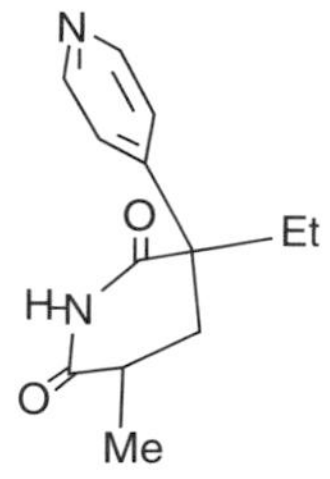

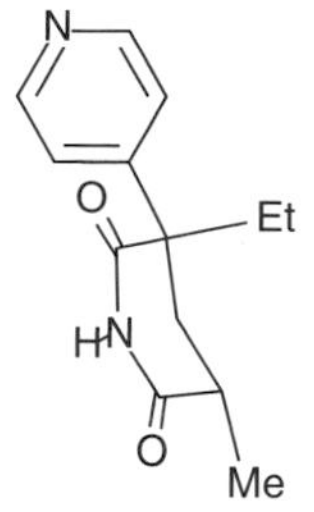

By far the largest area of competitive activity concerned steroidal inhibitors, particularly of the time-dependent variety, but similarly disappointing *in vivo* results generally applied here. Even the Brodie compound, 4-OHA (**8.8**), had been shown to have unwanted androgenic effects and an intramuscular depot formulation used for human dosage was not always well tolerated.

Also by this time, we had noted an association between azole-based antifungal activity and aromatase inhibition. And since many drug companies were or had been active in the antifungal area, we needed to give rapid attention to this potential source of leads. Fortunately for us, another team within ICI Pharmaceuticals had recently completed an antifungal program based on inhibition of the multienzyme-mediated biosynthesis of ergosterol, an essential constituent of fungal cell walls which is not synthesized in mammals. The specific target of that program had been fungal lanosterol-14-methyl demethylase. As part of a frequently applied 10-day teratology assessment, they had seen placental enlargement and effects on fetal development in pregnant rats, all potentially consistent with aromatase inhibition and a common property of the imidazole/triazole compounds they had explored. This work had heightened awareness and understanding of selectivity issues in the business, and placental enlargement in rats provided the Aromatase Team with a test (PE9) wherein chronic effects of estrogen depletion (compounds dosed once daily for 9 days) could be compared with similarly chronic effects on other systems in the same test animal. This overcomes problems of differential handling of compound between individuals, sexes, or species, all of which in retrospect can be seen to have misled us at some point of the program. As with all chronic tests, the accumulation of long half-life compounds (in this context, $t_{1/2}$>ca. 1 day) can present problems, but may also allow such compounds to be identified at an early stage.

We felt sure that time was not on our side, so we were well pleased when screening of our antifungal agents soon yielded a compound which was potent ($IC_{50} = 7$ ng/ml) in our aromatase screen (AR1: human placental microsomes; substrate, 40 nM [1,2-^{3}H]-androstenedione) and was inconsistently active in OI2/OI3 at 0.25–0.5 mg/kg. Poor aqueous solubility may have led to the inconsistency, but removal from that structure of an *ortho*-chloro substituent led to the bis-triazole (**8.17**), which was less active in AR1 ($IC_{50} = 40$ ng/ml), but consistently active in OI2, OI3, and PE9 at 0.2 mg/kg (approximately = ED_{50}). The compound is therefore 20 to 50 times as potent as AG. It was urgently subjected to as detailed an investigation as the perceived time pressure allowed.

The time pressure increased substantially during 1986 as the competitive situation grew still more intense. Schering was claiming long-lived oral effects for atamestane (1-methyl-androsta-1,4-diene-3,17-dione) dosed orally at 1 mg/kg to male volunteers while Ciba–Geigy disclosed that their racemic bicyclic imidazole derivative, CGS 16949A (**8.18**), is 1000 times as potent as AG, with ED_{50} in female rats of 30 μg/kg, and inhibition of aromatase was evident in human male volunteers even at 0.3 mg per man.

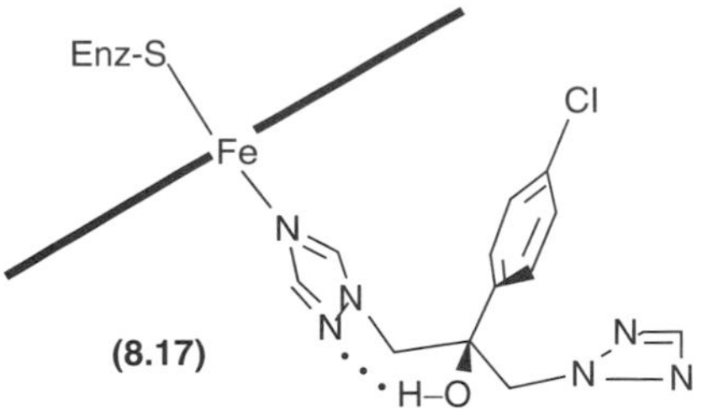

(8.17)

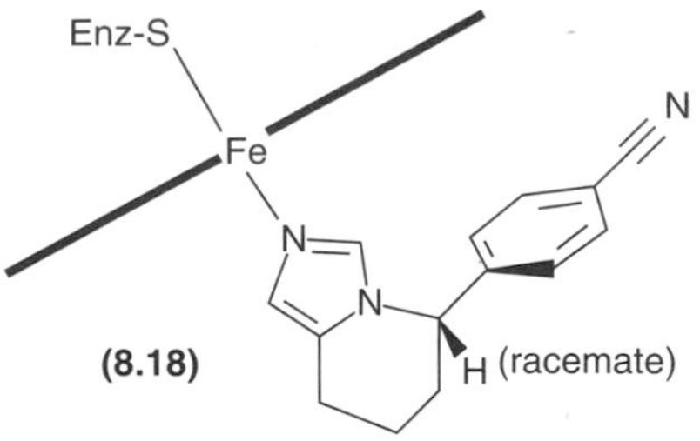

(8.18) H (racemate)

At that time, only in its duration of effect in volunteers, 4–10 h, did the Ciba–Geigy compound seem to present room for improvement. This placed still greater emphasis on the half-life requirement of our target drug.

8.4.4 Naphthol-lactones, Tight Binding and *in vitro–in vivo* Relationships

While much of the Team's early effort went on antifungal leads, we were also finding widespread activity with azoles attached to stilbenes related to (**8.3**), *cis*- and *trans*-2-aryl-tetrahydronapthalenes related to (**8.12**), and to more speculative frameworks based on computer modeling, such as the naphthol-lactone (**8.20**) whose synthesis and structure are shown in Scheme 8.3. *In vivo*, lactones and simple phenol esters such as pivalate esters (archetypal prodrugs) are almost always too rapidly hydrolyzed, via enzyme-mediated catalysis, to allow the longevity we demanded. But this lactone is a special case: it is impossible for it to be hydrolyzed at pH 7.4, the typical value for blood. It even stays ring closed in very dilute sodium hydroxide due to thermodynamics, not kinetics. The reason lies in the large increase in steric compression strain that accompanies ring opening. In contrast, the negligible problems generated in the ring-opened form of the synthetic intermediate (**8.19**) (Scheme 8.3) leave this compound highly sensitive to hydrolysis — $t_{1/2}$ in water at room temperature is ~120 sec at pH 9: this is 5000 times more reactive than ethyl acetate. The more hindered lactone (**8.20**) is however rapidly reduced by borohydride, but only to a lactol (hemiacetal); further reduction under the weakly basic conditions would require ring opening, which, like the ester hydrolysis, essentially does not occur.

A lactol ethyl ether, inadvertently produced in a reaction that had the cyclic ether as its target, was relatively poor *in vitro* but *in vivo* it had similar activity to the corresponding lactone. Rather efficient liver cytochrome P450-mediated reoxidation to the lactone seems a likely explanation.

The imidazole (**8.20**), X = Y = CH, is extremely potent *in vitro*. In a single AR1 test it inhibited the aromatization of tritium-labeled androstenedione by 74% at 1.25 ng/ml, the lowest concentration tested, while at higher concentrations the figures were: 2.5 ng/ml, 95.5% and 5.0 ng/ml, 99%. Such figures, if they can be relied on, are indicative of "tight binding," the condition in which, for the simplest case, *free* drug concentration is significantly depleted from the nominal value by binding to a site, usually the active site, which is present in the test medium at a concentration only somewhat less than or equal to twice the observed 50% inhibitory concentration. Ultimately, *half an equivalent* of inhibitor, essentially all bound to the target, is the absolute minimum required for 50% inhibition no matter how potent the agent might be (rare, catalytically active, irreversible inhibitors excepted). In most test situations one cannot rely on 95% inhibition being different from 100% inhibition, but here we are measuring release of tritiated water, which is easily and completely separable from the precursor, so very small amounts of reaction can quantitatively be measured. Other extremely potent inhibitors show similar responses, while weakly and moderately potent azole-based inhibitors all display classical inhibition curves consistent with the simplest outcome of 1:1 competition between substrate and inhibitor.

The sigmoidal appearance of linear-%inhibition versus log-concentration curves tends to obscure modest deviations from "normality," but the theoretical curve for the simplest case can be transformed to a linear function, or, more conveniently, the % inhibition axis can be transformed

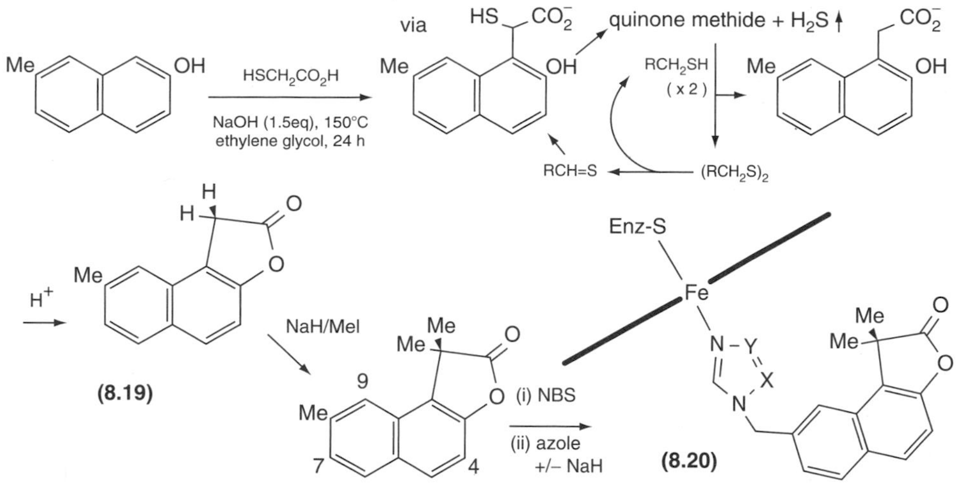

Scheme 8.3

so that experimental points for the simplest case should be linear and lie on a line which passes through 9.09, 50, and 90.9% inhibition at 0.1, 1, and 10 times the IC_{50}. Graph paper to this design was generated ''in house'' some years ago. Data point sets for two tight-binding inhibitors and another set for a borderline case are shown in this format in Figure 8.2, along with three theoretical ''curves'' (curved/inclined lines). The thinner central curve should fit observations when an enzyme present at 5 n*M*, acting on a negligible concentration of substrate, is inhibited by a compound with an equilibrium inhibition constant, K_i, equal to 5 n*M*: the IC_{50} of 7.5 n*M* is only a 1.5-fold underestimate of its true dissociation constant. If a second compound binds 1000-fold tighter, i.e., $K_i = 5$ p*M*, the thick curve on the left shows that, under the above conditions, the observed IC_{50} would be 2.5 n*M*, the limiting condition corresponding to half-an-equivalent referred to above. Tight binding thus limits the apparent potency advantage of the second compound, over the first, to threefold rather than the 1000-fold which would be observed with ''infinitely'' dilute enzyme solutions. The thick ''curve'' on the right for a compound with $K_i = 100$ n*M* differs only minutely from linearity. Experimental data points shown for three compounds, and other observations, pointed to the presence of roughly 3–5 n*M* binding sites (not necessarily active enzyme) in our typical AR1 test milieu, so compounds with IC_{50} values less than ~20 n*M* could, for the best comparisons, be corrected to nontight binding values, preferably nowadays by computed data-fitting techniques. Assuming a K_m for androstenedione of 40 n*M*, one can estimate pK_i ($-\log K_i$) values for the compounds in Figure 8.2; in sequence they are very approximately 10.3, approximately 9.3, and 8.1. For tight-binding corrected pIC_{50} ($-\log IC_{50}$) values subtract 0.3 from pK_i.

The data in Figure 8.2 show that the imidazole (**8.20**), X = Y = CH, may be the most potent inhibitor yet described. Modeling studies indicated that additional lipophilic substituents/fusions at C4 or C5 could produce yet further substantial improvements, but these ideas were not pursued because ICI 207658 had been identified as a compound with great potential and, more to the point, the improvements we most sought *in vivo* required an approach with reduced susceptibility to oxidative metabolism at its core. We returned to that task much later.

The *in vitro* binding/inhibitory sequence: imidazol-l-yl > triazol-l-yl > triazol-4-yl, with 10- to 30-fold gaps was a consistent finding in our work; 5-methylimidazol-l-yl may sometimes be superior to its parent, but tight binding usually makes this uncertain. Each of these azoles attached to the naphthol-lactone framework had good potency in OI3 (shorter-term test) with ED_{50}s of 250 to 500 μg/kg, but only the triazol-l-yl derivative retained its potency in OI2 (single dose given 20 h earlier than the time of dosing in OI3). In this series and in most others, the triazol-l-yl compounds were equal or superior to the imidazoles *in vivo* despite the order of magnitude disadvantage in

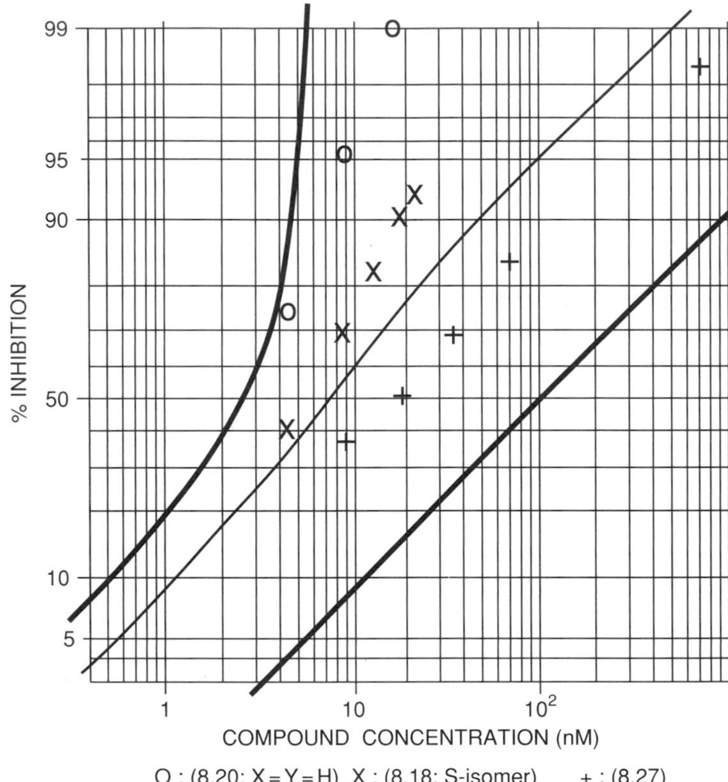

Figure 8.2 Effects of "tight binding" on % inhibition.

binding affinity. Advantages were most marked in the OI2 test. We believe relatively easy oxidation of the imidazole ring or the linking methylene group to be chiefly responsible for this disparity since the imidazole/triazole activity ratio is at its most extreme with the most robust frameworks. Robustness here is based partly on chemists' qualitative judgement and partly on observed plasma half-lives. In the naphthol-lactone case the imidazole had equal activity to the triazole but this still suggested easier than desired attack on the framework. Much later that idea was supported by a brief toxicology/pharmacokinetic once-per-day (u.i.d) oral dosing study on the N1-linked triazole, which returned a $t_{1/2}$ of less than 1 h in male rats.

The N4-linked triazole behaved worse than the imidazole in terms of its OI3/OI2 ratio (value for N4-linked triazole was =8), so it was expected to be rapidly cleared and to produce little toxicity in a similar study, which likewise involved u.i.d oral dosing. In fact it produced unacceptable liver toxicity. It seems likely that binding to heme-linked iron is disfavored by heme-N(δ-) to azole-N(δ-) repulsion, see (**8.21**), whereas when this heterocycle is a ligand to other metallo-enzymes (not just iron-based enzymes) such repulsion could be less or even attractive if an alcohol, water, or amide ligand is present, as shown for R–OH in (**8.22**).

(8.21) **(8.22)**

One should further reflect that the average initial unbound state of the compound *in vitro* is essentially equal to that of an aqueous solution [though perhaps not for the highly lipophilic *N*-''benzyl'' imidazole (**8.7**)], and the nitrogen lone pairs are solvated by hydrogen bonds to water, which is an excellent proton donor; so, since transfer to the enzyme-bound state involves loss of that solvation, the lack of a hydrogen bond or some equivalent in the bound state means that potency must suffer. This seems to apply in our case, but the SAR of the latest bis(4′-cyanophenyl)-methylazole Ciba–Geigy inhibitors (Lang et al., 1993), in particular the very high *in vitro* potency of the tetrazol-2-yl derivative CGS 45688, implies either that some H-bond replacement occurs or, perhaps more likely, there are marked conformational energetic/steric effects favoring the additional ring nitrogens in that particular compound.

8.4.5 The Design Principles Behind ICI 207658 — Later Named Arimidex™

The design of the naphthol-lactones, based on molecular modeling a wide selection of the large number of inhibitors then known, had *in vitro* potency as its dominant feature. But one has good reason to hope that selectivity will increase as potency increases. Testing for selectivity is time and resource consuming and can look at only a very small fraction of relevant P450s, let alone other enzymes. Effects due to receptor binding should also be considered, particularly if the drug is structurally similar to the substrate. The modeling approach by its nature has little, at least initially, to do with avoidance of metabolism, and ''adding on'' this feature at some later point is seldom likely to be easy. Substituting fluorine for hydrogen, especially in aromatic rings or as in CF_3 or CF_2 groups, is widely practised and comes closest to a panacea, usually tolerable changes in size and lipophilicity (Edwards, 1994), but synthetic difficulties and cost are often prohibitive and, like all supposed panaceas of the author's experience, it often seems least attractive when it is most needed.

Independent of modeling there are useful general principles that can be applied to seek extra selectivity. These may lead to achieving established selectivity requirements and perhaps also help avoid unexpected toxicity and nonpharmacologically related side effects:

- Without sacrificing much *in vitro* potency relative to a (notionally) iso-lipophilic comparator — flexibility, more specifically easily accessible conformational space, should be minimized; this is often a basis for improved potency.
- Without sacrificing much *in vitro* potency relative to a (notionally) iso-lipophilic comparator — one or more low-flexibility groups may be introduced which have high steric or solvation demands.
- Polar or more specifically hydrogen-bonding atoms and groups should be preferred to lipophilic isosteres and they should be well dispersed throughout the molecule; here the potency criterion is more complex due to competitive phase effects: reducing lipophilicity has to be consistent with maintaining adequate *in vivo* starting state to target-bound state energy differences, assuming steady state. For a set of analogues with positive $\log P_{\mathrm{octanol}}$ values this approximates, on a $\mathrm{pIC}_{50}/\log P$ plot, to being on the high-potency/low-$\log P$ side of a line of near unit slope (e.g., with slope near 0.7), drawn through data points for compounds of substantial current interest.

The problem remains how to achieve at least some of these aims while maintaining or preferably improving resistance to oxidative metabolism and other clearance processes.

In this regard, one perhaps widely useful group was first recognized as a result of the attempted synthesis of tetralone (**8.24**) (see Scheme 8.4). The tetralone and related naphthol-lactone targets had been conceived and worked on together, but a synthetic intermediate to the tetralone, (**8.23**), was early on converted to the imidazole (**8.25**). In view of the widespread activity in compounds of this type, it was not surprising to find good potency in AR1: $IC_{50} = 4$ ng/ml, but the ED_{50} in OI3 of 0.5 mg/kg was somewhat surprising. It was more surprising when compared to the results that had been obtained several years earlier with *N*-4′-cyanobenzylimidazole, one of the compounds that had shown unacceptable liver effects with multiple high doses. That very simple compound had a better IC_{50} (<1 ng/ml), but a slightly worse ED_{50} of 1 mg/kg. We expected the cyanophenyl compound to be metabolically the more robust and so probably should have been the more potent in OI3 and OI2, in line with its higher enzyme-inhibitory potency. In OI2 (single doses given 20 h prior to that in OI3), both were classified as inactive ($ED_{50} > 2.5$ mg/kg), so their relative potency in this test was unknown; further work at higher doses was not done because neither had the required ''duration'' characteristics.

Scheme 8.4

The unexpected potency reversal was rationalized by assuming that the 1-cyano-1-methyl-ethyl (CME) group was itself not easily attacked and in addition probably conferred some steric and electron-withdrawing protection to the benzenoid ring and its additional substituent. The modest difference in lipophilicity might then be invoked, through an increase in apparent volume of distribution and a lower rate of clearance of unchanged drug, to account for the potency reversal seen in OI3, where duration of action is moderately important. Regarding lipophilicity, the *f*-values for the CME group are +0.25 for octanol/water (measured) and −0.4 for hexane/water (estimated) compared to −0.40 and −0.99 (both averages of several measured values), respectively, for aromatic cyano. The modest octanol/hexane difference for the CME group indicates that its presence in a molecule should not itself seriously compromise penetration through lipid membranes; this bodes well for oral absorption and rapid distribution. Another factor assisting absorption, good solubility, could arise from the easy rotation of the strongly anisotropic CME group about the Ar−C bond in solution: this will help to keep melting points low since easy rotation in the crystal environment is highly unlikely and entropic factors therefore favor noncrystalline states, e.g., nonglassy melts and solutions. Melting points, together with $\log P_{\text{octanol}}$ values, are inversely correlated with log(aqueous solubilities).

That the geminal methyl groups, relative to most aliphatic groups, could have substantial resistance to oxidative attack follows from bond energetics (primary C−H bonds are the strongest) but also from a consideration of how the activation energy for hydrogen transfer is influenced by its surroundings. The forming H−OFe bond is highly polar and electron transfer to the extremely electrophilic $O=Fe^{IV}$ porphyrin$^{+\bullet}$ runs ahead of nuclear motion; this increases the fractional positive charge on the methylene group and the migrating H atom. The reaction-generated electric field and changes in fractional charges are opposed by the strong, electron-withdrawing field due to the cyano group ($\sigma_F \sim 0.57$) and also by the weaker electron-withdrawing field associated with the aryl ring ($\sigma_F \sim 0.13$); see (**8.26**).

The intervening quaternary carbon atom somewhat distances (insulates) the methyl groups from these field effects, but with an expected ''transmission coefficient'' of ~0.4, one could still expect substantial protection. Similar lines of argument apply to the benzylic methylene hydrogens with the more electron-withdrawing triazole, now with no ''insulating'' atom, inhibiting oxidation better than imidazole: $\sigma_F \sim 0.49$ and 0.35, respectively (azole σ_F data are from the PhD thesis of D.J. Hall, University of Wales at Bangor, 1990). The size and hydrophilicity of the azole rings also hinder oxidative metabolism of the methylene groups, but attack is speeded by their weak resonance stabilization of the transition state to the intermediate radicals. Fortunately the last-named effect is not dominant in these systems.

The effect of the 1-cyano-1-methyl-ethyl (CME) group on the ease of oxidation of the aryl rings is also expected to be substantial: thus for $CH_2{-}CN$, $\sigma_F \sim 0.23$ and $\sigma_{p+} \sim 0.16$ (more deactivating than an aromatic chloro-substituent: $\sigma_{p+} = 0.11$). The CME group is expected to be equally deactivating and even nonprotonated imidazolylmethyl, and more so the triazolylmethyl group, will further reduce rates of electrophilic (oxidative) aryl substitution.

Steric hindrance around the cyano gives confidence that hydrolytic or other nucleophilic attack at this group could be minimal, and no easy metabolic release of cyanide is predicted, unlike benzyl cyanide where oxidation at the relatively unhindered benzylic C−H bonds produces a cyanohydrin, which allows easy release of cyanide and potential acute toxicity.

Guided by the above ideas on selectivity and previous SAR, we thus set out urgently to synthesize the 3,5-bis-CME analogue of (**8.25**) and to make triazole equivalents. The triazol-1-yl analogue of (**8.25**) was disappointing but ICI 207658 (**8.27**) (see Table 8.1) was very potent *in vitro* and, more importantly at that time, *in vivo* it was equipotent with CGS 16949A in the demanding OI2 test, both having $ED_{50} \sim 15$ μg/kg. Clearly this was very exciting.

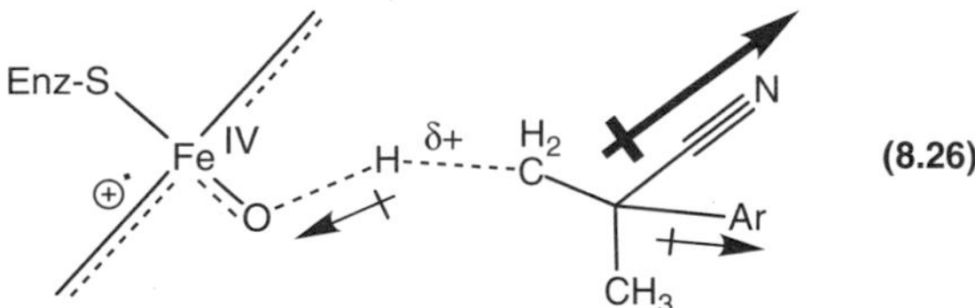

(8.26)

In preliminary studies in male rats, at extremely high multiples of the effective dose in females, ICI 207658 showed no untoward effect; in particular liver triglycerides remained normal. A small increase in liver weight was consistent with the observed induction of mixed function oxidases. Adrenal and other organ weights were the same as controls. The reason for using males for this and the many other compounds investigated is that estrogens indirectly regulate both adrenal weight and circulating triglyceride concentrations. We hoped that changes in the background levels in males would cause only minor changes in these parameters of central interest.

Problems arising from the use of different sexes are usually minor in most species except for rat: males frequently clear compounds faster than females and maximum plasma concentrations, C_{max}, are often lower. The effects are multiplicative on AUC (area under curve of plasma-concentration vs. time). Thus simple ratios of effective dose in one sex to side effect/toxic dose in the other can

Table 8.1 Potency of selected azoles

Compound number	1-(R^{α}-methyl)-3-CME-5-R^{m}-benzene: R^{α}	R^{m}	*In vitro* AR1 IC_{50} (ng/ml)	*In vivo* OI2 ED_{50} (mg/kg)	*In vivo* OI3 ED_{50} (mg/kg)
(8.27)	Triazol-l-yl	CME[a] (ICI 207658)	4	0.015	0.015
(8.28)	Imidazol-1-yl	CME	~0.2[b]		~0.5
(8.29)	Triazol-3-yl	CME	500		>1
(8.30)	Triazol-1-yl	$CH_2{-}S{-}CH_3$	5	>1	
(8.31)	Triazol-1-yl	$C(CH_3)_2{-}OH$	15	~0.5	
(8.32)	Triazol-1-yl	$C(=O)-CH_3$	2	>1	
(8.33)	Triazol-1-yl	$C(CH_3)_2{-}COCH_3$	~0.8[b]	>0.5	~0.2

[a]CME represents a 1-cyano-1-methyl-ethyl group.
[b]Estimated values (adjusted for tight binding).

sometimes mislead selectivity assessments. The same cautions and others are more widely known to apply to comparisons across species.

This discussion of selectivity ratios benefits greatly from hindsight and is relevant here mainly to the Team's first development compound described in Section 8.5. The retrospectroscope also indicates that our clamor for pharmacokinetic and preliminary toxicological studies, which exceeded the capabilities of the appropriate Safety of Medicines Department workgroup to respond, contributed to some insecure conclusions. Most such studies took place later than initial *in vivo* selectivity studies and were not chiefly driven by the need to better assess selectivity ratios.

In the case of ICI 207658, C_{max} was lower in males than females by two- to threefold, but half-lives of ~6 h, dropping to ~4 h at the end of the multiple-dose study were reported to be essentially the same in both sexes. The results of this preliminary study supported and expanded the basis for the Team's conclusion that ICI 207658 was a very promising compound.

Half-lives in rat usually are much shorter than in man so the above values, while being short of our target range, were not a cause for concern and it was predicted by our Safety of Medicines experts that induction of mixed function oxidases in liver was most improbable at the very low predicted human therapeutic dose. It was subsequent data from studies in dog and pigtailed macaque monkey, producing half-lives of ~8 h and 7 to 10 h, respectively, that seemed to indicate a remarkably uniform half-life across species and led to a majority view that the compound could not be relied on to achieve our target minimum half-life in patients.

Well before any of the pharmacokinetic data were available we had discovered that ICI 207658 occupies a pinnacle of *in vivo* SAR space; not a single analogue came within an order of magnitude in OI2 potency terms. This is not the place to go into detail so data on just a few compounds are shown in Table 8.1. Well over a hundred analogues were made with small and sometimes larger variations at every locus where change is possible. If you can think up a related structure, we probably made it or tried to make it (one rather obvious analogue is an exception, that involving changing cyano to nitro, we never did attempt to make it, despite it being one of our listed targets for a long time). Even such small changes as homologating one of the four methyl groups to an ethyl group, or converting two geminal methyls to cyclo-alkyl (three- or four-membered), or introducing an *ortho-* or *para-* bromo substituent, or changing the positions around the benzene ring, etc were markedly deleterious. Many active compounds were identified, but none was as supremely effective as ICI 207658.

Analogues of CGS 16949A (**8.18**) containing one or two *m*-CME groups had, consistent with modeling work, significantly inferior AR1-potency.

As can be seen even from the very small data set in Table 8.1, poor *in vivo* potency was rarely attributable to inadequate enzyme affinity. The early hypothesis concerning the properties of the CME group, now groups, with regard to *in vivo* handling of the drug stood the test of time, albeit one or two compounds with good and even excellent AR1 figures, but poor OI2 results, remain difficult to explain. Single test results may be wrong, we seldom had good reason to retest, or perhaps the anomalies relate to rat vs. human aromatase selectivity, or perhaps our analysis is flawed.

8.5 BIS-TRIAZOLE (8.17) — A TIMELY AROMATASE COMPOUND IN DEVELOPMENT

The antifungal lead had been converted to potential development candidate almost overnight, but that potential had to be assessed. At the time of the discovery of the bistriazole (**8.17**) no detail of the Ciba–Geigy compound was known, so AG was our yardstick and, because we were building up an ever-increasing body of data, it remained so for the first half of the resumed program. Against that yardstick the limited *in vitro* selectivity data was pleasing, particularly with regard to P450scc/ P450arom, where a 60-fold improvement over AG was seen. Most of the early efficacy data was very promising and not just in rats: in monkeys dosed at 0.1 mg/kg for 10 days a near maximum achievable reduction in estrogens was achieved. Similarly potent effects were not seen in dog, perhaps because levels of testosterone, a precursor of estrone and estradiol, increased five- to tenfold in a dose-related fashion with drug treatment. Literature reports ascribed the testosterone increases to hypothalamic aromatase inhibition so we tried to use this as a test system. Unfortunately, near maximal testosterone levels were seen only after multiple doses of 1 mg kg^{-1} day^{-1}. In view of the very high blood levels achieved in dogs this relatively massive required dose seemed unlikely to correspond simply to aromatase inhibition; we therefore placed little reliance on intraspecies dog selectivity ratio assessments. It is also possible that dog aromatase differs substantially from the rat and human enzymes.

In preliminary 7-day toxicity studies in rat, there were the expected increases in mixed function oxidases and increased liver weights at high doses, but, in the absence of effects on triglycerides, these were acceptable findings. The only slight concern expressed in the proposal for development was an increase in adrenal weights: in male rats at 50 mg/kg, 250 times the OI2 and placental enlargement ED_{50} doses; in dogs at 10 mg/kg.

The problem of rat sex differences in drug handling was substantial since both C_{max} and half-life in males were a third of those in females. AUC is therefore an order of magnitude lower and if tissue levels daily fall below some critical threshold for long enough, it is possible for body systems to largely recover from ''toxic'' effects. Small but significant reductions of the male rat accessory sex organ weights and testosterone and LH plasma levels at all doses down to and including the lowest tested, 0.1 mg/kg, were regarded as toxicologically inconsequential. Since other aromatase inhibitors tested had no such effects, these findings demonstrate a lack of selectivity, but in what way remains uncertain. It may be that, similar to AG, changes in P450-mediated rates and routes of hormone catabolism are occurring. Akin to this, interference with barbiturate metabolism in young male rats was evident at 1 mg/kg, but not at 0.1 mg/kg.

The Team's development proposal was accepted by higher management and, only 15 months after restarting the program, the Team had a compound in full development. With luck and rapid development we might still achieve commercial success, but the Ciba–Geigy compound, seemingly superior in potency and selectivity, was now clearly well ahead in the race. And there were still so many hurdles for the bis-triazole (**8.17**) to clear.

As the toxicity studies with (**8.17**) proceeded, the tally of adverse findings increased and our understanding of the unusual steroidogenesis in rat adrenal increased, leaving us with concerns about our ability to detect adverse changes relevant to other species, particularly humans. In dogs, hypokalemia was seen at modest doses and adrenal cortex vacuolation was slightly elevated from controls even at 0.5 mg/kg. It seemed certain that inhibition of 11-hydroxylation was to blame, but there was also no doubt that matters were made worse by the progressively higher C_{max} and AUC values which follow daily dosing of any long half-life compound. In our dogs the half-life was 2–4 days, so substantial accumulation would have occurred. This is the reverse situation to that described above for male rats and emphasizes the importance of temporal drug level profiles to safety/selectivity assessments. Such profiles are also very important to some chronic efficacy studies: (**8.17**) has a half-life in pigtailed monkeys of 1 day, so, barring enzyme induction, which is unlikely at the low doses used, there will be no gaps in drug cover and with chronic dosing a twofold elevation of C_{max} and AUC should occur. As stated previously, it is almost maximally effective with once-daily doses of 0.1 mg/kg. Contrast this with CGS 16949A: we found it to have a half-life of ~5 h in female rats but less than 2 h in monkeys; its large advantage over (**8.17**) in OI2 and still greater advantage (40-fold) in OI3 is reversed in monkeys: they require 0.1 mg/kg every 12 h to achieve near maximal reduction of estrogen levels.

This competitor compound mirrored our own in steadily revealing its weaknesses throughout the time of the bis-triazole (**8.17**) development. Our work in rats and dogs revealed increasing selectivity issues and the absence of, from our viewpoint, relevant selectivity data from both oral presentations and publications dealing with CGS 16949A (fadrozole hydrochloride), including human studies, encouraged us in due course to review our priorities.

As a business we had had many adverse experiences with long half-life compounds in chronic (6 months and more) toxicity studies. This was not to be an exception. In dogs, serum cholesterol and triglyceride levels were reduced by modest doses of the bis-triazole (**8.17**) and, by 6 months, cataracts were seen in the eyes of most dogs dosed at 7.5 mg/kg. The new fibre cells, which in mammals continuously enlarge the eye lens during life, need to synthesize their own cholesterol because they incorporate it in large amounts into their membranes and, being an avascular tissue, they cannot obtain it from the low-density lipoprotein (LDL) in plasma. A prolonged shortage of cholesterol in this tissue seems to lead to cataracts. We suspect that at the observed high plasma levels in dogs, (**8.17**) inhibits one or more of the P450-dependent enzymes that transform lanosterol to cholesterol.

The lead was born from a poor fungal lanosterol-14-methyl demethylase inhibitor and died, 18 months into development, due, probably, to inhibition of a canine lanosterol-demethylase.

In passing it is worth noting the large number of conformers easily accessible to (**8.17**) and its multiple chelation possibilities, bidentate and even tridentate. Perhaps these facts contribute to its inadequate selectivity.

8.6 THE SEARCH FOR THE IDEAL BACK-UP CANDIDATE

Inhibition of mammalian cholesterol synthesis had no precedent in aromatase inhibitors prior to the findings with (**8.17**), but we now urgently needed to look at possible successors to (**8.17**) in this new light. What had we found in that category during those 18 months? Not a lot. We found as expected that we could improve on the original naphthol-lactones (**8.20**) but not sufficiently to fall into the presently required category. Potency in OI2 and PE9 had been somewhat improved, those improvements being associated with electron-withdrawing substituents at C7. The better substituents, e.g., cyano, should hinder oxidative metabolism of the aromatic ring system and the proximal benzylic methylene. During this synthetic work we were surprised, following nitration or bromination, to observe amongst several products some substitution at the very sterically hindered C9 position; mostly reaction was at C7. As in the ICI 207658-like series, we made a large number of analogues but failed to make significant headway. We concluded, using an estimated σ_F value for the phenolic-lactone group, that extra protection of the gemdimethyls and the aromatic system was desirable. But as we saw no practical way to provide it we ceased work on this series. On reflection there were more changes we might have tried.

The only blemish on the profile of ICI 207658 was its projected ''inadequate'' half-life in man; in all *in vitro* selectivity tests conducted, including now against cholesterol synthesis, it performed superbly. None of a great many analogues was attractive. What else might be done?

There were clues in hand: androstenedione had been synthesized with deuterium or tritium in specific locations as part of the aromatase mechanistic studies. Kinetic isotope effects were seen. Of most relevance to us, the 19-trideutero compound in admixture with the nondeuterated parent showed an intermolecular isotope effect k_{H3}/k_{D3} approaching threefold, and the first hydrogen removal is known to be rate limiting. If this ratio, or even half the ratio, applied to the half-life of a deuterated ICI 207658, in at least two of the species used previously, might we not be home and dry? A quick back-of-the-envelope calculation showed that the additional cost of even a tetradeca-deutero compound could be trivial, probably only about one penny/mg, and, with an increased half-life, there was reason to believe that the daily dose might be only ~3 mg per patient.

We made the three deuterated compounds (**8.34**), (**8.35**), and (**8.36**), henceforth referred to as D2, D12, and D14. The hydrogens attached to the triazole ring were not changed because the SAR and metabolite studies pointed firmly against metabolism in this ring being relevant. Several antifungal triazoles had half-lives in male rats of about a week. Similarly the hydrogens directly attached to the benzene ring were not replaced because these are rarely subject to primary isotope effects: rate-determining attack takes place initially at carbon, on the π-system, and secondary isotope effects are typically too small for the present purpose.

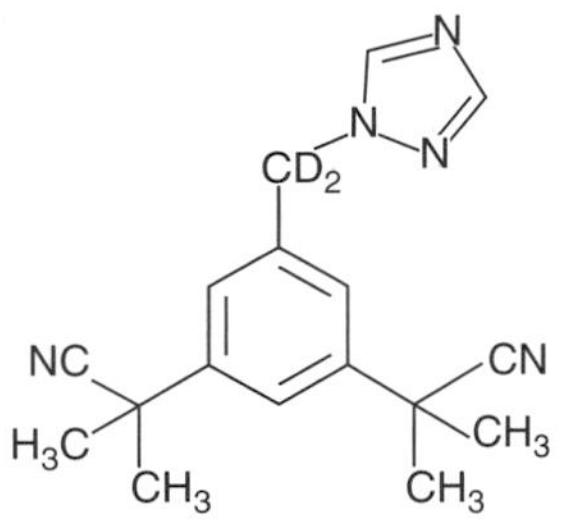

(8.34); D2 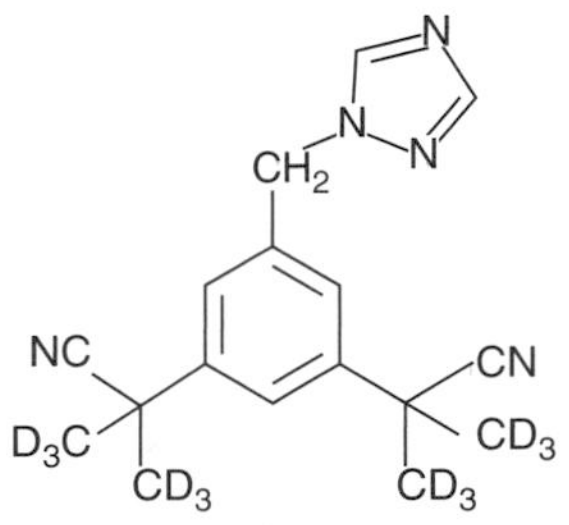 **(8.35);** D12 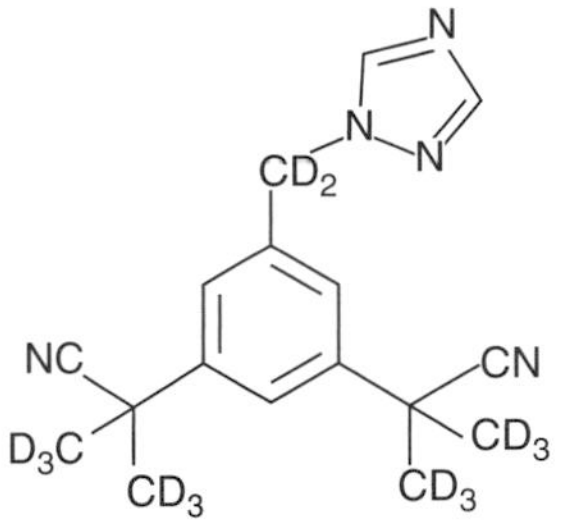 **(8.36);** D14

The D2 and D12 compounds were made despite concerns that wherever attack normally might occur, the partially caged substrate would still react at the remaining weakest point, so-called ''metabolic switching,'' and so reduce any advantage. The likelihood of metabolic switching in our target deuterated analogues also carried with it the risk of generating new, longer-lived, or more abundant metabolites with reduced selectivity or increased toxicity. Of course reduced toxicity is also possible: the subject has been reviewed by Pohl and Gillette (1984–1985).

Intramolecular isotope effects in P450-mediated oxidative reactions, as in nonenzymic heme-based model systems, can be very large: values around 20 are known and 5–10 are normal. In contrast, the isotope effect expressed in k_{cat} for metabolism of phenylethane or α,α-dideuterophenylethane with a rabbit liver-derived $P450_{LM2}$ enzyme is only 1.28. This indicates that at least one enzymic step with a large ''commitment to catalysis'' precedes hydrogen abstraction in this case (White et al., 1986). It can be argued that the effect is small in this case because of the activated nature of the secondary, benzylic C–H bonds. It is however by no means exceptional and ICI 207658 contains a related if much deactivated, more hindered and more hydrophilic part-structure. In contrast to this low value a related study on trideuteromethoxy anisole showed an *in vivo* isotope effect of 10.

Studies *in vivo* generally show less marked substrate dependence with smaller but still some substantial isotope effects. Increases in half-life of 1.5- to 2.5-fold are typical (Blake et al., 1975). In such clearance processes one is dealing with multistep events and the oxidative step is normally only partially rate determining. D.B. Northrop has developed a general equation, Equation (8.1), for the interpretation of isotope effects in multistep reactions. The maximum rate D_v is controlled by the ratio of catalysis R, which represents the ratio of the rate of the isotope-influenced catalytic step to the rate of the other forward steps contributing to the maximum rate:

$$D_V = (k_H/k_D + R)/(1 + R) \tag{8.1}$$

Octanol–water partition and *in vitro* aromatase inhibition studies showed the expected equivalence of all isotopic species. The effective size of the more slowly vibrating C–D fragment is on average very slightly smaller than a corresponding C–H fragment, but the difference in noncovalent binding properties is well below the detection limit in most biological systems.

In vivo potency and limited pharmacokinetic studies with the three compounds, mainly as single agents but sometimes as solid–solution mixtures (to avoid possible differential solubilization and absorption of individual samples) produced somewhat confusing results. OI2 tests (necessarily using females) in head to head comparisons with ICI 207658 (D0) showed a threefold potency improvement for D2 and improvements of 3.5- and twofold for D14 on separate occasions. The result for D12 was identical to that of D0. This indicated the benzylic methylene as the main site of oxidation in female rats at very low (2, 5, 10, and 20 μg/kg), near-therapeutic doses.

A very limited pharmacokinetic study compared the compounds at 1 mg/kg with historical data. Plasma levels of D12 in female rats were followed to 70 h postdosing and showed no detectable isotope effect. A similar result applied to D2 from samples taken at 1, 2, and 8 h postdosing in males; the timepoints at which data were available from the historical study of D0 in males. Only D14 showed some effect: in males followed to 24 h postdosing, the C_{max} and AUC increased by 60% (up to 8 h), but with no detectable change in half-life.

In part because of the limitations of the severely resource-constrained pharmacokinetic study, we tried in a very few animals to use mass spectrometric analysis to follow intraindividual handling of mixtures of compounds. In male rats dosed with a solid–solution of D0 with its D2 and D14 analogues, and using the historical data on D0 for comparisons, the apparent isotope effects interpreted as half-lives were 1.0 to 1.2 for the D2 compound (poor data due to plasma-related peaks) and 2.0 for D14. The same isotope effect, 2.0, was seen for a binary solid–solution of D0 and D14 compounds. A similar experiment in one dog yielded an apparent isotope-induced increase in half-life of 2.1-fold up to 12 h postdosing, but decreasing beyond this time to an average of 1.7-fold over the full 24 h of the experiment.

Being encouraged by these sighting experiments, but realizing their extreme limitations and the possible future need for more extensive work, we carried out a detailed analysis of the likely kinetic scheme for the overall process. Surprisingly, this revealed that results from these, at-first-sight, ideal experiments, involving intraindividual temporal changes in concentration ratios of compounds, cannot be *unambiguously* interpreted in terms of individual clearance rates or related half-lives. We were therefore left with insufficient solid evidence of benefit from deuteration and the undoubted penalties of increased compound costs, analysis costs, and uncertainties with regard to Registration Authority views and delays. The approach was abandoned.

The search for a better candidate continued mainly with *cis*-tetrahydronaphthalenes like (**8.12**), stilbenes related to stilboestrol (**8.3**), and corresponding reduced analogues with a 1,2-diarylethane framework. In many cases, compounds with a pyridine ring replacing one of the benzene rings, e.g., (**8.37**) (racemate), had excellent activity both in AR1 and in OI2. None of these compounds was satisfactory in all respects. In some the half-life was too long, longer than or similar to the bis-triazole (**8.17**) in the dog was now a near automatic bar to progression, while others had inadequate selectivity: like bis-triazole (**8.17**), the pyridine (**8.37**) substantially lowered serum cholesterol levels, by then a totally unacceptable encumbrance. Whether the effects on aromatase and cholesterol synthesis could be separated through resolution was not investigated. Time had almost run out. We had had to progress many compounds, first through larger scale synthesis, then often into semichronic and chronic tests before finding them unsatisfactory.

Janssen also was now forging ahead with the very impressive but racemic triazole R76713, (**8.38**). Published information showed potent activity in volunteers, so they too were now far ahead of us and our limited selectivity data on the compound gave us no comfort whatever. We had to make a choice now. That choice was by now almost inevitable: ICI 207658 was associated with only temporary accumulation following multiple large doses in dogs and it had an excellent selectivity profile. Selectivity had again come close to the top of the Team's priorities. Once again life comes pretty much full circle and ICI 207658 was entered into development under the number D1033.

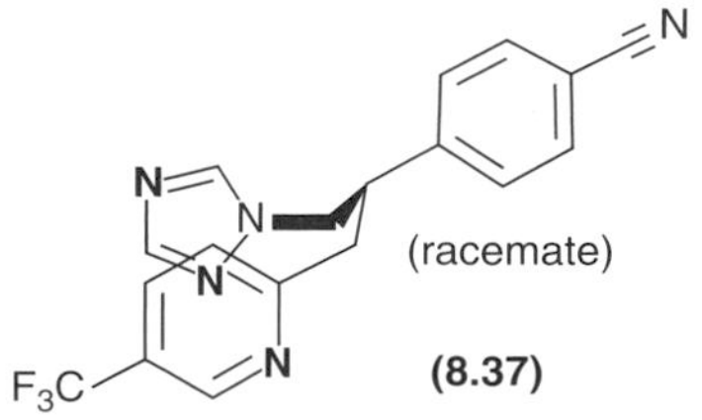

(racemate)

(8.37)

(racemate)

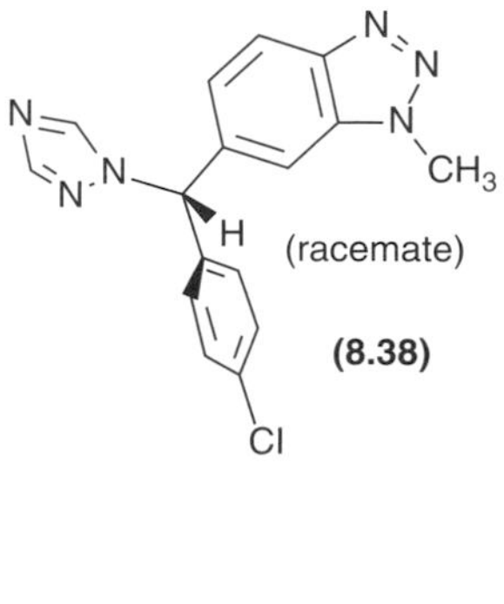

(8.38)

Long-term toxicity studies revealed no significant additional findings to those seen with shorter exposures, but comprehensive pharmacokinetic studies revealed that the preliminary half-life estimates in rat had been in error. Due to a plasma-associated material interfering with the assays, those estimates were generally too long; for example, the initial $t_{1/2}$ in males is now known to be ~2 h and a half-life of 2.3 h was observed after long-term dosing. It is therefore almost certain that the isotope work did achieve its objectives. But had we pursued this line into development we may have been faced with a supra-optimal half-life in patients (see below).

8.7 INTO THE CLINIC

Escalating dose studies in male volunteers confirmed our expectation of good absorption, rapid distribution, and high bioavailability, so the compound progressed into the first patients. The benefits we had confidently expected to find materialized and with a half-life of 2 days it fitted our target for optimum use long-term in postmenopausal women. No serious side effects, enzyme induction, or inhibition, barring aromatase, have been observed and no indications of a lack of selectivity have been seen at either the 10 mg or 1 mg u.i.d. doses investigated (Plourde et al., 1994). Since the lower dose gives >95% inhibition of aromatization in biochemical studies in patients, and equivalent anticancer efficacy to the higher dose, this smaller quantity, corresponding to approximately 15–20 μg/kg, was chosen as the recommended dose for use of Arimidex in postmenopausal breast cancer.

More impressive still are the results of the ATAC (Arimidex, Tamoxifen alone or in combination) study of over 9300 women, first reported in December 2001: this shows that Arimidex is significantly more effective and has important tolerability benefits compared with the current gold standard, tamoxifen, as an adjuvant treatment in postmenopausal women with early breast cancer. AstraZeneca now hope that Arimidex will achieve ''megabrand status'' with yearly sales exceeding one billion dollars.

FURTHER READING

Akhtar, M., Njar, V.C.O. and Wright, J.N. (1993) Mechanistic studies on aromatase and related C—C bond cleaving P-450 enzymes. *Journal of Steroid Biochemistry and Molecular Biology* **44**, 375–387.

Blake, M.I., Crespi, H.L. and Katz, J.J. (1975) Studies with deuterated drugs. *Journal of Pharmaceutical Sciences* **64**, 367–391.

Brodie, A., Brodie, H.B., Callard, G., Robinson, C., Roselli, C. and Santen, R. (eds.) (1993) Recent advances in steroid biochemistry and molecular biology. Proceedings of the Third International Aromatase Conference; Basic and Clinical Aspects of Aromatase. *Journal of Steroid Biochemistry and Molecular Biology* **44**(4–6).

Castagnetta, L., D'Aquino, S., Labrie, F. and Bradlow, H.L. (eds.) (1990) Steroid formation, degradation and action in peripheral tissues. *Annals of the New York Academy of Sciences* **595**.

Djerassi, C. (1992) *The Pill, Pygmy Chimps, and Degas' Horse: the Autobiography of Carl Djerassi*. New York: Basic Books.

Edwards, P.N. (1994) Uses of fluorine in chemotherapy. In Banks, R.E., Smart, B.E. and J.C. Tatlow (eds.) *Organofluorine Chemistry: Principles and Commercial Applications*. New York: Plenum Press, pp. 501–541.

Henderson, D., Philibert, D., Roy, A.K. and Teutsch, G. (eds.) (1995) Steroid receptors and antihormones. *Annals of the New York Academy of Sciences* **761**.

Lang, M., Batzl, Ch., Furet, P., Bowman, R., Hausler, A. and Bhatnagar, A.S. (1993) Structure–activity relationships and binding model of novel aromatase inhibitors. *Journal of Steroid Biochemistry and Molecular Biology* **44**, 421–428.

Plourde, P.V., Dyroff, M. and Dukes, M. (1994) Arimidex®: a potent and selective fourth generation aromatase inhibitor. In Brodie, A.M.H. and Santen, R.J. (eds.) *Breast Cancer Research and Treatment* (Special Issue: Aromatase and its Inhibitors in Breast Cancer Treatment) **30**(1), 103–111.

Pohl, L.R. and Gillette, J.R. (1984–1985) Determination of toxic pathways of metabolism by deuterium substitution. *Drug Metabolism Reviews* **15**(7), 1335–1351.

Schenkman, J.B. and Greim, H. (eds.) (1993) Cytochrome P450. *Handbook of Experimental Pharmacology*, Volume **105**.

Wakeling, A.E. (1990) Novel pure antioestrogens, mode of action and therapeutic prospects. *Annals of New York Academy of Sciences* **595**, 348–356.

White, R.E., Miller, J.P., Favreau, L.V. and Bhattacharyya, A. (1986) Stereochemical dynamics of aliphatic hydroxylation by cytochrome $P450_{LM2}$. *Journal of the American Chemical Society* **108**, 6024–6031.

9

Design of Enzyme Inhibitors as Drugs

Anjana Patel, H. John Smith, and Torsten Steinmetzer

CONTENTS

9.1 INTRODUCTION

Enzymes catalyze the reactions of their substrates by initial formation of a complex (ES) between the enzyme and substrate (S) at the active site of the enzyme (E). This complex then breaks down, either directly or through intermediary stages, to give the products (P) of the reaction with regeneration of the enzyme:

$$\mathrm{E + S} \rightleftharpoons \underset{\text{Enzyme–substrate complex}}{\mathrm{ES}} \xrightarrow{k_{cat}} \mathrm{E + products} \tag{9.1}$$

$$\mathrm{E + S} \rightleftharpoons \mathrm{ES} \xrightarrow{k_2} \underset{\text{Intermediate}}{\mathrm{E'}} + \mathrm{P_1} \xrightarrow{k_3} \mathrm{E + P_2} \tag{9.2}$$

k_{cat} is the overall rate constant for decomposition of ES into products, k_2 and k_3 are the respective rate constants for formation and breakdown of the intermediate E′ (i.e., $k_{cat} = k_2k_3/(k_2 + k_3)$). Chemical agents, known as inhibitors, modify the ability of an enzyme to catalyze the reaction of its substrates. The term inhibitor is usually restricted to chemical agents, other modifiers of enzyme activity such as pH, ultraviolet light, high salt concentrations, organic solvents, and heat, are known as denaturizing agents.

9.1.1 Basic Concept

The body contains several thousand different enzymes each catalyzing a reaction of a single substrate or group of substrates. An array of enzymes is involved in a metabolic pathway each catalyzing a specific step in the pathway, i.e.,

$$\mathrm{A} \xrightarrow{E_1} \mathrm{B} \xrightarrow{E_2} \mathrm{C} \xrightarrow{E_3} \cdots \xrightarrow{E_n} \text{Metabolite} \tag{9.3}$$

These actions are integrated and controlled in various ways to produce a coherent pattern governed by the requirements of the cell.

The basis for using enzyme inhibitors as drugs is that inhibition of a suitably selected target enzyme leads first to a build-up in concentration of substrates and then to a corresponding decrease in concentration of the metabolites, one of which leads to a useful clinical response. Where the substrate gives a required response (i.e., agonist) inhibition of a degradative enzyme leads to accumulation of the substrate and accentuation of that response. Build-up of the neurotransmitter acetylcholine by inhibition of acetylcholinesterase using neostigimine is used for the treatment of myasthenia gravis and glaucoma:

$$CH_3CO_2CH_2CH_2\overset{+}{N}(CH_3)_3 \xrightarrow{\text{Acetylcholinesterase}} CH_3CO_2H + HOCH_2CH_2\overset{+}{N}(CH_3)_3 \tag{9.4}$$

Anticholinesterases, e.g., donepezil, rivastigmine, galantamine capable of penetrating the blood–brain barrier and so exerting an effect on the central nervous system are used in the treatment of Alzheimer's disease (AD) for increasing the cognitive functions.

Where the metabolite has an action judged to be clinically undesirable or too pronounced, then enzyme inhibition reduces its concentration with a decreased (desired) response. Allopurinol is an inhibitor of xanthine oxidase and is used for the treatment of gout. The inhibition of the enzyme decreases conversion of the purines xanthine and hypoxanthine to uric acid, which otherwise deposits and produces irritation in the joints:

$$\text{Xanthine} \xrightarrow{\textit{xanthine oxidase}} \text{Uric acid} \tag{9.5}$$

In the above example an enzyme acting in isolation was targeted but several other strategies may be used with enzyme inhibitors to produce an overall satisfactory clinical response. The target enzyme may be part of a biosynthetic pathway consisting of a sequence of enzymes with their specific substrates and coenzymes (Equation (9.6)). Here the aim is to prevent, by careful selection of the target enzyme in the pathway (see Section 9.2.1), the overall production of a metabolite, which either clinically gives an unrequired response or is essential to bacterial or cancerous growth.

$$\mathrm{A} \xrightarrow{E_1} \mathrm{B} \xrightarrow{E_2} \mathrm{C} \xrightarrow{E_3} \mathrm{D} \xrightarrow{E_4} \mathrm{E\ (Metabolite)} \qquad \text{(Inhibitor acting on } E_1\text{)} \tag{9.6}$$

Inhibitor

Sequential chemotherapy involves the use of two inhibitors simultaneously on a metabolic chain (Equation (9.7)) and is employed with the aim of achieving a greater therapeutic effect than by application of either alone. This situation arises when dosage is limited by host toxicity or resistant bacterial strains have emerged. The best known combination is the antibacterial mixture co-trimoxazole, consisting of trimethoprim (dihydrofolate reductase (DHFR) inhibitor) and the sulfonamide sulfamethoxazole (dihydropteroate synthetase inhibitor) although the usefulness of the latter in the combination has been queried.

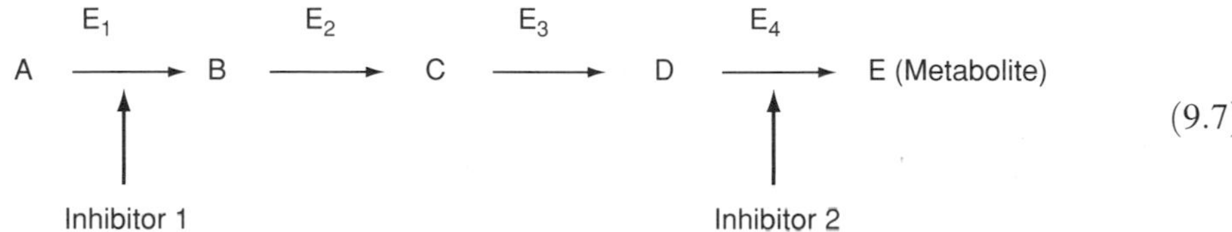

(9.7)

Inhibitors have been used (see Equation (9.8)) as co-drugs to protect an administered drug with a required action from the effects of a metabolizing enzyme. Inhibition of the metabolizing target enzyme permits higher plasma levels of the administered drug to persist, so prolonging its biological half-life and either preserving its effect or resulting in less frequent administration. Clavulanic acid, an inhibitor of certain β-lactamase enzymes produced by bacteria, when administered in conjunction with a β-lactamase-sensitive penicillin preserves the antibacterial action of the penicillin towards the bacteria.

$$\underset{\text{(Agonist)}}{\mathrm{A}} \xrightarrow{\textit{Metabolizing enzyme}} \text{Inert product(s)} \tag{9.8}$$

Co-drug (inhibitor)

Parkinson's disease is due to degeneration in the basal ganglia which leads to reduction in dopamine levels, which control muscle tension. Effective treatment for considerable periods involves administration of L-dopa, which is decarboxylated after passage into the brain by a central acting amino acid decarboxylase (AADC). Since L-dopa is readily metabolized by peripheral AADCs (see Figure 9.1) it is administered with a peripheral AADC inhibitor (which cannot penetrate the brain) to decrease this metabolism and reduce the administered dose necessary, i.e., benzserazide and carbidopa.

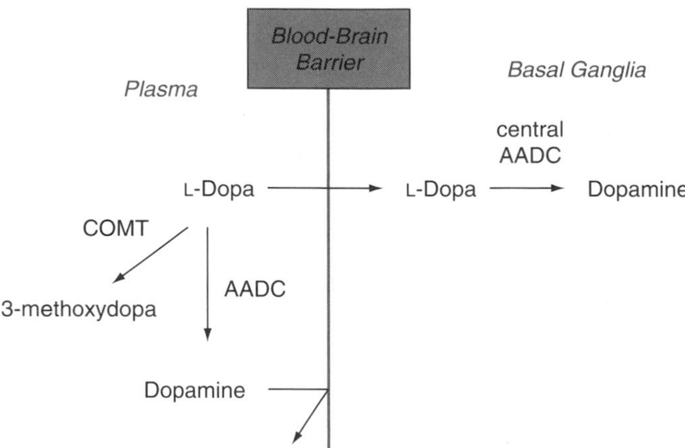

Figure 9.1 Peripheral and central metabolism of L-Dopa.

A further adjuvent to these combinations is a catechol-*O*-methyltransferase (COMT) inhibitor. COMT peripherally converts L-dopa to 3-methoxydopa with loss of potency. Entacapone (Comtess) is currently available for this purpose; tolcapone (Tasma) previously used led in a few instances to fatal hepatic effects and has been discontinued in the UK.

Other rare and thus less important *modus operandi* of enzyme inhibitors are as follows.

Inhibition of an enzyme on occasions leads to formation of a dead-end complex between the enzyme, coenzyme, and inhibitor rather than straightforward interaction between the inhibitor and the enzyme. 5-Fluorouracil inhibits thymidylate synthetase to form a dead-end complex with the enzyme and coenzyme, tetrahydrofolate, so preventing bacterial growth:

$$A \xrightarrow{E_1} B \xrightarrow{E_2} C \xrightarrow{E_3} D \xrightarrow{E_4} \text{E (Metabolite)}$$

Cofactor Z + Inhibitor → E_2Z' Inhibitor (Dead-end complex) (9.9)

Where product build-up progressively decreases the activity of an enzyme on its substrate, then enhancement of product inhibition (Equation (9.10)) can be achieved by inhibiting an enzyme which disposes of that product. *S*-adenosylhomocysteine (SAH), the product of methylating enzymes (e.g., catecholamine methyltransferase, COMT) using *S*-adenosylmethionine (SAM), and an inhibitor of these enzymes, is removed by the hydrolytic action of its hydrolase (SAH′ase). Inhibitors of SAH′ase should allow a build-up of the product, SAH, leading to a useful clinical effect.

A → $B\text{-}CH_3$ (Metabolite)
SAM → SAH (Inhibitionn)
SAH $\xrightarrow{\text{SAH'ase}}$ Products
Inhibitor (9.10)

9.1.2 Types Of Inhibitors

Enzyme inhibiting processes may be divided into two main classes, reversible and irreversible, depending upon the manner in which the inhibitor (or inhibitor residue) is attached to the enzyme.

Reversible inhibition occurs when the inhibitor is bound to the enzyme through a suitable combination of van der Waals, electrostatic, hydrogen bonding, and hydrophobic forces, the extent of the binding being determined by the equilibrium constant, K_I, for breakdown of the EI or EIS complex for classical inhibitors. However, on occasions a covalent bond may be formed with an active site residue, as in the case of a hemiacetal or hemiketal bond with the catalytic serine in serine proteases with a polypeptide aldehyde- or ketone-based inhibitor, but the EI complex readily dissociates back into free enzyme and inhibitor as the free inhibitor concentration falls due to excretion, metabolism, etc.

Reversible inhibitors may be competitive, noncompetitive, or uncompetitive depending upon their point of entry into the enzyme–substrate reaction scheme.

Competitive inhibitors, as their name suggests, compete with the substrate for the active site of the enzyme and by forming an inactive enzyme–inhibitor complex decrease the interaction between the enzyme and the substrate:

$$\begin{array}{ccccc} E + S & \underset{}{\overset{K_s}{\rightleftharpoons}} & ES & \underset{}{\overset{k_2}{\rightleftharpoons}} & E + P \\ I \Big\updownarrow K_i & & & & \\ EI & & & & \end{array} \qquad (9.11)$$

EI inactive enzyme–inhibitor complex

Michaelis–Menten equation for the rate (ν) of an enzyme-catalyzed reaction is given by

$$\nu = \frac{V_{max}[S]}{[S] + K_m}$$

which is modified in the presence of a competitive inhibitor,

$$\nu = \frac{V_{max}[S]}{[S] + K_m\left(1 + \frac{[I]}{K_i}\right)} = \frac{V_{max}}{1 + \frac{K_m}{S}\left(1 + \frac{[I]}{K_i}\right)} \qquad (9.12)$$

where K_m is the Michaelis constant which is the molar concentration of substrate at which $\nu = \frac{1}{2}V_{max}$. The extent to which the reaction is slowed in the presence of the inhibitor is dependent upon the inhibitor concentration [I], and the dissociation constant, K_I, for the enzyme–inhibitor complex. A small value for K_I ($\cong 10^{-6}$–10^{-8} M) indicates strong binding of the inhibitor to the enzyme. With this type of inhibitor the inhibition may be overcome, for a fixed inhibitor concentration, by increasing the substrate concentration. This fact can be readily established by examination of Equation (9.12) where it is seen that as the substrate concentration increases the second term decreases and the rate approaches V_{max}. With this type of inhibition only substrate binding, i.e., K_m is affected since the inhibitor competes with the substrate for the same binding site.

The type of inhibition and the value for K_I may be obtained by determining the initial rate of the enzyme-catalyzed reaction using a fixed enzyme concentration over a suitable range of substrate concentrations in the presence and absence of a fixed concentration of the inhibitor. Rearrangement of Equation (9.12) gives

$$\frac{1}{\nu} = \frac{1}{V_{\max}} + \frac{K_{\mathrm{m}}\left(1 + \frac{[\mathrm{I}]}{K_{\mathrm{i}}}\right)}{V_{\max}} \frac{1}{[\mathrm{S}]} \tag{9.13}$$

A plot of $1/\nu$ against $1/[\mathrm{S}]$, known as a Lineweaver–Burk plot for the two series of experiments, gives two regression lines which cut at the same point on the $1/\nu$ axis (corresponding to $1/V_{\max}$) but cut the $1/[\mathrm{S}]$ axis at values corresponding to $-1/K_{\mathrm{m}}$ and $-1/K_{\mathrm{m}}(1+[\mathrm{I}]/K_{\mathrm{I}})$ in the absence and presence of the inhibitor, respectively, from which K_{m} and K_{I} can be calculated (Figure 9.2). The manner of intersection of the two lines is characteristic of competitive inhibition. K_{I} is usually obtained from either a secondary plot of K'_{m} values vs [I] (Figure 9.3) or a Dixon plot of $1/\nu$ vs [I] at two (or more) substrate concentrations (Figure 9.4).

Very often the inhibitory potency within a series of inhibitors may be expressed as an IC_{50} value. The IC_{50} value represents the concentration of inhibitor required to halve the enzyme activity and this value should be used with care when comparing interlaboratory results since it is dependent on the concentration of substrate used:

$$\mathrm{IC}_{50} = K_{\mathrm{i}}\left(1 + \frac{S}{K_{\mathrm{m}}}\right) \tag{9.14}$$

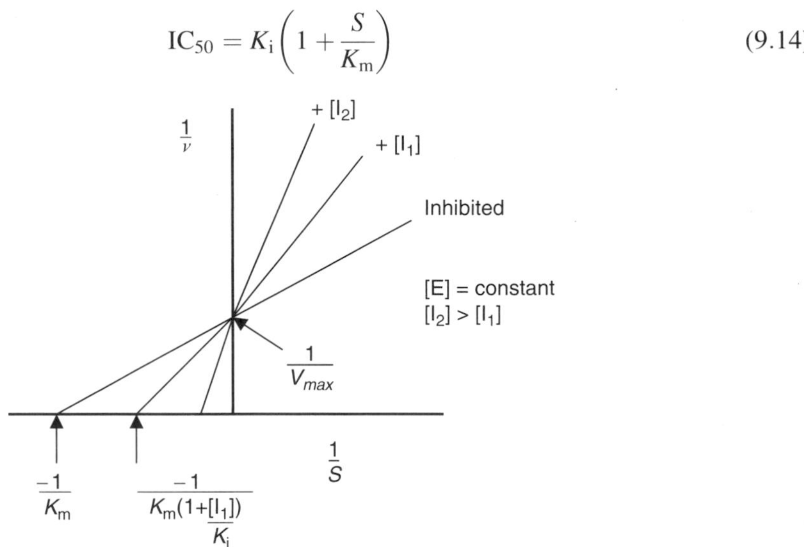

Figure 9.2 Lineweaver–Burk plot showing competitive inhibition.

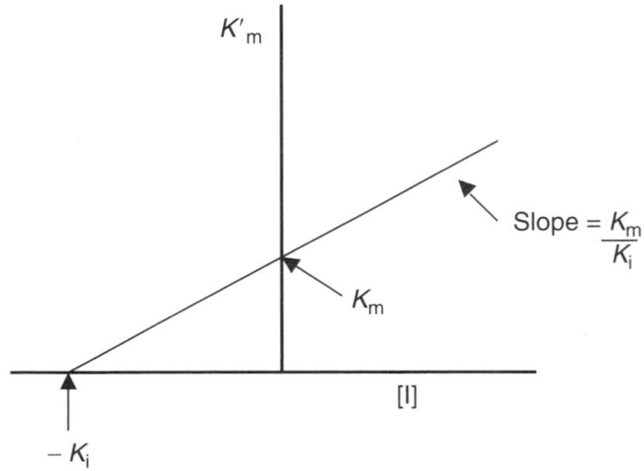

Figure 9.3 Secondary plot of K'_{m} values (from Figure 9.2) vs inhibitor concentration.

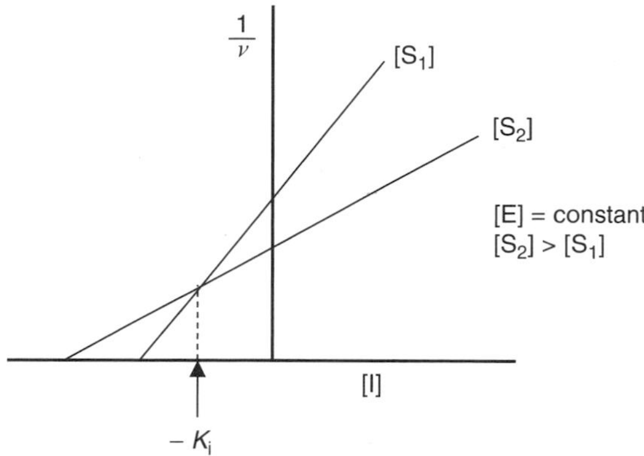

Figure 9.4 Dixon plot for competitive inhibition.

The relationship between percentage inhibition of an enzyme and inhibitor concentration ([S] = constant) is not linear (see Table 9.1). The relative concentration to inhibit enzyme activity by 50, 90, and 99% under standard conditions increases logarithmically, i.e., 10^0, 10^1, 10^2, respectively. In screening tests, although the potencies of different inhibitors may appear similar within the range 90–95% inhibition, their potencies may be very different when IC_{50} values from further experiments are compared.

Noncompetitive inhibitors combine with the enzyme–substrate complex and prevent the breakdown of the complex to products:

$$\begin{array}{ccccc} E + S & \rightleftharpoons & ES & \longrightarrow & E + P \\ {\scriptstyle +I}\downarrow\uparrow{\scriptstyle -I,\ K_i} & & {\scriptstyle +I,\ K_i}\downarrow\uparrow{\scriptstyle -I} & & \\ EI + S & \rightleftharpoons & EIS & & \end{array} \tag{9.15}$$

These inhibitors do not compete with the substrate for the active site and only change the V_{max} parameter for the reaction. The binding of the inhibitor to either E or ES is the same so that the value of K_I is identical. The kinetics for this type of inhibitor are given by

$$\nu = \frac{\dfrac{V_{max}}{1 + \dfrac{[I]}{K_I}[S]}}{[S] + K_m} = \frac{\dfrac{V_{max}}{(1 + [I]/K_I)}}{(1 + K_m/[S])} \tag{9.16}$$

Table 9.1 Relationship between percentage competitive reversible inhibition of an enzyme and relative inhibitor concentration

% Inhibition	Relative inhibitor concentrations
10.0	0.1
50.0	1.0[a]
67.0	3.0
76.0	5.0
90.0	10.0
99.01	100.0
99.90	1000.0
99.99	10000.0

[a] IC_{50} value.

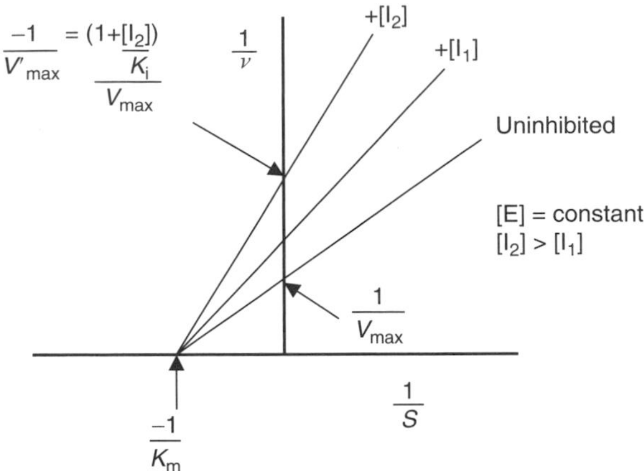

Figure 9.5 Lineweaver–Burk plot showing noncompetitive inhibition.

The extent of the inhibition, by a fixed concentration of inhibitor, cannot be reversed by increasing the substrate concentration (contrast competitive inhibition) since substrate and inhibitor bind at different sites. A Lineweaver–Burk plot of $1/\nu$ against 1/[S] gives a straight line which cuts the 1/[S] axis at $-1/K_m$ and the $1/\nu$ axis at $(1 + [I]/K_i)/V_{max}$. The shape of the plot (Figure 9.5) is typical of noncompetitive inhibition. The value for K_m is unchanged by the inhibitor as expected since it does not compete with the substrate for the substrate binding site, however catalytic activity is decreased (effects of V_{max}) by either binding of the inhibitor elsewhere on the catalytic site or by its binding producing a conformational change at the site. K_I values are usually determined from a secondary plot of $1/V'_{max}$ vs [I] (Figure 9.6) or a Dixon plot of $1/\nu$ vs [I] at two different substrate concentrations (Figure 9.7).

A third type of reversible inhibitor, rare for single substrate catalysis, is an uncompetitive inhibitor.

$$\begin{array}{ccccc} E + S & \rightleftharpoons & ES & \longrightarrow & E + P \\ & & {\scriptstyle +I}\downarrow\uparrow{\scriptstyle -I} \; {\scriptstyle K_i} & & \\ & & EIS & & \end{array} \tag{9.17}$$

This type of inhibitor only binds to the enzyme–substrate complex; perhaps substrate binding produces a conformation change in the enzyme which reveals an inhibitor-binding site. The modified Michaelis–Menten equation is shown in Equation (9.18) where it is seen that both K_m and V_{max} are modified.

$$\nu = \frac{\dfrac{V_{max}}{1 + \dfrac{[I]}{K_i}}}{1 + \dfrac{K_m}{[S]}\left(1 + \dfrac{[I]}{K_i}\right)} = \frac{V'_{max}}{1 + \dfrac{K'_m}{[S]}} \tag{9.18}$$

The Lineweaver–Burk equation is

$$\frac{1}{\nu} = \frac{K'_m}{V'_{max}}\frac{1}{[S]} + \frac{1}{V'_{max}} \tag{9.19}$$

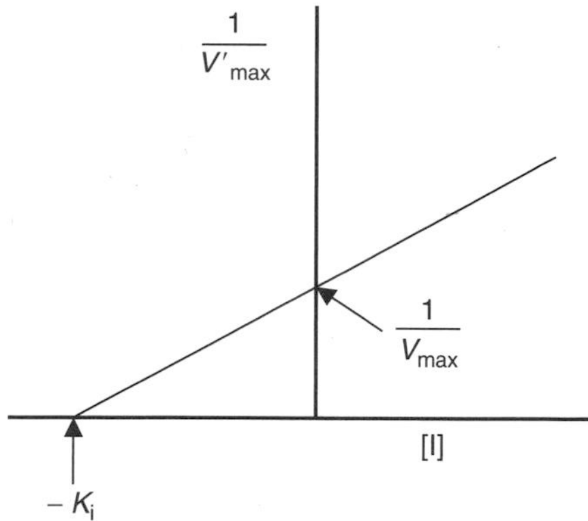

Figure 9.6 Secondary plot of $1/V'_{max}$ values (from Figure 9.5) vs inhibitor concentration.

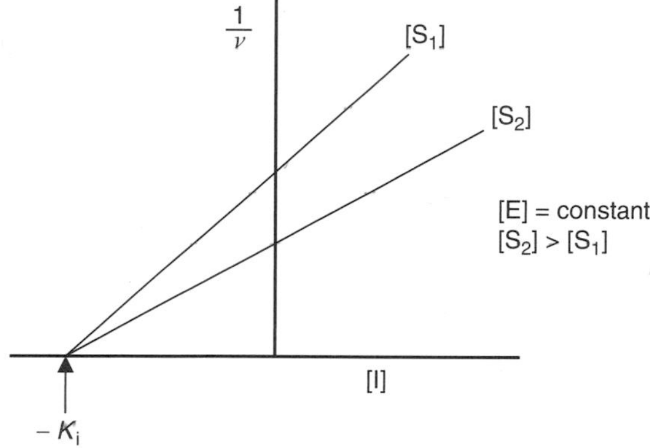

Figure 9.7 Dixon plot for noncompetitive inhibition.

A plot of $1/\nu$ vs $1/[S]$ in the absence and presence of an uncompetitive inhibitor is characterized by a series of parallel lines (equal slope since the slope is K_m/V_{max}) which cut the y-axis at $1/V'_{max}$ and the x-axis at $-1/K'_m$ (Figure 9.8). The value of K_I can be obtained more conveniently from a secondary plot of $1/V_{max}$ vs [I] where the intercept on the x-axis gives the value of $-K_i$ (Figure 9.9).

Occasionally the Lineweaver–Burk plot shows a pattern that can lie between either (a) competitive and noncompetitive inhibition or (b) noncompetitive and uncompetitive inhibition in that the regression lines intercept to the left of the y-axis and either above (i.e., (a)) or below (i.e., (b)) the x-axis (see example of (a) in Figure 9.10). This form of inhibition is termed mixed inhibition and arises because the inhibitor binds to both E and the ES complex but with different binding constants (K_i and K_I, respectively; note the different suffixes).

$$\begin{aligned} E + I &\underset{}{\overset{K_i}{\rightleftharpoons}} EI \\ ES + I &\underset{}{\overset{K_I}{\rightleftharpoons}} ESI \end{aligned} \tag{9.20}$$

The Michaelis–Menten equation for mixed inhibition is

$$\nu = \frac{V_{\max}}{\dfrac{(1+[\mathrm{I}]/K_{\mathrm{I}})}{1+\dfrac{K_{[\mathrm{I}]}}{[\mathrm{S}]}\dfrac{(1+[\mathrm{I}]/K_{\mathrm{i}})}{(1+[\mathrm{I}]/K_{\mathrm{I}})}}} = \frac{V'_{\max}}{1+\dfrac{K'_{[\mathrm{I}]}}{S}} \tag{9.21}$$

and the Lineweaver–Burk equation is

$$\frac{1}{\nu} = \frac{K'_{\mathrm{m}}}{V'_{\max}[S]} + \frac{1}{V'_{\max}} \tag{9.22}$$

Figure 9.10 shows a Lineweaver–Burk plot for the situation where $K_{\mathrm{I}} > K_{\mathrm{i}}$, i.e., K_{i} has the lower value and stronger binding to E occurs than ES. A secondary plot of $1/V'_{\max}$ vs [I] gives an intercept on the x-axis for $-K_{\mathrm{I}}$ (Figure 9.11(a)) and a plot of the slope of the primary plot vs [I] similarly gives the value for K_{i} (Figure 9.11(b)).

A special type of competitive inhibitor is a transition state analogue. This is a *stable compound* which resembles in structure the substrate portion of the enzymic transition state for chemical

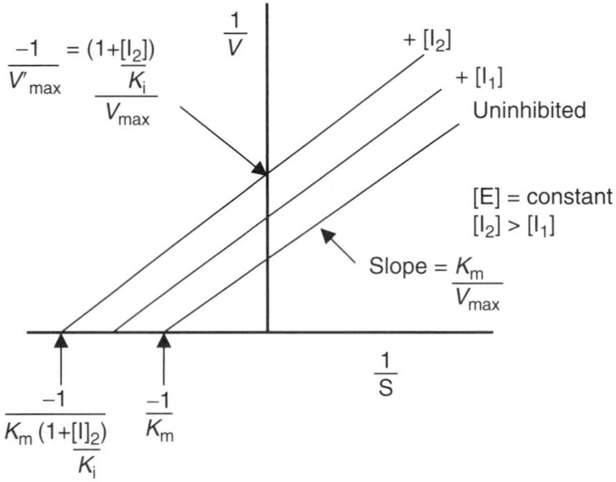

Figure 9.8 Lineweaver–Burk plot showing uncompetitive inhibition.

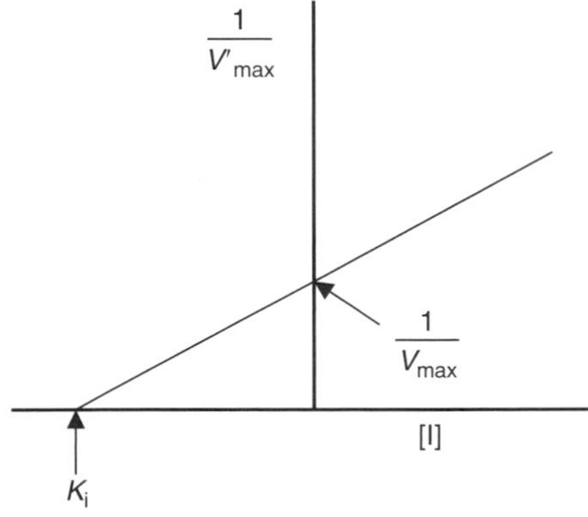

Figure 9.9 Secondary plot of $1/V'_{\max}$ values (from Figure 9.8) vs inhibitor concentration.

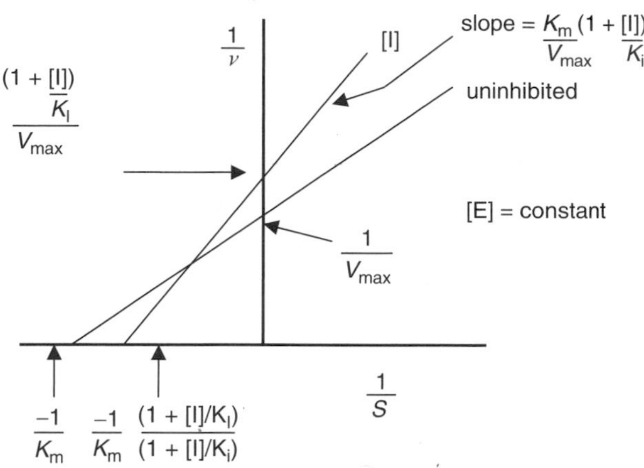

Figure 9.10 Lineweaver–Burk plot showing mixed inhibition where $K_I > K_i$ (strongest binding to enzyme).

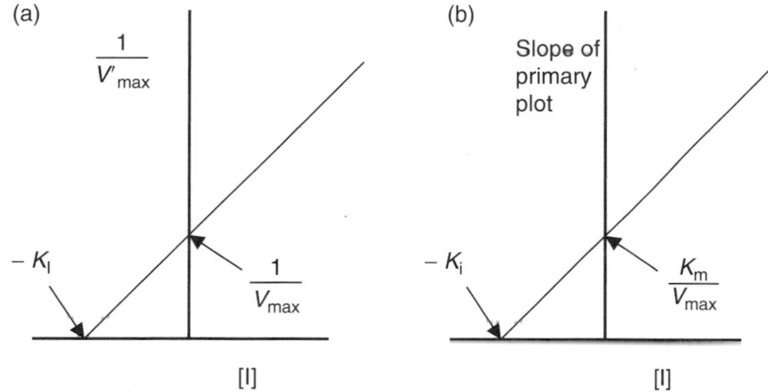

Figure 9.11 Secondary plot of (a) $1/V'_{max}$ values (from Figure 9.10) vs inhibitor concentration for determination of K_I and (b) slope (from Figure 9.10) vs inhibitor concentration for determination of K_i.

change; it differs in this respect from the transition state structure formed *after* reaction between, for example, a serine moiety at the active site of a serine protease with a peptidyl ketone inhibitor, i.e., the oxyanion-containing tetrahedral intermediate (see Section 9.7.1).

An organic reaction between two types of molecules is considered to proceed through a high energy activated complex known as the transition state which is formed by collision of molecules with greater kinetic energy than the majority present in the reaction. The energy required for formation of the transition state is the activation energy for the reaction and is the barrier to the reaction occurring spontaneously. The transition state may break down to give either the components from which it was formed or the products of the reaction. The transition state for the reaction between hydroxyl ion and methyl iodide is shown in Equation (9.23). The transition state shown depicts both commencement of formation of a C−OH bond and the breaking of the C−I bond. Enzymes catalyze organic reactions by lowering the activation energy for the reaction and one view is that they accomplish this by straining or distorting the bound substrate towards the transition state.

$$HO^- + CH_3{-}I \quad [HO\cdots CH_3 \cdots I] \quad HO{-}CH_3 + I^- \tag{9.23}$$

Equation (9.24) shows a single substrate–enzymatic reaction and the corresponding nonenzymatic reaction where $ES^{\neq}$ and $S^{\neq'}$ represent the transition states for the enzymatic and nonenzymatic reaction, respectively, and $K_N^{\neq}$ and $K_E^{\neq}$ are equilibrium constants, respectively, for their formation. K_s is the association constant for formation of ES from E and S, and K_T is the association constant for the hypothetic reaction involving the binding of $S^{\neq'}$ to E. Analysis of the relationships between these equilibrium constants shows that $K_T K_N^{\neq} = K_s K_E^{\neq}$. Since the equilibrium constant for a reaction is equal to the rate constant mutiplied by h/kT, where h is Planck's constant and k is Boltzmann's constant, then $K_T = K_s(k_E/k_N)$, where k_E and k_N are the first-order rate constants for breakdown of the ES complex and the nonenzymatic reaction, respectively. Since the ratio k_E/k_N is usually of the order 10^{10} or greater, it follows that $K_T \gg K_s$. This means that the transition state $S^{\neq'}$ is considered to bind to the enzyme at least 10^{10} times more tightly than the substrate.

$$\begin{array}{ccccc} E + S & \underset{}{\overset{K^{\neq}_N}{\rightleftharpoons}} & E' + S^{\neq'} & \longrightarrow & E + P \\ K_s \updownarrow & & \updownarrow K_T & & \updownarrow \\ ES & \underset{K_E^{\neq}}{\rightleftharpoons} & ES^{\neq} & \longrightarrow & EP \end{array} \tag{9.24}$$

A transition state analogue is a stable compound that structurally resembles the substrate portion of the unstable transition state of an enzymatic reaction. Since the bond-breaking and bond-making mechanism of the enzyme-catalyzed and nonenzymatic reaction is similar, the analogue will resemble $S^{\neq'}$ and have an enormous affinity for the enzyme compared to the substrate and consequently will be bound more tightly. It would not be possible to design a *stable* compound which mimics the transition state closely, since the transition state itself is unstable by possessing partially broken or made covalent bonds. Even crude transition state analogues of substrate reactions would be expected to be sufficiently tightly bound to the enzyme to be excellent reversible inhibitors. This expectation has been borne out in practice.

Design of a transition state analogue for a specific enzyme requires a knowledge of the mechanism of the enzymatic reaction. Fortunately, the main structural features of the transition states for the majority of enzymatic reactions are either known or can be predicted with some confidence.

Another class of competitive inhibitor which binds tightly to the enzyme is the slow, tight-binding inhibitor. These may be bound either noncovalently or covalently and are released very slowly from the enzyme because of the tight interaction. The slow binding is a time-dependent process and is believed to be due either to an enforced conformational change in the enzyme structure or reversible, covalent bond formation or, more probably, simply the very low inhibitor concentration used during measurement to allow observation of a residual activity. The rate constant which describes slow binding is

$$k_{\text{obs}} = \frac{k_{\text{on}}[I]}{\left(1 + \frac{[S]}{k_{\text{on}}}\right)} + k_{\text{off}} \tag{9.25}$$

where

$$E + I \underset{k_{\text{off}}}{\overset{k_{\text{on}}}{\rightleftharpoons}} EI \tag{9.26}$$

and k_{obs} describes the rate of formation of the EI complex. Since $k_{off} << k_{on}$ and $(1 + [S]/k_{on})$ is usually measured at a value of 10 or more, then, at an inhibitor concentration of 10^{-9} M and k_{on} limited to a ceiling value of 10^9 M^{-1} s^{-1} due to diffusion control,

$$k_{obs} = \frac{10^{-9} \cdot 10^9}{10} = 0.1\ s^{-1}$$

so that

$$t_{1/2} = \frac{\log 2}{0.1} = 6.9\ s$$

where $t_{½}$ is the time for half the slow binding to occur. At an inhibitor concentration of 10^{-10} M and $k_{on} = 10^8$ M^{-1} s^{-1} then $k_{obs} = 0.001\ s^{-1}$ and $t_{1/2} = 690$ s which gives rise to a readily observable time-dependent process. These longer half lives are only dependent on the used [I] and the value of k_{on}. Such examples have been observed for potent thrombin inhibitors where conformational change or covalent bond formation does not occur.

Coformycin, methotrexate, and allopurinol belong to this class and are useful drugs. Tight binding, where the dissociation from the complex takes days, is not distinguishable in effect from covalent bonding and this type of inhibitor may be classed as an irreversible inhibitor.

Compounds producing irreversible enzyme inhibition fall into two groups; active site-directed (affinity labelling) inhibitors and mechanism-based inactivators (k_{cat} inhibitors, suicide substrates).

Active site-directed irreversible inhibitors resemble the substrate sufficiently to form reversible enzyme–inhibitor complex, analogous to the enzyme–substrate complex, within which reaction occurs between functional groups on the inhibitor and enzyme. A stable covalent bond is formed with irreversible inhibition of the enzyme. Active site-directed irreversible inhibitors are designed to exhibit specificity towards their target enzymes, since they are structurally modeled on the specific substrate of the enzyme concerned.

In the previous discussion on reversible inhibitors, the potency of an inhibitor was shown to be reflected in the K_I value, which is characteristic of the inhibitor and independent of inhibitor concentration. However, the actual level of inhibition achieved in an enzyme system involves the use of equations into which inhibitor and substrate concentrations, as well as the K_m value for the substrate need to be incorporated. Similarly, the potency of an irreversible inhibitor is given by binding and rate constants which are both independent of inhibitor concentration. This allows a precise comparison of the relative potency of inhibitors, which is necessary in the design and development of more effective inhibitors of an enzyme.

Irreversible inhibition of an enzyme by an active site-directed inhibitor can be represented by

$$E + I \xrightarrow{k} EI \tag{9.27}$$

provided that complex formation between the inhibitor and the enzyme is ignored here for the present time. The reaction is bimolecular, however since the inhibitor is usually present in large excess of the enzyme concentration, the kinetics for inactivation of the enzyme follow a pseudo first-order reaction.

In the general case of a bimolecular reaction between two compounds A and B, the rate of reaction is given by

$$\frac{dx}{dt} = k_2(a - x)(b - x) \tag{9.28}$$

where k_2 is the second-order rate constant, a and b are the initial concentration of A and B, respectively, and the concentration of product is x at time t. Integration and rearrangement of Equation (9.28) gives

$$k_2 = \frac{2.303}{t(a-b)} \log \frac{b(a-x)}{a(b-x)} \tag{9.29}$$

In the situation where $a >> b$, this simplifies to

$$k_2 a = \frac{2.303}{t} \log \frac{b}{(b-x)} \tag{9.30}$$

Since $k_2 a = k_1$ where k_1 is the pseudo first-order reaction rate constant, then

$$k_1 = \frac{2.303}{t} \log \frac{b}{(b-x)} \tag{9.31}$$

A plot of $\log(b-x)$ vs t for the reaction as it proceeds, using a known concentration of the inhibitor, gives a regression line with slope $= -k_1/2.303$, from which k_1 and k_2 (i.e., $k_1 = k_2 a$) may be obtained. An alternative method for rapidly calculating k_1 is to determine the half-life ($t_{1/2}$) for the reaction by taking an interval of 0.3010 units (i.e., log 2) on the y-axis and reading off t_1 and t_2 from the x-axis. Then $t_{1/2} = (t_2 - t_1)$ and since from Equation (9.31) $b = 2(b-x)$ at the time for half the reaction to occur then

$$t_{1/2} = \frac{2.303 \log 2}{k_1} = \frac{2.303 \times 0.3010}{k_1} = \frac{0.693}{k_1} \tag{9.32}$$

from which k_1 can be calculated.

In practice in enzyme inhibition reactions it is sometimes found that k_1 is not directly proportional to a so that the value of k_2 is not constant with a change in the concentration of the inhibitor a. This is due to initial binding of the inhibitor to the active site of the enzyme before the irreversible inhibition reaction occurs.

$$\mathrm{E + I} \underset{}{\overset{K_\mathrm{I}}{\rightleftharpoons}} \underset{\text{Complex}}{\mathrm{(E)(I)}} \xrightarrow{k_{+2}} \underset{\text{Inhibited enzyme}}{\mathrm{EI}} \tag{9.33}$$

The rate of the inactivation reaction is given by

$$\frac{\mathrm{d}x}{\mathrm{d}t} = \frac{k_{+2}[\mathrm{E}]}{1 + K_\mathrm{I}/[\mathrm{I}]} \tag{9.34}$$

where x represents the concentration of the inhibited enzyme [EI], K_I is the dissociation constant for the enzyme–inhibitor complex and k_{+2} is the first-order rate constant for the breakdown of the complex into products. Integration of Equation (9.34) gives

$$k_1 t = 1\mathrm{nE} - 1\mathrm{n(E} - x) \tag{9.35}$$

where k_1 is the observed first-order rate constant and

$$k_1 = \frac{k_{+2}}{1 + K_I/[i]} \tag{9.36}$$

When Equation (9.36) is written in the reciprocal form

$$\frac{1}{k_1} = \frac{K_I}{k_{+2}[I]} + \frac{1}{k_{+2}} \tag{9.37}$$

a plot of $1/k_1$ against $1/[I]$ gives a regression line from which k_{+2} and K_I may be evaluated, since the intercepts on the $1/k_1$ and $1/[I]$ axes give the values for $1/k_{+2}$ and $-1/K_I$, respectively.

Many irreversible inhibitors of certain enzymes have previously been recognized in which the range of electrophilic centers normally associated with active site-directed irreversible inhibitors, e.g., $-COCH_2Cl$, $-COCHN_2$ $-OCONHR$, $-SO_2F$, are absent so that the means by which they inhibited the enzyme was not clear. The action of these inhibitors has now become understandable since they have been characterized as mechanism-based enzyme inactivators. Mechanism-based enzyme inactivators bind to the enzyme through the K_s parameter and are modified by the enzyme in such a way as to generate a reactive group, which irreversibly inhibits the enzyme by forming a covalent bond with a functional group present at the active site. On occasion, catalysis leads not to a reactive species but to an enzyme–intermediate complex which is partioned away from the catalytic pathway to a more stable complex by bond rearrangement (e.g., β-lactamase inhibitors).

These inhibitors are substrates of the enzyme, as suggested by their alternative name k_{cat} inhibitors, where k_{cat} is the overall rate constant for the decomposition of the enzyme–substrate complex in an enzyme-catalyzed reaction. Mechanism-based inactivators do not generate a reactive electrophilic center until acted upon by the target enzyme. Reaction may then occur with a nucleophile on the enzyme surface, or alternatively the species may be released and either react with external nucelophile or decompose:

$$E + I \underset{K_{-1}}{\overset{K_{+1}}{\rightleftharpoons}} EI \xrightarrow{k_{+2}} EI^* \xrightarrow{k_{+3}} E\text{–}I, \qquad EI^* \xrightarrow{K_{+4}} E + P \tag{9.38}$$

The ratio of the rate constants, i.e., k_{+4}/k_{+3} gives the partition ratio (r) for the reaction and where this approaches zero the mechanism-based inactivation will proceed with little turnover of the inhibitor and release of the active species as shown in Equation (9.39) where the noncovalent enzyme–inhibitor complex (EI) is transformed into an activated species (EI*) which then irreversibly inhibits the enzyme.

$$E + I \underset{K_{-1}}{\overset{K_{+1}}{\rightleftharpoons}} EI \xrightarrow{k_{+2}} EI^* \xrightarrow{k_{+3}} E\text{–}I \tag{9.39}$$

Consequently, the reactive electrophilic species, by not being free to react with other molecules in the biological media, has a high degree of specificity for its target enzyme and exhibits low toxicity.

The inactivation rate constant for a mechanism-based enzyme inactivation is termed k_{inact} and is a complex mixture of the rate constants k_2, k_3, and k_4 (Equation (9.38)). However, the kinetic form of Equation (9.38) and that for active site-directed inhibition are identical so that Equation (9.37) becomes

$$\frac{1}{k_1} = \frac{K_I}{k_{inact}[I]} + \frac{1}{k_{inact}} \tag{9.40}$$

which, since $t_{1/2} = 0.693/k_1$ (see Equation (9.32)) becomes

$$t_{1/2} = \frac{0.693K_I}{k_{inact}[I]} + \frac{0.693}{k_{inact}} \tag{9.41}$$

A plot of $t_{1/2}$ vs the reciprocal of the inhibitor concentration for the inactivation process using various concentrations of the inactivator gives a regression line which cuts the y-axis at $0.693/k_{inact}$ and the x-axis at $-1/K_I$. The meaning of K_I described here and K_I, the dissociation for the enzyme–reversible inhibitor complex, may not be the same under certain conditions, e.g., when k_3 becomes rate determining. Certain criteria need to be fulfilled before an irreversible inhibitor can be classified as a ''mechanism''-based enzyme inactivator (see Silverman, 1988).

9.2 GENERAL ASPECTS OF INHIBITOR DESIGN

9.2.1 Target Enzyme and Inhibitor Selection

Occasionally, drugs in current use for one therapeutic purpose have exhibited side effects indicative of potential usefulness for another, subsequent work establishing that the newly discovered drug effect is due to inhibition of a particular enzyme. Although the drug may possess minimal therapeutic usefulness in its newly found role, it does constitute an important ''lead'' compound for the development of analogues with improved clinical characteristics.

The use of sulfanilamide as an antibacterial was associated with acidosis in the body due to its inhibition of renal carbonic anhydrase (CA). This observation led to the development of the currently little used acetazolamide and subsequently the important chlorthiazide group of diuretics although these have a different mode of action.

The anticonvulsant aminoglutethide was withdrawn from the market due to inhibition of steroidogenesis and an insufficiency of 11β-hydroxy steroids. Aminoglutethimide, in conjunction with supplementary hydrocortisone, is now in clinical use for the treatment of estrogen-dependent breast cancer in postmenopausal women due to its ability to inhibit aromatase, which is responsible for the production of estrogens from androstenedione. Other more potent aromatase inhibitors have subsequently been developed (see Section 9.7.3).

Iproniazid, initially used as a drug in the treatment of tuberculosis, was observed to be a central nervous stimulant due to a mild inhibitory effect on monoamine oxidase (MAO). This observation eventually led to the discovery of more potent inhibitors of MAO, such as phenelzine, tranylcypromine, selegiline ((−)-deprenyl) and chlorgyline.

Many drugs introduced into therapy following detection of biological activity by pharmacological or microbiological screening experiments have subsequently been shown to exert their action by inhibition of a specific enzyme in the animal or parasite. This knowledge has helped in the development of clinically more useful drugs by limiting screening tests to involve only the isolated pure or partially purified target enzyme concerned and so introducing a more rapid screening protocol. However, translation of *in vitro* potency to the *in vivo* situation and finally the clinic is thwart with difficulties as will be seen later (also see Chapter 7).

The rational design of an enzyme inhibitor for a particular disease or condition in the absence of a lead compound presents a challenging task to the drug designer, since selection of a suitable target enzyme is a necessary first step in the process of drug design. *A priori* examination of the biochemical or physiological processes responsible for a disease or condition, where these are known or can be guessed at, may point to a suitable target enzyme in its biochemical environment, the inhibition of which would rationally be expected to lead to alleviation or removal of the disease or condition.

In a chain of reactions with closely packed enzymes in a steady state (see Equation (9.42)), where the initial substrate A does not undergo a change in concentration as a consequence of changes effected elsewhere in the chain, then any type of reversible inhibitor which inhibits the first step of the chain will effectively block that sequence of reactions.

$$\text{A} \xrightarrow[v_1]{\text{E}_1} \text{B} \xrightarrow[v_2]{\text{E}_2} \text{C} \xrightarrow[v_3]{\text{E}_3} \text{D} \xrightarrow[v_4]{\text{E}_4} \text{E metabolite} \qquad (9.42)$$

It is a general misconception that the overall rate in a linear chain can be depressed only by inhibiting the rate-limiting reaction, i.e., the one with lowest velocity at saturation with its substrate. Since individual enzymes will not be saturated with their substrates, the overall rate is determined largely by the concentration of the initial substrate, so that the first enzyme will often be rate-limiting, irrespective of its potential rate due to a low concentration of its substrate. Inhibitors acting at later points in the chain of closely bound enzymes may not block the metabolic pathway. If the reaction B $\rightarrow$ C (Equation (9.42)) is considered, competitive inhibition of E_2 will initially decrease the rate of formation of C but eventually the original velocity (ν_2) of the step will be attained as the concentration of B rises due to the difference between its rates of formation and consumption. However, selection of a target enzyme within a metabolic chain which does not inhibit the first step may lead successfully to translation of *in vitro* results, with the isolated target enzyme, to the *in vivo* situation due to additional changes. These changes relate to an increase in concentration of B which may have secondary effects on the chain due to product inhibition (B on E_1) or product reversal (A $\leftrightarrow$ B); either of these effects can slow ν_1, so leading to a slowing of the overall pathway.

This view is well illustrated by studies on inhibitors of the noradrenaline biosynthetic pathway (see ''Peripheral aromatic AADC inhibitors''). These were intended to decrease production of noradrenaline at the nerve–capillary junction in hypertensive patients, with an associated reduction in blood pressure. The selected target enzyme aromatic AADC catalyzes the conversion of dopa to dopamine in the second step of the biosynthesis of noradrenaline from tyrosine. Many reversible inhibitors, although active *in vitro* against this enzyme, fail to lower noradrenaline production *in vivo* although they, in an isolated scenario, may slow decarboxylation of dopa in peripheral tissues. Irreversible inhibitors of AADC successfully lower noradrenaline levels (see later).

However, competitive inhibitors have proved to be useful clinical agents, as examination of Table 9.2 illustrates, especially where the target enzyme has a degradative role on a substrate and is not part of the metabolic pathway in which the substrate is produced. Examples here are the anticholinesteases (Equation (9.4)) and AADC inhibitors as L-dopa protecting agents in the treatment of Parkinson's disease (Figure 9.1).

Irreversible inhibition progressively decreases the titer of the target enzyme to a low level and the biochemical environment of the enzyme is unimportant. For example α-monofluoromethyldopa is a mechanism-based inactivator of AADC and produces a metabolite which irreversibly inhibits and decreases the level of the enzyme by >99% (see ''Peripheral aromatic AADC inhibitors''). This leads to a near complete depletion of catecholamine levels in brain, heart, and kidney despite the occurrence of the enzyme in the second step of the noradrenaline biosynthetic pathway as discussed earlier.

The production of inhibited enzyme must be faster than the generation of new enzyme by resynthesis to maintain the target enzyme titer at a lower level so that dosing is infrequent. For

Table 9.2 Some reversible inhibitors used clinically (after Sandler and Smith 1989)

Drug	Enzyme inhibited	Clinical use
Allopurinol	Xanthine oxidase	Gout
Acetazolamide, methazolamide, dichlorphenamide, dorzolamide Ethoxzolamide, brinozolamide	Carbonic anhydrase II	Glaucoma, anticonvulsants
Trimethoprim, methotrexate, pyrimethamine	Dihydrofolate reductase	Antibacterial, anticancer, antiprotozoal agents
Cardiac glycosides	Na^+, K^+, -ATPase	Cardiac disorders
6-Mercaptopurine, azathioprine	Riboxyl amidotransferase	Anticancer therapy
Captopril, enalapril, cilazapril	Angiotensin-converting enzyme	Antihypertensive agent
Sulthiame	Carbonic anhydrase	Anticonvulsant (epilepsy)
Sodium valproate	Succinic semialdehyde dehydrogenase	Epilepsy
Idoxuridine	Thymidine kinase and thymidylate kinase	Antiviral agent
Cytosine arabinoside (Ara-C), 5-fluoro-2′5′-anhydro-cytosine arabinoside	DNA, RNA polymerases	Antiviral and anticancer agent
N-(Phosphonoacetyl)-L-aspartate (PALA)	Aspartate transcarbamylase	Anticancer agent
Indomethacin, ibuprofen, naproxen	Prostaglandin synthetase cyclooxygenase I and II	Antiinflammatory
Miconazole, clotrimazole, ketoconazole, ticonazole	Sterol 14α-demethylase of fungi fungi	Antimycotic
Benzserazide	AADC (peripheral)	Co-drug with L-dopa in Parkinson's disease
Aminoglutethimide, fadrozole, vorozole, letrozole, anastrozole	Aromatase	Oestrogen-mediated breast cancer
Saquinavir, ritonavir, indinavir, nelfinavir, amprenavir	HIV protease	HIV infections
Zidovudine, ddl, zacitabine, TIBO derivatives	HIV reverse transcriptase	HIV infections
Acyclovir, vidarabine, ganciclovir	Viral DNA polymerase	Herpes infections
Naftifine, terbinafine	Fungal squalene epoxidase	Antifungals
Finasteride	5α-reductase	Benign prostatic hyperplasia
Mevinolin, pravastatin, synvinolim	HMG-CoA reductase	Hyperlipidaemia
Adriamycin, etoposide	Topoisomerase II	Anticancer agents

mechanism-based inactivators, not only is the turnover rate of the enzyme important because of enzyme resynthesis, and this rate may be 10^3–10^5 slower than for natural substrates, but the partition ratio for the reaction should ideally be close to zero when every turnover should result in inhibition. A list of drugs which act by irreversible inhibition of the enzyme is given in Table 9.3.

9.2.2 Selectivity and Toxicity

Inhibitors used in therapy must show a high degree of selectivity towards the target enzyme. Inhibition of closely related enzymes with different biological roles (e.g., trypsin-like enzymes such as thrombin, plasmin, and kallikrein), or reaction with constituents essential for the well-being of the body (e.g., DNA glutathione, liver P-450 metabolizing enzymes) could lead to serious side effects. An inhibitor for potential clinical use is put through a spectrum of *in vitro* tests against other potential enzyme targets to ascertain that it is suitably selective towards the intended target. An inhibitor with high potency, e.g., $IC_{50} = 5$ nm would be screened at 1 μM against other targets and a small percentage inhibition would rate as a demonstration of acceptable selectivity. The aromatase inhibitor fadrazole (**9.106**) at higher doses than likely to be achieved clinically showed inhibition of the 18-hydroxyase in the steroidogenesis pathway, which could affect aldosterone

Table 9.3 Some irreversible inhibitors used clinical or in trial (after Sandler and Smith 1989)

Drug	Enzyme inhibited	Clinical use
Omeprazole	H^+, K^+-ATPase	Antiulcer agent
Sulfonamides	Dihydropteroate synthetase	Antibacterial
Iproniazid, phenelzine, isocarboxazid, tranylcypromine	MAO	Antidepressant
Neostigmine, eserine, dyflos, benzpyrinium, ecothiopate	Acetylcholinesterase	Glaucoma, myasthenia gravis Alzheimers disease
Penicillins, cephalosporins, cephamycins, carbapenems, monobactams	Transpeptidase	Antibiotics
Organic-arsenicals	Pyruvate dehydrogenase	Antiprotozoal agents
O-Carbamyl-D-serine	Alanine racemase	Antibiotic
D-Cycloserine	Alanine racemase	Antibiotic
Azaserine	Formylglycinamide ribonucleotide aminotransferase	Anticancer
γ-Vinyl GABA (Vigabatrin)	GABA transaminase	Epilepsy
Clavulanic acid, sulbactam	β-Lactamase	Adjuvant to penicillin antibiotic
α-Difuoromethylornithine (eflornithine)	L-Ornithine decarboxylase	Trypanosomal and other parasitic diseases
Selegiline ((-)-deprenyl)	MAO-B	Co-drug with L-dopa in Parkinson's disease
Coumate, 667-coumate	Estrone sulfatase	Estrogen-mediated breast cancer
4-hydroxyandrostendione, exemestane	Aromatase	Estrogen-mediated breast cancer
5-Fluorouracil	Thymidylate synthetase	Anticancer

production in the clinical setting. With the further developed compound letrozole (**9.107**) the observed selectivity between the two enzymes noted with fadrozole (tenfold) was widened by at least an order (100-fold).

Active site-directed irreversible inhibitors are alkylating or acylating agents and would be expected to react with a range of tissue constituents containing amino or thiol groups besides the target enzyme, with potentially serious side effects. They are mainly used in *in vitro* studies for labelling of amino acid residues at the active site (affinity labelling) of an enzyme for structural purposes.

Mechanism-based inactivators do not possess a biologically reactive functional group until after they have been modified by the target enzyme and, consequently, would be expected to demonstrate high specificity of action and low incidence of adverse reactions. It is these features which have encouraged their active application in inhibitor design studies.

In the situation where the target enzyme is common to the host's normal cells as well as to cancerous or parasitic cells, chemotherapy can be successful when host and parasitic cells contain different isoenzymes, e.g., DHFR, with that of the parasite being more susceptible to carefully designed inhibitors.

Alternatively, the target enzyme may be absent from the host cell. Sulfonamides are toxic to bacterial cells by inhibiting dihydropteroate synthetase, an enzyme on the biosynthetic pathway to folic acid. The host cell is unaffected, since it utilizes preformed folic acid while the susceptible bacterial cannot. Sulfonamides (**9.1**) are toxic to bacterial cells by inhibiting the utilization of *p*-aminobenzoate (**9.2**) by dihydropteroated synthetase, an enzyme in the biosynthetic pathway to dihydrofolic acid. Another example relates to the CA isoforms CA IX and CA XII, which predominate in cancer cells and are concerned in maintaining the acid–base balance and intercellular communication. Inhibitors of CAS IX and XII as antitumor agents would need to be very selective since up to 12 other CA isoforms are known in humans and are also involved in the interconversion between CO_2 and HCO_3^-, critical for many physiological processes (especially CA I, CA II, and CA IV).

Normal and cancerous cells contain the same form of the target enzyme, DHFR, but the faster rate of growth of the tumor cells makes them more susceptible to the effects of an inhibitor. Although side effects occur, these are acceptable due to the life-threatening nature of the disease.

9.3 RATIONAL APPROACH TO THE DESIGN OF ENZYME INHIBITORS

Once the target enzyme has been identified then usually a ''lead'' inhibitor has previously been reported or can be predicted from studies with related enzymes or has appeared, in more recent years, from the screening of industrial chemical collections or libraries. The design process is then initially concerned with optimizing the potency and selectivity of action of the inhibitor to the target enzyme using *in vitro* biochemical tests; nowadays the pure enzyme from recombinant DNA technology may be available for such studies. Candidate drugs are then examined by *in vivo* animal studies for oral absorption, stability to the body's metabolizing enzymes, and toxic side effects. Since many candidates may fall at this stage further design is then necessary to maintain desirable features and design out undesirable features from the *in vivo* profile. Since an *in vivo* profile in animal studies is not directly translatable to the human situation, studies with human volunteers are also required before a drug enters clinical trials.

Computerized molecular modeling is nowadays an essential part of the design process (see Chapter 4) but its relative importance in this process is determined by the state of knowledge concerning the target enzyme.

Ideally a high-resolution crystal structure of the target enzyme with the active site identified by co-crystallization with an inhibitor provides a knowledge of binding sites on the inhibitor and enzyme and their relative disposition; from these parameters chemical libraries can be searched for suitable lead compounds. Furthermore, an additional binding site may be identified so that a modified inhibitor using this additional site may be more potent or selective towards its target. Once the enzyme crystal structure is known the mode of binding of inhibitors fortuitously discovered earlier can be clarified (hindsight) to explain structural features responsible for their mechanism of action.

Usually for a newly discovered target the enzyme crystal structure is not known and the 3D-structure of the protein has to be less satisfactorily predicted from either NMR studies or by homology modeling from a related protein of known 3D-structure. For homology modeling the sequence similarity between the two proteins should be at least 30%. Either of these techniques can lead to the identification of prospective binding sites at the enzyme active site and on a lead inhibitor by ''docking'' the inhibitor at the active site using standard computer modeling software.

Observations can lead to further structural modification of the inhibitor to either improve fit (potency) in the model by taking advantage of additional binding areas such as hydrogen bonding groups or hydrophobic residues on the enzyme.

The relative positions of potential binding areas at the active site can provide a pharmacophoric pattern which can be used for *de novo* inhibitor design. Also, searching of 3D structural databases can provide novel structures previously designed for another purpose, with binding groups held in the correct 3D pattern through an appropriate carbon skeleton.

In the absence of a model of the enzyme active site then modeling with a series of inhibitors by superimposition (matching) of key functional groups, similar areas of electrostatic potential, and common volumes may identify areas, i.e., the pharmacophore, with similar physical and electronic properties in the more active members of a series. Whereas this approach is suitable for rigid structures it is less applicable to flexible molecules since the conformation in solution may be different to that required to efficiently bind to the enzyme active site. Alternatively the common conformational space available to a range of active inhibitors can be used to distinguish this from the space available to less active or inactive analogues which may lead to a defined model for the pharmacophore.

A more productive approach if a model of the enzyme active site does not exist, as is usual for a new target enzyme, is a design based on a knowledge of the substrate, a lead inhibitor (perhaps from a related enzyme), and of the mechanism of the catalytic reaction. Molecular modeling may enter into the design process at a later stage. A few selected examples are now given to illustrate this approach.

The antihypertensive drug captopril (**9.46**), an inhibitor of angiotensin I-converting enzyme (ACE), was designed from a knowledge of the substrate specificity and a known lead inhibitor of its target enzyme, together with a guess that the mechanism of action of ACE was similar to that of the zinc metalloprotease carboxypeptidase A about which much was known (see ''Metalloproteases''). Further structural modification gave the related enalaprilat and, from molecular modeling using inhibitor superimposition, cilazaprilat (**9.51**).

Many mechanism-based inactivators of pyridoxal phosphate-dependent enzymes are known, some of which were designed from a knowledge of the mechanism of action of their respective target enzymes. Inhibitors of AADC, histamine decarboxylase, ornithine decarboxylase (ODC), and GABA-transaminase (GABA-T) designed in this way have proved to be useful drugs (see Section 9.7.4).

Aspartate proteases, such as renin and human immunodeficiency virus (HIV)-protease catalyze the hydrolysis of their substrates by aspartate ion-catalyzed activation of the weak nucleophile water effectively to the strong nucleophile, hydroxyl ion. The hydroxyl ion attacks the carbonyl of the scissile amide bond in the substrate to give a tetrahedral intermediate which collapses to the products of the reaction (Figure 9.14).

HIV-protease is an aspartate protease which cleaves polyproteins formed in viral reproduction to the correct length for viral maturation. Inhibitors of HIV-protease have been designed based on the amino acid sequence around a scissile bond of the polyprotein substrate and the structure of the tetrahedral intermediate. Using the substrate sequence 165–9 (Leu-Asp-Phe-Pro-ILe) for a particular polyprotein a stable tripeptide analogue possessing a hydroxyethylamine moiety ($-CH(OH)-CH_2$) to resemble the tetrahedral intermediate ($-C(OH)_2-$) has been developed (see ''Metalloproteases''). This compound, saquinavir (Figure 9.21), has $IC_{50} = 0.4$ n*M* and is now in clinical use in drug combinations as an agent to prevent the spread of viral infection. Stable amino- and carboxy-terminal blocking groups are present and the hydrophobicity of the proline in the substrate has been increased in the perhydro isoquinoline residue.

Other HIV protease inhibitors have been developed for other scissile bonds in the polyprotein substrate using a variety of functions (see ''Aspartate proteases''), which simulate the tetrahedral intermediate formed during catalysis. Here, potent inhibitors ($IC_{50} = 0.4$ n*M*, saquinavir) have been designed without a knowledge of the crystal structure of the protease. However, this knowledge has become of paramount importance recently in the design of inhibitors against the drug-resistant mutant protease forms arising (see Section 9.6).

The quantitative structure–activity relationship (QSAR) for analogues within a series of structurally related inhibitors has been used to correlate potency with a wide range of physicochemical properties (see Chapter 6); missing members within the series with the required characteristics may be indicated for subsequent synthesis and study.

9.4 DEVELOPMENT OF A SUCCESSFUL DRUG FOR THE CLINIC

The development of an inhibitor from its inception through to clinical trials and then on to the market is thwart with difficulties. After satisfactory *in vitro* screening of a potent inhibitor for selectivity towards its target enzyme (i.e., little effect on related enzymes), *in vivo* studies in animals are undertaken to establish that the candidate drug is well absorbed when administered orally, has a low rate of metabolism (long biological half-life, $t_{1/2}$), and is free from toxic side effects. The *in vivo* studies present a formidable barrier to the development process and many candidate drugs can fall at this stage as has been described in Chapter 7 for the development of an aromatase inhibitor.

9.4.1 Oral Absorption

A drug needs to be absorbed through the gastrointestinal membrane and then carried by the plasma to its site of action. In the case of an inhibitor the site of action is the target enzyme and access may require further membrane penetration, e.g., cell membrane, bacterial cell, viral particle, cell nucleus, blood–brain barrier. Successful passage through the body's membranes requires the correct balance between the hydrophilic and hydrophobic properties of the drug, i.e., water: oil solubility ratio or partition coefficient (see relevant sections in Chapters 1, 3, and 6). Membranes are mainly lipid in nature and if the drug is too hydrophobic it will penetrate and remain in the membrane since it will have little tendency to pass through into the hydrophilic media of the plasma and alternatively if the drug is too hydrophilic it will have little tendency to penetrate into the gastrointestinal membrane from the aqueous media of the gut; either situation amounts to poor drug absorption.

Many potent inhibitors of thrombin are known from *in vitro* studies but few have the required hydrophilic:hydrophobic ratio to become useful clinical agents and this feature has dogged development of antithrombotic agents for a long time.

Oral absorption of a drug may be improved by chemical manipulation to a biologically inert but more absorbable form of a drug which after absorption is converted by the body's enzymes to the active parent drug, i.e., prodrug (see Chapter 8). This approach has proved particularly useful for drugs possessing a carboxylic acid group, which in the ionized form at pH 7, may not be well absorbed from the small intestine. Examples are ampicillin where well absorbed ester in the form of pivampicillin, bacampicillin, talampicillin release ampicillin in the plasma by initial hydrolysis by esterases to an intermediate which degrades in the aqueous media (see Section 7.4.1).

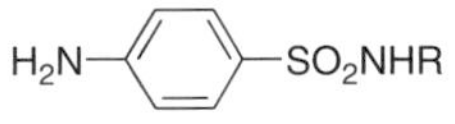

$H_2N-C_6H_4-SO_2NHR$

(**9.1**)

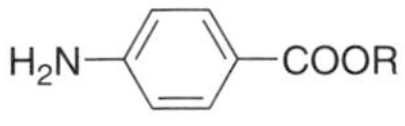

$H_2N-C_6H_4-COOR$

(**9.2**)

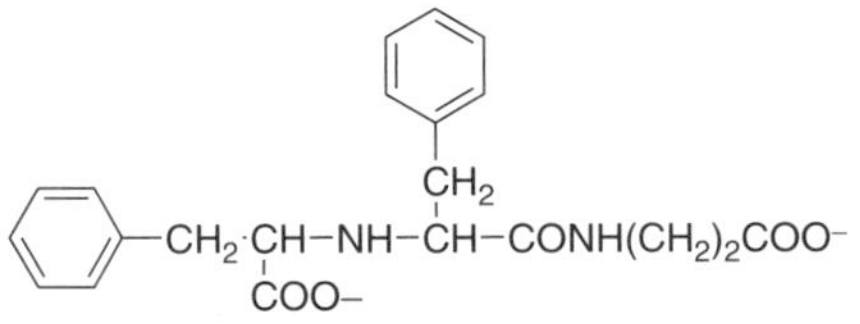

$C_6H_5-CH_2\cdot CH(COO^-)-NH-CH(CH_2C_6H_5)-CONH(CH_2)_2COO^-$

(**9.3**) SCH 32615

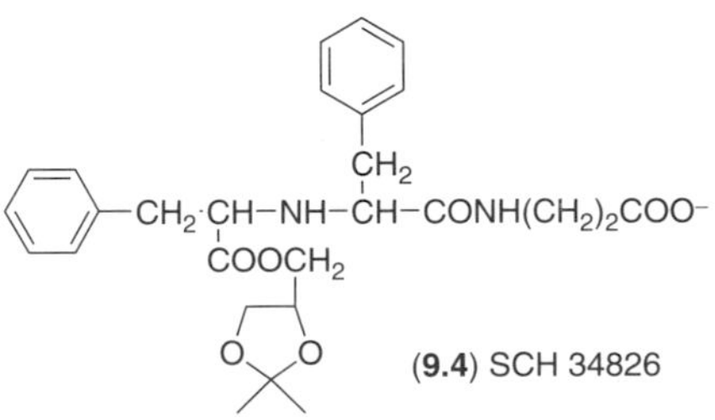

$C_6H_5-CH_2\cdot CH(COOCH_2\text{-dioxolane})-NH-CH(CH_2C_6H_5)-CONH(CH_2)_2COO^-$

(**9.4**) SCH 34826

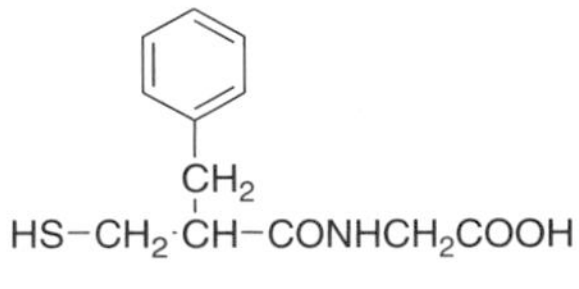

$HS-CH_2\cdot CH(CH_2C_6H_5)-CONHCH_2COOH$

(**9.5**) Thiorphan

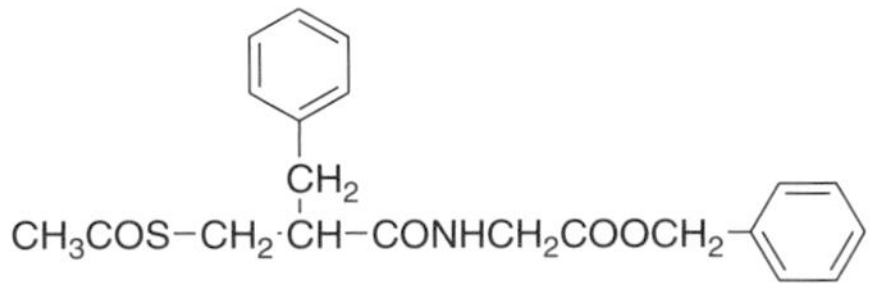

$CH_3COS-CH_2\cdot CH(CH_2C_6H_5)-CONHCH_2COOCH_2-C_6H_5$

(**9.6**) Acetorphan

The ACE inhibitor enalaprilat (see 9.7.1) is well absorbed as its ethyl ester, enalapril, and the enkephalinase A (MEP) inhibitor SCH 32615 (**9.3**), a dicarboxylic acid, is well absorbed as the acetonide of the glycerol ester, SCH 34826 (**9.4**).

The potent enkephalinase inhibitor thiorphan (**9.5**) is not active parenterally but the protected prodrug, acetorphan (**9.6**) is absorbed through the blood–brain barrier and subsequently converted by brain enzymes to the active drug. The absorption of the antiviral aciclovir (**9.7**) has been improved as the valine ester, valaciclovir (**9.8**) and other analogues penciclovir (**7.53**) and famciclovir (**7.52**) are further improvements.

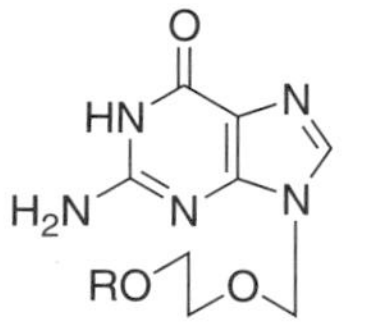

O
HN N
H_2N N N
RO O

(**9.7**) acyclovir: R = H

(**9.8**) valaciclovir: R = $—\overset{\text{O}}{\overset{\|}{\text{C}}}—\overset{\text{CH(CH}_3\text{)}}{\overset{|}{\text{CH}}}—NH_2$

Peptides as substrates of a peptide-degrading enzyme or agonists at a receptor site bind to the respective protein through a network of (>NH $\cdots$ O =C<) hydrogen bonds and consequently are hydrophilic molecules. Despite the introduction of terminal N and C hydrophobic residues or nonbinding ((>N$-CH_3$) modifications the oral administration of peptide-like enzyme inhibitors may lead to poor absorption due not only to the polar nature of the peptide back bone but also to degradation losses by intestinal proteases. Consequently high potency with an IC_{50} value in the low nanomolar range is required for such drugs. Saquinavir (Figure 9.21), a HIV protease inhibitor, has a low oral absorption (ca. 2%) but this is offset by a low IC_{50} of 0.4 n*M*.

9.4.2 Metabolism

For a reversible inhibitor to be a useful drug it must exist sufficiently long at the site of its target enzyme to exert its therapeutic effect. Since the level of the inhibitor at the site is a function of its plasma level, liver metabolism of the drug in the plasma to biologically inert products leads to progressive dissociation from the active site and, in time, reversal of the inhibition.

The biological half-life ($t_{1/2}$) of an inhibitor in humans is not directly related to that obtained from animal experiments although it is usually longer than that observed in the rat. A half-life of ca. 8 h is an acceptable figure in humans although for cancer chemotherapy a longer half-life 12–36 h is required to provide adequate drug cover in the event of patient noncompliance with the dose regimen.

The metabolic processes by which drugs are modified have been considered in Chapter 1 and most of these processes will lead to a shortening of the $t_{1/2}$ of the inhibitor. The most important of these involves hydroxylation by liver P 450 enzymes (CYP enzymes). The majority of drug metabolism in humans is carried out by CYP1, 2, and 3, particularly the CYP3A4 and CYP2C9 forms.

Several aromatase inhibitors used for the treatment of breast cancer (see Section 9.7.3) are imidazoles or triazoles. In general, within this group of inhibitors, replacement of imidazole by triazole may lead to a decrease in *in vitro* potency but this is reversed in the *in vivo* situation due to the greater metabolic stability of the triazole nucleus due to decreased hydroxylation and associated decreased potency. Further substitution of vulnerable $-CH_3$, and $-CH_2-$ groups with electron-withdrawing substituents decreases the chance of $-C^+$ development and subsequent hydroxylation (see Chapter 8). This approach is also illustrated in the development of fluconazole (**9.12**).

Ketoconazole (**9.9**), an antifungal agent, has a short $t_{1/2}$ when administered orally and is highly protein bound, due to its lipophilic nature, so that less than 1% of the unbound form exists at the site of action. Modification led to UK-46,245 (**9.10**), which had twice the potency in a murine candidosis model but further manipulation was required to improve metabolic stability and decrease lipophilicity. This was achieved in UK-47,265 (**9.11**) which has 100 times the potency of ketoconazole on oral dosing. Unfortunately this compound was hepatotoxic to mice and dogs and teratogenic to rats. Alteration of the aryl substituent to 2,4-difluorophenyl gave fluconazole (**9.12**) which is >90% orally absorbed and has $t_{1/2} = 30$ h. It is used for the treatment of candida infections and as a broad spectrum antifungal. The stability to metabolism of fluconazole could be attributed to possession of the stable triazole nucleus which is not hydroxylated unlike imidazole as well a protection of the $-CH_2-$ groups to hydroxylation by flanking electron-withdrawing groups (hydroxyl, triazole, triazole, difluorophenyl).

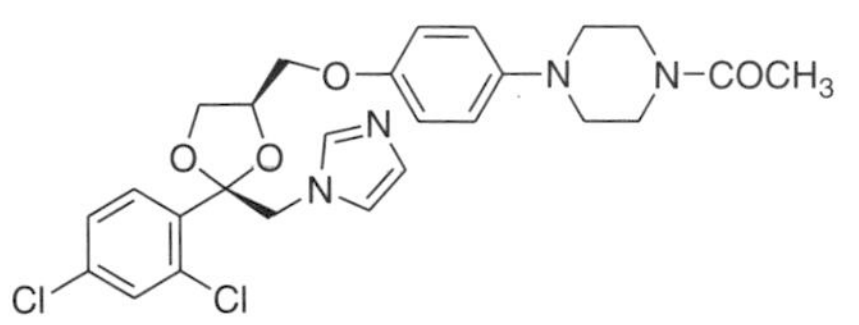

(**9.9**) ketoconazole

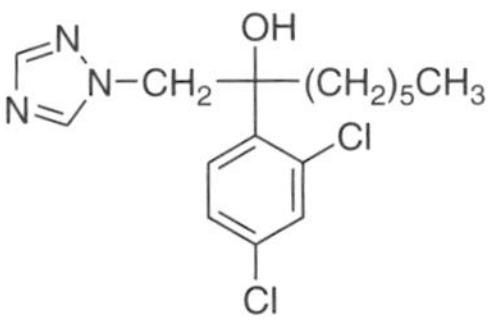

(**9.10**) UK-46,245

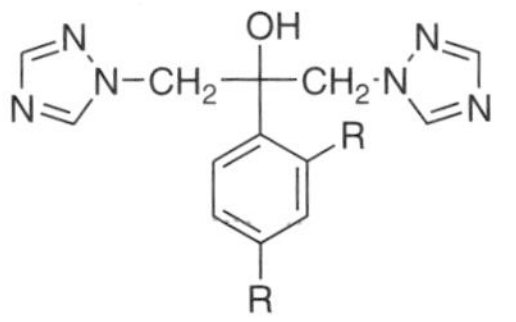

(**9.11**) R = Cl: UK-47,265
(**9.12**) R = F: fluconazole

Drug metabolism may be put to a beneficial use where, as described above, a poorly absorbable active drug may be chemically manipulated to a readily absorbable inert prodrug where the active drug is released on metabolism after passage through the gastrointestinal tract; esterases are the main class of enzymes for such conversions.

9.4.3 Toxicity

Toxic effects may become apparent on chronic dosing during animal pharmacology studies, clinical trials, or even after marketing. A well-known example is aminoglutethimide introduced as an antiepileptic and subsequently withdrawn due to effects on steroidogenesis enzymes leading to a ''medical adrenalectomy''. It was later re-introduced as an anticancer agent for the treatment of breast cancer by estrogen deprivation to capitalize on this toxic effect. The toxic side effects may merely be a matter of inconvenience or may be more severe. ACE inhibitor-induced dry cough affects about 40% of patients and could be due to build-up of bradykinin, a substrate of ACE, which increases NO generation through NO synthase with inflammatory effects on bronchial epithelial cells. The cough is said to be reduced by iron supplements ($FeSO_4$) where the activity of the NO synthase is reduced.

Many drugs, e.g., cimetidine, erythromycin, ketoconazole, choramphenicol, isoniazid, verapamil, including enzyme inhibitors are nonspecific inhibitors of liver cytochrome P-450 enzymes, i.e., inhibit many iso-enzyme forms. They consequently affect the metabolism of other drugs given concurrently leading to enhanced levels of these drugs and appearance of toxic effects. This interaction is particularly significant where enhanced drug levels are for a drug with a narrow therapeutic range between the therapeutically effective dose and the toxic dose, e.g., phenytoin. Further, in the normal adult population, there are ''slow'' and ''fast'' drug metabolizers so that the effect of one drug on the metabolism of another is not always predictable between patients.

Sildenafil (Viagra), a phosphodiesterase type 5 inhibitor is metabolized by CYPs 3A4 and 2C9 and should not be administered with drugs which inhibit CYP 3A4 such as erythromycin, ketoconazole, and HIV protease inhibitors (indinavir, nelfinavir, and saquinavir; rotonavir also inhibits CYP 2C9), which may increase sildenafil plasma levels.

Specific inhibitors of P-450 isoenzymes have similar effects but this effect is restricted to specific substrates of the particular isoenyme concerned. Examples include quinolone antibiotics (isoenzyme CYP1A2) and sulfaphenazole (CYP2C8/9).

9.5 STEREOSELECTIVITY

The stereochemistry of enzyme inhibitors possessing a chiral center is usually important in determining their potency towards a specific enzyme and this is a problem to be addressed in the early stages of drug design since it can sometimes be avoided by limiting the studies to achiral compounds.

Drug Registration Authorities worldwide are moving towards a requirement that for all new drugs the enantiomeric active form must be marketed unless for the racemate the activity of the separate enantiomers is available and enantioselective methods of chemical and biological analysis have been used in both animal and human studies. These requirements take into account the pharmacological consequences of the use of racemic drugs which has been previously described in Chapter 5.

Whereas the literature abounds with examples of activity residing mainly in one enantiomer of an inhibitor following *in vitro* studies, very few of these compounds have, as yet, reached the clinical or been subjected to registration requirements and *in vivo* information is not available from animal studies.

Aminoglutethimide (AG) (**9.13**), a long-established aromatase inhibitor, is used clinically as the racemate in the treatment of breast cancer in postmenopausal women (after surgery) to decrease their tumor estrogen levels. The (+) (*R*)-form is about 38 times more potent as an inhibitor than the (−) (*S*)-form. Aminoglutethimide is also an inhibitor of the side chain cleavage enzyme (CSCC) which converts cholesterol to pregnenolone in the adrenal steroidogenic pathway. Depletion of corticosteroids in this manner requires adjuvant hydrocortisone administration with the drug. Here the (+) (*R*)-form is about 2.5 times more potent than the (−) (*S*)-form.

For pyridoglutethimide (rogletimide) (**9.14**), an analogue of aminoglutethimide without the undesirable depressant effect, the inhibitor potency resides mainly in the (+) (*R*)-form (20 times that of the (−) (*S*)-form). 1-Alkylation improves potency *in vitro* but the activity for the most potent inhibitor in the series, the 1-octyl, resides in the (−)-(*S*)-form owing to a change in the mode of binding of inhibitor to enzyme.

A more selective inhibitor of aromatase than aminoglutethimide is the triazole vorozole (**9.15**) which is about 1000-fold more potent as an inhibitor. The (+) (*S*)-form is 32 times more active than the (−) (*R*)-form, but the very small inhibitory activity of the racemate towards other steroidogenic pathway enzymes, 11β-hydroxylase and 17,20-lyase, originates in the (−)- and (+)-forms, respectively.

It is of interest that in the benzofuranyl methyl imidazoles (**9.16**), some of which are 1000 times more potent as aromatase inhibitors in the racemic form than aminoglutethimide, comparable activity lies in both enantiomers. Homology modeling of the aromatase active site shows that the two aryl ring structures can fit equally well into the androstenedione (substrate) binding site.

MAO occurs in two forms, MAO-A and MAO-B. The use of MAO inhibitors as antidepressants is complicated by a dangerous hypertensive reaction with tyramine-containing foods (the "cheese-effect") which is due to inhibition of MAO-A located in the gastrointestinal tract which would otherwise remove the tyramine. L-(−) Deprenyl (selegiline) (**9.17**), a selective inhibitor of MAO-B, is widely employed to limit dopamine breakdown in Parkinson's disease in a selective inhibitory dosage. The (−)-isomer is much more potent than the (+)-isomer and, since the products of metabolism are (−)-metamphetamine and (+)-metamphetamine, respectively, the more potent (+)-metamphetamine side effects are removed from the racemate by use of L-deprenyl.

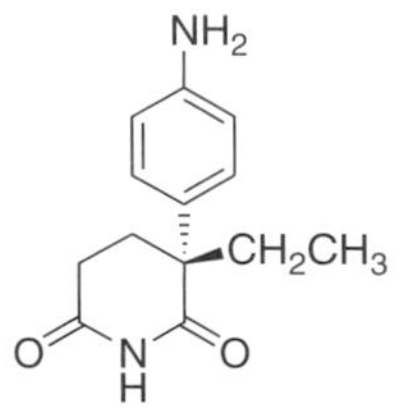

(**9.13**) (*R*)-Aminoglutethimide

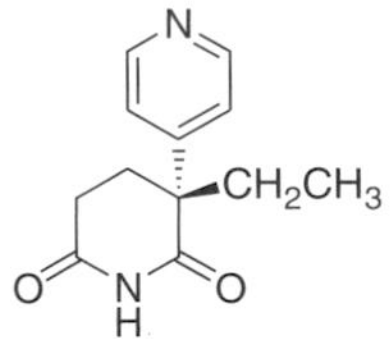

(**9.14**) (*R*)-Pyridoglutethimide

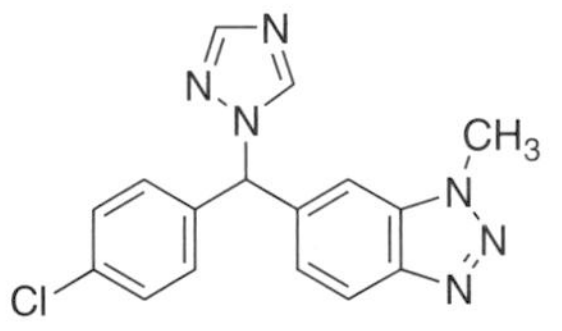

(**9.15**) Vorozole

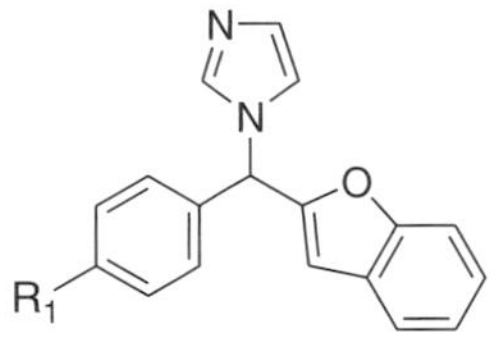

(**9.16**)

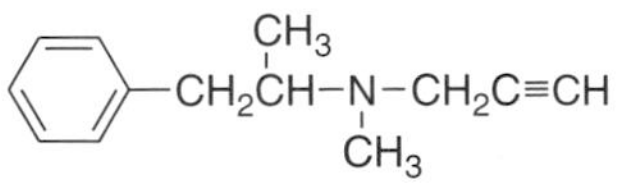

(**9.17**) Deprenyl

γ-Aminobutyric acid (GABA) transaminase inhibitors allow a build-up of the inhibitory neurotransmitter GABA and are potential drugs in the treatment of epilepsy. The inhibitory action of γ-vinyl GABA (vigabatrin, **9.118**), a drug now restricted in use in the treatment of this disease, resides mainly in the (*S*)-enantiomer (see Section ''GABA transaminase inhibitors'').

The HIV protease inhibitor saquinavir (Figure 9.21), used clinically in combination with other inhibitors directed at different targets to decrease resistance to the infection, has an (*R*)-configuration for the hydroxyl group since activity lies in this form ((*R*)-enantiomer $IC_{50} = 0.4$ n*M*, (*S*)-enantiomer = > 100 nm).

9.6. DRUG RESISTANCE

A setback to the use of an established inhibitor in the clinic in viral, bacterial, and parasitic diseases is the development of resistance to its action (see Chapter 16).

This may be caused by

1. Bypass of the antibiotic-sensitive step by duplication of the target enzyme in parasitic diseases, the second version being less susceptible to drug action, e.g., resistance to methicillin, trimethoprim, and sulfonamides.
2. Development of mutants under drug pressure in viral and parasitic diseases where an amino acid residue(s) of the natural (wild type) enzyme is changed in the transcription process. Suboptimal drug therapy selects for mutants which have a growth advantage over the wild type. Subsequent transmission of resistant variants to uninfected individuals may lead to infections that are drug resistant from the outset and require a new structural type of inhibitor for their suppression. Related drugs of a similar action and structure will show ''cross-resistance'' in that none will be superior in tackling the developed resistance. Resistance to drugs targeted at the reverse transcriptase (RT) of HIV-1 is due to mutational changes in the enzyme due to careless transcription so leading to failure, with time, of a drug to clear the virus. The nonnucleoside RT inhibitors (NNRTIs) nevirapine and efavirenz develop mutants which are resistant, point changes in the enzyme amino acid sequence observed at 103 (K → N), 101 (K → E), 188 (Y → L), 190 (G → S) in different mutants. In patients failing therapy with these drugs, cross-resistance to all available NNRTIs followed. Combination with other drugs attacking different targets is an approach to overcoming resistance to individual drugs, i.e., an RT inhibitor with a HIV aspartate proteinase.

Resistance to HIV aspartate proteinase inhibitors due to mutation is discussed in "Aspartate proteases."

3. Overproduction of target enzyme so that higher inhibitor concentrations are needed to inhibit growth, e.g., resistance to trimethoprim as well as developing pathways that bypass the inhibited process.
4. Reduced uptake of the drug or increased efflux (pushing out of the cell) of the drug by altering the number of transmembrane pumps in the cell membrane. These are general phenomena and can contribute with the other factors described above to overall drug resistance.

9.7 EXAMPLES OF ENZYME INHIBITORS AS DRUGS

9.7.1 Protease Inhibitors

The cleavage of peptide bonds by enzymes occurs by hydrolysis (proteolysis) and the enzymes are termed proteases. More recently the term peptidase has been used which is further divided into exopeptidase, where the action is near the end of a peptide chain, and endopeptidase, where the action occurs away from the termini. Berger and Schechter have proposed a scheme for identifying in an enzyme and substrate their common points of interaction between the amino acid residues (named P for peptide) and subsites (S for substrate) on the peptidase (Figure 9.12).

Peptidases are classified according to their catalytic mechanism of hydrolysis as serine, cysteine, aspartic, metallo, or threonine peptidases. The catalytic nucleophiles are OH (serine, threonine), SH (cysteine), H_2O (aspartic), and H_2O in conjunction with a metal (metallo).

More recently peptidases have been subclassified into clans where the members have evolved with similarities in tertiary folds and catalytic-site residues.

The selectivity of a peptidase for its substrates depends on the nature of the nucleophile, a specificity pocket for bonding between substrate and enzyme and an oxyanion hole to accommodate and distribute the negatively charged oxygen by attack of the catalytic nucleophile on the CONH bond (i.e., $>C(NH)-O^-$). Where the selectivity pocket is small, other sites in the proximity bind the individual residues of the substrate peptide chain to increase overall binding.

The mechanism for hydrolysis of peptide bonds by the respective serine, aspartate, and metallo peptidases will now be considered in detail since examples of inhibitors of these three classes of enzyme are described later where a knowledge of mechanism led to successful drug design.

Serine–OH as nucleophile

Chymotrypsin, trypsin, and subtilisin have a catalytic triad, Ser, His, and Asp at the catalytic site. After complex formation between the enzyme and substrate the Ser–OH nucleophile attacks the scissile carbonyl carbon atom with formation of a negatively charged intermediate (Figure 9.13). The nucleophilicity of the –OH group is increased by hydrogen bonding to the adjacent His. The

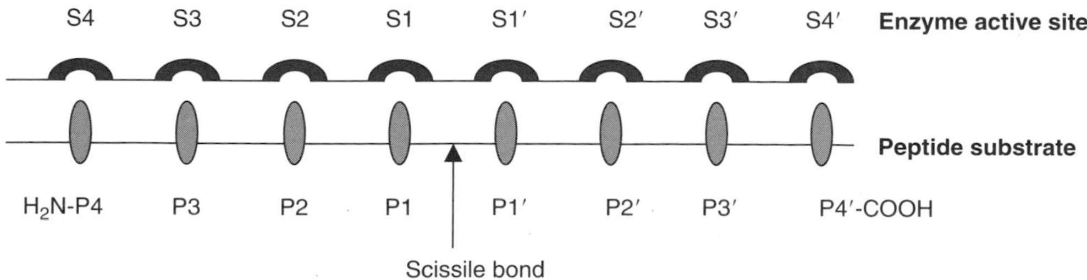

Figure 9.12 Terminology of specificity subsites of proteases and the complementary features of the substrate. (Adapted from Berger and Schechter, 1976.)

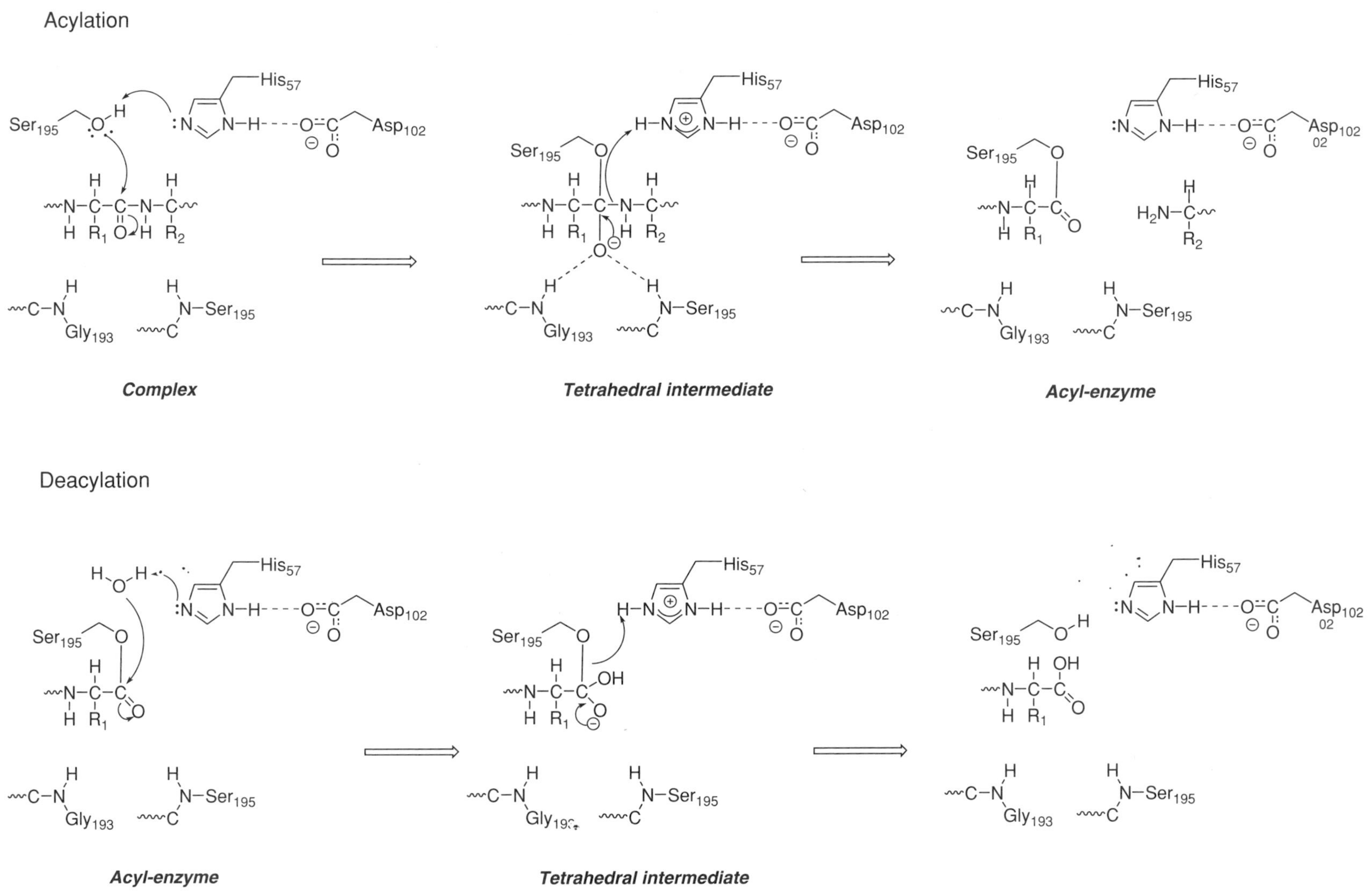

Figure 9.13 Catalytic mechanism proposed for serine peptidases with a catalytic triad consisting of Ser, His, Asp. (After Gerhatz et al., 2002; adapted from Beynon and Bond, 1989.)

intermediate is stabilized by the oxyanion hole formed by the backbone Gly_{193} NH and Ser_{195} NH residues (α-chymotrypsin) before collapse to a covalently bound acyl-enzyme with release of the initial cleaved peptide residue. The other cleaved residue is released with regeneration of the enzyme by a deacylation step where the catalytic nucleophile is a bound water molecule. The catalytic contribution of Asp is not clear but several different roles have been suggested (see Gerhartz et al.).

Water molecule as nucleophile

In aspartic peptidases hydrolysis of peptides is considered to proceed using a water molecule and a general acid–base mechanism utilizing an aspartate residue. In HIV-1 aspartic protease, a drug target, two aspartate residues from each monomer in the dimer are used, one in the ionized and the other in the unionized state (Figure 9.14).

Here, unlike in the serine peptidase-catalyzed reaction, a covalent acyl intermediate is not formed. The Asp residues act as general bases that activate the water molecule resulting in formation of a geminal diol. Subsequent protonation (general acid catalysis) of the leaving amino function in the cleaved residue (R_2) and de-protonation (general base catalysis) of the gem diol aiding peptide bond cleavage.

Water molecule as nucleophile and metal ion

The commonly occurring zinc peptidases among metallopeptidases, as illustrated by thermolysin but featuring as a drug target by angiotensin-converting enzyme (ACE), have a zinc ion tetrahedrally coordinated by two His, a Glu, and a water molecule in the resting state (Figure 9.15). On complex formation the carbonyl oxygen of the scissile peptide bond replaces the coordinated water molecule leading to activation of the carbonyl bond to nucleophilic attack by the general base

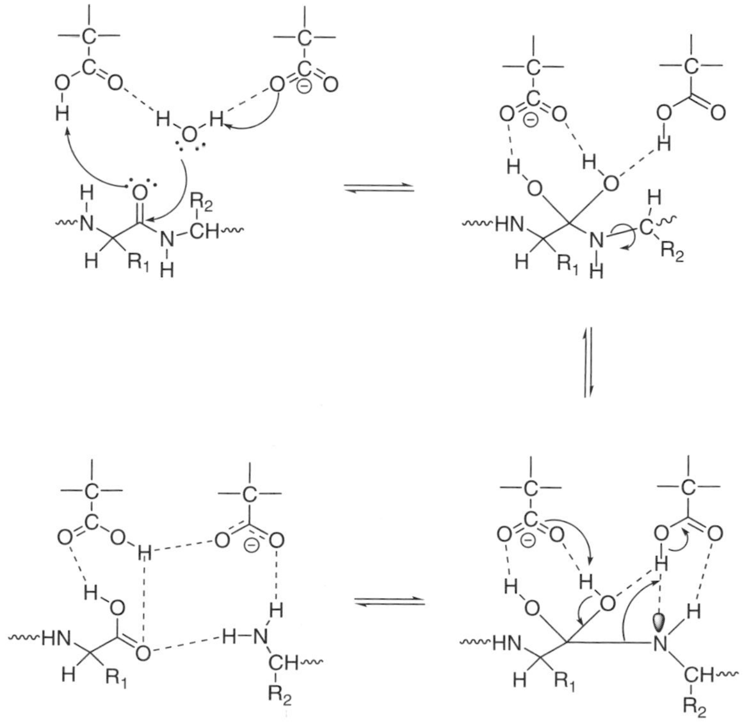

Figure 9.14 Catalytic mechanism proposed for aspartic peptidases. (After Gerhartz et al., 2002.)

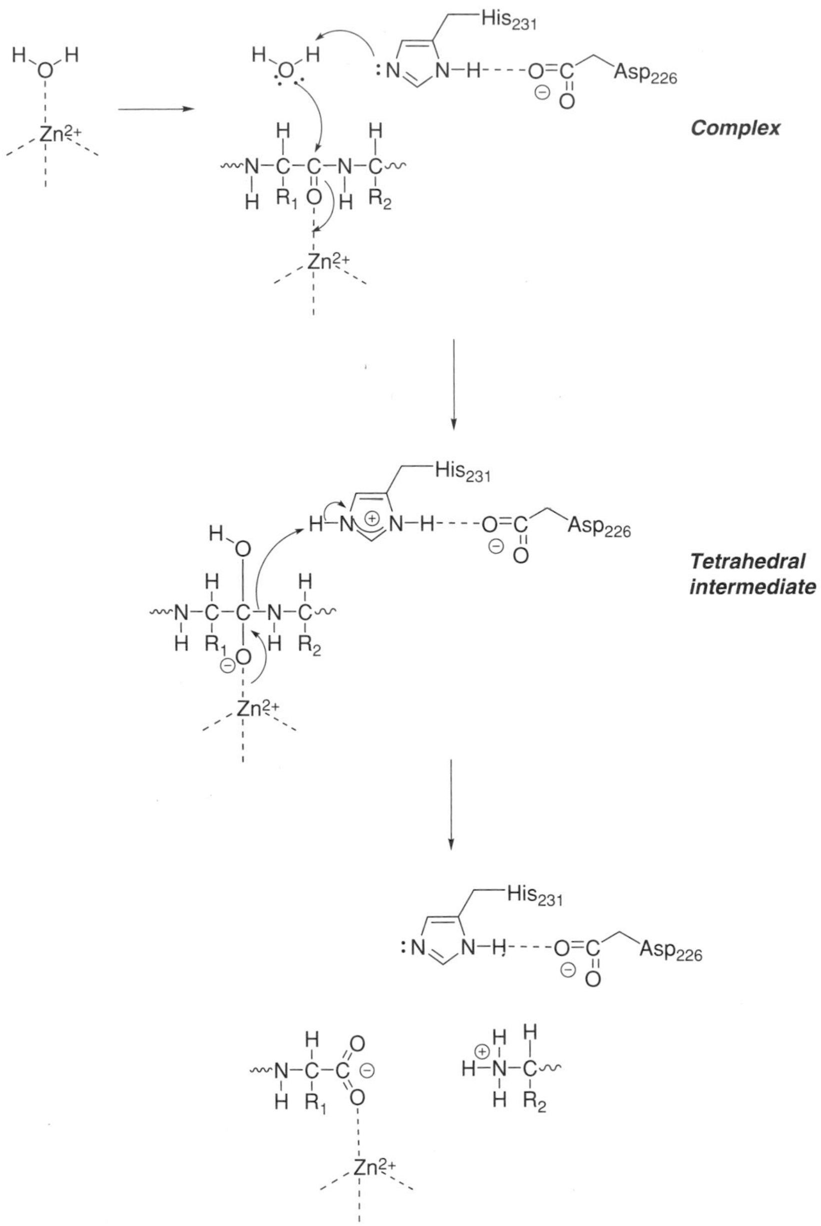

Figure 9.15 Catalytic mechanism for metallopeptidases with a water molecule bound to a single metal ion. (After Gerhartz et al., 2002; adapted from Mockand Standford, 1996.)

(His_{231}, thermolysin)-catalyzed water molecule. In the tetrahedral intermediate formed (Zn^{2+} with Arg as oxyanion hole) a proton is transferred from His_{231} to the leaving amino group in R_2.

Serine proteases

Thrombin inhibitors

Thrombin plays a central role within the coagulation cascade initiating not only fibrin clotting but also exerting several cellular effects. The serine protease thrombin is a member of the trypsin family which attacks peptide bonds following Arg or Lys residues; its catalytic mechanism is shown in Figure 9.13. Therefore, inhibitors occupying the active site must possess or imitate the basic amino- or guanidinoalkyl side chain of Lys and Arg. In extensive biochemical and pharmacological studies thrombin inhibitors were shown effective as anticoagulants and antithrombotics. The main criteria for a low-molecular-weight thrombin inhibitor to be an ideal anticoagulant are high selectivity and systemic bioavailability after oral application.

Thrombin is not present in an active form in blood but is formed from prothrombin after activation of the coagulation cascade, whereas its substrates (fibrinogen, thrombin activatable clotting factors, and protease activatable receptors) are permanently present. Consequently, inhibitors to be of therapeutic value must be present in the plasma at adequate concentrations to immediately neutralize the thrombin generated upon massive activation after vascular injury. It has been calculated that a pulse of thrombin is formed reaching a peak of about 200 nmol/L. However, immediately after initiation of the coagulation cascade the concentration of thrombin will be lowered by endogenous inhibitors, like the serpin (serine protease inhibitor) antithrombin. To be effective in anticoagulation the plasma concentration of a potent inhibitor (K_i in low nM range) should be at least 100 nmol/L or ≈ 0.05 μg/mL assuming an average molecular weight of 500 g/mol for a low-molecular-weight, synthetic thrombin inhibitor.

X-ray crystal structures of complexes between thrombin and several inhibitors and substrate analogues have been solved providing the basis for rational drug design (Figure 9.16 and Figure 9.17). Besides the primary specificity binding site to which the basic P1 amino acid of substrates or inhibitors is bound, there are located two further important hydrophobic binding pockets in the active site, which are called the proximal and distal binding site or abbreviated as P- and D-pocket, respectively. The D-pocket, often named also as aryl-binding site, is occupied by Phe at P9 of the fibrinopeptide A sequence and perfectly suited to accommodate phenyl or cyclohexyl rings in small, synthetic thrombin inhibitors. In contrast, the P-pocket is occupied by the side chain of Val in P2-position of the fibrinopeptide A and accepts even better the pyrrolidine ring of Pro in substrate-like inhibitor structures. An additional binding site, the anion-binding exosite 1, also called as fibrinogen recognition exosite (FRE), is located more than 20 Å far away from the active site and was discovered first from the crystal structure of the complex between thrombin and the naturally occurring thrombin inhibitor hirudin, originally isolated from the medicinal leech *Hirudo medicinalis*. Four Arg and five Lys residues but also hydrophobic amino acids contribute to this positively charged region, involved in both the recognition of the substrates fibrinogen and thrombin receptor but also in the binding of thrombomodulin and some natural inhibitors. The FRE is important for the design of synthetic analogues which are derived from the C-terminal sequence of hirudin, as will be described later.

Thrombin (Figure 9.16) possesses a second anion-binding exosite, which is named also the heparin-binding exosite. As can be derived from the name this site is important for the binding of heparin, a polysulfated polysaccharide, which is the most often clinically used anticoagulant. However, heparin is only an indirect inhibitor, because it enhances the inhibition of thrombin and some other coagulation factors by the serpin antithrombin.

Three main types of thrombin inhibitors were originally developed. These include peptide inhibitors based on natural substrates, arginine analogues, and benzamidine-derived compounds. However, in recent years the strategy was focused also on the design of compounds which contain a less basic P1 group to enhance their oral bioavailability. In addition, first successful examples of nonpolar prodrugs have been developed, which are converted into the active inhibitor form after oral absorption.

The first synthetic thrombin inhibitors, mainly esters and amides derived from the thrombin-sensitive Gly-Val-Arg sequence of the natural substrate fibrinogen and those resembling the Pro-Arg cleavage site of factors VIII, XIII, protein C, prothrombin, and the thrombin receptor showed relatively poor efficacy. However, extending the Pro-Arg sequence with a DPhe at P3 position gives effective inhibitors, such as the chloromethylketone H-DPhe-Pro-Arg-CH_2Cl (PPACK, **9.18**), the aldehyde H-DN(Methyl)Phe-Pro-Arginal (Efegatran, **9.19**) and the boronic acid derivative

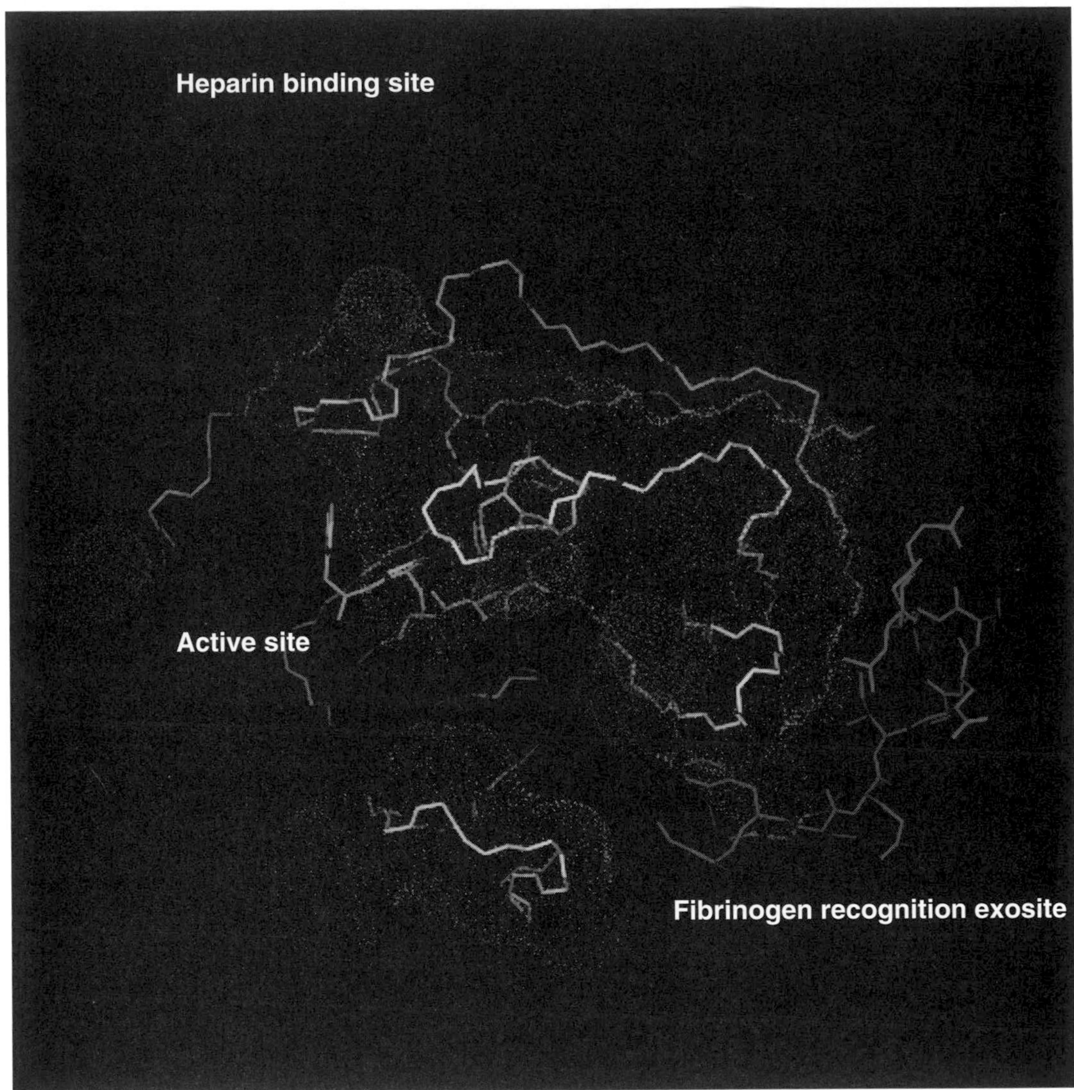

Figure 9.16 Front view of the thrombin molecule (backbone in yellow) in complex with the active site-directed inhibitor PPACK (green) and the C-terminal hirudin tail (residues 55–65 in pink) bound to the "fibrinogen recognition exosite." Thrombin is displayed with a Connolly dot surface in blue, red, and yellow for basic, acidic, or other residues, respectively. The "heparin-binding site" is located on the top of the thrombin molecule in this view defined as "standard" orientation.

Acetyl-DPhe-Pro-*boro*Arg (DuP 714, **9.20**). X-ray structures could prove that the DPhe at P3 in these inhibitors occupies the same aryl binding site as the Phe at P9 of the fibrinopeptide A sequence. Typically, all of these analogues (**9.18–9.20**) contain an activated carbonyl- or carbonyl-analogue group, which contributes to their high inhibitory potency and forms a covalent bond to the side chain hydroxyl group of thrombin's catalytic Ser195 residue.

The chloromethylketone PPACK (**9.18**) is the most powerful and most selective irreversible inhibitor of thrombin known, with a second-order rate constant three to five orders of magnitude higher than that for the inhibition of other trypsin-like proteases, such as factor Xa, plasmin, urokinase, plasma, and glandular kallikrein. The binding of **9.18** in complex with thrombin is shown in Figure 9.16 and Figure 9.17. After i.v. application the DPhe-Pro-Arg derived inhibitors

(**9.18–9.20**) exhibited anticoagulant effects in various animal experiments; however, their oral bioavailability is low (<10%). The methylation of the N-terminal amino group in efegatran (**9.19**) improved the stability and selectivity compared to the parent analogue H-DPhe-Pro-Arginal with a free amino group. The poor selectivity with respect to the trypsin-like enzymes of the fibrinolytic system and low oral bioavailability limited the clinical development of the boronic acid derivative (**9.20**) and prompted the development of analogues with neutral P1 side chains. Exchange of boroArg at P1 with the pinacol ester of methoxypropylboroGly in derivative TRI-50b (**9.21**) enhanced both oral bioavailability and selectivity of inhibition. This derivative (**9.21**), lacking the basic guanidino function of Arg which was thought to be essential for binding to Asp189 in the P1 specificity pocket of trypsin-like proteases, is still a potent thrombin inhibitor with nanomolar K_i-value (7 nM) and entered clinical development.

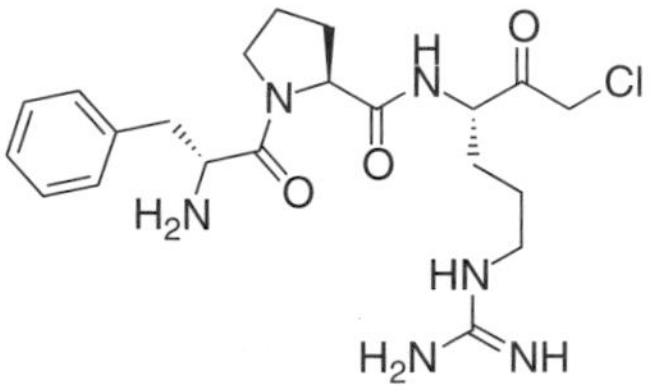

(**9.18**) PPACK 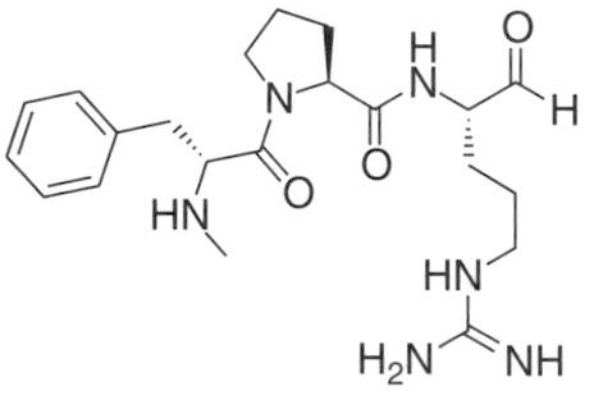 (**9.19**) Efegatran 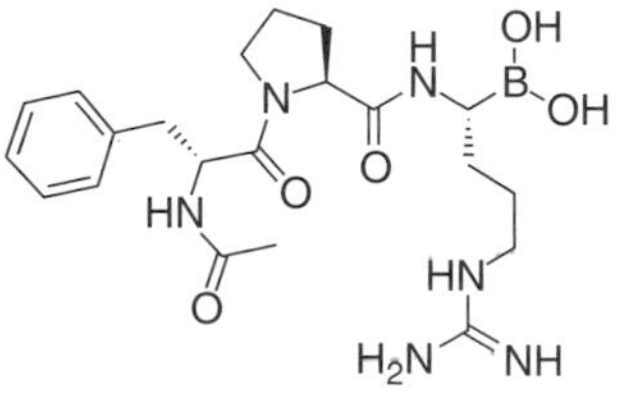 (**9.20**) DUP-714

An important strategy in the design of substrate analogue inhibitors is the replacement of the cleavable P1–P1′–peptide bond by stable analogues. In addition to PPACK (**9.18**), which irreversibly alkylates His57 in the active site of thrombin, several potent reversible inhibitors with an Arg ketone structure have been developed. Although these inhibitors also form a covalent hemiketal bond with Ser195 of thrombin, the enzyme–inhibitor complex can dissociate under certain conditions (e.g., reduction in free inhibitor concentration) back into free enzyme and intact inhibitor. Such Arg α-ketoheterocyles were attached to bicyclic lactams, which have been developed as DPhe-Pro-mimetics to reduce the peptidic character of the inhibitors. MOL-144 (**9.22**) is one example for a highly potent thrombin inhibitor (K_i 0.64 nM), which showed also a modest oral bioavailability of ≈ 25% both in rats and nonhuman primates.

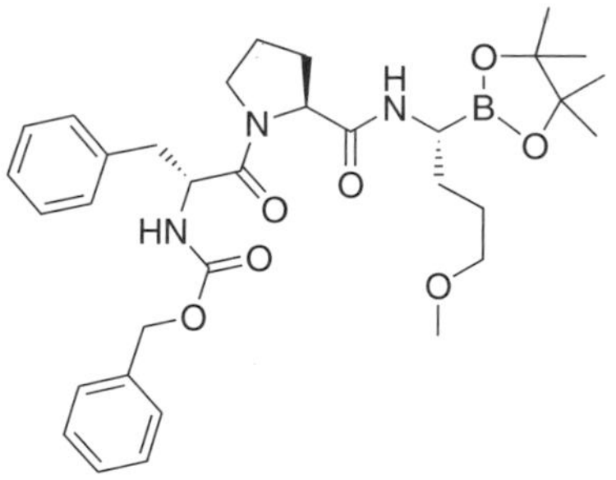

(**9.21**) TRI-50b 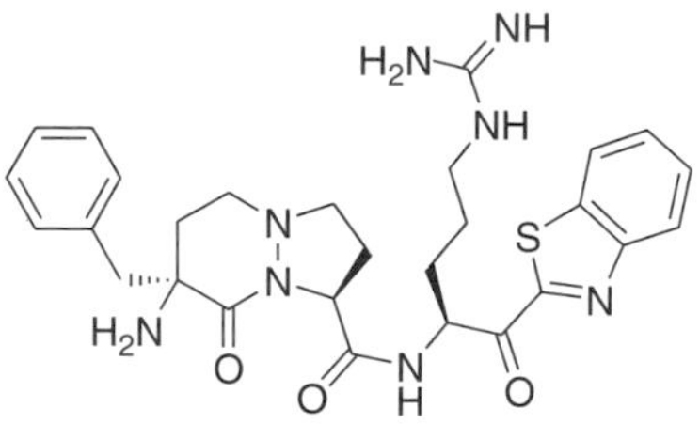 (**9.22**) MOL-144

A remarkable improvement of the pharmacokinetic properties, both absorption and half-life in circulation, was obtained with decarboxylated P1-mimetics of Arg. The first analogue was inogatran (**9.23**, K_i 15 nM) containing an noragmatine at P1 and an N-carboxymethylated amino acid at P3-position. This structure was further improved using a 4-amidinobenzylamine as P1 residue in melagatran (**9.24**, K_i 2 nM), which has a half-life of approximately 1.7 h after i.v. treatment in humans. In addition, a double-prodrug approach of melagatran has been successfully applied in case of ximelagatran (**9.25**), which contains a less basic hydroxyamidino group and an esterified carboxyl group. In humans, **9.25** is rapidly absorbed and biotransformed to **9.24**, the bioavailability ranged from 18 to 24%. In December 2003, after successful clinical phase III studies, ximelagatran has been approved in France as the first orally active thrombin inhibitor for the prevention of venous thromboembolic events in major orthopaedic surgery.

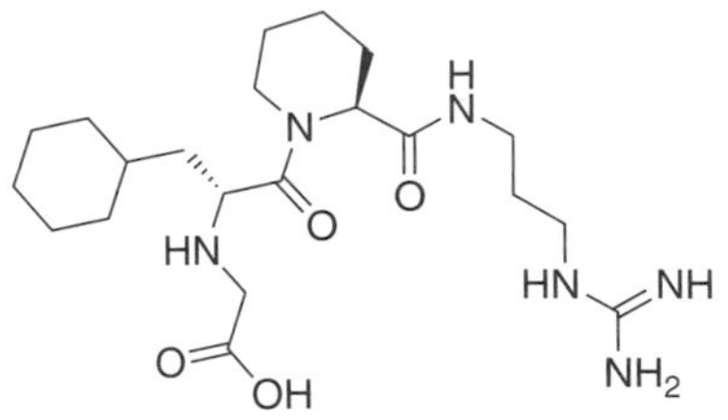

(**9.23**) Inogatran

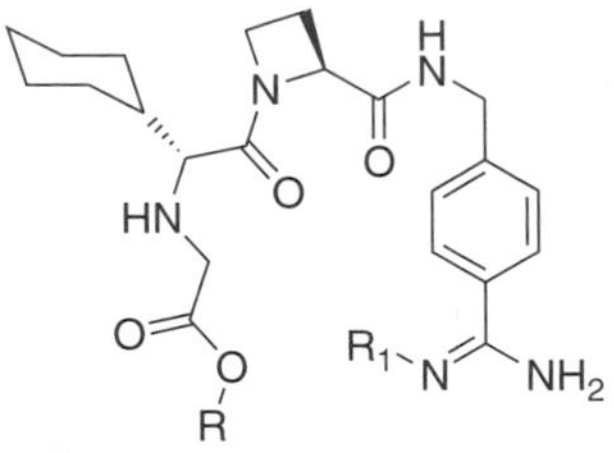

(**9.24**) Melagatran (R = H, R_1 = H)

(**9.25**) Ximelagatran (R = CH_2-CH_3, R_1 = OH)

Another type of thrombin inhibitors has been developed from synthetic Nα-arylsulfonylated arginine ester- and amide-type substrates. Argatroban (**9.26**, K_i 19 n*M*) is very selective and the only small molecular weight thrombin inhibitor used in clinical practice so far. The drug has been successfully used in humans as an antithrombotic agent instead of heparin. However, argatroban is not sufficiently absorbed after oral administration, therefore it is only used in a very limited number of patients who cannot be treated with heparin.

A very similar analogue with improved efficacy is UK-156406 (**9.27**, K_i 0.46 n*M*), in which the guanidinoalkyl side chain of Arg is replaced by a.benzamidine-moiety (see later discussion for basis of this change). **9.27** is well tolerated in rats; the oral bioavailability in dogs was found to be 45% with a plasma half-life of 48 min. In addition, data were presented that **9.27** may be potentially useful in the treatment of lung fibrosis in which activation of protease activatable receptors (thrombin receptor) has been implicated. This was one of the first reports on therapeutic *in vivo* effects of thrombin inhibitors not related to blood clotting.

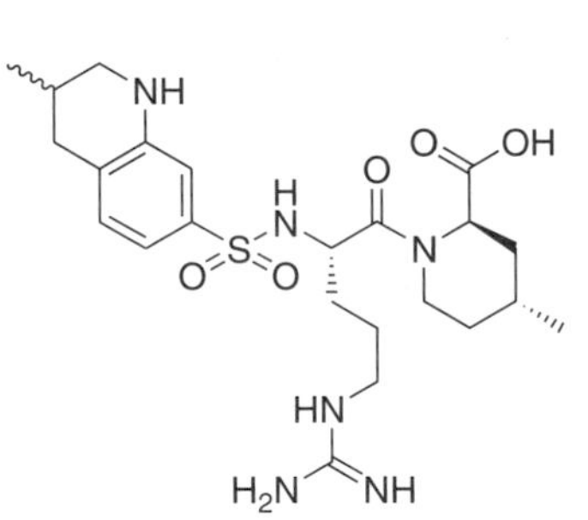

(**9.26**) Argatroban

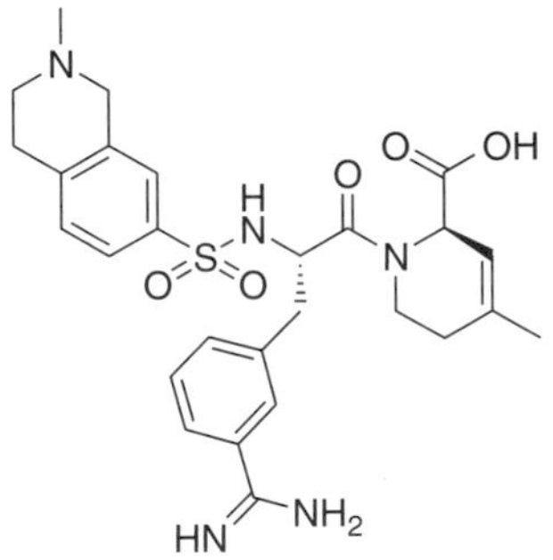

(**9.27**) UK-156406

As has already been seen in **9.24**, **9.25**, and **9.27** the conformationally constrained benzamidine group is a useful replacement for the flexible arginine side chain in different types of thrombin inhibitors. In addition to the interaction of the amidino group with Asp189 the aryl ring makes several hydrophobic contacts in the S1-binding site and therefore contributes to the high affinity of these analogues.

One of the first benzamidine-based inhibitors was 4-amidinophenylpyruvic acid (APPA, **9.28**), which is an outstanding inhibitor with respect to oral bioavailability (up to 80%). Despite its low selectivity and moderate affinity (K_i 6.5 μ*M*) anticoagulant and antithrombotic effects could be demonstrated *in vivo*. Other benzamidine derivatives, like the Nα-tosylated piperidides of 3-amidinophenylalanine (3-TAPAP, **9.29**) and of 4-amidinophenylalanine (4-TAPAP, **9.30**) had potencies and binding properties similar to those of the arginine derivative argatroban **9.26**. NAPAP

(**9.31**, Nα-naphthylsulfonyl-Gly-D-4-amidinophenylalanyl-piperidide) with a glycine spacer between the arylsulfonyl group and the D-4-amidinophenylalanine was the first thrombin inhibitor with a nanomolar K_i of 2 nM. Both anticoagulant and antithrombotic effects correspond to its pronounced antithrombin activity, however, NAPAP is not enterally absorbed and is rapidly eliminated from the circulation via a hepatic uptake.

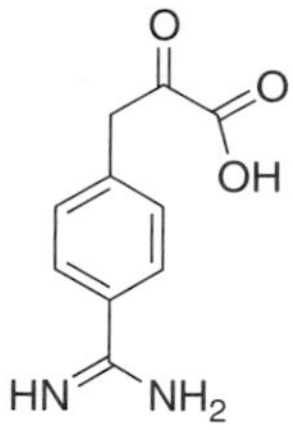

(**9.28**) APPA

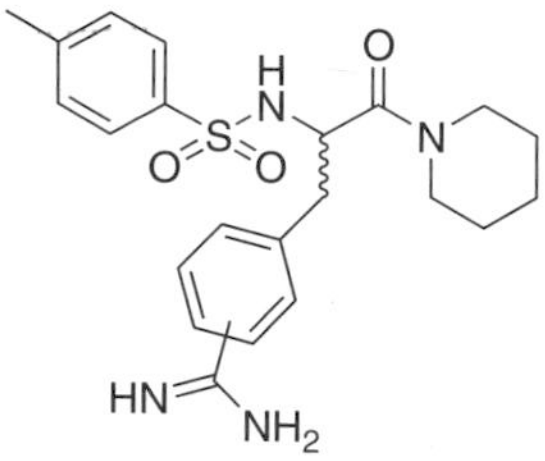

(**9.29**) 3-position, 3-TAPAP

(**9.30**) 4-position, 4-TAPAP

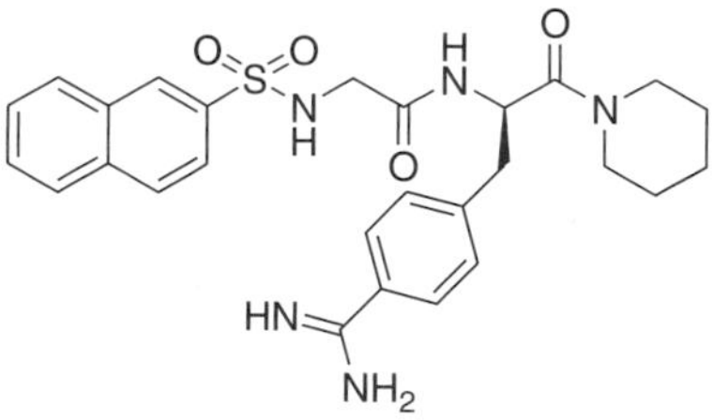

(**9.31**) NAPAP

Despite NAPAP (**9.31**) having a quite different structure compared to PPACK (**9.18**) the key segments of both inhibitors occupy identical binding sites in thrombin (Figure 9.17). For example, the naphthyl and piperidine rings in NAPAP are superimposable with the phenyl and pyrolidine rings of the P3-DPhe and P2-Pro in PPACK, the basic amidino- and guanidine-groups are located at the same position and make a strong salt bridge with Asp189 at the bottom of thrombin's S1-binding pocket.

From the x-ray crystal structure of the NAPAP–thrombin complex (Figure 9.17) it was obvious that the NAPAP molecule is bound nearly perfectly to thrombin, so that there is only limited space for additional substituents. Despite this, the NAPAP–thrombin structure was used for modeling of several derivatives. Changing Gly at P2 to Asp and replacement of the N-terminal 2-naphthylsulfonyl group by a 4-methoxy-2,3,6-trimethyl-phenylsulfonyl residue (CRC-220, **9.32**) enhanced the antithrombin potency but not bioavailability. A very potent compound was found by changing the benzamidine moiety to a 3-aminomethyl-N-(amidino)piperidine. Ro 46–6240 (**9.33** or Napsagatran) inhibits thrombin with a K_i of 0.3 nM and has outstanding selectivity, but no improved pharmacokinetic properties.

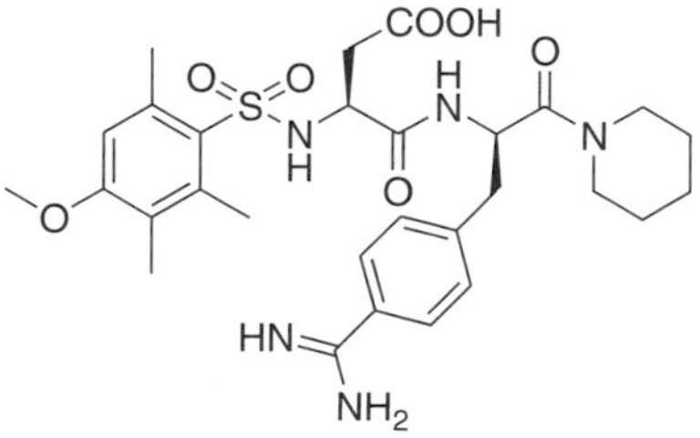

(**9.32**) CRC 220

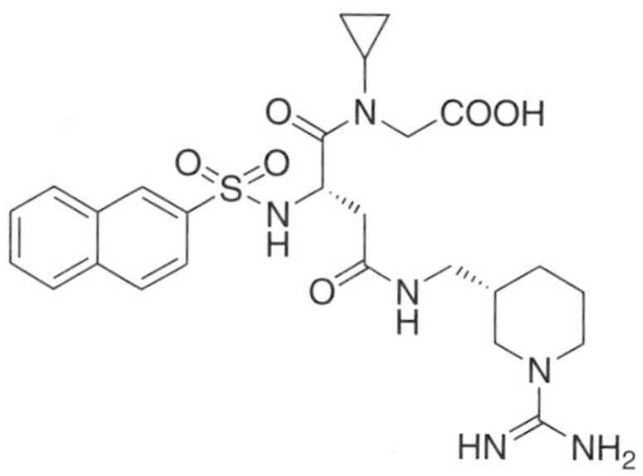

(**9.33**) Napsagatran

Nearly all of the thrombin inhibitors described above have a poor oral bioavailability due to their basic guanidino- and amidino-groups with pK_a values of ≈13 and ≈11.5, respectively. In addition to prodrugs (e.g., ximelagatran **9.25**), which contain a masked, and therefore less polar P1-function that is easily converted in their active form in blood, several new thrombin inhibitors containing less basic P1-residues have been developed in recent years based on long known lead structures.

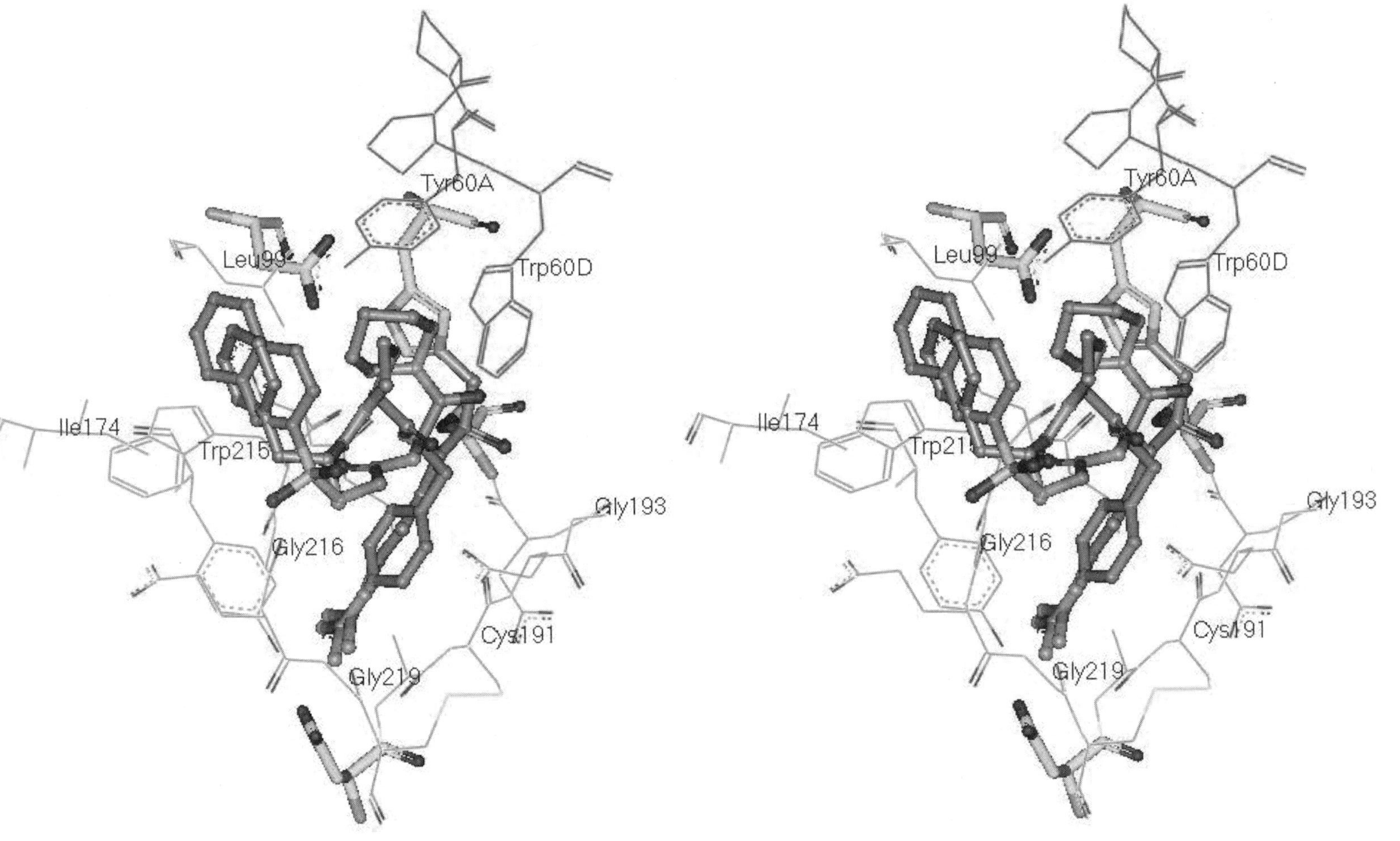

Figure 9.17 Stereo view of the active-site region of PPACK–thrombin superimposed with the experimentally determined inhibitors PPACK (**9.18**, ball and sticks with grey carbon atoms) and NAPAP (**9.31**, ball and sticks with orange carbon atoms). The residues of the catalytic triad (Ser195, His57, Asp102) and Asp189 at the bottom of the S1-binding pocket are shown as sticks with yellow carbon atoms. All other thrombin residues are shown as lines, the thrombin-specific insertion loop (Tyr60A-Pro60B-Pro60C-Trp60D) is colored in magenta. The irreversible inhibitor PPACK (**9.18**) forms a covalent, hemiketal-like bond to the thrombin residue Ser195 and alkylates His57, whereas the high potency of NAPAP (**9.31**) is based only on noncovalent interactions. Selected thrombin residues are labeled.

The DPhe-Pro-Arg structure of PPACK (**9.18**) was transformed into analogues with a C-terminal aminopyridyl structure (**9.34**, K_i 0.82 n*M*), having pK_a values between 6.2 and 6.5. In the course of work the P3–P2 segment of **9.34** was replaced by a nonpeptidic pyridinon-core in case of **9.35** (L-374,087, K_i 0.5 n*M*), which was further modified into a pyrazinon structure (**9.36**, L-375,378, K_i 0.8 n*M*). L-375,378 (**9.36**) is a very specific thrombin inhibitor, after oral treatment using an inhibitor dose of 0.5 mg/kg a maximal concentration of 1 μ*M* was observed in dogs, the half-life was prolonged to 230 min.

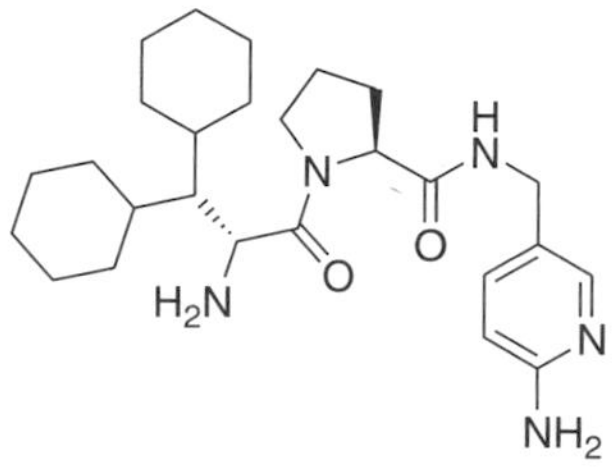

(**9.34**)

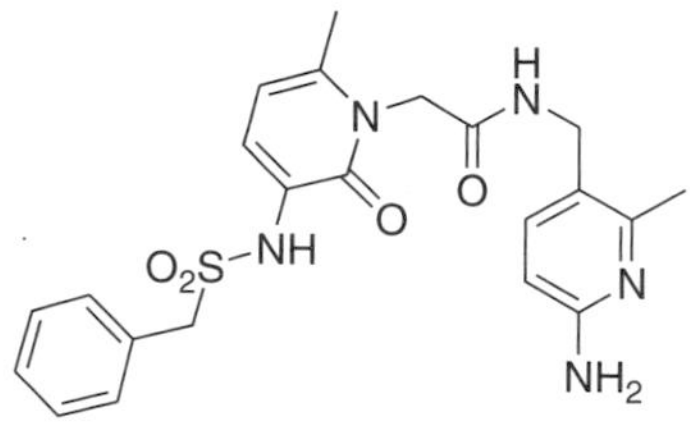

(**9.35**) L-374,087

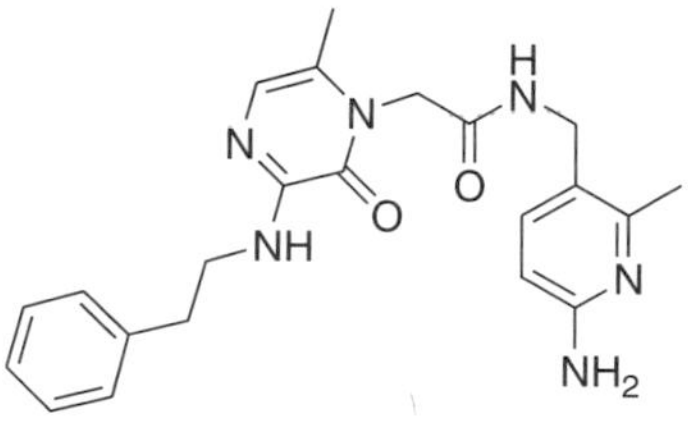

(**9.36**) L-375,378

LB-30057 (**9.37**, K_i 0.38 n*M*) is closely related to the long known inhibitor 4-TAPAP (**9.30**) but shows a significantly improved affinity and selectivity profile. The oral bioavailability in dogs after a dose of 10 mg/kg is 58% with an elimination half-life of 112 min. **9.37** is very safe; a statistically significant elevation of surgical blood loss and bleeding time was detected only at dose of 30 mg/kg, therefore **9.37** was selected for further clinical development. A significant oral bioavailability was found also for the related analogues **9.38** and **9.39**, which also contain less basic P1-amino acids.

Nowadays, high-throughput screening of available compound databases is a very useful approach in medicinal chemistry. In the case of success, completely new lead structures can be identified in this way. The lead identification is normally followed by an extensive process of lead optimization, as done also in case of the aminopyridine inhibitor **9.40** (BM 51.1081, K_i 3.2 n*M*) and the hydroxy-substituted benzothiophene compound **9.41** (K_i 0.3 n*M*). Both analogues are additional examples for very potent and selective thrombin inhibitors missing a strong basic P1-residue.

In the meantime two recombinant variants of the thrombin inhibitor hirudin, a 65 amino acid long polypeptide, have been approved for the prevention of deep venous thrombosis after hip replacement surgery and as heparin replacement for patients with heparin-induced thrombocytopenia. r-Hirudin is an extremely potent (K_i 230 f*M*) and very specific thrombin inhibitor because it exerts a concerted binding of its N-terminal and C-terminal domains to the active site and the FRE of thrombin.

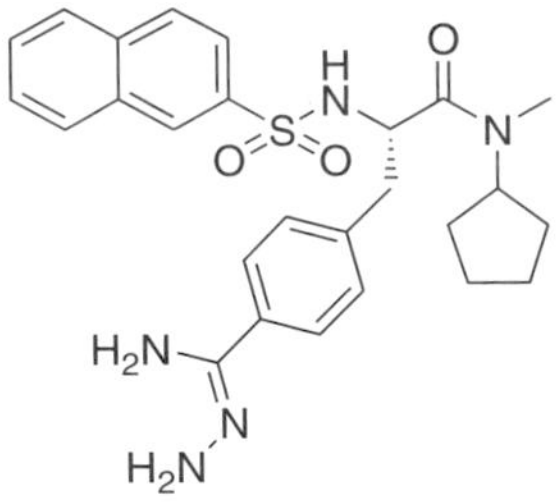

(**9.37**)LB-30057

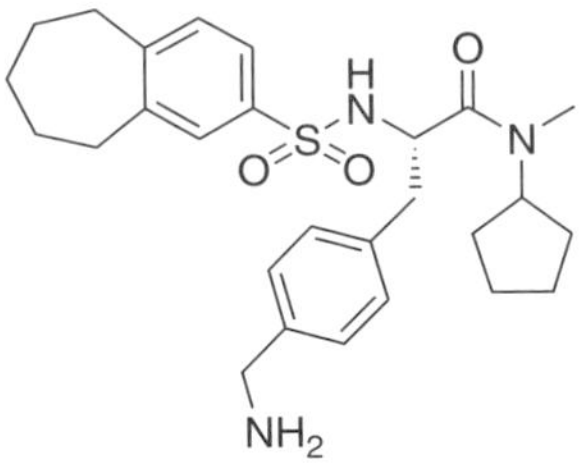

(**9.38**)

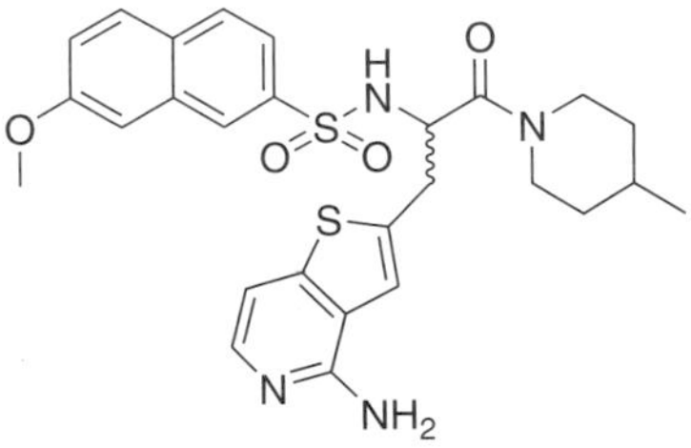

(**9.39**)

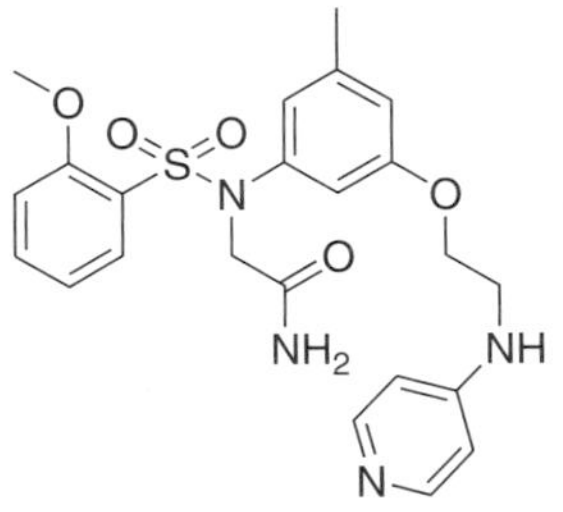

(**9.40**) BM 51.1081

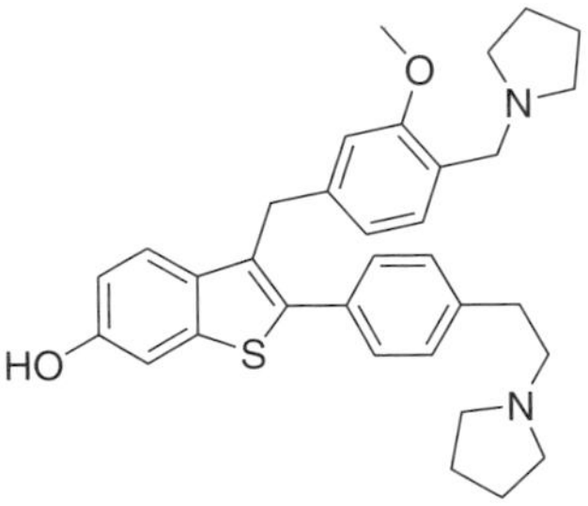

(**9.41**)

Two classes of synthetic peptides were designed after solving the x-ray crystal structure of the thrombin–hirudin complex. To the first group belong analogues of the C-terminal anionic hirudin tail (residues 55–65), which bind only to the FRE of thrombin. Therefore, the so-called hirugens (e.g., **9.42**, MDL 28,050, IC_{50} 0.15 μ*M*) inhibit thrombin-induced fibrinogen clotting in micromolar concentrations but not the cleavage of synthetic peptide substrates in the active site. Very similar to hirudin, the second class of peptidic thrombin inhibitors consist of two inhibitor parts, which simultaneously block the active site and the FRE of thrombin. The first analogue of these bifunctional inhibitors was originally named as hirulog, later as bivalirudin (**9.43**, K_i 1.9 n*M*). The active site-directed inhibitor segment of bivalirudin consists of a DPhe-Pro-Arg-Pro sequence which is connected to the hirudin sequence 53–64 by a pentaglycine linker. The P1′-Pro is important for the stabilization of the peptide bond after the P1-Arg residue. Therefore, bivalirudin can be considered as a poor thrombin substrate with inhibitory potency, which tightly binds to thrombin but is very slowly cleaved. In October 1999 bivalirudin (**9.43**) was approved for use as an anticoagulant in patients undergoing coronary angioplasty in New Zealand, in the meantime it is also approved in the US under the name Angiomax. Several highly potent analogues of bivalirudin have been developed, which contain proteolytically stable active site-directed inhibitor segments derived from argatroban (**9.44**, K_i 17 f*M*) or NAPAP (**9.45**, K_i 290 f*M*). Although these newer analogues were also very effective antithrombotics in animal models, none of them reached clinical development. The major drawback of hirudin and all of the hirulog-like oligopeptides is the lack of oral bioavailability; therefore they can be only used in the clinic.

Succinyl-Tyr-Glu-Pro-Ile-Pro-Glu-Glu-Ala-Cha-DGlu-COOH

(**9.42**) MDL 28,050

H-DPhe-Pro-Arg-Pro-(Gly)$_5$-Asn-Gly-Asp-Phe-Glu-Glu-Ile-Pro-Glu-Glu-Tyr-Leu-OH

(**9.43**) Bivalirudin

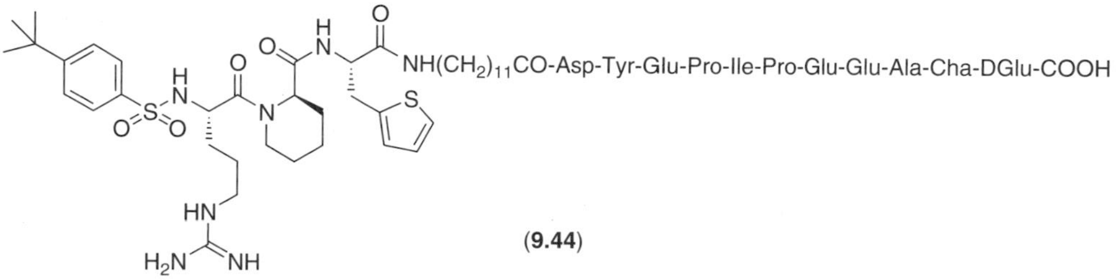

NH(CH$_2$)$_{11}$CO-Asp-Tyr-Glu-Pro-Ile-Pro-Glu-Glu-Ala-Cha-DGlu-COOH

(**9.44**)

(Gly)$_5$-Asn-Gly-Asp-Tyr-Glu-Pro-Ile-Pro-Glu-Glu-Ala-Cha-DGlu-COOH

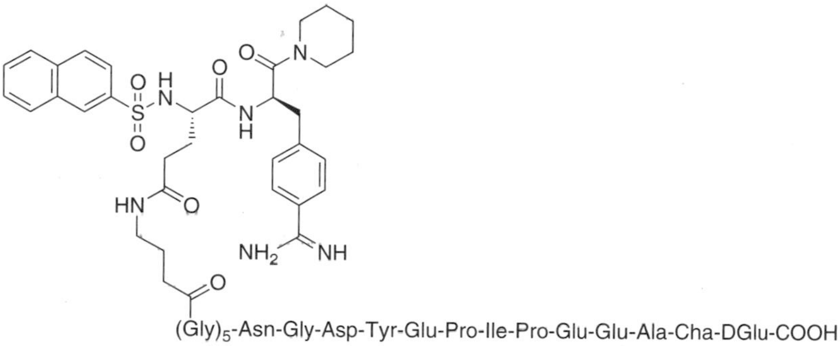

(**9.45**)

The knowledge on thrombin inhibitors accumulated so far shows that the desirable goal in the development of new analogues for therapeutic use should not only be the design of more active compounds. The work has to be focused on the improvement of the pharmacokinetic properties of known types of inhibitors, especially with regard to oral bioavailability. However, it still has to be awaited, whether the extensive clinical use of the recently launched ximelagatran can demonstrate for the first time that orally available thrombin inhibitors offer a real benefit in terms of efficacy and safety over established antithrombotic therapies.

Metalloproteases

Metalloproteases are a group of enzymes which possess a catalytically essential zinc atom at the active site. Much of the information regarding the catalytic mechanism (see Figure 9.15) of metalloproteases has been based on the active site models of carboxypeptidase A and thermolysin. Decomposition of the tetrahedral intermediate results in bond cleavage. The structures of the zinc–inhibitor complexes, determined using thermolysin–inhibitor complexes, have shown thiol (2-benzyl-3-mercaptopropanoyl-L-alanylglycinamide), carboxylate (L-benzylsuccinic acid), and phosphinyl (phosphoramidon) inhibitors to be tetrahedrally coordinated to zinc with displacement of a water molecule. The structure of the thermolysin–hydroxamic acid inhibitor complex differs in that the hydroxamate moiety (CONHOH) forms a bidentate complex with zinc through the carbonyl oxygen and the hydroxyl group, so that a penta-coordinate complex is formed between three ligands from the enzyme and two from the hydroxamate group. The active sites of the metalloproteases also contain a varying number of residues responsible for substrate recognition which differ significantly between individual metalloproteases.

Examples of metalloproteases include leucocyte collagenase, membrane metalloendopeptidase (neutral endopeptidase), and ACE. Approaches used in the design and development of inhibitors of zinc metalloproteases include leads from biproduct inhibitors (incorporating features of both cleavage products from the enzyme-catalyzed reaction); introduction of a zinc-binding ligand (e.g., thiol, hydroxamate) and modification of substrates.

Angiotensin 1-converting enzyme inhibitors

ACE cleaves a dipeptide from angiotensin I (a decapeptide) at the carboxyl terminal to generate angiotensin II (an octapeptide), which has hypertensive activity and stimulates the release of aldosterone (Equation (9.43)). ACE also catalyzes the hydrolysis of the vasodilator bradykinin (a nonapeptide). ACE inhibitors are therefore important antihypertensive agents.

$$\underset{\textbf{Angiotensin I}}{\text{Asp-Arg-Val-Tyr-Ile-His-Pro-Phe-His-Leu}} \xrightarrow{\text{ACE}} \underset{\textbf{Angiotensin II}}{\text{Asp-Arg-Val-Tyr-Ile-His-Pro-Phe}} \tag{9.43}$$

During the initial stages of development of ACE inhibitors, little was known concerning its structure and characteristics other than that it was a zinc-dependent enzyme. In the early work a model of the active site of ACE was tentatively constructed based on information from the well-known structure of carboxypeptidase A, where a zinc moiety at the active site forms a complex with the scissile amide bond of the peptide substrate and a positively charged residue binds to the negatively charged C-terminal carboxyl group of the substrate. Unlike carboxypeptidase A, which cleaves a single amino acid from the C-terminus, ACE cleaves off a dipeptide. ACE does not show specificity for C-terminal hydrophobic amino acids, indicating that its active site does not have a hydrophobic pocket corresponding to that possessed by carboxypeptidase A. Two sites, corresponding to side chain substituents of the two terminal amino acid residues of the substrate (AA_1 and AA_2, see **9.46**), were included in the model and have since been validated by structure–activity studies.

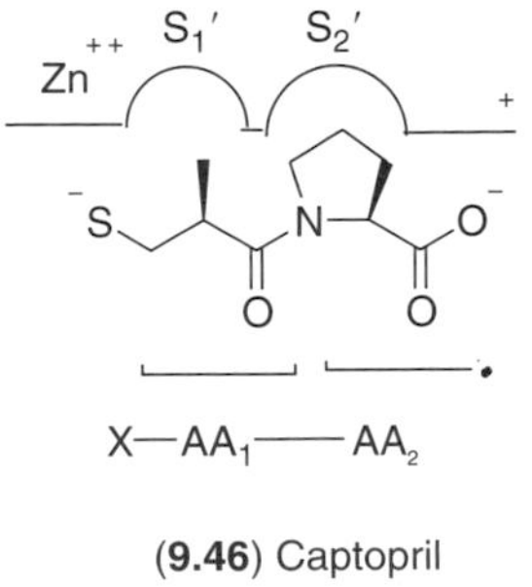

(**9.46**) Captopril

D-Benzylsuccinic acid, a biproduct inhibitor of carboxypeptidase A, was introduced into the structure of a peptide and became the starting point for the design of ACE inhibitors. Proline, the C-terminus amino acid in naturally occurring peptide inhibitors such as the nonapeptide SQ 20881 (Glu-Trp-Pro-Arg-Pro-Gln-Ile-Pro-Pro), became the natural choice as the C-terminal amino acid in synthetic inhibitors. Structural requirements for binding to the active site were determined by using simple peptide inhibitors.

Zinc-binding ligand. Succinyl-L-proline, the first biproduct inhibitor, has a weak inhibitory activity for ACE. Replacement of the carboxyl group of succinyl-L-proline by a sulfhydryl function as a zinc chelating group produced an inhibitor, captopril (**9.46**), which was 1650-fold more potent and orally active. An alkyl chain length of two carbons, between the amide carbonyl and zinc-complexing ligand, is optimum for efficient binding. Other zinc-binding ligands such as *N*-carboxylate and phosphate ions are also effective.

AA_1 substituent. The presence of an α-substituent contributes to inhibitory potency but is not essential. Methyl or benzyl substituents increase the inhibitory potency of succinyl-L-proline, but a cyclohexyl residue results in loss of potency. These substituents may contribute to increased binding, but the increase in inhibitory potency is more likely to be due to the restriction in conformation introduced at the chiral center.

AA_2 substituent. Binding is enhanced by a free C-terminal carboxyl group and esterification reduces inhibitory potency. Succinyl-L-proline is the most effective, although other C-terminal aromatic amino acids and leucine residues are also acceptable but are less efficient than proline. The superiority of L-proline is probably due to its rigid structure which may lock the carboxyl group into a favorable conformation for interaction with the positively charged residue (probably protonated arginine) at the active site of the enzyme.

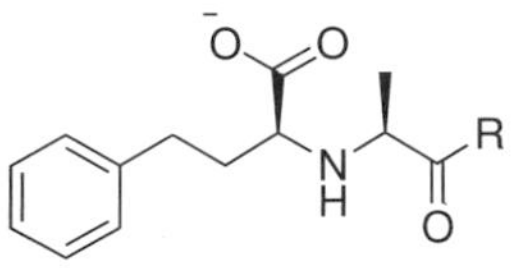

(**9.47**) Enalaprilat R=

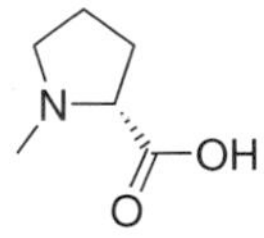

(**9.48**) Quinaprilat R =

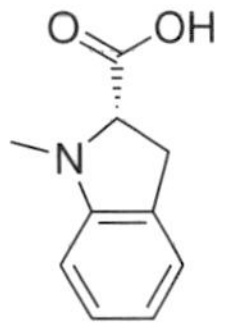

(**9.49**) Ramiprilat R =

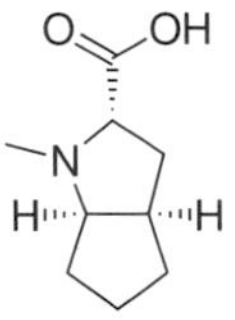

(**9.50**) Spiraprilat R =

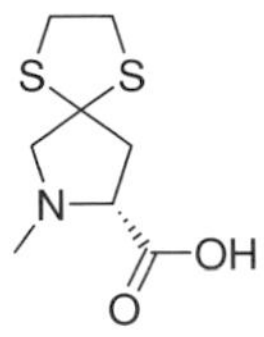

AA_1–AA_2 amide bond. The presence of an amide bond in the correct position is critical for binding of the inhibitor. Analogues which either lack the amide bond or in which the amide bond is displaced have significantly reduced inhibitory potency.

Development of other inhibitors has also been based on the concept of biproduct inhibition. Optimization of orientations of the C-terminus carboxyl and amide carbonyl for binding to the enzyme by incorporating some features of succinyl-L-proline into an Ala-Pro ''backbone'' led to the development of thiol-free inhibitors such as enalaprilat (**9.47**). This had a better side effect profile than captopril but was not active when given orally. The ethyl ester enalapril is well absorbed and is subsequently hydrolyzed *in vivo* to the active inhibitor enalaprilat. Enhancement of enzyme–inhibitor interactions by introduction of bulky hydrophobic groups at the C-terminal produced inhibitors such as quinaprilat (**9.48**), ramiprilat (**9.49**), and spiraprilat (**9.50**).

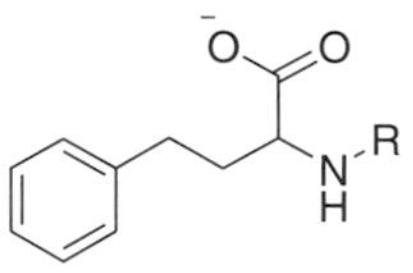

(**9.51**) Cilazaprilat R =

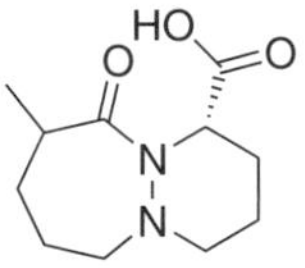

(**9.52**) Lisinopril R =

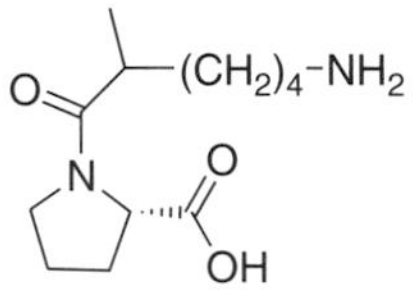

(**9.53**) Benzeprilat R =

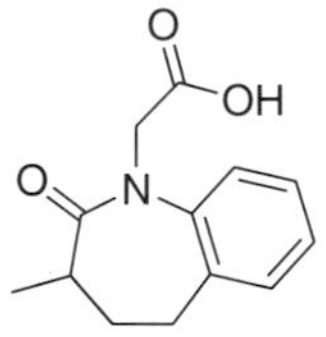

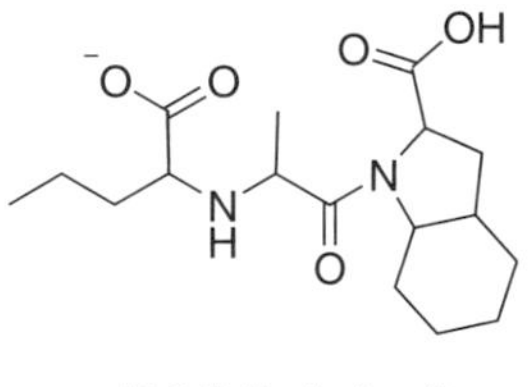

(**9.54**) Perindoprilat

A methyl substituent on the P_1' residue enhances potency of inhibitors with proline at the C-terminal (e.g., enalaprilat), and is built into compounds with large hydrophobic residues at the C-terminus such as cilazaprilat (**9.51**), lisinopril (**9.52**), benzeprilat (**9.53**), and perindoprilat (**9.54**). With these inhibitors, the larger group appears to be sufficient to fill the $S_{1'}$–$S_{2'}$ pockets and gives the right orientation to the remaining part of the molecule for interaction with the enzyme. The phosphonate-containing inhibitor, fosinoprilat (**9.55**), as well as ceranapril (**9.56**), has features of both a biproduct inhibitor and the transition state of the enzyme-catalyzed reaction. With the exception of captopril and lisinopril, all inhibitors in clinical use have to be given as ester prodrugs for oral bioavailability. Apart from their use as antihypertensives, ACE inhibitors are proving to be particularly useful as an adjunct with diuretics or digitoxin in the treatment of heart failure. They also appear to have a particular role in reducing blood pressure in patients with diabetic nephropathy.

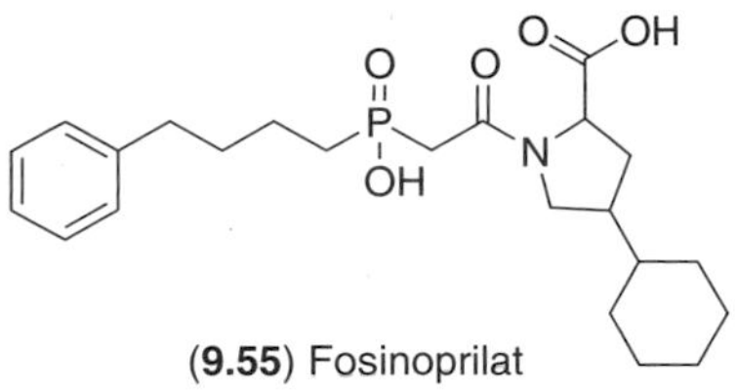

(**9.55**) Fosinoprilat

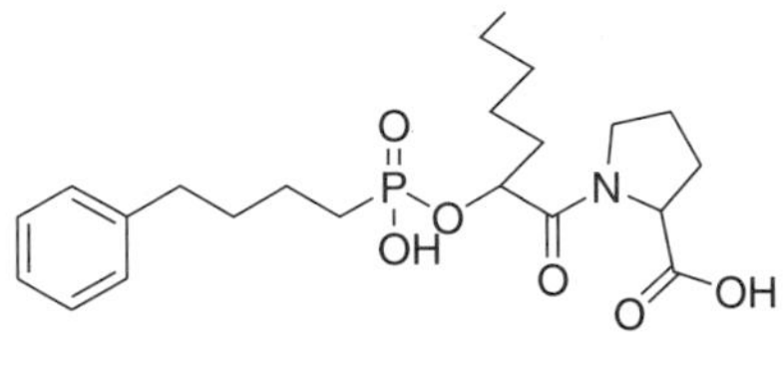

(**9.56**) Ceranapril

Captopril (**9.46**) is slightly N-selective, whereas lisinopril (**9.52**) and enalaprilat (**9.47**) are more C-selective. Recent ACE inhibitors, such as the phosphinic tetrapeptides RXPA380 and RXP407, (Figure 9.18) are 3000 and 1000-fold more C- and N-selective, respectively. C-domain selectivity is enhanced by the presence of a bulky P_1 group and a larger P_1 side chain promotes C-domain selectivity, e.g. $-(CH_2)_4-NH_2$ in lisinopril which displays greater C-selectivity than enalaprilat which contains the smaller methyl side chain. The enhanced C-domain selectivity observed with ACE inhibitors containing a bulky P_2' group, such as the tetrahydroisoquinoline group in quinaprilat (**9.48**) and the hexahydroindoline group in perindoprilat (**9.54**), was confirmed by radioligand-binding studies. The bulky methylindole P_2' substitutent of the phosphinic tetrapeptide RXPA380 would also lend weight to this requirement. The requirements for N-domain selectivity are an amidated (blocked) terminal carboxyl at P_2' and an acidic group at the P_2 position.

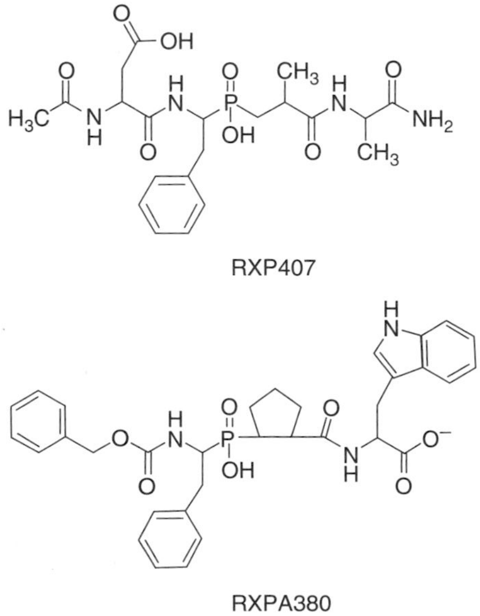

RXP407

RXPA380

Figure 9.18 C- and N-domain selective phosphinic tetrapeptide ACE inhibitors.

Membrane metalloendopeptidase inhibitors

Membrane metalloendopeptidase (MEP, also known as enkephalinase and neutral endopeptidase) is involved in the deactivation of the enkephalin pentapeptides and other peptides and hormones including atrial natriuretic peptide (ANP), Substance P, cholecystokinin, bradykinin, and chemotactic peptide. MEP cleaves the enkephalin pentapeptides (Tyr-Gly-Gly-Phe-Leu/Met) at the Gly_3–Phe_4 bond and much earlier work has focused on inhibition of enkephalin degradation (to allow a build-up of the pentapeptides *in vivo*) in search for compounds as potential nonaddictive analgesics. Although many compounds which have been developed are potent MEP inhibitors *in vitro*, the hope of a therapeutically useful analgesic remains to be realized. Attention has now shifted to another pharmacological aim. The loss of biological activity of ANP is the result of cleavage by MEP at the Cys_7–Phe_8 bond. Inhibitors which prolong the biological activity of ANP have a potential therapeutic role in the treatment of hypertension and congestive heart failure.

A model of the MEP-binding site, based on information from the earlier development of inhibitors of related zinc metalloproteases such as carboxypeptidase A, thermolysin, and ACE has been developed. Using MEP inhibitors with the general structure X-AA_1-AA_2 (X = zinc complexing ligand, AA_1 and AA_2 = amino acid corresponding to P_1' and P_2' positions, respectively, see **9.57**), the optimal amino acid requirements for S_1' and S_2' subsites, zinc complexing ligands of varying affinity together with the importance of other binding sites have been determined.

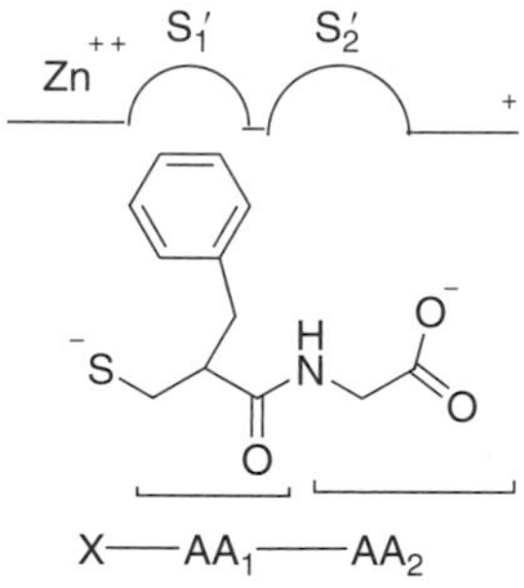

(9.57) Thiorphan

Zinc binding ligands. Suitable amino acids or short peptides containing terminal zinc liganding groups such as thiol, carboxyl, phosphoramidite, or hydroxamate all show inhibitory activity for MEP. From studies on hydroxamic acids, the position of the zinc-binding group seems critical, the optimal inhibitory potency being obtained when the zinc-binding ligand is separated by a single carbon from the chiral center of the AA_1 residue.

S_1' subsite. This site can accommodate large groups such as cyclohexyl and biphenyl moieties, but optimum inhibitory activity has been observed when benzyl is the side chain substituent on AA_1. Introduction of a methyl-, methoxy- or amino-substituent on the phenyl ring does not affect inhibitory potency but nitro- and dimethylamino substituents reduce potency. The presence, rather than the absolute configuration, of the AA_1 is important, indicating flexibility within the region of the active site containing the S_1' subsite and zinc. However, the (*S*)-isomers exhibit greater inhibitory potency.

S_2' subsite. The binding requirements of this subsite have been established using the dipeptides Phe-Y or Tyr-Y. Compounds without a side chain on AA_2 (Phe-Gly, Tyr-Gly) or substitution on the alpha carbon with a methyl group (Phe-Ala, Tyr-Ala), an aromatic or a large hydrophobic residue all show good inhibitory activity. β-Alanine or GABA in the AA_2 position also increases inhibitory activity. The (*S*)-isomer is the preferred configuration at this subsite.

Positively charged arginine residue. This binds with the C-terminal ionized carboxyl of AA_2. The presence of a free terminal carboxylate group in AA_2 therefore increases binding between the enzyme and the substrate or inhibitor. Inhibitors containing a sulfonic acid instead of a carboxy group, e.g., *m*-aminobenzenesulfonic derivatives show a high degree of inhibitory potency and selectivity for MEP.

Hydrogen bond donor group. This binds with the terminal amide (peptide) linkage. Evidence that the amide group of the peptide bond between AA_1 and AA_2 is hydrogen bonded to the active site of the enzyme comes from the observation that *N*-methylation of the peptide link in the dipeptides Phe-Gly, Phe-Ala, or Phe-Leu, leads to 100-fold reduction in inhibitory activity.

The amino acid sequence of MEP has now also been determined. It consists of 749 amino acids spanning the cell membrane and includes a 27-amino acid residue cytoplasmic domain, a 13-amino acid residue hydrophobic domain, and a large extracellular domain containing the active site. The three zinc-coordinating residues have been identified as His-583, His-587 and Glu-646, and Glu-584 as the residue involved in the acid–base catalytic mechanism occurring at the active site.

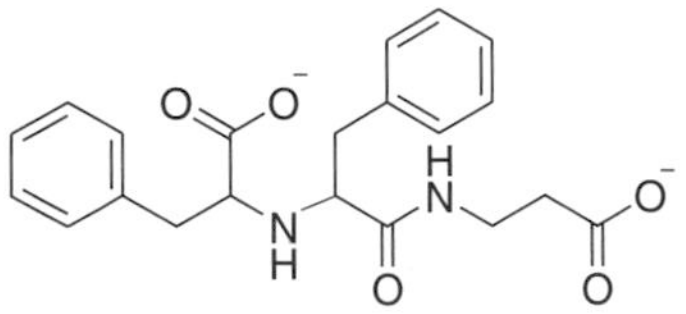

(**9.58**) SCH 32615

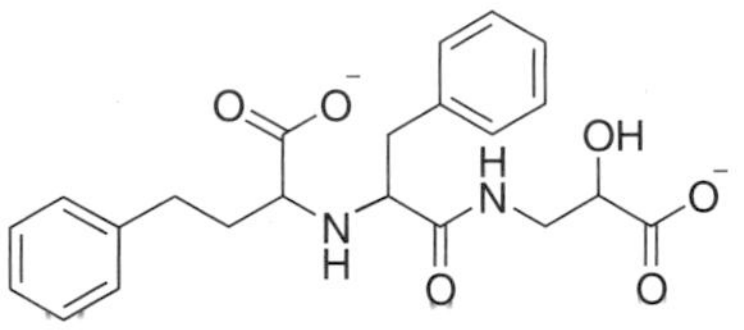

(**9.59**) SCH 39370

The use of the sulfydryl group as the zinc-binding ligand inserted into dipeptides was shown to be optimal for binding with the active site and led to the development of the first potent inhibitor of MEP, thiorphan (**9.57**, $K_i = 4.7$ n*M*). Thiorphan was also found to be a relatively efficient inhibitor of ACE ($K_i = 150$ n*M*), retroinversion of the amide bond (retrothiorphan) increased selectivity for MEP (MEP, $K_i = 6$–10 n*M*, ACE $K_i > 10$ μ*M*). The *N*-carboxyalkyl-based MEP inhibitors SCH 32615 (**9.58**) and SCH 39370 (**9.59**) were developed from concepts similar to those used in the development of *N*-carboxyalkyl ACE inhibitors. Two aromatic amino acid residues occupying the S_1 and S_1' subsites combined with β-alanine or GABA at AA_2 enhanced MEP inhibitory potency and selectivity over ACE. It has been proposed that the *N*-carboxyalkyl group, which serves to bind the zinc and the β-alanine residue, is a critical component in determining selectivity for MEP as significant ACE inhibitory activity is observed when alanine is present as AA_2. More conformationally restrained molecules, based on GABA in the AA_2 position combined with cycloleucine at AA_1 led to the development of candoxatrilat (**9.60**, UK 69578), where the (+)-enantiomer is 30-fold more potent than the (−)-enantiomer.

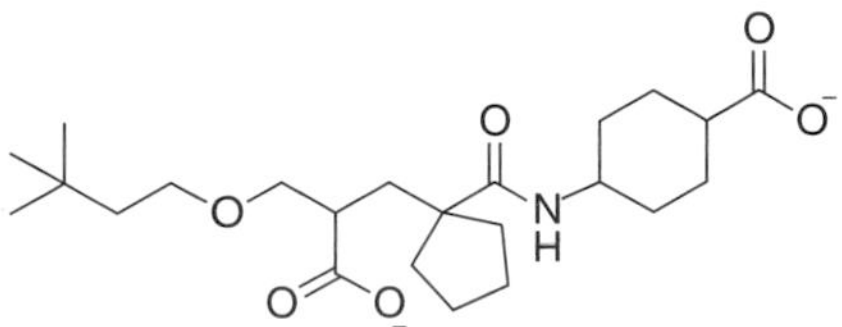

(**9.60**) UK 69578, Candoxatrilat

Phosphoramidon, a phosphoryl dipeptide of microbial origin, inhibits both thermolysin and MEP and has formed the basis for the development of specific phosphoryl inhibitors of MEP. A phosphonic acid dipeptide containing a β-alanine residue (**9.61**) has shown selectivity for MEP. *N*-Phosphonomethyl dipeptide inhibitors such as CGS 24592 (**9.62**) were based on the observation that the ACE inhibitors fosinopril (**9.55**) and ceranapril (**9.56**) tend to be longer acting than other carboxylic acid or thiol-containing analogues. It was noted that CGS 24592 (**9.62**) underwent a very slow hydrolysis in bicarbonate solution to the derivative (**9.63**), which exhibited unexpected inhibitory potency for MEP (IC_{50} = 15 n*M*). The structure represented a significant departure from other MEP inhibitors which contain a modified di- (or tri-) peptide backbone, with a critical secondary amide bond and a zinc-chelating ligand. Modification of the C-terminal carboxylic acid functionality of (**9.63**) to a tetrazole led to a highly potent, nonpeptide MEP inhibitor CGS 26303 (**9.64**).

As with ACE inhibitors, these MEP inhibitors are not well absorbed orally which limits their potential therapeutic usefulness. To improve pharmacokinetic profiles, the inhibitors have been further developed as prodrugs such as sinorphan (prodrug of (*S*)-thiorphan), SCH 34826 (a lipophilic ester of **9.57**), UK 79300 (an indanyl ester of (+)-isomer of **9.60**) and CGS 25462 and CGS 26393 (the aminomethyl phosphonate derivatives of **9.62** and **9.64**, respectively).

A different approach to improving potential therapeutic efficacy in the development of non-addictive analgesics has been the realization of combined inhibitors of more than one enzyme in a single inhibitor. Kelatorphan (**9.65**) inhibits MEP, aminopeptidase N (APN), and dipeptidylaminopeptidase, the enzymes involved in inactivation at different points of the enkephalin pentapeptides in the CNS. A variation of this approach has been the concept of covalently linking two different types of inhibitor in a ''prodrug.'' An APN inhibitor and a MEP inhibitor have been linked by a thioester or a disulfide bond in order to increase the hydrophobicity, and so absorption, of each molecule (**9.66**). Hydrolysis or reduction, respectively, leading to the release of two active inhibitors, occurs once the compound has passed the blood–brain barrier.

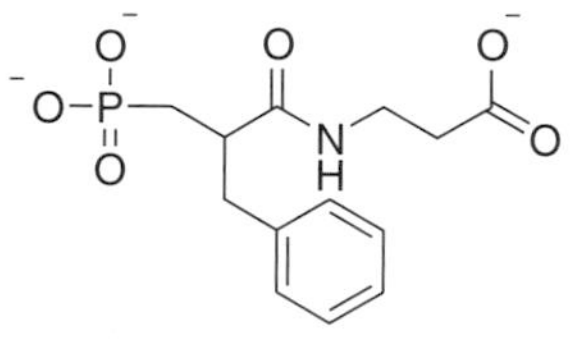

(**9.61**)

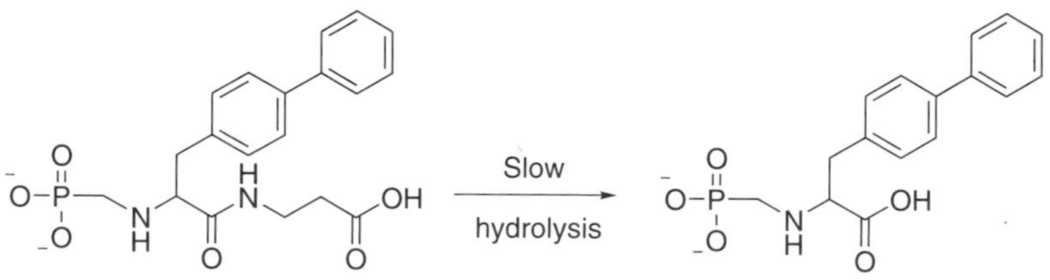

Slow hydrolysis

(**9.62**) CGS 24592

(**9.63**)

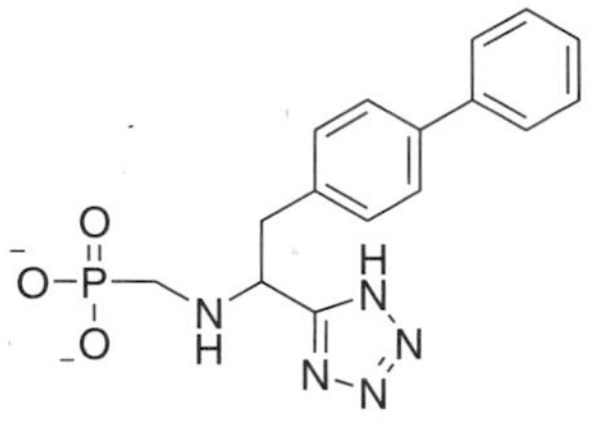

(**9.64**) CGS 26303

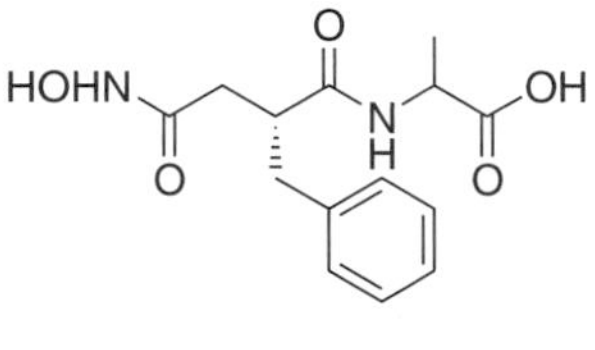

(**9.65**) Kelatorphan

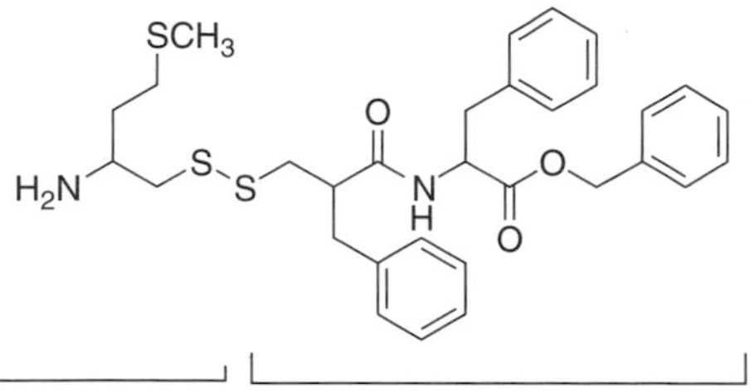

(**9.66**) APN inhibitor MEP inhibitor

Combined inhibitors, such as the mercaptoalkyl derivatives alatrioprilat (**9.67**) and glycoprilat (**9.68**) display both MEP and ACE inhibitory activity and are being assessed for their therapeutic potential in the treatment of cardiovascular diseases.

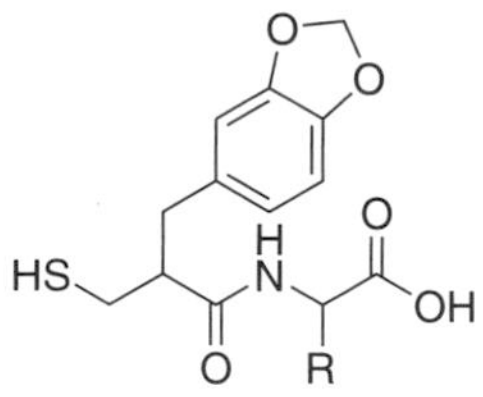

(**9.67**) R = CH_3, Alatrioprilat
(**9.68**) R = H, Glycoprilat

The promising activities displayed by these combined inhibitors, known as vasopeptidase inhibitors, has resulted in the generation of a new class of potent therapeutics, a number of which are in clinical trials (Figure 9.19). Omapatrilat, in Phase III clinical trials has been shown to have improved natriuretic and humoral effects compared with the ACE inhibitor lisinopril (**9.52**). The OCTAVE trial, which compared omapatrilat with enalapril (**9.47**) also confirmed that omapatrilat was a more effective antihypertensive agent but with a comparable side effect profile; however, other trials with omapatrilat have indicated a benefit from the mixed inhibitor approach.

Vasopeptidase inhibitor	Structure	Clinical status
Omapatrilat	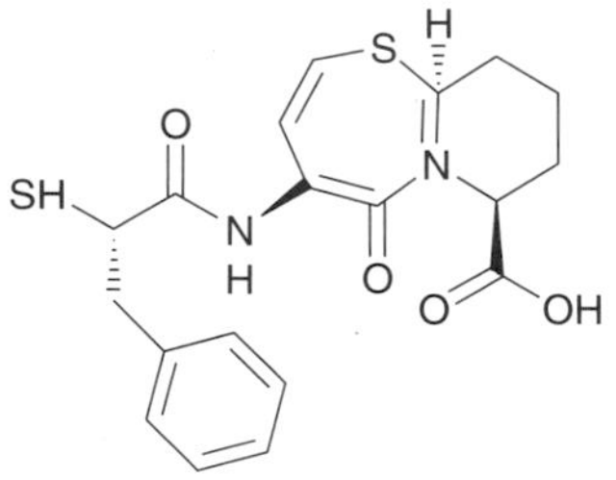	Phase III
Sampatrilat	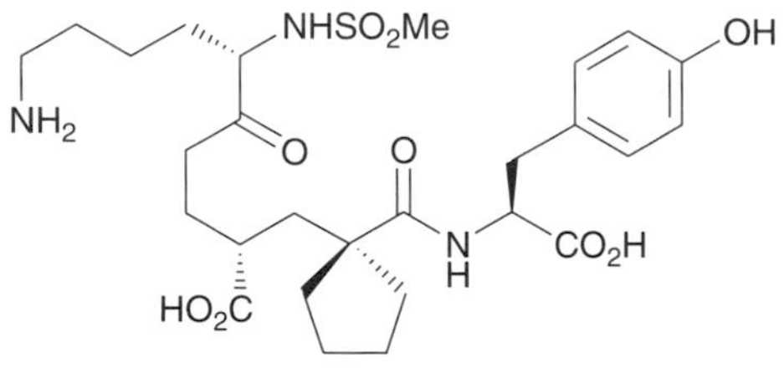	Phase II
Fasidotril	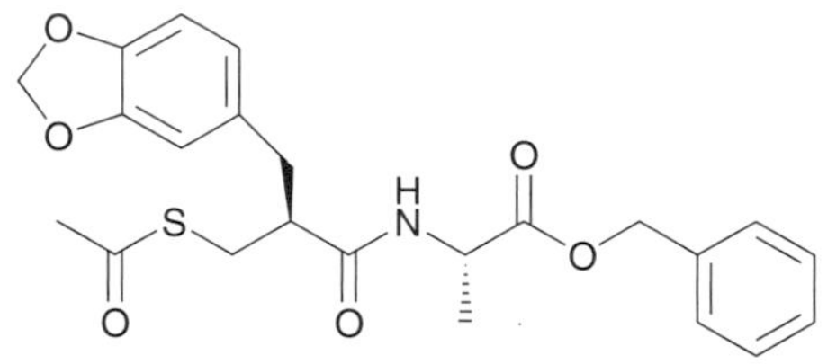	Phase II
BMS 189921	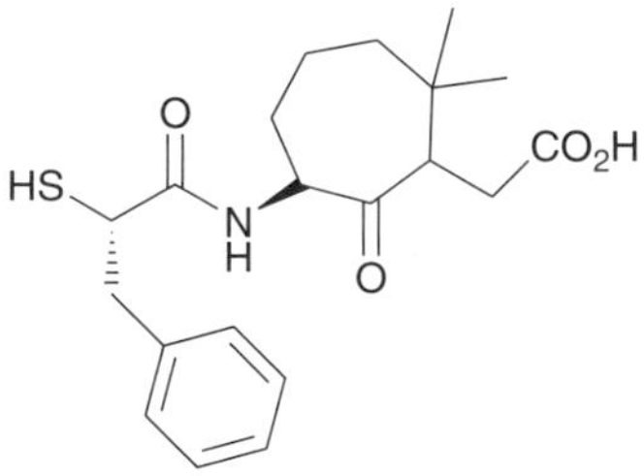	Phase II

Figure 9.19 Combined MEP/ACE inhibitors in Phase II and III clinical trials.

Matrix metalloproteinases

Matrix metalloproteinases (MMPs) are a family of enzymes that are secreted by inflammatory cells and connective tissue cells and are involved in the turnover and remodeling of extracellular matrix proteins in the normal wound healing process. This family of enzymes are capable of breaking down most components in the extracellular matrix including collagen, laminin, fibronectin, elastin, and serpin. Examples of MMPs include collagenase-3 (MMP-13), gelatinase-A (MMP-2), gelatinase-B (MMP-9), interstitial collagenase (MMP-1), matrilysin (MMP-7), metalloestalase (MMP-12), MT-MMP (MMP-14), MT2-MMP (MMP-15), MT3-MMP (MMP-16), MMT4-MMP (MMP-17), neutrophil collagenase (MMP-8), stromelysin-1 (MMP-3), stromelysin-2 (MMP-10), and stromelysin-3 (MMP-11). Under normal physiological conditions, the activity of MMPs is tightly regulated, both at the level of synthesis and secretion and extracellularly by the need for activation and through the presence of general inhibitors such as α2-macroglobulin and specific inhibitors known as tissue inhibitors of metalloproteinases (TIMPS). Overexpression and activation of MMPs has been linked with a range of diseases including arthritis, cancer, and multiple sclerosis.

Initially, the interest in developing matrix metalloproteinase inhibitors (MMPIs) arose because of the possible role of MMPs in cartilage and bone degradation in rheumatoid arthritis. Experimental models of arthritis provided evidence that cartilage destruction may result from an imbalance of MMPs over TIMPs. However, greater interest has been shown in the evidence for excessive MMP activity in a range of different tumors and a correlation between the level of MMPs and the invasiveness of the tumor. MMPIs have been developed because of the potential importance of MMPs for tumor progression and metastases. MMPs contain a zinc atom in a highly conserved active site. The majority of synthetic MMPIs are substituted peptide derivatives in which the zinc binding group is attached (Figure 9.20). The rank order of potency for the zinc binding group has been determined for inhibition of fibroblast collagenase as: hydroxamate >> formylhydroxylamine > sulfydryl > phosphinate > aminocarboxylate > carboxylate. There is considerable variation between different MMPs in the residues that line the S′ pocket. Therefore the S_1' subsite offers the greatest opportunity for selective inhibitor design. The S_2' subsite is a shallow cleft and various amino acid residues can be tolerated at this site suggesting that the S_2' subsite does not play a dominant role. The methyl group is the preferred P_3' substituent for binding at the S_3' subsite. The amide backbone is also involved in the hydrogen bonding interactions of MMP enzymes.

Both endogenous and synthetic inhibitors have been evaluated, although the endogenous MMPIs (TIMP-1 and TIMP-2) have not progressed to pharmaceutical products due to a lack of oral bioavailability. Two broad spectrum MMPIs have been developed with activity against most of the major MMPs listed above. Batimastat and marimastat (Figure 9.20) are competitive reversible inhibitors that work by mimicking the MMP substrate. Batimastat is insoluble so it is not possible to achieve therapeutic plasma concentrations after oral administration, although it can be administered

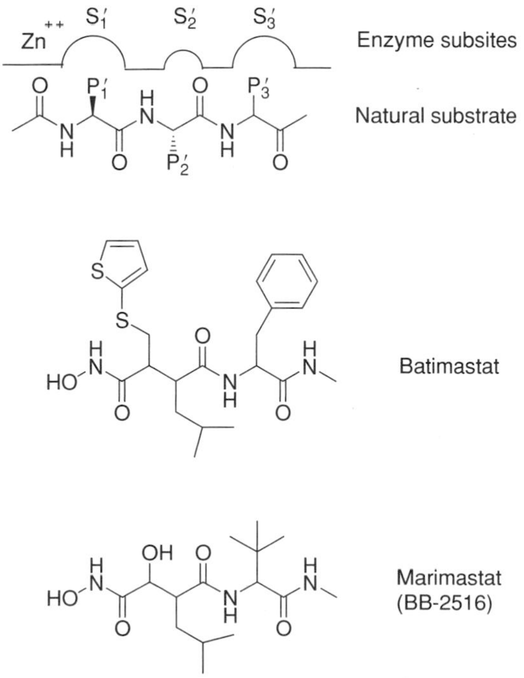

Figure 9.20 Proposed binding of Batimastat and Marimastat at MMP active site.

directly into peritoneal and pleural cavities. Marimastat has high oral bioavailability. Marimastat has been investigated in various malignant disorders, but with little success.

Aspartate proteases

HIV protease inhibitors

Two genetically distinct subtypes, HIV-1 and HIV-2, of HIV have been identified. RT inhibitors such as AZT have had limited success because of emergence of viral resistance and drug toxicity. Blockade of the virally encoded protease, which is critical for viral replication, has become a major target in the search for an effective antiviral agent and several inhibitors have been approved for use in patients.

HIV-1 protease catalyzes the conversion of a polyprotein precursor (encoded by *gag* and *pol* genes) to mature proteins needed for the production of an infectious HIV particle. A highly conserved triad, Asp-Thr(Ser)-Gly, in the viral enzyme which is also found in mammalian proteases belonging to the aspartic acid family, suggested a similar mechanistic class for HIV protease. This has now been confirmed by elucidation of the crystal structure of the native HIV protease and the HIV protease complexed with aspartyl protease inhibitors. There are however significant structural differences between the retroviral and classical aspartyl proteases such as renin. Mammalian and fungal aspartyl proteases generally are comprised of 200 amino acids and consist of two homologous domains with the key catalytic triad occurring twice.

The structure of HIV protease has been identified by x-ray crystallographic methods as a homodimer comprising of two identically folded subunits (each comprising of 99 amino acids). Each subunit contributes one of the two conserved aspartates (Asp 25 and Asp 25^1) to the single hydrophobic active site cavity. It is believed that during hydrolysis, a structural water molecule attacks the carbonyl carbon of the peptide bond of the substrate while the carbonyl oxygen accepts the proton from one of the catalytic aspartic acid residues leading to the formation of a tetrahedral transition state (Figure 9.14). Catalytic studies have suggested that in the transition state, one of the aspartic acid residues exists in the neutral form whereas the other residue is negatively charged. However, the protonation state of the protease aspartic acid residues in the complex with its inhibitors remains controversial. After the formation of the transition state, two conformationally flexible flaps (one per subunit) close around the substrate.

HIV-protease cleaves the polyprotein precursor at eight different sites, of which Tyr-Pro and Phe-Pro residues (occurring as P_1-P_1' at three of the cleavage sites of HIV-1), are of particular interest in relation to the development of inhibitors. The amide bonds N-terminal to proline are not hydrolyzed by mammalian aspartic proteases and therefore offer selectivity for the viral enzyme. Leu-Ala, Leu-Phe, Met-Met, and Phe-Leu are also found at HIV-1 cleavage sites. The amino acid sequences flanking the cleavage have been divided into three classes.

Class 1: Phe-Pro or Tyr-Pro at P_1-P_1'
Class 2: Phe-Leu at P_1-P_1' and Arg at P_4
Class 3: Gln or Glu at P_2'

Studies using oligopeptides have shown that seven residues spanning P_4-P_3' are required for specific and efficient hydrolysis of the P_1-P_1' amide bond and crystallographic data suggest multiple hydrogen bonding to the backbone of inhibitors spanning this site and close van der Waals contact for the P_3-P_3' side chains.

Incorporation of a transition-state mimic into substrate analogues has been one of the strategies used in the development of enzyme inhibitors. Substitution of the scissile amide bond with nonhydrolyzable dipeptide isosteres in the appropriate sequence context has also proved to be successful in the development of potent renin inhibitors. A number of such dipeptide isosteres (inserted into a heptapeptide template spanning P_4-P_3' and which mimic the tetrahedral intermediate of peptide hydrolysis) have been evaluated. Hydroxyethylene (**9.69**), dihydroxyethylene (**9.70**), and hydroxyethylamine (**9.71**) isosteres provide the greatest intrinsic affinity for the enzyme. The order of affinity of other isosteres for HIV-1 protease has been established as difluoroketones (**9.72**) = statine (**9.73**) > phosphinate (**9.74**) > reduced amide isostere (**9.75**). The principle structural feature in most transition state analogues designed to inhibit HIV protease is the critical hydroxyl group shown by x-ray analysis to bind both aspartic acid groups.

Many inhibitors of HIV protease have been designed and discovered but have not progressed to the clinical stage and these are discussed below to bring out the design aspects.

Pepstatine A (Iva-Val-Val-Sta-Ala-Sta), a natural product, contains two residues of the amino-acid statine. It is a nonspecific inhibitor of aspartic acid proteases and inhibits several retroviral proteases, including the hydrolysis of both polyprotein and oligopeptide substrates by HIV-1 protease. The concentration of inhibitor required to inhibit HIV-1 protease is significantly higher than those required for mammalian or fungal aspartic proteases. The structure of H-261 (**9.76**) mimics the cleavage sequence of the renin substrate angiotensinogen (Leu-Val). It is also non-specific and inhibits both HIV-1 (K_i = 5 nM) and HIV-2 (K_I = 35 nM) protease. Analogues incorporating the cyclohexyalanine-Val hydroxyethylene isostere, U-81749 (**9.77**, K_I = 70 nM) and the dihydroxyethylene isostere of cyclohexylalanine-Val, U-75875 (**9.78**, K_I < 1 nM) both show potent antiviral activity in cell cultures.

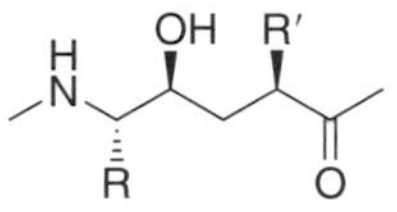

(**9.69**) Hydroxyethylene isostere

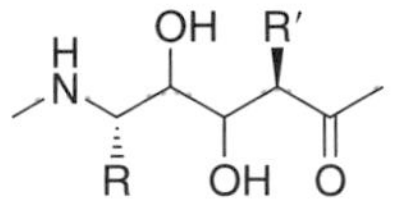

(**9.70**) Dihydroxyethylene isostere

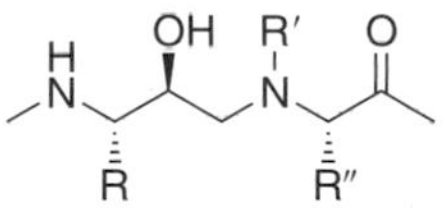

(**9.71**) Hydroxyethylamine isostere

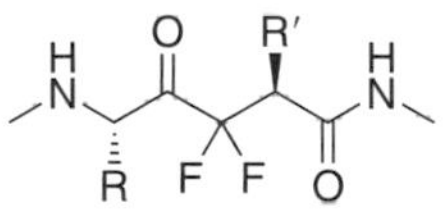

(**9.72**) Difluoroketone

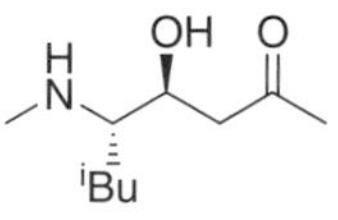

(**9.73**) Statine

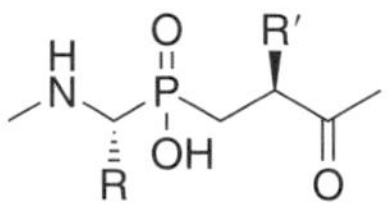

(**9.74**) Phosphinate

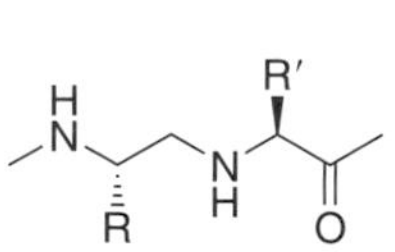

(**9.75**) Reduced amide

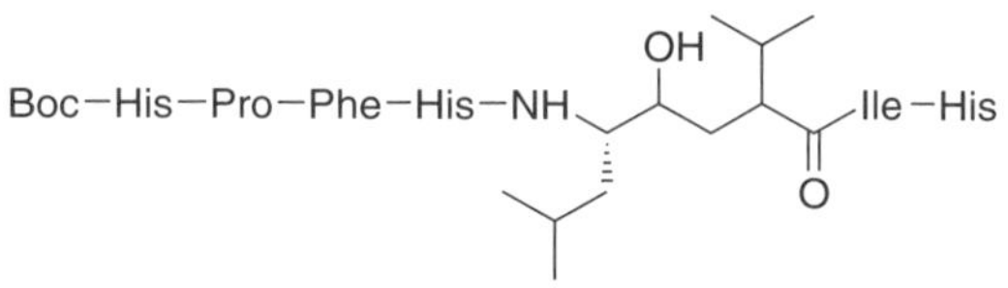

(**9.76**) H-216

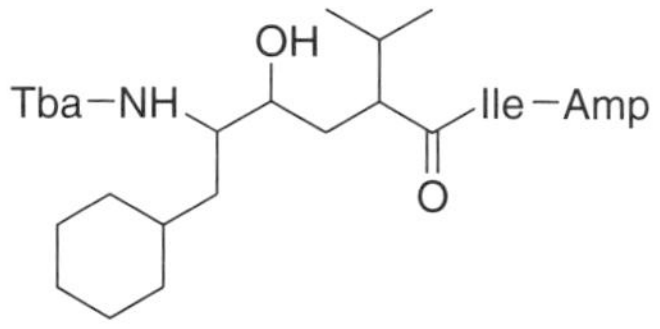

(**9.77**) U-81749

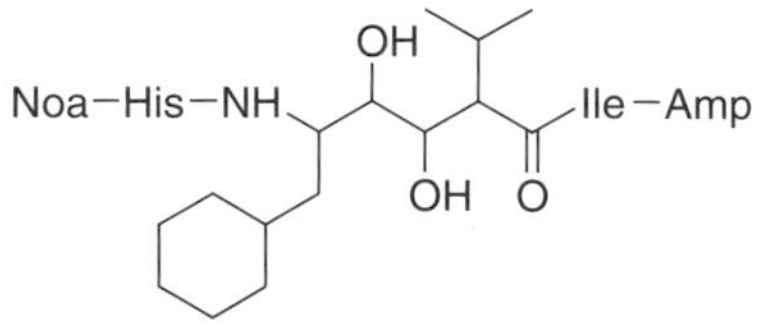

(**9.78**) U-75875

Adaption of the hydroxyethylamine dipeptide isostere to mimic the Phe-Pro site has produced inhibitors with selectivity for the retroviral protease (**9.79**, $K_I = 0.66\,\text{n}M$). Conversion of the proline to a decahydroisoquinoline nucleus has been very successful in the development of the potent selective HIV protease inhibitor saquinavir (**9.80**, $K_I < 0.12\,\text{n}M$), which has licensing approval for clinical use. Unlike compound JG-365 (**9.81**), where the crystal structure has shown a preference for the (*S*)-hydroxyl enantiomer of the isostere fragment of the molecule, the (*R*)-configuration is preferred for saquinavir (*R*-enantomer $IC_{50} = 0.4\,\text{n}M$, *S*-enantomer $IC_{50} = >100\,\text{n}M$).

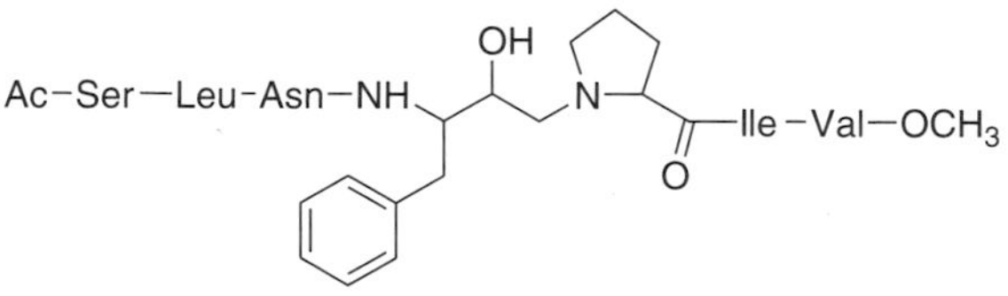

(**9.79**)

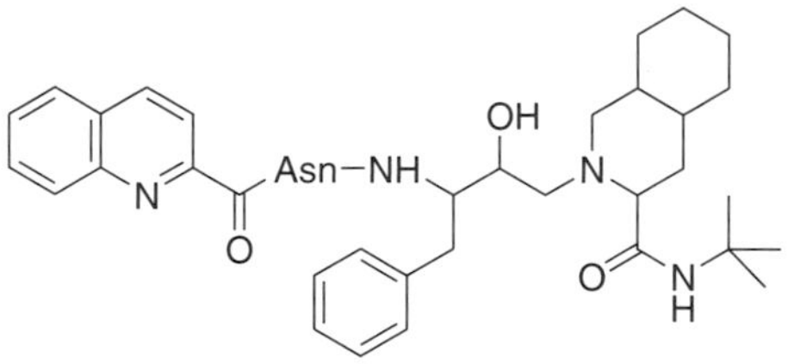

(**9.80**) Saquinavir

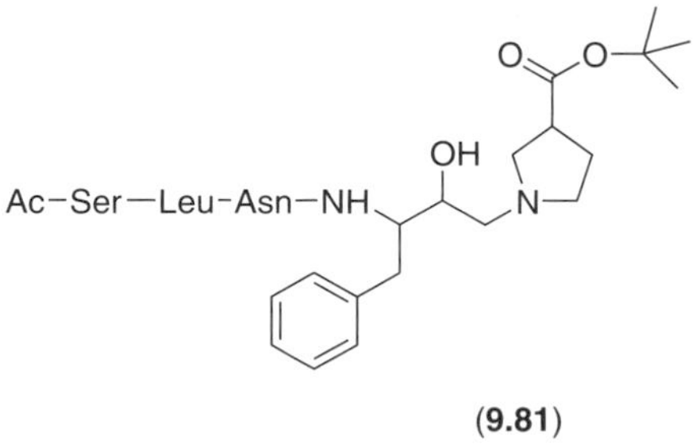

(**9.81**)

X-ray crystallography studies have shown that the hydroxyl group is located between the aspartic acids in both JG-365 and saquinavir, but the adjacent methylene groups fit in a different manner into the active site.

SC 52151 (**9.82**, $IC_{50} = 6.3$ n*M*), based on hydroxyethylurea isostere has oral bioavailability. L 735 524 (**9.83**, $IC_{50} = 0.36$ n*M*), which is a combination of a hydroxyethylene isostere and a hydroxyethylamine isostere, is also orally active. The sulfonamido moiety, in the novel (*R*)-hydroxyethyl sulfonamides isostere (**9.84**), has also been used to replace the $P_1'P_2'$ amide linkage of the inhibitor (**9.85**, $K_I = 1$ n*M*).

Symmetrical inhibitors (**9.86**, $IC_{50} = 0.2$ n*M*; **9.87**, $K_I = 0.8$ μ*M*) capitalize on the unique symmetry of the homodimeric enzyme. Unlike transition-state analogues, the stereochemistry of the two hydroxyl groups is not significant. Modifications to evaluate the effect of polar heterocyclic end groups led to the nonsymmetrical inhibitor A77003 (**9.88**, $IC_{50} < 1$ n*M*). Improved oral bioavailability was obtained with A 80987 (**9.89**, $K_I = 0.25$ n*M*) where the methylamide groups had been replaced by esters.

Penicillin-derived symmetrical dimers (**9.90**) have been identified as good lead structures from screening programs and symmetrical cyclic dihydroxy ureas (**9.91, 9.92**) have played a large part in experimental studies on resistance. The antipsychotic agent haloperidol (**9.93**, $K_I = 100$ μ*M*) was identified as a weak inhibitor through a computational search of a structural database based on a complementary shape of the HIV-1 protease active site. The 1,3-dithiolane analogue (**9.94**, $K = 15$ μ*M*) exhibits greater inhibitory potency.

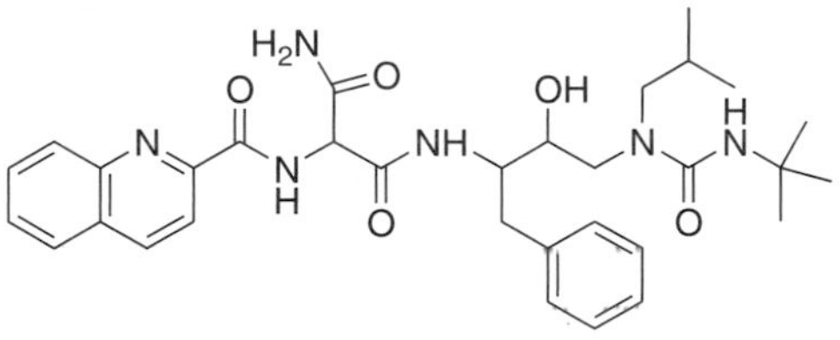

(**9.82**) SC52151

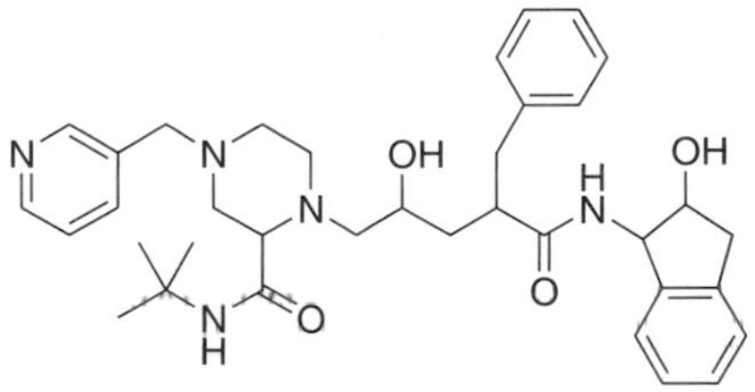

(**9.83**) L735524

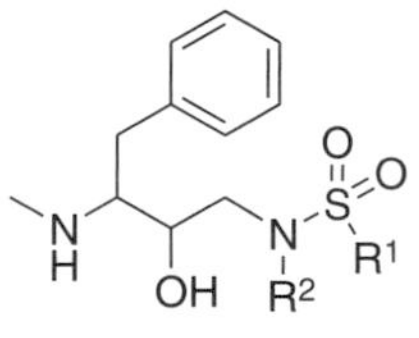

(**9.84**)

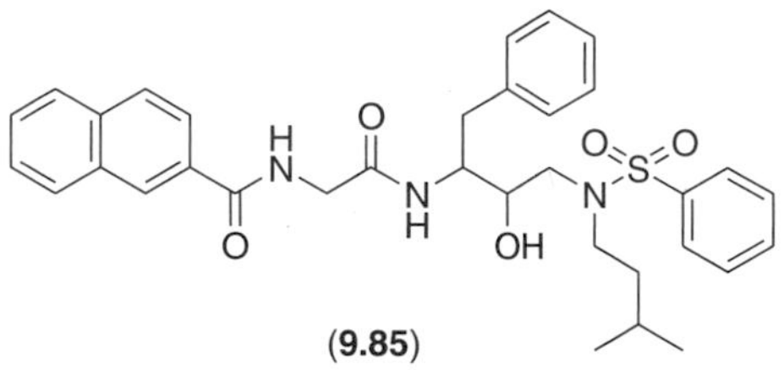

(**9.85**)

Of the many HIV protease inhibitors described to date only five, saquinavir (Invirase/Fortovase), ritonavir (Norvir), indinavir (Crixivan), nelfinavir (Viracept), and amprenavir (Agenerase) (Figure 9.21) have been approved for use in the treatment of HIV. The problem in the treatment of HIV by inhibition of the RT or the protease is the development of resistance due to the ability of the virus to mutate as a result of a rapid rate of replication and the high error rate of RT (one error per HIV replication cycle). Variants occurring as a minor population together with the wild-type strain (i.e., unaltered form) in patients encode mutant proteases which have a reduced affinity for inhibitors but are sufficiently enzymatically active to produce viral precursors having the ability

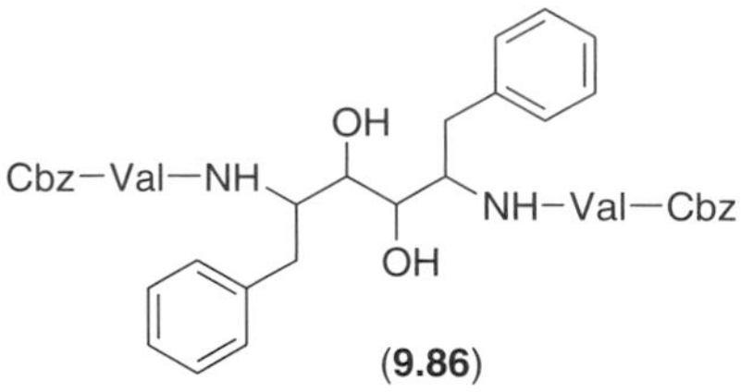

(**9.86**)

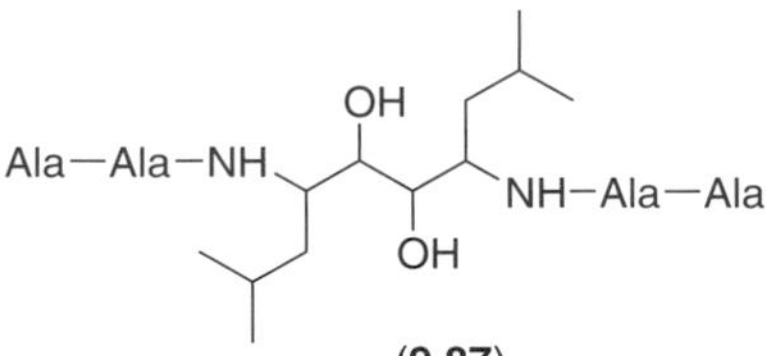

(**9.87**)

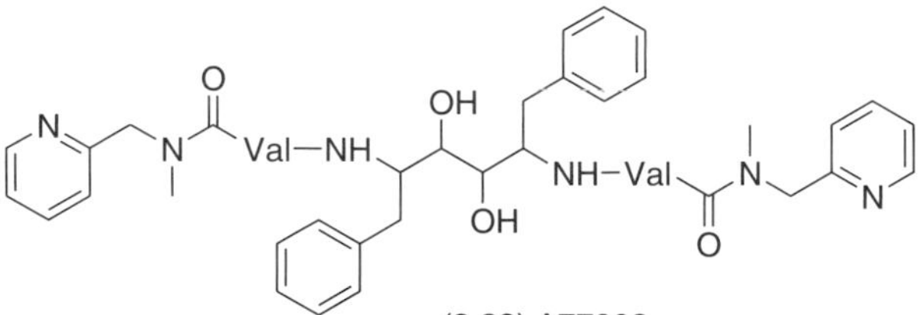

(**9.88**) A77003

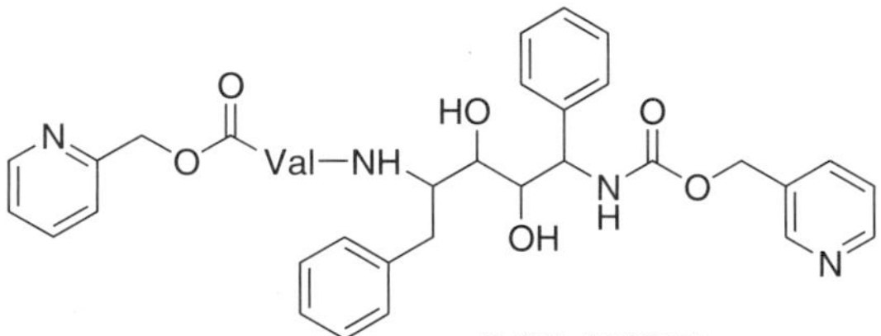

(**9.89**) A80987

to continue HIV replication where the wild type is partially suppressed. Combinations of RT/protease inhibitors have been used in an attempt to prevent the virus acquiring resistant mutations. The need for complete suppression of the disease became clear with the discovery of a latently infected reservoir of HIV in some patients.

Examination, by sequence analysis and in some cases x-ray crystallography, of the HIV protease from drug-resistant HIV strains has shown 20 amino acid substitutions in the active site, flap pivot point and elsewhere, half of which by being located in the active site reduce binding of the inhibitor as a result of reduction of van der Waals contacts, increased steric hindrance, and generation of unfavorable charge interactions. Elsewhere the mutations may compensate for the effect of the inhibitor in reducing enzyme activity by increasing the catalytic efficiency of the active site through conformational changes.

New design strategies have originated in an attempt to counter the effects of mutations in the active site on inhibitor binding. Inhibitors that interact with residues that are involved in substrate binding or the catalytic process are less likely to be affected by mutation in these areas since the variant protease could be catalytically impaired or inactive. Many potent inhibitors hydrogen bond to the catalytic Asp 25 and Asp 25′ and Ile 50 and Ile 50′ on the flaps via the structural water so taking advantage of this situation. Asymmetric compounds would avoid interaction with the double mutant symmetrically positioned in the homodimer to a lesser degree than symmetrical compounds. Increase in the flexibility of an inhibitor usually reduces the potency but here with variants the movement of inhibitor-binding residues to other enzyme-binding areas may reduce loss of binding affinity. Increase in the size of an inhibitor and thus increase in the number of interactions between inhibitor and enzyme could minimize the overall loss of binding affinity due to mutations as à consequence of any reduction in binding contributing only a small amount to the overall binding energy of the inhibitor.

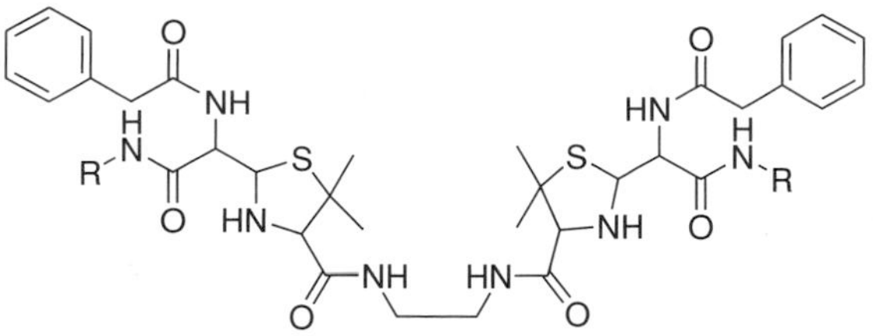

(**9.90**)

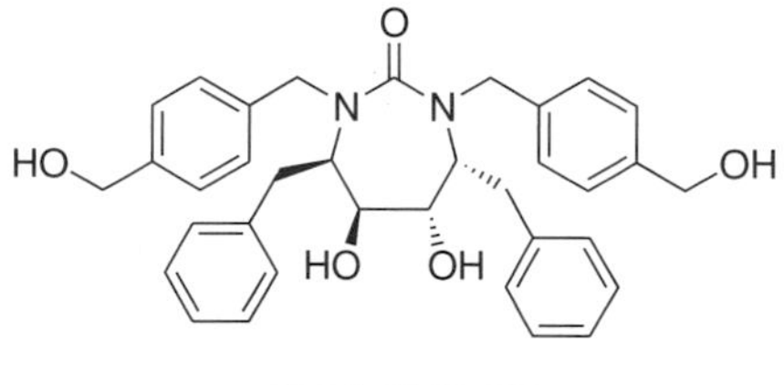

(**9.91**) DMP-323

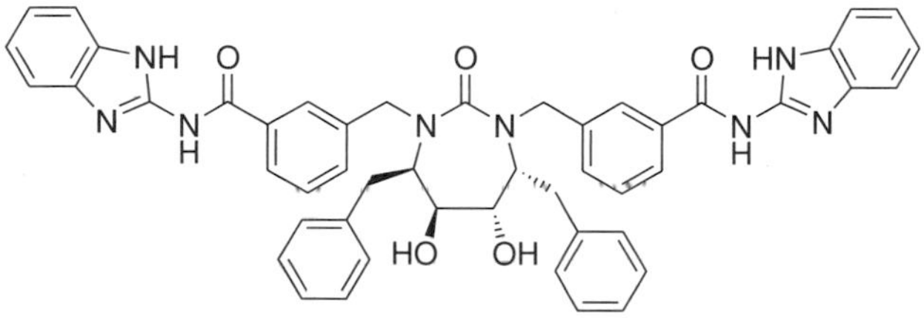

(**9.92**) SD-146

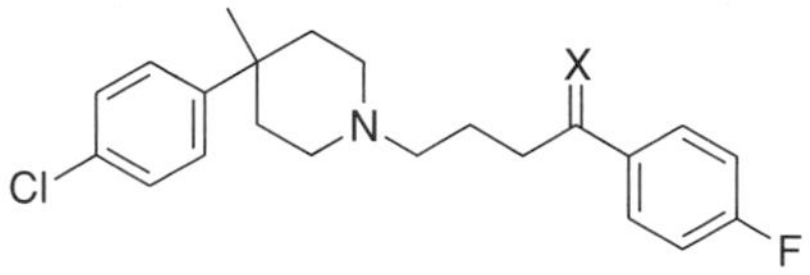

(**9.93**) X = O : Haloperidol

(**9.94**) X =

The currently available inhibitors are potent, asymmetric, flexible inhibitors that interact with the catalytic aspartates and flaps but have a similar linearity, substituents (OH, benzyl group at P1 or P1′) and subsite binding so that different mutants of the protease emerging are unlikely to show a decrease in resistance to any particular combination of these protease inhibitors. The design of new structural types is required so that by combination of these with existing protease inhibitors cross-resistance is eliminated due to different binding profiles on the enzyme and a wider selection of mutant variants eliminated from viral replication.

Saquinavir
(Hoffmann-La Roche)

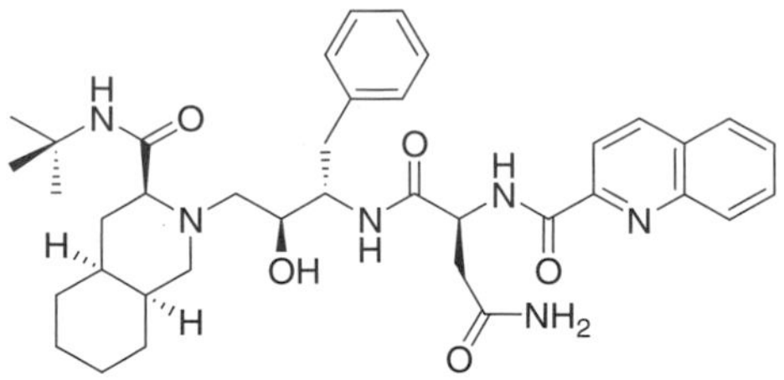

Indinavir
(Merck)

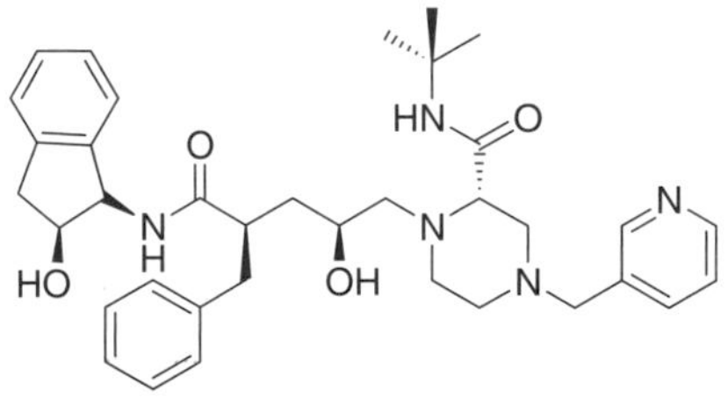

Ritonavir
(Abbott Laboratories)

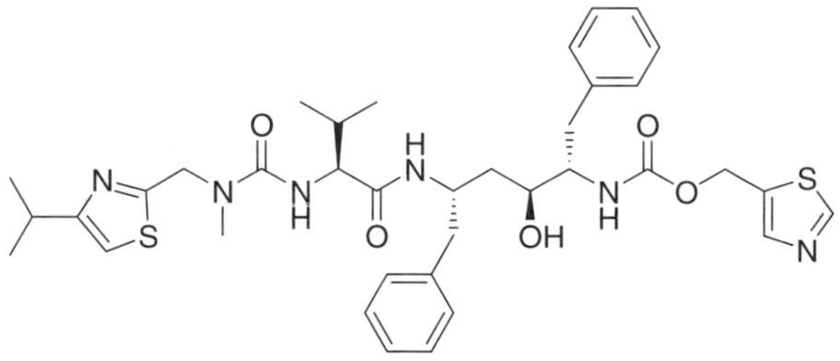

Nefinavir
(Agouron)

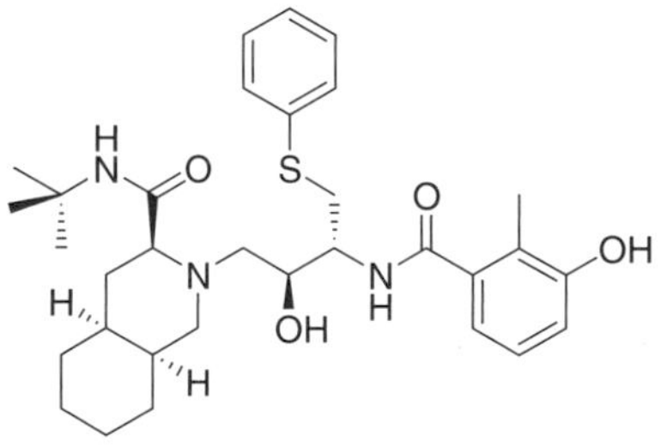

Amprenavir
(Vertex/GlaxoWellcome)

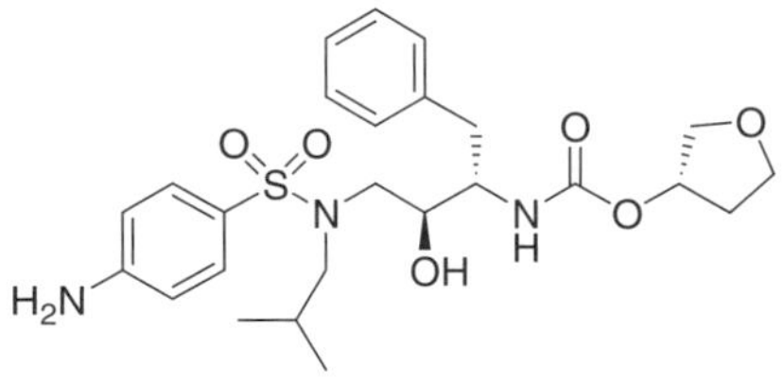

Figure 9.21 HIV protease inhibitors approved for the treatment of HIV.

9.7.2 Acetylcholinesterase Inhibitors

Acetylcholine is the chemical transmitter released at the nerve endings in the parasympathetic and motor nervous systems following a nervous impulse. After a response from the tissue the acetylcholine is removed by hydrolysis to inert products by acetylcholinesterase (see Equation (9.44)) in the proximity. Inhibitors of acetylcholinesterase allow a build-up of acetylcholine at the nerve endings so that a more prolonged effect is produced which is useful in the treatment of myasthenia gravis, a disease associated with the rapid fatigue of muscles, as well as in the treatment of glaucoma where stimulation of the ciliary body improves drainage from the eye and decreases intraocular pressure. A more recent potential use has been in the treatment of AD and senile dementia of the Alzheimer's type (SDAT).

$$E + H_3C-\overset{O}{\overset{\|}{C}}OCH_2CH_2N^+(CH_3)_3 \rightleftharpoons \underset{\text{Complex}}{E\text{-}(CH_3-\overset{O}{\overset{\|}{C}}OCH_2CH_2N^+(CH_3)_3)} \longrightarrow \underset{\text{Acyl enzyme}}{E-\overset{O}{\overset{\|}{C}}-CH_3} \xrightarrow{H_2O} E + CH_3COOH \quad (9.44)$$

$$+\ HO-CH_2CH_2N^+(CH_3)_3$$

Inhibitors of acetylcholinesterase fall into two groups: the reversible carbamate inhibitors such as eserine (physostigmine (**9.95**), neostigmine (**9.96**), and benzylpyrinium (**9.97**)) and the irreversible organophosphorous inhibitors, dyflos (**9.98**) and ecothiopate (**9.99**).

$N(CH_3)_3^+$ Br^- — $-O-C(=O)-N(CH_3)_2$

(**9.96**) Neostigmine

$N^+-CH_2-C_6H_5$ Br^- — $-O-C(=O)-N(CH_3)_2$

(**9.97**) Benzylpyrinium

The carbamates carry a positive charge and are bound at the anionic site (carboxylate ion) of the enzyme and correctly positioned to form a carbamyl enzyme with the serine hydroxyl group at the esteratic site (see Equation (9.45)). The carbamyl enzyme is only slowly decomposed ($t_{1/2}$ = ~20 min) and in the presence of excess inhibitor the enzyme is partially locked up in this form so that its activity towards the substrate acetylcholine is decreased. Dilution or removal of excess inhibitor leads to a shift in the steady-state inhibition level with an increase in activity of the enzyme.

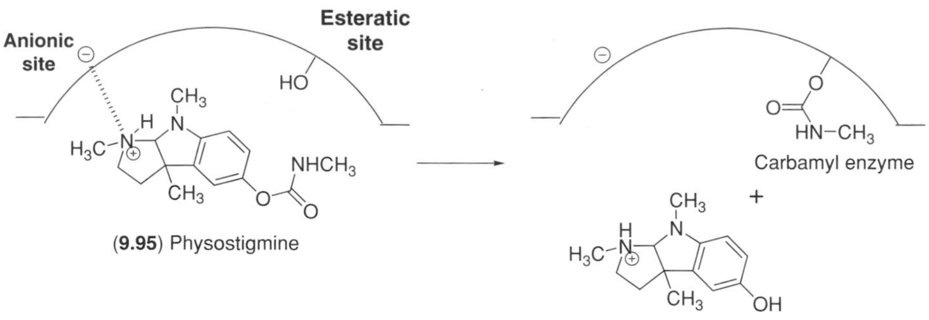

Anionic site
Esteratic site

(**9.95**) Physostigmine

Carbamyl enzyme (9.45)

The organophosphorus compounds rapidly react with the enzyme to form a stable phosphoryl enzyme and the enzyme is irreversibly inhibited (see Equation (9.46)).

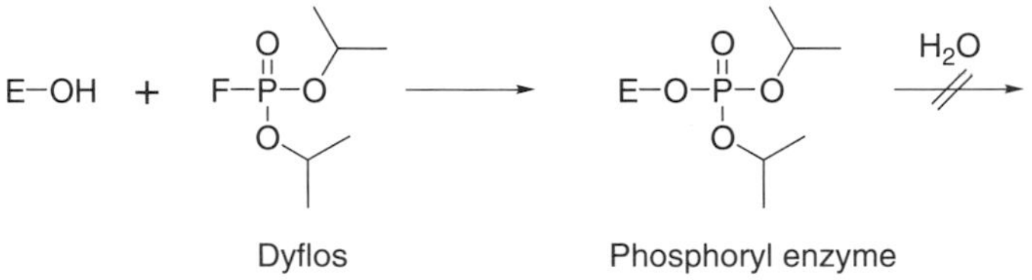

$$E{-}OH + \text{Dyflos} \longrightarrow \text{Phosphoryl enzyme} \overset{H_2O}{\not\longrightarrow} \qquad (9.46)$$

Dyflos   Phosphoryl enzyme

$R_2{-}P(=O)(R){-}R_1$

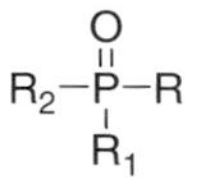

(**9.98**) $R = R_1 = -OCH(CH_3)_2$, $R_2 = F$; dyflos
(**9.99**) $R = R_1 = -OCH_2CH_3$, $R_2 = -S{-}CH_2{-}CH_2{-}N^+Me_3$; ecothiopate
(**9.100**) $R = -CH_3$, $R_1 = -OCH(CH_3)_2$, $R_2 = F$; sarin
(**9.101**) $R = -OCH_2CH_3$, $R_1 = N(CH_3)_2$, $R_2 = -CN$; tabun

The organophosphorus compounds have a long duration of action in the body after a single dose of the drug and enzyme activity only returns after synthesis of fresh enzyme. Due to dangers of overdosage, as well as handling, they are little used except for treatment of glaucoma where the other less toxic cabamate drugs have not proved satisfactory in a particular therapy.

Volatile organophosphorus compounds such as sarin (**9.100**) and tabun (**9.101**) have been prepared for use as nerve gases in war and other less volatile compounds have been used as insecticides for the spraying of crops. Inhibition of the mammalian or insect enzyme leads to a build-up of acetylcholine and death from accumulated acetylcholine poisoning.

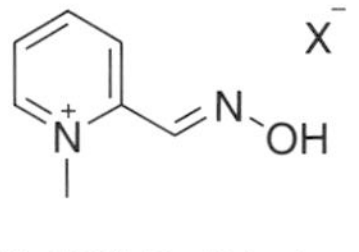

(**9.102**) Pralidoxime

Much research has been carried out to find antidotes, for nerve gas poisoning, which could be distributed to the population in the event of war. One of these discoveries, pyridine-2-aldoxime mesylate (pralidoxime (**9.102**)) has been successfully used, in conjunction with atropine to block the action of acetylcholine on receptors, in the treatment of accidental poisoning during crop spraying. Pralidoxime is considered to complex at the anionic site where it is firmly held by electrostatic attraction in the correct spacial configuration for attack by the oxime anion on the phosphorus atom with displacement of the inhibitor residue from the enzyme.

There is evidence that AD and SDAT are associated with dysfunction of normal cholinergic neurotransmission in the brain leading to learning and memory deficiencies. Examination of patients with these diseases has shown reduced levels of ChAT (acetyl-Co A: choline *O*-transferase), acetylcholinesterase and the muscarinic receptor sub type M_1. ChAT is responsible for the synthesis of acetylcholine in the cerebral cortex and it has been postulated that by inhibiting acetylcholinesterase in the brain the associated build-up of acetylcholine will enhance the cognitive function. Several acetylcholinesterase inhibitors, capable of penetrating into the CNS (i.e., not quaternary compounds), have been introduced into trials/clinic as one of few treatments to date for mild or moderately severe AD.

Tacrine (Cognex, tetrahydroaminoacridine, **9.103**) was one of the first acetylcholinesterase inhibitors to be used for AD and has a noncompetitive action. It showed modest efficacy in the disease but its usefulness was limited by its frequent dosing, the cholinergic side effects that occur to varying degrees (gastrointestinal symptoms, sweating, bradycardia) with all inhibitors of the enzyme and also by a specific, reversible hepatotoxicity; in general it has been replaced by more recently discovered inhibitors.

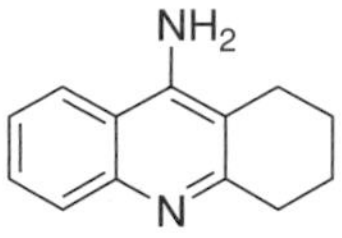

(**9.103**) Tacrine

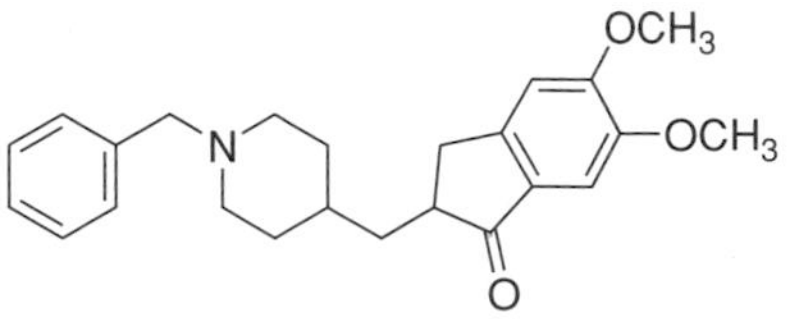

(**9.104**) Donepezil

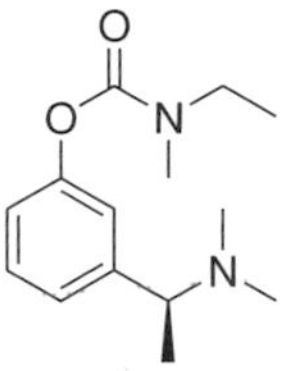

(**9.105**) Rivastigmine

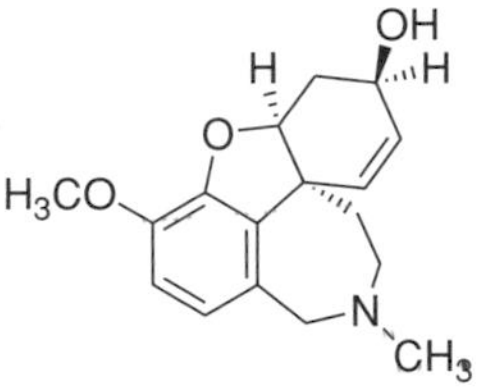

(**9.106**) Galantamine

Donepezil (**9.104**, Aricept) introduced for the symptomatic treatment of mild or moderately severe AD is a highly selective, reversible acetylcholinesterase inhibitor with much less activity against butyrylcholine, an enzyme mainly existing in peripheral tissues rather than the CNS. Oral absorption is complete and the half-life is 70–80 h allowing a single daily dosing. Overall, it seems that 40% of patients will respond positively to the drug. Rivastigmine (Exelon, **9.105**) carbamylates the enzyme, and, despite a short plasma half-life, inactivates it for about 10 h (pseudoirreversible inhibition). It is rapidly metabolized to a metabolite with little activity. Benefits in cognition, all-round functioning, and daily living activities have been shown in trials with a suitable dosage. Adverse effects attributable to cholinesterase inhibitors as a class are apparent as well as weight loss. Galantamine, hydrobromide (Reminyl, **9.106**), the most recently introduced of the inhibitors, is an alkaloid originally isolated from snowdrop and daffodil bulbs but now prepared synthetically. It is a reversible competitive inhibitor and is more active against acetylcholinesterase than butylcholinesterase. It may have a different profile to the other agents described here, although this requires clinical confirmation, since it is an allosteric modulator of nicotinic cholinergic receptors (i.e., enhances effect of acetylcholine).

9.7.3 Aromatase and Steroid Sulfatase Inhibitors

Aromatase inhibitors

Aromatase belongs to a group of cytochrome P-450 enzymes responsible for hydroxylation processes in the body. It contains a Fe^{3+}–heme catalytic site which, after reduction to Fe^{2+}, binds and activates oxygen, leading to initial insertion of two hydroxyl groups on the C-19 (methyl)carbon of its substrates androstenedione and testosterone. A further hydroxylation occurs, followed by aromatization to estrone and estradiol, respectively, accompanied by elimination of water and formate by a mechanism only partially understood.

The steroidogenic pathway (see Figure 9.22) from cholesterol to the substrates of aromatase commences in the adrenals with the action of the cytochrome P-450 enzyme, cholesterol side chain cleavage (CSCC) enzyme, producing pregnenolone which is then isomerized by another enzyme to progesterone. Progesterone is converted by 17α-hydroxy: 17,20-lyase (P-450–17), another P-450 enzyme, to androstenedione which can be reduced by a dehydrogenase to testosterone. Aromatase is located mainly in fatty tissue in postmenopausal women and mainly in ovarian tissue in pre-menopausal women.

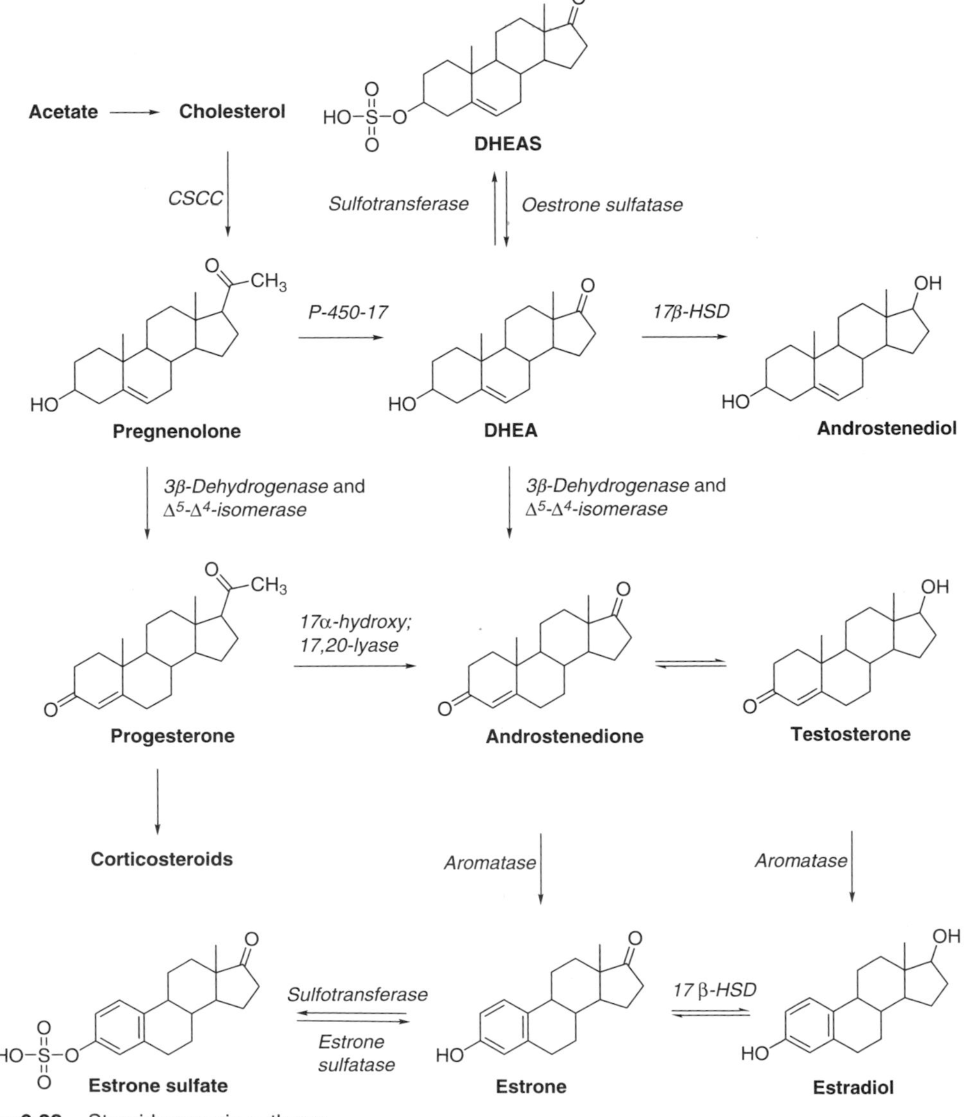

Figure 9.22 Steroidogenesis pathway.

After diagnosis of a breast tumor, it is removed by surgery and this is followed by a course of chemotherapy to reduce new tumor growth or suppress metastasis (microtumor spread) in other parts of the body. Mammary tissue contains estrogen receptors (see Chapter 14), and depending on their concentration the patient can be categorized as either estrogen receptor-positive (ER+) or receptor-negative (ER−). About one third of the cases of breast cancer in women are hormone-dependent, the major hormone involved in supporting the growth of the tumors being estradiol. The categorization can determine the type of chemotherapeutic treatment employed.

The first line drug for use in the treatment of mammary cancer in postmenopausal women with (ER+) and (ER−) tumors is tamoxifen. This is an estrogen receptor antagonist which, by competing with estradiol for the receptor, can reduce the ability of estradiol to stimulate tumor growth. Estrogen has important physiological effects on reproductive tissues (breast, uterus, ovaries) as well as preserving bone mineral density (prevention of osteoporosis) and protecting the cardiovascular system by reducing cholesterol levels.

Tamoxifen has weak estrogenic activity (agonist) as an antagonist and whereas this is beneficial in decreasing cardiovascular effects and risk of osteoporosis on blocking estrogen action it may lead to a small risk of endometrial cancer. Selective estrogen receptor modulators (SERMs) have been designed to increase the antagonist potency of agents and remove their undesirable agonist effect on the endometrium and these are discussed in Chapter 14. Tamoxifen-resistant tumors (ER+) are sometimes amenable to treatment with a second line drug which is an aromatase inhibitor. This reduces the plasma level of circulating estradiol available to the tumor tissue by inhibiting the action of aromatase, present in the fatty tissue, on androstenedione.

The nonsteroidal aromatase inhibitor aminoglutethimide (**9.13**) was in clinical use until recently for the treatment of (ER+) breast cancer in postmenopausal women. On chronic administration of the drug, the already low plasma estrogen level present in elderly women is further rapidly lowered and maintained, enabling a success rate in terms of remission or stabilization of about 33% (unselected patients) or 52% (ER+ patients).

Aminoglutethimide was initially introduced into therapy as an antiepileptic drug, but after initial withdrawal due to noted side effects of adrenal insufficiency it was re-introduced into cancer chemotherapy due to its potential effect for interrupting the steroidogenic pathway to estrogen production. Subsequent work showed that it was a potent, competitive, reversible inhibitor of aromatase with a weaker effect on the CSCC enzyme (which accounts for its effects on adrenal hormone production).

Aminoglutethimide is co-administered with hydrocortisone to supplement decreased production of 11β-hydrosteroids due to its effect on CSCC. Side effects associated with use of the drug are ataxia, dizziness, and lethargy due to its sedative nature. These effects, which can lead to patient noncompliance, decrease after several weeks of administration of the drug. Consequently, more specific inhibitors without these side effects have been sought.

Several antifungal agents based on imidazole, e.g., ketoconazole (**9.9**), econazole were known at this time which inhibit the fungal P-450 14α-demethylase enzyme. They are inhibitors of aromatase but have a wide spectrum of activity against other P-450 enzymes in the steroidogenic chain. Several potent specific inhibitors of aromatase containing an imidazole or triazole nucleus (increased *in vivo* stability) have subsequently been developed. Fadrozole (**9.107**), (+)-vorozole (**9.15**, Rivizor), letrozole (**9.108**, Femara, achiral), and anastrozole (**9.109**, Arimidex, achiral) have proved successful in clinical trials. These compounds are 400- to 1000-fold more potent than aminoglutethimide and have no CNS effects. Fadrozole also inhibits the 18-hydroxylase enzyme responsible for aldosterone production at doses much higher than used clinically; this side effect has been designed out in the more selective letrozole.

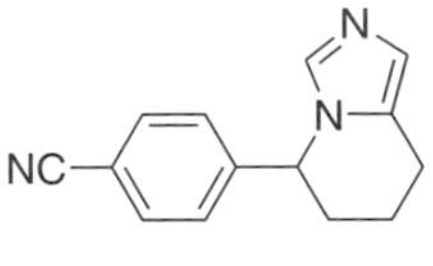

(**9.107**) Fadrazole 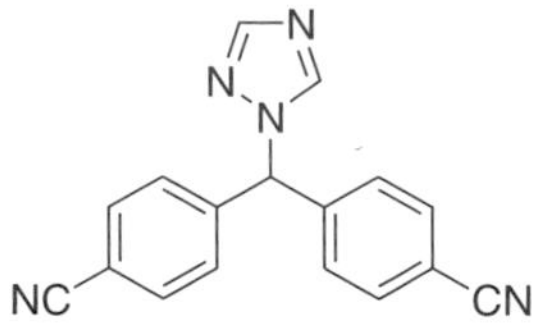 (**9.108**) Letrozole 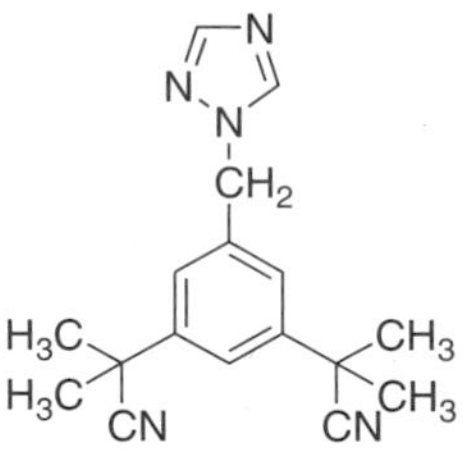 (**9.109**) Anastrazole

Mechanism-based inactivators of aromatase are known and these are based on the androstenedione (substrate) skeleton. 4-Hydroxyandrostenedione (**9.110**, formestane) is in clinical use as an intramuscular injection (formestane) given once weekly and is a specific irreversible inhibitor of the enzyme although the mechanism is not clear. It has to be administered parenterally since it is rapidly metabolized by first-pass metabolism following oral administration. Other steroidal irreversible inhibitors include plomestane (**9.111**) and the orally active exemestane (**9.112**).

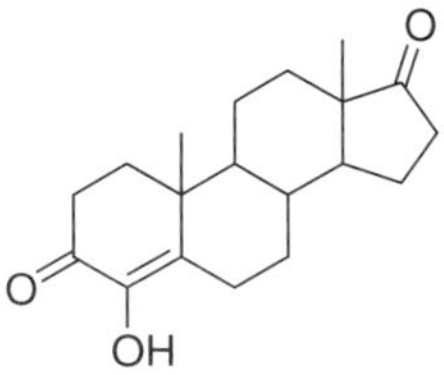

(**9.110**) 4-Hydroxyandrostenedione 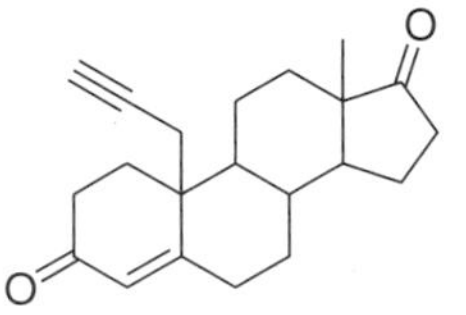 (**9.111**) Plomestane 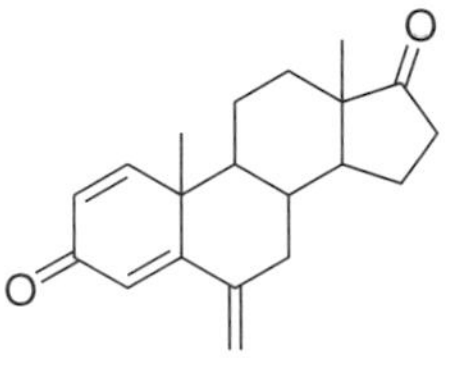 (**9.112**) Exemestane

Anastrozole (1 mg day^{-1}) and letrozole (2.5 mg day^{-1}) reduce serum levels of estrogens by 97 and 99%, respectively, which is beyond the limit of detection in many patients and greater than that for aminoglutethimide (1000 mg day^{-1}, 90%) but comparable with exemestane (25 mg day^{-1}, 97%), which is more effective than formestane (250 mg every 2 weeks, 85%). The advantages of the third generation aromatase inhibitors over aminoglutethimide are their improved tolerance and fewer side effects rather than their improved response rates (11–24%) and durations (18–23 months). (cf. aminoglutethimide, 12–30%, 13–24 months). Anastrazole is to be used in a large trial (I bis II) as a prophylactic drug for postmenopausal women considered to be at higher risk due to a family history of breast cancer.

Recent views are that breast tissue is capable of synthesizing estrogens from the action of a steroid sulfatase on estrone sulfate as well as the action of aromatase on androstenedione. Inhibitors of the steroid sulfatase have been developed as potential adjuvants to aromatase inhibitors in the treatment of postmenopausal breast cancer patients, as described in the following section.

Estrogen sulfatase inhibitors

In postmenopausal patients four- to sixfold higher concentrations of estrogens have been found in breast tissue compared to the plasma. This is considered to be due to local synthesis of estrone (and then estradiol by the action of 17β-hydroxysteroid dehydrogenase (17β HSD)) from estrone sulfate by estrone sulfatase (see Figure 9.22). Estrone sulfate is an inactive storage and transport form of estrogen, its concentration in plasma being up to 20-fold greater than estrone. Similarly dehydroepiandrosterone sulfate (DHEAS) is converted by the action of the sulfatase to dehydroepiandrosterone (DHEA). DHEA is convertible to androstenedione, a substrate for estrogen production by aromatase (present in breast tissues), and by a 17β-HSD to androstenediol, an androgen having low affinity for ER (Figure 9.22). However, androstenediol's 100-fold higher plasma concentration than estradiol produces estrogenic effects. It is considered that local production of estrogen in the breast due to the actions of aromatase and estrone sulfatase may play a greater part in breast tumor promotion than circulating plasma estrogens.

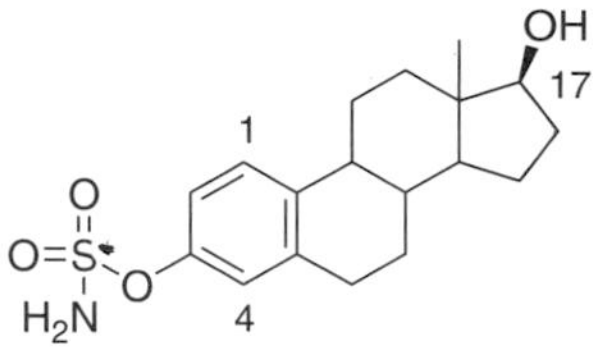

(**9.113**) EMATE

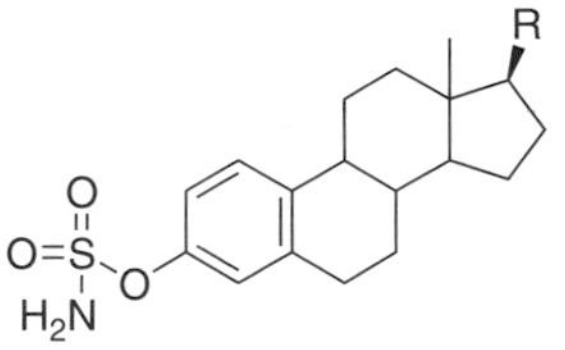

(**9.114**) R = $NHCO(CH_2)_6CH_3$
(**9.115**) R = $CONH(CH_2)_6CH_3$

Inhibitors of estrone sulfatase are being developed as potential adjuvants to aromatase inhibitors in the treatment of breast cancer to prevent the use of estrone sulfate and DHEAs as a source of estrogen and the ER agonist androstenediol. Estrone-3-sulfamate (**9.113**, EMATE) is a potent irreversible inhibitor of estrone sulfatase. Unfortunately it has estrogenic effects due to the release of estrone (estradiol) as the leaving group during the reaction with the enzyme. Potent non-estrogenic analogues of EMATE were obtained by substitution of the OH group in the 17β-position with a long hydrophilic alkylamido chain (**9.114**) and (**9.115**), which improved binding through a hydrophobic area on the enzyme in the vicinity of the 17β-position and led to the required loss of estrogenic activity in the leaving group on reaction.

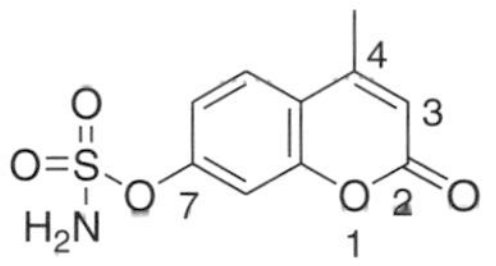

(**9.116**) COUMATE

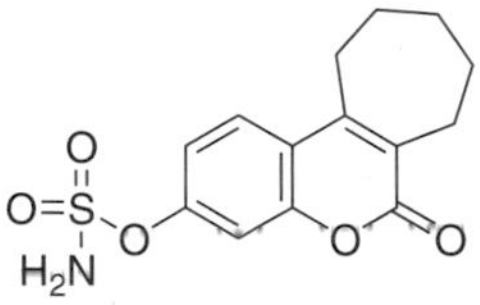

(**9.117**) 667 COUMATE

Considerable success has been achieved with substituted coumarins as mimics of the steroidal A/B rings with a correctly positioned 7-*O*-sulfamate function and alkyl residues overlapping the hydrophobic binding areas for the C/D rings of the steroid. COUMATE (**9.116**) is a potent, orally active, non-estrogenic irreversible inhibitor of the enzyme and extension of the 3,4-alkyl chains with ring formation gave the optimum heptene tricyclic 667 COUMATE (**9.117**). This was more potent than EMATE (IC_{50} = 8 n*M* vs 25 n*M*) with estrone sulfate as substrate or with DHEAS as substrate (IC_{50} 4.5 n*M* vs 110 n*M*). 667 COUMATE is an irreversible inhibitor of the enzyme, and in animal experiments with rats showed no estrogenicity and caused regression of implanted tumor growth; it is now in preclinical development for the treatment of ER-(+) postmenopausal patients.

9.7.4 Pyridoxal Phosphate-Dependent Enzyme Inhibitors

Enzymes using pyridoxal phosphate as coenzyme catalyze several types of reactions of amino acid substrates, such as (1) transamination to the corresponding α-ketoacid; (2) racemization; (3) decarboxylation to an amine; (4) elimination of groups on the β- and γ-carbon atoms; (5) oxidative deamination of ω-amino acids. The coenzyme is bound to the enzyme by formation of an aldimine (Schiff base) with the ω-amino group of a lysine residue. The first step in the reaction with the amino acid substrate is an exchange reaction to form an aldimine with the α-amino group of the amino acid (see Equation (9.47)). Either by hydrogen abstraction (transamination, racemization) or by decarboxylation, a negative charge is developed on the α-carbon atom, and this is distributed over the whole conjugated cofactor system. Protonation then occurs on either the α-carbon atom (decarboxylation, racemization) or on the carbon atom adjacent to the pyridine ring (transamination) as shown in Equation (9.47). The direction of the fission which occurs is dictated by the nature of the protein at the active site so that a specific enzyme catalyzes a particular type of reaction. Information has recently become available on the crystal structure of several of these enzymes and the role of their active site residues.

R–CH–COO⊖ ⊕NH₃ + CHO ⇌ R–CH–COO⊖ HN⊕=CH ⇌ (− H⁺, α-Transaminase) R–C⊖–COO⊖ HN⊕=CH ⟷ R–C–COO⊖ =HN⊕ ⊖CH

+ H⁺ → R–C–COO⊖ =HN⊕ CH₂

R–CH⊖ HN⊕=CH ⟷ R–CH=HN⊕ CH

+ H⁺ → R–CH₂ HN⊕=CH

(9.47)

At one time, several irreversible inhibitors of several pyridoxal phosphate-dependent enzymes were known but their mechanism of action was not clear since they did not possess the electrophilic centers present in the active site-directed irreversible inhibitors known at that time. Later, when a new class of inhibitor, the mechanism-based enzyme inactivator, became known, their inhibition mechanism became predictable from the well-established mechanism of action of these enzymes. The next step for design was to manipulate the amino acid substrate structure of a suitable target enzyme in such a manner as to obtain maximal exploitation of the enzyme's machinery.

The inhibitors act as substrates of the enzyme but their structure is such that they either (1) divert the electron flux from the α-carbanion formed away from the coenzyme moiety, or (2) using the normal electron flux either give rise to reactive species or generate a stable substrate-cofactor, which binds strongly to the enzyme active site. All these mechanisms can lead to irreversible inhibition of the enzyme.

Mechanism-based inactivators of many pyridoxal phosphate-dependent enzymes are known but only a few target enzymes and their inactivators of therapeutic interest will be discussed here.

GABA transaminase inhibitors

GABA is considered as the main inhibitory neurotransmitter in the mammalian central nervous system. There has been much interest recently in the design of inhibitors of the pyridoxal phosphate-dependent enzyme, α-ketoglutarate-GABA transaminase. This enzyme governs the levels of GABA in the brain (see Equation (9.48)). Inhibitors of the enzyme would allow a build-up of GABA and could be used as anticonvulsant drugs for the treatment of epilepsy.

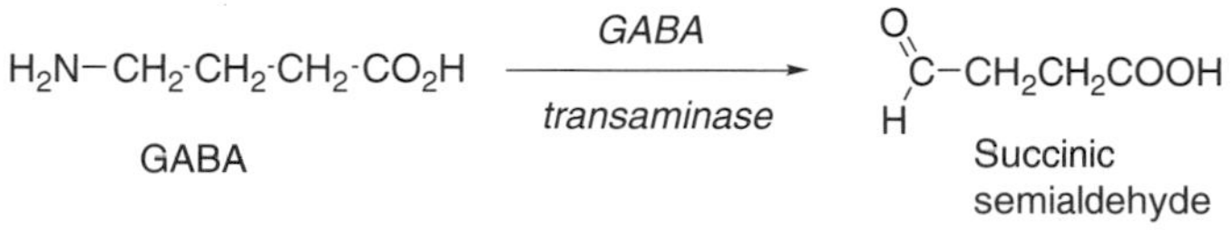

$$H_2N-CH_2\cdot CH_2\cdot CH_2\cdot CO_2H \xrightarrow[\textit{transaminase}]{\textit{GABA}} OHC-CH_2CH_2COOH \quad (9.48)$$

GABA → Succinic semialdehyde

γ-Acetylenic GABA (**9.118**) is a time-dependent inhibitor of GABA-T but also inhibits other pyridoxal phosphate-dependent enzymes. γ-Vinyl GABA (vigabatrin, **9.119**) acts in a similar manner through its (*S*)-enantiomer but has a more specific action.

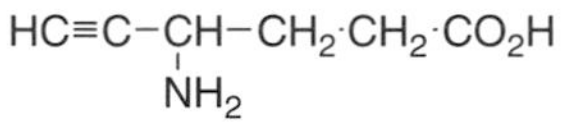

$HC{\equiv}C-CH(NH_2)-CH_2\cdot CH_2\cdot CO_2H$

(**9.118**)

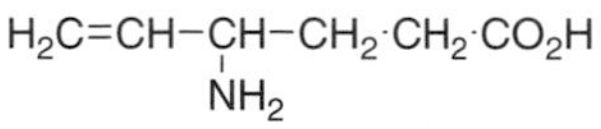

$H_2C{=}CH-CH(NH_2)-CH_2\cdot CH_2\cdot CO_2H$

(**9.119**) Vigabatrin

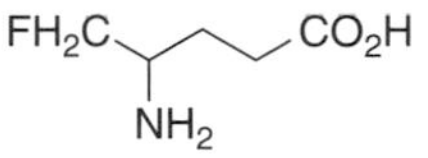

$FH_2C-CH(NH_2)-CH_2-CH_2-CO_2H$

(**9.120**)

$R-CH(NH_2)-CH_2-CO_2H$

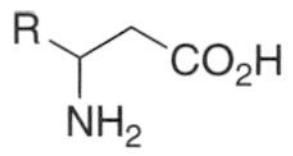

(**9.121**); R = CH_2F
(**9.122**); R = CHF_2

Vigabatrin (Sabril) has shown promise as a drug for the treatment of epilepsy. Studies on drug-resistant epileptic patients have indicated that additional therapy with vigabatrin reduces seizures by over 50% in more than half the population studied without development of tolerance. However, since about one third of patients using the drug have visual field defects (irreversible) restrictions on its prescribing and new monitoring recommendations have been introduced.

Halomethyl derivatives of GABA have also been described as inhibitors of GABA-T. The fluoromethyl derivative (**9.120**) is the best time-dependent inhibitor and the inactivation is accompanied by elimination of fluoride ion. Shortening of the chain of (**9.120**) to give the β-alanine derivatives (**9.121**) and (**9.122**) produced inhibitors with similar kinetic constants. However, *in vivo* (**9.122**) was almost 100-fold more active than vigabatrin but showed unexplained delayed toxicity after a single administration and was not further developed.

The mechanism of action of the GABA-T inhibitors based on GABA and bearing either an unsaturated function or a leaving group has not yet been clearly elucidated. The initial postulated mechanism was that these two groups of inhibitors formed a Schiff-base with pyridoxal phosphate, followed by loss of the α-carbon proton.

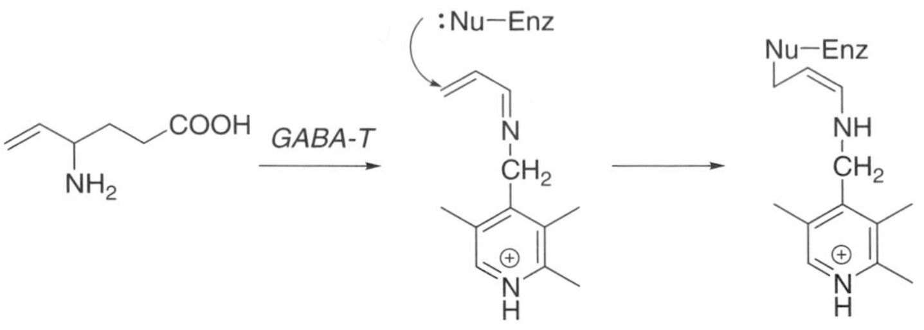

(9.49)

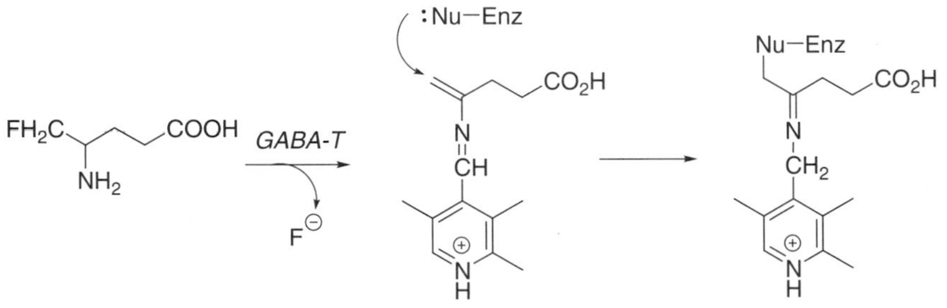

(**9.123**)

(9.50)

(**9.124**)

With the unsaturated derivatives the electron flow that followed was towards the coenzyme moiety to give the vinylimine (**9.123**), whereas with the fluoromethyl derivatives the electron flow was away from the coenzyme and accompanied by loss of fluoride ion to give the enimine (**9.124**) (see Equations (9.49) and (9.50)). The electrophilic centers developed in the conjugated systems by the normal or abnormal electron flow react with a nucleophile at the active site of the enzyme. More recent work with other pyridoxal phosphate-dependent enzymes has suggested an alternative mechanism, which is illustrated in Equation (9.51) for the fluoromethyl derivatives. Here the enimines dissociate from the pyridoxal phosphate to give an enamine which then recombines with the lysine of the active site. The cofactor is then attacked by the electrophilic center of the enamine to yield a stable complex at the active site, which leads to irreversible inhibition of the enzyme.

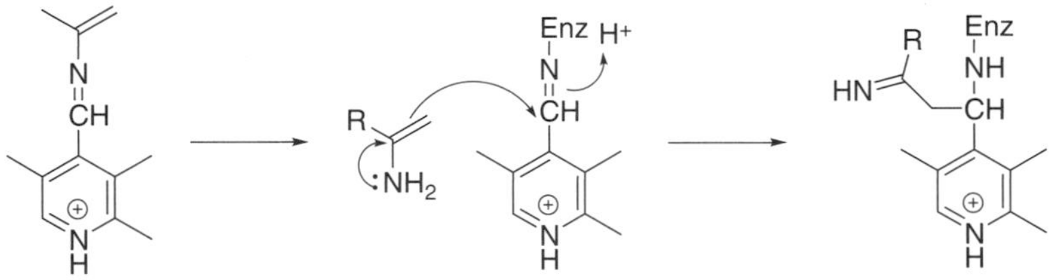

(9.51)

Another potent irreversible inhibitor of GABA transaminase is gabaculine (**9.125**), which is a naturally occurring neurotoxin isolated from *Streptomyces toxacaenis*. Although not of clinical application, this inhibitor is interesting since it is considered to inhibit the enzyme by a different mechanism to that previously described for suicide inactivators. Gabaculine acts as a substrate and is converted in the normal manner to the ketimine (**9.126**) (Equation (9.52)). This then aromatizes under the influence of a basic group to form a stable enzyme-bound pyridoxamine derivative and the enzyme is inactivated.

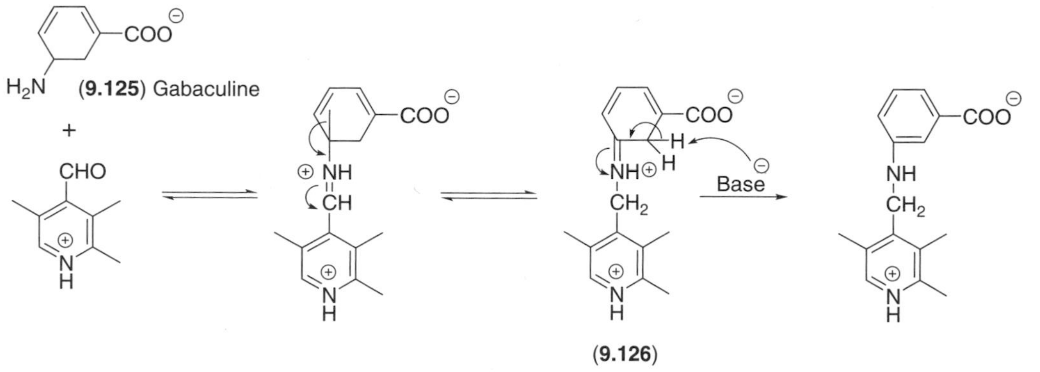

(9.125) Gabaculine

(9.126)

(9.52)

Peripheral aromatic AADC inhibitors

Noradrenaline is synthesized at the nerve endings of the postganglionic fiber by a series of reactions from tyrosine (see Equation (9.53)). Inhibitors of AADC have been synthesized as potential antihypertensive drugs on the basis that a decrease in the biosynthesis of noradrenaline would deplete noradrenaline stores at the nerve endings and lead to a decrease in blood pressure. Although many reversible inhibitors of the enzyme are known from *in vitro* studies (e.g., methyldopa (**9.127**)) only a few exert an antihypertensive action *in vivo* and probably by an alternative mechanism since inhibition of the first step in the pathway is not involved (see Section 9.2.1). However, this work led to the discovery of the inhibitors carbidopa (**9.128**) and serazide (**9.129**) which have proved useful as adjuvants in the treatment of Parkinson's disease with L-dopa. L-Dopa penetrates into the basal ganglia of the brain where it is decarboxylated to the active agent, dopamine. Large doses of L-dopa are required in therapy since it is depleted in the plasma by *peripheral* AADC to dopamine which is readily removed by monoamine oxidase. Combination of L-dopa with serazide or carbidopa leads to decreased metabolism of L-dopa so that smaller effective doses can be used in therapy, which have fewer side effects than large doses. Necessary features of these inhibitors are that they do not penetrate the blood–brain barrier and interfere with the decarboxylation of L-dopa to dopamine in the brain and, for the reason given above, neither do they reduce the synthesis of endogenous amines in the peripheral tissues.

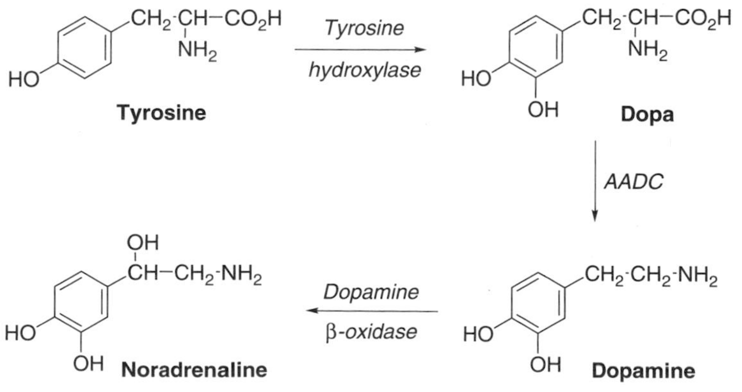

Tyrosine —(Tyrosine hydroxylase)→ Dopa —(AADC)→ Dopamine —(Dopamine β-oxidase)→ Noradrenaline

(9.53)

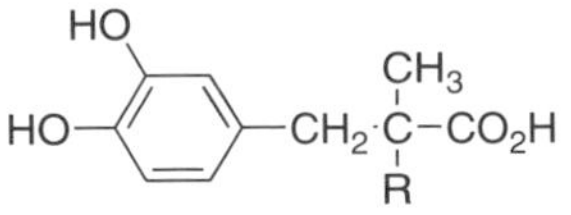

(**9.127**) R = NH_2 ; Methyldopa
(**9.128**) R = $NHNH_2$; Carbidopa

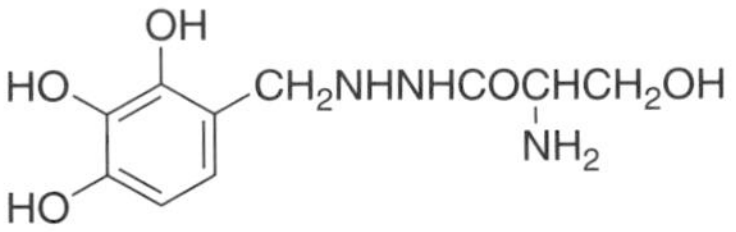

(**9.129**) Serazide

AADC is a pyridoxal phosphate-dependent enzyme and serazide and carbidopa are potent pseudoirreversible inhibitors of the enzyme. They probably function by binding to pyridoxal phosphate as carbonyl group reagents.

Several mechanism-based inactivators of AADC have been described, including the α-monofluoromethyl (**9.130**) and α-difluoromethyl (**9.131**) derivative of dopa. These compounds are time-dependent irreversible inhibitors of the enzyme. During one turnover of the inhibitor by the enzyme, one equivalent each of CO_2 and F^- is released and the inhibitor binds in a 1:1 ratio to the enzyme–cofactor complex.

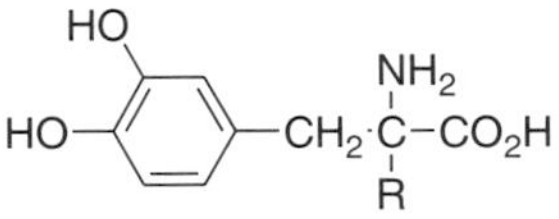

(**9.130**) R = CH_2F
(**9.131**) R = CHF_2

α-Difluoromethyldopa is comparable to carbidopa and effectively protects exogenous dopa against decarboxylation. It inhibits brain AADC only at high concentrations. α-Monofluoromethyldopa effectively inhibits AADC centrally as well as peripherally and the resulting depletion of peripheral catecholamines produces antihypertensive effects which can be reversed by i.v. infusions of dopamine.

Ornithine decarboxylase inhibitors

Naturally occurring polyamines such as putrescine, spermidine, and spermine are required for cellular growth and differentiation. Spermidine and spermine are derived in human-type cells from putrescine. Putrescine is synthesized by decarboxylation of ornithine, catalyzed by the pyridoxal phosphate-dependent enzyme ODC (Equation (9.54)). ODC has a very short biological half-life and its synthesis is stimulated ''on demand'' by trophic agents and controlled by putrescine and spermidine levels. ODC has been considered a suitable target enzyme for the control of growth in tumors and disease caused by parasitic protozoa.

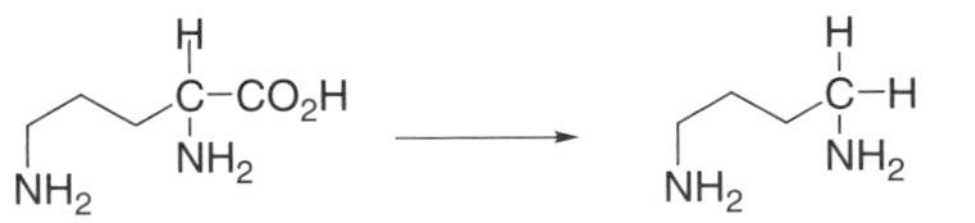

Ornithine → **Putrescine** (9.54)

α-Difluoromethylornithine (eflornithine; (**9.132**)) is a mechanism-based inactivator of the enzyme and irreversibly inhibits the enzyme by the general mechanism previously depicted with elimination of a single fluoride ion to produce a conjugated electrophilic imine (cf. **9.124**) which reacts with the nucleophilic thiol of Cys-390. A further fluoride ion is then eliminated which, following transaldimation with Lys-69, leads to the species $Cys-^{390}$-S−CH=C(NH_2)− $(CH_2)_3$−NH_2, which loses ammonia and cyclizes to (2-(1-pyrroline)methyl) cysteine.

$$\mathrm{H_2N-CH_2CH_2CH_2-C(CHF_2)(NH_2)-CO_2H}$$

(**9.132**) Eflornithine

Eflornithine has low toxicity in animals and has shown antineoplastic and antiprotozoal actions in clinical trials. The methyl or ethyl esters are effectively hydrolyzed at the higher cellular concentrations attained due to improved absorption and are ten times more effective than eflornithine in decreasing ODC activity in animal tissues, an effect which is long lasting.

Eflornithine is used for the treatment of trypanosomiasis (sleeping sickness) as an alternative to the arsenic-containing drug melarsoprol with its dire side effects. More recently it has been used as a topical prescription treatment (Vaniqua) for women with unwanted facial hair by slowing hair growth.

FURTHER READING

Acharya, K.R., Sturrock, E.D., Riordan, J.F. and Ehlers, M.R.W. (2003) ACE revisited: a new target for structure-based drug design. *Nature Reviews Drug Discovery* **2**, 891–902.

Ala, P.J. and Chang, C.-H. (2002) HIV aspartate proteinase: resistance to inhibitors. In H.J. Smith and C. Simons (eds.), *Proteinase and Peptidase Inhibition: Recent Potential Targets for Drug Development*. Taylor and Francis, London, pp. 367–382.

Aldridge, W.N. (1989) Cholinesterase and esterase inhibitors and reactivation of organophosphorus inhibited esterase. In M. Sandler and H.J. Smith (eds.), *Design of Enzyme Inhibitors as Drugs*, Vol. 1. Oxford University Press, Oxford, pp. 294–313.

Ali, S. and Coombes, R.C. (2002) Endocrine-responsive breast cancer and strategies for combating resistance. *Nature Reviews Cancer* **2**, 101–112.

Basury, I. (2003) Neutral peptidase inhibitors: new drugs for heart failure. *Indian Journal of Pharmacology* **35**, 139–145.

Beckett, R.P., Davidson, A.H., Drummond, A.H., Huxley, P. and Whittaker, M. (1996) Recent advances in matrix metalloproteinase inhibitor research. *Drug Discovery Today* **1**, 16–26.

(a) Bode, W., Huber, R., Rydel, T.J. and Tulinsky, A. (1992) X-ray crystal structures of human α-thrombin and of the human thrombin–hirudin complex, pp. 3–61. (b) Powers, J.C. and Kam, C.-M. (1992) Synthetic substrates and inhibitors of thrombin, pp. 117–159. (c) Stone, S.R. and Maraganore J.M. (1992) Hirudin interactions with thrombin, pp. 219–256. In L.J. Berliner (ed.), *Thrombin — Structure and Function*. Plenum Press, New York, London.

Edwards, P.D. (2002) Human neutrophil elastase inhibitors. In H.J. Smith and C. Simons (eds.), *Proteinase and Peptidase Inhibition: Recent Potential Targets for Drug Development*. Taylor and Francis, London, pp. 154–177.

Fisher, J.F., Tarpley, W.G. and Thaisrivongs, S. (1994) HIV protease inhibitors. In M. Sandler and H.J. Smith (eds.), *Design of Enzyme Inhibitors as Drugs*, Vol. 2. Oxford University Press, Oxford, pp. 226–289.

Fournie-Zaluski, M.-C., Coric, P., Turcaud, S., Lucas, E., Noble, F., Maldonado, R. and Roques, B.P. (1992) ''Mixed inhibitor-prodrug'' as a new approach towards systemically active inhibitors of enkephalin-degrading enzymes. *Journal of Medicinal Chemistry* **35**, 2473–2481.

Gerhartz, B., Niestroj, A.J. and Demuth, H.-U. (2002) Enzyme classes and mechanism. In H.J. Smith and C. Simons (eds.), *Proteinase and Peptidase Inhibition: Recent Potential Targets for Drug Development*. Taylor and Francis, London, pp. 1–20.

Hlasta, D.J. and Pagani, E.D. (1994) Human leukocyte elastase inhibitors. In J.A. Bristol (ed.), *Annual Reports in Medicinal Chemistry*. Academic Press, New York, pp. 195–204.

Hooper, N.M. (2002) Zinc metallopeptidases. In H.J. Smith and C. Simons (eds.), *Proteinase and Peptidase Inhibition: Recent Potential Targets for Drug Development*. Taylor and Francis, London, pp. 352–366.

John, R.A. (1995) Pyridoxal phosphate-dependent enzymes. *Biochimica et Biophysica Acta, Protein Structure and Molecular Enzymology* **1248**(2), 81–96.

Katzenellenbogen, B.S., Montano, M.M. and Ekena, K., et al. (1997) Antioestrogens: mechanism of action and resistance in breast cancer. *Breast Cancer Research and Treatment* **44**, 23–38.

Leonetti, G. and Cuspidi, C. (1995) Choosing the right ACE inhibitor. *Drugs* **49**, 516–535.

Patel, A., Smith, H.J. and Sewell, R.D.E. (1993) Inhibitors of enkephalin-degrading enzymes as potential therapeutic agents. In G.P. Ellis and D.K. Luscombe (eds.), *Progress in Medicinal Chemistry*, Vol. 30. Elsevier Science, Amsterdam, pp. 327–378.

(a) Sandler, M. and Smith, H.J. (1989) Introduction to the use of enzyme inhibitors as drugs, pp. 1–18; (b) Frick, L. and Wolfenden, R. (1989) Substrate and transition-state analogue inhbitors, pp. 19–48; (c) Shaw, E. (1989) Active-site-directed irreversible inhibitors, pp. 49–69; (d) Tipton, K. (1989) Mechanism-based inhibitors, pp. 70–93. In M. Sandler and H.J. Smith (eds.), *Design of Enzyme Inhibitors as Drugs*, Vol 1. Oxford University Press, Oxford.

Schwartz, J.C., Gros, C., Duhamel, P., Duhamel, L., Lecomte, J.M. and Bralet, J. (1994) ''Atriopeptidase'' (EC 3.4.24.11) inhibition and protection of atrial natriuretic factor. In M. Sandler and H.J. Smith (eds.), *Design of Enzymes Inhibitors as Drugs*, Vol. 2. Oxford University Press, Oxford, pp. 739–754.

Silverman, R.C. (1988) *Mechanism-Based Enzyme Inactivation: Chemistry and Enzymology*, Vol. 1. CRC Press, Boca Raton, Florida, pp. 3–23.

(a) Slater, A.M., Timms, D. and Wilkinson, A.J. (1994) Computer-aided molecular design of enzyme inhibitors, pp. 42–64; (b) Luscombe, D.K., Tucker, M., Pepper, C.M., Nicholls, P.J., Sandler, M. and Smith, H.J. (1994) Enzyme inhibitors as drugs: from design to the clinic, pp. 1–41. In M. Sandler and H.J. Smith (eds.), *Design of Enzyme inhibitors as Drugs*, Vol. 2. Oxford University Press, Oxford.

Smith, H.J., Nicholls, P.J., Simons, C. and Le Lain, R. (2001) Inhibitors of steroidogenesis as agents for the treatment of hormone-dependent cancers. *Expert Opinion on Therapeutic Patents* **11**(5), 789–824.

Sturzebecher, J., Hauptmann, J. and Steinmetzer, T. (2002) Thrombin. In H.J. Smith and C. Simons (eds.), *Proteinase and Peptidase Inhibition: Recent Potential Targets for Drug Development.* Taylor and Francis, London, pp. 178–201.

The Royal College of Ophthalmology (2001) Annual Report. The Ocular Side-Effects of Vigabatrin (Sabril). Information and Guidelines for Screening.

Vanden Bossche, H. (1992) Inhibitors of P450-dependent steroid biosynthesis: from research to medical treatment. *Journal of Steroid Biochemistry and Molecular Biology* **43**, 1003–1021.

10

Peptide Drug Design

David J. Barlow

CONTENTS

10.1 INTRODUCTION

The recent advances in biotechnology and solid-phase synthetic techniques have resulted in a major expansion of interest in the development of peptides as pharmaceuticals. Peptides make ideal candidates as therapeutic agents in that they are usually very potent and exhibit a high degree of specificity for their target receptors. The anticoagulant peptide hirudin, for example, isolated from the medicinal leech, *Hirudo medicinalis*, is a highly selective inhibitor of the blood clotting enzyme thrombin and has an equilibrium dissociation constant of 0.2–1.0 p*M*.[1] Likewise, the vasoactive peptide endothelin is highly selective for its receptor on vascular endothelial cells and has an EC_{50} of about 1 n*M*.[2]

10.1.1 Problems with Peptides as Drugs

Despite the high potency and selectivity of peptides, their physical and chemical properties are generally such that they are not readily developed for *direct* use as drugs. The majority of peptides of interest have relatively high molecular weights and are often highly charged. As a consequence they do not readily traverse epithelial cell membranes and so are poorly absorbed when administered parenterally. There are complications too that are caused by the fact that they are always highly susceptible to enzymic degradation. As a general rule, therefore, peptides suffer very low bioavailabilities (in the case of orally administered compounds, typically $<1\%$[3,4]). In many cases the molecules are also difficult to formulate because they exhibit relatively poor aqueous solubility and are sensitive to heat, pH, and oxidation. For their effective exploitation as drugs, therefore, most peptides must be chemically modified so that they are made smaller, more lipophilic, more readily soluble, and more stable to enzymes and denaturants.

10.1.2 Scope of this Chapter

In the following sections each of the problems associated with peptide drug development is considered in turn, along with the various solutions that have been found to circumvent these difficulties. There are a few specific case studies presented to illustrate how particular peptides were modified to provide some of the peptide drugs currently used clinically, but there are no *detailed* considerations given to the mechanisms of action of individual peptide drugs. The challenges posed in development of *orally active* peptide drugs are considered in some depth, and there are discussions too of the use of peptides as antimicrobial agents and absorption aids. To begin, however, we present a brief summary of peptide structure and its associated terminology.

10.2 PEPTIDE STRUCTURE

Peptides are unbranched polymers (with predefined size and composition) formed from the monomer units, α-amino (carboxylic) acids (Figure 10.1a). All eukaryotic (and most prokaryotic) peptides are formed from amino acids that have the L configuration (Figure 10.1b) but prokaryotes frequently also use amino acids that have the D configuration (Figure 10.1c). Most natural peptides involve a basic set of 20-amino acids, each distinguished by the nature of its side chain (R) (see Table 10.1). The different amino acids are classified according to the polarity of their side chains. The hydrophilic amino acids include those with uncharged side chains (such as serine) and those with cationic and anionic side chains (the latter including aspartic acid and the former, lysine). The hydrophobic amino acids include those with aromatic side chains (like phenylalanine) as well those with aliphatic side chains (such as valine).

The polymerization of the monomeric units proceeds by condensation reactions, which link each amino acid (residue) to its neighbors by secondary amide (or so-called peptide) bonds (Figure 10.1d). The covalent structure of the resulting chain is commonly abbreviated as shown in Figure 10.2, with an ordered list of the amino acids presented using standard (IUPAC) three- or one-letter codes (Table 10.1).

These data, together with details of any chemical modifications (such as glycosylation or amidation) and inter- or intrachain disulfide bonds (formed through the reduction of cysteine

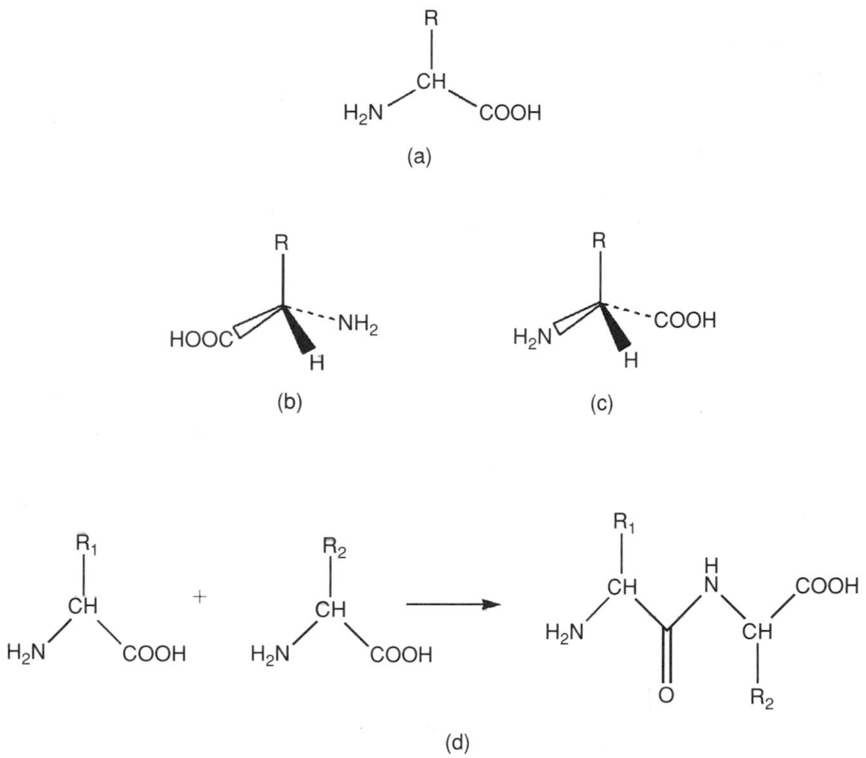

Figure 10.1 The structure of α-amino (carboxylic) acids (a), their L- (b) and D- (c) enantiomeric forms, and their condensation via peptide bond formation (d).

residues to form cystine, Figure 10.3), are referred to as the primary structure of the peptide, with the synonymous descriptions: amino acid sequence or peptide sequence.

Among the natural small peptides, and even more commonly in the synthetic analogs designed specifically for use as pharmaceuticals, there are frequently ''nonprotein'' amino acids incorporated, in the latter case to improve their *in vivo* activity. Examples of these include: ornithine (three-letter code, Orn) (**10.1**), which is found in the antibiotic peptide, gramicidin S (see below); penicillamine (three-letter code, Pen) (**10.2**), which is used to limit the conformational freedom of a peptide in the immediate vicinity of a disulfide bond (as in the case of enkephalin analogs, see below); α-aminoisobutyric acid (three-letter code, Aib) (**10.3**), which favors helical conformation (as in the membrane permeabilizing antibiotic peptide, alamethacin[5]); and pyroglutamate (three-letter code, Glp) (**10.4**), which at the N-terminus of peptide can limit degradation by aminopeptidases (as with luteinizing hormone-releasing hormone, see below).

Those peptides that involve only a relatively small number of residues are usually described explicitly, e.g., dipeptide (two residues), tripeptide (three residues), etc., whereas those with 10 to 30 residues are termed oligopeptides, and those with more than 30 residues are termed polypeptides. (Molecules having more than 50 residues are generally classed as proteins.)

Table 10.1 Structures and properties of the most common L-α-amino acids found in peptides[a]

Alanine Ala (A) $—CH_3$	Cysteine Cys (C) SH Able to form disulfide bonds (see Figure10.3)	Aspartic acid Asp (D) COOH pK_a: 3.86 *N*-glycosylation site residue; deamidated to Asp	Glutamic acid Glu (E) COOH pK_a: 4.25	Phenylalanine Phe (F) UV absorption: λ_{max}, 257 nm; ε, 200 cm^2 mol^{-1}
Glycine Gly (G) —H Optically inactive; confers conformational flexibility to peptide main chain	Histidine His (H) H N N pK_a: 6.0	Isoleucine Ile (I) CH_3 H_3C	Lysine Lys (K) H_2N pK_a: 9.67	Leucine Leu (L) H_3C CH_3
Methionine Met (M) CH_3 S Readily oxidized to sulfoxide	Asparagine Asn (N) $CONH_2$	Proline Pro (P) N H COOH An α-imino acid	Glutamine Gln (Q) $CONH_2$ Deamidated to Glu	Arginine Arg (R) NH H_2N NH pK_a: 12.48
Serine Ser (S) OH *O*-glycosylation site residue	Threonine Thr (T) OH CH_3 *O*-glycosylation site residue	Valine Val (V) CH_3 CH_3	Tryptophan Trp (W) N H A fluorophore; UV absorption: λ_{max}, 280 nm; ε, 5600	Tyrosine Tyr (Y) OH UV absorption: λ_{max}, 274 nm; ε,1400 cm^2 mol^{-1}

[a]Amino acid name, IUPAC three-letter (and one-letter) code, side chain structure, side chain pK_a.

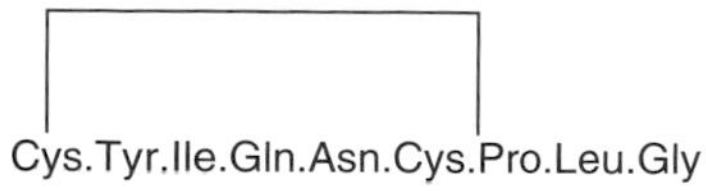

Cys.Tyr.Ile.Gln.Asn.Cys.Pro.Leu.Gly

Figure 10.2 Primary structure of the nonapeptide hormone, oxytocin, with amino acid residues indicated using the IUPAC three-letter codes (see Table 10.1). Half-cystine residues 1 and 6 are linked by a disulfide bond (Figure 10.3).

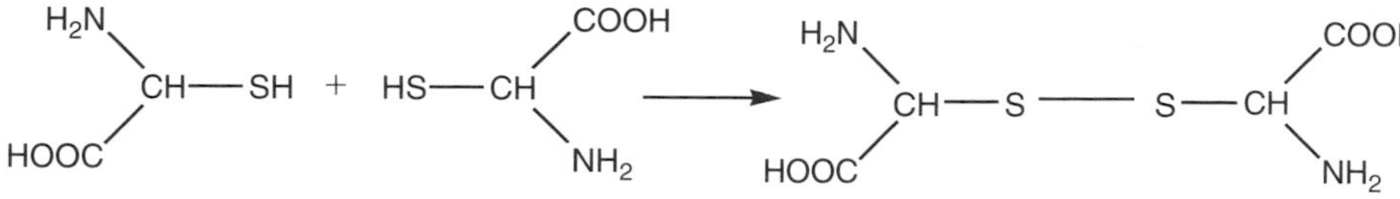

Figure 10.3 The formation of cystine through the reaction of two thiol groups on a pair of cysteine residues.

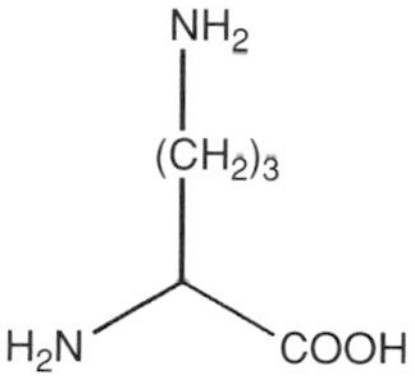

(10.1)

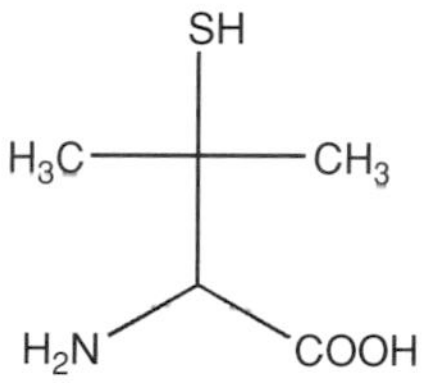

(10.2)

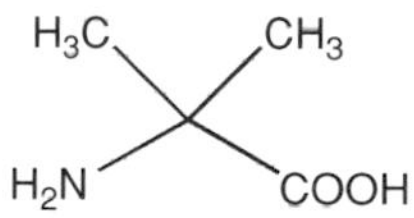

(10.3)

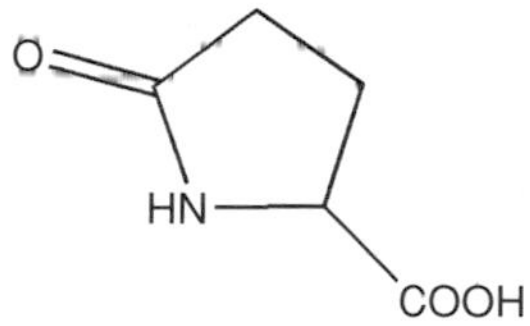

(10.4)

10.3 PEPTIDE DRUGS

The majority of peptide pharmaceuticals, which are currently marketed or undergoing clinical trials, derive from natural molecules, although the generation and screening of combinatorial peptide libraries now provides an increasing number of lead compounds. Table 10.2 gives a catalog of some representative examples, summarizing their clinical indications, intended routes of administration, and status. It will be noted that almost all of the peptides listed are intended for intravenous administration, with very few molecules formulated for use as oral pharmaceuticals. As will become apparent over the following sections, this is due to the problems that arise as a consequence of their physical and chemical properties.

10.3.1 Issues of Peptide Size

The majority of peptides of interest have molecular weights that are much greater than those of the more conventional pharmaceuticals (Table 10.3), which means that they are poorly absorbed and must be made smaller to ensure reasonable bioavailabilities.

Table 10.2 Representative examples of peptide pharmaceuticals approved or undergoing clinical trials 2002–2003

Peptide	Drug (company)	Indications	Status	Route of administration
Insulin	Exubra (Inhale; Aventis; Pfizer)	Diabetes types I and II	Phase III	Pulmonary
	Oralin (Generex)		Phase II/III	Oral
Insulin LisPro	Humalog (Eli Lilly)	Diabetes types I and II	Approved	Injection
Calcitonin	Oratonin (NOBEX))	Osteoporosis	Phase I	Oral
	Calcimar (Aventis)		Approved	Injection
	Miacalcin (Novartis)		Approved	Nasal (spray)
Glucagon-like peptide	Exendin-4 (Amylin)	Diabetes type II	Phase II	Injection
Luteinizing hormone-releasing hormone (LH-RH)	Lupron depot (Takeda Abbott)	Prostate cancer	Approved	Intramuscular
	Plenaxis (Praecis)	Prostate cancer	Approved	Intramuscular
Interferon-α N	Alferone N (Interferon Sciences)	Genital warts	Approved	Injection; topical (gel)
Factor IX (antihemophilia factor B)	BeneFIX (Genetics Institute)	Hemophilia	Approved	Injection
Interferon-β	Betaserone (Chiron, Berlex)	Multiple sclerosis	Approved	Injection
Growth hormone	Biotropin (BTG)	Growth hormone deficiency	Approved	Buccal (spray)
Erythropoietin α	EPOGEN (Amgen)	Anemia associated with renal failure in dialysis patients	Approved	Injection
FSH (Follitropin α/β)	Follistim (Organon)	Female infertility	Approved	Injection
	Fertinex (Serono)		Approved	Injection
Somatropin	GenoTropin (Pharmacia & Upjohn)	Growth hormone deficiency	Approved	Injection
	Humatrope (Eli Lilly)		Approved	Injection
	Serostim (Serono)		Approved	Injection
Growth hormone-releasing factor (GH-RF)	Geref (Serono Lab)	Growth hormone deficiency	Approved	Injection
Thyrotropin α	Thyrogen (Genzyme)	Diagnostic tool for serum thyroglobulin in patients with thyroid cancer	Approved	Injection
Oxytocin	Syntocinon (Novartis)	Induction of labor	Approved	Injection
	Oxytocin (Am Pharm Partners)		Approved	Injection
Vasopressin	Desmopressin acetate (Abbott)	To control the frequent urination, increased thirst and water loss associated with diabetes insipidus	Approved	Injection
	Desmopressin acetate (Bausch & Lomb)		Approved	Nasal (spray)

In order to reduce the size of a potential peptide drug, first it is necessary to determine which of its amino acid residues can be omitted and which ones must be retained because of their importance for the molecule's activity. There are two simple strategies that can be adopted to secure this information. The first involves testing the activities of shorter versions of the peptide in which residues have been truncated from its N- or C-termini. By such means, for example, it proved possible to trim the 14 residues of somatostatin (also known as growth hormone release inhibitory factor) (**10.5**), down to just six residues, truncating five residues from the molecule's N-terminus, and three from its C-terminus (**10.6**), reducing the MW from 1638 to 878. (Note here, however, that it was also found necessary to fix the conformation of the analog molecule, not only maintaining the

Table 10.3 Typical peptide drug sizes and molecular weights

	MW
Oligopeptides	
Thyrotropin-releasing hormone (3 residues)	362
Luteinizing hormone-releasing hormone (10 residues)	1182
Somatostatin (14 residues)	1638
Polypeptides	
Insulin (51 residues)	5733
Glucagon (29 residues)	3485
"Traditional" pharmaceuticals	300–400

disulfide linkage, but also adding two further conformational constraints, in the form of a D-Trp at position 3 and a second N- to C-terminal linkage provided by aminoheptanoic acid, Aha.)

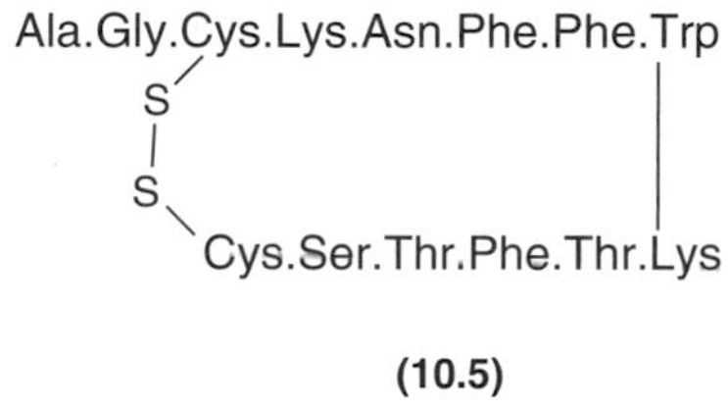

(10.5)

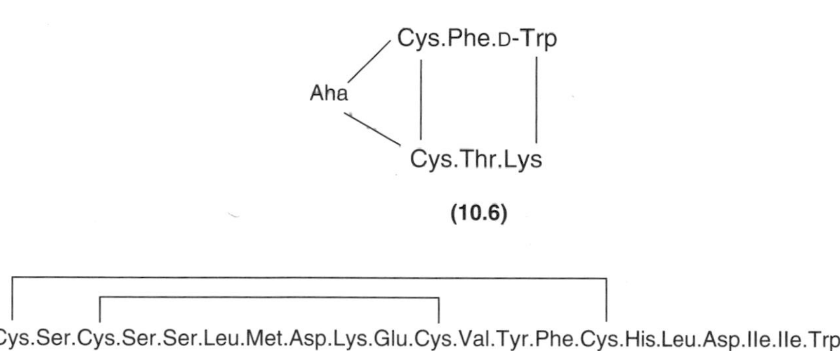

(10.6)

(10.7)

A second, complementary strategy, which provides information on the relative importance of residues within the body of a peptide involves performing an L-alanine scan. Here, full-length variants of the peptide are synthesized and tested, with each one having one of its amino acids substituted by L-alanine. If a given alanine-substituted peptide shows no significant loss in activity, then the replaced residue is identified as functionally nonessential. Alanine is chosen as the replacing residue because of its relatively small and neutral side chain; the use of glycine, which is also small and neutral, is not usually favored because of its tendency to disrupt regular secondary structures in the peptide.

For the case of the vasoactive peptide hormone, endothelin (**10.7**), the performance of an L-alanine scan revealed (Table 10.4)[6] that the hydrophilic residues Ser 2, Ser 4, Ser 5, and Lys 9 are relatively unimportant as regards the peptide's activity, with the L-Ala substitutions at these positions having little effect either on receptor binding or vascular smooth muscle contraction. Residues Tyr 13, Phe 14, Leu 17, and Trp 21 on the other hand are seen to be critical to activity: the L-Ala substitutions at these positions give rise to analogs with $<10\%$ of the vasoconstrictor activity of native endothelin. Since the Ala 13, Ala 14, and Ala 21 analogs show correspondingly low

Table 10.4 Biological properties of L-Ala monosubstituted analogs of endothelin 1 (10.7)[a]

L-Ala-substituted residue	Receptor binding[b]	Contraction[c]
–	100.0	100.0
Ser 2	100.0	70.0
Ser 4	200.0	51.5
Ser 5	142.0	24.8
Leu 6	101.0	73.0
Met 7	350.0	88.5
Asp 8	103.0	0.8
Lys 9	504.0	56.0
Glu 10	162.0	121.0
Val 12	16.2	29.4
Tyr 13	0.8	1.1
Phe 14	1.8	8.4
His 16	339.0	382.0
Leu 17	25.0	0.9
Asp 18	50.0	14.0
Ile 19	406.0	120.0
Ile 20	22.0	22.4
Trp 21	0.8	< 0.1

[a]Data taken from Tam et al.[6]
[b]Binding to A617 cells incubated in serum-free medium, expressed as a percentage of endothelin 1 binding ($IC_{50} = 0.20 \times 10^{-9}$ *M*).
[c]Contraction of rabbit vena cava, expressed as a percentage of endothelin 1 contraction (EC_{50} for endothelin $= 2.89 \times 10^{-10}$ *M*).

receptor binding, the residues at these positions are implicated as receptor contact residues. The Ala 17 analog, however, shows a reasonable level of receptor binding, and so its low vasoconstrictor activity would seem to suggest that this residue is only peripherally involved in receptor binding but is crucial for signal transduction.

When nonessential residues have been identified within the body of a peptide they cannot simply be excised from the structure because this is likely to alter the molecule's conformation and so may adversely affect its activity. It may be possible, however, to substitute such residues using amino acids that have smaller side chains (thus reducing the size of the molecule) or else using amino acids that have more nonpolar side chains (thereby enhancing the peptide's lipophilicity).

10.3.2 Peptide Conformation

Most small peptides, and in particular those that are highly polar, do not adopt well-defined conformations in aqueous solution: they have a limited capacity for establishing the intramolecular interactions which help to stabilize folding intermediates, and so their structures, which although perhaps exhibiting a strong conformational bias[7] vary dynamically over a range of conformations.[8] In order to carry out their biological functions, however, such molecules will adopt a particular conformation on binding to their target receptor. With some peptides this ''active'' conformation may be selected as the result of interaction with a specific macromolecule, and with others it may be selected as the result of self-association or because of the amphipathic nature of their folded structure at a lipid–water interface. The former class of peptide is thought to include molecules like atrial natriuretic factor[9] and in the latter category are those such as glucagon and vasoactive intestinal polypeptide.[8]

In the case of small peptides that do adopt well-defined conformations in aqueous solution, these usually have conformational constraints imposed either by intramolecular disulfide bonds or by the cyclization of their peptide main chains.

The disulfide-linked molecules include α-helix-containing peptides like endothelin (Figure 10.4) as well as β-sheet peptides like transforming and epidermal growth factors (Figure 10.5). The peptides having cyclic main chains are illustrated by the antibiotic ionophore, gramicidin S

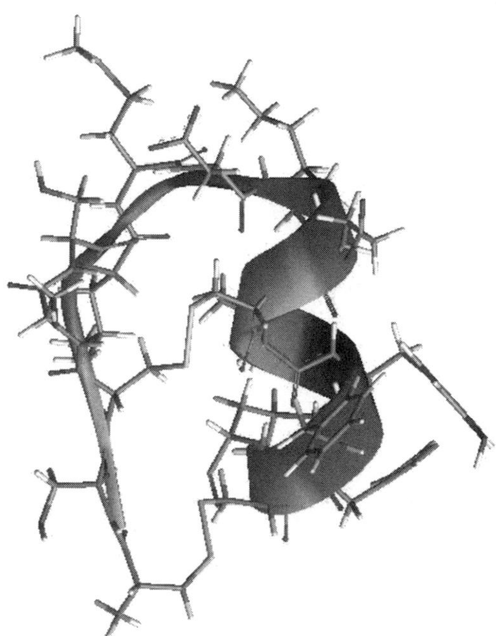

Figure 10.4 (See color insert after p. 368) The three-dimensional structure of human endothelin-1, as revealed by 2-D ^{1}H NMR studies.[13] Residues 9–15 form an irregular α-helix, which is tethered to the N-terminal half of the molecule through disulfide bonds linking Cys 1 to Cys 15, and Cys 3 to Cys 11.

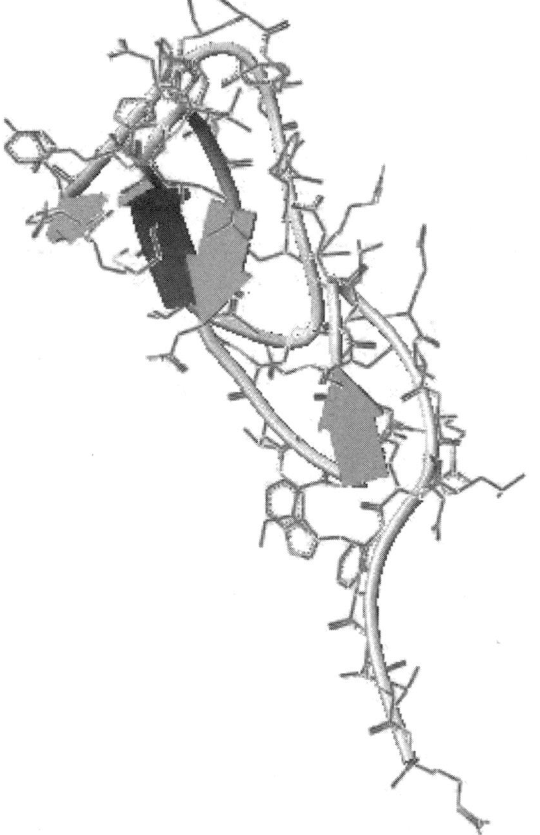

Figure 10.5 (See color insert after p. 368) The three-dimensional structure of mouse epidermal growth factor.[12] β-Strands are shown schematically as blue arrows.

(**10.8**), and the immunosuppressor, cyclosporin A (**10.9**). Note here that gramicidin *S* involves the amino acid, ornithine, and also features both D- and L-amino acids. Note too that cyclosporin is rather unusual in that it contains the nonprotein amino acids sarcosine (or *N*-methyl glycine) and (2*S*,3*R*,4*R*,6*E*)-3-hydroxy-4-methyl-2-methylamino-6-octenoic acid, and has 7 of its 11 peptide bonds methylated (with NCH_3 rather than the usual NH).

L-Val·*L*-Orn·*L*-Leu·*D*-Phe·*L*-Pro

L-Pro·*D*-Phe·*L*-Leu·*L*-Orn·*L*-Val

(10.8)

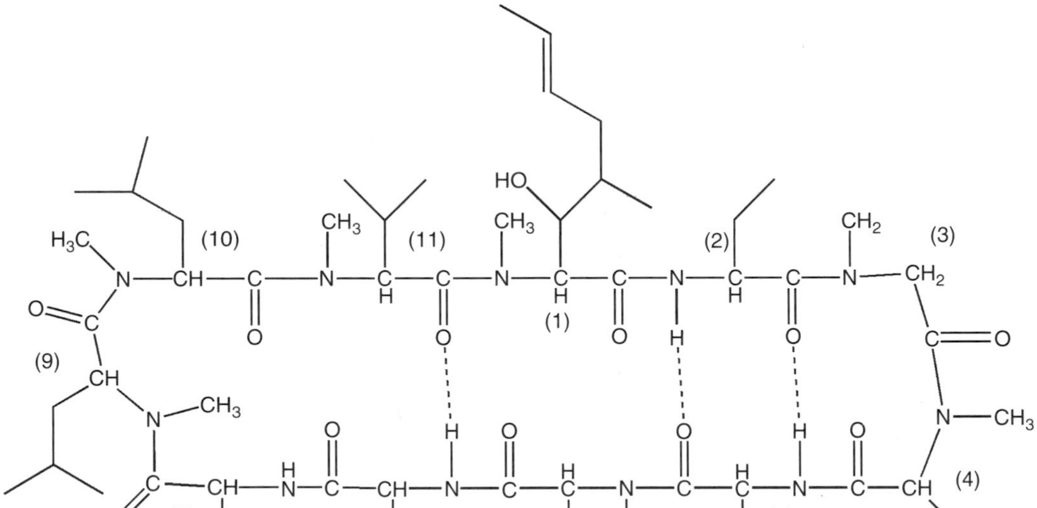

(10.9)

The details of the three-dimensional structures of peptides are sometimes obtained by x-ray crystallography (as in the case of oxytocin[10] and vasopressin[11]), but more frequently are derived from two-dimensional NMR studies (e.g., epidermal growth factor[12] and endothelin[13]). In some instances circular dichroism studies are also instructive,[14] and since this technique is more straightforward and yet provides a highly sensitive means of detecting conformational changes, it is very often used in confirming the three-dimensional structures of peptide pharmaceuticals.

When tailoring the design of a peptide so that it has a conformation appropriate for its required pharmacological activity, the key consideration is the conformation adopted by the peptide when it binds to its target receptor. As noted earlier, when most peptides are in free solution they have conformations that change constantly over time, and so their chain entropy is relatively high. When they bind to their receptors, they effectively become locked in a particular conformation and their chain entropy decreases. This loss of entropy incurred on binding is unfavorable and reduces the free energy of complexation between the peptide and its receptor. In order to produce analogs with greater potency, therefore, one strategy is to arrange that the peptide be conformationally constrained. The peptide will then have lower chain entropy in free solution, and will suffer less of a loss of entropy on binding to its receptor. (Of course, exactly the same principle applies to all other types of drug; it is just that the loss of entropy on receptor binding is particularly punitive for peptides because of their inherently high conformational flexibility in free solution.)

For an acyclic peptide, the conformational constraints required to enhance the free energy of receptor binding can easily be engineered through the introduction of one or more disulfide bonds or by arranging for ring closure mediated through reaction between its N- and C-terminal groups. The former approach is exemplified by the case of gramicidin S (**10.8**), and the latter by the case of somatostatin analogs (**10.6**).

Such strategies are not always successful, however, and it is important to note that the cyclic and acyclic forms of a peptide may show differing activities. For example, the acyclic Met-enkephalin (**10.10**) binds fairly nonselectively to μ, δ, and κ opioid receptors, and when the molecule is cyclized as (**10.11**) the resulting peptide is also nonselective. When the cyclization is engineered through D-penicillamine residues, however, the resulting analog (**10.12**) is found to be a potent and selective δ-opioid receptor agonist, with strong analgesic activity.

Tyr.Gly.Gly.Phe.Met

(10.10)

Tyr.Cys.Gly.Phe.Cys

(10.11)

Tyr.*D*-Pen.Gly.Phe.*D*-Pen

(10.12)

As a subtler means of arranging conformational constraints in an acyclic peptide, researchers have often substituted key residues with their D-enantiomers. Indeed, as a means to gain insights into the conformational requirements of a peptide's activity, researchers often perform a D-residue scan — analogous to an L-Ala scan, but with each residue in turn replaced by its D-enantiomer. The introduction of D-residues into a peptide can stabilize certain types of reverse (or β-) turn (Figure 10.6) may destabilize α-helices, or otherwise indicate parts of the molecule where conformation appears critical to biological activity. For example, it was discovered that when Phe-7 in α-melanocyte-stimulating hormone (α-MSH) (**10.13**) was replaced by D-Phe, the resulting analog proved to be more potent and longer acting than the parent hormone. It was thus proposed

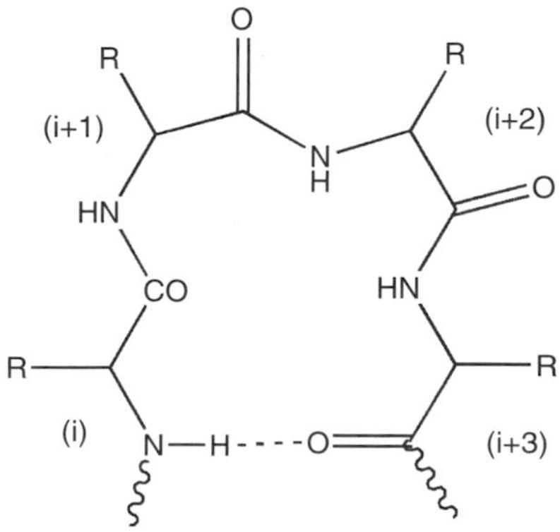

Figure 10.6 A four-residue peptide reverse turn (also referred to as a β-turn). The stabilizing i to $i+3$ hydrogen bond is shown as a dashed line.

that residues 6 to 9 could be involved in the formation of a β-turn — because the introduction of a D-amino acid at position $i+1$ in a turn serves to increase the peptide's propensity for turn formation. It is taken, therefore, that the introduction of the D-Phe-7 in α-MSH biases the solution equilibrium of the peptide toward the conformation required for receptor binding, and so leads to an analog with increased potency.

A partial confirmation of this hypothesis was obtained through the preparation and testing of a cyclic analog of α-MSH (**10.14**) in which the β-turn over residues 6–9 was stabilized by a disulfide created between residues at positions 4 and 10. The resulting analog was found to be superpotent, but was not as long acting as the D-Phe-7 analog. Note here that Cys residues required for the formation of the disulfide were engineered by replacement of the Met-4 and Gly-10 residues — the combined side chains of these two residues are roughly isosteric with the disulfide moiety.

Ac.Ser.Tyr.Ser.Met.Glu.His.Phe.Arg.Trp.Gly.Lys.Pro.Val.NH_2

(10.13)

Ac.Ser.Tyr.Ser.Cys.Glu.His.Phe.Arg.Trp.Cys.Lys.Pro.Val.NH_2

(10.14)

As an alternative way of forcing β-turn formation, it is also possible to use the imino acid, proline. Again, the introduction of the proline residue at position $i+1$ in a turn region increases the propensity for formation of a (type II) β-turn.

When attempting to design a conformationally constrained peptide analog, it is too important to allow for the fact that its conformation also influences its absorption (and hence its bioavailability). A number of researchers have shown how the introduction of conformational constraints can have a detrimental effect on a peptide's absorption. Both cysteinyl and penicillinoyl analogs of enkephalin are three to seven times less permeable than the corresponding acyclic peptides.[15]

10.3.3 Increasing Peptide Lipophilicity

Most natural peptides are highly polar and have very low lipophilicities. Their octanol–buffer (pH 7.4) partition coefficients generally lie in the range 0.01–0.1 (see Table 10.5), and so they do not readily cross lipid bilayers, and suffer very low bioavailabilities. For the most part, therefore, it is necessary to prepare analogs of the molecules that have significantly increased lipophilicities.

When the size of a peptide has been reduced to give the minimum length fragment that retains biological activity, the opportunities for further enhancing its absorption across epithelial membranes by increasing its lipophilicity may be rather limited. Any side chain substitutions made must be confined to those residues remaining that are not critical to its pharmacological activity, and care must be taken not to compromise the progress already made in reducing molecular weight. With these restrictions, therefore, it is generally difficult to provide any significant improvement in lipophilicity using the naturally occurring amino acids (like leucine and phenylalanine), and researchers have consequently resorted to using synthetic amino acids that are far more lipophilic. Examples of these include *t*-butyl-glycine (**10.15**), β-naphthylalanine (**10.16**), and *p*-phenylphenylalanine (**10.17**).

Alternative strategies employed to increase lipophilicity include cyclization, and the conjugation of fatty acids to the peptide termini. With a cyclized peptide, the increase in lipophilicity results from the removal of the charged N- and C-terminal groups, reduction in the overall solvent accessible surface area, and (in certain cases at least) the occupation of some of the peptide's main chain NH and CO groups in intramolecular hydrogen bonding. Since cyclization will also

Table 10.5 Log *P* values for representative peptides (with *P* measured for partitioning between octanol and aqueous buffer pH 7.4)

Peptide	log *P*
Thyrotropin-releasing hormone (TRH) (**10.18**)	−1.43
δ-Sleep inducing peptide Trp.Ala.Gly.Gly.Asp.Ala.Ser.Gly	−2.06
Leu-enkephalin Tyr.Gly.Gly.Phe.Leu	0.05
Met-enkephalin (**10.10**)	−1.52
Substance P (Figure 10.7)	−0.56
Luteinizing hormone-releasing hormone (LH-RH) Glp.His.Trp.Ser.Tyr.Gly.Leu.Arg.Pro.Gly.NH_2	−1.35
Arginine vasopressin (**10.20**)	−2.15
Cyclosporin (**10.9**)	3.00

reduce the enzymatic degradation of the peptide (see below) this approach has much to commend it, but it is important to note that the conformational constraints imposed might have detrimental effects on pharmacological activity.

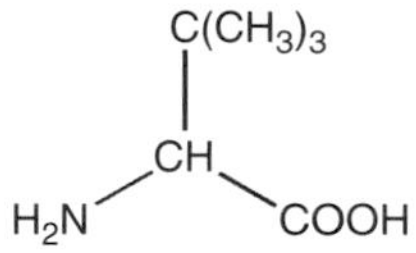

$C(CH_3)_3$–CH(H_2N)–COOH

(10.15)

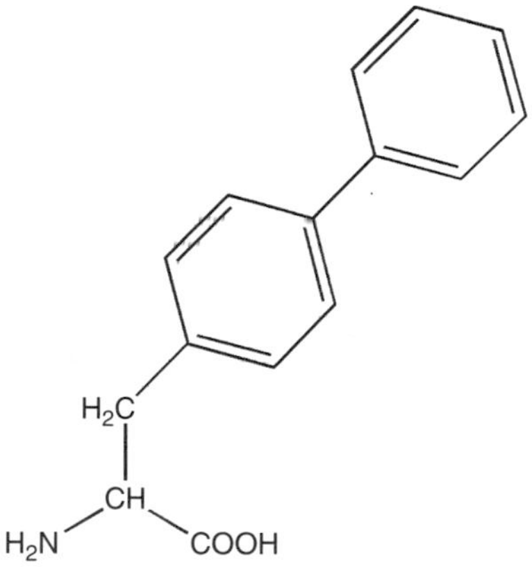

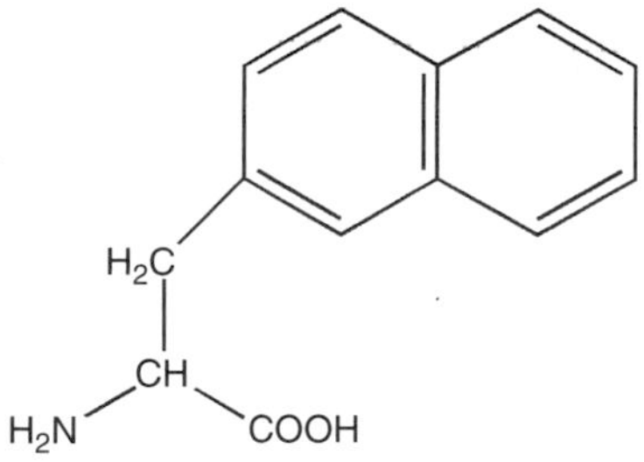

(10.16)

(10.17)

The same disadvantage may also apply in the case of fatty acid conjugation. It has been shown, for example, that although lauroyl (dodecanoyl) group acylation of thyrotropin-releasing hormone (TRH) (**10.18**) results in an analog (**10.19**) with much improved log (octanol–water partition coefficient) (1.91 vs. 0.07), the acylated molecule only has 64% of the activity of the parent peptide.[16] In experiments with C_6–C_{16} chain alkylation of TRH, tetragastrin, and insulin, however, it has been shown that particular analogs may give >70% activity.[17]

(10.18) R = H

(10.19) R = $CO(CH_2)_{11}CH_3$

10.3.4 Susceptibility of Peptides to Enzymic Degradation

By far the most significant factor limiting the use of peptides as drugs is their extreme susceptibility to hydrolytic enzymes. Although some routes of administration cause fewer problems than others, it must be remembered that all living systems have a biochemistry that is based around proteins and peptides, and thus they contain a plethora of proteinases and peptidases that will readily digest the molecules down to inactive products. The scale of the problem can be illustrated by the case of the peptide hormone, substance P, for which there are five different inactivating enzymes (so far) identified, with six different scissile (i.e., hydrolysable) peptide bonds (Figure 10.7).

Enzymes that attack at the ends of a peptide are termed exopeptidases, and are subclassified as aminopeptidases or carboxypeptidases, according to whether they hydrolyse peptide bonds at the molecules' N- or C-termini, respectively. Those enzymes that hydrolyse the peptide bonds within the body of a peptide are referred to as endopeptidases.

Proteases like trypsin and chymotrypsin do not pose too much of a problem for peptide drug stability because these enzymes are much more active against (large) proteins than they are against (small) peptides. The major problems arise because of the peptidase activities associated with epithelial cell membranes and intracellular components.

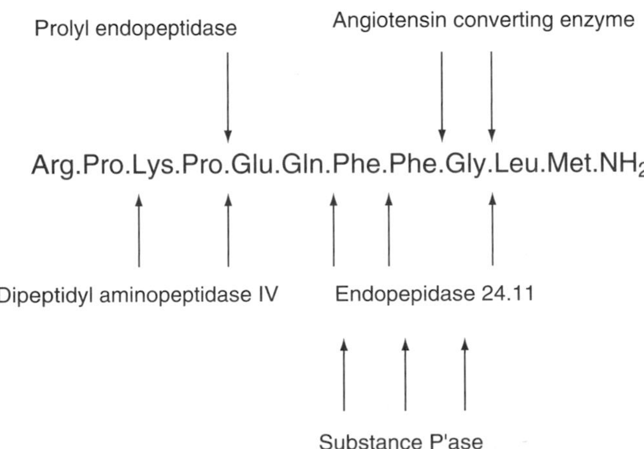

Figure 10.7 Principle enzymatic cleavage sites for the peptide hormone, substance P.

In order to protect against exopeptidase activity, the N- and C-termini must be blocked, preferably by some means that renders them electrically neutral. At the N-terminus this can be accomplished using acetylation or an N-terminal pyroglutamate residue (**10.4**). At the C-terminus, the same job can be performed by amidation or the reduction of the C-terminal carboxyl to an alcohol. (Note that the protection of the C-terminus by means of esterification is a poor option, because of the very high esterase activity *in vivo*.)

Both types of exopeptidase activity can also be limited by synthesizing molecules with D- rather than the usual L-amino acids at the termini, always providing of course that the conformation of the terminal residues is not critical to the peptide's activity. By far the best option, however, is provided by cyclization. As noted earlier, this also helps to increase the lipophilicity of a peptide. In addition, it leads to a decreased surface area and decreased flexibility of the peptide, and so hinders its degradation by endopeptidases.

Endopeptidase activity is otherwise limited by modifying the side chains of amino acid residues that flank the scissile bonds in the peptide, by substituting D- instead of L-amino acids at these positions, or by blocking or using alternative linkages to replace the peptide groups. In various angiotensin-converting enzyme inhibitors, for example, the scissile peptide bond is replaced by a ketomethylene group. Other such *pseudo-peptide* bonds include a reduced amide bond (CH_2NH), and a retro-peptide bond (NH·CO). The reduced amide bond is not particularly useful since it is not isosteric with the peptide bond, and its incorporation in a peptide may give a detrimental increase in conformational flexibility, and lead to the generation of a potential charged group on the peptide main chain, which may cause pH effects. Various researchers have also proposed the use of unsaturated ethene and ethyne linkages, but these are unlikely to be acceptable in peptide pharmaceuticals because of their inherently high reactivity.

10.4 PHARMACEUTICAL PEPTIDES

10.4.1 Vasopressin

Arginine vasopressin (**10.20**) is a cyclic nonapeptide, which has a rather short-lived antidiuretic effect. The analog 1-desamido-8-D-arginine vasopressin (ddAVP, desmopressin) (**10.21**) is more than 1000 times more potent than the parent peptide and is used in the treatment of severe diabetes insipidus. The clinical dose of desmopressin (10–20 μg intranasally, twice daily) makes it one of most potent of all pharmacologically active agents in man. The deletion of the parent hormone's N-terminal amino group acts to protect the peptide against aminopeptidase degradation, and the substitution of Arg-8 by D-Arg-8 protects it against hydrolysis by endopeptidases.

Cys.Tyr.Phe.Gln.Asn.Cys.Pro.Arg.NH_2 (disulfide bridge Cys–Cys)

(10.20)

$S(CH_2)_2$COTyr.Phe.Gln.Asn.Cys.Pro.*D*-Arg.Gly.NH_2 (bridge S–Cys)

(10.21)

10.4.2 Somatostatin

Somatostatin (also known as growth hormone release inhibitory factor) is a 14-residue peptide (**10.5**) produced in the hypothalamus. It was first identified as an inhibitor of pituitary growth hormone secretion but was subsequently shown to inhibit a number of other endocrine functions as well. As with vasopressin, the parent hormone suffers a very short plasma half-life, which severely restricts its usefulness as a pharmaceutical. A more potent and longer acting analog, sandostatin (**10.22**) was developed for symptomatic treatment of acromegaly and carcinoid syndrome. In sandostatin, the size of the molecule has been reduced by around 50% (which improves its absorption), and the N- and C-termini are both protected against exopeptidase attack, the former by means of a D-amino acid residue, and the latter by the reduction of the carboxyl group to an alcohol. The further use of a D-Trp residue at position 4 helps to constrain the central residues in the disulfide-linked loop as a β-turn, and this serves to increase the molecule's biological potency.

D-Phe.Cys.Phe.*D*-Trp.Lys.Thr.Cys.Thr(ol)

(10.22)

10.4.3 Luteinizing Hormone-releasing Hormone

Luteinizing hormone-releasing hormone (LH-RH) (**10.23**) (also known as gonadotrophin-releasing hormone) is a ten-residue peptide produced by the hypothalamus. Its normal mode of action is to stimulate pituitary secretion of gonadotrophin (or luteinizing hormone), and this in turn stimulates production of sex steroids by the sex organs. The hormone initially found use in fertility control, and it is still used today as an injectable for inducing ovulation in women suffering from hypogonadotrophic hypogonadism. Once again, however, the natural hormone has an extremely short plasma half-life, highly susceptible to enzymatic degradation, and this deficiency, together with a general desire to explore new modes of administration, prompted the development of more potent, and more stable synthetic analogs.

Glp.His.Trp.Ser.Tyr.Gly.Leu.Arg.Pro.Gly.NH_2

(10.23)

Glp.His.Trp.Ser.Tyr.*D*-Ser(tBu).Leu.Arg.Pro.NEt

(10.24)

Superpotent analogs such as buserelin (**10.24**) have been developed for use in treating prostatic cancers whose growth is dependent upon testosterone. The utility of such peptides for this purpose arises as a consequence of the fact that LH-RH is normally released in a pulsatile manner, with the pituitary gland's normal response dependent upon this pulsatility. Chronic administration of either LH-RH or an agonist like buserelin leads to a downregulation of the LH-RH receptors in the pituitary, and this then causes a *suppression* of gonadotrophin production, and to a concomitant decrease in sex hormone production by the gonads. Ultimately, therefore, there is less testosterone produced, and so the prostatic tumor growth is checked.

The incorporation of the modifications at positions 6 and 10 in the sequence provides increased resistance against endo- and exopeptidases, and increases potency by around 100-fold.

10.4.4 Insulin

Insulin (**10.25**), a twin-chain 51 residue polypeptide hormone produced in the pancreas, has long been used in the treatment of diabetes to control blood glucose levels. The hormone is conventionally administered by injection, and is active in its monomeric form (the oligomers resulting from its self-association are inactive). The modified peptide, LysPro (Humalog), developed by Eli Lilly & Co., has a sequence inversion in the C-terminal portion of the B-chain, with residues Pro-B28·Lys-B29 replaced by Lys-B28·Pro-B29. This inversion results in a reduced self-association, perturbing the equilibrium of the hormone in solution in favor of the monomeric species, thereby altering the pharmacokinetic properties of the hormone, giving a more rapid onset and earlier peak plasma concentrations with shorter duration of action.

(**10.25**)

10.5 ANTIMICROBIAL PEPTIDES

10.5.1 Activity and Origin

Antimicrobial peptides, sometimes also referred to as antibiotic peptides, are peptides that show activity against microorganisms. Such peptides have now been found in a whole host of different organisms including insects, fish, amphibians, and mammals. They form important components of the organisms' innate defense systems, many of them showing good activity against a range of pathogenic organisms, including both Gram-positive and Gram-negative bacteria, protozoa, and fungi. Some of them also show activity against cancer cells. In consequence, these types of peptides are attracting considerable interest as leading candidates for drug development. In addition to the work centered on the natural molecules, there are also research efforts expended in the development of synthetic variants. A catalog of all of the natural peptides discovered (now exceeding 800) is maintained at http://www.bbcm.univ.trieste.it/~tossi/antimic.html, and an equivalent repository for details relating to the synthetic variants can be found at http://oma.terkko.helsinki.fi:8080/~SAPD/.

10.5.2 Antimicrobial Peptide Structures and Classification

The structures of many of the natural antimicrobial peptides have now been determined both in the solid state (via single crystal x-ray diffraction) and also in solution (using ^{1}H-NMR). The peptides can be broadly classified according to their size, conformation, or predominant amino acid content (Table 10.6).

The defensins are small cationic peptides produced by mammals. They are subclassified as α- or β-defensins according to their disulfide topology. Six α-defensins and two β-defensins have so far been found in humans. Four of the six α-defensins are stored in secretory granules in neutrophil cells, and are sometimes referred to as neutrophil peptides (hNPs 1–4); the other two α-defensins (α-defensins 5 and 6) are secreted by Paneth's cells in the ileum. The microbicidal activity of the α-defensins is apparent only at relatively high concentrations and they exhibit a broad spectrum of activity, acting on Gram-positive and Gram-negative bacteria, fungi, and enveloped viruses. The β-defensins, by contrast, are inducible, much more highly potent, and are selective for Gram-negative

Table 10.6 Classification and nomenclature of some antimicrobial peptides

Peptide	Species (organ)
Group I: α-helical peptides without cysteines	
Bombinins	Frog (skin)
Cecropins	Insects (hemocytes)
LL-37	Human (neutrophils, epithelia)
Magainins	Frog (skin)
Styelins	Tunicates (hemocytes)
Clavanins	Tunicates (hemocytes)
Melittin	Bee (venom)
Group II: β-sheet structures stabilized by disulfides	
Protegrins	Pig (intestine)
Tachyplesins	Horseshoe crab (hemocytes)
α- and β-defensins	Vertebrates (immune cells, epithelia)
Insect defensins	Insects (hemocytes)
Plant defensins	Plants (seeds, leaves)
Drosomycin	Insects (hemocytes)
Group III: peptides with a predominance of one or more amino acids	
PR-39	Pig (intestine, neutrophils)
Bac5, Bac7	Cow (neutrophils)
Drosocin	Fruit fly (hemolymph)
Metchnikowin	Fruit fly (hemolymph)
Group IV: peptides with loop structures	
Bactenicin	Cow (neutrophils)
Ranalexin	Frog (skin)
Nigrocin	Frog (skin)

bacteria and yeast. Both the α- and β-defensins also act to mobilize cells involved in the adaptive immune response.

The human α-defensins, like hNP-3, are 29–35 residue peptides with three intrachain disulfides. They have a triple-stranded β-sheet structure, with a β-turn region that is rich in cationic residues. In the crystal structure, hNP-3 packs as a dimer.[18]

The β-defensins are much the same length as the α-defensins and they too have a triple-stranded β-sheet structure, but in the case of β-defensin-2 (**10.26**) at least, there is also an N-terminal α-helix (involving residues 4–10) which packs against the β-sheet and plays a key role in the aggregation of the peptide to form octamers.

(10.26)

The Group IV, ''loop-containing'' antimicrobial peptides, like ranalexin and nigrocin-1 (**10.27**) are α-helical peptides that have a single disulfide at their C-terminus. In many of these peptides, the α-helix is interrupted by a central proline residue, and the resulting kink in the structure is thought to serve as a hinge to facilitate ion channel formation in the microbial membrane.

Proline-kinked α-helices are also seen in the Group I antimicrobial peptides such as cecropin A, caerin 1.1, and maculatin 1.1 (**10.28**).

G.L.L.D.S.I.K.G.M.A.I.S.A.G.K.G.A.L.Q.N.L.L.K.V.A.S.C.K.L.D.K.T.C

(10.27)

G.L.F.G.V.L.A.K.V.A.A.H.V.V.P.A.I.A.E.H.F

(10.28)

10.5.3 Mechanisms of Action

The microbicidal activity of the antimicrobial peptides derives from their permeabilization of the microbial membranes and the subsequent release of cellular contents. The initial association between the peptides and the membranes is governed mainly by electrostatic forces — involving Coulombic interactions between their numerous cationic residues and the anionic lipid bilayer. As to the mechanism by which the peptides then cause permeabilization of the membrane, a variety of models have been proposed.

One mechanism of permeabilization is suggested to involve the formation of (fixed) ion pores in the microbial membrane. The existence of such pores and estimates of their likely dimensions have been based on measurements of membrane ion conductance and the passage of molecules of various sizes through model lipid vesicles and bilayers. A model of such a pore has been constructed on the basis of the x-ray crystal structure reported for hNP-3.[19] In this ''barrel stave'' or ''transmembrane helix bundle'' model, 12 hNP-3 monomers are arranged in the membrane with their nonpolar parts facing the lipid bilayer and their hydrophilic parts directed inwards, forming hydrophilic channels of around 20 Å diameter which span the membrane. Similar types of channels are suggested to account for the microbicidal activities of the peptides: maculatin, caerin, and alamethacin.

An alternative mechanism of membrane permeabilization has been proposed to account for the activities of other antimicrobial peptides including the magainins and cecropins. According to this ''carpet'' model, the antimicrobial peptides aggregate to form positively charged patches on the microbial membrane surface, effectively neutralizing the anionic lipid head groups over a wide area of the membrane, disrupting the integrity of the membrane, and causing transient gaps (or ''worm holes'') to be formed which allow the indiscriminate passage of ions and larger molecules. In the case of human β-defensin 2, the aggregates formed in the membrane may be the octamers seen in the peptide crystal structure.[20]

10.5.4 Development of Antimicrobial Peptides as Therapeutic Agents

Both natural and synthetic antimicrobial peptides have been evaluated as therapeutic agents for a variety of clinical conditions, including stomach ulcers caused by *Helicobacter pylori*, oral mucositis, pneumonia, and septic shock. Many of the early studies in this area were carried out by small Biotech companies, but the large pharmaceutical companies are now also showing interest. The antimicrobial peptides isolated from frog skin (magainins) were amongst the first to be developed for use as pharmaceuticals, with the magainin derivative, pexiganan, submitted for phase III clinical trials in the USA in 2003, with the intended application involving the treatment of polymicrobic diabetic foot ulcers. Brief details of these and other antimicrobial peptides currently in development as drugs are given in Table 10.7.

10.6 PEPTIDES AS ABSORPTION AIDS

In addition to their direct use as therapeutic agents, peptides can also be employed as vectors to improve delivery of other types of drug. In this role the peptide (typically a dipeptide) acts to improve the absorption of the parent drug, most often serving as a substrate for a membrane transporter. As the peptide moiety binds to the transporter, so the entire peptide–drug conjugate

Table 10.7 Some antimicrobial peptides currently under development as drugs

Peptide	Company	Target disease	Clinical status
Pexiganan (MSI-78)	Genaera	Infected diabetic foot ulcers	Phase III trials
Iseganan (protegrin)	Intrabiotics	Mucositis	Phase II trials
Daptomycin	Cubist Pharmaceuticals	Sepsis	Phase III trials
Lactoferricin-B	AM Pharma	Antifungal	Preclinical tests
Heliomycin	Entomed	Antibacterial	Preclinical tests

is taken across the cell membrane. For example, addition of dipeptide moieties to the antiviral nucleoside, acyclovir (ACV), yielding the prodrugs Val.Val-ACV (**10.29**) or Val.Tyr-ACV (**10.30**), not only enhances the drug's corneal permeability (probably through exploitation of the oligopeptide transporter on the cornea), but also increases the drug's solubility, and reduces its cytotoxicity, providing a superior means of treating herpes simplex viral infections (HSV-1).[21]

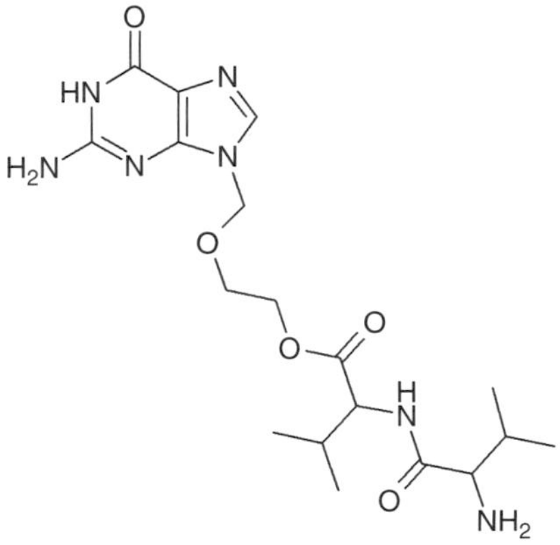

(10.29)

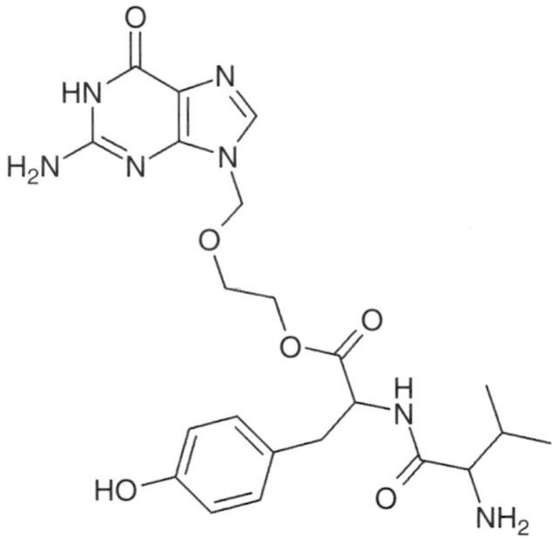

(10.30)

Likewise, it has been shown that prodrugs (**10.31**), prepared through the conjugation of peptide vectors to the anticancer agent, doxorubicin, can give a sixfold improvement in the uptake of the drug across the blood–brain barrier, leading to a 20-fold increase in doxorubicin levels in the brain parenchyma.[22]

One of the peptide vectors employed in this case was the so-called Trojan peptide, penetratin-1 (**10.32**). This peptide is a synthetic 16-residue peptide whose amino acid sequence derives from the third α-helix in the homeodomain of the *Drosophila melanogaster* transcription factor, Antennapedia.

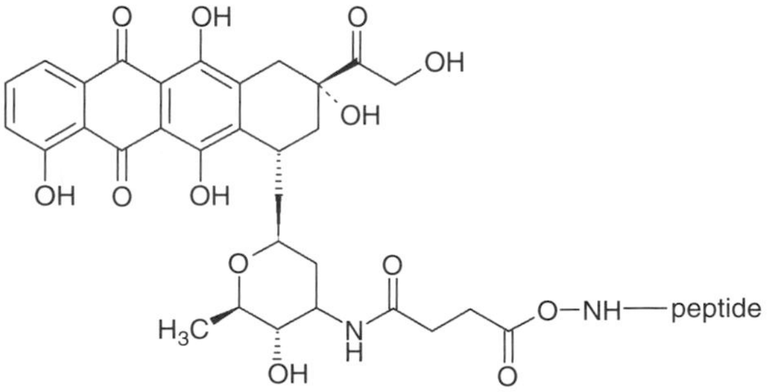

(10.31)

Arg.Gln.Ile.Lys.Ile.Trp.Phe.Gln.Asn.Arg.Arg.Met.Lys.Trp.Lys.Lys

(10.32)

The third α-helix in the homeodomain of Antennapedia has been shown to be responsible for the internalization of the protein into cells across the plasma membrane, and its synthetic peptide mimic, penetratin-1, has been shown to have similar translocation properties.[23] Intriguingly, it seems that the helicity of the penetratin peptide is unimportant, and that its translocation mechanism does not involve recognition by any kind of membrane receptor. Rather, it is suggested that the peptide interacts with the charged phospholipids in the outer leaflet of the membrane, causing the bilayer to become destabilized, and forming inverted micelles that then traverse the membrane and open into the cytoplasm.

Large peptides (identified through the use of phage display libraries) also show promise as a means to engineer increased uptake of peptides into tumors in the treatment of cancer. For example, by linking the peptide (**10.33**) to an antisense phosphorothioate oligonucleotide against the ErbB2 receptor, Shadidi and Sioud[24] achieved a highly specific delivery of the oligonucleotide to breast cancer cells, thereby effecting an inhibition of ErbB2 gene expression, which was not achieved using the free antisense oligonucleotide.

Leu.Thr.Val.Ser.Pro.Trp.Tyr

(**10.33**)

10.7 CONSIDERATIONS IN DESIGN OF ORALLY ACTIVE PEPTIDES

Given all of the foregoing discussions of the problems attendant upon the design of peptide drugs, one is naturally led to wonder whether the pharmaceutical potential offered by this class of molecule is ever likely to be exploited to any significant extent. As was noted earlier (Table 10.2), very few of the peptide drugs currently marketed or undergoing clinical trials are intended for oral use, and yet this route of administration remains the most favored. So, are orally active peptide drugs a real possibility, and what will be their general structural characteristics?

10.7.1 Considerations of Peptide Size and Potency

Let us first deal with the issue of peptide size and consider how small a peptide can be made and yet still provide an acceptable biological activity. The relationship between peptide potency and size has been conveniently quantified through analyses of the enzyme kinetic and x-ray crystallographic data available for complexes involving natural and synthetic peptidyl inhibitors of metallo-, serine, and aspartyl proteases.[25] The potency of each peptide is described simply by the equilibrium dissociation constant for its complexation with its target enzyme (K_i), and the sizes of the molecules are described with reference to their molecular surface areas. The peptide inhibitors' K_i values are correlated with the amounts of their surface areas that are buried upon complexation with their target enzymes. These *buried* surface areas (A_b, in Å^2) are calculated simply as the differences in the surface areas of the molecules in their complexed and uncomplexed (or free) states (respectively, A_c and A_f). A plot of the buried surface areas of peptide inhibitors against their $\log(K_i)$ values gives a respectable correlation (Figure 10.8), with a linear least-squares fit giving the relation:

$$-\log(K_i) = 1.02 + 0.04(A_b), \quad r = 0.87$$

Although the regression line thus accounts for only 76% of the variance in the $\log(K_i)$ data, this is not unexpected, because the parameter of buried surface area measures only the van der Waals and

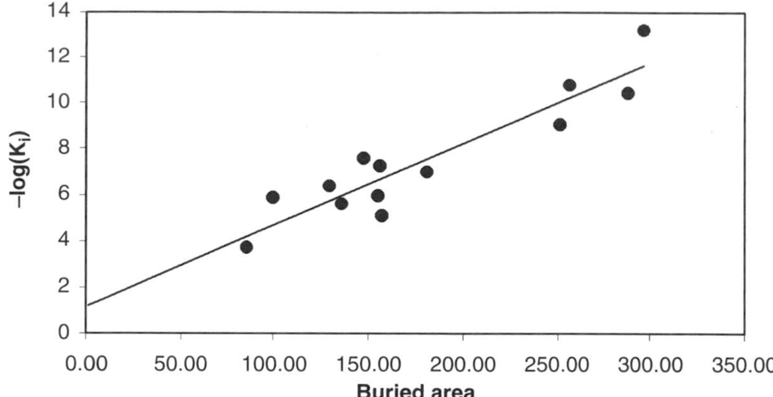

Figure 10.8 Peptide inhibitor potency vs. buried surface area. K_i are the inhibitors' molar equilibrium dissociation constants, and the buried areas are the areas of the molecules (in Å^2) that are made inaccessible to solvent when they form complexes with their target enzymes.

hydrophobic contributions to the stabilization of a complex, and does not account for the entropic, hydrogen bonding, and electrostatic components.

In order to relate the buried surface area of a peptide to its size, we simply relate A_b and A_f (Figure 10.9). Here, we find that for peptides with a molecular weight <700, there is a fairly good linear relationship, with a least-squares fit giving:

$$A_b = 57.7 + 0.4A_f, \quad r = 0.9$$

where A_b and A_f are again measured in Å^2.

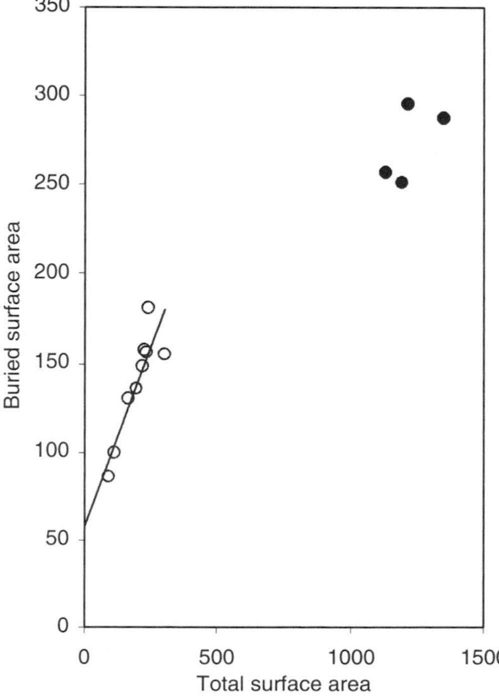

Figure 10.9 Peptide inhibitor total surface area vs. buried surface area (both areas measured in Å^2). Data for peptides with molecular weights >5000 are plotted as filled symbols; those for the smaller peptides (with molecular weights <700) are shown with open symbols.

With the larger peptides (all having molecular weights >5000), however, it is apparent that this relation no longer holds, and that a much smaller fraction of their total surface area is buried when they complex with their target enzymes. From this we can deduce that much of the structure of a large peptide can be considered ''redundant'' (at least as regards the recognition of its target receptor), and so it is concluded that a potent analog of the peptide could be produced which (given an appropriate sequence and conformation) may be made considerably smaller than the parent molecule.

From these relationships, therefore, it is clear that if we wish to develop a *small* peptide drug, the minimum acceptable size would seem to be around four to six amino acid residues. This size of molecule would have a molecular weight in the range 400–800, and could be expected to have a potency in the nanomolar range; the surface area of such a molecule buried upon complexation with its receptor target would be around 250 $Å^2$, and its surface area in its free state would thus be ~480 $Å^2$.

10.7.2 Considerations of Peptide Size and Specificity

Having thus determined the minimum size of peptide likely to exhibit a respectable affinity for its target receptor, it is necessary to establish whether so small a molecule will retain target specificity. Inspection of the various data presented in the literature shows that the size reduction is unlikely to present a problem. Table 10.8 shows some representative data, for two competitive reversible inhibitors of human renin: A-72517 (**10.34**) and CGP 38560 (**10.35**). Both of these inhibitors show good oral activity, and involve roughly four amino acid residues. A-72517 has a molecular weight of 706, and CGP 38560 a molecular weight of 730. Despite their relatively small size, however, it can be seen that the two peptides exhibit a remarkable degree of both species and enzyme specificity.

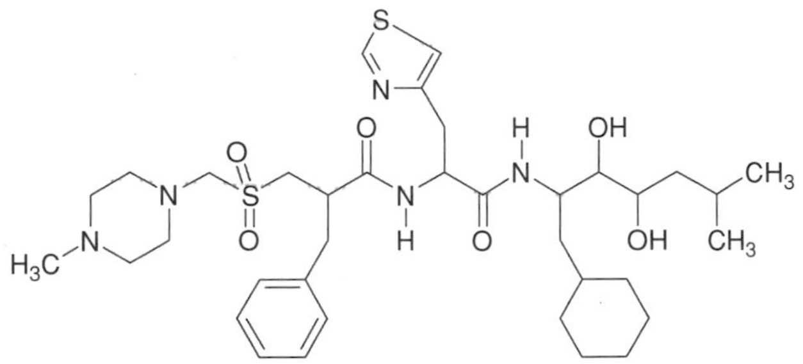

(10.34)

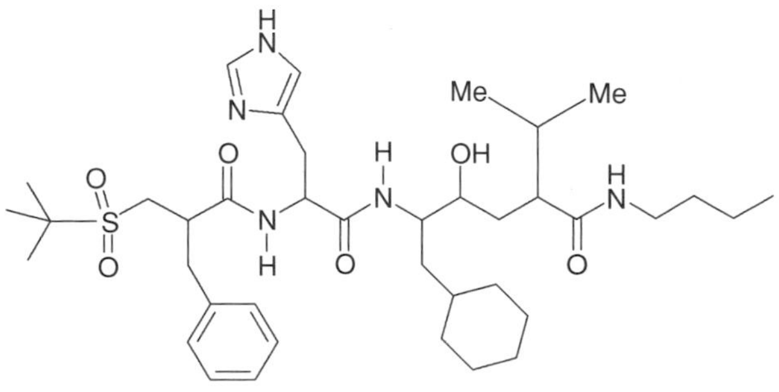

(10.35)

Table 10.8 In vitro specificities of orally active peptidyl inhibitors of human renin

Inhibition of various species forms of plasma renin by A72517 (10.34)[26]		Potency of CGP 38560 (10.35) against human aspartyl proteases[27]	
Species	**IC_{50} (n*M*)**	**Enzyme**	**K_i (n*M*)**
Human	1.1	Renin	0.4
Guinea pig	9.4	Cathepsin D	600
Dog	110	Gastriscin	3000
Rat	1400	Pepsin	6000

A-72517 is approximately ten times less effective against guinea pig renin as against human renin, is about 100 times less potent against dog renin, and shows negligible activity against rat renin.[26] CGP 38560 is a potent inhibitor of human renin, but displays insignificant activity against the related enzymes cathepsin D, gastricsin, and pepsin.[27]

It is concluded, therefore, that in striving for a potent and yet fairly small peptide drug, there will be no loss of specificity (and hence no increase in side effects) if the molecule involves only four amino acids.

10.7.3 Considerations of Peptide Lipophilicity

From the above observations, it is apparent that any peptide drug, which exhibits a good biological potency and specificity, will necessarily involve more than three amino acid residues. Since molecular modeling shows that the minimum diameter of such a molecule will exceed 10–12 Å, it seems unlikely that it would be readily able to traverse epithelial barriers via the paracellular route. It is also unlikely that it could cross transcellularly by active transport, since there is no evidence of any peptide transporter catering for molecules larger than three residues. There is a possibility that the peptide may undergo mediated absorption if it can be conjugated to a suitable carrier molecule, or that small amounts of it may gain access into cells by endocytosis. The most straightforward option to exploit, however, is simple, transcellular diffusion. With this mechanism of uptake the rate of absorption will be governed by the size of the peptide, and also by its lipophilicity: it has been shown that the rate of transport of peptides across Caco-2 cells decreases by roughly 50–70% per residue in the progression through di-, tri-, and tetrapeptides, and that molecules of comparable sizes exhibit a two- to sevenfold increase in absorption as their log(octanol−water partition coefficient) ($\log P$) is increased by 2 to 3 units.[28] For the purposes of this discussion, the conservative estimate is made that a successful (orally active) peptide drug must have a $\log P$ in the range 2–3.

As was noted earlier (Section 10.3.3), most natural peptides have a $\log P$ of the order of −1 and it is clear, therefore, that analogs of the parent compounds must be produced which are much less hydrophilic. In order to determine the extent of the modifications required, and to identify the sorts of modifications that are most appropriate, we need some way to predict the $\log P$ of a peptide based purely on consideration of its three-dimensional structure. Such predictions may be achieved in a relatively straightforward manner, by relating the $\log P$ values for peptides to the proportions of their molecular surface areas contributed by polar groups (that is, their % polar surface areas).[25] Figure 10.10 shows the correlation between $\log P$ and % polar surface area for a set of 15 peptides involving 3–11 residues. A linear least-squares fit to these data gives the relation:

$$\log P = 3.3 - 0.15\ (\%\ \text{polar area}), \quad r = 0.86$$

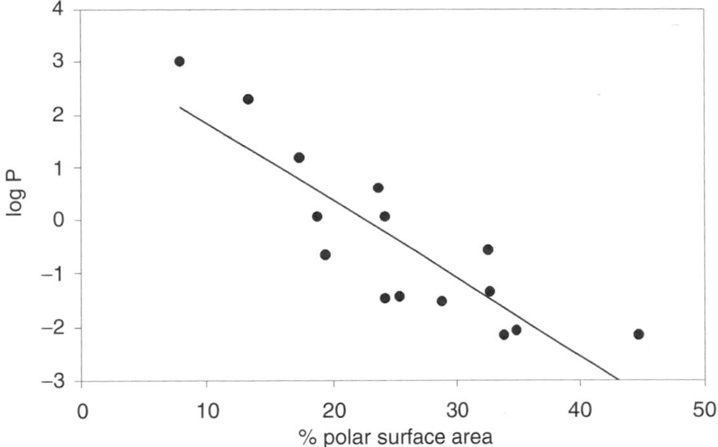

Figure 10.10 Peptide log *P* vs. % polar surface area. *P* are the octanol/buffer (pH 7.4) partition coefficients of the peptides. Surface areas measured in Å^2.

Quite simply, therefore, we see that the more lipophilic peptides like cyclosporin (**10.9**) have very few of their polar groups exposed on their surface, whereas the highly hydrophilic ones, like vasopressin (**10.20**), have the majority of these groups accessible to solvent (see Figure 10.11).

By extrapolation from Figure 10.10, it is apparent that for a peptide to have a log $P > 2$, it must have <10% polar surface area. If we thus combine this requirement with the earlier observation that a biologically potent and specific peptide drug must have a total surface area of around 480 Å^2, we learn that our readily absorbed (orally bioavailable) peptide drug must have a polar surface area of <48 Å^2.

10.7.4 Summary of Structural Requirements for Orally Active Peptide Drugs

On the basis of these simple analyses, therefore, we are able to define the essential structural features for a small, potent, target specific, and absorbable peptide drug: the required characteristics are found to be provided by a molecule involving four to six amino acid residues, with a molecular

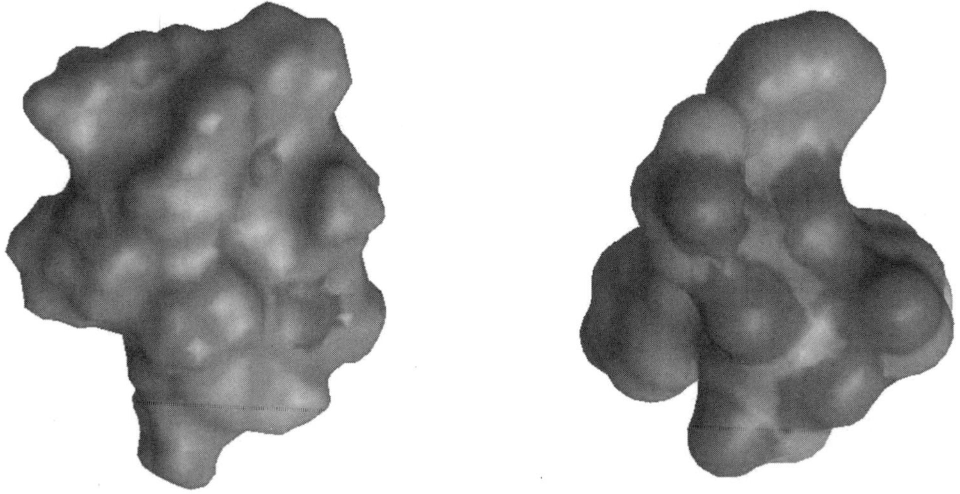

Figure 10.11 (See color insert after p. 368) Solvent accessible surfaces of the peptides cyclosporin (left) and lysine vasopressin (right). Regions of the surfaces that are contributed by polar atoms are colored red (O atoms) and blue (N atoms), the nonpolar parts of the surfaces being colored green.

weight in the range 400–800, a total surface area of around 480 Å^2, and a polar surface area of <48 Å^2.

In order to test these hypotheses, we may again refer to the structures of the two peptidyl inhibitors of human renin, CGP 38560 and A-72517. Both of these molecules have IC_{50}/K_i values in the nanomolar range (Table 10.8) and, in line with the observations recorded above, it is calculated that they have total surface areas of ~350 Å^2. The two compounds have molecular weights of roughly 700, and involve the equivalent of four amino acid residues. As regards their lipophilicity, since both molecules are calculated to have around 13% polar surface area, they should have log *P* values somewhat greater than 2. Accordingly, it would be predicted that both of these compounds should show a significant rate of membrane permeation by simple (transcellular) diffusion and, by extension, could be expected to have respectable oral bioavailabilities. In fact the reported log *P* for A-72517 is 4.57,[26] and both this compound and CGP 38560 are reported to show better than 10% oral bioavailability.[26,27]

It would thus seem that small, orally active peptide drugs are perfectly attainable. A closer inspection of CGP 38560 and A-72517, however, shows that "successful" peptide drugs are likely to have very limited peptide character; both of these renin inhibitors involve almost no natural amino acids, and incorporate a variety of nonpeptide moieties. The reasons for this are made apparent when we consider the requirement that absorbable peptides of this size must have <48 Å^2 polar area. Given that the charged N- and C-termini and the peptide bonds contribute, respectively, 16, 20, and 8 Å^2 to a peptide's surface area, in a true peptide of four to six residues, the total amount of polar surface area contributed by its two termini and peptide bonds will be 60–75 Å^2. Even without considering any polar groups involved in the amino acid side chains, therefore, this is already greater than our limit of 48 Å^2. If we then consider an analog that has the N- and C-termini removed or blocked in some way, the polar area due to the peptide's peptide bonds will be just below our 48 Å^2 limit, at 20–40 Å^2. There is still little scope for introducing polar groups into the peptide's side chains, therefore, and so our analog must be modified yet further. As was noted earlier, the necessary reduction in polar surface area could be achieved through methylation of some or all of the peptide's NH groups, or by cyclizing the molecule so that its peptide CO and NH groups are occupied in intramolecular hydrogen bonding. Both of these modifications would also serve to protect the molecule against attack by exo- and endopeptidases.

As the culmination of all the changes needed to produce our clinically useful peptide drug, we thus find that our "peptide" drug now bears very little resemblance to a peptide.

REFERENCES

1. Stone, S.R. and Hofsteenge, J. (1986) Kinetics of inhibition of thrombin by hirudin. *Biochemistry* **25**, 4622.
2. Yanagisawa, M., Kurihara, H., Kimura, S. and Tomobe, Y. et al. (1988) A novel potent vasoconstrictor peptide produced by vascular endothelial cells. *Nature* **332**, 411.
3. Lee, V.H.L. and Yamamoto, A. (1990) Penetration and enzymatic barriers to peptide and protein absorption. *Advanced Drug Delivery Reviews* **4**, 171.
4. Ganderton, D. The development of peptide and protein pharmaceuticals. In: R.C. Hider and D.J. Barlow (eds), *Polypeptide and Protein Drugs*. Ellis Horwood, London, 1991, 211 pp.
5. Fox, R.O. and Richards, F.M. (1982) A voltage-gated ion channel model inferred from the crystal structure for alamethicin at 1.5 Å resolution. *Nature* **300**, 325.
6. Tam, J.P., Liu, W., Zhang, J.-W. and Galantino, M. *et al.* (1994) Alanine scan of endothelin: importance of aromatic residues. *Peptides* **15**, 703.
7. Marqusee, S. and Baldwin, R.L. (1987) Helix stabilisation of $Glu^- \ldots Lys^+$ salt bridges in short peptides of de novo design. *Proceedings of the National Academy of Sciences USA* **84**, 8898.
8. Blundell, T.L. and Wood, S.P. (1982) The conformation, flexibility and dynamics of polypeptide hormones. *Annual Reviews of Biochemistry* **51**, 123.

9. Theriault, Y., Boulanger, Y., Weber, P.L. and Reid, B.R. (1987) Two-dimensional ^{1}H-NMR investigations of the water conformation of atrial natriuretic factor (ANF 101–126). *Biopolymers* **26**, 1075.
10. Wood, S.P., Tickle, I.J., Treharne, A.M., Pitts, J.E. et al. (1986) Crystal structure analysis of deamino-oxytocin: conformational flexibility and receptor binding. *Science* **232**, 633.
11. Rose, J.P. and Wang, B.-C. (2001) Structures of an uncomplexed neurophysin and its vasopressin complex. *Protein Science* **10**, 1869.
12. Montelione, G.T., Wuthrich, K., Burges, A.W., Scheraga, H.A. et al. (1992) Solution structure of murine epidermal growth factor determined by NMR spectroscopy and refined by energy minimisation with restraints. *Biochemistry* **31**, 236.
13. Andersen, N.H., Chen, C., Marschner, T.M., Krystek, S.R. et al. (1992) Conformational isomerism of endothelin in acidic aqueous media. *Biochemistry* **31**, 1280.
14. Perkins, T.D.J., Hider, R.C. and Barlow, D.J. (1990) Proposed solution structure of endothelin. *International Journal of Peptide and Protein Research* **36**, 128.
15. Boguslavsky, V., Hruby, V.J., O'Brien, D.F., Misicka, A. et al. (2003) Effect of peptide conformation on membrane permeability. *Journal of Peptide Research* **61**, 287.
16. Muranishi, S., Sakai, A., Yamada, K., Murakami, M. et al. (1991) Lipophilic peptides — synthesis of lauroyl thyrotropin releasing hormone and its biological activity. *Pharmaceutical Research* **8**, 649.
17. Yamamoto, A., Yamada, K., Tenma, T., Douen, T. et al. (1992) Improvement of intestinal absorption of peptides and proteins by chemical modification with fatty acids. *Journal of Pharmacobiodynamics* **15**, 28.
18. Hill, C.P., Yee, J., Selsted, M.E. and Eisenberg, D. (1991) Crystal structure of defensin HNP-3, an amphiphilic dimer: mechanisms of membrane permeabilization. *Science* **251**, 1481.
19. Fujii, G., Selsted, M.E. and Eisenberg, D. (1993) Defensins promote fusion and lysis of negatively charged membranes. *Protein Science* **2**, 1301.
20. Hoover, D.M., Rajashankar, K.R., Blumethal, R., Puri, A. et al. (2000) The structure of human β-defensin-2 shows evidence of higher order oligomerization. *Journal of Biological Chemistry* **275**, 32911.
21. Anand, B.S., Nashed, Y.E. and Mitra, A.K. (2003) Novel dipeptide prodrugs of acyclovir for ocular herpes infections. *Current Eye Research* **26**, 151.
22. Rouselle, C., Clair, P., Lefauconnier, J.M., Kaczorek, M. et al. (2000) New advances in the transport of doxorubicin through the blood–brain barrier by a peptide vector-mediated strategy. *Molecular Pharmacology* **57**, 679.
23. Derossi, D., Chaissaing, G., Prochaintz, A. (1998) Trojan peptides: the penetratin system for intracellular delivery. *Trends in Cell Biology* **8**, 84.
24. Shadidi, M. and Sioud, M. (2002) Identification of novel carrier peptides for the specific delivery of therapeutics into cancer cells. *FASEB Journal* **16**, U478.
25. Barlow, D.J. and Satoh, T. (1994) The design of peptide analogues for improved absorption. *Journal of Control Release* **29**, 283.
26. Kleinert, H.D., Rosenberg, S.H., Baker, W.R., Stein, H.H. et al. (1992) Discovery of a peptide-based renin inhibitor with oral bioavailability and efficacy. *Science* **257**, 1940.
27. Wood, J.M., Criscione, L., Degasparo, M., Buhlmayer, P. et al. (1989) CGP 38 560 — orally active, low molecular weight renin inhibitor with high potency and specificity. *Journal of Cardiovascular Pharmacology* **14**, 221.
28. Conradi, R.A., Higers, A.R., Ho, N.F.H. and Burton, P.S. (1991) Influence of peptide structure on transport across Caco-2 cells. *Pharmaceutical Research* **8**, 1453.

FURTHER READING

Hider, R.C. and Barlow, D.J. (1991) *Polypeptide and Protein Drugs*. Ellis Horwood, London, 1991.

Hruby, V.J. (2002) Designing peptide receptor agonists and antagonists. *Nature Reviews — Drug Discovery* **1**, 847.

Koczulla, A.R. and Bals, R. (2003) Antimicrobial peptides. *Drugs* **63**, 389.

Lee, V.H.L. (1988) Enzymatic barriers to peptide and protein absorption. *Critical Reviews in Therapeutic Drug Carrier Systems* **5**, 69.

Lien, S.L. and Lowman, H.B. (2003) Therapeutic peptides. *Trends in Biotechnology* **21**, 556.

Sood, A. and Panchagnula, R. (2001) Peroral route: an opportunity for protein and peptide drug delivery. *Chemical Reviews* **101**, 3275.

Witt, K.A., Gillespie, T.J., Huber, J.D., Egleton, R.D. et al. (2001) Peptide drug modifications to enhance bioavailability and blood–brain barrier permeability. *Peptides* **22**, 2329.

11

Combinatorial Chemistry: A Tool for Drug Discovery

Barrie Kellam

CONTENTS

11.1 INTRODUCTION

The extraordinary advances in biology and biotechnology over recent decades have revealed an unprecedented number of new drug targets. This has been intrinsically coupled to major advances in high-throughput screening (HTS), allowing pharmaceutical companies to biologically evaluate large numbers of potential drug candidates extremely rapidly as these new targets emerge. When presented alongside the efforts of the pharmaceutical companies to lower the cost of drug development, it has become increasingly important for discovery chemists to be able to address these issues and ultimately provide greater chemical diversity in a more cost-effective and time-efficient manner. Traditional approaches utilized ''primary screening'' of collections of chemical entities, either maintained by the pharmaceutical company or acquired from external sources, to ascertain structural information on molecules that bind to a novel target. Such compound libraries are potentially expensive to maintain, and may present long-term stability problems. In addition, identification of a new lead molecule does not always allow the rapid synthesis of numerous analogs to further explore structure–activity relationships (SAR).

Put in its simplest terms, combinatorial chemistry is a technique for producing the large number of required compounds in a short time period. Depending on the technique employed, anywhere from dozens to hundreds of thousands of compounds can be generated in a resource-efficient manner. This chapter will explore the fundamentals of combinatorial library synthesis and its impact on modern drug discovery.

11.2 THE DRUG DISCOVERY PROCESS

The drug discovery process comprises a number of different stages, each requiring different numbers, quality, and quantities of synthesized materials. The first stage is often referred to as the ''Lead Discovery Stage'' and the primary goal of this exercise is to identify chemical substances that bind to the pharmacological target. As previously mentioned, this often involves the development of a HTS assay to analyze in-house compound libraries. Some of these collections contain anywhere from 100,000 to over 500,000 individual members and are often somewhat historically biased toward chemical classes developed or acquired by the company. Very small quantities of material are required for primary screening; however this does result in an increasing number of samples becoming depleted. Usually at the end of the lead discovery stage, several compounds that bind to the target have been identified and these serve as the basis for a medicinal chemistry program that will enable the development of a new drug entity; the so-called ''lead optimization stage.'' Thus, the ability to rapidly synthesize numerous analogs of the identified leads is a critical next stage. Combinatorial chemistry can play a significant part in all of these early stages of drug discovery. In brief, combinatorial chemistry allows the rapid and inexpensive synthesis of large numbers of compounds, and these ''combinatorial libraries'' can significantly enhance the chemical diversity of established collections. In principle, analogs of identified leads can also be readily synthesized using combinatorial techniques such as parallel synthesis, enabling the simultaneous preparation of hundreds to thousands of compounds in quantities sufficient for the lead optimization process.

Combinatorial chemistry has already made a significant impact on drug discovery and all pharmaceutical companies have established programs in this arena. Indeed, there is a continuing emergence of many small companies whose focus is upon various aspects of combinatorial technology and this combined focus of efforts has initiated the development of numerous complementary technologies in high-throughput purification and analysis. There have also been significant advancements in terms of automation, information technologies, and other related areas.

11.3 THE CONCEPTS OF COMBINATORIAL SYNTHESIS

11.3.1 Libraries

Let us initially consider how a range of simple chemical building blocks can be utilized to generate libraries of diverse compounds. Let us start with the simple example of two building blocks ''X'' and ''Y'' reacted together to produce a single product ''XY'' (Figure 11.1). If we assume that there are five derivatives of ''X'' available to react with ''Y,'' then it is obvious that we could produce five discrete compounds, namely X^1Y to X^5Y. Making the situation a little more complex, let us consider what would happen if we also had five derivatives of ''Y'' available for our synthesis. In this situation X^1 could be reacted with all ''Y'' derivatives to give five discrete products. Likewise this could also be achieved with X^2, X^3, X^4, and X^5 giving us $5 \times 5 = 25$ new compounds in total (Figure 11.1).

Combinatorial chemistry as a means of generating large libraries of compounds heralded from peptide and oligonucleotide chemistry and the former provides a perfect illustration of the power of numbers that can be potentially generated. If you were to employ the 20 natural amino acids for the synthesis of a pentapeptide library, you would produce a massive 3,200,000 compounds (20^5). The realization of this synthetic achievement was greatly facilitated by the employment of solid-phase peptide synthesis, initially developed by Bruce Merrifield in 1963. As the technology was further embraced and moved away from the peptide arena, a renaissance in solid-phase organic chemistry was palpable (Figure 11.2).

Solid-phase synthesis (SPS) was an extremely powerful technique for the synthesis of polypeptides and oligonucleotides, since their construction necessitated the repetitive cycle of a high yielding, room temperature coupling reaction, washing, protecting group removal, washing, and subsequent coupling. This laborious process therefore lent itself extremely well to automation on peptide and oligonucleotide synthesizers. However, the chemistry required for the synthesis of these biopolymers was quite limited and a huge volume of research was therefore undertaken during the 1990s to enhance the scope and application of SPS to the construction of more ''drug-like'' molecules. As such, Jonathan Ellman's seminal paper on the SPS of a 1,4-benzodiazepine library provided a clear demonstration that this was a tractable proposition and confirmed Merrifield's 1969 vision that ''a goldmine awaits discovery by organic chemists . . . . . .'' We shall return to this important area later in the chapter.

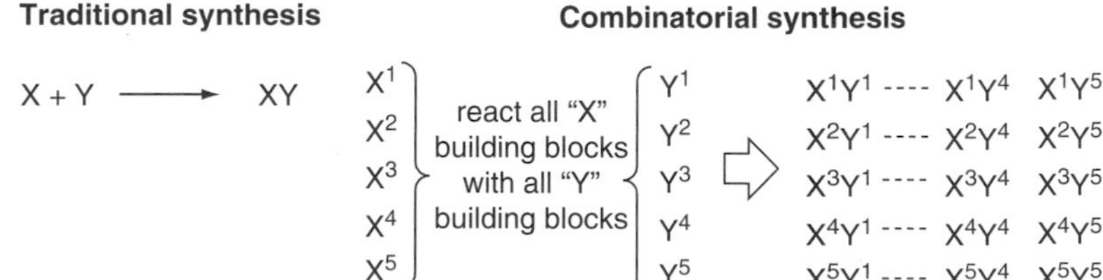

Figure 11.1 A representation of traditional and combinatorial synthesis.

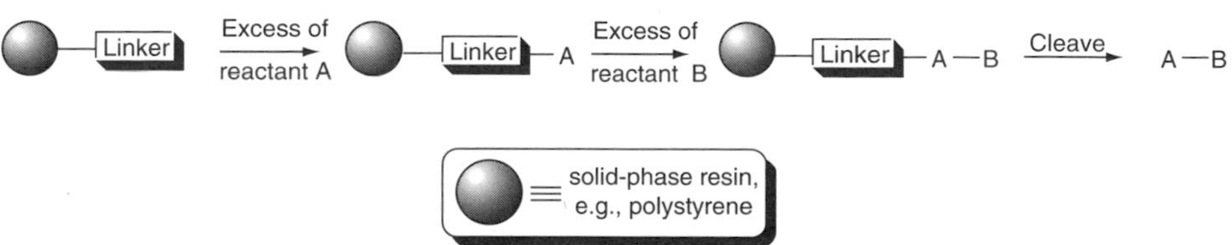

Figure 11.2 A representation of the general principles of solid-phase synthesis.

11.3.2 The Mix and Split Method

This simple and elegant method of generating chemical diversity was an enormous factor in the developmental history of combinatorial chemistry and further empowered the use of solid-phase organic chemistry. The Hungarian chemist Arpad Furka first reported the technique in 1988, focusing then on peptide synthesis. However, the independent research efforts of the groups of Richard Houghten and Kit Lam truly pioneered the concept and resulted in an explosion of the research and development of combinatorial techniques throughout the late 1980s and the 1990s.

It is probably easiest to illustrate the concepts of ''mix and split'' synthesis using peptide chemistry as the exemplar. Let us imagine that we wanted to make all the possible tripeptides of the three amino acids leucine (Leu), phenylalanine (Phe), and glycine (Gly). Initially we would take a batch of resin and divide it into three equal portions, each of which would then be placed in an individual reaction vessel (Figure 11.3). The resin in reaction vessel 1 would then be acylated by the amino acid leucine, that in reaction vessel 2 with phenylalanine, and that in reaction vessel 3 with glycine.

After washing the individual batches of resin beads to remove excess reagents and by-products, the beads are pooled together in one pot and thoroughly mixed before getting split into equal portions again and placed in three fresh reaction vessels. You can now see how the name ''mix and split'' was coined. As before, the beads in reaction vessel 1 are acylated with leucine, reaction vessel 2 with phenylalanine, and reaction vessel 3 with glycine. It is important to realize that because the original three resin batches were mixed and split, after the second coupling step, each reaction vessel will contain three different dipeptides (Figure 11.4).

Altogether, we have generated nine new compounds in just two steps. If we then repeat the process again we will produce a total of 27 tripeptides ($3^3 = 27$) in just three steps (Figure 11.5).

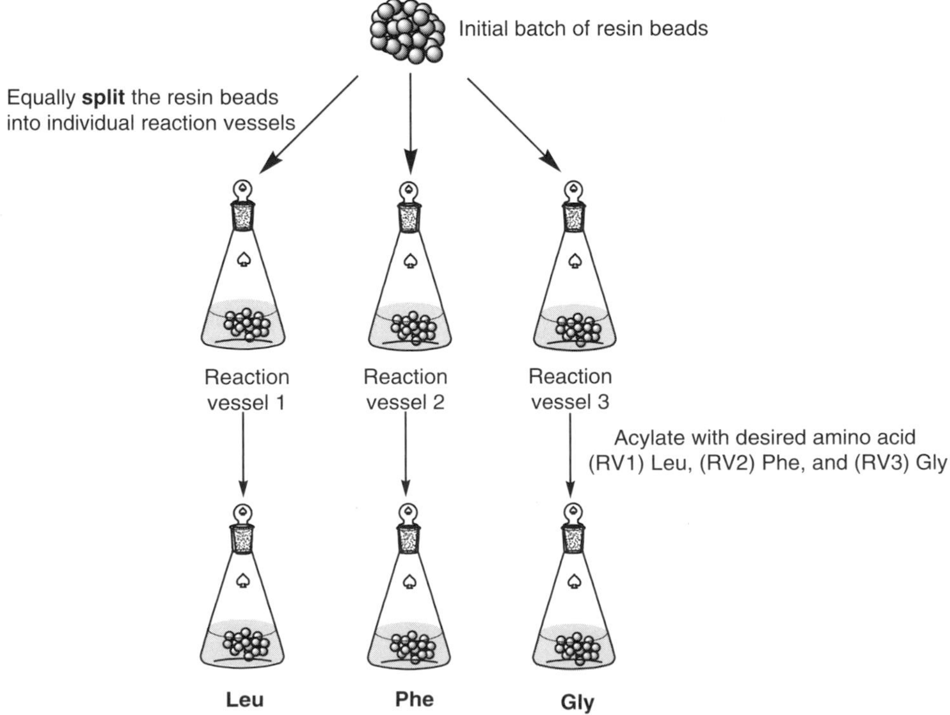

Figure 11.3 Stage 1 of the mix and split method.

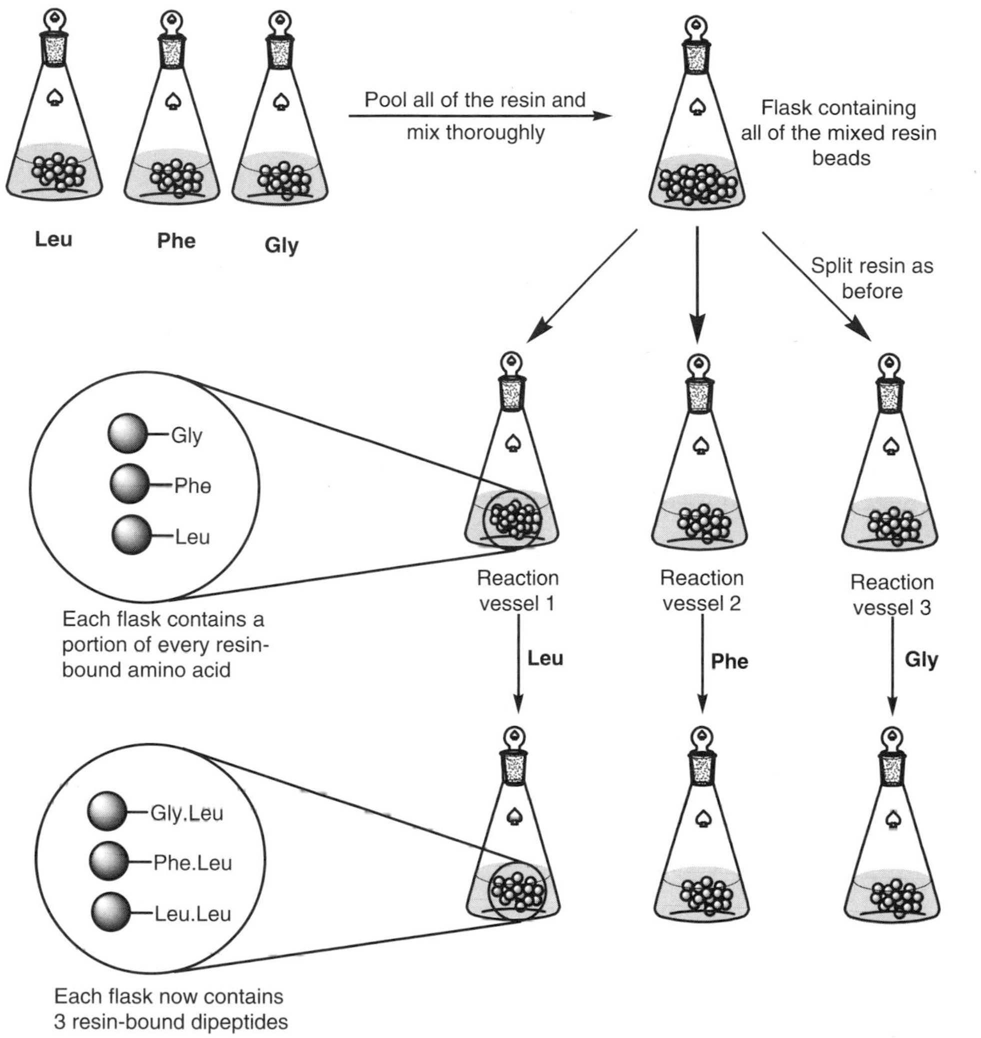

Pool all of the resin and mix thoroughly
Flask containing all of the mixed resin beads
Leu
Phe
Gly
Split resin as before
Gly
Phe
Leu
Each flask contains a portion of every resin-bound amino acid
Reaction vessel 1
Reaction vessel 2
Reaction vessel 3
Leu
Phe
Gly
Gly.Leu
Phe.Leu
Leu.Leu
Each flask now contains 3 resin-bound dipeptides

Figure 11.4 Stage 2 of the mix and split method.

Gly.Leu.Leu
Phe.Leu.Leu
Leu.Leu.Leu
Gly.Gly.Leu
Phe.Gly.Leu
Leu.Gly.Leu
Gly.Phe.Leu
Phe.Phe.Leu
Leu.Phe.Leu
Gly.Leu.Phe
Phe.Leu.Phe
Leu.Leu.Phe
Gly.Gly.Phe
Phe.Gly.Phe
Leu.Gly.Phe
Gly.Phe.Phe
Phe.Phe.Phe
Leu.Phe.Phe
Gly.Leu.Gly
Phe.Leu.Gly
Leu.Leu.Gly
Gly.Gly.Gly
Phe.Gly.Gly
Leu.Gly.Gly
Gly.Phe.Gly
Phe.Phe.Gly
Leu.Phe.Gly

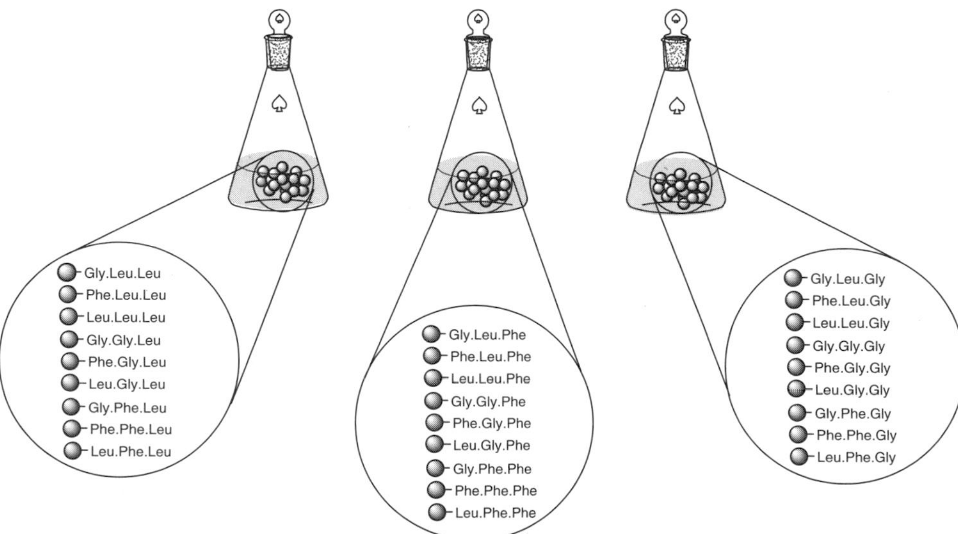

Figure 11.5 The final outcome of a 3 amino acid tripeptide mix and split library synthesis ($3^3 = 27$ tripeptides).

While we have illustrated the concept of "mix and split" synthesis using a peptide example, one can envisage that any set of monomer units can be assembled into a diverse library using the same strategy.

Once synthesized, one now faces the issue of finding biological activity from screening such library mixtures. This process can be extremely rapid, however deconvoluting the identity of a single active molecule provides more of a challenge. One has to realize that using this technique, all combinations of the building blocks are present in the test mixture and the most potent compound, in this example, is just one of 27 discrete molecules. One successful approach to identifying the structure of the active compound is to undertake a recursive deconvolution, i.e., perform further rounds of synthesis and screening. Another approach comprises tagging the resin bead to afford information on the structure. These and other techniques will be discussed further in Section 11.4.3.

11.3.3 Position Scanning Libraries

This is a variation on the mix and split approach whereby the same library compounds are prepared several times as so-called "sublibraries," and for each of the sublibraries discrete mixtures are synthesized where a different residue is held constant. Let us illustrate this by considering the synthesis of a complete tripeptide library utilizing all 20 proteinogenic amino acids; this would therefore contain 8000 members in total. The whole library could be broken down into three sublibraries containing 8000 tripeptides, where one of the amino acids in the sequence is held constant. The first sublibrary could also be further divided into 20 discrete mixtures (consisting of 400 compounds per mixture) in which the first amino acid was constant within each individual mixture yet was different between mixtures. The second sublibrary would also consist of 20 mixtures in which the second amino acid was constant within each mixture but different between mixtures, with the same principle applied to the third sublibrary mixtures, only keeping the third amino acid constant. By then testing the activity of the 60 individual mixtures one obtains a readout of which amino acid is favored in each of the three positions. One cannot assume that the best three amino acids when combined as a tripeptide will produce a highly active compound since the activity measured is the combined effect of 400 tripeptide structures in the mixture. It is therefore much better to identify the best three or four amino acids for each position and synthesize all the possible variations these could produce.

Of course, the example given provides the entire positional scan for a tripeptide sequence. Often, peptide libraries are constructed where only a sample of the 20 proteinogenic amino acids are included, making the size and complexity of the library considerably easier to manage.

11.4 SPS OF "DRUG-LIKE" LIBRARIES

11.4.1 Resins and Linkers

The use of peptides to illustrate the simple solid-phase processes involved in library generation was fine, however for combinatorial chemistry to benefit the pharmaceutical industry, the expeditious synthesis of small organic "drug-like" molecules was required. This significantly raises the complexity of the tasks since the required synthetic protocols are potentially far more varied and demanding, depending on the class of molecule the research team are interested in pursuing.

At the inception of a project one has to choose an appropriate solid support. Solid-phase organic synthesis can take place on a variety of these including polymeric resin beads, polyethylene pins, crowns, tubes, photoresponsive chips, paper, glass, cotton, and membranes. The term "solid-phase" is actually rather misleading, since reactions do not take place in, or on the surfaces of the solid-phases, which are in heterogeneous contact with the solutions. In fact, the reactions occur in swollen gel systems formed by the penetration of solvent and solute molecules into the polymeric matrix. To date, the most commonly used resins are: (i) polystyrene cross-linked with 1–2% divinylbenzene and (ii) TentaGel — 80% polyethylene glycol (PEG) grafted to cross-linked polystyrene.

While the former remains the more physicomechanically and thermally robust, the PEG "spacer arms" facilitate projection of the resin-bound substrate into the bulk solvent, generating a reactivity environment closer to solution-phase chemistry. It is important to remember that in addition to the choice of an appropriate solid support, choice of the linker, solvent, reagent concentrations, and temperature are also the key to establishing a successful solid-phase library synthesis.

The linker acts as a "handle" for the attachment of the initial building block to the solid support. The linker needs to be cleavable if one is to remove the completed molecule from the solid support at the end of the synthesis. In this respect one can liken the linker to a type of protecting group (extensively used in traditional solution-phase organic synthesis) as the principles of what make a good protecting group are virtually identical to what make a good linker.

Linkers cleaved under acidic conditions remain popular, principally because they already possessed a well-established track record in solid-phase peptide synthesis. Generally these linkers release carboxylic acids and amides upon cleavage (Table 11.1).

While the linkers illustrated all release acids or amides, it is important to note that there are now many commercially available variants which release a multitude of alternative functional groups and the reader is directed to some of the excellent suggestions in the "further reading" section to explore this area in more detail.

11.4.2 Monitoring Reactions "On-Bead"

This objective remains a major headache for all chemists undertaking solid-phase organic chemistry. The pursuit of traditional solution-phase synthesis allows the routine employment of simple analytical techniques such as thin-layer chromatography (TLC) to monitor the progress of a reaction. However, this luxury is not an option for the solid-phase chemist since the molecule of interest remains linked to the insoluble support. One can envisage that cleavage of a small sample of resin, followed by analysis could provide the required information, yet this relies upon the resin-bound intermediate being stable to the cleavage conditions. In addition, the process adds a potentially time-consuming operation, which may not be appropriate for a chemist attempting to monitor the path of a transformation in the middle of a long synthetic route. To this end, much research has been focused on surmounting these issues and there now exist a range of options with varying degrees of advantages and disadvantages. For example, one can employ traditional elemental analysis through combustion of the molecule–resin conjugate. Alternatively one can use colorimetric tests for the presence of key functional groups, e.g., the purple color produced by ninhydrin in the presence of a primary amine. More recently the introduction of certain spectroscopic methods such as FTIR, gel phase, and magic angle spinning (MAS) NMR have become more popular.

11.4.3 Parallel Synthesis

Having examined the basics of the materials and methods that can be applied to facilitate synthesis on solid phase, we shall now explore some key methods employed in combinatorial library generation. We looked at the "mix and split" approach earlier. If you think in more detail about this technique one soon realizes that each individual resin bead, despite having undertaken a potentially vast array of manipulations and chemistries, can only contain one sequence of building blocks. If one extrapolates this observation we can start to think about parallel synthesis techniques whereby a single reaction product is produced in each reaction vessel.

Table 11.1 Some commonly employed linkers for the release of acids and amides (these linkers must be acylated by a carboxylic acid to afford the above derivatives)

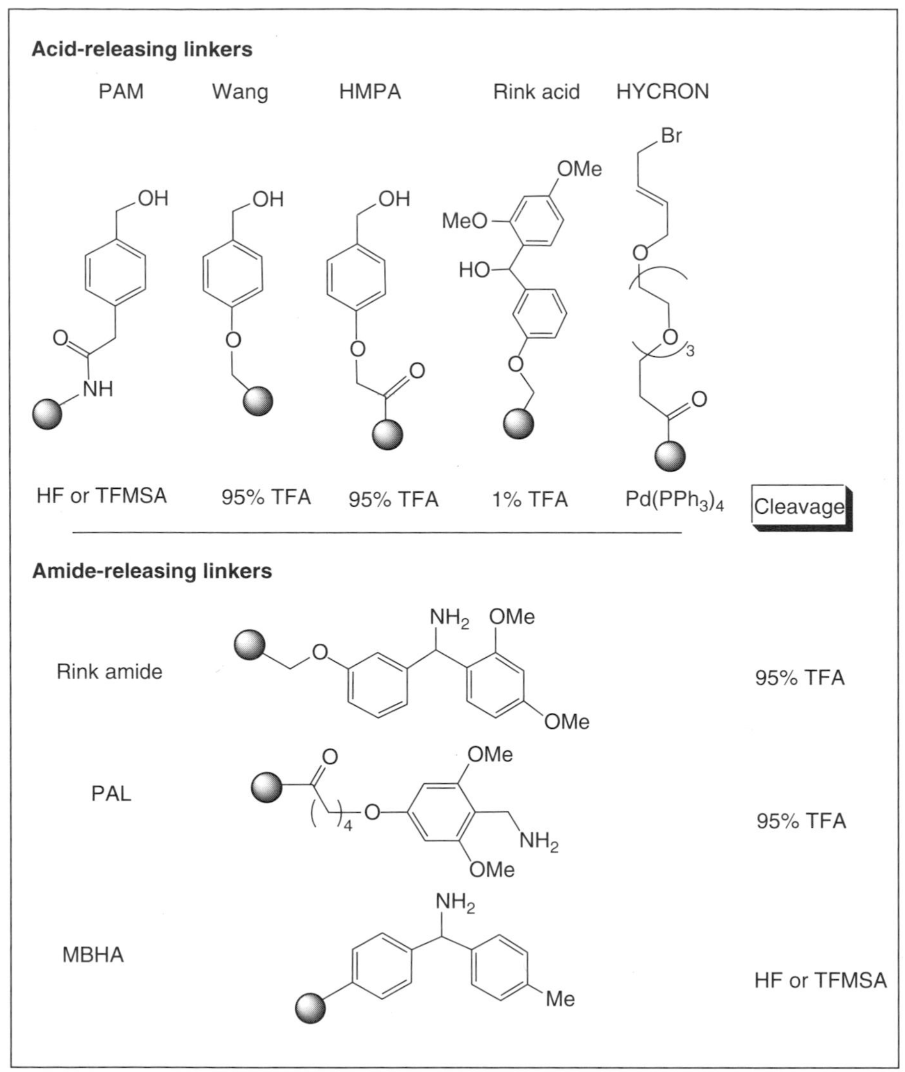

Houghten's "tea bag" methodology

This is an elegant manual approach to parallel synthesis that has been employed for the generation of numerous libraries. We will again use the synthesis of a peptide-based library to illustrate the concept. Initially, the resin (~100 mg) is distributed into individual polypropylene meshed bags (3 × 4 cm) (these look like miniature tea bags — hence the name associated with this technique) and each bag is sealed and labeled (Figure 11.6).

The tea bags are then distributed into individual reaction vessels and the resin is acylated with a specific amino acid. The bags can then be washed and pooled for communal deprotection and washing. The tea bags can then be redistributed into fresh reaction vessels for the addition of the next amino acid, with the cycle repeating until the desired peptide length is achieved. This technique is extremely cheap and as such can be easily undertaken within any research laboratory.

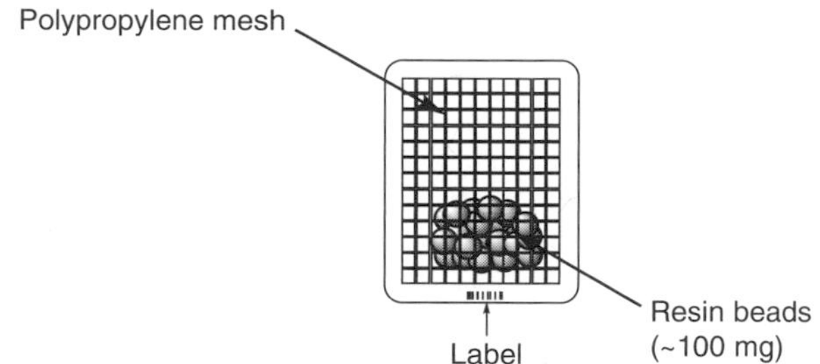

Figure 11.6 Houghten's tea bag.

However, it is quite labor intensive thereby limiting the quantity and speed at which new compounds can be synthesized. It was primarily the latter reason, which meant that this process was not fully embraced by the pharmaceutical industry, which instead preferred to use automation (or at least semiautomation) for their parallel synthesis methods.

Automated parallel synthesis

An excellent example of a technique that can be used in both manual and automated methodologies is the solid-phase support in the form of sticks or pins (4 mm diameter, 40 mm length). As such, these pins can be immersed into individual reaction vessels or well plates (Figure 11.7) .

Common operation such as washing or protecting group removal can be performed in a reaction bath, whereby all of the pins are immersed in the same environment. Each pin generally supports between 80 and 300 nmol of material.

11.4.4 Deconvolution and Encoding Libraries

If we now move forward to the situation where the synthesis of a library has been successfully accomplished and a compound mixture has proven to be biologically active, we now face the daunting task of identifying the active molecule (or molecules). This process is called *deconvolution* and there are several approaches to achieving this goal.

Recursive deconvolution

Let us think back to our tripeptide library synthesized in Section 11.3.2, which served to illustrate the mix and split approach. If we were to test the three final mixtures of tripeptides we could quickly identify which one produced the best biological response and thereby identify the N-terminal

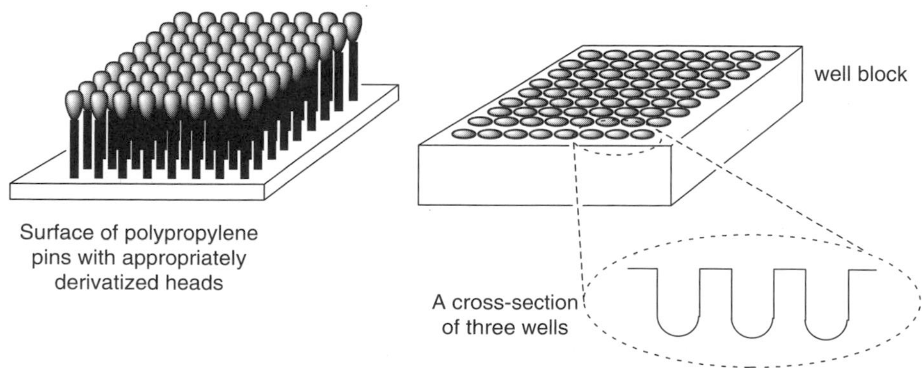

Figure 11.7 Representation of solid-phase pins.

residue associated with it. From Figure 11.5, if the most active mixture were the middle one, then the favored N-terminal amino acid would be phenylalanine. With this information in hand we have two options open to us. Firstly, we could go away and synthesize all nine possible tripeptides terminating with Phe, but this seems a little long-winded (and imagine if we were dealing with a hexapeptide library of these three amino acids, we would then have to synthesize 243 individual peptides). Alternatively, if we had kept small samples of the three dipeptide mixtures produced during the library synthesis (i.e., those illustrated in Figure 11.4), we could now acylate phenylalanine to each mixture, affording the nine tripeptides in three separate mixtures (Figure 11.8). If these mixtures are then tested, we can identify both the second and third amino acids, which give us the best biological activity. Now it is a straightforward task to synthesize the three possible tripeptides from this mixture and ultimately identify the most active member.

Clearly this example serves to illustrate a point, and the appropriate use of the mix and split approach, coupled with the judicious retention of useful intermediate mixtures for recursive deconvolution can significantly help in economizing the effort involved in dealing with much larger libraries.

Multiple release resins

Linkers have been designed which allow the release of a certain amount of the target molecule from the insoluble support. This way, a consecutive series of cleave and analyze steps can be undertaken until the active bead is identified. At this stage final cleavage will allow spectroscopic identification of the active compound. This approach can be illustrated by a double cleavage linker originally

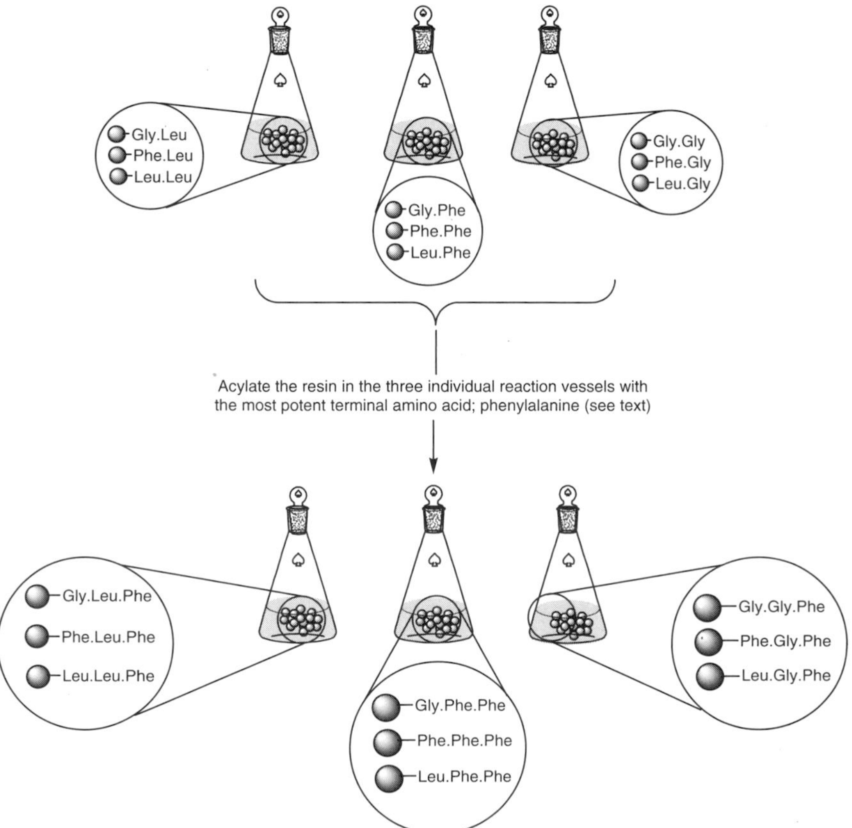

Figure 11.8 Recursive deconvolution.

designed for peptides, where the first cleavage is effected by TFA-mediated removal of a Boc group on a secondary amine. Neutralization results in the formation of a diketopiperidine with concomitant release of the first amount of compound. The second cleavage is effected by base (e.g., sodium hydroxide) saponification, with the third batch of the active compound remaining resin bound for subsequent on-bead sequencing using the Edman degradation (Figure 11.9).

Chemical tagging

An excellent example of this was devised by Still and coworkers and commercially developed by Pharmacopeia. The technique utilizes haloaromatic tags with varying length hydrocarbon chains. These were installed onto the polymeric resin by means of a rhodium-catalyzed carbene insertion following the addition of a building block (Figure 11.10).

This basically provided a chemical ''bar-code'' for the series of reactions involved in the library synthesis. At the end of the synthesis, the tagging molecules can be oxidatively released, silylated, and analyzed by their retention times using electron-capture GC.

There are numerous examples of alternative chemical tagging procedures that have been successfully employed in combinatorial library synthesis and the reader is directed to the ''further reading'' section at the end of this chapter to explore this enormous area.

Radio frequency tagging

This elegant method employs a microchip tag, which acts like a bar-code, identifying each library member. These tags are robust, encapsulated in a glass vessel, and can withstand a wide variety of chemicals and synthetic conditions. The radio frequency tag accompanies a particular portion of resin on its synthetic journey in a tea bag type vessel (Figure 11.11).

Encoded sheets

Pfizer developed a methodology to allow separation of the individual solid-phase products while including an in-built coding system to deconvolute the synthetic history. The resin beads were sandwiched between two sheets of inert woven polypropylene, and the sheets were then fused together so that the beads were immobilized. If one thinks back to the earlier example, this is really just an extension of Houghten's tea bag concept, but in a sheet format. The individual squares generated on the sheets were given simple three-letter codes (Figure 11.12).

Let us consider an example of three sheets measuring 6 cm $\times$ 6 cm, divided into a 3 $\times$ 3 array to produce nine individual squares each of which is given a three-letter code (27 squares spread across three individual sheets). If we were to separate the individual sheets and place them in a reaction vessel and acylate the resin beads with a single amino acid, the top sheet would now have phenylalanine attached, the second sheet glycine, and the bottom sheet valine. These sheets could be easily washed and deprotected in a universal reaction vessel. The cleaver maneuver was to now restack the sheets and cut them into strips of three to afford three sets of columns. If each of the sets of columns were now acylated with a new amino acid, a unique dipeptide would be generated on each column of the material (Figure 11.13).

Again, global washing and deprotection precedes the final maneuver of restacking the strips and cutting them into individual squares. If these are divided by rows this time to form three sets of squares and each set is acylated with the third amino acid, the end product would be the generation of all possible 27 tripeptides in well-defined polypropylene bags ready for cleavage and biological screening (Figure 11.14).

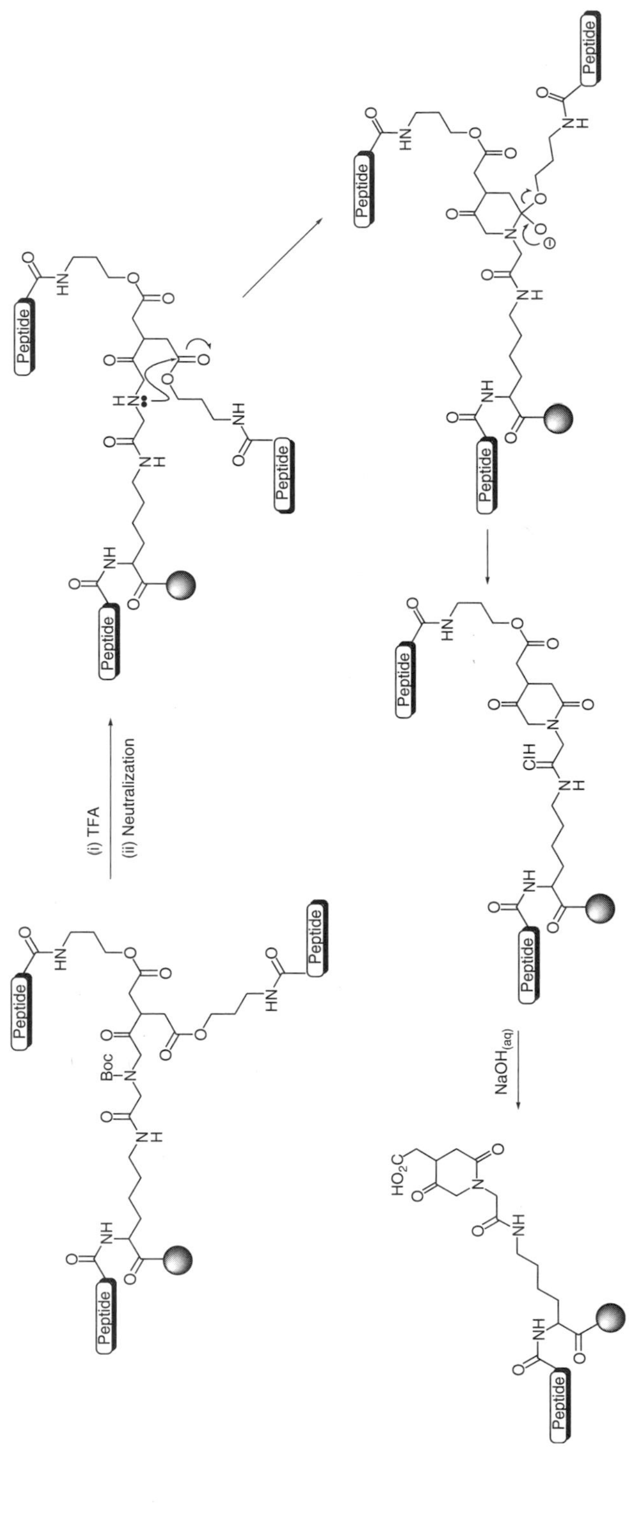

Figure 11.9 Mechanistic representation of a double cleavable linker for solid-phase library synthesis.

1. Couple building block A
2. Install tagging molecule X
A—●—Tag X → C—B—A—●—Tag X, Tag Y, Tag Z

Tags: N_2-diazoketone aryl tag, OMe, O(CH$_2$)$_n$OAr; Ar = 2,6-dichloro-4-fluorophenyl; 2,4,6-trichlorophenyl; pentachlorophenyl

Figure 11.10 Haloaromatic tagging for binary encoded solid-phase synthesis.

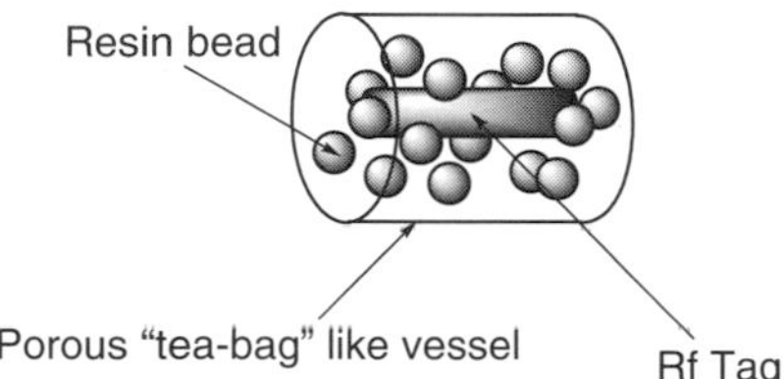

Figure 11.11 A cartoon illustration of an RF tagging vessel with encased resin beads.

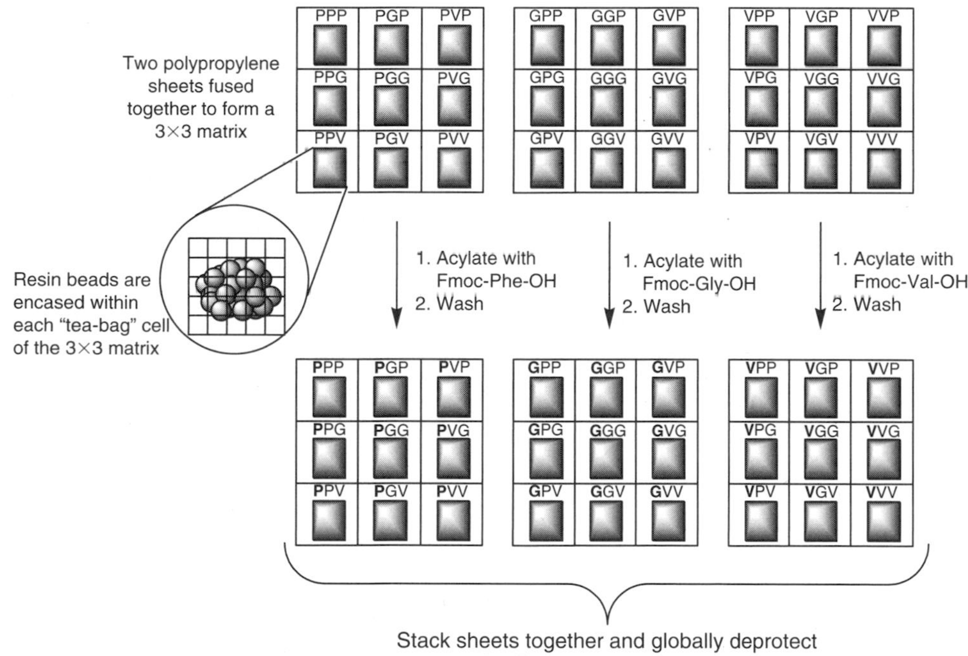

Figure 11.12 Stage 1 of an encoded sheet tripeptide library synthesis.

Light-directed synthesis

This technology was developed by scientists at the Affymax Research Institute in the early 1990s and combined the techniques of photolithography, photochemistry, and SPS in a process termed "light-directed, spatially addressable parallel synthesis." If we once again use peptide synthesis as an example of the technology, the solid support in this instance is not a resin bead, but instead a

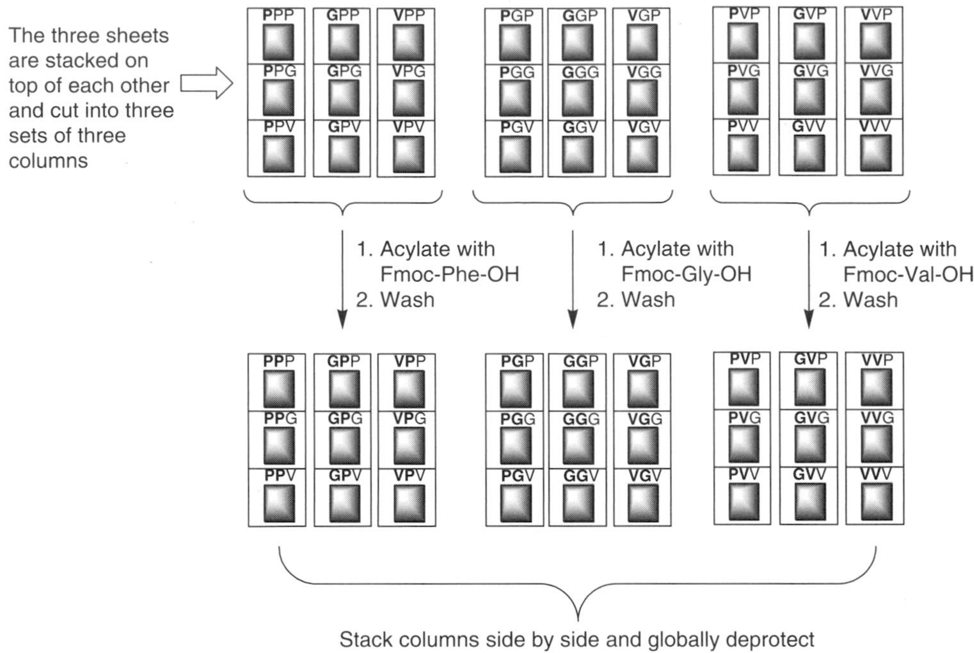

Figure 11.13 Stage 2 of the encoded sheet library synthesis.

two-dimensional glass surface (a reaction plate) displaying amines masked by the photolabile protecting group nitroveratryloxycarbonyl (Figure 11.15).

Part of the solid-phase surface can be covered by a mask, which does not permit transmission of UV light. Exposure of this set-up to UV light results in selective deprotection of amines that are not covered by the masked area. Incubation of the plate with an activated N-photolabile protected amino acid will then afford a protected dipeptide only on the area that was not covered by the UV mask. After washing, one can then envisage using a new mask to deprotect a new area on the surface for coupling to an alternative amino acid, building up a ''checker-board'' library of peptides, where the sequence can be ascertained by a grid reference approach.

Once the library is complete and all side-chain protecting groups removed, then activity was often ascertained by incubating the plate with a fluorescently tagged protein target of interest. Thus, only those regions of the plate which contained bioactive peptides bound the fluorescent protein and these could be readily determined using fluorescence microscopy. In addition, the intensity of the measured fluorescence provided a means of determining the affinity of the peptide for the protein.

11.5 COMBINATORIAL CHEMISTRY IN THE 21ST CENTURY

From the descriptions so far, one can see that a lot of excitement, hope, and high-quality research and development went into the area of combinatorial chemistry during the last two decades of the 20th century. Has this impetus been carried over into the new millennium? If one looks at the statistics so far, then one can reasonably say yes to this question. For example, in 2002 a total of 388 chemical libraries were published in the open scientific literature. One must also not lose sight that this figure is probably a gross underestimate of the actual number of libraries produced, since many combinatorial or pharmaceutical companies will understandably not release technical details of their libraries into the public domain in order to protect their intellectual property. However, the figure of 388 was in fact a 25% increase on the number of published libraries reported in 2001. This

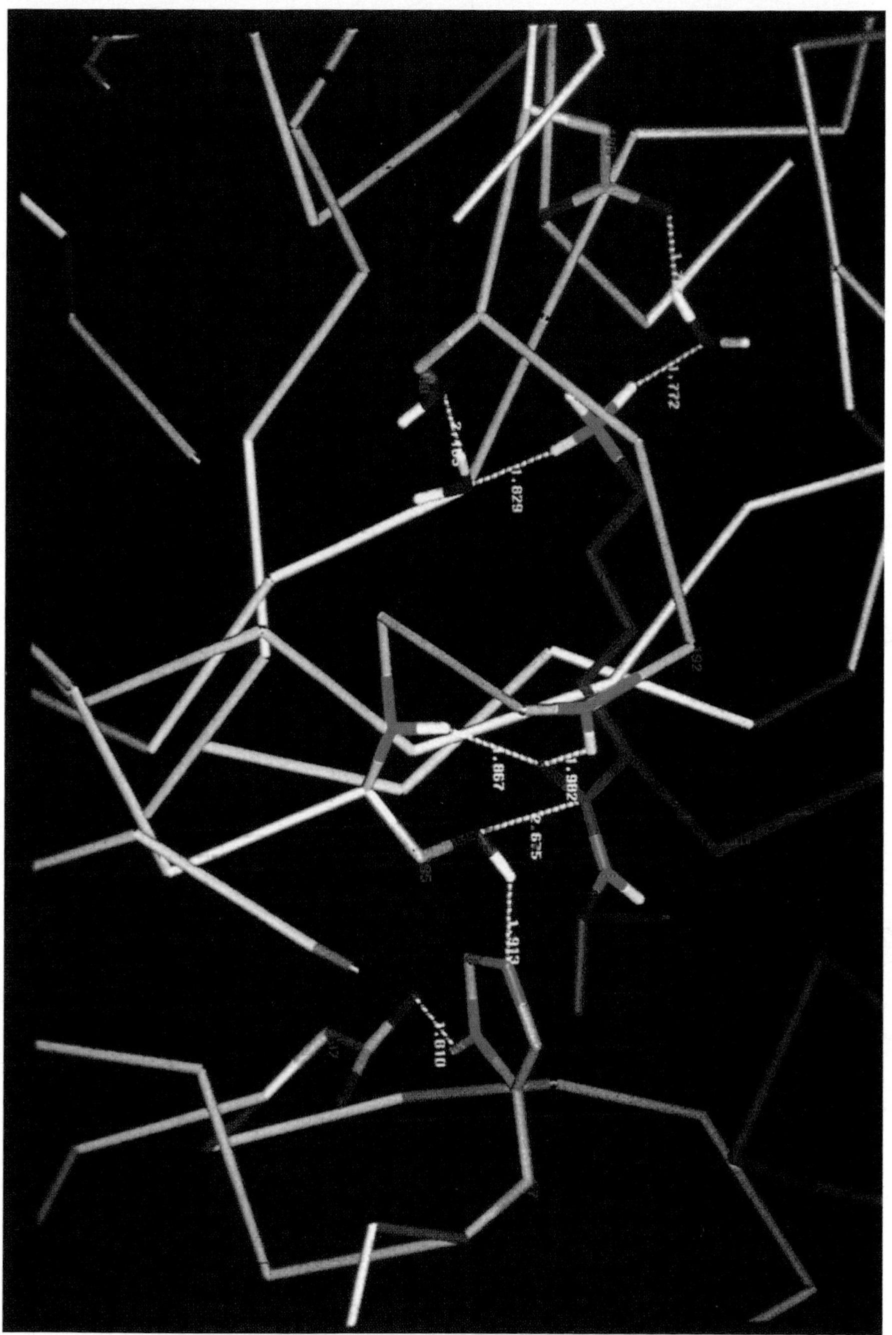

Color Figure 4.8 Serine proteases. Proton movement and enzymatic cleavage of the peptide bond.

Color Figure 4.8 Continued

$\xrightarrow{+\,H_2O}$ Cleavage of peptide bond and reformation of catalytic triad

Color Figure 4.9 Inhibition of a serine protease and protein–protein recognition. The natural ligand inhibitor, Hirudin binding to the catalytic Asp-His-Ser triad within the serine protease α-Thrombin (Vitali et al., 1992).

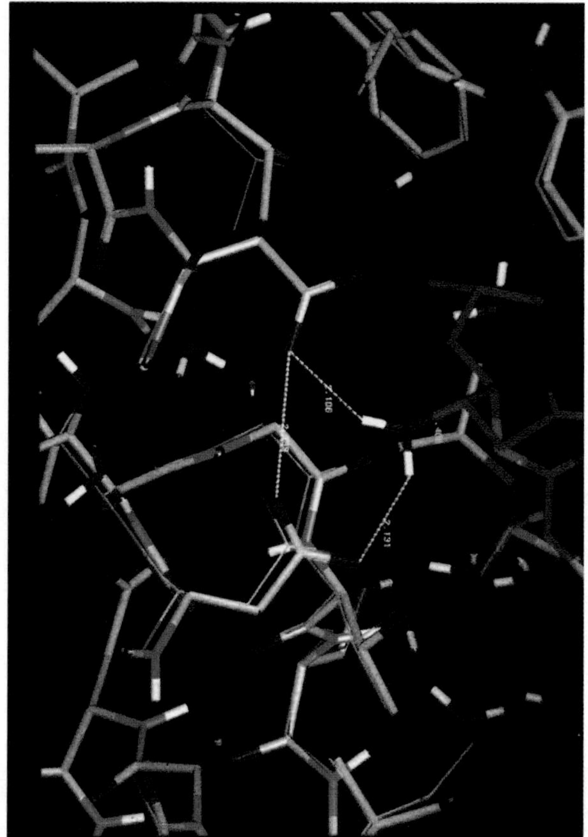

Color Figure 4.10 Aspartate proteases.

Asp^{32} — Asp^{215}

'inner'

→

Asp^{32} — Asp^{215}

'inner'

→

Cleavage of peptide bond and resetting of Aspartate protonation

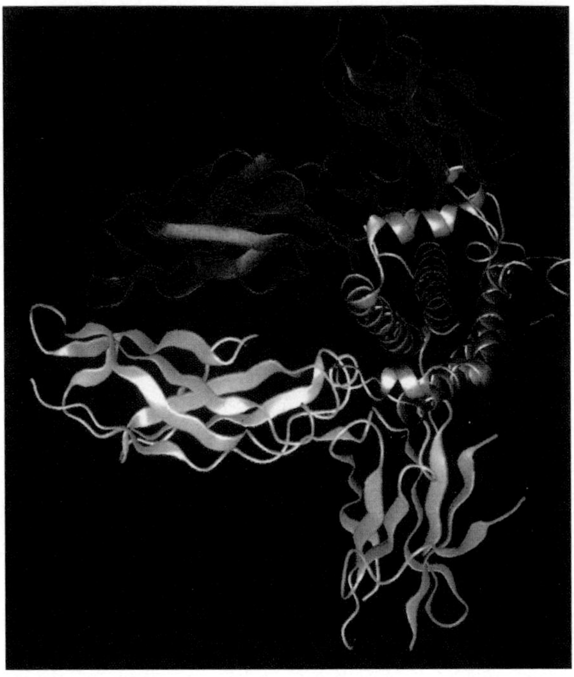

Color Figure 4.11 Protein–protein recognition. The influence of a hormone on protein dimerization. Human growth hormone (hGH) binding to the extracellular domain of its receptor (de Voset al., 1992).

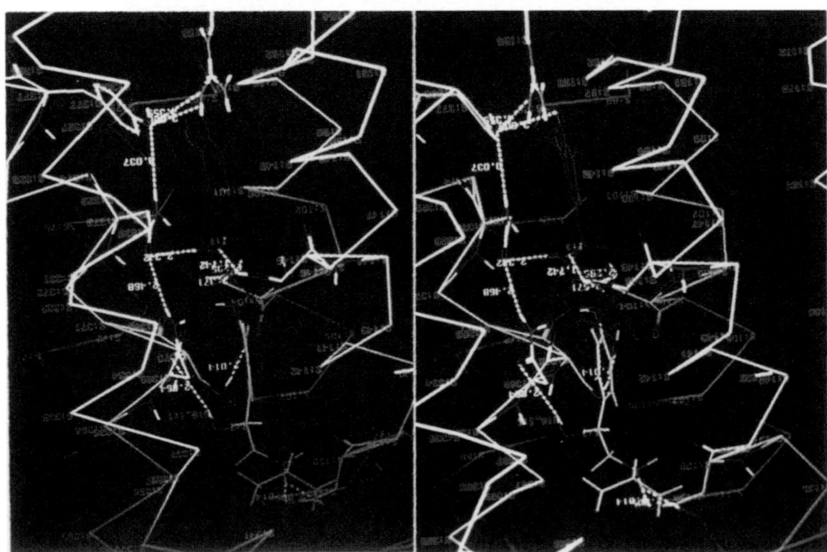

Color Figure 4.12 A segment of the β_1-adrenergic receptor adapted from the crystallographic structure of the mammalian G protein-coupled receptor, bovine rhodopsin (Palczewski et al., 2000), receptors characteristically represented by seven *trans*-membrane α-helices and activated by small ligand hormones.

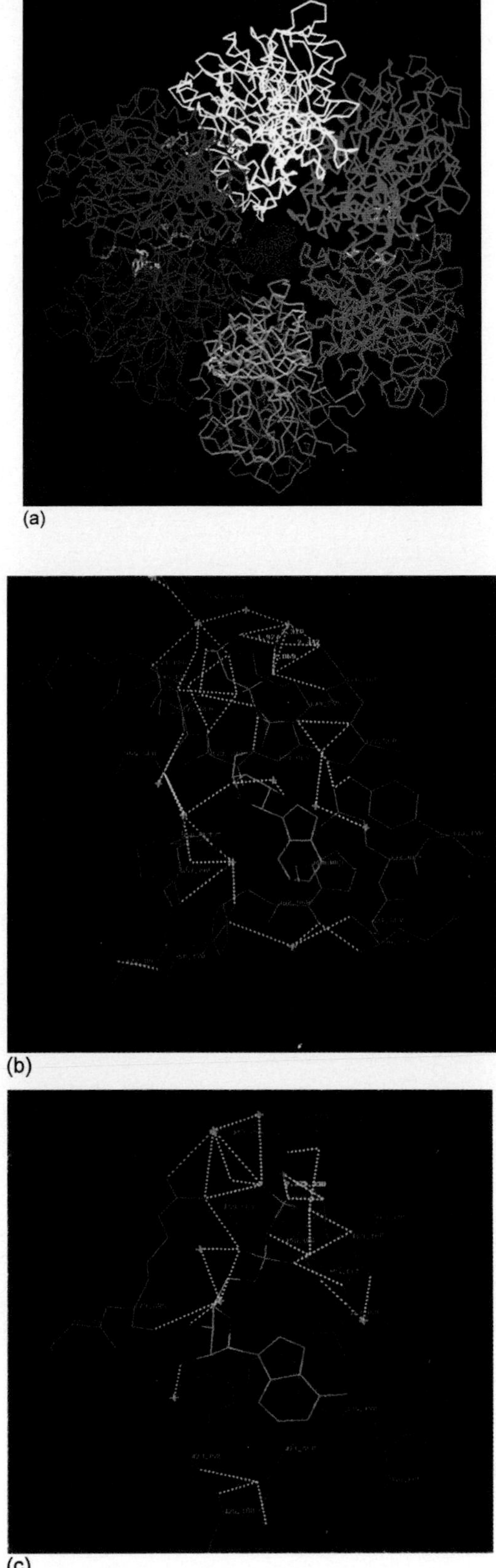

Color Figure 4.13 (a) Adenosine triphosphate synthase (ATP synthase, F_1 F_0 synthase) is the central enzyme in energy conversion in mitochondria, chloroplasts, and bacteria. and uses a proton gradient across the membrane to synthesis ATP from the diphosphate, ADP, and inorganic phosphate.

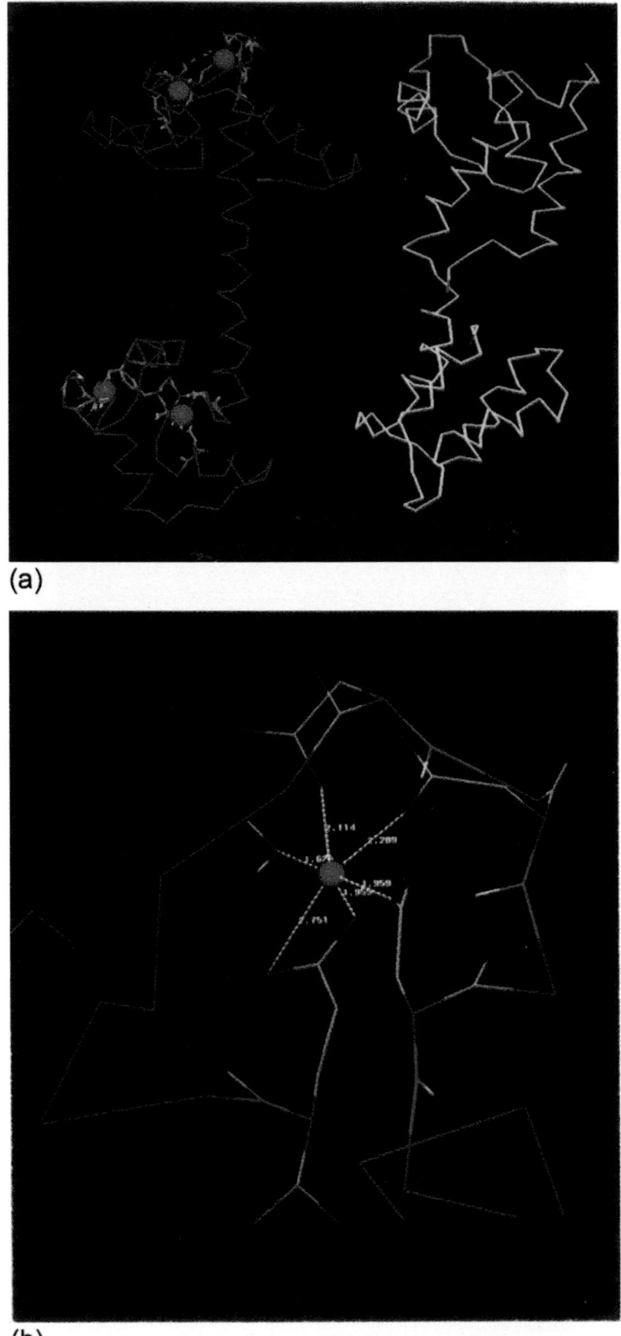

Color Figure 4.14 The influence of strong charge on conformation. The structure of calmodulin with and without the interaction of four calcium ions.
(a) *Top*: Calcium-bound calmodulin from *Drosophila melanogaster* (2.2 Å resolution; Taylor et al., 1991) has a seven-turn α-helix connecting the two calcium-binding domains. **(b)** *Bottom*: The six-coordination octahedral form of a binding site is shown in plate (b) where the Ca^{2+} ion is held by four acidic residues.

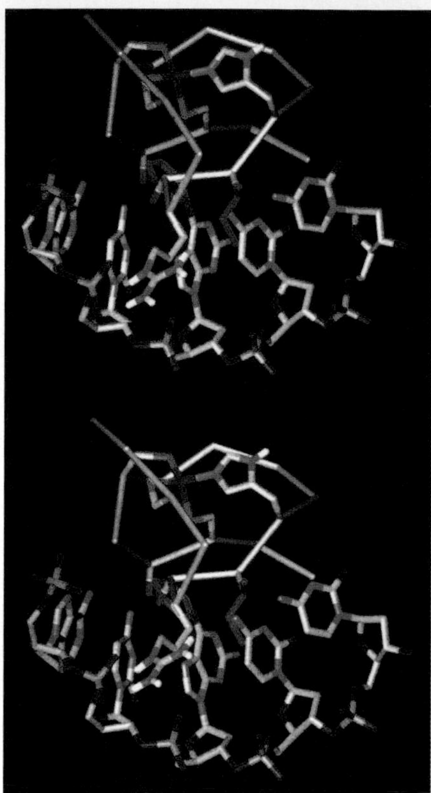

Color Figure 4.15 Protein–single strand DNA recognition. A zinc finger domain binding to a single-stranded DNA sequence. Interaction of an NMR-determined zinc finger domain in the HIV–1 nucleocapsid protein (South et al., 1991).

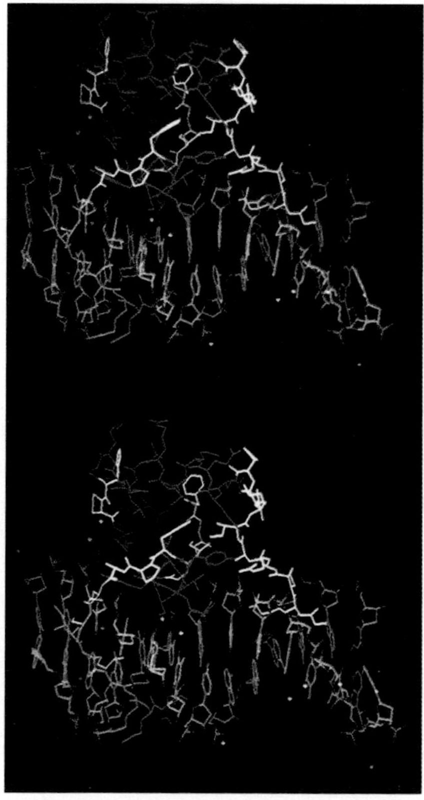

Color Figure 4.16 Protein–double strand DNA recognition. The selectivity of protein binding in the major and minor grooves of the DNA. The binding of the prokaryotic enzyme Hin recombinase to DNA in the *Salmonella* chromosome (Feng et al., 1994).

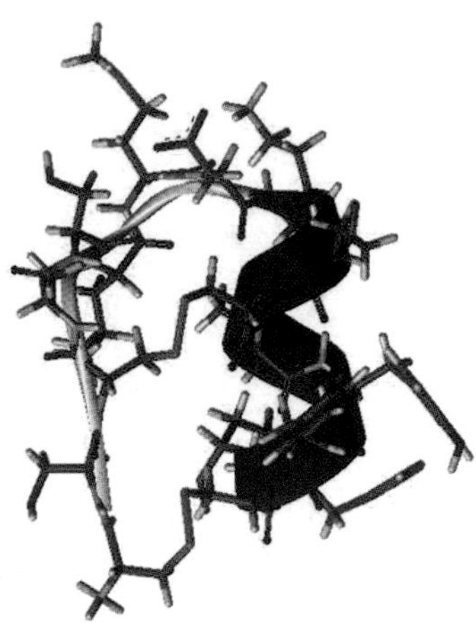

Color Figure 10.4 The three-dimensional structure of human endothelin-1, as revealed by 2-D ^{1}H NMR studies.[13] Residues 9–15 form an irregular α-helix, which is tethered to the N-terminal half of the molecule through disulfide bonds linking Cys 1 to Cys 15, and Cys 3 to Cys 11.

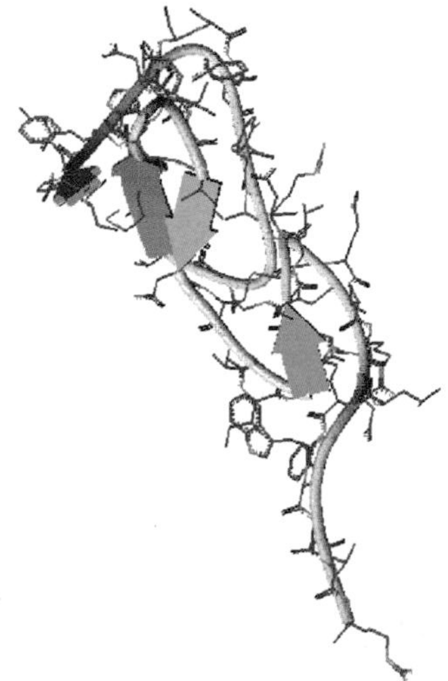

Color Figure 10.5 The three-dimensional structure of mouse epidermal growth factor.[12] β-strands are shown schematically as blue arrows.

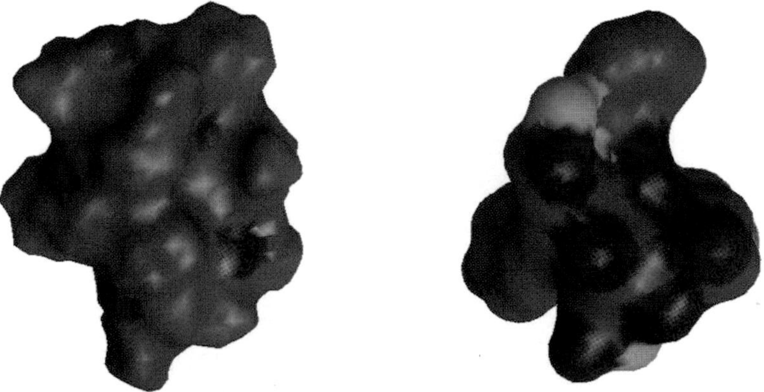

Color Figure 10.11 Solvent accessible surfaces of the peptides cyclosporin (left) and lysine vasopressin (right). Regions of the surfaces that are contributed by polar atoms are colored red (O atoms) and blue (N atoms), the nonpolar parts of the surfaces being colored green.

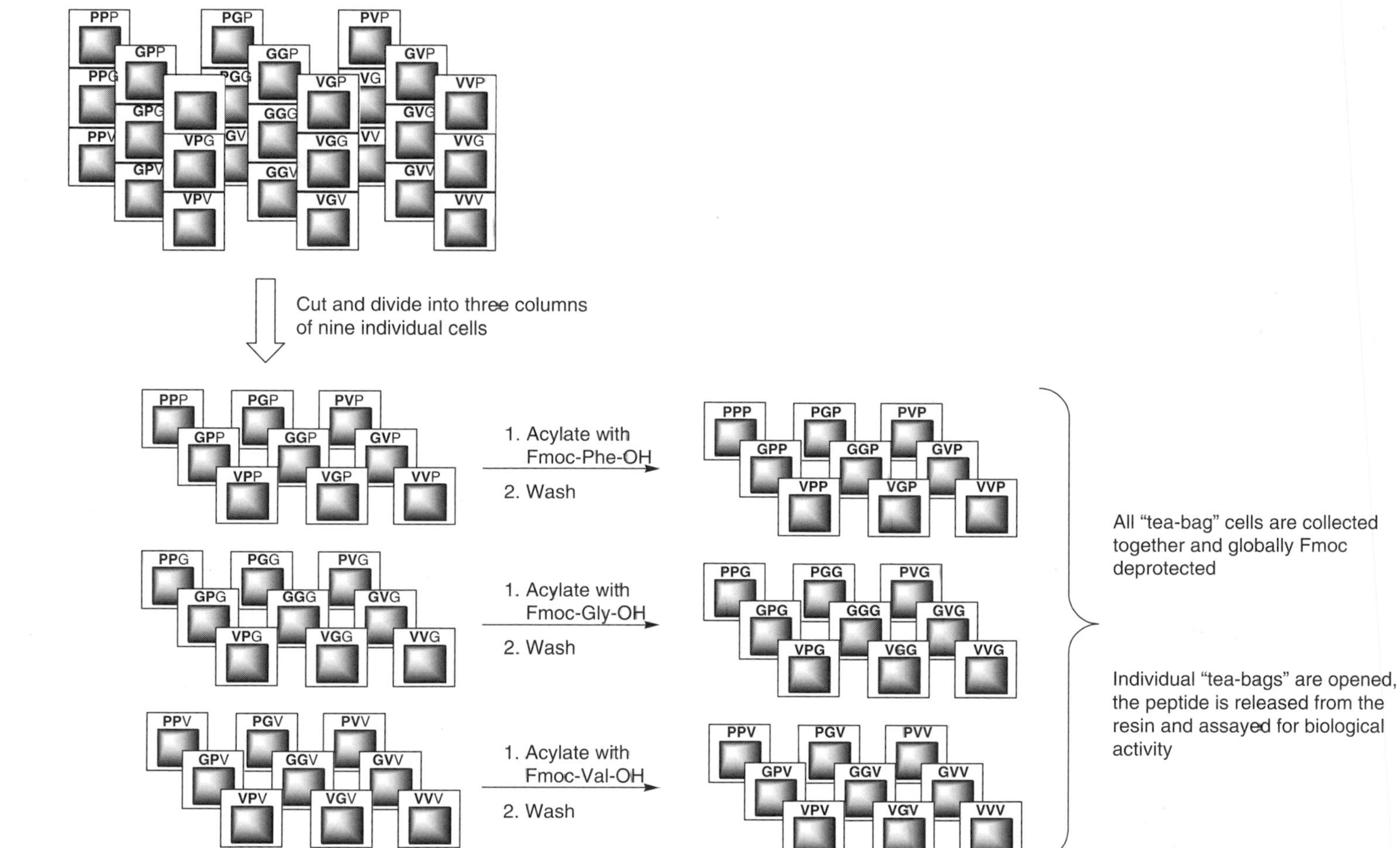

Figure 11.14 Final stage of the encoded sheet library synthesis.

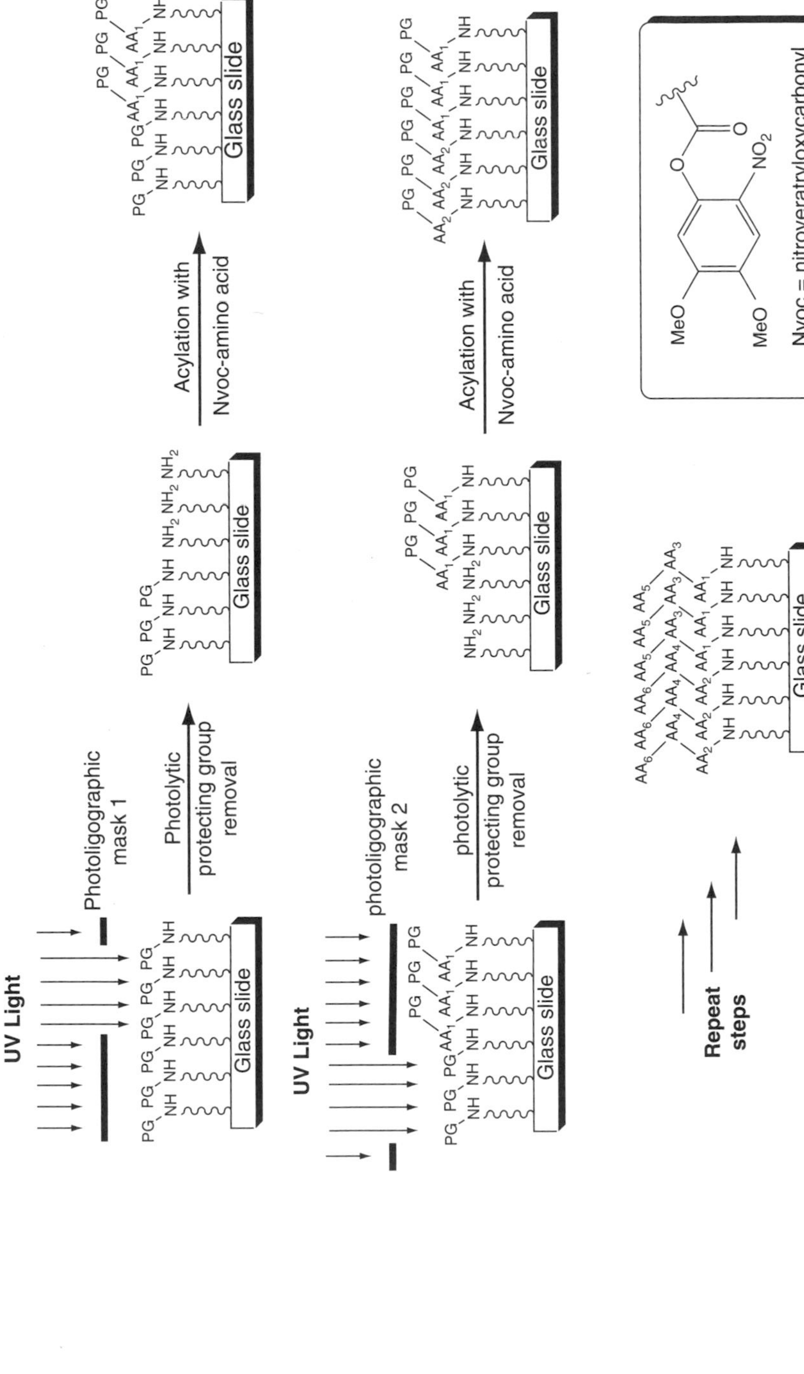

Figure 11.15 Light-directed spatially addressable parallel synthesis.

would seem to suggest that combinatorial chemistry is still a highly populated field of scientific endeavor pursued at both the industrial and academic levels.

Throughout this chapter we have used peptide synthesis as an excellent strategy to illustrate many of the concepts associated with combinatorial chemistry and library construction, however, it is of worth to note that the majority of libraries described in the modern literature are in fact nonpeptide based. Instead the vast majority describe small molecule, heterocyclic libraries, and when one thinks about this it is not very surprising as to the reasons for this observation. Firstly, peptides are not ideal drug candidates, since they often display poor oral activity, usually as a result of their rapid metabolism by digestive enzymes. Secondly, the history of drug discovery and development is literally brimming with examples of low molecular weight (typically less than 500) heterocyclic molecules with proven track records as clinically effective and safe drug entities.

Therefore, it is highly appropriate to provide a brief account of some important small molecule libraries, which have been developed. As previously discussed in this chapter, one of the earliest examples of a nonpeptide library was reported by Jonathan Ellman and coworkers, and described the construction of a 1,4-benzodiazepine library. This was an ideal molecular scaffold to investigate since it is formed by covalently linking three discrete units, ultimately affording a final bicyclic heterocyclic core with five positions of structural diversity, two of which can additionally be varied on the aromatic ring (Figure 11.16).

Since this seminal work was reported, huge efforts have been made to enhance the repertoire of chemistries, which can be applied to SPS, and has ultimately facilitated the generation of enormously diverse small molecule chemical libraries. Again, it would be impossible to cover the enormous amount of material that has been reported, however the reader is encouraged to explore some of the ''further reading'' texts highlighted at the end of the chapter.

It is worth, however, dedicating some space to serve as a case study and draw together some of the key aspects discussed within this chapter. To illustrate this, an excellent and recent example was published by Merck chemists, who generated a 128,000-member library of 2-arylindoles. What made this an interesting piece of work revolved around the fact that it described a combinatorial mixture synthesis, with the authors claiming that this type of library could still play a vital role in the discovery of new leads for a variety of targets. This was in stark contrast to the vast majority of published libraries, which instead favored the synthesis of single, purified compounds using parallel synthesis, and purification techniques.

The Merck chemists chose the 2-arylindole scaffold due to its presence in many biological compounds. They reasoned that by tethering an amine to the 3-position of the indole ring, while making minor modifications to the 4-, 5-, 6-, and 7-positions and placing a variety of aryl substituents in the 2-position they would be able to generate potent and selective compounds for many G-protein coupled receptors (GPCRs) (Figure 11.17).

The solid-phase linker strategy employed the Kenner safety-catch linker; an excellent linker that has been utilized in countless combinatorial library syntheses. It is based on an aryl sulfonamide structure, which can be readily acylated by an appropriately activated carboxylic acid. This acylsulfonamide product is very stable and can be exposed to many chemistries without deleterious side reactions. However, on completion of the synthesis, alkylation of the acylsulfonamide NH generates a highly electrophilic species, which can undergo nucleophilic displacement at the carbonyl group with a variety of nucleophiles. This is where the linker gets the term ''safety catch'' from; the acylated linker is stable (safety catch ''on'') until a specific chemical reaction generated the N-alkylated derivative (safety catch ''off'') thereby triggering the resin for subsequent cleavage and release of the desired library member (Figure 11.18).

The synthesis of the 128,000-member 2-arylindole library involved just two synthetic steps on solid phase and then one further step on half of the cleaved product (Figure 11.19).

Initially, 20 arylalkyl keto acids were immobilized on the sulfonamide resin using the corresponding symmetrical anhydrides (termed the ''X subunits''). A portion of resin for each X subunit was archived for later deconvolution. The 20 pools were then mixed and split into 20 equal portions

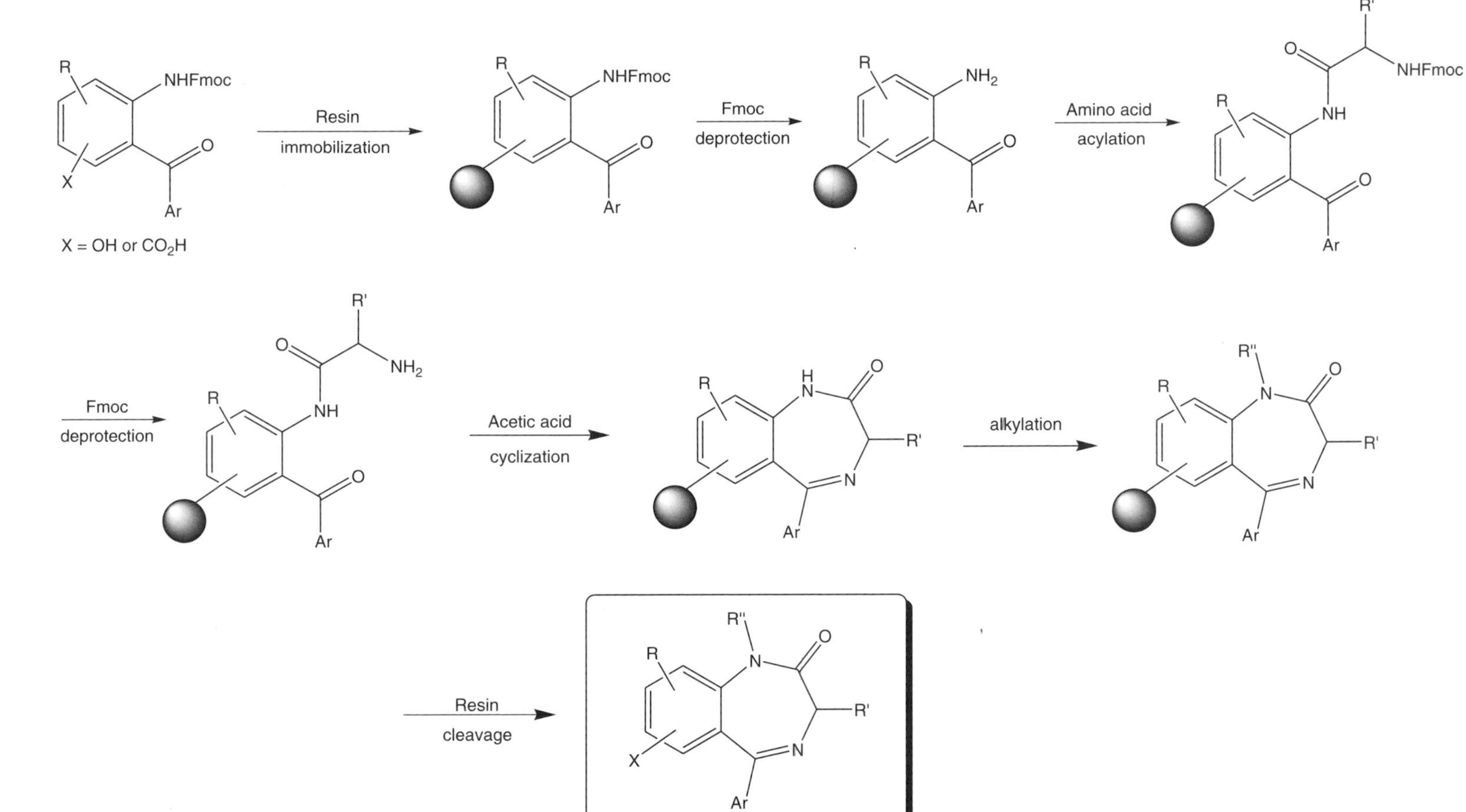

Figure 11.16 Ellman's solid-phase 1,4-benzodiazepine library synthesis.

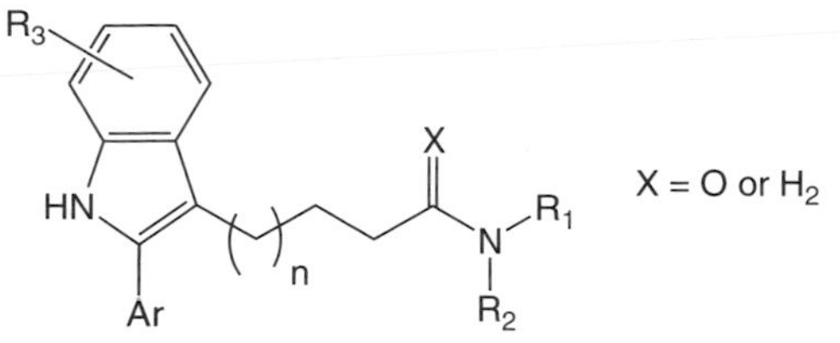

Figure 11.17 The generalized structure of Merck's proposed 2-arylindole library.

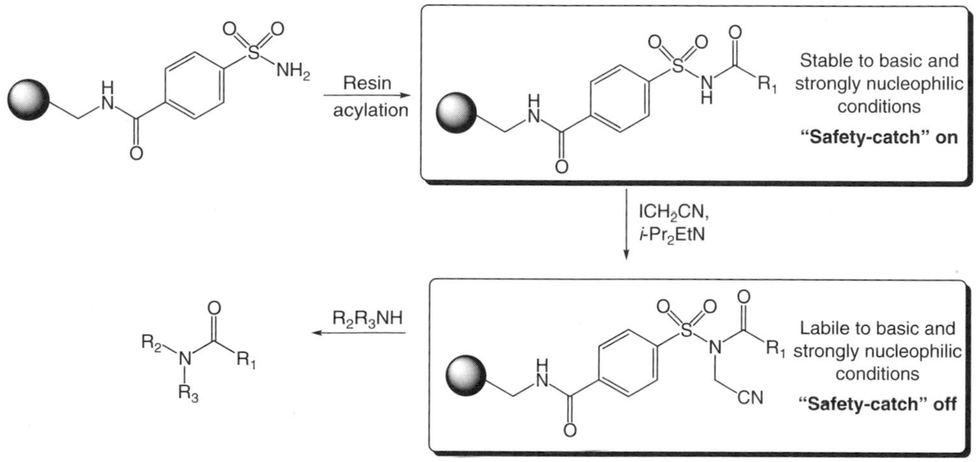

Figure 11.18 Ellman's modification of Kenner's "safety-catch linker."

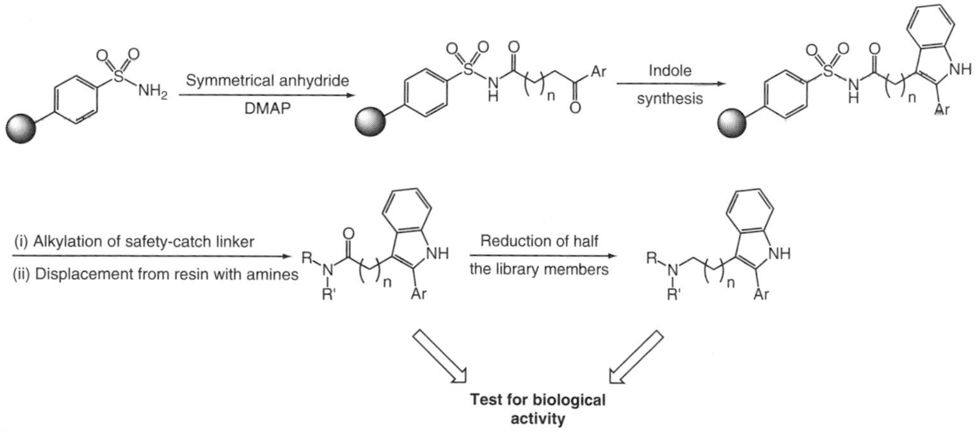

Figure 11.19 Synthetic strategy for the Merck 2-arylindole combinatorial library.

and exposed to indole cyclization with the necessary arylhydrazine (the Y subunit) in the presence of zinc chloride and acetic acid. Therefore at this stage, there were 20 pools of 20 different indole derivatives, and again these were archived, mixed, and split into 80 equal portions. At this stage the sulfonamide linker was alkylated and then displaced using 80 different amines (the Z subunit) to afford 80 pools of compounds containing 400 compounds/pool. At this juncture, the individual pools were split into half. One half was left as the amide derivative, while the other half was reduced with borane–methylsulfide complex to the corresponding amine derivatives. This split/reduction step doubled the library from 32,000 to 64,000 members in 160 pools of 400 compounds.

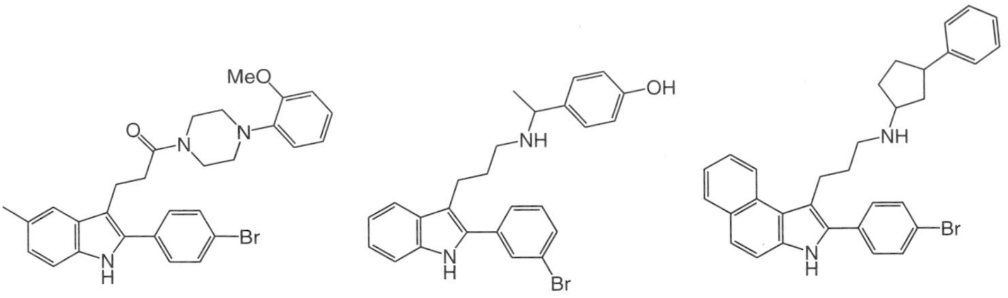

Highest activity at the NK1 receptor Highest activity at the NPY5 receptor Highest activity at the CCR3 receptor

Figure 11.20 Three highlighted molecules from the Merck library and their major receptor targets.

In parallel to this, a second pair of libraries was also prepared using different X, Y, and Z subunits to ultimately generate 128,000 new molecules as 320 pools of 400 compounds.

The resulting 320 mixtures were screened in a wide variety of GPCR binding assays, with activity observed in many families of receptor. These included neurokinin, chemokine, and serotonin receptors. The Z subunit pools of interest were subsequently deconvoluted in two stages by the resynthesis of the compounds from the previously archived resin pools. Once achieved, a series of individual compounds were then prepared and screened to determine their absolute pharmacology (Figure 11.20).

11.6 DETERMINING ACTIVITY

11.6.1 HTS

With all that has been described to date, it is quite clear that combinatorial synthesis produces large quantities of molecular structures in a much quicker time than traditional solution-phase synthesis. The biological testing of these discrete compounds or mixtures must also therefore be carried out quickly to meet the demands of the pharmaceutical industry to shorten the time and cost of drug development. The overarching terminology to describe this process is HTS and as mentioned earlier, it was developed before combinatorial chemistry, thereby actually acting as a major driver in the latter's development. In fact the whole arena has turned full circle, since combinatorial techniques have matured to such a level, there are now so many compounds produced each year, the main drive in the pharmaceutical industry is to make current HTS strategies even more efficient. In the early HTS screens, assays were typically performed using 96-well plates, with each well capable of holding 100 μl. This has now moved forward in quite a considerable manner, whereby plates containing 1536 wells are now routinely employed in HTS screens. In this latter example, the individual wells hold between 1 and 10 μl, and this in itself has raised issue over liquid handling techniques and their reproducibility and ease of automation. Alongside these developments are the introduction of fluorescence and chemiluminescence based read-out systems in both 96 and 1536 based formats. Not only are these beneficial in moving away from radioligand-based assays, in terms of both safety and disposal issues, but they also open the possibilities of high content screening (HCS). This technique is again driven by the needs of the pharmaceutical industry to streamline its R & D efforts and make the whole drug discovery process a most cost-effective operation. In this arena, the key driver is the amount of data one can obtain from a single well on a HCS plate. It does not take much to realize that if in a single experiment one could ascertain not only whether a novel compound binds to a certain receptor or not, but also if it is an agonist, and antagonist or an inverse agonist, then the cost to the company per well becomes enormously more efficient. These advances coupled with the developments in microfluidics and ''chip''-based technologies certainly indicate a period of further exciting and rapid change. While it is important to be aware of these issues, an in-depth discussion of the various HTS and HCS techniques is beyond the scope of this chapter, however, the reader is again directed to some excellent reviews in the ''further reading'' section which deal with various aspects associated with this rapidly expanding area of drug discovery.

11.7 CONCLUSIONS

Irrespective of the nature of the library members, the greatest potential of combinatorial chemistry lies in the numbers. The technologies highlighted both within this brief introduction to the area and the ''further reading'' articles clearly illustrate that the science has advanced to the stage where numerically large, quality libraries can be constructed on a practical scale. While we have concentrated on solid-phase technologies within the confines of this chapter, it is important to the reader that there has also been an enormous amount of research effort focused upon solution-phase combinatorial libraries. In these examples, the concept of solid-phase technologies is not lost, however, since there are now vast numbers of resin-immobilized reagents that are commercially available and good review articles encompassing this arena are cited within the ''further reading'' section.

Over the past 20 years medicinal chemists have exploited all of the available techniques in both lead identification and lead optimization to great success. However, one must never lose sight of combinatorial chemistry's place within the whole drug discovery field and therefore appreciate that to be at its most effective and enable meaningful drug discovery to forge into the new millennium, medicinal chemists must utilize combinatorial chemistry in an integrated fashion with both well-established traditional methods and the rapidly shifting arena of modern instrumentation, detection methods, and biological screening platforms. A truly exciting and multidisciplinary era therefore awaits.

FURTHER READING

Reviews on Library Synthesis

Dolle, R.E. (2000) Comprehensive Survey of Combinatorial Library Synthesis: 1999. *J. Comb. Chem.* **2**, 383–433.

Dolle, R.E. (2001) Comprehensive Survey of Combinatorial Library Synthesis: 2000. *J. Comb. Chem.* **3**, 477–517.

Dolle, R.E. (2002) Comprehensive Survey of Combinatorial Library Synthesis: 2001. *J. Comb. Chem.* **4**, 369–418.

Dolle, R.E. (2003) Comprehensive Survey of Combinatorial Library Synthesis: 2002. *J. Comb. Chem.* **5**, 693–753.

Dolle, R.E. (2004) Comprehensive Survey of Combinatorial Library Synthesis: 2003. *J. Comb. Chem.* **6**, 623–679.

Reviews on Linkers

Gordon, K. and Balasubramanian, S. (1999) Solid phase synthesis — designer linkers for combinatorial chemistry: a review. *J. Chem. Technol. Biotechnol.* **74**, 835–851.

Qiang, H., Quan, L. and Zheng, B.Z. (2004) Linkers for solid-phase organic synthesis. *Prog. Chem.* **16**, 236–242.

Reviews on Combinatorial Chemistry and Tagging

Geysen, H.M., Schoenen, F., Wagner, D. and Wagner, R. (2003) Combinatorial compound libraries for drug discovery: an ongoing challenge. *Nature Rev. Drug Discov.* **2**, 222–230.

Ed, N.J. and Wu, Z. (2003) Beyond Rf tagging. *Curr. Opin. Chem. Biol.* **7**, 374–379.

Reviews on High Throughput and High Content Screening

Walters, W.P. and Namchuk, M. (2003) Designing screens: how to make your hits a hit. *Nature Rev. Drug Discov.* **2**, 259–266.

Bleicher, K.H., Böhm, H.-J., Müller, K. and Alanine, A.I. (2003) Hit and lead generation: beyond high-throughput screening. *Nature Rev. Drug Discov.* **2**, 369–378.

Reviews on Solid-Phase Reagents

Ley, S.V. and Baxendale, I.R. (2002) New tools and concepts for modern organic synthesis. *Nature Rev. Drug Discov.* **1**, 573–586.

Hodge, P. (2003) Organic synthesis using polymer-supported reagents, catalysts and scavengers in simple laboratory flow systems. *Curr. Opin. Chem. Biol.* **7**, 362–373.

Additional Book Chapters Concerning Solid-Phase And Combinatorial Chemistry

Easson, M.A.M. and Rees, D.C. (2002) Combinatorial chemistry: tools for the medicinal chemist, Chapter 16, in *Medicinal Chemistry: Principles and Practice*, 2nd Edition, Ed. King, FD. Royal Society of Chemistry, pp. 359–381.

Thomas, G. (2000) Combinatorial chemistry, Chapter 2, in *Medicinal Chemistry: an Introduction*, 1st Edition, Wiley, pp. 69–90.

Patrick, G.L. (2001) Combinatorial synthesis, Chapter 12, in *An Introduction to Medicinal Chemistry*, 2nd Edition, Oxford University Press, pp. 289–318.

12

Recombinant DNA Technology: Monoclonal Antibodies

Frederick J. Rowell

CONTENTS

12.1 RECOMBINANT DNA TECHNOLOGY

12.1.1 Introduction

For hundreds of years humankind has utilized microorganisms to produce a whole range of natural products which we can use (e.g., ethanol, organic acids, dextrans, and antibiotics). Microorganisms are extremely easy to cultivate and large-scale culture results in high yields of the product required which can then be purified and utilized. Natural products can also be extracted from plant tissue. Biologically active compounds from animals can be isolated from the appropriate organ or tissue but as these are extremely potent compounds, they are normally present only in small quantities and large amounts of the appropriate tissue are required to obtain useful quantities of the product. This is a particular problem with compounds of human origin due to lack of cadavers and to the possibility of contamination of the resulting product with human viruses such as hepatitis and the AIDS virus.

For proteins extracted from animals and used in humans such as insulin derived from pigs, since the protein is not chemically identical to the equivalent human protein, its use may evoke an immune response leading to sensitization of the patient. Somatostatin, a hormone that inhibits the secretion of pituitary growth hormone, required half a million sheep brains to be processed to yield about 5 mg of the product. Today, using recombinant DNA technology, the same amount of hormone corresponding to the human protein can be harvested from 9 L of a culture medium in which has been grown a microorganism possessing the inserted human somatostatin gene. It is, therefore, now possible to produce, in large quantities, a whole range of biologically active polypeptides of identical composition to those found naturally in humans or any other living organism using recombinant DNA technology (often termed genetic engineering).

The success of recombinant DNA technology is exemplified by the substantial and increasing impact of this class of medicine is making. Since the inception of the new centralized European drug approval system in 1995, some 88 recombinant proteins/monoclonal antibody-based products have gained approval. This number accounts for some 36% of all new EU drug approvals during the period 1995–2003. Of the approved drugs, recombinant proteins constitute the major fraction (64%) with the major types like cytokines (23%), recombinant hormones (21%), blood products and blood-related products (12%), and therapeutic enzymes (7% of the total). The remaining classes comprise vaccines (21%) and monoclonal antibodies (15%). In contrast to this success, to date, no gene therapy product has gained EU or US approval. As of 2002, according to the American Association of Pharmaceutical Researchers and Manufacturers (PhRMA), of the 371 medicines based on biotechnology in development, only 9 are of the antisense type and 16 are gene therapy-based. The bulk of these products are of the types listed above with vaccines and antibodies constituting the major classes of product.

12.1.2 Principles of Recombinant DNA Technology

The insertion of a human gene into, say, a bacterial cell can only be achieved through techniques that enable the gene to replicate within the cell so that all the progeny derived from the original cell possess the inserted gene and that the product defined by the genetic code from the inserted gene is produced via the normal processes of transcription and translation within the cells of the recipient or host cells. The technique is to insert the gene into extranuclear DNA molecules such as plasmids (extrachromosomal loops of DNA) and bacteriophages (viruses that utilize bacteria as their hosts). They are easily isolated from cells and opened up so that the new gene can be covalently attached to the open strand of DNA, the loop reformed, and the plasmid or bacteriophage containing the new gene reinserted into the bacterium or other host cell. The agents, which open and reform the DNA loops, are specific enzymes termed endonucleases and ligases, respectively.

The six main steps in the genetic engineering process (see Figure 12.1) are as follows:

1. Isolating the gene for the protein to be synthesized
2. Opening the cloning vector (the plasmid or bacteriophage)
3. Covalently attaching the DNA corresponding to the new gene into the open loop of the cloning vector
4. Reforming the bonds within the enlarged DNA to regenerate the loop
5. Reinserting the enlarged vector into the host

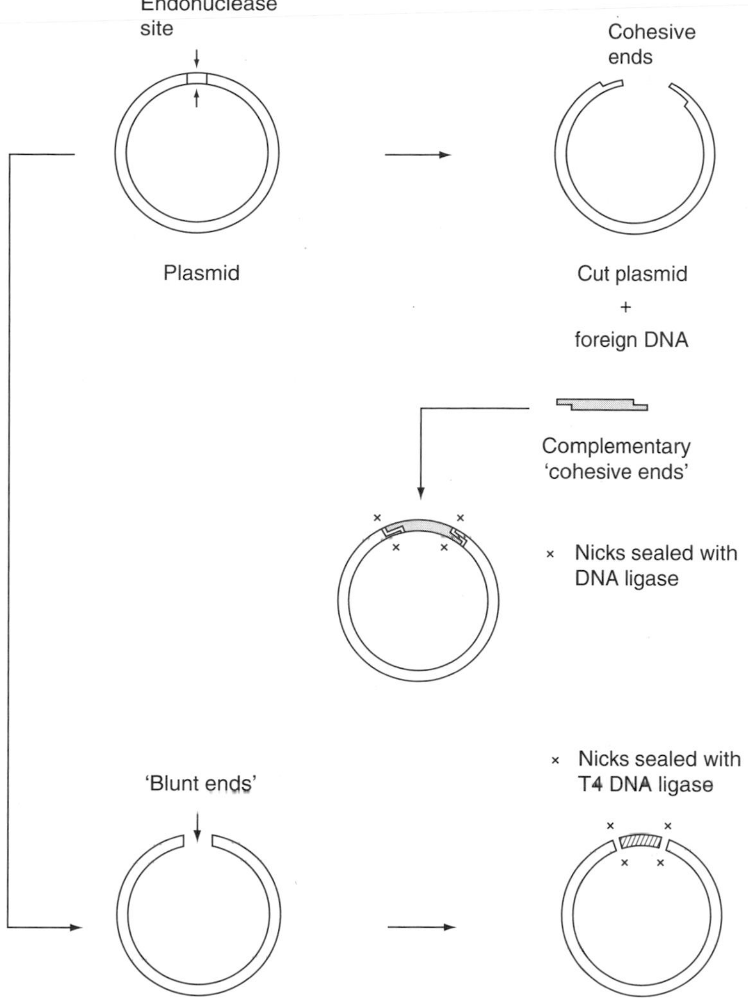

Figure 12.1 Insertion of foreign DNA into a bacterial plasmid.

6. Culturing the mutant or chimeric host cell to enable it to replicate and produce the required protein, which is isolated from the culture medium.

Identification and isolation of the required gene

If the amino acid sequence of the protein to be synthesized by recombinant DNA technology is known then its complementary sequence of codons (triplet sequences of nucleotides, the sequence of each codon corresponding to the amino acid sequence in the protein) can be synthesized. If the protein consists of many amino acids, as is the usual case, then this approach is impractical and the approach is limited to synthesis of the sequences of codons unique to the gene for the required protein. In addition the synthesis incorporates a radioactive tag, usually in the form of ^{32}P into the sugar–phosphate backbone of the synthetic DNA strips. These radioactive fragments will now bind to the complementary codons on the DNA corresponding to unique codon sequences of the required gene. This provides us with the means of identifying the location of the required gene within a multitude of DNA molecules obtained from synthesis of mRNA mixtures using the enzyme reverse transcriptase and nucleotides.

It is assumed that cells or tissues which are producing the required protein will also have a high concentration of the messenger RNA (mRNA) coding for the protein since this contains the transferred genetic message which is read at the ribosome during the synthesis of the protein. mRNA from these target cells or tissues is extracted and the minute amounts of mRNA thus isolated can be transformed into the genetic DNA code for the protein and finally larger quantities of this key intermediate product can be synthesized using a second enzyme called DNA polymerase.

In practice the process described is more complicated since firstly, single-stranded mRNA must be used and is formed from the double stranded naturally occurring form and secondly, the mRNA coding for the required gene will be embedded within a much larger segment of mRNA coding for a variety of other genes. Hence, this large mRNA fragment has to be cut into smaller pieces using specific enzymes and this process may cut the required gene into smaller segments in the process whereas only the complete intact sequence corresponding to the code for the protein is required.

Modifications of the isolated gene before insertion

Having isolated the DNA sequence coding for the required gene it is necessary to ensure that it is in a form which can be successfully incorporated into the DNA of the cloning vector. This requires four key preliminary modifications of the gene.

Firstly, sequences of DNA known as introns which are interruptions of the code found in mammalian genes must be removed enzymatically since their presence leads to incorrect translation of the spliced gene by the bacterial or viral cloning vector during protein synthesis.

Secondly, a signal has to be incorporated at the beginning of the DNA code for protein to signal to the enzyme RNA polymerase to initiate the transformation of mRNA from the synthetic DNA code for the protein. Likewise the correct signals must be attached to the gene's DNA code to instruct the vector's ribosome to start and stop the gene's synthesis during the translation process. A diagrammatic representation of the resulting vector plasmid is illustrated in Figure 12.2.

Thirdly, an ancillary DNA code for a marker gene (e.g., the gene for resistance to tetracycline) is spliced adjacent to the gene coding for the required protein. This serves as a means of detecting whether the total sequence has been successfully incorporated into the host DNA.

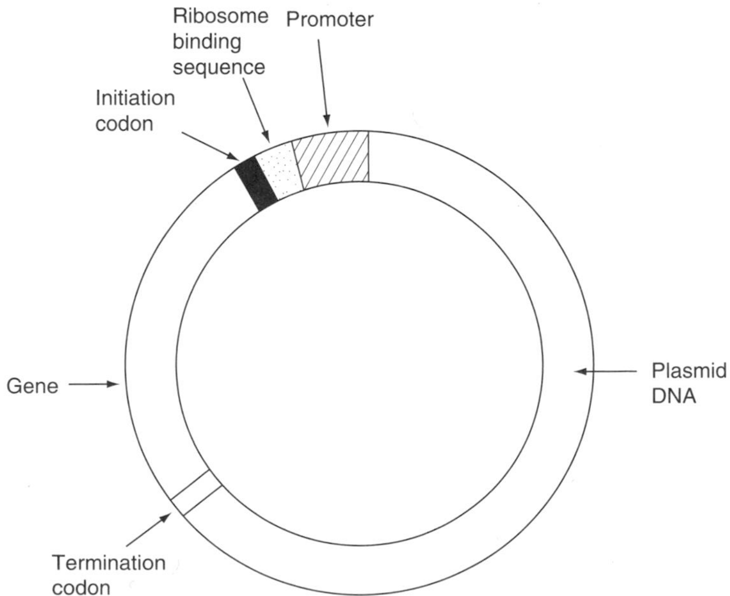

Figure 12.2 Diagrammatic representation of a vector plasmid.

Finally, the synthesized double-stranded DNA carrying all the modifications listed must be treated with enzymes to expose protrusions of bases at the ends of the DNA duplex that can form bonds with complementary protrusions on the ends of the opened circle of the bacterial or viral DNA to which the foreign DNA is to be attached.

Insertion into the cloning vector

The extrachromosomal circular DNA molecules found in bacteria can be separated from the rest of the chromosomal material in the cell by agarose gel electrophoresis. They can be opened up by breaking bonds between specific pairings of nucleotides in the DNA molecule using enzymes termed restriction enzymes (or endonucleases). Different restriction enzymes break bonds selectively between different pairs of nucleotides as shown in Table 12.1. In order that the correct orientation of the bases on the end of the opened plasmid and the complementary bases on the ends of the gene to be inserted occurs, it is necessary to use the appropriate endonuclease. Production of the mutually complementary sequences at the exposed ends of the strands of vector and foreign DNA produces cohesive or ''sticky ended'' strands since they will tend to aggregate together due to the formation of complementary hydrogen bonds between them. The nicks in the sugar–phosphate backbone are now sealed using a DNA ligase so that the foreign gene becomes an integral part of the plasmid's genetic material (Figure 12.1).

The next process is the insertion of the chimeric plasmid into the host bacteria. Bacteria can take up free extracellular DNA by a process termed transformation. The rate of uptake is slow but this can be enhanced by allowing the process to take place at low temperatures (0–5°C) in the presence of calcium ions. It is necessary to identify those cells which have taken up the enlarged plasmid. This is achieved by the use of the marker gene such as that for antibiotic resistance. Successful incorporation of this gene together with the gene for the protein should have produced a bacterium which exhibits resistance to the antibiotic tetracycline. Hence, bacteria carrying this resistance gene will survive in a culture medium containing quantities of antibiotic which is fatal to non-resistant bacteria.

Resistant cells can then be isolated, propagated, and tested to determine whether the required protein is synthesized by the cells. This is achieved by exposing dishes on which cultures are growing to cellulose nitrate filters. If the cultures are producing the required protein this protein will be adsorbed onto the surface of the filter. Subsequent exposure of the filter to protein-specific antibodies carrying a radioactive label produces radioactive patches corresponding to the location of cultures producing the required protein.

A major problem that can occur is the proteolytic degradation of the synthesized polypeptide. This can be prevented by fusing the synthetic gene to the gene of a larger protein associated with the plasmid, e.g., beta-galactosidase or beta-lactamase. The fusion protein is resistant to proteolysis and the required polypeptide can then be cleaved off and isolated. A number of proteins undergo post-translational modification such as modifications of signal or leader amino acid sequences and

Table 12.1 Recognition sites and end products of endonuclease activity

Enzyme	Recognition site	Cleavage product
EcoR I	↓[a] — G A A T T C — C T T A A G ↑	— G A A T T C — — C T T A A G — "Sticky ends"
Hae III	↓ — G G C C — — C C G G — ↑	— GG C C — — C C G G — "Blunt ends"

[a]↓ = Cleavage points.

glycosylation of the protein. An example of how this is achieved for insulin is discussed in a later section on insulin. Choice of the host cell is important since different host organisms have differing capacities to perform post-translational changes and may differ in their efficiency of recombinant protein production. Currently the most popular hosts are *Escherichia coli*, *Bacillus subtilis*, yeast, and cultured cells of higher eukaryotes such as insect and mammalian cells. For proteins such as insulin, which require post-translational modifications and require formation of disulfide bonds to achieve the active product, *E. coli* is not the vector of choice. In that case the yeast *Saccharomyces cerevisiae* has been successfully used as the host. Also in contrast to *E. coli*, coproduction of pyrogens and endotoxins is not a problem with *S. cerevisiae*.

12.1.3 Production of Polypeptides Using Recombinant DNA Technology

Somatostatin

Somatostatin is a small polypeptide (14 amino acids long) and the gene is relatively easily synthesized. The synthetic gene is illustrated in Figure 12.3. It should be noted that the initiation amino acid, methionine, preceded the NH_2 terminal amino acid of somatostatin and that the COOH-terminal amino acid is followed by two-step codons. An Eco RI cleavage site is at one end of the gene and a Bam HI cleavage site at the other, thereby providing cohesive termini to facilitate its insertion at these sites in the plasmid vector.

The plasmid used is the artificially created plasmid pBR 322 which has been completely sequenced. About 20 endonucleases cleave this plasmid at known sites. Two antibiotic resistance markers are associated with this plasmid: tetracycline resistance (Tc^r) and ampicillin resistance (Ap^r). Small peptides like somatostatin are rapidly degraded by *E. coli* and it is necessary to fuse the somatostatin gene to the beta-galactosidase gene for protection. This is achieved by inserting the beta-galactosidase gene together with the lac control region adjacent to the somatostatin gene. The lac control region comprises a promoter site, an operator site which ''switches on'' the adjacent structural genes, and the ribosomal binding site. Thus we have all the elements necessary for successful transcription and subsequent translation. The initial and final plasmids are illustrated in Figure 12.4. Note that the Bam HI site is in the Tc^r marker and cutting and gene insertion results in the loss of tetracycline resistance.

The hybrid protein produced at the ribosome consisting of the beta-galactosidase protein fused to the somatostatin is treated with cyanogen bromide (CNBr). This cleaves at the methionine residue, which lies between the two molecules, yielding somatostatin plus beta-galactosidase fragments. The use of CNBr cleavage is limited to those peptides not possessing methionine as part of their sequence. The somatostatin is detected by immunoassay. Similar techniques have been used for the synthesis of other smaller peptides such as endorphins and enkephalins, which are considered to be opioid peptides.

Insulin

Insulin is an excellent example of how the problems of post-translational modification have been overcome. The protein secreted contains a signal sequence of amino acids at the N-terminus. During passage through the membrane these are cleaved off so that the pro-insulin formed consists

Eco RI																
	Met	*Ala*	*Gly*	*Cys*	*Lys*	*Asn*	*Phe*	*Phe*	*Trp*	*Lys*	*Thr*	*Phe*	*Thr*	*Ser*	*Cys*	Stop Stop
5' AATTC	ATG	GCT	GGT	TGT	AAG	AAC	TTC	TTT	TGG	AAG	ACT	TTC	ACT	TCG	TGT	TGA TAG
G	TAC	CGA	CCA	ACA	TTC	TTG	AAG	AAA	ACC	TTC	TGA	AAG	TGA	AGC	ACA	ACT ATCCTAG 5
																B am HI

Figure 12.3 The synthetic somatostatin gene. (Reproduced by kind permission from Old and Primrose (1985), p. 60.)

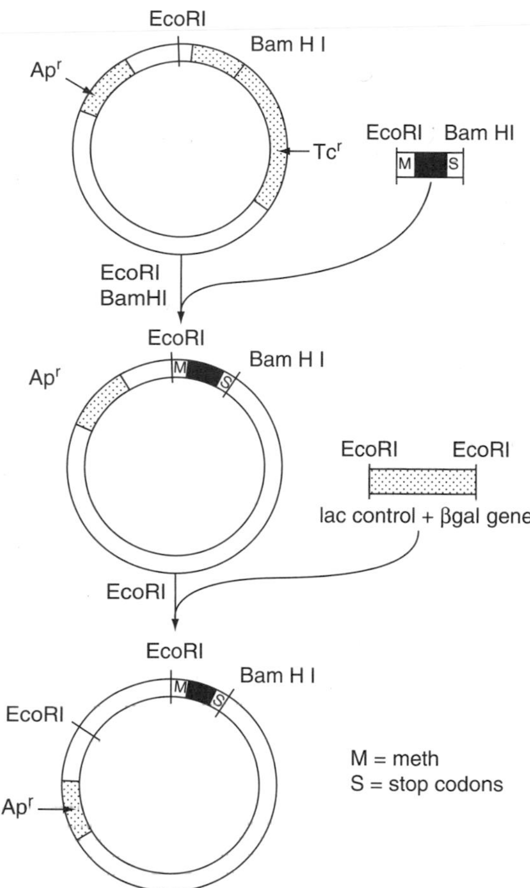

Figure 12.4 Diagrammatic representation of the steps involved in generating a recombinant plasmid for the bacterial synthesis of somatostatin. (Reproduced by kind permission from Emery (1984), p. 163.)

of three chains (A, B, and C). Insulin is formed by the removal of the C chain by proteases. This leaves the A and B chains of insulin in a stable tertiary structure held together by the two disulfide bonds formed when the molecule originally folded as pro-insulin. Bacteria cannot bring about these processes. Thus the approach used is to construct separately the genes for the A and B chain, insert them separately into pBR plasmids and then add the elements of the lac control and beta-galactosidase gene, followed by cloning into the two separate bacterial strains. Thus one culture is producing a hybrid A chain and another a hybrid B chain. Separate cleavage with CNBr frees the two polypeptide chains which, after purification, can be joined by disulfide bonds. Genetically engineered human insulin has now replaced porcine insulin in use.

Human growth hormone

Initially, the only source of human growth hormone (HGH) was human pituitary tissue which was removed at autopsy. HGH from this source was insufficient for clinical treatment. The peptide contains 191 amino acids and although chemical synthesis of the gene is feasible, it is not an easy process. It was therefore necessary to prepare copy DNA (cDNA) by extracting mRNA from pituitary tissue. It was found that there was a Hae III cleavage site in the sequence coding for amino acids 23 and 24. Treatment with Hae III yielded the larger fragment (amino acids 24–191) which could then be combined with the smaller chemically synthesized fragment (amino acids 1–23) which was preceded by the initiating amino acid methionine. This was now inserted into the plasmid next to an appropriate promoter, etc., that is required for successful transcription and translation. The product, however, is not completely identical to the natural hormone as it contains an extra NH_2-terminal, methionine, which could induce an immunological reaction. An alternative approach is to insert the entire cDNA sequence into a Simian virus (SV) 40, using monkey kidney cell tissue culture as the host. The HGH excreted by this method is identical to that found in the pituitary gland.

Lymphokines and monokines

These are families of proteins that have been shown to exhibit antiviral effects, together with the enhancement of elements of the immune system with resultant anticancer effects. These proteins regulate the cellular parts of the immune system. Macrophages produce monokines and the T cells and B cells produce lymphokines. For example alpha, beta, and gamma interferons are produced by the leukocytes, fibroblasts, and activated T cells, respectively. Natural yields of lymphokines and monokines are low but genetically engineered human versions are now available.

Purity of genetically engineered proteins

A variety of analytical methods are used to ensure that products resulting from genetic engineering are fit for human use. These include tests for:

1. *Identity*: Polyacrylamide gel electrophoresis, isoelectric focusing, chromatography (particularly reverse phase HPLC), peptide mapping (in which the protein is digested under controlled conditions by protease enzymes and the HPLC profile of the resulting polypeptide fragments serves as a fingerprint for the parent protein), amino acid analysis, and spectroscopy.
2. *Purity*: Chromatography, spectroscopy, assays for host DNA, assays for pyrogens, and other residual cellular proteins derived from the outer membranes of the host organism, particularly for products to be administered chronically or in high doses.
3. *Potency of the product*: Bioassay against a national or international reference preparation. This ensures that the product has the required biological activity.

Gene therapy

The above applications aim to counteract the deficiency of a natural protein through its substitution by injection of the protein synthesized outside the body through genetic engineering. Some diseases are due to defects in the patient's genes, and examples of such diseases are listed in Table 12.2, together with the deficient gene. This deficiency may be manifested in the lack of production of a hormone or factor, synthesis of an inactive enzyme, or synthesis of a malfunctioning receptor. It is now possible in some cases to synthesize the normal gene and to insert it into a vector using the processes described above, to produce vectors which express the functioning human gene. Transplantation of this vector into the patient so that the vector produces the required gene product in the patient has been attempted with some success for cystic fibrosis. Alternatively the DNA for the normal gene could be introduced into the patient's own cells through genetic engineering.

12.2 MONOCLONAL ANTIBODIES

12.2.1 Introduction

Antibodies are proteins that are designed to bind specifically to foreign or antigenic molecules or microorganisms, which invade higher living organisms. This specific binding initiates a range of *in vivo* processes designed to neutralize the adverse biological activity of the invading molecules or

Table 12.2 Some diseases possibly amenable to treatment by gene therapy

Disease state	Defective gene
Cancer-melanoma	HLA-B7
Cystic fibrosis	Cystic fibrosis transmembrane regulator
Growth hormone deficiency	Growth hormone
Hemophilia	Factor VIII and factor XI
Hypercholesteremia	Low density lipoprotein receptor
Phenylketonuria	Phenylalanine hydroxylase

microorganisms and expedite their elimination from the body. It is their specific binding with relatively high affinity to antigens, which has found exploitation in many areas of biological sciences. It has been the ease of production of antibodies to a wide variety of antigenic species coupled with the ability to produce a single type of antibody of constant specificity and composition through monoclonal antibody technology, which has led to the use of antibodies as potent biological targeting agents for *in vivo* use and as diagnostic agents.

The immune system

Higher animals possess a highly sophisticated immune system. Substances that activate the immune system are known as antigens. Two kinds of effector mechanisms mediate the immune response to antigens. One response is mediated by a population of lymphocytes known as T lymphocytes (T cells). These T cells act in conjunction with a second set of lymphocytes termed B lymphocytes (B cells) to ensure that antibodies are only produced to foreign molecules and microorganisms. It is the function of activated B cells to produce antibodies and of the T cells to police the antibody production process so that antibodies are only produced to invading foreign molecules and microorganisms.

Each B cell has on its surface a unique receptor. A small fraction of the B cells, which normally circulate in the blood and lymph, will fortuitously bind to patches on the surface of the foreign molecule or microorganism (called epitopes). The T cell screens the resulting B cell-molecular complex and if the foreign molecule does not possess the marker flags on its surface which identify it as ''self'' then the T cells signal to the complexed B cells to activate division of these B cells through the release of activator molecules (particularly interleukins). The activated B cells now rapidly increase in number and secrete antibodies which possess antibody-binding sites, the structure of which is identical to the receptors on the preactivated cells which themselves bind to the foreign molecules or microorganisms. Hence these antibodies will bind to the same foreign molecules or epitopes on the surface of larger invading microorganisms. This is the basis of the humoral immune response.

It should be noted that each activated B cell will produce a series of identical daughter cells or clones each of which secretes the same unique antibody. In practice a number of B cells are activated for each antigen and consequently a range of different cloned B cells is generated and hence a variety of antibodies each recognizing different epitopes will be present in the antiserum of the animal exposed to the foreign antigen. The resulting antiserum is termed as polyclonal antiserum.

Antigens

An antigen is any substance which can elicit an immune response in an animal which possesses a functional immune response. Proteins, which are foreign to the animal, are generally highly immunogenic and will stimulate the production of a range of different antibodies, each specific for a particular determinant (or epitope) associated with the surface of the protein. These antibodies will bind to that particular portion of the protein only. Polysaccharides and nucleic acids are less immunogenic than proteins, even though they may have a high molecular weight. Low molecular weight molecules (M_r < about 2000) do not elicit the formation of antibodies but when they are covalently coupled to a foreign protein molecule which acts as a carrier for these smaller molecules, a range of antibodies may be produced when the carrier complex is used as an immunogen, some of which recognize the smaller molecules on the surface of the carrier complex. These antibodies can now also bind to the uncomplexed small molecules in solution. In this way antibodies to a range of smaller molecules including drugs, steroids, vitamins, and pesticides, can be raised.

Antibody structure and classes

All antibodies have the same general structure and are symmetrical molecules made up of four polypeptide chains. Two chains contain identical sequences of 400–500 amino acid residues and are termed heavy (H) chains. The two other chains are termed light (L) chains and these contain over 200 amino acid residues. The H chains are joined together by disulfide bonds and each L chain is joined to a H chain by a disulfide bond. The sequence of amino acids in the amino terminal half of both H and L chains differs substantially between antibodies stimulated by different antigens. This region of the chain is termed the variable region (VH and VL) and within each of these variable regions lie three hypervariable segments. The VH and LH are folded in such a way as to bring together the hypervariable regions together to form a groove or cavity into which the epitope fits. The antigen–antibody binding is highly specific due to the stereochemical complementarity which is required to couple complementary hydrophobic and/or ionic interactions between the amino acids in the binding site and the contact groups on the surface of the epitope.

The carboxyl terminal shows far less sequence variation between antibodies and is termed the constant region (C) of the antibody molecule. Each species of animal will produce identical constant regions for each subclass of antibody. Cleavage of the antibody molecule can be achieved with proteolytic enzymes such as pepsin and trypsin into fragments that retain antigen-binding properties (termed F_{ab} fragments) and fragments that do not (termed F_c fragments). The major humoral antibody is IgG. Its structure is illustrated in Figure 12.5.

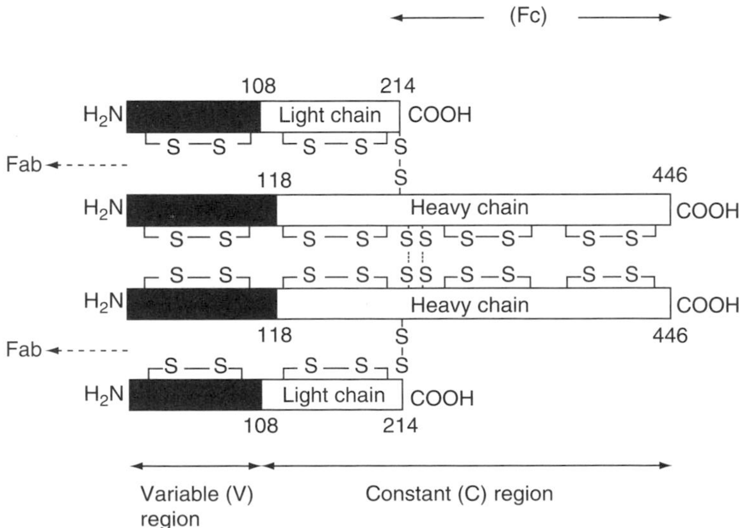

Figure 12.5 Diagrammatic representation of a molecule of human IgG.

12.2.2 Monoclonal Antibodies

It was noted in an earlier section (the immune system) that an activated B cell will subdivide to produce a clone of identical daughter cells each of which secretes identical antibodies. If a single activated B cell could be isolated and cloned then the resulting antibodies derived from a single clone of daughter cells is termed as monoclonal antiserum. Unfortunately it is not feasible to produce monoclonal antibodies via this route as the quantities of antibodies thus obtained are limited since culture and growth of the cell line will rapidly result in cell death due to the mortality of B cells. It is therefore necessary to render activated B cells immortal and this is achieved by fusing the genetic material from the required B cells with a cancerous B cell from the same species since cancerous cells are immortal and the resulting hybrid cells should contain the genes for production of the required antibody and for immortality. Large-scale tissue culture of the resulting fused cells should enable production and harvesting of monoclonal antibodies on commercial scales.

In practice mice or rats are used and the most common type of mouse used for monoclonal antibody production is an inbred strain known as BALB/C. The cancerous B cells required for fusion are myeloma cells.

Kohler and Milstein in 1975 demonstrated that mouse myeloma cells could be fused with B cells taken from the spleen of immunized mice and the resultant hybrids produced antibody. The technique is called somatic cell hybridization and the cell product is termed hybridoma (Figure 12.6).

The five major steps involved in monoclonal antibody formation through this process are as follows:

1. Immunization of the selected animals (usually BALB/C mice) with immunogen
2. Isolation of the spleen from a hyperimmunized animal and fusion of B cells from the spleen with myeloma cells from the same species
3. Culture of the resulting hybridoma cells in (HAT) medium
4. Selection of single clones of immortal cells secreting the required antibody
5. Scale up of the culture process to produce the required monoclonal antibody

The initial stage is to repeatedly inject the antigen into a group of about six animals, often in conjunction with immunostimulants such as Freund's adjuvant (complete for the first immunization followed by incomplete for subsequent ones). Immunizations take place at intervals of about a week and with successive immunizations there is an increased stimulation of B cell clones within the animal responding to the antigen. The presence of high concentrations of antibody to the antigen

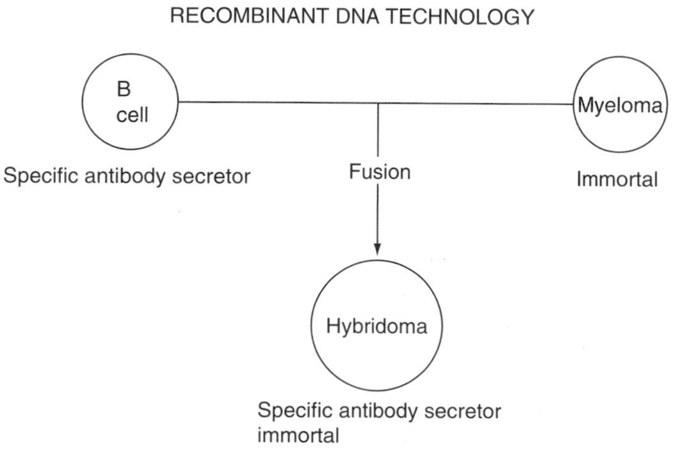

Figure 12.6 Principle of hybridoma formation.

can be demonstrated by taking blood samples from the animal and analyzing the sample. In this way the animal giving the best response to the antigen can be identified.

A suitable source of these B cells is the spleen from which they are harvested after the animal showing the best response has been sacrificed. The fusion partners (the myeloma cells) are now well-established cell lines and many of these cell lines have mutated to non-secretors or better still, non-synthesizers of antibody. The latter is the ideal partner for the antigen-stimulated B cell. The cell fusion is normally carried out using large numbers of the two cell types, as the rate of successful fusion is low (about 1 in 10^5 cells). The inclusion of 50% polyethylene glycol (PEG, M_r 140–4000) and about 5–10% dimethyl sulfoxide (DMSO) in the solution creates a favorable medium for membrane–membrane fusion between cells to occur.

Due to the low fusion rate and of the probability that fusion will occur between identical cells in addition to that between B cells and myeloma cells, the fused cells are transferred to a medium which only allows the required successful fusions to flourish. The standard technique is to use myeloma cells, which have lost the capacity to synthesize hypoxanthine–guanine phosphoribosyl-transferase (HGPRT), one of the two enzymes required for the eventual synthesis of DNA in the cell.

The other pathway leading to DNA synthesis is the salvage pathway and it can be blocked by the addition of chemical aminopterin to the cell culture medium. If the transferred cells from the fusion step are placed in a medium containing aminopterin (A) then only cells containing HGPRT can survive. The gene for this enzyme will be derived from the activated B cells and use of this pathway for DNA production is also encouraged by the presence of hypoxanthine (H) and thymidine (T) in the medium since these compounds are utilized in the synthesis by the enzyme. The medium is termed HAT medium due to the presence of these three additives in it.

Unfused cells and fused cells containing genetic material from only the B cells will not survive in the HAT medium as the unfused myeloma cells die due to inability to synthesize DNA while the other cells eventually die out. Hybridoma cells are able to proliferate as they contain the genes for immortality and for HGPRT.

As antigens contain many epitopes, it follows that the resulting hybridoma cells will secrete a number of antibodies corresponding to these epitopes. It is therefore necessary to screen for clones, which produce antibodies to the required epitope. This is achieved in two steps; firstly single hybridoma cells are isolated and then these are screened to identify which are producing the required antibodies.

Two strategies can be used for cloning cells:

a. *Limiting dilution*: The hybrid suspension is diluted so that the volume put into each culture well contains on average a single cell. No growth will occur in those wells that receive no cells and those wells that receive more than one cell may result in antibodies of two specificities being found. The clones that develop may be broken up and the dilution process and cloning repeated several times to ensure monoclonality.
b. *Solid medium*: A solid medium based on agarose gel has been developed which permits the cells to divide on the surface to form visible colonies. These can be broken up and the process repeated to obtain pure clones of hybridoma cells which can then be transferred to a suitable culture medium to test for antibody production.

The clones are screened for antibody production by using a suitable assay technique, which can initially detect the class of Ig produced, since IgG or IgM are commonly produced. This test uses a second antibody reagent which specifically binds to IgG or IgM molecules and this binding interaction can be monitored. The next step is to determine whether any of the clones are secreting antibody to any of the epitopes associated with the antigen. Once again the use of a second antibody reagent is employed (see Figure 12.7). The test uses the antigen or epitope immobilized on a convenient surface such as the wells of microtitre plates. If the culture supernatant contains antibodies from the antigen then these will bind to the epitope on the surface. Addition of a labeled

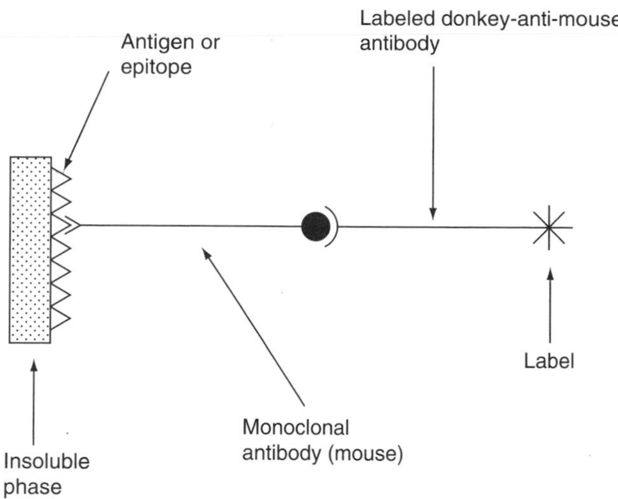

Figure 12.7 Use of anti-antibodies to detect synthesis of antibody by hybridoma cells.

antibody, which specifically binds to the class of antibody identified in the first screen, to the wells will result in the formation of a second antibody complex linked to the first murine antibody attached to the epitope on the surface. After washing the plate to remove excess second reagent, the presence of the bound second reagent in the complex can be determined from the presence of label in the wells.

Once the clones secreting the required antibody have been identified these cells are transferred to large culture vessels where their propagation can take place and monoclonal antibodies subsequently harvested. A variety of culture systems can be employed ranging from conventional animal cell culture in bottles to use of hollow fiber systems onto which cells adhere and through which the culture medium is pumped. The latter systems offer advantages since the secreted antibodies can be collected in the medium emerging from the fibers and higher yields of antibody can be obtained (g/L compared with mcg/L for conventional culture systems). Alternatively, the murine hybridoma can be propagated in BALB/C mice where the concentration in ascites fluid can reach 5–20 mg/mL. A diagrammatic representation of monoclonal antibody production is shown in Figure 12.8.

Monoclonal antibodies are chemically homogeneous but may not be truly monospecific as they can react with antigens that share an epitope or can react with epitopes that are structurally closely related.

12.2.3 Application of Monoclonal Antibodies

There has been a dramatic increase in the use of monoclonal antibodies and the greatest expansion has been in the field of therapy where they have been used to target drugs to specific tissues. However, they will continue to find an ever-increasing role as analytical and diagnostic reagents due to their reproducible properties of specificity and affinity for specific antigens, and from the ability to produce them in commercial quantities.

Analytical

Because of their specificity, monoclonal antibodies are used extensively in the assay of serum and urine levels of hormones (e.g., detection of human chorionic gonadotrophin in the urine of pregnant women), drugs, enzymes, etc. They are also finding application in assays for environmental contaminants such as pesticides and dioxins, in soil, water, and air samples. A wide range of

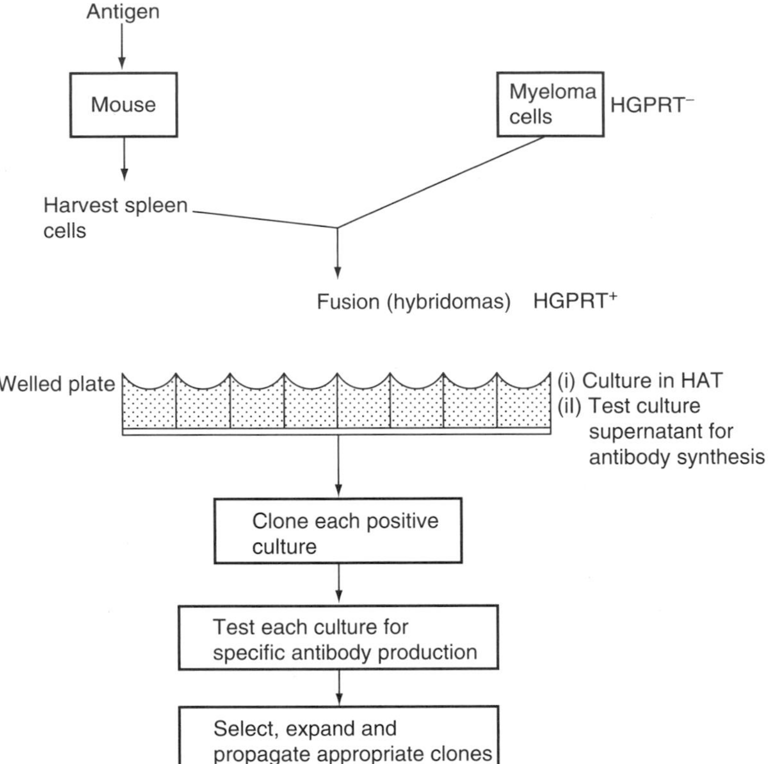

Figure 12.8 Diagrammatic representation of monoclonal antibody production.

immunoassay procedures has been developed which employ a variety of end points. Of particular importance are non-competitive, non-isotopic assays based on fluorescence and spectrophotometric end points, the latter resulting from use of enzyme labels to rapidly convert substrates to colored products. Such assays are used for therapeutic drug monitoring where fully automated assays provide results within seconds or minutes per sample.

Diagnostic

Monoclonal antibodies have been used to identify specific antigens associated with cell surfaces. This has led to the development of rapid tissue typing techniques before transplantation surgery, the classification of cells and sub-populations (particularly the T cell family), cell–cell interactions and differentiation, the biochemistry of cell surfaces, the classification of microorganisms, and the detection of tumor markers associated with certain types of tumor.

Therapeutic

Therapeutic antibodies have now established themselves as one of the most successful classes of therapeutic drugs. Currently over 25% of new therapeutic agents under development are based on antibodies and this percentage is predicted to increase with estimated sales of US$10–20 billion by 2010. What has led to this remarkable success? The major factors have been the ability to produce pure antibodies on a large scale that are highly specific for a therapeutic target with good therapeutic outcomes and without unacceptable side effects. The main application has been in the fields of cancer and autoimmune disease. Table 12.3 lists the 14 FDA-approved therapeutic antibodies and Table 12.4 indicates the major areas of therapeutic development.

One of the prime developments has been the application of genetic engineering to antibody production. All the therapeutic antibodies introduced since 1997 are based on chimeric or humanized antibodies. In these antibodies, the majority of the murine-derived amino acids have been replaced by the corresponding human sequences. This renders the resulting antibody more "human" and hence less likely to evoke an immune response when given to the patient. This is an essential prerequisite since therapeutic antibody therapy generally requires repeated administration, and generation of antibodies by the patient to the administered antibodies would lead to a quenching of their activity and possibly produce an anaphylactic reaction in such patients.

Figure 12.5 shows the structure of an IgG antibody molecule. The antibody-binding site possesses a unique sequence of amino acids that results in the high affinity and specific binding between the antibody and the antigen. The two sites are located at the tips of the four peptide chains. The constant regions possess amino acid sequences that are unique for a particular type of IgG molecule of a given species. The same applies to the major part of the light chain — it is only at the ends of the chains that variations in sequences are found. Since the unique sequences are known for human IgG and the corresponding amino acid sequences associated with the variable regions of the

Table 12.3 Monoclonal antibodies approved by the FDA as of April 2003

Name	Indication	Antibody	Company (Year)
Orthoclone OKT®3	Acute kidney transplant rejection	Murine IgG2a	Johnson & Johnson (1986)
ReoPro®	Prevention of blood clotting refractory unstable angina	Chimeric Fab	Centor/Lilly (1994); (1997)
Panorex®	Colorectal cancer	Murine	Centooor (1997)
Rituxan®	Non-Hodgkin's lymphoma	Chimeric IgG1	IDEC/Genentech/ Roche (1997)
Zenepax®	Acute kidney transplant rejection	Humanized Ig1	Roche (1997)
Remicade®	Crohn's disease rheumatoid arthritis	Chimeric IgG1	Centocor (1998); (1999)
Simulect®	Acute kidney transplant rejection	Chimeric IgG1	Novartis (1998)
Synagis®	Respiratory syncytial virus (RSV)	Humanized IgG1	Medimmune (1998)
Herzeptin®	Metastatic breast cancer	Humanized IgG1	Genentec (1998)
Mylotarg®	Relapsed acute myeloid leukemia	Humanized IgG4-toxin conjugated	Wyeth Ayerst (2000)
Campath®	Chronic lymphocytic	Humanized IgG1	Millenium (2001)
Zevalin®	Non-Hodgkin's lymphoma	Mouse IgG1-radionuclide	IDEC (2002)
Xolar®	Allergy	Humanized	Tanox/Genentech/ Novartis (2002)
Humira®	Rheumatoid arthritis	Human IgG1	Abbot/CAT (2003)

Table 12.4 Examples of classes of therapeutic antibodies under development[a]

Cancer and related conditions	38
Autoimmune disorders	10
Respiratory disorders	7
Digestive conditions	6
Skin disorders	6
Infectious disease	4
Neurological disorders	4
Transplantation	3
Diabetes and related conditions	1
Eye conditions	1
Heart disease	1
Others	3

[a]http://www.phrma.org/newmedicines/biotech (2002).

murine monoclonal antibody can be determined. Hence the corresponding DNA codon sequences can be inferred and genes synthesized or isolated that when inserted via a vector into an appropriate host, will result in expression of hybrid peptides possessing the human constant region with the murine variable region grafted on. These are the chimeric antibodies referred to in Table 12.3.

The process can be further refined so that the expressed antibody consists of human sequences of the constant region and most of the variable region with only the rodent amino acids responsible for antigen binding left in the final structure. Such antibodies are those referred to as ''humanized'' antibodies in Table 12.3. Such antibodies contain less than 10% murine-derived amino acids and consequently have lower antigenicity. It is now possible to eliminate all murine amino acids by using human antibody libraries that are expressed in phage particles or to use transgenic mice that possess human genes for the heavy and light antibody chains. Such antibodies are termed ''human.'' If the transgenic mouse is immunized then it will produce a range of ''human'' antibodies to the antigen. The standard procedure for hybridoma generation described above can then be applied to the responding animals. The first fully human monoclonal antibody gained approval by the FDA in 2003. It is adlimumab (Humira®; Table 12.3) and is designed for the treatment of rheumatoid arthritis. It is a human IgG1 that binds with high affinity to tumor necrosis factor-alpha (TNF-α) and it was obtained from a phage antibody library.

The majority of the 300 or so therapeutic antibodies currently in clinical trails are based on ''humanized'', or, ''human'' F_{ab} or similar fragments (such as single chain F_v, scF_v). Isotypes of IgG are also chosen (e.g., IgG1, IgG2, IgG4) to produce the required response following binding to the target antigen. For cancer therapy this antigen is often a cell surface protein that is highly expressed or uniquely expressed in the cancer cell. Thus IgG1 is a good mediator of phagocytosis and antibody dependent cellular cytotoxicity and binding of this antibody isotype leads to elimination of the target cell via these mechanisms. In contrast if antibody binding is required to block or inhibit the functioning of the target cells, then IgG4 or F_{ab} fragments are preferred.

Another approach to enhance the effector activity of lymphocytes or phagocytes toward the target cells is to use bi-specific antibodies. These consist of two different antibodies that are joined together so that they will recognize and bind to two different target antigens. One can be for an epitope on the surface of the target cell such as the cancer cell. The second is specific for an epitope on an effector cell such as T cells — such binding activates the effector cell and the response is directed against the target cell. Again such bi-specific antibodies can be generated as single protein molecules by genetic engineering.

A variation of the above approach involves a hybrid protein consisting of an antibody and an enzyme. The antibody again directs the enzyme to the target cell. A prodrug can then be administered that is inactive. Once it reaches the target cell it will be converted into an active form, by the enzyme, that will kill cells in the immediate vicinity. This is known as the antibody-directed enzyme prodrug therapy (ADEPT) approach (see Chapter 7).

In a final alternative strategy, the target cell can be eliminated by attaching a molecule to it so that on antibody binding the cell is destroyed. This can be achieved by attaching a radioactive atom or a toxic molecule. In the former approach a radionuclide such as iodine-131 or yttrium-90 is attached to the antibody. Once the antibody accumulates on the surface antigen, the target cell is exposed to a fatal dose of ionizing radiation. An example is Zevalin® (Table 12.3).

In contrast, in order to be effective, once the toxin has been transported to the target site via piggy-backing on an antibody, the antibody must become internalized within the target cell. An example is Mylotarg®, which utilizes the toxin calicheamicin (Table 12.3). Another advantage of using antibody–isotope conjugates is that the target cell and surrounding cells will be subjected to the ionizing radiation whereas antibody–toxin conjugates will only eliminate the single target cell to which it becomes attached and is internalized.

12.2.4 Recombinant Vaccines

As noted in the introduction, vaccines based on recombinant vaccines now constitute one of the major categories of pharmaceutical products. These have proved beneficial as they *provide us with non-pathogenic vectors*, which have been genetically modified to express on their surface, one or more antigenic epitopes from viruses, bacteria, and organisms which currently produce widespread illness and mortality such as measles, tuberculosis, diphtheria, poliomyelitis, and so on. Thus a single vaccination to a polyvaccine from such an engineered vector would be highly cost-effective and safer to use. Alternatively, instead of using live recombinant vectors, it is possible to produce and isolate recombinant surface antigen proteins for use as vaccines. Examples of recombinant vaccines include recombinant BCG, an avirulent bovine tubercle bacillus, and the recombinant surface antigen protein for hepatitis B.

FURTHER READING

Brekke, O.H. and Loset, G.A. (2003) New technologies in therapeutic antibody development. *Current Opinion in Pharmacology* **3**, 544–550.

Brown, T.A. (1991) *Essential Molecular Biology, Vol. 1. A Practical Approach*. Oxford: IRL Press.

Emery, A.E.H. (1984) *An Introduction to Recombinant DNA*. Chichester: John Wiley.

Fanger, M.W. and Guyre, P.M. (1991) Bispecific antibodies for targeted cytotoxicity. *Trends in Biotechnology* **9**, 375–380.

Glennie, M.J. and van de Winkel, J.G.L. (2003) Renaissance of cancer therapeutic antibodies. *Drug Discovery Today* **8**, 503–510.

Holliger, P. and Winter, G. (1993) Engineering bispecific antibodies. *Current Opinion in Biotechnology* **4**, 446–449.

Hudson, L. and Hay, F.C. (1989) *Practical Immunology*, 3rd edn. Oxford: Blackwell Scientific Publications.

Hurle, M.R. and Gross, M. (1994) Protein engineering techniques for antibody humanisation. *Current Opinion in Biotechnology* **5**, 428–433.

Lerner, R.A., Kang, A.S., Bain, J.D., Burton, D.R. and Barbas, C.F. (1992) Antibodies without immunisation. *Science* **258**, 1313–1315.

Old, R.W. and Primrose, S.B. (1985) In: *Principles of Gene Manipulation*, N.G. Carr, J.L. Ingraham, and S.C. Rittenberg (eds). Oxford: Blackwell Scientific Publications.

Paliwal, S.K., Nadler, T.K. and Regnier, F.E. (1993) Rapid process monitoring in biotechnology. *Trends in Biotechnology* **11**, 95–101.

Pezzuto, J.M., Johnson, M.E. and Manasse, H.R. (1993) *Biotechnology and Pharmacy*. New York: Chapman & Hall.

Tomlinson, E. (1992) Impact of new biologies on the medical and pharmaceutical sciences. *Journal of Pharmacy and Pharmacology* **44** (Suppl. 1), 147–149.

Walker, J.M. and Gingold, E.B. (1993) *Molecular Biology and Biotechnology*, 3rd edn. Cambridge: The Royal Society of Chemistry.

Walsh, G. (2003) Pharmaceutical biotechnology products approved within the European Union. *European Journal of Pharmaceutics and Biopharmaceutics* **53**, 3–10.

13

The Human Genome and its Impact on Drug Discovery and Therapy

Frederick J. Rowell

CONTENTS

13.1 THE HUMAN GENOME

13.1.1 Introduction

The unraveling of the human genome and that of other organisms have provided scientists with a powerful new tool to gain a greater insight into relationships between disease and genetic makeup. Such insight has raised the possibility of finding new targets for drug action based on new sites of action. This approach implies an ability to identify both new target genes and the way that these are expressed *in vivo* via proteins, leading to new pharmacological pathways. This in turn has led to the introduction of new technologies to rapidly identify genes and the proteins that they express. The former is termed, ''genomics,'' and the latter, ''proteomics.'' In this chapter we will review the current potential and limitations of genomics and proteomics with regard to drug discovery. We will also describe how such information can be used to screen for diseases and used to predict how genetic makeup can affect pharmacokinetics, via ''pharmocogenetics,'' and affect drug metabolism leading to possible toxicological problems, via, ''toxicogenetics.'' Finally the new technologies that are leading to high-throughput screening in these emerging areas will be reviewed.

13.1.2 Genomics

The genome describes the identity and sequence of the genes of an organism. Much progress has been made in elucidating genomes for a number of organisms starting with those for simple organisms such as bacteria and yeast, those of more complex life forms such as plants, roundworms, the fruit fly, and the mouse. For humans the draft sequence was completed in 2000 with the complete sequence announced to coincide with the 50th anniversary of the discovery of the structure of DNA, in 2003. This information enables us to compare our genetic makeup with that of other organisms, a study termed, ''comparative genomics.'' Such studies help to provide interspecies comparisons of evolution, biochemistry, genetics, metabolic, and physiological pathways. It appears that there are some 32,000 genes, which constitute the human genome but that we share about 50% of these with the banana plant, over 80% with the mouse, and about 99% with the chimpanzee.

The information that has accrued following the elucidation of the genome for each species has been analyzed to identify genes that are implicated in certain disease states. This information is available on, ''The Ensemble,'' website where the human genome data together with that for a number of other species are stored and regularly updated and expanded.[1] This site therefore provides information regarding disease states and implicated genes, and their locations on specific chromosomes. Recent work, for example, following the sequencing of human chromosome 20, has implicated genes involved in diabetes, childhood eczema, and leukemia. Other examples of diseases that are thought to be associated with one or more malfunctioning genes include heart disease, cystic fibrosis, and prostrate cancer.

13.1.3 Pharmacogenetics

Compound discovery in the pharmaceutical industry is both expensive and time-consuming. It takes on average $500 million to bring a new drug to market and this process requires about 15 years. However, only 1% of new chemical entities emerging from drug discovery are successful. The major reasons for this large drop out rate are poor pharmacokinetics (about 40%) and preclinical toxicity (11%). The concept of *in vitro* screening for ADME/Tox (drug absorption, distribution, metabolism, and excretion/toxicology) profiles for new drug candidates is now well established. The effect that differences in the genetic makeup and the resulting expression of proteins of individuals play on ADME/Tox is now beginning to be understood and used as a predictive tool in the screening of new drug entities. This new approach may reduce both the cost and time for such screening.

It may also play an important future role in tailoring drug prescribing and dispensing to those individuals of greatest genetic compatibility, thereby reducing adverse drug effects. The urgent need for such an approach stems from the fact that in the USA alone, there are currently over 100,000 avoidable deaths each year with adverse drug reactions and are the major cause of hospitalization and death.

13.1.4 Proteomics

It is now clear that organisms are more complex than their simple genetic composition indicates. Thus it is often observed that there is no correlation between mRNA expression levels and protein expression, and secondly, it is the case that genes are able to express more than one protein per gene. Hence in bacteria the average number of proteins per gene is one to two, three in yeast, and three to six or more in humans. Hence the whole gamut of proteins expressed by the organism's genome also needs to be determined. The process by which this is achieved is termed proteomics.

The implications of this are obvious and profound since the ability to identify protein expression and relate it with healthy or diseased states in humans offers us the potential to gain insight into the mechanisms leading to diseases and of ways of correcting such faulty biochemical/physiological pathways. This may be via designing drugs to directly correct the defect by binding say to a specific protein, or by producing the correct protein via recombinant DNA technology (see Chapter 12) and using this as a therapeutic agent as has been the case with insulin. A similar approach can be applied to pathogenic organisms since knowledge of their genomics and proteomics can lead to better insights into their mode of action within the host leading to disease. As an example, this approach is used to develop new vaccines for malaria and the AIDS virus, and the development of drugs to new targets.

13.2 SCREENS FOR GENES AND EXPRESSED PROTEINS

13.2.1 Genomes for Species

If the information that is potentially available from analysis of genes and their expressed proteins is to be harnessed and used to diagnose disease progression, develop new drugs, therapeutic proteins, and new vaccines, then rapid analytical methods based on high throughput of samples are needed. Such methods have emerged partly as a result of programs set up to elucidate the genome for each species. This involved the refinement of existing analytical methods and development of new, more rapid analytical methods due to the magnitude of the task.

The processes involved included the isolation of each chromosome, its unraveling, and determination of the base sequence associated with each DNA strand within it. In practice two main strategies were used: the "top down" and "bottom up" approaches. In the former, the DNA was chopped into relatively large fragments using the types of restriction enzymes described in Chapter 12. This resulted in common or overlapping sequences in some fragments. The precise base sequences in these overlapping fragments were then determined so that the whole DNA sequence could be worked out using computer algorithms. In the second approach, the whole genome was randomly chopped up and then following sequencing of the fragments, the sequence was established. This so called, "shot gun" approach did not rely on hierarchical sequencing and hence is more direct. However, it is best suited to compact genomes such as those of bacteria while the ''top down'' approach is best suited to diffuse genomes which contain ''junk DNA,'' that is sequences associated with regulatory or structural information rather than for genes. This applies to the mammalian genome where only about 5% of the genome codes for genes.

13.2.2 Electrophoresis

The main method used to separate the large number of smaller fragments was gel electrophoresis. The mixture is placed at one end of a gel, usually made from agarose, or polyacrylamide where the mixture consists of small fragments or where increased resolution is desired such as in sequencing. These gels are thinly spread over the surface of a planar support. This plate is located within an electrophoresis system so that a high voltage can be applied across the ends of the gel. Under these conditions the DNA fragments move according to their relative size and charge through the cross-linked gel matrix. Since the backbone of DNA contains negatively charged phosphate groups net migration is toward the anode. Hence, separation of the fragments is achieved in a linear fashion across the gel. For protein mixtures the plate can then be turned through 90°, and the process is repeated when further resolution is then achieved. This two-step process is known as 2-D gel electrophoresis or 2-D PAGE (polyacrylamide gel electrophoresis when polyacrylamide gel is used).

13.2.3 Staining and Blotting

Having separated the fragments it is then necessary to identify the base sequence of the individual fragments. This can be achieved by transferring some of the material associated with the spots associated with each resolved component, from the surface of the gel to a new ''sticky'' membrane. The process involves placing a membrane, often made of nylon or cellulose nitrate, onto the gel and applying pressure to ensure good contact between the two surfaces. After several hours the upper membrane is removed when the composition of the transferred spots can be determined. When DNA fragments are analyzed, this process is often termed ''Southern'' blotting after its inventor.

Before this can commence it is first necessary to know the position of the spots on the surface of the original gel template following its development. This can be determined using specific dyes that bind to DNA or by using complementary strands of DNA that carry either radioactive labels such as ^{32}P or fluorescent label, and which will stick to DNA carrying specific codon sequences. Knowing the location of the spots on the gel surface, the complementary pattern on the membrane can be deduced. It is now possible to interrogate each spot on the membrane by removing some of the DNA material and to subject it to analysis. Spot removal is generally achieved by either physically removing the spot, extracting the DNA, and then analyzing the base sequence.

13.2.4 Analysis of Sequences

The Sanger method is still used for genetic sequencing processes. However, mass spectrometry (MS) is applied to several areas of analysis and two approaches are under development. The first is termed as matrix-assisted laser desorption ionization–time of flight–mass spectrometry (MALDI-TOF-MS) and the second electrospray-MS. The former is best suited for direct use with samples on membranes and provides information on the molecular mass and the mass fingerprint — the unique pattern of peaks produced in the mass spectrum resulting from the fragments produced from the fragmentation of the parent DNA strand. It is then applied to high-resolution SNP screening although currently most SNPs are still screened by PCR-based methods (see Chapter 12) using real time technology. The second method is best suited for samples in solution and provides information for *de novo* sequencing since the accurate mass of larger fragments can be determined from its accurate mass and hence the composition and sequence of the bases can be deduced.[2]

However, both approaches have found greater application for protein sequencing and identification as discussed below.

13.3 DNA MICROARRAYS

13.3.1 Basis of the Method

The above approach using 2-D PAGE displays admirable resolution, i.e., ability to separate genes differing by only single bases, the so-called single nucleotide polymorphisms (SNPs). Only 0.1% of the genome is different between individuals, and it is these SNPs that constitute the major source of these differences. Over 1 million such SNPs have already been identified. However, the PAGE process followed by blotting, digestion, and MS analysis is a relatively slow process involving expensive equipment and skilled operators. It also generally requires an amplification step (via the polymerase chain reaction described in the previous chapter) at the outset to obtain sufficient DNA to undertake the subsequent PAGE-blotting and MS analysis. It is therefore unsuited as a simple method for screening for genes on a routine basis.

To overcome these limitations an alternative approach has been developed which has the potential as a high-throughput technology for automated genome analysis. The basis of this approach is outlined in Figure 13.1. The target cell (A) is cultured or removed from the tissue and its total mRNA (B) is extracted (Step 1). These are converted to cDNA via reverse transcriptase (Step 2). The cDNA can be rendered fluorescent during this step as shown in Figure 13.2 when a fluorescently tagged nucleotide is incorporated into the cDNA chain. Alternatively a biotinylated nucleotide can be used when biotin molecules are incorporated into unlabeled cRNA chains via transcription.

A planar surface that is commonly a silica glass slide or a nonporous membrane such as cellulose nitrate or polyvinylidene difluoride (PDVF) is prepared so that nucleotides of defined base sequences can be attached as small spots on the surface via covalent binding or strong adsorption (Figure 13.1C). In this way an array of microspots is constructed upon the surface producing a so-called gene chip. The nucleotide sequence on the spots is chosen to be complementary to a particular nucleotide sequence specific to a particular gene so that if that gene is present in the labeled extract from the cell then that gene will become anchored at that spot (Figure 13.3).

Following incubation of the tagged cDNAs with the chip, specific binding takes place via hybridization (Figure 13.1, Step 3). Subsequently, following washing and drying of the chip, the positions of gene attachment can be determined by scanning the surface of the chip with a fluorescence scanner tuned to the label used to label the cRNA (Figure 13.1, Step 4). If cRNA tagged with biotin is used in the hybridization step, then a labeled form of avidin, which rapidly and irreversibly binds to biotin on cDNA, is added to stain the spots before scanning. In this way the genes present and their expression levels/abundance in the initial extract can be identified. An example of such a screen is shown in Figure 13.4. This was produced using three different cDNA samples each labeled with a different fluorescent dye.

13.3.2 Uses In Gene Expression Studies

This information is currently most widely used for gene expression profiling where it is providing useful information on differentiating tumor and normal cells.[3] For example, following microarray analysis of 17,856 genes in samples from patients with non-Hodgkin's lymphomas, two subtypes of the disease have been distinguished for the first time. Similarly different expression profiles have been observed in other cancers such as melanoma, epithelial, breast, colon, renal, and ovarian cancer. In addition DNA microarrays can be used for diagnostic purposes as screens for diseases,

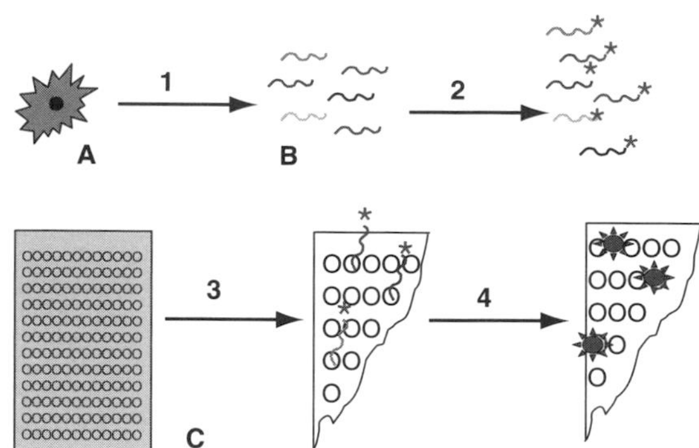

Figure 13.1 The basic steps in DNA microarray analysis.

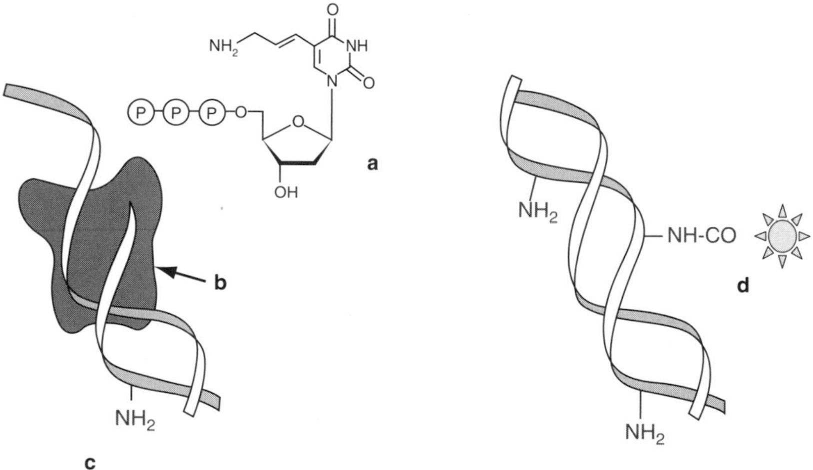

Figure 13.2 Labeling DNA molecules. A modified nucleotide such as aminoallyl deoxyuridine triphosphate is incorporated into the DNA chain via DNA polymerase (a). An activated derivative of a fluorescent dye now reacts with amino groups on the new DNA chain producing a fluorescently labeled DNA chain.

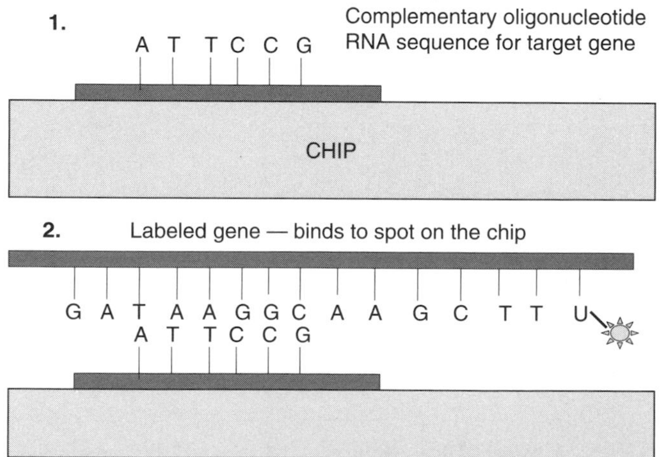

Figure 13.3 Illustration of the hybridization step in which a labeled DNA strand attaches to a complementary oligonucleotide chain attached as a single spot to the surface of a microarray plate.

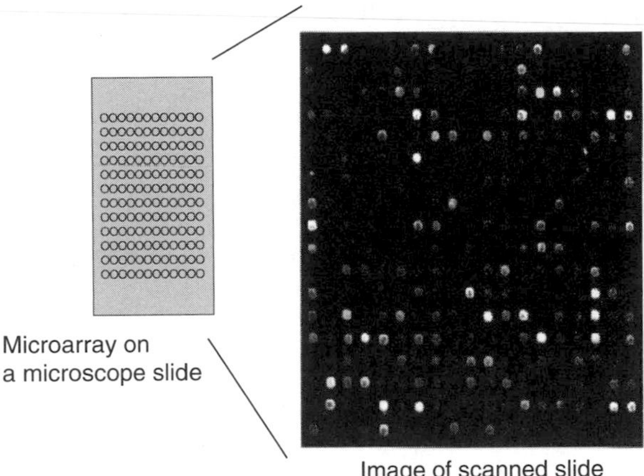

Figure 13.4 Example of a scanned cDNA microarray using a mixture of three different fluorescent tags.

for polymorphism analysis in studies on pharmacogenetics for example, and when coupled to mass spectrometry it can lead to novel gene identification. Useful summaries of DNA microarray technology that describe the main features of some hybridization microarray formats that are currently commercially available can be found on the Gene-Chips.com website[4] and in the reviews of Li et al.[5] and Weng and DeLisi.[6]

This approach presupposes that the genes present in the cell are known. However, the recent elucidation of the entire human genome and some of the major pathogenic organisms means that this approach is now a feasible way of screening for the presence or absence of specific genes associated with disease or for the presence of genes from potentially hazardous pathogens.

13.3.3 Limitations of Genome Information

There are still some limitations to this approach. The major restriction is not simply a technology-related one, since as noted above, commercial systems are now available that can screen for the entire 32,000 genes of the human genome on a single chip, but limitations due to the interpretation of the data.

As noted above for humans, one gene can result in the expression of an average of six different proteins. This stems from the way that the genetic code can be read during translation and the way that the resulting polypeptide sequences are spliced together following transcription. Also, although genes may be identified via the above methods, they may be mutated and not always transcribed. Conversely, some mRNA messengers are transcribed but not translated, and even if they are, the number of mRNA copies may not reflect the actual numbers of functional protein molecules produced. In addition, the resulting proteins can then be biochemically modified *in vivo* in a number of ways such as by phosphorylation, acetylation, glycosylation, and other posttranslational changes. Finally there is a sampling difference in disease, proteins may be secreted into bodily fluids such as urine or saliva, whereas biopsy samples may be needed to obtain DNA or mRNA. Due to these ambiguities, it is becoming apparent that screening for the protein products following translation, i.e., proteomics, may provide better information than that provided by the methods described above, i.e., by genomics. Consequently simple methods for determining the presence and amounts of proteins using 2-D gel electrophoresis, capillary electrophoresis, protein chips analogous to gene chips, and other methods have been and are being developed.

13.4 SCREENS FOR EXPRESSED PROTEINS

13.4.1 2-D PAGE

Conventionally the processes for the separation and identification of proteins have mirrored those used for genes. Hence the standard method has been a modified form of PAGE, blotting, and analysis. The strategy is to apply the mixture of proteins derived from the tissue or cells of interest to one end of a PAGE system and to undertake electrophoresis. In this case proteins will migrate according to their net charge. This will depend on their amino acid composition and the pH of the medium used to run the electrophoresis. Hence some proteins will carry positive charges, some negative charges, and others no charges during the development. Hence the direction and rate of migration is likely to differ for each protein with some remaining at the origin where they were applied to the membrane, and others migrating according to their overall charges and molecular sizes as they negotiate their way through the cross-linked coating of the membrane. Hence two developments are generally performed; in the first, separation is achieved according to the proteins' p*I* values during isoelectric focusing, followed by SDS-PAGE for the 2-D when further resolution takes place due to size.

13.4.2 Staining and Scanning

Following this development it is necessary to locate the position of the proteins on the membrane. This is achieved by staining the membrane with reagents that stick to proteins. Examples of stains include Coomassie blue, silver, and fluorescent dyes. Scanning systems are then used to locate the positions of the stained proteins. The resulting scans can be exported to an image analysis system and the patterns analyzed for expression of known proteins in health and disease.

13.4.3 Identification of Known Proteins

An example of a 2-D PAGE electrophoretogram of proteins separated by isoelectric focusing is shown in Figure 13.5. The black crosses correspond to known proteins. This analysis was achieved by accessing a 2-D PAGE proteomic database on the World Wide Web and logging onto the database

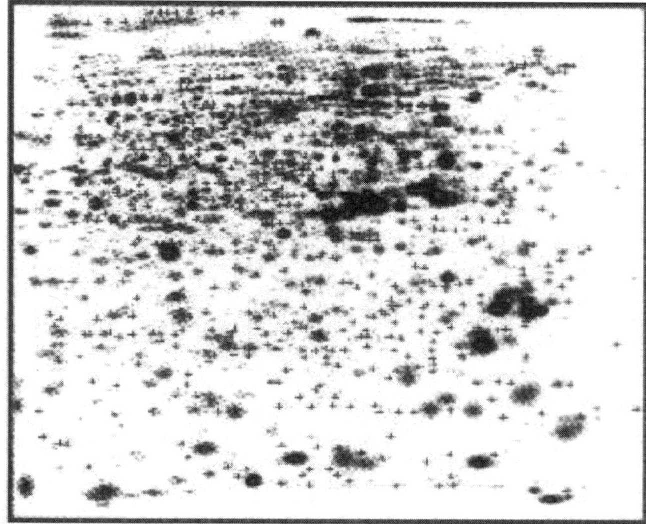

Figure 13.5 An example of a 2-D electrophoretogram of a protein mixture for identification of proteins expressed in cancer.

for proteins of that class and species (e.g., Ref. [7] — which also gives links to other databases). This example illustrates a common application of this approach, namely for monitoring expression of proteins during disease progression, in this case cancer. The general objective is to use proteome expression profiles to identify protein biomarkers for early diagnosis, and to obtain information that may lead to a better understanding of the molecular mechanisms responsible for progression of the disease. This information may be coupled with genomic data and the knowledge may then be used to identify targets for designing new candidate drugs.

13.4.4 Characterization of Unknown Proteins

Alternatively these areas or spots can be removed from the gel surface in order to identify unknown proteins. This may be via an automated spot-picking robot that transfers them to the wells of a microtiter plate where they are partially digested when a proteolytic enzyme such as trypsin is added to each well. This step is used to break the protein into smaller fragments. This digestion often takes place overnight and under standard conditions so that each protein will produce a mixture of fragments of exactly the same composition when digested under these conditions. Hence this unique mixture of fragments produces a "fingerprint" of peaks in the mass spectrum on subsequent analysis, that is characteristic of the parent protein. Analysis of the fragments by MALDI–TOF–MS can be used to characterize the parent protein but use of a nanospray MS–MS instrument or its equivalent may also be needed for the identification of novel proteins since these provide an accurate mass of the intact protein and of its amino acid sequence.

13.4.5 Example of Analysis for Plasma Proteins

At the end of the digestion process the resulting peptide fragments are transferred as spots onto metal plates that will be used as the target plates for MALDI–TOF–MS analysis.[8] Figure 13.6 shows an example of enzyme digestion of plasma proteins. In this example, following a process of chemical reduction, alkylation, and then 2-D electrophoresis, the resolved proteins were transferred via electroblotting onto a PVDF membrane. The proteins were visualized by staining and the blot was attached to a metal MALDI–TOF–MS plate using double-sided conducting tape to ensure good electrical conductivity. Minute droplets of the enzyme (100 pl) were applied to the stained spots on the membrane, in this case by a chemical inkjet printer, causing white spots due to bleaching of the blue stain. Multiple enzymes can be applied to a single protein spot and the peptide fragments are then analyzed by direct interrogation on each blot. Examples of the resulting MALDI–MS of tryptic peptides from spots 22 to 24 are shown in Figure 13.7.[9,10] The resulting mass spectrograms enable identification of the parent protein from matches in the fragmentation patterns and the accurate

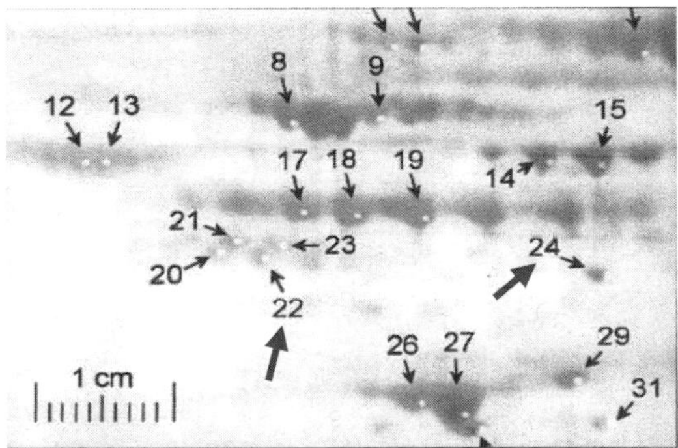

Figure 13.6 Example of a 2-D electrophoretogram of plasma proteins showing tryptic digests.

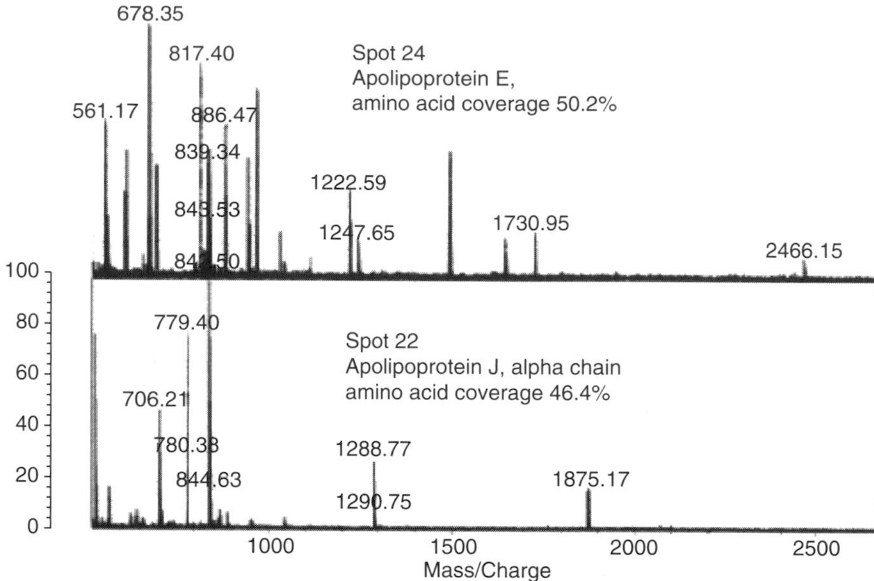

Figure 13.7 Example of a MALDI–TOF–MS of digests from spots 22 to 24 in Figure 13.6, illustrating identification of proteins from their amino acid sequences.

masses of fragments. This is performed automatically through direct comparison of the acquired data with information in a database provided with the instrument. Figure 13.8 shows the layout and main components of a typical MALDI–TOF–MS system.

It should be noted that automated equipment is now available for each of the processes described, so that high throughput of samples is possible but at a considerable cost with respect to the initial investment. However, commercial services are available and suppliers of these services can be found on the World Wide Web.

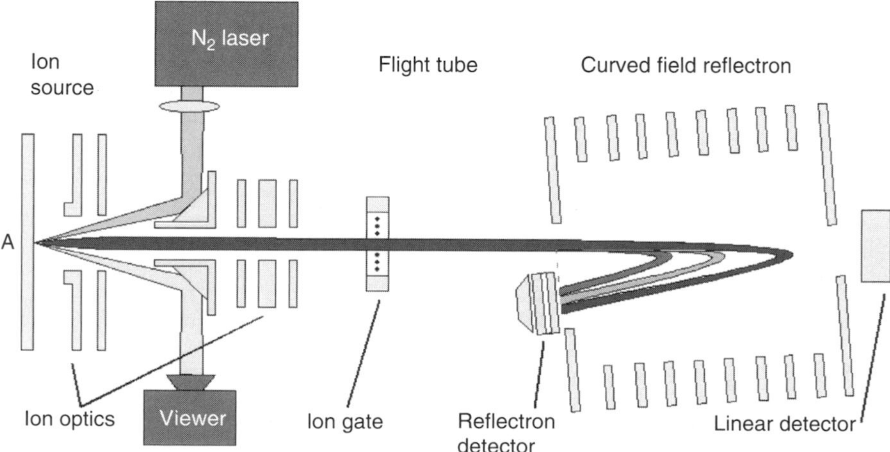

Figure 13.8 Schematic representation of the components of a MALDI–TOF–MS system. The laser beam is focused onto the surface of the plate at A and this results in desorption of molecules and their ionization. Once released they enter the MS and impact on the detector.

13.5 MINIATURIZED ANALYTICAL SYSTEMS FOR PROTEINS

13.5.1 Microarray Systems Using Antibodies

It was noted above that microarrays of spots can be produced on small surfaces such as chips. These spots can each be of a unique complementary DNA/oligonucleotide sequence so that when the chip is exposed to mixtures of genes, specific docking of genes will take place at the sites of the specific spots. Scanning of the spots will then identify which genes are present. It is currently possible to use commercial systems that can rapidly probe for the presence of whole genome sequences on a single chip.[11] In a similar manner, antibodies each raised to a specific protein can also be immobilized as discrete spots on these chips (Figure 13.9A).[12] As with the analysis of DNA mixtures on the chip, the proteins are covalently labeled with a fluorescent molecule and then the mixture is added to the protein chip. Specific binding takes place only on the surface between the immobilized antibodies and the antigenic fluorescent-labeled protein molecules (Figure 13.9B). The location of spots on which binding has occurred can again be achieved by scanning the surface for fluorescence when the identities of the proteins in the sample and their abundance (which is proportional to the intensity of the spots) can be determined (Figure 13.9C).

As an alternative strategy, the target proteins can be immobilized as spots on the surface of the chip and the protein mixture from the cell or organism is introduced onto the chip together with a mixture of tagged antibodies each labeled with a fluorescent molecule. If a protein is present in the mixture, the labeled antibody will bind to it and not to the corresponding protein on the surface spot. Conversely, if the protein is absent in the mixture then binding will take place and a fluorescent spot will appear on scanning. In the former case it is assumed that the labeling process does not alter the protein's immunogenicity (reduction of the subsequent ability of the antibody to recognize the chemically changed protein). In the second it is assumed that the process of protein immobilization does not produce conformational changes reducing the subsequent recognition of the protein on the spot. In practice it is necessary to check that no loss of immunogenicity occurs for both approaches.

13.5.2 Non-Antibody-Based Systems

Alternatives to antibodies have recently been investigated and the most promising developments include analyte-specific reagents such as aptamers and molecular imprint polymers, and more general non-specific affinity reagents such as spots of strong cation exchange material, hydrophobic material, and immobilized metal affinity material. Aptamers are single-stranded oligonucleotides that have been synthesized in a random sequence of bases to produce a large number of chains that are then screened for their binding to target analytes such as proteins. The advantage of these binding molecules is that they are fully compatible with the technology behind the DNA arrays described earlier.

The other alternative, molecular imprints are cavities left in plastic-binding surfaces following polymerization in the presence of the target analyte molecule and subsequent removal of that target species. The cavities that remain then have surface contours and charges that are complementary to the original target molecules. The non-specific binding materials have been used as multiple spots on the surface of a chip. Following incubation with the protein mixture, binding takes place according to the affinity for the surface spot and the resulting profile can be analyzed by a MALDI–TOF–MS.[13]

13.5.3 Applications of Protein Chips

In practice volumes as small as 100 pL can be applied to such chips so that literally thousands of spots can be used to identify thousands of proteins. Automatic systems are again available to

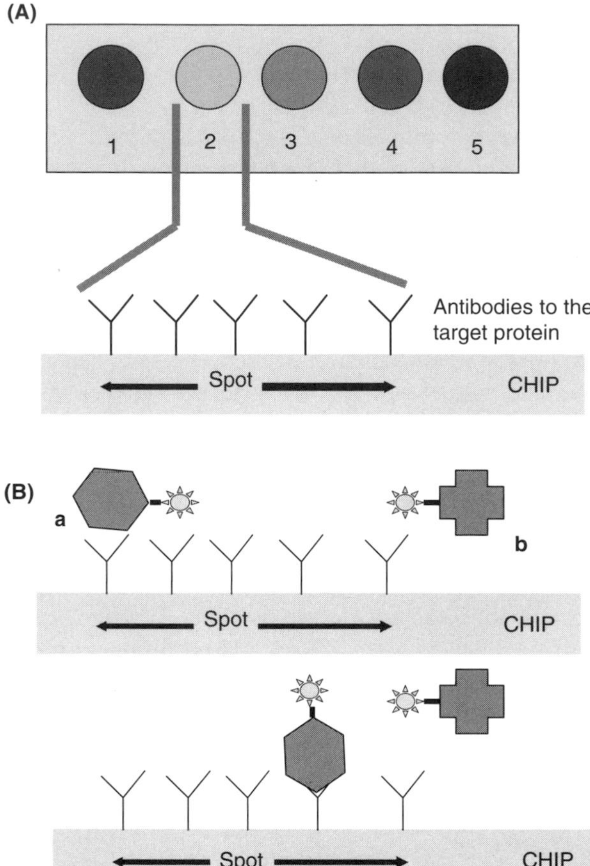

(C) Chip is then exposed to fluorescent light

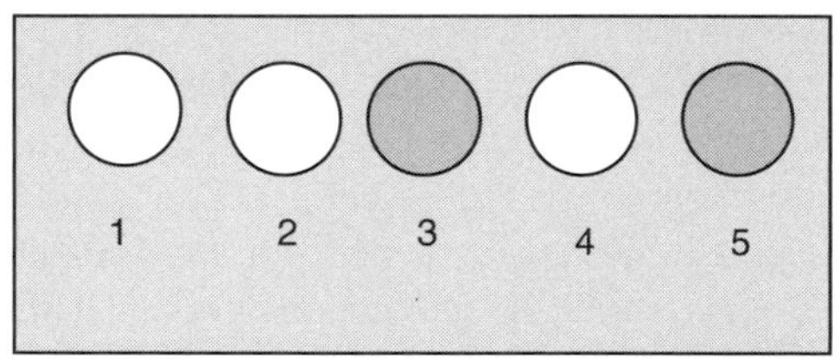

Proteins 3 and 5 are present in the protein mixture

Figure 13.9 Steps in producing microarrays for proteins using immobilized antibodies. Antibody molecules of different specificities are immobilized on the surface of the chip as discrete spots (A). Only complementary antigenic molecules (a) bind to the antibody molecules attached as a spot on the surface of the chip (B). Other molecules (b) do not bind and are washed away. Fluorescent tags are covalently attached to each protein molecule in the mixture applied to the chip and their binding produces fluorescent spots (C).

perform addition of reagents and samples to the chips and commercial chips are available, designed for use with such systems.[14,15]

As with the DNA microarray chip technologies, the antibody array systems give a pattern of fluorescent spots, which can be used to identify the presence or absence of known proteins and their relative abundance. Hence this approach can be used as rapid screen for diagnostic purposes since the presence of specific proteins can be used as biomarkers for disease. Its advantage lies in the ability to screen for up to thousands of proteins in a single assay and that the process can be largely automated.

13.6 MICROFLUIDIC SYSTEMS

Another approach aims to perform the separation of components in a liquid chromatography analogous to gel electrophoresis but performed within microchannels cut/etched into silica chips.[16] This requires only nanoliters of reagents and separations are achieved within seconds/minutes rather than the hours needed for the current gel electrophoresis systems. The separated components can then be analyzed using their spectroscopic properties such as UV, fluorescence, NMR, and MS.[17] Such systems are commonly called, ''labs on a chip.'' One of the first commercial examples of a system that is based on this approach is the 2100 Bioanalyser produced by Agilent Technologies.[18]

13.7 EXAMPLES OF APPLICATIONS OF THE NEW TECHNOLOGIES

13.7.1 Screens for Gene Biomarkers in Disease

As noted above, the discovery of SNPs in the human genome is of great interest to the pharmaceutical industry and the genome maps currently available[1] enable identification of regions within genes of linkage disequilibrium (where some genetic markers such as SNPs occur more or less frequently than expected in the population) that may be important in disease. Methods used to discover these SNPs involve repetitive sequencing of cloned DNA fragments from pools of 10–50 subjects followed by computer-assisted analysis of the sequences to identify sites where significant variations occur. It is then necessary to correlate such sites with diseases. Once this has been achieved the information can be used for diagnostic purposes, using DNA-chips for example, which use immobilized oligonucleotides complementary to sequences around the identified SNPs. Alternatively, the information can be used as a starting point to elucidate the physiological consequences following expression of the genes containing these SNPs and, once this has been established, to identify potential biological sites as targets for new drugs.

13.7.2 *ApoE4* Gene and Alzheimer's Disease

An example of this approach has been demonstrated for the *ApoE4* gene, the presence of which is implicated in earlier onset of Alzheimer's disease. Since not all individuals with this gene develop the disorder, scanning for its presence has little diagnostic value. It was not until a high-density map of SNPs was constructed on chromosome 19 from a region of 4,000,000 bp on either side of *ApoE*, that a highly significant picture emerged. When the allele (the different forms of DNA sequences at identical locations on a chromosome) frequency of each SNP was compared for Alzheimer's patients and controls, three SNP peaks were identified that were associated with the disease.[19]

13.7.3 Pharmacogenetics and Toxicoproteomics

Pharmacogenetics

Genetic makeup is likely to determine an individual's inherent response to a particular drug. This response may affect the efficacy of the drug and its toxicity. Such different responses result from the net effects of their genes on drug absorption, distribution, metabolism, and excretion, in addition to other factors such as age, concurrent drug medication, diet, ethnicity, and so on.

Factors affecting pharmacokinetics

The origin of genetically determined differences in drug pharmacokinetics and pharmacodynamics in individuals can stem from a variety of sources. These include variations in transport proteins, variations in the response to drugs at the receptor level, and in changed rates of drug metabolism due to differences in the enzymatic efficiency of cytochrome P450 enzymes in phase I metabolism or phase II enzymes such as *N*-acetyltransferase or thiopurine *S*-methyltransferase.[20]

It is assumed that differences in the genes will result in structural differences in the proteins for which they code and this may affect the functioning of the proteins as transporting agents, as receptor sites, and as metabolizing enzymes. SNPs constitute the main cause of genetic differences between individuals that lead to the differences in pharmacokinetics and pharmacodynamics described.

In order to apply the technologies described, further identification and cataloging of SNPs within the human genome will again be necessary. This will be undertaken using the DNA chip technologies similar to those described. It will then be necessary to correlate defined SNPs with variations in drug response using extensive clinical studies.

Genetically derived variations in enzyme structure may not only result in changed pharmacokinetics due to altered rates of drug metabolism, but also to altered kinetics and dynamics if prodrugs are involved. Another possible consequence of such variations in metabolism is drug toxicity resulting from a greater sensitivity of an enzyme to either the parent drug or one or more of its metabolites and resulting in adverse drug reactions.

Adverse drug reactions

Adverse drug reactions may be due to type A (dose-related) and type B (idiosyncratic).[21] Genetic factors play an important role in type A reactions and polymorphisms (variations with a frequency of above 1% in the normal population of DNA sequence between individuals in a particular region of the genome) involved in phase I and phase II drug metabolizing enzymes may put individuals at a higher risk of dose-related toxicity than those without such polymorphisms. Type B reactions are subject to even greater influence of genetic influence than type A. Examples include variations in enzyme structure leading to toxicity due to greater sensitivity toward the drug (glucose-6-phosphate dehydrogenase), and poorer detoxification of drugs (dihydropyrimidine dehydrogenase which catabolizes 5-fluorouracil).

Other examples include adverse reactions leading to immunological sensitization. These result from chemically reactive drug metabolites that form conjugates with the patient's proteins *in vivo* thereby initiating an immune response to the resulting immunogen. Finally drug targets such as membrane-bound receptors may exhibit differing sensitivities to the bound drug due to genetic variations. Thus the dopamine D3 receptor has been shown to be subject to genetic polymorphisms leading to substitution of serine by glycine. This may result in higher incidence of tardive dyskinesia in such patients following neuroleptic therapy.

Unfortunately it is not generally possible to currently predict idiosyncratic adverse reactions and they are only manifested following widespread use of the drug concerned. It is hoped that screening will lead to preclinical tools that can be used to predict the toxicity profile with respect to hepatotoxicity, nephrotoxicity, cardiotoxicity, and carcinogenicity for candidate drugs. Such screens, termed toxicoproteomics, are based on exposing target cells or tissues to the drug and monitoring the protein expression profile. Candidate protein biomarkers can then be identified that can be used as early predictors of drug-induced toxicity. Conventional 2-D electrophoresis can be used following the exposure experiments to discover new proteins and to detect specific isoforms but antibody-based and affinity-based chips with rapid sample throughput can also be used.[22] Alternatively microfluidic devices can be used with the cells immobilized within microchannels and exposed to the drug when the expressed proteins can be separated in flow and identified within the device.

Drug resistance

Another related area concerns the induction of drug resistance following therapy. This is a major problem in cancer chemotherapy where resistance to antineoplastic drugs often limits effective treatment. Again, it is now possible using DNA microarrays to monitor how DNA is expressed on exposure to different anticancer drugs and thus to identify genes responsible for chemosensitivity and/or drug resistance in specific tissues. This approach has been explored using cDNA arrays and Affymetrix arrays on the same cell lines, and with tissues from biopsies for cDNA and tissue array systems. In addition proteomic expression has been performed using extracts from exposed cell lines and 2-D PAGE. A number of genes that are differentially expressed on drug exposure or in drug-resistant cell lines have been identified, as have proteins that are overexpressed in chemo-resistant cell lines. These studies also showed up deficiencies in the sensitivity and accuracy of current microarray technologies, in the extraction of data and in its interpretation, and its relevance to complex *in vivo* interactions.[23] However, further studies will hopefully lead to the identification of genetic biomarkers and to a better understanding of mechanisms involved in drug resistance.

13.8 FINAL WORD

The last section illustrates how genomics and proteomics can be used in conjunction with advanced technological methods to further our understanding of disease and of the mechanism of drug action at the cellular and subcellular levels. Although we now have powerful new tools for use in drug discovery and therapy, we should be aware of their limitations since *in vitro* studies on ex *in vivo* samples are likely to provide only limited insights to the complex web of interacting systems that pertain *in vivo*.

REFERENCES

1. www.ensembl.org
2. Graber, J.H., O'Donnell, M.J., Smith, C.L. and Cantor, C.R. (1998) Advances in DNA analysis. *Current Opinion in Biotechnology* **9**, 14–18.
3. Blohm, D. and Guiseppi-Elie, A. (2001) New developments in microarray technology. *Current Opinion in Biotechnology* **12**, 41–47.
4. www.Gene-Chis.com
5. Li, X., Gu, W., Mohan, S. and Baylink, D.J. (2002) DNA microarrays: their use and misuse. *Microcirculation* **9**, 13–22.
6. Weng, Z. and DeLisi, C. (2002) Protein therapeutics: promises and challenges for the 21st century. *Trends in Biotechnology* **20**, 29–35.
7. http://proteomics.cancer.dk
8. Rappsilber, J., Moniatte, M., Nielsen, M.L., Podtelejnikov, A.V. and Mann, M. (2003) Experience and perspectives of MALDI MS and MS/MS in proteomic research. *International Journal of Mass spectrometry* **226**, 223–237.
9. Sloane, A.J., Duff, J.L., Wilson, N.L., Gandhi, P.S., Hill, C.J., Hopwood, F.G., Smith, P.E., Thoma, M.L., Cole, R.A., Packer, N.H., Breen, E.J., Cooley, P.W., Wallace, D.B., Williams, K.L. and Gooley, A.A. (2002) High throughput peptide mass fingerprinting and protein macroarray analysis using chemical printing strategies. *Molecular and Cellular Proteomics* **1**, 490–499.
10. http://shimadzu-biotech.net
11. www.affymetrix.com

12. Cahill, D.J. (2001) Protein and antibody arrays and their medical applications. *Journal of Immunological Methods* **250**, 81–91.
13. www.ciphergen.com
14. http://home.appliedbiosystems.com
15. http://www.genomicsolutions.com
16. Khanddurina, J. and Guttman, A. (2002) Bioanalysis in microfluidic devices. *Journal of Chromatography A* **943**, 159–183.
17. Weigl, B.H., Bardell, R.L., and Cabrera, C.R. (2003) Lab-on-a-chip for drug development. *Advanced Drug Delivery Reviews* **55**, 349–377.
18. www.agilent.com/chem./pharma3
19. Roses, A.D. (2000) Pharmacogenetics and the practice of medicine. *Nature* **405**, 857–865.
20. Meyer, U.A. (2000) Pharmacogenetics and adverse drug reactions. *The Lancet* **356**, 1667–1671.
21. Pirmohamed, M. and Park, B.K. (2001) Genetic susceptibility to adverse drug reactions. *Trends in Pharmacological Sciences* **22**, 298–305.
22. Bandara, L.R. and Kennedy, S. (2002) Toxicoproteomics — a new preclinical tool. *Drug Discovery Today* **7**, 411–418.
23. Huang, Y. and Sadee, W. (2003) Drug sensitivity and resistance genes in cancer chemotherapy: a chemogenomics approach. *Drug Discovery Today* **8**, 356–363.

14

The Chemotherapy of Cancer

David E. Thurston

CONTENTS

14.1 INTRODUCTION

Cancer is a disease in which the control of growth is lost in one or more cells leading to a solid mass of cells known as a tumor. A growing (primary) tumor will often become life-threatening by obstructing vessels or organs. However, death is most often caused by spread of the primary tumor to one or more other sites in the body (metastasis) making surgical intervention impossible. Other types of cancers such as leukemia involve a build-up of large numbers of cells in the blood.

In the first three decades of the last century cancer accounted for less than 10% of all UK deaths, infectious diseases being the main cause of mortality. While dramatic progress has been made in controlling infections, similar progress has not been made with the treatment of cancer. Improved diet, living conditions, and healthcare have now increased the average life span to the point where cancer, which is a disease of advanced years (70% of all new cases of cancer in the UK occur in those over 60 years), has become more prevalent. Consequently, about 300,000 new cases of cancer occur each year in the UK. The annual number of deaths from cancer of all types is approximately 160,000, which constitute approximately 25% of all UK deaths. Statistics show that approximately one in three of the population will suffer from some form of cancer during their lifetimes, and one in four will die from the disease. Furthermore, 1 in 10,000 children will be diagnosed annually as suffering from cancer, which means that there are 1300 new cases each year.

It is thought that exposure to an ever-increasing number of chemicals (carcinogens) in the environment and the diet may be a significant contributing factor. Occupational factors are thought to account for 6% of cancers, while lifestyle and diet may account for up to 30%. Recognition of this is currently leading to major changes in social behavior and legislation (e.g., overall reduction in cigarette consumption, curbing of smoking in public places, healthier diets, more exercise, more restrictive health, and safety rules in working environments). Genetic predisposition is also a factor in some types of the disease.

14.2 TERMINOLOGY

A tumor (or neoplasm) is an abnormal tissue mass or growth which results from neoplasia, a state in which the control mechanisms governing cell growth become deficient, leading to cell proliferation. In the early stages of tumor growth, cancer cells often resemble the original cells from which they derive. Later, they lose the appearance and function of these tissues. Normal adult tissues tend not to grow but to maintain a steady number of cells. In some tissues (e.g., liver) this is achieved without proliferation because there is little cell loss. In the bone marrow, however, a steady number of cells are maintained by a fast rate of cell division balanced by the rate of cell loss. Often it requires only a slow increase in the rate of proliferation to gradually outgrow normally controlled cellular populations.

Cancers are generally named in relation to the type of tissue from which they arise. For example, *sarcoma* describes those tumors occurring in mesodermal tissue which includes connective tissue, bone, and muscle. *Osteosarcoma* refers to bone cancer, and *carcinomas* to tumors of the epithelial tissues such as the mucous membranes and glands (including cancers of the breast, ovary, and lung).

Cancers of the blood or hemopoietic tissue are generally known as *blastomas*. These can involve lymphoid, erythroid, or myeloid cells which generally fall into the *sarcoma* category. *Leukemias* describe those cancers which originate in leucocytes, and may be *myeloid*, *lymphatic*, or *monocytic*. In addition, these particular cancer types may be described as chronic or acute. Bone marrow cell tumors are referred to as *myelomas*, and in *multiple myeloma* (the most common bone marrow cancer) a clone of plasma cells is involved. Neoplasia of erythroid stem cells is known as *primary polycythaemia*.

The reticuloendothelial system is also susceptible to tumorogenesis. *Lymphosarcoma* is cancer of the lymphoid cells, whereas Hodgkin's disease is an example of a lymph adenoma which, although it mainly affects reticulum cells, can extend to eosinophils, fibroblasts, and lymphocytes.

14.3 METASTASES

Metastasis is the ability of solid tumors to spread to new sites in the body thus establishing *secondary* cancerous growths. Tumor cells may easily penetrate the walls of lymphatic vessels, distribute to draining lymph nodes, and hence move to distant sites. They can also invade blood vessels directly, since capillaries have weak thin walls which offer little resistance. A tumor may also spread across body cavities from one organ to another; e.g., stomach to ovary. Most patients who die of cancer do so as a consequence of metastatic spread to vital organs. Furthermore, both primary and secondary tumors not only expand in size but also infiltrate surrounding tissue. When nerve-endings are affected then pain and discomfort is experienced.

At the point of diagnosis of cancer, curative surgical or radiological treatment is usually only possible if metastasis of the primary tumor has not occurred. Therefore, early diagnosis is essential. Since about 50% of malignant tumors have already metastasized prior to diagnosis, the condition is often beyond the reach of curative surgery or radiotherapy alone. It is in these cases that systemic chemotherapy can often help to reduce the risk of further formation and growth of secondary tumors. Recognition of the significant benefits of early diagnosis has led to an emphasis on mass screening programs. For example, cervical and breast cancer screening in woman has become routine during the last decade, and prostrate screening for men is presently being introduced. Rapid advances in the development of highly sensitive analytical techniques to detect tumor markers (e.g., proteins or glycoproteins) in easily obtained body fluids such as blood, urine, and saliva are likely to revolutionize early diagnosis during the next decade.

14.4 CAUSES OF CANCER

Cancer is now known to be a disease that involves dynamic changes to the genome caused by both internal and external factors (e.g., environmental factors; see Section 14.5.2). This has been established by the discovery of mutations that produce *oncogenes* with dominant gain of function and *tumor suppressor genes* with recessive loss of function. Weinberg and others (see Further Reading) have proposed a set of simple rules that may govern the transformation of normal human cells into malignant tumor cells. They have postulated that there are a small number of molecular, biochemical, and cellular traits (so-called *acquired* or *hallmark* capabilities) that are shared by most and perhaps all types of human cancers.

Several lines of evidence now suggest that tumorogenesis in humans is a multistep process with each step reflecting genetic alterations that drive the progressive transformation of normal cells into highly malignant derivatives. These observations have been supported by a large body of work indicating that the genomes of tumor cells are invariably altered at multiple sites, having suffered disruption as subtle as point mutations and as obvious as changes in chromosome complement. For example, transformation of cultured cells is itself a multistep process with rodent cells requiring at least two introduced genetic changes before tumorigenic competence is acquired. Human cells are even more difficult to transform.

Observations of human cancers and animal models suggest that tumor development proceeds via a process analogous to Darwinian evolution, in which a succession of genetic changes, each conferring one or another type of growth advantage, leads to the progressive conversion of normal cells into cancer cells. Weinberg has suggested that the vast catalogue of cancer cell genotypes is a manifestation of just six essential alterations in cell physiology that collectively induce malignant growth. These are:

- Self-sufficiency in growth signals
- Insensitivity to growth-inhibitory signals
- Evasion of programmed cell death (apoptosis)
- Limitless replicative potential
- Sustained angiogenesis
- Tissue invasion and metastasis

Weinberg proposes that these six capabilities are shared in common by most and perhaps all types of human tumor cells. Furthermore, he suggests that this multiplicity of defences may explain why cancer is relatively rare during an average human lifetime.

Our ability to analyze individual human cancers at the genetic and biochemical levels has undergone dramatic changes during the last 10 years (e.g., DNA arrays, genome sequencing, proteomics), and is likely to benefit from further developments in the future at an increasingly rapid pace. At present, description of a recently diagnosed tumor in terms of its underlying genetic lesions remains in its infancy. However, in 10 or 20 years time the diagnosis of all somatically acquired lesions in a tumor cell genome will become a routine procedure along with routine gene expression or proteomics profiling of tumor cells. Rather than the relatively primitive approaches to chemotherapy available now, in the future it should be feasible to produce ''customized'' anticancer drugs targeted to each of these hallmarks of cancer. It may be possible to use these in appropriate combinations and in concert with sophisticated diagnostic technologies to detect and treat all stages of disease progression. This may ultimately allow the prevention of incipient cancers from developing as well as the cure of pre-existing cancers.

14.5 MECHANISMS OF GENOMIC DAMAGE

As discussed above, it is now accepted that cancer is a ''genetic'' disease resulting from changes to DNA sequence information in specific genes. There are a number of ways in which these changes can occur through internal, external, or hereditary factors, and these are described below.

14.5.1 Internal Factors

Tumor formation may result from changes to DNA sequence brought about by malfunction of the normal DNA processing systems within the cell.

Mutation

Genetic mutations can take several forms. In a *point* mutation only one base is altered, and the resulting new codon can cause insertion of an incorrect amino acid into the corresponding position of the protein. Should the protein be critical for proliferation, e.g., a tumor suppressor protein, then tumorogenesis may result. In a *translocation* mutation, an entire segment of DNA may be moved from one part of a gene or chromosome to another. In this case, loss of the proteins corresponding to the two original DNA sequences or the production of a new protein corresponding to the novel fusion sequence may lead to tumorogenesis. The original genes are known as ''proto-oncogenes''; that is, genes which will not cause cancer unless suitably activated (i.e., by translocation to form an ''oncogene''). The concept of proto-oncogenes and oncogenes has been validated in cancers such as Burkitt's lymphoma and chronic myelogenic leukemia (CML) where the precise sequences involved in the translocations have been identified.

Addition or loss of genetic material

During normal DNA handling processes such as repair, DNA bases may be inadvertently added or deleted. This can have a similar effect to a point mutation as it can alter the codon reading frame and lead to the production of faulty proteins critical for control of cell growth.

Changed gene expression

A problem with gene expression may occur, such as uncontrolled expression or amplification. Should, for example, growth factors or proteins responsible for receptor formation be involved, then tumorogenesis may result.

14.5.2 External Factors

Viruses

A link between viruses and cancer was first recognized in 1911 when Peyton Rous demonstrated that avian spindle cell cancer could be transmitted from one bird to another by a cell-free filtrate containing the virus (which now carries his name, the Rous sarcoma virus). Since then, other viruses have been identified as linked to human cancers. Well-known examples include the involvement of the Epstein–Barr virus (EBV) in Burkitt's lymphoma, and the human papilloma virus (HPV) in cervical cancer.

Viruses may be either RNA retroviruses such as human-T-cell leukemia virus (HTLV-1) or DNA viruses. RNA viruses contain DNA polymerases which facilitate the production of double-stranded viral DNA. On being incorporated into the host DNA, the viral DNA may cause tumorogenesis via a number of different mechanisms including production of an oncogene from an existing proto-oncogene, damage to a tumor suppresser gene, or insertion of a completely new gene. For example, HTLV-1 introduces a gene known as *tax*, which results in the overexpression of interleukin-2 (IL-2). This can lead to adult T-cell lymphomas and leukemias with an increase in the number of activated lymphocytes, although these may take years to develop in susceptible individuals. HTLV-1 is endemic in Southeast Asia and the Caribbean; in the Far East it is also associated with nasopharyngeal cancers.

EBV infects 90% of the world's population and is considered to be an ''initiator'' of cancer as opposed to a specific cause. For example, 90% of Burkitt's lymphoma cells test positive for EBV which allows infected lymphocytes to become immortal, leading to a potentially cancerous state. Burkitt's lymphoma is endemic in those parts of Africa with chronic malaria suggesting that the latter may be a cofactor in lymphoma development.

Hepatocellular carcinoma has been linked to the hepatitis B virus (HBV), and is endemic in Southeast Asia and tropical Africa. The risk of tumor formation is greatest in those who are infected from an early age, and males are four times more likely to develop the cancer than females. It is believed that the X-gene in HBV codes for proteins which promote transcription.

Finally, there are over 50 different types of HPV, and HPVs 16 and 18 have been linked to cervical cancer. The virus produces several proteins, some of which enhance mitosis while others interfere with P53 (a tumor suppresser gene) or modify the interaction between cellular proteins and transcription factors.

Bacterial infections

Examples are known of bacterial infections that can lead to tumorogenesis. The best known example is *Helicobacter pylori*, which colonizes the stomach of some individuals and is associated with the formation and development of stomach ulcers. It has now been established that, in some cases, the infection can lead to stomach tumors although the precise mechanism is still unclear.

Chemicals

There is firm evidence that certain chemicals in the environment, and some encountered through the diet, lifestyle, or occupation, can lead to tumorogenesis. For example, the link between cigarette smoke and lung cancer is now well established, and it is also known that carcinogenic polycyclic aromatic hydrocarbons (PAHs) are formed when red meat is overcooked (especially through frying or barbecuing) and can lead to colon cancers. Carcinogenic amines are formed in the stomach as a result of the bacterial degradation of nitrites used as preservatives in meat and fish, and potent carcinogens (the aflatoxins) are found in peanut butter due to their production as secondary metabolites by a fungus infecting the peanuts during growth.

Occupation-associated cancer is not just a feature of recent times as, in 1775, Sir Percival Pott noted the high frequency of scrotal skin cancer in young chimney sweeps. Poor hygiene meant that tarry deposits from coal fires were in contact with the sensitive skin of the scrotum for long periods of time. More recently, vinyl chloride used by workers in the plastics industry has been associated with angiosarcoma of the liver, and furniture industry workers have been prone to nasopharyngeal malignancies induced by the inhalation of small particles carrying organic compounds that arise from leather and wood polishing processes. Many of these organic carcinogens work by covalently modifying DNA (either before or after metabolism).

In addition to these organic carcinogens, certain dusts and minerals are known to cause cancer. For example, the link between asbestos and pleural and peritoneal tumors (particularly mesothelioma) is well established and is thought to be due to physical damage of chromosomes by microscopic fibers of asbestos which are needle-like and can penetrate the cellular and nuclear membranes.

Radiation

Malignancies have been linked to exposure to α or β particles or λ- or x-rays which are known to damage DNA by fragmentation through the formation of free radicals. A link between nuclear fall-out and cancer was firmly established after the Hiroshima bombing in World War II. It has also been postulated that children living close to nuclear power stations are at a higher risk of contracting leukemias and brain tumors, although statistical analyses of these data are often controversial. However, the danger of escape of radioactive materials from nuclear reactors was highlighted by the Chernobyl incident in Russia which caused widespread contamination of the food chain and led to a variety of different cancers in the exposed population. It is also known that a build-up of radon gas produced by certain types of granite can endanger the occupants of houses built from this material. Radon is a naturally occurring radioactive gas that, once inhaled, enters the bloodstream and delivers radiation to all tissues; bone marrow is particularly sensitive and so leukemias predominate. In the UK, local councils have been obliged to offer grants to affected householders so that buildings can be structurally modified to improve ventilation.

More recently there has been speculation in the lay press that electromagnetic radiation from overhead power lines may be associated with childhood leukemias. Similarly, microwaves emanating from mobile phone masts have come under suspicion, and there has been speculation that mobile telephones held to the ear may cause brain tumors. However, these associations remain controversial in scientific circles due to the difficulties of interpreting data produced from a small number of cases. In the case of overhead power lines, one scientific study has suggested that, rather than the electromagnetic radiation itself causing cancer, the high electric fields generated by power lines may concentrate radioactive radon gas into local pockets. The UK Government has been sufficiently concerned about these possibilities to fund more detailed studies.

14.5.3 Hereditary Factors

A number of genes have now been identified that, if inherited, can predispose individuals to certain types of cancer. For example, two genes, *BRCA1* and *BRCA2*, have been identified and sequenced by UK and US researchers. These genes are inherited and closely associated with breast cancer. Other genes associated with colon and bowel tumors are known to be inherited. This has lead to the introduction of diagnostic screening with subsequent genetic counseling for affected individuals. In some cases, women who discover that they are carrying *BRCA1* or *BRCA2* and thus have a high risk of developing cancer choose to have their breasts removed at an early age as a prophylactic measure.

14.6 TREATMENT

Cancer treatment often encompasses more than one approach, and the strategy adopted is largely dependent on the nature of the cancer and how far it has progressed. The main treatments are still surgery, radiotherapy, and chemotherapy. However, other approaches such as photodynamic therapy (PDT), antibody- and vaccine-related approaches and gene therapy are in development.

14.6.1 Surgery

If a tumor is small and reasonably well defined it can sometimes be removed by surgery. However, additional treatment with chemotherapy or radiotherapy is often required to attempt to eliminate any cells that may have remained behind or metastasized. Alternatively, radiotherapy or chemotherapy may be administered prior to surgery, in order to shrink the tumor and facilitate its removal. Where possible, a large area of surrounding healthy tissue is also removed in addition to the tumor in order to ensure complete removal of cancer cells from the site.

14.6.2 Radiation Therapy (radiotherapy)

Radiation therapy utilizes x-rays or radiopharmaceuticals (radionuclides) which act as sources of γ-rays. X-rays are delivered locally in a highly focused beam to avoid damage to healthy tissue. Although radiotherapy is a well-established technique, there is still ongoing research into the most effective treatment regimes in terms of the duration and frequency of exposure. Latest developments include multiple beams under computer guidance that can converge and focus on a tumor with a high degree of accuracy, thus sparing surrounding healthy tissue from damage.

The main radionuclides in use include cobalt-60, gold-198, and iodine-131. Gold-198 concentrates in the liver, and iodine-131 is used to treat thyroid cancers as iodine accumulates in this gland. A significant proportion of tumor cells are hypoxic (i.e., have a low oxygen level) and are thus less sensitive to damage by irradiation. Therefore, prior to and during radiation therapy, oxygen is sometimes given to sensitize the tumor cells to treatment. Radiosensitizing drugs such as metronidazole have also been administered prior to treatment in an attempt to improve therapeutic outcome. A process known as high linear energy transfer (HLET) has been applied to the irradiation of hypoxic cells. In this procedure, the tumor is irradiated with neutrons (heavier than x-rays or γ-rays) which decay to α-particles, the latter causing cell damage in an oxygen-independent manner.

14.6.3 Photodynamic Therapy

PDT is a relatively new form of cancer treatment involving the initial systemic administration of a photosensitizer such as the porphyrin derivative Photofrin (see Section 14.13.4), which is selectively retained by malignant cells. After the agent has localized, the tumor is irradiated with an intense light source of an appropriate wavelength which excites the Photofrin. Upon decay to its ground state, available oxygen is transformed into the singlet form which is highly cytotoxic and damages tumor cells. There is also some evidence that, by damaging endothelial cells, PDT restricts blood flow to the tumor. Laser light sources are now well developed, and the use of flexible optical fibers means that tumors in inaccessible organs and areas of the body including the GI tract, bladder, and ovaries can be easily reached through endoscopy and key–hole surgery techniques. As other less-expensive non-laser light sources become widely available and new types of photosensitizers are developed, the use of PDT is likely to escalate in the future.

14.6.4 Immunotherapy, Antibody Targeting and Vaccines ("Biologicals")

There are a number of so-called *biological agents* or therapies either in use or in development. For example, several tumor types, including some types of breast cancer, have been found to produce specific tumor antigens on their cell surfaces, and this has led to the development of monoclonal antibodies specific for these tumors (e.g., Herceptin, see "Development of inhibitors of receptor and protein kinases"). Tumor-specific antibodies can also be used to selectively deliver a cytotoxic agent (e.g., Mylotarg) or a radionucleotide (see Section 14.13.1). An antibody–enzyme conjugate designed to release an active form of a cytotoxic agent from a non-toxic prodrug form selectively at the tumor site (e.g., ADEPT therapy, see Section 14.13.2) is presently being evaluated in clinical trials. Finally, research is ongoing into the development of vaccines that may either prevent tumor formation or modify the growth of established tumors.

14.6.5 Chemotherapy

Chemotherapy is the use of small low-molecular-weight drugs to selectively destroy a tumor or limit its growth. Nitrogen mustards were the first agents to be used clinically; this resulted from the accidental discovery that the mustard gas used in World War II had antileukemic properties. Since then, significant advances have been made in the development of new anticancer drugs. For example, cisplatin, which was also discovered serendipitously, has provided a major advance in the treatment of testicular and ovarian carcinomas, with a very high cure rate for the former with early diagnosis.

One advantage of chemotherapy is that low-molecular-weight drugs normally pervade all tissues of the body and so can destroy cells in protected areas (e.g., the brain) or in the process of metastases. Disadvantages include the unpleasant side effects of bone marrow suppression, hair loss, and nausea common for many agents, and the rapid development of clinical resistance in most tumor types. However, some of the newer, more selective agents tend to have better side effect profiles. The following sections summarize the major issues relating to chemotherapeutic agents, and sections describe the major families of agents in detail.

Discovery of anticancer drugs and their preclinical evaluation

Most clinically used anticancer drugs were discovered either through chance (serendipity) (e.g., cisplatin and the nitrogen mustards) or through screening programs (e.g., vinblastine and taxol). Only recently has a more detailed knowledge of the fundamental biochemical differences between normal and tumor cells allowed a more rational approach to the drug design process (e.g., the kinase inhibitors Gleevec and Iressa). A combination of the power of screening techniques and rational drug design has been partially realized with the widespread trend towards the use of combinatorial chemistry and compound libraries to provide large numbers of molecules of diverse structures for screening (see Chapter 11).

New agents are nearly always evaluated initially in *in vitro* tumor cell lines. In fact, this only measures the cytotoxicity (i.e., cell killing ability) of an agent and provides no indication of whether it is likely to have useful *in vivo* antitumor activity. However, by studying panels of different tumor cell types, an indication can be obtained of whether the agent has selective cytotoxicity towards a particular tumor type which may in turn suggest suitable *in vivo* experiments. Initial *in vivo* experiments may involve observation of the effect of systemically administered novel agents on tumor cells growing in porous fibers inserted subcutaneously or intraperitoneally in mice or rats (the so-called ''hollow fiber assay'' pioneered by the National Cancer Institute [NCI]). However, human tumor xenograft assays represent the most successful animal model to date in which human tumor fragments are transplanted into immunosuppressed rodents. Effect on tumor growth is then observed after administration of the novel agent. Even so, these models do not reliably relate to equivalent tumors in humans, as numerous differences exist including the integrity of the blood supply to the transplanted tumor and general biochemical species differences including metabolism.

This means that, sometimes, drugs that are active in humans show no effect in animal models. For example, hexamethylmelamine [2,4,6-tris(dimethylamino)-1,3,5-triazine] was the lead compound for the ''melamine'' class of antitumor agents which reached clinical evaluation. This agent works in humans because it is metabolized to a carbinolamine species which has some activity against carcinomas of the bronchus, ovary, and breast. However, the parent compound exhibits only minimal activity when tested in mouse models, as mice fail to carry out the crucial metabolic step.

Despite these problems, most drugs in clinical use today (with the exception of some hormonal agents) were introduced as a result of activity demonstrated in animal models. The NCI in the USA has carried out a worldwide screening cascade of the type described above for many years. Compounds sent to the NCI from academic and industrial sources worldwide are initially screened in a 3- and then 60-cell line panel, and compounds with interesting activity are progressed through hollow fiber and then human tumor xenograft assays to establish efficacy. The NCI has now established a substantial database of information and uses a computer algorithm known as ''COMPARE'' to establish whether the characteristics of a novel agent evaluated in the 60-cell line panel are similar to any existing families of agents or individual compounds thus predicting the likely mechanism of action.

Accessibility of tumor cells to drugs

The accessibility of anticancer drugs to tumor cells varies greatly. While leukemia cells are fully exposed to drugs in the blood stream, solid tumors have a less reliable blood supply. Small tumors are usually reasonably well supplied and more susceptible to drug action than large tumors which often have poor capillary access, particularly in their centers which can be hypoxic (oxygen deficient) and even necrotic. The degree of accessibility of a chemotherapeutic agent is therefore one of a number of reasons for the greater sensitivity of small primary and early metastatic tumors to chemotherapy and highlights the importance of early diagnosis and treatment. It is noteworthy that brain tumors are particularly resistant to chemotherapy as few drugs are capable of crossing the blood–brain barrier.

Achieving selective toxicity

The development of more-effective chemotherapeutic agents is critically dependent upon the discovery of exploitable biochemical differences between normal and tumor cells. Such differences should allow a more rational approach to drug design rather than relying on the empirical manner in which many of the present day drugs have been discovered. Drugs resulting from such a fundamental change in the discovery process are presently limited, but the kinase inhibitors Gleevec and Iressa are good examples of the trend in this direction. An older example is the discovery that some lymphoid malignancies are dependent on an exogenous supply of asparagine whereas healthy cells can synthesize their own; this led to the clinically useful agent asparaginase (see Section 14.14.3).

There is hope that complete selectivity may one day be achieved through gene targeting in which genetic differences between tumor cells and healthy cells are exploited (see ''Gene targeting''). There is speculation that some of the presently used DNA-binding agents such as the mustards that bond covalently to GC sequences of DNA may already exploit the fact that some oncogenes are particularly GC-rich. However, they may simply be more damaging to faster growing cells. Although it is true that certain types of cancer cells grow faster than cells in healthy tissues, in the majority of tumors the rate of cell division is still slower than that of normal bone marrow, skin epithelium, or the mucosa of the GI tract. This explains the consistent pattern of side effects accompanying chemotherapy which are dose-limiting in practice.

Limiting the toxicity of chemotherapeutic agents

The cell cycle is a target for many chemotherapeutic agents. Competition between cells within the tumor favors those that continuously progress through the cell cycle. As a result, tumor cells generally proliferate and differentiate at a faster rate than immediately surrounding normal cells. However, any healthy rapidly dividing cells will also be affected by chemotherapy and this gives rise to side effects such as myelosuppression, alopecia, dermatitis, and GI, liver or kidney toxicities. In addition, chemotherapeutic agents which penetrate the blood–brain barrier are not well-tolerated by the central nervous system (CNS) and usually induce side effects such as nausea and vomiting. In order to gain selectivity, new agents must target features that are unique to tumor cells.

It has proved possible to limit bone marrow toxicity by exploiting cell kinetic differences between normal and tumor stem cells. Stem cells constitute the smallest, yet most important, compartment in a proliferating system. They are capable of an indefinite number of divisions and are responsible for maintaining the integrity and survival of a cell population. Thus, the increase in size of a lymphoma cell population compared with a normal stem cell population and the toxicity of chemotherapeutic agents to the bone marrow can be explained by the fact that only 20% of the bone marrow stem cells are usually in active cycle at any one time, the remaining 80% being in the resting phase (G_0). This small dividing marrow stem cell population may be significantly reduced in size by chemotherapy, leading to the observed toxicity. However, within 3–4 days the remaining stem cells can move from G_0 into active cycle. This means that very high doses of drugs may be given for 24–36-h periods provided these are co-ordinated with interdispersed recovery periods.

Studies of the cell kinetic patterns of tumor growth have suggested a classification for cytotoxic agents based on their ability to reduce the stem cell population of normal bone marrow and lymphoma cells in mice. Two classes of antitumor agents have been distinguished:

Class 1: Cell-cycle-specific agents, which kill cells in only one phase of the cycle; e.g., S-phase (the period of DNA synthesis) (see Figure 14.1): 6-mercaptopurine, cytosine arabinoside, and methotrexate; or M phase (mitosis): vinblastine and vincristine. An increased dose of these drugs will not kill more bone marrow stem cells than are killed by the initial dose.

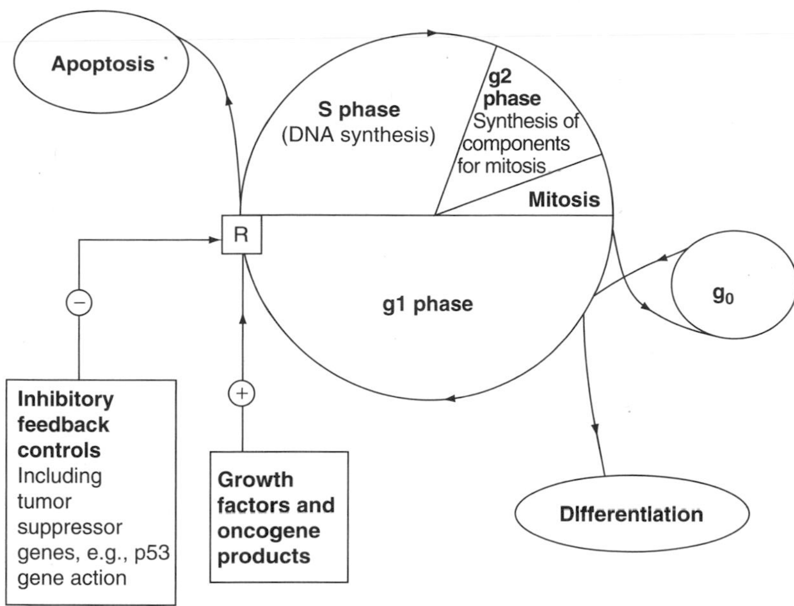

Figure 14.1 The cell cycle (R = restriction/check point).

Class 2: Non-cell-cycle-specific agents that kill cells at all phases of the cell cycle; e.g., cyclophosphamide, melphalan, chlorambucil, cisplatin, BCNU, CCNU, 5-fluorouracil (5-FU), actinomycin D, and daunorubicin. An increased dose of these drugs will increase the number of bone marrow stem cells killed. Although cell killing can occur in all phases, some of these agents are still more active in a given phase of the cycle.

With few exceptions, such as the cumulative toxicities associated with adriamycin (cardiac), bleomycin (pulmonary), and cisplatin (renal), the common toxicities of anticancer drugs are usually reversible within 2–3 weeks. Mucositis (e.g., associated with actinomycin D, adriamycin, bleomycin, methotrexate, 5-FU, and daunorubicin) is reversible over a period of 5–10 days, reflecting the rapid recovery of normal tissues. Nausea and vomiting, which accompany the administration of many drugs, may be partially overcome with modern antiemetic agents such as the 5HT3 antagonists (see Section 14.17.1).

Overview of the different mechanisms of action of chemotherapeutic agents

Different mechanisms of action include interference with metabolism (e.g., antimetabolites such as methotrexate) and interruption of cell division (e.g., antitubulin agents such as vinblastine and taxol). Other classes of agents work as antihormonal agents (e.g., the action of tamoxifen on the estrogen receptor [ER], or the inhibition of steroid biosynthesis via the aromatase inhibitors). A large number of clinically useful antitumor target agents DNA. Some block the synthesis of DNA (e.g., 6-mercaptopurine), while others act by becoming incorporated into DNA and then interfering with its function (e.g., 6-thioguanine). However, many interact directly with DNA by mechanisms including intrastrand (e.g., cisplatin) or interstrand (e.g., nitrogen mustards) cross-linking, intercalation (e.g., mitoxantrone), or interaction in the minor (e.g., ET-387 and SJG-136) or major (e.g., temozolomide) grooves. Others drugs interact with DNA (by one of the above mechanisms) and then cause strand scission (strand breakage or cleavage) at the binding site. The recently developed signal transductase inhibitors work by selectively blocking the enzymes involved in, for example, growth factor signaling (e.g., the kinase inhibitors such as Gleevec and

Iressa), and the Vascular targeting agents (VTAs) work by either blocking new capillary growth (e.g., the antiangiogenic agents), or reducing the capacity of existing tumor vasculature. There is also a broad family of agents known as the ''Biologicals'' which includes antibodies targeted to markers on the surface of tumor cells (e.g., Herceptin) or involved with prodrug strategies (e.g., ADEPT or GDEPT), immune response modifiers (e.g., the interferons), and vaccines. Many new mechanisms of selective cell kill are being explored, and these are described in Section 14.16.2.

Drug resistance

The development of drug resistance may seriously interfere with treatment. Initial selective cytotoxicity towards a tumor can sometimes be followed by rapid recovery in rate of tumor growth, and the degree of resistance may increase after each subsequent administration to the point where the chemotherapeutic agent becomes completely ineffective. Drug resistance has been observed in most drug-sensitive tumor types and with most types of drugs. In some resistant tumor cells, an upregulation of particular proteins can be observed to occur. For example, glutathione transferase production may increase which is usually responsible for resistance to alkylating agents through catalysis of their covalent interaction with glutathione rather than DNA. Another mechanism of resistance involves the active transport of drugs out of cells. The discovery of the multidrug resistance (MDR) gene has led to an understanding of how tumors can become resistant to a wide variety of agents. Future therapies may include strategies to downregulate the MDR gene as a means to enhance the effectiveness of existing chemotherapeutic agents. This problem of resistance has been overcome in several tumor types by using a combination of different cytotoxic agents as described below.

Combination chemotherapy

Attempts to treat tumors with single agents are often disappointing. A single drug usually kills the most sensitive population of cells in a tumor leaving a resistant fraction unharmed and still dividing. This led, in 1960, to the first use of a combination of drugs for treating testicular tumors. This ''cocktail'' principle was then rapidly extended to other tumor types. As a rule, each drug included in a particular combination should be active as a single agent and have different toxic (dose-limiting) side effects compared to the others. Multiple drug therapy also enables the simultaneous attack of different biological targets thus enhancing the effectiveness of treatment. The successful application of this approach to the treatment of acute lymphoblastic leukemia is illustrated below with the comparative response rates for both single agents and various combinations:

Drugs	Complete remission (%)
Methotrexate (M)	22
Mercaptopurine (MP)	27
Prednisone (P)	63
Vincristine (V)	57
Daunorubicin (D)	38
P, V	94
P, V, M, MP	94
P, V, D	100

A variety of combination schedules using different drugs and dosage regimes are now available and, in many cases, are accepted as superior to single drug therapy. Many cancers are already disseminated at the time of clinical diagnosis, and so combination chemotherapy may be commenced concurrently with local treatment (e.g., radiotherapy, surgery) to maximize benefit for the patient. Furthermore, the micrometastases associated with the primary tumor are often very sensitive to combination chemotherapy since they have a good blood supply that facilitates drug access. They are also less likely to develop drug resistance than an older tumor.

Adjuvant chemotherapy

In treating cancer, it is sometimes necessary to co-administer other agents that can either enhance the activity of the anticancer drug or counteract any side effects produced. An example of the former is the recent clinical evaluation of the co-administration of the 6-alkylguanine transferase inhibitor Patrin with temozolomide (Temodal) in an attempt to enhance efficacy by slowing removal of the DNA adducts formed. In terms of adjuvants to reduce side effects, antiemetics will often be administered to counteract the nausea commonly associated with many chemotherapeutic agents. Myelosuppression is more problematic in that it can lead to an increased risk of infection, and so antibiotic or antifungal therapy may be required. Finally, steroids such as prednisolone may be co-administered with some anticancer agents to reduce the severity of side effects.

14.7 ANTIMETABOLITES

Antimetabolites function by blocking crucial metabolic pathways essential for cell growth. Selectivity is based on the fact that tumor cells are often faster growing compared to surrounding normal cell populations with the exception of the bone marrow or parts of the GI tract which can lead to the well-known side effects. This growth differential is significant in some leukemias explaining why antimetabolites can lead to remissions or even cures. However, older solid tumors usually have only a small fraction of cells in active growth and so only partially respond. All agents of this type in clinical use, including the antifolates and the purine and pyrimidine antimetabolites, interfere with DNA synthesis.

14.7.1 DHFR Inhibitors (Antifolates)

Methotrexate (Maxtrex)

Tetrahydrofolic acid is produced by the action of the enzyme dihydrofolate reductase (DHFR) on dihydrofolic acid, and is required for the synthesis of thymine which becomes incorporated into DNA (Equation (14.1)). Slight modification of the structure of folic acid produced the lead antimetabolite aminopterin. Methotrexate, which was shown to be more selective, followed in the 1950s. It binds more strongly to the active site of DHFR than the natural substrate by a factor of 10^4 due to the presence of an amino rather than a hydroxyl moiety, which increases the basic strength of the pyrimidine ring.

The most basic center in the methotrexate molecule is at the N1 and adjacent C2-NH_2 position as confirmed by ^{13}C nuclear magnetic resonance (NMR) measurements at C2. Examination of the drug–enzyme complex by x-ray diffraction has shown that the pyrimidine ring is situated in a lipophilic cavity with the cation of N1/C2-NH_2 binding to an aspartate-26 anion of the enzyme. Other binding points revealed by x-ray include hydrogen bonding between C4-NH_2 and the carbonyl groups of both Leu-4 and Ala-97, and ionic interactions between the α-COOH of the glutamate residue and the basic side chain of Arg-57. The *p*-aminobenzoyl residue lies in a pocket formed on one side by the lipophilic side chains of Leu-27 and Phe-30, and, on the other side, by Phe-49, Pro-50, and Leu-54. A neighboring pocket lined by Leu-4, Ala-6, Leu-27, Phe-30, and Ala-97 accommodates the pteridine ring. The nicotinamide (NADPH) portion of the fully extended coenzyme lies sufficiently close to the pteridine ring to facilitate transfer of a hydride anion from the pyridine nucleus to the C6-position.

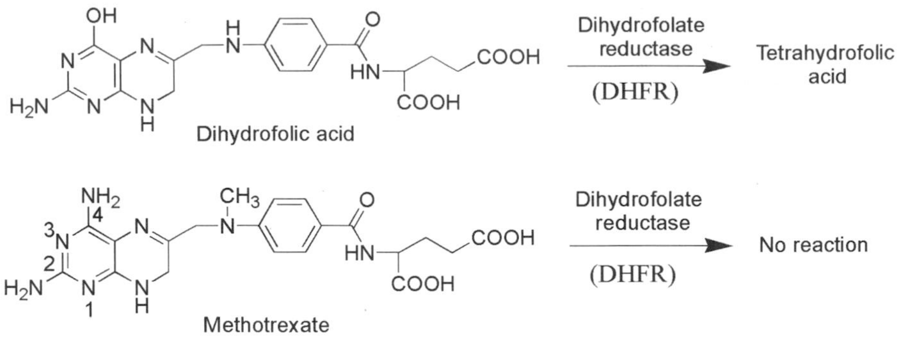

(14.1)

Surprisingly, methotrexate occupies the reverse position at the active site of the enzyme compared to the substrate. Although the *p*-aminobenzoyl and glutamate portions of both are identically bound, with dihydrofolate N-1 is unbound, C2-NH_2 and C4-OH bind only to water molecules, N3 is hydrogen-bonded to Asp-26, N5 is unbound and N8 interacts with Leu-4 via van der Waals forces. This results in the substrate being more loosely bound, a consequence of the differences in position and strength of the most basic centers in the substrate and inhibitor molecules.

Methotrexate is used as maintenance therapy for childhood acute lymphoblastic leukemia, where it can be given intrathecally for CNS prophylaxis. It is also useful in choriocarcinoma, non-Hodgkin's lymphoma, and a number of solid tumors. Side effects include myelosuppression, mucositis, and gastrointestinal ulceration with potential damage to the kidneys and liver which may require careful monitoring. Resistance is also a problem with the tumor cells eventually increasing production of DHFR.

In high-dose intermittent schedules the adverse effect on bone marrow can be relieved by the periodic administration of the calcium salt of N5-formyltetrahydrofolic acid (folinic acid or Leucovorin) which enables blockade of tetrahydrofolic acid production to be bypassed (i.e., folinic acid ''rescue'' therapy).

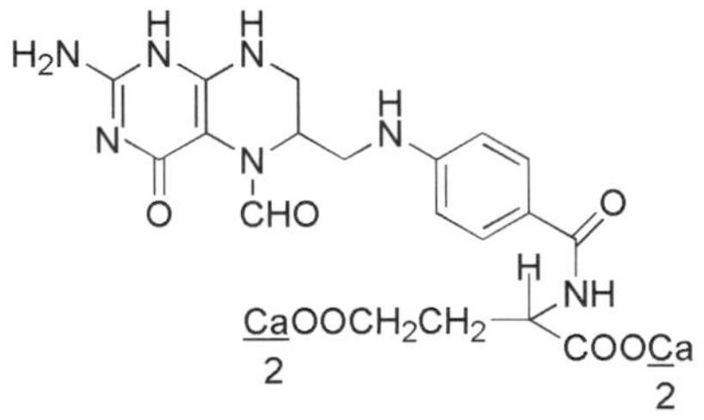

Leucovorin

(14.1)

Many derivatives of methotrexate have been synthesized with a view to reducing its toxicity, however it is still the major DHFR inhibitor in clinical use. Analogues containing a fluorine atom have also been synthesized so that their interaction with the DHFR enzyme can be studied by F-NMR both *in vitro* and *in vivo*.

14.7.2 Purine Antimetabolites

Purine analogues inhibit a later stage in DNA synthesis compared to DHFR inhibitors. Their major problem is a lack of selective toxicity since purines are involved in many other cellular processes in addition to nucleic acid synthesis. Mercaptopurine (Puri-Nethol; 6-MP) is used almost exclusively as maintenance therapy for acute leukemias. The free base form is converted by sensitive tumor cells to the ribonucleotide 6-mercatopurin-9-yl (MPRP) which results from interaction of the base with 5-phosphoribosyl transferase. Resistance to 6-MP usually arises due to loss of production of this enzyme within tumor cells.

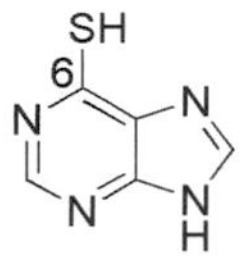

6-Mercaptopurine
(Puri-Nethol, 6-MP)
(14.2)

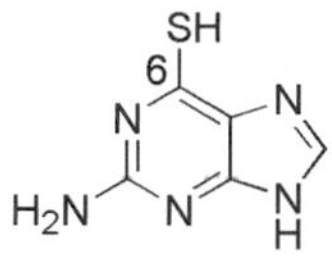

6-Tioguanine
(Lanvis)
(14.3)

Although MPRP inhibits several enzymatic pathways in the biosynthesis of purine nucleotides including the conversion of inosine-5′-phosphate to adenosine-5′-phosphate, the main inhibitory action appears to occur at an earlier stage when 5′-phosophoribosylpyrophosphate is converted into phosphoribosylamine by phosphoribosylpyrophosphate amido-transferase. Allopurinol is contraindicated during treatment as it interferes with the metabolism of 6-MP by inhibiting xanthine oxidase-mediated degradation of 6-MP to thiouric acid which can lead to renal damage. Interestingly, the immunosuppressant agent azathioprine (Imuran) is metabolized to mercaptopurine.

Another cytotoxic drug used for treating myeloblastic leukemia, tioguanine (previously thioguanine or 6-thioguanine), is metabolized to the 9-(1′-ribosyl-5′-phosphate) by tumor cells. However, in contrast to MPRP, this does not inhibit an enzyme but is further phosphorylated to the triphosphate and then incorporated into DNA as a ''false'' nucleic acid. The lack of selectivity towards tumor cells is due to rapid incorporation of 6-thioguanine into the DNA of bone marrow cells. It is used orally to induce remission in acute myeloid leukemia (AML).

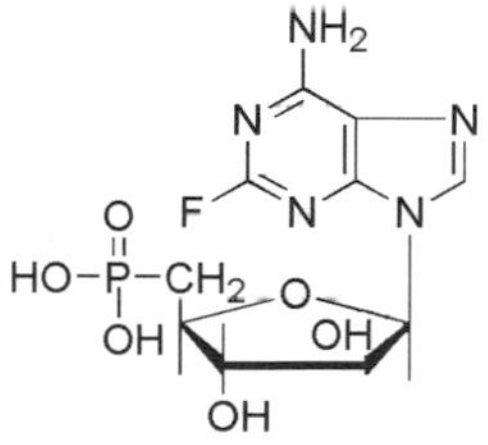

Fludarabine phosphate
(Fludara)
(14.4)

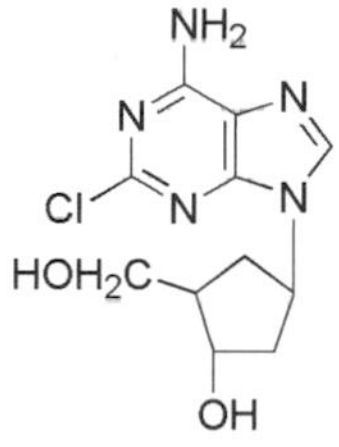

Cladribine (Leustat)
(14.5)

Fludarabine (Fludara) and Cladribine (Leustat) are more recent agents in this family which retain the purine nucleus but have sugar-like moieties already attached. Fludarabine is recommended for patients with B-cell chronic lymphocytic leukemia (CLL) after initial treatment with an alkylating agent has failed. In addition to myelosuppression, fludarabine can also cause immunosupression. Cladribine is an effective but potentially toxic drug given by intravenous infusion for the first-line treatment of hairy cell leukemia and the second-line treatment of CLL patients who have failed on standard regimes of alkylating agents. Its usefulness is limited by both myelosuppression and neurotoxicity.

14.7.3 Pyrimidine Antimetabolites

Cytarabine (Cytosar; ARA-C) and 5-FU are the two prototypical pyrimidine antimetabolites that work by interfering with pyrimidine synthesis. Cytarabine is still one of the most effective single agents available for treating acute myeloblastic leukemia, the major side effect being myelosuppression. A disadvantage of cytarabine therapy arises from its rapid hepatic deamination by cytosine deaminase to give an inactive uracil derivative. These agents work by inhibiting DNA synthesis (S-phase) and by blocking the progression of cells through the G1/S-phase boundary.

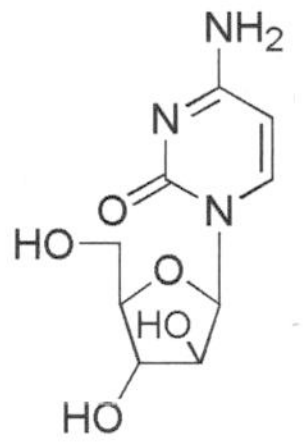

Cytarabine (Cytosar)
(14.6)

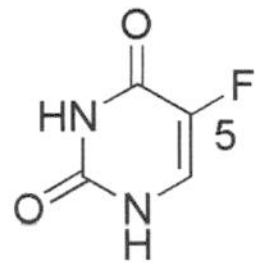

5-Fluorouracil
(14.7)

This short half-life can be counteracted by continuous infusion methods of administration, although the agent can also be administered subcutaneously and intrathecally. This rapid deamination has led to a quest for deaminase-resistant agents or pyrimidine nucleoside deaminase inhibitors which might be co-administered.

Of the many halogenated analogues investigated, only fluoro derivatives have any appreciable antitumor activity. 5-FU is used for the treatment of breast tumors and cancers of the GI tract including advanced colorectal. It is also highly effective as a 5% cream (Efudix) in treating skin cancer. The main side effects include myelosuppression and mucositis. 5-FU is initially metabolized to the 2′-deoxyribonucleotide, 5-fluoro-2′-deoxyuridylic acid (FUdRP), which is a potent inhibitor of thymidylate synthetase. The latter causes the transfer of a methyl group from the coenzyme methylenetetrahydrofolic acid to deoxyuridylic acid which is converted to thymidylic acid and incorporated into DNA. 5-FU has been shown to have an affinity for thymidylate synthetase several thousand times greater than that of the natural substrate. This remarkable property is associated with the unique characteristic of the fluorine atom whose van der Waals radius compares favorably with that of hydrogen although the bond strength is considerably greater. Additionally, the high electronegativity of fluorine affects the electron distribution, conferring a lower pK_a on the molecule compared to uracil. These two features combine to enable FUdRP to fit the active site of the enzyme extremely well although the fluorine cannot be removed thus effectively inhibiting the enzyme. Further studies have suggested that a nucleophilic sulfydryl group at the active site forms a covalent bond to FUdRP leading to a ''dead end'' adduct of the enzyme, coenzyme and 5-FudRP. Structure–activity studies have shown that the increased size but lower electronegativity of other types of halogen atoms reduces activity. It has been postulated that the high selectivity of 5-FU, especially in skin treatments, may reflect the fact that certain types of cancer cells lack the relevant enzymes to degrade it.

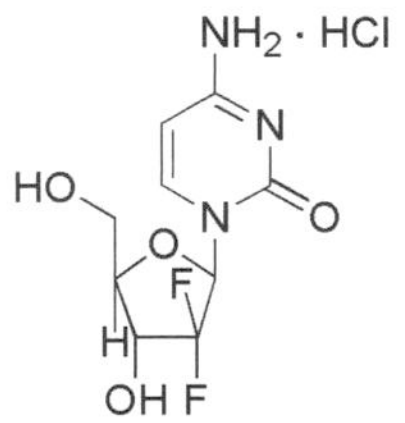

Gemcitabine
(Gemzar)
(14.8)

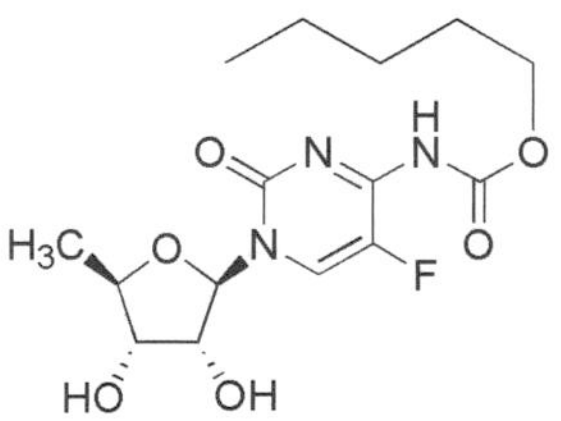

Capecitabine
(Xeloda)
(14.9)

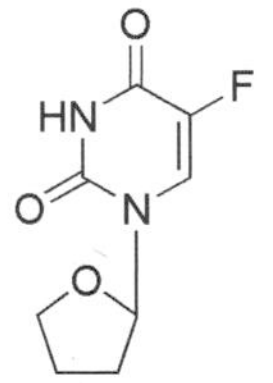

Tegafur (Uftoral)
(14.10)

Other newer antimetabolites in this family include Gemcitabine (Gemzar), which is used intravenously in combination with cisplatin for the treatment of metastatic non-small cell lung, pancreatic, and bladder cancer. It is generally well tolerated but can cause GI disturbances, renal impairment, pulmonary toxicity, and influenza-like symptoms. It is metabolized intracellularly to active diphosphate and triphosphate nucleosides which are responsible for inhibiting DNA synthesis and inducing apoptosis. Capecitabine (Xeloda) is metabolized to 5-FU and is useful as oral monotherapy for metastatic colorectal cancer. Similarly, tegafur is a prodrug of fluorouracil given orally in combination with uracil (Uftoral), which inhibits degradation of the fluorouracil. It is used in combination with calcium folinate for the treatment of metastatic colorectal cancer.

14.7.4 Thymidylate Synthase Inhibitor

Although working through the same mechanism as fluorouracil by inhibiting thymidylate synthase, ralitrexed (Tomudex) is a recently introduced potent inhibitor that represents a new structural class.

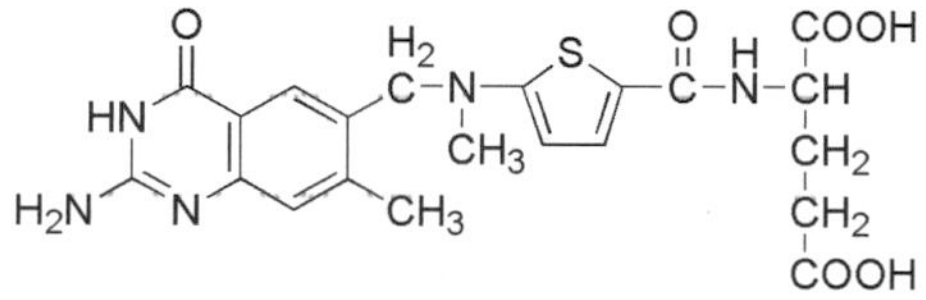

Raltitrexed (Tomudex)
(14.11)

It is given intravenously for palliation of advanced colorectal cancer in cases where 5-FU and folinic acid cannot be used. It is generally well tolerated but can cause myelosuppression and GI toxicity.

14.7.5 Adenosine Deaminase Inhibitor

There is one example of an adenosine deaminase inhibitor in clinical use. Pentostatin (Nipent) was initially isolated from *Streptomyces antibioticus* in the mid-1970s, and a full synthesis was reported in 1982.

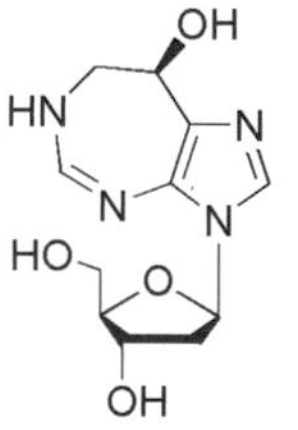

Pentostatin (Nepent)
(14.12)

Administered intravenously on alternative weeks, pentostatin is highly active in hairy cell leukemia and is capable of inducing prolonged remission. It is potentially toxic, causing myelosuppression, immunosupression, and a number of other side effects which may be severe.

14.8 DNA-INTERACTIVE COMPOUNDS

A large number of anticancer drugs exert their effect by interacting with DNA, and are among the most widely used in cancer chemotherapy. In broad terms they act by damaging DNA, thus interfering with cell replication. However, although the more simple agents in this class (e.g., the nitrogen mustards) may cause relatively non-specific DNA damage (i.e. interstranel cross-linking) at numerous sites on DNA, some of the newer experimental agents (e.g., SJG-136) are much more specific, recognizing and binding to specific DNA sequences.

Some agents in this class intercalate between the base pairs of DNA, whereas others alkylate in either the minor or major grooves. Other drugs cross-link the DNA strands together in either an "intra" or "interstrand" manner, and some exert their effect by cleaving the DNA strands. As with other families of anticancer drugs, the selective toxicity towards cancer cells may arise solely from the difference in growth rate of populations of cancer cells compared to normal cells, which also explains their toxicity towards bone marrow and cells of the GI tract. Alternatively, selectivity may arise through a reduced capability of cancer cells to repair the DNA lesions compared to normal cells thus triggering apoptosis instead. However, there is increasing evidence that some DNA-interactive agents selectively target specific DNA regions (for example, GC-rich sequences) which may be prevalent in some tumor cells.

As a class there are often short-term side effects with DNA-interactive drugs such as alopecia (hair loss), GI toxicity (often mucositis and diarrhoea) and, more seriously, reversible bone marrow suppression. However, there are also two problems associated with prolonged usage. First, gametogenesis is often severely affected and sperm storage is now recommended in young male patients prior to treatment. Second, prolonged use, particularly in combination with radiotherapy, can increase the incidence of acute nonlymphocytic leukemia occurring later in life presumably due to the DNA damage caused by these treatments. The agents described below are categorized according to their mechanism of action.

14.8.1 Alkylating Agents

Dacarbazine and temozolomide are known as methylating agents because they work by methylating guanine bases within the major groove of DNA predominantly at O6 positions. However, the experimental agent Ecteinascidin-743 (ET-743) alkylates the N2 of guanine in the minor groove thus forming a bulky adduct which blocks DNA processing.

Methylating agents

Dacarbazine (DTIC-Dome)

Dacarbazine (DTIC-Dome) was one of several triazenes originally evaluated as potential inhibitors of purine biosynthesis. Although it was found to have a wide spectrum of activity ranging from malignant lymphomas to melanomas and sarcomas, it was later established that its mechanism of action is not associated with inhibition of purine biosynthesis. Instead it was demonstrated that N-demethylation occurs *in vivo* to afford 5-aminoimidazole-4-carboxamide and a transient methanediazonium ion (Equation (14.2)). Through radiolabeling experiments it has been shown that the latter methylates DNA at guanine N7 positions.

$$\text{Dacarbazine (DTIC-Dome)} \longrightarrow \text{5-Aminoimidazole-4-carboxamide} + CH_3N_2^+ \text{ (Methanediazonium ion, DNA-methylating species)} \longrightarrow CH_3^+ \quad (14.2)$$

It is given intravenously since the irritant properties of dacarbazine preclude contact with skin and mucous membranes. Also, since triazenes are liable to photochemical decomposition, the intravenous infusion bag containing dacarbazine must be protected from light. It is used as a single agent to treat metastatic melanoma and in combination with other agents for soft-tissue sarcomas (STSs). It has also been used as a component of a combination therapy for Hodgkin's disease known as ABVD (doxorubicin [i.e., Adriamycin], bleomycin, vinblastine, and dacarbazine). The predominant side effects are myelosuppression and intense nausea and vomiting.

Temozolomide (Temodal)

Temozolomide (Temodal) is a newer alkylating agent licensed for the second-line treatment of malignant glioma, and also used experimentally for melanoma. It is structurally related to dacarbazine but is similar in its spectrum of activity and cross-resistance profile to the nitrosoureas. As with dacarbazine and the nitrosoureas, its major dose-limiting toxicity is bone marrow suppression.

Like dacarbazine, temozolomide is a prodrug and serves to transport a methylating agent to guanine sequences within the major groove of DNA. The mechanism of activation involves hydrolytic cleavage of the tetrazinone ring at physiological pH to give the unstable monomethyl triazene (MTIC), which then undergoes cleavage to liberate 5-aminoimidazole-4-carboxamide and the highly reactive methanediazonium methylating species (Equation (14.3)). After methylating DNA (or otherwise decomposing) the latter forms N_2, and it is the generation of the small stable molecules 5-aminoimidazole-4-carboxamide, CO_2, and N_2 that provides the driving force for the mechanism of action of temozolomide.

$$\text{Temozolomide (Temodal)} \xrightarrow{H_2O} \text{Monomethyl triazine (MTIC)} + CO_2 \xrightarrow{H^+}$$

$$\text{5-Aminoimidazole-4-carboxamide} + \text{Methanediazonium ion (DNA-methylating species)} \longrightarrow CH_3^+ + N_2 \qquad (14.3)$$

Temozolomide (Temodal) — Monomethyl triazine (MTIC) — 5-Aminoimidazole -4-carboxamide — Methanediazonium ion (DNA − methylating species)

After treatment of patients with labeled temozolomide (^{11}C label at the methyl group), the relative abundance of covalently modified DNA in brain tumors relative to normal tissue can be observed by positron emission tomography (PET) imaging. It is thought that this selectivity may be attributable to the slightly different pH environments of normal vs malignant tissues in the brain, coupled with differential capacities to repair the methylation lesions by O6-alkyl-DNA alkyltransferase (ATase) or other repair processes. It is also known that methylation is favored in guanine-rich regions of DNA (for electrostatic reasons), and so an over-representation of guanine-rich regions in the tumor genome may also play a role.

Procarbazine

Procarbazine, *N*-(l-methylethyl)-4-[(2-methylhydrazino)methyl]benzamide, is a hydrazine derivative first synthesized as a mono-amine oxidase inhibitor. It was only later discovered to have significant activity in lymphomas and carcinoma of the bronchus.

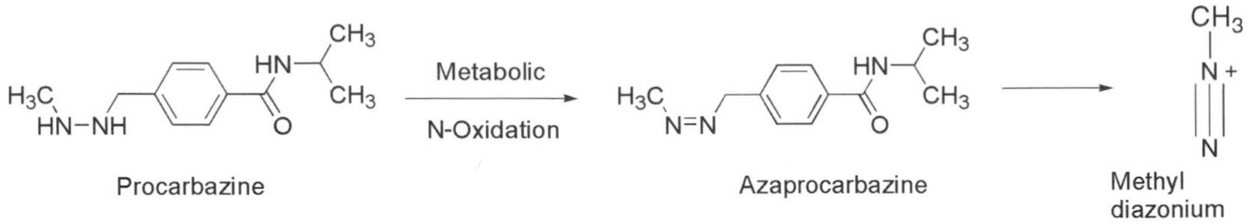

(14.4)

The mechanism of action is thought to involve metabolic N-oxidation to an azoprocarbazine species, followed by subsequent rearrangement to produce either methanediazonium or methyl radicals which act as DNA methylating agents towards guanine residues (Equation (14.4)).

Procarbazine is most often used in Hodgkin's disease, for example in the ''MOPP'' combination therapy which includes Mustine (chlormethine), Oncovin™ (vincristine), Procarbazine, and Prednisolone. It is administered orally and toxic effects include nausea, myelosuppression, and a hypersensitivity rash that prevents further use of the drug. The mild monoamine-oxidase inhibition effect of Procarbazine does not require any dietary restriction, however alcohol ingestion may cause a disulfiram-like reaction.

Ecteinascidin-743

ET-743 is a novel DNA-binding agent derived from the marine tunicate *Ecteinascidia turbinata* that is presently under evaluation in the clinic. Rather than methylating DNA in the major groove as is the case with dacarbazine and temozolomide, ET-743 alkylates the N2 of guanine in the minor groove thus forming a bulky adduct which blocks DNA processing. It has significant *in vitro* activity against melanoma, breast, ovarian, colon, renal, non-small cell lung, and prostate carcinoma cell lines and is under evaluation in the clinic against STS (sarcomas) which are some of the most difficult forms of cancer to treat. The majority of chemotherapeutic agents have only marginal activity against STS, with the most active agents (doxorubicin and ifosfamide) having objective response rates of ~20% as first-line treatment with their use limited by serious toxicities and resistance.

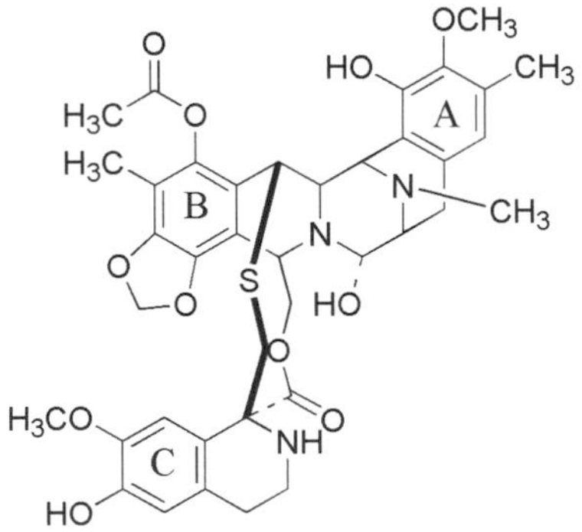

Ecteinascidin-743 (ET-743)

(14.13)

The drug has a unique mechanism of action which involves covalent binding to the N2 position of guanine in the minor groove of DNA causing the double helix to bend towards the major groove. This is a unique feature distinguishing ET-743 from all currently available DNA-binding agents, which cause structural perturbation of the DNA molecule by bending it towards their site of interaction with DNA rather than away from it. The molecule consists of three fused tetrahydroisoquinoline rings, two of which (subunits A and B) provide the framework for covalent interaction within the minor groove of the DNA double helix. The third ring (subunit C) protrudes from the

DNA duplex and interacts with adjacent nuclear proteins, which is thought to account for the cytotoxicity.

The cytotoxicity of ET-743 appears to be associated with DNA repair capability. *In vitro* studies have demonstrated inhibition of transcription-dependent nucleotide excision repair pathways and inhibition of cell cycle progression leading to p53-independent apoptosis. The transcription-coupled nucleotide excision repair process (TC-NER) involves recognition of DNA damage and recruitment of various nucleases at the site of DNA damage. At micromolar concentrations, ET-743 has been shown to trap these nucleases in a malfunctioning nuclease–(ET-743)–DNA adduct complex, thereby inducing irreparable single-strand breaks in the DNA. This is supported by the fact that mammalian cell lines deficient in TC-NER show resistance to ET-743. *In vitro* exposure of human colon carcinoma cells to clinically relevant (i.e., low nanomolar) concentrations of ET-743 induces strong perturbation of the cell cycle with delay of cell progression from G1 to G2 phase, an inhibition of DNA synthesis and cell cycle arrest in G2 phase resulting in p53-independent apoptosis. Furthermore, there is evidence that nanomolar concentrations of ET-743 cause promoter-selective inhibition of transcriptional activation of genes involved in cell proliferation (e.g., c-jun, c-fos). Also, unlike other DNA-damaging drugs (e.g., doxorubicin) which cause rapid induction of MDR1 gene expression in human sarcoma cells, ET-743 selectively inhibits transcriptional activation of the multidrug-resistance (MDR1) gene in human sarcoma cells *in vivo*.

ET-743 is generally well tolerated; the most common adverse events in clinical practice are non-cumulative hematological and hepatic toxicities. Transient and reversible elevation of hepatic transaminases, nausea, vomiting, and asthenia are common but seldom severe or treatment-limiting. Other side effects commonly associated with cytotoxic agents such as mucositis, alopecia, cardiac or neurotoxicities are not observed.

Pyrrolobenzodiazepine (PBD) monomers

The Pyrrolobenzodiazepine (PBD) family of antitumor agents is based on the natural product anthramycin which was originally isolated from a *Streptomyces* species in the 1960s. The PBD structure contains an electrophilic carbinolamine moiety at the N10–C11 position which can also exist in an equivalent methyl ether form, or under some conditions in equilibrium with the highly electrophilic imine species (Equation (14.5)).

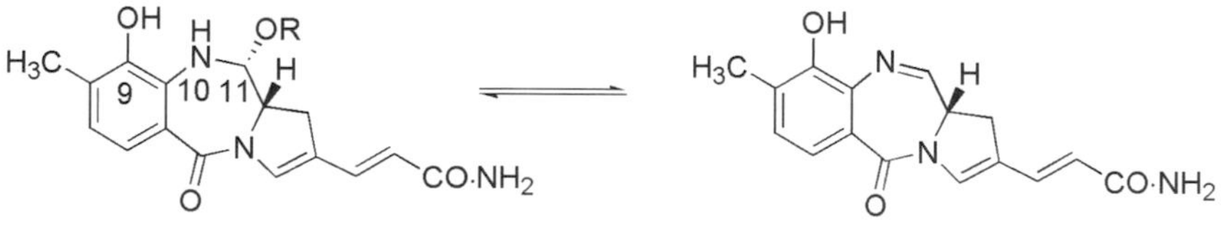

OH H OR H$_3$C N H 9 10 11 N O CO·NH$_2$ ⇌ OH H$_3$C N H N O CO·NH$_2$ (14.5)

Anthramycin (R = H)
Anthramycin methyl ether (R = CH_3)

Anthramycin imine

The PBDs are perfectly shaped to fit into the minor groove of DNA where they covalently attach to the N2-position of guanine through an aminal linkage to their C11 position. They are sequence-specific, recognizing a purine–guanine–purine motif before bonding to the central guanine. The PBDs have been shown previously to possess antitumor activity in the clinic although they have been associated with severe side effects including cardiotoxicity and bone marrow suppression. Through numerous structure–activity relationship (SAR) studies it is now known that the cardiotoxicity is due to the C9 phenolic −OH group that can be converted to a quinone species capable of producing free radicals with the potential to damage heart muscle. A number of second generation non-cardiotoxic PBD monomers lacking a C9-phenolic group but structurally modified to enhance antitumor activity are presently in pre-clinical development. A related family of PBD compounds known as the PBD dimers, where two PBD monomeric units are joined together to produce DNA interstrand cross-linking agents, are described in ''Sequence-selective cross-linking agents.''

14.8.2 Cross-Linking Agents

Cross-linking agents contain two alkylating moieties separated by various distances. By alkylating two nucleophilic functional groups either on the same or opposite strands of DNA, the adducts formed represent either ''intrastrand'' or ''interstrand'' cross-links, respectively (Figure 14.2).

The principle that a DNA cross-linking agent can have a potential beneficial effect in treating leukemia was discovered serendipitously during World War II when in 1943 allied military personnel were accidentally exposed to sulfur mustard gas [$S(CH_2CH_2Cl)_2$] when a US ship, the USS Liberty, carrying 100 tons of this material was attacked in the port of Bari in Italy. Clinicians observed that the leukocyte count dropped in those surviving exposure which suggested a possible use in the treatment of leukemia-type diseases. However, while some leukemias proved responsive, any benefit was greatly outweighed by the toxic effect of the compound. It is now known that sulfur mustard and its derivatives, the nitrogen mustards, are highly reactive and relatively unselective, forming multiple cross-linked sites on DNA that led to their high degree of toxicity. This view has driven the design of new experimental mustard analogues such as tallimustine (a nitrogen mustard conjugated to a netropsin analogue) which has enhanced DNA sequence-recognition properties. It binds in the minor groove and spans five base pairs, recognizing a 5′-GAAAT sequence. It has also led to the design of experimental agents such as SJG-136 (see ''Sequence-selective cross-linking agents'') which are much more selective, forming cross-linked adducts at specific DNA sequences. Through this approach it is hoped to develop agents more selective for specific DNA sequences, thus reducing the frequency of cross-linked sites in genomic DNA and potentially reducing the associated toxicity.

Nitrogen mustards

The nitrogen mustards were developed as bio-isosteric derivatives of sulfur mustard gas that was first synthesized in 1886 but made its debut as a war gas in 1922. They had an improved therapeutic index compared to sulfur mustard, and were introduced into the clinic by Goodman in 1946. The

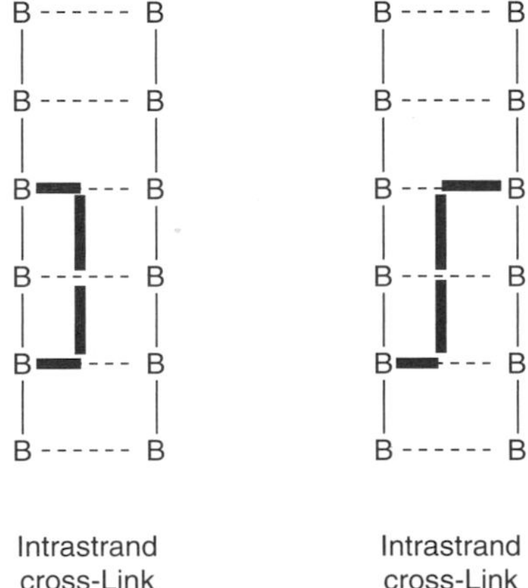

Figure 14.2 Two types of DNA cross-linked adducts (B = DNA base, adducts may span two or more base pairs).

first aliphatic example was mechlorethamine hydrochloride (now called chlormethine or mustine hydrochloride) which was found to effectively depress the white blood cell count and was used for treating certain leukemias.

Despite much research and an in-depth understanding of the medicinal chemistry of the nitrogen mustards, their precise mechanism of action at the molecular level remains to be properly explained. It is clear that these molecules ''staple'' the two strands of DNA together via covalent interactions in the major groove, and presumably block replication. It can also be demonstrated that enzymes such as RNA polymerase are blocked by mustard-DNA adducts. It can be shown from both electrophoresis-based experiments and molecular modeling that mustard adducts cause distortion of the DNA around the binding site which can be transmitted through a number of base pairs (the ''teleomeric'' effect). Therefore, DNA processing may be affected at a point distant from the adduct site. Apart from the ''kinetic'' explanation (i.e., killing faster growing cells) for the antitumor activity of the mustards, another possibility is that their GC selectivity may be relevant. For example, it is known that some of the gene sequences associated with Burkitt's lymphoma are particularly GC-rich, and that this disease is highly responsive to cyclophosphamide. An alternative explanation is that some cancer cells are less proficient at repairing the mustard adducts compared to healthy cells and so opt for the apoptotic route.

In addition to MDR-related resistance, the effect of nitrogen mustards can be significantly reduced by an increase in concentration of glutathione in the cell. The highly nucleophilic glutathione forms adducts with mustards which are then no longer electrophilic and able to react with DNA. Some cancer cells become resistant to nitrogen mustards by carrying out repair processes whereby the mustard adducts are excised and the damaged DNA re-synthesized. Repair inhibitors that might be co-administered with mustards to enhance their clinical effectiveness are under development.

Aliphatic nitrogen mustards

The only example of a simple aliphatic nitrogen mustard in clinical use is chlormethine (mustine). The high chemical reactivity of the aliphatic nitrogen mustards towards DNA and also their vulnerability to attack by a wide range of other nucleophilic centers is thought to account for the observed toxicities.

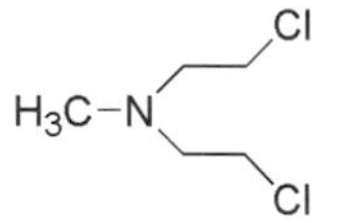

Cl

H_3C-N

Cl

Chlormethine (mustine)

(14.14)

Under physiological conditions, chlormethine and related compounds undergo initial internal cyclization through elimination of chloride to form a cyclic aziridinium (ethyleneiminium) ion. This cyclization process involves intramolecular catalysis through neighboring group participation and results in a positive charge on the nitrogen which is delocalized over the two adjacent carbon atoms. This cation, although relatively stable in aqueous biological fluids, is highly strained and reacts readily with any nucleophile. The clinically useful antitumor adduct appears to result from attack of the N7-atoms of guanine residues in the major groove of DNA. The process is then repeated with the second chloroethyl group and a second guanine N7-atom located on the opposite DNA strand. This gives rise to an interstrand cross-link that effectively locks the two strands of DNA together (Equation (14.6)).

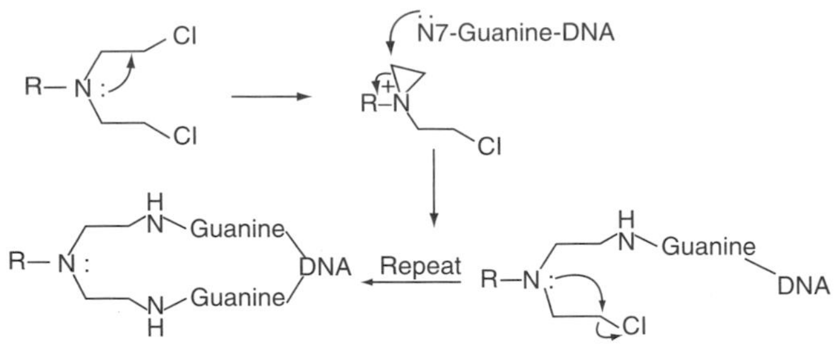

(14.6)

Theoretically, this type of alkylation resembles an S_N2 process since the rate-controlling step is the bimolecular reaction between the cyclic aziridinium ion and the nucleophile which involves simultaneous bond formation and breakage. The preceding step, involving iminium ion formation, is a fast unimolecular process. However, in practice, it is difficult to draw sharp distinctions between the contributions of S_N1- and S_N2-type mechanisms.

Chlormethine is used in some treatment regimens for the management of Hodgkin's disease although it causes severe vomiting. Due to its chemical reactivity the freshly prepared injection must be given into a fast-running intravenous infusion, as any local extravasation causes severe tissue necrosis.

Aromatic nitrogen mustards

The aromatic nitrogen mustards were introduced in the 1950s and are milder alkylating agents. The aromatic ring acts as an electron-sink, withdrawing electrons from the nitrogen atom. Thus it is thought that, in comparison to aliphatic mustards, the central nitrogen atom of an aromatic mustard is not sufficiently basic to form a cyclic aziridinium ion since the nitrogen electron pair is delocalized by interaction with the π electrons of the aromatic ring. Therefore, alkylation most likely proceeds via an S_N1 mechanism, with normal carbocation formation (resulting from chloride ion ejection) providing the rate-determining step (Equation (14.7)).

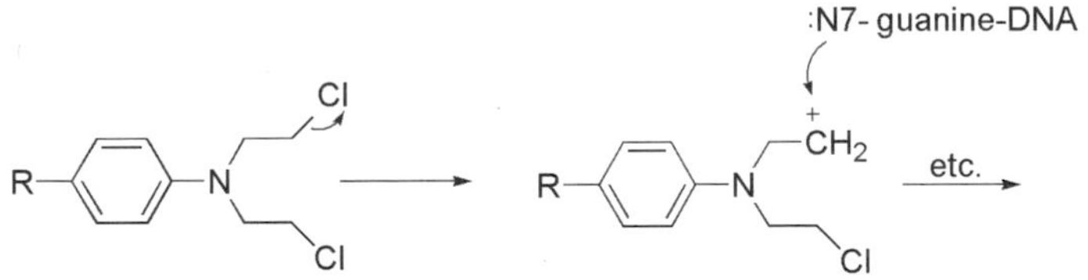

(14.7)

These analogues are sufficiently deactivated that they can reach their target DNA sites before being degraded by reaction with collateral nucleophiles. This means that the aromatic mustards can be taken orally which is a significant advantage.

Chlorambucil (Leukeran)

During early SAR studies it was found that inclusion of a carboxylic acid group greatly improved the water solubility of aromatic mustard analogues. However, direct attachment of a carboxyl group to the aromatic ring reduced chemical and biological activity to an unsatisfactory level due to the additional electron-withdrawing effect. Therefore, the carboxyl group was electronically insulated from the aromatic ring by a number of methylene groups (three proving optimal) to give rise to chlorambucil (Leukeran) which is one of the slowest acting and least toxic nitrogen mustards. It is useful in the treatment of CLL, the indolent non-Hodgkin's lymphomas, Hodgkin's disease, and ovarian cancer.

Chlorambucil (Leukeran™)
(14.15)

The lower chemical reactivity is reflected in the relative lack of side effects apart from bone marrow suppression which is reversible. Occasionally patients develop widespread rashes which can develop into serious dermatological conditions unless treatment is stopped.

Melphalan (Alkeran)

Melphalan (phenylalanine mustard, Alkeran), which also contains a carboxylic acid, was designed with a view to introducing greater selectivity based on the hypothesis that attachment of an amino acid residue (i.e., phenylalanine) might facilitate selective uptake by tumor cells in which rapid protein synthesis is occurring.

Melphalan (Alkeran)
(14.16)

Since melphalan, prepared from D-phenylalanine, is much less active than that prepared from the L-form, it has been postulated that melphalan may be conveyed into cells by the L-phenylalanine active transport mechanism. It is used to treat myeloma and occasionally solid tumors (e.g., breast and ovarian) and lymphomas. Like chlorambucil it is given orally but can also be given intravenously. Bone marrow toxicity is delayed with melphalan and so it is usually given at intervals of 6 weeks.

Oxazaphosphorines

A further attempt to produce more selective mustards was based on the hypothesis that some tumors may over-produce phosporamidase enzymes. Cyclophosphamide (Endoxana) is the most successful mustard to result from this work and it has been used safely for many years. Ifosfamide (Mitoxana) is a related analogue that has been more recently introduced.

Cyclophosphamide (Endoxana)

The design of this mustard prodrug was based on the concept that the P = O group should decrease the availability of the nitrogen lone pair in an analogous manner to the phenyl ring of the aromatic mustards, thus deactivating the molecule to nucleophilic attack. However, it was postulated that the P = O group might be cleaved by phosphoramidases which were thought to be over-produced in some tumor cells, thus releasing the nitrogen lone pair and restoring the electrophilicity of the molecule. Hence a degree of tumor selectivity could be achieved if the phosporamidase enzymes were confined mainly to tumor cells.

Cyclophosphamide (Endoxana)
(14.17)

$HS{-}CH_2CH_2{\cdot}\overset{-}{S}O_3\overset{+}{Na}$

MESNA (Uromitexan)
(14.18)

Although this was a reasonable design concept, and cyclophosphamide has become an important clinical agent, it was later demonstrated that activation *in vivo* is not due to enzyme-catalyzed hydrolysis of the P = O group but rather to oxidation by liver microsomal enzymes. After 4-hydroxylation, the molecule fragments to give phosphoramide mustard which is assumed to be the biologically active species, along with highly electrophilic acrolein which is toxic and can cause hemorrhagic cystitis (Equation (14.8)). The production of normustine [$HN(CH_2CH_2Cl)_2$] has also been observed.

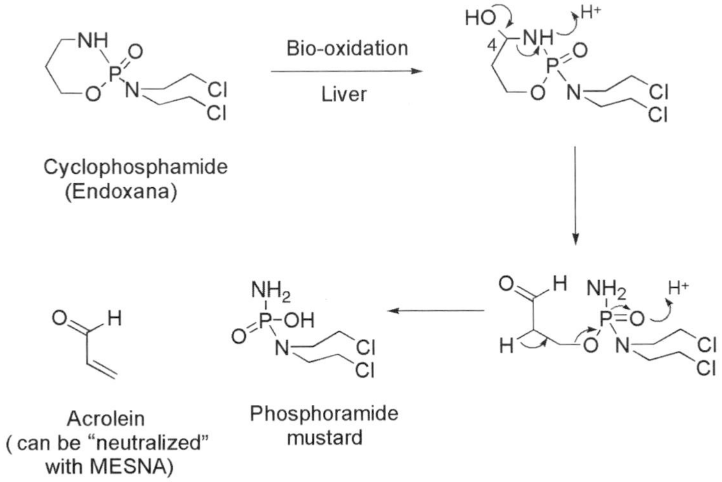

Bio-oxidation

Liver

Cyclophosphamide (Endoxana)

Acrolein (can be "neutralized" with MESNA)

Phosphoramide mustard

(14.8)

Cyclophosphamide has a wide spectrum of clinical activity. It is widely used in the treatment of CLL, the lymphomas and solid tumors including carcinomas of the bronchus, breast, ovary, and various sarcomas. It is given by mouth or intravenously but is inactive until metabolized by the liver. The acrolein metabolite is excreted in the urine, where it can cause hemorrhagic cystitis in susceptible patients, a rare but very serious complication. This is because acrolein is a potent electrophile and can react with nucleophiles on the surface of cells lining the bladder. An increased fluid intake after intravenous injection can help to avoid this problem. However, after high-dose cyclophosphamide therapy (>2 g intravenously), MESNA (Uromitexan) is routinely co-administered to ''neutralize'' the effect of the acrolein. MESNA, which is 2-mercaptoethanesulfonate, acts as a ''sacrificial'' nucleophile, forming a non-reactive water soluble adduct that is eliminated safely in the urine (Equation (14.9)).

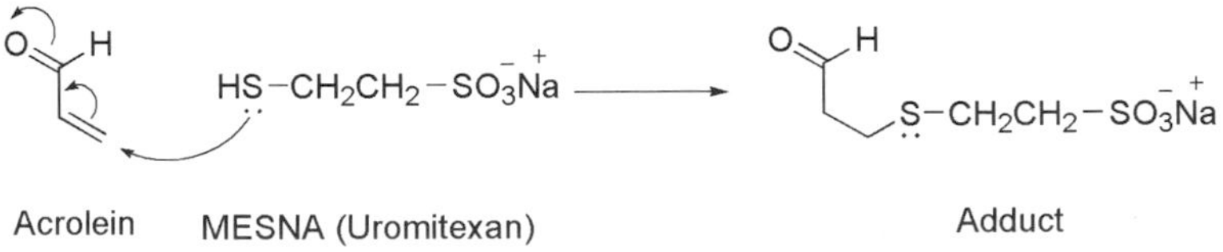

$HS-CH_2CH_2-SO_3^-\overset{+}{Na}$ → $S-CH_2CH_2-SO_3^-\overset{+}{Na}$ (14.9)

Acrolein MESNA (Uromitexan) Adduct

MESNA is given before and after treatment with cyclophosphamide when oral therapy is used, but is co-administered with cyclophosphamide and given afterwards with intravenous therapy.

The slow rate of *in vivo* hydroxylation of cyclophosphamide in humans has led to the synthesis of a number of experimental 4-hydroxy derivatives (i.e., the 4-hydroperoxy analogue) designed to spontaneously cleave *in vivo* to yield the active 4-hydroxy metabolite. However, none of these have shown any advantage over cyclophosphamide.

Ifosfamide (Mitoxana)

Ifosfamide (Mitoxana) is an analogue of cyclophosphamide that is not technically a nitrogen mustard due to the translocation of one chloroethyl moiety to another position within the molecule (i.e., to a nitrogen attached to the P = O group). However, it has a similar spectrum of activity to cyclophosphamide but can only be administered intravenously.

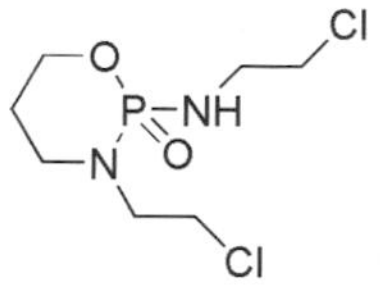

Ifosfamide (Mitoxana)
(14.19)

MESNA is routinely co-administered with Ifosfamide and given afterwards to reduce urethelial toxicity.

Conjugated nitrogen mustards

Given the significant antitumor activity of the mustards, considerable medicinal chemistry research has been devoted to attaching them to other therapeutic moieties either to achieve a combination therapy, or for the purposes of targeting the mustard to a particular organ or cell type. The most commercially successful of these approaches has been the combination of a mustard and an estrogen which appears to provide both an antimitotic effect and (by reducing testosterone levels) a hormonal effect.

Estramustine phosphate (Estracyt)

Estramustine, which is given orally, is a combination of an estrogen and chlormethine (mustine). It is used as primary and secondary therapy in patients with metastatic prostate cancer. At one time, estrogens represented the primary treatment for metastatic prostate cancer, but with the discovery of their serious cardiovascular side effects in the early 1970s, this treatment fell from favor. However, because of their multiplicity of actions, estrogens are still attractive for the treatment of prostate cancer. Not only do they lower luteinizing hormone (and therefore testosterone level), block 5-α-reductase (and the conversion of testosterone to dihydrotestosterone [DHT]), and increase the concentration of sex hormone-binding globulin (thereby preventing free testosterone from attaching to prostate cancer cells), they have recently been shown to have a direct cytotoxic effect on prostate cancer cells in culture.

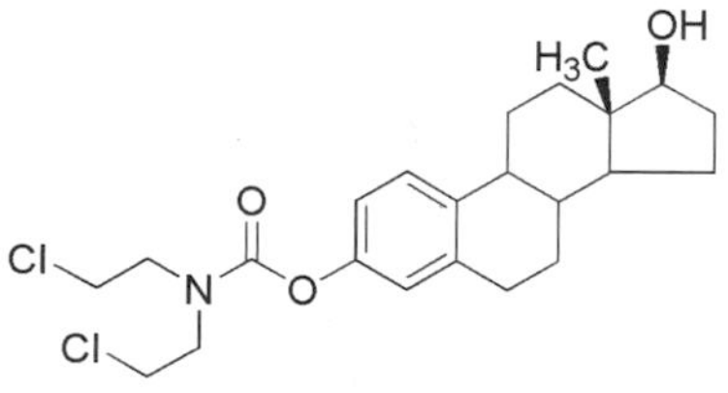

Estramustine (Estracyt)
(14.20)

Estracyt, the combination of an estrogen and a nitrogen mustard through a carbamate linkage, was first introduced in 1976. Although it is feasible that the intact molecule may still alkylate and cross-link DNA, it is more likely that the carbamate linker is cleaved by plasma or cellular carbamases, and that the two released components act at their different sites.

From a number of clinical studies it appears that Estracyt must be used in high doses at the start of therapy if it is to have both estrogenic and cytotoxic effects. At a dose level of 10 mg/kg in patients with poorly differentiated prostate cancer, Estracyt is slightly superior to orchiectomy.

Aziridines

Rather than form aziridinium ions as reactive intermediates, thiotepa and related analogues such as the experimental agents AZQ and BZQ have an aziridine ring already incorporated in their structure. Ring-opening of the aziridines with nucleophiles is slower compared to the fully charged transient aziridinium ions of the mustards. However, depending upon the pK_a of the aziridine nitrogen, there is likely to be significant protonation at physiological pH, meaning that, in practice, the aziridinium ion may be the reactive species.

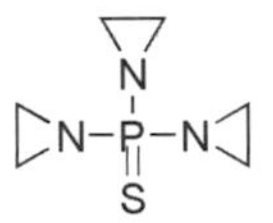

Thiotepa

(14.21)

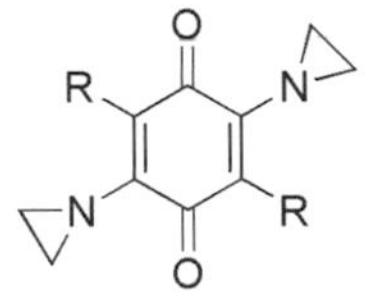

AZQ: R = HNCOOEt
BZQ: R = $HNCH_2CH_2OH$

(14.22)

Thiotepa is usually used as an intracavitary drug (e.g., by intrapleural, intrabladder, or peritoneal infusion) for the treatment of malignant effusions, ovarian, or bladder cancer. It is also occasionally used to treat breast cancer but requires parenteral administration for this purpose. A substituted benzoquinone ring has also been employed as an ''anchor'' for the aziridine groups in the experimental agents AZQ and BZQ. The aziridine moieties are deactivated by the withdrawal of electrons from the nitrogens into the quinone carbonyl groups via the ring. These molecules have been employed as experimental bioreductive prodrugs, as reduction of the quinone ring to either the semiquinone or the hydroxyquinone species reverses the electron flow and raises the pK_a of the aziridine nitrogens thus allowing them to be activated via protonation.

It should be noted that the cross-linking agent Mitomycin C also contains an aziridine moiety and this agent is discussed in ''Mitomycin C.''

Epoxides

Epoxide functionalities are similar to aziridines in their propensity to alkylate DNA. Although some carcinogens such as the aflatoxins are known to alkylate DNA via an epoxide, there are currently no clinically used drugs that contain preformed epoxide moieties. However, treosulfan (L-threitol 1,4-bismethanesulfonate, Ovastat) is a prodrug that converts non-enzymatically to L-diepoxybutane via the corresponding monoepoxide under physiological conditions (Equation (14.10)).

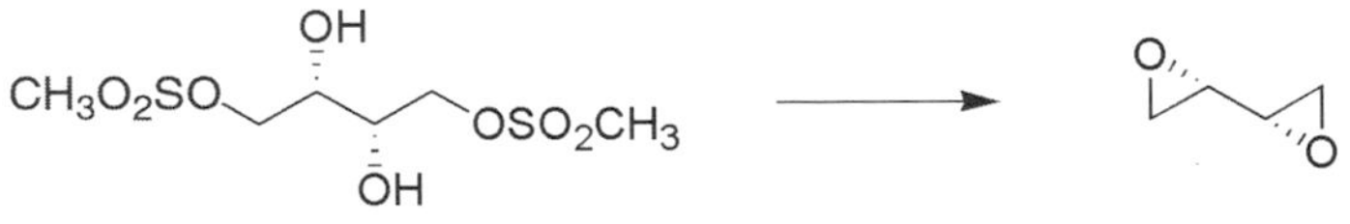

CH_3O_2SO–CH_2–CH(OH)–CH(OH)–CH_2–OSO_2CH_3 ⟶ L-Diepoxybutane (14.10)

Treosulfan (Ovastat)

L-Diepoxybutane

In vitro studies have shown that this conversion is required for the alkylation and interstrand cross-linking of plasmid DNA, and it is assumed that it also occurs *in vivo*. Alkylation occurs at guanine bases with a sequence selectivity similar to other alkylating agents such as the nitrogen mustards. In treosulfan-treated K562 cells growing *in vitro*, cross-links form slowly, reaching a peak after approximately 24 h. However, incubation of K562 cells with preformed epoxides such as L-diepoxybutane provides faster and more efficient DNA cross-linking, supporting the proposed prodrug conversion.

Treosulfan is mainly used to treat ovarian cancer. It can be administered either orally or by intravenous or intraperitoneal injection. The major side effects are similar to those for the nitrogen mustards and include bone marrow suppression and skin toxicities. Nausea, vomiting, and hair loss occur less frequently.

Methanesulfonates

Busulfan [1,4-di(methanesulfonyloxy)butane] is the best known example of an alkyl dimethane-sulfonate with significant antitumor activity. It is known to form DNA cross-links, with the methanesulfonyloxy moieties acting as leaving groups after attack by nucleophilic sites on DNA. From a mechanistic viewpoint, the methanesulfonate groups should participate in S_N2-type alkylation reactions. SAR studies have revealed that unsaturated analogues of known stereochemistry such as the corresponding butyne and *trans*-butene derivatives are inactive, while the *cis*-butene derivative retains activity. It is thought that the activities of the *cis*-analogue and the more flexible saturated busulfan depend on their ability to form cyclic derivatives by 1,4-bisalkylation of suitable nucleophilic groups. For example, 1,4-di(7-guanyl)butane has been identified as a product of the reaction between busulfan and DNA suggesting that this agent acts as an interstrand cross-linking agent in a similar manner to the nitrogen mustards. However, studies of the structure of urinary metabolites suggest that cysteine residues in certain proteins are also alkylated.

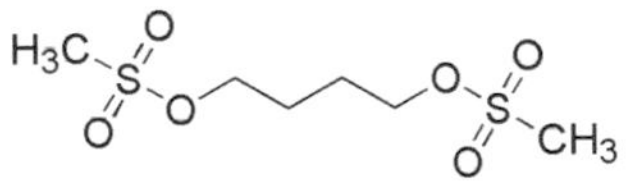

Busulphan (Myleran)
(14.23)

Busulfan is administered orally and causes significantly less nausea and vomiting than other DNA cross-linking agents, thus being more acceptable to patients. It is used almost exclusively for the treatment of CML in which it is highly effective and can keep patients almost symptom-free for long periods of time. Unfortunately, it causes excessive myelosuppression which can result in irreversible bone marrow aplasia, and this requires careful monitoring. Hyperpigmentation of the skin is another common side effect and, more rarely, pulmonary fibrosis may occur.

Nitrosoureas

The nitrosoureas alkylate DNA and lead to both mono-adducts and interstrand cross-links at a number of different sites. The study of a large number of nitrosourea analogues has established the structural unit for optimal activity as the 2-chloroethyl-*N*-nitrosoureido moiety, and the likely mechanism of both monoalkylation and cross-linking is shown below. Although these molecules possess chloroethyl fragments, their activity is not associated with aziridinium ion formation as in the mustards, since the corresponding nitrogen atom is part of a urea structure and so the electron pair on the nitrogen is not available to participate in a cyclization reaction. Instead it is thought that the alkylation of nucleic acids proceeds via generation of a chloroethyl carbonium ion. The alkyl isocyanate fragment also formed is thought to carbamoylate the amino groups of proteins (Equation (14.11)).

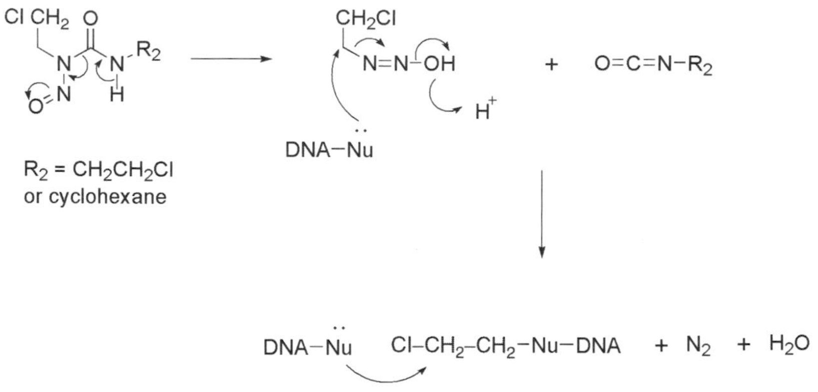
Cl CH2 O
R2
N
N
O=N
H
R2 = CH2CH2Cl
or cyclohexane
CH2Cl
N=N-OH
H+
DNA-Nu
+
O=C=N-R2
DNA-Nu
Cl-CH2-CH2-Nu-DNA
+ N2
+ H2O

(14.11)

The most significant property of the nitrosoureas is their activity towards cancer cells in the brain and cerebrospinal fluid, the so-called sanctuary sites. This is due to the relatively high lipophilicity of these molecules compared to other agents such as the nitrogen mustards. However, they are also active in malignant lymphomas and carcinomas of the breast, bronchus, and colon. Carcinoma of the GI tract, which is notably intractable to drug treatment, can also respond to the nitrosoureas. Unfortunately, these agents cause severe bone marrow toxicity which is usually dose-limiting. Carmustine and lomustine are two examples of clinically-useful nitrosoureas.

Lomustine

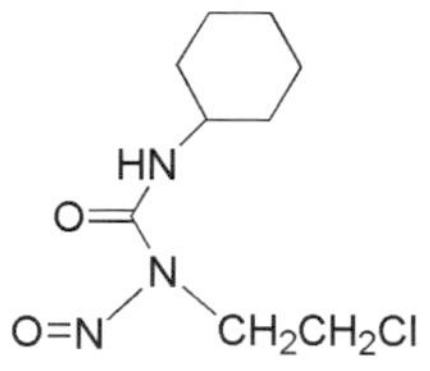
HN
O
N
O=N
CH2CH2Cl

Lomustine (CCNU)
(14.24)

Lomustine [*N*-(2-chloroethyl)-*N'*-cyclohexyl-*N*-nitrosourea, CCNU] is a lipid soluble nitrosourea which is administered orally. It is mainly used to treat Hodgkin's disease and certain solid tumors. Bone marrow toxicity is delayed, and the drug is therefore given at intervals of 4 to 6 weeks. Nausea and vomiting are common and moderately severe, although permanent bone marrow damage can occur with prolonged use.

Carmustine (BiCNU)

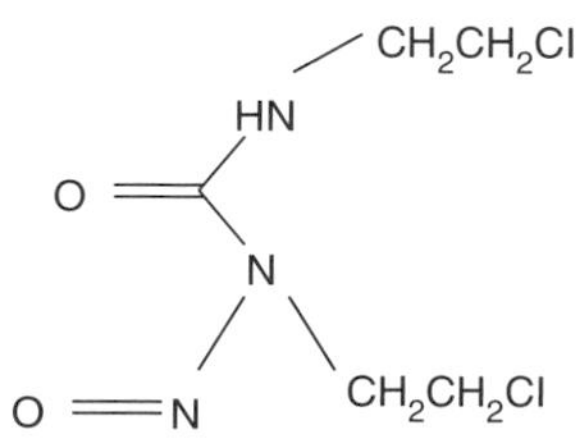
CH2CH2Cl
HN
O
N
O
N
CH2CH2Cl

(4.25)

Carmustine [*N,N'*-bis(2-chloroethyl)-*N*-nitrosourea, BiCNU] has similar activity and toxicity profiles to lomustine but is administered intravenously. It is most commonly used to treat patients with myeloma, lymphoma, and brain tumors. A major problem with carmustine is that cumulative renal damage and delayed pulmonary fibrosis may occur.

Platinum complexes

The discovery and development of the platinum complexes which include cisplatin, carboplatin (Paraplatin), and oxaliplatin (Eloxatin) is often quoted as one of the great successes of cancer chemotherapy due to the pronounced activity of cisplatin, particularly in testicular and ovarian cancers. However, as with many other clinically useful drugs, the first member of the family, cisplatin, was discovered by serendipity rather than by design.

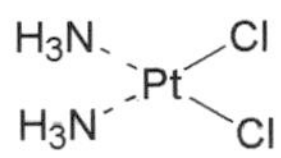

Cisplatin **(14.26)**   Carboplatin (Paraplatin) **(14.27)**   Oxaliplatin (Eloxatin) **(14.28)**

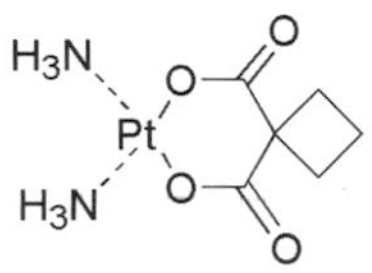

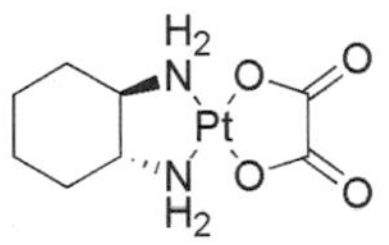

Cisplatin, also known as Peyrone's salt or Peyrone's chloride, is a co-ordination complex and was originally prepared and reported in 1845. In the 1960s, Rosenberg and coworkers observed that passing an alternating electric current through platinum electrodes in an electric cell containing *Escherichia coli* led to arrest of cell division without killing the cells. Continued growth without division led to unusually elongated cells with a spindle-like appearance. The cause of the cytostatic effect was eventually traced to platinum complexes formed electrolytically at concentrations of only 10 parts per million in the presence of ammonium salts and light. *cis*-Diaminedichloroplatinum (cisplatin) was identified as one of the most active complexes, and Rosenberg then went on to show that it possessed significant cytotoxicity towards cancer cells growing *in vitro* and had antitumor activity *in vivo*. As a result, cisplatin was introduced into the clinic in the UK in 1979. The related analogues, carboplatin and oxaliplatin, were later developed in an attempt to reduce the problematic side effects of cisplatin. It is now known that the platinum complexes work by interacting in the major groove of DNA and forming intrastrand cross-links. The platinum complexes are all administered intravenously although an orally active experimental analogue (JM 216) has been studied in the clinic but not yet commercialized. One disadvantage of the metal complexes is their high cost due to the platinum content.

Cisplatin

Cisplatin produces intrastrand cross-links in the major groove of DNA with preferential interaction between (guanine N7)-(guanine N7), (guanine N7)-(adenine N7) (in both cases with the bases adjacent to one another), and (guanine N7)-X–(guanine N7) (with one base ''X'' in-between the alkylated guanines). Based on gel electrophoresis and NMR studies, these intrastrand cross-links are known to kink the DNA at adduct sites, a phenomenon that can now be confirmed by techniques such as high-field NMR, x-ray crystallography, and atomic force microscopy. Interestingly, the configurational isomer, *trans*-platin, has significantly less antitumor activity which relates to a less efficient interaction with DNA.

Cisplatin is of significant value in patients with metastatic germ cell cancers (i.e., seminoma and teratoma). For example, clinical trials of cisplatin in combination with vinblastine have produced complete remission in 59% of patients with testicular cancer, and 30% remission in ovarian carcinoma. In some cisplatin-resistant cell lines the adducts are rapidly repaired, and it is thought that the DNA repair surveillance enzymes recognize the distortion around the adduct site. One possible explanation for the significant activity of cisplatin in testicular cancer is that the germ cells have a limited ability to repair cisplatin adducts compared to other cells. Other tumors which respond to cisplatin include squamous cell carcinoma of the head and neck, bladder carcinoma, refractory choriocarcinoma, and lung, upper GI, and ovarian cancer. However, carboplatin is now preferred for ovarian cancer.

Cisplatin has a number of dose-limiting side effects which include nephrotoxicity (monitoring of renal function is essential), ototoxicity, peripheral neuropathy, hypomagnesaemia, and myelosuppression. It requires intensive intravenous hydration during treatment and progress may be complicated by severe nausea and vomiting. It is, however, increasingly given in a day care setting.

Carboplatin (Paraplatin)

Carboplatin is an analogue of cisplatin that incorporates a cyclobutyl-substituted dilactone ring. It is better tolerated than cisplatin in terms of nausea and vomiting, GI toxicity, nephrotoxicity, neurotoxicity, and ototoxicity, although myelosuppression is more pronounced. It is widely used in the treatment of advanced ovarian cancer and lung cancer (particularly the small cell type). Carboplatin is better tolerated than cisplatin and the dose is determined according to renal function rather than body surface area. It is often used on an outpatient basis.

Oxaliplatin (Eloxatin)

Oxaliplatin is licensed for the treatment of metastatic colorectal cancer in combination with fluorouracil and folinic acid. Neurotoxic side effects which include sensory peripheral neuropathy are dose-limiting. Other side effects include gastrointestinal disturbances, ototoxicity, and myelosuppression. Renal function is normally monitored during treatment.

Carbinolamines

The carbinolamine group [−NR−CH(OH)−] (R = alkyl or aryl) is an electrophilic moiety found in a number of synthetic and naturally occurring compounds that can form a covalent bond with various nucleophilic sites on DNA bases. Trimelamol, which contains three carbinolamine moieties, is an experimental agent developed from the clinically-active hexamethylmelamine (and the closely related pentamethylmelamine) which do not contain carbinolamines themselves but are converted during oxidative metabolism. Trimelamol was designed to have the carbinolamine moieties already in place. Phase II trials carried out in the early 1990s indicated that trimelamol is active in refractory ovarian cancer, and is less emetic and neurotoxic than pentamethylmelamine.

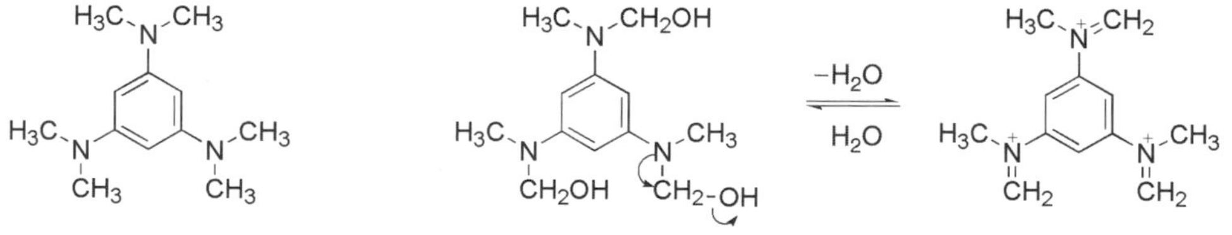

H_3C N CH_3; H_3C N CH_3; N CH_3 CH_3 | H_3C N CH_2OH; H_3C N CH_2OH; N CH_3 CH_2-OH | $-H_2O$ / H_2O | H_3C N^+ $=CH_2$; H_3C N^+ CH_2; N^+ CH_3 CH_2

Hexamethylmelamine **(14.29)** Trimelamol iminium form **(14.30)**

The precise mode of action of trimelamol has not yet been established. However, it is known to be a reasonably potent DNA interstrand cross-linking agent although the exact sites of alkylation or cross-linking have not been established. It is known that the carbinolamine moieties can dehydrate to give iminium ions, and it is likely that it is these highly electrophilic species that alkylate DNA bases. In practice, there is likely to be a pH-dependent equilibrium mixture of the carbinolamine and imine forms in aqueous solution.

One reason why trimelamol has not yet been commercialized is its poor solubility in a number of physiologically compatible solvents, thus making it difficult to formulate. It also suffers from stability problems in that amine-containing degradation products couple with trimelamol itself to form dimeric and higher order polymers which are extremely insoluble. Some analogues of trimelamol that partially overcome these problems have been developed, and so a new version of this agent may be studied in clinical trials in the future.

The PBD dimer cross-linking agents also possess carbinolamine moieties or the equivalent. However, due to their unique mechanism of action (i.e., DNA sequence-selectivity), they are discussed separately in ''Sequence-selective cross-linking agents.''

Cyclopropanes

Cyclopropane moieties are the homocarbon analogues of aziridines and epoxides, and are also electrophilic and capable of alkylating DNA. Part of the driving force for such an alkylation reaction is the release of strain within the three-membered ring. The experimental agent Bizelesin is the only known example of a cyclopropane-containing DNA interstrand cross-linking agent. However, as this agent is also an example of a sequence-selective cross-linking agent, it is discussed in ''Sequence-selective cross-linking agents.''

Mitomycin C

Mitomycin C is a naturally occurring antitumor antibiotic rich in chemical functional groups. The three components of the molecule essential for its mode of action are the quinone, aziridine, and carbamate moieties. The mechanism of action involves a bioreductive step, and so mitomycin is regarded as a ''bioreductive agent'' (see Section 14.13.5). However, it is discussed in this section due to its DNA cross-linking properties.

1. Enzymatic reduction
2. Reaction with DNA

(14.12)

Bioreduction of Mitomycin C (Mitomycin C Kyowa) followed by DNA cross-linking

It is thought that initial reduction of the quinone (one-electron reduction yields a semiquinone, while a two-electron reduction gives the hydroquinone) leads to transformation of the heterocyclic nitrogen from a conjugated amido to an amino form thus making it more electron-rich. This facilitates elimination of the ring junction methoxy group. Tautomerization of the resulting iminium ion and loss of the carbamate group then creates an electrophilic center, which is susceptible to attack by a nucleophilic DNA base. Nucleophilic attack of the aziridine moiety by a nucleophile on the opposite strand of DNA also occurs, leading to an interstrand cross-link. The predominant adducts appear to form between two guanine-N2 groups within the minor groove (Equation (14.12)).

The most important feature of mitomycin is the bioreductive ''trigger'' that is required before cross-linking can take place. It is known that the centers of some tumors, particularly older ones of larger size, are hypoxic due to a poor blood supply. The bioreductive conditions that exist at the center of these tumors are thought to account, in part, for the tumor selectivity of mitomycin. The drug is administered intravenously to treat upper GI and breast cancers, and by bladder instillation for superficial bladder tumors. It causes delayed bone marrow toxicity and so is usually administered at six-weekly intervals. Prolonged use may result in permanent bone marrow damage, and it may also cause lung fibrosis and renal damage.

The concept of bioreductive activation continues to attract interest due to the hypoxic conditions in certain tumors, and a number of other compounds have been designed based on this mechanism of action (see Section 14.13.5).

Sequence-selective cross-linking agents

Most of the cross-linking agents described in the previous sections are relatively small in size and rarely span more than one or two base pairs. There has been an ongoing effort to develop agents that can span longer sequences of DNA (e.g., six or more) and that can recognize specific sequences of base pairs within the stretch spanned. The reasoning is, for agents with lower sequence selectivity, a large number of adducts are formed across the genome, many of which will lead to the toxic side effects (e.g., myelosuppression) observed with these agents. If longer, more sequence-selective agents can be developed, then there will be fewer adducts formed across the genome with a potential reduction in the occurrence and intensity of side effects and a potential improvement in therapeutic index. The ultimate design objective is an agent that recognizes and cross-links a DNA sequence that only occurs in tumor cells (e.g., oncogenes) but not in healthy cells. Two families of agents have so far been developed based on this principle: the CBI dimers which interstrand cross-link adenine bases via their N3 positions, and the PBD dimers such as SJG-136 which interstrand cross-link guanine bases via their N2 positions.

Pyrrolobenzodiazepine dimer (SJG-136)

The experimental agent SJG-136 (NCI-694501) is a PBD dimer that consists of two monomeric PBD units joined through their C8/C8′-positions via a propyldioxy linker.

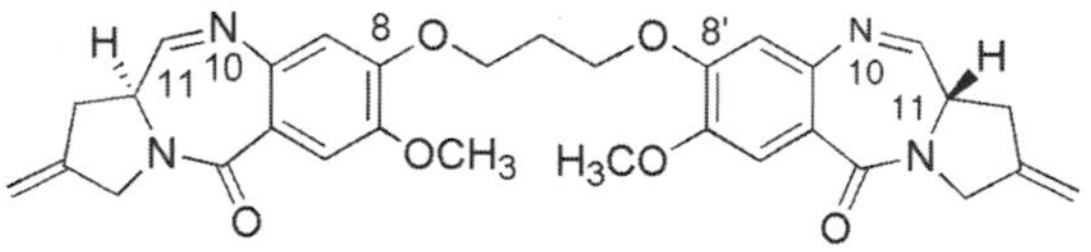

H N 8 O O 8′ N H 11 10 10 11 N OCH$_3$ H$_3$CO N O O

SJG-136
(14.31)

The two imine moieties (i.e., N10–C11/N10′–C11′) in each seven-membered ring bind covalently in the minor groove of DNA to the N-2 positions of guanines on opposite strands of DNA thus forming an interstrand cross-link. The molecule spans six DNA base pairs and occupies half a complete turn of the DNA helix (Figure 14.3). SJG-136 has a preference for binding to purine–GATC–pyrimidine sequences while recognizing the central GATC sequence.

The molecule has shown significant activity in the NCI 60 cell line screen and other cell lines, and in follow-up hollow fiber and human tumor xenograft assays in both mice and rats where it provides a dose-dependent effect on tumor growth in a number of tumor models and is curative in some. SJG-136/DNA adducts are resistant to repair compared to those of other DNA-interactive agents, and the agent retains full potency in a number of drug resistant cell lines. SJG-136 also has significant antitumor activity in a mouse xenograft model based on a cisplatin-resistant tumor

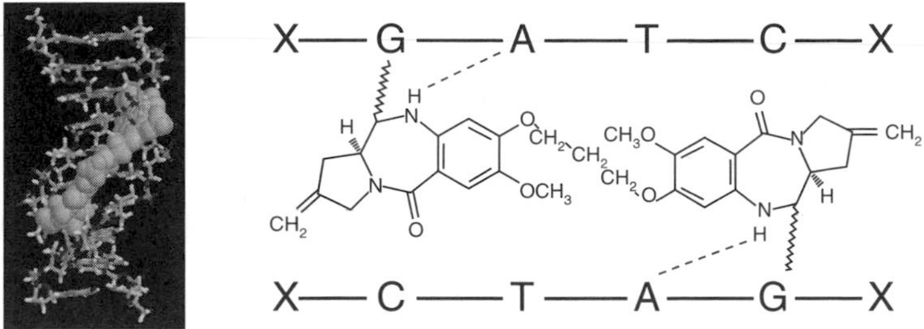

Figure 14.3 SGJ-136 binding covalently in the minor groove of DNA.

suggesting that it may have potential use in drug resistant disease. SJG-136 is presently being evaluated in Phase I clinical trials.

Cyclopropanebenzindole (CBI) dimer (Bizelesin)

The experimental agent bizelesin is similar to SJG-136 in binding in the minor groove of DNA and spanning approximately six base pairs. However, it cross-links two adenines via their N3 positions as opposed to the G–G cross-link caused by SJG-136. In the case of bizelesin, the alkylating moieties are cyclopropane rings which are formed *in situ*. Thus bizelesin itself is a prodrug which undergoes chemical transformation into the electrophilic dicyclopropane species. As shown below, this involves spontaneous formation of the quinone and cyclopropane moieties through loss of chloride ion (Equation (14.13)).

Bizelesin

(14.13)

N3-Adenine

Dicyclopropane active species

Cyclopropane moieties are the homocarbon analogues of aziridines and epoxides, and are thus electrophilic and capable of alkylating DNA. However, in the case of bizelesin each cyclopropane moiety is attached to the *para*-position of a dihydroquinone and so is further activated. Attack of the cyclopropane ring by an adenine N3 is driven by the energy released in re-aromatizing the dihydoquinone to a phenol. The central indole units also recognize AT base pairs but noncovalently, and so overall bizelesin recognizes a run of AT base pairs.

Unfortunately, bizelesin itself has been shown to cause severe myelosuppression and so has not progressed to the clinic.

14.8.3 Intercalating Agents

The DNA intercalating agents are one of the most widely used families of antitumor agents. They are flat in shape, consisting of three or four fused aromatic rings. Their mechanism of action involves insertion between the base pairs of DNA, perpendicular to the axis of the helix. Once in position, they are held in place by interactions including hydrogen bonding and van der Waals forces. In addition, the side chains are often rich in hydrogen-bonding functionalities (e.g., the amino sugar of doxorubicin or the pentapeptide rings of dactinomycin), and are positioned in the DNA minor or major grooves where they can further stabilize the adduct formed. Some intercalators with arrays of functional groups at either end of the molecule protrude into both the minor and major grooves after insertion. In this case they are known as ''threading'' agents.

Intercalation can be detected in naked DNA by an increase in helix length which can be evaluated as an increase in viscosity using sedimentation values or as a change in mobility of DNA fragments in an electrophoresis gel. Adducts of intercalating agents with DNA can also be studied by such techniques as NMR and x-ray crystallography.

A number of different modes of action have been ascribed to intercalating-type drugs. For example, it can be demonstrated in the laboratory that some intercalators and threading agents block transcription and interfere with other DNA processing enzymes. Many intercalating agents have a preference for GC-rich sequences and, as with the alkylating agents, this may play a part in their mechanism of action. However, some intercalators are known to ''trap'' complexes between topoisomerase enzymes and DNA thus leading to strand cleavage. Others are known to chelate metal ions and produce DNA-cleaving free radicals, or to interact with cell membranes.

The largest family of intercalating agents in clinical use is the anthracyclines, members of which contain four fused aromatic rings and include the naturally occurring antibiotics doxorubicin, daunorubicin and aclarubicin, and the related semisynthetic analogues epirubicin and idarubicin. Members of the anthracene family possess three aromatic rings, with mitoxantrone (also known as mitozantrone) being the only example in common use in the UK. The third group is the phenoxazine family, members of which contain three fused six-membered rings but with the central ring containing oxygen and nitrogen heteroatoms. The best known member of this group is dactinomycin which also contains two cyclic peptide side chains which stabilize the drug–DNA adduct by interacting in the minor groove of DNA.

It should be noted that many intercalating agents act as radiomimetics, and so simultaneous use of radiotherapy should be avoided as it may result in enhanced toxicity.

Anthracyclines

The anthracyclines (sometimes known as the anthraquinones) are a group of antitumor antibiotics, the first compound of which was isolated from *Streptomyces peucetius*. It is the best known family of intercalating agents, and members consist of a planar anthraquinone nucleus attached to an amino sugar. Although doxorubicin, daunorubicin, and aclarubicin are natural products, semisynthetic analogues such as epirubicin and idarubicin have also been developed.

A number of different mechanisms have been suggested to explain the biological activity of the anthracyclines, and there is still controversy about the relative importance of each one. The first centers on the fact that the planar ring system inserts between DNA base pairs perpendicular to the long axis of the double helix with the amino sugar conferring stability on the adduct through hydrogen bonding interactions with the sugar phosphate backbone. It is known that this intercalation process can interfere with DNA processing including transcription. Second, the anthracyclines are known to form complexes with topoisomerase enzymes and DNA, and this can lead to strand breaks. Third, binding to cell membranes has been observed, and it has been suggested that this might alter membrane fluidity and ion transport, and disturb various biochemical equilibria in the cell. Lastly, generation of semiquinone species can lead to free radical or hydroxyl radical production which may cause DNA and other cellular damage. Radical formation may be mediated by chelation of divalent cations such as calcium and ferrous ions by the phenolic and quinone functionalities, and this is thought to be responsible for the cardiotoxicity observed with the anthracyclines.

Doxorubicin (Caelyx and Myocet)

Doxorubicin was first isolated from *S. peucetius*, and is presently one of the most successful and widely used anticancer drugs because of its broad spectrum of activity. It is used to treat acute leukemias, lymphomas, and a variety of solid tumors including carcinoma of the breast, lung, thyroid, and ovary, as well as soft-tissue carcinomas. Doxorubicin is given by injection into a fast-running infusion, usually at 21-day intervals, although care must be taken to avoid local extravasation which can cause severe tissue necrosis. Common toxic effects include nausea and vomiting, myelosuppression, alopecia, and mucositis. The drug is largely excreted by the biliary tract, and an elevated bilirubin concentration is an indication for reducing the dose. Higher cumulative doses are associated with cardiomyopathy, and potentially fatal heart failure can occur. Thus cardiac monitoring is sometimes used to assist in safely limiting total dosage. Liposomal formulations are also available which are thought to reduce the incidence of cardiotoxicity and lower the potential for local necrosis. For example, Caelyx is licensed for advanced AIDS-related Kaposi's sarcoma, and for advanced ovarian cancer when platinum-based chemotherapy has failed. Myocet is licensed for use with cyclophosphamide for metastatic breast cancer. Doxorubicin can also be given by bladder instillation for superficial bladder tumors.

Daunorubicin

Daunorubicin was originally isolated from fermentation broths of *S. peucetius*. It is an important agent in the treatment of acute lymphocytic and myelocytic leukemias. Administered intravenously, it has general properties similar to those of doxorubicin. An intravenous liposomal formulation (DaunoXome) is licensed for advanced AIDS-related Kaposi's sarcoma.

Aclarubicin Aclarubicin is produced by *S. galilaeus* and has general properties similar to doxorubicin. It is used for acute nonlymphocytic leukemia in patients who have relapsed or are resistant or refractory to first-line chemotherapy.

Epirubicin (Pharmorubicin) Epirubicin is a semisynthetic analogue of doxorubicin differing only in the stereochemistry of the C-4 hydroxy group of the sugar moiety. Clinical trials suggest that it is as effective as doxorubicin in the treatment of breast cancer, however it is necessary to carefully monitor the maximum cumulative dose to help avoid cardiotoxicity. Like doxorubicin, it is given either intravenously or by bladder instillation.

Idarubicin (Zavedos) Idarubicin is a semisynthetic analogue of daunorubicin with general properties similar to doxorubicin. It is the only anthracycline that can be given orally as well as intravenously. Idarubicin is used in advanced breast cancer after failure of first-line chemotherapy (not including anthracyclines), and also in acute nonlymphocytic leukaemia.

Naturally Occurring Anthraquinone Antitumor Antibiotics

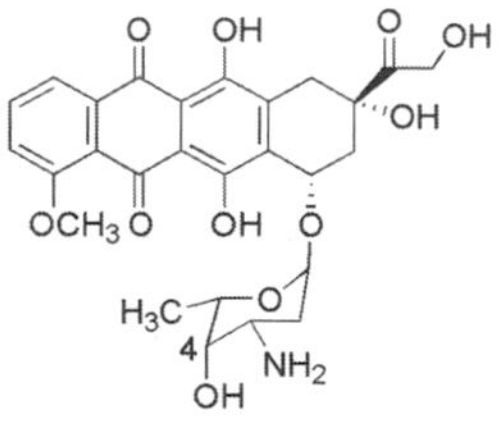

Doxorubicin (Caelyx or Myocet)

(14.32)

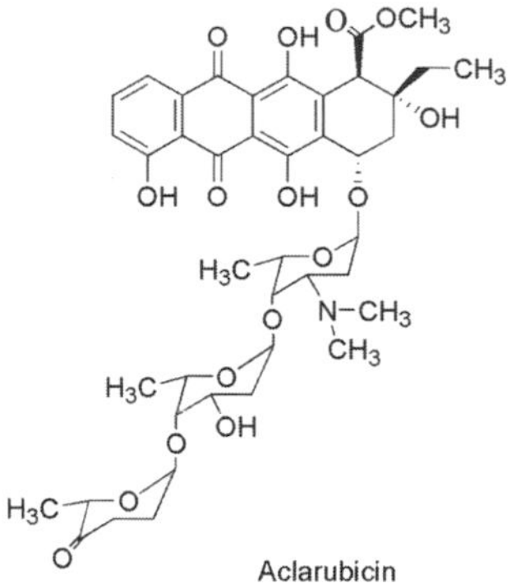

Aclarubicin

(14.33)

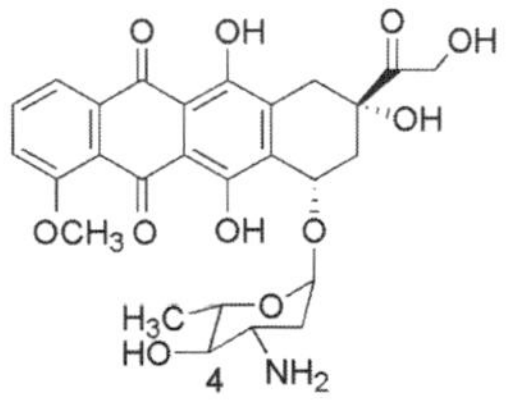

Epirubicin (Pharmorubin)

(14.34)

Semi-synthetic anthraquinone antitumor antibiotics

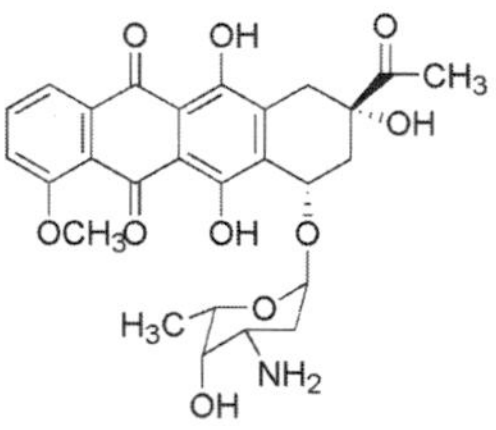

Daunorubicin (DaunoXome)

(14.35)

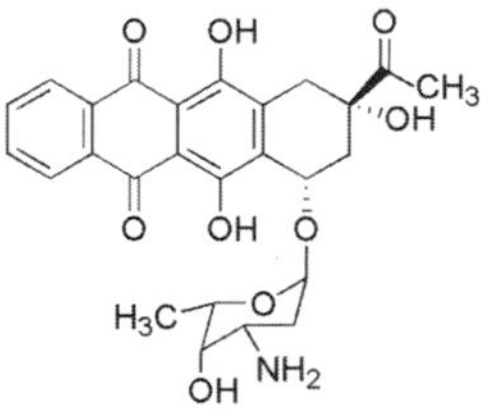

Idarubicin (Zavedos)

(14.36)

Anthracenes

These compounds are based on the anthracene nucleus and have three fused rings rather than the four present in the anthracyclines. Mitoxantrone (also known as mitozantrone) is the most important anthracene agent used in the UK.

Mitoxantrone (Novantrone or Onkotrone)

Mitoxantrone (mitozantrone) is rich in oxygen- and nitrogen-containing substituents and side chains, and these allow extensive stabilization of the intercalated adduct by hydrogen bonding interactions. The molecule is known to have a preference for binding to GC-rich sequences and, like the anthracyclines, there is evidence that DNA is cleaved although the mechanism is not thought to be linked to the generation of reactive oxygen species.

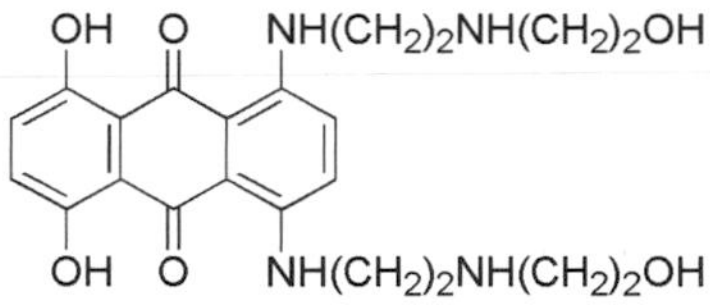

Mitoxantrone (Novantrone or Onkotrone)
(14.37)

Mitoxantrone is used for metastatic breast cancer and is also licensed for use in the treatment of non-Hodgkin's lymphoma and adult non-lymphocytic leukaemia. It is given intravenously and is generally well tolerated; however, myelosuppression and dose-related cardiotoxicity are major side effects. Although the cardiotoxicity is less prominent than with the anthracyclines, cardiac examinations and monitoring are still recommended after a certain cumulative dose.

Phenoxazines

Members of the phenoxazine family are similar to the anthracenes in containing three fused six-membered rings but differ in that the central ring contains oxygen and nitrogen heteroatoms. The best known member is dactinomycin that contains two cyclic peptide side chains which stabilize the drug–DNA adduct by interacting in the minor groove of DNA.

Dactinomycin (Cosmegen Lyovac)

Dactinomycin, known as a chromopeptide antibiotic, was isolated from *S. parvulus* in the 1950s and developed initially as a potent bacteriostatic agent although it was found to be too toxic for general use. The antitumor activity of dactinomycin was not observed until 10 years later when it was tried with great success in the treatment of Wilm's tumor (a kidney tumor in children) and a type of uterine cancer.

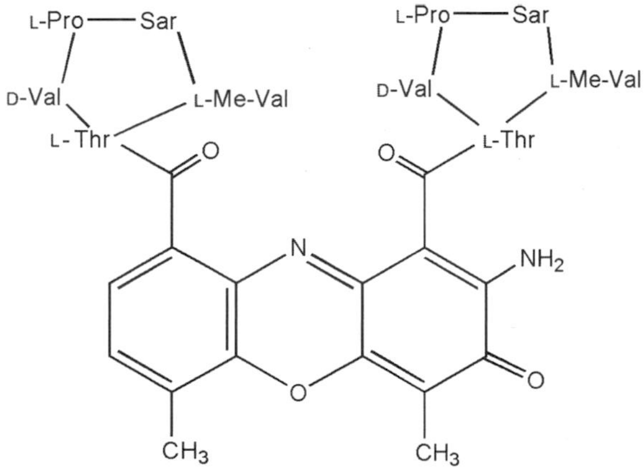

Dactinomycin (Cosmegen Lyovac)
(14.38)

The molecule consists of a tricyclic phenoxazine-3-one chromophore with two identical cyclic pentapeptide side chains. The mechanism of action appears to be dependent on its concentration, with either blockade of DNA synthesis occurring or inhibition of DNA-directed RNA synthesis thus preventing chain elongation. X-ray crystallography studies have shown that the phenoxazone ring intercalates preferentially between GC base pairs where it can interact with the N2-amino groups. The cyclic peptide moieties position themselves in the minor groove and participate in extensive hydrogen bonding and hydrophobic interactions with functional groups in the floor and walls of the groove, thus providing significant stabilization of the adduct. It is the stability of the adduct that is important for the blockage of RNA polymerase. Adduct formation also leads to single-strand DNA breaks in a similar manner to doxorubicin, either through radical formation or by interaction with topoisomerase.

Given intravenously, dactinomycin is principally used to treat pediatric cancers. For example, it is used in combination with vincristine in the treatment of Wilm's tumor, and combined with methotrexate for gestational choriocarcinoma. It is also used in some testicular sarcomas and in AIDS-related Kaposi's sarcoma. The side effects of dactinomycin are similar to those of doxorubicin except that cardiac toxicity is less prominent. Tumor resistance to dactinomycin is common, and this is thought to be due to both reduced uptake and active transport of the drug out of the tumor cells.

14.8.4 Topoisomerase Inhibitors

DNA topoisomerases are a family of enzymes responsible for the cleavage, annealing, and topological state (e.g., supercoiling) of DNA. In particular, these enzymes are responsible for the winding and unwinding of the supercoiled DNA composing the chromosomes. If the chromosomes cannot be unwound, transcription of the DNA message cannot occur and the corresponding protein cannot be synthesized, ultimately leading to cell death.

There are two types of topoisomerase enzymes, known as I and II. Topoisomerase I enzymes are capable of removing negative supercoils in DNA without leaving damaging nicks. They work by breaking only one strand of DNA, and attaching the free phosphate residue of the broken strand to a tyrosine residue on the enzyme. The complex then rotates, relieving the supercoiled tension of the DNA, and the two ends are re-sealed. Topoisomerase II enzymes cleave both strands of double-stranded DNA simultaneously, passing a complete duplex strand through the cut, followed by re-sealing of both strands. The mode of action of intercalators such as the anthracyclines, anthracenes, and acridines (e.g., Amsacrine) is thought to be partly associated with inhibition of these enzymes. However, other families of agents such as the camptothecins and ellipticines have been discovered that specifically bind to and inhibit topoisomerase enzymes.

Topoisomerase I inhibitors

Topoisomerase I (Topo I) is an essential enzyme controlling DNA topology that works by transiently breaking one DNA duplex and passing the second strand through the break followed by re-sealing. This is an absolute requirement of many nuclear processes including replication, transcription, and recombination. Topo I inhibitors work by keeping the chromosomes wound tight, and so the cell cannot make proteins which leads to cell death. Because some cancer cells grow and reproduce at a much faster rate than normal cells, they are thought to be more vulnerable to topoisomerase inhibition than normal cells, perhaps explaining the selective toxicity.

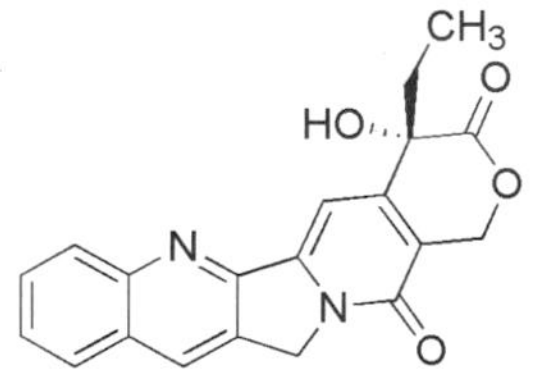

Camptothecin

(14.39)

The lead structure for topo I inhibitors is the natural product camptothecin, which is a cytotoxic quinoline-based alkaloid with a unique five-ring system extracted from the barks of the Chinese *camptotheca* tree (e.g., *Camptotheca acuminate*) and the Asian *nothapodytes* tree. Camptothecin and its derivatives are unique in their ability to inhibit DNA topoisomerase by stabilizing a covalent reaction intermediate termed the ''cleavable complex'' which ultimately leads to tumor cell death.

The use of camptothecin in the clinic was found to be limited due to serious side effects and poor water solubility. However, a number of derivatives of camptothecin were produced to address these issues and are now used in the treatment of breast and colon cancers, malignant melanoma, small-cell lung cancer, and leukemia. Topotecan (Hycamtin) was approved by the FDA in 1996 for the treatment of advanced ovarian cancers resistant to other chemotherapeutic agents. The injectable irinotecan HCl (Camptosar) was also approved in 1996 as a treatment for metastatic cancer of the colon or rectum, although it is normally only prescribed in colorectal cancer cases that have not responded to the standard treatment with fluorouracil.

The major side effects of camptothecin-based drugs include severe diarrhoea, nausea, and lowered leukocyte (white blood cell) counts. The drug may also damage bone marrow.

Topotecan (Hycamtin)

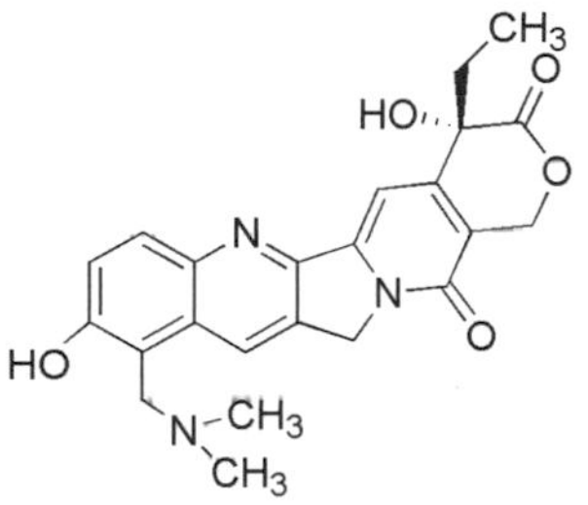

Topotecan (Hycamtin)
(14.40)

The solubility problems of camptothecin were solved by the addition of hydroxyl and dimethylaminomethyl substituents to the benzenoid ring to give the derivative topotecan. This analogue has the same mechanism of action as camptothecin in inhibiting the action of topoisomerase I. It binds to the topoisomerase I–DNA complex and prevents re-ligation of the DNA strand, resulting in double-strand DNA breakage and cell death. Unlike irinotecan, topotecan is found predominantly in the carboxylate form at neutral pH (i.e., resulting from opening of the lactone ring). As a result, topotecan has different antitumor and toxicity profiles compared to irinotecan. It is cell cycle phase-specific (S-phase), and is also a radiation-sensitizing agent.

Topotecan is given by intravenous infusion for the treatment of metastatic ovarian cancer when first-line or subsequent therapy has failed. In addition to dose-limiting myelosuppression, side effects of topotecan include GI disturbances including delayed diarrhoea, asthenia, alopecia and anorexia.

Irinotecan (Campto)

Irinotecan represents another attempt to improve the water solubility of camptothecin by the addition of a di-piperidine carbamate functionality to the benzenoid ring. It is known that irinotecan is hydrolyzed *in vivo* to the active metabolite, 7-ethyl-10-hydroxycamptothecin (SN-38), so in this sense it is a prodrug. Both irinotecan and SN-38 bind to the topoisomerase I–DNA complex and prevent re-ligation of the DNA strand, resulting in double-strand DNA breakage and cell death.

However, the precise contribution of SN-38 to the activity of irinotecan in humans is not known. Like topotecan, irinotecan is cell cycle phase-specific (S-phase).

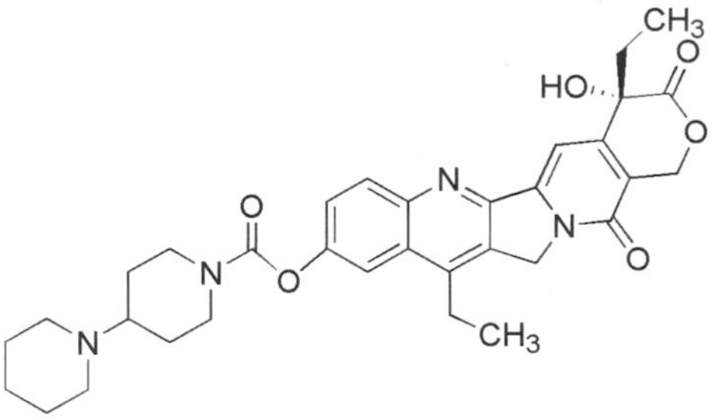

Irinotecan (Campto)
(14.41)

It is licenced for metastatic colorectal cancer in combination with fluorouracil and folinic acid or as monotherapy when treatment containing fluorouracil has failed. Irinotecan is given by intravenous infusion, and the side effects are the same as for topotecan.

Topoisomerase II inhibitors

Human cells express two genetically distinct isoforms of topo II known as topo IIα and topo IIβ. The lead structure for drugs such as etoposide and teniposide that inhibit these enzymes is podophyllotoxin, a plant alkaloid isolated from the American mandrake rhizome.

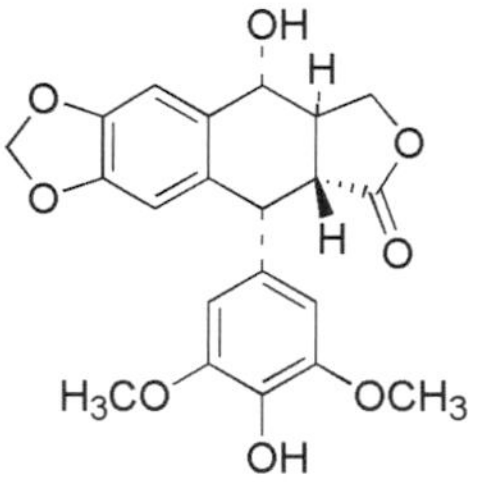

Podophyllotoxin
(14.42)

Other semisynthetic derivatives, such as podophyllic acid ethyl hydrazide, have been synthesized and studied, but none have been commercialized. As a class, topo II inhibitors appear to have activity in cancers including leukemias and lymphomas, oat cell carcinoma of the bronchus, malignant teratomas, and testicular cancer.

Etoposide (Etopophos or Vepesid)

Etoposide is a semisynthetic glucoside of epipodophyllotoxin, and is used clinically to treat small cell bronchial carcinoma for which it is claimed to be one of the most effective compounds known. Etoposide works by inhibiting the ability of topo IIα and IIβ to re-seal cleaved DNA duplexes. Therefore, normally reversible DNA strand breaks are converted into lethal breaks by processes such as transcription and replication.

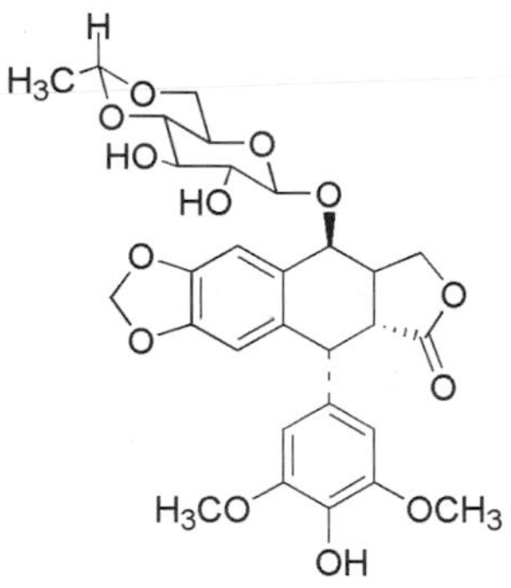

Etoposide (Etopophos or Vepesid)
(14.43)

Etoposide may be given orally or by slow intravenous infusion, the oral dose being double the intravenous dose. A preparation containing etoposide phosphate can be given by intravenous injection or infusion. It is usually given daily for 3–5 days, but courses should not be repeated more frequently than at intervals of 21 days. Etoposide has clinical activity in small cell carcinoma of the bronchus, the lymphomas, and testicular cancer. Toxic effects include alopecia, myelosuppression, nausea, and vomiting.

Teniposide (Vumon)

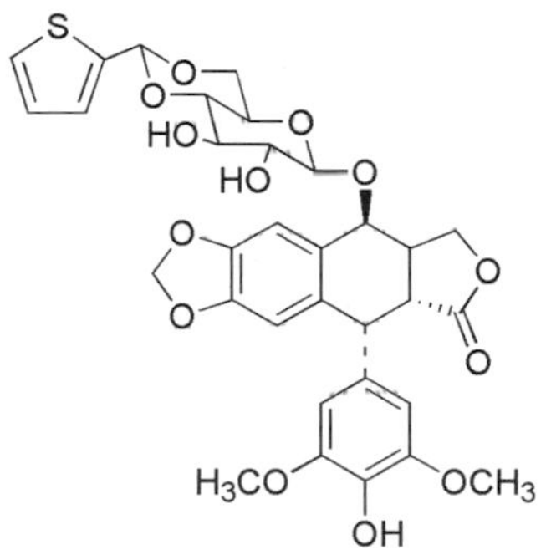

Teniposide (Vumon)
(14.44)

Teniposide is an analogue of etoposide that differs in structure only in that the methyl substituent on the sugar moiety is exchanged for a thiophene ring. It has broadly similar clinical activity to etoposide.

Ellipticine

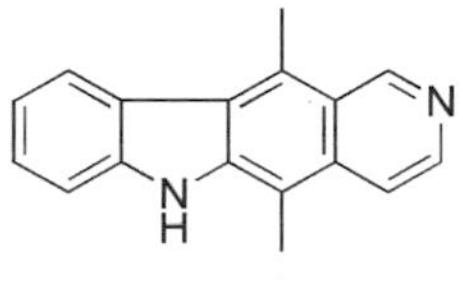

Ellipticine
(14.45)

The experimental agent ellipticine is a plant alkaloid isolated from *Ochrosia elliptica* that exerts its antitumor action through intercalation with DNA and inhibition of the topoisomerase II enzyme. In *in vitro* cytotoxicity studies, ellipticine has significant activity in nasopharyngeal carcinoma cell lines. Despite much research on the ellipticines, no successful clinical candidates based on the ellipticine nucleus have yet emerged.

Amsacrine

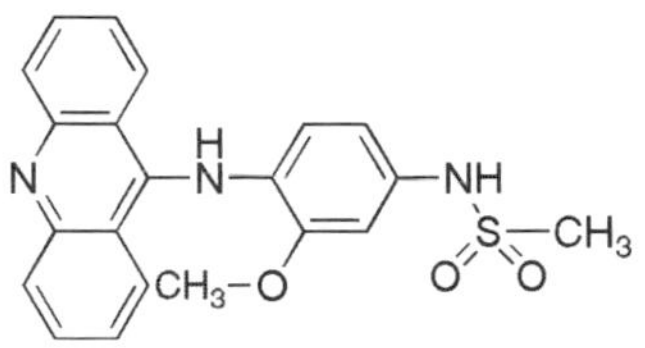

Amsacrine (Amsidine)
(14.46)

Amsacrine is an acridine derivative that was first synthesized in 1974 and entered clinical trials in 1977. The mechanism of action is not completely understood, but it is known to intercalate with DNA, inhibit topoisomerase II, and cause double-strand DNA breaks. Cytotoxicity is greatest during the S phase of the cell cycle when topoisomerase levels are at a maximum. Clinically, amsacrine has activity and toxicity profiles similar to doxorubicin. It is administered intravenously and is used in the treatment of advanced ovarian carcinomas, myelogenous leukemias, and lymphomas. Side effects include myelosuppression and mucositis. Electrolytes are normally monitored during treatment as fatal arrhythmias have occurred in association with hypokalaemia.

14.8.5 DNA-Cleaving Agents

These agents work by binding to the DNA helix, usually in a sequence-selective manner, and then by cleaving the double strand through radical production. The best known example of this class is the natural product bleomycin which is used clinically. The experimental enediyne agent calicheamicin has proved too toxic to be used in its own right as an antitumor agent, however it has been successfully used as the cytotoxic ''warhead'' in the antibody–drug conjugate gemtuzumab ozogamicin (Mylotarg).

Bleomycins

The bleomycins are a group of closely related natural products that exert their antitumor activity by binding to DNA in a sequence-selective manner followed by strand cleavage. The preparation known as bleomycin sulfate consists of a mixture of the glycopeptide bases (e.g., A_2, A_2I, B_{1-4}, etc.) with A2 as the predominant component. The mixture is obtained from *Streptomyces verticillus*, and the individual molecular weights are in the region of 1300. Despite the size and complexity of the molecule, particularly with regard to the number of chiral centres, the first total synthesis of bleomycin was reported in 1982. The bleomycins tend to accumulate in squamous cells and are therefore suitable for treatment of tumors of the head, neck, and genitalia, although it has also been used in Hodgkin's disease and testicular carcinomas. Unlike most anticancer drugs, it is only slightly myelosuppressive, and dose-limiting toxicity is confined to the skin, mucosa, and lungs.

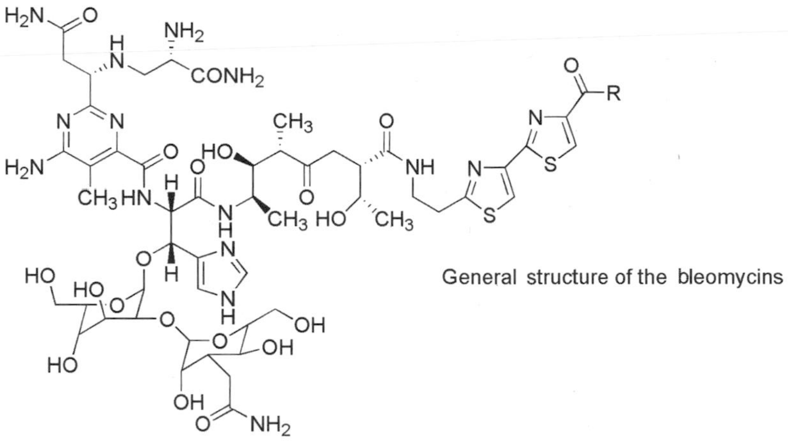

General structure of the bleomycins

(14.47)

Within the bleomycin molecule there are three distinct regions which are believed to contribute to its mechanism of action. First, the heterocyclic bithiazole moiety (top right as drawn in 14.47) is thought to interact with DNA. Electrostatic attraction of the highly basic sulfonium ion (A_2, R = $NH(CH_2)_3S^+Me_2$) to the phosphate residues in DNA stabilizes the adduct. Once bound, the second domain (top left) which consists of a β-hydroxyhistidine, a β-aminoalanine, and a pyrimidine forms an iron (II) complex which interacts with oxygen to generate free radicals, leading to single- and double-strand breaks. Currently, it is not clear whether the activation of this complex is self-initiating or the result of enzyme catalysis. The third region of bleomycin (bottom left) is glycopeptidic in nature and, while having no direct antitumor activity of its own, may contribute to drug uptake by tumor cells or provide additional stabilizing hydrogen bonding interactions with DNA or associated histone proteins.

Bleomycin is given intravenously or intramuscularly to treat metastatic germ cell cancer, squamous cell carcinoma, and, in some regimens, non-Hodgkin's lymphoma. Although it causes little bone marrow suppression, the most serious side effect is progressive pulmonary fibrosis which occurs in 5–15% of patients. This is dose related, and patients who have received extensive treatment with bleomycin may be at risk of respiratory failure under some conditions. Dermatological toxicity is also common, and increased pigmentation (particularly affecting the flexures) and subcutaneous sclerotic plaques may occur. As a result, ulceration in these areas and pigmentation of the nails may occur. Mucositis is also relatively common and an association with Raynaud's phenomenon has been reported. Hypersensitivity reactions manifesting as by chills and fevers commonly occur a few hours after administration and can be counteracted by intravenous hydrocortisone. Enzymes in most tissues rapidly deactivate the bleomycins, probably as a result of deamination or peptidase activity.

Enediynes

The enediynes are a family of approximately 15 to 20 antitumor antibiotics produced by *Micromonospora echinospora* ssp. *calichensis*, a bacterium isolated from chalky soil or caliche clay, a soil found in Texas. Naming of the individual compounds is based both on thin-layer chromatography (TLC) mobility using Greek letters with subscripts, and on the halogen substitution pattern which is indicted by the superscript. Some members of the enediyne family are also known as esperamycins.

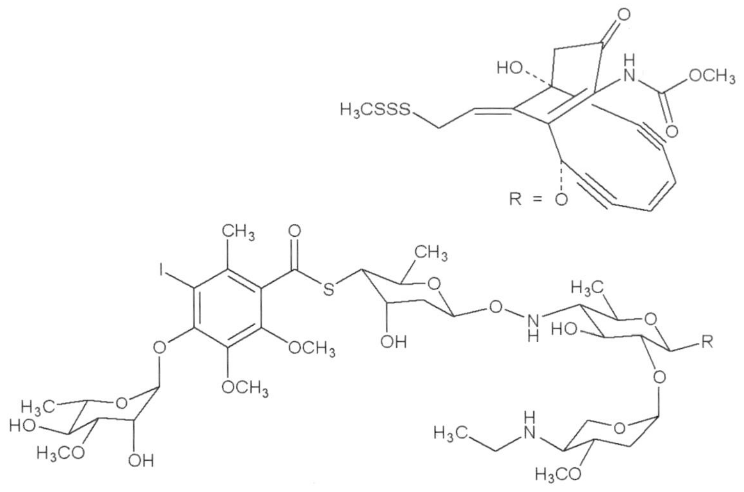

Calicheamicin γ_1
(14.48)

These molecules are unique in structure in containing two triple bonds in close proximity (i.e., separated by an ethylene unit) as part of a large ring. They bind in the minor groove of DNA and initiate double-stranded DNA cleavage via a radical abstraction process which involves a unique thiol-mediated cyclization (the Bergman cyclization). During this process, the triple bonds rearrange to form an aromatic ring which provides the driving force for the reaction. This causes a proton to be abstracted from a sugar in the DNA backbone thus leading to strand cleavage. This type of DNA damage is difficult for the cell to repair and thus explains the significant *in vitro* cytotoxicity of these compounds.

Although the enediynes are exquisitely potent *in vitro*, there does not appear to be any basis for tumor cell selectivity *in vivo*, and no analogues have so far progressed to the clinic. However, a calicheamicin analogue has been used as the cytotoxic ''warhead'' in the antibody–drug conjugate gemtuzumab ozogamicin (Mylotarg), and this is described in Section 14.13.1. The antibody portion of Mylotarg binds specifically to the CD33 antigen found on the surface of leukemic myeloblasts and immature normal cells of myelomonocytic lineage, and the calicheamicin then kills the cells by the mechanism described above.

14.9 ANTITUBULIN AGENTS

Plants and trees have been an invaluable source of medicinal compounds through the centuries, and this includes a number of anticancer agents such as the vinca alkaloids and the taxanes. These families of agents both work by interfering with spindle formation although by different mechanisms. Binding of the vinca alkaloids to the tubulins interferes with microtubule assembly, causing damage to the mitotic spindle apparatus and preventing chromosomes from traveling out to form daughter cell nuclei. This is similar to the mechanism of action of colchicine, but different from paclitaxel, which interferes with cell division by promoting the assembly of microtubules and inhibiting the tubulin disassembly process.

14.9.1 Vinca Alkaloids

In the late 1950s a screening program for agents with potential hypoglycemic properties led to the isolation of two alkaloids, vinblastine and vincristine, as minor constituents of the Madagascar periwinkle (*Vinca rosea*). These agents were shown to reduce white blood cell count, and have subsequently played an important role in cancer chemotherapy. Vinblastine, vincristine, and the related semisynthetic vindesine and vinorelbine are used to treat acute leukemias, lymphomas, and some solid tumors (notably breast and lung). All of these agents are given by intravenous administration, and side effects include neurotoxicity, myelosuppression, and alopecia.

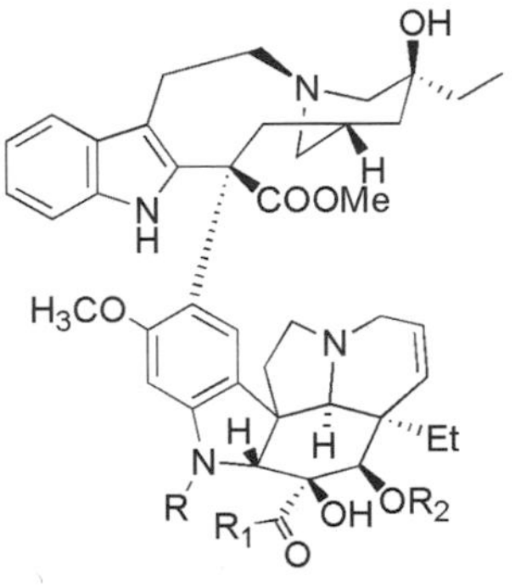

	R	R_1	R_2
Vinblastine (Velbe™)	Me	OMe	COMe
Vincristine (Oncovin™)	CHO	OMe	COMe
Vindesine (Eldesine™)	Me	NH_2	H

(14.49)

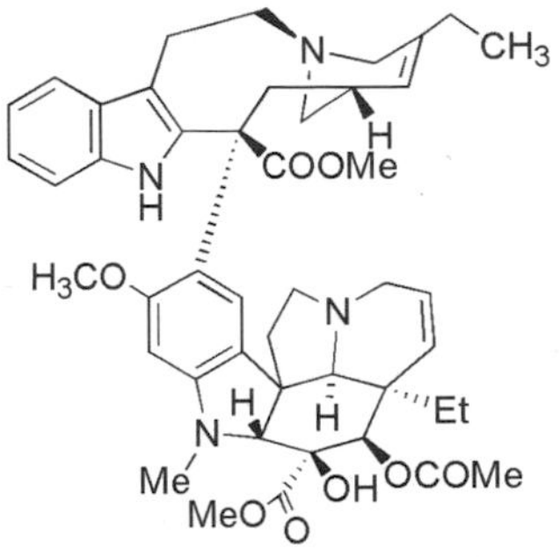

Vinorelbine (Navelbine™)

(14.50)

The naturally occurring vinca alkaloids have complex but similar chemical structures. Vincristine is more widely used than vinblastine but the plant produces the latter in approximately 100-fold greater quantity. Fortunately, vinblastine may be converted to vincristine by a simple chemical step involving oxidation of the methyl to a formyl group. Since these alkaloids have proved so useful in therapy, efforts have been directed towards the design of new analogues with reduced toxicity (generally neurotoxicity), and this has resulted in the semisynthetic analogues vindesine and vinorelbine. Research has also been carried out into cell culture techniques and the use of immobilized plant enzymes as a means to produce the alkaloids more efficiently. Compounds representing the two halves of the structure of the dimeric Vinca molecules occur in much higher proportions in the plant extract, and the possibility of linking these at the appropriate positions and with the correct stereochemistry has also become feasible.

The vinca alkaloids are cell-cycle-specific agents that block mitosis by metaphase arrest. Their cytotoxic effects result from binding to the microtubules. These structures were first characterized in the cytoplasm over 25 years ago and comprise two main proteins, the α and β tubulins (M_r ca. 55,000), which form the microtubule scaffolding upon which many of the dynamic internal processes in living cells, including cell division, depend. Microtubules are long tubular structures of about 25 nm in diameter which form the major component of the mitotic spindle apparatus responsible for the movement of chromosomes during cell division. Binding of the vinca alkaloids to the tubulins interferes with microtubule assembly, causing damage to the mitotic spindle apparatus and preventing chromosomes from traveling out to form daughter cell nuclei. This is similar to the action of colchicine, but is different from the action of paclitaxel, which interferes with cell division by keeping the spindles from being broken down. There is also some evidence that these alkaloids interfere with a tumor cell's ability to synthesize DNA and RNA. However, overall, the basis for the tumor cell selectivity of these compounds is not clear.

Neurological toxicity, usually manifested as peripheral or autonomic neuropathy, is a feature of treatment with all vinca alkaloids and is a dose-limiting side effect of vincristine. Interestingly, it occurs less often with vindesine, vinblastine, and vinorelbine despite their similar chemical structures. Conversely, myelosuppression is the dose-limiting side effect of vinblastine, vindesine, and vinorelbine whereas vincristine causes negligible myelosuppression. The vinca alkaloids commonly cause reversible alopecia. They also cause severe local irritation, and so extravasation must be avoided. Patients with neurological toxicity often experience peripheral paraethesia, loss of deep tendon reflexes, abdominal pain, and constipation. Motor weakness can also occur and is an indication that treatment should be discontinued. Generally, recovery of the nervous system is slow but complete.

Vinblastine (Velbe)

The isolation of vinblastine from *Vinca rosea* and its structural identification were reported in 1959 and 1960, respectively, and a synthesis starting from catharanthine and vindoline units was reported in 1979. As the sulfate salt, the agent is included on a weekly basis in several drug regimens for treating lymphocytic and histiocytic lymphomas, Hodgkin's disease, advanced disseminated breast carcinoma, choriocarcinoma, advanced testicular carcinoma, Kaposi's sarcoma, and Letterer–Siwe disease. There is some evidence that the antitumor activity of vinblastine may be due, in part, to interference with glutamic acid metabolism (specifically, the pathways leading from glutamic acid to the Krebs cycle and to urea formation).

Vincristine (Oncovin)

The isolation and structural elucidation of vincristine were reported in 1961 and 1964, respectively. Despite its similarity in structure to vinblastine, vincristine sulfate has a different spectrum of both antitumor activity and side effects. Notably it is more neurotoxic than vinblastine although it causes significantly less myelosuppression. Several drug combinations include vincristine for the treatment of acute lymphoblastic and myeloblastic leukemias, Hodgkin's disease, Wilm's tumor, rhabdomyosarcoma, neuroblastoma, retinoblastoma, STSs, and disseminated cancer of the breast, testes, ovaries, and cervix. The relatively low bone marrow toxicity renders it suitable for combination with drugs that cause greater bone marrow depression.

Vindesine (Eldisine)

Vindesine sulfate is a semisynthetic product derived from vinblastine, and its toxicity and side effects are therefore similar to those of the parent compound. It was first reported in 1974, and is mainly used to treat melanoma and lung cancers (carcinomas) and, in combination with other drugs, to treat uterine cancers.

Vinorelbine (Navelbine)

Vinorelbine was discovered serendipitously when it was observed that, when vinblastine is placed in strong acid conditions, dehydration of the tertiary alcohol group in the piperidine ring of the A subunit occurs to introduce a double bond. In the UK, vinorelbine is presently recommended as an option for the second-line or subsequent treatment of advanced breast cancer where anthracycline-based regimens have failed or are unsuitable. However, it is currently in Phase II clinical trials as a treatment for ovarian cancer and has shown some activity in combination with cisplatin for the treatment of non-small-cell lung cancers. The side effects include diarrhoea, nausea, and hair loss, although it appears to be significantly less neurotoxic than vindesine.

14.9.2 The Taxanes

Paclitaxel (Taxol) is a highly complex tetracyclic diterpene found in the bark and needles of the Pacific yew tree *Taxus brevifolia*. The cytotoxic nature of extracts of *Taxus brevifolia* was first demonstrated in 1964. Pure paclitaxel was isolated in 1966 and its structure published in 1971. However, it has only recently appeared in the clinic, over 30 years since its discovery. Docetaxel (Taxotere) is a more recently introduced semisynthetic analogue with similar therapeutic and toxicological properties.

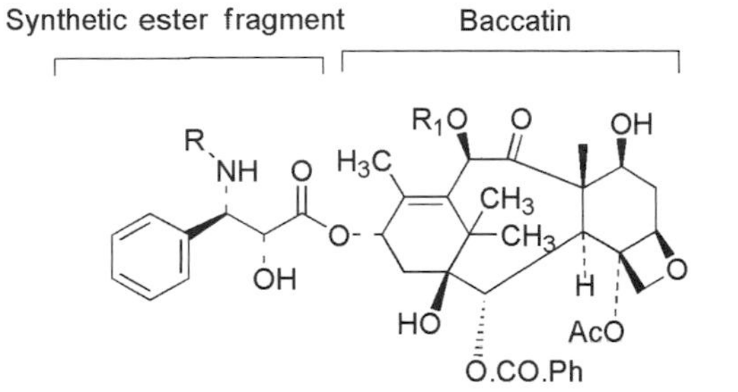

Synthetic ester fragment Baccatin

	R	**R_1**
Paclitaxel (Taxol)	CO.Ph	$CO.CH_3$
Docetaxel (Taxotere)	$CO.O.C(CH_3)_3$	H

(14.51)

Like the vinca alkaloids, the mechanism of action of these agents involves the microtubules but in a different way. In the cell there is an equilibrium between the microtubules and tubulin dimers, and the assembly and disassembly of dimers is governed by cell requirements. Paclitaxel is thought to promote microtubule assembly and inhibit their depolymerization to free tubulin, thus shifting the equilibrium in favor of the polymeric form of tubulin and reducing the critical concentration of the non-polymerized form by stabilizing the microtubule complex. This interferes with the ability of the chromosomes to separate during cell division. There are also some reports of paclitaxel acting as an immunomodulator and activating macrophages to produce IL-1 and tumor necrosis factor (TNF).

The barks from a large number of yew trees would be required to provide a single course of treatment of paclitaxel for an ovarian cancer patient, and the problems associated with producing sufficient quantities of drug account, in part, for its delay in being introduced to the clinic. Fortunately, semisynthesis is now possible by extracting baccatin (paclitaxel without the ester fragment) in large amounts from the leaves (a renewable resource) of a related species, *Taxus baccata*. The ester side chain can be made synthetically and then joined to baccatin to provide paclitaxel. Similarly, docetaxel can be prepared from the natural precursor 10-deacetylbaccatin III (i.e., $R_1 = H$ instead of $COCH_3$) extracted from the needles of the European yew tree, *Taxas baccata*, through attachment of the synthesized ester fragment. Although both of these agents are produced commercially by semisynthesis, the first total synthesis of paclitaxel was reported in early 1994. One of the reasons for pursuing new analogues is that paclitaxel has relatively poor water solubility.

Paclitaxel (Taxol)

Given by intravenous infusion, there is increasing evidence that paclitaxel in combination with cisplatin or carboplatin is the treatment of choice for ovarian cancer. This combination is also used for women whose ovarian cancer is metastatic and initially considered inoperable, and where standard platinum therapies have failed. It is also used for metastatic breast cancer where standard anthracycline therapy has failed, and for non-small cell lung cancer (NSCLC) when no further treatment options including surgery or radiotherapy remain. Routine pre-medication with a corticosteroid, an antihistamine and a histamine H2-receptor antagonist is recommended to prevent severe hypersensitivity reactions. More commonly, only bradycardia or asymptomatic hypotension occur. Other side effects include myelosuppression, peripheral neuropathy, and cardiac conduction defects with arrhythmias. Paclitaxel can also cause alopecia and muscle pain. Nausea and vomiting is mild to moderate.

Docetaxel (Taxotere)

Docetaxel is licensed for use in combination with doxorubicin for initial chemotherapy of advanced breast cancer, or alone where adjuvant cytotoxic chemotherapy has failed. Docetaxel is also used in the treatment of advanced or metastatic NSCLC where first-line therapy has failed. Its side effects are similar to paclitaxel but persistent fluid retention (generally leg oedema) can be resistant to treatment. Hypersensitivity reactions also occur, and prophylactic dexamethasone is recommended to reduce the fluid retention and possible hypersensitivity reactions.

14.10 SIGNAL INHIBITION

Conventional anticancer drugs such as alkylating agents, antimetabolites, topoisomerase inhibitors, and antimicrotubule agents have been traditionally focused on targeting DNA synthesis and cell division. Although these drugs show efficacy, their lack of selectivity for tumor cells over normal cells usually leads to severe adverse effects such as bone marrow suppression and cardiac, hepatic, and renal toxicities which limits their use. In an attempt to circumvent these unpleasant side effects, a new class of anticancer agents known as signal transduction or secondary message inhibitors has been developed. Cells use a wide variety of both intra- and intercellular mechanisms to signal for processes including growth, apoptosis, and intracellular protein degradation. Due to up-regulation or greater dependence on some of these pathways in tumor cells, inhibition should lead to an anticancer effect. Research in the protein kinase area has recently led to Glivec which is considered a landmark drug for the treatment of leukaemia. A second protein kinase inhibitor, Herceptin, has also been successfully commercialized. Although not yet commercially available, inhibitors of the D-type cyclins and their kinase partners such as CDK4 and CDK6 are also being developed and are described below. Finally, the ubiquitin–proteasome pathway is the principal pathway for intracellular protein degradation and Velcade has been developed to interfere with this pathway.

14.10.1 Kinase Inhibitors

These agents act by inhibiting the actions of protein kinases that modulate the signaling systems necessary for cellular division, growth, and migration. Two such agents, Gleevec and Herceptin, have led the field in this area, and many new agents of this type are poised for commercialization.

Classification of protein kinases

Protein kinases are enzymes within cells that act by attaching phosphate groups to amino acid residues. This process of phosphorylation serves two primary roles; as a molecular on–off switch to trigger a cascade of cellular events and as a connector that binds proteins to each other. Protein kinases play a primary role in the complex signaling system that transfers information between and within cells. They are generally classified into two groups:

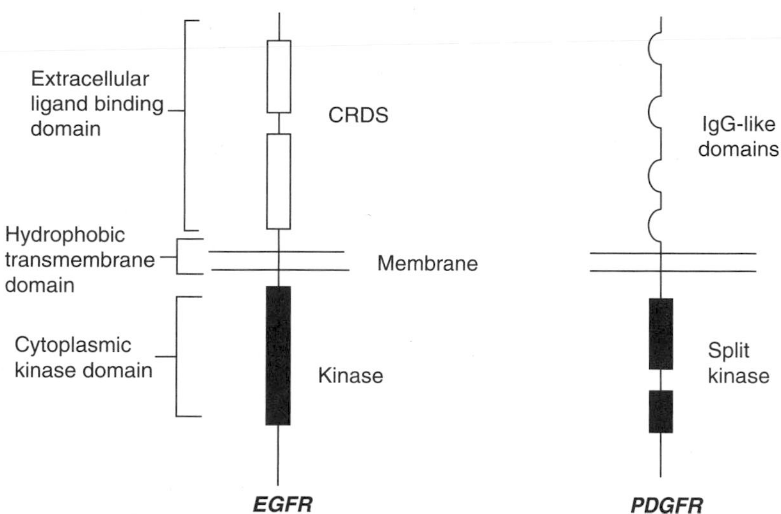

Figure 14.4 Schematic structures of receptor kinases EGFR and PDGFR.

A. *Based on specificity for target amino acids*:

- The *serine* or *threonine specific kinases* — catalyzing the phosphorylation of serine and threonine residues
- The *tyrosine-specific kinases* — catalyzing the phosphorylation of tyrosine residues
- The *mixed function kinases* — catalyzing both serine/threonine and tyrosine phosphorylation

B. *Based on structure and cellular localization:*

- *Receptor kinases* — defined by a hydrophobic transmembrane domain which passes through the plasma membrane, an extracellular ligand-binding domain, and a cytoplasmically-oriented kinase domain (Figure 14.4). The extracellular ligand-binding domain is typically glycosylated and conveys ligand specificity. Examples include epidermal growth factor receptor (EGFR), platelet-derived growth factor receptor (PDGFR), transforming growth factor (TGF)-β receptor, and ErbB.
- *Non-receptor kinase* — these have no transmembrane or extracellular domains unlike the receptor kinases and may be associated with the cytoplasmic surfaces by membrane localization via a lipid modification which anchors them to the phospholipid bilayer or by non-covalent binding to a membrane receptor. Examples include *ABL*, JAK, FAK, and SRC kinase.

Functions of protein kinase

The function of protein kinases is to provide a mechanism for transmitting information from a factor outside a cell to the interior of the cell without a requirement for the initiating factor to cross the cell membrane. For example, growth factors or polypeptide hormones in the extracellular milieu exert a regulatory effect on cells by activating specific gene transcription in the nucleus of a target cell without passing through the cytoplasmic membrane. By this mechanism, protein kinases help regulate cellular functions such as proliferation, cell–matrix adhesion, cell–cell adhesion, movement, apoptosis control, transcription, and membrane transport.

Mechanism of signal transfer

The transfer of a signal from an extracellular factor through the membrane via a kinase usually occurs by one of two mechanisms. In the first, the receptor kinase is solely responsible for the

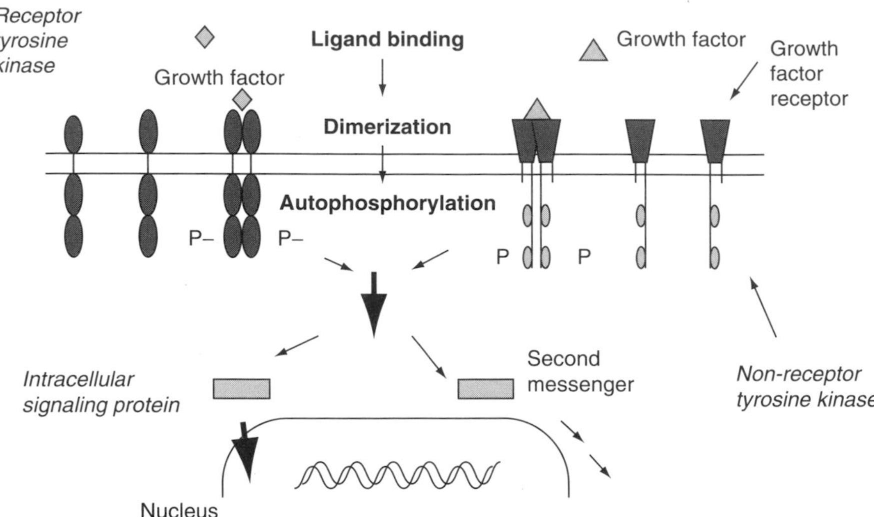

Figure 14.5 Signaling through tyrosine kinase.

transfer of signals across the membrane. Ligand binding to a receptor kinase induces receptor dimerization or oligomerization and this leads to interactions between adjacent cytoplasmic domains with consequent activation of the kinase (Figure 14.5). Activation of a non-receptor kinase is similarly induced in response to the appropriate extracellular signal, however, dimerization may or may not be necessary for activation. In either case it is the activation of the cytoplasmic kinase domain that is the key step in transferring the signal across the membrane. The activated kinase then initiates a cascade of phosphorylation reactions resulting in the activation of other proteins, as well as the production of second messenger molecules leading to the transmission of the signal initiated by the extracellular factor to the nucleus.

Regulation of kinase activity

In order to attempt to control the cellular effects of kinase activation such as proliferation, a number of mechanisms of regulation of kinase signaling might be envisaged. First, use could be made of the opposing effects of protein kinases. For example, activation of TGF-β receptor signal transmission can modulate cell cycle arrest. Second, activation of protein phosphatases that remove phosphates from kinase residues could switch off kinase signaling. Third, reducing the activity of protease enzymes involved in the formation of ligands or the activation of specific enzymes that degrade the ligands could modify signaling. Finally, the availability of protein kinases and their relative localization within cells could be manipulated.

Role of protein kinase in cancer

Protein kinases can play major etiologic roles in the initiation of malignancy and may contribute to the uncontrolled proliferation of cancer cells, tumor progression, and development of metastatic disease. It is thought that cancer cells depend almost entirely on signaling by protein kinases for their continued proliferation as compared to normal cells which rarely invoke these pathways. Protein kinases can be affected in two main ways in tumor cells:

a. *Mutations of protein kinases*: Several studies have implicated a mutation of the protein kinase ABL (i.e., BCR-ABL) as the etiologic agent in CML. BCR/ABL is a constitutively active cytoplasmic tyrosine kinase present in virtually all patients with CML and 15–30% of adult patients with acute lymphoblastic leukemia. Mutations in RET tyrosine kinase have been observed in several multiple endocrine neoplasia (Type 2) patients and are presumed to be responsible for the development of the disease. Similarly, EGFR mutations with enhanced kinase activity have been detected in several human tumor types.

b. *Over-expression of protein kinases*: Expression of EGFR, and its primary ligands EGF and TGFα, have been studied in several human malignancies with co-expression of the EGFR and EGF observed to have both a prognostic significance and a possible role in the pathogenesis of several human cancers. It has been established that over-expression of EGFR and EGF in several tumor types significantly reduces patient prognosis. Members of the EGFR kinase family (EGFR, ErbB-2, HER-2/NEU, ErbB-3, and ErbB-4) are reported to be overexpressed in some types of breast tumors. HER2/neu receptor tyrosine kinase is amplified up to 100-fold in tumor cells of approximately 30% of patients with invasive breast cancer and its presence is associated with poor prognosis. Over-expression of PDGF and PDGFR have been reported in meningioma, melanoma, and neuroendrocrine cancers, and tumors of the ovary, pancreas, stomach, lung, and prostrate. Elevated levels of SRC kinase activity have been noted in colon cancer specimens implying that over-expression may be important in this disease.

Development of inhibitors of receptor and protein kinases

There is presently significant research activity in the design of small molecule inhibitors of the various kinases thought to be important in the initiation and promotion of tumors. As protein kinases use ATP as a source of phosphate, one of the main approaches to date has been to design inhibitors that bind to the ATP-binding site of the protein. This makes it impossible for the kinase to phosphorylate proteins and thus signaling halts. This approach has led to Glivec. For receptor kinases, one of the main approaches has been the development of antibodies that bind to the extracellular portion of the kinase thus blocking its function. This approach has led to Herceptin. Many other kinase inhibitors are presently in development.

HER2/neu inhibition: Trastuzumab (Herceptin)

Trastuzumab is a recombinant humanized anti-P185 monoclonal antibody targeted against HER2/neu receptors. It is a potent inhibitor of the HER2/neu signaling pathway and was developed for use in breast cancer. It is believed to induce receptor internalization and inhibition of cell cycle progression through mechanisms that include up-regulation of p27, an intracellular inhibitor of cyclin-dependent kinase 2 (CDK2). Several clinical trials have shown an antitumor response. For example, one trial has shown that standard chemotherapy plus trastuzumab compared to standard chemotherapy alone gives a longer median time-to-disease progression (7.4 months vs 4.6 months $P<0.001$), a significantly greater rate of overall tumor response (50% vs 32%, $P<0.001$, NNT 6), a longer median duration of response (9.1 months vs 6.1months, $P<0.001$), a longer median time to treatment failure (6.9 months vs 4.5 months, $P<0.001$) and a significantly lower rate of death at 1 year (22% vs 33% $P = 0.008$). The most common adverse events include fever, chills, pain, asthenia, nausea, vomiting, and headache.

Trastuzumab is licensed in the UK, in combination with paclitaxel, for metastatic breast cancer in patients with tumors over-expressing HER2 who have not received chemotherapy for metastatic breast cancer and in whom anthracycline treatment is inappropriate. It is also licensed as monotherapy for metastatic breast cancer in patients with tumors that over-express HER2 who have received at least two chemotherapy regimens including, where appropriate, an anthracycline and a taxane. Women with estrogen-receptor-positive breast cancer should also have received hormonal therapy before being given trastuzumab.

The side effects of trastuzumab, which is given by intravenous infusion, include chills, fever, and possible hypersensitivity reactions such as anaphylaxis, urticaria, and angiodema. Cardiotoxic, pulmonary events, and GI disturbances can also occur. Concomitant use of trastuzumab with anthracyclines is associated with cardiotoxicity and should be avoided. Cardiac function should be carefully monitored if there is a need for co-administration.

Trastuzumab is the first example of an anticancer drug used in so-called ''personalized medicine'' regimes. A HER2 test kit has been developed and women must be screened for HER2 overexpression before trastuzumab can be administered. It is likely that new agents of this type will be co-developed with a test kit as there is a move towards personalized medicine regimes.

BCR/ABL inhibition: Imatinib (Glivec)

Imatinib (Glivec), also known as STI-571 or Gleevec in the USA, is a low-molecular-weight inhibitor of the tyrosine kinase activity of BCR/ABL as well as the ABL, PDGFR, and c-kit kinases. It acts by blocking the ATP-binding site of the kinases. Clinical trials have been carried out in patients in chronic phase Philadelphia-chromosome-positive CML refractory to interferon alfa. All patients treated with 140 mg/day dose or more had a hematological response, and 98% of patients treated at a dose of 300 mg/day or more had a complete hematological response which was maintained in 96% of these patients for a median follow-up of 265 days. Total cytogenetic response was observed in 54% of patients treated at doses of 300 mg/day or more with 13% having complete cytogenetic remissions.

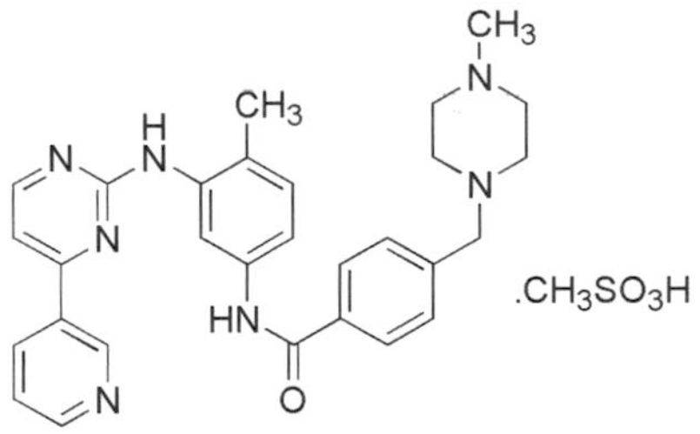

Imatinib mesylate (Glivec)
(14.52)

One of the main advantages of imatinib is that it is administered orally which is convenient for patients. It is now licensed in the UK for the treatment of Philadelphia-chromosome-positive chronic myeloid leukaemia in chronic phase after failure of interferon alfa, or in accelerated phase, or in blast crisis. It is also licensed for Kit-positive unresectable or metastatic malignant gastro-interstitial stromal tumors (GISTs). The most frequent side effects of imatinib are nausea, vomiting, diarrhoea, oedema, muscle pain (myalgia), and headache all of mild to moderate severity. Anemia, thrombocytopenia, neutropenia, and elevation of liver enzymes have also been reported in some patients but do not show any dose correlation.

EGFR inhibition: Gefitinib (Iressa), Erlotinib (Tarceva), and Cetuximab (Erbitux)

The EGFR receptor is part of a subfamily of four closely related receptors: EGFR (ErbB-1), Her 2/neu (ErbB-2), Her 3 (ErbB-3), and Her 4 (ErbB-4). Increased EGFR-mediated signaling can contribute to a cell moving into a state of continuous, uncontrolled cell division therefore expanding the population of malignant cells and increasing the tumor mass. The process by which this occurs begins when the receptors, which are inactive single units or monomers, become activated by the binding of the appropriate ligand. This causes the receptors to pair together forming a dimer. The dimers may be formed either from two identical receptors, for example the EGF-1 receptor (EGFR) can pair with another EGFR-1 receptor to form a homodimer; or from two non-identical receptors, for example an EGFR can pair with a different receptor such as Her2/neu, forming an asymmetrical heterodimer. The pairing of the receptors activates the tyrosine kinase enzyme located in the intracellular domain of the receptor. This causes both intracellular domains to become transphosphorylated which, in turn, initiates a cascade of events that eventually results in the signal reaching the nucleus.

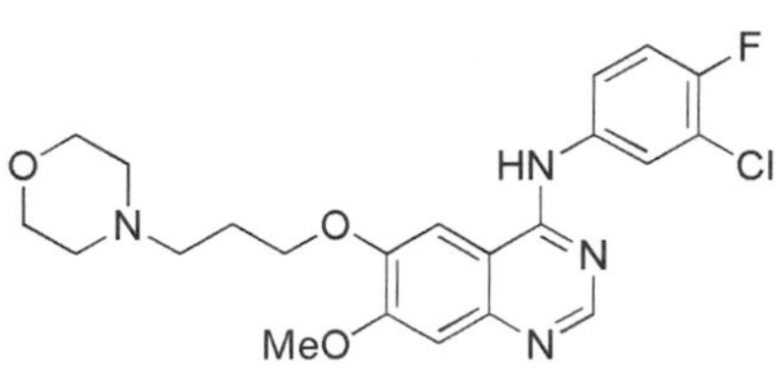

Gefitinib (Iressa)

(14.53)

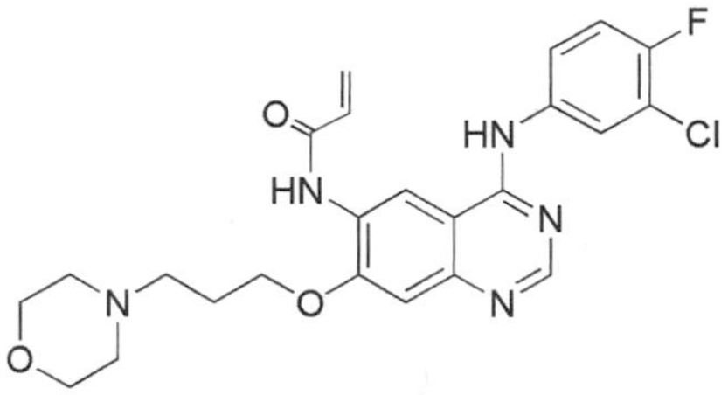

Canertinib (CI-1033)

(14.54)

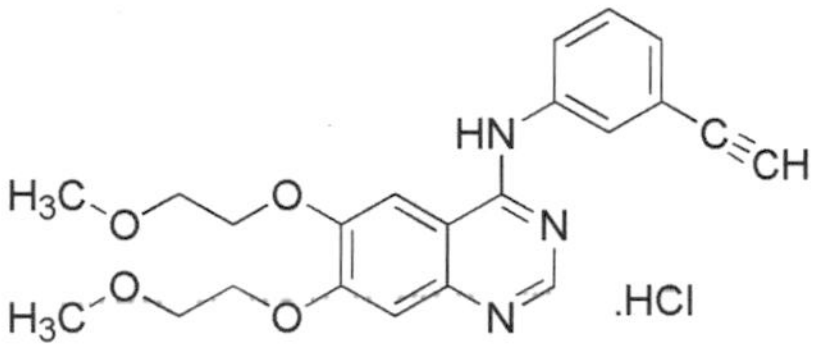

Erlotinib (Tarceva)

(14.55)

Gefitinib (Iressa) is the first in a new class of small molecule agents that target the EGFR receptor by blocking the ATP-binding site. Taken orally, it is used for the treatment of advanced NSCLC. At present there is no standard therapy for NSCLC once first-line therapy has failed. In clinical trials, gefitinib monotherapy has been shown to have significant activity in NSCLC with objective responses and disease stabilization in around 50% of patients. Gefitinib has also been shown to bring rapid symptom relief, in some cases within 8–10 days of starting treatment.

Erlotinib (Tarceva) has been granted orphan drug status by the FDA for patients with malignant glioma. It acts to block tumor cell growth by targeting human epidermal growth factor-1 (HER-1). It is the first HER1-inhibitor to receive orphan drug classification for brain cancer treatment. Other EGFR inhibitors being developed include Cetuximab (Erbitux) for colorectal cancer. Erbitux is a chimerized (part mouse, part human) monoclonal antibody that targets and inhibits epidermal growth factor receptor (EGFR), which is associated with tumor cell growth in a number of EGFR-positive solid malignant tumors. EGFR is over-expressed in more than 35% of all solid malignant tumors. Results from pre-clinical studies demonstrated that Erbitux acts by inhibiting angiogenesis associated with human transitional cell cancer, including bladder cancer. Other pre-clinical data indicated that Erbitux was effective in suppressing growth of malignant tumors, and that combined administration of Erbitux with radiation improved efficacy of local tumor irradiation.

VEGFR inhibition

VEGF and its receptors stand out as the growth factor consistently associated with angiogenesis. Inhibiting the function of VEGFR2 in endothelial cells has been shown to be sufficient to prevent tumor growth in experimental models. VEGFR inhibitors being developed include Semaxanib

(SU5416), a selective VEGFR2 inhibitor which completely blocks ATP binding to the kinase domain of the VEGFR2. Avastin is being developed by Genentech to bind to the VEGF receptor.

PDGFR inhibition

A large proportion of tumor cells express both PDGF and PGGFR. A number of experimental PDGFR inhibitors are presently under evaluation including CD-P860 and SU-101.

Other potential targets

Other potential targets for inhibition are the MET and SRC kinases and all other protein kinases implicated in cancer. The attraction of MET kinase is that the process of metastasis could be blocked thus allowing the primary tumor to be treated.

14.10.2 Cell Cycle Inhibitors

The D-type cyclins and their kinase partners CDK4 and CDK6 phosphorylate the tumor suppressor protein, pRb during the G1-phase of the cell cycle and contribute to its inactivation. They also act as integrators of extracellular signals. Cyclin D1 expression can be stimulated by the ras signaling pathway, and most human cancers contain mutations that affect either cyclins, CDK4, CDK6, their regulators or pRB.

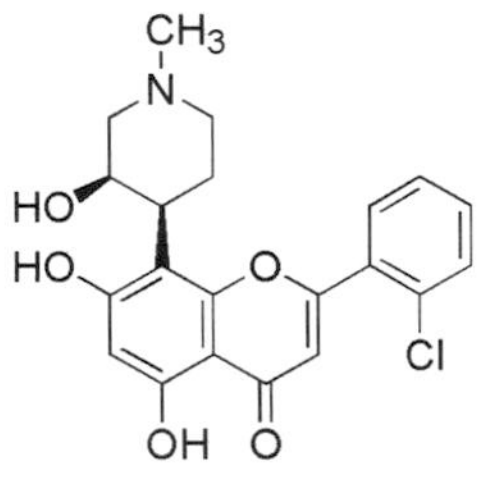

Flavopiridol
(14.56)

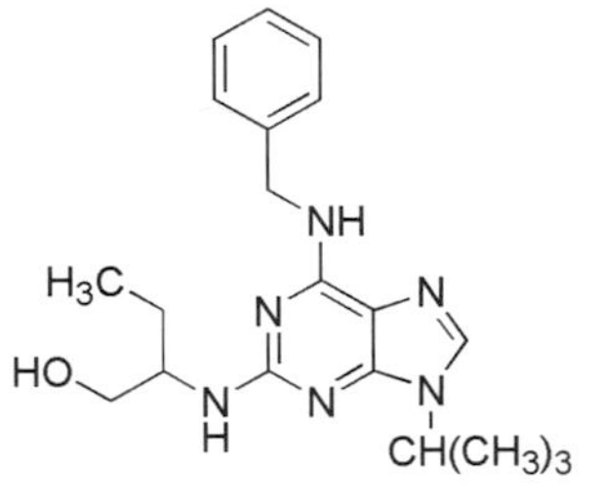

Roscovitine
(14.57)

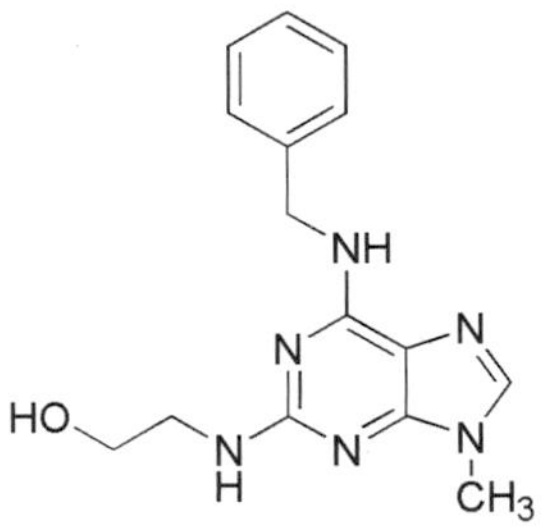

Olomucine
(14.58)

Flavopiridol, a flavone, is the prototype inhibitor of cyclin-dependent kinases and is capable of inducing apoptosis in some tumor cells. However, further development of this agent was terminated in early 2004 after poor results from intermediate-stage Phase II clinical trials conducted by Aventis. Several other CDK inhibitors have been described such as olomoucine (14.58) and roscovitine which induce apoptosis in several cell lines. The Paullones inhibit various CDKs including CDK1 (cyclin B), 2 (cyclin A), 2 (cyclin E), and 5 (p25) with different IC_{50} values (0.4, 0.68, 7, and 0.85 μM). Examination of these compounds and their chemical derivatives in the NCI drug screen has shown that there is generally a lack of correlation between CDK inhibitory and antitumor activity. Although one derivative, alsterpaullone (14.59), has shown a much higher activity against CDK1 which is also associated with a higher antitumor activity, the molecule has been discovered to potently inhibit glycogen synthase kinase-3β and CDK5/p25, thus demonstrating that this class of inhibitors also targets other proteins. These families have acted as leads for later generations of inhibitors, and one such compound, CYC202, has just completed phase I clinical evaluation.

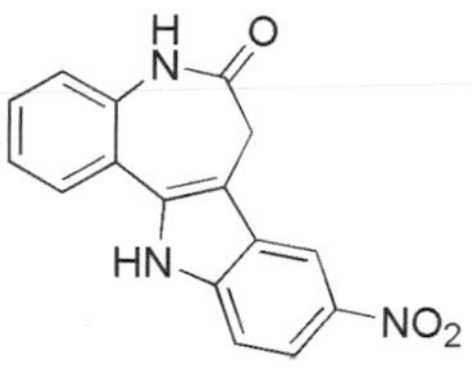

Alsterpaullone
(14.59)

One problem with agents of this class is that there is often a discrepancy between *in vitro* and *in vivo* data. For example, *in vitro* IC_{50} values for regulating CDK2 in the low nM range typically translate into *in vivo* IC_{50} values in the micromolar range for inhibition of progression of cells from G1- to S-phase. Although poor bioavailability is a likely explanation, it is also possible that the molecules do not inhibit CDK2 in the cell as they do *in vitro*. Most of the lead molecules under development bind at the ATP-binding site of kinases. However, the ATP concentration within the cell is between 5 and 15 mM while the ATP concentration in CDK2 assays is usually between 10 and 400 μM. Thus it is possible that inhibition of CDKs might be occurring by a different mechanism. Another possible use of CDK inhibitors is to protect cells from the lethal effect of cytotoxic drugs.

14.10.3 Proteasome Inhibitors

The ubiquitin–proteasome pathway is the principal pathway for intracellular protein degradation. Protein substrates are ''marked'' with a polyubiquitin chain and then degraded to peptides and free ubiquitin by a large multimeric protein known as the proteasome which exists within all eukaryotic cells. Numerous examples of regulatory proteins have been found to undergo ubiquitin-dependent proteolysis. Many of these proteins function as important regulators of physiological as well as pathophysiological cellular processes. Importantly, the ubiquitin pathway plays a significant role in neoplastic growth and metastasis. For example, the ordered and temporal degradation of numerous key proteins such as cyclins and tumor suppressors is required for cell cycle progression and mitosis.

Aberrant regulation of cell cycle proteins can result in accelerated and uncontrolled cell division, leading to tumorogenesis, cancer growth, and spread. Hence, proteasome inhibitors should arrest or retard cancer progression by interfering with the ordered, temporal degradation of these regulatory molecules. For example, inhibition of proteasome-mediated IκB degradation may limit metastasis via the attenuation of NF-κB-dependent cell adhesion molecule expression and make dividing cancer cells more sensitive to apoptosis. Thus proteasome inhibitors could act through multiple mechanisms to arrest tumor growth, spread, and angiogenesis. However, dosing regimens must be optimized to limit the effects of proteasome inhibition in healthy cells.

A number of boron-containing proteasome inhibitors have been developed that are potent, selective, and reversible. These compounds inhibit the chymotryptic activity of the proteasome, thereby blocking the activity of the enzyme. By attenuating the degradation of cell cycle regulatory proteins, such agents elicit multiple effects leading to the inhibition of tumor cell growth and to apoptosis. Importantly, in these boronic series, the inhibitor potency (K_i) data correlate with the cytotoxicity profile *in vitro* and also with *in vivo* activity thus supporting the proposed mechanism of action.

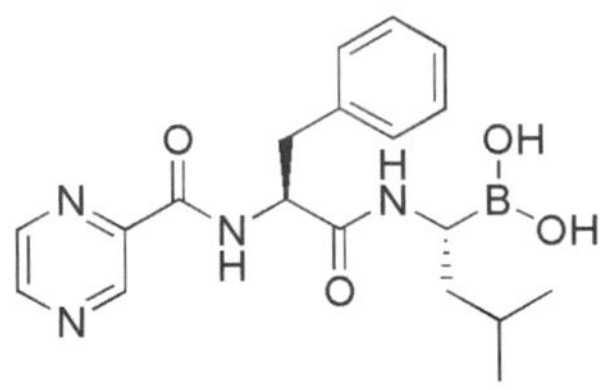

Bortezomib (Velcade)
(14.60)

Velcade (bortezomib) is the first in a new class of cancer drugs known as proteasome inhibitors that will be used to treat multiple myeloma, the second most common blood cancer to non-Hodgkin's lymphoma. It is a cancer of the plasma cell, which is an important part of the immune system that makes antibodies to fight off disease. The FDA recently approved bortezomib injection as a treatment for multiple myeloma. The FDA reviewed the license application in less than 4 months under the accelerated approval program.

The proteasome is also required for activation of NF-κB by degradation of its inhibitory protein IκB. NF-κB is required, in part, to maintain cell viability through the transcription of inhibitors of apoptosis, in response to environmental stress or cytotoxic agents. Stabilization of the IκB protein and blockade of NF-κB activity has been demonstrated to make cells more susceptible to apoptosis.

14.11 HORMONAL THERAPIES

It has long been recognized that tumors derived from hormone-dependent tissues such as the breast, endometrium, and testes are themselves dependent on the same hormones. This is demonstrated by the remissions observed in premenopausal breast cancer following ovariectomy (surgical removal of the ovaries) and in prostatic cancer following orchidectomy (surgical removal of the testes). Hormonal treatments thus play a large role in the treatment of these tumors. While not curative, they may provide excellent palliation of symptoms in selected patients, sometimes for many years. As with all treatments, tumor response and toxic side effects should be carefully monitored and treatment changed if progression occurs or side effects exceed benefit.

Hormonal treatment of advanced prostatic cancer is based on the concept that hormone-dependent tumors will regress when deprived of hormone stimulation. Furthermore, in prostatic carcinoma which is usually androgen-dependent, estrogens are often effective but their use is limited by cerebro- and cardiovascular side effects. A different approach involves the use of an antiandrogen such as cyproterone acetate which can block production of the adrenal and testicular androgens that provide the androgenic stimulus to the prostate. This is a more rational form of treatment for prostatic cancer and has produced responses in patients who have relapsed after estrogen therapy and orchidectomy. With antiandrogens gynecomastia (enlargement of the male breast) is less of a problem and the cardiovascular risk is decreased. Flutamide is also an antiandrogen which works by inhibiting androgen uptake or nuclear binding of androgens in target tissue. In advanced prostatic cancer, flutamide produces a response comparable to that achieved with estrogen therapy, and when used in combination with a luteinizing hormone-releasing hormone (LH-RH) agonist survival rate is prolonged. Bicalutamide is an antiandrogen used to block androgens formed in the adrenal glands as an adjunct to LH-RH agonists or orchidectomy in advanced prostatic carcinoma. It binds to androgen receptors without activating gene expression, thus inhibiting androgenic stimulus and ultimately leading to tumor regression. The gonadotrophin releasing hormone (Gn RH) analogues such as buserelin, gôserelin, leuprorelin, and triptorelin are also effective and devoid of the limiting side effects of the estrogens. They produce a downregulation of the receptors in the pituitary, thus preventing LH release and thereby reducing serum testosterone.

The antiestrogens tamoxifen and toremifine bind selectively to ERs preventing estrogen-induced stimulation of DNA synthesis and hence cell growth in breast cancer. Progestogens are second- or third-line therapies in breast carcinoma and are also used in endometrial carcinoma. Aromatase inhibitors such as anastrozole and letrozole can be used in the treatment of advanced postmenopausal breast cancer and are licensed for first-line use. They prevent the conversion of androgens to estrogens in nonovarian sites thereby removing the stimulus for growth in estrogen-dependent tumors. The aromatase inhibitor aminoglutethimide also inhibits adrenal steroid production, so corticosteroid replacement therapy must be given concomitantly. Anastrazole, exemestane, and letrozole have a greater specificity and do not affect adrenal steroid synthesis.

14.11.1 Breast Cancer

The management of patients with breast cancer involves surgery, radiotherapy, drug therapy, or a combination of these. The most common cytotoxic chemotherapy regimen for both adjuvant use and metastatic disease has been cyclophosphamide, methotrexate, and fluorouracil. However, anthracycline-containing regimens are now increasingly used and should be regarded as standard therapy unless contraindicated (e.g., in cardiac disease). In metastatic disease, the choice of chemotherapy regimen will be influenced by whether the patient has previously received adjuvant treatment and the presence of any co-morbidity. For women who have not previously received chemotherapy, either cyclophosphamide, methotrexate, and fluorouracil, or an anthracycline-containing regimen is the standard initial therapy for metastatic breast disease. Patients with anthracycline-refractory or resistant disease should be considered for treatment with a taxane either alone or in combination with trastuzumab if they have tumors that overexpress human epidermal growth factor-2 (HER-2). Other drugs licensed for breast cancer include capecitabine, gemcitabine, ralitrexed, mitoxantrone, mitomycin, and vinorelbine. In cancers that overexpress HER-2, trastuzumab is an option for cytotoxic chemotherapy resistant disease.

Early breast cancer

In early breast cancer all women should be considered for adjuvant therapy following surgical removal of the tumor. Adjuvant therapy is used to eradicate the micrometastases that cause relapses. The choice of adjuvant treatment is determined by the risk of recurrence, the estrogen-receptor status of the primary tumor, and menopausal status.

Tamoxifen is an estrogen-receptor antagonist and is the adjuvant hormonal treatment of choice in all women with estrogen-receptor-positive breast cancer. It is supplemented in selected cases by cytotoxic chemotherapy. Premenopausal women may also benefit from treatment with a gonadorelin analogue or ovarian ablation. Treatment with tamoxifen delays the growth of metastases and increases survival. Therefore, if tolerated, treatment should be continued for 5 years. Tamoxifen also decreases the risk of developing cancer in the other breast.

Anastrozole is licensed for the adjuvant treatment of estrogen-receptor-positive early breast cancer in postmenopausal women who are unable to take tamoxifen because of high risk of thromboembolism or endometrial abnormalities. Cytotoxic chemotherapy is preferred for both premenopausal and postmenopausal women with estrogen-receptor-negative breast cancer.

Advanced breast cancer

Tamoxifen is used in postmenopausal women with estrogen-receptor-positive tumors, patients with long disease-free intervals following treatment for early breast cancer, and those with disease limited to bone or soft tissues. However, aromatase inhibitors such as anastrozole or letrozole may be more effective and are regarded as preferred treatment in postmenopausal women. Ovarian ablation or a gonadorelin analogue should be considered in premenopausal women. Progestogens such as medroxyprogesterone acetate continue to have a role in postmenopausal women with advanced breast cancer. They are as effective as tamoxifen but are not as well tolerated. They are less effective than the aromatase inhibitors. Cytotoxic chemotherapy is preferred for advanced estrogen-receptor-negative tumors and for aggressive disease, particularly where metastases involve visceral sites (e.g., the liver) or where the disease-free interval following treatment for early breast cancer is short.

Role of estrogen in tumor growth

Estrogen has effects on the growth and function of reproductive tissues and also preserves bone mineral density, thus reducing the risk of osteoporosis and protecting the cardiovascular system by reducing cholesterol levels. Estrogens act as promoters rather than initiators of breast tumor development and can also facilitate tumor invasiveness by stimulating the production of proteases which can degrade the extracellular matrix.

Estrogen action is conducted through ERs of which there are two types, ERα and ERβ, although the role of the latter in breast cancer is unclear. ERs can be detected in 60–80% of human breast cancers and their presence (i.e., ER(+)) is related to the sensitivity of the tumor to antiendocrine treatment with 60% of ER(+) breast cancer patients responding.

Estradiol (E2) the principal estrogen is synthesized in the ovaries by aromatase from androstenedione but in postmenopausal women, where ovarian synthesis has ceased, local synthesis occurs in adipose and other tissue, especially the breast where E2 levels are 20-fold higher than in the plasma.

After entering the cell, E2 binds to the ER which leads to a conformational change allowing the ER dimer, through its DNA-binding domain, to bind to the promoter (activator) region of the gene (Figure 14.6). Gene transcription is controlled by two areas on the ER; AFI whose activity is regulated by phosphorylation and AF2 where ligand-binding of E2 (activating) and antiestrogens (e.g., tamoxifen-deactivating) occurs. Cofactors complex to ER which have either activating (CoA) or deactivating (Co-repressor, CoR) effects on transcription.

The dormant CoR–ER complex recruits histone deacetylases (HDACs) which maintain the histones making up the chromatin in a nonacetylated state such that transcription from the associated DNA cannot occur. The activated CoA–E2–ER complex formed upon E2 binding recruits histone acetyltransferases (HATs) which acetylate histones leading to chromatin decondensation and transcription. E2 binding at the ligand-binding pocket produces a conformational change which allows the CoA binding surface on ER to appear. Estrogen binding facilitates binding of CoA at the AF2 site whereas antiestrogen (e.g., tamoxifen) binding prevents CoA binding thus, in part, mediating their anticancer effects.

Several growth factors can stimulate ER activity in the absence of estrogen, thus providing a possible pathway for endocrine-resistant breast tumors where, upon prolonged treatment with endocrine agents, the patient relapses with disease progression. This ER stimulation is considered to be due to phosphorylation of the AFI region of the ER by a cascade of events in at least two pathways, some involving kinases, initiated at the receptor tyrosine kinase external to the cell (Figure 14.7). One of these kinases, HER-2, a member of the EGFR family, on overexpression in tumors leads to resistance to tamoxifen. Trastuzumab (Herceptin), a recombinant humanized mouse monoclonal antibody against HER-2, acts as an inhibitor and is used in metastatic breast cancer in combination with paclitaxel in situations where HER-2 is overexpressed.

It should be noted that other drugs are used in the treatment of breast cancer. Trilostane is licensed for postmenopausal breast cancer. It is quite well tolerated but diarrhoea and abdominal discomfort may be a problem. Trilostane causes adrenal hypofunction, and so corticosteroid

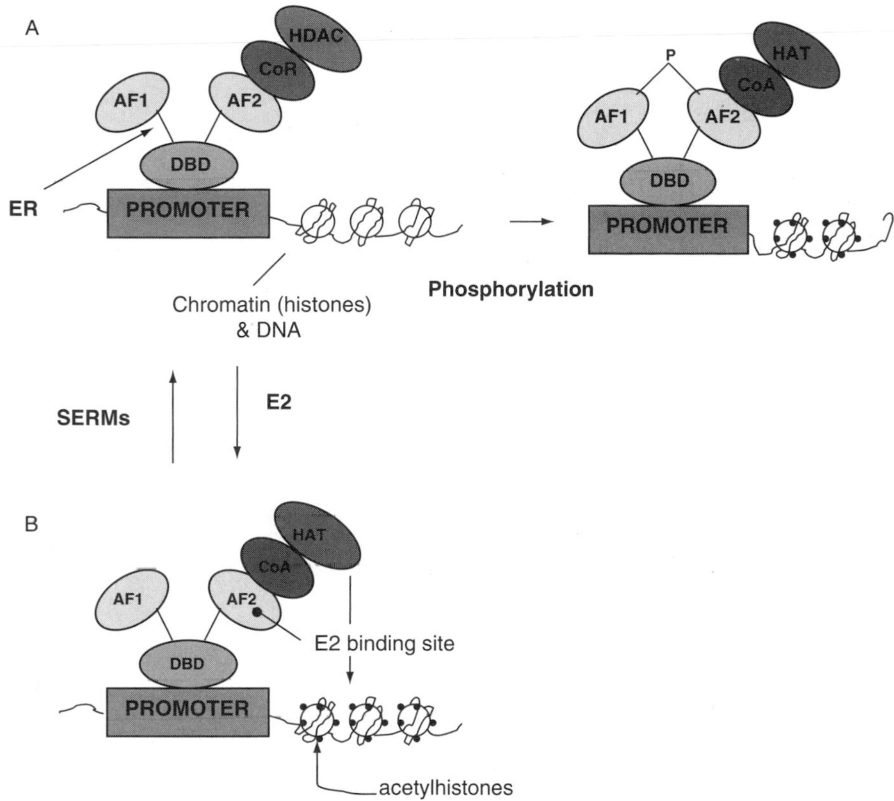

Figure 14.6 ER–gene binding showing (A) resting state, and (B) estrogen-activated state. (After S. Ali and R.C. Coombs (2002) *Nature Reviews Cancer* **2**, 101–112. With permission.)

replacement therapy is needed. The use of biphosphonates in patients with metastatic breast cancer may prevent skeletal complications of bone metastases.

Antiestrogens

After surgery with associated radiation therapy to remove a tumor mass, adjuvant therapy with antiestrogens is usually initiated to prevent the growth of metastases (some of which are ER-(+)).

Tamoxifen (Nolvadex)

Tamoxifen (Nolvadex) is a nonsteroidal estrogen antagonist used as a first-line antiestrogen. It is licensed in the UK for breast cancer and also anovulatory infertility. The structure was first reported by (the then) ICI in 1964, and the separated geometrical isomers 2 years later. The first stereospecific synthesis was reported in 1985. Tamoxifen competes with estrogen for the ER (at AF2) thus preventing estrogen activation and subsequent tumor growth. One third of nonselected postmenopausal patients respond and the rate is higher (50–60%) for patients with ER-(+) tumors.

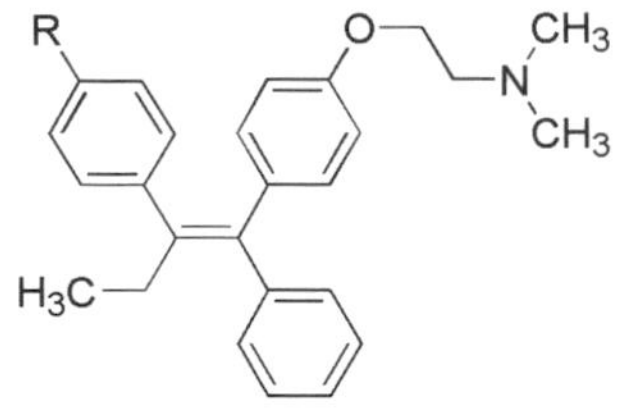

Tamoxifen: R = H (Nolvadex)
Hydroxytamoxifen: R = OH
(14.61)

Whereas tamoxifen has a beneficial effect due to its agonist actions on preserving bone mineral density and protection of the cardiovascular system by reducing plasma lipids, its agonist effect on the gynecological tract presents a small but significant risk of generating endometrial cancer, possibly due to nonprevention of activation of AF1 since this is more significant in the uterus. Pure antiestrogens (e.g., ICI 164384 and Faslodex) prevent activation of both AF2 and AF1 and have only antagonist effects, as well as reducing the half-life of ER.

The side effects of tamoxifen can include hot flushes, vaginal discharge and bleeding, suppression of menstruation in some premenopausal women, pruritic vulvae, GI disturbances, headache, light-headedness, tumor ''flare,'' and decreased platelet count. Occasionally cystic ovarian swellings in premenopausal women can occur. Most importantly, tamoxifen can increase the risk of thromboembolism particularly during and immediately after major surgery or periods of immobility. Patients should be made aware of the symptoms of thromboembolism and advised to report sudden breathlessness and any pain in the calf of one leg. Increased endometrial changes can also occur including hyperplasia, polyps, cancer, and uterine sarcoma. Clinical trials have shown that 20 mg daily substantially increases survival in early breast cancer. Despite the small risk of endometrial cancer, the benefits generally outweigh the risks.

Recent evidence suggests that tamoxifen prophylaxis can reduce breast cancer in women at high risk of the disease. However, the adverse effects of this agent preclude its routine use in most healthy women. Research is underway to find new agents with the beneficial chemopreventive effects of tamoxifen but without the accompanying side effects.

Toremifene (Fareston)

Toremifine is a nonsteroidal antiestrogen structurally similar to tamoxifen and first reported in 1983 by Farmos. It is licensed in the UK for the treatment of hormone-dependent metastatic breast cancer in postmenopausal women, although it is not often used due to the side effect profile as outlined below.

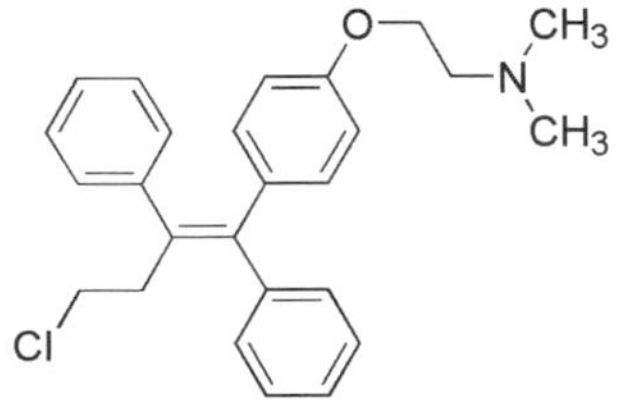

Toremifene (Fareston)
(14.62)

Given orally, a large number of side effects have been reported including hot flushes, vaginal bleeding or discharge, dizziness, edema, sweating, nausea, vomiting, chest or back pain, fatigue, headache, skin discolorization, weight increase, insomnia, constipation, dyspnoea, paresis, tremor, vertigo, pruritis, anorexia, reversible corneal opacity, and asthenia. Thromboembolic events have also been reported. Rarely, dermatitis, alopecia, emotional lability, depression, jaundice, and stiffness can occur. Hypercalcemia may also appear especially if bone metastases are present. Finally, there is also an increased risk of endometrial changes including hyperplasia, polyps and cancer, and so any abdominal pain or vaginal bleeding, etc. should be immediately investigated. Therefore, toremifene should not be used in women with a history of endometrial hyperplasia, severe hepatic impairment, or thromboembolytic events.

Novel selective ER modulators

There has been much research into the discovery of novel selective ER modulators (known as ''SERMs'') as replacements for tamoxifen. Ideally, SERMs should possess increased potency at AF2, have agonist effects on bone and plasma lipids but should lack the unwanted agonist effects on the uterus. They fall into three general categories: (1) nonsteroidal compounds resembling tamoxifen in having a triphenylethylene structure; (2) nonsteroidal compounds based on a benzothiophene structure; and (3) steroidal pure antiestrogens.

The first group includes compounds such as toremifene, droloxifene, and idoxifene some of which have shown limited individual advantages over tamoxifen in preclinical evaluations. For example, some compounds have a greater potency, lack of hepatocarcinogenicity, have a shorter half-life or have a reduced estrogenic effect in rat uterus. However, in clinical trials there were no major differences in efficacy and safety in comparison with tamoxifen.

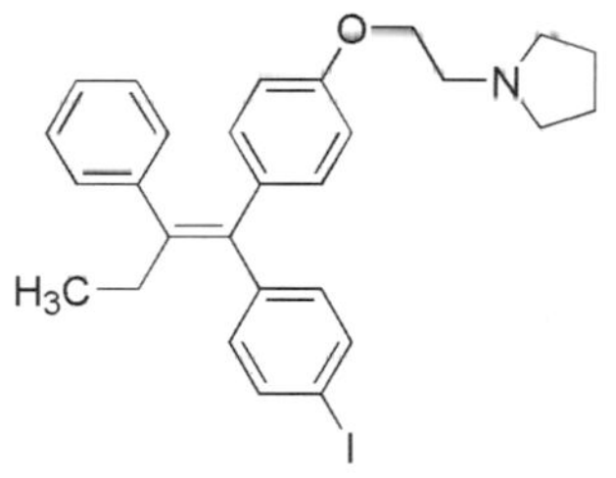

Idoxifene (CB-7432)

(14.63)

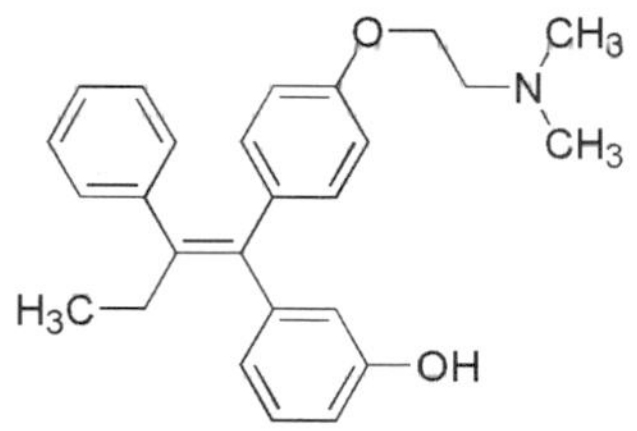

Droloxifene

(14.64)

First Generation SERMs: Tamoxifen-Like Triphenylethylene Structures

In the benzothiophene group, raloxiphene initially developed for treatment of osteoporosis significantly reduced the incidence of breast cancer during development for osteoporosis and has been in trials for breast cancer. Arzoxifene (LY353381) is more potent than raloxiphene, has an improved profile and is also in development.

EM-800, an orally active prodrug of the active benzopyrene EM-652 (SCH 57068) in preclinical studies had the required profile with no agonist activity on the uterus and, like the steroid antiestrogen fulvestrant, downregulated ER levels; they are undergoing further examination as first-line adjuvant treatment for breast cancer.

The steroidal antiestrogen fulvestrant (ICI 182, 780, Faslodex), which is estradiol with a long 7β-hydrophobic chain, not only impairs the necessary ER dimerization for agonist action but also downregulates the ER, so acting as a potent pure antiestrogen in all tissues (breast, uterus, and probably bone).

Raloxifene (LY 156,758)

(14.65)

Arzoxifene (LY 353,381)

(14.66)

Second Generation SERMs: Benzothiophene Structures

X =

Fulvestrant (Faslodex)

(14.67)

EM-652 (SCH 57068)

(14.68)

Third Generation SERMs: Steroidal Pure Anti-Oestrogens

Other analogues include the orally active SR-16234 and ZK-191703. Whether the latest generation of SERMs has an important future role in the treatment of breast cancer remains to be seen.

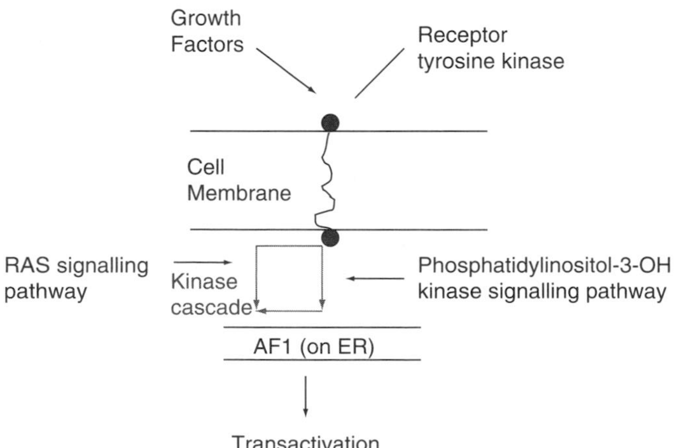

Figure 14.7 Simplified view of growth factor activation of ER by phosphorylation of AF1. (After S. Ali and R.C. Coombs (2002) *Nature Reviews Cancer* **2**, 101–112. With permission.)

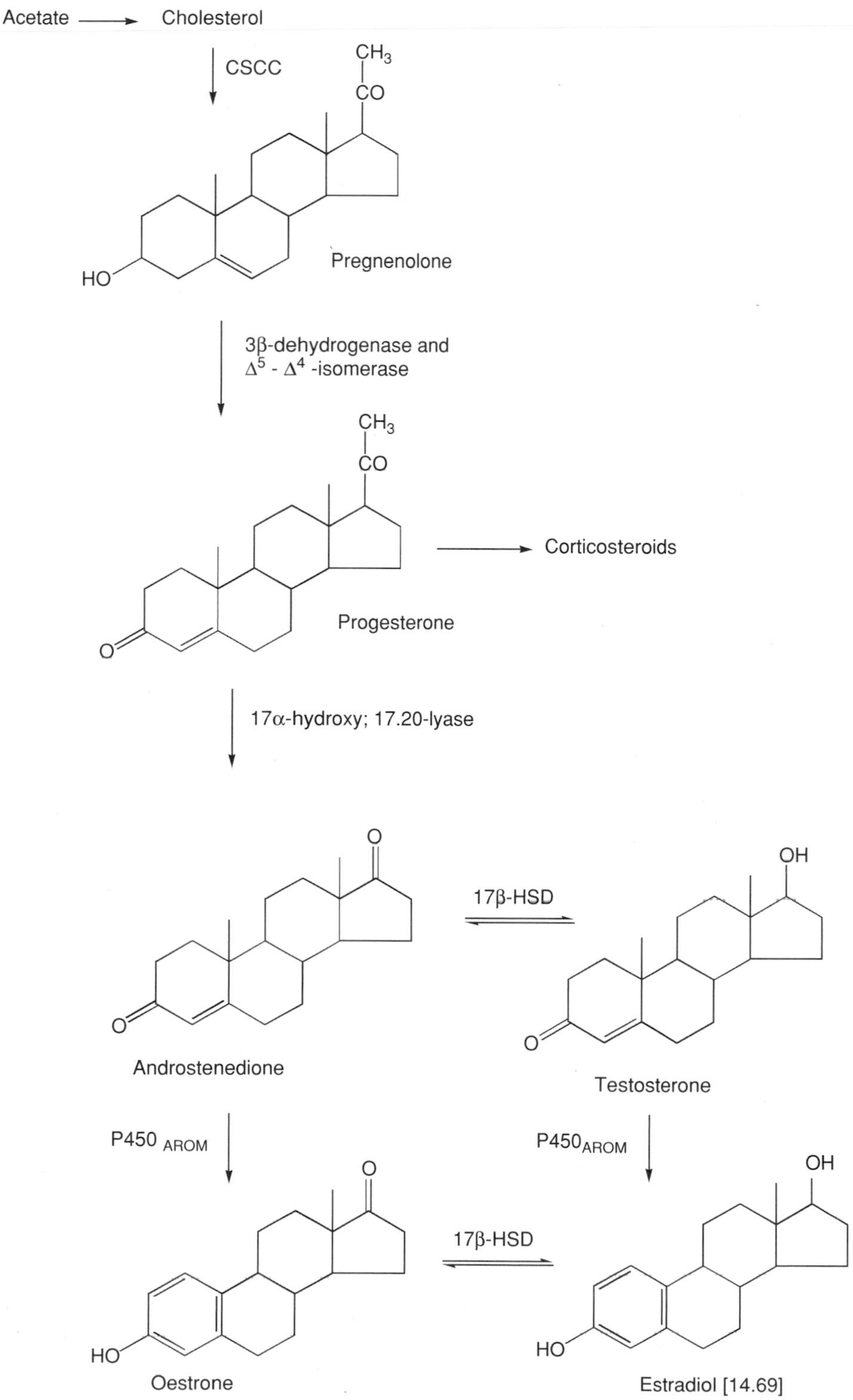

Acetate → Cholesterol

CSCC

CH_3 CO

Pregnenolone

HO

3β-dehydrogenase and
Δ^5 - Δ^4 -isomerase

CH_3 CO

→ Corticosteroids

Progesterone

O

17α-hydroxy; 17.20-lyase

O

17β-HSD

OH

O

Androstenedione

O

Testosterone

$P450_{AROM}$

$P450_{AROM}$

O

OH

17β-HSD

HO

Oestrone

HO

Estradiol [14.69]

Figure 14.8 Steroidogenesis pathway.

Aromatase inhibitors

Aromatase inhibitors act predominantly by blocking the conversion of androgens to estrogens in the peripheral tissues. They do not inhibit ovarian estrogen synthesis and thus should not be used in premenopausal women. Androstenedione and testosterone are converted by the cytochrome P450 enzyme aromatase ($P450_{AROM}$) to estrone and estradiol, respectively, as the final step in the steroidogenesis pathway from cholesterol (Figure 14.8).

Selective inhibition of aromatase leads to reduced estrogen plasma levels without affecting other hormones produced by the steroidogenesis pathway. In postmenopausal women where estrogen synthesis has ceased in the ovaries and is carried out mainly in adipose and other tissues, the breast has 20-fold greater concentration of estrogen compared to the plasma due to local synthesis by the action of sulfatase enzyme on estrone sulfate (see Chapter 9).

First-generation inhibitors

Aminoglutethimide, the first aromatase inhibitor to be discovered, was initially developed as an anticonvulsant but was withdrawn from the clinic after reports of adrenal insufficiency. It was subsequently found to inhibit several cytochrome P-450 enzymes involved in adrenal steroidogenesis and was then re-developed for ''medical adrenalectomy'' against advanced breast cancer. Side effects including drowsiness and rash limited its use, but the discovery that its efficacy was mainly due to aromatase inhibition stimulated the development of numerous new inhibitors during the 1980s and early 1990s. They are sometimes described as first-, second-, and third-generation inhibitors according to the chronologic order of their clinical development. They are further classified as type 1 or type 2 inhibitors according to their mechanism of action. Type 1 inhibitors are steroidal analogues of androstenedione and bind to the same site on the aromatase molecule, but unlike androstenedione they bind irreversibly because of their conversion to reactive intermediates by aromatase. Therefore, they are now commonly known as enzyme inactivators. Type 2 inhibitors are nonsteroidal and bind reversibly to the heme group of the enzyme by way of a basic nitrogen atom. For example, anastrozole and letrozole, both third-generation inhibitors, bind via their triazole groups.

Second-generation inhibitors

The second-generation aromatase inhibitors include formestane (4-hydroxyandrostenedione), a type 1 compound, and fadrozole, a type 2 imidazole. Each has been found to have clinical efficacy, but formestane has the disadvantage of requiring intramuscular injection, and fadrozole at high dose causes aldosterone suppression, limiting its use to doses that produce only about 90% inhibition. Other second-generation aromatase inhibitors have been investigated clinically but have never been approved for clinical use.

Third-generation inhibitors

The third-generation inhibitors, developed in the early 1990s, include the nonsteroidal triazoles anastrozole, letrozole, and vorozole (Type II — reversible), and the steroidal agent exemestane (Type I — irreversible) which can be administered orally. In contrast to aminoglutethimide and fadrozole, these third-generation inhibitors have not had any selectivity issues at clinical doses and are 10^3- to 10^4-fold more potent than aminoglutethimide, with little or no effect on basal levels of cortisol or aldosterone. Anastrozole and letrozole are at least as effective as tamoxifen for first-line treatment of metastatic breast cancer in postmenopausal women.

The nonsteroidal 1,2,4-triazoles such as anastrozole and letrozole are usually less active *in vitro* than the imidazoles but are more stable to metabolism *in vivo* so that overall clinical activity is in favor of the triazoles. They are very potent *in vivo* with anastrozole (1 mg daily) and letrozole (2.5 mg daily) reducing serum levels of estrogens by 97 and 99%, respectively (beyond the limit of detection in many patients). This is superior to the performance of aminoglutethimide (1000 mg daily, 90% reduction) but comparable with exemestane (25 mg daily, 97% reduction). The advantages of these third-generation inhibitors include improved tolerance with fewer side effects rather than their improved response rates (11–24%) and durations (18–23 months) in comparison to aminoglutethimide (12–30%, 13–24 months) or the progestin, megestrol (8–17%, 12–18 months). However, in conclusion, it is not yet known whether the benefits of aromatase inhibitors persist over the long term.

Aminoglutethimide (Orimeten)

$P450_{AROM}$ inhibition was discovered serendipitously in the late 1950s when the experimental anticonvulsant agent aminoglutethimide was observed to block adrenal steroidogenesis. Follow-up clinical trials in advanced breast cancer showed that a 20–40% response rate and a duration of remission of 6–12 months could be achieved. However, the agent caused a number of serious side effects including adrenal hypofunction (especially under conditions of stress such as surgery, trauma, or acute illness) which required corticosteroid replacement therapy. Aminoglutethimide has now been largely replaced by the newer, more specific, aromatase inhibitors such as anastrozole which are much better tolerated.

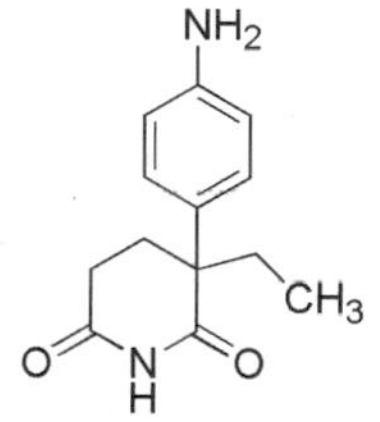

Aminoglutethimide (Orimeten)
(14.70)

Aminoglutethimide lacks selectivity for $P450_{AROM}$ and side effects include CNS depression (e.g., drowsiness, lethargy), rash, and sometimes rash with fever. Occasionally, dizziness, nausea, and other minor side effects can occur. It inhibits several other P450 enzymes on the steroidogenesis pathway, thus blocking production of various adrenal steroids and requiring co-administration of glucocorticoid. Given orally, it is used to treat advanced breast or prostate cancer. In prostate cancer it is often co-administered with a glucocorticoid and sometimes with a mineralocorticoid. It is also used to treat Cushing's syndrome arising from malignant disease.

Anastrozole (Arimidex)

Anastrozole was first reported in 1989 by (the then) ICI. Given orally, it is used for the adjuvant treatment of estrogen-receptor-positive early breast cancer in postmenopausal women who are unable to take tamoxifen because of high risk of thromboembolism or endometrial abnormalities. It is also used for advanced breast cancer in postmenopausal women. The drug is recommended for a duration of treatment of 5 years only, and a laboratory test for menopause is recommended if there is doubt. Caution is required in women susceptible to osteoporosis, and bone mineral density should be assessed before and after treatment when necessary.

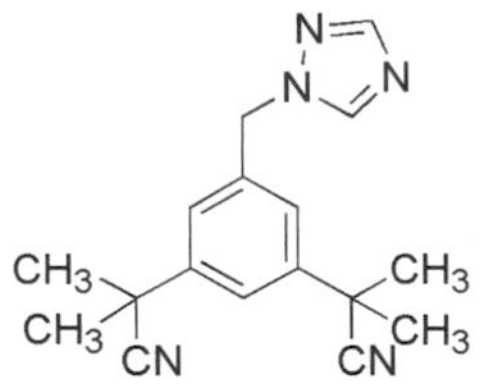

Anastrazole (Arimidex)
(14.71)

Possible side effects of anastrozole include hot flushes, vaginal dryness, vaginal bleeding, hair thinning, anorexia, nausea, vomiting, diarrhoea, arthralgia, bone fractures, rash (including Stevens–Johnson syndrome), asthenia, and drowsiness. A slight increase in total cholesterol has also been reported.

Letrozole (Femara™)

Letrozole, a nonsteroidal aromatase inhibitor structurally related to fadrozole, was first reported in 1987 by (the then) Ciba-Geigy, and is used mainly for advanced breast cancer in postmenopausal women. It is also used for preoperative treatment of postmenopausal women with localized hormone-receptor-positive breast cancer to allow subsequent breast conserving surgery. Letrozole is not recommended for use in premenopausal women or those with severe hepatic impairment.

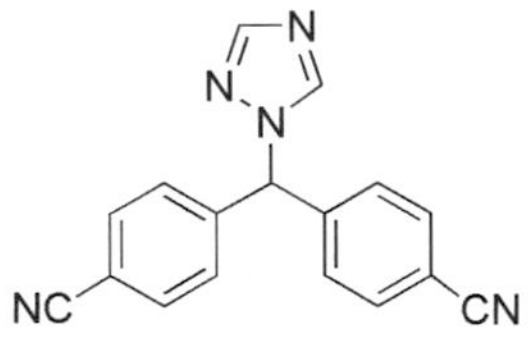

Letrozole (Femara)
(14.72)

Given orally, the drug can cause a range of side effects including hot flushes, nausea, vomiting, dyspepsia, constipation, diarrhoea, abdominal pain, anorexia (and sometimes weight gain), dyspnoea, chest pain, coughing dizziness, fatigue, headache, infection, muscular skeletal pain, peripheral oedema, rash, and pruritis.

Vorozole (Rivizor™)

Vorozole is a nonsteroidal aromatase inhibitor closely related to anastrozole and letrozole that was first reported in 1988 by Janssen. Resolution of the enantiomers of vorozole was reported in 1994.

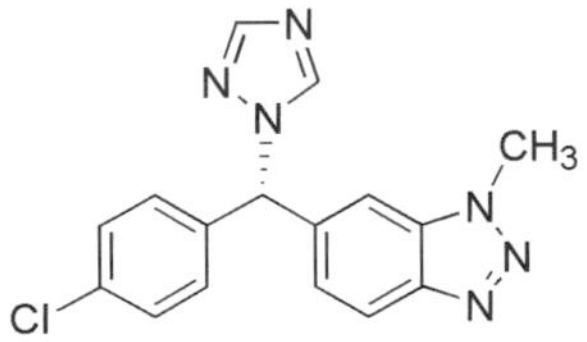

Vorozole (Rivizor)
(14.73)

This drug is not currently licensed for use in the UK.

Exemestane (Aromasin™)

Exemestane is the first example of an irreversible aromatase inhibitor and was reported in 1987 by Farmitalia. It is used for advanced breast cancer in postmenopausal women in whom antiestrogenic therapy has failed.

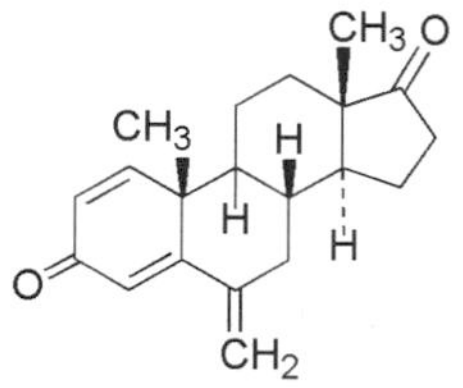

Exemestane (Aromasin)
(14.74)

Given orally, the main side effects include hot flushes, nausea, dizziness, fatigue, and sweating. Less frequently, vomiting, dyspepsia, constipation, abdominal pain, anorexia, peripheral edema, headache, depression, insomnia, alopecia, and rash have been observed. More rarely, thrombocytopenia and leucopenia can occur. It should be used with caution in women with hepatic and renal impairment, and is not indicated for use in premenopausal women.

14.11.2 Prostatic Cancer

Prostatic cancer is mainly hormone-dependent, being promoted by the androgen DHT which is derived from testosterone by the action of 5α-reductase. Thus, one obvious form of treatment involves removal of the DHT stimulus. Prostatic cancer is usually well developed on presentation so that survival rates are low and treatment is aimed at increasing the time of survival and the quality of life. Removal of the DHT stimulus to tumor growth can be achieved either by blocking its synthesis or its action at its receptor. Surgical removal of the prostate or testes (orchidectomy) is now less prevalent treatments having been largely replaced by endocrine therapy (estrogens) and, in more recent years, by treatment with antiandrogen and LH-RH analogues.

Hormonal therapy does not provide a cure for prostate cancer because the cancer usually becomes resistant (refractory) to treatment by a process that is not completely understood. A prostate cancer cell population usually contains both androgen-sensitive (i.e., dependent on DHT for growth and viability) and androgen-insensitive cells (i.e., not dependent on DHT). It has been proposed that, over time, the hormonal treatment suppresses the sensitive cells leaving the insensitive cells to flourish. Resisting programmed cell death, they gradually become the majority in the population.

Androgen receptors are proteins with a molecular weight of 120 kDa, and their synthesis is guided by genes on the X chromosome. In cells, the androgen-binding monomer (4.4S) is bound to receptor-associated protein and to other small molecules to produce a 9S inactive oligomer. DHT binds to the androgen receptor (AR), as does testosterone but with less affinity. The AR–DHT complex then binds to the nuclear androgen response element sequence of the gene, so promoting transcription of mRNAs leading to protein synthesis and maintenance of cellular growth of the prostate (Figure 14.9).

The zinc fingers of the androgen receptor, each consisting of a zinc atom and four cysteine amino acids, are separately responsible for recognition of the response element and the androgen receptor dimer formation area. The fingers insert into the DNA at two six base-pair runs of the nucleotide sequence separated by a three base spacer. Two have been identified as binding to ATAGCAtctTGTTCT or AGTACTccaAGAACC sequences (Figure 14.10 and Figure 14.11).

The binding of testosterone/DHT to the hormone-binding site near the terminal COOH activates the receptor complex so that it can bind to the response element. This is achieved by disaggregation of the macromolecular complex and release of an accessory protein, so exposing the DNA-binding site of the 4.4S receptor. Antiandrogens competitively prevent DHT from binding to the androgen receptor, and the androgen receptor–(antiandrogen) complex formed cannot become correctly aligned to bind to the response element so preventing dimerization of the monomer androgen receptor and subsequent promotion of transcription.

Metastatic cancer of the prostate usually responds to hormonal treatment aimed at androgen depletion. Standard treatments include bilateral subcapsular orchidectomy or use of gonadorelin analogues such as buserelin, goserelin, leuprorelin, or triptorelin. In these cases responses can last for 12 to 18 months. No entirely satisfactory therapy exists for disease progression (i.e., hormone-refractory prostate cancer), but occasionally patients respond to other methods of hormone manipulation such as use of an antiandrogen. Bone metastases can often be palliated with radiation or, if widespread, with strontium, aminoglutethimide, or prednisolone.

Gonadorelin analogues (LH-RH analogues; agonists and antagonists)

Luteinizing hormone-releasing factor (also known as LH-RF, LH-RH, LRF, LRH, gonadorelin, gonadotropin-releasing factor, gonadoliberin, luliberin, or LH-RH/FSH-RH) is a neurohumoral

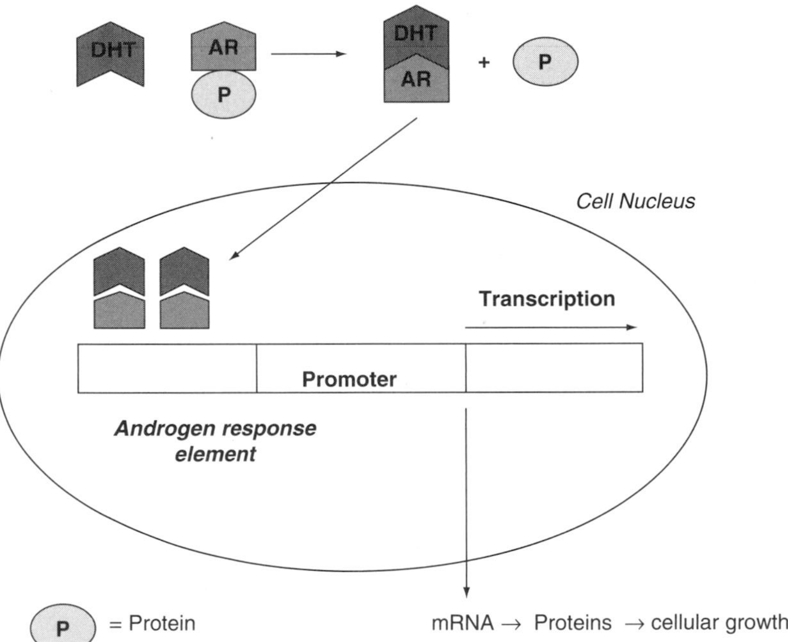

Figure 14.9 DHT activation of the androgen response.

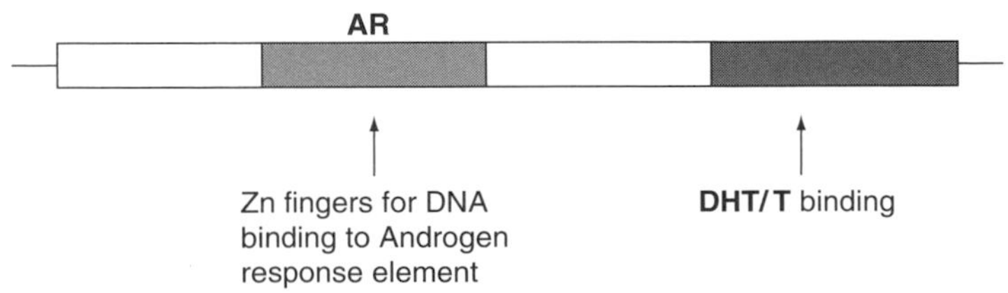

Figure 14.10 Schematic view of the androgen receptor (AR).

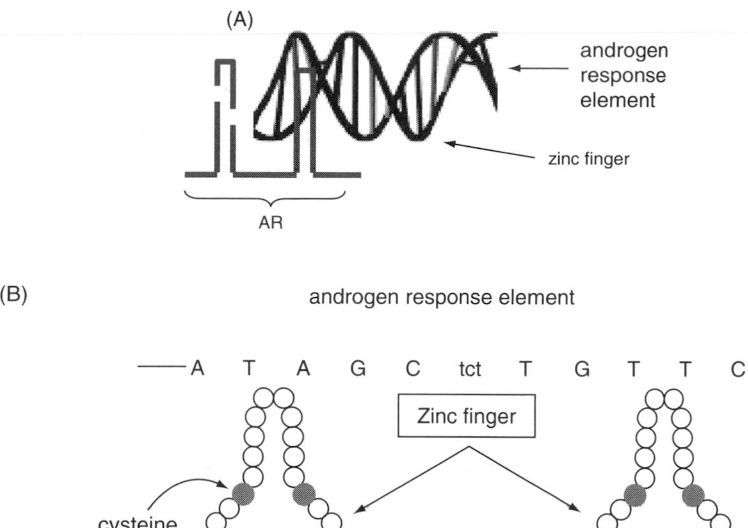

Figure 14.11 (A) Binding of AR through zinc fingers to androgen response element. (B) Zinc finger binding to nucleotide sequence of androgen response element.

hormone of molecular weight 1182.29 produced in the hypothalamus. It stimulates the secretion of the pituitary hormones, luteinizing hormone (LH), and follicle-stimulating hormone (FSH), which in turn produce changes resulting in the induction of ovulation. It was first isolated from porcine hypothalamic extracts, then structurally elucidated and synthesized in the early 1970s. It has been previously commercialized under the names Fertagyl (Intervet), Fertiral (HMR), Kryptocur (HMR), Relefact LH-RH (HMR), Cystorelin (Abbott), Hypocrine (Tanabe Seiyaku), Lutrelef (Ferring), and Lutrepulse (Ortho).

$$\text{5-oxoPro–His–Trp–Ser–Tyr–Gly–Leu–Arg–Pro–GlyNH}_2$$

Luteinizing Hormone - Releasing Factor (LH-RH)
(14.75)

Gonadorelin analogues are as effective as orchidectomy or diethylstilbestrol but they are expensive and require parenteral administration, at least initially. These agents work by causing initial stimulation, then depression of luteinizing hormone release by the pituitary. During the initial stages of treatment (1–2 weeks), increased production of testosterone may cause progression of the cancer. In susceptible patients this tumor ''flare'' may cause spinal chord compression, ureteric obstruction, or increased bone pain. When such problems are anticipated, alternative treatments (e.g., orchidectomy) or concomitant use of an antiandrogen such as cyproterone acetate or flutamide are recommended. In this case, antiandrogen treatment should be started 3 days before the gonadorelin analogue and continued for 3 weeks. For this reason, gonadorelin analogues should be used with caution in men at risk of tumor flare and these patients should be monitored closely during the first month of therapy.

The introduction of LH-RH analogues has provided an alternative to estrogens and orchidectomy in the treatment of advanced prostate cancer but without the significant side effects. Naturally occurring LH-RH (Pyr-Glu-His-Trp-Ser-Tyr-Gly-Leu-Arg-Pro-Gly-NH_2) has a short half-life (and a pulsitile action on the receptor), but by substituting the amino acid at the 6-position, deleting the amino acid at the 10-position, and adding an ethylamide group to the proline residue at the 9-position, the first synthetic analogue was produced (leuprolide, Lupron™). This early agent had a greatly increased potency together with prolonged activity and a nonpulsitile action on the receptor.

The initial effect of LH-RH agonists is to stimulate the secretion of LH and FSH, leading to increased testosterone synthesis. This can provide elevation of serum testosterone to 140–170% of basal levels within several days and an accompanying resurgence of symptoms (i.e., ''tumor flare''). Continuous administration, however, leads to dramatic inhibitory effects through a process of downregulation of LH-RH pituitary membrane receptors, and a reduction in gonadal receptors for LH and FSH, thus resulting in suppression of testosterone secretion comparable to surgical castration. Thus, chronic administration causes the pituitary gland to become refractory to additional stimulation by endogenous LH-RH, and so testicular androgen production is prevented.

The most commonly used LH-RH agonists in the UK are buserelin (Suprefact), goserelin (Zoladex), leuprorelin acetate (Prostap SR), and triptorelin (De-capeptyl SR). A specific side effect of LH-RH analogue treatment is a transient worsening of symptoms (including increased bone pain) during the first week of therapy as a result of the initial testosterone surge, with approximately 3–17% of patients affected. This has led to co-administration of antiandrogens before or with the first-line LH-RH analogue to prevent this effect. Other side effects include atrophy of the reproductive organs, loss of libido, and impotence. As they are only able to cause a decrease in testicular androgens while leaving adrenal androgen production unaffected (i.e., incomplete androgen clearance), the use of antiandrogens in combination with LH-RH agonists has been studied and found to give greater survival rates than LH-RH agonists alone.

Gonadorelin analogues are also used for breast cancer in women and other related indications. Caution is required in women with metabolic bone disease because decreases in bone mineral density may occur. The gonadorelin analogues cause side effects similar to the menopause in women and orchidectomy in men. These include hot flushes and sweating, sexual dysfunction, vaginal dryness or bleeding, and gynaecomastia or changes in breast size. Other side effects include hypersensitivity events (rashes, pruritis, asthma, and rarely anaphylaxis), injection site reactions, headache (rarely migraine), visual disturbances, dizziness, arthralgia and myalgia, hair loss, peripheral oedema, gastrointestinal disturbances, weight changes, sleep disorders and mood changes.

Buserelin (Suprefact)

Buserelin is a synthetic nonapeptide agonist analog of LH-RH, first reported by Hoechst in 1976. It is used for the treatment of advanced prostate cancer.

5-oxoPro–His–Trp–Ser–Tyr–D-Ser(*t*-Bu)–Leu–Arg–Pro$NHCH_2CH_3$

Buserelin (Suprefact)
(14.76)

Side effects can include worsening hypertension, palpitations, glucose intolerance, altered blood lipids, thrombocytopenia, leucopenia, nervousness, fatigue, memory and concentration disturbances, anxiety, increased thirst, hearing disorders, and musculoskeletal pain. It is normally given by subcutaneous injection for the first 7 days and this is followed by administration by nasal spray, although the latter can cause nasal irritation, nose bleeds, and altered sense of taste and smell. Buserelin should not be given to patients with depression.

Goserelin (Zoladex)

Goserelin is a synthetic peptide agonist gonadorelin analogue of LH-RH first synthesized by ICI in 1977. It is licensed in the UK for the treatment of prostate cancer, estrogen-receptor-positive early breast cancer and also advanced breast cancer in premenopausal women.

5-oxoPro–His–Trp–Ser–Tyr–D-Ser(*t*-Bu)–Leu–Arg–Pro–$NHNHCONH_2$

Goserelin (Zoladex)
(14.77)

For either breast or prostate cancer, a 3.6 mg (as acetate) implant (Zoladex) is administered by subcutaneous injection into the anterior abdominal wall every 28 days. A longer acting version (Zoladex LA) containing 10.8 mg of the acetate is licensed for prostate cancer and can be administered every 12 weeks. Side effects include the usual ones experienced with agonist analogues of LH-RH, but other adverse events include transient changes in blood pressure, paraesthesia, and rarely hypercalcemia (in patients with metastatic breast cancer). Goserelin is contraindicated in pregnancy and in women with undiagnosed vaginal bleeding.

Leuprorelin acetate (Prostap SR)

Leuprorelin (also known as leuprolide) is a synthetic nonapeptide agonist analog of LH-RH first reported by Takeda in 1975. It is licensed for the treatment of advanced prostate cancer.

5-oxoPro–His–Trp–Ser–Tyr–D-Leu–Leu–Arg–ProNHC_2H_5

Leuprorelin Acetate (Prostap SR)
(14.78)

It is administered as a 3.75 mg dose in a microsphere formulation (Prostap SR) by subcutaneous or intramuscular injection every 4 weeks. A larger 11.25 mg subcutaneous formulation can be given every 3 months. Apart from the usual side effects common with agonist analogues of LH-RH, other adverse reactions include fatigue, muscle weakness, paraesthesia, hypertension, palpitations, and alteration of glucose tolerance and blood lipids. Thrombocytopenia and leucopoenia have also been reported.

Triptorelin (De-capeptyl SR)

Triptorelin is a synthetic peptide agonist analog of LH-RH, first reported in 1976. It is used to treat advanced prostate cancer and also endometriosis.

5–oxoPro–His–Trp–Ser–Tyr–D-Trp–Leu–Arg–Pro–$GlyNH_2$

Triptorelin (De-capeptyl SR)
(14.79)

It is administered in the form of a 3 mg aqueous copolymer microsphere suspension (De-capeptyl SR) by intramuscular injection every 4 weeks. In addition to the usual side effects of agonist analogs of LH-RH, other reported adverse events include transient hypertension, dry mouth, excessive salivation, paraesthesia, and increased dysuria.

LH-RH receptor antagonists

LH-RH receptor antagonists can suppress LH release from the pituitary from the outset, thus circumventing the ''flare effect'' observed with the gonadorelin analogues. However, the relatively low potency of antagonists and their adverse effects due to histamine release has delayed their introduction into the clinic.

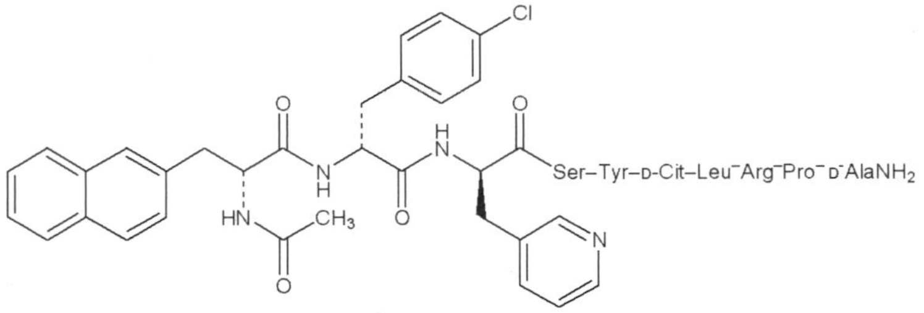

Ser–Tyr–D-Cit–Leu–Arg–Pro–D-AlaNH$_2$

Cetrorelix
(14.80)

Several peptide antagonists with low adverse effects have been evaluated in clinical trials (e.g., Cetrorelix [SB 75300] and Abarelix), however many have the drawback that, due to insufficient bioavailability, they need to be administered by either daily injection, intranasal spray, or a sustained release delivery system.

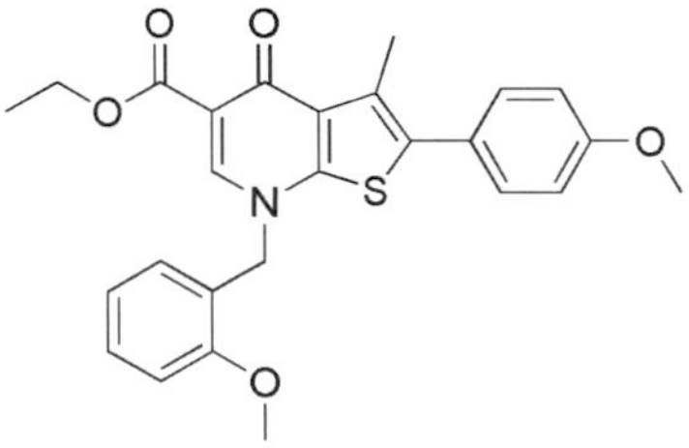

(14.81) 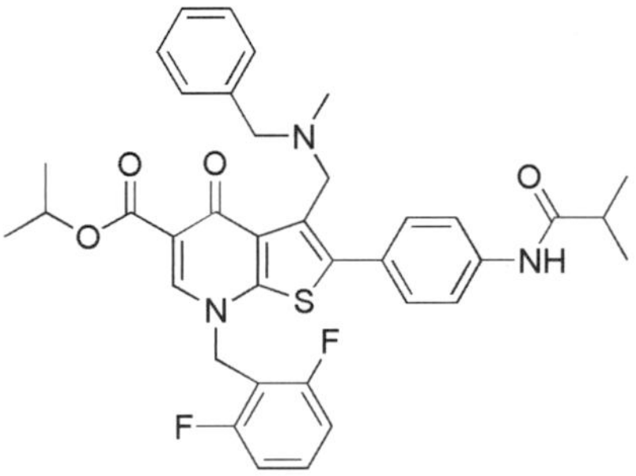 **(14.82)**

Potent orally active nonpeptide LH-RH antagonists (Takeda). The fluorinated analogue (right) has 50-times the affinity of LH-RH for the receptor.

Potent orally active nonpeptide antagonists have been described by Takeda (see above), designed to mimic the shape of the Type II β-turn involving residues 5 to 8 (Tyr-Gly-Leu-Arg) of LH-RH (*p*-Glu-His-Trp-Ser-Tyr-Gly-Leu-Arg-Pro-Gly-NH_2). These were discovered by directed screening of a library of compounds active at G protein receptors. Interestingly, the fluorinated analogue has 50 times the affinity of LH-RH for its receptor and is equiactive with the agonist leuprolide.

Merck have also described several series of nonpeptide antagonists (see below). Chemical library screening led to a lead quinoline agent that was optimized by changes to the molecule including conversion of the pyridine ring to a piperidine, and the addition of nitro and methyl groups.

Lead quinolone LH-RH receptor antagonist (Merck)
(14.83)

Optimized Lead
(14.84)

Other structures with LH-RH antagonist properties have also been described. For example, the indole- and pyrrolopyridine-based analogues shown below have significant antagonist properties.

Indole-based LH-RH receptor antagonists
(14.85)

Pyrrolopyridine-based LH-RH Receptor Antagonists
(14.86)

Antiandrogens

Antiandrogens inhibit the binding of DHT and other androgens to the androgen receptors in target tissues. Target cells are located in all areas of the body that depend on androgens such as the male genital skin, the seminal vesicles, the prostate, fatty tissues, and breast tissue as well as the hypothalamus and the pituitary. Antiandrogens bind to the androgen receptor, creating a receptor–antiandrogen complex, which is unable to elicit a response at the response-element, and so androgen-dependent gene transcription and protein synthesis are not stimulated. Therefore, antiandrogens can block the tropic effect of all androgens, not only peripherally (e.g., the prostate) but also centrally (e.g., hypothalamus and pituitary). They are used in conjunction with LH-RH agonists to reduce the effects of androgens produced by the adrenals which are not under LH control.

There are two main families of antiandrogen agents based on either steroidal or nonsteroidal structures. The most extensively studied steroidal agent is cyproterone acetate (Cyprostat, Androcur). In prostatic cells it acts as a competitive inhibitor of the binding of DHT to androgen receptors but also acts centrally, suppressing the corticotropic axis and therefore reducing plasma testosterone levels due to its progestin-like activity as well as its agonist activity (approximately 1/10 testosterone) (Figure 14.12).

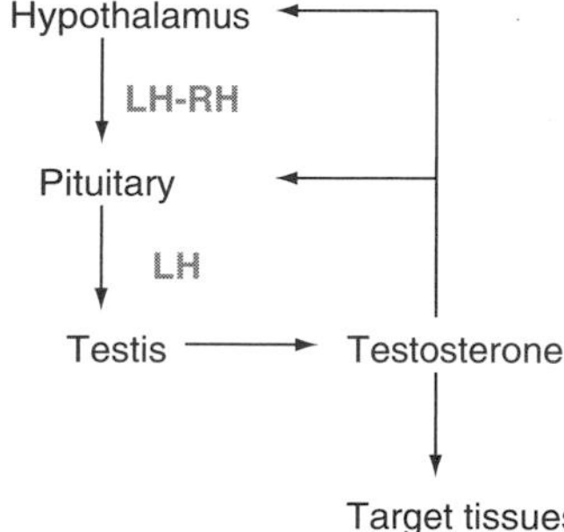

Figure 14.12 Control of androgen levels by the hypothalamus–pituitary axis.

Nonsteroidal antiandrogens have more limited central activity (depending on their individual structures) but all have significant and potent peripheral effects. They displace testosterone and DHT from the androgen receptor in the prostate but can also do the same in the hypothalamus. If such agents are used alone, this central blocking effect can potentially lead to an increase in the release of LH-RH by feedback control and subsequent LH production, thus causing a slow but gradual rise in serum testosterone levels to overcome the blockade. This is undesirable as it can stimulate prostatic tumor growth.

Cyproterone acetate, flutamide, and bicalutamide can be used to inhibit the tumor ''flare'' that may occur after commencing treatment with gonadorelin analogues. Cyproterone acetate and flutamide are also licensed in the UK for use alone in patients with metastatic prostate cancer refractory to gonadorelin analogue therapy. Bicalutamide is used for prostate cancer either alone or as an adjunct to other therapy according to the clinical circumstances.

The nonsteroidal antiandrogens are interesting from a medicinal chemistry standpoint in that bicalutamide and the active metabolite of flutamide (hydroxyflutamide) both contain a pseudo five-membered ring stabilized by a cyclic NH $\cdots$ O hydrogen bond. However, nilutamide contains a fully formed bioisosterically equivalent imidazolidinedione ring of similar three-dimensional shape.

Cyproterone Acetate (Cyprostat)

Cyproterone is the main steroidal antiandrogen in clinical use. The free alcohol form was first reported in 1965 by Schering AG. Interestingly, the alcohol is an antiandrogen whereas the acetate (e.g., Cyprostat) is both an antiandrogen and a progestogen and has an effect on both hormone secretion and spermatogenesis in humans. In target prostatic cells it acts as a competitive inhibitor of the binding of DHT to androgen receptors. It also acts centrally, suppressing the corticotropic axis and therefore reducing plasma testosterone levels due to its progestin-like activity as well as its agonist activity (approximately 1/10 testosterone).

Cyproterone is licensed in the UK for the treatment of prostate cancer, and is given orally for the tumor flare that occurs with initial gonadorelin therapy. It is also used for long-term palliative therapy where gonadorelin analogues or orchidectomy are either contraindicated or not well tolerated, or where oral therapy is preferred. It can also be used to control the hot flushes that sometimes occur with gonadorelin therapy or after orchidectomy. Clinical trials have shown that, with cyproterone acetate, objective responses can be obtained in 33% of patients with advanced prostate cancer and stabilization of the disease in 40%.

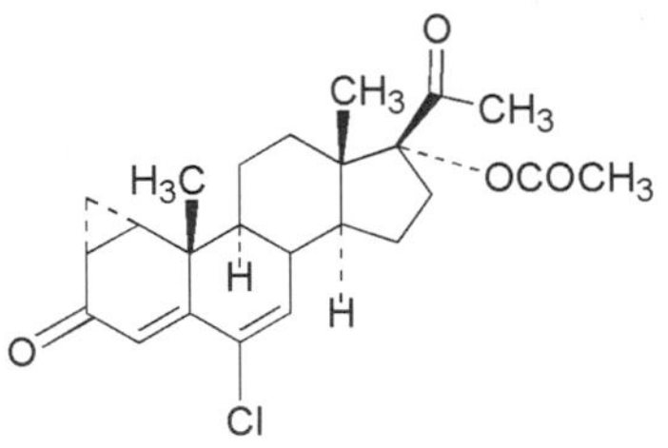

O CH3 CH3 OCOCH3 H3C H H O Cl

Cyproterone acetate (Cyprostat)
(14.87)

One of the most serious side effects of cyproterone is direct hepatotoxicity including jaundice, hepatitis, and hepatic failure which have been reported usually after several months of therapy in patients on large dose of 200–300 mg daily. For these patients, careful monitoring is necessary with liver function tests before and after treatment. Furthermore, any patient showing symptoms should be monitored, and the drug should not be given to patients with any type of liver problem unless the benefits are judged to outweigh the risks. Lethal cardiovascular events can also occur with cyproterone acetate, and at a slightly higher rate than with diethylstilboestrol (DES). Blood counts should also be monitored initially and throughout the treatment. Other side effects include fatigue and lassitude which may impair performance of skilled tasks such as driving. There is also a risk of recurrence of thromboembolic disease, diabetes mellitus, sickle-cell anemia, and severe depression. In addition, adrenocortical function should be monitored regularly throughout the treatment period.

Flutamide (Drogenil or Eulexin)

Flutamide was the first nonsteroidal antiandrogen to be developed, and was reported by Sherico in 1967. It is a prodrug, with the active metabolite hydroxyflutamide acting by inhibiting the nuclear binding of testosterone and DHT to the androgen receptor. Hydroxyflutamide has both peripheral and central activity on all androgen target cells.

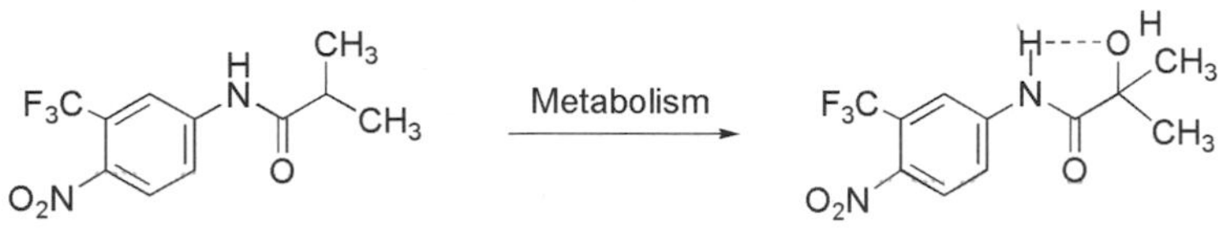

Metabolism

Flutamide (Drogenil)
(14.88)

Hydroxyflutamide (active metabolite)
(14.89)

Molecular modeling studies of hydroxyflutamide have attributed its greater binding affinity compared to flutamide to the dominant conformation in which the NH-proton is hydrogen-bonded to the hydroxyl function. The related agent bicalutamide is able to form a comparable hydrogen-bonded conformation and thus take up a similar three-dimensional shape. In the case of nilutamide, the structure contains a fully formed imidazolidinedione ring of similar three-dimensional shape.

Given orally, flutamide is licensed in the UK for use in advanced prostate cancer, although it should not be given to patients with cardiac disease or hepatic impairment. Due to the potential for hepatotoxicity (hepatic injury with occasional resulting deaths have been reported), periodic liver function tests are important for those on long-term therapy or when the first sign of liver disorder appears. One of the most common side effects is gynecomastia with approximately 60% of patients being affected, 10% of whom suffer severely. Nausea, vomiting, diarrhoea, increased appetite, insomnia, and tiredness can also be significant. Other reported side effects include decreased libido, inhibition of spermatogenesis, gastric and chest pain, headache, dizziness, edema, blurred vision, rash, pruritis, hemolytic anaemia, systemic lupus erythematosus-like syndrome, and lymphoedema.

Nilutamide (Anandron)

Nilutamide (Anandron) was first reported in 1977 by Roussel-UCLAF. It is a nonsteroidal anti-androgen with structural similarities to the hydroxyflutamide-active metabolite of flutamide. Unlike the pseudo five-membered rings of hydroxyflutamide and bicalutamide, which are held in place by hydrogen bonds, nilutamide (which does not require metabolic activation) contains a fully formed five-membered imidazolidinedione ring of similar three-dimensional shape.

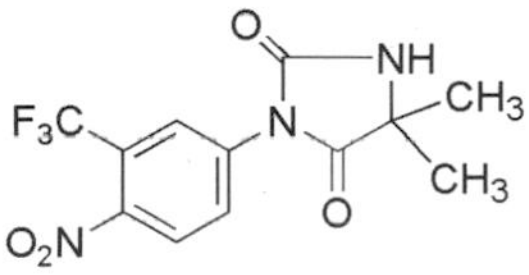

Nilutamide (Anandron)
(14.90)

Nilutamide is well absorbed and has a much longer half-life than flutamide (45 h compared to 5–6 h for flutamide), permitting a once-daily dosage. It is both centrally and peripherally active, and is therefore associated with a rise in plasma LH and subsequently plasma testosterone. Nilutamide is particularly useful in patients who are intolerant to flutamide.

Bicalutamide (Casodex)

Bicalutomide (Casodex) is a nonsteroidal peripherally active antiandrogen first reported by (the then) ICI in 1984. Interestingly, the structure allows a hydrogen-bonded pseudo five-membered ring to form in a similar manner to the active hydroxyflutamide metabolite, and this is thought to contribute to its activity.

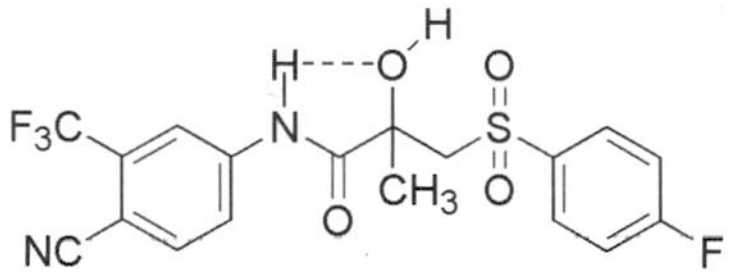

H H----O O F_3C N S CH_3 O O NC F

Bicalutamide (Casodex®)
(14.91)

Licensed in the UK for the treatment of prostate cancer, bicalutamide is well absorbed orally, and has a half-life of 5–7 days. It selectively blocks peripheral androgen receptors which is a potential advantage over centrally acting agents in that testosterone (DHT) synthesis will not be stimulated by LH-RH release from the hypothalamus. Side effects include dyspepsia, nausea, vomiting, diarrhoea, abdominal pain, asthenia, depression, gynaecomastia, breast tenderness, decreased libido, hot flushes, alopecia, hirsutism, pruritis, dry skin, hematuria, cholestasis, and jaundice. Other less common side effects include cardiovascular events and GI disturbances. Bicalutamide should not be given to patients with hepatic impairment.

14.11.3 Neuroendrocrine Tumors: Somatostatin Analogues

Neuroendrocrine (particularly carcinoid) tumors (and acromegaly) are usually stimulated by various hormone-releasing factors. Thus, agonist analogues of the hypothalamic release-inhibiting hormone Somatostatin (e.g., octreotide and lanreotide) can slow down the growth of such tumors.

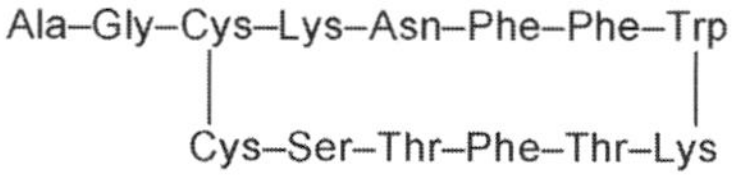

Ala–Gly–Cys–Lys–Asn–Phe–Phe–Trp
Cys–Ser–Thr–Phe–Thr–Lys

Somatostatin (hypothalamic release-inhibiting hormone)
(14.92)

Somatostatin (also known as hypothalamic release-inhibiting hormone (HR-IH), growth hormone-release inhibiting factor (GH-RIF), or somatotropin release inhibiting factor (SRIF or SRIF-14)) is a widely occurring cyclic tetradecapeptide with a molecular weight of 1637.91. It mediates, together with other peptides (e.g., somatoliberin), the neuroregulation of somatotropin secretion. In particular, it inhibits release of growth hormone, insulin and glucagons, and is also a potent inhibitor of a number of systems, including central and peripheral neural, gastrointestinal, and vascular smooth muscle. It was first isolated from ovine hypothalamic extracts and then structurally elucidated and synthesized in the early 1970s.

As somatostatin is a naturally occurring neuropeptide that is degraded rapidly in the bloodstream, attempts have been made to produce agonist analogues with drug-like properties. This approach has been successful, and octreotide (Sandostatin) and lanreotide (Somatuline LA) are licensed for the relief of symptoms associated with neuroendrocrine (particularly carcinoid) tumors and acromegaly. Octreotide is also licensed for the prevention of complications following pancreatic surgery, and can also be useful in reducing vomiting in palliative care and in stopping variceal bleeding.

It should be noted that growth hormone-secreting pituitary tumors can expand during treatment with somatostatin analogues leading to serious complications. Thus patients should be carefully monitored for tumor expansion. The gall bladder can also be affected, and the duration and depth of hypoglycemia produced should be monitored. Side effects with these agents mainly involve GI disturbances including anorexia, nausea, vomiting, abdominal pain and bloating, flatulence, diarrhea, and steatorrhea. In addition, postprandial glucose tolerance may be impaired, gall stones have been reported after long-term treatment and more rarely, pancreatitis has been reported. Pain and irritation may occur at the injection site.

Octreotide (Sandostatin)

Administered subcutaneously, octreotide is licensed in the UK for carcinoid tumors with features of carcinoid syndrome, VIPomas and glucagonomas. It is a long acting octapeptide analog of somatostatin first reported in 1981.

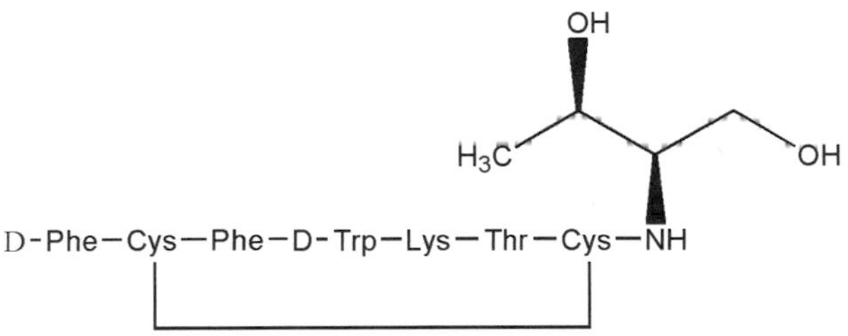

OH
H_3C OH
D-Phe—Cys—Phe—D-Trp—Lys—Thr—Cys—NH

Octreotide (Sandostatin)
(14.93)

A microsphere powder for reconstitution to an aqueous suspension depot preparation known as Sandostatin LAR is also available and is licensed as a longer acting preparation for neuroendrocrine tumors and acromegaly. Sandostatin LAR is designed for deep intramuscular injection into the gluteal muscle.

Lanreotide (Somatuline LA)

Also known as angiopeptin, lanreotide is an octapeptide disulfide analogue of somatostatin, and was first reported in 1987. Lanreotide in the form of copolymer microparticles for reconstitution as an aqueous suspension for intramuscular injection (i.e., Somatuline) is licensed in the UK for the treatment of neuroendrocrine tumors. It is also available as the acetate in a gel form suitable for injection (Somatuline Autogel).

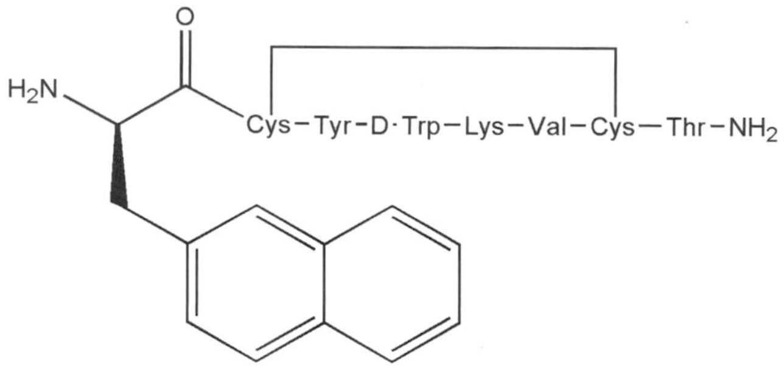

Lanreotide (Somatuline LA)
(14.94)

Side effects of lanreotide include asthenia, fatigue, and raised bilirubin. Other less common symptoms include skin nodules, hot flushes, leg pains, malaise, headache, tenesmus, decreased libido, drowsiness, pruritis, increased sweating and, rarely, hypothyroidism.

14.11.4 Estrogen Therapy

Estrogen therapy is occasionally used in postmenopausal women to treat breast cancer. It is rarely used to treat prostate cancer because of its side effects but will be considered if no other options are available. In prostate cancer, an estrogenic agent such as DES works by inhibiting the hypothalamic–pituitary system through a negative feedback (shut off) mechanism. A shortage of testosterone/DHT reaching the cells is signaled which results in increased production of testosterone by the testis (some of which will reach cells and occupy receptors) thus resulting in a fall in the secretion of luteinizing hormone (LH) from the pituitary and a subsequent decrease in testosterone synthesis by the testes (an outcome known as ''chemical castration'').

However, toxicity with this treatment is common, and dose-related side effects include nausea, fluid retention, and venous and arterial thrombosis. Impotence and gynecomastia always occur in men, and withdrawal bleeding may be a problem in women. Estrogen therapy may also cause hypercalcemia and bone pain when used to treat breast cancer. Ethinylestradiol (ethinylestradiol) is the most potent estrogen available and is used to treat breast cancer.

Diethylstilbestrol

Diethylstilbestrol (also known as stilboestrol or DHS) is a synthetic nonsteroidal estrogen first reported in 1938. Its structure is interesting from the medicinal chemistry perspective, as the two ethyl and two phenolic substituents arrange themselves around the central double bond to mimic a steroidal four-ring system in a similar manner to the estrogen inhibitor tamoxifen.

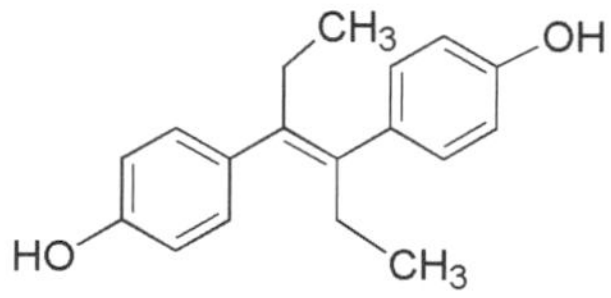

Diethylstilbestrol
(14.95)

Its use in prostate cancer has now lost favor due to the cardiovascular complications and the feminizing side effects associated with estrogen. Other side effects include sodium retention with edema, thromboembalism, and jaundice. Diethylstilbestrol is well absorbed orally, and is licensed in the UK for use in both breast and prostate cancers.

Ethinylestradiol

Ethinylestradiol is a synthetic steroid with high estrogenic potency which is well absorbed orally. Its synthesis from estrone was first reported in 1938.

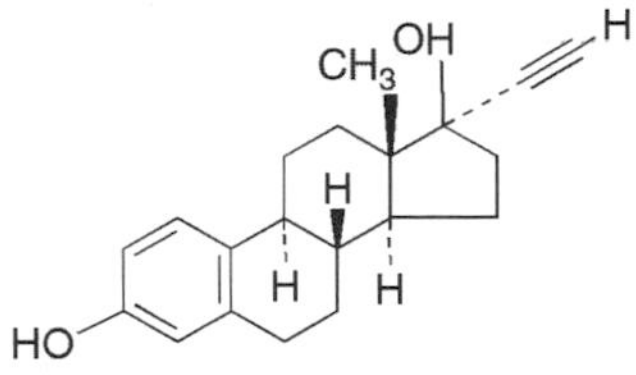

Ethinylestradiol
(14.96)

Ethinylestradiol is the most potent estrogenic agent available. Unlike other estrogens, it is only slowly metabolized in the liver. In the UK it is licensed for use in breast cancer, however it also appears in a large number of contraceptive products worldwide.

14.11.5 Progestogen Therapy

Progestogens are used in the treatment of breast carcinoma and other hormone-dependent cancers including endometrial cancer, and renal and prostatic adenocarcinomas. Four agents, gestonorone caproate, medroxyprogesterone acetate, megestrol acetate, and norethisterone are licenced in the UK for these uses. Medroxyprogesterone and megestrol are the agents of choice and can be given orally. Side effects of progestogen therapy in general are mild but may include nausea, fluid retention, and weight gain.

Gestonorone caproate (Depostat)

Also known as gestronol hexanoate or Depostat, gestonorone is a steroidal progestogen first reported in 1969. It is used in the treatment of prostatic hypertrophy.

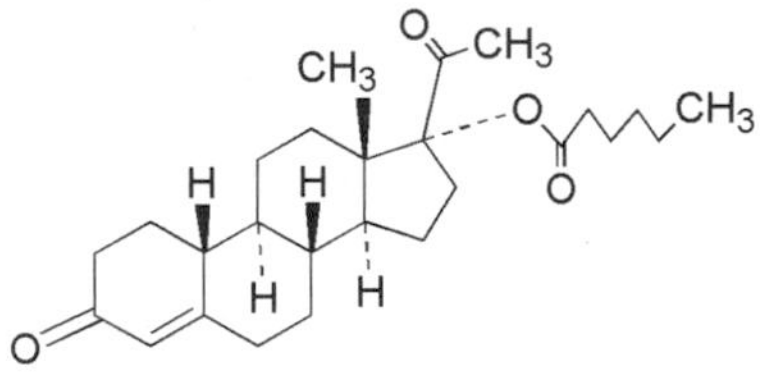

Gestonorone caproate
(14.97)

Given by intramuscular injection as an oily formulation, it is licenced in the UK for the treatment of benign prostatic hypertrophy and endometrial cancer.

Medroxyprogesterone acetate (Farlutal, Provera)

Medroxyprogesterone acetate is an orally active progestogen formerly used in combination with other steroids as an oral contraceptive. It was first reported by the Upjohn company in 1958.

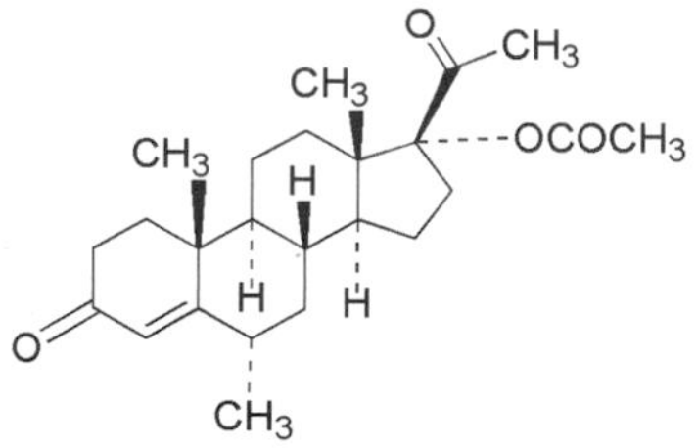

Medroxyprogesterone acetate (Farlutal, Provera)
(14.98)

In addition to the general adverse events relating to progestogens, glucocorticoid effects at higher dose levels may lead to a cushingoid syndrome. Given orally or by deep intramuscular injection into the gluteal muscle, medroxyprogesterone acetate is licensed in the UK for the treatment of breast carcinoma, and other hormone-dependent cancers including endometrial cancer, and renal and prostatic adenocarcinomas.

Megestrol acetate (Megace)

Megestrol acetate is an orally active progestin, formerly used in combination with other steroids as an oral contraceptive, and first reported by Searle in 1959. It is mainly used for the palliative treatment of breast and endometrial cancers.

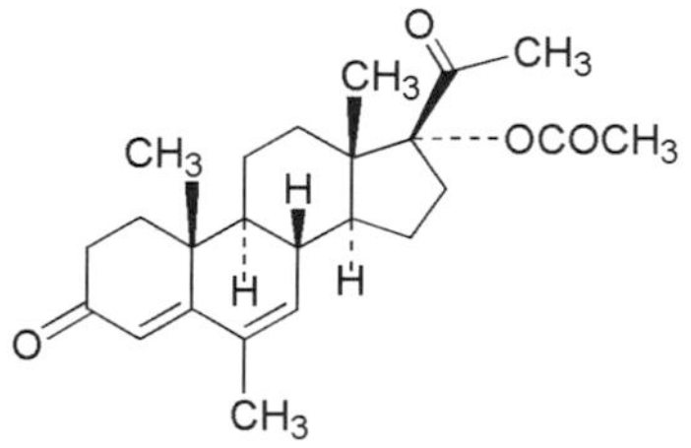

Megestrol acetate (Megace)
(14.99)

Taken orally, it is used for in both breast and endometrial cancers. It causes the usual side effects associated with progestogens.

Norethisterone

Norethisterone, also known as norethindrone in the USA, was first reported by Syntex in the mid-1950s.

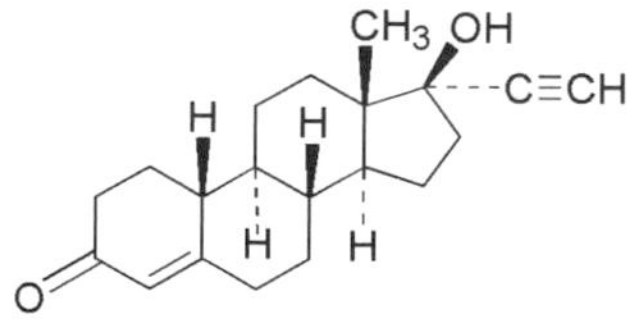

Norethisterone (Norethindrone)
(14.100)

Given orally, norethisterone is licenced in the UK for use in breast cancer.

14.12 VASCULAR TARGETING AGENTS

A new class of agents known as vascular targeting agents (VTAs) are being developed which may be useful in pathologies such as cancer where an abnormal growth of blood vessels is an essential component of the disease and its progression. VTAs differ from antiangiogenic agents in that they are designed to cause rapid and selective vascular shutdown in tumors in a matter of minutes to hours, rather than preventing new blood vessels forming. Such agents are designed to be used in intermittent doses to synergise with conventional treatments rather than be used chronically. VTAs are divided into two main classes: (1) small molecules which do not specifically localize to tumor endothelium but exploit the known differences between tumor and normal endothelium to induce selective vascular dysfunction, and (2) the biologics such as antibodies and peptides used to deliver toxins, and pro-coagulant and pro-apoptotic effectors to tumor endothelium. The former class of agent will be discussed below.

14.12.1 Combretastatin

The combretastatins are a family of *cis*-stilbene compounds discovered at the University of Arizona in the 1990s. Combretastatin A-4 (CA-4) is a tubulin-binding agent isolated from the root bark of the African shrub *Combretum caffrum*, also known as the Cape Bushwillow. Interestingly, Zulu warriors utilized a substance made from this shrub to prepare poison arrow tips and as a charm to ward off enemies.

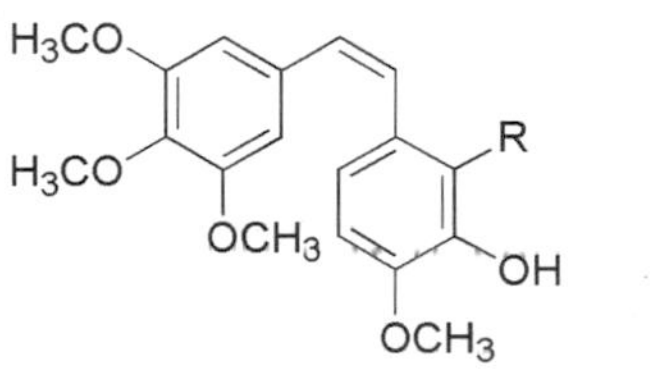

H_3CO H_3CO OCH_3 R OH OCH_3

Combretastatin A-1[1] (Oxi4500): R = OH
Combretastatin A-4[1]: R = H
(14.101)

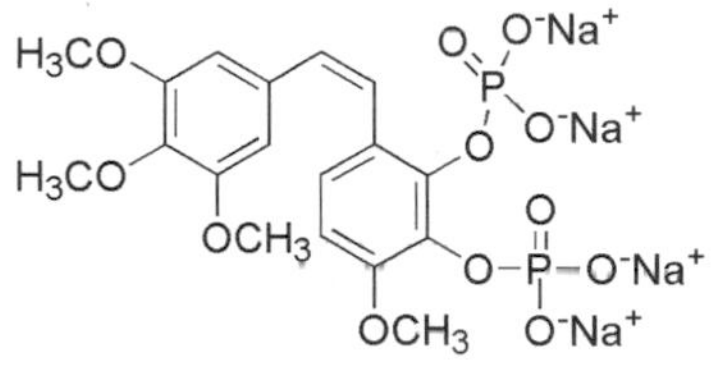

H_3CO H_3CO OCH_3 O^-Na^+ O^-Na^+ O^-Na^+ O^-Na^+ OCH_3

Oxi4503
(14.102)

The interesting feature of the combretastatins is the selectivity for tumor vasculature compared to that of healthy tissue. CA-1 is known to bind to tubulin with a higher affinity for β-tubulin than colchicine at or near colchicine-binding sites thus causing destabilization of the tubulin cytoskeleton. When this tubulin structure is disrupted, without an internal skeleton to maintain its elongated shape, the endothelial cells change from a flattened streamlined profile to a round-bloated profile, effectively plugging the capillaries and preventing blood flow to the tumor. This starves the tumor of nutrients, causing tumor cell death, exposing the basement membrane and, as a result, inducing hemorrhage and coagulation. The selectivity arises because in tumors the blood vessels are newly formed and immature compared to the mature endothelial cells that line blood vessels in normal tissues. It is only in newly formed endothelial cells that combretastatin is able to disrupt the internal skeleton that gives the endothelial cell its characteristic flat shape because tubulin alone is responsible for maintaining physical structure. The disruption cannot occur in mature endothelial cells because they are protected by the protein Actin (not present in immature endothelial cells) which protects the tubulin responsible for maintaining cell shape as it matures. Actin is not present until days after the formation of new endothelial cells. In addition to these fundamental differences between immature and mature endothelial cells, characteristics of the tumor microcirculation such as high interstitial fluid pressure, pro-coagulant status, vessel tortuosity, and heterogeneous blood flow distribution assist the selectivity process. The reduced blood flow results in significant hypoxia within the tumor, causing tumor cell death and regression. Tumor blood flow can be measured by a variety of methods including PET and magnetic resonance imaging (MRI).

SAR studies reveal that the *Z*-configuration of the combretastatins is an essential structural requirement for cytotoxic activity, and the appropriate distance between the two rings is also critical. Other tubulin depolymerizing agents are in clinical development (e.g., ZD6126 and AVE8062), along with DMXAA which has a different mode of action.

Oxi4503

Oxi4503 is a diphosphate prodrug of combretastatin A-1 that is ten times more potent than the monophosphate of CA-4, and has superior *in vivo* activity compared to CA-4. It is active against various cell lines including those that express the MDR phenotypes. Oxi4503 is presently in clinical trials and was designed to improve the pharmaceutical properties of the parent molecule, in particular by enhancing its water solubility. After injection, it is hydrolyzed to the parent CA-1 by enzymes in the blood stream and then quickly enters the endothelial cells that line blood vessels. It then damages the tumor vasculature thus starving tumors of nutrients.

14.13 TUMOR-TARGETING THERAPIES

This type of therapy involves the administration of a cytotoxic agent attached to an antibody. Although, in principle, the antibody should guide the drug selectively to the tumor cells, little success has been achieved in practice. One problem with this approach is that, once bound to the tumor cell, the drug–antibody conjugate may not be internalized and so little cytotoxic effect is achieved. For example, early studies with vinblastine–antibody conjugates demonstrated that although doses of the agent could be lowered thus reducing toxicity to some degree, the overall efficacy of the treatment was not a significant improvement over vinblastine treatment alone.

A number of different strategies are now available for targeting cytotoxic agents to tumor sites or for activating them inside or near a tumor, and some examples of these are described below. Gene targeting should also be considered as a tumor-targeting strategy and this is described in ''Gene targeting.''

14.13.1 Antibody-Targeting Therapies

Gemtuzumab ozogamicin (Mylotarg)

Gemtuzumab ozogamicin (Mylotarg) is the first chemotherapeutic agent targeted directly to cancerous cells using monoclonal antibody technology. It has a molecular weight of 151–153 kDa, and is composed of a recombinant humanized IgG_4, kappa antibody conjugated to a cytotoxic antitumor antibiotic, calicheamycin (a natural product isolated from the bacterium, *Micromonospora echinospora* sp. *calichensis* discovered by Wyeth-Ayerst researchers in caliche clay, a soil found in Texas). The antibody portion of Mylotarg binds specifically to the CD33 antigen, a sialic acid-dependent adhesion protein found on the surface of leukemic myeloblasts and immature normal cells of myelomonocytic lineage, but not on normal hematopoetic stem cells.

Gemtuzumab ozogamicin (Mylotarg)
(14.103)

Approximately 98.3% of the amino acid sequence of the anti-CD33 hP67.6 antibody is of human origin. The constant region and framework regions contain human sequences while the complementarity-determining regions are derived from a murine antibody (p67.6) that binds CD33. It is produced by mammalian cell suspension culture using a myeloma NS0 cell line and is purified under conditions which remove or inactivate viruses. The antibody was developed by the Fred Hutchinson Cancer Research Center in Seattle, and subsequently licensed to Wyeth-Ayerst. The anti-CD33 antibody was humanized by the Celltech Group, and then linked to *N*-acetyl-gamma calicheamycin via a bifunctional linker with a loading of 4–6 moles of calicheamycin per mole of antibody over 50% of the surface of the antibody.

Mylotarg works by targeting and binding to the CD33 antigen which is expressed on the surface of hematopoietic cells. For example, CD33 is expressed on leukemic blasts in more than 80% of patients with AML. It is also expressed on normal and leukemic myeloid colony-forming cells, including leukemic clonogenic precursors, but is not expressed on pluripotent hematopoietic stem cells or on nonhematopoietic cells. Binding of the anti-CD33 antibody portion of Mylotarg with the CD33 antigen results in the formation of a complex that is subsequently internalized. The calicheamycin derivative is then released inside the lysosomes of the myeloid cell. The calicheamycin binds to DNA in the minor groove, resulting in DNA double-strand breaks and cell death. In terms of side effects, the cytotoxic effect on normal myeloid precursors leads to substantial myelosuppression, but this is reversible because pluripotent hematopoietic stem cells are spared. Mylotarg has been approved for the treatment of patients aged 60 years or older with CD33-positive relapsed AML.

14.13.2 ADEPT

ADEPT is an acronym for "antibody-directed enzyme prodrug therapy," and is an advance over the use of antibody–drug conjugates alone. This treatment is based on the concept that a prodrug is usually converted into its active form by some enzymatic process. In ADEPT, an antibody–enzyme conjugate is initially administered to the patient which leads to the localization of enzyme at the tumor site. The prodrug is then administered so that conversion to the active cytotoxic agent only occurs when it comes into contact with enzyme at the tumor site. Significant progress has been made with the development of this approach, despite the fact that the choice of enzyme is limited to those not generally found in the body (e.g., bacterial enzymes). Carboxypeptidase and nitro reductase enzymes are presently being studied, and a number of potential prodrugs have been developed. Phase I clinical trials with a carboxypeptidase-release system have so far produced encouraging results.

14.13.3 GDEPT

Gene-directed enzyme prodrug therapy (GDEPT) is related to ADEPT in that an enzyme localized at the tumor site releases a cytotoxic agent from its prodrug form. However, instead of delivering the enzyme to the tumor via an enzyme–antibody conjugate, it is delivered using a gene therapy approach. For example, research is carried out at present into the use of viral vectors to introduce genes coding for specific enzymes into tumors and organ systems.

14.13.4 Photoactivated Drugs

PDT involves the administration of a non-toxic prodrug that can be activated selectively at the tumor site by light of a specific wavelength. This general strategy has been in use for a number of years for the treatment of psoriasis using 8-methoxypsoralen (PUVA treatment). This agent is relatively non-toxic until exposed to UV light when it cross-links DNA at thymine sites causing distortion of the DNA helix with consequent toxicity towards psoriatic cells.

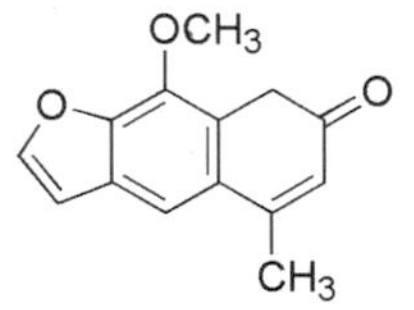

OCH_3 O O CH_3

8-Methoxypsoralen
(14.104)

This prodrug photoactivation concept was extended into cancer therapy in the mid-1990s when it was discovered that porphyrin-type molecules can be selectively taken up by some tumors. This led to the development of porfimer sodium (Photofrin) and temoporfin (Foscan) which are now used in the photodynamic treatment of various tumors. These drugs accumulate in malignant tissue and are activated by laser light to produce a cytotoxic effect. Although indications are presently limited to obstructing esophageal and NSCLCs, the rapid development of surgical lasers with flexible optical fibers has allowed experimental use of these agents for other tumors in inaccessible places such as other parts of the GI tract and the ovaries. An intense non-laser light source is also being marketed for this type of therapy, and many new types of prodrugs are under development.

Porfimer sodium (Photofrin)

Porfimer sodium is not a single entity but is a dark reddish brown mixture of oligomers formed by up to eight porphyrin units joined through ether and ester linkages. As part of a two-stage treatment process, it is administered by intravenous infusion over 3 to 5 min and this is followed 40–50 h later by illumination of the tumor with laser light of 630 nm wavelength. Porfimer sodium is licensed for use in obstructing esophageal and endobronchial NSCLCs which can be reached with laser endoscopy methods. It is also in use experimentally for other tumors of the skin and body cavities (e.g., stomach, colon, ovarian) which can be reached with a laser.

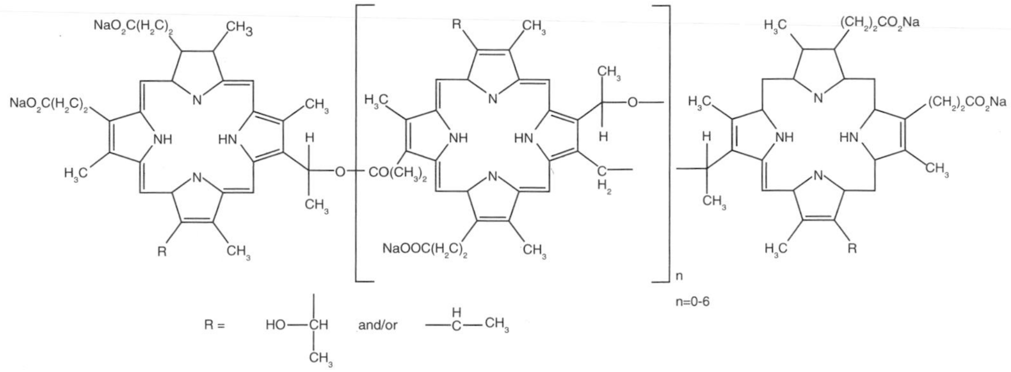

Photofrin (porfimer sodium)
(14.105)

The cytotoxicity and antitumor efficacy of porfimer sodium are both light- and oxygen-dependent. After intravenous injection, clearance from a variety of tissues occurs over 40–72 hours. Tumors, skin, and organs of the reticuloendothelial system (including liver and spleen) retain the agent for a longer time period. Therefore, selectivity in treatment occurs through a combination of selective retention of agent by the tumor and selective delivery of laser light which induces a photochemical rather than a thermal effect. Cellular damage caused by porfimer sodium is a consequence of the propagation of radical reactions which are initiated after the agent absorbs light to form singlet oxygen. Subsequent radical reactions can form superoxide and hydroxyl radicals. Tumor shrinkage can also occur through ischemic necrosis secondary to vascular occlusion which appears to be partly mediated by thromboxane A_2 release. The necrotic reaction and associated inflammatory responses may evolve over several days.

Side effects of porfimer sodium include severe photosensitivity with sunscreens offering no protection. Thus, patients are advised to avoid exposure of skin and eyes to direct sunlight or bright indoor light for at least 30 days. Some patients remain photosensitive for up to 90 days or more due to residual drug in all parts of the skin. Therefore, careful monitoring of treated patients is required and it is important to expose small test areas of skin to sunlight immediately after treatment to assess the degree of photosensitivity likely to occur. Constipation can also be a problematic but treatable side effect.

Temoporfin (Foscan)

The main advantage of temoporfin compared to porfimer sodium is that it is a single chemical entity rather than a mixture of oligomers. It is licensed for the PDT of advanced head and neck cancers.

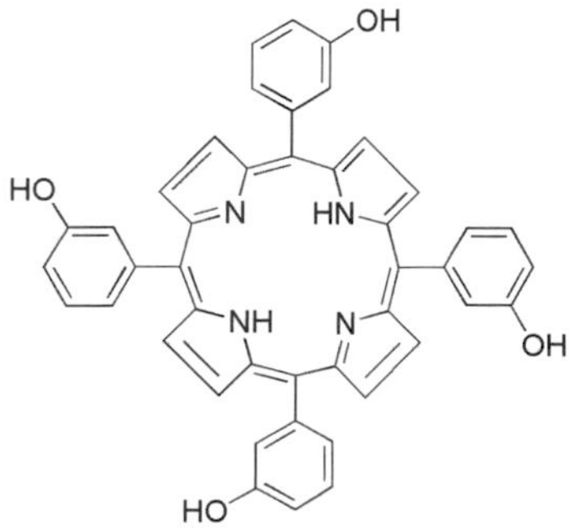

Temoporfin (Foscan)
(14.106)

Temoporfin is administered by intravenous injection over at least 6 min. Side effects include photosensitivity (as with porfirin sodium, sunscreens offer no protection), and so exposure of skin and eyes to direct sunlight or bright indoor light for at least 15 days after administration should be avoided. Other side effects include constipation, local hemorrhage, facial pain and oedema, possible scarring, and dysphagia. Temoporfin is contraindicated in patients with porphyria or other diseases exacerbated by light. In addition, ophthalmic slit-lamp examination cannot be carried out for 30 days after administration.

14.13.5 Bioreductive Agents

There have been numerous attempts to capitalize on the fact that the center of a large tumor is often deficient in oxygen due to a poor blood supply. It is possible that these ''bioreductive'' conditions can be used to trigger conversion of a suitable agent into an active form. The best known example of this is Mitomycin C which is discussed in ''Mitomycin C.'' It is thought that initial reduction of the quinone leads to transformation of the heterocyclic nitrogen from a conjugated amido to an amino form thus making it more electron-rich. This facilitates elimination of the ring junction methoxy group. Tautomerization of the resulting iminium ion and loss of the carbamate group then creates an electrophilic center which is susceptible to attack by a nucleophilic DNA base. Nucleophilic attack of the aziridine moiety by a nucleophile on the opposite strand of DNA also occurs, leading to an interstrand cross-link. The predominant adducts appear to be between two guanine-N2 groups within the minor groove.

14.14 BIOLOGICALS

Biological agents are large macromolecular agents, usually proteins or glycoproteins, that fall in to three categories: biological response modifiers (BRMs), immunotherapy agents, and enzymes. They are produced either by extraction from human cells or by genetic engineering processes. Biological agents may be used to stop, control, or suppress processes that permit cancer growth. They can also make cancer cells more recognizable and therefore more susceptible to destruction by the immune system. This can be achieved by boosting the killing power of immune system cells, such as T cells, NK cells, and macrophages. Biological agents can also encourage cancer cells to adopt the growth patterns of healthy cells, or block or reverse processes that change normal or precancerous cells into cancerous ones. Finally, biological agents may enhance the body's ability to repair or replace normal cells damaged or destroyed by other forms of cancer treatments such as chemotherapy or radiation, and prevent cancer cells from spreading to other parts of the body. Other agents in this category include enzymes such as asparaginase and vaccines. One of the major concerns when administering a biological product is hypersensitivity to the protein or glycoprotein and the possibility of severe anaphylaxis.

14.14.1 Biological Response Modifiers

Antibodies, interferons, ILs, and other immune system substances can be produced in the laboratory for use in cancer treatment and are known as BRMs. They can alter the interaction between the immune defenses and cancer cells to boost, direct, or restore the body's ability to fight the disease. Growth factors, colony-stimulating factors, and vaccines may also be considered as BRMs but these are presently experimental and are discussed elsewhere. This is a highly active area of research and many more agents are likely to become available in the near future.

Monoclonal antibodies

Monoclonal antibodies (see Chapter 12) can now be produced in large quantities to GMP standards. Their uses include targeting receptors on cell surfaces (e.g., Alemtuzumab), transporting cytotoxic agents to specific cells (e.g., Gemtuzumab), or for targeted prodrug therapies (e.g., ADEPT therapies).

Rituximab (MabThera)

Rituximab is an antibody that causes lysis of B lymphocytes, and is licensed for the treatment of chemotherapy-resistant advanced follicular lymphoma and for diffuse large B-cell non-Hodgkin's lymphoma in combination with other chemotherapy. It should be used with caution in patients with a history of heart disease as it can exacerbate angina, arrhythmias, and heart failure. Infusion-related side effects including cytokine release syndrome (characterized by severe dyspnoea) are common with rituximab and occur mainly during the first infusion. They include fever, chills, nausea, vomiting, allergic reactions such as rash, etc., flushing, and tumor pain. Transient hypotension can also occur during infusion. Premedication with an analgesic, antihistamine, and a corticosteroid can help with these side effects. More rarely, pulmonary infiltration and tumor lysis syndrome can occur.

Alemtuzumab (MabCampath)

Alemtuzumab is a monoclonal antibody that causes lysis of B lymphocytes. It is licensed for use in chronic lymphocytic leukaemia which has failed to respond to treatment with an alkylating drug, or which has remitted for only a short period (i.e., less than 6 months) following fludarabine treatment. In common with rituximab, it causes infusion-related side effects including cytokine release syndrome.

Interferons

Interferons are types of cytokines that occur naturally in the body, and were the first cytokines produced in the laboratory for use as BRMs. There are three major types of interferons: interferon alpha, beta, and gamma. Interferon alpha is the type most widely used in cancer treatment. It has been established that interferons can improve the way the immune system acts against cancer cells. For example, some interferons can stimulate NK cells, T cells, and macrophages, thus boosting the anticancer activity of the immune system. They can also act directly on cancer cells by slowing their growth or promoting their differentiation into cells with more normal behavior.

The first interferon was discovered in 1957 by scientists at the National Institute for Medical Research (UK). Interferons are glycoproteins normally induced in response to viral infections, and usually only effective in the species in which they are produced. Although initially studied for their antiviral activity, in 1981 doctors in Yugoslavia reported substantial improvements or total remissions in head and neck cancers after human leukocyte interferon preparations were injected directly into the tumors. Despite the intense interest after this discovery, a considerable time elapsed before agents were brought into clinical use, the major difficulty being commercial production of pure interferons in sufficient quantities. Relatively large doses are required for treatment, and initially only minute amounts of varying levels of purity were available from human tissue culture methods. However, in the early 1980s there were significant advances in production techniques, including the development of recombinant DNA technologies which allowed biosynthetic interferons of high purity to be made available in clinically useful quantities (see Chapter 12).

Combinations of interferon alpha with doxorubicin, cisplatin, vinblastine, melphalan, and cyclophosphamide have been evaluated in ovarian, cervical, colorectal, and pancreatic carcinomas with some promising results. However, dose-related side effects of interferon alpha include influenza-like symptoms, lethargy, and depression. Myelosuppression, affecting granulocytes, may occur along with cardiovascular problems such as hypotension or hypertension and arrhythmias. Interferon beta is available as interferon beta-1a (Avonex and Rebif) or as beta-1b (Betaferon); however, in the UK it is licensed only for various stages of multiple sclerosis and not cancer.

Interferon alpha (IntronA, Roferon-A, Viraferon)

Interferon alfa has demonstrated limited antitumor activity in certain lymphomas and solid tumors. It is also used in the treatment of chronic hepatitis B and C. Side effects are dose-related but commonly include anorexia, nausea, influenza-like symptoms, and lethargy. Ocular side effects and depression including suicidal behavior have also been reported. Myelosuppression may occur particularly affecting granulocyte counts, and cardiovascular problems, nephrotoxicity, and hepatotoxicity have also been reported. Other side effects include hypersensitivity, thyroid abnormalities, hyperglycemia, alopecia, psoriasiform rash, confusion, coma, and seizures.

The following preparations are licensed in the UK and are all administered by subcutaneous injection. IntronA which is interferon alfa-2b (rbe) is licensed for the treatment of hairy cell leukaemia, follicular lymphoma, chronic myelogenous leukaemia, lymph, or liver metastases of carcinoid tumor, and as an adjunct to surgery in malignant melanoma and maintenance of remission in multiple myeloma. It is also used for chronic hepatitis B and C. Roferon-A, which is interferon alfa-2a (rbe), is used for the treatment of AIDS-related Kaposi's sarcoma, hairy cell leukemia, chronic myelogenous leukemia, recurrent or metastatic renal cell carcinoma, progressive cutaneous T-cell lymphoma, follicular non-Hodgkin's lymphoma, and as an adjunct to surgery in malignant melanoma. It is also used for chronic hepatitis B and C. Viraferon, which is interferon alfa-2b (rbe), is licenced for chronic hepatitis B and C. A polyethylene glycol-conjugated (''pegylated'') derivative of interferon alfa-2b, known as peginterferon alfa-2b (Pegasis, PegIntron, and ViraferonPeg), has been introduced recently. Pegylation increases the persistence of the interferon in the blood. It is licensed for use in the treatment of chronic hepatitis C, ideally in combination with ribavarin.

Interleukins

ILs are proteins that occur naturally in the body. There are several classes which are being studied for their role in immunomodulatory and inflammatory processes. IL-1 has been shown to possess both direct and indirect antitumor effects. It has been investigated for its ability to protect bone marrow cells from the deleterious effects of radiation and chemotherapy. It can also release a cascade of hemopoietic growth factors. A recombinant IL, IL-2, is used clinically in metastatic renal cell carcinoma. However, the response rate is less than 50% and it causes capillary leakage leading to hypotension and pulmonary edema. Studies are underway to determine whether IL-2 can enhance the efficacy of tumor vaccines.

Interleukin-2 (Aldesleukin, Proleukin)

Aldesleukin, which is recombinant IL-2, is licensed for metastatic renal cell carcinoma. It is usually given by subcutaneous injection. It is now rarely given by intravenous infusion because of an association with capillary leak syndrome, which can cause pulmonary edema and hypotension. Aldesleukin produces tumor shrinkage in a small proportion of patients, but it has not been shown to increase survival. Bone marrow, hepatic, renal, thyroid, and CNS toxicities are common.

14.14.2 Immunotherapy

The immune system is a complex network of cells and organs that work together to defend the body against attacks by foreign or ''nonself'' invaders. This network is one of the body's main defenses against diseases including cancer. For example, the immune system may recognize the difference between healthy and pre-cancerous or cancerous cells and eliminate the latter before the disease fully developed. It follows that cancer may develop when the immune system breaks down or is not functioning adequately, a good example being Kaposi's sarcoma that is common in patients with AIDS.

Immunotherapies are designed to repair, stimulate or enhance the immune system's responses. Non-specific immunomodulating agents are substances that broadly stimulate or indirectly augment the immune system. More specific agents target key immune system cells and cause secondary responses such as increased production of cytokines and immunoglobulins. BCG is a non-specific immunomodulating agent commonly used in the treatment of superficial bladder cancer either before or after surgical removal of the tumor.

BCG bladder instillation (ImmunoCyst and OncoTICE)

Bacillus Calmette-Guérin (BCG) is a live attenuated strain derived from *Mycobacterium bovis*. It is licensed as a bladder instillation for the treatment of primary or recurrent bladder carcinoma, and for the prevention of recurrence following transurethral resection. It is contraindicated in patients with active tuberculosis, an impaired immune response, or with other infections (e.g., HIV, urinary tract infection). Side effects include cystitis, dysuria, urinary frequency, hematuria, malaise, fever, influenza-like syndrome, and potential systemic BCG infection. More rarely, hypersensitivity reactions can occur (e.g., arthralgia and rash), orchitis, transient urethral obstruction, bladder contracture, renal abscess, and ocular disturbances. The two commercial products, ImmunoCyst and OncoTICE are derived from the Connaught and TICE strains, respectively.

14.14.3 Enzymes

One successful enzyme-based therapeutic strategy involves the use of an enzyme to break down, and reduce systemic concentrations of a nutrient required by tumor cells. The selectivity for tumor cells is achieved because healthy cells are usually able to synthesize supplies of certain nutrients for themselves whereas, due to genetic abnormalities, tumor cells may be deficient in some aspect of the biosynthetic pathway and thus become starved of the nutrient. The best known example of this strategy is the therapy utilizing the enzyme asparaginase which is used to break down the amino acid asparagine, thus lowering its concentration in blood and tissues.

Asparaginase (Crisantaspase; Erwinase)

Crisantaspase (Erwinase) is the enzyme asparaginase produced by *Erwinia chrysanthem*. It is a 133,000 molecular weight tetrameric protein used almost exclusively in the treatment of acute lymphoblastic leukemia. The mechanism of action is based on the fact that these particular tumor cells have very low levels of asparagine which is required for cell growth; instead they must obtain this amino acid exogenously. Healthy cells, on the other hand, can synthesize their own asparagine. The treatment involves the intramuscular or subcutaneous administration of crisantaspase which reduces the concentration of asparagine in the body by converting it to aspartic acid and ammonia, thereby removing it from the protein synthesis cycle (Equation (14.14)). Whereas healthy cells rapidly synthesize their own supply of asparagine, the tumor cells succumb to the reduced levels in their environment.

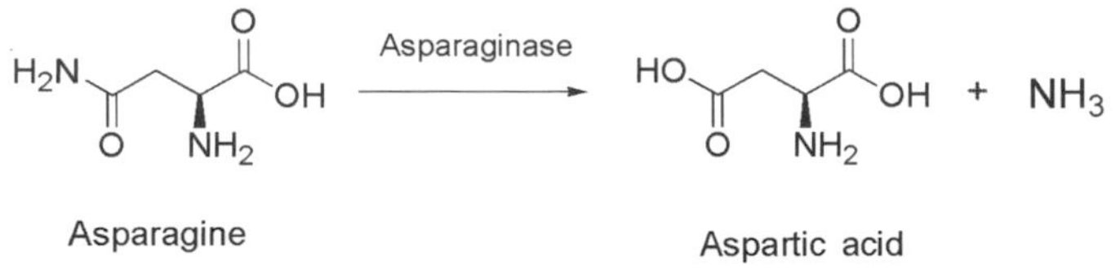

Asparagine —Asparaginase→ Aspartic acid + NH_3 (14.14)

Side effects include nausea, vomiting, CNS depression, and liver function and blood lipid changes. As crisantaspase is a large protein, facilities for management of anaphylaxis should always be available during administration. Resistance to the drug develops when the tumor cells begin to synthesize their own asparagine.

14.15 MISCELLANEOUS AGENTS

14.15.1 Hydroxycarbamide (Hydroxyurea; Hydrea)

Hydroxycarbamide (Hydrea), also known as hydroxyurea, is an orally active drug first synthesized in the 1860s. Its antitumor activity, which was discovered much later, is thought to be due to inhibition of ribonucleotide reductase with a resulting effect in the S-phase of the cell cycle. The drug causes a depletion of the deoxynucleoside triphosphate pool and blocks DNA synthesis and repair.

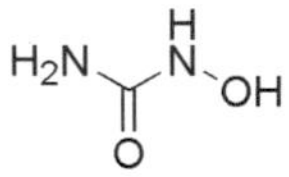

Hydroxycarbamide (hydroxyurea; Hydrea)
(14.107)

Hydroxycarbamide is used mainly in the treatment of chronic myeloid leukaemia, often in combination with other drugs, but has also been used (off-label) in melanoma. It is occasionally used for polycythemia (the usual treatment is venesection). It induces fetal hemoglobin production in patients with sickle-cell anemia and so is also used in the treatment of this disease. The most common toxic effects are myelosuppression, nausea, and skin reactions.

14.5.2 Razoxane

Razoxane is a cyclized analog of EDTA, first synthesized in the late 1960s. It has limited activity in leukemias, and is little used in the clinic.

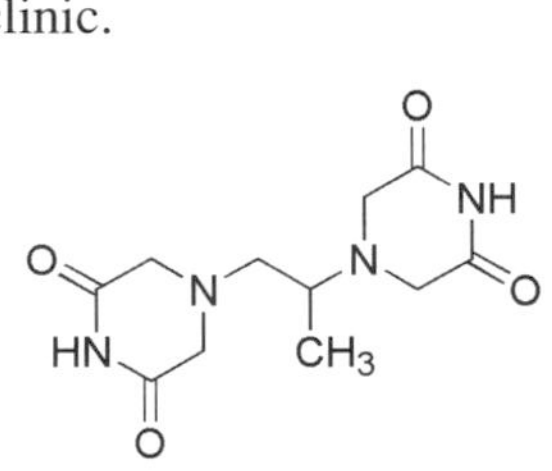

Razoxane
(14.108)

Not surprisingly, this agent exhibits intracellular iron-chelating activity. It also plays a role in preventing the formation of doxorubicin–iron complexes which generate radical oxygen species that are associated with cardiotoxicity.

14.16 THE FUTURE

Cancer research is funded mainly through pharmaceutical companies, but also by sources including government bodies and initiatives, and charitable organizations and trusts. It is sometimes argued that the current lack of efficacious drugs and therapeutic strategies to treat cancer is a reflection of the complexities of the disease and the shortage of ideas for novel agents and therapies rather than a lack of funding for basic research. Given the resources that have been channeled into cancer research worldwide since President Nixon declared his famous ''War on Cancer'' in 1971, the present lack of effective drugs and treatments would tend to support this view. Some examples of new areas currently under development are briefly described below. New research tools available to scientists in cancer research are also discussed, and the developing area of chemoprevention briefly described.

14.16.1 New Therapies

Gene targeting

Once a gene has been identified and associated with a particular disease such as cancer (e.g., BCR/ABL in the case of CML), it should be theoretically possible to develop agents capable of selectively targeting either the gene itself or the equivalent messenger RNA. This should prevent production of only the protein relating to the ''faulty'' gene. Selectivity is the most important issue to consider with this strategy, as it has been calculated that it may be necessary to actively recognize up to 15 to 20 base pairs of DNA or RNA in order to selectively target one gene in the entire human genome (approximately >30,000 genes). However, in practice, sufficient selectivity to minimize side effects might be acceptable.

A large number of oncogenes have now been identified and sequenced, and technologies to design and produce molecules with DNA/RNA-recognition properties are rapidly advancing. Agents that target DNA or the corresponding messenger RNA are known as antigene or antisense agents, respectively. Ribozymes are a special family of oligonucleotides that target and bind to mRNA but then induce cleavage. One nucleic acid targeting technology (e.g., antisense) takes advantage of the hydrogen bonding interactions that allow nucleic acid strands to pair together. However, the same objective is achievable with lower molecular weight ligands (e.g., hairpin polyamides).

Antigene

One overriding advantage of the antigene strategy is that inhibition of just one gene will prevent the formation of numerous copies of the mRNA transcript and thus the corresponding protein. The experimental macromolecular antigene approach utilizes oligonucleotides that interact with double-stranded DNA to form a so-called ''triple helix.'' The oligonucleotide, which is typically 15–20 base pairs in length, lays in the major groove and is held in place mainly by hydrogen bonding interactions. Although a great deal of effort has been put into this area, two major problems remain. First, recognition of DNA is presently limited to runs of cytosines or thymines, and this level of selectivity is presently insufficient to successfully target clinically important gene sequences. Secondly, in practice, oligonucleotides do not make ideal drugs as they suffer from a number of problems including poor stability, pharmacokinetics, and cellular penetration. A great deal of effort has been made to stabilize oligonucleotides towards chemical or enzymatic degradation, and this has been achieved by chemically modifying the backbone phosphate groups. Strategies to enhance the stability of oligonucleotide–DNA complexes have also been pursued; for example, by tethering oligonucleotides to intercalating moieties.

In the small molecule area (e.g., molecules of less than 1000 in molecular weight), agents have now been produced that can recognize more than 10 base pairs of DNA, and this is an intense area of research. For example, the hairpin polyamides are polymers of five-membered heterocycles that fold back on themselves to form hairpin structures that bind in the minor groove of DNA. The intriguing aspect of hairpin polyamides is that the five-membered heterocycles can be chosen to specifically recognize individual base pairs. However, although it has been possible to design hairpin polyamides with impressive sequence-selectivity for stretches of naked DNA, they do not appear able to penetrate the nuclear membrane in most cells in *in vitro* experiments. Other approaches to the design of antigene agents that interact in the minor groove of DNA are currently underway that utilize molecules based on the naturally occurring PBD antitumor antibiotics that are known to penetrate the nucleus of cells. Such molecules have been shown to inhibit transcription *in vitro* in a sequence-selective manner, and it is anticipated that this activity will translate the *in vivo* situation.

Antisense

An antisense oligonucleotide is a relatively short length of single-stranded nucleic acid (e.g., 15–20 base pairs) with a sequence complementary to a region of the target mRNA. Hybridization with the RNA then interferes with the process of translation. However, a similar set of problems exist to those described above with regard to poor stability, pharmacokinetics, and cellular penetration. In addition, this approach is not likely to be as efficient as the antigene strategy in which inhibition of just one copy of the gene should prevent many copies of mRNA from being produced. Despite disappointments with clinical evaluation of antisense technology in the mid-1990s, there is currently a resurgence of interest, and a number of agents are presently in clinical trials. For example, oblimersan sodium (Genesense), an antisense agent targeted to the Bcl-2 oncogene is in clinical trials for the treatment of melanoma in combination with dacarbazine. Similarly, GEM-231, an agent targeted against Protein Kinase A R1α (PKA) is being evaluated against a number of solid tumors in combination with irinotecan (Camptosar). AVI-4126, an antisense agent targeted to the c-myc oncogene is being evaluated in prostate cancer. Research is also underway to find small molecule ligands (e.g., molecular weight <1000) that might interact with messenger RNA in a sequence-selective manner, thus avoiding the pharmaceutical problems apparent with antisense oligonucleotides.

Ribozymes

Ribozymes are similar to antisense oligonucleotides in that they target and bind to mRNA. The base-pair sequence of the ribozyme allows it to adopt a hairpin structure when associating with RNA. Once bound, ribozymes can elicit cleavage of the bound mRNA fragment by invoking an enzyme known as RNase H. Ribozyme agents have been in development since the mid-1990s, but so far no clinical candidate has emerged.

RNA interference

RNA interference (RNAi) is a recently identified phenomenon in which the introduction of short double-stranded RNA (dsRNA) fragments into a cell can downregulate or ablate the production of protein for the gene associated with the sequence of the dsRNA introduced. This natural process that occurs in all cells is thought to act as protection from infections with RNA viruses. Critically, the cell responds to the introduction of a foreign double-stranded form of small interfering RNA (siRNA) by destroying all internal mRNA with the same sequence. The phenomenon was first observed in the *C. elegans* worm, and later in *Drosophila*, trypanosomes, and planaria. The post-transcriptional gene silencing observed in plants is also thought to be due to a related RNAi mechanism.

In the natural pathway, dsRNA present in a cell through infection is recognized by the cell as foreign. First, it is bound by the enzyme DICER which cuts the dsRNA into nucleotide duplexes each containing 21 bases and known as small interfering RNAs or siRNAs. These in turn become bound by a complex of proteins called the RNA-induced silencing complex (RISC). One strand of the siRNA duplex is then stripped off, leaving the remaining ''complimentary'' strand associated with the protein complex. The RISC complex then locates and binds a target RNA using the complimentary siRNA strand as a guide. Only those RNAs that can form a complete set of Watson–Crick base pairs with the siRNA complimentary strand will bind, thus ensuring that only the specific target RNA is bound and destroyed. Once the RISC complex correctly matches up the complimentary siRNA strand with its specific target, the target RNA is cleaved and destroyed enzymatically.

This natural phenomenon can be potentially utilized to downregulate a viral RNA, a normal cellular RNA, or a disease-related RNA. The process can be initiated by introducing into the cell a 21 base-pair dsRNA with a sequence matching the mRNA of the gene intended for downregulation or ablation. siRNAs are remarkably potent and only a few dsRNA molecules per cell are required to produce effective interference. The reason for this is that a single dsRNA molecule can mark hundreds of target mRNAs for destruction before it is ''spent,'' and, in addition, the RISC complexes work catalytically with many mRNA molecules being destroyed by one complex. If the target RNA is a viral genome, then viral replication will be blocked. Similarly, if the target RNA is disease-related then protein synthesis by that gene will be prevented and the disease may be ameliorated.

As with the other antigene and antisense approaches that utilize oligonucleotides, the major problem with developing a therapy based on this technology relates to the difficulties associated with delivering dsRNA constructs to cells, which are even more severe than delivering single-stranded oligonucleotides as in the antisense approach. Therefore, although the RNAi phenomenon would appear to have great promise for the *in vivo* downregulation or ablation of oncogenes critical to cancer cells, at present potential therapies are based on ''gene therapy'' approaches where a viral construct is used to deliver a fragment of DNA that can be inserted in the host genome. This insertion is then used to synthesize an RNA that is designed to fold back on itself to produce a double-stranded structure as well as containing the critical targeting sequence. However, as with all gene therapy approaches, the problem is that a viral vector will typically deliver the DNA fragment to all cells of the body with the consequent risk of collateral genomic damage. Therefore, as with antisense, it is likely that the first applications of this technology will be in topical therapies where systemic drug delivery issues may be avoided.

Gene therapy

Gene therapy for the treatment of cancer usually involves the use of a viral vector to deliver a working version of a mutated gene known to be associated with the tumor. One example is replacement of a mutated tumor suppressor gene such as p53. Another strategy is to deliver a non-human gene that can produce an enzyme capable of activating a prodrug. For example, the experimental GDEPT involves selective delivery to the tumor of genes capable of expressing bacterial carboxypeptidase or nitroreductase enzymes (see Section 14.13.3). Systemic administration of enzyme-activatable prodrugs then leads to local drug action at the tumor site.

Delivery of genes to the tumor site is often attempted using viral vectors such as the adenovirus. However, the main problem is that the viral constructs are rarely tumor or organ specific and so the DNA insert may be delivered, in the worst case scenario, to all healthy cells of the body, thus leading to concerns about genomic damage in healthy cells. Much work is underway to develop other delivery systems such as liposome-based vehicles. However, selective delivery to tumor cells remains a significant challenge in this area of research.

Other therapeutic strategies

Resistance inhibitors

Drug resistance provides a major obstacle to successful cancer chemotherapy, particularly as it can develop to most classes of agents. Mechanisms of resistance include the induction of the multiple drug resistance (MDR), breast cancer resistance protein (BCRP), and related proteins which can actively transport different classes of drugs out of a cell, an increase in glutathione production which can "neutralize" alkylating agents, and an enhancement of metabolism or DNA repair.

It is postulated that agents capable of blocking any of these processes might not only enhance the action of most anticancer drugs, but may also allow them to be used for longer periods of time without resistance developing. One early lead in this area came from the observation that co-administration of verapamil (a calcium-channel blocker) can enhance the effect of some anticancer drugs, and several agents of this type are now being evaluated in the clinic. A related strategy is to develop "chemoprotecting" agents that can protect healthy cells during chemotherapy. There is parallel research into the development of "radiosensitizing" agents to increase the sensitivity of tumors to radiation during radiotherapy, and "radioprotecting" agents that can protect surrounding healthy tissue during treatment.

DNA repair inhibitors

The efficiency of DNA-interacting agents such as the nitrogen mustards and cisplatin is known to be reduced by the numerous repair systems that normally protect DNA from damage by carcinogens, radiation, and viruses. DNA repair is mediated by a remarkable set of enzymes that first recognize the damage incurred and then set out to repair the lesion, usually by excising the damaged segment of DNA and re-synthesising a new one. The mechanism by which these enzymes work is an active area of research, and the first crystal structure of a repair enzyme has only just become available. There is presently interest in developing agents capable of inhibiting DNA repair as, theoretically, they could enhance the therapeutic value of the DNA-binding class of anticancer drugs. However, presumably there is an associated risk that carcinogens and other sources of DNA damage (e.g., radiation) might be rendered more dangerous during the treatment period.

The poly(ADP) ribosylation (PADPR) of damaged DNA is thought to be associated with the attraction of repair enzymes to the site, and so there has been a considerable effort to develop PADPR inhibitors. So far, a number of potential inhibitors have been identified, some based on simple benzamide analogues. One such agent is presently in clinical trials.

Telomerase inhibitors

Telomeres are repeat sequences of DNA found at the end of chromosomes. They are thought to be crucial for cell division during which they signify the end of a chromosome molecule. A number of telomeres are lost during each cycle of cell division, and this is thought to act as a type of "biological clock" leading to a natural cell death when the telomeres have been depleted. Telomerase is an enzyme that can add new telomeric repeats to the ends of chromosomes thus extending the life of cells. Although telomerase is not expressed in normal cells which have a defined life cycle, cancer cells exploit the telomerase enzyme to attain a state of immortality. Hence, over 80% of different tumor types have been shown to express this enzyme. Telomerase has now been purified, assays have been established, and a number of telomerase inhibitors have been identified and are in development.

Histone deacetylase (HDAC) inhibitors

Transcription in cells is controlled by a number of mechanisms including the degree of acetylation of the lysine residues of the histone N-terminal tails. This is achieved by two families of enzymes known as HATs and HDACs. These enzymes dictate the pattern of histone acetylation and deacetylation which is associated with transcriptional activation and repression, respectively.

HATs, of which there are five classes, catalyze the addition of an acetyl group to the lysine residues of histone N-terminal tails thereby masking the residues' charge. This causes a decreased affinity of the histones for DNA, which is generally associated with transcriptional activity. Deacetylation by HDACs removes these charge-neutralizing acetyl groups. The protein then becomes more positively charged, causing an increased affinity of the histones for DNA and leading to transcriptional repression. HATs and HDACs do not work independently but exist in multiprotein complexes which control transcriptional activation or repression.

HDAC inhibitors inhibit the process of histone deacetylation, thus initially leading to transcriptional up-regulation. However, it is not clear why this should exert a selective cytotoxicity in tumor cells as opposed to healthy cells. It is thought that they must initially activate a selected number of genes whose expression causes inhibition of tumor cell growth and induction of apoptosis. There are three classes of HDACs, Classes I, II, and III. The current HDAC inhibitors inhibit Class I and II, but not III. Although valproic acid has been shown to inhibit HDACs of Class I with more efficiency than Class II, a great deal of research effort is currently underway to identify novel inhibitors capable of selectively targeting specific classes or individual members of each family.

Several structural classes of HDAC inhibitors exist which include the hydroxamates (e.g., Trichostatin A [TSA] and suberoylanilide hydroxamic acid [SAHA]); the cyclic tetrapeptides (e.g., FR901228 and apicin); the aliphatic acids (e.g., sodium/phenyl butyrate and valproic acid); the benzamides (e.g., MS-275); and the ketones (e.g., trifluoromethyl ketones, α-ketomides).

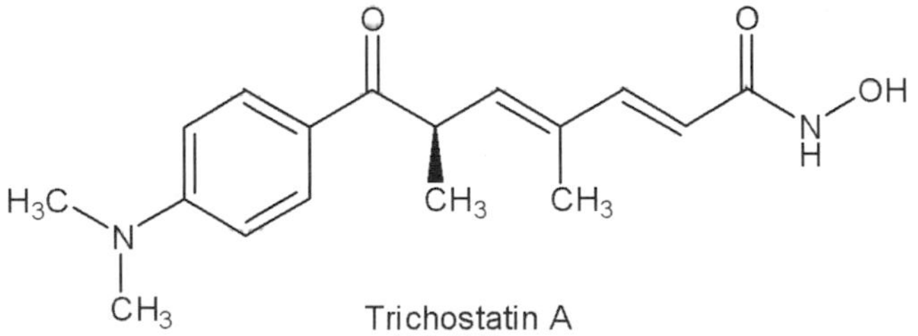

Trichostatin A
(14.109)

The epigenetic inactivation of genes via histone modification through HATS and HDACs has been studied in a variety of tumor types *in vitro*, where effects can include growth arrest, activation of differentiation, or induction of apoptosis. For example, it has been demonstrated that some HDAC inhibitors can reverse the malignant phenotype of lung cancer cells growing *in vitro*. Furthermore, altered HDAC expression in lung cancer cells leads to unresponsiveness to TGF-β, which normally inhibits epithelial cell proliferation, suggesting that HDAC inhibitors should be useful in the treatment of lung cancers. Interestingly, some HDAC inhibitors such as the experimental agent FR901228 have overcome both multidrug-resistance protein (MRP)- and BCRP-mediated chemotherapy resistance in small cell lung cancer cell lines.

A number of Phase I and II clinical trials are presently underway to evaluate HDAC inhibitors as single agents against various tumor types, including lung cancers and hematological malignancies. The FDA has already granted orphan drug status to SAHA, a drug candidate which is in Phase I and II clinical trials for a variety of cancers including multiple myeloma. However, it is likely that the optimal use of HDAC inhibitors will be in combination with established cytotoxic drugs.

Antimetastatic agents

A primary tumor is not often the direct cause of death of a patient as it can be removed by surgery or treated with radio- or chemotherapy. It is usually the secondary tumors (metastases) that lead to death as they become too dispersed throughout the body to make further treatment, particularly surgery, effective. Attention has recently been focused on the mechanism by which tumor cells move around the body, either via the blood or the lymph system, and establish themselves in new locations. There is an intense effort to develop drugs that can interfere with the metastatic process. For example, the metalloproteinase enzyme family has been identified as important for the metastatic process, and one novel agent, Marimastat (a metalloproteinase inhibitor), has been evaluated in clinical trials but not progressed due to lack of efficacy. However, research in this area continues, and there is now an effort to find inhibitors of the MET kinase enzyme which is also associated with the metastatic process.

Blood-flow modifying agents

A good blood supply is crucial for the growth of a primary tumor and can be even more important for the establishment of a metastatic tumor at a distant location. The enzyme angiogenin has been identified as important for establishing new microvasculature at the site of a metastatic tumor. The structure of angiogenin has now been established from x-ray crystallographic studies, and an effort is currently underway to design inhibitors.

Vaccines

Vaccination against cancer is an active area of research, with the objective of either activating or inducing a host response to tumor-associated antigens. For example, patients with melanoma have been treated with a melanoma vaccine using BCG as an adjuvant. While there was evidence to suggest an immunological response in some patients, efficacy was poor overall. Tumors associated with viruses such as Hepatitis B and HPV are also targets for a vaccination approach to cancer treatment or prophylaxis. With hepatitis, a combination of hepatitis immunoglobulin and heat-inactivated Hepatitis B vaccine has produced greater protection than vaccination with hepatitis immunoglobulin alone. A number of clinical trials are presently underway to evaluate HPV vaccines in the prophylaxis of cervical cancer.

Growth factors

Growth factors or cytokines are proteins which affect cell growth and maturation. Recombinant technology (see Chapter 12) has allowed the production of large amounts of cytokines and there are several being evaluated in clinical trials. For example, hemopoietic growth factors have found a use in counteracting the myelosuppressive side effects associated with many anticancer agents. Granulocyte colony stimulating factor (G-CSF) and granulocyte-macrophage colony stimulating factor (GM-CSF) increase the circulating number of neutrophils, eosinophils, and macrophages by inducing inflammation, and it has been shown that some tumor cells possess receptors for these CSFs. The dosing regimen is usually once daily, with side effects including influenza-like symptoms. Erythropoietin has also been shown to be clinically useful in treating certain types of malignant anemia, and is naturally produced by the body in response to hypoxia. Recombinant technology has now been applied to the production of erythropoietin.

Inhibiting growth factors can lead to useful antitumor activity, and known inhibitors include octreotide (Sandostatin®) which is in clinical use, and suramin (a polysulfonated naphthylurea). These two compounds are analogues of somatostatin, a naturally occurring growth hormone.

Octreotide is administered subcutaneously and is useful in controlling symptoms, but does not always cause tumor reduction. Suramin binds proteins extensively due to its polyanionic nature. Early clinical trials with suramin were hampered by severe toxicities (renal and liver dysfuntion, adrenal insufficiency, and peripheral neuropathy), but with suitable dosage adjustments clinical trials in breast and prostate cancers appear promising. The clinical use of lymphokine (a protein or glycoprotein produced by a lymphocyte) and natural killer cells is also under investigation.

Tumor necrosis factor

TNF is a glycoprotein produced by macrophages, monocytes, and natural killer cells, and is partly responsible for tumor cell lysis. Phase I and II studies, administering the agent either intravenously or intramuscularly, have produced mixed results. Part of the problem is ensuring that a high enough concentration of TNF reaches the tumor site, but this has to be balanced against adverse effects which include hypotension and cardiotoxicity. Transient fever has also been observed in some patients, as have hematological disturbances, the latter being reversible on cessation of treatment. Studies on the potential clinical applications of TNF are continuing.

Immunomodulation

Thalidomide is a chiral glutamic acid derivative first reported in 1957 by Chemie Grünenthal. It is very poorly soluble and exists as a racemic mixture of two enantiomers. Thalidomide was approved in various European countries as a sedative in 1957 and became widely used in the rest of the world, specifically marketed to pregnant women. By late 1961, the association between fetal abnormalities and thalidomide became recognized. Although it was withdrawn commercially at this time, thalidomide has continued to be investigated for other conditions including erythema nodosum leprosum, AIDS wasting, non-microbial aphthous ulcers of the mouth and throat associated with AIDS, multiple myeloma, and other malignancies. There is presently a resurgence of interest in thalidomide, and it is being further evaluated for the treatment of resistant or relapsing multiple myeloma.

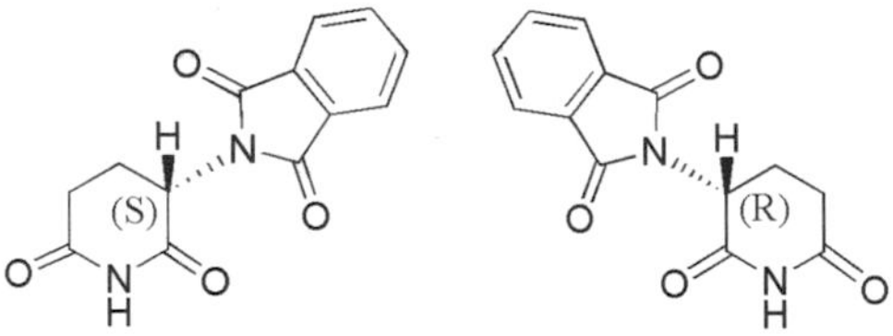

Enantiomers of thalidomide

(14.110) **(14.111)**

The main pharmacological activity of thalidomide is as an immunomodulator. However, the actual mechanism of action in multiple myeloma has not been fully established. Potential mechanisms include direct inhibition of myeloma cell growth and survival, anti-angiogenesis, suppression of the production of TNF-α, inhibition of selected cell surface adhesion molecules that assist leucocyte migration, shifts in the ratio of $CD4^+$ lymphocytes (helper T cells) to $CD8^+$ lymphocytes (cytotoxic T cells), and effects on ILs and interferon-γ. Interestingly, the molecule can also exhibit partial immunostimulatory as well as anti-inflammatory effects.

The most commonly observed adverse reactions in humans include somnolence, constipation, and polyneuropathia leading to painful tingling and numbness. Furthermore, hypersensitivity with signs such as maculo-papulous rash, possibly associated with fever, tachycardia, and hypotension may be observed.

Thalidomide is administered as a racemic mixture. The enantiomers differ from the racemate in each having higher water solubility by a factor of about three, and there is indirect evidence that enantiomerically pure thalidomide is absorbed more readily and undergoes faster hydrolytic cleavage. *In vitro* and *in vivo* studies have shown that, under physiological conditions, there is very rapid racemization, so that whichever enantiomer is administered, a similar equilibrium mixture will quickly be established. The (+)-(*R*)-form is the predominant enantiomer at equilibrium in a ratio of 1:1.6.

14.16.2 New Research Tools and Methodologies

During the last two decades many advances have been made in chemical and biological technologies that have been applied to elucidating the molecular basis of a number of tumor types and to the development of new agents and therapeutic strategies to treat cancer. These incremental advances led to the recent introduction of Imatinib (Glivec), which is widely heralded as the first agent to be rationally designed based on a detailed knowledge of the molecular pharmacology of the cancer cell. Arguably the arrival of Glivec in the early 2000s represented a landmark advance in the design and discovery of new anticancer agents, contrasting with the empirical approaches that characterized the previous decade. Modern cancer researchers now have a number of powerful scientific tools and methodologies at their disposal, and some of the more significant ones are briefly described below.

Gene hunting and sequencing

Advances in molecular biology allow new genes to be identified and then sequenced with great rapidity. Modern methods of DNA sequencing are based on the technology and sequencing machines developed for the Human Genome Project. An astonishing number of cancer-related genes, such as the breast cancer related BRCA1 and BRCA2, have been discovered and identified during the last decade. Hunting for new cancer-related genes (oncogenes) is a major activity by both commercial and academic research groups throughout the world, and the list of known oncogenes is rapidly expanding. Once a cancer-related gene has been identified, then the resulting messenger RNA and protein can be elucidated and novel agents can be designed to interact with them.

Genomics and proteomics

In their broadest sense, the terms genomics and proteomics refer to the recently developed technologies that allow gene expression and protein production to be studied in living cells and tissues including tumors (see Chapter 13). Genomics studies are based on DNA arrays (or DNA microchips), which have large numbers of cDNAs on their surface corresponding to significant genes in the human genome. A typical research program might involve the study of gene expression in tumor cells which can have their mRNA extracted and used to produce fluorescently labeled cDNA probes. The probes corresponding to those genes being highly expressed will hybridize to the cDNA on the chip and this will produce a signal at locations where a match occurs. The over- or under-expression of individual genes can thus be measured. The major potential problem with this approach is that, just because a cell expresses a gene and makes the corresponding mRNA, it does not necessarily mean that it would produce the corresponding protein. For example, due to RNA splicing, single genes can lead to multiple proteins. This means that DNA array studies may give a false impression of which proteins are important in the cell.

Proteomics techniques are designed to directly measure protein rather than mRNA production, and thus circumvent the problem described above. This is achieved by extracting the entire population of proteins from a cell, separating them using an electrophoretic method, and then identifying them using mass spectrometry. The important point to note is that the two technologies (i.e., genomics and proteomics) are complementary and as research tools can be used experimentally to identify genes that are overexpressed in cancer cells, and to observe the effects of experimental agents on gene expression.

Protein structural studies

Thanks to advances in x-ray crystallography and high-field NMR instrumentation and techniques, it is now possible to elucidate the three-dimensional structure of proteins to high resolution provided they can be isolated from cells, purified (for NMR and x-ray) and crystallized (for x-ray). Even if a protein will not crystallize, techniques have been developed that will allow a lower resolution x-ray map to be constructed. Once the structure of a cancer-related protein has been established, it is possible to feed its co-ordinates into molecular modeling software packages where it can be used for drug design purposes. For example, using this approach it is possible to design ligands that will fit into the active site of an enzyme as inhibitors (see Chapters 4 and 9).

Chemical technologies

During the last decade, combinatorial chemistry methodologies have been developed that allow large libraries of novel compounds to be rapidly produced to take advantage of high-throughput screening (HTS) (see Chapter 11). With combinatorial techniques, molecular fragments are joined together in a sequentially random fashion to provide a large number of molecules (sometimes more than a million) within a short time period. The use of a large number of fragments provides ''molecular diversity'' in which libraries of molecules of widely different three-dimensional shapes are produced. The key to success with this technology is some form of tracking or tagging procedure in which ingenious techniques are used to either ensure that each new compound has a unique physical position within a two-dimensional array (or multiwell plate), or contains a traceable ''label'' which may take a variety of forms ranging from chemical to radiofrequency tags. These libraries can be passed through high-throughput screens and the structure of any one compound in the mixture can be traced through the tag or by the positional history of the synthesis. The active molecule can then be re-synthesized on a larger scale for further evaluation.

Screening technologies

During the last decade significant advances have been made in screening technologies, mainly based on the robotic handling of multiwell plates. Provided a robust assay can be set up for the interaction of an inhibitor or agonist with an enzyme or receptor, then large libraries of compounds can be rapidly and automatically screened with minimum human intervention. Assays of this type are often light-based with a ''hit'' leading to the emission of light of a certain wavelength which can be easily monitored and quantitated by automated detection systems. Bioinformatics systems are required to record, manipulate, and report the vast amount of data produced by these screens.

There have also been dramatic developments in *in vivo* models. For example, there is now a commercially available mouse strain known as Oncomouse® that has various (according to the model) oncogenes incorporated into its genome. These mice will typically succumb to the gene-related tumor within a defined period of time, and so the model can be used to study the relationship of the gene to the course of the disease and also to the effect of novel therapeutic agents on the time-to-disease and life span etc. A more recent mouse model involves the expression of green jellyfish protein (GFP) by tumor cells. The entire mouse can be viewed in a scanner and tumor sites imaged by measuring the extent and intensity of fluorescence. This provides an extremely accurate measure of tumor growth, shrinkage, or spread, thus allowing novel agents and therapies to be accurately evaluated.

14.16.3 Chemopreventive Agents

In Section 14.11.1 the prophylactic use of tamoxifen and related SERMS by women to reduce the risk of breast cancer was discussed, and there is much research activity in this area to identify compounds with improved toxicity profiles. However, with 10.3 million new cancer cases diagnosed each year worldwide, medical investigators are taking a closer look at how diet may help to reduce cancer risk.

Experts believe that as many as two-thirds of these cancers may be caused by diet and lifestyle factors. A number of studies indicate that diets rich in fruits and vegetables contribute to decreased risk of incurring many types of cancers. Researchers are now investigating specific phytochemicals (natural plant chemicals) found in fruits and vegetables that may serve to prevent cancer and other diseases. Phytochemicals number in the thousands, and only a small number of them have been identified and studied. In addition, it is now thought that aspirin may reduce the risk of breast and bowel cancers, and clinical trials are underway.

Various different mechanisms can be envisaged for agents of this type. For example, it should be possible to ''neutralize'' carcinogens in the GI tract or prevent their absorption. Alternatively, it may be possible to selectively enhance the metabolism of some carcinogens. For example, it is known that some compounds can enhance the effectiveness of Phase 2 detoxification enzymes that help neutralize cancer-causing chemicals and free radicals before they can damage DNA and initiate the development of cancer. A further mechanism may relate to enhanced repair of DNA lesions formed from the interaction of carcinogens with the genome.

Although highly attractive as a concept, there are numerous practical problems with identifying and evaluating potential chemopreventive agents. One problem with the scientific validation of chemoprevention is that clinical trials would need to be conducted over many years, as it is possible that cancer deaths in old age might be associated with carcinogens ingested in the first few decades of life. It is also difficult to devise *in vitro* screens for chemopreventive agents. For example, although it is possible to screen for compounds that enhance metabolism, there is no guarantee that faster metabolism, even for selected carcinogens, will lead to prevention of carcinogenesis. Indeed, some carcinogens are activated by metabolism.

A number of lead compounds of diverse structure (e.g., flavonoids, terpenes) have originated from epidemiological studies relating to diet. For example, soy beans are known to contain chemopreventive agents, and a soya-rich diet can lead to a lower incidence of bowel cancer. Similarly, broccoli, curry powder, and orange peel are known to contain chemopreventive agents, and compounds thought to be the active agents have been isolated and characterized. A selection of four such compounds are discussed below.

Resveratrol

Resveratrol is one of a group of compounds known as phytoalexins which are produced in plants during times of environmental stress such as insect, animal, or pathogenic attack, or adverse weather conditions. It has been identified in more than 70 plant species examples of which include red grapes, mulberries, peanuts, white hellebore, and fescue grass. Resveratrol is an active ingredient of the Asian folk medicine ''Kojo-Kon'' which is the powdered root of the Japanese knotweed.

The best known source of resveratrol is the red grape where it is found in the skin and not the flesh. Fresh grape skin contains about 50–100 μg of resveratrol per gram, while red wine concentrations range from 1.5 to 3.0 mg/L. It is thought to be responsible, in part, for the blood fat and cholesterol-lowering effects of red wine and may explain why those consuming a Mediterranean-type diet (which includes red wine) may have a reduced risk of heart disease despite sometimes consuming higher levels of saturated fat (e.g., the ''French Paradox'').

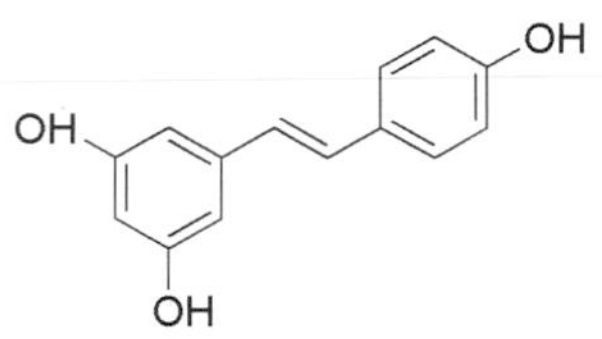

Resveratrol
(14.112)

There is some evidence that resveratrol can retard all three phases of the cancer process: initiation, promotion, and progression. It is known to have antioxidant and antimutagenic activity, and can also increase levels of the Phase II drug-metabolizing enzyme quinone reductase. The latter is particularly significant as this enzyme is capable of metabolically detoxifying carcinogens, thereby removing them from the body and decreasing exposure of the genome. All three of these physiological effects suggest that resveratrol may help to prevent cancer initiation, the initial, irreversible stage of the tumor-forming process. In support of this, resveratrol has been shown to inhibit the development of pre-neoplastic lesions in mouse mammary glands treated with a carcinogen with no toxic effects observed.

Resveratrol also has anti-inflammatory and antiplatelet aggregating activity, and inhibits the activity of the cyclooxygenase and hydroperoxidase enzymes (suggestive of anti-promotion activity), in addition to causing the differentiation of human promyelocytic leukemia cells. These additional pharmacological activities suggest that it may also depress the progression phase of cancer.

Given that the toxicity profile of resveratrol is so positive, and that it is already ingested by the public in food and drink products, future research to establish dose levels and schedules may lead to purified resveratrol being taken by healthy individuals on a routine basis to prevent the initiation of cancer. A number of clinical studies are presently underway, and the NCI have been involved in studies using resveratrol isolated from a crude extract of the roots of a tree collected in Peru. The implications for resveratrol in coronary heart disease are also being studied.

Phytoestrogens

The phytoestrogens are naturally occurring phenolic plant compounds found in foods such as beans, cabbage, soya bean, grains, and hops, and are part of a wider family of structures found in all plants known as polyphenols. They are structurally similar to the mammalian estrogen, estradiol, and thus have estrogenic properties. However, their estrogenic activity is generally much less than that of human estrogens (estrogenic activity ranges from 1/500 to 1/1000 of the activity of estradiol). Therefore, in a relative sense, phytoestrogens can act as antiestrogens by blocking the ERs and exerting a much weaker estrogenic effect compared with the natural hormone. As a consequence, it has been suggested that they might partly suppress or inhibit normal estrogenic activity in estrogen-responsive tissues such as breast, and may thus reduce the risk of breast cancer. In addition to these endocrine effects, they may also have estrogen-independent action on other cellular targets thereby complicating the prediction of their properties in humans.

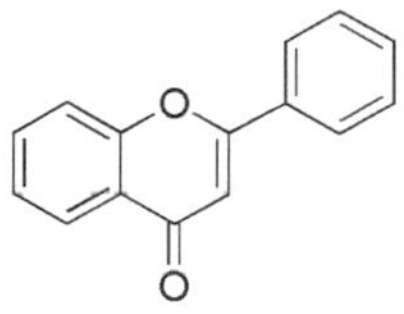

Flavone structure
(14.113)

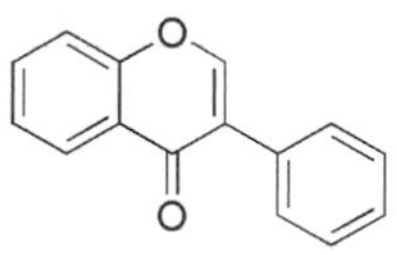

Isoflavone structure
(14.114)

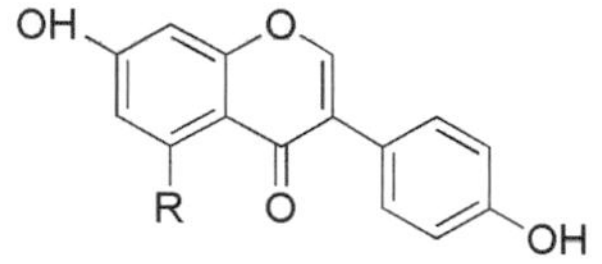

Genistein: R = OH
Daidzein: R = H
(14.115)

Phytoestrogens are found in the seeds, stems, roots, or flowers of legumes, where they serve as natural fungicides and act as part of the plant's defense mechanism against microorganisms. They also act as the molecular signals emanating from the roots of leguminous plants to attract specific nitrogen-fixing soil bacteria. The main classes of phytoestrogens are the isoflavones, coumestans, and lignans. Isoflavones are receiving a great deal of commercial interest at present with the soya bean being the most abundant source. The most important soya isoflavones are genistein and daidzein. The latter has also been isolated from red clover, and has been shown to have specific protein kinase inhibition properties. Lignans, however, are also an important source of phytoestrogens in the UK diet as they are present in most fiber-rich foods.

Research into the possible benefits of phytoestrogens has focused on cancer (breast and prostate in particular), menopause, osteoporosis, and heart disease (antioxidant activity). Other potential areas of benefit include diabetes and cognitive function. It has been observed that women in Japan have one of the lowest incidences of breast cancer in the world, and that Asian men have even lower rates of prostate cancer. These rates are generally lower in the homeland, but when Asian women migrate to America for example, the breast cancer incidence significantly increases. Therefore, it has been suggested that environmental factors, especially the diet, play an important role. For example, there is a marked difference in dietary consumption of phytoestrogens between Japan and the UK; the intake of phytoestrogens in Japan is 30 times higher. The total isoflavone intake of the Japanese population was documented in 1994 as being between 150 and 200 mg/day, although more recent studies suggest it to be more in the region of 32–50 mg/day. However, an intake of 50 mg of isoflavones per day achieves plasma isoflavone concentrations of 50–800 ng/mL, and it is the high consumption of soya isoflavones in Japan that is thought to be a major factor in the reduced incidence of breast cancer amongst the Asian population.

With regard to the possible beneficial role of phytoestrogens in reducing the risk of prostate cancer, the only available data have come from animal studies, and there are no significant positive findings from human clinical trials. In addition, phytoestrogens are thought to be an alternative to estrogen replacement therapy during the menopause due to their positive estrogenic activity. This is supported by the observation that the menopause is less of a problem in soya-consuming countries. For example, postmenopausal Japanese women who eat large amounts of phytoestrogens reportedly have fewer hot flushes and night sweats.

In conclusion, it is possible that dietary lignan and isoflavones may have a role in the prevention of several types of cancer, but at present the evidence is not sufficient to recommend particular dietary practices or changes. For cancers of the breast and prostate, the evidence is most consistent for a protective effect resulting from a high intake of plant foods, including grains, legumes, fruit, and vegetables. At present it is not possible to definitively identify particular food types or components that may be responsible.

Curcumin

Curcumin (turmeric yellow) is the bright yellow ingredient in turmeric, a spice used in curry dishes. It has been known since the early 1990s that curcumin can slow the growth of new cancers and, furthermore, that it can arrest angiogenesis. In 2003 it was discovered that curcumin can irreversibly inhibit aminopeptidase N (APN), an enzyme that spurs tumor invasiveness and angiogenesis (blood vessel growth). APN is a membrane-bound, zinc-dependent metalloproteinase that breaks down proteins at the cell surface and helps cancer cells invade the space of neighboring cells.

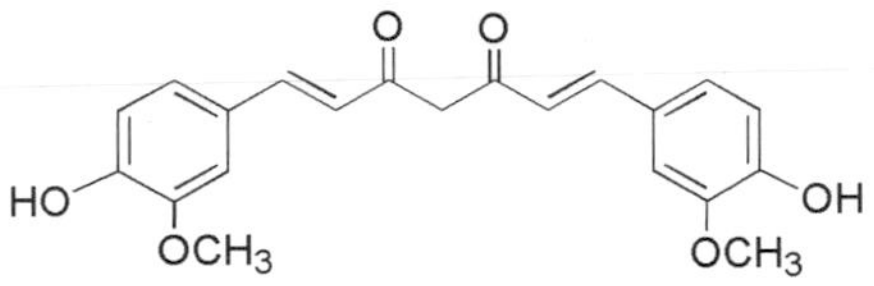

Curcumin
(14.116)

This discovery was made when researchers at Sejong University, Seoul, were screening libraries to find inhibitors of APN as potential antiangiogenic agents. After screening approximately 3000 molecules for hits, curcumin showed up as one of the most potent inhibitors. Through a combination of *in vitro* and *in vivo* enzyme assays and surface plasmon resonance studies, it was established that curcumin's inhibition of APN is direct and irreversible. Although the exact mode of binding is not yet established, it has been postulated that the α,β-unsaturated ketone moieties may covalently link to two nucleophilic amino acids in APN's active site.

This finding suggests the possibility of developing more potent APN inhibitors based on the structure of curcumin which may be taken orally with minimum side effects. Curcumin itself is presently in Phase I clinical trials for colon cancer.

Sulforaphane

Sulforaphane is a member of a class of chemicals known as isothiocyanates that are found in broccoli, cabbage, cauliflower, kale, other cruciferous vegetables and some fruits, and may help to explain why populations that consume a diet rich in fruits and vegetables appear to have a reduced risk of developing several types of cancers. Although most research to date has been conducted in animals, a number of clinical studies are presently underway to attempt to extend these observations to humans.

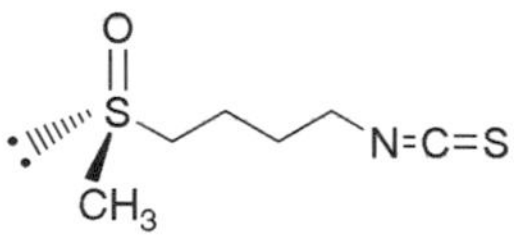

L-Sulforaphane
(14.117)

In 1992, researchers at Johns Hopkins University in the USA isolated and identified sulforaphane, and found it to be a potent selective inducer of Phase II detoxification enzymes. In its precursor form, sulforaphane glucosinolate (SGS) functions as an indirect antioxidant. As such, it does not directly neutralize free radicals as is the case with direct antioxidants such as vitamins E and C and beta-carotene. Instead, indirect antioxidants induce (or boost) the activity of the Phase II detoxification enzymes. These enzymes act as a defense mechanism, triggering broad spectrum antioxidant activity that neutralizes many free radicals before they can cause the cell damage that may cause mutations leading to cancer. More importantly, it is thought that the indirect antioxidant effects are long-lasting, triggering an ongoing process that continues to be effective and may last for days after the sulforaphane has left the body. This contrasts with the direct antioxidants, which neutralize only one molecule of a radical at a time, and are destroyed in the process.

In vivo experiments have shown that sulforaphane can block the formation of mammary tumors in rats treated with a potent carcinogen. For example, the number of rats developing tumors can be reduced by as much as 60%. In addition, the number of tumors developing in each animal can be reduced by 80%, and the size of the tumors decreased by 75%. The formation of tumors can also be delayed and they may grow more slowly. It has also been shown that sulforaphane can inhibit the formation of premalignant lesions in the colons of rats, and that it can induce cell death in human colon carcinoma cells growing *in vitro*. These results suggest that, in addition to the activation of detoxifying enzymes, induction of apoptosis (cell death) may also be involved in the chemoprevention properties of sulforaphane.

A wide range of broccoli plants have been examined to determine which have the highest levels of sulforaphane. It has been shown that many varieties of fresh and frozen broccoli differ significantly in the amounts of sulforaphane they contain, and thus in their ability to enhance protective Phase II enzymes. It is also known that as plants grow older, the concentration of SGS decreases. Conversely, young plants (e.g., three-day-old sprouts) have greater enzyme-inducing activity. For example, it has been shown that three-day-old broccoli sprouts can contain between 20 and 50 times the concentration of SGS than the mature cooked vegetable.

It is noteworthy that commercial broccoli is highly variable in SGS content. However, it has been shown that young broccoli sprouts, grown from the seeds of certain varieties of broccoli, can be produced under carefully standardized conditions to contain consistently high concentrations of SGS. For example, a variety of broccoli known as BroccoSprouts™ has been patented that contains, on average, 20 times the concentration of SGS than mature cooked broccoli.

14.17 ADJUNCT THERAPIES

The role of antiemetics in cancer chemotherapy is of great importance since many of the cytotoxic agents in clinical use cause profound nausea and vomiting. Uncontrolled vomiting may outweigh the benefits of treatment and can lead to poor patient compliance. Traditionally, drugs such as metoclopramide were used, but with the addition of the $5HT_3$ serotonin antagonists such as ondansetron and granisetron, the incidence and severity of emesis have been substantially reduced. Steroids also play a role in the palliation of symptomatic end-stage malignant disease where prednisolone, for example, can enhance appetite and produce a sense of well-being.

14.17.1 Anti-emetic Agents

Nausea and vomiting cause considerable distress to many patients receiving chemotherapy, and to a lesser extent abdominal radiotherapy, and may be so severe as to lead to refusal of further treatment. Symptoms may be acute (i.e., occurring within 24 h of treatment), delayed (i.e., first occurring more than 24 h after treatment), or anticipatory (i.e., occurring before subsequent doses). Delayed and anticipatory symptoms are more difficult to control than acute ones and require different management.

Patients vary in their susceptibility to drug-induced nausea and vomiting. Those affected more often include women, patients under 50 years of age, anxious patients, and those who experience motion sickness. Susceptibility also increases with repeated exposure to the drug. Drugs may be classified according to their emetogenic potential. However, symptoms vary according to the dose, the type and dose of other drugs co-administered, and the individual susceptibility.

Treatments with a mild emetogenic effect include the drugs 5-FU, etoposide, methotrexate, and the vinca alkaloids, and also abdominal radiation. Drug treatments classed as having a moderate emetogenic effect include doxorubicin, intermediate and low doses of cyclophosphamide, mitoxantrone, and high doses of methotrexate. Highly emetogenic treatments include cisplatin, dacarbazine, and high doses of cyclophosphamide.

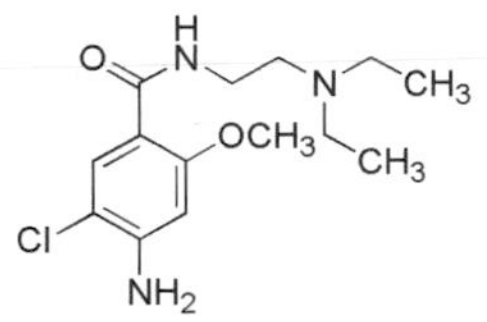

Metaclopramide
(14.118)

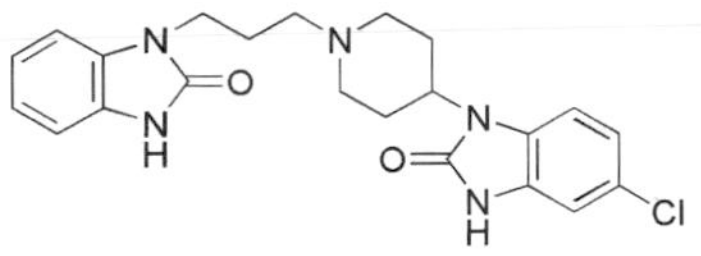

Domperidone (Motilium)
(14.119)

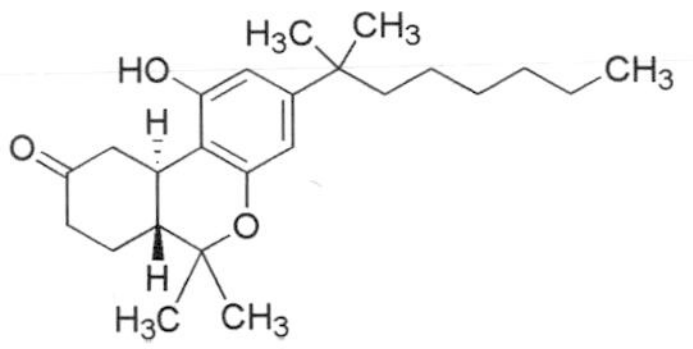

Nabilone
(14.120)

For prevention of acute symptoms, patients at low risk can be pretreated with metoclopramide or domperidone continued for up to 24 h after chemotherapy. Dexomethasone may also be added to enhance effectiveness. For patients at high risk of emesis or when other anti-nauseant treatments have proved inadequate, a specific $5HT_3$ serotonin antagonist such as ondansetron, granisetron, or tropisetron usually given by mouth is often highly effective, particularly when used in conjunction with dexamethasone.

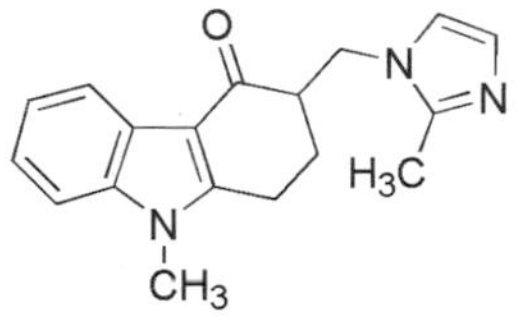

Ondansetron (Zofran)
(14.121)

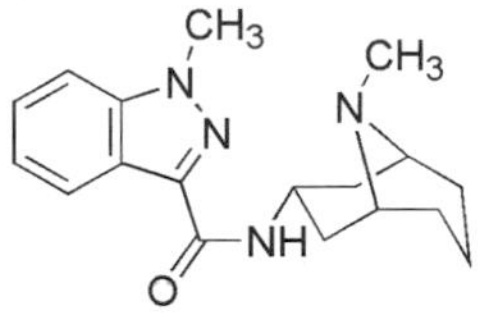

Granisetron (Kytril)
(14.122)

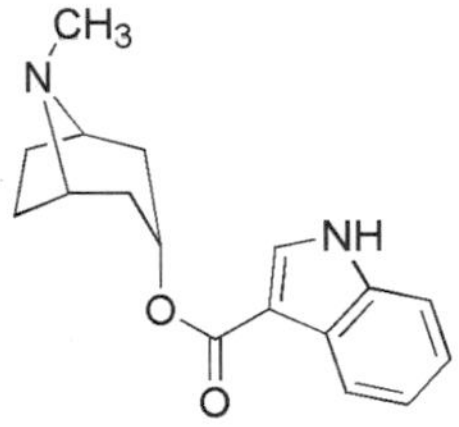

Tropisetron (Navoban)
(14.123)

Delayed symptoms of nausea may be treated with oral dexamethasone, sometimes in combination with metoclopramide or prochlorperazine. The $5HT_3$ antagonists are less effective for this purpose.

Good symptom control is the best way to prevent anticipatory nausea. The addition of lorazepam to antiemetic therapy is helpful in this regard because of its amnesic, sedative, and anxiolytic effects. The cannabinoid derivative navalbine is also occasionally used to control emesis associated with chemotherapy.

14.17.2 Immunosuppressive Agents

Prednisolone

Prednisolone is an orally administered synthetic corticosteroid that is widely used in oncology for its significant antitumor effect in acute lymphoblastic leukaemia, Hodgkin's disease, and the non-Hodgkin's lymphomas.

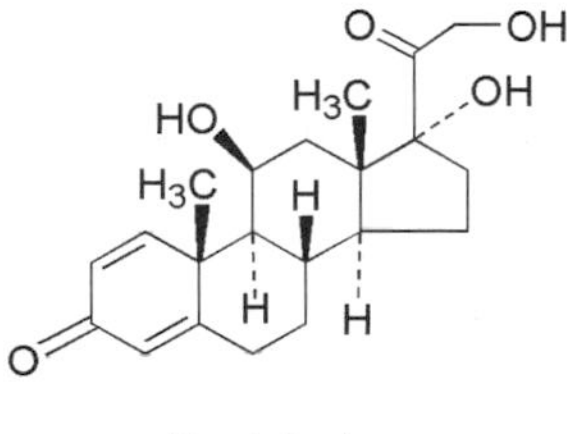

Prednisolone
(14.124)

It also has a role in the palliation of symptomatic end-stage malignant disease where it can enhance appetite and produce a sense of well-being.

FURTHER READING

Ali, S. and Coombs, R.C. (2002) Endocrine-responsive breast cancer and strategies for combating resistance. *Nature Reviews Cancer* **2**, 101–112.

Anon (1996a) Nature (Supplement to issue 6604) Intelligent Drug Design **384**, 1–26.

Anon (1996b) Scientific American (Special Issue) What you need to know about cancer **275**, 4–167.

Baguley, B.C. and Kerr, D.J. (eds) (2002) *Anticancer Drug Development*. San Diego, California: Academic Press, 397 pp.

Bishop, J.M. and Weinberg, R.A. (eds) (1996) *Molecular Oncology*. New York: Scientific American, Inc.

Browne, M.J. and Thurlby, P.L. (eds) (1996) *Genomes, Molecular Biology and Drug Discovery*. London: Academic Press.

Cassidy, A. (1999) Dietary phytoestrogens — potential anti-cancer agents? *British Nutrition Foundation Bulletin* **24**, 22–30.

Chung, F.L., Conaway, C.C., Rao, C.V., and Reddy, B.S. (2000) Chemoprevention of colonic aberrant crypt foci in Fischer rats by major isothiocyanates in watercress and broccoli. *Proceedings of the American Association for Cancer Research* **41**, 660.

Culver, K.W., Vickers, T.M., Lamsam, J.L., Walling, H.W., and Seregina, T. (1995) Gene therapy of Solid Tumours. *British Medical Bulletin* **51**, 192–204.

Dalgleish, A.G. (1994) Viruses and cancer. *British Medical Bulletin* **47**, 21–46.

Dobrusin, E.M. and Fry, D.W. (1992) Protein tyrosine kinases and cancer. *Annual Reports in Medicinal Chemistry* **27**, 169–178.

Druker, B. (2002) Perspectives on the development of a molecularly targeted agent. *Cancer Cell* **1**, 31–36.

Hanahan, D. and Weinberg, R.A. (2000) The hallmarks of cancer. *Cell* **100**, 57–70.

Hochhauser, D. and Harris, A.L. (1991) Drug resistance. *British Medical Bulletin* **47**, 178–196.

Johnstone, R.W. (2002) Histone–deacetylase inhibitors: novel drugs for the treatment of cancer. *Nature Reviews Drug Discovery* **1**, 287–299.

Larson, E.R. and Fischer, P.H. (1989) New approaches to antitumour therapy. *Annual Reports in Medicinal Chemistry* **24**, 121–128.

Lee, M.D., Ellestead, G.A., and Borders, D.B. (1991) Calicheamicins: discovery, structure, chemistry and interaction with DNA. *Accounts of Chemical Research* **24**, 235–243.

Lemoine, N.R. and Cooper, D.N. (eds) (1996) *Gene Therapy*. Oxford: BIOS Scientific Publishers Ltd.

Macdonald, F. and Ford, C.H.J. (1997) *Molecular Biology of Cancer*. Oxford: BIOS Scientific Publishers Ltd.

Malcolm A. (2002) *The Cancer Handbook*, Vol. 1. London: Nature Publishing Group, Macmillan Publishers Ltd. pp. 161–177.

Miller, A.D. (1992) Human gene therapy comes of age. *Nature (London)* **357**, 455–460.

Mulligan, G.C. (1993) The basic science of gene therapy. *Science* **260**, 926–932.

Neidle, S.J. and Waring, M.J. (eds) (1993) *Molecular Aspects of Anticancer Drug-DNA Interactions*. London: The Macmillan Press.

Parker Hughes Cancer Center (2000) Tyrosine kinase inhibitors: molecules with an important mission. *http://www.ih.org/pages/tyrosine_kinase_inhibitors.html*.

Pratt, W.B. and Ruddon, R.W. (eds) (1979) *The Anticancer Drugs*. Oxford: Oxford University Press.

Pullman, B. (1991) Sequence specificity in the binding of antitumour anthracyclines to DNA. *Anti-cancer Drug Design* **7**, 95–105.

Schlessinger, J. (2002) Cell signalling by receptor tyrosine kinases. *Cell* **103**, 211–225.

Shawver L., Slamon, D., and Ullrich, A. (2002) Smart drugs: tyrosine kinase inhibitors in cancer therapy. *Cancer Cell* **1**, 117–123.

Silverman, R.B. (1992) *The Organic Chemistry of Drug Design and Drug Action*. London: Academic Press.

Summerhayes, M. and Daniels, S. (2003) *Practical Chemotherapy: A Multidisciplinary Guide*. London: Radcliffe Medical Press, 398 pp.

Suffness, M. (1993) Taxol: from discovery to therapeutic use. *Annual Reports in Medicinal Chemistry* **27**, 305–314.

Teicher, B.A. (ed.) (1997) *Anticancer Drug Development; Preclinical Screening, Clinical Trials, and Approval*. Totowa, New Jersey: Humana Press, 311 pp.

Thurston, D.E. (1999) Nucleic acid targeting: therapeutic strategies for the 21st century. *British Journal of Cancer* **80**(Supplement 1), 65–85.

Vousden, K.H. and Farrell, P.J. (1994) Viruses and human cancer. *British Medical Bulletin* **50**, 560–581.

Workman, P. (ed.) (1992) *New Approaches in Cancer Pharmacology. Drug Design and Development*. London: Springer-Verlag.

Yarnold, J.R., Stralton, M., and McMillan, T.J. (eds) (1996) *Molecular Biology for Oncologists*. 2nd edition, London: Chapman and Hall.

15

Neurotransmitters, Agonists, and Antagonists

Robert D. E. Sewell, H. John Smith, Holger Stark, Walter Schunack, and Philip G. Strange

CONTENTS

15.1 OVERVIEW

Robert D. E. Sewell

15.1.1 Introduction

Over recent years, our concepts about neurotransmitters have evolved firstly, because the number of putative transmitter candidates has increased and secondly, because our understanding of neurotransmitter function has widened to embrace a more diverse range of actions. The aim of this chapter is not to describe the medicinal chemistry of all known potential neurotransmitters and their agonists and antagonists, but more to concentrate on selected candidates, namely histamine, dopamine, L-glutamate (Glu), and γ-aminobutyric acid (GABA). Among these examples, there are corresponding receptor types, which typify both of the major functional classes of neuronal receptor (i.e., metabotropic receptors and transmitter-gated ion channels which are also known as ionotropic receptors).

Neurotransmitters

Neurotransmitters are a varied assortment of substances implicated in the transfer of signals across chemical synapses. These neuronal elements consist of narrow clefts which separate presynaptic nerve terminals from receptors located on postsynaptic membranes found on subsequent neurons, muscles, or glands. Neurotransmitters are released by presynaptic terminals and evoke rapid excitatory or inhibitory responses in postsynaptic neurons. In addition to postsynaptic receptors, there are receptors sited presynaptically on the nerve terminals themselves. These are sometimes designated as autoreceptors, capable of responding to released neurotransmitter, providing a negative feedback function concerned with regulation of transmitter release. Other neuronal synapses exist which rely on electrical rather than chemical transmission and this represents a very fast mode of communication between cells.

The chemical neurotransmitters (Figure 15.1) may be classed into three major categories:

1. Simple amino acids like glutamate, GABA, and glycine account for transmission at a high proportion of central nervous system (CNS) synapses. They are relatively fast communicators and occur at the highest concentrations of all the neurotransmitter groups in brain tissue (micromoles per gram of tissue).

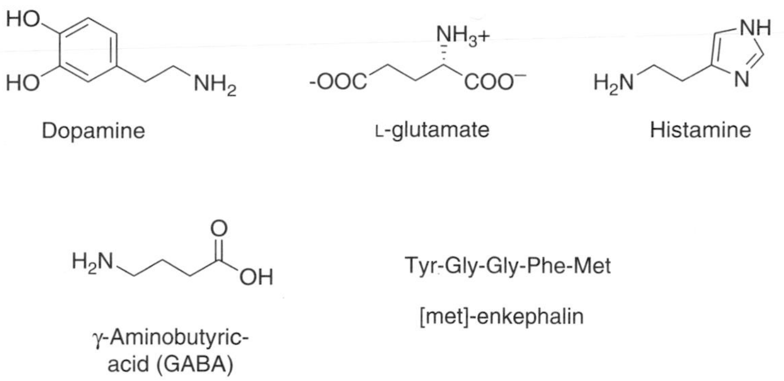

HO HO NH$_2$

Dopamine

NH_3+ -OOC COO$^-$

L-glutamate

NH H$_2$N N

Histamine

O H$_2$N OH

γ-Aminobutyric-acid (GABA)

Tyr-Gly-Gly-Phe-Met

[met]-enkephalin

Figure 15.1 Examples of neurotransmitters.

2. The amine neurotransmitters are composed of ''classical'' transmitters such as acetylcholine, serotonin (5-hydroxytryptamine (5-HT)) histamine, and the catecholamines (dopamine and noradrenaline). Also included on this group are purine neurotransmitters like adenosine and adenosine triphosphate (ATP). All the neurotransmitters in the group are found in moderate concentrations in neuronal tissue (nanomoles per gram of whole brain tissue) and tend to exert a slower modulatory type of action.
3. Neuropeptides (e.g., Met-enkephalin) occur at the lowest neuronal concentrations of all three groups (nanomoles per gram of whole brain tissue). In addition to release at synapses, they may also be released at nonsynaptic locations to diffuse more freely onto receptor sites to produce slow long-lasting effects and, in such circumstances, might be more appropriately considered as local neuromodulators.

It is not uncommon for neurons to contain more than one neurotransmitter, a typical example is neuropeptides occurring in combination with amines in one and the same neuron.

Receptors

Neurotransmitter effects are mediated via receptors located in neuronal membranes and these are thought to be of two functional classes.

Metabotropic receptors

Metabotropic receptors are linked to intracellular proteins, which transduce signals across the cell membrane. These proteins are known as G-proteins (so-called because they are coupled to the guanosine nucleotides GDP or GTP). They are comprised of three subunits (α, β, and γ) of which the α subunit possesses GTPase activity. The DNA sequences coding for the majority of G-protein-coupled metabotropic receptors do have some sequence homology so they are viewed as a superfamily and over 100 constituent members have been cloned.

Binding of a neurotransmitter or an agonist with a metabotropic receptor often stimulates the formation of an intracellular second messenger by means of an effector enzyme. The second messengers include adenosine 3′,5′-cyclic phosphate (c-AMP), inositol phosphates, diacylglycerol (DAG), and arachidonate. The overall mechanism invariably involves an amplification process since a single neurotransmitter/agonist–receptor complex may activate several G-protein molecules in turn to generate many secondary messenger molecules intracellularly. There appear to be subtypes of virtually all metabotropic receptors that differ either in location or second messenger coupling, but they all possess seven transmembrane hydrophobic spanning regions in their peptide sequences, with an extracellular amino terminus and an intracellular C-terminus. The membrane spanning portions of receptor proteins form α-helices and possess similarities within each receptor group so it is probable that the agonist-binding site resides at least partly in the membrane-spanning region.

Transmitter-gated ion channels (Ionotropic receptors)

Transmitter-gated ion channels (ionotropic receptors) are linked directly to ion channels in neuronal membranes. They are responsible for transient (millisecond) increases in the conductance of specific ions in each instance and this gives rise to rapid synaptic transmission. There are several examples, which include the nicotinic acetylcholine receptor, the 5-HT_3 receptor, and receptors for the amino acids such as $GABA_A$, glycine, and *N*-methyl-D-aspartate (NMDA). Ionotropic receptors are comprised of four to five protein subunits in a complex linked to an ion channel. The basic structure for each subunit consists of a protein, which loops in and out of the neuronal membrane. In the case of the nicotinic receptor, which has been comprehensively studied, there are five subunits (α_2, β, γ, δ) each traversing the membrane a total of five times. Four of the spanning segments are hydrophobic in nature and are considered as truly integral transmembrane domains. The fifth segment has only one face of its α-helix structure which is hydrophobic while the other face is hydrophilic and, along with counterparts in the other four subunits, constitutes the lining of the ion channel interfacing with an aqueous environment. When an agonist binds to this pentameric receptor, a conformational change occurs in the complex, which allows the passage of ions through the channel.

Neuronal signal effectors

Both metabotropic receptors and transmitter-gated ion channels are coupled to effectors involving enzymes or ion channels and the modulation of cytoplasmic Ca^{2+} concentration features in several of these systems. Thus ionotropic receptors can regulate cytoplasmic Ca^{2+} concentration either directly by conducting Ca^{2+} itself, or via permeability to monovalent cations which depolarize the membrane as Na^+ influxes and this subsequently activates Ca^{2+} channels in the proximity of the receptor.

Metabotropic receptors may activate slower biochemical effector processes which modulate Ca^{2+} concentrations intracellularly through the following G-proteins:

1. G_s which catalyzes the conversion of ATP to c-AMP by stimulating adenylate cyclase while concomitantly activating Ca^{2+} channels
2. G_i/G_o which inhibits both adenylate cyclase and Ca^{2+} channels while simultaneously activating K^+ channels
3. G_q which activates phospholipase C (PLC) to catalyze the hydrolysis of phosphatidylinositol 4,5-biphosphate (PIP2) to inositol 1,4,5-triphosphate (IP3), and DAG. IP3 may cause Ca^{2+} release from such intracellular storage sites as the endoplasmic reticulum. DAG on the other hand, stimulates the enzyme protein kinase-C (PKC), which is capable of phosphorylating other cellular or membrane proteins, enzymes, and ion channels. If ion channel proteins are phosphorylated, this may alter ion fluxes albeit in a slower manner than that generated by ionotropic receptors

FURTHER READING

Alexander, S.P.H., Mathie, A., and Peters, J.A. (2004) Guide to receptors and channels. *British Journal of Pharmacology* **141** (Suppl. 1), 1–126.

Liu, F., Wan, Q., Pristupa, Z., Yu, X.-M., Wang, Y.T., and Miznik, H.B. (2000) Direct protein–protein coupling enables cross-talk between dopamine D_5 and γ-aminobutyric acid A receptors. *Nature* **403**, 274–280.

Schoneberg, T., Schultz, G., and Gudermann, T. (1999) Structural basis of G-protein coupled receptor function. *Molecular Cell Endocrinology* **151**, 181–193.

Schwartz, T.W. (1996) Molecular structure of G-protein-coupled receptors. In: *Textbook of Receptor Pharmacology*, J.C. Foreman and T. Johanesen (eds). CRC Press, Boca Raton, FL.

Simonds, W.F. (1999) G-protein regulation of adenylate cyclase. *Trends in Pharmacological Science* **20**, 66–72.

15.2 CNS GLUTAMATE AND GABA RECEPTORS AS POTENTIAL DRUG TARGETS

Robert D. E. Sewell and H. John Smith

15.2.1 Introduction

Considerable attention has been directed toward understanding neuronal communication within the CNS. Disturbances in neurotransmitter processing and function make a significant contribution toward explaining possible underlying causes of a variety of conditions/diseases such as stroke, Parkinson's disease, epilepsy, Alzheimer's disease, anxiety, and schizophrenia. Imbalances between neurotransmitter systems producing a predominance of either excitatory or inhibitory activity have been suggested as a possible causal factor for at least some of these disease states. L-Glutamate (Glu) is the major, ubiquitous excitatory neurotransmitter in the CNS whereas GABA is the principal inhibitory brain transmitter. Much is known about these two transmitters, especially their respective receptors, composition, subtypes, structure, complexity, and distribution in the brain. This knowledge has been gained using sophisticated neurochemical and molecular biological techniques. Where these transmitters have been associated with particular disease states, compounds developed to target their specific receptors have lacked selectivity due to small structural differences within either Glu or GABA receptor subtypes. Here, Glu and GABA receptors, their agonists/antagonists and their function is described and progress in the development of drugs aimed to modify these systems is discussed in the search for ''quality of life'' — enhancing drugs for the diseases previously mentioned.

15.2.2 Glutamate Receptors

Glutamate released at excitatory synaptic junctions acts on a mixed population of pentameric ionotropic receptors termed either NMDARs (after the agonist *N*-methyl-D-aspartate) or non-NMDARs, which fall into two distinct families; AMPA-receptor subunits (after the agonist α-amino-3-hydroxy-5-methyl-4-isozole propionic acid) and kainate-receptor subunits (after the agonist kainate). NMDARs operate through ligand-gated cation channels with slow kinetics and high Ca^{2+} permeability, whereas the non-NMDARs operate via ligand-gated cation channels with fast kinetics and low Ca^{2+} permeability. Glutamate also stimulates monomeric metabotropic receptors mentioned later.

15.2.3 NMDA Receptors

Three families of the NMDA receptors are known each consisting of a multiassembly (NR1/2) of several isoforms (NR1 = 1a–4b, NR2 = A–D, NR3 = A–B) with differential distribution in the CNS. The subunit assemblies (isoforms) possess an extracellular N-peptide terminus containing a ligand-binding region, three membrane-spanning regions (M1, M3, and M4) with an extracellular loop between M3 and M4, a loop (M2) from the cytoplasm partially spanning the membrane and forming an ion channel lining domain as well as an additional ligand-binding site and an intracellular C-peptide terminus.

NMDAR agonists

NMDA was discovered from a structure–activity study on aspartic- and glutamic acid analogs, the D-enantiomer being the active form. The search for selective analogs led to the agonists α,ε-diamino-pimelate (DAP, **15.1**) and 3-amino-hydroxy pyrolidin-2-one (HA-966, **15.2**). Modification of the terminal COOH to give planar analogs has improved potency in the 4-methylglutamic isomers but selectivity has been lost. Conformational restriction of the interacidic moiety into a ring structure or the α-amino function and interacidic chain into a heterocyclic ring leads to compounds comparable in potency to aspartic- and glutamic acids, e.g., (*RS*)-ibotenic acid (**15.3**), HQA (**15.4**), (*2S, 1′R, 2′S*)-CCG (**15.5**), and (*RS*)-α-Tetgly (**15.6**). The enhanced potency of (**15.5**) and others suggest that Glu binds in a folded conformation at the receptor.

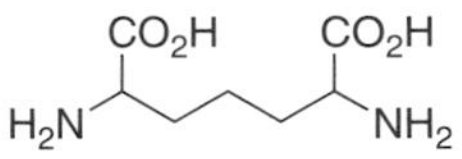

(**15.1**) DAP

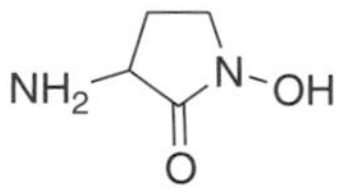

(**15.2**) (*R*)-HA-966

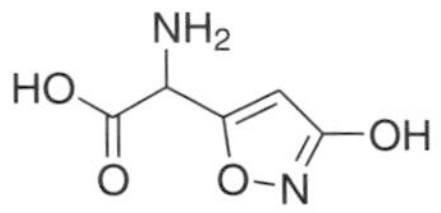

(**15.3**) (*RS*) - ibotenic acid

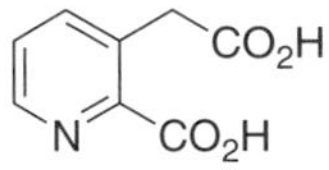

(**15.4**) HQA

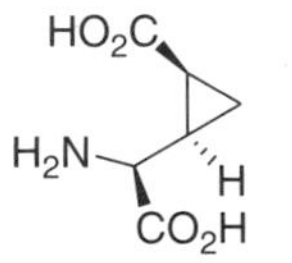

(**15.5**) (2*S*, 1′*R*, 2′*S*) - CCG

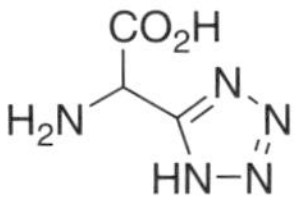

(**15.6**) (*RS*)-α-Tetgly

NMDAR competitive antagonists

(*R*)-AP5 (**15.7**), a chain extended analog of Glu bearing a ω-phosphono group was a highly selective antagonist at NMDA receptors. Compounds (*R*)-AP5 and (*R*)-AP7 (**15.8**) have been used to investigate NMDA receptor function and its role in epilepsy and neurodegeneration but they lacked the blood–brain barrier penetration required for effective drugs. Restriction of the conformation of AP5 by a double bond (CGP 37849 (**15.9**); CGP 40116 (**15.10**); CGP 39653 (**15.11**)) or in AP5 a keto group (MDL 100, 453, **15.12**) gave potent antagonists.

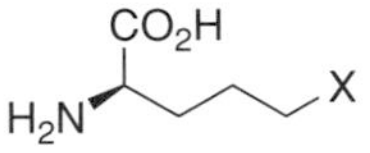

(**15.7**) X=PO_3H_2, (*R*)-AP5

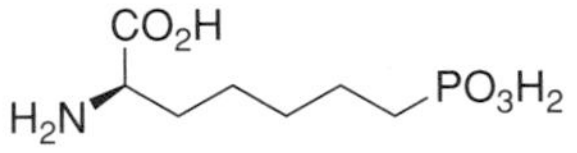

(**15.8**) (*R*)-AP7

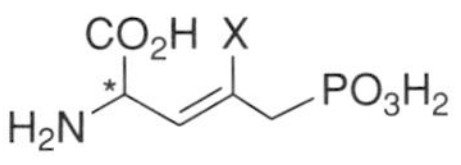

(**15.9**) X = CH_3 CGP37849
(**15.10**) * = *R*, CGP40116
(**15.11**) X = n-C_3H_7 CGP39653

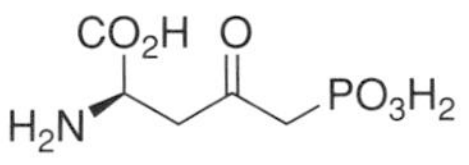

(**15.12**) MDL 100453

The ethyl ester (**15.13**) prodrug of (**15.9**) is an orally active anticonvulsant. Similarly, restriction of AP5 and AP7 by incorporation of the α-amino group and interacidic chain into piperazine (CPPP (**15.14**)) or piperidine (CGS 19755 (**15.15**)) has given potent analogs. The biphenyl analog of AP7 (SDZ 220 581 (**15.16**)) is neuroprotective and orally active. Molecular modeling studies indicate that the increased binding of the double-ionized phosphono group (PO_3^{2-}) compared to the COO^- analog is due to double interaction in the folded conformation of the former group at different sites on the receptor near the ligand-binding site. A degree of subtype selectivity for some antagonists has been noted. A number of side affects of these agents have been observed in animal models, which include motor impairment (ataxia), psychotomimetic effects, and learning impairment.

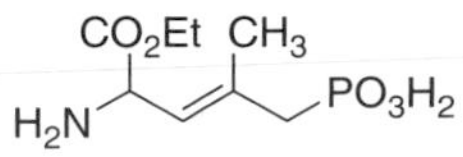

(**15.13**) CGP 39551

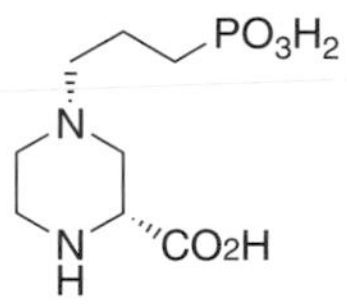

(**15.14**) CPPP

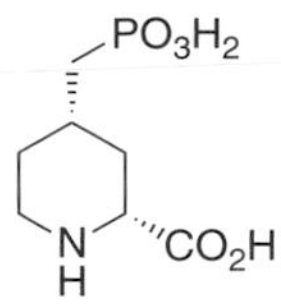

(**15.15**) CGS 19755

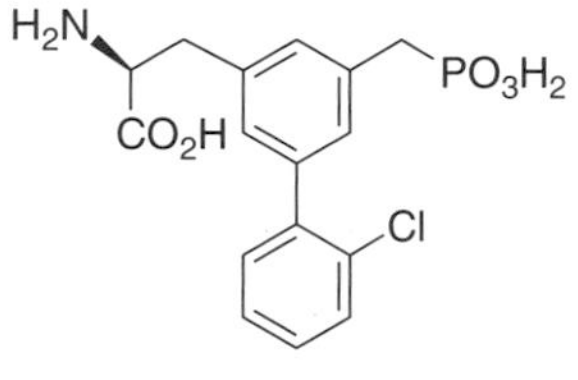

(**15.16**) SDZ-220581

Glycine-binding site antagonists

Glycine is required as a co-agonist for activation of NMDA receptors and binding site antagonists have been described, e.g., kynurenic acid (**15.17**), quinolin-2-ones, quinoxaline-2,3-diones, and 2-carboxyindoles. A number of these agents have been shown to possess *in vivo* anticonvulsant activity.

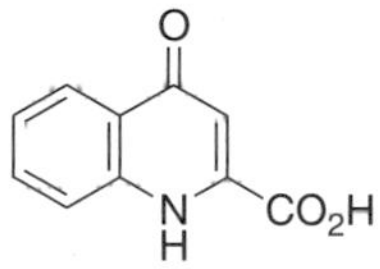

(**15.17**) Kynurenic acid

Noncompetitive NMDA receptor antagonists

Several noncompetitive NMDA receptor antagonists are recognized, but their mechanism of action is complex. In this context, endogenous polyamines such as spermine and spermidine act on an accessory site in the receptor assembly to facilitate cation channel opening (Na^+, Ca^{2+}), and the experimental compounds ifenprodil (**15.18**) and its analog eliprodil block this polyamine activity. Eliprodil led to potent NR2B-selective antagonist analogs such as CP-283,097 (**15.19**). Ifenprodil also reacts with other receptors (serotonin receptors, σ-receptors, and α-adrenoceptors) but the analog Ro25-6981 (**15.20**) is more selective.

15.2.4 Non-NMDA Receptors (AMPA/Kainate)

AMPA receptor agonists

A number of naturally occurring agonists at AMPA/kainate receptors are potent neurotoxins, e.g., quisqualate (**15.21**, *Quisqualis* sp.), ibotenic acid (**15.3**, *Amanita muscaria*), TAN-950A (**15.22**, *Streptomyces* sp.), willardine (**15.23**, *Acacia and Mimosa* sp.), kainate (**15.24**, *Digenia simplex*).

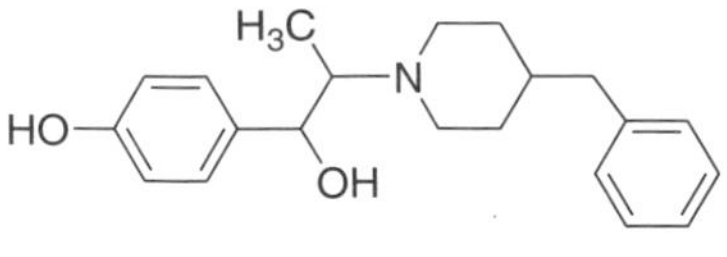

(**15.18**) Ifenprodil

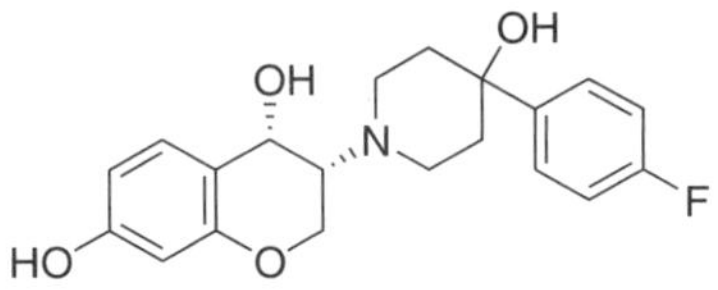

(**15.19**) CP-283,097

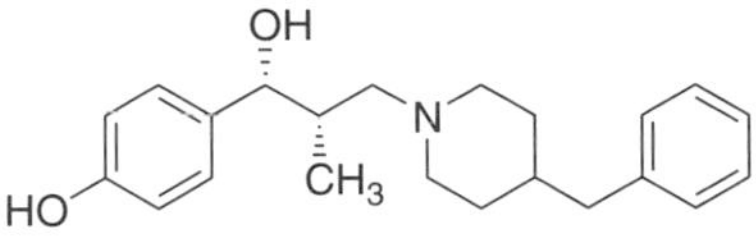

(**15.20**) Ro 25-6981

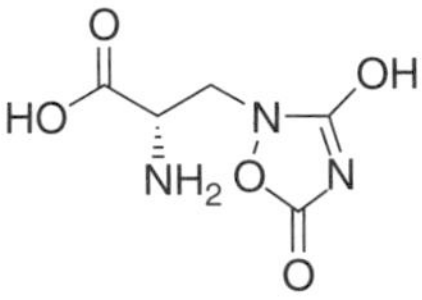

(**15.21**) Quisqualate

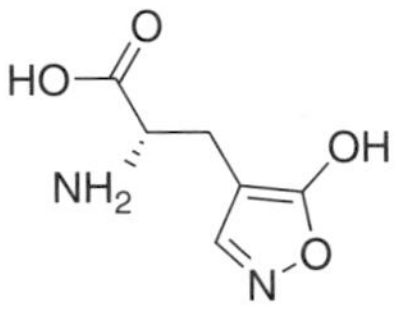

(**15.22**) TAN-950A

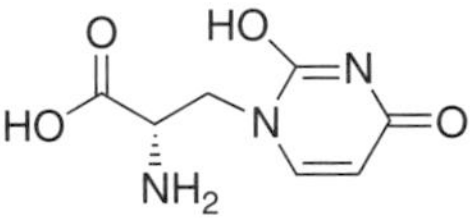

(**15.23**) Willardine

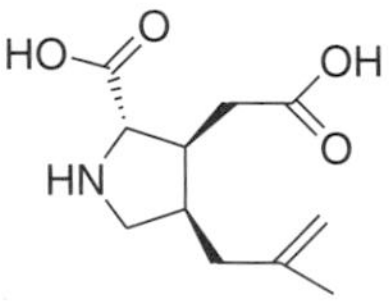

(**15.24**) Kainate

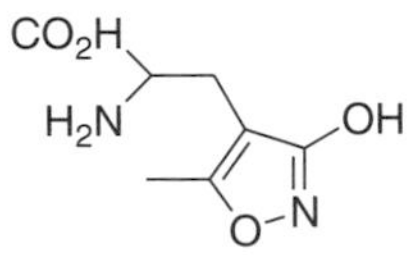

(**15.25**) AMPA

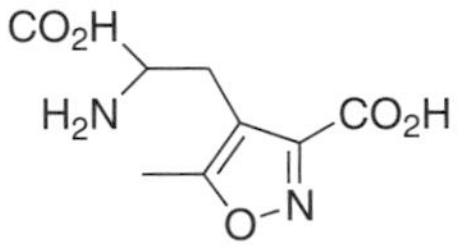

(**15.26**) ACPA

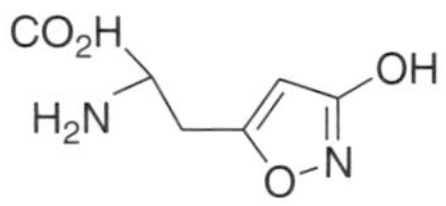

(**15.27**) HIBO

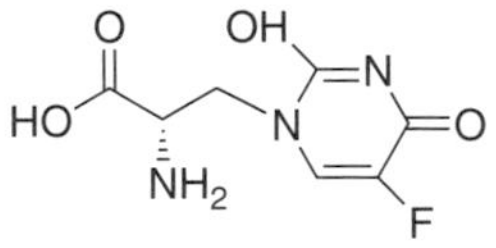

(**15.28**) (*S*)-Fluorowillardine

The 3-isoxazolol bioisostere of Glu, ibotenic acid (**15.3**), has been used as a lead compound in the design of AMPA agonists with different pharmacological profiles, e.g., AMPA (**15.25**), ACPA (**15.26**), and HIBO (**15.27**). Similar bioisosteres but based on the pyrimidine-2,4-dione ring are present in the potent agonists (*S*)-fluorowillardine (**15.28**) and (*S*)-chloroazowillardine.

AMPA receptor antagonists

The quinoxalinedione CNQX (**15.29**) is a potent nonselective AMPA receptor antagonist, but its analog NBQX (**15.30**) possesses improved selectivity and has been further chemically modified to improve water solubility (e.g., NS 257, **15.31**), which is systemically active as an anticonvulsant). Lengthening the carbon backbone in the ligand AMPA (**15.25**) gives the antagonists AMOA (**15.32**) and AMPO (**15.33**). The *tert*-butyl analog of AMPO is selective for AMPA receptors compared to NMDA/kainate receptors, the activity residing in the (*S*)-form.

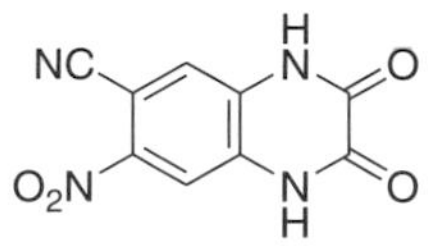

(**15.29**) CNQX

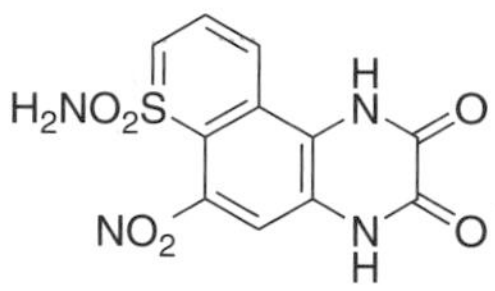

(**15.30**) NBQX

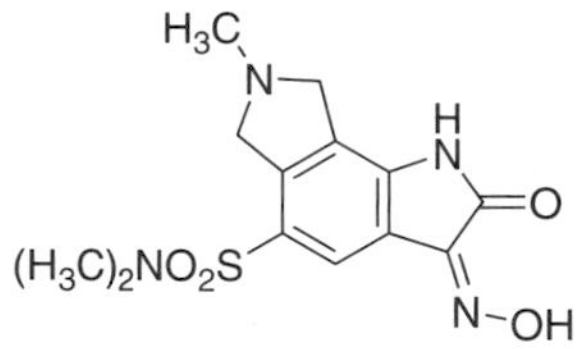

(**15.31**) NS 257

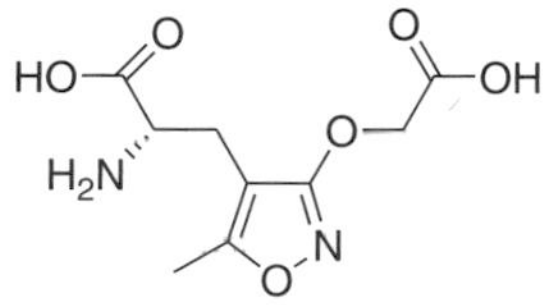

(**15.32**) (*S*)-AMOA

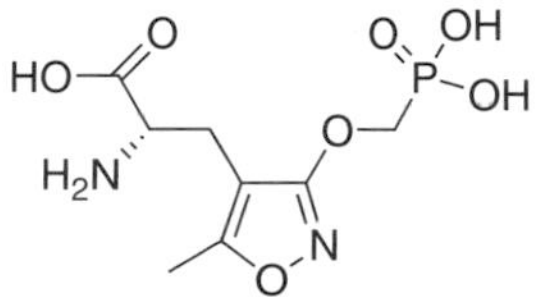

(**15.33**) AMPO

Kainate receptor agonists

There is a distinct scarcity of selective agonist compounds available for kainate receptors and even kainate (**15.24**) itself also acts as an agonist at AMPA sites. Other kainate receptor agonists include domoic acid (**15.34**) and acromelic acid (**15.35**), which have been isolated from seaweed and a poisonous mushroom, respectively. These naturally occurring agonist compounds are more potent than kainate in their neurotoxic actions.

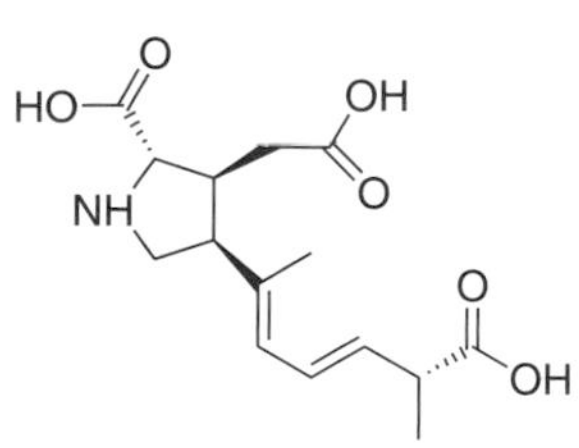

(**15.34**) Domoic acid

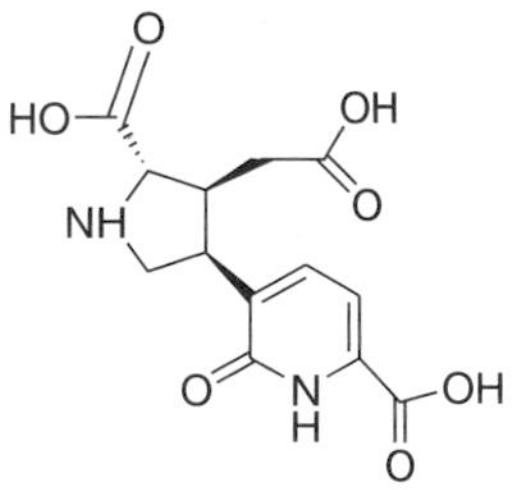

(**15.35**) Acromelic acid A

15.2.5 Metabotropic Glutamate Receptors

The NMDA and non-NMDA receptors described previously are ionotropic Glu (iGlu) receptors in that Glu-mediated excitatory transmission involves cation membrane flux. The metabotropic Glu receptor (mGluR) involved in Glu-mediated transmission, and localized at pre- and postsynaptic sites of Glu-ergic (and GABA-ergic) neurons, consists postsynaptically of Group I mGluRs (subtypes of $mGluR_1$ and $mGluR_5$) coupled to PLC, Group II mGluRs ($mGluR_2$ and $mGluR_3$), and Group III mGluRs ($mGluR_4$ and $mGluR_{6\text{–}8}$) occurring presynaptically and negatively coupled to adenylyl cyclase (AC). These receptors possess a different structure compared to the majority of other G-protein-coupled receptors. The mGluRs modulate Glu (and GABA) neurotransmission through second messenger production, kinases, iGluR activities, and their synaptic location.

mGlu Agonists and antagonists have been used as high-affinity compounds at Group I ($mGlu_{1/5}$) receptors and Group II ($mGlu_{2/3}$) receptors in radioligand binding studies and as experimental tools for investigating their pharmacology, expression, and regulation in tissues.

Group I mGlu receptor agonists

Quisqualate (**15.36**) is the most potent agonist at mGluRs 1 and 5 but also has high affinity at AMPA receptors and therefore lacks selectivity. Moreover, the related conformationally restrained CBQA isomers (**15.37**) activate $mGluR_5$. Likewise, restrained Glu analogs such as 1*S*, 3*R*-ACPD (**15.38**) exhibit nonselective agonist actions at $mGluR_{1-6,8}$. The phenylglycine analog *S*-3,5-DHPG (**15.39**) is selective for Group I mGlu receptors and is devoid of activity at Group II or III receptors.

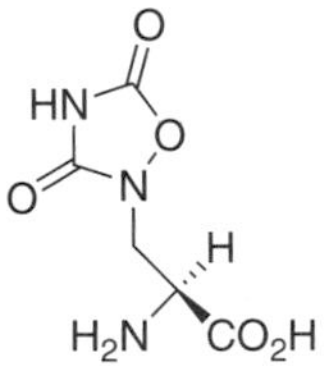

(**15.36**) Quis 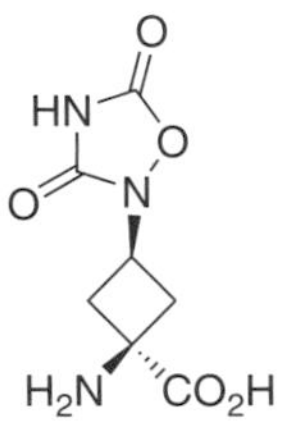 (**15.37**) Z-CBQA 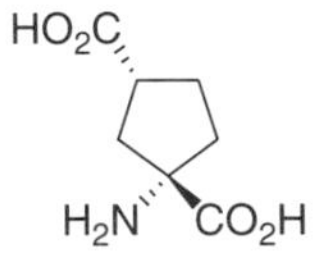 (**15.38**) 1*S*, 3*R*-ACPD 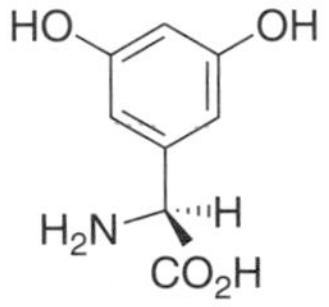 (**15.39**) *S*-3, 5-DHPG

Group II mGlu receptor agonists

Conformationally restrained glutamate analogs based on cyclopropylglycine in the extended conformation tend to be selective agonists for mGlu receptors without activating iGlu receptors. L-CCG-1 (**15.40**) is the most potent agonist for Group II ($mGluR_2$/$mGluR_3$), but loses selectivity at higher concentrations (i.e., binds to $mGluR_{1/5,4/6/8}$). In this respect, a greater selectivity for Group II receptors is shown by DCG-IV (**15.41**).

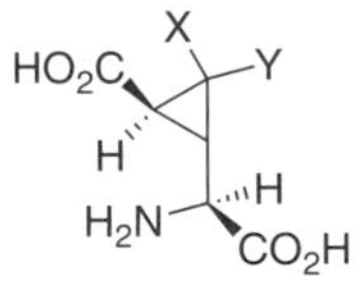

	X	*Y*
(**15.40**) L-CCG-I	H	H
(**15.41**) DCG-IV	H	CO_2H

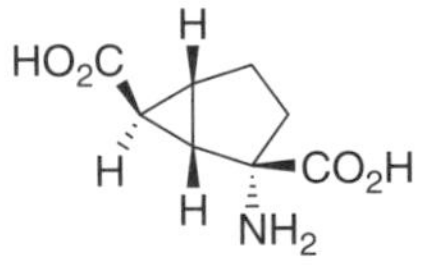

(**15.42**) LY354740

The Group II mGlu receptor agonists described lack CNS activity following systemic administration. LY354740 (**15.42**) exhibits systemic pharmacological effects at $mGluR_{2/3}$ with 10^2–10^4-fold selectivity for other mGlu subtypes or iGlu receptors. Oral administration of this compound in rats induced anxiolytic activity without CNS depression (unlike diazepam) and blocked withdrawal effects of morphine, nicotine, and diazepam.

Group III mGlu receptor agonists

(*S*)-AP4 (**15.43**) is a selective agonist for Group III receptors and can distinguish between Groups I–III receptors but it is not an ideal diagnostic agent and other compounds are under examination in this expanding area.

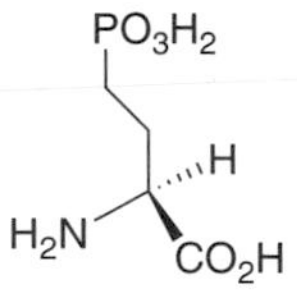

($\mathbf{15.43}$) S-AP4

Group I mGlu receptor antagonists

(*S*)-4-carboxyphenylglycine ((*S*)-4-CPG, **15.44**), an antagonist at $mGluR_1$ with weak agonist activity at $mGluR_2$ was converted to an antagonist ((*S*)-MCPG, **15.45**) at both receptors by an α-methyl substitution. Homologation of the α-alkyl group gave (*R*,*S*)-PeCPG (**15.46**), which had improved potency at $mGluR_5$. Aryl substitution of the parent compound (**15.44**) led to (*S*)-4C3HPG (**15.47**), which is an antagonist at $mGluR_1$ and a potent agonist at $mGluR_2$. This is an interesting combination of pharmacological properties which yielded an expected anticonvulsant profile in models of epilepsy.

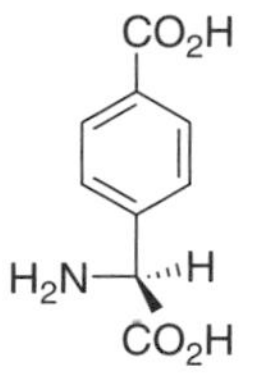

(**15.44**) *S*-4-CPG 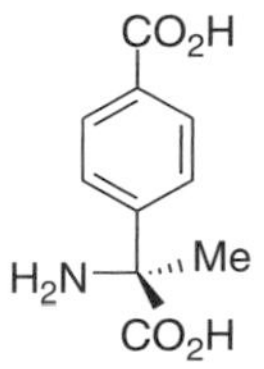 (**15.45**) *S*-MCPG 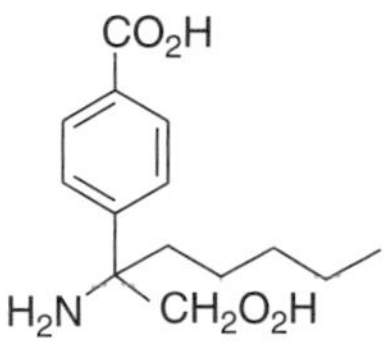 (**15.46**) *R,S*-PeCPG 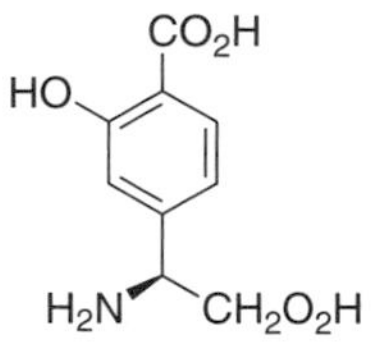 (**15.47**) *S*-4C3HPG

Group II mGlu receptor antagonists

Some 4-substituted glutamate analogs display antagonist activity. ADBD (LY307452, **15.48**) is selective at $mGluR_2$ and $mGluR_3$ and potency is increased in ADED (LY310225, **15.49**). LY341495 (**15.50**), a cyclopropyl glutamate analog, is selective for Group II receptors at low concentration but at higher concentrations its antagonist action spreads to other receptor groups rendering it as a useful pharmacological tool for studying multiple mGlu receptors.

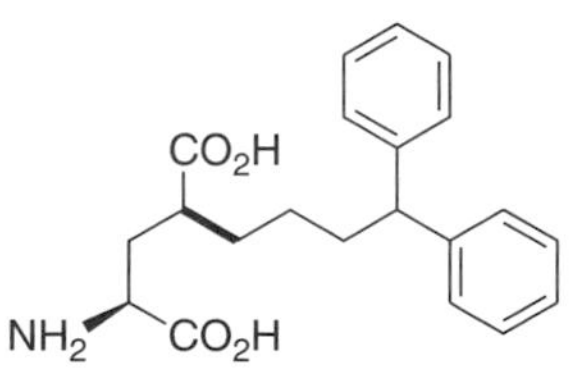

(**15.48**) LY307452 (ADBD) 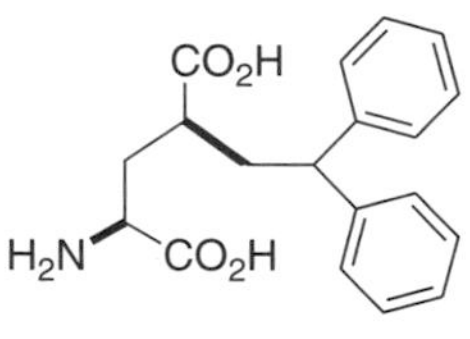 (**15.49**) LY310225 (ADED) 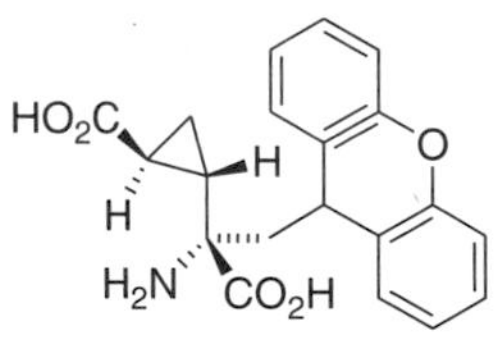 (**15.50**) LY341495

Group III mGlu receptor antagonists

Several antagonists are known based on phosphono amino acids (MAP4, **15.51**; MSOP, **15.52**) or phenyl glycine (MPPG, **15.53**; CPPG, **15.54**) but may have activity towards Group II receptors as agonists or antagonists, respectively.

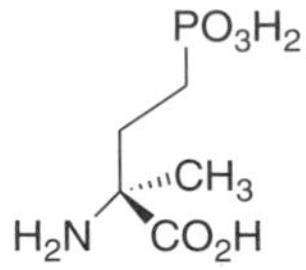

(**15.51**) MAP4 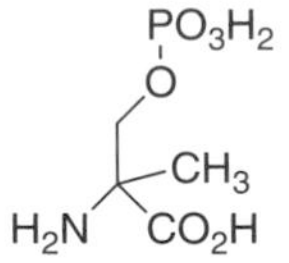 (**15.52**) MSOP 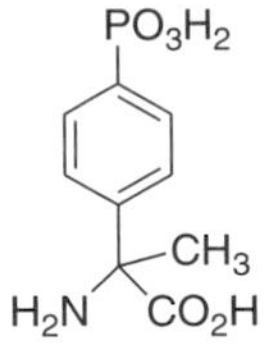 (**15.53**) MPPG 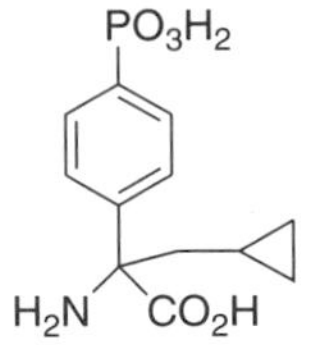 (**15.54**) CPPG

15.2.6 Drug Development

The discovery of the importance of Glu neurotransmission in CNS pathways and the intricacies of the systems involved at the molecular level (revealed by recent strides in molecular biology) has prompted interest in the basis of diseases arising from neuronal malfunction and possible treatments using carefully designed small molecular drugs.

Ionotropic Glu (iGlu) receptors (NMDA, AMPA, and kainate) are modulated by metabotropic (mGlu) receptor activation. Glu not only exerts its excitatory activity through iGlu receptors, but also neurotoxic effects resulting from excessive NMDA receptor stimulation which has been implicated in CNS injury, e.g., hypoxia, ischemia, and trauma. Under excitation of NMDA receptors is likely to occur in schizophrenia and Alzheimer's disease as illustrated using receptor antagonists such as ketamine, phencyclidine (PCP, **15.55**), CPP (**15.56**), and CPP-ene (**15.57**) to prevent the action of Glu on NMDA receptors. The antagonists probably block receptor channels by binding to a site located deeply within it and thus have a potential therapeutic use as neuroprotective agents for the treatment of ischemia caused by hyperactivation of NMDA channels and excessive Ca^{2+} entry. However, PCP and other high-affinity channel blockers have unwanted side affects, i.e., as for competitive antagonists, they exhibit motor impairment (ataxia), learning impairment, and psychotomimetic effects in laboratory models. Hence, low-affinity blockers (amantadine and analogs) may prove to be more clinically useful drugs.

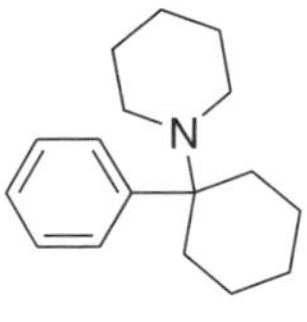

(**15.55**) Phencyclidine

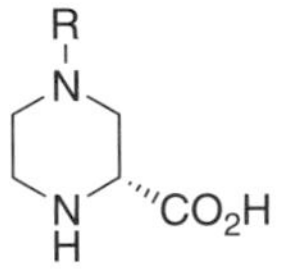

(**15.56**) R = $(CH_2)_3PO_3H_2$ (*R*)-*CPP*
(**15.57**) R = CH_2-CH=CH-PO_3H_2 (*R*)-*CPP*-ene

The noncompetitive NMDA receptor antagonist ifenprodil (**15.18**) and its analogs possess both anticonvulsant and neuroprotective properties without the attendant psychotomimetic side effects of competitive antagonists and some of the channel blockers mentioned above. Ifenprodil interacts with the NR2B-containing receptor but also binds nonspecifically to other sites such as neuronal calcium channels. More selective analogs have been developed (Ro 25–6981, **15.20**) and other drugs acting on iGlu receptors are under development for diseases associated with excitatory neuronal degeneration as previously mentioned. In parallel with these developments, the potential for developing mGluR agonist/antagonists, which have a regulatory role, has been examined. In future, this may lead to the development of useful drugs for pain, schizophrenia, epilepsy, and neurodegenerative conditions such as Parkinson's disease.

15.2.7 $GABA_A$ Receptors

GABA is the primary inhibitory neurotransmitter found mainly at local interneurons in the CNS and it exerts effects through at least three different receptor subtypes. The quaternized alkaloid bicuculline (**15.68**) is a competitive $GABA_A$ antagonist, which aided the early characterization of this receptor subtype. A subpopulation of GABA receptors ($GABA_B$) was subsequently identified using the $GABA_B$ agonist baclofen and a $GABA_C$ subtype has also been defined. $GABA_{A/C}$ receptors are ionotropic while $GABA_B$ is a metabotropic site that modulates the opening of K^+ channels through second messengers involving a G-protein. The $GABA_A$ subtype is the most well characterized of the GABA receptors and it is composed of a macromolecular complex (heteropentameric glycoprotein assembly) of five subunits clustered around a membrane channel that is preferentially permeable to Cl^- ions. Several variants of each of these subunits have been cloned to date and there are different conformations of the $GABA_A$ receptor. Probably the most prevalent mammalian arrangement of constituent subunits in the $GABA_A$ receptor is: $2\alpha_1$, $2\beta_2$, and γ_2 since mRNAs encoding these subunits are commonly localized throughout the brain. The particular combination determines affinity for the neurotransmitter, the rates of activation and deactivation, and ion conductance through the channel. Each receptor unit consists of hydrophobic peptide membrane-spanning domains (α-helices), and an extracellular region with an N-terminal glycosylated domain carrying the GABA-binding site and a short C-terminal tail as well as an intracellular loop. The unit is an allosteric protein such that ligand binding affects the channel ion pore opening (gating). Other allosteric modulators such as benzodiazepine agonists bind to an accessory site (the benzodiazepine receptor) on the $GABA_A$ receptor thus facilitating GABA-binding and consequently its agonist activity. Barbiturates and some anesthetics behave in a likewise fashion. The $GABA_A$ receptor also has a less well-defined binding site for neurosteroids. These compounds are related to steroid hormones (e.g., pregnenolone and its metabolites) and are synthesized in the brain from cholesterol. The synthetic agent alphaxolone (**15.58**) is a neurosteroid, which modulates GABA agonist activity and has been developed as an anesthetic.

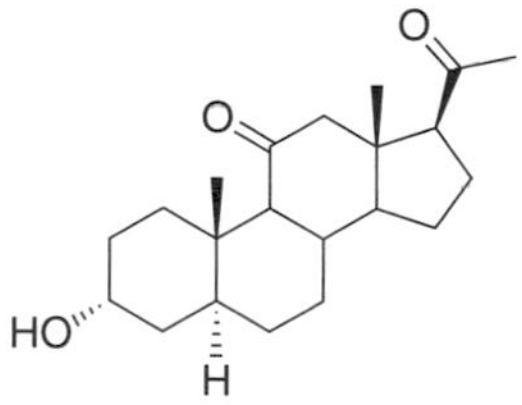

(**15.58**) Alphaxolone

$GABA_A$ receptor agonists

Different classes of agonists have been designed using the powerful $GABA_A$ agonist muscimol (**15.59**), a constituent of the mushroom *A. muscaria*, as lead compound. This led to the potent agents thiomuscimol (**15.60**) and dihydromuscimol (**15.61**), the *S*-form of the latter being the most potent agonist known. Ring closure of muscimol to THIP (**15.62**) and the isomeric THPO (**15.63**) gave an agonist and a GABA reuptake inhibitor, respectively. Further development of THIP gave the agonists isoguvacine (**15.64**) and isonipecotic acid (**15.65**). Development of the GABA uptake inhibitor THPO led to nipecotic acid (**15.66**) and guvacine (**15.67**) with similar actions.

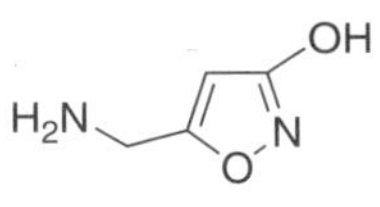

(**15.59**) Muscimol 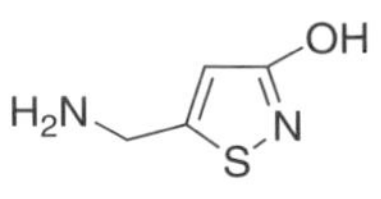 (**15.60**) Thiomuscimol 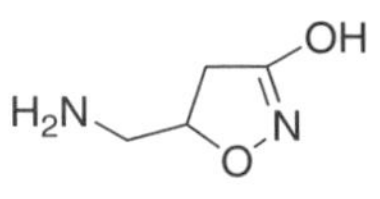 (**15.61**) Dihydromuscimol

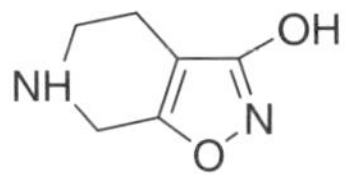

(**15.62**) THIP 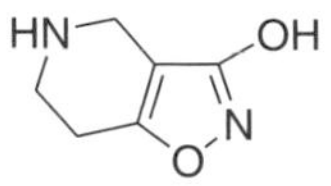 (**15.63**) THPO 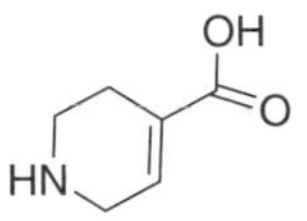 (**15.64**) Isoguvacine

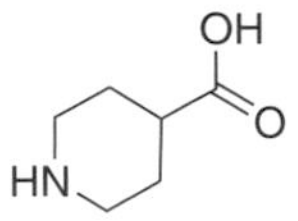

(**15.65**) Isonipecotic acid 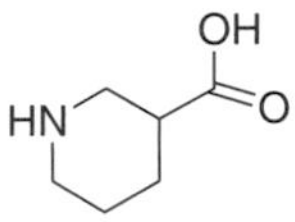 (**15.66**) Nipecotic acid 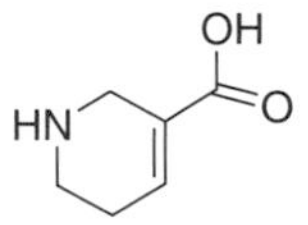 (**15.67**) Guvacine

$GABA_A$ receptor antagonists

Bicuculline and its methochloride (**15.68**) are well-known $GABA_A$ receptor antagonists having been used as tools for determining the pharmacological role of such receptors. Potent effects have also been shown by the THIP analog Iso-THAZ (**15.69**), gabazine (**15.70**), and its thiomuscimol isostere (**15.71**).

Glutamate and GABA transporters

After stimulating Glu receptors, glutamate is rapidly removed from the synapse and extracellular fluid to regenerate the receptors for further stimulation and also ensure that the damaging effects of prolonged neuronal exposure to glutamate are terminated. Removal of glutamate by reuptake into surrounding cells is carried out by protein transporters in the cell membrane which convey glutamate intracellularly thus preventing neuronal damage and recycling the neurotransmitter.

GABA transporters have been cloned from several species and four are known in the mouse (GAT1-4). A number of compounds are known which inhibit GABA uptake by these transporters, some showing selectivity for a specific transporter protein and these include nipecotic acid (**15.66**), guvacine (**15.67**), and tiagabine (**15.72**).

15.2.8 Drug Development

Since the early pharmacological observation that GABA produced CNS depression, the mechanisms underlying this action have been examined at the molecular level. The structural picture of individual receptors has indicated a diversity of actions resulting from stimulation of $GABA_A$ receptors, the ongoing unraveling process aided by selective pharmacological tools (agonists, antagonists, modulators, transporter-directed agents, etc.) in animal model systems, tissue radioligand binding studies, and site-directed mutagenesis in reconstituted receptor systems.

All this basic research is necessary before selective agents, free of undesirable side effects, can be developed and processed to market for diseases arising through malfunctioning of the $GABA_A$ system. Sufficient progress has not been made to date for this to have been achieved due to the complex and interdependent nature of the systems involved. Nevertheless, in this section, the prospective outlook for clinical therapies is explored.

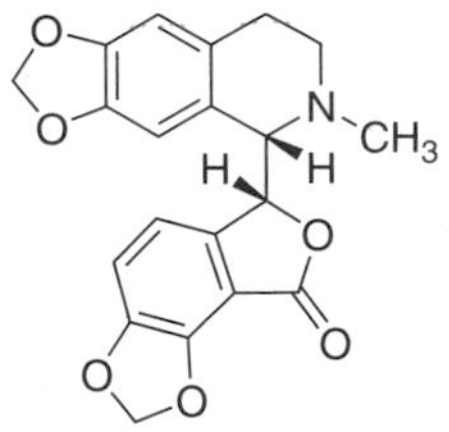

(**15.68**) Bicuculline

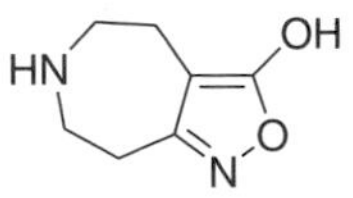

(**15.69**) iso-THAZ

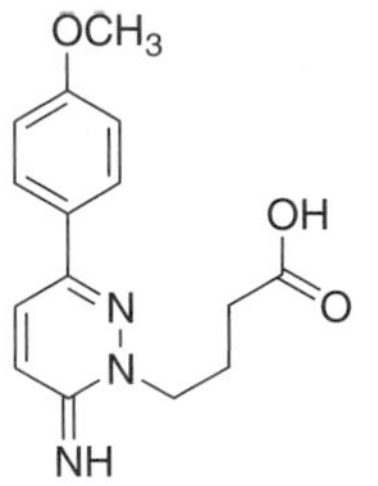

(**15.70**) Gabazine

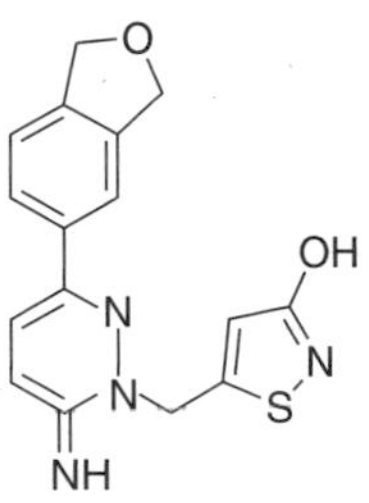

(**15.71**) Thiomuscimol isostere of gabazine

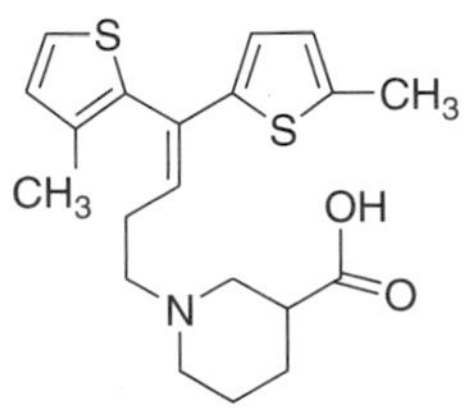

(**15.72**) Tiagabine

Hypoactivity of the GABA system is considered to be at least partially implicated in epilepsy, anxiety, pain, and sleep disorders whereas hyperactivity may be conducive to schizophrenia. Enhancement of GABA-ergic function may be achieved by administration of (1) GABA agonists, or modulators which increase GABA-like effects, (2) GABA reuptake inhibitors or GABA-metabolizing enzyme inhibitors (see Chapter 9) which maintain extracellular GABA concentrations. The benzodiazepines produce sedation, muscle relaxation, antianxiety, and anticonvulsant effects as well as memory impairment. The actions of currently available drugs are not wholly selective for the benzodiazepine-binding site on different $GABA_A$ receptor combinations or cell types in the CNS and this may conceivably account for unwanted side effects. It has been shown in genetically modified mice that α_1-containing $GABA_A$ receptors are responsible for the sedative effects of diazepam while the anxiolytic and muscle relaxant effects are mediated through α_2-containing receptors. Moreover, the α_3-containing receptor is not responsible for any of the above features of diazepam. α_5-Containing receptors are mainly restricted to the hippocampal area of the CNS, which is known to be concerned with memory. It has also been shown, using transgenic models, that α_5-containing receptors suppress learning and memory and hence α_5-receptor selective inhibitor would be useful for improving memory in the elderly or in Alzheimer's disease patients.

Tiagabine (Gabitril, **15.72**), a GABA reuptake inhibitor is available for the treatment of certain types of epileptic seizures, and by potentiating GABA levels, it shifts the Glu/GABA balance of excitatory/depressant activity in favor of a CNS depressant effect. Tiagabine selectively targets the GABA transporter protein GAT1 (see earlier). A mixed GAT1/3 inhibitor may therefore be useful against a wider variety of seizure types where tiagabine may not be so effective.

Tiagabine is also effective in preclinical models of anxiety, but this finding has not yet been exploited in the clinic. $GABA_A$ modulators such as barbiturates, neurosteroids, and benzodiazepines have been used to produce sleep (sedative hypnotics). In this respect, the $GABA_A$ agonist THIP (Gaboxadol, **15.62**) induces sleep without the drug ''hangover'' reported for benzodiazepines and similar results have been obtained with other agonists like muscimol (**15.59**) and the GABA reuptake inhibitor tiagabine (**15.72**).

FURTHER READING

Egebjerg, J., Schousboe, A., and Krogsgaard-Larsen, P. (eds) (2002). *Glutamate and GABA Receptors and Transporters: Structure, Function and Pharmacology*. Taylor & Francis, London. (1) Jane, D. Pharmacology of NMDA receptors, pp. 69–98. (2) Madsen, U., Johansen, T.N., Stensbol, T.B., and Krogsgaard-Larsen, P. Pharmacology of AMPA/kainate receptors, pp. 99–118. (3) Schoepp, D.D. and Monn, J.A. Pharmacology of metabotropic glutamate receptors, pp. 151–186. (4) Krogsgaard-Larsen, P., Frolund, B., Kristiansen, U., and Ebert, B. Ligands for the $GABA_A$ receptor complex, pp. 236–274.

Whiting, P.J. (2003) $GABA_A$ receptor subtypes in the brain: a paradigm for CNS drug discovery. *Drug Discovery Today* **8**(10), 445–450.

15.3 HISTAMINE RECEPTORS

Holger Stark and Walter Schunack

15.3.1 Introduction

The story of histamine (**15.73**, specific histamine nomenclature is shown on this structure) began in the early 1900s with classical pharmacological investigations concerning its physiological and pathophysiological effects. Histamine has been recognized as an important chemical messenger communicating information from one cell to another. A large variety of cell types including smooth muscles, endocrine and exocrine glands, blood cells, and cells of the immune system respond to histamine stimulus with large differences from species to species mainly investigated in vertebrates. New aspects were brought into this field of research by the discovery that histamine does not only act as a local hormone, but also acts as a neurotransmitter. Although histamine is widely distributed within mast cells in almost all mammalian peripheral tissues it plays an important role in the mammalian brain displaying powerful neuromodulatory, immunomodulatory, and neurotransmitter effects. Histamine itself does not cross the blood–brain barrier. Physiologically it is produced by decarboxylation of L-histidine, mainly by the specific histidine decarboxylase. This enzyme could be selectively inactivated by (*S*)-α-fluoromethylhistidine (**15.74**) a ''suicide'' substrate.

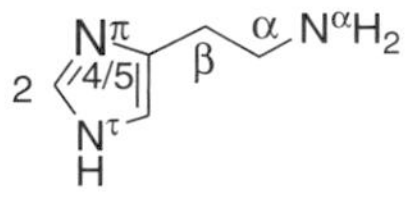

(**15.73**) Histamine

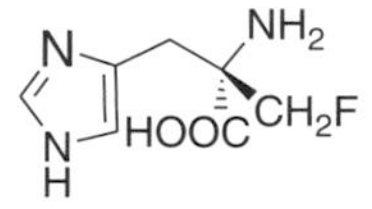

(**15.74**) Fluoromethylhistamine

Histamine catabolism occurs along two alternative pathways. One metabolic pathway is the methylation by the specific histamine *N*-methyltransferase to *N*′-methylhistamine, the other one is the oxidative deamination by diamine oxidase to imidazole acetaldehyde and further oxidative and coupling products. The first pathway seemed to be the only one to operate in the mammalian brain. Therefore, the methylated metabolite is the most important catabolite for investigations on central histamine levels for neurotransmitter aspects.

According to their chronological order of discovery four subtypes of histamine receptors have been described so far: histamine H_1, H_2, H_3, and H_4 receptors. The effects following direct stimulation of these subreceptors belonging to the superfamily of G-protein-coupled receptors are manifold and depend on species and tissues. H_1 receptors are coupled to the phosphatidylinositol cycle, H_2, and H_3 receptors to the adenylate cyclase; the signaling system for H_4 receptor is unknown so far. Meanwhile all receptors have been cloned and characterized by molecular biology methods. H_1 and H_2 receptors are postsynaptically located. The H_3 receptor was identified at first as a presynaptically located autoreceptor controlling the synthesis and release of histamine in histaminergic neurons. Later on, the newly found function of H_3 heteroreceptors modulating the release of a number of different neurotransmitters (noradrenaline, acetylcholine, dopamine, serotonin, neuropeptides, etc.) gave further hints for therapeutic indications of H_3 receptor ligands. The recently found H_4 receptors are expressed in human blood leukocytes and bone marrow suggesting their function on immune modulation.

15.3.2 H_1 Receptors

Histamine possesses two basic moieties: the primary nitrogen on the side-chain (N^{α}) protonated under physiological conditions ($pK_{a1} = 9.73$) and the protonatable aromatic imidazole nucleus ($pK_{a2} = 5.91$). The first developed H_1 receptor agonists followed the minimal structural requirements of having an aromatic ring and an ethylamine side chain. This approach was more or less successful with *N*-methyl-2-(2-pyridyl)ethanamine (betahistine, **15.75**). Although it shows less than 30% intrinsic activity (i.a.) compared to the endogenous ligand (histamine i.a. = 100%) betahistine is still used as a drug in Menièr's syndrome. Recent developments in the class of 2-substituted histamine derivatives resulted in 2-(3-(trifluoromethyl)phenyl)histamine (**15.76**) and methylhistaprodifen (**15.77**). These compounds have 128% and ≈400% relative activity (histamine = 100%), respectively. Compounds (**15.76**) and (**15.77**) make experimental characterization of H_1 receptor-mediated biological functions much more precise than former tools.

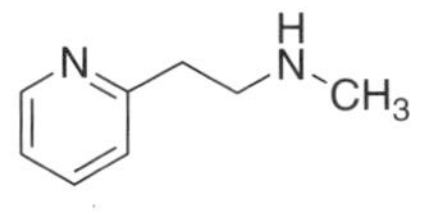

(**15.75**) Betahistine 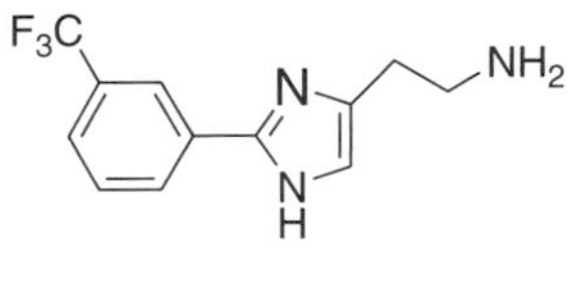 (**15.76**) Trifluoromethylphenylhistamine 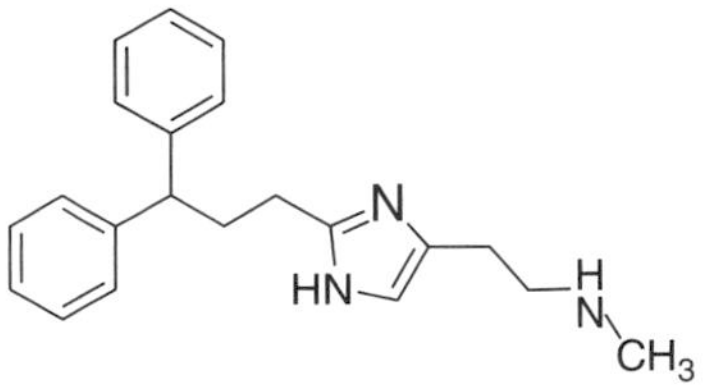 (**15.77**) Methylhistaprodifen

Most people have allergic and inflammation reactions in mind when thinking of histamine. Compounds preventing this response to histamine are the first so-called "antihistamines." These histamine H_1 receptor antagonists are among the most widely used medications in the world. In contrast to histamine these agents are generally lipophilic compounds due to their aromatic moieties. This aromatic ring structure is connected by a small chain to a protonatable basic center, in most cases a tertiary amino group. One of the most potent and selective antagonists pharmacologically used is mepyramine (pyrilamine, **15.78**). In radiolabeled form (^{3}H, ^{11}C) it can also be used for binding characterizations *in vitro* as well as *in vivo*. One of the most useful [^{125}I]iodinated H_1 receptor antagonists is iodobolpyramine (**15.79**) which could be easily prepared in a reaction with a radiolabeled Bolton–Hunter precursor. Many other compounds have been developed since then. Chlorpheniramine (**15.80**) and diphenhydramine (**15.81**) are only two examples of a large number. Chlorpheniramine shows a stereoselective effect presenting higher antagonist activity with its (*S*)-(+)-enantiomer (eutomer). These H_1 receptor antagonists of the first generation have a number of side effects, which led to further developments of neuroleptics (chlorpromazine), antidepressants (amitriptyline, doxepin), or anticholinergic agents (trihexyphenidyl, biperiden). The H_1 receptor antagonist hydroxyzine is used as a tranquillizer. The most common central side effects of lipophilic H_1 receptor antagonists are sedation and avoiding nausea with kinetose. Both indications are now given with diphenhydramine and promethazine (**15.82**). Sedation and atropine-like side effects led to the next generation of H_1 receptor antagonists. Ketotifen (**15.83**) is a potent H_1 receptor antagonist and possesses antagonist activity to other allergy-inducing agents like bradykinin or serotonin. This is an example of an antihistamine with additional antiallergic activity. Terfenadine (**15.84**, R $= CH_3$) and astemizole are not capable of easily penetrating the blood–brain barrier. Due to this lack of central effects in therapeutic dosage, as a consequence of pharmacokinetic differences, terfenadine, astemizole, and other newer antihistamines (loratidine, desloratidine, (levo)cetirizine, ebastine, (levo)cabastine, mizolastine (**15.85**)) may have advantages in their side-effect profile. Due to the cardiac side effects of terfenadine and astemizole (prolonged QT_c interval and severe arrhythmias) especially in higher dosage or in combination with cytochrome P_{450} (CYP450) inhibitors like ketoconazole or erythromycin, these compounds were withdrawn from the market and the active metabolite of (**15.84**) (fexofenadine (**15.84**), R $=$ COOH) with improved side-effect profile and without metabolism by CYP450 was introduced.

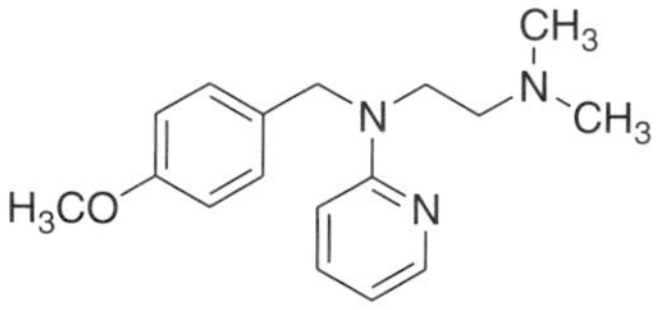

(**15.78**) Mepyramine

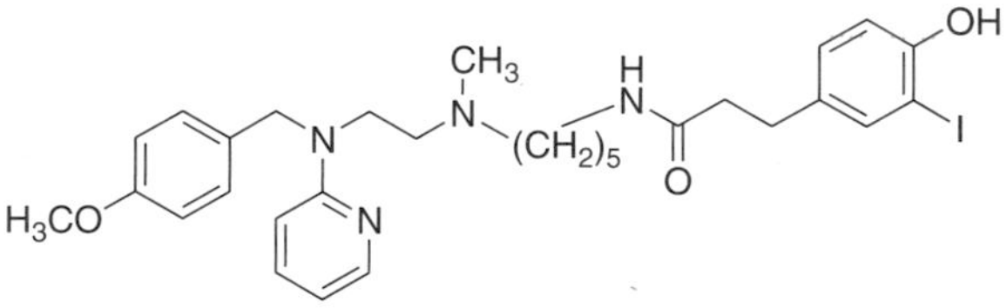

(**15.79**) Iodobolpyramine

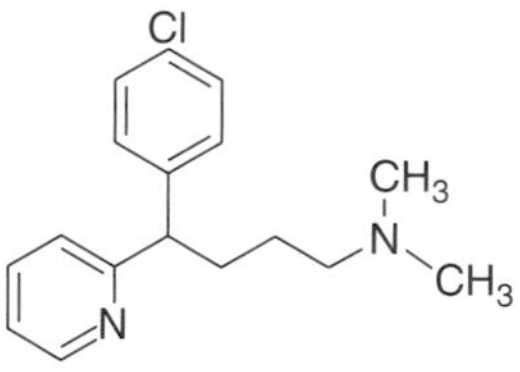

(**15.80**) Chlorpheniramine

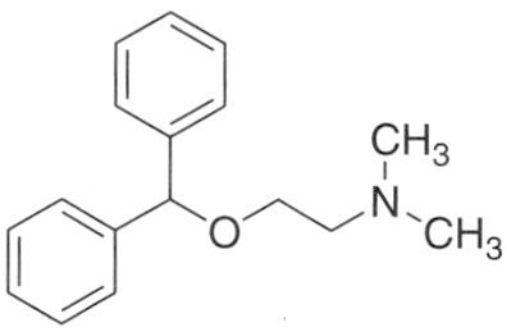

(**15.81**) Diphenhydramine

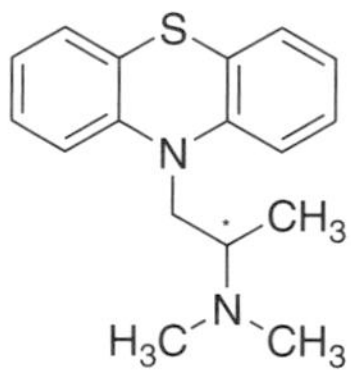

(**15.82**) Promethazine

15.3.3 H_2 Receptors

Structural similarities of the H_2 receptor agonists dimaprit (**15.86**) and amthamine (**15.87**) to histamine (**15.73**) are obvious. In addition to the protonated side-chain nitrogen the structures are capable of making hydrogen bonds and of undergoing a 1,3-prototropic tautomerism like the imidazole moiety. The isothiourea and the aminothiazole moieties may be considered as bioisosteres of imidazole in this particular case. These compounds are roughly in the same activity range as histamine. Prolongation of the alkyl chain and variation of the amino functionality to a strongly basic substituted guanidine group led to a strong increase in H_2 receptor agonist activity. Impromidine (**15.88**), which contains two imidazole fragments, is 48 times more potent than histamine. The imidazolylpropylguanidine group seems to be responsible for the agonist binding, whereas the other part of the molecule (cimetidine part, cf. **15.90**) contributes additional binding. This binding is improved by a diarylalkyl structure as in arpromidine (**15.89**). On cardiac H_2 receptors this compound is about 100 times more potent than histamine, but shows different activity with different preparations of H_2 receptors giving a hint for H_2 receptor subpopulations. Thereby, the ''fluoropheniramine'' partial structure of (**15.89**) (cf. **15.80**) also incorporates H_1 receptor antagonist activity.

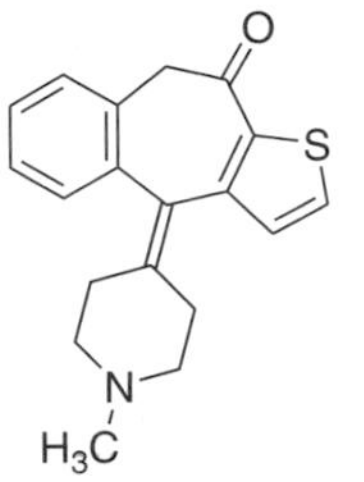

(**15.83**) Ketotifen

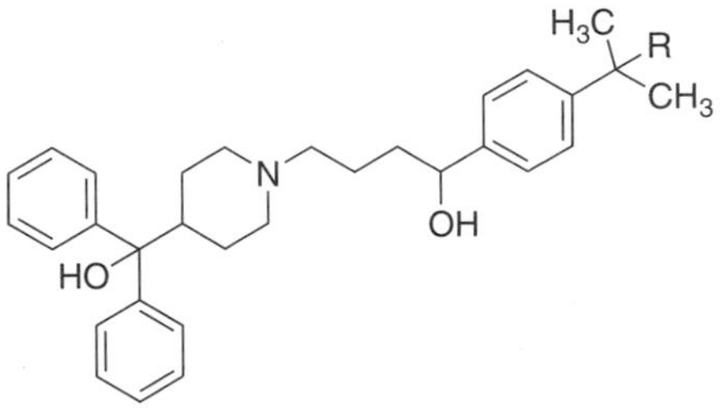

(**15.84**) R=CH_3, Terfenedine; R=COOH, Fexofenidine

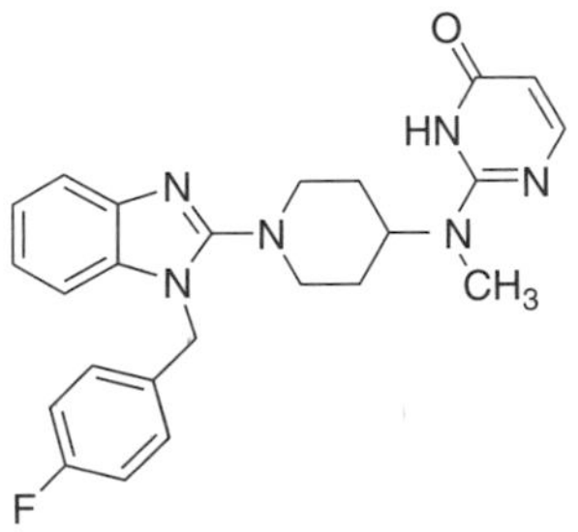

(**15.85**) Mizolastine

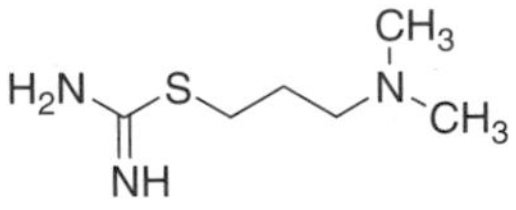

(**15.86**) Dimaprit

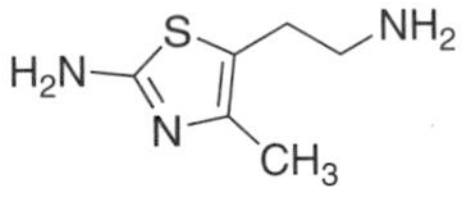

(**15.87**) Amthamine

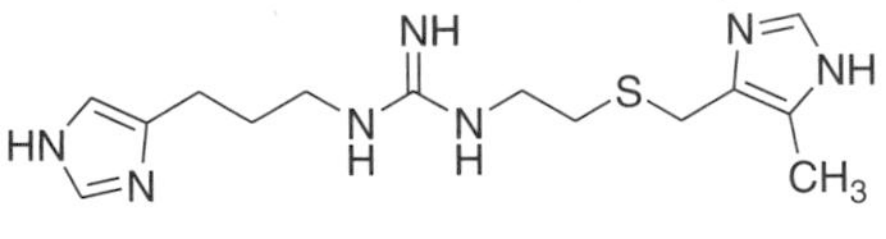

(**15.88**) Impromidine

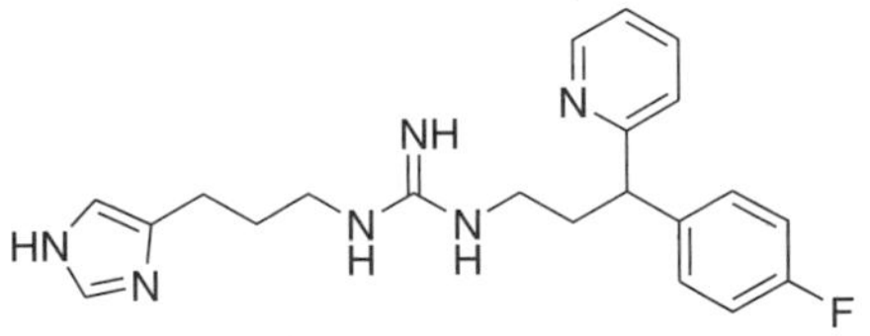

(**15.89**) Arpromidine

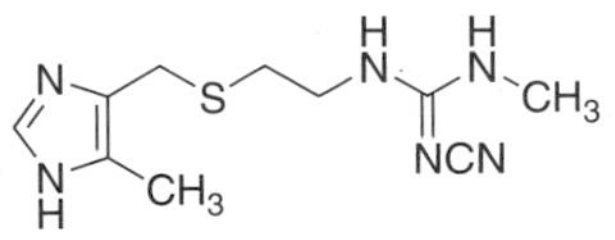

(**15.90**) Cimetidine

The partial structure of impromidine that enhances activity is the main structural element of cimetidine (**15.90**), the first worldwide marketed H_2 receptor antagonist. The guanidine group is substituted by a cyano group that reduces the basic properties of this moiety and leads to a highly polar group. The basic imidazole ring connected by a flexible chain to a polar group, which is uncharged under physiological conditions leads to H_2 receptor antagonists. The imidazole moiety seems to be responsible for the inhibition of CYP450-dependent reactions leading to unwanted side effects with comedications in man. Therefore, new developments replaced the imidazole ring bioisostere with a basic substituted furan (ranitidine (**15.91**)) and thiazole (famotidine (**15.92**)) rings, or a basic substituted phenoxyalkyl moiety (roxatidine acetate (**15.93**); iodoaminopotentidine (**15.94**); zolantidine (**15.95**)). Other nonclassical bioisosteres for the thiourea group (cf. the early H_2 receptor antagonist burimamide (**15.103**)) are the cyanoguanidine, the nitroethendiamine, and aminosulfonyl amidine moieties. Histamine H_2 receptor antagonists are used for the treatment of conditions associated with gastric hyperacidity like peptic ulcer disease or reflux esophagitis.

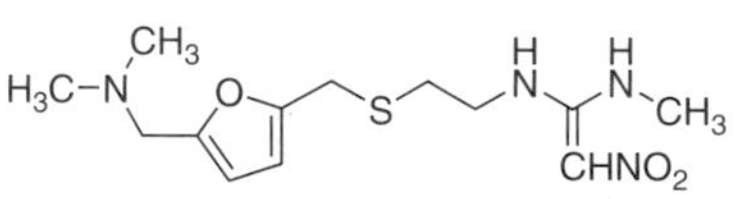

(**15.91**) Ranitidine

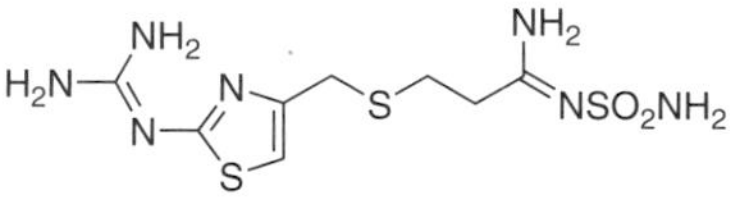

(**15.92**) Famotidine

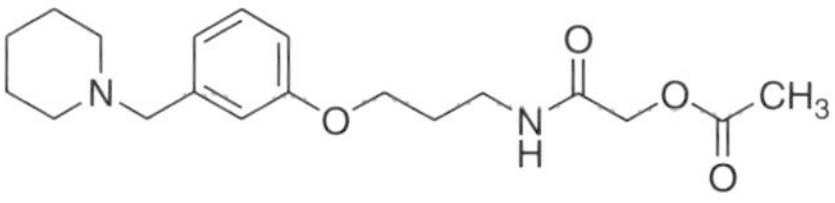

(**15.93**) Roxatidine acetate

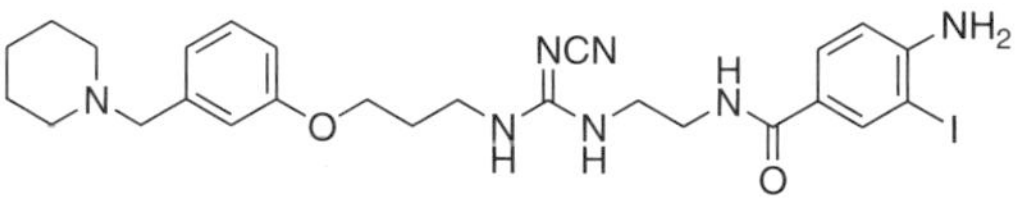

(**15.94**) Iodoaminopotentidine

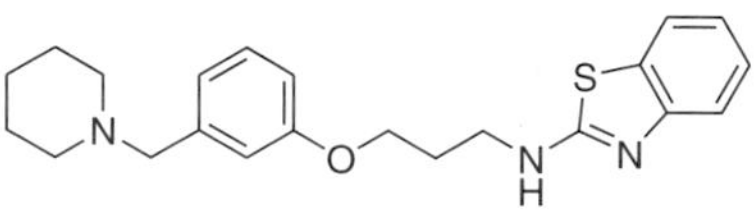

(**15.95**) Zolantidine

The new compounds have a high therapeutic index showing low incidence of side effects. One of the most potent H_2 receptor antagonists is iodoaminopotentidine. This compound and a related azido derivative were used in radiolabeled form for autoradiographic localization, specific binding studies, and photoaffinity labeling of cerebral H_2 receptors. All these compounds possess a cyanoguanidine, nitroethenediamine, or comparable ''urea equivalents'' as a structural alternative for the polar group. Therefore, they do not readily cross the blood–brain barrier. No important central effects could be detected with these drugs *in vivo*. Their physicochemical properties have to be taken into account and drug design to optimize the partition coefficient, ionization constant, and molecular size with maintenance of H_2 receptor antagonist activity led to zolantidine, which penetrates into the CNS in a reasonable concentration. Hallucinatory effects of some H_2 receptor antagonist when given in high dosage and the control of nociceptive responses may be clarified by behavioral experiments with zolantidine. Although proton-pump inhibitors (inhibitors of H^+/K^+-ATPase) like omeprazole, lansoprazole, or pantoprazole have replaced the H_2 receptor antagonists as first regime therapeutics in peptic ulcer disease the H_2 receptor antagonists still have their place as reliable and safe therapeutics.

15.3.4 H_3 and H_4 Receptors

The function of histamine as a neurotransmitter in addition to its autacoid function was strengthened by the discovery of the histamine H_3 receptor. Histamine (**15.73**) shows higher affinities for the H_3 and H_4 receptors than for the H_1 and H_2 receptors (H_3, pD_2 7.2; H_4, $pK_i \approx 8$; H_2, pD_2 6.0; H_1, pD_2 6.85). Most H_3 and H_4 receptor ligands with known high affinity so far possess an imidazole ring. Many replacement studies in histamine for H_3 receptor ligands led to a dramatic decrease in affinity. Even substitution by small methyl groups on the imidazole ring decreases H_3 receptor affinity (cf. **15.90**). On the other hand, methylation on the side chain increases agonist activity in particular cases. Methylation on the side-chain nitrogen resulted in the H_3 receptor agonist N^{α}-methylhistamine (**15.96**). Although this compound is often used in tritium-labeled form in H_3 receptor binding studies it shows remarkable activity at H_1 and H_2 receptors. The use of the chiral methylated histamine derivative [^{3}H](*R*)-α-methylhistamine should be favored over other commercially available [^{3}H]labeled H_3 receptor agonists for these investigations. The reason for this preference is that the selectivity of side-chain methylated histamine derivatives was increased in (*R*)-α-methylhistamine (**15.97**), now the standard agonist for histamine H_3 receptors, and was furthermore increased with (α*R*,β*S*)-α,β-dimethylhistamine (**15.98**). Despite their structural similarity these compounds show low activity at H_1 and H_2 receptors, but compared to histamine they have 15 to 18 times the H_3 receptor agonist activity, respectively, displaying impressive receptor selectivity. The branched side-chain histamine derivatives show a high degree of stereoselectivity. In all cases the enantiomer with the same relative configuration in the α-position as L-histidine is the eutomer. The eudismic ratio of (**15.97**) compared to its (*S*)-enantiomer (distomer) is about 130. Even the imidazolylmethylpyrrolidine derivative (immepyr, **15.99**) shows this stereoselectivity. A comparable compound containing an achiral piperidinylmethyl moiety (immepip) has also been developed recently. One of the most potent H_3 receptor agonist so far is the imidazolylethylisothiourea derivative (**15.100**) (imetit). The replacement of the side-chain nitrogen by the polar isothiourea moiety, cationic under physiological conditions, leads to improved receptor activity. The sulfur atom in imetit is not critically important for activity and may be replaced by oxygen or methylene. Recent studies with molecular modeling methods showed that all these H_3 receptor agonists could be superimposed to one pharmacophore model displaying similar molecular interaction patterns. Unfortunately all these ligands display similar physicochemical properties. They are extremely hydrophilic compounds, which could not easily cross the blood–brain barrier and reach the CNS, the area with the highest H_3 receptor density. Especially the well-investigated (*R*)-α-methylhistamine is a good substrate for the inactivating enzyme histamine methyltransferase. Clinical trials were not as promising as the first pharmacological experiments.

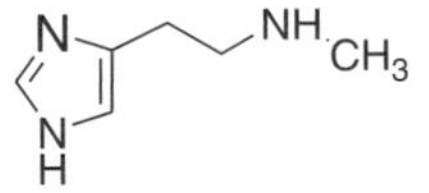

(15.96) Nα - Methylhistamine

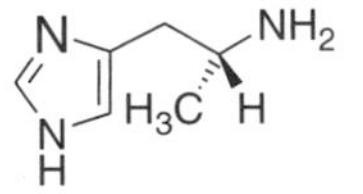

(15.97) (*R*) α-Methylhistamine

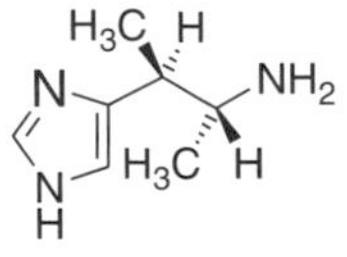

(15.98)

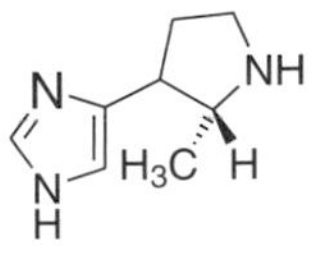

(15.99) Immepyr

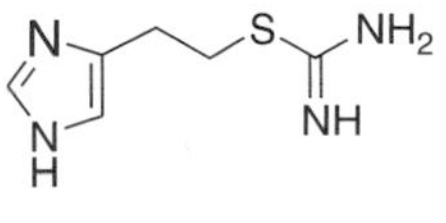

(15.100) Imetit

Prodrugs of (**15.97**) were designed to prepare lipophilic compounds, which could easily penetrate biological membranes and were no longer a substrate for the inactivating enzyme. Compound (**15.101**) (BP-294) is the lead structure for prodrugs of (**15.97**). The liberation of the active drug (**15.97**) depends on chemical hydrolysis of the azomethine bond. Depending on the substitution pattern of the benzophenone promoiety the compounds could be targeted to central or peripheral tissues. High plasma and CNS levels of the biologically highly active (*R*)-α-methylhistamine could be achieved by this approach for a prolonged duration. Compound BP-294 (**15.101**) has already been introduced into clinical trials. This compound shows extraordinary cytoprotective and anti-inflammatory effects in experiments on animals with lesions in the gastrointestinal tract. Newer developments on nonaminergic imidazole-containing (partial) agonists suggest similar results without the prodrug principle.

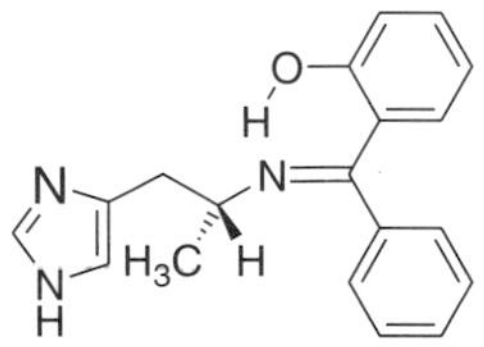

(**15.101**) BP-294

Histamine derivatives with larger substituents on the side-chain N^{α} nitrogen lead to partial agonists, and further increase in size of these substituents leads to histamine H_3 receptor antagonists. This transition from agonist to antagonist was extensively shown on histamine derivatives, but the same is true for derivatives of imetit (**15.100**). Clobenpropit (**15.102**) is *in vitro* a highly potent H_3 receptor antagonist obtained by this approach. The first compounds detected as H_3 receptor antagonists were compounds from the line of H_2 receptor antagonists. Burimamide (**15.103**), which was used as the first ''selective'' H_2 receptor antagonist for the characterization of H_2 receptors, was ironically also used for the characterization of H_3 receptors later on. The H_1 receptor agonist betahistine (**15.75**) shows remarkable H_3 receptor antagonist activity and the H_2 receptor agonists impromidine (**15.88**) and arpromidine (**15.89**) show high H_3 receptor antagonist activity. Once more the imidazole structure like that in histamine seems to be an essential part of highly potent H_3 receptor ligands. With the optimization of the thiourea derivative (**15.103**) a new rigid piperidino and a cyclohexyl moiety were introduced. The resulting thioperamide (**15.104**) was the first highly potent and selective H_3 receptor antagonist. This reference antagonist was replaced by ciproxifan (**15.105**) due to higher *in vitro* and *in vivo* potency and higher selectivity.

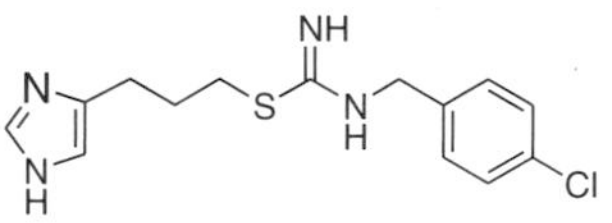

(**15.102**) Clobenpropit

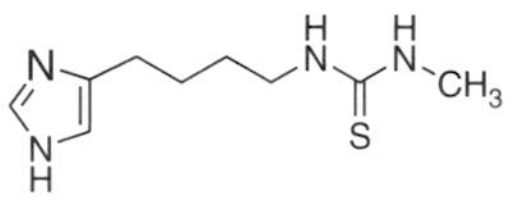

(**15.103**) Burimamide

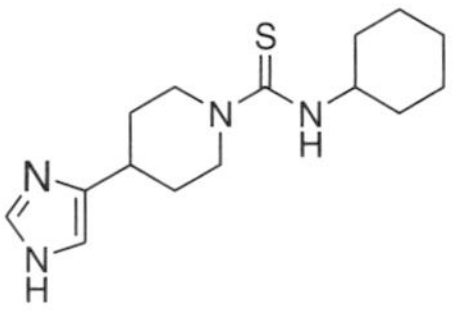

(**15.104**) Thioperamide

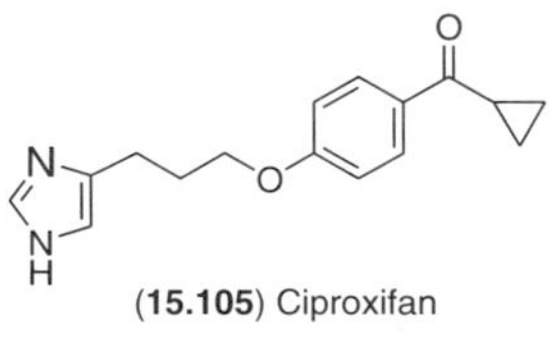

(**15.105**) Ciproxifan

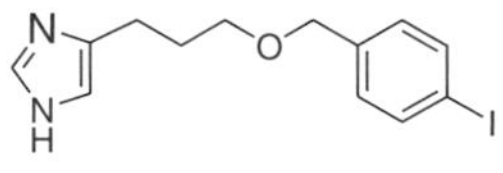

(**15.106**) Iodoproxifan

A general structure–activity pattern was developed by different series of antagonists. A nitrogen-containing heterocycle (mostly imidazole and now also piperidino and related moieties) is connected to a polar group through a chain. These structures seem to be essential for a potent antagonist interaction with H_3 receptors. A lipophilic residue, linked to the polar group by a spacer, seems to enable the molecule to reach additional binding areas, e.g., a hydrophobic pocket of the receptor and thereby the H_3 receptor antagonist activity of the resulting molecule increases. This structural pattern was used to design a new H_3 receptor radioligand with high potency and selectivity. In contrast to the former antagonists possessing a polar group easily able to form hydrogen bonds the resulting iodoproxyfan (**15.106**) has an ether functionality. This structure is suited for hydrogen bonding to a limited extent only. This seems necessary to lower the unspecific binding of a useful radioligand. [^{125}I] Iodoproxyfan fulfills all the criteria for a radioligand such as high activity, selectivity, and specificity as well as saturable and reversible binding. Although radioligands are useful for different pharmacological experiments the development of therapeutically acceptable drugs is a different type of investigation. In this respect it may be important to know that H_3 receptors display a high value of constitutive activity and meanwhile numerous antagonists have been described as inverse agonists rather than as true antagonists whereas only one compound (proxyfan) is described as behaving as a neutral antagonist in some test models.

Recently, new classes of nonimidazole alkyl ethers were developed possessing high H_3 receptor antagonist potency *in vivo* after oral administration. FUB 649 (**15.107**) is an important lead in this series. Depending on the substitution pattern of the phenyl ring and the length of the spacer between this phenyl and the ether moiety, pharmacodynamic and pharmacokinetic properties could be largely varied with retention of antagonist potency. The first antagonist introduced into clinical trials is cipralisant (**15.108**, GT-2331). This compound is tested in attention-deficit hyperactivity disorder (ADHD), which is a severe disorder in children and young adults that is from a therapeutic point of view not satisfactorily countered at the moment. Other different central disorders and diseases may be potential targets for H_3 receptor antagonists, e.g., epilepsy, stress, food intake, sleeping, vertigo, drug abuse, or dementias such as Alzheimer's disease.

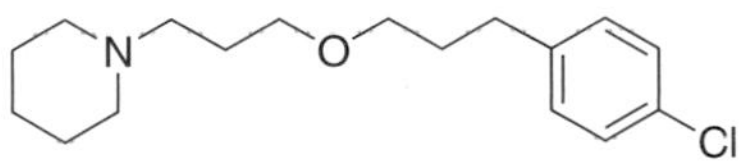

N O Cl

(**15.107**) FUB 649

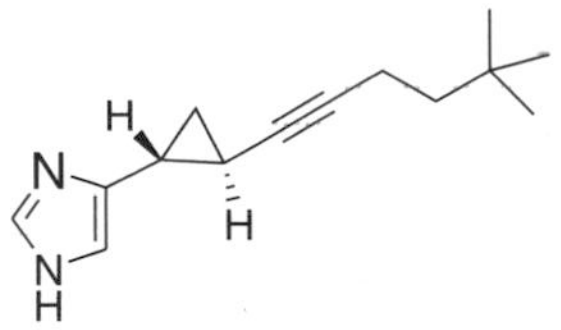

H N N H H

(**15.108**) Cipralisant

Up to now no selective ligands have been described for the H_4 receptor. Most imidazole-containing compounds, which are active at H_3 receptors, also show comparable affinity for H_4 receptors. Interesting developments are expected in this field within the next years.

Influencing the histaminergic neurotransmitter system seems to be an attractive new approach for numerous diseases. The new ligands should improve our knowledge of the physiological and pathophysiological interactions of different neurotransmitters and show new possibilities for the treatment of different diseases.

FURTHER READING

Hill, S.J., Ganellin, C.R., Timmerman, H., Schwartz, J.-C., Shankley, N.P., Young. J.M., Schunack, W., Levi, R., and Haas, H.L. (1997) International union of pharmacology. XIII. Classification of histamine receptors. *Pharmacological Reviews* **49**, 253–278.

Cooper, D.G., Young, R.C., Durant, G.J., and Ganellin, C.R. (1990) Histamine receptors. In: *Comprehensive Medicinal Chemistry: The Rational Design, Mechanistic Study & Therapeutic Application of Chemical Compounds*, C. Hansch (ed.). Pergamon Press, Oxford and New York, pp. 323–421.

Buschauer, A., Schunack, W., Arrang, J.-M., Garbarg, M., Schwartz, J.-C., and Young, J.M. (1989) Histamine receptors. In: *Receptor Pharmacology and Function*, M. Williams, R.A. Glennon, and P.B.M.W.M. Timmermans (eds). Marcel Dekker, New York, pp. 293–347.

Schwartz, J.-C., Arrang, J.-M., Garbarg, M., Pollard, H., and Ruat, M. (1991) Histaminergic transmission in the mammalian brain. *Physiology Reviews* **71**, 1–51.

Hill, S.J. (1990) Distribution, properties, and functional characteristics of three classes of histamine receptor. *Pharmacological Reviews* **42**, 45–83.

Lipp, R., Stark, H., and Schunack, W. (1992) Pharmacochemistry of H_3-receptors. In: *The Histamine Receptor, Series of Receptor Biochemistry and Methodology*, J.-C. Schwartz and H.L. Haas (eds). Wiley-Liss, New York, Vol. 16, pp. 57–72.

Zingel, V., Leschke, C., and Schunack, W. (1995) Developments in histamine H_1-receptor agonists. In: *Progress in Drug Research*, E. Jucker (ed.). Birkhäuser Verlag, Basel, Vol. 44, pp. 49–85.

Leurs, R., Blandina, P., Tedford, C., and Timmerman, H. (1998) Therapeutic potential of histamine H_3 receptor agonists and antagonists. *Trends in Pharmacological Sciences* **19**, 177–183.

Krause, M., Stark, H., and Schunack, W. (2001) Azomethine prodrugs of (*R*)-α methylhistamine, a highly potent and selective histamine H_3-receptor agonist. *Current Medicinal Chemistry* **8**, 1329–1340.

Schwartz, J.-C., Morisset, S., Rouleau, A., Tardivel-Lacombe, J., Gbahou, F., Ligneau, X., Heron, A., Sasse, A., Stark, H., Schunack, W., Ganellin, C.R., and Arrang, J.-M. (2001) Application of genomics to drug design: the example of the histamine H_3 receptor. *European Neuropsychopharmacology* **11**, 441–448.

Hough, L.B. (2001) Genomics meets histamine receptors: new subtypes, new receptors. *Molecular Pharmacology* **95**, 415–419.

Stark, H., Arrang, J.-M., Ligneau, X., Garbarg, M., Ganellin, C.R., Schwartz, J.-C., and Schunack, W. (2001) The histamine H_3 receptor and its ligands. *Progress in Medicinal Chemistry* **38**, 279–308.

15.4 DOPAMINE RECEPTORS

Phillip G. Strange

15.4.1 Introduction

Dopamine receptors have been very important targets for drug design by medicinal chemists partly because of the involvement of dopamine systems in important physiological functions and partly because dopamine receptors are important targets for drug action. In the brain, dopamine systems are involved in the control of movement and certain aspects of behavior, in the pituitary dopamine is important in the control of the secretion of prolactin and melanocyte-stimulating hormone (α-MSH), in the cardiovascular system dopamine is important in the control of blood pressure and heart rate, and in the eye dopamine is important for the control of certain aspects of visual function. In these systems the actions of dopamine are mediated by binding to receptors and blockade or activation of these receptors can offer therapy for certain disorders. For example, dopamine antagonists have been shown to be important in the treatment of schizophrenia, whereas dopamine agonists have been shown to be of use in the treatment of the brain disorder Parkinson's disease, in the therapy of excessive prolactin secretion and certain prolactin-secreting tumors, and in the therapy of cardiovascular disorders. For these reasons a very wide range of compounds has been synthesized that bind to receptors for the neurotransmitter dopamine. The development of the concept of multiple dopamine receptors has added further impetus to this drug discovery program. Based on studies of the actions of dopamine using pharmacological and biochemical techniques it became apparent in the late 1970s that the concept of a single receptor for dopamine was insufficient to explain the information emerging and it was suggested that there were two receptors for dopamine that were termed D_1 and D_2. These had different pharmacological and biochemical properties some of which are summarized in Table 15.1. The concept of two dopamine receptors survived until the techniques of molecular biology were applied to the dopamine receptors. This showed that there were at least five dopamine receptor subtypes (D_1–D_5), which have different structural and functional properties and different localizations in tissues. Some of their properties

Table 15.1 Dopamine receptor subtypes defined on the basis of biochemical and pharmacological studies

	D_1 (D_1-like)	D_2 (D_2-like)
Selective agonists	SKF 38393 (**15.127**)	Quinpirole (**15.126**)
Selective antagonists	SCH 23390[a]	Sulpiride[a]
Biochemical response	cAMP↑	cAMP↓, K^+ channel ↑, Ca^{2+} channel ↓

The data in the table are based upon the suggestion of Kebabian and Calne (1979) but have been expanded to include more recent information (Vallar and Meldolesi, 1989). The original classification was into D_1 and D_2 receptor subtypes but, as discussed in the text, with the advent of the isoforms defined by molecular biology (Table 15.2) these should be termed D_1-like receptors.

[a]Formula in Table 15.3.

are summarized in Table 15.2. This rather complicated picture can be simplified by the realization that on the basis of structural and functional properties these five receptor subtypes can be grouped into two subfamilies: D_1/D_5 which have properties similar to those of the pharmacologically defined D_1 receptor and $D_2/D_3/D_4$ which have properties similar to those of the pharmacologically defined D_2 receptor. The two subfamilies are therefore now termed the D_1-like and D_2-like subfamilies. In the subsequent discussion when a receptor is referred to as D_1 this will imply that this is the receptor subtype defined by gene cloning whereas when the receptor has only been defined by pharmacological or biochemical analyses the nomenclature D_1-like/D_2-like will be used.

This emerging understanding of the multiple subtypes of dopamine receptors has importance for the way the activities of potential new dopamine receptor-directed drugs are assayed. In early studies of these compounds animal behavioral tests were used, e.g., the induction of stereotyped behavior or the induction of turning in rodents; these tests detect dopamine receptor activity but do not distinguish compounds with selectivity for different receptor subtypes. The definition of D_1- and D_2-like receptors from biochemical and pharmacological studies enabled compounds to

Table 15.2 Dopamine receptor subtypes defined on the basis of molecular biological studies

	D_1-like		D_2-like		
Receptor isoform	D_1	D_5	D_2	D_3	D_4
Amino acids in human receptors	446	477	414/443	400	387
Pharmacological properties					
Agonist binding (K_i nM)					
(−)-Apormorphine (**15.109**)	0.7	—	0.7	32	4
Dopamine	0.9	0.9	7	4	30
Quinpirol (**15.126**)	1900	—	4.8	24	30
7-OH DPAT (**15.119**)	5000	—	10	1	650
SKF 38393 (**15.127**)	1	0.5	150	5000	1000
Antagonist binding (K_i nM)					
Haloperidol*	80	100	1.2	7	2.3
Chlorpromazine*	90	130	3	4	35
Clozapine*	170	330	230	170	21
Raclopride	18,000	—	1.8	3.5	2400
Remoxipride	24,000	—	300	1600	2800
(−)-Sulpiride*	45,000	77,000	15	13	1000
SCH 23390*	0.2	0.3	1100	800	3000

The different dopamine receptor isoforms can be distinguished structurally on the basis of the sizes of the predicted third intracellular loops and C-terminal tails (Strange, 1991; Civelli et al., 1993). The D_1-like receptors both have short third intracellular loops and long C-terminal tails whereas the D_2-like receptors each have long third intracellular loops and short C-terminal tails. The D_1- and D_2-like subgroups can also be distinguished on the basis of amino acid homologies. The data for K_i values shown in the table are derived from Seeman and Van Tol (1994) and Hacksell et al. (1995) using ligand binding. There is some variability in the values derived from different studies and this is a particular problem for agonists where the complexities of agonist binding studies need to be considered. The selectivity of clozapine for D_4 over D_2 receptors (Seeman and Van Tol, 1994) has not been found in all subsequent studies (Strange, 1994; Hacksell et al., 1995).

[a]Formula in Table 15.3.

be assayed for their interaction at the two subclasses using activity-based assays, e.g., the stimulation of adenylate cyclase for the D_1-like receptors or using ligand-binding assays. More recently the availability of cloned genes for the five receptor subtypes has offered the prospect of the assay of selective substances for their activities against each subtype expressed in a suitable cell host using ligand-binding assays or activity-based assays.

The use of these different assay systems poses certain problems in the definition of the selectivity of compounds directed at the different receptors. This is particularly acute for the agonists where there are two qualities of an agonist that are potentially of interest; its affinity for the receptor and its ability to stimulate a response. In defining selectivity between receptors in activity-based assays, these two quantities are not always defined or separated and this can lead to confusion. Even in ligand-binding assays there is potential confusion in extracting the relevant parameters for defining selectivity. Therefore in the discussion below ''activity'' will be referred to for agonists which is a broad definition of selectivity based on the tests used in the particular publication cited and should be taken only as a guide to actual selectivity. For antagonists there are fewer problems of this nature although the emerging phenomenon of ''inverse agonism'' may complicate matters.

15.4.2 Dopamine Agonist

It was in the late 1950s that an independent role for dopamine as a neurotransmitter was postulated and this led to the development of models for the assessment of dopamine agonism mostly based on animal behavioral tests. From these tests it became clear that a number of naturally occurring or semisynthetic substances possessed dopamine agonist activity. Notable among these were the aporphine alkaloids, e.g., apomorphine (**15.109**) and the ergot alkaloids, e.g., ergotamine (**15.110**). At the same time synthetic programs were initiated to obtain dopamine agonists with greater potency or activity and some of the important chemical classes of agonists will be considered below.

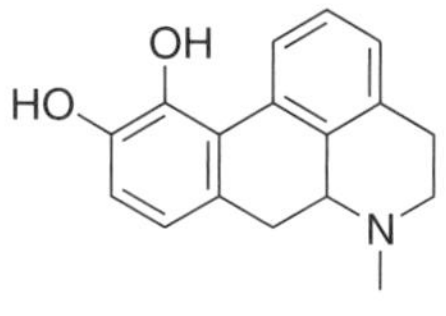

(**15.109**) Apomorphine

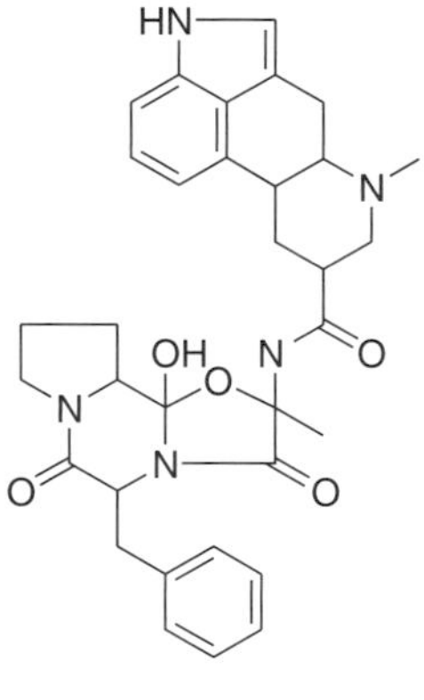

(**15.110**) Ergotamine

Aminotetralins and related compounds

A major synthetic effort has been devoted to compounds related to dopamine whereby the dopamine molecule is ''locked'' into a rigid structure. The best-known examples of these are the aminotetralins. These have been synthesized in a variety of analogs with different hydroxyl substitution patterns on the aromatic ring. The 5,6- and 6,7-dihydroxy aminotetralins (ADTNs) (**15.111**, **15.112**) may be considered to be equivalent to the dopamine molecule, frozen into one or other of its two principal conformations (termed α and β conformations, **15.113**, **15.114**). The possibility of using these compounds to determine the active conformation of dopamine attracted much interest but it has not been proven to draw any firm conclusion from the results of the studies.

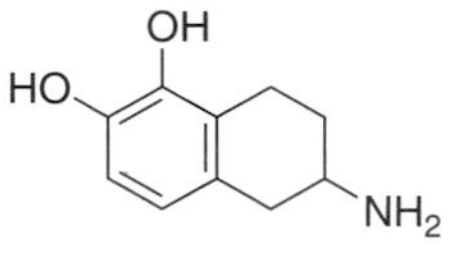

(**15.111**) 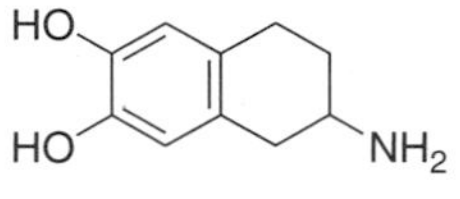 (**15.112**) 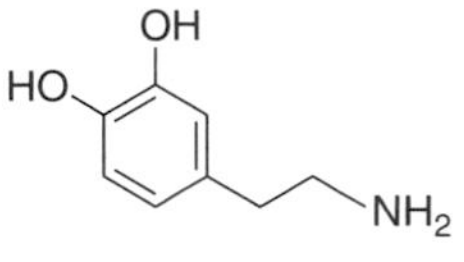 (**15.113**) 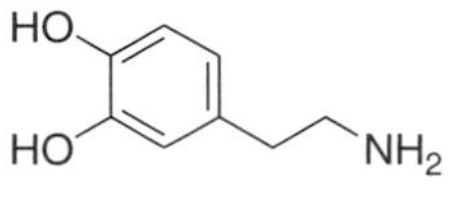 (**15.114**)

The aminotetralins have activities at both the D_1- and D_2-like subfamilies of dopamine receptors and it has been possible to draw some broad conclusions about the structure–activity relationships involved. Compounds with two hydroxyl groups show the highest activity and the 5,6- and 6,7-congeners both have high activity. Substantial activity is retained in the monohydroxy compounds, for example, the 5-hydroxy aminotetralins (e.g., **15.115**). The monohydroxy equivalents of dopamine (tyramines) also have some activity as dopamine agonists supporting this idea. Compounds lacking hydroxyl groups, e.g., *N,N*-dipropyl-2-aminotetralin and phenylethylamine, also exhibit weak agonist activity so these hydroxyl groups are not essential for agonist action. The amino group of the aminotetralins has been derivatized in a number of cases and this has shown that the addition of two alkyl groups enhances activity both at D_1- and D_2-like receptors, e.g., **15.116**. Activity increases with increasing alkyl chain length up to the di *N*-propyl derivatives, which have the highest activities and the di *N*-butyl compounds which have less activity. It seems that there may be some additional site on the receptor with steric constraints which is occupied by these alkyl groups and which enhances agonist activity.

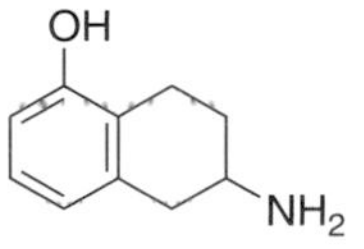

(**15.115**) 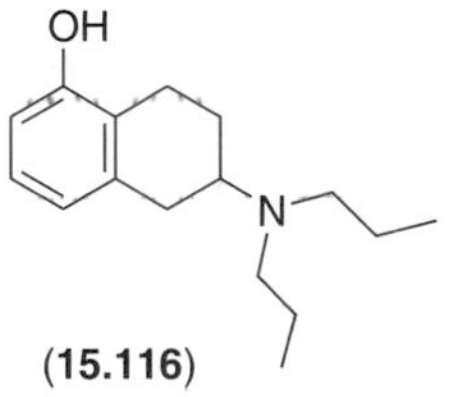 (**15.116**)

A series of 5-hydroxy aminotetralins has been synthesized where the *N*-*n*-propyl, *N*-phenylethyl, and *N*-*n*-propyl, *N*-thenylethyl congeners are extremely potent agonists with substantial D_2-like selectivity (**15.117**, **15.118**). This suggests that particular groups larger than *n*-propyl can enhance potency further when attached to the aminotetralin structure. Generally the aminotetralins do not show great selectivity among the different D_2-like receptors but 7-hydroxy-*N,N′* di *n*-propyl aminotetralin (**15.119**) has some selectivity for the D_3 receptor mostly based on data from ligand-binding assays.

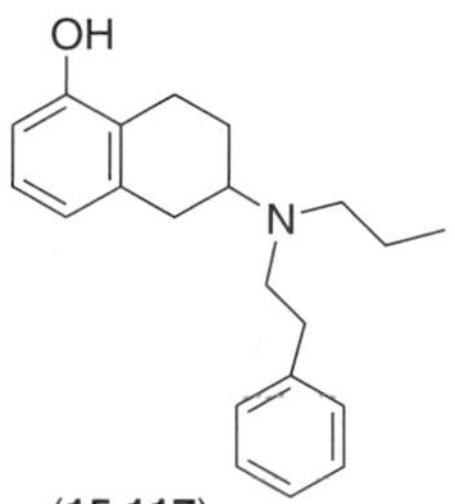

(**15.117**) 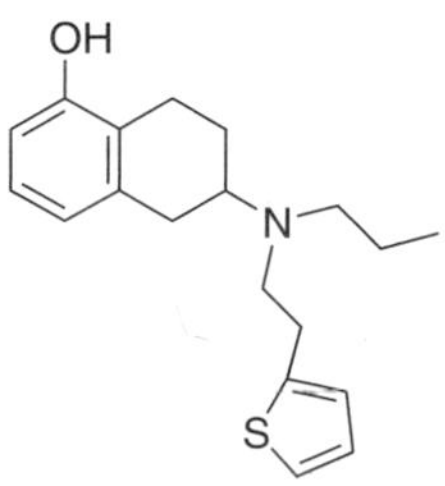 (**15.118**) 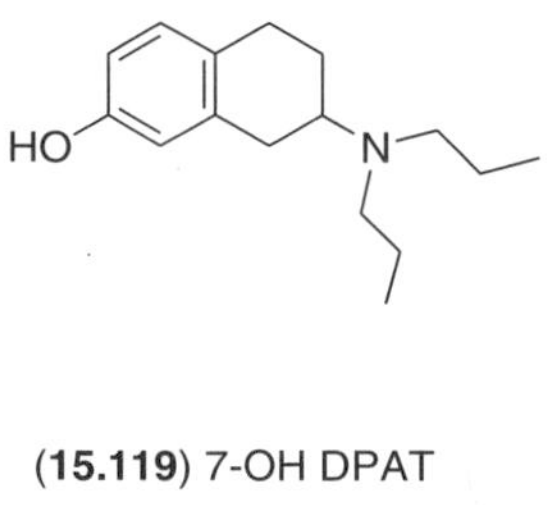 (**15.119**) 7-OH DPAT

Aporphine alkaloids

The aporphine alkaloids contain the dopamine structure in a rigid conformation. *R*(−)- apomorphine (**15.109**) has been shown to have D_1- and D_2-like agonist activity although it tends to be a partial agonist at the D_1-like receptors. The replacement of the *N*-methyl group with an *N*-propyl group in *R*(−)-*N*-propyl norapomorphine (**15.120**) increases the affinity for the D_2-like receptors relative to the D_1-like receptors.

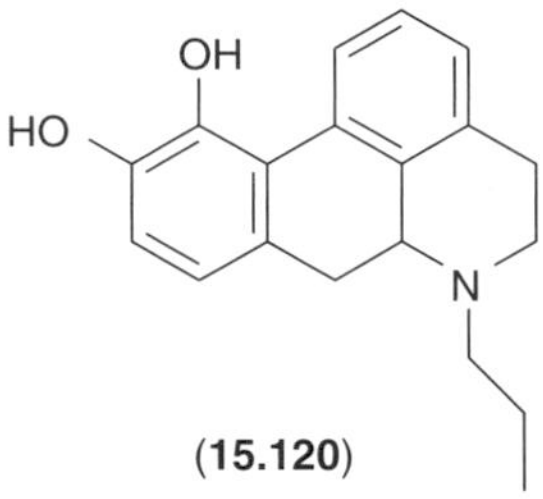

(**15.120**)

Ergot alkaloids

A number of ergot alkaloids with dopamine agonist activity have been isolated from the crude mixture of natural products known as ergot. Typical examples of these are ergotamine (**15.110**) and α-ergocriptine (**15.121**), which contain the D-lysergic acid structure linked by an amide bond to a cyclic peptide moiety and these have come to be called ''ergopeptines.'' Modification of these natural products by chemical synthesis has provided substances with better selectivity for dopamine receptors and notable here is bromocriptine (**15.122**), which has potent D_2-like receptor activity but D_1-like receptor antagonistic properties. It is used in the treatment of Parkinson's disease and excessive prolactin secretion.

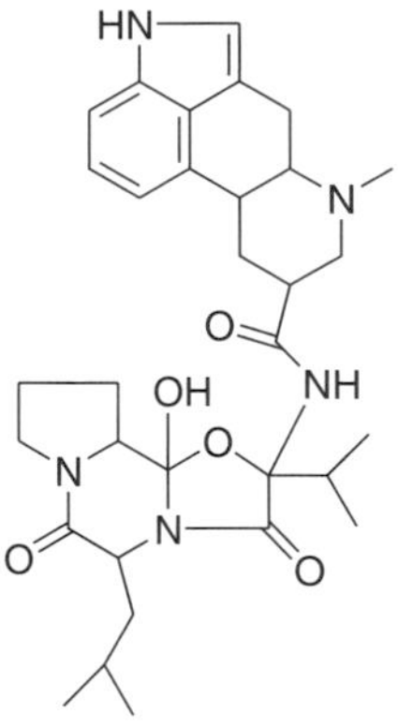

(**15.121**) Ergocriptine

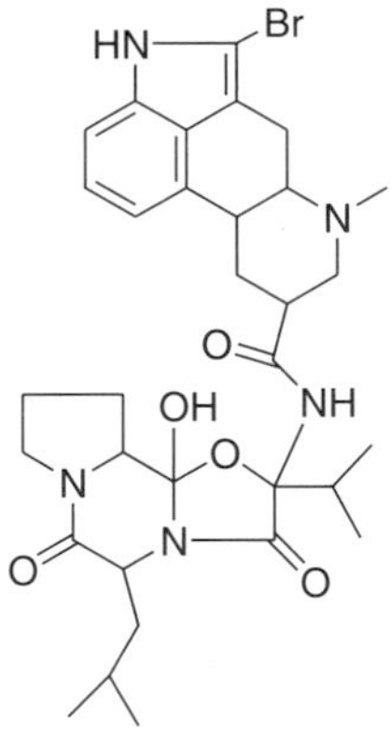

(**15.122**) Bromocriptine

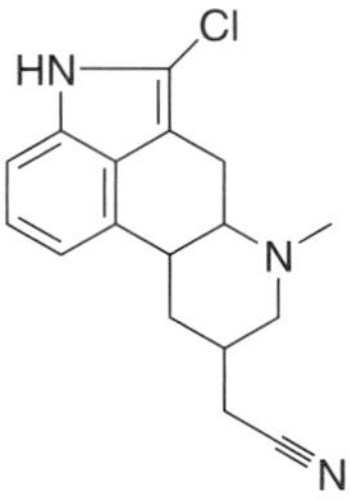

(**15.123**) Lergotrile

A large number of semisynthetic ''ergolines'' exist where the structure is based on the lysergic acid structure and the peptide side chain of the ergopeptines has been eliminated. These have been shown to possess potent agonist activity at the D_2-like receptors, e.g., lergotrile (**15.123**), pergolide (**15.124**), and lisuride (**15.125**). Some of these compounds also possess significant agonist activity at the D_1-like receptors, e.g., pergolide, whereas others possess little or no agonist activity and in fact may be D_1 partial agonists or antagonists depending on the test system, e.g., lisuride (**15.125**).

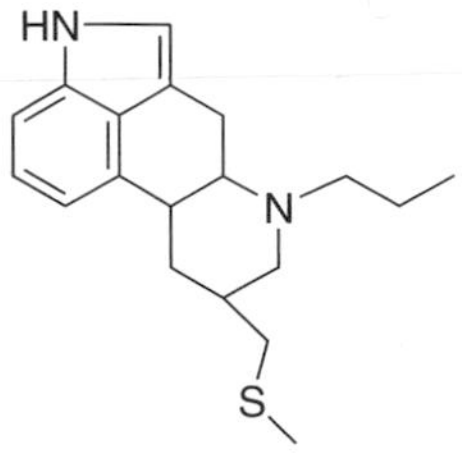

(**15.124**) Pergolide 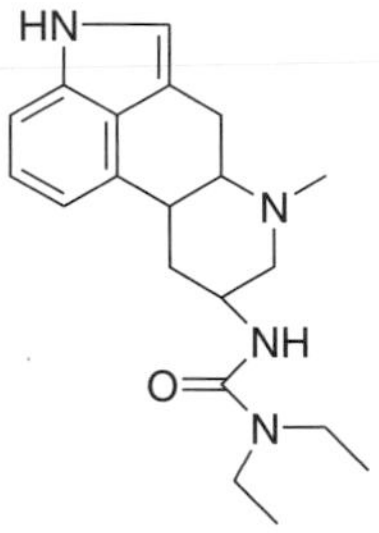 (**15.125**) Lisuride 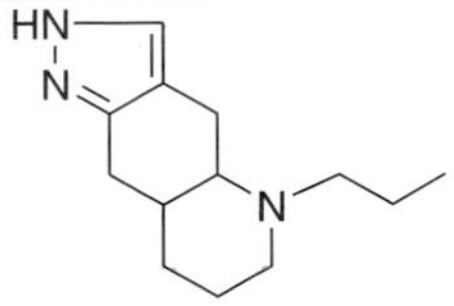 (**15.126**) Quinpirole

Much synthetic work has ensued in order to identify the part of ergoline that is responsible for the dopaminergic activity. Among the compounds synthesized are a group of partial ergolines including quinpirole (**15.126**), which is a very selective D_2-like agonist.

Benzazepines

The benzazepine nucleus has been used to provide another series of molecules some of which are potent and selective D_1-like agonists, e.g., SKF 38393 (**15.127**), fenoldopam (**15.128**). Other analogs exhibit activity at both the D_1-like receptors. Benzazepines also provide selective D_1-like antagonists (see below).

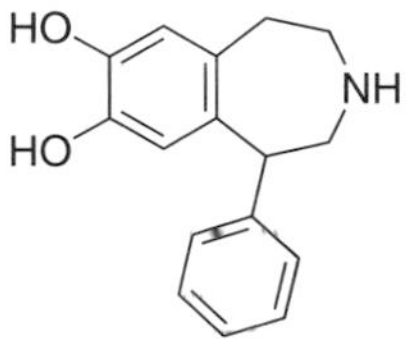

(**15.127**) SKF 38393 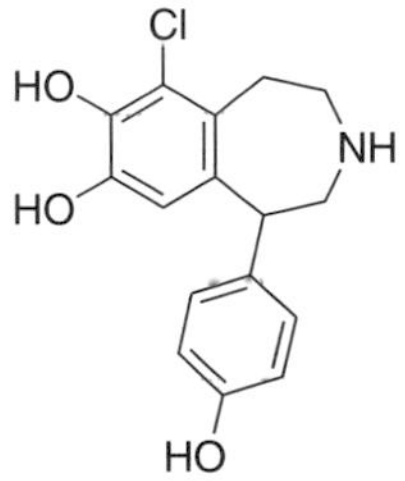 (**15.128**) Fenoldopam

Miscellaneous structures

The naphthoxazine PHNO (**15.129**) has been synthesized and is one of the most potent D_2-like receptor agonists available with little ability to bind to or activate D_1-like receptors. *N*-*n*-Propyl-3-(hydroxyphenyl) piperidine (3-PPP) (**15.130**) is an example of a compound where the 3(*R*)-stereoisomer has D_2-like receptor agonist activity whereas the 3(*S*)-isomer has variable intrinsic activity depending on the receptor preparation. This can be seen as a preferential ability to act as an agonist at presynaptic autoreceptors but to behave as an antagonist at postsynaptic receptors. The compounds have little D_1-like receptor activity.

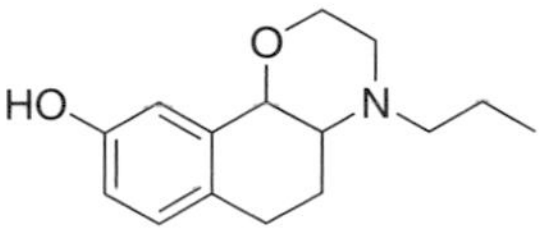

(**15.129**) PHNO 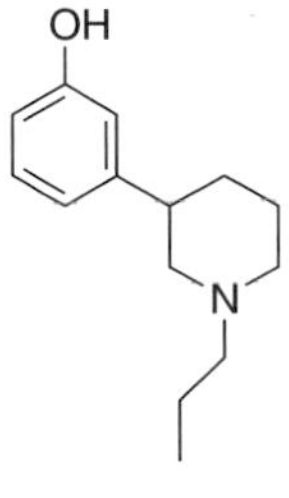 (**15.130**) 3-PPP

Selective agonists

Compounds exist, as indicated above, that have the ability to act selectively as agonists on D_1-like receptors, e.g., SKF 38393 (**15.127**) or D_2-like receptors, e.g., quinpirole (**15.126**), PHNO (**15.129**). These compounds do not, however, show clear selectivity for the different isoforms comprising the two subfamilies with the exception of 7-OH DPAT (**15.119**) as mentioned earlier. In Table 15.2 some data are given for agonists from ligand binding studies illustrating this point. A second area where selectivity has been claimed for dopamine agonists is at autoreceptors. These are the receptors on dopamine nerve terminals or cell bodies that mediate inhibition of neurotransmitter synthesis, release, or cell firing. It was found that although these receptors exhibited a pharmacological profile consistent with a D_2-like receptor the autoreceptors were more sensitive to agonists compared to postsynaptic receptors. This would be consistent with autoreceptors being D_2-like receptors but with a larger amplification (spare receptor ratio). This can at least in part explain the apparent autoreceptor selectivity of 3 (*S*)-3-PPP (**15.130**). If this compound is a partial agonist at D_2-like receptors rather than at receptors with a large amplification, clear agonist activity will be seen, while at receptors with a lower amplification factor antagonist activity may be displayed. More recently, compounds have also been synthesized with apparent autoreceptor antagonist selectivity (see below).

15.4.3 Dopamine Antagonists

In the early 1950s the phenothiazine, chlorpromazine (Table 15.3) was discovered to have the ability to induce in humans a state of indifference without loss of consciousness and it was used as an antipsychotic drug. In the late 1950s the butyrophenone series of drugs was discovered, e.g., haloperidol (Table 15.3), and these were shown to have antipsychotic activity. It was eventually found that a prominent action of these drugs was to inhibit various actions of dopamine and it became clear that one of their principal activities was as dopamine antagonists. The phenothiazines and butyrophenones are D_2-like antagonists and have varying abilities as D_1-like antagonists. They also show varying abilities to act as serotonergic, muscarinic, histaminergic, and adrenergic antagonists. Extensive synthetic programs have been performed with the aim of developing more selective drugs with different structures and in Table 15.3 some of the key structural classes of dopamine antagonist are shown together with an indication of their selectivity. Of note here, are the benzazepines which are selective D_1-like antagonists and the substituted benzamides which are selective D_2-like antagonists.

While some of the compounds shown in Table 15.3 may show some selectivity between D_1- and D_2-like receptors, they do not in general have any marked abilities to discriminate the individual members of the two subfamilies. Table 15.2 gives some data from ligand binding studies on the dissociation constants for some antagonists at the different receptor subtypes defined by gene cloning. It can be seen that there are some drugs such as raclopride that show low affinity for some receptor isoforms but there are few substances available with clear selectivity for a single receptor isoform. It was claimed that clozapine had a clear selectivity for the D_4 receptor subtype but subsequent work has not reported the same selectivity.

Behavioral evidence for preferential actions of the compounds (+)-UH 232 (**15.131**) and (+)-AJ 76 (**15.132**) as antagonists at dopamine autoreceptors has been presented but the relation of these findings to the different dopamine receptor isoforms is unclear. These compounds do show some selectivity for binding to the D_3 dopamine receptor but the selectivity is not great.

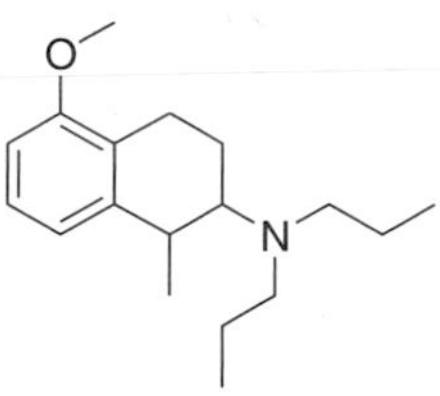

(**15.131**) (+)-UH 232 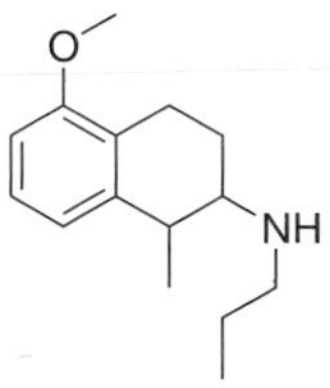 (**15.132**) (+)-AJ 76

Extensive synthetic work has led to the development of some compounds with limited selectivity for different D_2-like receptors. For example, the following compounds have been

Table 15.3 The major classes of dopamine antagonists[a]

	Selectivity for dopamine receptor subtypes
Phenothiazines chlorpromazine (aliphatic side chain)	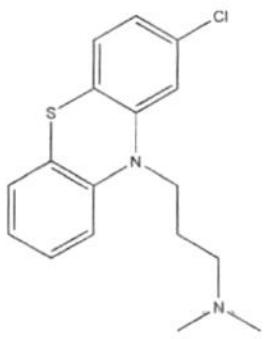 D_1-like/D_2-like
Fluphenazine (piperazine side chain)	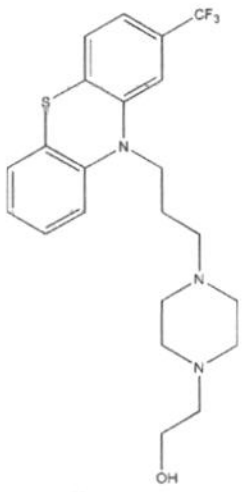 D_1-like/D_2-like
Thioridazine (piperidine side chain)	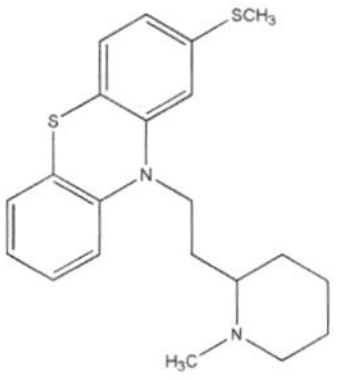 D_1-like/D_2-like
Thioxanthines flupenthixol	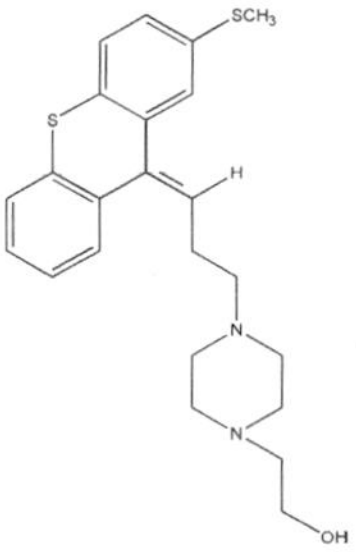 D_1-like/D_2-like

[a]Some selectivity data derived from ligand binding studies are given in Table 15.2.

Table 15.3 The major classes of dopamine antagonists — Continued

	Structure	Selectivity for dopamine receptor subtypes
Butyrophenones (haloperidol)	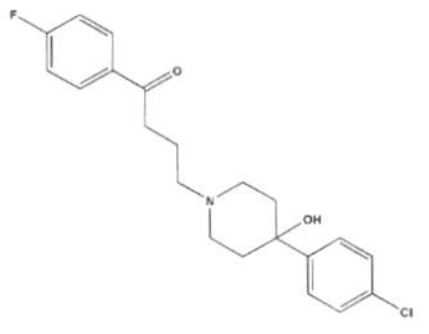 F, O, N, OH, Cl	D_1-like/D_2-like
Dibenzazepines (clozapine)	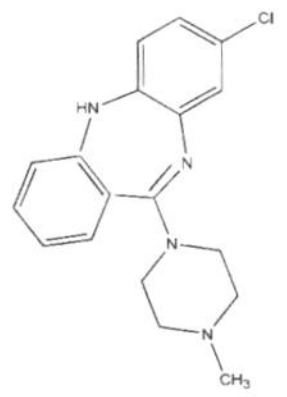 Cl, HN, N, N, N, CH_3	D_1-like/D_2-like
Substituted benzamides (sulpiride)	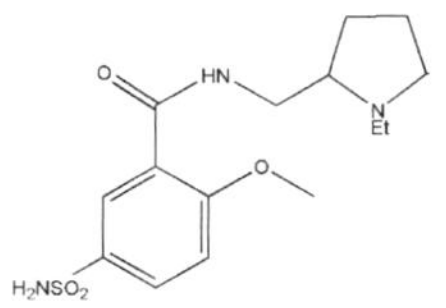 O, HN, N, Et, O, H_2NSO_2	D_2-like
Benzazepines (SCH23390)	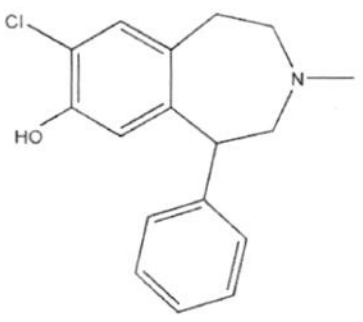 Cl, HO, N	D_1-like
Miscellaneous (butaclamol)	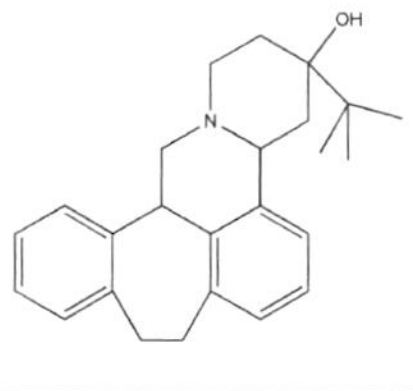 OH, N	D_1-like/D_2-like

described as selective antagonists for D_2 receptors (L741626 (**15.135**), ~40-fold), D_3 receptors (S33084 (**15.133**), SB-277011-A (**15.134**), ~100-fold), and D_4 receptors (L745870 (**15.136**), ~2000-fold) (Kulagowski et al., 1996; Bristow et al., 1997; Millan et al., 2000; Reavill et al., 2000).

An emerging phenomenon for some of the compounds described as dopamine antagonists is the observation that they possess inverse agonist activity in some assay systems. This means that rather than being neutral antagonists, these compounds possess an efficacy opposite to that of dopamine. Although this has been described for many dopamine antagonists, the importance of the phenomenon for *in vivo* effects of the drugs is unclear.

(**15.133**) S33084

(**15.134**) SB-277011-A

(**15.135**) L741626

(**15.136**) L745870

FURTHER READING

Beaulieu, M., Itoh, Y., Tepper, P., Horn, A.S. and Kebabian, J.W. (1984) *N, N*-disubstituted 2-aminotetralins are potent D_2 dopamine receptor agonists. *European Journal of Pharmacology* **105**, 15–21.

Bristow, L.J., Kramer, M.S., Kulagowski, J., Patel, S., Ragan, C.I. and Seabrook, G.R. (1997) Schizophrenia and L-745,870, a novel dopamine D4 receptor antagonist. *Trends in Pharmacological Sciences* **18**, 186–188.

Cannon, J.G. (1983) Structure–activity relationships of dopamine agonists. *Annual Reviews of Pharmacology and Toxicology* **23**, 103–130.

Cavero, I., Massingham, R. and Lefevre-Borg (1982) Peripheral dopamine receptors, potential targets for a new class of antihypertensive agents. *Life Sciences* **31**, 939–948; 1059–1069.

Civelli, O., Bunzow, J.R. and Grandy, D.K. (1993) Molecular diversity of the dopamine receptors. *Annual Reviews of Pharmacology and Toxicology* **32**, 281–307.

Hacksell, U., Jackson, D.M. and Mohell, N. (1995) Does the dopamine receptor subtype selectivity of antipsychotic agents provide useful leads for the development of novel therapeutic agents? *Pharmacology Toxicology* **76**, 320–324.

Hauth, H. (1979) Chemical aspects of ergot derivatives with central dopaminergic activity. In: *Dopaminergic Ergot Derivatives and Motor Function*, K. Fuxe and D.B. Calne (eds.). Pergamon Press, Oxford, pp. 23–31.

Hogberg, T. (1991) Novel substituted salicylamides and benzamides as selective D2 receptor antagonists. *Drugs of the Future* **16**, 333–357.

Johansson, A.M., Arvidsson, L.E., Hacksell, U., Nilsson, J.L.G., Svensson, K., Hjorth, S. et al. (1985) Novel dopamine receptor agonists and antagonists with preferential action on autoreceptors. *Journal of Medicinal Chemistry* **28**, 1049–1053.

Kebabian, J.W. and Calne, D.B. (1979) Multiple receptors for dopamine. *Nature (Lond.)* **277**, 93–96.

Kulagowski, J.J., Broughton, H.B., Curtis, N.R., Mawer, I.M., Ridgill, M.P., Baker, R., Emms, F., Freedman, S.B., Marwood, R., Patel, S., Patel, S., Ragan, C.I. and Leeson, P.D. (1996) 3-((4-(4-Chlorophenyl)-piperazin-1-yl)-methyl)-1H-pyrrolo-2,3-b-pyridine: an antagonist with high affinity and selectivity for the human dopamine D4 receptor. *Journal of Medicinal Chemistry* **39**, 1941–1942.

Leff, P. (1995) Inverse agonism: theory and practice. *Trends in Pharmacological Sciences* **16**, 256.

Leysen, J.E. and Niemegeers, C.J.E. (1985) Neuroleptics. *Handbook of Neurochemistry*, vol. 9, pp. 331–361.

Martin, G.E., Williams, M., Pettibone, D.J., Yarborough, G.G., Clineschmidt, B.V. and Jones, J.H. (1984) Pharmacologic profile of a novel potent direct-acting dopamine agonist, (+)-PHNO. *Journal of Pharmacology and Experimental Therapeutics* **230**, 569–576.

Millan, M.J., Gobert, A., Newman-Tancredi, A., Lejeune, F., Cussac, D., Rivet, J.M., Audinot, V., Dubuffet, T. and Lavielle, G. (2000) S33084, a novel, potent, selective, and competitive antagonist at dopamine D(3)-receptors: I. Receptorial, electrophysiological and neurochemical profile compared with GR218,231 and L741,626. *Journal of Pharmacology and Experimental Therapeutics* **293**, 1048–1062.

Reavill, C., Taylor, S.G., Wood, M.D., Ashmeade, T., Austin, N.E., Avenell, K.Y., Boyfield, I. et al. (2000) Pharmacological actions of a novel, high-affinity, and selective human dopamine D(3) receptor antagonist, SB-277011-A. *Journal of Pharmacology and Experimental Therapeutics* **294**, 1154–1156.

Seeman, P. (1980) Brain dopamine receptors. *Pharmacological Reviews* **32**, 229–313.

Seeman, P. (1987) Dopamine receptors in brain and periphery. *Neurochemistry International* **10**, 1–25.

Seeman, P. and Van Tol, H.H.M. (1994) Dopamine receptor pharmacology. *Trends in Pharmacological Sciences* **15**, 264–270.

Seeman, P., Watanabe, M., Grigoriadis, D., Tedesco, J.L., George, S.R., Svensson, U. et al. (1985) Dopamine D_2 receptor binding sites for agonists. *Molecular Pharmacology* **28**, 391–399.

Seiler, M.P. and Marksteine, R. (1982) Further characterisation of structural requirements for agonists at the striatal dopamine D_1 receptor. *Molecular Pharmacology* **22**, 281–289.

Sibley, D.R. and Creese, K. (1983) Interaction of ergot alkaloids with anterior pituitary D_2 dopamine receptors. *Molecular Pharmacology* **23**, 585–593.

Sorensson, C., Waters, N., Svensson, K., Carlsson, A., Smith, M.W., Piercey, M.F. et al. (1993) Substituted 3-phenyl piperidines: new centrally acting autoreceptor antagonists. *Journal of Medicinal Chemistry* **36**, 3188–3196.

Strange, P.G. (1991) Interesting times for dopamine receptors. *Trends in Neurosciences* **14**, 43–45.

Strange, P.G. (1992) *Brain Biochemistry and Brain Disorders*. Oxford University Press, Oxford.

Strange, P.G. (1994) Dopamine D_4 receptors: curiouser and curiouser. *Trends in Pharmacological Sciences* **15**, 317–319.

Strange P.G. (2001) Antipsychotic drugs: importance of dopamine receptors for mechanisms of therapeutic actions and side effects. *Pharmacological Reviews* **53**, 119–133.

Strange, P.G. (2001) Dopamine Receptors, published by Tocris Cookson Chemicals (revised version). http://tocris.com/reviews.html

Strange P.G. (2002) Mechanisms of inverse agonism at G protein coupled receptors. *Trends in Pharmacological Sciences* **23**, 89–95.

Vallar, L. and Meldolesi, J. (1989) Mechanisms of signal transduction at the dopamine D_2 receptor. *Trends in Pharmacological Sciences* **10**, 74–77.

Waddington, J.L. and O'Boyle, K.M. (1987) The D_1 dopamine receptor and the search for its functional role: from neurochemistry to behaviour. *Reviews in the Neurosciences* **1**, 157–184.

Wikstrom, H., Sanchez, D., Lindberg, P., Hacksell, U., Arvidsson, L.E., Johansson, A.M. et al. (1984) Resolved 3-(3-hydroxyphenyl) *N-n*-propylpiperidine and its analogues: central dopamine receptor activity. *Journal of Medicinal Chemistry* **27**, 1030–1036.

Wolf, M.E. and Roth, R.H. (1987) Dopamine autoreceptors. In: *Dopamine Receptors*, I. Creese and C.M. Fraser (eds). Alan R. Liss, New York, pp. 45–96.

16

Design of Antibacterial, Antifungal, and Antiviral Agents

Claire Simons and A. Denver Russell*

CONTENTS

*Deceased.

16.1 INTRODUCTION

Human infections are caused by a variety of microorganisms including bacteria, fungi, viruses, and parasites. There is a considerable arsenal of drugs currently available for the treatment of such infections and many more compounds in development. Microbes present a number of therapeutic targets. Current drugs either target the integrity of the microbial structure (e.g., cell wall inhibitors) or target key points during the life cycle of the particular microorganism (e.g., inhibitors of protein/RNA/DNA synthesis and nucleic acid biosynthesis) (Table 16.1).

Table 16.1 Mechanism of action of antimicrobial agents

Effect	Examples	Comments
Inhibition of cell wall synthesis	D-Cycloserine	Competitive inhibition of alanine racemase and synthetase
	β-Lactam antibiotics	Binding to PBPs (specific)
	Glycopeptides	Binding to peptidoglycan precursor
Membrane active agents	Polymyxins	Affect outer membrane Gram-negative bacteria
	Ionophores	Specific cation conductors: nonselective
	Ethambutol, isoniazid, PA-824, Isoxyl	Interfere with mycolic acid synthesis
Inhibition of protein synthesis	Aminoglycosides	Inhibit initiation stage
	Tetracyclines	Inhibit binding of aminoacyl-tRNA to 30*S* ribosomal subunit
	Oxazolidinones	Inhibit before initiation
	Macrolides	Inhibit translocation
	Streptogramins	Bind to bacterial ribosomes
Inhibition of nucleic acid synthesis	Actinomycin D	Binds to double-stranded DNA
	Rifampicin	Inhibits DNA-dependent RNA polymerase
	Mitomycin C	Covalent linking to DNA
	Quinolones	Effect on DNA gyrase
Inhibition of tetrahydrofolate synthesis	Sulfonamides	Competitive inhibitors of dihydropteroate synthetase
	Trimethoprim	Inhibits dihydrofolate reductase
Inhibitors of sterol biosynthesis	BM212	Probable inhibition of P450 MT1

Drug resistance is a serious concern as it has major implications in the use of chemotherapeutic agents and the control of infectious disease. Resistance occurs primarily due to the emergence of acquired resistance, which arises after exposure of a microbial population to a drug, and is caused either by acquisition of plasmids and transposons or by chromosomal mutation.

There are three main biochemical mechanisms by which acquired drug resistance can occur: (1) enzymatic alteration of the drug resulting in inactivation, often enzyme inactivation, e.g., β-lactams, chloramphenicol, aminoglycosides, and fosfomycin; (2) alteration of the target site within the antimicrobial cell — the microbes become insensitive to the drug but are still able to carry out their normal function, e.g., β-lactams, aminoglycosides, quinolones, rifampicin, tetracyclines, glycopeptides, and macrolides; (3) decreased intracellular accumulation of the drug, caused by impaired uptake or enhanced efflux, observed with β-lactams, aminoglycosides, quinolones, tetracyclines, macrolides, sulfonamides, trimethoprim, and mupirocin.

Intrinsic resistance associated with the cell wall is also observed with Gram-negative bacteria and mycobacteria. The cell envelopes of these two types of organisms are considerably more complex than the envelopes of Gram-positive organisms (Figure 16.1), and act as efficient barriers to the entry of many high molecular weight hydrophilic molecules.

Chemotherapeutic agents suitable for the treatment of human infections must possess two key elements, high antimicrobial activity, and low toxicity. Ongoing drug design and development is aimed at enhancing these two key elements and, with the increasing problem of drug resistance, the development of drugs capable of counteracting antimicrobial resistance mechanisms is a primary goal. Additionally, any drug must ideally possess a broad spectrum of activity, with a rapid microbiocidal activity.

Pharmacological properties such as drug uptake (high plasma and tissue levels), rate of excretion, and half-life are important considerations. Likewise chemical and physical properties of a drug must also be addressed with issues such as aqueous solubility and stability of importance in the drug design process.

A considerable amount of information is now available for individual compounds or drug classes. This information is summarized in this section and, where possible, ways in which this knowledge can be used to further the design of new antimicrobial agents will be discussed.

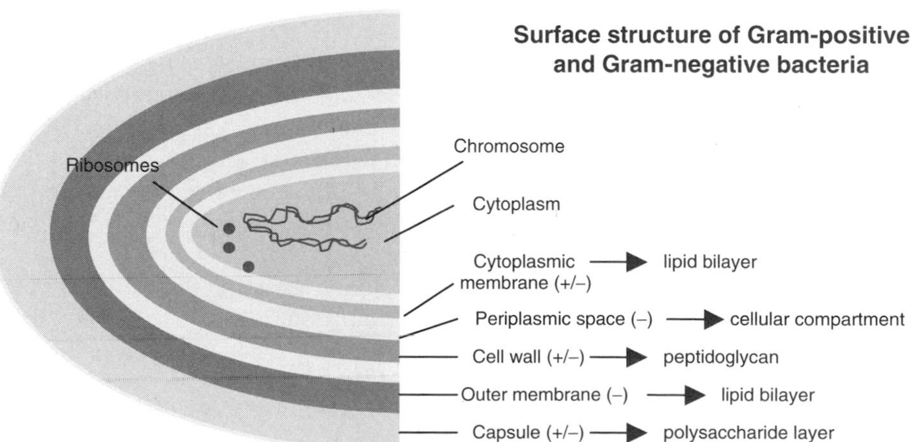

Figure 16.1 Surface structure of bacteria. Layers found in Gram-positive and Gram-negative bacteria are indicated (+/–), those generally only found in Gram-negative bacteria are indicated with (–).

16.2 MECHANISMS OF ACTION OF ANTIBACTERIAL AGENTS

In 1929, Fleming published the results of his chance finding that a *Penicillium* mould caused lysis of staphylococcal colonies on an agar plate. Extraction of the active principle, the antibiotic benzylpenicillin, from the culture medium was later achieved by Florey, Chain, and their colleagues at Oxford University and in the USA. The development of antibiotics with enhanced therapeutic potential has been possible due to further discoveries in antibiotic production, mechanisms of antibacterial action, resistance, and pharmacokinetics. The major targets of antibacterial agents are cell wall, protein, RNA/DNA, and tetrahydrofolate syntheses (Table 16.1).

Resistance to current antibacterial agents is a major health problem. Of particular concern are infections caused by multidrug-resistant (MDR) Gram-positive pathogens, including methicillin-resistant *Staphylococcus aureus* (MRSA), methicillin-resistant *S. epidermidis* (MRSE), vancomycin-resistant *Enterococcus faecalis* and *Enterococcus faecium* (VREF), and penicillin- and cephalosporin-resistant *S. pneumoniae*. MDR *Mycobacterium tuberculosis* (MDRTB), the pathogen responsible for tuberculosis, which is the leading cause of death worldwide from a single infectious agent, has increased with the seriousness of this advance declared a global emergency by the WHO in 1993.

The following sections (16.2.1–16.2.6) discuss key therapeutic targets of antibacterial agents and, where relevant, mechanisms of resistance of the identified drug classes.

16.2.1 Inhibitors of Cell Wall Synthesis

The bacterial wall is a complex structure, with differences observed between the walls of Gram-positive and Gram-negative bacteria. Figure 16.1 outlines the key components of bacterial surface structure. The capsule, outer membrane, and periplasmic space are generally found only in Gram-negative bacteria.

The outer membrane of the mycobacterial cell wall is very hydrophobic, composed of long fatty acids (C_{60}–C_{90}), known as mycolic acids, resulting in an efficient barrier to a range of antimicrobial agents. A layer of arabinogalactan and peptidoglycan is found between the outer mycolic acid membrane and the lipid bilayer of the plasma membrane.

However, both Gram-positive and Gram-negative bacteria and mycobacteria contain a basal peptidoglycan (murein, mucopeptide) cell wall, which is composed of chains of alternating aminosugars *N*-acetylglucosamine (GlcNAc) and *N*-acetylmuramic acid (MurNAc) to which amino acids, some in the unnatural D-configuration, are attached. In brief, peptidoglycan synthesis involves (1) the stepwise addition of amino acids to MurNAc, (2) linking to GlcNAc to form a linear polymer, and finally (3) a cross-linking (transpeptidation, via a transpeptidase enzyme) of the linear polymers to form a rigid structure, in which the degree of cross-linking varies (Figure 16.2). The cross-links are either direct cross-links between the tetrapeptide on one strand of peptidoglycan and the tetrapeptide of a second strand of peptidoglycan, formed via the diamines, lysine, or diaminopimelate, or indirect cross-links where two tetrapeptides are joined by a pentglycyl peptide. The cross-links confer cell wall resistance to lysis by intracellular osmotic pressure. Peptidoglycan in Gram-negative bacteria consists mainly of direct cross-links whereas Gram-positive bacteria contains both direct and indirect cross-links.

Compounds that inhibit enzymes involved in the biosynthesis of the cell wall components include D-cycloserine (**16.1**), a structural analog of D-alanine (**16.2**) used in the second-line therapy of tuberculosis, which inhibits two enzymes, a racemase and a synthetase, involved in the synthesis of the D-alanyl-D-alanine dipeptide (Figure 16.2). More recent enzyme targets include those involved in the biosynthesis of lysine (**16.3**) and diaminopimelic acid (DAP, **16.4**).

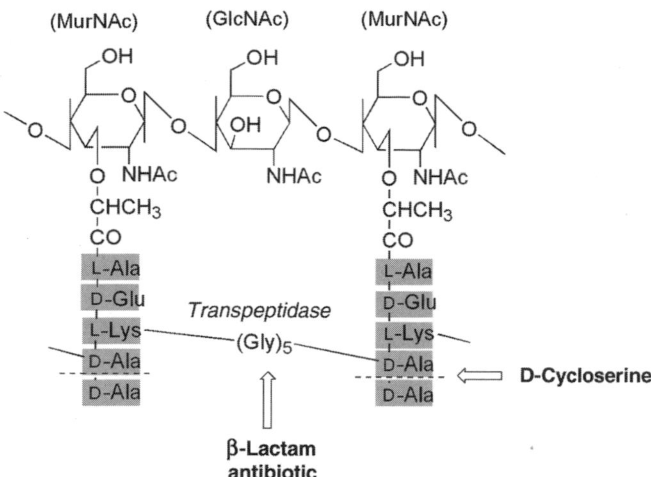

Figure 16.2 Peptidoglycan showing cross-links and role of transpeptidase. (MurNAc, *N*-acetylmuramic acid; GlcNAc, *N*-acetylglucosamine; Ala, alanine; Glu, glutamine; Lys, lysine; $(Gly)_5$, 5 molecules of glycine). Variability in the amino acids present occurs, e.g., the $(Gly)_5$ bridge is found in *S. aureus* but not in *E. coli.*

(**16.1**) D-Cycloserine (**16.2**) D-Alanine (**16.3**) L-Lysine (**16.4**) LL-DAP

The assembly process of the peptidoglycan layer has long been a target for antibiotic action with the β-lactam penicillins, cephalosporins, and glycopeptides being the major drug classes in clinical use.

β-Lactam penicillins

Bacterial cell membranes (inner membranes) contain proteins known as penicillin binding proteins (PBPs), which are associated with specific enzyme activity involved in cell wall synthesis, e.g., transglycosylase, transpeptidase, and carboxypeptidase enzyme activity (Table 16.2).

β-Lactam antibiotics combine with PBPs and induce morphological effects. PBPs 1B, 2, and 3 appear to be the most important in *Escherichia coli*, with binding to PBP1 resulting in rapid cell lysis or spheroplasts (osmotically fragile forms). With PBP2 binding, spherical, osmotically stable bacterial forms are produced, and with PBP3 binding filamentation is induced (Table 16.2).

The original penicillins such as benzylpenicillin (penicillin G) (**16.5**) suffered clinically as a result of their narrow spectrum, susceptibility to β-lactamases (see β-lactamase inhibitors section), and acid lability resulting in administration by injection (Table 16.3). Bacterial resistance to β-lactam antibiotics results primarily from β-lactamases, which cause cleavage of the lactam ring in susceptible members (Figure 16.3).

There are four groups of β-lactamases (over 250 different β-lactamase enzymes are known). The most clinically important β-lactamases occur in Group 2, including β-lactamases produced by *S. aureus* which favor penicillin as substrates. The design of the semisynthetic penicillins has had one common goal: to achieve, through the introduction of different substituents, a new antibiotic with an improved spectrum of activity or enhanced stability to β-lactamases. The following examples illustrate the substitution at the 6-position (Table 16.3).

Table 16.2 Function of penicillin-binding proteins (PBPs) in *Escherichia coli* and examples of β-lactam inhibitors

PBP	Enzyme activity	Function	Inhibitory effect	Inhibitors
1A	Transglycosylase, transpeptidase	Cell wall growth	Lysis	Benzylpenicillin, cephalosporins
1B (a,b,d)	Transglycosylase, transpeptidase	Cell wall growth	Lysis	Benzylpenicillin, cephalosporins
2	Transglycosylase, transpeptidase	Initiation of cell wall growth	Oval cells	Mecillinam, imipenem
3	Transglycosylase, transpeptidase	Septum formation, cell division	Filaments	Many cephalosporins, piperacillin, aztreonam
4	Carboxypeptidase, endopeptidase	Regulation of cross-linking	Unknown	Benzylpenicillin, ampicillin, imipenem
5	Carboxypeptidase	Regulation of cross-linking	Unknown	Cefoxitin
6	Carboxypeptidase	Regulation of cross-linking	Unknown	Cefoxitin
7,8	Unknown	Unknown	Unknown	Penems

Table 16.3 β-Lactam penicillin antibiotics

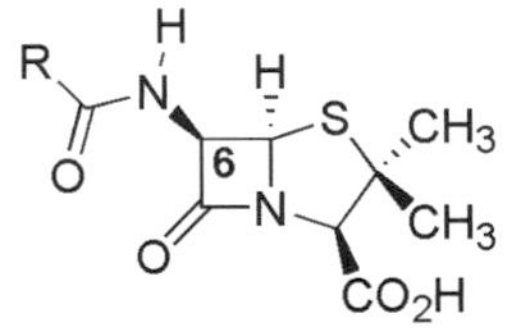

β-Lactam	R	Gram-positive activity	Gram-negative activity	Oral formulation
Benzylpenicillin (**16.5**)	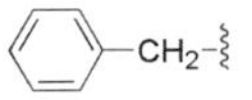 CH_2	Yes	No[a]	No, acid-labile
Methicillin (**16.6**)	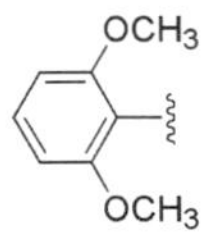 OCH_3, OCH_3	Yes (also active against β-lactamase staphylococcal producers)	No	No, acid-labile
Ampicillin (**16.7**)	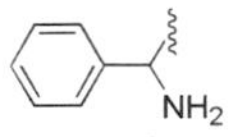 NH_2	Yes	Yes (with the exception of *P. aeruginosa*)	Yes (also given by injection)
Cloxacillin (**16.8**)	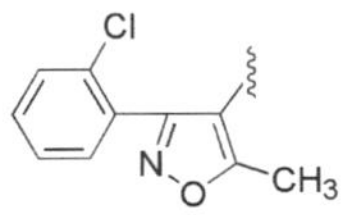 Cl, N, O, CH_3	Yes	No	Yes
Flucloxacillin (**16.9**)	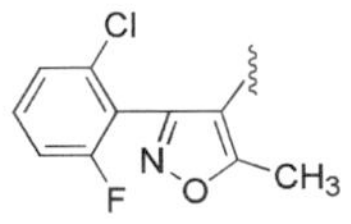 Cl, F, N, O, CH_3	Yes	No	Yes

Table 16.3 β-Lactam penicillin antibiotics — Continued

β-Lactam	R	Gram-positive activity	Gram-negative activity	Oral formulation
Carbenicillin (**16.10**)	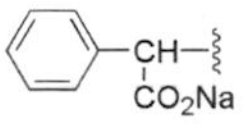 Ph–CH(CO$_2$Na)–	Yes	Yes[b]	Not absorbed
Ticarcillin (**16.11**)	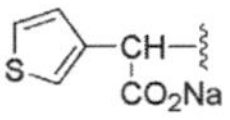 3-thienyl–CH(CO$_2$Na)–	Yes	Yes[b]	No
Amoxycillin (**16.12**)	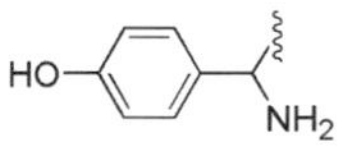 HO–C$_6$H$_4$–CH(NH$_2$)–	Yes	Yes[b]	Yes
Piperacillin (**16.13**)	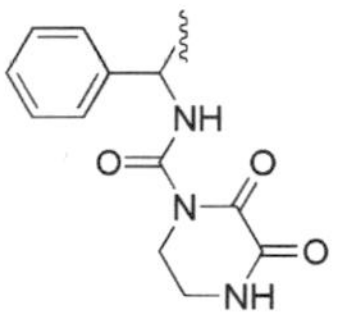 Ph–CH(NH–CO–N(dioxopiperazine, NH))–	Yes	Yes[b]	No
Azlocillin (**16.14**)	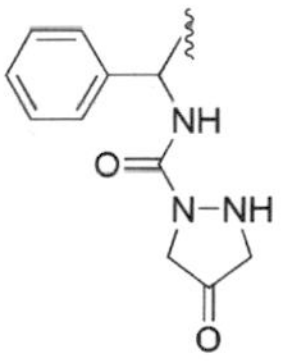 Ph–CH(NH–CO–N–NH pyrazolidinone)–	Yes	Yes[b]	No
Mezlocillin (**16.15**)	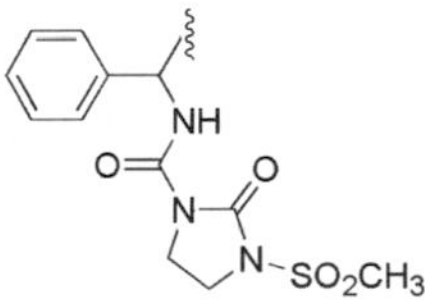 Ph–CH(NH–CO–N(imidazolidinone)–N–SO$_2$CH$_3$)–	Yes	Yes[b]	No

[a]Except gonococci and meningococci.
[b]Includes activity *vs. P. aeruginosa*.

The first semisynthetic penicillin of any clinical consequence was methicillin (**16.6**), which although inactive against Gram-negative bacteria, possesses significant activity against β-lactamase staphylococcal producers. Ampicillin (**16.7**) was the next advance due to its marked activity against Gram-negative organisms (excluding *P. aeruginosa*) and oral bioavailability. Ampicillin, however, is susceptible to β-lactamases of *S. aureus* and most Gram-negative bacteria. Many other important semisynthetic penicillins followed including cloxacillin (**16.8**), the first oral, β-lactamase-stable penicillin, although still lacking significant action on Gram-negative cells; its derivative flucloxacillin (**16.9**) carbenicillin (**16.10**) the first penicillin with activity against

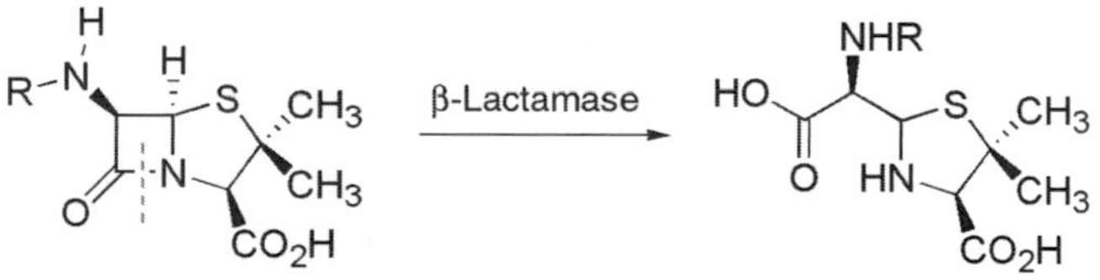

Figure 16.3 Mode of action of β-lactamase.

P. aeruginosa, and its derivative ticarcillin (**16.11**); amoxycillin (**16.12**) with a similar profile to ampicillin but with improved oral absorption; piperacillin (**16.13**), azlocillin (**16.14**), and mezlocillin (**16.15**) which combine the spectra and degree of activity of ampicillin and carbenicillin (Table 16.3).

The 3-position has also been important in the development of the penicillins, primarily for the introduction of ester moieties to enhance uptake and oral bioavailability (prodrugs, see Chapter 7). Esters of ampicillin (**16.7**), such as pivampicillin (**16.16**), talampicillin (**16.17**), and bacampicillin (**16.18**), are metabolized *in vivo* to produce higher blood levels of the parent antibiotic. Ester formation at the 6-position has also produced some valuable prodrugs. Of note are the prodrugs of carbenicillin (**16.10**), carfecillin (**16.19**), and carindacillin (**16.20**), which are metabolized *in vivo* to give a similar blood level of the parent antibiotic to that obtained with an equivalent dose of carbenicillin given intramuscularly (Table 16.4).

Cephalosporins

Cephalosporins (extended-spectrum β-lactams) contain a six-membered dihydrothiazine ring fused to a β-lactam ring. Cephalosporins are susceptible to enzymatic hydrolysis by β-lactamases (cephalosporinases) produced by some Gram-negative bacteria resulting in cleavage of the β-lactam ring to give the inactive cephalosporoic acid with release of the side-chain at position 3 (unless $R^2 = H$) and general breakdown of the molecule (Figure 16.4).

Research on cephalosporins and cephamycins (7α-methoxycephalosporins) has been to improve antibacterial activity and increase resistance to β-lactamases ("cephalosporinases") produced by

Table 16.4 β-Lactam penicillin antibiotics modified at the 3- and 6-positions

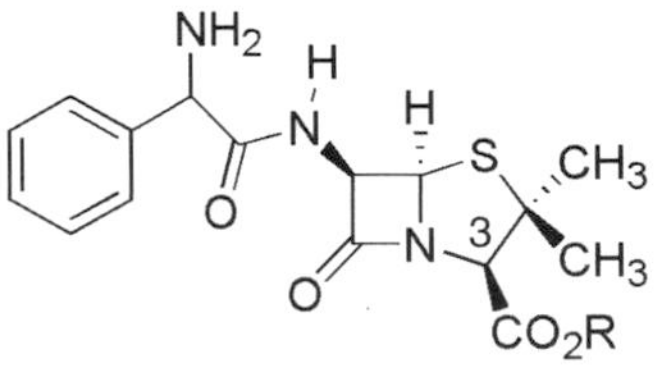

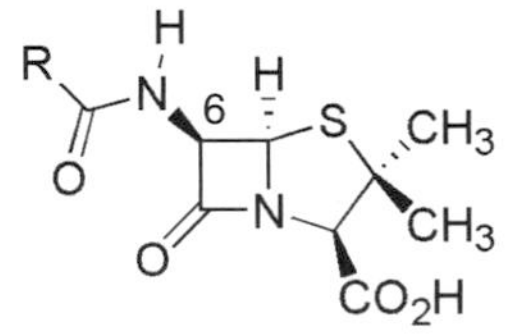

3-position esters (ampicillin, CO_2R)	6-position esters (CO_2H at 3)
(**16.16**) Pivampicillin; R = $-H_2C-O-C(=O)-C(CH_3)_3$	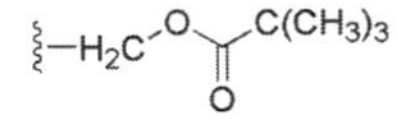 (**16.19**) Carfecillin; R = 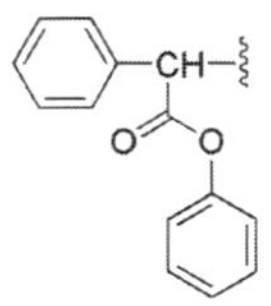
(**16.17**) Talampicillin; R =	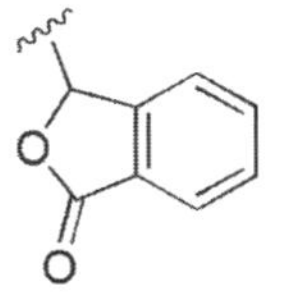 (**16.20**) Carindacillin; R = 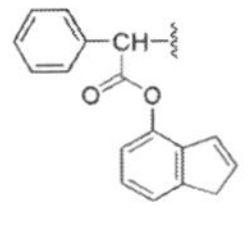
(**16.18**) Bacampicillin; R = $-CH(CH_3)-O-C(=O)-O-CH_2CH_3$	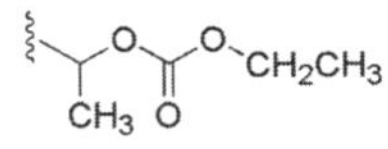

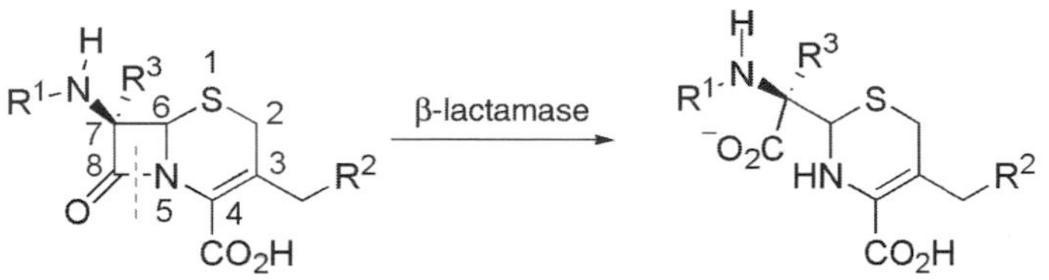

Figure 16.4 Enzymatic hydrolysis of cephalosporins with β-lactamase.

Gram-negative bacteria, and to improve pharmacokinetic properties by varying substituents on the molecule.

Structure–activity relationships

Structure–activity relationship (SAR) studies have been extensive and some of the key findings are described. The cephamycins, such as cefoxitin (**16.21**), display high antibacterial activity and stability to most β-lactamases. Antibacterial activity is increased due to greater penetration through the outer membrane of Gram-negative bacteria, but increasing the size of the ether group results in reduced activity.

Isosteric replacement of the sulfur at position 1 of the cephalosporins with oxygen gives the oxacephems. The 1-oxacepham antibiotics, such as latamoxef (**16.22**), have high antibacterial activity and β-lactamase resistance. Normally the presence of the ring oxygen would make the compound both chemically and enzymatically unstable (due to susceptibility to cleavage of the more polar O−C bond). However, the presence of the 7α-methoxy group stabilizes the molecule. Removal of either sulfur or oxygen from position 1 (carbacephem antibiotics), as in loracarbef (**16.23**), results in increased stability to β-lactamase.

The side-chain is of importance not only for increasing stability towards β-lactamases but also for enhancing binding affinity. Introduction of an imine–ether moiety in the side-chain enhances stability towards β-lactamases, with the additional steric bulk making the cephalosporin molecule less favorable as a substrate for the β-lactamases.

Side-chains containing a 2-aminothiazolyl group, such as cefotaxime (**16.24**), ceftriaxone (**16.25**), and ceftazidime (**16.26**, Figure 16.5), display increasing affinity for PBPs of Enterobacteriaceae and Streptococci. The interplay of various substitutions is important for antibacterial activity and β-lactamase stability as illustrated for ceftazidime in Figure 16.5.

Pharmacokinetic properties

3-Acetoxymethyl cephalosporins, such as cephalothin (**16.27**) and cephacetrile (**16.28**), are active against Gram-positive and β-lactamase-negative Gram-negative bacteria. However, they

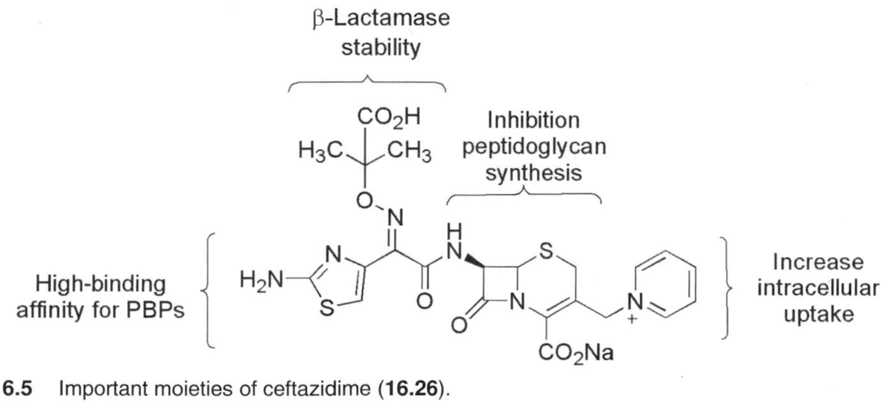

Figure 16.5 Important moieties of ceftazidime (**16.26**).

are readily converted *in vivo* by esterases to the less active 3-hydroxymethyl derivatives, which are rapidly excreted resulting in a short half-life (Figure 16.6).

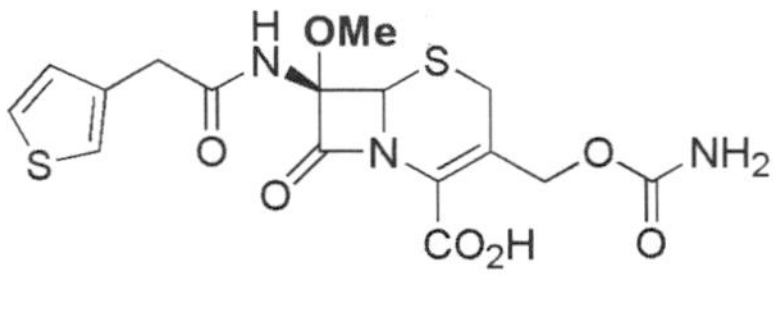

(**16.21**) Cefoxitin

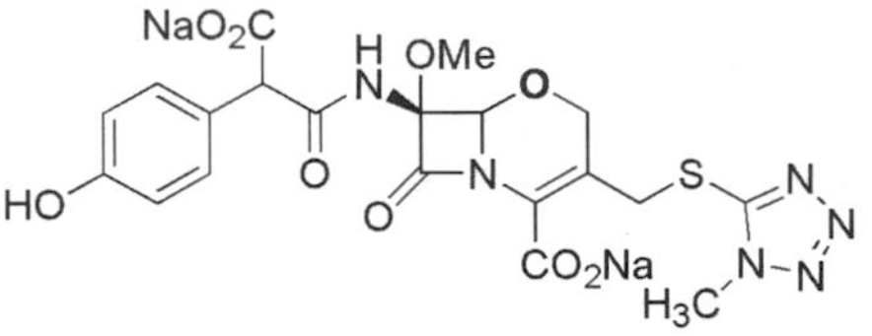

(**16.22**) Latamoxef

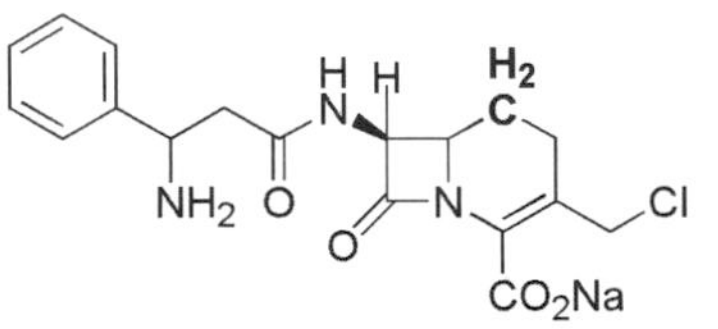

(**16.23**) Loracarbef

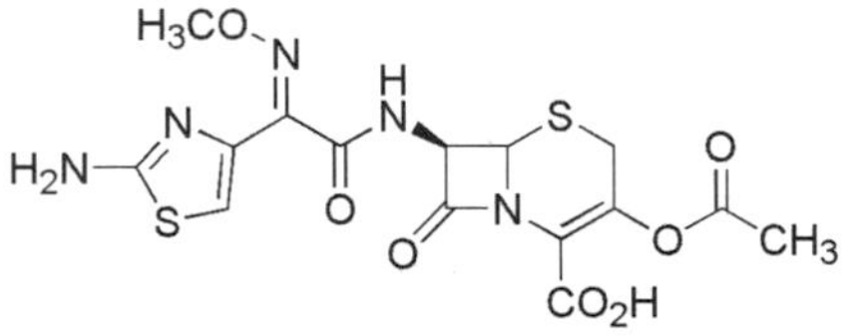

(**16.24**) Cefotaxime

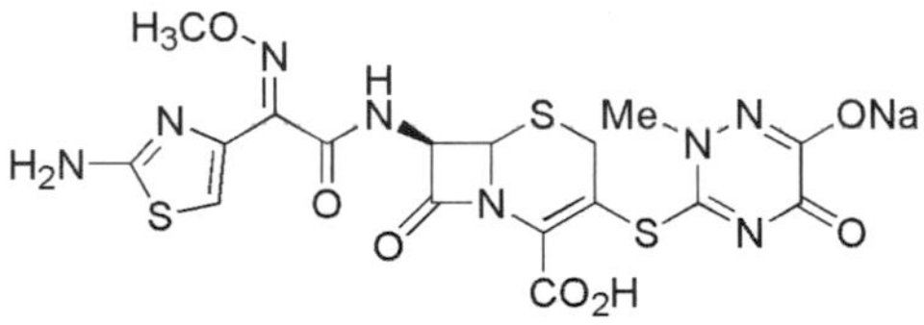

(**16.25**) Ceftriaxone

In addition to rapid metabolism, cephalothin (**16.27**) and cephacetrile (**16.28**) are also inactive when given orally. More recent oral cephalosporins, such as loracarbef (**16.23**), cefixime (**16.29**), and cefpodoxime (**16.30**), show increased stability to Gram-negative β-lactamases. Continued development of cephalosporins has produced more selective and stable drugs, such as the narrow-spectrum antibiotic cefsulodin (**16.31**) used for the treatment of pseudomonal infections.

Other classes of β-lactam antibiotics of note are the monobactams, for example, aztreonam (**16.32**). Aztreonam is highly active against most Gram-negative bacteria, with a predominant effect on PBP3, and stable to most β-lactamases. Interestingly aztreonam is inactive against *S. aureus* strains, probably due to the absence of PBP3 from staphylococci. The penems contain a double bond in the thiazolidine ring and a methylene group in place of the S-atom at position 1 in the

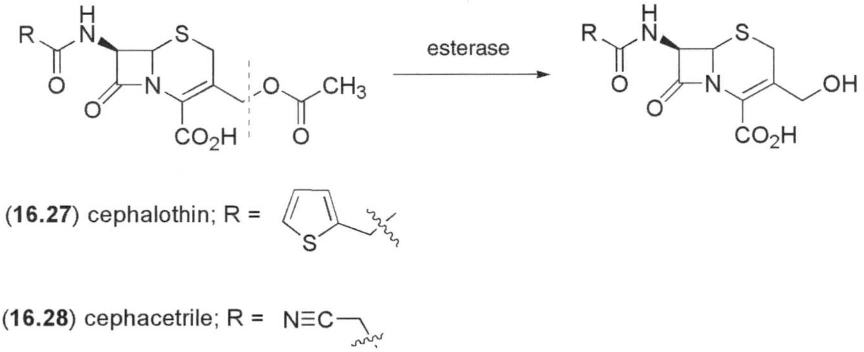

esterase

(**16.27**) cephalothin; R =

(**16.28**) cephacetrile; R = N≡C–

Figure 16.6 Esterase cleavage of 3-acetoxymethyl cephalosporins.

penicillin molecule. Imipenem (**16.33**) is a broad-spectrum antibiotic that is highly resistant to β-lactamases. It is, however, metabolized in the kidney by the enzyme dehydropeptidase-I and must be administered intraveneously with cilastatin, an inhibitor of this enzyme.

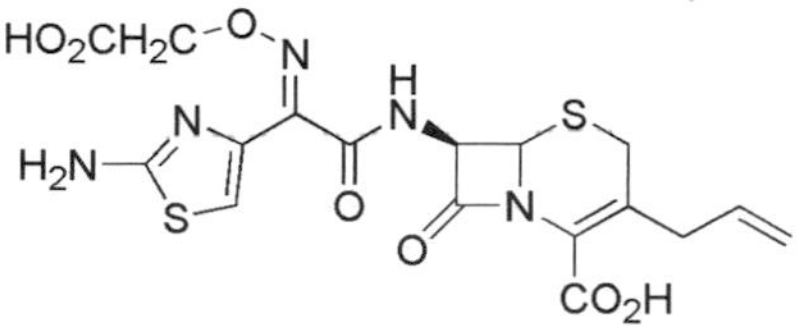

(**16.29**) Cefixime

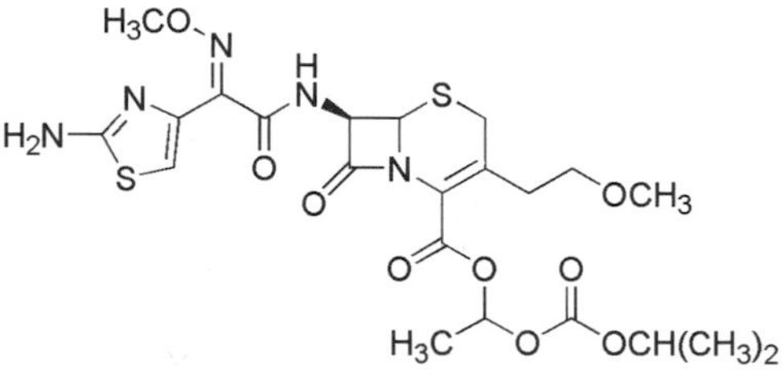

(**16.30**) Cefpodoxime

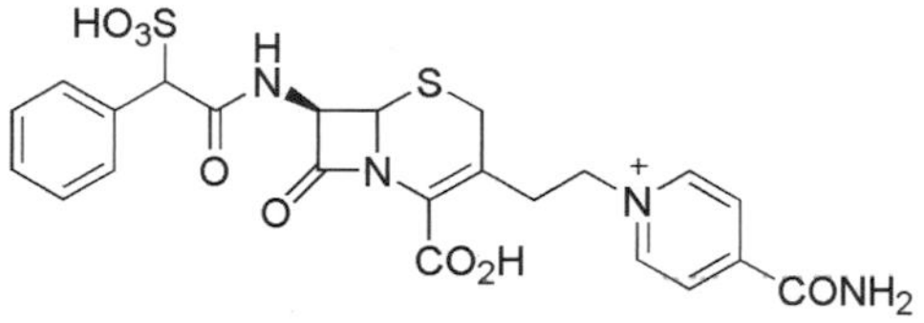

(**16.31**) Cefsulodin

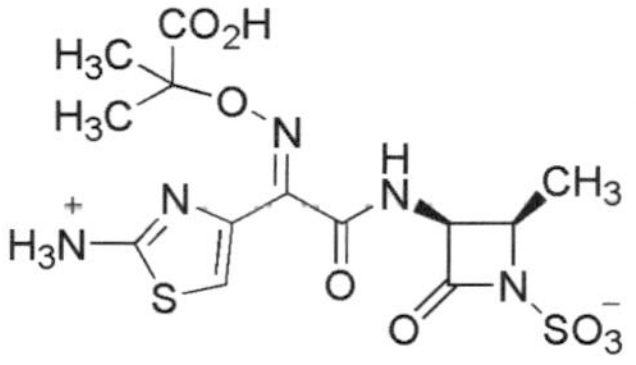

(**16.32**) Aztreonam

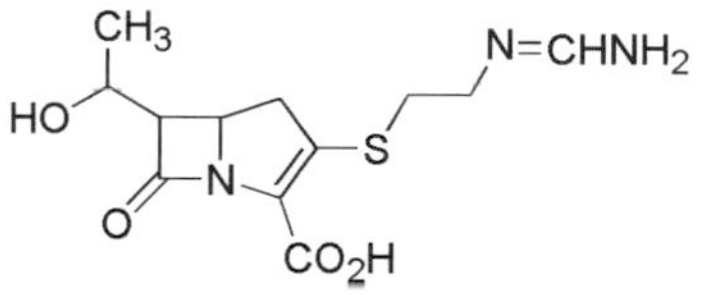

(**16.33**) Imipenem

β-Lactamase inhibitors

Inhibition of β-lactamases potentiates β-lactam activity. The preferred inhibitors are irreversible (or suicide) inhibitors, which inactivate the enzyme for extended periods and so prolong the effective life of the β-lactam drugs.

Three suicide inhibitors are currently licensed for clinical use in combination with specific β-lactam antibiotics (Table 16.5). These are clavulanic acid (**16.34**) (a clavam), sulbactam (**16.35**), and tazobactam (**16.36**). Inhibitory efficiency can be measured by determining the number of inhibitor molecules required to inhibit one molecule of enzyme (the higher the number, the less effective).

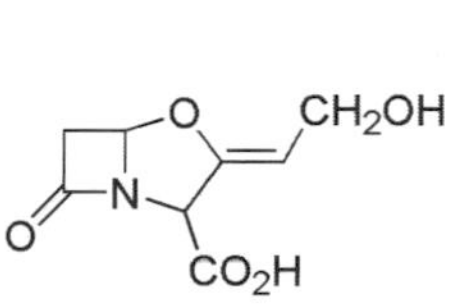

(**16.34**) Clavulanic acid

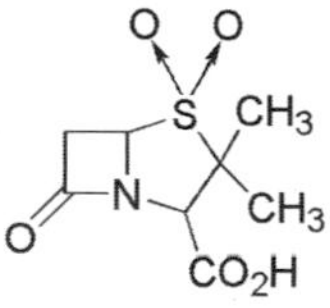

(**16.35**) Sulbactam

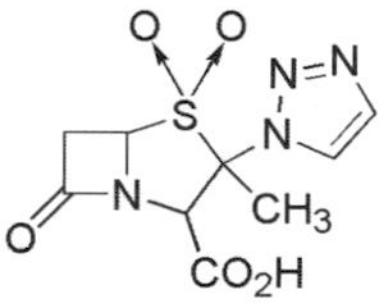

(**16.36**) Tazobactam

The mechanism of action of the mechanism-based inactivators (irreversible inhibitors) of β-lactamase, such as clavulanic acid and sulbactam, involves initial acylation of the β-lactamases by the inhibitors, with reaction at a serine moiety in the active site (Figure 16.7).

Table 16.5 β-Lactamase inhibitor/β-lactam combinations

Inhibitor	Ratio [I]:[E]	Combination	Formulation
Clavulanic acid (**16.34**)	115:1	Amoxycillin (**16.12**)	Oral/parenteral
		Ticarcillin (**16.11**)	Parenteral
Sulbactam (**16.35**)	3100:1	Ampicillin (**16.7**)	Oral/parenteral
Tazobactam (**16.36**)	125:1 → 50:1[a]	Piperacillin (**16.13**)	Parenteral

[a]Dependent on β-lactamase.
[I], β-lactamase inhibitor; [E], enzyme.

The active site serine hydroxyl group attacks the β-lactam carbonyl to form a tetrahedral intermediate **A**, which then collapses to the acyl-enzyme **B**. **B** can follow three different reaction paths: (1) The normal course of deacylation results in the formation of the imine **C** which is then hydrolyzed spontaneously to give the monosemialdehyde and the sulfinate of penicillamine. (2) The second reaction of **B** is as a transient inhibitor of the enzyme, the enamine **D** predominating as the more stable tautomer of the imine **B**; this inhibition is slower than the hydrolytic process. (3) The third reaction is the slowest, with the acyl-enzyme **B** undergoing a transimination reaction by an enzyme lysine residue. This gives rise to the new chromophore **E** and an inactivated enzyme in which two active site residues have been cross-linked; half of the suicide inhibitor's structure is lost during this process. Interaction of the enzyme with a suicide inhibitor ultimately leads to irreversible inhibition.

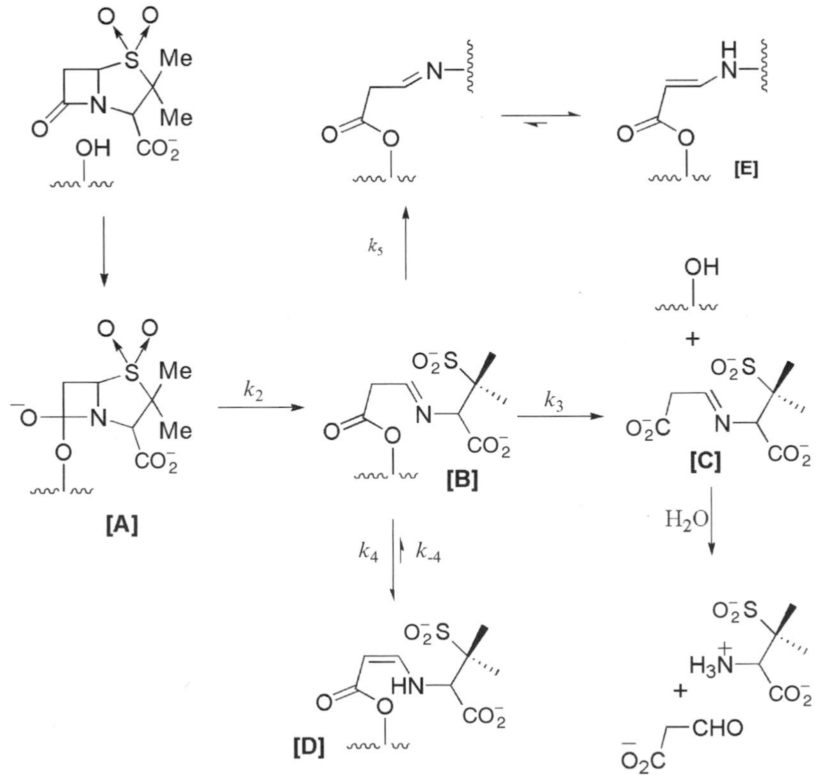

Figure 16.7 Interaction of sulbactam (**16.35**) with β-lactamase.

Glycopeptides

The glycopeptides vancomycin (**16.37**) and teicoplanin (**16.38**) are very important for the treatment of infections caused by staphylococci (in particular *S. aureus*) and *E. faecium*. However, both vancomycin and teicoplanin are potentially nephrotoxic (fewer adverse effects associated with teicoplanin), and the emergence of vancomycin-resistant *E. faecium* (VREF) and other MDR Gram-positive organisms is of increasing concern.

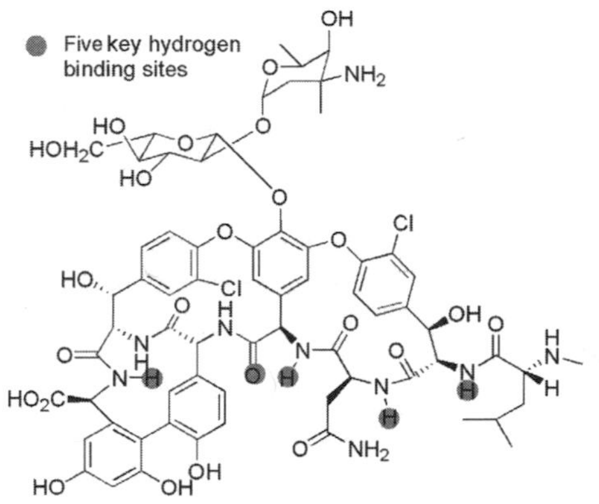

● Five key hydrogen binding sites

(**16.37**) Vancomycin

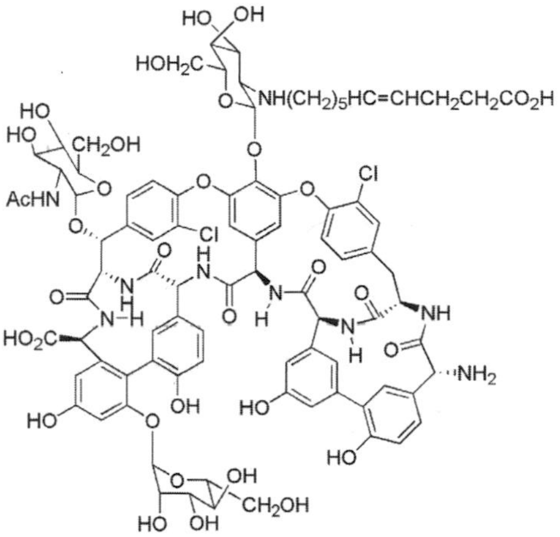

(**16.38**) Teicoplanin

Glycopeptides inhibit cell wall biosynthesis by binding to the terminal D-alanine-D-alanine sequence of the peptidoglycan precursor. The complex of vancomycin with the dipeptide, which is stabilized by multiple hydrogen bonding, prevents the substrate from interacting with the active site of the enzyme (Figure 16.8). As a result, transglycosylase and transpeptidase enzymes are prevented from constructing and cross-linking the rigid bacterial cell wall that maintains the structural integrity of the bacterium.

Natural and semisynthetic analogs of vancomycin have been described; however, vancomycin and teicoplanin are the only clinically used glycopeptides.

Glycopeptide resistance

Resistance is mediated by the ligase, VanA, which catalyzes the synthesis of cell-wall precursors with a D-alanine–D-lactate terminus instead of the usual D-alanine–D-alanine terminus. The transformation of an amide bond to an ester replaces an attractive carbonyl–NH hydrogen bond with a repulsive interaction between the ester oxygen from the D-Lac and the same carbonyl group within the binding pocket of the antibiotic (Figure 16.9).

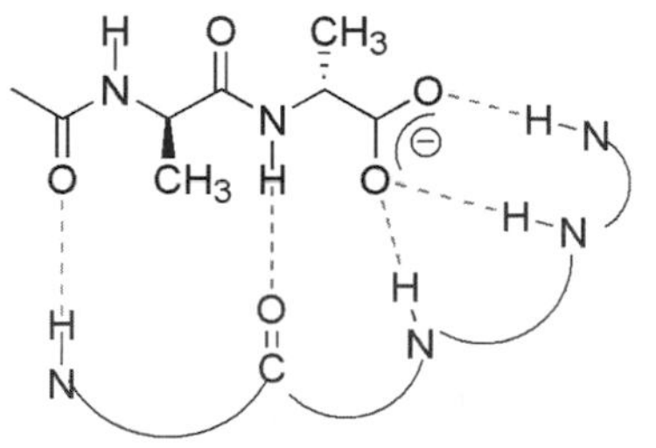

Figure 16.8 Binding of the terminal D-Ala-D-Ala with vancomycin via five hydrogen bonds.

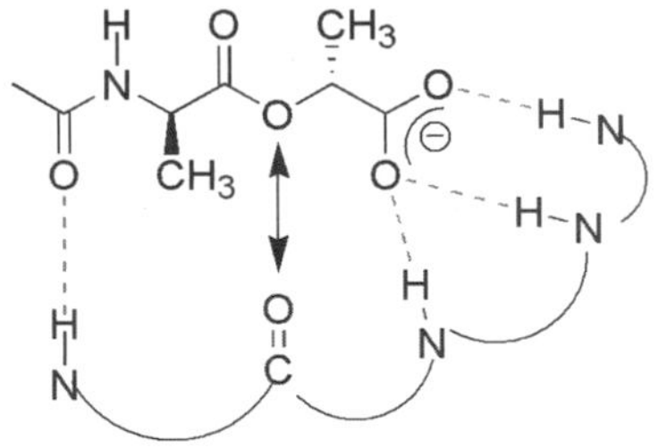

Figure 16.9 Binding of the terminal D-Ala–D-Lac with vancomycin showing four hydrogen bonds and one repulsive interaction.

The inability of glycopeptides to tightly bind the D-alanine–D-lactate peptidoglycan precursor is attributed to the loss of one critical hydrogen bond in the complex (cf. Figure 16.8 and Figure 16.9), a change sufficient to diminish binding and reduce enzyme inhibition (and therefore interruption of cell wall synthesis) to ineffective levels.

There are three approaches to counteracting resistance to glycopeptides: (1) inhibition of VanA ligase; (2) derivatives of glycopeptides with increased binding affinity to the modified target site; and (3) new antibiotics active against vancomycin-resistant Gram-positive bacteria (see ''Streptogramins'').

VanA-type VRE VanA-type VRE is the most frequently identified in clinical isolates, and expresses resistance through five plasmid-born genes, *vanS*, *vanR*, *vanH*, *vanA*, and *vanX*, resulting in the formation of the D-alanine–D-lactate peptidoglycan precursor. Weak inhibitors of both VanH, an α-keto dehydrogenase, and VanA, a depsipeptide ligase, are known, but greater success has been achieved with the development of inhibitors against VanRS and VanX. The tyramines (e.g., RWJ-49815, **16.39**) are active against VRE and are believed to act by inhibition of the VanRS system. VanX is a novel zinc-dependent D,D-dipeptidase for which a number of inhibitors are known, including dithiols (e.g., 2,3-dimercaptopropan–1-ol), phosphinates (**16.40**), and fluorodipeptides (**16.41**).

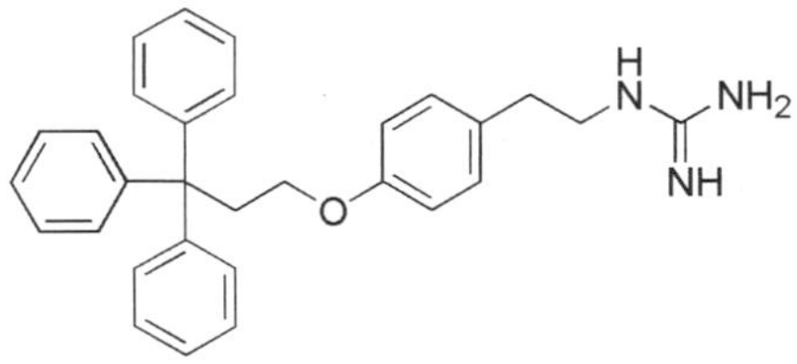

(**16.39**) RWJ-49815

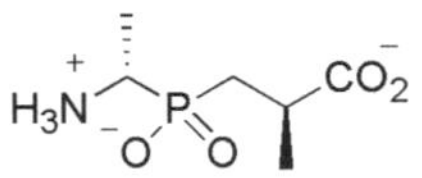

(**16.40**)

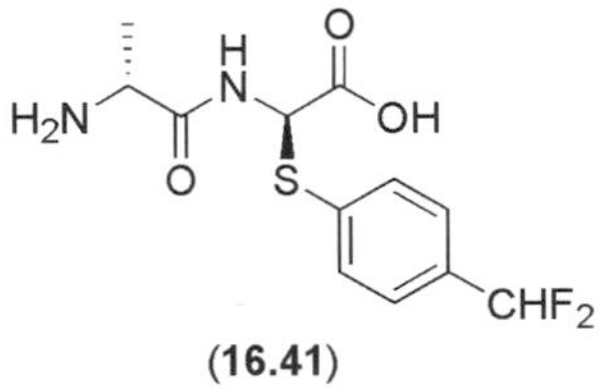

(**16.41**)

Derivatives of glycopeptides Several semisynthetic glycopeptides have recently been produced which show activity against resistant strains of bacteria. These antibiotics can either (a) dimerize (e.g., eremomycin), a property which occurs once the antibiotic has bound to the ligand and which leads to higher binding constants, or (b) are capable of membrane anchoring (e.g., teicoplanin) (Figure 16.10).

Biphenylchloromomycin (BCE) (Figure 16.11), a derivative of eremomycin, is ca. 500 times more active than vancomycin. Like vancomycin, BCE dimerizes. This enhancement in activity results from the ability of the hydrophobic biphenyl functionality to locate the antibiotic at its site of

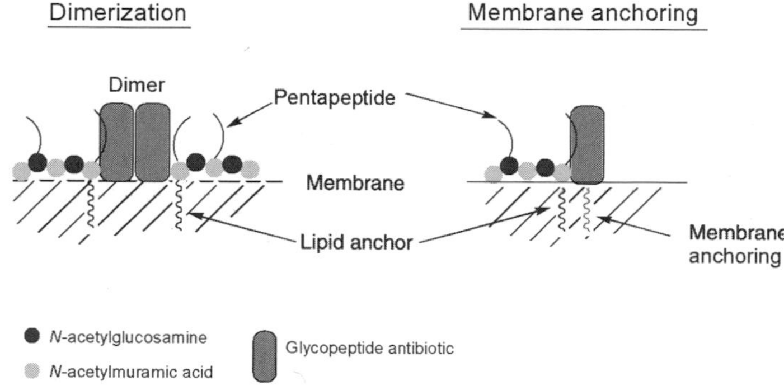

Figure 16.10 Antibiotic dimerization and membrane-anchoring properties.

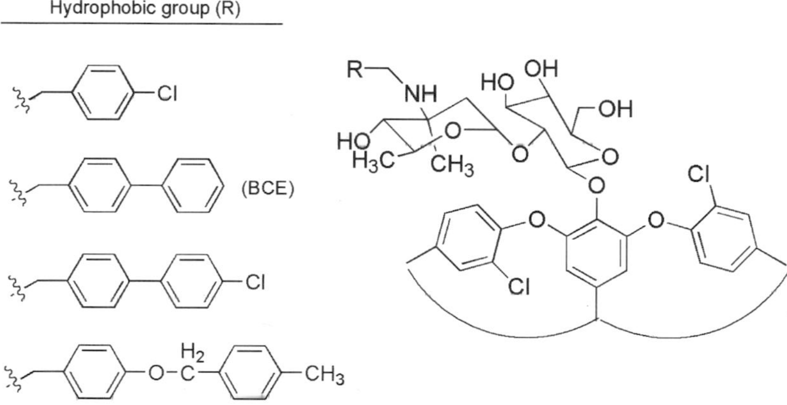

Figure 16.11 Aromatic membrane-anchoring moieties of BCE and derivatives.

action, i.e., the biphenyl moiety acts as a membrane anchor, which also results in an increase of the effective concentration of the antibiotic at the cell surface.

The activity of BCE and other derivatives is attributed to membrane anchoring and dimerization acting cooperatively to enhance the intrinsic affinity for -D-Lac terminating ligands (i.e., the membrane anchoring compensates for the absence of a H-bond in the antibiotic–cell wall complex).

16.2.2 Membrane-Active Agents

The term ''membrane-active agent'' is generally taken to mean an agent that affects the cytoplasmic membrane in microorganisms. Gram-negative bacteria (Figure 16.1) and mycobacteria also possess an outer membrane, which acts as a penetration barrier to some drugs. The polymyxins, such as polymyxin B (**16.42**, a mixture of polymyxin B1 and B2) and colistin (**16.43**, polymyxin E), cause leakage of intracellular constituents by damaging the cytoplasmic membrane of Gram-negative bacteria and disruption of the outer membrane lipopolysaccharide. The polymyxins are used as intestinal antibiotics, though their use is limited due to toxicity to mammalian cells.

Ionophoric drugs facilitate the passage of inorganic cations across the cytoplasmic membrane of Gram-positive bacteria. Valinomycin, a cyclododecadepsipeptide isolated from *Streptomyces fulvissimus*, is a K^+-conducting ionophore. Because of the lack of selectivity, these drugs are toxic to mammalian cells.

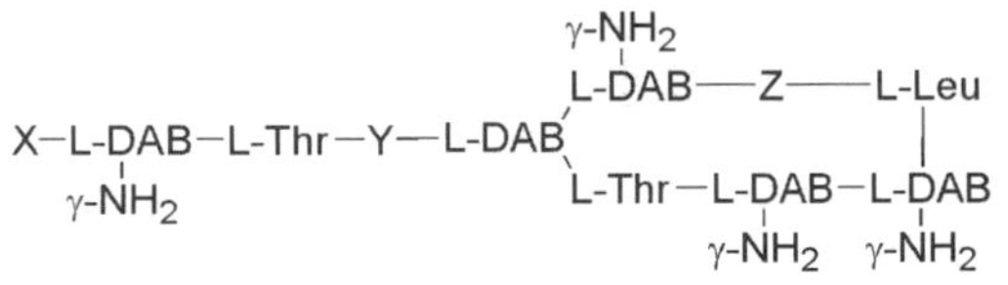

DAB = α,γ-diaminobutyric acid

Polymixin	X	Y	Z
(**16.42**) polymyxin B1	(+)-6-methyloctanoyl	L-DAB	D-Phe
(**16.42**) polymyxin B2	6-methylheptanoyl	L-DAB	D-Phe
(**16.43**) Colistin	(+)-6-methyloctanoyl	L-DAB-γ-NH_2	D-Leu

Inhibitors of mycolic acid biosynthesis

The mycobacterial cell wall contains a unique permeability barrier composed of 3-hydroxy fatty acids, known as mycolic acids, covalently bound to arabinogalactan. The biosynthesis of mycolic acids, alpha-, keto-, and methoxymycolates, provides a number of enzyme targets for the development of therapeutic agents.

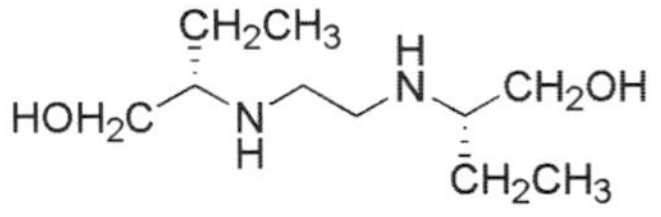

(**16.44**) Ethambutol 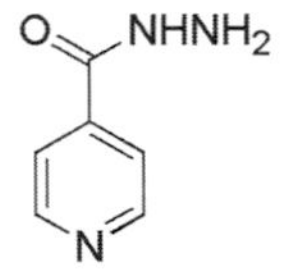 (**16.45**) Isoniazid 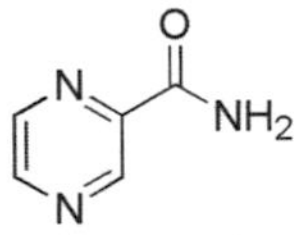 (**16.46**) Pyrazinamide

The exact mechanism of action of ethambutol (**16.44**), used in both the first- and second-line therapy of tuberculosis (TB), is unknown. However, its structural similarity to trehalose monomycolate suggests that it may interfere with mycolic acid biosynthesis in mycobacteria. Mycolic acid biosynthesis would also appear to be a major target for the anti-TB drugs isoniazid (Nydrazid®, **16.45**) and pyrazinamide (**16.46**). Isoniazid is a prodrug which is first metabolized to *N*-acetylisoniazid and then to isonicotinic acid which is incorporated into aberrant NAD^+ (Figure 16.12). Acetyl radicals are also generated which may be responsible for the hepatotoxicity of isoniazid. The nicotinamide derivative, pyrazinamide (**16.46**), is bactericidal at pH 5.5 but inactive at pH 7, i.e., the activity of pyrazinamide is pH-dependent.

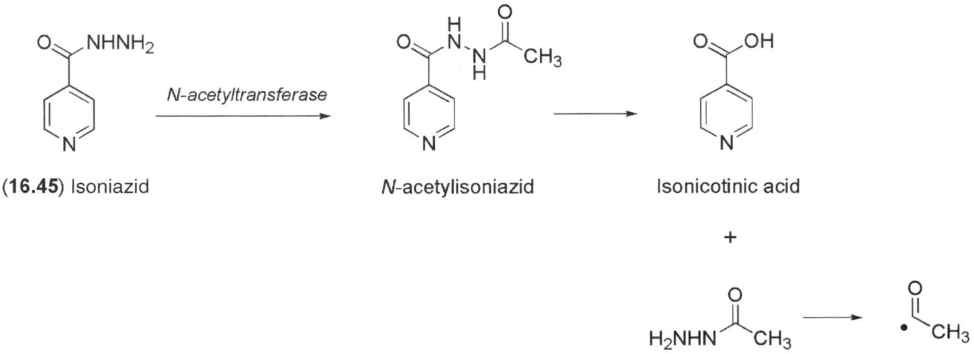

(**16.45**) Isoniazid — *N-acetyltransferase* → *N*-acetylisoniazid → Isonicotinic acid + $H_2NHN\text{-}CO\text{-}CH_3$ → $\bullet CO\text{-}CH_3$

Figure 16.12 Metabolism of the anti-TB drug isoniazid (**16.45**).

The nitroimidazopyran derivative PA-824 (**16.47**), currently in Phase I clinical trials, is also a prodrug requiring activation, possibly a nitro-reduction step, by *M. tuberculosis* F420 cofactor. PA-824 displays bactericidal activity comparable with isoniazid and is active against MDR strains. PA-824 was designed from the parent compound CGI-17341 (**16.48**), which exhibited mutagenic properties in addition to antimycobacterial activity. SAR studies of over 300 nitroimidazopyran derivatives determined that optimal activity was obtained with a lipophilic substituent at the 3-position, with activity generally residing in the (*S*)-enantiomer. PA-824 inhibits the biosynthesis of cell wall ketomycolate from hydroxymycolate, either by inhibition of an enzyme or cofactor responsible for the oxidation reaction.

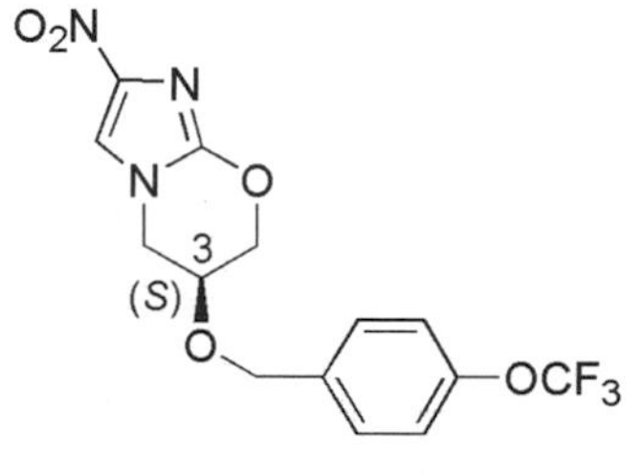

(**16.47**) PA-824 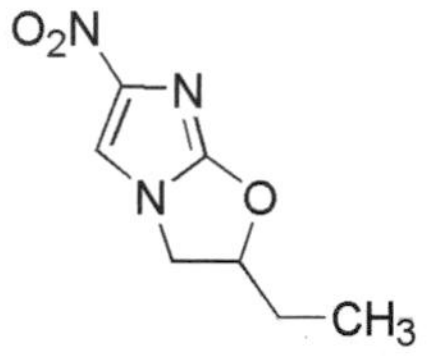 (**16.48**) CGI-17341

Isoxyl (**16.49**), a substituted diacyl thiourea, was first developed in the 1960s but was found to exhibit only modest activity against *M. tuberculosis*. However, more recent studies using either a longer treatment schedule or isoxyl/isoniazid combination showed isoxyl to be an effective drug. Isoxyl inhibits mycolyl transferase, a key enzyme in mycolic acid biosynthesis. Trehalose monomycolate, the natural substrate, binds at the active site of the transferase enzyme, which contains a hydrophilic pocket to accommodate the trehalose structure and two hydrophobic pockets to accommodate the mycolate chains. The structure of the thiourea inhibitors reflects this architecture containing a central hydrophilic thiourea unit flanked by hydrophobic alkyl-, alkoxy-, or alkylthio-chains.

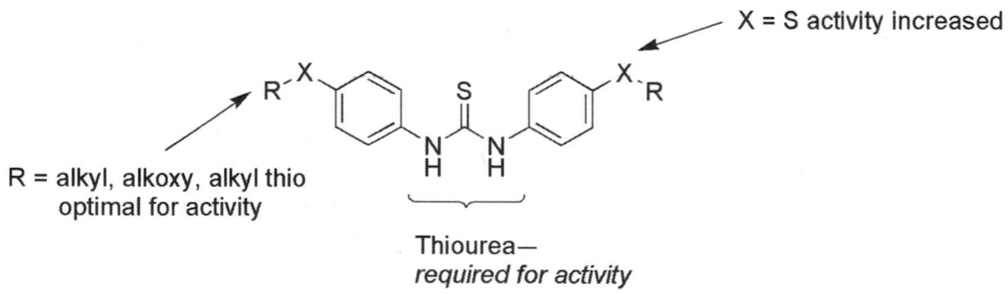

X = S activity increased

R = alkyl, alkoxy, alkyl thio optimal for activity

Thiourea—*required for activity*

(**16.49**) X = S, R = $CH_2CH_2CH(CH_3)_2$; isoxyl

16.2.3 Inhibitors of Protein Synthesis

Bacterial ribosomes have different characteristics (sedimentation coefficients of 70, i.e., are 70*S*, with 50*S* and 30*S* subunits) from those of mammalian ribosomes (80S, with 60*S* and 40*S* subunits). Most antibacterial antibiotics that are clinically important inhibitors of protein synthesis show selectivity for the 70*S* ribosomes.

For example, chloramphenicol affects 70*S* ribosomes but not 80*S*. By contrast, the tetracyclines inhibit protein synthesis on isolated 70*S* and 80*S* ribosomes. The tetracyclines and aminoglycoside antibiotics bind to the 30*S* subunit whereas macrolides and the oxazolidinones bind to the 50*S*

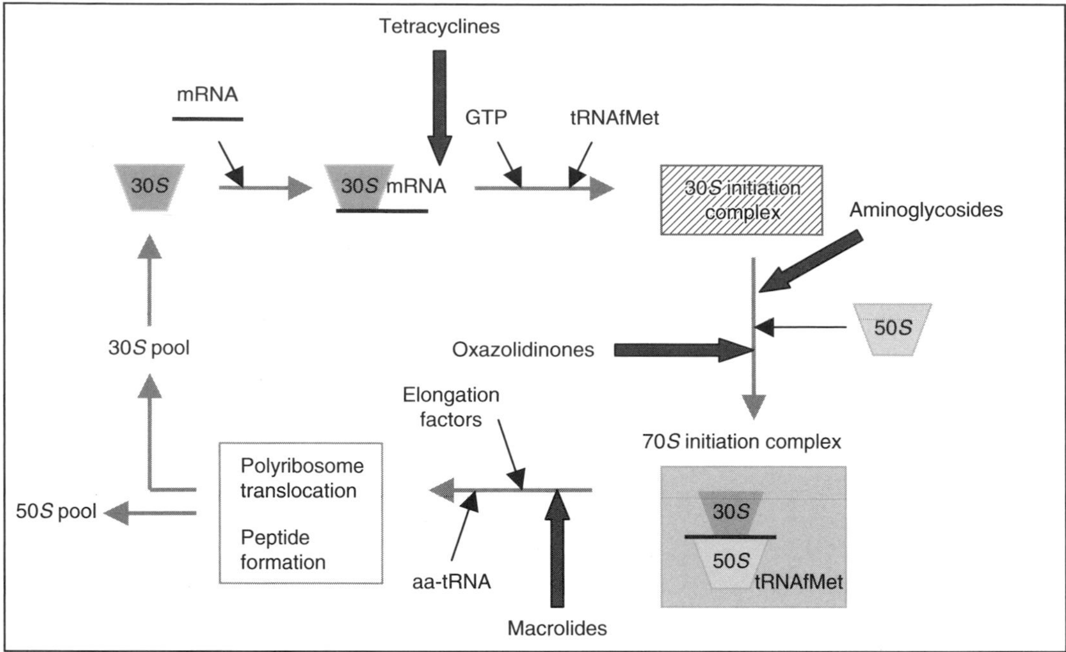

Figure 16.13 Inhibition sites of protein biosynthesis by antibiotics that bind to the 30*S* and 50*S* ribosomes. tRNA-fMet = *N*-formylmethionine–transfer RNA complex.

subunit (Figure 16.13). The various ribosomal preferences result from the energy-dependent active transport system present in bacterial but not mammalian cells which allow transport of these drugs into bacteria.

Aminoglycoside–aminocyclitol antibiotics

The aminoglycoside antibiotics are bactericidal to Gram-negative bacteria, staphylococci, and mycobacteria. The more important aminoglycoside antibiotics are streptomycin (**16.50**), kanamycin (**16.51**), gentamicin (**16.52**), tobramycin (**16.53**), and sisomicin (**16.54**). The development of newer aminoglycoside antibiotics has resulted in antibiotics with increased antimicrobial activity against resistant bacterial strains, enhanced pharmacokinetic properties, and reduced toxicity.

(**16.50**) Streptomycin

(**16.51**) Kanamycin

(16.52) Gentamicin C1 **(16.53)** Tobramycin **(16.54)** Sisomicin

Aminoglycoside-modifying enzymes

The aminoglycoside antibiotics are not completely inactivated by plasmid-encoded enzymes but modified in the outer regions of the resistant cell and do not bind to the ribosome. Only a small proportion of the external aminoglycoside need to be modified for resistance to be expressed. Aminoglycoside-modifying enzymes are of three types:

i. acetyltransferases, which transfer an acetyl group from acetylcoenzyme A to susceptible $-NH_2$ groups in the antibiotic;
ii. adenylyltransferases, which transfer adenosine monophosphate (AMP) from adenosine triphosphate (ATP) to susceptible $-OH$ groups in the antibiotic;
iii. phosphotransferases, which phosphorylate susceptible $-OH$ groups, with ATP acting as the source of phosphate.

Kanamycin (**16.51**) can be modified in at least six different sites (Figure 16.14), thus emphasis has been the development of aminoglycosides, such as amikacin (**16.55**), with greater resistance to these enzymes (Figure 16.14).

Tetracyclines

Tetracyclines are active against both Gram-positive and Gram-negative bacteria, but their widespread use in human and animal health care has had a major effect on the acquisition and spread of resistance to these agents. The most clinically used tetracyclines inhibit bacterial growth by blocking protein synthesis and are bacteriostatic. It has been hypothesized that tetracyclines interfere with aminoacyl-tRNA binding to the 30*S* subunit of ribosomes, resulting in inhibition of protein synthesis (Figure 16.13).

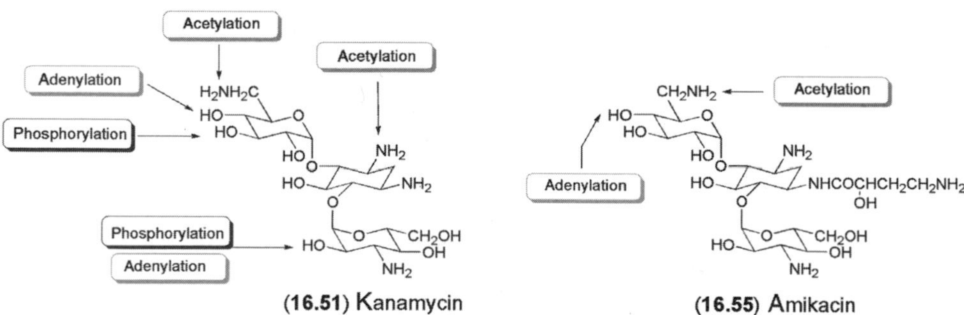

Figure 16.14 Points of enzymatic modification of kanamycin and amikacin.

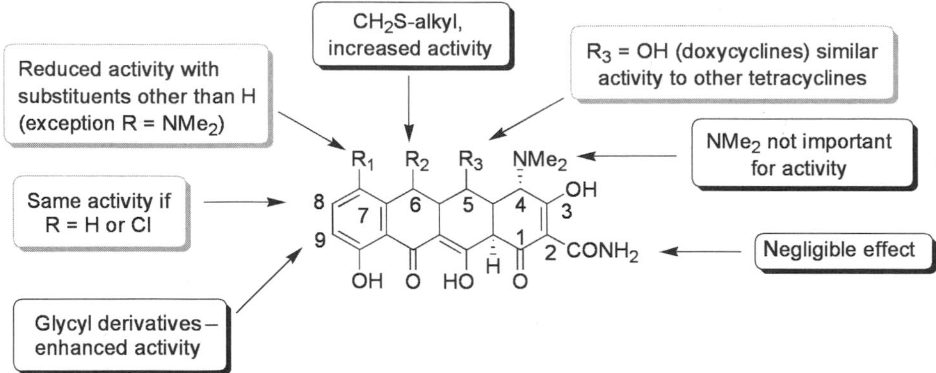

Figure 16.15 Summary of the effects of substituent modifications (SAR) on the activity of tetracyclines. No effect, minor effect, major effect.

Table 16.6 Clinically important tetracyclines

	R^1	R^2	R^3
(**16.56**) Doxycycline	H	CH_3	OH
(**16.57**) Clomocycline	Cl	CHOH	H (At 2:$CONHCH_2OH$)
(**16.58**) Chlorotetracycline	Cl	CHOH	H
(**16.59**) Demethyltetracycline	Cl	CHOH	H
(**16.60**) Methacycline	H	$= CH_2$	OH
(**16.61**) Minocycline	NMe_2	H2	H

SAR studies of the tetracyclines have shown that alterations in the hydrophilic domain reduce activity, whereas modifications in the hydrophobic area are usually tolerated, and sometimes lead to enhanced potency. The effects of specific modifications are shown in Figure 16.15. Important tetracyclines to arise from SAR studies are doxycycline (**16.56**), clomocycline (**16.57**), chlorotetracycline (**16.58**), demethyltetracycline (**16.59**), methacycline (**16.60**), and minocycline (**16.61**) (Table 16.6).

The major breakthrough in the development of potent tetracycline antibiotics active against resistant bacterial strains was the development of the glycyclines (**16.62**). The 9-glycyl-derivatives have been designed with a peptidic attachment to enhance permeation and ribosome binding. These compounds exhibit potent activity against a broad spectrum of Gram-negative and Gram-positive bacteria. Importantly, they have excellent activity against strains that are resistant to tetracycline because of efflux or ribosomal protection mechanisms (see ''Tetracycline resistance'').

Glycyl unit

(**16.62**) Glycycline

The basic nitrogen of the glycyl unit is essential and derivatives with a small alkylamino or cyclic amino groups in the glycyl unit display optimal activity. Replacement of the alkyl-substituted glycyl group with other amino acids, such as leucine, alanine, and phenylalanine, results in significantly reduced activity.

Tetracycline resistance

Nearly all clinically relevant resistant bacterial strains have acquired dedicated tetracycline resistance genes, usually encoded on mobile genetic elements (plasmids, transposons) which function by one of the two distinct mechanisms: efflux or ribosomal protection.

Efflux mechanism Efflux pumps belong to the major facilitator family of proteins in the cytoplasmic membrane, and work as antiporters driven by the proton motive force (PMF), with a proton imported into the cell in exchange for a monocationic magnesium–tetracycline complex. Genes involved include *tet* (A), (B), (C), (D), (E), (G), (H), (K), (L), *tetA* (B), and *otr* B.

Ribosomal protection Resistance due to ribosomal protection is encoded by the *tet* (M) gene and its homologs, the *tet* genes O, Q, P(B), S, and T. These genes are most commonly found in Gram-positive bacteria but are becoming more widespread.

The exact mechanism has not been elucidated. However, it appears that TetM associates with the ribosome and either blocks the binding of the tetracycline or causes dissociation of tetracycline from the ribosome. Importantly, the glycyclines (**16.62**) possess activity against organisms expressing the *tet*M and *tet*O determinants. Presumably, the affinity, or mechanism of binding of the glycyclines to ribosomes modified by the TetM or TetO proteins, is sufficient to prevent aminoacyl-tRNA binding despite the expression of the resistant determinant in the cell.

Macrolides

The macrolides inhibit bacterial protein synthesis by binding to the ribosomal 50*S* subunit, stimulating the dissociation of peptidyl-tRNA from ribosome during the translocation process. Development of the macrolide antibiotics has involved semisynthetic compounds derived from erythromycin (**16.63**), active generally against Gram-positive bacteria and some Gram-negative bacteria including *Legionella pneumophila*.

The macrolides consist of a large lactone ring containing 12–16 atoms, which are attached to one or more sugar residues. Recent compounds of clinical importance are clarithromycin (**16.64**) and azithromycin (**16.65**) used in the treatment of respiratory tract infections. The acid sensitivity of the macrolides requires administration in enteric-coated tablets or as the more acid-stable esters and ester salts, such as miocamycin (**16.66**), the diacetyl derivative of midecamycin (**16.67**), and rokitamycin (**16.68**), the butyryl ester of leucomycin A_5 (**16.69**) (Table 16.7).

Macrolide resistance

The clinically important resistance mechanism is by plasmid or transposon target modification mediated by erythromycin-resistant methylases, *Erm* enzymes. These *Erm* enzymes catalyze the $N^{6,6}$-dimethylation of an adenosine residue of bacterial 23*S* ribosomal RNA, which results in a reduced affinity of the ribosome for the macrolide. Other resistant mechanisms include efflux (PMF) and enzymatic modification of the macrolide (lactone hydrolysis, phosphorylation, and glycosylation).

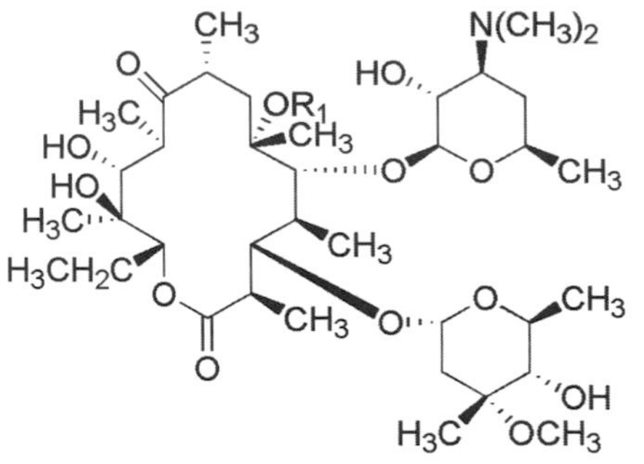

(**16.63**) R = H; Erythromycin
(**16.64**) R = CH_3; Clarithromycin

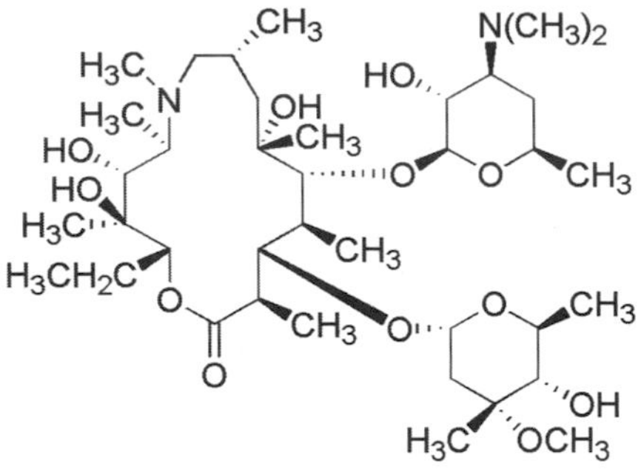

(**16.65**) Azithromycin

Telithromycin (**16.70**), a ketolide derivative which contains a 3-keto group instead of an L-cladinose on the aglycone A, is a new addition to the macrolide–lincosamide–streptogramin B (MLSb) class of antimicrobial agents with good activity against macrolide-resistant pathogens *in vitro* and *in vivo*. As well as the 3-keto group, the presence of the aryl-substituted 11,12-carbamate moiety was shown to be an important structural feature for potent activity against resistant strains. SAR studies determined the maximum chain length between the aryl and carbamate substituents to be C_3–C_4. Likewise the 4′-carbamate macrolides, such as CP-544372 (**16.71**), were shown to have potent *in vitro* activity against macrolide-resistant strains including *S. pneumoniae*. The aryl-containing ketolide derivatives appear to protect specific binding domains of the 23*S* ribosomal RNA against chemical modifications such as ribosomal methylation.

Table 16.7 Esterified macrolides

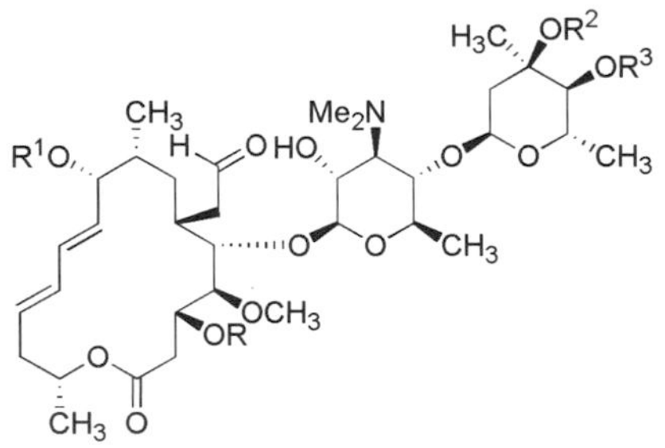

	R	R^1	R^2	R^3
(**16.66**) Miocamycin	$COCH_2CH_3$	$COCH_3$	$COCH_3$	$COCH_2CH_3$
(**16.67**) Midecamycin	$COCH_2CH_3$	H	H	$COCH_2CH_3$
(**16.68**) Rokitamycin	H	H	$COCH_2CH_3$	$CO(CH_2)_2CH_3$
(**16.69**) Leucomycin A_5	H	H	$COCH_2CH_3$	H

Aryl group C_3-C_4 side chain

11,12-cyclo carbamate

3-keto

(**16.70**) Telithromycin

(**16.71**) CP-544372

Oxazolidinones

The oxazolidinones represent a new class of antibacterial agents with activity against MRSA, MRSE, and VREF bacterial strains. The oxazolidinones bind to the 50*S* subunit, distorting the binding site for tRNAfMet. This prevents the formation of the tRNAfMet-mRNA-70*S* or -30*S* initiation complex, that is, protein synthesis is inhibited before initiation (Figure 16.13).

Extensive SAR studies have identified key pharmacophores (Figure 16.16). From this research, two oxazolidinones were advanced for clinical trials, linezolid (**16.72**) and eperezolid (**16.73**), with linezolid (Zyvox™) approved by the FDA for clinical use in 2000.

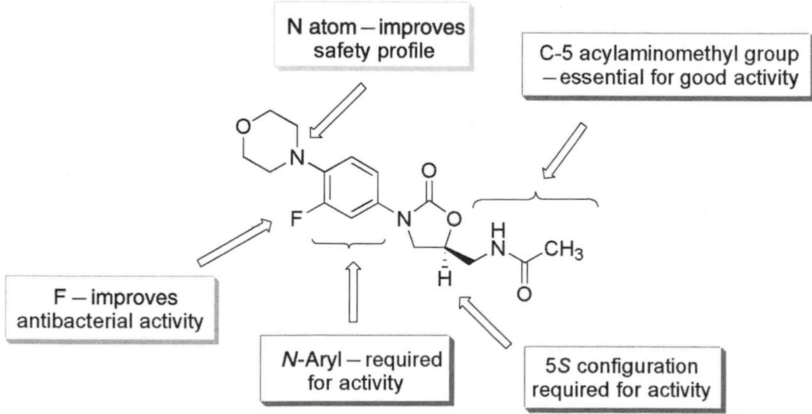

Figure 16.16 Important pharmacophores for antibacterial activity of oxazolidinones.

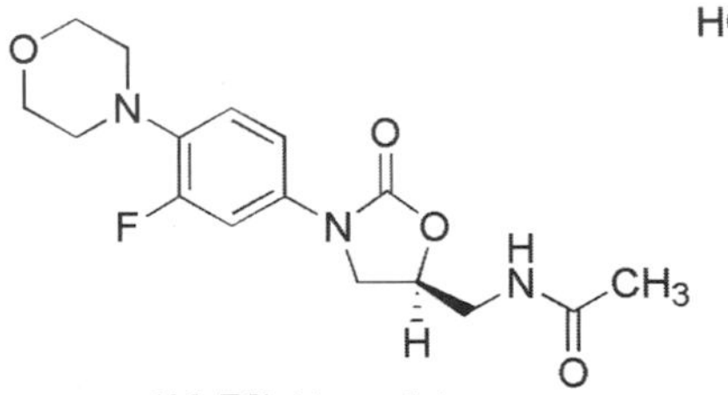

(**16.72**) Linezolid

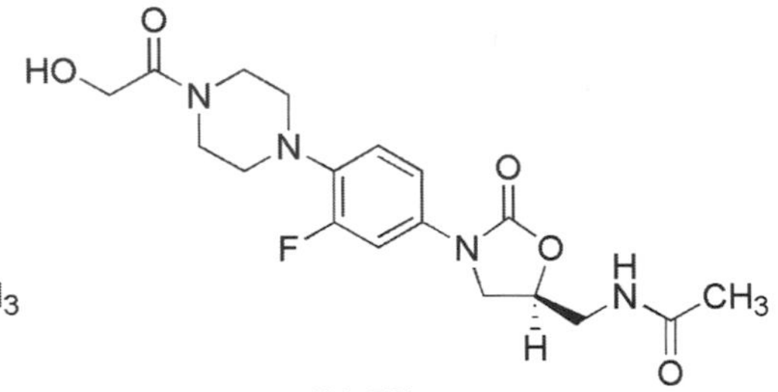

(**16.73**) Eperezolid

Despite the early enthusiasm for these novel antibacterial agents, linezolid-resistant *S. aureus* and enterococci have emerged during treatment and, in addition, the high incidence of myelosuppression in patients is cause for concern.

Streptogramins

The streptogramins bind to the bacterial ribosomes resulting in inhibition of protein synthesis. An unusual feature of this family is its mix of two structurally unrelated cyclic depsipeptides: a substituted 23-membered macrolactone (type A, e.g., quinupristin (**16.74**)) and a more diverse group of cyclic hexadepsipeptides (type B, e.g., dalfopristin (**16.75**)). Although type B have greater bacteriostatic activity, the combination of type A and B creates a synergistic bactericidal effect. These combinations are orally active against Gram-positive and some Gram-negative bacteria.

Synercid®, which belongs to the streptogramin family, is a mixture of two compounds — quinupristin (**16.74**)/dalfopristin (**16.75**) (administered as a 3:7 mixture) (Figure 16.17). The Synercid® emergency-use program was initiated as a result of the increasing number of infections caused by VREF, and has been shown to be as effective as vancomycin in non-VREF. Synercid® has recently been licensed (1999) under the FDA accelerated approval program, and represents the first alternative in three decades to vancomycin.

16.2.4 Inhibitors of Nucleic Acid Synthesis

Inhibitors of nucleic acid synthesis have several modes of action (Table 16.1) and fall into two categories: (a) those that inhibit the synthesis of purine and pyrimidine nucleotides; (b) those that inhibit synthesis at the polymerization level. This involves the appropriate polymerase, with nucleic acid as a template, and is the stage where nucleoside triphosphates are incorporated, through a 3′-5′-phosphodiester linkage, into the polynucleotide chain. Azaserine (**16.76**), a glutamine analog,

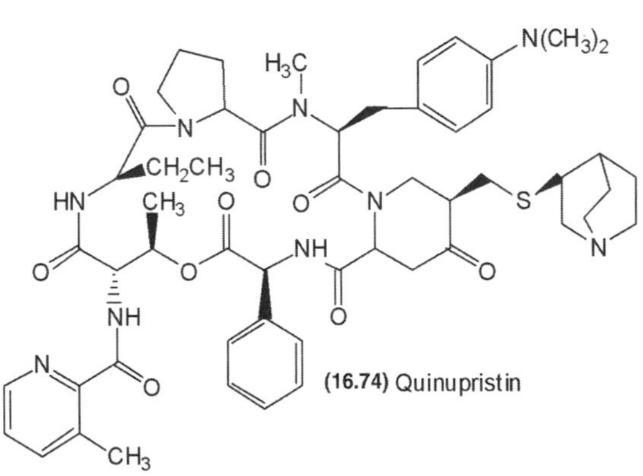

(**16.74**) Quinupristin

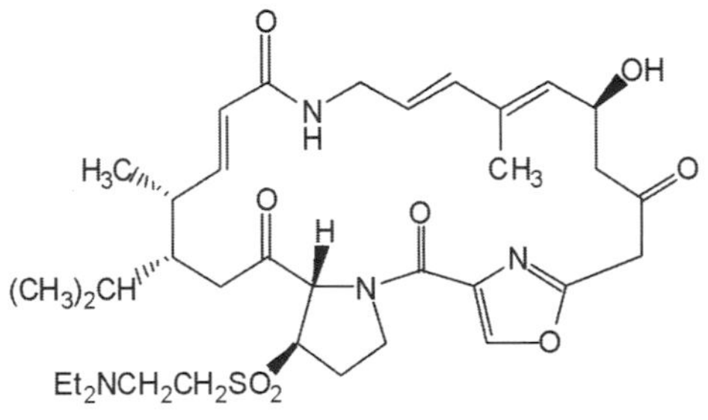

(**16.75**) Dalfopristin

Figure 16.17 Structures of the components of Synercid, quinupristin (**16.74**), and dalfopristin (**16.75**).

inhibits purine and pyrimidine nucleotide biosynthesis but is toxic to mammalian cells; actinomycin D and mitomycin C are intercalating agents which are also not selectively toxic; rifampicin (**16.77**), used in the treatment of tuberculosis and leprosy, binds and inhibits mycobacterial DNA-dependent RNA polymerase with a consequent-specific action on RNA synthesis. Rifampicin does not inhibit mammalian DNA polymerase and therefore has a selective toxic action.

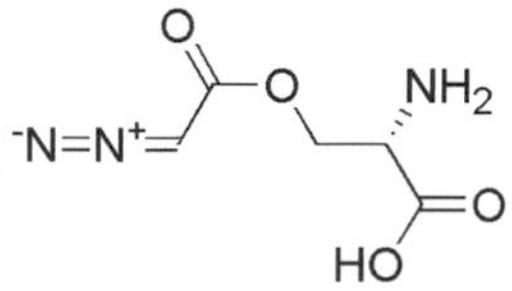

(**16.76**) Azaserine

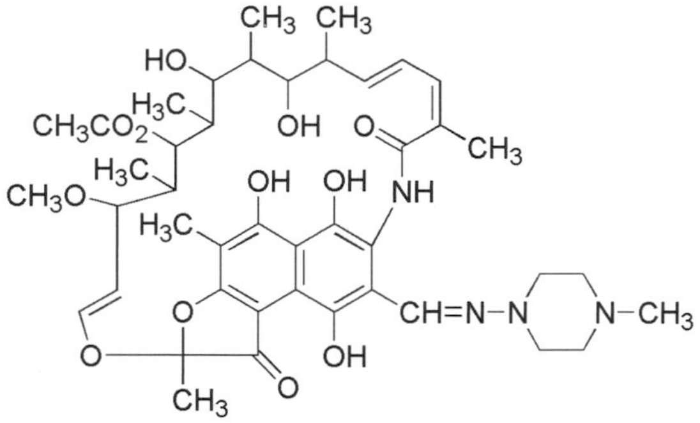

(**16.77**) Rifampicin

Perhaps the most clinically used group in this category are the quinolones, which usually inhibit the supercoiling of bacterial DNA by interfering with the activity of DNA gyrase, an essential type II topoisomerase that exists only in bacteria. However, in certain bacteria such as *S. aureus* DNA topoisomerase IV is thought to be the major target of the quinolones. DNA gyrase is an enzyme that nicks double-stranded DNA, introduces negative supercoils, and then seals the nicked DNA. Mammalian topoisomerase II does not possess supercoiling activity, suggesting that the action of mammalian DNA gyrase is substantially different from bacterial DNA gyrase, which is a satisfactory explanation for the selective action of the quinolone antibacterial agents.

Quinolones

The quinolone antimicrobial agents are synthesized chemically and have been extensively studied with over 10,000 derivatives synthesized. Nalidixic acid (**16.78**), which has a limited use as a drug due to a narrow antibacterial activity (Enterobacteriaceae only), short half-life, and high protein binding, is regarded as the progenitor of the newer quinolones, the most important of which are the fluoroquinolones, for example norfloxacin (**16.79**) and the market leader ciprofloxacin (**16.80**).

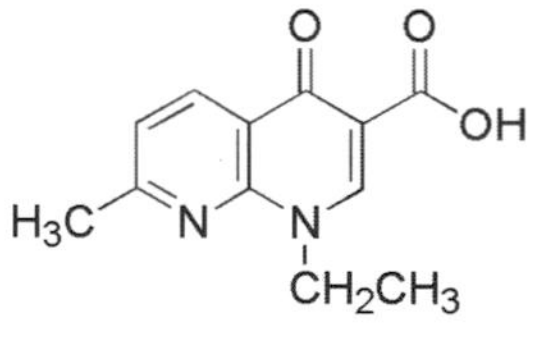

(**16.78**) Nalidixic acid

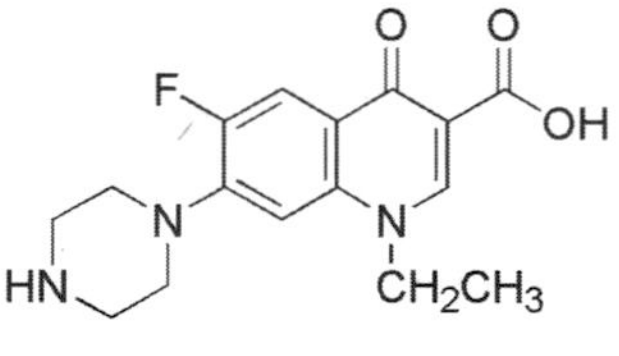

(**16.79**) Norfloxacin

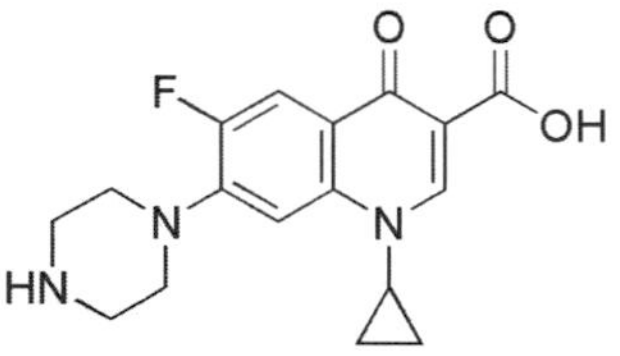

(**16.80**)Ciprofloxacin

The pharmacophore common to all fluoroquinolones has three key domains, enzyme binding, DNA binding, and an auto-assembling domain (Figure 16.18). Introduction of a fluoro substituent at the C-6 position resulted in enhanced potency and spectrum of antibacterial activity (Gram-negative and some Gram-positive bacteria, but not anaerobes) of these first generation fluoroquinolones, compared with nalidixic acid (**16.78**). Removal of the pyridine nitrogen atom reduced protein binding and introduction of a piperazine at the C-7 position resulted in an increased half-life.

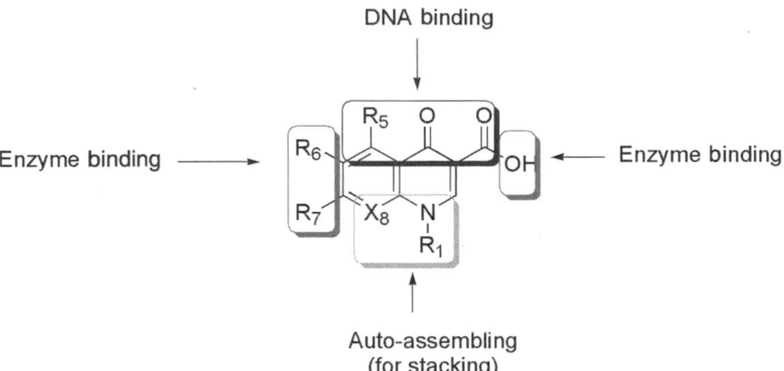

Figure 16.18 Key binding domains of the fluoroquinolones.

The fluoroquinolones are used for the treatment of bacterial infections including respiratory tract, skin, and urinary tract infections. However, the low potency of fluoroquinolones against a number of clinically important Gram-positive bacteria (e.g., *S. aureus* and *S. pneumoniae*) and the development of resistance mechanisms limit the use of these first generation fluoroquinolones.

Quinolone resistance

Resistance to fluoroquinolones results from mutations leading to altered amino acids in the resistance-determining region of DNA gyrase. A single amino acid substitution in the structural genes, *gyrA*, *gyrB*, and *grlA*, has been shown to result in high-level resistance to fluoroquinolones. Efflux and mutations that reduce the permeability of the outer bacterial membrane also influence quinolone resistance in staphylococci. It has also been found that concentrations of fluoroquinolones well in excess of inhibitory levels reduce the bactericidal activity of nalidixic acid (**16.78**) and some fluoroquinolones. Paradoxically, this is believed to result from inhibition of RNA/protein synthesis at these high drug concentrations. Second generation fluoroquinolones were designed to have a broader spectrum of Gram-positive activity with anaerobe and MRSA activity introduced into the third generation fluoroquinolones.

Second and third generation fluoroquinolones

The second generation fluoroquinolones, temafloxacin (**16.81**), lomefloxacin (**16.82**), and sparfloxacin (**16.83**), displayed broad activity against Gram-positive bacteria, while retaining high activity against Gram-negative bacteria. Introduction of a halogen at C-8 (X_8 in Figure 16.18) and the presence of a bulky group (cyclopropyl (**16.83**) or difluorophenyl (**16.81**)) resulted in enhanced activity (reduced minimum inhibitory concentration (MIC)). Substitution at C-5 (e.g., NH_2 in **16.83**) improved Gram-negative activity as did a methyl substitution in the piperazine ring. The methyl substitution in the piperazine ring was also found to increase the half-life. It should be noted that a number of adverse effects have been recognized with fluoroquinolone antibiotics including photosensitivity, hemolytic, and nephrotoxic reactions and CNS toxicity, resulting in their withdrawal from the market or termination of clinical trials.

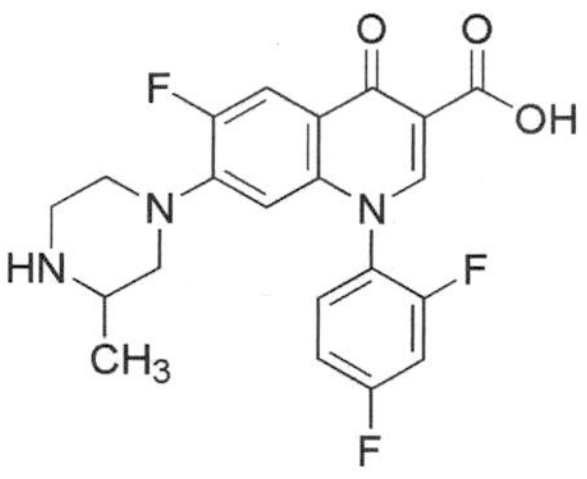

(**16.81**) Temafloxacin 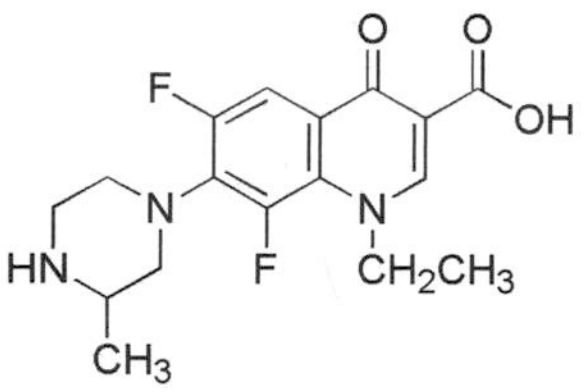 (**16.82**) Lomefloxacin 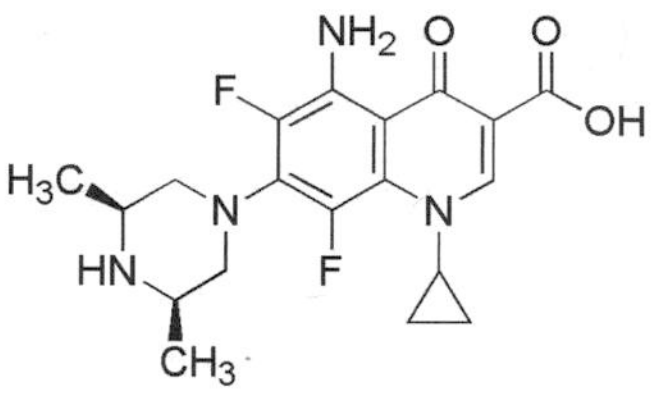 (**16.83**) Sparfloxacin

The newer third generation fluoroquinolones display broad-spectrum activity against Gram-negative and Gram-positive bacteria and anaerobes. Moxifloxacin (**16.84**) possesses potent activity against Gram-positive bacteria and *M. tuberculosis*, but is not active against MRSA. Clinafloxacin (**16.85**) has broad Gram-positive activity and is active against ciprofloxacin-resistant MRSA. Sitafloxacin (**16.86**) is more active than other currently available quinolones against quinolone-resistant isolates of *P. aeruginosa*, as well as possessing activity against quinolone-resistant strains of MRSA and MRSE. It has been suggested that the enhanced activity of sitafloxacin results from potent inhibition of mutant quinolone-resistant DNA gyrase.

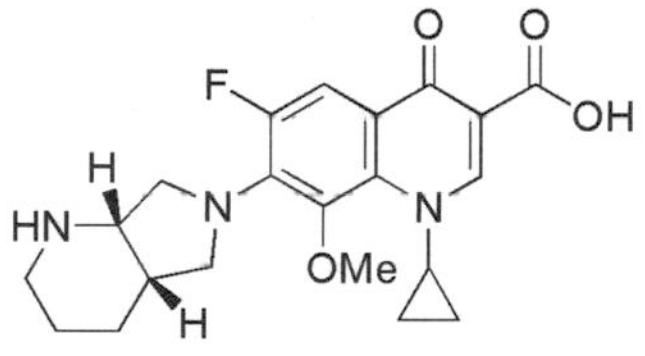

(**16.84**) Moxifloxacin 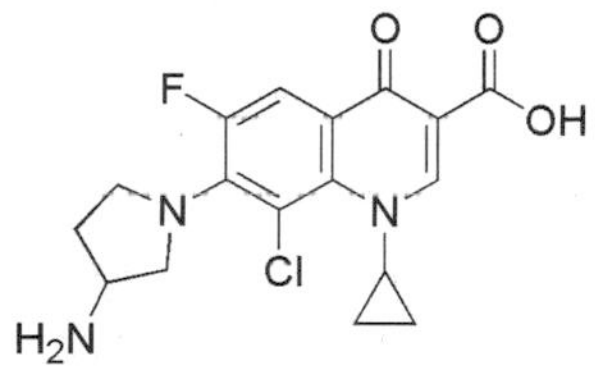 (**16.85**) Clinafloxacin 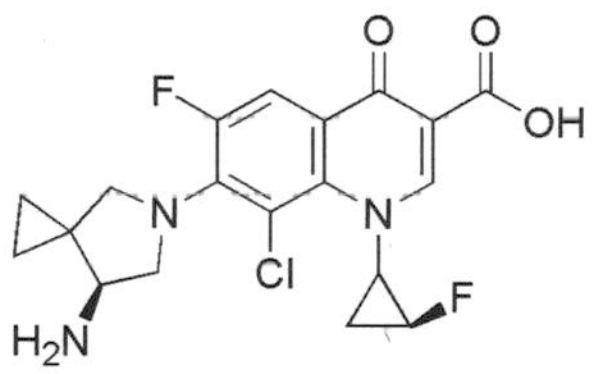 (**16.86**) Sitafloxacin

SAR studies have indicated sites for modification leading to either enhanced bactericidal activity (improved MIC) or improved Gram-positive/Gram-negative/anaerobe spectrum of activity (Figure 16.19).

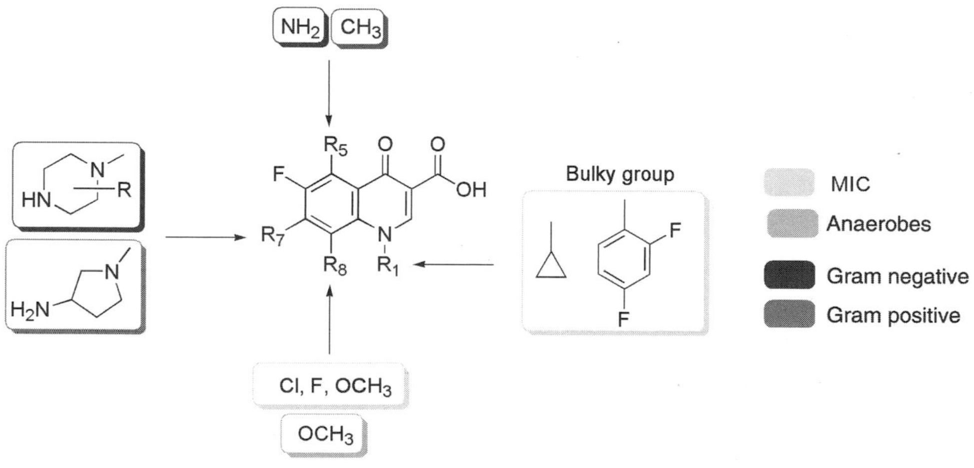

Figure 16.19 Modifications of the fluoroquinolone pharmacophore affecting MIC, Gram-negative, Gram-positive, and anaerobe antibacterial activity.

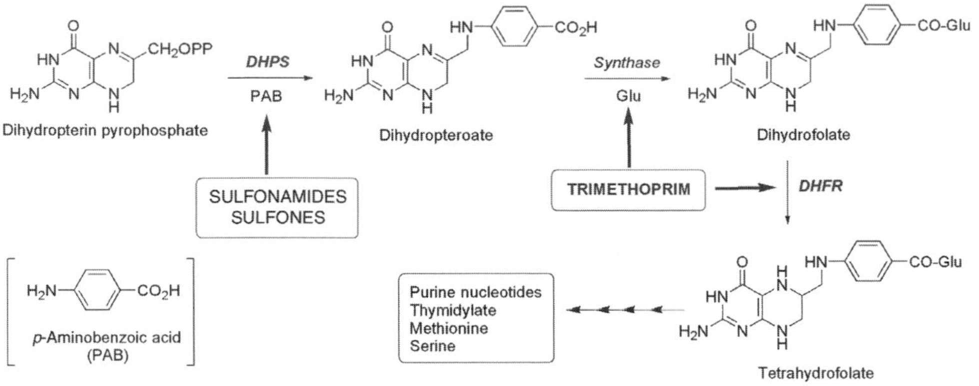

Figure 16.20 Folate metabolism in microorganisms and target enzymes of sulfonamides, sulfones, and trimethoprim.

16.2.5 Antibacterial Folate Inhibitors

The discovery that alkyl- or phenyl-substituted diaminopyrimidines had an antifolate action led to the development of compounds that were active against bacteria or protozoa and also to agents that were selectively toxic for the parasite rather than for the host (mammalian) cells. Bacteria must synthesize folate as they are unable to absorb preformed folate, therefore enzymes involved in folate metabolism are targets for inhibition (Figure 16.20).

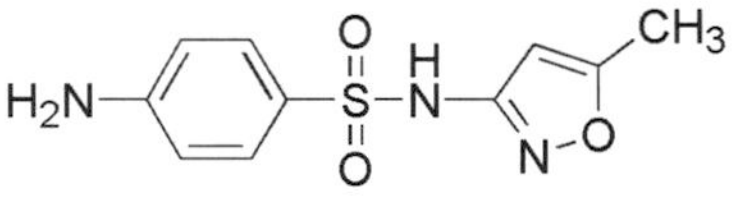

(**16.87**) Sulfamethoxazole

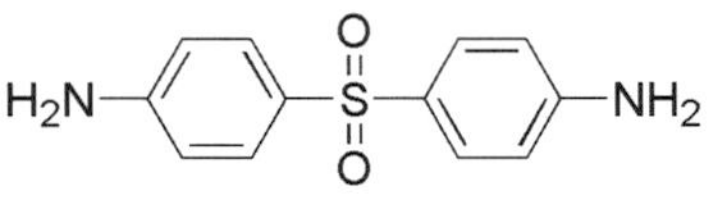

(**16.88**) Dapsone

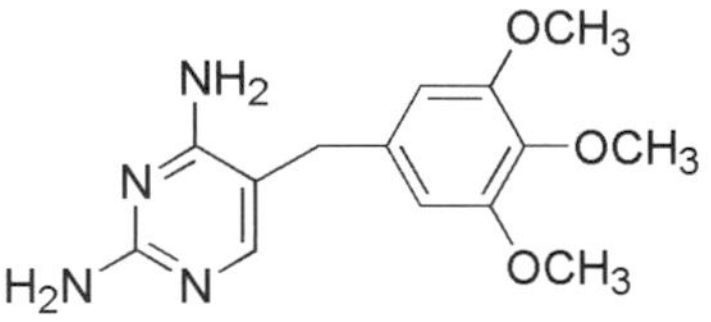

(**16.89**) Trimethoprim

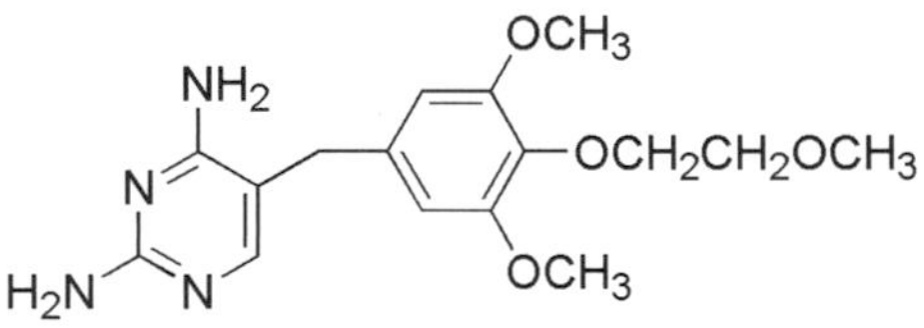

(**16.90**) Tetroxoprim

Sulfonamides, such as sulfamethoxazole (**16.87**), act by competitively inhibiting dihydropteroate synthetase (DHPS), an enzyme involved in the production of dihydropteroate from dihydropterin pyrophosphate and *p*-aminobenzoate (PAB). These drugs can become incorporated in an aberrant dihydropteroate or dihydrofolate. The sulfones, such as dapsone (**16.88**), have a similar mechanism of action to that of the sulfonamides. Dapsone is active against bacteria, protozoa, and mycobacteria, with exceptionally high activity observed against *M. leprae*. SAR studies of the sulfones have shown that the presence of the *p*-aminobenzenesulfonyl pharmacophore is essential for maintaining good activity, with small electron donating groups in the *para*-position tolerated in the second benzene ring. No major new antimicrobial sulfonamide has been introduced into clinical practice since 1970 and with the sulfones, dapsone is still the most frequently used in the clinic.

The diaminopyrimidines, such as trimethoprim (**16.89**), are inhibitors of dihydrofolate reductase (DHFR), the enzyme responsible for the conversion of dihydrofolate to tetrahydrofolate. The presence of three alkoxy substituents in the phenyl ring, as in trimethoprim, produces highly selective and potent antibacterial agents, which bind much less strongly to human DHFR than to the bacterial enzyme. One of the most active folate inhibitors is tetroxoprim (**16.90**), which contains a methoxyethoxy *para*-substitution.

The rationale for combining trimethoprim with a sulfonamide (as in co-trimoxazole) is based upon the *in vitro* assertion that the mixture has a markedly increased activity, a synergistic effect, in comparison with either drug alone, i.e., a ''sequential blockade'' in which a sulfonamide acts as a competitive inhibitor of DHPS and trimethoprim inhibits DHFR (Figure 16.20). Although such a combination reduces the risk of the emergence of resistance, the sequential blockade antibacterial mechanism has been questioned with results indicating that the sulfonamide, sulfamethoxazole (**16.87**) binds to purified *E. coli* DHFR.

Resistance and cross-resistance are common and are plasmid mediated. Changes in both DHPS and DHFR enzymes result in a decreased affinity for sulfonamides/sulfones and diaminopyrimidines, respectively.

16.2.6 Inhibitors of Sterol Biosynthesis

M. tuberculosis encodes for a large number of enzymes, most notable cytochrome P450 enzymes, normally associated with synthesis and metabolism of membrane lipids and sterols. Inhibition of sterol 14α-demethylase (P450 DM, CYP51) is already a successful strategy in antifungal therapy (see ''Azole antifungal agents'') and known antifungal CYP51 inhibitors econazole and clotrimazole have shown good inhibitory activity against *M. smegmatis*. Thus, inhibition of the P450 mycobacterial enzymes may provide a useful target for anti-TB agents. BM212 (**16.91**) is a pyrrole derivative with inhibitory activity against *M. tuberculosis* and *M. avium* as well as marked antifungal activity against *Candida albicans* and *Cryptococcus neoformans*. It is likely that BM212 inhibits sterol biosynthesis by inhibition of the P450 enzymes.

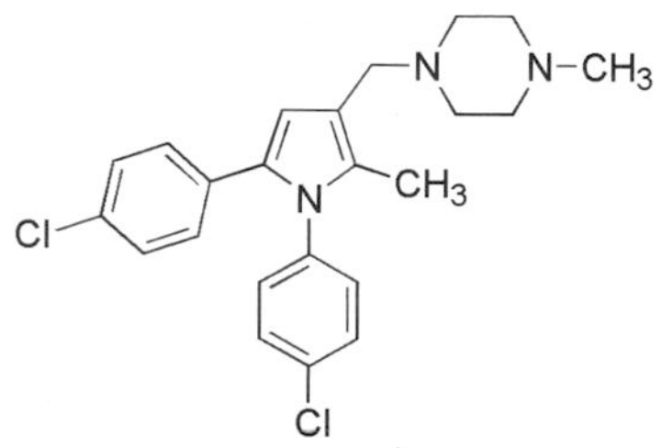

(**16.91**) BM212

16.2.7 Summary Figure

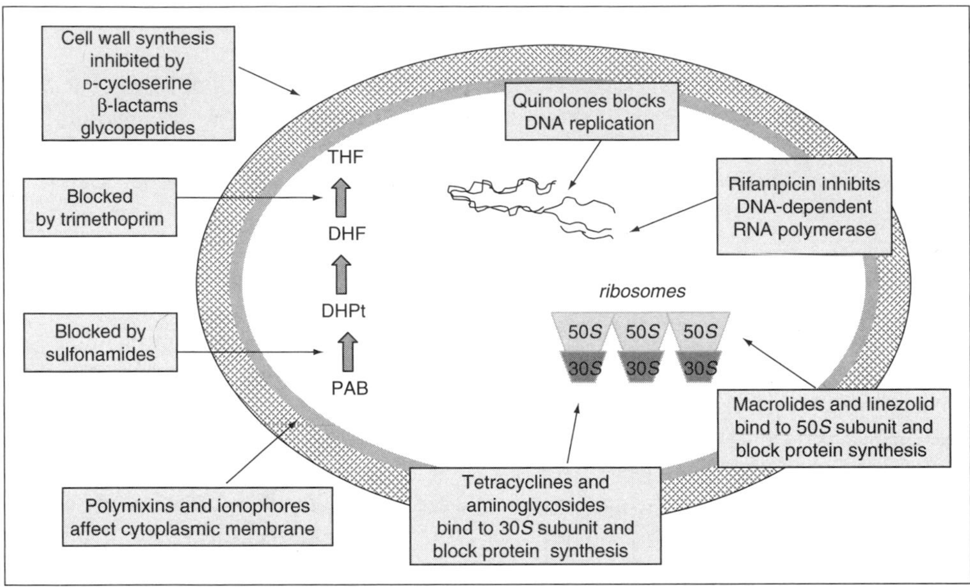

Figure 16.21 Summary diagram indicating the sites of action of antibacterial agents.

16.3 MECHANISM OF ACTION OF ANTIFUNGAL AGENTS

The design of antifungal agents poses different problems from those associated with antibacterial agents. Owing to the difference in structure and some biosynthetic processes between bacteria and fungi, antibacterial agents are generally ineffective against fungal infections. Both fungi and mammalian cells are eukaryotic, therefore selectivity between pathogen and host is also challenging. Fungal diseases were only recognized as important pathogens in the late 1980s with the significant increase in fungal infections, most notably candidiasis. The widespread use of antibacterial agents and immunosuppressive drugs combined with the advent of AIDS all contributed to the increase in fungal infections. Prior to the 1980s, the nephrotoxic amphotericin B was the only drug available for the treatment of serious fungal infections.

In the 1980s and 1990s a number of safer antifungal drugs, such as the azoles, were introduced into the clinic. However, the widespread use of the newer antifungal agents has been accompanied by increasing reports of resistance. The important drugs in clinical use are the macrolide polyenic antibiotics, imidazole derivatives, allylamines, flucytosine, and griseofulvin. The main targets of these antifungal drugs are either the cell wall, fungal sterol biosynthesis, or nucleic acid synthesis (Table 16.8).

The ideal antifungal agent should be fungicidal with a broad spectrum of activity, be suitable for oral and intravenous administration and possess good pharmacodynamic properties without development of resistance during therapy. At present none of the clinically used drugs satisfies all these criteria.

16.3.1 Inhibitors of Cell Wall Synthesis

The fungal cell wall is a complex structure (Figure 16.22) and contains a number of components that are unique to fungi including chitin, β-glucans, and mannoprotein, all of which have a structural role. A cell wall is absent in mammalian tissue, therefore the cell wall is an attractive

Table 16.8 Mechanism of action of antifungal agents

Effect	Examples	Comments
Inhibition of cell wall synthesis	Polyoxins, nikkomycins	Competitive inhibition of chitin synthetase
	Echinocandins	Inhibition of β-glucan synthetase
Inhibition of sterol biosynthesis	Polyenes	Bind to membrane sterols
	Azoles	Competitive inhibition of 14α-demethylase (CYP51)
	Allylamines, thiocarbamates, butenafine	Inhibition of squalene epoxidase
	Amorolfine	Inhibition Δ^{14}-reductase and $\Delta^{7,8}$-isomerase
Inhibition of protein synthesis	Sordarins	Bind to translational elongation factor 2 (EF2), stabilizing the complex formed by EF2 and the ribosome
Inhibition RNA/DNA synthesis	Griseofulvin	Inhibits fungal mitosis by interaction with polymerized microtubules
	Flucytosine	Replacement of uracil with 5-flurouracil in fungal RNA. Inhibits thymidylate synthetase and interferes with fungal DNA

target as these unique components provide the possibility of achieving fungal rather than host (mammalian) toxicity, i.e., a selective action. Antifungal agents have been developed to target two key enzymes, chitin synthetase and β-glucan synthase.

Chitin synthetase inhibitors

The biosynthesis of chitin, which is composed of linear chains of β-(1,4)-linked *N*-acetylglucosamine residues and constitutes approximately 1% of the cell wall of *C. albicans*, is catalyzed by chitin synthetase. The peptido-nucleosides, the polyoxins, and nikkomycins, are structural analogs of a cell-wall precursor, uridine diphosphate *N*-acetylglucosamine, and act as competitive inhibitors of chitin synthetase.

The polyoxins and nikkomycins (neopolyoxins (**16.92**)), such as polyoxin D (**16.93**) and nikkomycin Z (**16.94**), were isolated from *Streptomyces cacoi* and *S. tendae*, respectively, in the 1960s and 1970s. The nikkomycins are generally more active against whole cells, presumably due to better transport into the cell. Although chitin synthetase inhibitors have been extensively studied, none has reached the market. The discovery that many fungi produce multiple chitin synthetases of

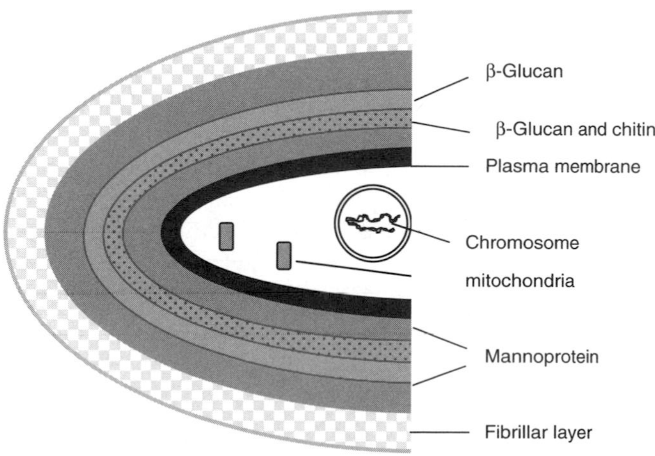

Figure 16.22 The fungal cell.

varying susceptibility to these compounds may account for the failure of chitin synthetase inhibitors as therapeutic agents.

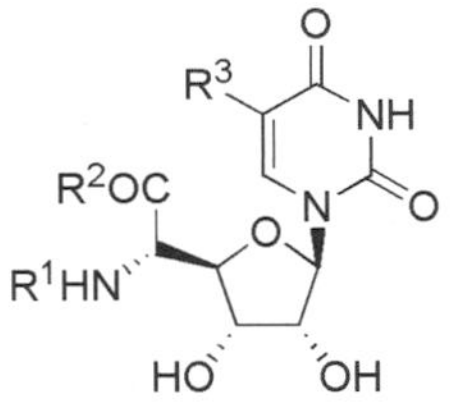

(**16.92**) neopolyoxins
R^1 = C-terminal amino acid
R^2 = N-terminal amino acid or OH
R^3 = H, CH_3, CH_2OH, CO_2H

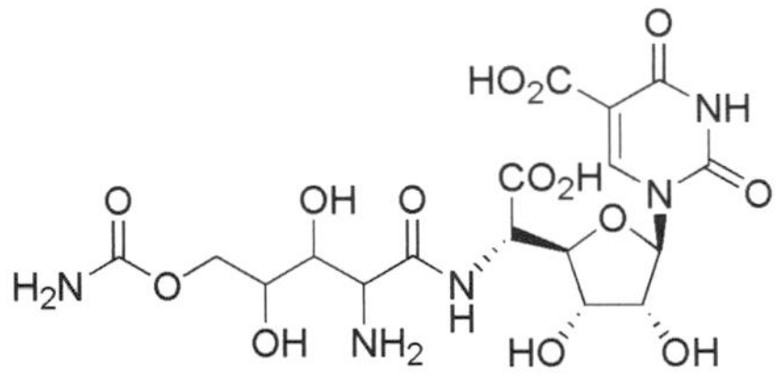

(**16.93**) Polyoxin D

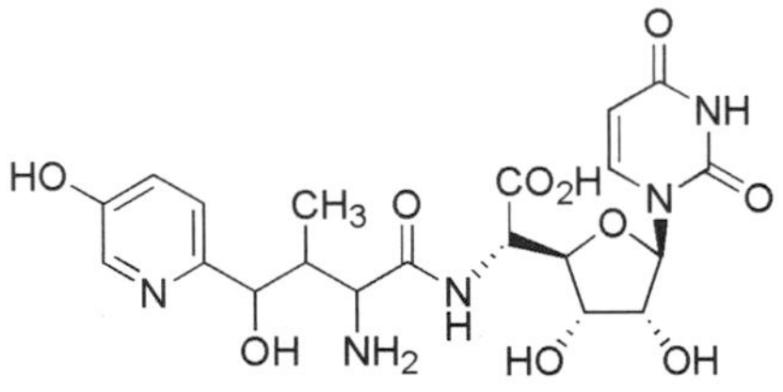

(**16.94**) Nikkomycin Z

β-Glucan synthase inhibitors

β-Glucan, which is composed of long chains of β-(1,3)-linked glucose residues with occasional side-chains involving β-(1,6)-linkages, and mannoprotein constitute 80% of the *C. albicans* cell wall. β-Glucan plays the dominant role in cell-wall architecture and is important in maintaining cell wall rigidity. The echinocandins, which contain the echinocandin B nucleus (**16.95**), are acetylated cyclic hexapeptides in which all the amino acid residues contain hydroxyl groups and all possess a lipophilic side-chain derived from linoleic acid. These compounds selectively inhibit β-1,3- and β-1,6-linked glucans and are therefore highly active against *Candida*, which contains predominantly β-linked glucans, but are inactive against *Cryptococcus*, which contains mainly α-1,3-glucans. Cilofungin (**16.96**), the parent agent, was developed in the 1970s, but because of its toxic effects was withdrawn during Phase II clinical trials. A structurally similar class of compounds based on the pneumocandin A0 nucleus has also been identified. Because of their similar mechanism of action, both structure classes are often referred to as echinocandins.

The echinocandins have two major drawbacks as potential clinical agents: (a) a narrow spectrum of activity and (b) poor solubility in pharmacologically acceptable solvents. Development of novel echinocandins has concentrated on modification of the lipid side-chain to resolve solubility and toxicity problems associated with the parent cilofungin.

Anidulafungin (LY303366) (**16.97**) is in Phase II clinical trials for intravenous use. This alkoxypolyaryl derivative has been shown to be more effective than cilofungin in experimental *C. albicans* infection models with a broader, although still rather limited, spectrum of activity.

Caspofungin (Cancidas™) (**16.98**) is a polypeptide antifungal related to pneumocandin B0 licensed for intravenous use. Caspofungin has fungicidal activity for *C. albicans* and is approved for invasive aspergillosis in patients refractory to, or intolerant of, other therapies. It has limited activity against other fungal species. Micafungin (FK463) (**16.99**) is in Phase II trials for intravenous use and is similar to caspofungin and anidulafungin in terms of susceptibility profile. It has a broad spectrum of activity against *Candida* spp., including azole-resistant *C. albicans*, is active against *Aspergillus*, and has moderate activity against some other species.

R = $-(CH_2)_7CH{=}CHCH_2CH{=}CH(CH_2)_4CH_3$; (**16.95**) Echinocandin B

R = $-C_6H_4-C_6H_4-O(CH_2)_7CH_3$; (**16.96**) Cilofungin

R = $-C_6H_4-C_6H_4-C_6H_4-O(CH_2)_4CH_3$; (**16.97**) Anidulafungin

(**16.98**) Caspofungin (Cancidas)

(**16.99**) Micafungin

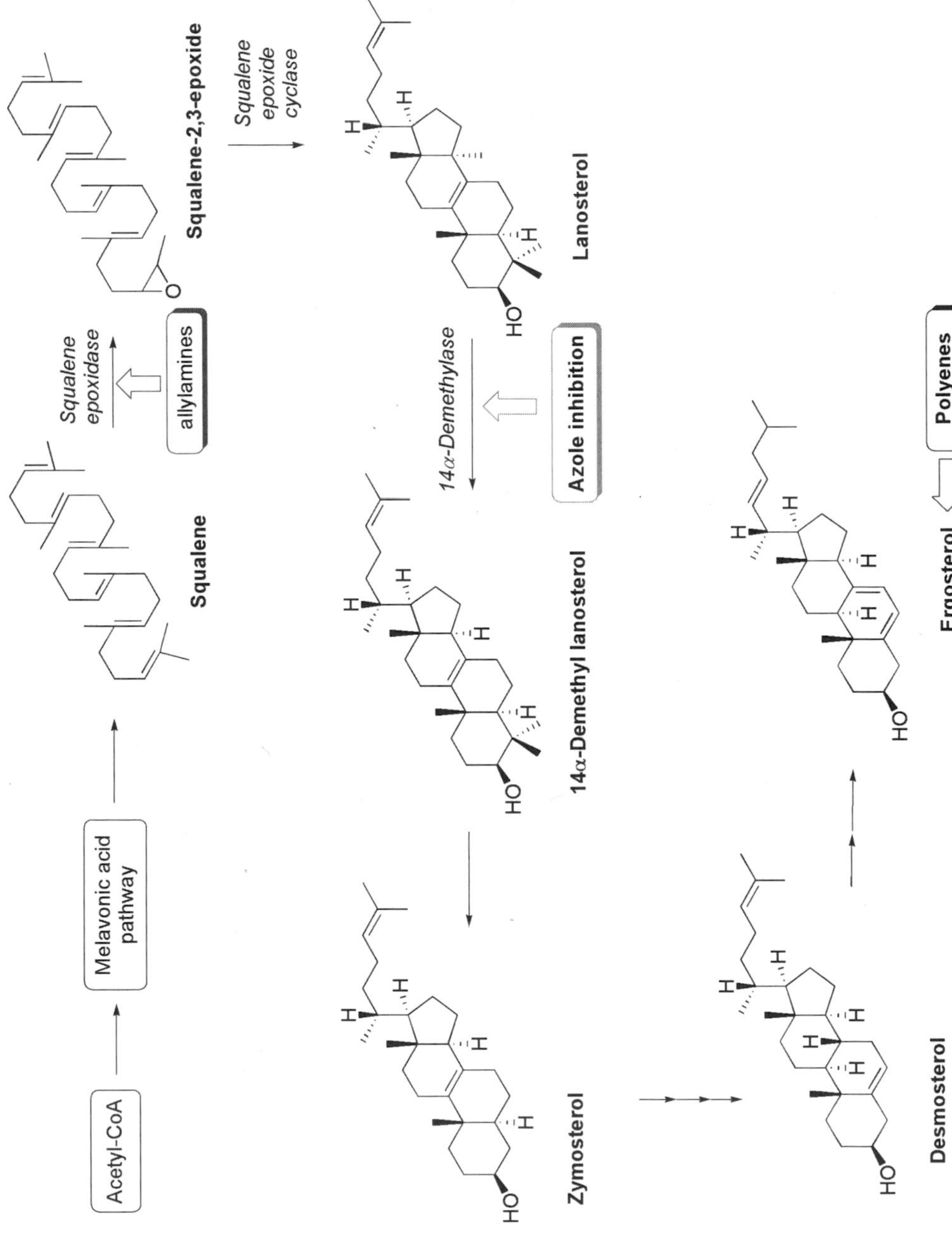

Figure 16.23 Ergosterol biosynthesis and antifungal targets.

As the echinocandins have only recently been introduced into the clinic, drug resistance is not an issue.

16.3.2 Inhibitors of Sterol Biosynthesis

Sterol biosynthesis is an essential metabolic pathway in animals, fungi, and plants (Figure 16.23). Importantly for drug development, differences between metabolic pathways and sterol products exist, allowing selective targeting. Ergosterol, which is absent from animals and higher plants, is the main component of the fungal cell membrane and is the target for the majority of clinically licensed antifungal agents namely polyenes, allylamine/thiocarbamates, and azoles.

Polyene antibiotics

The majority of polyene antibiotics are produced by *Streptomyces* species, with nystatin (Mycostatin™) the first to be isolated. The macrolide ring of the polyenes is larger than that of the other macrolide group and contains a series of conjugated double bonds. Generally, antifungal activity increases with the number of conjugated olefinic bonds, although solubility decreases from the tetraenes to the heptaenes.

Large polyenes, such as amphotericin B, interact with membrane sterol, resulting in the production of aqueous pores in which the polyene hydroxyl groups face inward. These aqueous pores result in altered permeability, leakage of vital cytoplasmic components, and death of the microorganism. Polyenes show activity against yeast and fungi but not bacteria and resistance to polyene antibiotics is rare. The therapeutic usefulness of polyenes is limited by their poor solubility and stability; however, the most pronounced limitation is their toxicity. Nephrotoxicity is the major adverse reaction and the majority of patients suffer from acute side effects including fever, myalgias, hypotension, and bronchospasm.

The low water solubility of amphotericin B has led to the introduction of four formulations: a colloidal suspension, which uses deoxycholate as the solubilizing agent, amphotericin B deoxycholate (Fungizone™), and three lipid preparations, amphotericin B colloidal dispersion (ABCD; Amphocil™ or Amphotec™), amphotericin B Lipid Complex (ABLC; Abelcet™), and liposomal amphotericin B (L-AMB; Ambisome™) (Table 16.9). The lipid formulations have reduced toxicity compared with the deoxycholate bile salt. Due to the improved therapeutic index of the lipid preparations of amphotericin B, liposomal nystatin (Nyotran™), containing dimyristoyl phosphatidyl choline and dimyristoyl phosphatidyl glycerol, was developed and is currently in late Phase III clinical trials (Table 16.9).

The reasons for the enhanced therapeutic index of the lipid-based formulations are not yet known. It has been hypothesized that, once the polyene drug is incorporated into the liposomes, it may participate in a selective transfer mechanism. This mechanism involves transfer of drug from the ''donor'' liposome to the ergosterol-containing ''target'' in the fungal cell membrane aided by either fungal or host phospholipases. Pimaricin, an ophthalmic suspension, is the only other polyene used clinically (Table 16.9).

Antifungal azoles

The major advance in the treatment of fungal infections was the introduction of the azole antifungals. The main target of azole antifungals is the cytochrome P450-dependent 14α-demethylation of lanosterol (Figure 16.23). Inhibition of sterol 14α-demethylase (CYP51) results in the depletion of ergosterol and accumulation of sterol precursors including lanosterol. Accumulation of the sterol precursors causes formation of a plasma membrane with altered structure and function. The azoles contain either an imidazole or triazole heterocyclic moiety that interacts with the haem of CYP51.

Table 16.9 Polyenes licensed/under development

Polyene	Delivery/application	
Fungizone ABCD; Amphocil/Amphotec ABLC; Abelcet L-AMB; Ambisome	IV, topical IV IV IV	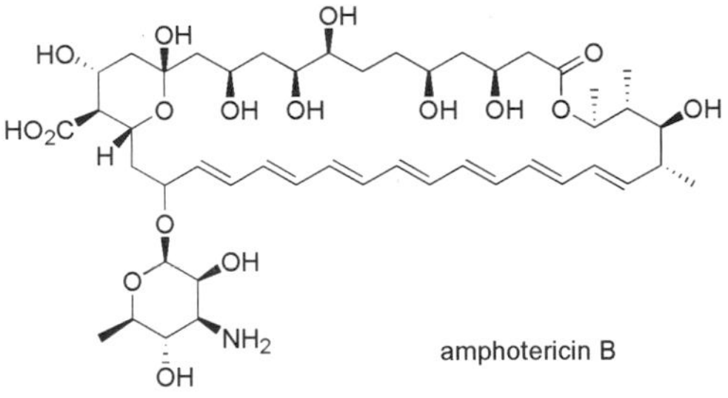 amphotericin B
Mycostatin Nyotran	Topical IV	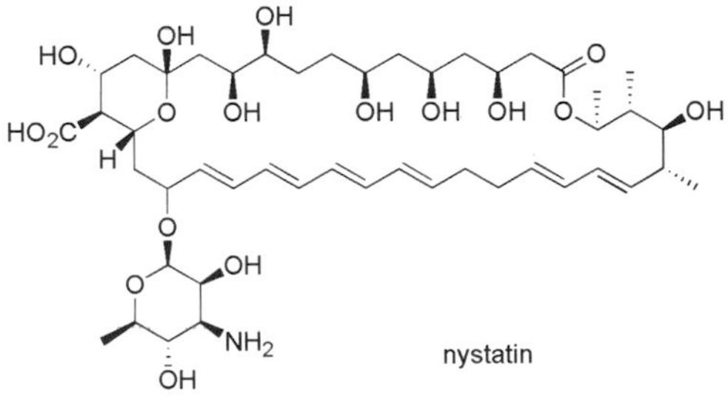 nystatin
Pimaricin	Ophthalmic	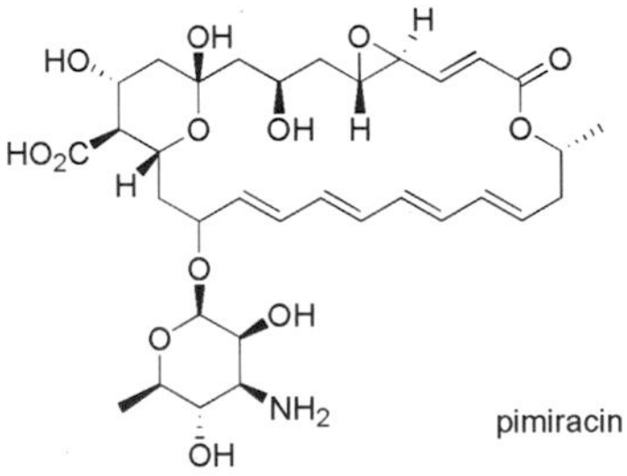 pimiracin

The first azoles identified as antifungal agents were the imidazoles clotrimazole (**16.100**), miconazole (**16.101**), and ketoconazole (**16.102**), all of which are fungistatic rather than fungicidal. Miconazole and ketoconazole provided the initial breakthrough in systemic antifungal agents. Ketoconazole (Nizoral™) is now the only imidazole azole used for the treatment of systemic infections. However, toxicity concerns, especially hepatotoxicity, and the lipophilic property of ketoconazole, which results in high drug concentrations in fatty tissues and poor distribution into cerebrospinal fluid, have resulted in less frequent use as a first-line therapy.

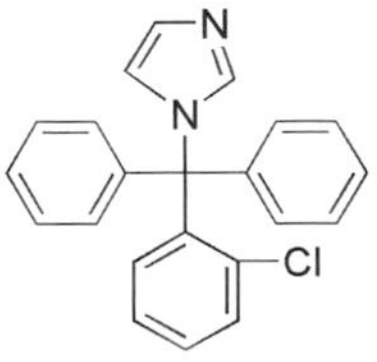

(**16.100**) Clotrimazole

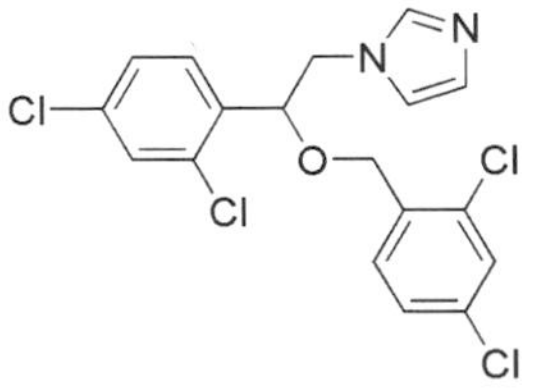

(**16.101**) Miconazole

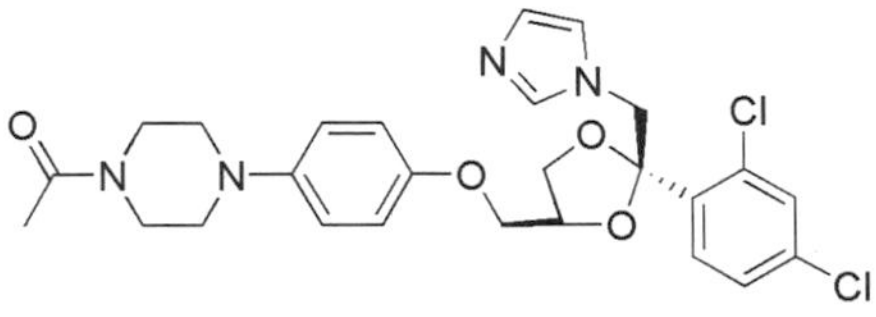

(**16.102**) Ketoconazole

The triazoles, fluconazole (Diflucan™) (**16.103**), and itraconazole (Sporanox™) (**16.104**), were found to have much improved therapeutic profiles. Fluconazole is fungistatic with activity against *Candida* and *Cryptococcus* spp., although some *Candida* species are resistant to fluconazole. Fluconazole has no activity against *Aspergillus*, unlike itraconazole (**16.104**), which was the first azole antifungal to be licensed for the treatment of aspergillosis. Itraconazole is a broader-spectrum and more potent antifungal agent than fluconazole, but the high lipophilicity of itraconazole requires the use of a cyclodextrin formulation for both oral and intravenous administration.

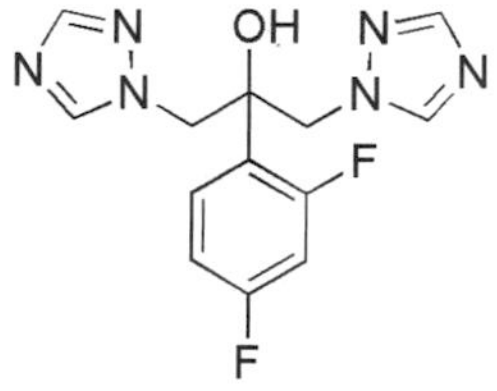

(**16.103**) Fluconazole

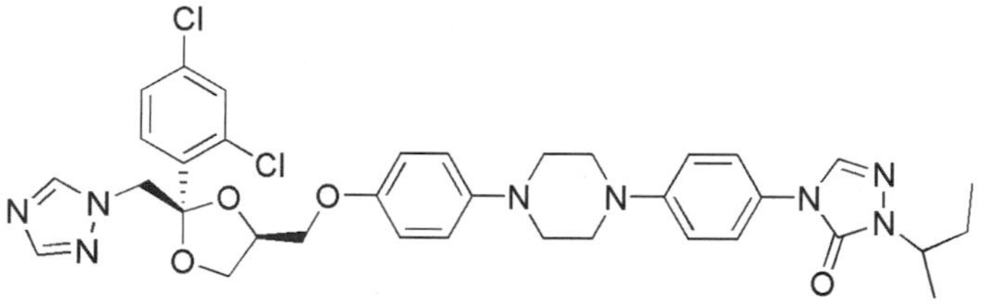

(**16.104**) Itraconazole

The success and widespread use of the azole antifungals has been accompanied by increasing issues of drug resistance (see azole antifungal resistance section). This has resulted in the need for broader-spectrum and potent azole antifungals with activity against fluconazole- and itraconazole-resistant fungal strains. Ideally new azoles should possess fungicidal activity.

Voriconazole (Vfend™) (**16.105**), approved by the FDA in 2002, is structurally related to fluconazole. Voriconazole is generally considered to be a fungistatic agent against *Candida* spp. and *C. neoformans* and may be fungicidal against *Aspergillus* spp. Voriconazole is more potent than fluconazole (**16.103**) and itraconazole (**16.104**) with a broad spectrum of activity including activity against fluconazole-resistant *Candida* spp. However, dose-related, transient visual disturbances, skin rash, and elevated hepatic enzyme levels have been reported which may limit administration in susceptible patients.

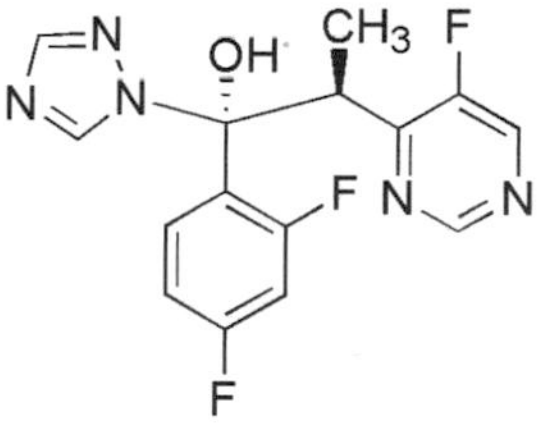

(**16.105**) Voriconazole

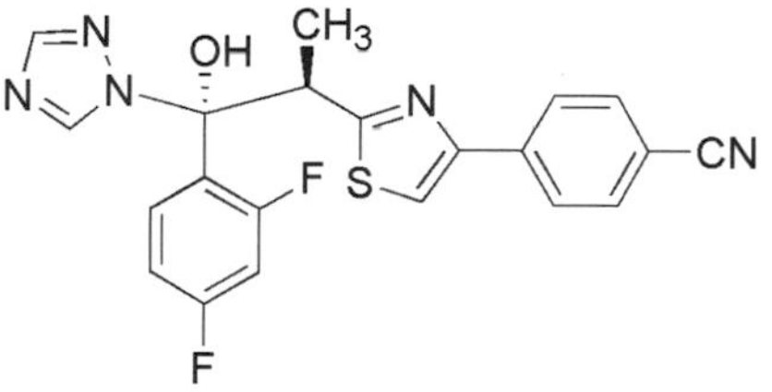

(**16.106**) Ravuconazole

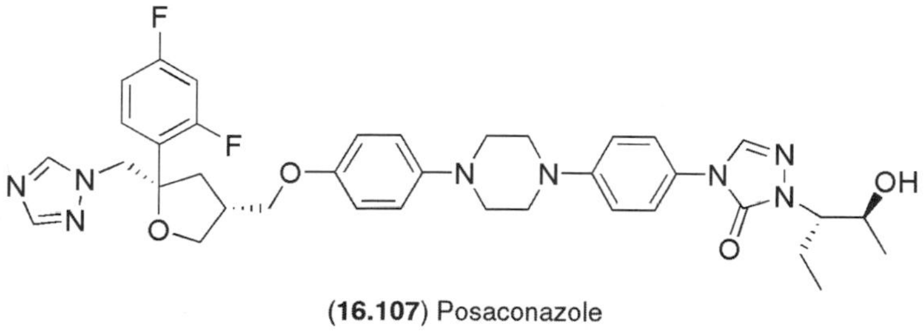

(**16.107**) Posaconazole

Ravuconazole (**16.106**), which is in Phase II clinical trials as an oral antifungal agent, is also structurally related to fluconazole. Ravuconazole is comparable with itraconazole (**16.104**) in its inhibition of CYP51 with a broad spectrum of activity, although activity is limited against some *Candida* spp. and cross-resistance with itraconazole and fluconazole is indicated. Posaconazole

(**16.107**) is structurally related to itraconazole and is currently in Phase III trials in oral and suspension forms. Posaconazole is a significantly more potent inhibitor of sterol C14 demethylation, particularly in *Aspergillus*. As with the newer triazoles, posaconazole displays a broad spectrum of activity and is fungistatic against *Candida* spp. with fungicidal activity noted against *neoformans* and *Aspergillus* species.

Azole antifungal resistance

There are several mechanisms of azole resistance including overexpression of 14α-demethylase, drug efflux, alteration in sterol biosynthesis, and active site mutation. Overexpression of 14α-demethylase causes increased ergosterol content accompanied by a decrease in susceptibility to azoles. The overexpression of the target enzyme also contributes to cross-resistance between fluconazole and itraconazole. An important mechanism of resistance to azoles is active drug efflux. Two efflux pumps have been found in fungi, viz. major facilitator superfamily (MFS) and ATPase efflux transporters. Overexpression of these efflux pumps results in reduced penetration across the fungal cell membrane and consequently reduced uptake and accumulation.

Alteration in sterol biosynthesis has been shown in *Candida* and *Saccharomyces* spp. where a change in sterol metabolism caused by mutation in sterol $\Delta^{5,6}$-desaturase results in accumulation of 14α-methylfecosterol rather than ergosterol 14-methyl-3,6-diol. Unlike ergosterol 14-methyl-3,6-diol, 14α-methylfecosterol allows continued cell growth, therefore the cell becomes resistant to the azole CYP51 inhibitors.

Active site mutation results in reduced drug binding. Investigations of CYP51 from *C. albicans* clinical isolates implicated a single amino acid substitution in the heme-binding domain of the active site of CYP51 and a similar point mutation, which alters susceptibility of the target enzyme, close to the active site in resistance development. Observed resistance/cross-resistance to the azole antifungals is generally a result of several resistance mechanisms.

Other sterol biosynthesis inhibitors

Allylamines, such as terbinafine (**16.108**), act as competitive inhibitors of squalene epoxidase, the enzyme involved in the conversion of squalene to squalene-2,3-epoxide in the early steps of ergosterol biosynthesis (Figure 16.23). Inhibition of this enzyme results in accumulation of squalene in the fungal cell membrane resulting in increased membrane permeability and disruption of cellular organization. Terbinafine (Lamisil™) (**16.108**) is the best-selling nonazole drug used in the treatment of skin, nail, and hair infections, and is more effective and considerably less toxic than griseofulvin (**16.115**) in the treatment of dermatophytosis. Recent reports indicate that terbinafine has a broader spectrum of activity, suggesting a possible use in azole-resistant oropharyngeal infections.

The benzylamine derivative butenafine (Mentax™) (**16.109**) also inhibits squalene epoxidase and has the same therapeutic indications as terbinafine. Importantly, butenafine displays anti-inflammatory activity as well as antifungal activity, which is a useful property in the treatment of dermatophytosis accompanied by an inflammatory reaction. Other inhibitors of squalene epoxidase include the thiocarbamate tolnaftate (**16.110**), also used in the treatment of dermatophyte infections.

The morpholines, such as amorolfine (Loceryl™) (**16.111**), which has fungicidal activity against dermatophytes, are characterized by a 2,6-dimethylmorpholine ring with a bulky *N*-substituent. The mechanism of action of the morpholines involves inhibition of two enzymes involved in sterol biosynthesis, Δ^{14}-reductase and $\Delta^{7,8}$-isomerase. Inhibition results in depletion of ergosterol and accumulation of ignosterol in the fungal cytoplasmic membrane, causing hyperfluidity of the membrane and deposition of chitin.

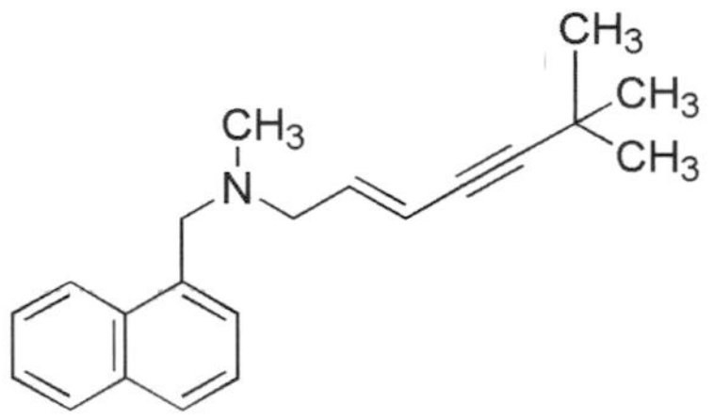

(**16.108**) Terbinafine

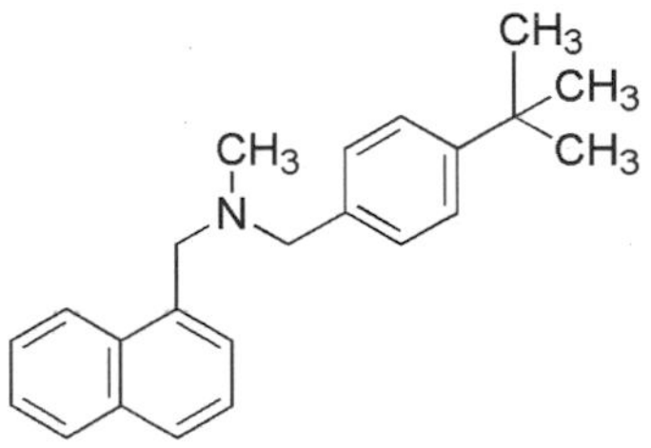

(**16.109**) Butenafine

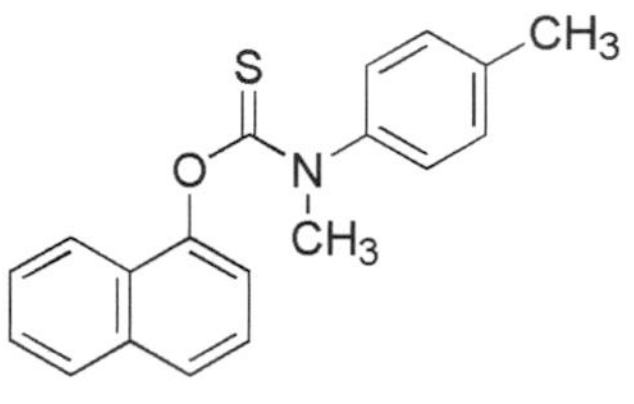

(**16.110**) Tolnaftate

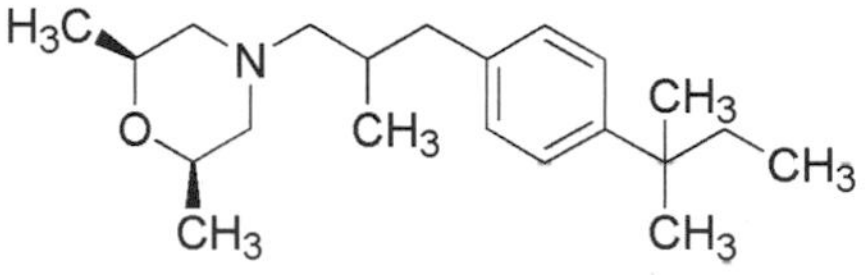

(**16.111**) Amorolfine

16.3.3 Inhibitors of Protein Synthesis

The sordarins, which are tetrahydropyran derivatives of the naturally occurring diterpene sordaricin (**16.112**), are a novel class of antifungal agents that inhibit fungal protein synthesis. The sordarins specifically target translation elongation factor 2 (EF2), binding to EF2, and stabilizing the complex formed by EF2 and the ribosome. EF2 is unique to fungi therefore sordarins are highly selective to fungal cells and demonstrate a broad spectrum of activity *in vivo*, including activity against azole-resistant *C. albicans* and *Pneumocystis carinii*.

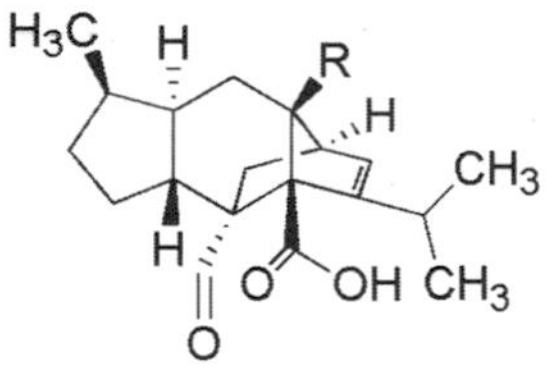

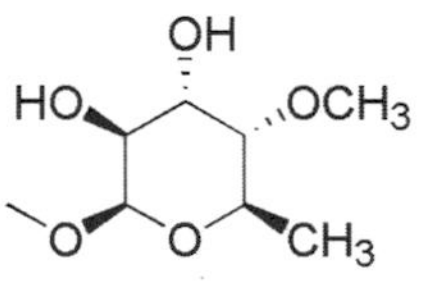

(**16.112**) Sordaricin

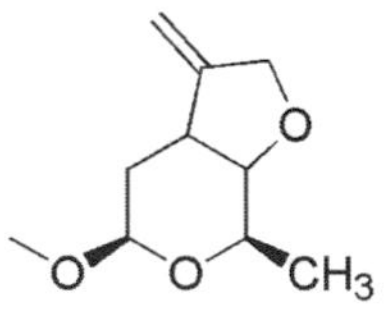

(**16.113**) GM 237354

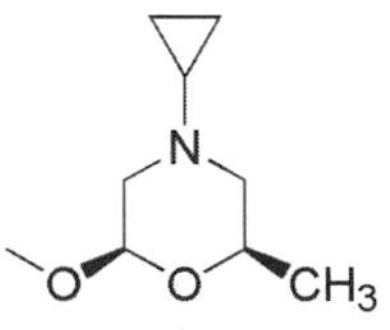

(**16.114**) GM 471552

The sordarins, of which the lead compound is GM 237354 (**16.113**), are currently undergoing evaluation in oral, subcutaneous, and intravenous formulation in animal models. Further drug development has produced the azasordarins, for example GM 471552 (**16.114**), which contain a morpholine ring in place of the tetrahydropyranyl ring of the sordarins, have also shown promise in *in vitro* and *in vivo* studies.

16.3.4 Inhibitors of Nucleic Acid Synthesis

Griseofulvin (**16.115**) was isolated from *Penicillium griseofulvum* in 1939 and was later discovered to exhibit potent fungistatic activity against dermatophytes. Griseofulvin inhibits fungal mitosis by disrupting the mitotic spindle through interaction with polymerized microtubules. By interfering with tubulin polymerization, griseofulvin stops mitosis at metaphase. The destruction of cytoplasmic microtubules interferes with the transport of secretory materials to the cell periphery, which may inhibit cell wall synthesis. Griseofulvin was the first-line drug for treatment of dermatophytosis. However, newer antifungal agents such as itraconazole (**16.104**) and terbinafine (**16.108**), which have reduced toxicity, enhanced efficacy, and shorter duration of therapy, are now the preferred antifungal agents for dermatophytosis.

(**16.115**) Griseofulvin

Flucytosine (Ancobon™) (**16.116**) is the only available antimetabolite drug having antifungal activity and is activated by deamination within fungal cells to 5-fluorouracil (5-FU). Flucytosine inhibits fungal protein synthesis by replacing uracil with 5-FU in fungal RNA and also inhibits thymidylate synthetase via 5-fluoro-2′-deoxy-uridine monophosphate (5-FdUMP) interfering with fungal DNA synthesis (Figure 16.24).

Flucytosine is selectively toxic to fungi as mammalian cells lack cytosine deaminase, although side effects, including bone marrow depression, may be observed. Flucytosine has a narrow spectrum of activity, but importantly it is active against *Cryptococcus* spp. and *C. albicans* and is used clinically in combination with amphotericin B (**16.98**) for the treatment of cryptococcal meningitis and systemic *Candida* infections.

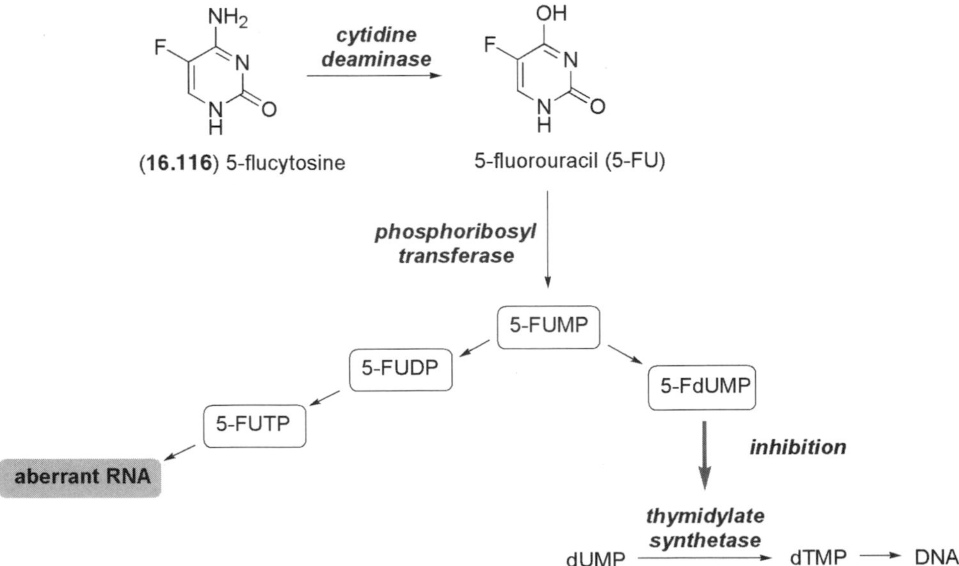

Figure 16.24 Mechanism of action of antimetabolite flucytosine.

16.3.5 Summary Figure

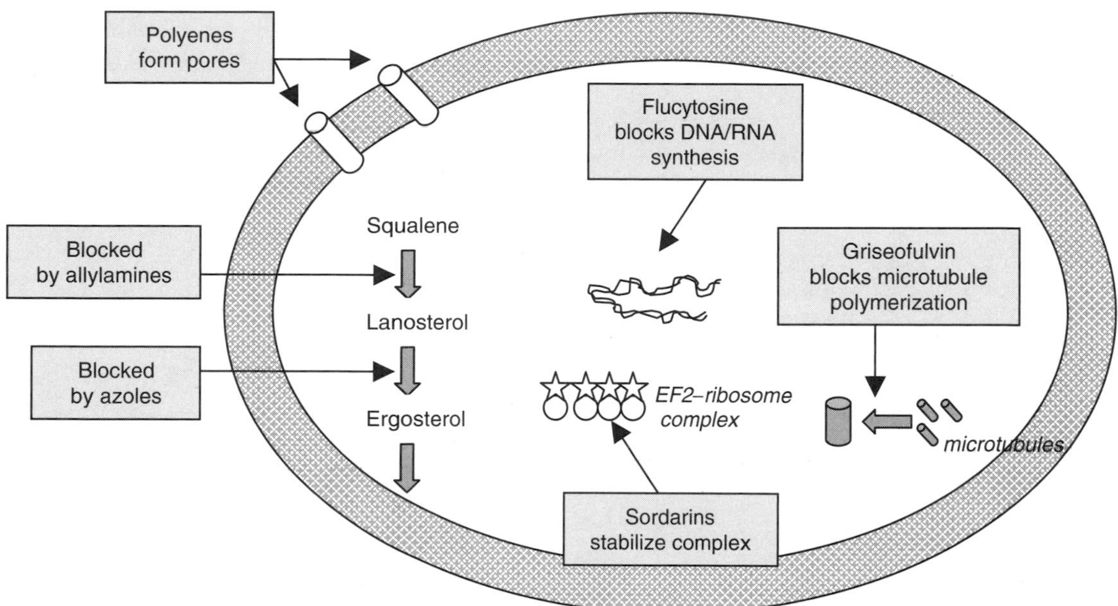

Figure 16.25 Summary diagram indicating the sites of action of antifungal agents.

16.4 ANTIVIRAL AGENTS

Viruses are classified into seven groups based on the nature of their genome/replication strategy (Table 16.10). However, regardless of the class category, all viruses use the host (human) cell machinery in order to replicate and hide their own DNA/RNA in the DNA of the host cell, so that when the host cell attempts to replicate it inadvertently produces more viruses.

Outside of host cells, viruses exist as a protein coat or capsid, sometimes enclosed within a membrane. The capsid encloses either DNA or RNA, which codes for the virus elements. Viruses cause a number of diseases in humans, including smallpox, the common cold, chickenpox, influenza, shingles, herpes, polio, rabies, ebola, hanta fever, and AIDS. Many forms of cancer have also been linked to viruses, such as liver cancer (hepatitis B and C virus), Burketts lymphoma (Epstein–Barr virus), and cervical and skin cancer (human papillomavirus).

As viral replication is linked with host cell replication, the major difficulty in designing an antiviral drug is selective toxicity for the virus over the host. However, very often the more selective an antiviral agent the more prone it is to the development of viral resistance. Development

Table 16.10 Viral classification

Class	Genome	Virus
I	Double-stranded DNA	Adenoviruses, herpes viruses, poxviruses
II	Single-stranded (+)sense DNA	Parvoviruses
III	Double-stranded RNA	Reoviruses, birnaviruses
IV	Single-stranded (+)sense RNA	Picornaviruses; togaviruses
V	Single-stranded (−)sense RNA	Orthomyxoviruses, rhabdoviruses
VI	Single-stranded (+)sense RNA with DNA intermediate	Retroviruses
VII	Double-stranded DNA with RNA intermediate	Hepadnaviruses

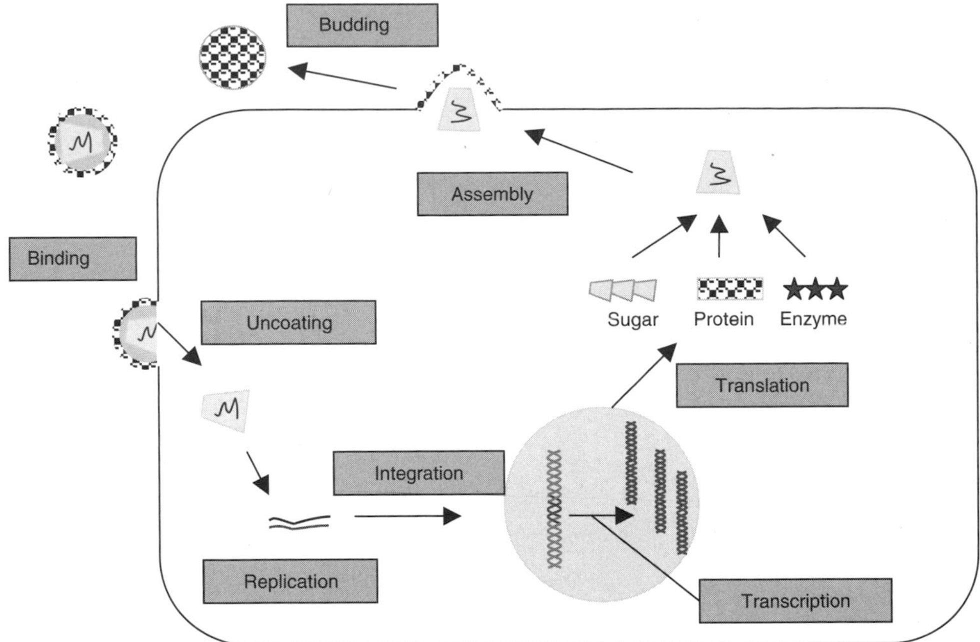

Figure 16.26 General viral life cycle indicating possible drug targets.

of viral resistant strains is rapid against certain drug classes requiring the use of multidrug therapies (with up to three different drug classes), a therapeutic strategy termed highly active antiretroviral therapy (HAART) in the treatment of human immunodeficiency virus (HIV) infection, to delay progression of viral resistance and disease state.

The viral life cycle provides numerous targets for therapeutic agents including virus–cell fusion, virus uncoating, transcription, integration, translation (addition of proteins, sugars, etc.), and budding from the cell (Figure 16.26).

The major considerations in the development of antiviral agents are good selectivity, negligible toxicity, and low potential for the development of drug resistance. There is an abundance of literature describing antiviral compounds with potential as therapeutic agents. In the following sections, however, only licensed antiviral drugs and those in Phase II/III clinical trials will be discussed.

16.4.1 Capsid Binders

Capsid binders such as disoxaril (**16.117**) and pleconaril (**16.118**) target the three-dimensional structure of picornaviruses, such as the cold virus rhinovirus. Both disoxaril and pleconaril contain a 3-methylisoxazole group that fits into a hydrophobic pocket (capsid VP1) of the outer protein coat. Once bound, structural changes required for uncoating are blocked and the virus is unable to enter its life cycle. Pirodavir (**16.119**), which is in Phase II clinical trials for the treatment of rhinovirus infection, contains a 3-methyldiazine heterocyclic ring that would be expected to bind in a similar manner as the 3-methylisoxazole ring.

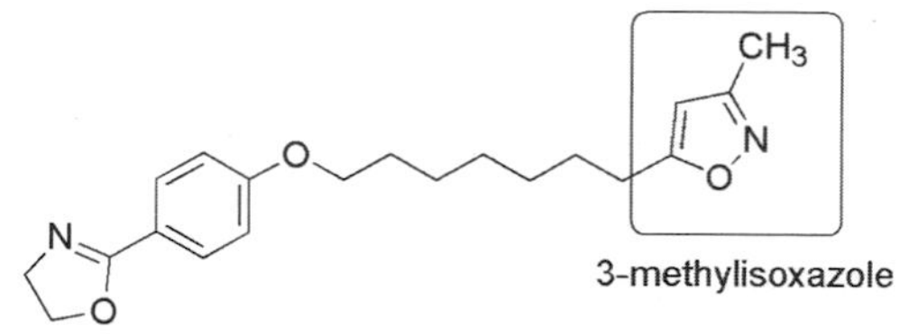

3-methylisoxazole

(**16.117**) Disoxaril

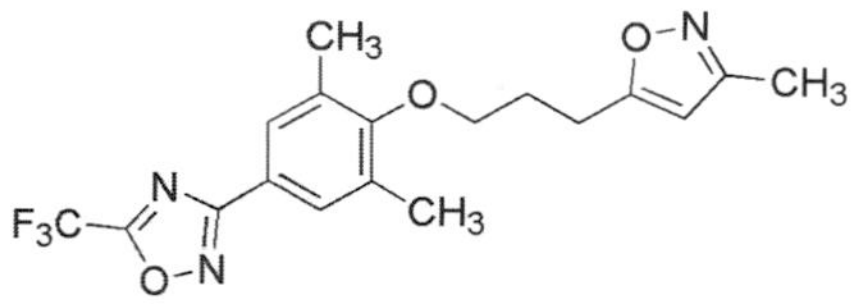

(**16.118**) Pleconaril

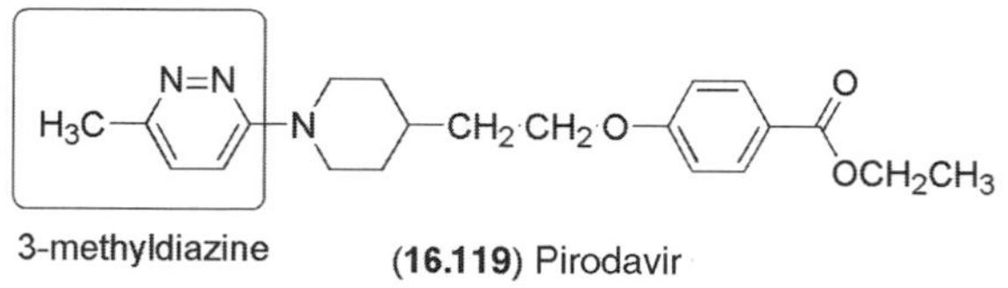

3-methyldiazine

(**16.119**) Pirodavir

16.4.2 Receptor and Fusion-Binding Inhibitors

Receptor binding and fusion have proved valuable targets for the development of drugs for the treatment of influenza virus and HIV infections. In the treatment of influenza virus, neuraminidase and M2 inhibitors are licensed for clinical use. Fusion inhibitors have recently been introduced for the treatment of HIV, with coreceptor (CCR5 and CXCR4) inhibitors currently under development (preclinical and Phase I clinical trials).

Neuraminidase inhibitors

Influenza virus binds via hemagglutinin residues to glycoprotein host cell residues. Once bound, the virus infects the host cell and replicates. Release of new virus requires cleavage from the host, which is achieved by the action of neuraminidase. Neuraminidase catalyzes the cleavage of terminal sialic acid residues on host glycoprotein and glycolipid surface residues allowing release of the new virus from the cell. Neuraminidase plays a pivotal role in the spread of virus to new cells and is also involved in the introduction of apoptosis to the infected cell.

Zanamivir (Relenza®) (**16.120**), a sialic acid derivative, was the first neuraminidase inhibitor licensed for clinical use. Zaminivir is active against influenza A virus only and is delivered by inhalation due to poor oral bioavailability. The more recent inhibitor oseltamivir phosphate (Tamiflu®) (**16.121**) is active against both influenza A and B with good oral bioavailability. Oseltamivir phosphate is an ethyl ester prodrug, which undergoes rapid hepatic conversion to the active form (**16.122**).

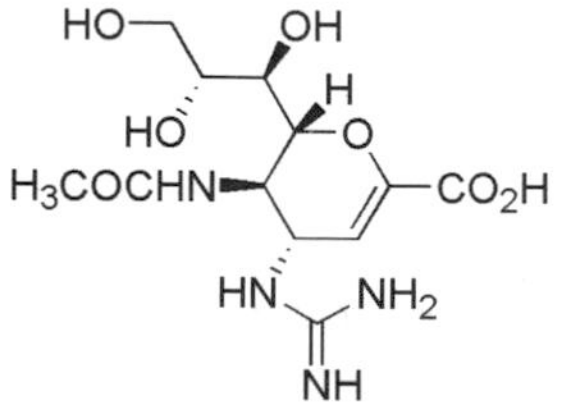

ester hydrolysis

(**16.120**) Zanamivir

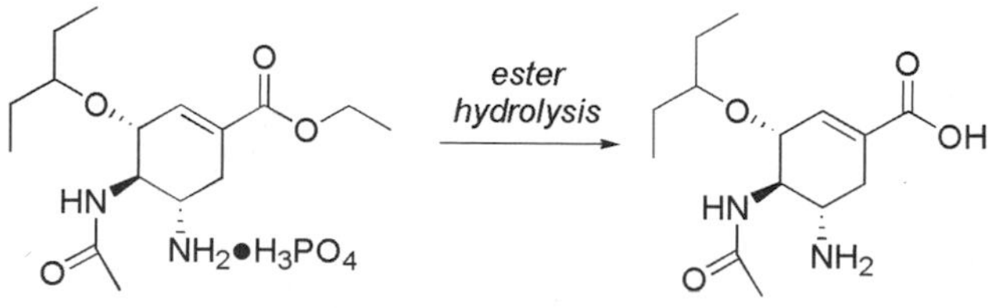

(**16.121**) Oseltamivir phosphate

(**16.122**)

M2 inhibitors

Amantadine (Symmetrel®) (**16.123**) and rimantadine (Flumadine®) (**16.124**) are chemically related adamantane derivatives active against influenza A virus. After influenza A viruses enter cells, these drugs inhibit their uncoating by affecting the ion-channel activity of the viral M2 protein, thereby inhibiting viral replication. The exact mechanism of action is not known, but it has been suggested that amantadine and its derivative rimantadine may interact with residues either outside or inside the channel and alter the conformation of the channel preventing normal functioning of the M2 protein. Resistance can be rapid and arises from mutations in the RNA sequence coding for the structural M2 protein.

(**16.123**) Amantadine (**16.124**) Rimantadine

Fusion inhibitors

Fusion inhibitors are the most recent addition to combination therapy of HIV, with the peptide enfuvirtide (Fuzeon™) (**16.125**) approved for clinical use by the FDA in 2002 (USA) and 2003 (Europe). The outer lipid bilayer of HIV-1 contains gp120 surface glycoproteins and gp41 transmembrane glycoproteins, which are involved with binding of the virus to the host cell, and which are derived from gp160. The main target of HIV-1 is T4-lymphocyte or ''T-helper cells,'' which contain CD4 receptors with which the virus can interact.

The initial stage in virus–host cell fusion is interaction of gp120 with the CD4 host receptor. Once attached to CD4, gp120 binds to the chemokine receptors (CCR5 or CXCR4) on the host cell surface. A conformation change in each gp120 unit then unfolds the gp120 complex, allowing the gp41 units to bind to the host cell membrane. The final stage is gp41-mediated membrane fusion and entry (Figure 16.27). The binding/fusion process provides a number of stages for inhibition, including CD4–gp120 receptor binding, chemokine receptor binding, and gp41-mediated fusion. Fuzeon (**16.125**) is a 36-amino acid synthetic peptide (Ac-YTSLIHSLIEESQNQQEKNEQEL-LELDKWASLWNWF-NH_2) representing amino acid sequence 642–678 of gp160, which interferes with gp41-mediated fusion. It has been suggested that Fuzeon disrupts the formation of a

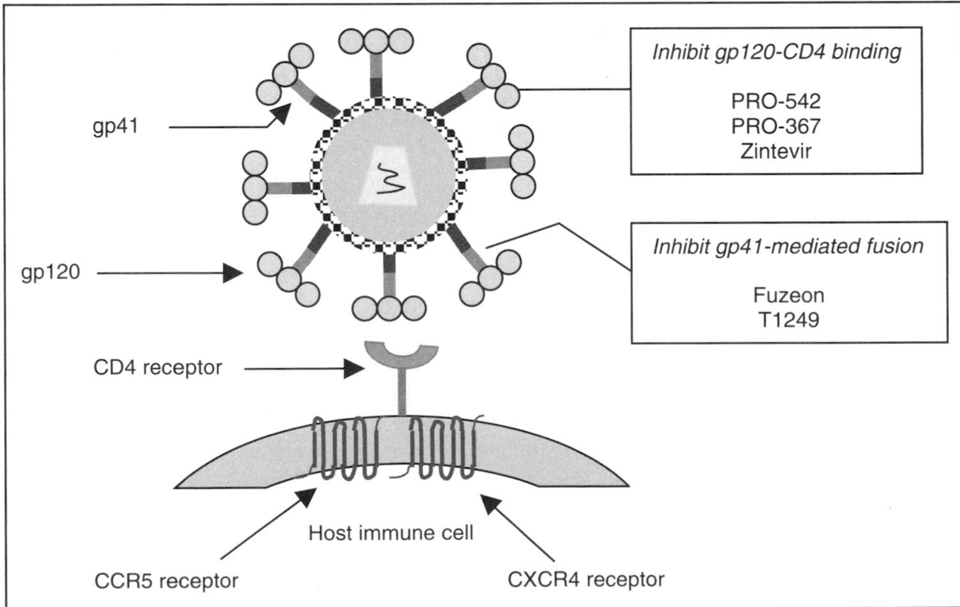

Figure 16.27 Surface and membrane glycoproteins and receptors involved in viral binding and fusion.

stable gp41 configuration, which is required for membrane fusion. T1249 (**16.126**) is a 39-amino acid peptide in Phase I/II clinical trials, which binds to a different region of the gp160 receptor.

PRO-542 (**16.127**), which consists of a fusion protein of the HIV-binding region of CD4 incorporated into gamma globulin, and a similar drug PRO-367, neutralize HIV by binding to gp120, thereby preventing interaction with host receptors. Both drugs are currently in Phase II clinical trials (Figure 16.27). Zintevir (**16.128**), a 17mer phosphothiorate oligonucleotide, was originally designed as an integrase inhibitor, in clinical studies (currently Phase II/III), however, it has been shown to act as an inhibitor of virus–cell binding by preventing gp120–CD4 interaction. *n*-Docosanol (Abreva®) (**16.129**) ($CH_3(CH_2)_{20}CH_2OH$), a fusion inhibitor approved for the topical treatment of recurrent herpes simplex virus infection, has also been shown to inhibit HIV fusion, although the mechanism of action is unclear and no further advance in clinical studies has been reported.

16.4.3 Inhibitors of Replication

The major targets of replication are the enzymes DNA polymerase and reverse transcriptase (RT). DNA-polymerase inhibitors are used for the treatment of herpes simplex virus and hepatitis B virus infections, whereas RT inhibitors are important drugs in anti-HIV therapy. The majority of inhibitors for both enzymes are nucleoside derivatives, that is, derivatives of the natural substrates. Non-nucleoside RT inhibitors (NNRTIs) are a newer class of anti-HIV drugs which are used in combination with an NRTI and a protease inhibitor (see Section 16.4.4) in HAART.

DNA-polymerase inhibitors

Acyclic nucleosides, which differ from conventional nucleosides in that the sugar ring is replaced by an acyclic moiety, are used widely as DNA-polymerase inhibitors. The first acyclic nucleoside to show selective inhibition of herpes simplex virus replication was acyclovir (Zovirax®, ACV) (**16.130**), a guanosine-based nucleoside whose clinical effectiveness stimulated considerable interest in acyclic nucleosides. Acyclovir is used in the treatment of mucosal, cutaneous, and systemic herpes simplex virus-1 and herpes simplex virus-2 infections and in the first-line therapy of varicella zoster virus, which causes chickenpox and shingles. Triphosphate-derived acyclovir, ACV-TP, is the active form, which targets viral DNA polymerase (Figure 16.28).

The poor bioavailability of ACV and the emergence of ACV-resistant strains of herpes simplex and varicella zoster virus has led to the development of drugs with improved efficacy. These are ganciclovir (Cymevene®) (**16.131**), penciclovir (Vectavir®) (**16.132**), and the prodrugs famciclovir (Famvir®) (**16.133**) and valacyclovir (Valtrex®) (**16.134**), which demonstrate improved oral bioavailability, and more recently the acyclic nucleoside phosphonates, such as the *bis*(POM)-PMEA prodrug (Adefovir dipivoxil®) (**16.135**) and Cidofovir (Vistide®) (**16.136**) which are active against hepatitis B virus and ACV-resistant virus strains, respectively (Table 16.11)

Ganciclovir (GCV) (**16.131**) is active against all the herpes viruses and Epstein–Barr virus, targeting DNA polymerase as described for ACV. However, GCV can be distinguished from other guanosine acyclic analogs in that it demonstrates pronounced activity against human cytomegalovirus, and as such it is used for the treatment of human cytomegalovirus retinitis. Penciclovir (PCV) (**16.132**) is the carboacyclic equivalent of ganciclovir, with the ether oxygen replaced by a methylene unit. The antiviral activity of PCV is similar to ACV, as its mode of action. The improved potency of PCV, compared with ACV, is attributed to the increased half-life of the major metabolite (*S*)-PCV-TP. The poor oral bioavailability of PCV led to the design of the orally available prodrug famciclovir (FCV) (**16.133**).

The mode of action of famciclovir involves initial hydrolysis of the 4′-acetyl group with subsequent removal of the remaining ester group in the liver to give 6-deoxy-PCV. The 6-deoxy-PCV is then oxidized in the liver by the molybdenum hydroxylase, aldehyde oxidase, to give PCV (Figure 16.29).

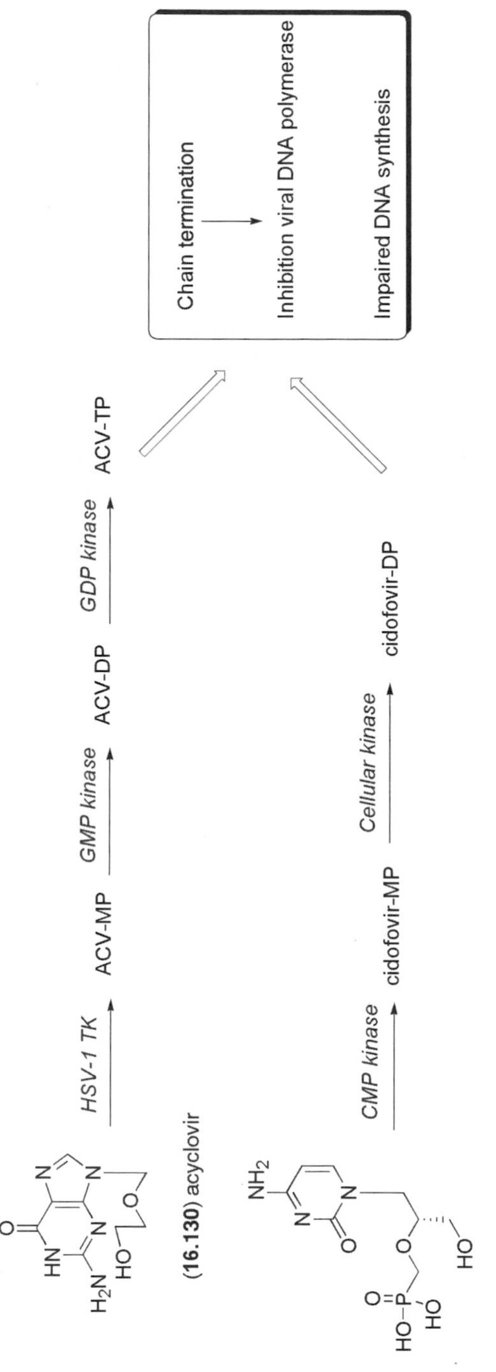

Figure 16.28 Mechanism of inhibition of DNA polymerase. HSV-1, herpes simplex virus type 1; TK, thymidine kinase; GMP, guanosine monophosphate; GDP, guanosine diphosphate; CMP, cytidine monophosphate.

Table 16.11 Acyclic nucleoside DNA-polymerase inhibitors licensed for use and in Phase II/III development

Inhibitors	Chemical structure	Therapeutic indication
Acyclovir (Zovirax®) **16.130**	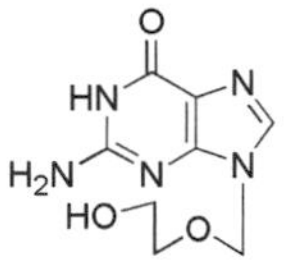	HSV-1, HSV-2, and VZV
Ganciclovir (Cymevene®) **16.131**	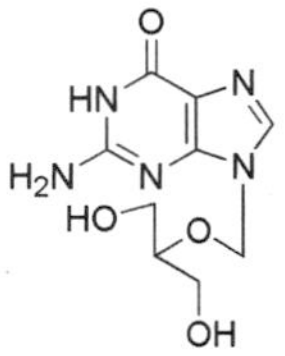	HCMV retinitis
Penciclovir (Vectavir®) **16.132**	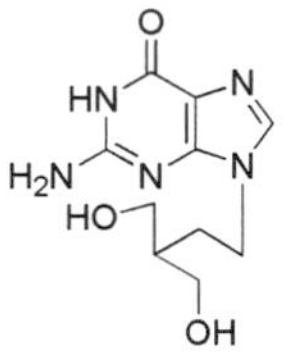	Topical HSV
Famciclovir (Famvir®) **16.133**	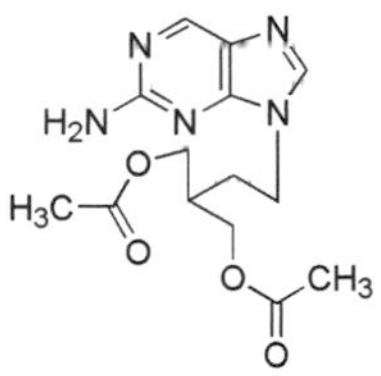	VZV, HSV-1, HSV-2. Phase III trials forHBV
Valacyclovir (Valtrex®) **16.134**	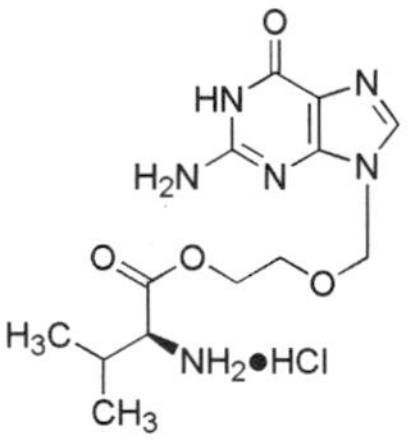	HSV-1, HSV-2, VZV
Valomaciclovir stearate (RP-606) **16.137**	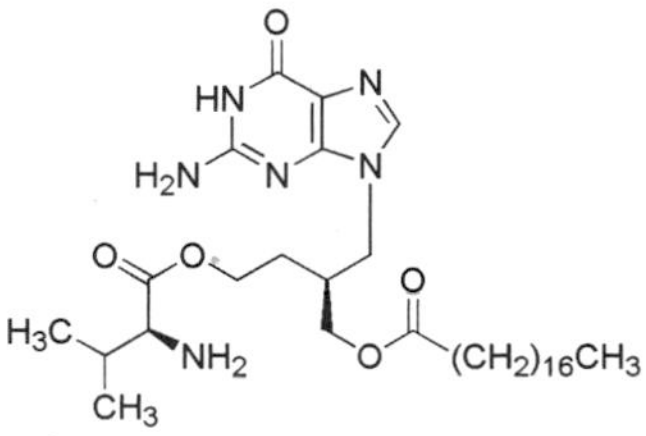	Phase II for VZV completed

Table 16.11 Acyclic nucleoside DNA-polymerase inhibitors licensed for use and in Phase II/III development — Continued

Inhibitors	Chemical structure	Therapeutic indication
bis (POM)-PMEA (Adefovir DIPIVOXIL®) **16.135**	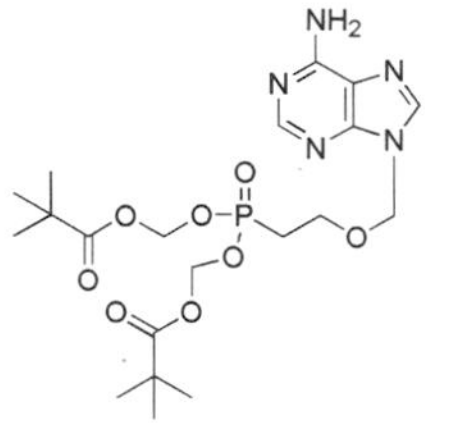	Phase III trials for HBV
Cidofovir (Vistide®) **16.138**	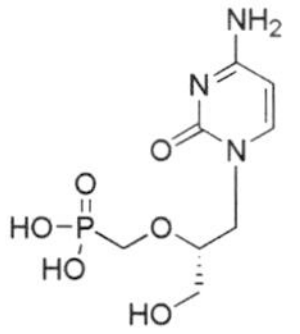	HCMV retinitis, topical HSV-1

Abbreviations: herpes simplex virus (HSV), type 1 (HSV-1) and type 2 (HSV-2); varicella zoster virus (VZV); human cytomegalovirus (HCMV); hepatitis B virus (HBV).

The most promising prodrug of ACV is the L-valyl ester valacyclovir (Valtrex®) (**16.134**), an oral prodrug which undergoes rapid first-pass metabolism to produce the parent nucleoside, ACV, and L-valine. Valacyclovir administration is particularly beneficial where high concentrations of ACV are required, as in the treatment of the less sensitive varicella zoster virus. Improved uptake has been achieved with valomaciclovir stearate (RP-606) (**16.137**), currently undergoing clinical trials (Table 16.11).

The acyclic nucleoside phosphonates are stable mimics of acyclic nucleosides, which display both a broad spectrum of antiviral activity and prolonged antiviral action. The unique feature of the acyclic nucleoside phosphonates is their ability to bypass the first rate-limiting phosphorylation step, which is normally carried out by virus-encoded kinases (Figure 16.28).

The diphosphorylated acyclic nucleoside phosphonate (a triphosphate mimic) has a long intracellular half-life and selectively targets the viral DNA polymerase. However, other enzymes may be inhibited by these compounds, such as herpes simplex virus type 1 encoded ribonucleotide reductase, and certain compounds also target RT and act as chain terminators. The acyclic nucleoside phosphonates suffer from poor oral bioavailability and also have a number of toxicity implications (e.g., hepatic toxicity, neutropenia, and nephrotoxicity). To improve both the bioavailability and safety profile of the acyclic nucleoside phosphonates, a number of prodrugs have been developed, most notably, bis(POM)-PMEA (Adefovir dipivoxil®, **16.135**), which is in Phase III trials for the treatment of hepatitis B virus infections.

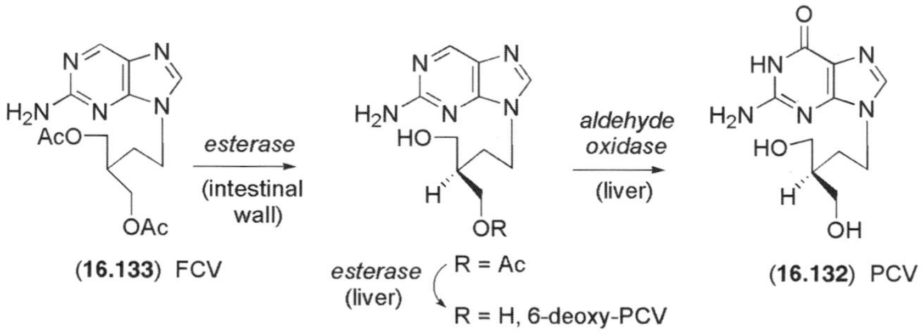

(**16.133**) FCV — *esterase* (intestinal wall) → R = Ac; *esterase* (liver) → R = H, 6-deoxy-PCV — *aldehyde oxidase* (liver) → (**16.132**) PCV

Figure 16.29 Metabolism of the prodrug FCV.

The more conventional nucleoside compounds, vidarabine (Vira-A®) (**16.138**), idoxuridine (Herpid®) (**16.139**), and trifluridine (Viroptic®) (**16.140**), were the drugs of choice before the introduction of the acyclic nucleoside antiviral agents (Table 16.12). Vidarabine (**16.138**), which is composed of the purine base adenine coupled to an arabinofuranosyl sugar moiety, is phosphorylated to its active triphosphate form by cellular kinases rather than viral thymidine kinase (TK), i.e., activation is independent of TK, so that vidarabine is active against TK-deficient virus. Vidarabine acts as a competitive inhibitor of herpes simplex virus DNA polymerase and also inhibits ribonucleotide reductase, and the overall effect is inhibition of DNA synthesis. Clinically, vidarabine is used for the treatment of herpes simplex virus keratinitis and has now been superseded by acyclovir (**16.130**) in the treatment of herpes simplex virus encephalitis. Idoxuridine (**16.139**) and trifluridine (**16.140**) are 5-substituted derivatives of 2′-deoxyuridine. The iodo- and trifluromethyl-substituents are isosteres of methyl, i.e., both compounds are mimics of the pyrimidine nucleoside thymidine. Idoxuridine and trifluridine are competitive inhibitors of herpes simplex virus DNA polymerase used as topical ophthalmic preparations (Table 16.12).

A more recent 5-substitued 2′-deoxyuridine nucleoside, brivudin (Helpin, Zostrex®) (**16.141**) has good oral bioavailability and displays slightly enhanced anti-herpes simplex virus type 1 activity compared with acyclovir. Brivudin is considerably more active than acyclovir against varicella zoster virus and is licensed for varicella zoster virus therapy.

There is no recommended therapy for hepatitis B; the two most commonly used treatments are alpha interferon (IFN) and lamivudine (Epivir®) (**16.142**) (Table 16.12). Lamivudine, an analog of L-dideoxycytidine, is a DNA-polymerase inhibitor used in patients with chronic hepatitis B virus infection and also in HIV therapy. Emtricitabine (Coviracil®) (**16.143**) is the 5-fluoro-derivative of lamivudine, with a similar mechanism of action, currently in Phase II trials for hepatitis B virus and Phase III for HIV therapy. A novel carbocyclic nucleoside, entecavir (**16.144**), is in Phase II trials; importantly entecavir has been shown to be effective in patients who have developed resistance to lamivudine.

Resistance to DNA-polymerase inhibitors

Three main mechanisms have been identified for the acyclic antiviral nucleosides: (i) altered TK substrate specificity, (ii) absence of thymidine kinase (TK-deficient strains), and (iii) point mutations resulting in altered viral DNA polymerase. TK deficiency is the main contributor to acyclovir, penciclovir, famciclovir, and valacyclovir resistance, whereas for ganciclovir alterations in the viral enzyme encoded by the human cytomegalovirus gene UL97, which is involved in the first phosphorylation step to form ganciclovir monophosphate, is the main contributing resistance mechanism. Mutations in the DNA polymerase of hepatitis B virus and human cytomegalovirus have been shown to mediate resistance to *bis*(POM)-PMEA and cidofovir, respectively. The prevalence of drug-resistant virus to the acyclic nucleosides is, however, very low. For example, acyclovir and penciclovir resistance is ~0.3% in immunocompetent patients and 4–7% in immunocompromised patients. Mutations in the DNA polymerase of herpes simplex virus have been shown to mediate resistance to vidarabine and lamivudine. In the case of lamivudine, a single amino acid point mutation in the YMDD (tyrosine, methionine, aspartate, aspartate) locus of the polymerase gene results in resistance.

RT inhibitors

The virally encoded enzyme RT plays a pivotal role in the replicative process of HIV (Figure 16.26). RT is a multifunctional enzyme with RNA-dependent DNA polymerase, DNA-dependent DNA polymerase, and RNase H activity. RT is responsible for transcribing the genetic material contained within the two copies of (+)-strand RNA, located within the core of the virus, into DNA

Table 16.12 Nucleoside DNA-polymerase inhibitors licensed for use and in Phase II/III development

Inhibitors	Chemical structure	Therapeutic indication
Vidarabine (Vira-A®) **16.138**	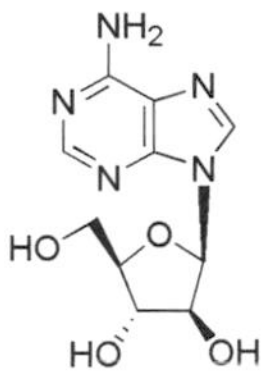	HSV keratitis
Idoxuridine (Herpid®) **16.139**	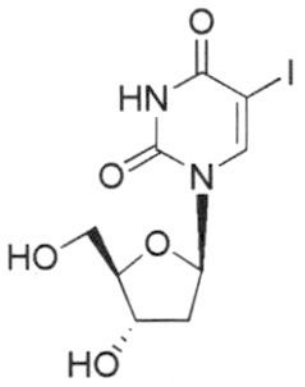	Topical treatment of HSV/VZV derived keratoconjunctivitis
Trifluridine (Viroptic®) **16.140**	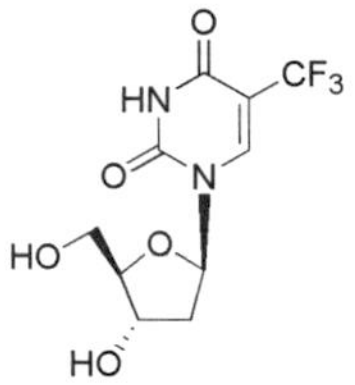	Topical HSV
Brivudine (Helpin/Zostrex®) **16.141**	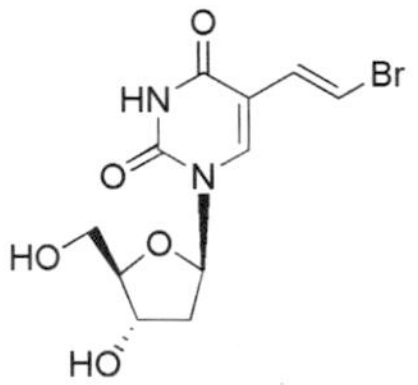	HSV-1, VZV
Lamivudine (Epivir®) **16.142**	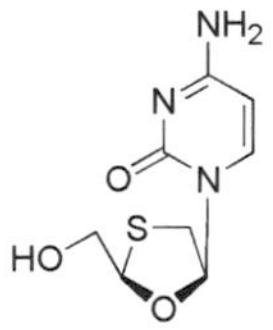	HBV, HIV
Emtricitabine (Coviracil®) **16.143**	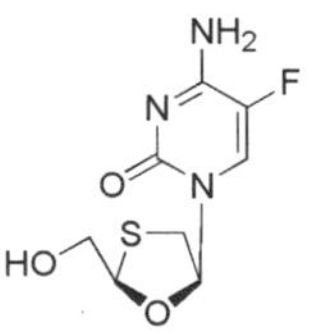	Phase II trials for HBV Phase III trials for HIV
Entecavir **16.144**	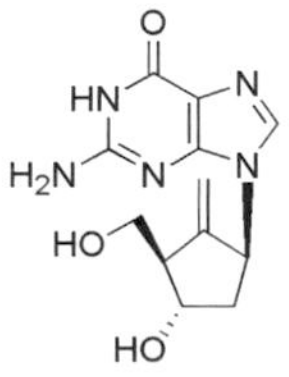	Phase II trials for HBV

Abbreviations: herpes simplex virus (HSV), type 1 (HSV-1); varicella zoster virus (VZV); human immunodeficiency virus (HIV); hepatitis B virus (HBV).

once inside the host cell. Many of the antiviral nucleoside mimetics interact at the substrate-binding site of the HIV RT where they act as competitive inhibitors/alternate substrates (chain terminators). Non-nucleoside RT inhibitors interact at an allosteric-binding site and are specific for HIV-1.

Nucleoside RT inhibitors

The first compounds approved for the treatment of AIDS were the nucleoside mimetics 3′-azido-2′,3′-dideoxythymidine (AZT, Retrovir®) (**16.145**), 2′,3′-dideoxyinosine (DDI, Videx®) (**16.146**), and 2′,3′-dideoxycytidine (DDC, Hivid®) (**16.147**). More recent additions are the L-nucleosides Epivir (**16.142**) and Emtricitabine® (Emtriva®, **16.143**). This class of compounds is known collectively as NRTIs. These NRTIs are active in their triphosphate form and interact at the substrate-binding site of the HIV RT where they act as competitive inhibitors/alternate substrates. The NRTIs lack a 3′-hydroxy group, which is essential for the extension of the viral DNA chain, therefore once they are incorporated into the growing DNA chain, RT is inhibited by premature chain termination (Figure 16.30).

The NRTIs show differences in their ability to suppress HIV, and this has been found to be directly related to their intracellular metabolism to the 5′-triphosphate form. For 2′,3′-dideoxy-2′,3′-didehydrothymidine (D_4T, Zerit®) (**16.148**), the rate-limiting step is the formation of D_4T-MP, whereas for AZT the rate-limiting step is the formation of AZT-DP from AZT-MP. This results from the ability of AZT-MP to inhibit deoxythymidine monophosphate (dTMP) kinase, the enzyme which catalyzes the conversion of AZT-MP to AZT-DP. All the NRTIs are bioprecursor prodrugs, that is, they are activated by intracellular metabolism to their active triphosphate forms, which would be too polar to traverse the lipophilic cell membranes.

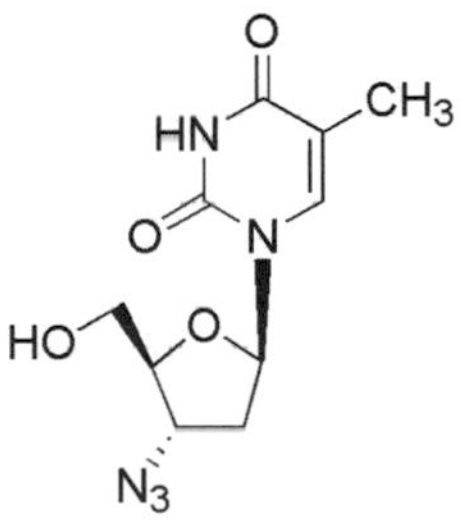

(**16.145**) Retrovir

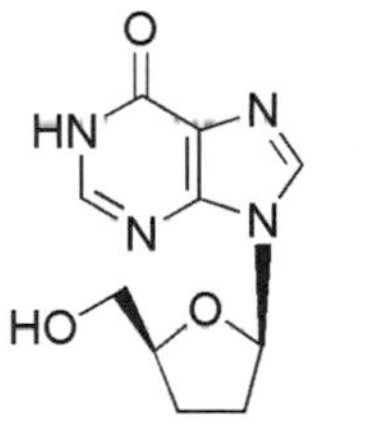

(**16.146**) Videx

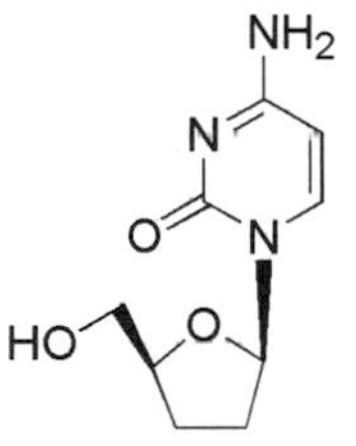

(**16.147**) Hivid

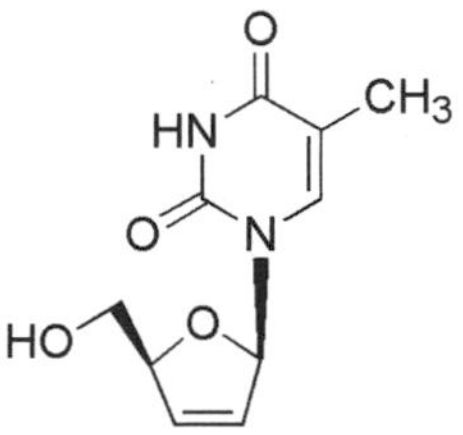

(**16.148**) Zerit

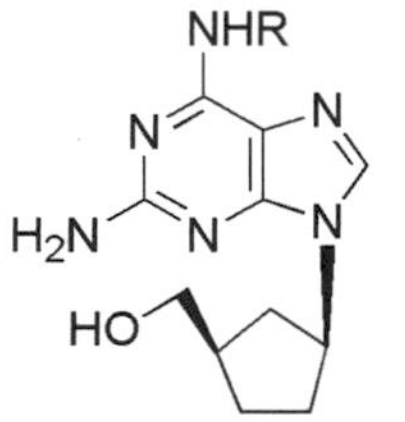

(**16.149**) R = ᶜPr, Abacavir
(**16.150**) R = H, Carbovir

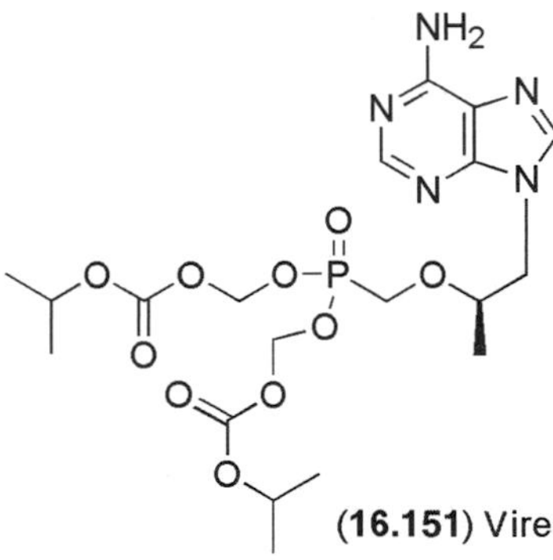

(**16.151**) Viread

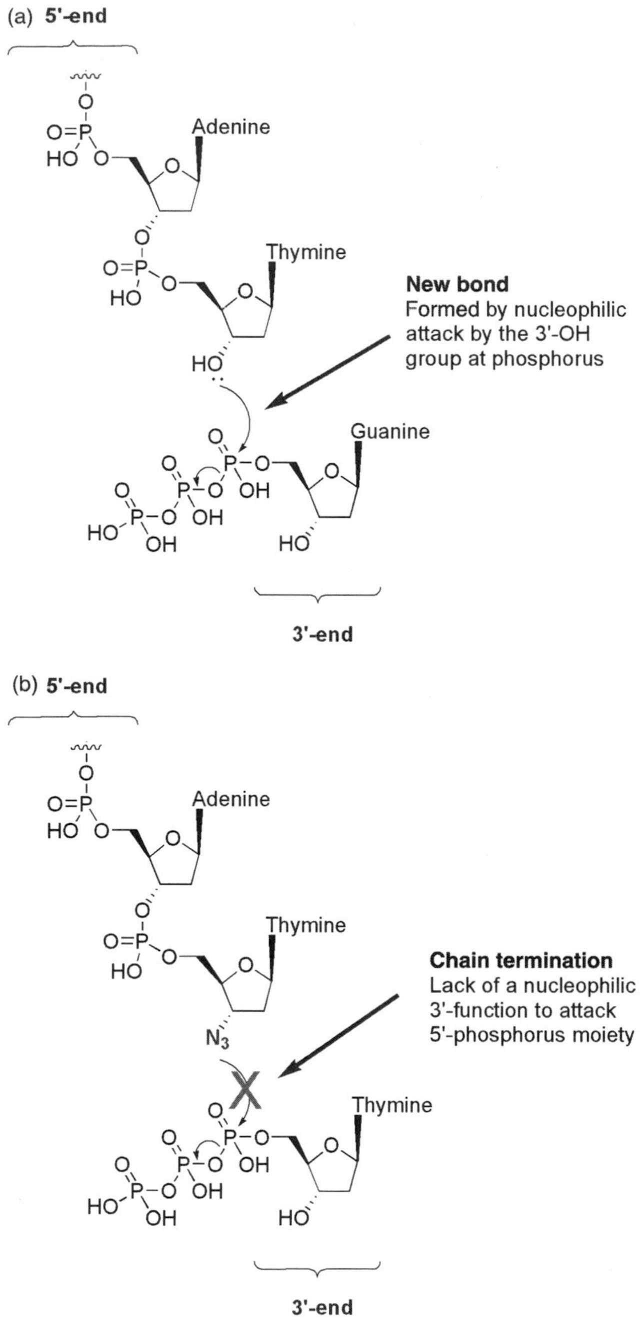

(a) **5'-end**

Adenine

Thymine

New bond
Formed by nucleophilic attack by the 3'-OH group at phosphorus

Guanine

3'-end

(b) **5'-end**

Adenine

Thymine

Chain termination
Lack of a nucleophilic 3'-function to attack 5'-phosphorus moiety

Thymine

3'-end

Figure 16.30 (a) DNA polymerization and (b) chain termination.

Carbocyclic nucleosides arise from isosteric replacement of the oxygen of the furanose ring with a methylene unit. This replacement has been shown to result in enhanced biostability. However, carbocyclic nucleosides are synthetically more challenging than the more conventional nucleosides. Carbocyclic nucleosides are also bioprecursor prodrugs. In the case of the carbocyclic nucleoside, Abacavir (**16.149**), initial monophosphorylation is followed by metabolism by cytosolic enzymes to the parent nucleoside, carbovir monophosphate, which is subsequently converted to the active carbovir triphosphate. This design allows the delivery of carbovir triphosphate without the pharmacokinetic and toxicological deficiencies associated with direct administration of carbovir (**16.150**). Numerous prodrugs have been developed to overcome the problems of delivery and metabolism limitations. One such prodrug is Viread® (**16.151**), which uses the lipophilic prodrug moiety isopropoxycarbonyl-oxymethyl (POC) to enhance the oral bioavailability of the acyclic nucleoside phosphonate PMPA (Tenofovir). Viread is the only nucleotide currently licensed for the treatment of HIV.

NRTI combinations, for example Combivir® (Retrovir + Epivir) and Trizivir® (Abacavir + Retrovir + Epivir), are also commonly used. They are, however, always given in combination with at least one other anti-HIV drug, usually a non-nucleoside RT inhibitor (NNRTI, see non-nucleoside RT inhibitors section) or a protease inhibitor (see Section 16.4.4).

Non-nucleoside RT inhibitors (NNRTIs)

The NNRTIs are specific inhibitors of HIV-1 RT and do not affect the replication of other retroviruses. They are described as NNRTIs as they do not bind at the substrate (deoxynucleotide triphosphate) binding site with the selectivity of the NNRTIs resulting from a specific interaction with a nonsubstrate (allosteric) binding site of the HIV-1 RT.

The interest in NNRTIs commenced with the discovery of 1-(2-hydroxyethoxymethyl)-6-phenylthiothymidine (HEPT, **16.152**) and tetrahydroimidazo[4,5,1-jkj][1,4]benzodiazepin-2-one and thione (TIBO, **16.153**). Over 30 structural classes of NNRTIs have been described, although only three are currently in clinical use: efavirenz (Sustiva®, **16.154**), delaviridine (Rescriptor®, **16.155**), and nevirapine (Viramune®, **16.156**). A few candidate NNRTIs are under Phase II clinical evaluation including the imidazole capravirine (**16.157**) and the natural product calanolide A (**16.158**).

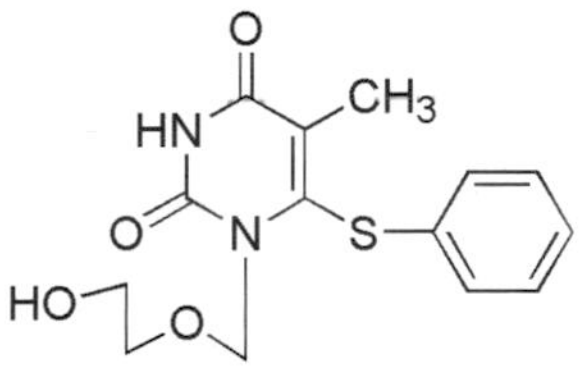

(**16.152**) HEPT

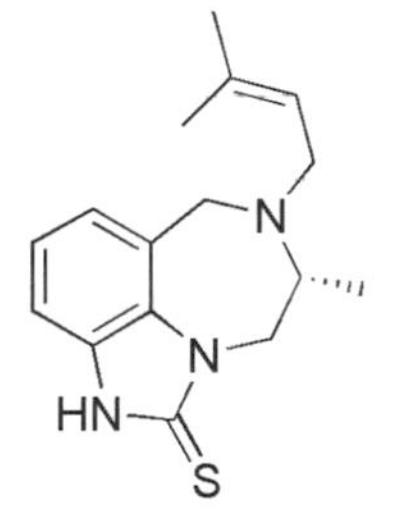

(**16.153**) TIBO

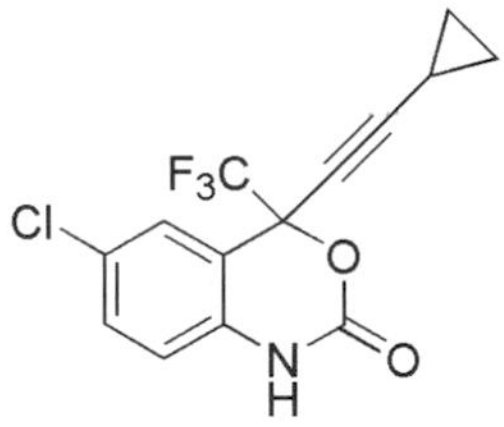

(**16.154**) Efavirenz

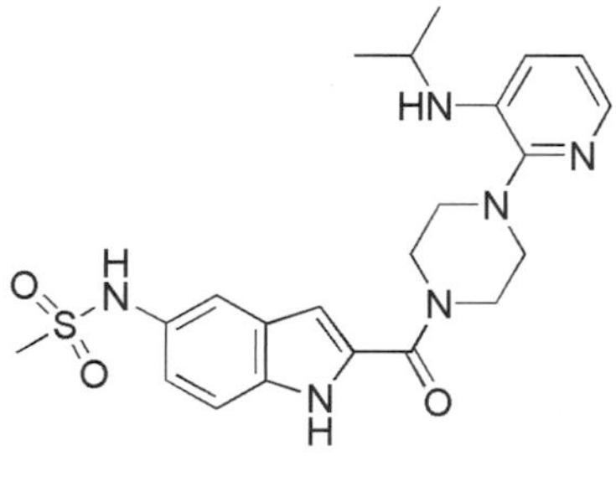

(**16.155**) Delaviridine

Although the NNRTIs appear very diverse in structure, a common "butterfly-like" conformation has been identified: a (thio) urea, (thio) carboxamido, or (thio) acetamide moiety flanked by two bulky hydrophobic groups. The bulky hydrophobic, frequently aromatic, groups serve as π-electron donors for aromatic side-chain residues in the vicinity of the active site.

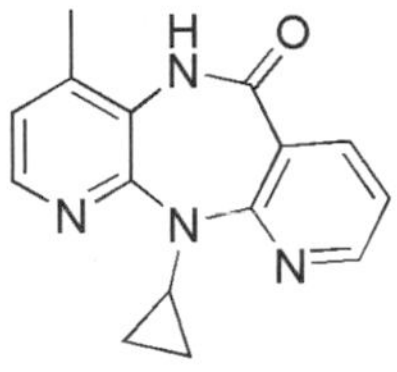

(16.156) Nevirapine

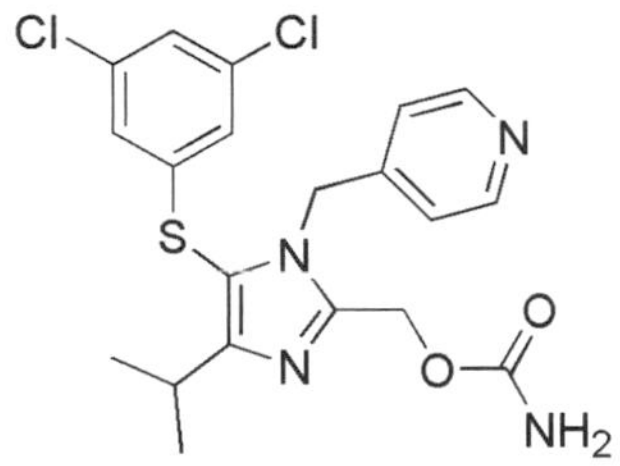

(16.157) Capravirine

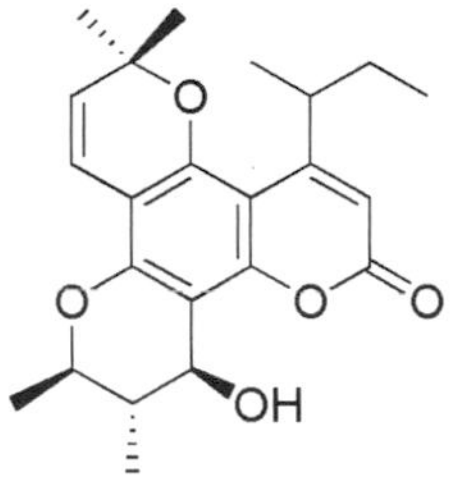

(16.158) Calanolide A

Resistance to RT inhibitors

The emergence of drug-resistant viral strains after exposure to RT inhibitors is a consequence of both high turnover of HIV-1 in patients and low fidelity of viral RT. In untreated individuals, HIV-1 generates approximately 10 billion copies per day and in the process makes an error of one base per cycle mainly due to the lack of proofreading ability of its RT. Resistance specifically results from target gene mutation (point mutations) within the *pol* gene, which encodes the viral protease, RT, and integrase. The slower progress of NNRTIs, compared with NRTIs, into the clinic results from rapid emergence of viral resistance, which arises from point mutations (K103N, Y181C, Y188C/L/H, and M230L identified as key mutations) at the NNRTI-binding site resulting in reduced fit/interactions of the NNRTIs.

16.4.4 Protease Inhibitors

HIV protease plays a pivotal role in the maturation step of viral particles by processing the polyprotein gene products of *gag* and *gag-pol* into active structural and replicative proteins. Inhibition of HIV protease results in the production of immature, noninfectious viral particles. HIV protease interacts with an octapeptide region (sequentially labeled P4–P1, P1′–P4′) of the polyprotein substrate and cleaves the viral polyprotein precursor at the scissile bond site, which contains a phenylalanine or proline sequence at P1 and P1′. Mammalian proteases do not generally cleave peptide bonds that contain a proline at the C-terminal side of the scissile bond, therefore this clip site provided an important distinction for the design of novel peptidomimetic drugs. Currently licensed inhibitors mimic both the P1–P1′ dipeptide clip site and the P2 (Asn) and P2′ (Ile) residues of the RT-protease clip site.

Saquinivir (Invirase®, Fortovase®, **16.159**), the first protease inhibitor licensed for clinical use, is a transition-state isostere containing Asn and Phe at P2 and P1, and a decahydroisoquinoline-*tert*-butylamine group which mimics the Pro–Ile dipeptide at P1′–P2′ (Figure 16.31). Saquinivir (**16.159**), nelfinavir (Viracept®, **16.160**), and indinavir (Crixivan®, **16.161**) are described as class IA protease inhibitors as they contain similar Pro–Ile replacements. The introduction of a *S*-phenyl cysteine moiety in nelfinavir (**16.160**) resulted in improved enzyme–inhibitor interaction and therefore improved inhibitory activity. Nelfinavir also has reduced peptide character compared with saquinivir (**16.159**), which results in improved oral bioavailability. Improved oral bioavailability was achieved with indinavir (**16.161**) by the inclusion of a basic pyridylmethylpiperazine moiety.

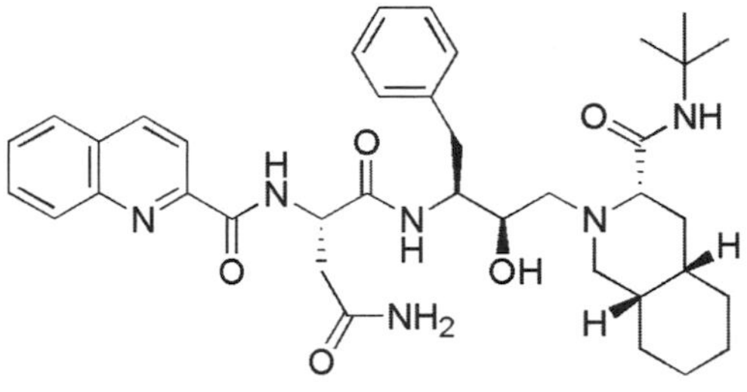

(**16.159**) Saquinivir

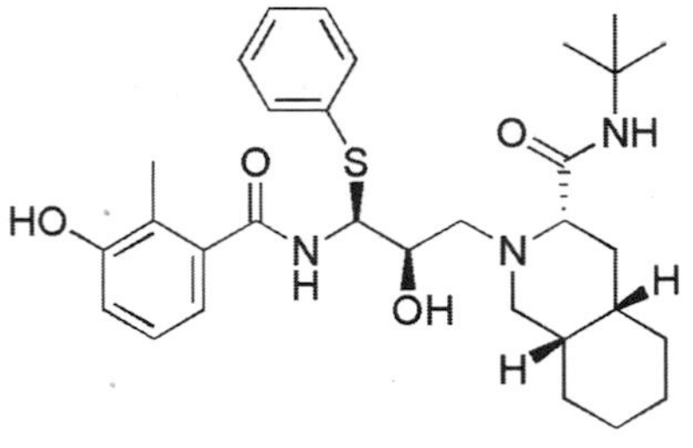

(**16.160**) Nelfinavir

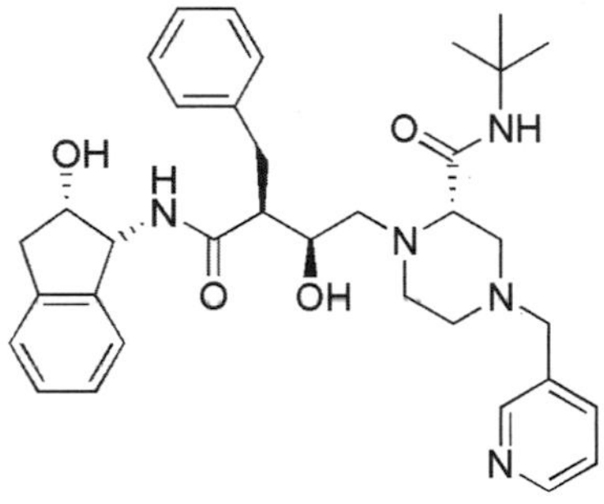

(**16.161**) Indinavir

With high molecular weight structures, class IA inhibitors are subject to rapid hepatic metabolism and clearance. This has led to the design of amprenavir (Agenerase®, **16.162**), a smaller molecular weight class IB protease inhibitor, which is less susceptible to hepatic clearance. The presence of a sulfonamide group at the P2′-site of amprenavir results in retained activity due to

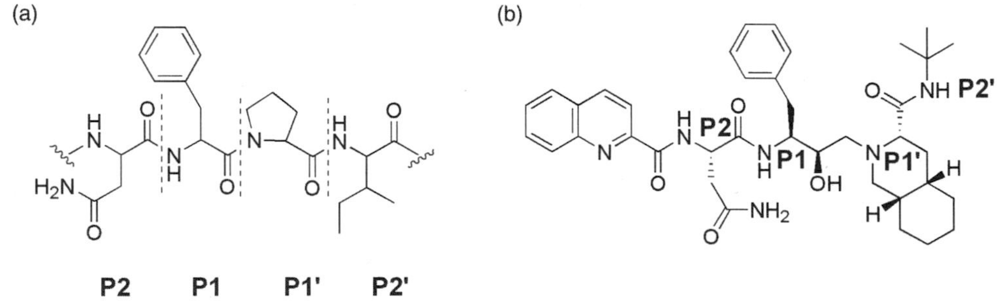

(a)

P2 P1 P1' P2'

—Asn—Phe—Pro—Ile—

(b)

(**16.159**) saquanivir

Figure 16.31 (a) P2–P1–P1′–P2′ clip site of HIV-1 protease and (b) peptodomimetic protease inhibitor saquinivir (**16.162**) with sites of comparison indicated.

strong interactions between this moiety and the protease. Other class 1B inhibitors fosamprenavir (**16.163**) and TMC114 (**16.164**) are almost identical to amprenavir (**16.162**).

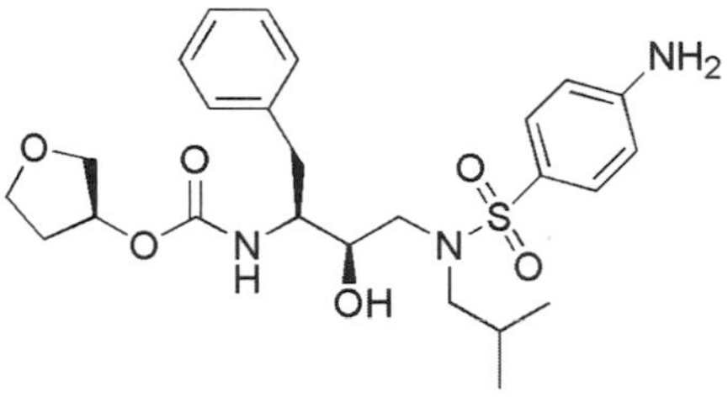

(**16.162**) Amprenavir

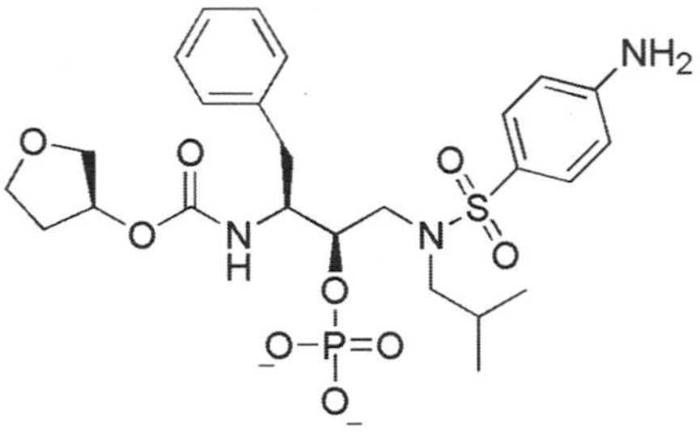

(**16.163**) Fosamprenavir

The most recently licensed protease inhibitor is atazanavir (Rayataz®, **16.165**) which is taken as a once daily oral tablet, an improvement compared with the 2 to 3 times daily dosing associated with earlier protease inhibitors. It can also be used as a booster in dual-PI regimens for patients who have failed previously on protease inhibitor-based regimens.

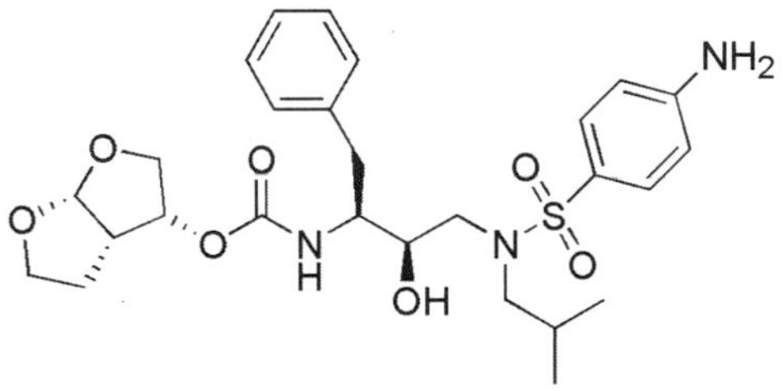

(**16.164**) TMC114

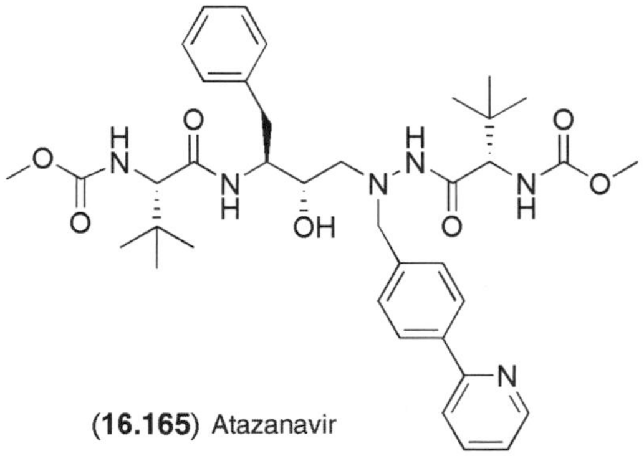

(**16.165**) Atazanavir

The unique symmetry of the HIV protease dimer structure has been exploited for the design of ritonavir (Norvir®, **16.166**) and lopinavir (**16.167**). These class ID inhibitors are designed to superimpose their axis of symmetry, positioned in the P1–P1′ region, on that of HIV protease, to maximize enzyme–inhibitor interactions. The Phe–Phe core, thiazole, and tetrahydro-pyrimidin-2-one moieties of ritonavir and lopinavir, respectively, decrease the rate of metabolism. The heterocyclic moieties also serve to enhance oral bioavailability.

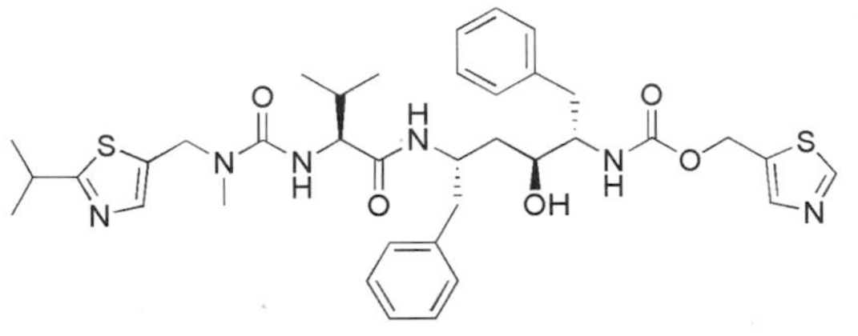

(**16.166**) Ritonavir

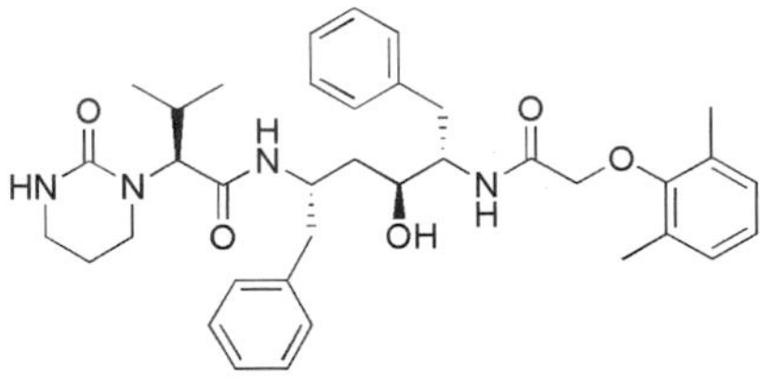

(**16.167**) Lopinavir

Nonpeptidic, class II, inhibitors are attractive as drug candidates as they are generally more readily synthesized and potentially cheaper to manufacture; however, no class II protease inhibitors have reached the market. Tipranavir (**16.168**), which is in Phase III clinical trials, is the closest to clinical use. It has, however, been found that tipranavir must be boosted with ritonavir (**16.166**) to compensate for its short half-life and poor bioavailability. Mozenavir (**16.169**) showed early promise but is no longer in development.

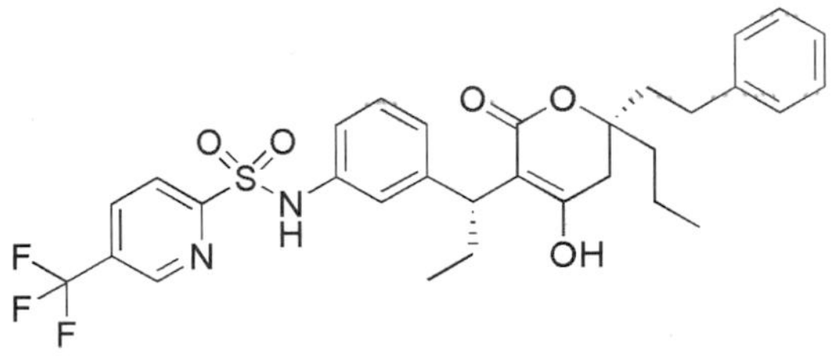

(**16.168**) Tipranavir

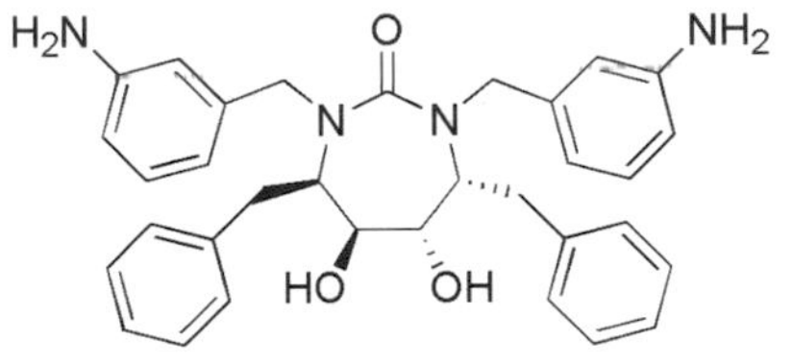

(**16.169**) Mozenavir

Resistance to protease inhibitors

All protease inhibitors have similar resistance profiles because they interact with the same residues in the active site. Loss of sensitivity to protease inhibitors mainly arises in resistant viral strains that encode protease molecules containing specific amino acid mutations. These mutations result in protease molecules with lower affinity for the inhibitors but maintain sufficient affinity for the natural polyprotein substrate. Four major mutations have been identified: (i) active site mutations which directly interferes with binding; (ii) nonactive site mutations which lead to modification of the active site conformation; (iii) mutations in the cleavage sites of HIV-1 protease polyprotein substrate, and (iv) mutations that reduce the stability of the protease dimer structure resulting in decreased drug affinity.

16.4.5 Translation Inhibitors

One other inhibitor of note is the antisense oligonucleotide fomivirsen (Vitravene®, **16.170**). Fomivirsen is a 21-mer phosphorothiate antisense oligonucleotide that prevents viral replication by binding to a complementary human cytomegalovirus mRNA sequence, inhibiting translation of several cytomegalovirus early stage proteins. Fomivirsen has a unique mode of action and is active against GCV, foscarnet, and cidofovir-resistant virus strains. Administration is as an intraocular injection in the salvage therapy of AIDS patients with human cytomegalovirus retinitis.

16.4.6 Summary Figure

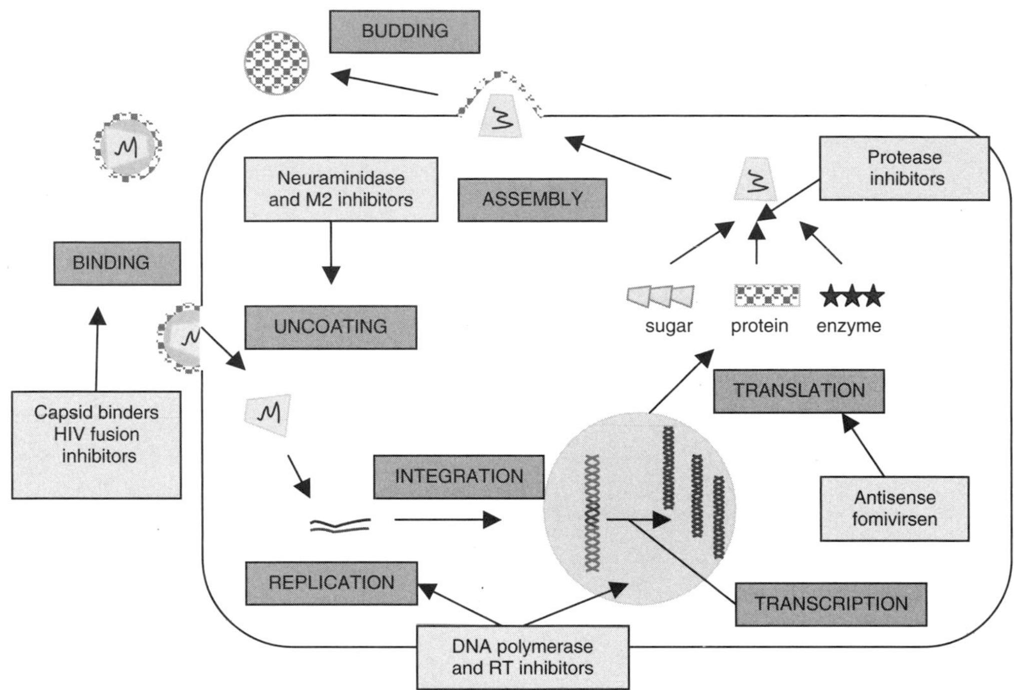

Figure 16.32 Summary diagram indicating the sites of action of antiviral agents.

FURTHER READING

Antibacterial

Chu, D.T.W. (1999) Recent progress in novel macrolides, quinolones, and 2-pyridones to overcome bacterial resistance. *Medicinal Research Reviews* **19**, 497–520.

Cox, R.J., Sutherland, A. and Vederas, J.C. (2000) Bacterial diaminopimelate metabolism as a target for antibiotic design. *Bioorganic and Medicinal Chemistry* **8**, 843–871.

Gao, Y. (2002) Glycopeptide antibiotics and development of inhibitors to overcome vancomycin resistance. *Natural Products Reports* **19**, 100–107.

Knowles, J.R. (1985) Penicillin resistance: the chemistry of β-lactamase inhibition. *Accounts of Chemical Research* **18**, 97–104.

Massova, I. and Mobashery, S. (1997) Molecular bases for interactions between β-lactam antibiotics and β-lactamases. *Accounts of Chemical Research* **30**, 162–168.

Moellering, R.C. (1999) Quinupristin/dalfopristin: therapeutic potential for vancomycin-resistant enterococcal infections. *Journal of Antimicrobial Chemotherapy (Suppl.)* **44**, 25–30.

Phetsuksiri, B., Baulard, A.R., Cooper, A.M., Minnikin, D.E., Douglas, J.D., Besra, G.S. and Brennan, P.J. (1999) Antimycobacterial activities of isoxyl and new derivatives through the inhibition of mycolic acid synthesis. *Antimicrobial Agents and Chemotherapy* **43**, 1042–1051.

Russell, A.D. and Chopra, I. (1996) *Understanding Antibacterial Action and Resistance*, 2nd edn. Chapman & Hall, London.

Sensi, P. and Grassi, G.G. (1996) Antimycobacterial agents. In: *Burger's Medicinal Chemistry and Drug Discovery. Vol. 2: Therapeutic Agents*, 5th edn, M.E. Wolff (ed.). John Wiley & Sons, New York, pp. 575–635.

Sharman, G.J., Try, A.C., Dancer, R.J., Cho, Y.R., Staroske, T., Bardsley, B., Maguire, A.J., Cooper, M.A., O'Brien, D.P. and Williams, D.H. (1997) The roles of dimerization and membrane anchoring in activity of glycopeptide antibiotics against vancomycin-resistant bacteria. *Journal of the American Chemical Society* **119**, 12041–12047.

Souter, A., McClean, K.J., Ewen Smith, W. and Munro, A.W. (2000) The genome sequence of *Mycobacterium tuberculosis* reveals cytochromes P450 as novel anti-TB targets. *Journal of Clinical Technology and Biotechnology* **75**, 933–941.

Storer, C.K., Warrener, P., VanDevanter, D.R., Sherman, D.R., Arain, T.M., Langhorne, M.H., Anderson, S.W., Towell, J.A., Yuan, Y., McMurray, D.N., Kreiswirth, B.N., Barry, C.E., and Baker, W.R. (2000) A small-molecule nitroimidazopyran drug candidate for the treatment of tuberculosis. *Nature* **405**, 962–966.

Sum, P.-E., Sum, F.-W. and Projan, S.J. (1998) Recent developments in tetracycline antibiotics. *Current Pharmaceutical Design* **4**, 119–132.

Tenover, F.C. (2001) Development and spread of bacterial resistance to antimicrobial agents: an overview. *Clinical Infectious Diseases* **33** (Suppl. 3), S108–S115.

Antifungal

Arikan, S. and Rex, J.H. (2000) New agents for the treatment of systemic fungal infections. *Emerging Drugs* **5**, 135–160.

Fostel, J.M. and Lartey, P.A. (2000) Emerging novel antifungal agents. *Drug Discovery Today* **5**, 25–32.

Ghannoum, M.A. and Rice, L.B. (1999) Antifungal agents: mode of action, mechanisms of resistance, and correlation of these mechanisms with bacterial resistance. *Clinical Microbiology Reviews* **12**, 501–517.

Zotchev, S.B. (2003) Polyene macrolide antibiotics and their application in human therapy. *Current Medicinal Chemistry* **10**, 211–223.

Antiviral

Ala, P.J. and Chang, C.-H. (2002) HIV aspartate proteinase: resistance to inhibitors. In: *Proteinase and Peptidase Inhibition: Recent Potential Targets for Drug Development*, H.J. Smith and C. Simons (eds). New York: Taylor and Francis, London, pp. 367–382.

De Clercq, E. (1999) Perspectives of non-nucleoside reverse transcriptase inhibitors (NNRTIs) in the therapy of HIV-1 infection. *Il Farmaco* **54**, 26–45.

Kilby, J.M. and Eron, J.J. (2003) Novel therapies based on mechanism of HIV-1 cell entry. *New England Journal of Medicine* **348**, 2228–2238.

Moore, J.P. and Stevenson, M. (2000) New targets for inhibitors of HIV-1 replication. *Nature Reviews* **1**, 40–49.

Naesens, L. and De Clercq, E. (2001) Recent development in herpes virus therapy. *Herpes* **8**, 12–16.

Simons, C. (2001) *Nucleoside Mimetics: their Chemistry and Biological Properties*, *Advanced Chemistry Texts*. Gordon and Breach, Amsterdam.

Tupin, J.A. (2003) The next generation of HIV/AIDS drugs: novel and developmental anti-HIV drugs and targets. *Expert Review in Anti-Infective Therapy* **1**, 97–128.

Useful Websites

Sigma-Aldrich antibiotic explorer website, http://www.sigmaaldrich.com/Area_of_Interest/Biochemicals___Reagents/Antibiotic_Explorer.html

Dr Fungus website, *Medical Antifungal Agents*, http://www.webillustrated.com/site_pre_endnote/thedrugs/medical.htm

Mediscover website, Infectious diseases; *Antiviral Agents FactFile 2000–2001*. http://www.mediscover.net/hivbycat.cfm

17

Pharmaceutical Applications of Bioinorganic Chemistry

David M. Taylor and David R. Williams

CONTENTS

17.1 INTRODUCTION

Bioinorganic chemistry may be defined loosely as the study of the interactions of inorganic ions or radicals with the components of living cells and tissues. However, perhaps the large area of interest lies in the interactions of metals with the plethora of metal-binding compounds (ligands) that exist *in vivo*. The human body contains about 50 of the elements in the periodic table and about half of these are either *essential* or *beneficial* to life, while the remainder are adventitious having been introduced from the environment. Ten of these essential elements, oxygen, carbon, hydrogen, nitrogen, calcium, phosphorus, sulfur, potassium, sodium, and chlorine, are present in amounts ranging from a few kilograms to a few tens of grams, and these so-called *bulk* elements are contained in the proteins, fats, and carbohydrates that are the building blocks for the organs and tissues.[1a–f] The major emphasis of this chapter will rest on the remaining 20 or so of the essential and beneficial elements, which, because they are generally present in very small quantities, are often called *trace* elements. In addition the pharmaceutical applications of some nonessential, nonbeneficial elements will also be considered.

The total mass of the essential or beneficial elements in the human body is less than 50 g, yet they must be present within specific limits of concentration and in particular chemical forms if the individual is to enjoy a healthy life. Of the adventitious elements, some can be described as pollutants arising from human activities, especially from the industrial developments of the past two centuries, while the remainder enter the human body simply because they are present naturally in drinking water or in the plant and animal tissues which make up our food. The concentrations of the adventitious elements vary from person to person, depending on their geographical environment and their eating habits.

There are strict criteria that must be fulfilled before a trace element is classified as essential — it must be present in all healthy tissues and it must cause reproducible symptoms of ill health if it is excluded from the diet. The essential and beneficial trace elements and their locations in the periodic table are shown in Figure 17.1. Table 17.1 lists the amounts of the 30 essential and beneficial elements found in the male and female human body.

Life on this planet probably began around 5 billion ($\sim 5 \times 10^9$) years ago from primitive cells that evolved in the ancient oceans utilizing biochemical substrates synthesized on the surfaces of sand particles on tidal beaches.[1a] Such evolution was based on the elements readily available in the ancient seabed and the primitive oceans; and any life forms dependent on less readily available elements would have been bred out millions of years ago. Thus, the composition of the human body

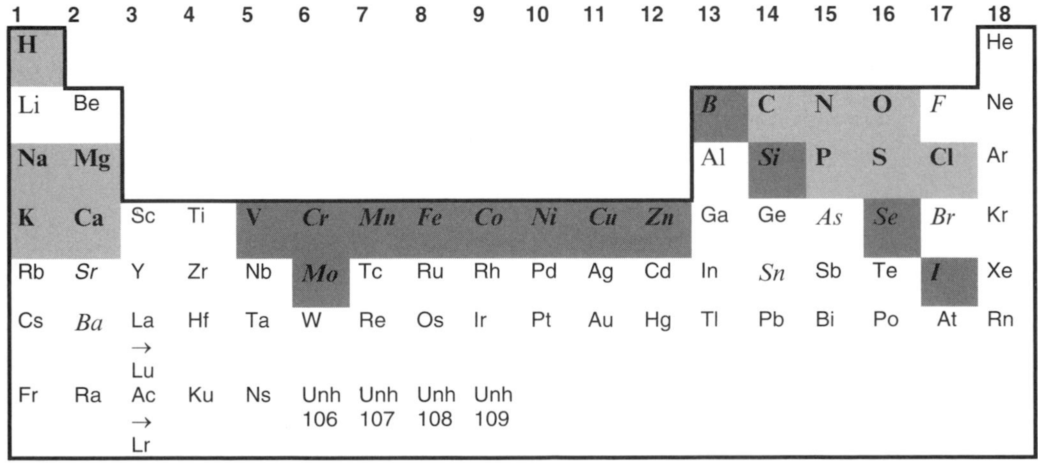

Figure 17.1 The periodic table indicating the **bulk**, ***essential***, and *possibly essential*, elements.

Table 17.1 The approximate masses of the essential and beneficial elements occurring in 73 kg male and 60 kg female reference persons. Data are from Refs. [8a–c]

	Male		Female	
Element	**Grams**	**Moles**	**Grams**	**Moles**
Hydrogen	7,300	7,300	6,000	6,000
Carbon	16,500	1,375	13,700	1,140
Nitrogen	1,880	134	1,545	110
Oxygen	44,900	2,810	38,900	2,310
Phosphorus	730	23.5	600	19.3
Sulfur	146	4.6	120	3.8
Chlorine	104	2.9	85	2.4
Sodium	104	4.5	86	3.7
Potassium	144	3.7	103	2.6
Calcium	1,180	29	860	22
Lithium	0.0007	0.0001	0.0006	0.00009
Boron	0.01	0.0009	0.009	0.00008
Fluorine	0.8	0.04	0.7	0.036
Magnesium	36	1.5	30	1.3
Silicon	1.5	0.05	1.3	0.046
Vanadium	0.02	0.0004	0.017	0.00034
Chromium	0.005	0.0001	0.0043	0.00008
Manganese	0.021	0.00038	0.017	0.00031
Iron	4.4	0.079	3.6	0.064
Cobalt	0.0007	0.00001	0.0006	0.00001
Nickel	0.01	0.0002	0.01	0.0002
Copper	0.12	0.0018	0.094	0.0015
Zinc	2.4	0.037	2.0	0.030
Arsenic	0.015	0.0002	0.012	0.00016
Selenium	0.02	0.0025	0.017	0.00022
Bromine	0.2	0.0026	0.17	0.0021
Molybdenum	0.005	0.00005	0.004	0.00004
Tin	0.03	0.00026	0.026	0.00022
Iodine	0.013	0.0001	0.011	0.00008
Barium	0.017	0.00012	0.015	0.00011

broadly resembles the elemental composition of the ancient seabed and oceans. These are the lighter elements of the periodic table, because, assuming that the planet was formed from a cloud of dust particles, the middleweight and heavier elements would have been compacted by gravity into the mantle and core of the Earth, respectively.[1a,b]

The evolution of life has not been a smooth process. The earliest cells evolved under the highly reducing atmosphere of water vapor, hydrogen sulfide, ammonia, and methane that was present at the beginning of terrestrial time. Around 2 billion years ago the early cell probably contained only about 100 different protein molecules, as compared with the million or so proteins found in modern cells; it also contained a range of metal ions, some of which fulfilled structural or osmotic roles while others acted as catalysts.

The heaviest metals used in evolution were those from the first transition series of the periodic table, and metals such as manganese, iron, cobalt, and copper all existed in their lower oxidation states in primitive cells. Then, relatively suddenly, perhaps about 1.8 billion years ago, blue-green algae in the oceans began to produce oxygen in sufficient quantities to slowly convert the Earth's reducing atmosphere into the present oxygen-containing one. This had the effect of raising the oxidation states of the aforementioned transition metals and of releasing the previously insoluble cuprous ores into the biosphere as the more soluble cupric salts. At the same time iron and manganese were immobilized as their higher oxidation states, Fe_3O_4, FeOOH, Mn_3O_4, and MnO_2.

The atmospheric ozone content also began to rise, reaching ~1%, which was sufficient to screen out the harmful effects of the sun's ultraviolet radiation which appears to be more destructive to aerobic than to anaerobic systems. As a result of these dramatic environmental changes, organisms that had evolved using ferrous ions as oxygen carriers became vulnerable to oxidation and probably died out. However, biochemical systems involving cupric complexes and oxygen were now possible. Thus, because of the chequered history of the four metals, iron, manganese, cobalt, and copper, in their different oxidation states — some originally rejected by nature and only relatively recently incorporated into essential biochemical mechanisms — it is to be expected that in complex multicellular species, such as *Homo sapiens*, serious biochemical disturbances may arise from changes in metal balances, and from challenges to optimal concentrations of such metals, caused by environmental factors or by pharmacological intervention.

17.2 THE IMPORTANCE OF TRACE ELEMENTS IN HUMANS

Homo sapiens is a complex multicellular species that depends for its health upon the correct functioning of ~10^{15} cells of many different, yet interdependent, types.[1e,f] Thus, the maintenance of good health will also depend on the supply of all 30 essential or beneficial elements in adequate, but not excessive, quantity and in a chemical form that is utilizable within the body.

In all human tissues there are complicated interactions between the spectrum of metals and of proteins and other ligands within the cells. The spectrum of proteins within a tissue or tissue compartment — the *proteome* — is controlled by the genes that make up the *genome*; Fausto de Silva and Willams[1h] have therefore proposed the concept of the *metallome* to describe the concentration patterns of essential metals in human or animal tissues. Figure 17.2 compares the metallomes for the total metal concentration in normal human blood plasma, liver, and brain.

The principle of the biphasic response[1c] reminds us that while an insufficiency of an element may lead to ill health by preventing the optimal function of some biochemical process, an excess may cause serious, even life-threatening, toxic reactions. In disease it might be expected that the metallome of one or more specific tissues would be altered; and indeed there is much evidence to show that the concentrations of zinc, copper, iron, and other metals may be changed in some disease states. However, the extent to which the spectrum of metals which make up the metallome, Na, K,

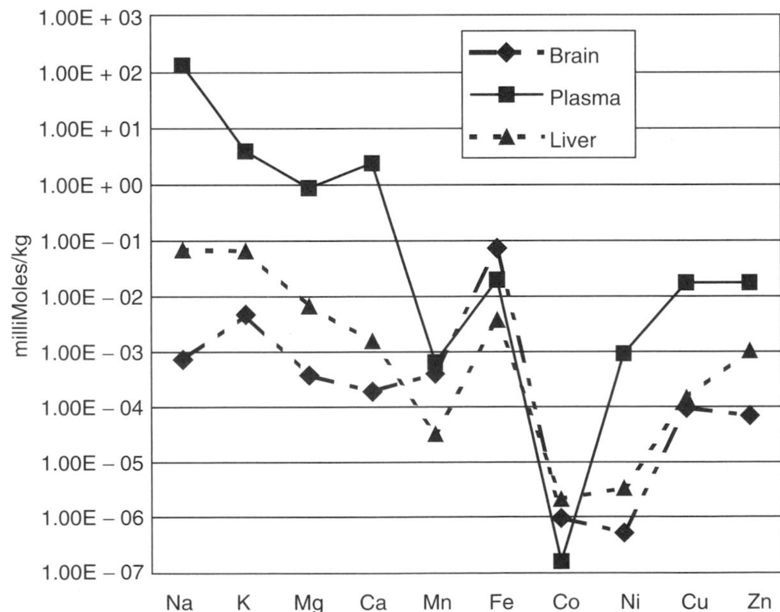

Figure 17.2 The metallomes (total concentrations of metals) of normal human plasma, liver, and brain (data taken from Ref. [1 h]) (1.0E + 3 = 1000; 1.0E – 7 = 1.0×10^{-7}).

Mg, Ca, Mn, Fe, Co, Ni, Cu, Zn, is altered by disease[1h] is at present little understood, thus the systematic investigation of changes in metallome composition, and of methods for correcting them, are potentially exciting areas for future research.

Trace element insufficiency, or imbalances of elemental intake, may arise because: the diet itself is deficient in the element; the element is in a chemical form that cannot readily be absorbed from the gastrointestinal tract into the plasma, or disease causes excessive loss from the body. Poor bioavailability may arise because the element is present in a highly insoluble form, or as an electrically charged complex with a ligand present within the gastrointestinal tract that cannot readily be transferred across the intestinal mucosa. These ligands may arise from food residues or from intake of one or more pharmaceuticals.

The transition metals, which collectively weigh less than 10 g, are particularly vulnerable to interactions that may produce the effects of a deficiency. For example:

1. Each of these trace elements may be present in an organ or body fluid in only microgram quantities, yet most of the drugs which could interact with them will be administered in milligram or greater doses — the same pharmaceuticals may have little influence upon the 1.2 kg of calcium or the 36 g of magnesium but they may easily render a few micrograms of copper or zinc biochemically inert.
2. Poorly controlled industrial activities have resulted in the release of elements not normally present in our environment or diet, for example lead, cadmium, and mercury. The chemical properties of such metals may be so similar to those of an essential metal that they can compete for important biochemical-binding sites *in vivo*.
3. Modern food processing, preservation, and packaging techniques, as well as the advent of convenience foods, have all influenced the spectrum of trace elements and metal complexing ligands that enter the human body.
4. Advances in medicine and pharmacy have extended human life span toward 90 or more years and long-continued dietary or therapeutic practices that cause even very minor disturbance of trace elements may, over a period of years, lead to serious deficiency or imbalance. The importance of this can be seen in the treatment of conditions such as arthritis, hypertension, diabetes, gout, or minor cardiac disorders with medicaments that may need to be taken for perhaps 20 or more years.
5. As has already been mentioned, our basic biochemistry evolved under anaerobic conditions, and even in today's aerobic atmosphere often we still need divalent metals such as manganese, iron, cobalt, and copper in their lower oxidation states. Thus, drugs that interfere with oxidation–reduction (redox) mechanisms may lead to biochemical inactivation of such metals.

Fortunately, trace element balances can usually be relatively easily restored if deficiencies or excesses are recognized early enough. However, the correction of trace element deficiency is not a simple task and a series of criteria need to be satisfied. The preparation must ensure that the element will be present at the absorptive surfaces in the small intestine in solution, in the appropriate oxidation state, and in a form that can be absorbed into the blood. It may be the community pharmacist–patient relationship that first discusses such trace element-dependent side effects. Thus, it is important for pharmacists to have some understanding of trace element biochemistry.

There is no known aspect of biochemistry in which trace metals are not involved. Table 17.2 contains a partial listing of the more serious conditions that are associated with trace element imbalance — a full list would require many pages.[1g,h] For a more detailed discussion of the normal and pathological biochemistry of trace elements, the reader is referred to some of the general texts listed in Refs. [1a–l] but a few of the more important roles of such elements are mentioned here:

- Some elements are essential as cations or anions for the maintenance of osmotic pressures and to neutralize species of opposite charge. Sodium, potassium, and hydrogen ions fall into this category. Nature has evolved very specific mechanisms to maintain the concentrations of these elements at just the correct levels. For example, in blood the pH is buffered at ~7.4 despite severe challenges from the metabolic production of bicarbonate ions, etc.; changes of even 0.1 of a pH unit may be

Table 17.2 Examples of some more serious human disorders associated with bulk or essential trace metals[1g,h]

Element	Disorder associated with a deficiency	Disorder associated with an excess
Bulk metals		
Calcium	Bone deformities (Rickets), tetany	Atherosclerosis, cardiac arrhythmia, cardiac arrest
Potassium	Cardiovascular and renal symptoms	
Magnesium	Irritability of the central nervous system (CNS)	Muscular paralysis
Essential metals		
Chromium	Defective glucose metabolism, hyperlipidemia, skeletal deformities, testicular dysfunction	Allergy, lung cancer
Manganese		Ataxia, liver cirrhosis, psychiatric disorders
Iron	Anemias	Primary and secondary hemochromatosis.
Cobalt	Anemia, anorexia	Cardiomyopathy, hypo-thyroidism, polycythemia
Copper	Anemia, anorexia	Wilson's disease
Zinc	Anorexia, anemia, hypogonadlsm, depressed immune response, CNS disturbance	Hyperchromic anemia, changes in HDL/LDL cholesterol ratio
Selenium	Endemic cardiomyopathy (Kesham syndrome), osteoarthropathy (Kashin–Beck syndrome)	Hepatic and renal damage, teratogenesis, fetal toxicity
Molybdenum	Growth depression, hyperpurinemia	Anemia

life-threatening. Similarly, the sodium and potassium concentrations in blood are strictly maintained at 140 and 4 mmol dm^{-3}, respectively, whereas inside cells the concentrations are 20 and 95 mmol dm^{-3}. Thus, specific mechanisms have evolved to keep sodium out of cells while encouraging the influx of potassium. Similarly, chloride is kept outside the cells, whereas intracellular fluid is rich in HPO_4^{2-}.

- More than 1000 metalloenzymes are known to date; some 900 or more contain zinc, either as part of the active site or in a structural role, while others require manganese, molybdenum, copper, nickel, or selenium for their activity.
- Cobalt, once a very important element in primitive cells, now appears to play an important role only in vitamin B_{12} and in allowing free radical chemistry at low redox potentials.
- Copper, although less than 1 g is found in the average person, has been linked to many low molecular mass metal complexes and to ceruloplasmin, metallothionein, and other proteins in blood plasma and in tissues. Copper is now recognized as playing many roles *in vivo*; for example, copper oxidases, such as superoxide dismutase, play important roles in scavenging free radicals, removing excess neurotransmitter amines, and in the development of hormonal messengers such as adrenaline.
- Several metals play important roles through redox reactions in facilitating electron flow to enable high-speed transport of signals in nerves, and in storing energy or information.

17.3 METAL COORDINATION CHEMISTRY

17.3.1 General Aspects

Since many biochemicals and pharmaceuticals are often near ideal ligands for transition metal ions, it is appropriate here to summarize briefly the basic aspects of the chemistry of metal complexation. For a full discussion of this topic the reader is referred to the works given in Refs. [4a–d]. A *ligand* may be defined as a chemical that has a pair of electrons that can be donated to a vacant orbital in a metal ion. Important biochemical/pharmacological donor groups are RS•, $-NH_2$, COO^-, and $-O^-$, in addition to other groups such as $-PO_4^{3-}$ $-NO_4^{3-}$ etc. can form dative covalent bonds by donating electrons to vacant orbitals in a metal ion. Metal coordination is biochemically important because it may mask the normal chemical properties of the metal or alter the properties of the ligand. The classic example of this phenomenon is the relatively nontoxic substance potassium ferrocyanide ($K_4Fe(CN)_6$) that is made by mixing two very toxic solutions potassium cyanide and ferrous cyanide. The underlying chemistry of this was explained more than a century ago by Werner, who postulated the following principles.

Metals have two types of valency; first, a primary and ionizable valency, that is satisfied by negative ions such as Cl^-, NO_3^-, SO_3^{2-}, CN^-, and a secondary and nonionizable (covalent) one. For each metal there is a fixed number of secondary valencies — called the coordination number; for Fe^{2+} the coordination number is 6, while for Cu^{2+} it may be either 4 or 6, and coordination numbers ranging up to 8 are found with other metals; these secondary valencies are directional in nature radiating out from the central metal ion toward the corners of a regular tetrahedron or a regular octahedron with coordination numbers of 4 and 6, respectively. In potassium ferrocyanide, the complex iron-cyanide moiety, $[Fe(CN)_6]^{4-}$, ionizes as a complete unit and no highly toxic CN^- ions are produced.

Large ligands can be attached to the metal ion by two or more bonds providing that their spacing is sufficient to accommodate the spatial distribution of the secondary valencies of the metal ion. The terms bi-, tri-, hexa-, or polydentate are used to describe ligands with 2, 3, 6, or many points of attachment. Often rings are formed which are called *chelate* rings, 5- and 6-membered rings usually are most stable and hence are those most commonly encountered. Chelation may increase bond strengths by factors even as great as 1 millionfold; some polydentate ligands, such as the porphyrin rings in hemoglobin or vitamin B_{12}, form exceptionally stable complexes which fix Fe^{2+} and Co^{2+}, respectively, in their lower oxidation states and facilitate the biochemical functions of the molecules. The directional nature of the secondary bonds of a metal ion enables them to act as templates to hold specific configurations; such templates probably played a major role in biochemical evolution and they are important today in, for example, the production of chiral isomers.

In biochemistry and pharmacology complexing ligands can be used to hold a metal in an unfavorable valence state, to neutralize a charge to enable a metal complex to pass through a cell membrane, to produce exceptionally stable bonding by multiple chelation and to assist in the design of ligand drugs designed to react with a specific ion *in vivo*, for example to detoxify and remove a polluting metal.

17.3.2 Metal–Ligand Specificity

Complex stability

Metal ions and ligands show a definite order of affinity for each other. One method of quantifying the degree of tightness of the binding is to use mass action equilibrium data for the reactions between the metal and a ligand, or ligands, to calculate a formation constant, K, or β. Thus for the reaction of one atom of a metal M with one molecule of a ligand L, the equilibrium is:

$$K_1 = \frac{[\mathrm{ML}]}{[\mathrm{M}][\mathrm{L}]}$$

where K_1 is the formation constant. A second constant K_2 describes the stability of the 2:1 complex ML_2. A cumulative constant, $\beta = K_1 \cdot K_2$, is used to describe the overall reactions:

$$\mathrm{M}^{n+} + \mathrm{L}^- \rightleftharpoons \mathrm{ML}^{(n-1)+} \quad \text{and} \quad \mathrm{ML}^{(n-1)} + \mathrm{L}^- \rightleftharpoons \mathrm{ML}_2^{(n-2)+}$$

When considering whether the *modus operandi* of a new or existing pharmaceutical may involve metal ion complexing it is helpful to be able to predict which metal ion–drug interactions may occur and how strong they will be. Such predictions can be made using the hard and soft acid and base approach.

The hard and soft acid and base approach

This approach[4a] assumes: (a) that if a bond exists between two atoms, one will play the role of an acid and the other a base, and (b) that electrons hold the bonded atoms together.

The acid is taken to be a Lewis acid-type species (atom, molecule, or ion) that has vacant accommodation for electron pairs and the base has the tendency to give up electron pairs to the acid. A typical acid–base reaction is:

$$\underset{\text{Acid}}{A} + \underset{\text{Base}}{B} \rightleftharpoons \underset{\text{Complex}}{AB}$$

The bond A−B may be any chemical bond, e.g., $[(H_3N)_5Co{-}NH_3]^{3+}$, $C_2H_5{-}OH$, or H−OH. Thus, we may theoretically dissect any species into an acid and a base fragment, irrespective of whether either fragment could exist in isolation.

Softness arises from the electron mobility or polarizability of a species. If the electrons are easily moved, the species is *soft*, if they are firmly held it is *hard*. Other descriptions of a soft base could include such terms as polarizable, easily oxidized, loosely held valence electrons: these are all associated with a low charge density on the base. A hard acid has a high charge density, is of small size and, usually, does not contain unshared electrons in its valence shell. Naturally, a hard base and a soft acid are the converse of these descriptions.

The single principle underlying the hard and soft acid and base (HSAB) approach is that a *strong bond* is formed by interaction of a *hard acid* with a *hard base*, or a *soft acid* with a *soft base*. *Hard–soft* bonds are *weak* or do not form at all. Some examples of physiologically or pharmacologically important hard and soft acids and bases are shown in Table 17.3.

The HASB principle in biochemistry and pharmacy

The HSAB concept[4a] that strong bonds are formed only between hard–hard or soft–soft components is widely seen in chemistry, biochemistry, and pharmacology and many examples are to be

Table 17.3 Hard and soft acid and base (HSAB) classification of some physiologically or pharmacologically important metals and ligands[4a]

Hard	Soft	Borderline
Acids		
H^+, Li^+, Na^+, K^+	Cu^+, Tl^+, Hg^+, CH_3Hg^+	
Mg^{2+}, Ca^{2+}, Mn^{2+}	Cd^{2+}, Pt^{2+}, Hg^{2+}, Pt^{4+}, $Co\,(CN)_5^{2-}$	Fe^{2+}, Co^{2+}, Ni^{2+}, Cu^{2+}, Zn^{2+}, Pb^{2+}, Sn^{2+}, Bi^{2+}, SO_2, NO^+
N^{3+}, Cl^{3+}, Cr^{3+}, Co^{3+}, Fe^{3+}, As^{3+}, Si^{4+}	Tl^{3+}, BH_3, RS^+, RSe^+, I^+, Br^+, HO^+, RO^+	SO_2, NO^+
VO^{2+}, MoO^{3+}	I_2, Br_2, chloranil, quinones, O, Cl, Br, I, N, RO, RO_2	
HX (hydrogen-bonding molecules)	M^0 (metal atoms)	
Bases		
H_2O, OH^-, F^-	R_2S, RSH, RS^-	$C_6H_5NH_2$, N_3^-, Br^-, NO_3^{2-}, SO_3^{2-}, N_2
COO^-, CH_3CO^-, PO_4^{2-}, SO_4^2	I^-, SCN^-, $S_2\,O_3^{2-}$	
Cl^-, CO_3^{2-}, $(CN)_5^{2-}$ NO_3^-	CN^-, RNC, CO,	
ROH, RO^-, R_2O	C_6H_6, H^-, R^-	
NH_3, RNH_2,		

found in the suggestions for further reading. One of the earliest examples of this concept at work is found in the observations made by Berzelius (1779–1848) who, in the early years of the 19th century, noted that some metal ores occurring on the Earth's surface were carbonates or oxides, while others were sulfides. This can be explained on the HSAB principle, since hard acids, e.g., Mg^{2+}, Ca^{2+}, Al^{3+}, or Fe^{3+} form strong bonds with hard bases such as O^{2-}, CO_3^{2-}, or SO_3^{2-}, whereas softer acids, e.g., $Cu^{+/2+}$, Pb^{2+}, or Ag^{+}, prefer soft bases such as S^{2-}. Any hard acid–soft base or soft acid–hard base compounds would have been so unstable that they would have hydrolyzed away many millions of years ago.

Many biochemically essential transition metal ions are shown as ''borderline'' in Table 17.3, but through chemical symbiosis they can exhibit both hard and soft properties. This is especially prevalent when a metal has more than one oxidation state. Chemical symbiosis is a process whereby one hard, or soft, base on a metal ion attracts other hard, or soft, bases to join it. Metal ions are often in a state of dynamic equilibrium between two oxidation states and the lower state can be stabilized by adding soft ligands and *vice versa*. Thus, if *very hard* or *very soft* ligands are added the metal will be completely anchored in one oxidation state thereby preventing the natural biochemical process (e.g., a redox reaction). This, of course, ''poisons'' the system and many of the best-known poisons are acids or bases that are so strongly held to the active sites of an enzyme that the site is effectively blocked. Cadmium ions and organic mercurials are good examples of soft acid poisons. In addition to blocking active sites, poisoning by heavy metal ions may result in deleterious structural changes in the protein, or even its precipitation. Soft acids, such as the heavy metal ions, bind strongly to sulfur groups and thus may effectively rob the organism of important sulfur-containing proteins, for example those containing cysteine residues. Very soft base poisons may effectively deprive the body of metal ions. For example, cyanides, sulfides, and trivalent arsenic compounds exert their toxic effect by attaching to the metals in metalloporphyrins and metalloenzymes and at high concentrations they may remove the metal ion entirely from the protein.

Conversely, nonpoisonous or inert materials are needed for artificial prostheses that have to be inserted surgically. Thus, pure metals, or pure metal alloys, which, if dissolved, would give soft acids, are often chosen, e.g., gold, silver, and tantalum. Because such metals yield soft ions, there is a negligible tendency for them to ''corrode'' in the body by forming complexes with endogenous hard bases, such as water, carbonate, or biological amines.

The principle of HSAB matching of acids and bases to achieve strong bonds also underlies the clinical use of chelating, or sequestering, agents to promote metal excretion in cases of excessive metal ion accumulation, thereby reducing or preventing toxic effects. This principle is illustrated in Table 17.4.

In vivo *complexing and metabolic specificity*

Replacement of zinc in carboxypeptidase with the isomorphous cobalt(II) ion leads to increased peptidase activity. Why then does the human body contain only zinc carboxypeptidase and no cobalt carboxypeptidase? The answer is in two parts, firstly, the natural abundance of the elements and, hence, the dietary intake of our ancestors and ourselves excludes some elements which may

Table 17.4 Some of the earliest sequestering agents used for treating metal excesses (from Fiabane and Williams[1f] with permission)

Ligand	Donor atoms	Metals mobilized	HSAB classification
Dimercaptopropanol [BAL]	$2S^{2-}$	As(III,V), Au(I), Hg(I,II)	Soft
EDTA	$4O^{2-}$, 2N	Pb(II), Co(II), Ca(II)	Borderline
D-Penicillamine	S, N, O^{2-}	Cu(I,II)	Borderline/soft
Desferrioxamine	Several O	Fe(III)	Hard

appear to be feasible on HSAB grounds and on the basis of strict isomorphous replacement. The importance of this natural abundance factor is further illustrated by the Fe^{2+}-containing porphyrins. The order of preference for the central metal ion as determined (a) thermodynamically is $Ni^{2+} < Cu^{2+} > Co^{2+} > Fe^{2+} > Zn^{2+}$ but (b) kinetic measurements have shown the order to be $Cu^{2+} > Co^{2+} > Fe^{2+} > Ni^{2+}$. Thus human hemoglobin could equally well have contained cobalt in place of iron, but iron is more than 1000 times more abundant in the hydrosphere than cobalt, thus mammalian hemoglobins are iron-based.

Secondly, the whole process by which an element, or an energy-rich compound of it, is absorbed from the diet, incorporated in the correct oxidation state into a metalloenzyme or other active center and, finally, excreted from the body involves many complexing reactions. These involve many steps including the absorption of the metal from the gastrointestinal tract, its transportation across many membranes, and the buffering and redox reactions necessary to ensure that its concentration and oxidation states are appropriate at each stage. Although cobalt can apparently replace zinc in carboxypeptidase *in vitro*, it certainly cannot follow through all the other biochemical processes that are necessary to insert it into the enzyme *in vivo*. If this chain of metal insertion into enzymes is broken, the organism cannot thrive, or multiply. Sometimes, a contaminating metal can survive through the chain and be inserted into the enzyme but it cannot be excreted and accumulates in the body as a poison. Cadmium is one such example; it can be incorporated into metalloenzymes and other metalloproteins but accumulates there as a poison.

Two key concepts run through the discussions in this section, *selectivity* and *specificity*. Normal human biochemistry is usually very specific. However, when man seeks to disturb normal biochemistry, for example by introducing a ligand drug designed to mobilize a metal ion, or to disturb some aberrant metal–ligand interaction, in order to treat disease, the best that can be hoped for is selectivity. There will inevitably be some side reactions (not necessarily manifesting themselves clinically) because more than the one desired metal will be complexed or because less than 100% of the drug will be absorbed from the intestine and transported to the site of action. Examples of this will be discussed later.

17.4 PHARMACOLOGICAL AND PHARMACEUTICAL CONSIDERATIONS

The concentrations of many essential and beneficial metal ions *in vivo* are controlled within narrow limits, some complexed to high molecular mass species such as proteins and others to low molecular mass ligands such as amino- or carboxylic acids.[1a–d] In some metalloenzymes the metal ion is so tightly bound that it is chemically inert and cannot be removed without dismantling and rebuilding the whole molecule, while in others the metal ion is reversibly complexed and is in equilibrium with low molecular mass species and aquated metal ions. Table 17.5 shows some examples of these metalloproteins. The noninertly bound portion of the metal is in a state of labile equilibrium that can readily be disturbed by internal or external causes so that the organism can no longer function normally.

Table 17.5 Inert and labile metal–protein systems *in vivo* (from May and Williams[3a] with permission)

Inert and/or thermodynamically nonreversible	Labile and thermodynamically reversible
Iron	
Hemoglobin, myoglobin (Ferritin)	Transferrin $\rightleftharpoons$ lmm Fe^{3+} complexes $\rightleftharpoons$ $[Fe(H_20)_6]^3$
Copper	
Ceruloplasmin (Metallothionein)	Serum albumin $\rightleftharpoons$ lmm Cu^{2+} complexes $\rightleftharpoons$ $Cu(H_2O)_6]^{2+}$
Zinc	
α-2-Macroglobulin, metallothionein	Serum albumin $\rightleftharpoons$ lmm Zn^{2+} complexes $\rightleftharpoons$ $[Zn(H_2O)_6]^{2+}$

lmm = Low molecular mass complexes.

As is indicated in Table 17.2 a number of disease states have been associated directly with changes in the concentrations of trace metals in tissues and body fluids. Drugs whose *modus operandi* is through electron donation may well complex metals, either as their central mechanism or as a side effect. Invariably the matching of ligands to metals is in keeping with the principles of the HSAB concept.

The late development of knowledge of this subject has arisen, in part, from the fact that most of these trace metal complexes are normally present at concentrations below those measurable by even the most sophisticated analytical techniques. Further, any attempt to concentrate the species up to analytical levels completely upsets the equilibrium shown in Table 17.5 so that unrepresentative species (usually the least soluble) precipitate or are extracted.

During the last four decades, computer simulation has been used to analyze metals in biological fluids (biofluids) at normal concentrations. In its simplest form, computer speciation accepts that constants are invariant and just as the bioavailability of aspirin can be calculated from its protonation constant, so the computer can combine the effects of literally thousands of such constants, solubility products, etc. along with total metal and ligand concentrations to compute the distribution of species in biofluids at steady state.

The analysis of the concentration and distribution of the different chemical forms (species) of a metal in a system is known as *speciation*. As an illustration Table 17.6 shows the principal species in which zinc(II) exists in normal blood plasma and in the fluid draining from a surgical wound.[7a,b] Computer simulations of metal speciation in biofluids, such as plasma, saliva, wound fluids, intestinal juices, or milk, often run to more than 10,000 species. They require a databank of well-validated constants and analytical data and a medium-sized computer; nowadays many speciation programs can be run on a high-performance personal computer. It is important that as far as possible, computer simulations should be validated by experimental data. Many of the principles discussed earlier in this chapter have been revealed, or verified retrospectively, using computer speciation.

17.5 METALS AS THE *MODUS OPERANDI* OF DRUGS

Some drugs are metal compounds in which the metal is essential for their desired pharmacological effect; others are ligands that exert their effects by interaction with an endogenous metal at their site of action. The use of metals as drugs can be traced back for thousands of years, for example zinc oxide ointments for wound healing or solutions of rust in acid wine for the treatment of anemia were used in pre-Christian times. Paracelsus in the 16th century made the general introduction of some heavy metals into the *materia medica*. Today many metals are known to play important roles, or to

Table 17.6 Predominant zinc complexes identified in the low molecular mass fractions (lmm) of blood plasma and wound fluid by computer speciation analysis[7a,b]

		% Total Zn(II) in lmm form	
Complex species	**Charge**	**Plasma**	**Wound fluid**
Zn(II)-cysteinate-citrate	3−	43	53
Zn(II)-H-cysteinate-phosphate	3−		23
Zn(II)-*bis*-cysteinate	2−	19	6
Zn(II)-H_2-cysteinate	2+		4
Zn(II)-histidine-cysteinate	−	12	3
Zn(II)-histidinate	+	3	
Zn(II)-cysteinate	0	3	
Others (each <1%)		20	11

show great potential for a future role, in the treatment of cancer and other serious diseases; many of these aspects are reviewed in Refs. [1i–j].

17.5.1 Metals as Drugs

More than a dozen metals have been used as drugs during the past few hundred years, including highly toxic elements such as arsenic, mercury, and lead; Table 17.7 lists some of those which still find applications in healthcare. In addition, vanadium has recently been shown to be a potential drug for the treatment of Type II diabetes, and radioactive isotopes of a number of elements ranging in diversity from technetium to rhenium have found applications as diagnostic or radiotherapeutic agents in nuclear medicine.[6a,b] Some of these applications are discussed below.

Bismuth in the treatment of peptic ulcer

The heavy metal bismuth has been widely used for many years for the treatment of gastric or duodenal ulceration.[1j] One widely prescribed preparation is tri-potassium dicitratobismuthate (bismuth chelate). Intragastric fiber-optic color photography suggested that the bismuth acted as a cytoprotectant by coating the ulcerated area with a precipitate that protected the raw surface from further attack from the gastro- or intestinal juices. Computer speciation analysis subsequently showed that in the acid environment of the stomach the soluble charged $Bi(citrate)_2^{3-}$ is converted into insoluble precipitates of BiOCl and insoluble bismuth citrate.

Computer simulation also suggested other bismuth formulations that could produce cytoprotective bismuth patches over ulcers at sites having more neutral pH values. However, optimization of such ulcer therapies is difficult because of the vagaries of ulcer origins and their response to therapy.

A new aspect to ulcer therapy with bismuth compounds was created by the discovery in 1982 that acid secretion by the intestinal bacterium *Helicobacter pylori* may well play a major role in the induction of gastric ulcers, and even gastric cancer. Bismuth is itself bactericidal, but its effect is weak, however, combination of ranitidine bismuth citrate with antibiotics such as clarithromycin plus metronidazole or amoxicillin or tetracycline has raised the success rate for ulcer therapy to about 90%, i.e., to about four times that achievable with bismuth alone.

Bismuth oxide and/or bismuth subgallate are also used in combination with other agents in pain-relieving preparations for the treatment of hemorrhoids, e.g., Anusol-HC®.

Table 17.7 Some metals present in pharmaceuticals administered to humans

Element	Compound	Proprietary name — and use
Aluminum	Aluminum hydroxide	Aludrox® — antacid
	Aluminum silicate	Kaolin — antidiarrheal
Antimony	Sodium stibogluconate	Pentostam® — antilieshmaniasis
Bismuth	Tripotassiumcitrato-bismuthate; ranitidine bismuth citrate	De-Nol® — antiulcer; Pylorid® — antiulcer
Cobalt	Vitamin B_{12}	Cyanocobalamin — pernicious anemia
Iron	Ferrous glycine sulfate; iron sorbitol	Plesmet® — iron deficiency; Jectofer® — parenteral iron therapy
Gold	Sodium aurothiomalate	Myocrisin® — antiarthritic
Magnesium	Magnesium sulfate	Epsom salts — laxative
Platinum	*cis*-Dichlorodiammine-platinum(II)	Cisplatin — anticancer agent
	cis-(1,1-Dicarboxycyclo-butane) diammine-platinum(II)	Carboplatin — anticancer agent
Selenium	Selenium sulfide	Selsun — seborrheic dermatitis
Silver	Silver sulfadiazine	Flamazine® — infected burns; leg ulcers
Zinc	Zinc sulfate	Solvazinc® — proven zinc deficiency

Metals as anticancer agents — the "platinum drugs"

A number of metals, such as gallium, hafnium, palladium, have been shown in animal experiments to be able to slow down the growth of tumors, but none has yet proved sufficiently effective to justify clinical trials.[1] However, an important, but accidental, discovery in 1964 led to the development of an exciting new class of inorganic cytotoxic drugs for cancer chemotherapy. Studies of the effects of electric currents on the growth of microorganisms led to the discovery that the simple presence of a platinum electrode in a culture, without any electric current applied, led to severe growth disturbance. The active species were recognized to be tiny concentrations of platinum complexes formed by interactions between the culture medium and the pure platinum metal electrodes.

Tests of one such complex, especially *cis*-dichlorodiammineplatinum(II) (cisplatin) (F1), in animal tumor systems showed that this was a very potent cytostatic agent. Unfortunately, it also exhibited such high general toxicity, especially nephrotoxicity, that, initially, any clinical application appeared unlikely. However, despite this disadvantage, a Phase I clinical trial was commenced which confirmed the high antitumor effectiveness of cisplatin and showed that some of the more general toxic effects, especially the nephrotoxicity, could be avoided by intensive hydration of the patient.

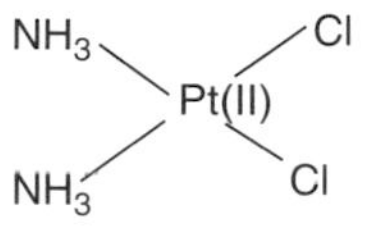

F1 Cisplatin 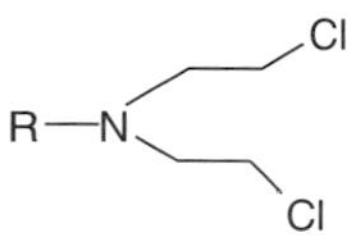 **F2 Nitrogen mustard** 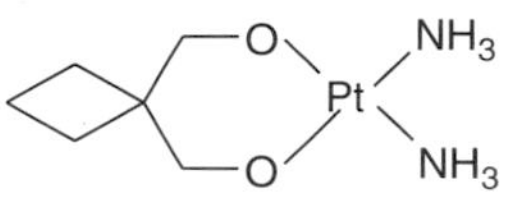 **F3 Carboplatin**

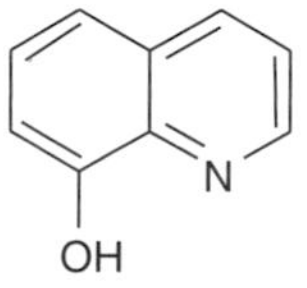

F4 8-Hydroxyquinoline (Oxine) 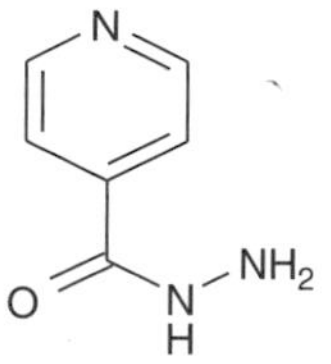 **F5 Isoniazid** 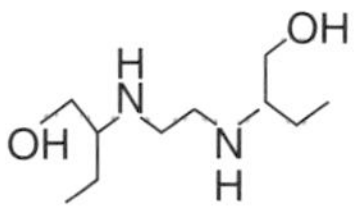 **F6 Ethambutol**

Cisplatin is a planar, electrically neutral complex that is able to cross cell membranes. Its cytotoxic action arises because *in vivo* the two adjacent chloride ions are lost, permitting the formation of platinum chelates with two nitrogens from purine or pyrimidine bases in the deoxyribonucleic acid (DNA) chain to create an intrastrand link within the DNA; this then interferes with the replication of the DNA when the cell next attempts to divide. The formation of the intrastrand link by cisplatin in a single strand of the DNA contrasts to the formation of an interstrand links between the two DNA strands of the α-helix that commonly occur with other types of alkylating agent, such as those based on nitrogen mustards (F2). This difference in the mode of action of cisplatin and the nitrogen mustards probably lies in the separation between the two chloride ions in the molecules. In cisplatin the distance between the two chloride ions is 0.33 nm, while the separation between the chloride ions at the ends of the arms of the nitrogen mustards is 0.80 nm; these distances correspond perfectly to the space required to form inter- and intrastrand linkages, respectively.

Cisplatin was introduced clinically in the UK in the late 1970s and rapidly became a first line drug for use, either alone or in combination with vinblastine or other cytotoxic agents, in the treatment of ovarian and testicular cancer and also for the treatment of bladder, lung, and upper gastrointestinal tract cancer. However, cisplatin suffers from serious disadvantages; it must be infused intravenously; it causes severe nausea and vomiting and is one of the most powerful emetics known; it also causes leukopenia, and renal dysfunction cannot be entirely avoided. Further, some tumors develop resistance to the drug. Later research showed that replacement of the chloride groups by 1,1-dicarboxycyclobutane to yield the *cis*-(1,1-dicarboxycyclobutane)diammineplatinum(II), carboplatin (F3), derivative removed some of these problems and introduced a second-generation agent into clinical use. Carboplatin is now widely used in the treatment of advanced ovarian and lung cancer, especially small cell type tumors. Carboplatin, like cisplatin, must be administered intravenously but it is better tolerated with nausea and vomiting being less severe and renal, neuro-, and ototoxicity being much less of a problem than with cisplatin. However, carboplatin has a greater effect on the bone marrow (myelosuppression) than cisplatin.

A third platinum drug, containing an oxalic acid residue, oxaliplatin, has now entered clinical practice for the treatment of metastatic colorectal cancer in combination with fluorouracil and folinic acid (leucovorin). Like the other platinum drugs oxaliplatin must be given intravenously and it exhibits side effects such as gastrointestinal disturbance, ototoxicity, myelosuppression, and neurotoxicity, the latter being the dose-limiting factor.

Another fascinating development arose from the discovery in the 1990s that several *trans*-platinum complexes, for example *trans*-ammine(cyclohexylamine)-dichlorodihydroxyplatinum(IV), showed *in vitro* and *in vivo* sensitivity against tumor cells that were resistant to cisplatin. These complexes appear to form interstrand cross-links in DNA and have forced a reevaluation of the structure–activity relationships for platinum antitumor drugs, because the factors determining the cytotoxic activity of *trans*-platinum complexes are not the same as those of the *cis*-complexes. This may provide an exploitable rationale for the development of a range of *trans*-platinum antitumor drugs.[11]

Copper and rheumatoid arthritis

Rheumatoid arthritis afflicts about 5% of the UK population and is particularly prevalent among the elderly. The origins of the disease are unknown but it appears to involve disturbances in the chain of reactions that control the patient's autoimmune response and a role for copper has been suggested.[3a,b] However, although such ''folklore'' remedies as wearing a copper bracelet have been widely used, to date no successful copper-containing drug for rheumatoid arthritis has been developed. In part this may reflect the fact that orally administered copper compounds can cause extreme gastrointestinal irritation.

Vanadium compounds in the treatment of diabetes

In recent years several vanadium complexes of various oxidation states (III–V) have shown considerable promise as potential orally active therapeutic agents for Type II diabetes mellitus. One of the compounds of great interest is bis(maltolato)oxovanadium(IV). The mechanism of action is, as yet, uncertain, and the results of clinical trials are awaited.[11]

Zinc and the immune system

Zinc is known to play an important role in preserving the integrity of the immune system.[5a] It has been reported that daily administration of zinc gluconate could alleviate the symptoms of the common cold.[5a] This claim was based on studies in thousands of volunteers who have been tested in double blind clinical trials of the reduction in symptoms of the common cold, after administration of zinc gluconate. Studies of the correlation of the apparent biological response measured as a reduction in common cold symptoms with (a) the total amount of zinc gluconate administered and (b) the salivary concentration of zinc species as determined by computer speciation modeling, showed that there appeared to be a significant link between the biological response and the concentration of *free* ${Zn_{aq}}^{2+}$ ions in saliva. However, there was no obvious correlation between biological response and the *total* amount of zinc ingested as zinc gluconate lozenges. Optimal activity appears to require salivary *free* ${Zn_{aq}}^{2+}$ ion concentrations of 2 and 4 mmol dm^{-3} at pH = 6.75 and pH = 5.5, respectively.

17.6 DRUGS THAT EXERT THEIR EFFECTS *VIA* METAL COMPLEXATION OR CHELATION

17.6.1 Metal Chelation in Antimicrobial Activity

Many microorganisms are critically dependent on one or more metals for their growth and a number of the more effective antimicrobial drugs act by denying the organism the use of such metals.[1i,j]

The chelating agent 8-hydroxyquinoline, oxine (F4), a rapidly acting antimicrobial and fungicidal agent, was one of the first such agents to be shown to exert its action *via* metal chelation, since oxine-derivatives in which the ligand donor groups were blocked with −O− methyl, ≡N−methyl, or isomeric hydroxy groups showed no antimicrobial activity. Traces of Fe^{2+} or Fe^{3+} were shown to be required for activity and evidence suggests that the site of action lies inside the cell, or at least within the plasma membrane, since the derivative hydroxyquinoline-5-sulfonic acid, that has increased hydrophilic properties and the ability to chelate iron is not antibacterial. The oxine–iron complex, as a result of rearrangement of the orbitals of the Fe^{3+} ion, is able to catalyze the oxidation of thiol groups in lipoic acid, an essential coenzyme required by bacteria for the oxidative decarboxylation of pyruvic acid. The importance of the lipophilic properties is illustrated by the activity of halogenated derivatives such as 5,7-diiodo-8-hydroxyquinoline and 5-chloro-8-hydroxy-7-iodoquinoline against organisms causing bacterial dysentery.

The mode of action of the 4-quinolonecarboxylic acid family of antibiotics are now known to act by the inhibition of bacterial DNA synthesis, perhaps with the enzyme DNA-gyrase as the target. It is suggested that the inhibition of this enzyme is due to the formation of a reversible ternary DNA–drug–enzyme complex and that iron or copper ions are involved in the inhibition.

The entry of the antitubercular agent isonicotinic acid hydrazide (isoniazid) (F5) into the tubercle bacillus is mediated by the formation of a lipid-soluble copper chelate. The activity of other antitubercular drugs, including ethambutol (F6), is also dependent on copper(II) chelation.

The tetracyclines (F7) constitute a group of important agents for treating systemic bacterial infections. High values for stability constants and the presence of hard basic groups, such as hydroxyl anions and tertiary amino moieties, indicate a readiness to complex Ca^{2+} and Mg^{2+}. Much evidence suggests that tetracycline owes its antibacterial activity to its ability to complex Mg^{2+} in the bacterial cell membrane. The increased lipophilicity of the Mg^{2+}–tetracycline complex facilitates concentration in the bacterial cell where it blocks protein biosynthesis by interfering with the binding of aminoacyl-*t*-RNA to ribosomal receptors.

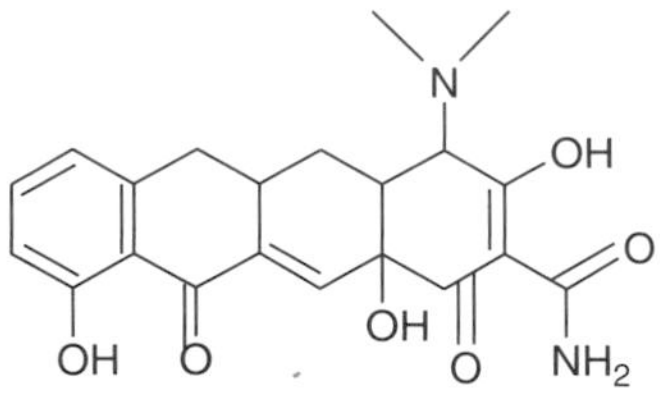

F7 Tetracycline

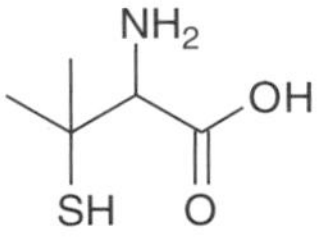

F8 D-Penicillamine

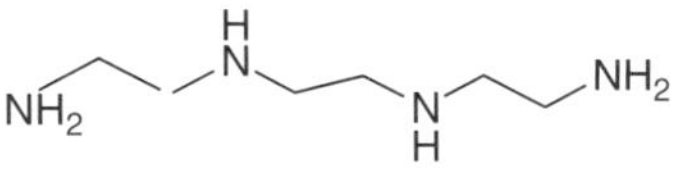

F9 Triethylenetetramine (TREN or Trientine)

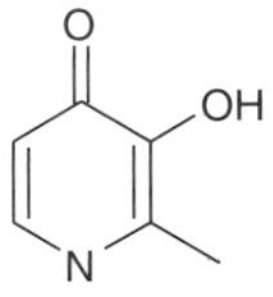

F11 Deferiprone

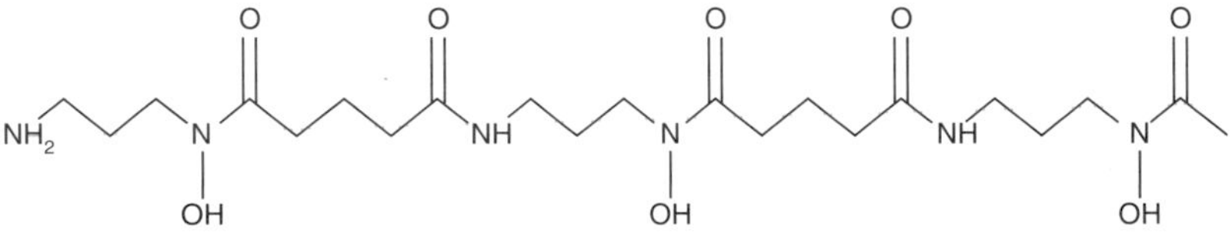

F10 Desferrioxamine (Desferal or DFOA)

17.6.2 Metal Ion Removal

In metal storage diseases, such as Wilson's disease and hemochromatosis, the symptoms can often be alleviated and the progress of the disease slowed down by treatment with a chelator with a high affinity for the metal concerned. Similarly, poisoning by exogenous metals can also be treated with appropriate chelators. The use of chelating agents to accelerate the elimination of, or to *decorporate*, unwanted metal from the body is sometimes called *decorporation therapy*.

Removal of copper in Wilson's disease

Wilson's disease (hepatolenticular degeneration) is an idiopathic condition characterized by an inability of the body to use copper for ceruloplasmin synthesis that results in massive overloading of albumin and the low molecular mass ligands with copper. Dietary copper becomes deposited in excessive amounts in the brain, liver, eyes, and other tissues causing neurological symptoms and cirrhosis of the liver leading to death relatively early in life. The progression of the disease can be markedly slowed by treatment with the chelating agent D-penicillamine (F8), a hydrolysis product of some penicillins. This agent does not release copper from serum albumin neither does it degrade ceruloplasmin nor release its vast stores of copper; however, it may possibly liberate the metal from liver metallothionein-binding sites, and it certainly does markedly increase the urinary excretion of copper.

Although D-penicillamine does not apparently disturb serum albumin copper, it does mobilize zinc from this protein and this together with bone marrow depression are recognized side effects of D-penicillamine therapy. These are potentially serious side effects and D-penicillamine is now often replaced by the tridentate ligand triethylenetetramine (TREN or Trientine) (F9) in the treatment of patients with Wilson's disease who are intolerant toward penicillamine.

D-Penicillamine is also used for the treatment of severe active rheumatoid arthritis, in this application it is interesting to speculate that the mode of action of the drug could be a local release of copper in areas of severe disease.

Removal of iron in hemochromatosis

The massive storage of iron encountered in either primary or secondary hemochromatosis is another condition that is amenable to treatment with chelators.[2a,b] The genetic disease thalassemia (sickle cell anemia) causes severe anemia that has to be treated with repeated blood transfusions and this leads in time to severe secondary hemochromatosis. This disease creates a major clinical problem in some tropical countries. For the last four decades the fungal siderophore desferrioxamine (Desferal) (F10) has been the only selective iron chelator available for clinical use. This is an extremely powerful ligand for the chelation of Fe^{3+} and it has the advantage of having little affinity for other essential metals such as copper, zinc, calcium, or magnesium. Desferal forms had acid–hard base complexes with iron in which the iron(III) is bound more strongly than in the Fe–EDTA complex. X-ray diffraction studies suggest that the complex formed involves iron(III) bound to three hydroxamic acid–N(OH)CO groups.

Desferal is relatively expensive and must be administered by slow subcutaneous infusion. Thus, the cost of the drug and the need for infusion pumps are serious disadvantages when the pressing clinical need is to treat large numbers of young people in poor countries. There is great interest in developing inexpensive, orally active chelators for iron removal and derivatives of 3-hydroxypyridin-4-one appear to offer the desired properties. One such derivative, originally code named L1 (F11), has recently been licensed for clinical use as a second line agent under the name deferiprone.[2c] This agent is able to mobilize iron from the main iron-storage protein ferritin, possibly because the molecule is small enough to enter into the so-called tunnels in the ferritin molecule and to directly chelate the ferric iron deposited there. The hydroxypyridones have also been shown to mobilize iron from hemosiderin, the degradation product of ferritin that accumulates in the tissues of thalassemia patients. The iron so mobilized is excreted from the body mainly through the urine. Unfortunately, deferiprone is readily metabolized and this leads to undesirable side effects, in particular gastrointestinal disturbances, agranulocytosis, neutropenia, and arthropathy.[2d]

Treatment of exogenous metal poisoning

Acute or chronic poisoning by lead, or other heavy metals, remains an important clinical problem. The polyaminopolycarboxylic acid chelator ethylenediamine-N,N,N',N'-tetraacetic acid (EDTA) (F12) has a high affinity for lead and this agent has been shown to be a reasonably efficient chelator for lead *in vivo*. However, EDTA suffers from some disadvantages. First, as a charged ion $ETDA^{4-}$, the form in which the drug exists in plasma, it is unable to pass through lipid membranes to reach lead deposited in cells; second EDTA is also able to complex essential ions such as Mn^{2+} and Zn^{2+} and its prolonged use can lead to a deficiency of these metals.

For some purposes EDTA is administered as the disodium salt because the tetrasodium salt is too alkaline. The disodium salt of EDTA, disodium edetate, is also used to reduce blood calcium levels in hypercalcemia. The hard basic groups, $-COO^-$ and $-NH_2$, of the tetradentate EDTA form a stable water-soluble complex with the hard Ca^{2+} ions which is excreted via the kidney. Disodium edetate has also been used for treating lime burns on the cornea and for restoring the K^+/Ca^{2+} balance in cardiac arrhythmias accidentally induced by digoxin. ''Decorporation therapy'' by regular infusions of disodium edetate is also used as a rather equivocal form of treatment for patients with atherosclerosis; the rationale that the chelator mobilizes calcium from the atherosclerotic plaques thus improving general blood flow; some spectacular results have been claimed by protagonists of this form of therapy.

For the treatment of poisoning with lead or mercury, the administration of the disodium calcium–EDTA complex is preferred as this prevents the chelation and excretion of the essential endogenous metal calcium. An analog of EDTA, diethylenetriamine pentaacetic acid (DTPA) (F13), is employed in the nuclear industry for the treatment of the infrequent accidents involving human contamination with plutonium or americium.

The problems created by the charged nature of the $EDTA^{4-}$ and $DTPA^{5-}$ species, which limits their transport across cell membranes and, thus, their access to intracellular metal deposits may, in theory at least, be circumvented by treatment with two ligands. If a ligand that exists in electrically neutral form in the blood, for example D-penicillamine, is administered first, followed by EDTA administration, then intracellular deposits of a metal such as lead may be mobilized and transported into the blood stream as neutral D-penicillamine complexes and then quantitatively eliminated in

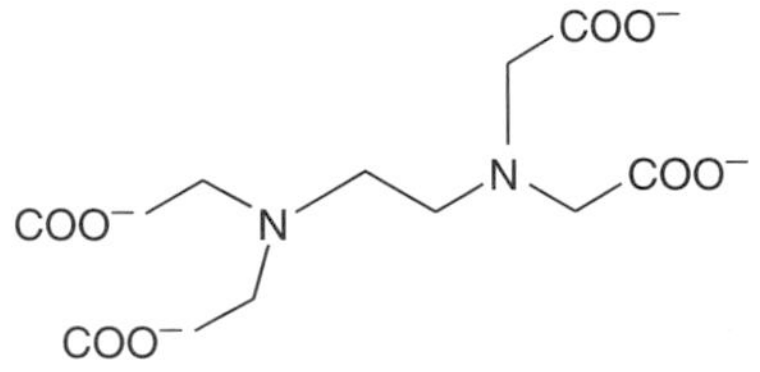

F12 Ethylenediamine tetra-acetate (EDTA)

F13 Diethylenetriamine-penta-acetate (DTPA)

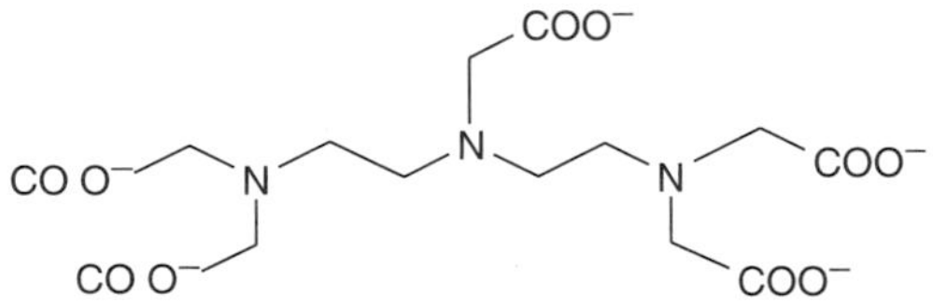

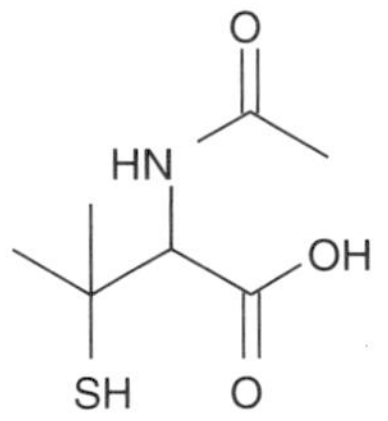

F14 *N*-acetyl-D-penicillamine

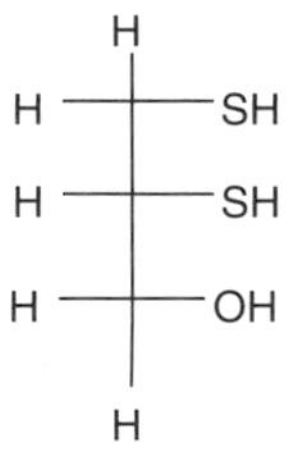

F15 Dimercapto-propanol (BAL)

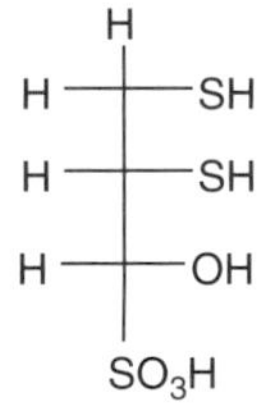

F16 Dimercatopropane sulfonate (DMPS or Dimeval)

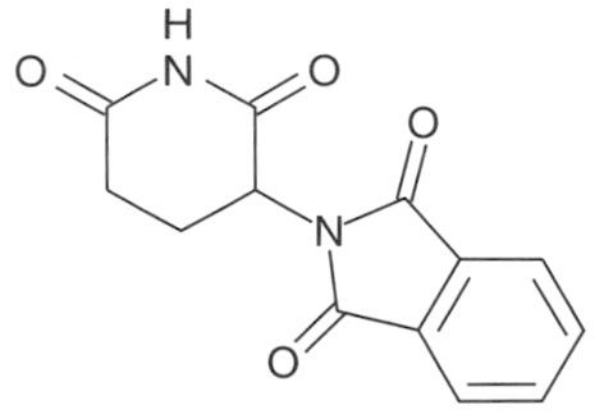

F17 Thalidomide

the urine following interaction with EDTA — such a regimen of using two chelators is known as synergistic chelation therapy.

Mercury and lead poisoning have also been treated successfully with D-penicillamine, although the *N*-acetyl derivative (F14) that has a softer basic group is considered to be more effective.

Poisoning by mercury, arsenic, gold, or antimony may be treated with dimercaptopropanol (dimercaprol) (F15) or British Anti-Lewisite (BAL), one of the first chelators to be used clinically. Dimercaprol provides soft sulfydryl groups that bind these soft metals forming water-soluble complexes. Dimercaprol must be injected in an oil suspension which has a number of disadvantages, and the water-soluble derivative dimercaptopropane sulfonate, dimeval (F16), has been used as an alternative drug for mercury and arsenic poisoning.

17.6.3 Drugs as Metalloenzyme Inhibitors

As mentioned in Section 17.2, the body contains more than 1000 metal-containing enzymes and in most cases the metal is essential for the specific enzymatic function. Thus, the metal ions in these enzymes may form important targets for drug action: this is discussed in more detail in Chapter 9.[1m] Increasingly many drugs are said to be metal ion-seeking since their active sites are electron donors, which inorganic chemists commonly term "ligand donor groups." As many as 900 of the metalloenzymes contain zinc as their prosthetic group. A very important example of these Zn-metalloenzymes is carbonic anhydrase (CA), which exists in most human tissues as one or more of 14 different isozymes or CA-related proteins. CA catalyzes a single reaction that is essential for the function of most human tissues, the interconversion of a molecule of carbon dioxide with a water molecule to form a bicarbonate ion:

$$O{=}C{=}O + H_2O \rightleftharpoons HCO_3^- + H^+$$

The active site of the enzyme is the zinc atom itself, thus if this element is removed or blocked by a drug the enzyme itself is inhibited. Sulfonamides, such as acetazolamide, inhibit CA and produce a diuretic function, but the effect is weak and the drug is little used clinically for this purpose. The CA-inhibitors, acetazolamide administered orally, or dorzolamide and brinzolamide, applied as eye drops, reduce intraocular pressure by decreasing aqueous humor production and are important drugs for the treatment of glaucoma. Another important Zn-metalloenzyme is the angiotensin-converting enzyme (ACE) (see Chapter 9) that has two main actions. Firstly, it cleaves a dipeptide from angiotensin I to form the hypertensive agent angiotensin II that in turn stimulates the release of aldosterone. Secondly, ACE catalyzes the hydrolysis of another peptide, the vasodilator bradykinin. Thus, as discussed in Chapter 9, the Zn-binding drugs such as captopril, enalapril, and lisinopril that inhibit ACE play important roles in the treatment of high blood pressure. A further important group of metalloenzymes are the so-called membrane metalloendopeptidases (MEP) which also contain Zn; these enzymes are involved in the deactivation of a range of peptides and hormones, including cholecystokinin, bradykinin, and atrial natriuretic peptide (ANP). Inhibitors of MEP could potentially play a role in analgesia, however, no clinically useful agent has yet appeared. Substances, which could prolong the activity of ANP, have a potential role in the treatment of hypertension and congestive heart and failure; this is an area of active drug research.[1m]

The matrix metalloproteinases (MMP) are another group of Zn-metalloproteases that play major roles in many areas of cellular activity, including embryogenesis, nerve growth, ovulation, angiogenesis, wound healing, bone remodeling, apoptosis, arthritis, and cancer invasion and metastasis.[1n] Inhibitors of MMPs could thus prove useful in the treatment of a variety of diseases including some cancers. Hydroxamates, binding preferentially to the catalytic Zn atom, are potent inhibitors of MMP and a range of hydroxamate derivatives, currently under study, may well have potential as therapeutic agents in the treatment of various diseases including cancer.

17.7 METAL-DEPENDENT SIDE EFFECTS OF DRUGS

A *prima facie* case can usually be made linking a metal ion interaction with almost any drug in the pharmacopeias and several therapeutic examples of such links have already been discussed. However, the interaction with the metal ion is not necessarily part of the desired effect and may lead to unwanted side effects. Examples of such effects would be a drug that in the presence of an endogenous metal ion is extensively inactivated, or a substance that produces a metabolite that complexes with, and inactivates, an essential metal, or an agent designed to accelerate the excretion of an unwanted metal but which also enhances the excretion of an essential metal.

It has been postulated that thalidomide (F17), the drug which when taken by pregnant females as a tranquillizer caused so many tragic birth abnormalities in the 1960s, could have produced its effects of limb-shortening through hydrolysis in the embryo of the peptide-like bonds in the molecule, to produce complexing moieties that sequestered the Ca^{2+} ions that were essential for the normal development of the limb buds. Thalidomide, although unlicensed, may also be used in the treatment of men and postmenopausal women suffering from leprosy and who have become corticosteroid-dependent, in this case also Ca^{2+} or other metal sequestration may be a *modus operandi*.

17.7.1 Metal Ion Sequestration by a Metabolite

Most antitubercular drugs are potential metal ion chelators and, as mentioned earlier, this can enhance their biological activity by rendering them more bioavailable. However, whether such metal chelation is a characteristic of their activity or not, the administration of a ligand drug is likely to interfere with normal trace element behavior. Such interactions need to be characterized and quantified lest topping-up therapy is necessary.

Ethambutol ((+)-2,2′-ethanediyldiimino)-*bis*-1-butanol)) (F6), an antitubercular drug, is one such example. While it has been postulated that the mode of action of ethambutol may involve the formation of a ternary complex involving copper(II) ions and RNA, clinical studies have shown that it does lead to an increase in urinary zinc excretion.

Solution studies have been used to measure the formation constants for the complexing of ethambutol and its principal metabolite, 2,2′-(ethanediyldiimino)-*bis*-1-butane carboxylic acid (EBDA), with a range of metal ions essential to humans. Insertion of these data in a computer simulation of the interactions in blood plasma showed that a dose of 25 mg ethambutol/kg body weight, which produces an EBDA concentration of 1.25×10^{-6} mol dm^{-3}, raised the concentrations of low molecular mass zinc complexes by up to one third, the new complex Zn–EBDA^0 is electrically neutral. However, up to ten times this dose of ethambutol has no effect on the metal ion speciation. Clearly, the clinical disturbances reported during ethambutol therapy are not due to the drug itself, but to its metabolite EBDA. Further, such undesirable side effects as peripheral vision loss may be explained by the ability of the neutral zinc complexes formed causing migration of the metal from relatively zinc-rich areas in the eye.

17.7.2 Sequestering Drug-Nonspecific Metal Ion Interactions

EDTA-H_4 is the most widely used synthetic chelating agent *in vivo* and also finds extensive use in pharmaceutical formulations as well as in industrial and domestic applications.

As mentioned in Section 17.6, EDTA is used in medicine for the treatment of lead or calcium excess. The disodium salt is usually used because it increases the solubility of the compound, intravenous infusion is necessary since EDTA^{4-} as a charged ion is poorly absorbed from the gastrointestinal tract making oral treatment largely ineffective. For the removal of excess metal from the body (*decorporation therapy*) quite large doses are administered, up to 40 mg kg^{-1} body weight or ~3 g day^{-1} and the treatment may need to be repeated many times.

The first recorded clinical use of EDTA was as the nickel complex, which was administered as a potential treatment for adenocarcinoma, but unfortunately, the agent was almost quantitatively excreted in unchanged form in the urine. This presumably reflects the very high formation constant of the Ni^{2+}–$EDTA^{4-}$ complex and the fact that no metal ions could be released *in vivo* to exert any antitumor effect. Over the last 30 years or so more than 2 million treatments with EDTA have been administered in the USA alone, either for calcium removal ($EDTAHNa_3$ — Limclair) or lead removal ($EDTAH_2Ca$ — Ledclair).

The agent is not specific for calcium, lead or any similar metal, and it is not without side effects, including renal damage at high doses. It is not yet known if all the undesirable effects of EDTA *in vivo* are due to metal chelation, but prolonged therapy does lead to hypocalcemia and, to a lesser extent, depletion of essential metals like zinc and manganese. Speciation analysis suggests that a ratio of 3:1 $EDTAH_2Ca$:$EDTAH_2Zn$ ought to reduce any need for topping up therapy. In the case of DTPA, which has found use in the treatment of human contamination with plutonium or americium, $DTPAHZn_2$ has been shown to be less toxic than $DTPAHCa_2$, presumably because there is no depletion of essential metals such as Zn, Mn, or Mo.

17.8 TRACE ELEMENT SUPPLEMENTATION

It has been mentioned several times in the foregoing discussion that there may be occasions when treatment with a ligand drug may produce a depletion of essential metals that requires the introduction of a topping-up therapy.

Metal supplementation sounds deceptively simple, but, in fact, many factors must be taken into consideration in selecting or developing a satisfactory preparation. Oral administration is the most commonly used route of supplementation, but it is also the one presenting the most difficulties. Most of the essential and beneficial metals required by the human body are absorbed from the upper part of the small intestine, a region where the pH ranges from ~6 to ~8; further, the metals must reach the absorptive sites on the intestinal mucosa in a soluble and absorbable form. However, at this pH in the aqueous environment of the small intestine most multivalent metal ions react almost quantitatively with water to form hydrolyzed, insoluble, and thus nonabsorbable hydroxides and oxides. Other reactions occur with the numerous complexing ligands present in the intestinal contents to form metal complexes, the major fraction of which may be electrically charged and also nonabsorbable; the fraction of electrically net neutral and thus absorbable species may well represent only a tiny fraction of the total metal which enters the gastrointestinal tract.

In formulating metal supplements a suitable complexing ligand may be added with a view to enhancing the proportion of soluble, neutral metal complexes that are formed near the absorptive surfaces in the intestine. However, because such complex formation is usually very pH-dependent the choice of ligand is not always easy; for example, when ferrous iron reacts with ascorbic acid the total percentage of neutral complexes formed in the pH range 5 to 7.5 is $>60\%$, whereas with galacturonic acid that percentage is $<30\%$, thus ascorbic acid appears to be the better ligand for facilitating iron absorption.

17.8.1 Iron, Zinc, and Copper Supplementation

These are the most prevalent metals *in vivo,* as mentioned earlier, and supplementation treatments can be traced back thousands of years into pre-Christian times when solutions of rust in acid wine were used for anemia and zinc oxide ointments for wounds or skin conditions. Today, the *Pharmacopeias* list more than 40 preparations for iron, but only three for zinc, and none for copper.

Modern oral supplementation therapy involves a combination of scientific approaches to which certain psychological and commercial factors have to be added. These considerations:

a. aim to increase the flow of metal complexes from intestine to blood by increasing the concentration of lipid-soluble, low molecular mass complexes present in the intestinal fluids; for example, iron preparations may contain Fe(II) in association with complexing ligands such as ascorbate, malate, fumarate, gluconate, or amino acids that promote the formation of neutral complexes at pH ~6 to 7; gluconate may be used for the same purpose for metals such as Cu, Zn, or Co;
b. tend to favor iron(II) rather than iron(III) compounds, since the former can be up to ~17 orders of magnitude more soluble (K_{sp} $Fe(OH)_2 = 10^{-15.1}$, $Fe(OH)_3 = 10^{-38.7}$)[1]; K_{sp} is the concentration solubility product $[Fe^{2+}][OH^-]^2$, or $[Fe^{3+}][OH^-]^3$;
c. cause least irritation to the gastrointestinal tract; for this reason ferrous sulfate is not an agent of choice.

Irritation of the gastrointestinal tract can be a problem, especially in iron supplementation. Such irritation tends to follow the Irving–Williams series of complex stability for divalent ions, i.e., $Mn<Fe<Co<Ni<Cu>Zn$. This indicates why oral supplementation with copper is exceedingly difficult.

Chelation approaches have been used to overcome the problems of gastrointestinal irritation by iron; thus, in addition to ferrous sulfate, preparations containing ferrous fumarate, ferrous gluconate, ferrous succinate, or ferrous glycine sulfate are available.

In cases where oral iron preparations are not tolerated by the patient, or there is a need to rapidly increase iron levels, the metal may be administered by injection. A widely used injection form of iron is an iron(III) sorbitol (a reduction product of glucose) citric acid complex (Jectofer), but ferric gluconate and $Fe(OH)_3$–dextran complexes are also used.

In contrast to iron, zinc supplementation presents few problems, administration of daily doses of 150 mg Zn^{2+} as zinc sulfate is well tolerated. However, this dose is some ten times the recommended daily intake of zinc and prolonged supplementation at this level may induce deleterious changes such as alteration of the high-density lipoprotein (HDL)/low-density lipoprotein (LDL) cholesterol ratio and an increased risk of heart disease.

17.9 BIOINORGANIC CHEMISTRY AND RADIOPHARMACY

Radioactive metal complexes and other preparation have been widely used during the past four decades for the diagnosis of a wide spectrum of diseases and the ability of large amounts of certain radioactive isotopes (radionuclides) to destroy tissue has been exploited for the treatment of cancer and some other diseases.[9] A great variety of complexes of the short radioactive half-life isotope ^{99m}Tc ($t_{1/2} = 6.0$ h) are used extensively in diagnostic nuclear medicine (the half-life of a radionuclide is the time required for the radioactivity to decay to one-half of its original value). Less frequently complexes, or simple salts, of longer-lived metallic radionuclides such as ^{67}Ga ($t_{1/2} = 3.26$ days), ^{111}In ($t_{1/2} = 2.8$ days), and ^{201}Tl ($t_{1/2} = 3.04$ days) are also used. Some examples of diagnostic ''bioinorganic'' radiopharmaceuticals are listed in Table 17.8. The development of complexes containing metallic radionuclides involves the same principles of bioinorganic chemistry as those discussed above. The preparation of such ''radiopharmaceuticals'' has now become one of the responsibilities of many hospital pharmacists.

17.9.1 Radionuclide Therapy

Radionuclides are also of increasing interest for therapy, two of the earliest examples of this were the use of colloidal solutions of the γ-ray and β-particle emitting isotope ^{198}Au ($t_{1/2} = 2.7$ days) for the treatment of intracavitary accumulations of fluid (ascites) in cancer patients; and of the use of the nonmetallic radionuclide ^{131}I ($t_{1/2} = 8.0$ days), administered as iodide, for the treatment of thyrotoxicosis and thyroid cancer. More recently ^{131}I-labeled metaiodobenzylguanidine (MIBG) has been introduced for the treatment of pheochromocytoma, neuroblastoma, and other tumors that

Table 17.8 Examples of some typical "bioinorganic" radiopharmaceuticals used in diagnostic nuclear medicine

Radiopharmaceutical	Half-life	Administered activity [MBq]	Imaging application
^{99m}Tc-Pertechnetate	6.02 h	50	Thyroid function
^{99m}Tc-1,2-bis[bis(2-ethoxyethyl) phosphino]–ethane (Myoview)	6.02 h	200	Heart (myocardium)
^{99m}Tc-Methyl-oxy-isobutyl-isonitrile (MIBI)	6.02 h	500–700	Heart
^{99m}Tc-Diphosphonate	6.02 h	600	Skeletal metastases
^{99m}Tc-Hexamethylpropyleneamineoxine [HM-PAO]	6.02 h	700	Brain in stroke
^{99m}Tc-Dimercaptosuccinic acid [DMSA]	6.02 h	75	Kidney function
^{99m}Tc-Hydroxyiminodiacetate [HIDA]	6.02 h	150	Liver/gall bladder
^{67}Ga citrate	3.26 days	200	Tumor localization
^{111}In-DTPA-D-Phe-1-octreotide (Somatostatin analog)	2.8 days	110	Tumors with somatostatin receptors
^{131}Iodide	8.04 days	0.4	Thyroid function
^{201}Tl Chloride	3.05 days	75	Heart

are relatively rich in chromaffin granules. Some examples of therapeutic radiopharmaceuticals are shown in Table 17.9. The doses required for effective therapy are up to 1000 times greater than those normally used in diagnostic procedures.

Systemic treatment with radionuclides is increasingly used for the management of the pain caused by secondary tumors (metastases) arising in bone, particularly in patients suffering from prostate or breast cancer. Currently $^{89}SrCl_2$ is used as an effective first-line agent for treating skeletal pain; strontium, a chemical analog of calcium, deposits in areas of bone resorption and remodeling in the vicinity of the metastases.

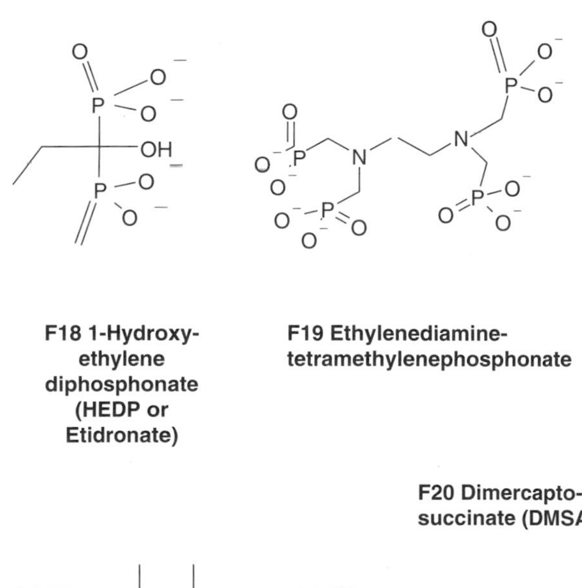

F18 1-Hydroxy-ethylene diphosphonate (HEDP or Etidronate)

F19 Ethylenediamine-tetramethylenephosphonate

F20 Dimercapto-succinate (DMSA)

Recently, complexes of ^{186}Re ($t_{1/2}$ = 3.78 days) with 1-hydroxy-ethylene diphosphonate, etidronate, or HEDP (F18), a substance that deposits in bone and is widely used for the treatment of Paget's disease, have been introduced for the palliation of bone pain. Complexes of the β-γ-emitting lanthanide radionuclide ^{153}Sm ($t_{1/2}$ = 1.95 days) with ethylenediamine–tetramethylene-phosphonate acid (EDTMP) (F19) have also been introduced for the same purpose.

Table 17.9 Examples of some "bioinorganic" radiopharmaceuticals for radionuclide therapy

Radiopharmaceutical	Half-life	Therapeutic application
[^{89}Sr]-Chloride	50.5 days	Pain relief from skeletal metastases
[^{117m}Sn](IV)-Diethylenetriaminepentaacetate	13.6 days	Pain relief from skeletal metastases
[^{131}I]-Iodide	8.04 days	Thyrotoxicosis, thyroid cancer
[^{131}I]-Metaiodobenzylguanidine (MIBG)	8.04 days	Pheochromocytoma; neuroblastoma
[^{153}Sm]-Ethylenediaminetetramethylphosphonate (EDTMP) (Quadramet™)	1.95 days	Pain relief from skeletal metastases
[^{186}Re]-Hydroxyethylenediphosphonate (HEDP)	3.78 days	Pain relief from skeletal metastases
[^{188}Re]-Dimercaptosuccinate (DMSA)	17.0 h	
[^{198}Au]-Colloid	2.7 days	Malignant ascites
[^{211}At]-Monoclonal antibody	7.1 h	Radioimmunotherapy of tumors
[^{213}Bi]-Monoclonal antibody	47 min	Radioimmunotherapy of tumors
[^{225}Ac]-Monoclonal antibody	10.0 days	Radioimmunotherapy of tumors

This is an active area of radiopharmaceutical research and other radiolanthanide diphosphonate complexes, and ^{188}Re ($t_{1/2}$ = 17.0 h) dimercaptosuccinate (DMSA) (F20) and ^{117m}Sn(IV) ($t_{1/2}$ = 13.6 days) DTPA complexes are entering into clinical trials and practice. In this work HSAB principles and computer-assisted speciation analysis play important roles in identifying complexes with good clinical potential.

There is also wide interest in the development of lanthanide and other metallic radionuclide complexes with monoclonal antibodies, or antibody fragments, that will bind specifically to tumor cells and thus specifically irradiate the tumor mass, with minimal effects on surrounding normal tissues. In this area of radioimmunotherapy, in addition to β-γ-emitting radionuclides, such as ^{47}Sc ($t_{1/2}$ = 4.54 days), ^{67}Cu ($t_{1/2}$ = 2.58 days), ^{153}Sm and ^{199}Au ($t_{1/2}$ = 3.14 days), α-particle emitting radionuclides, such as ^{213}Bi ($t_{1/2}$ = 47 min), are of special interest in view of the fact that α-particles decaying in an organ or a tumor deposit large amounts of radiation energy in a very small volume of tissue.

Complexes of α-particle emitting metallic radionuclides that would localize in or close to the nucleus of a tumor cell are of considerable interest at the research level. This is because the short range of the α-particle means that all the large amount of radiation energy is deposited in a tiny volume of tissue, causing severe damage to the DNA. Monoclonal antibody complexes with ^{211}At-labeled ($t_{1/2}$ = 7.1 h) have entered Phase I and Phase II clinical trials as tumor-specific α-particle emitting therapeutic radiopharmaceuticals. Astatine is the fifth, and heaviest, member of the halogen series and its chemical properties show a number of similarities with those of iodine. All astatine isotopes are radioactive and their half-lives are too short for them to exist in nature, thus, ^{211}At must be produced in a cyclotron close to the site at which it is to be used. This makes ^{211}At-radiopharmaceuticals both expensive and of limited availability. Similar complexes with ^{212}Bi, ^{213}Bi, and ^{225}Ac are also undergoing clinical trials. These latter radionuclides are of special interest since they can be produced relatively easily from a generator.[6b]

17.10 CONCLUDING REMARKS

The aim of this chapter has been to indicate, hopefully in a convincing manner, that all agents, in one way or another, interact with metal ions — as *modus operandi*, as side effects, as counter ions (anions or cations), as competing electron acceptors for drug active groups, as possible impurities or even as trace element supplements.

The late development of this subject has been due to the scarcity of analytical techniques capable of measuring the very low metal concentrations involved, and also due to the predominance of organic chemistry in pharmacological and pharmaceutical research with the result that the fact that life is an inseparable combination of inorganic and organic chemical reactions is often overlooked.

One might comment on the success, or otherwise, of isosteric modifications in drug design. Bearing in mind the odds against evolutionary processes, unassisted by modern technology, selecting *Homo sapiens* to be the advanced organism that we are with perfectly balanced biochemistry, it would appear that drug designers with all their armamentarium of computers and laboratory equipment directed at putting the right isoelectronic group in the right place for ideal drug activity, would be well placed to achieve perfection. What then is missing?

The fact is that each active group is not moved from one position to a nearer optimal one without knock-on effects. The same electrons on the groups that interlock with the active sites in order to produce the desired pharmacological effects also attract HSAB bases such as protons and metal ions, each being quantifiable in terms of a formation constant, K. Indeed the target site for a drug's activity can only react with the active groups if the protons or metal ions are displaced. The pK values, however, vary with the position, the energetics of the molecule, and the solvent.

It is humbling to note that in nature biochemical reactions do not occur as the discrete sequences that we depict as equations involving letters, numbers, or other symbols, but rather that every reaction between elements or molecules is an interaction of packets of electron energy. Since these packets are often quite diffuse in their nature, it is not surprising that biochemical interactions are influenced, or are under the influence of other packets of energy in their immediate environment. Future successes in drug design, especially in designing agents for treating the more intransigent human diseases, will depend not only on molecular modeling and creative synthetic developments, but also on a sound understanding of all the interactions of the drug with both the inorganic and the organic components of the body.

FURTHER READING

1. ***General aspects*** — (a) Miller, S. and Orgel, L.E. (1973) *The Origins of Life*. London, Chapman & Hall. (b) Williams, D.R. (1976) *Introduction to Bio-Inorganic Chemistry*. Illinois, C.C. Thomas. (c) Albert, A. (1979) Metal-binding substances. In: *Selective Toxicity*, 6th edn. London, Chapman & Hall, pp. 385–442. (d) Daniels, T.C. and Jorgensen, E.C. (1977) Physicochemical properties in relation to biological action. Chelation and biological action. (e) Schultz, H.W. (1977) Surfactants and chelating agents. In: Wilson, C.O., Gisvold, O., and Doerge, R.F. (eds) *Textbook of Organic, Medicinal and Pharmaceutical Chemistry*, 7th edn. Philadelphia, Lippincott, pp. 50–56, pp. 222–246. (f) Fiabane, A.M. and Williams, D.R. (1977) In: *Principles of Bio-Inorganic Chemistry*. Royal Society of Chemistry, London, Monographs for Teachers No. 31, pp. 82–107. (g) Fraústo da Silva, J.J.R. and Williams, R.J.P. (1991) *The Biological Chemistry of the Elements*. Oxford, Clarendon Press. (h) Fraústo da Silva, J.J.R. and Williams, R.J.P. (2001) *The Biological Chemistry of the Elements. The Inorganic Chemistry of Life*, 2nd edn. Oxford, Oxford University Press. (i) Williams, D.R. (1983) Historical outline of the biological importance of trace metals. *Journal of Metabolic Diseases*, (Suppl. **1**), 1–4. (j) Taylor, D.M. and Williams, D.R. (1995) *Trace Element Medicine and Chelation Therapy*. London, The Royal Society of Chemistry. (k) Hay, R.W. (1984) *Bio-Inorganic Chemistry*. Chichester, Ellis Horwood, p. 210. (l) Sigel, A. and Sigel, H. *Metal Ions and their Complexes in Medication and in Cancer Diagnosis and Therapy. Metal Ions in Biological Systems*, Vol. 41. New York, Marcel Decker, pp. 1–39 and 139–183. (m) Patel, A., Smith, H.J., and Steinmetzer, T. (in press) Design of enzyme inhibitors as drugs, Chapter 9. In: Smith, H.J. (ed.) *Introduction to the Principles of Drug Design and Action*, 4th edn, Chapter 9. (n) Smith, H.J. and Simons, C. (eds) (2005) *Enzymes and Their Inhibition: Drug Development*. London, Taylor and Francis, London.
2. ***Chelation of iron*** — (a) May, P.M. and Williams, D.R. (1980) The inorganic chemistry of iron metabolism. In: Jacobs, A. and Worwood, M. (eds) *Iron in Biochemistry and Medicine*, Vol. II. London, Academic Press, pp. 1–27. (b) Hider, R.C. and Liu, Z.D. (2003) Emerging understanding of the advantage of small molecules such as hydroxypyridones in the treatment of iron overload.

Current Medical Chemistry **10**(12), 1051–1064. (c) Nick, H., Acklin, P., Lattmann, R., Buehlmayer, P., Haufle, S., Schupp, J., and Alberti, D. (2003) Development of tridentate iron chelators: from desferrithiocin to ICL670. *Current Medical Chemistry* **10**(12), 1065–1076. (d) Kontoghiorghes, G.J., Neocleous, K., and Kolnagou, A. (2003) Benefits and risks of deferiprone in iron overload in thalassaemia and other conditions: comparison of epidemiological and therapeutic aspects with desferrioxamine. *Drug Safety* **26**(8), 553–584. (e) Piga, A., Gaglioti, C., Fogliacco, E., and Tricta, F. (2003) Comparative effects of deferiprone and desferrioxamine on survival and cardiac disease in patients with thalassaemia major: a retrospective analysis. *Haematologica* **88**(5), 489–496.

3. ***Copper*** — (a) May, P.M. and Williams, D.R. (1981) Role of low molecular weight copper complexes in the control of rheumatoid arthritis. In: Sigel, H. (ed.) *Metal Ions in Biological Systems*, Vol. 12. New York, Marcel Decker, pp. 283–317. (b) Sorenson, J.R. J. (1982) *Inflammatory Diseases and Copper*. New Jersey, Humana press, p. 662.
4. ***Chemistry of metal complexes*** — (a) Pearson, R.G. (1963) Hard and soft acids and bases. *Journal of the American Chemical Society* 85, 3533–3539. (b) Huheey, J.E. (1978) Coordination chemistry — theory, structure and mechanisms. In: *Inorganic Chemistry, Principles of Structure and Reactivity*, 2nd edn. New York, Harper & Row, pp. 332–488. (c) Stenlake, J.B. (1979) Metal chelation. In: *Foundations of Molecular Pharmacology*, Vol. 2. *The Chemical Basis of Drug Action*. University of London, Athlone Press, pp. 86–99. (d) Williams, D.R. and Halstead, B.W. (1982–1983) Chelating agents in medicine. *Journal of Toxicology & Clinical Toxicology* **19**(10), 1081–1115.
5. ***Zinc*** — Abu Bakar, N.K., Taylor, D.M., and Williams, D.R. (1999) Chemical speciation of zinc in saliva containing zinc gluconate lozenge correlated with reduction in symptoms of the common cold. *Chemical Speciation and Bioavailability* **11**(3), 95–102.
6. ***Radionuclide therapy*** — (a) Atkins, H.L. (1998) Overview of nuclides for bone pain palliation. *Applied Radiation and Isotopes* **49**, 277–283. (b) Srivastava, S. and Dadachova, E. (2001) Recent advances in radionuclide therapy. *Seminars in Nuclear Medicine* **31**, 330–341.
7. ***Wounds*** — (a) Jones, P.W., Taylor, D.M., Williams, D.R., Finney, M., Iorwerth, A., Webster, D., and Harding, K.G. (2001) Using wound fluid analyses to identify trace element requirements for efficient healing. *Journal of Wound Care* **10**, 205–208. (b) Jones, P.W., Taylor, D.M., and Williams, D.R. (2000) Analysis and chemical speciation of copper and zinc in wound fluid. *Journal of Inorganic Biochemistry* **81**, 1–10.
8. ***Human data*** — (a) International Commission on Radiological Protection (1975) ICRP Publication 23 Reference Man. Oxford, Pergamon Press. (b) International Commission on Radiological Protection (2002) ICRP Publication 89 Basic anatomical and physiological data for use in radiological protection: reference values. *Annals of the ICRP* 32(3–4). (c) Iyengar, G.V., Kollmer, W.E., and Bowen, H.J.M. (1978) *The Elemental Composition of Human Tissues and Body Fluids*. Weinheim, Verlag Chemie.

Index